GW01606125

NUCLEAR CHEMICAL ENGINEERING

McGraw-Hill Series in Nuclear Engineering

Thomas H. Pigford, *Consulting Editor*

Ash: *Nuclear Reactor Kinetics*
Benedict, Pigford, and Levi: *Nuclear Chemical Engineering*
Bonilla: *Nuclear Engineering*
Ellis: *Nuclear Technology for Engineers*
El-Wakil: *Nuclear Power Engineering*
Glower: *Experimental Reactor Analysis and Radiation Measurements*
Hoisington: *Nucleonics Fundamentals*
Knief: *Nuclear Energy Technology*
Meghreblian and Holmes: *Reactor Analysis*
Price: *Nuclear Radiation Detection*
Schultz: *Control of Nuclear Reactors and Power Plants*
Stephenson: *Introduction to Nuclear Engineering*

NUCLEAR CHEMICAL ENGINEERING

Second Edition

Manson Benedict

Professor Emeritus of Nuclear Engineering
Massachusetts Institute of Technology

Thomas H. Pigford

Professor of Nuclear Engineering
University of California, Berkeley

Hans Wolfgang Levi

Hahn-Meitner-Institut für Kernforschung Berlin
and apl. Professor of Nuclear Chemistry
Technische Universität Berlin

McGraw-Hill Book Company

New York St. Louis San Francisco Auckland Bogotá Hamburg
Johannesburg London Madrid Mexico Montreal New Delhi
Panama Paris São Paulo Singapore Sydney Tokyo Toronto

This book was set in Press Roman by Hemisphere Publishing Corporation.
The editor was Diane D. Heiberg;
the production supervisor was Rosann E. Raspini.
Kingsport Press, Inc. was printer and binder.

NUCLEAR CHEMICAL ENGINEERING

4 5 6 7 8 9 0 KPKP 8 9 8 7 6 5 4 3

Library of Congress Cataloging in Publication Data

Benedict, Manson.
Nuclear chemical engineering.

(McGraw-Hill series in nuclear engineering)
Includes bibliographies and index.
1. Nuclear engineering. 2. Nuclear chemistry.
I. Pigford, Thomas H., joint author. II. Levi, Hans Wolfgang, joint author. III. Title.
TK9350.B4 1981 621.48 80-21538
ISBN 0-07-004531-3

This text is dedicated to the authors' wives,
Marjorie Allen Benedict, Catherine Cathey Pigford,
and Ruth Levi,
whose assistance, encouragement, and patience
made this book possible.

CONTENTS

PREFACE

The development of nuclear fission chain reactors for the conversion of mass to energy and the transmutation of elements has brought into industrial prominence chemical substances and chemical engineering processes that a few years ago were no more than scientific curiosities. Uranium, formerly used mainly for coloring glass and ceramics, has become one of the world's most important sources of energy. Thorium, once used mainly in the Welsbach gas mantle, promises to become a nuclear fuel second in importance only to uranium. Zirconium and its chemical twin hafnium, formerly always produced together, have been separated and have emerged as structural materials of unique value in reactors. New chemical engineering processes have been devised to separate these elements, and even more novel processes have been developed for producing deuterium, ^{235}U, and the other separated isotopes that have become the fine chemicals of the nuclear age. The processing of radioactive materials, formerly limited mainly to a few curies of radium, is now concerned with the millions of curies of radioactive isotopes of the many chemical elements that are present in spent fuel discharged from nuclear reactors.

The preceding introduction to the preface of the first edition of this book can still serve as the theme of this second edition. Since 1957 nuclear power systems have become important contributors to the energy supply of most industrialized nations. This text describes the materials of special importance in nuclear reactors and the processes that have been developed to concentrate, purify, separate, and store safely these materials. Because of the growth in nuclear technology since the first edition appeared and the great amount of published new information, this second edition is an entirely new book, following the first edition only in its general outline.

Chapter 1 lists the special materials of importance in nuclear technology and outlines the relationship between nuclear reactors and the chemical production plants associated with them. Chapter 2 summarizes the aspects of nuclear physics and radioactivity that are pertinent to many of the processes to be described in later chapters. Chapter 3 describes the changes in composition and reactivity that occur during irradiation of fuel in a nuclear reactor and shows how these changes determine the material and processing requirements of the reactor's fuel cycle. Chapter 4 describes the principles of solvent extraction, the chemical engineering unit operation used most extensively for purifying uranium, thorium, and zirconium and reprocessing irradiated fuel discharged from reactors.

Chapters 5, 6, and 7 take up uranium, thorium, and zirconium in that order. Each chapter discusses the physical and chemical properties of the element and its compounds, its natural occurrence, and the processes used to extract the element from its ores, purify it, and convert it to the forms most useful in nuclear technology.

The next four chapters take up processing of the highly radioactive materials produced in reactors. Chapter 8 describes the isotopic composition and radioactive constituents of spent fuel discharged from representative types of reactors and deals briefly with other radioisotopes resulting from reactor operation. Chapter 9 describes the physical and chemical properties of the synthetic actinide elements produced in reactors: protactinium, neptunium, plutonium, americium, and curium, and their compounds. Chapter 10 describes the radiochemical processes that have been developed for reprocessing irradiated fuel to recover uranium, plutonium, and other valuable actinides from it. Chapter 11 describes conversion of radioactive wastes from reactor operation and fuel reprocessing into stable forms suitable for safe, long-term storage, and systems to be used for such storage.

The last three chapters deal with separation of stable isotopes. Chapter 12 lists the isotopes of principal importance in nuclear technology, discusses their natural occurrence, and develops the chemical engineering principles generally applicable to isotope separation processes. Chapter 13 describes processes useful for separating deuterium and isotopes of other light elements, specifically distillation, electrolysis, and chemical exchange. Chapter 14 describes processes used for separating uranium isotopes, specifically gaseous diffusion, the gas centrifuge, aerodynamic processes, mass and thermal diffusion, and laser-based processes.

Four appendixes list fundamental physical constants, conversion tables, nuclide properties, and radioactivity concentration limits for nuclear plant effluents.

As may be seen from this synopsis, this text combines an account of scientific and engineering principles with a description of materials and processes of importance in nuclear chemical technology. It aims thus to serve both as a text for classroom instruction and as a source of information on chemical engineering practice in nuclear industry.

Problems at the end of each chapter may prove useful when the text is used for instruction. References are provided for readers who wish more details about the topics treated in each chapter. Extensive use has been made of information from the *Proceedings* of the four International Conferences on the Peaceful Uses of Atomic Energy in Geneva, Switzerland, sponsored by the United Nations, which are listed as *PICG,* followed by the number of the conference, in the references at the ends of chapters.

This book was written in a transition period when U.S. engineering and business practice was changing from English to SI units. When the references cited used English units, these have been retained in the text in most cases. Equivalent SI values are also provided in many passages, or conversion factors are given in footnotes. In addition, conversion tables are provided in App. B. The multiplicity of units is regrettable, but it is unavoidable until the world's technical literature has changed over completely to the SI system.

In preparing this text the authors have been blessed with assistance from so many sources that not all can be mentioned here. We are grateful to our respective institutions, Massachusetts Institute of Technology, University of California (Berkeley), and Hahn-Meitner-Institut (Berlin), for the freedom and opportunity to write this book. For help with calculations, illustrations, and typing, thanks are due Marjorie Benedict, Ellen Mandigo, Mary Bosco, Sue Thur, and many others. Editorial assistance from Judith B. Gandy and Lynne Lackenbach is acknowledged with gratitude. To the many generations of students who used the notes on which this book is based and helped to correct its mistakes we are greatly indebted. Among our more recent students we wish to thank Allen Croff, Charles Forsberg, Saeed Tajik, and Cheh-Suei Yang.

Among our American professional colleagues we are greatly indebted to Don Ferguson and his associates at Oak Ridge National Laboratory; Paul McMurray and others of Exxon Nuclear Company; James Buckham and Wesley Murbach of Allied General Nuclear Services; James Duckworth of Nuclear Fuel Services, Inc.; Joseph Megy of Teledyne Wah Chang Albany Company; Paul Vanstrum and Edward Von Halle of Union Carbide Corporation; Lombard Squires, John

Proctor, and their associates of E. I. duPont de Nemours and Company; Marvin Miller of MIT; and Donald Olander of the University of California (Berkeley). In Germany, we wish to thank Hubert Eschrich of Eurochemic, Richard Kroebel of Kernforschungszentrum Karlsruhe, Erich Merz of Kernforschungsanlage Jülich, Walther Schüller of Wiederaufarbeitungsanlage Karlsruhe, and Eckhart Ewest of Deutsche Gesellschaft für Wiederaufarbeitung von Kernbrennstoff.

Assistance provided to one of the authors (MB) by a fellowship from the Guggenheim Foundation is acknowledged with gratitude.

Despite the valued assistance the authors have had in preparing this text, it doubtless still contains many errors and omissions. We shall be grateful to our readers for calling these to our attention.

Manson Benedict
Thomas H. Pigford
Hans Wolfgang Levi

CHAPTER

ONE

CHEMICAL ENGINEERING ASPECTS OF NUCLEAR POWER

1 INTRODUCTION

The production of power from controlled nuclear fission of heavy elements is the most important technical application of nuclear reactions at the present time. This is so because the world's reserves of energy in the nuclear fuels uranium and thorium greatly exceed the energy reserves in all the coal, oil, and gas in the world [H1], because the energy of nuclear fuels is in a form far more intense and concentrated than in conventional fuels, and because in many parts of the world power can be produced as economically from nuclear fission as from the combustion of conventional fuels.

The establishment of a nuclear power industry based on fission reactors involves the production of a number of materials that have only recently acquired commercial importance, notably uranium, thorium, zirconium, and heavy water, and on the operation of a number of novel chemical engineering processes, including isotope separation, separation of metals by solvent extraction, and the separation and purification of intensely radioactive materials on a large scale. This text is concerned primarily with methods for producing the special materials used in nuclear fission reactors and with processes for separating isotopes and reclaiming radioactive fuel discharged from nuclear reactors.

This chapter gives a brief account of the nuclear fission reaction and the most important fissile fuels. It continues with a short description of a typical nuclear power plant and outlines the characteristics of the principal reactor types proposed for nuclear power generation. It sketches the principal fuel cycles for nuclear power plants and points out the chemical engineering processes needed to make these fuel cycles feasible and economical. The chapter concludes with an outline of another process that may some day become of practical importance for the production of power: the controlled fusion of light elements. The fusion process makes use of rare isotopes of hydrogen and lithium, which may be produced by isotope separation methods analogous to those used for materials for fission reactors. As isotope separation processes are of such importance in nuclear chemical engineering, they are discussed briefly in this chapter and in some detail in the last three chapters of this book.

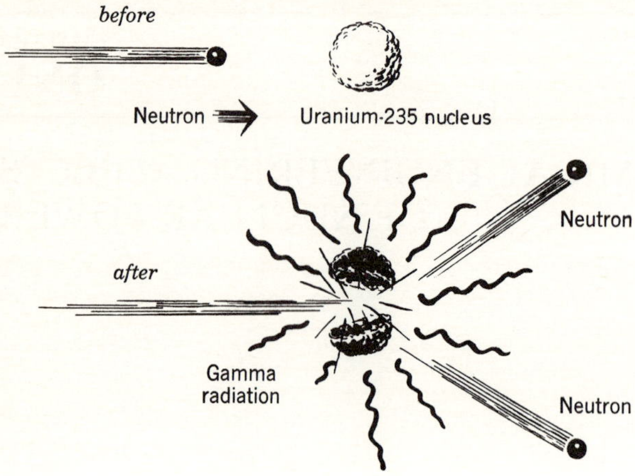

Figure 1.1 Fission of ^{235}U nucleus by neutron.

2 NUCLEAR FISSION

The nuclear fission process utilized in today's power-producing reactors is initiated by interaction between a neutron and a fissile nucleus, such as ^{235}U.† The nucleus then divides into two fragments, with release of an enormous amount of energy and with production of several new neutrons. Under proper conditions, these product neutrons can react with additional ^{235}U atoms and thus give rise to a neutron chain reaction, which continues as long as sufficient ^{235}U remains to react. Fission of a single nucleus of ^{235}U is represented pictorially in Fig. 1.1, and a fission chain reaction is shown in Fig. 1.2. To keep the rate of the chain reaction constant, neutrons are allowed to leak from a nuclear reactor or are absorbed in boron, ^{238}U, or other nonfissionable materials placed in the reactor. A steady chain reaction is depicted in Fig. 1.3.

The fission of ^{235}U can take place in a number of ways, one of which is shown in Fig. 1.4. The nucleus of ^{235}U, which contains 92 protons and 143 neutrons, divides into two fragments, plus some extra neutrons, in such a manner that the total number of protons and neutrons in the product nuclei equals the total number in the reactant neutron and ^{235}U nucleus. In the example of this figure, the fission fragments are ^{144}Ba, containing 56 protons and 88 neutrons; ^{89}Kr, containing 36 protons and 53 neutrons; and three extra neutrons. The fission fragments are unstable and subsequently undergo radioactive decay. In the radioactive decay some of the neutrons of the nucleus are converted into protons, which remain in the nucleus, and into electrons, which fly out as beta radiation. In this example, four neutrons in ^{144}Ba are successively converted into protons, resulting in ^{144}Nd as end product, and three neutrons in ^{89}Kr are converted into protons, resulting in ^{89}Y as end product.

The numbers assigned to each reactant or end product represent its mass in atomic mass units (amu). This unit is defined as the ratio of the mass of a neutral atom to one-twelfth the mass of an atom of ^{12}C. In the present instance the mass of the products is less than that of the reactants‡:

†In this text each nuclide, such as uranium-235, is referred to by its chemical symbol, in this case ^{235}U.

‡The mass of the electrons is not included in this calculation because the electrons emitted from the nucleus in radioactive decay ultimately return as orbital electrons surrounding the nucleus of a neutral atom.

Reactants		Products		Difference
^{235}U	235.043915	^{144}Nd	143.910039	
Neutron	1.008665	^{89}Y	88.905871	
		3 neutrons	3.025995	
Total	236.052580		235.841905	0.210675

A fraction 0.210675/235.043915 = 0.0008963 of the mass of the ^{235}U atom disappears in this fission reaction. This reduction in mass is a measure of the amount of energy released in this fission reaction. The Einstein equation (1.1) expressing the equivalence of energy and mass,

$$\Delta E = c^2 \, \Delta m \tag{1.1}$$

predicts that when Δm kilograms of mass disappears, ΔE joules of energy appears in its place. In this relation, c is the velocity of light, 2.997925×10^8 m/s.† The energy released in this fission reaction thus is

$$(0.0008963)\,(2.997925 \times 10^8)^2 = 8.06 \times 10^{13} \text{ J/kg } ^{235}\text{U} \tag{1.2}$$

or 3.46×10^{10} Btu/lb.

Energy changes associated with a single nuclear event are commonly expressed in terms of millions of electron volts (MeV), defined as the amount of energy acquired by an electronic charge (1.602×10^{-19} C) when accelerated through a potential difference of 1,000,000 V. One MeV therefore equals $1.602 \times 10^{-19} \times 10^6 = 1.602 \times 10^{-13}$ J.

The energy released when one atom of ^{235}U undergoes fission in the above reaction is

$$\frac{(8.06 \times 10^{13} \text{ J/kg})(235.04 \text{ g/g-atom})}{(1.602 \times 10^{-13} \text{ J/MeV})(6.023 \times 10^{23} \text{ atoms/g-atom})(1000 \text{ g/kg})} = 196 \text{ MeV/atom} \tag{1.3}$$

†Fundamental physical constants are listed in App. A. A table of mass and energy equivalents is given in App. B.

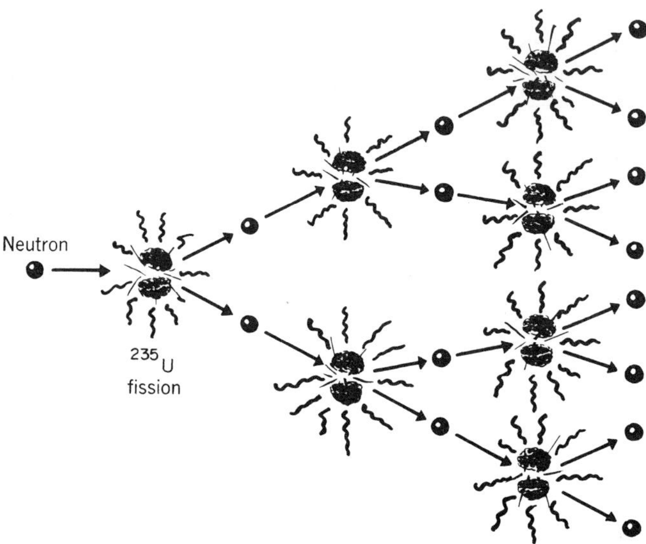

Figure 1.2 Fission chain reaction.

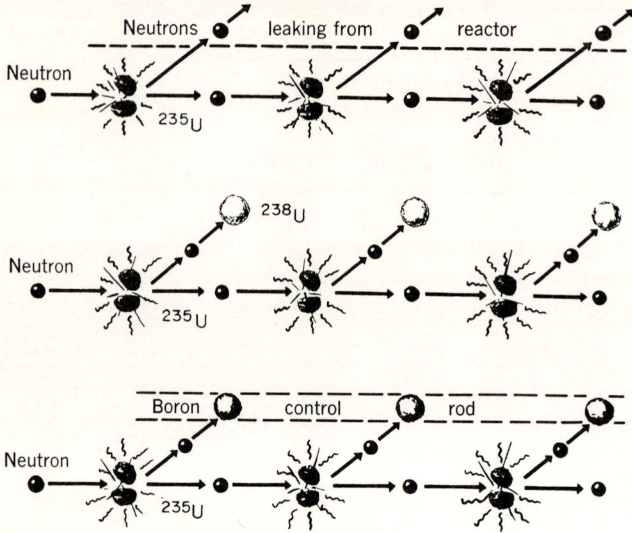

Figure 1.3 Steady fission chain reaction.

Atoms of ^{235}U may undergo fission in a variety of ways, of which the reaction shown in Fig. 1.4 is only one. The average yield of particles and energy from fission of ^{235}U in all possible ways is shown in Fig. 1.5. In the primary fission reaction shown at the top of this figure, ^{235}U splits into two parts, the radioactive fission products, while at the same time giving off several fast neutrons (2.418 on the average) and gamma radiation. One of these neutrons is used to maintain the fission reaction. The remaining neutrons may either be used to bring about other desired nuclear reactions or be lost either through leakage from the reactor or through capture by elements present in the reactor to produce unwanted or waste products.

Following the primary fission reaction, the radioactive fission products undergo radioactive disintegration, yielding beta particles and delayed gamma rays and ending up as stable fission products. Since the radioactive fission products have half-lives ranging from fractions of a

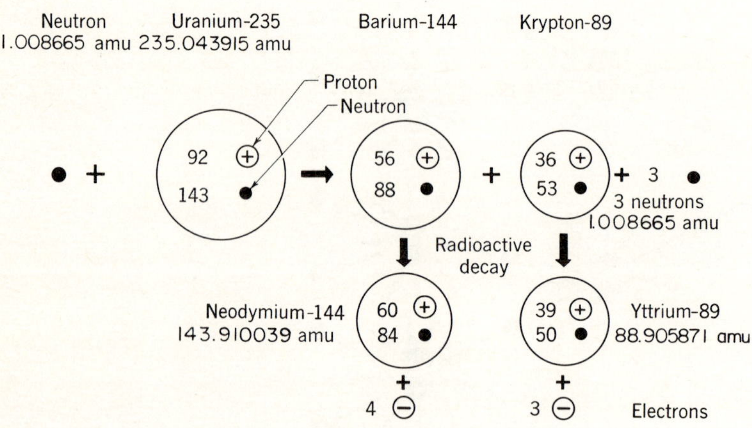

Figure 1.4 Example of fission of ^{235}U.

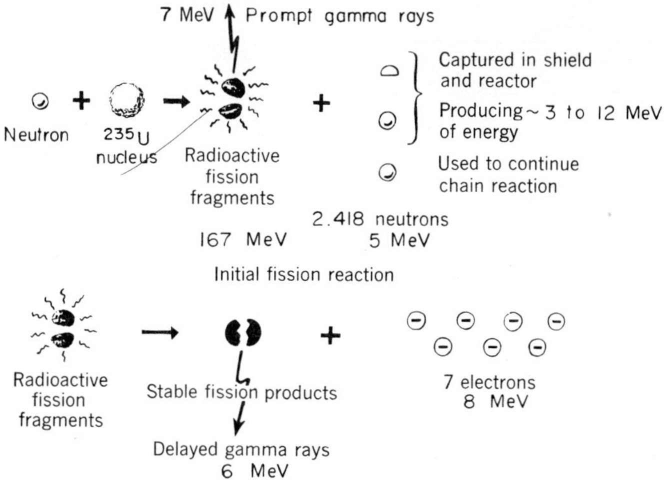

Figure 1.5 Average yields in fission of ^{235}U.

second to millions of years, the emission of beta particles and delayed gamma rays takes place over a long period of time after a reactor has been shut down, but at a diminishing rate.

The total energy released in fission is the sum of the energies associated with the different particles shown in this figure, 196 to 205 MeV. As up to 5 MeV of gamma energy escapes from a typical power reactor and is not utilized, a nominal figure for the energy released in fission is 200 MeV. This corresponds to around 35.2 billion Btu of energy per pound or 0.95 MWd of energy per gram of ^{235}U undergoing fission. In addition, some ^{235}U is consumed without undergoing fission by reacting with neutrons to form ^{236}U. When this reaction is taken into account, the energy released is around 29 billion Btu per pound, or 0.78 MWd per gram of ^{235}U consumed. This is about 2 million times the energy released in the combustion of an equivalent mass of coal.

3 NUCLEAR FUELS

In addition to ^{235}U, two other isotopes can be used as fuel in nuclear fission reactors. These are plutonium-239, ^{239}Pu, produced by absorption of neutrons in ^{238}U; and ^{233}U, produced by absorption of neutrons in natural thorium. The reactions by which these isotopes are made are as follows:

$$\begin{array}{l} {}^{238}\mathrm{U} + \underset{\text{Neutron}}{{}^{1}n} \rightarrow {}^{239}\mathrm{U} \rightarrow {}^{239}\mathrm{Np} + e^- \\ \qquad\qquad\qquad\qquad\qquad\quad \downarrow \\ \qquad\qquad\qquad\qquad\qquad {}^{239}\mathrm{Pu} + e^- \\ \qquad\qquad\qquad\qquad\qquad\qquad\quad \text{Beta particles} \\ {}^{232}\mathrm{Th} + {}^{1}n \rightarrow {}^{233}\mathrm{Th} \rightarrow {}^{233}\mathrm{Pa} + e^- \\ \qquad\qquad\qquad\qquad\qquad\quad \downarrow \\ \qquad\qquad\qquad\qquad\qquad {}^{233}\mathrm{U} + e^- \end{array}$$

Properties of these three fissile fuel nuclides are listed in Table 1.1.

The number of neutrons produced per neutron absorbed by fissile material is less than the number of neutrons produced per fission because some of the neutrons absorbed produce the higher isotopes ^{236}U, ^{240}Pu, or ^{234}U rather than causing fission.

Table 1.1 Nuclear fuels

Isotope	^{235}U	^{239}Pu	^{233}U
Obtained from	0.7% of natural uranium	Absorption of neutrons by ^{238}U	Absorption of neutrons by ^{232}Th
Neutrons produced per			
Fission	2.418	2.871	2.492
Thermal† neutron absorbed	1.96	1.86	2.2
Absorption cross section, b:			
Thermal† neutrons	555	1618	470
Fast neutrons	1.5	2	2

†In a typical reactor for power production.

The fact that the number of neutrons produced per neutron absorbed exceeds 1.0 for each fuel indicates that each will support a nuclear chain reaction. Neutrons in excess of the one needed to sustain the nuclear chain reaction may be used to produce new and valuable isotopes, for example, to produce ^{239}Pu from ^{238}U or ^{233}U from thorium by the reactions cited earlier.

When the number of neutrons produced per neutron absorbed in fissile material is greater than 2.0, it is theoretically possible to generate fissile material at a faster rate than it is consumed. One neutron is used to maintain the chain reaction, and the second neutron is used to produce a new atom of fissile material to replace the atom that is consumed by the first neutron. This process is known as breeding. The reactions taking place in breeding ^{239}Pu from ^{238}U are shown in Fig. 1.6. ^{238}U is the only material consumed over all; ^{239}Pu is produced from ^{238}U and then consumed in fission.

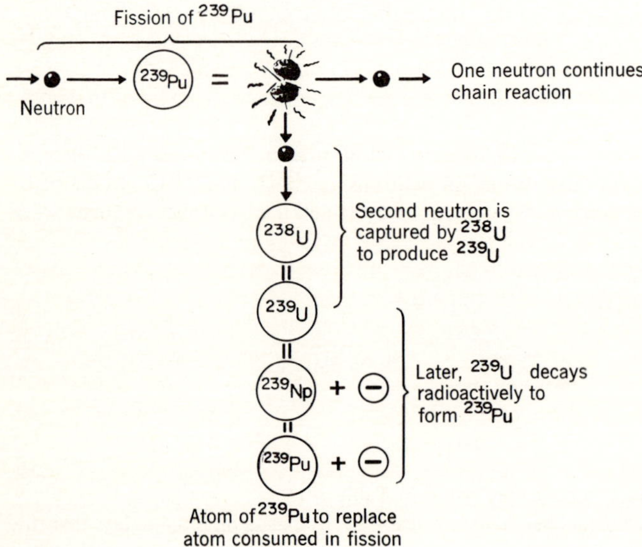

Figure 1.6 Breeding of ^{239}Pu.

In thermal reactors fueled with plutonium, the number of neutrons produced per neutron absorbed is less than 2.0 and breeding is impossible. For ^{233}U, on the other hand, this number is substantially greater than 2.0, and breeding is practicable in a thermal reactor. In fast reactors, the number of neutrons produced per neutron absorbed is close to the total number of neutrons produced per fission, so that breeding is possible with both ^{233}U and plutonium. Breeding as here defined is not possible with ^{235}U, because there is no naturally occurring isotope from which ^{235}U can be produced.

A fast reactor is one in which the average speed of neutrons is near that which they have at the moment of fission, around 15 million m/s. At these high speeds the probability of a neutron's being absorbed by a fissionable atom is low, and the neutron-absorption cross section, which is a measure of this probability, is small.

A thermal reactor is one in which the neutrons have been slowed down until they are in thermal equilibrium with reactor materials; in a typical power reactor, thermal neutrons have speeds around 3000 m/s. At these lower speeds, the neutron-absorption cross sections are much larger than for fast neutrons.

The critical mass of fissile material required to maintain the fission process is roughly inversely proportional to the neutron-absorption cross section. Thus the critical mass is lowest for plutonium in thermal reactors, larger for the uranium isotopes in thermal reactors, and much greater in fast reactors. For this reason, as well as others, thermal reactors are the preferred type except when breeding with plutonium is an objective; then a fast reactor must be used.

4 NUCLEAR REACTOR TYPES

In addition to classifying nuclear reactors as thermal or fast, they may be characterized by their purpose, by the type of moderator used to slow down neutrons, by the type of coolant, or by the type of fuel. The principal purposes for which reactors may be used are for research, testing, production of materials such as radioisotopes or plutonium, or power generation. This text is concerned mainly with power reactors.

The most effective substances for slowing down neutrons are those elements of low molecular weight that have low probability of capturing neutrons, namely, hydrogen, deuterium (the hydrogen isotope of atomic mass 2, chemical symbol D), beryllium, or carbon. Examples of moderators containing these elements are light water (H_2O), heavy water (D_2O), beryllium oxide, and graphite.

In many types of thermal power reactors, moderator, fuel, and coolant are kept separate in the reactor. Figure 1.7 is a schematic diagram of a nuclear power plant utilizing such a reactor. Table 1.2 lists five examples of reactors with separate moderator, fuel, and coolant and gives references where more detailed information about these reactors may be obtained. In this type of reactor, fuel and moderator ordinarily remain in place in the reactor and only coolant flows through the reactor to remove the heat of fission. Hot coolant flows from the reactor to a steam generator, where it is cooled by heat exchange with feedwater. The feedwater is converted to steam, which drives a steam turbine. The steam then is condensed, preheated, and recirculated as feedwater to the steam generator. Coolant, after being cooled in the steam generator, is returned to the reactor by the coolant circulator. The steam turbine drives an electric generator.

When H_2O is used as coolant, the same material serves also as moderator, so that the reactor structure can be simplified. Figure 1.8 is a schematic diagram of a pressurized-water reactor, in which the coolant and moderator consist of liquid water whose pressure of 150 bar (2200 lb/in^2) is so high that it remains liquid at the highest temperature, around 300°C (572°F), to which it is heated in the reactor. The main difference in principle from Fig. 1.7 is

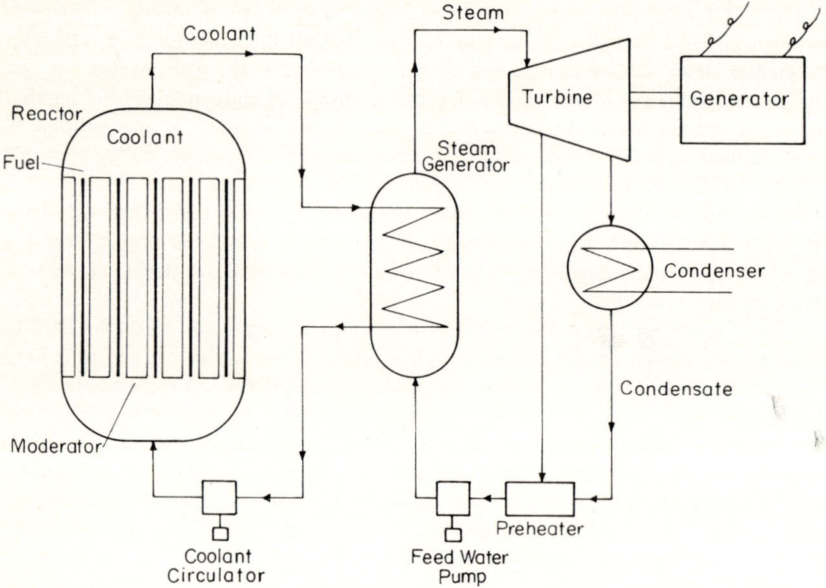

Figure 1.7 Schematic of nuclear power plant with separate fuel, moderator, and coolant.

that there is no separation of coolant from moderator in the reactor. The pressurized-water reactor is one of the two types of power reactor in most common use in the United States. More information about it is given in Chap. 3.

The boiling-water reactor is the other type of power reactor in common use in the United States that uses H_2O as coolant and moderator. In this type the water in the reactor is at a lower pressure, around 70 bar (1000 lb/in^2), so that it boils and is partially converted to steam as it flows through the reactor. Coolant leaving the reactor is separated into water, which is recycled, and steam, which is sent directly to the turbine as illustrated in Fig. 1.9. Comparison with Fig. 1.8 shows that the boiling-water system differs from the pressurized-water system in having no external steam generator, the reactor itself providing this function.

In a fast-breeder reactor it is impractical to use water as coolant because it is too effective a moderator for neutrons. Liquid sodium is the coolant most extensively investigated for fast

Table 1.2 Examples of nuclear power reactors with separate fuel, moderator, and coolant

	Gas-cooled reactor	Advanced gas-cooled reactor	High-temperature gas-cooled reactor	Heavy-water reactor	Heavy-water organic-cooled reactor
Fuel form	U alloy	UO_2	ThC_2+UC_2	UO_2	UO_2
Enrichment	Natural U	2% ^{235}U	93% ^{235}U	Natural U	0.7–2% ^{235}U
Cladding	Mg alloy	Stainless	Graphite	Zircaloy	Zircaloy
Moderator	Graphite	Graphite	Graphite	D_2O	D_2O
Coolant	CO_2	CO_2	He	D_2O	Terphenyl
Control material	B	B	B_4C	B_4C	B_4C
Reference	[L1]	[C2]	[S1]	[C1]	[E2]

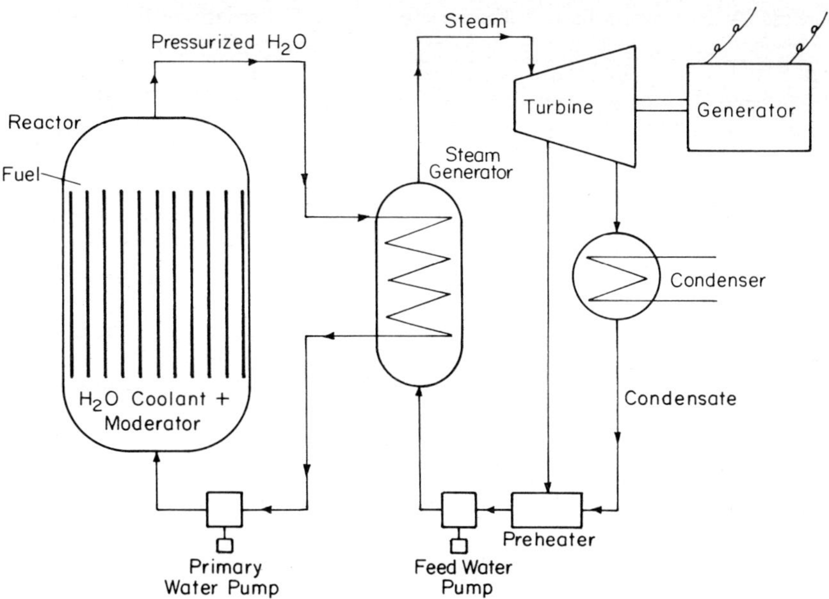

Figure 1.8 Schematic of pressurized-water nuclear power plant.

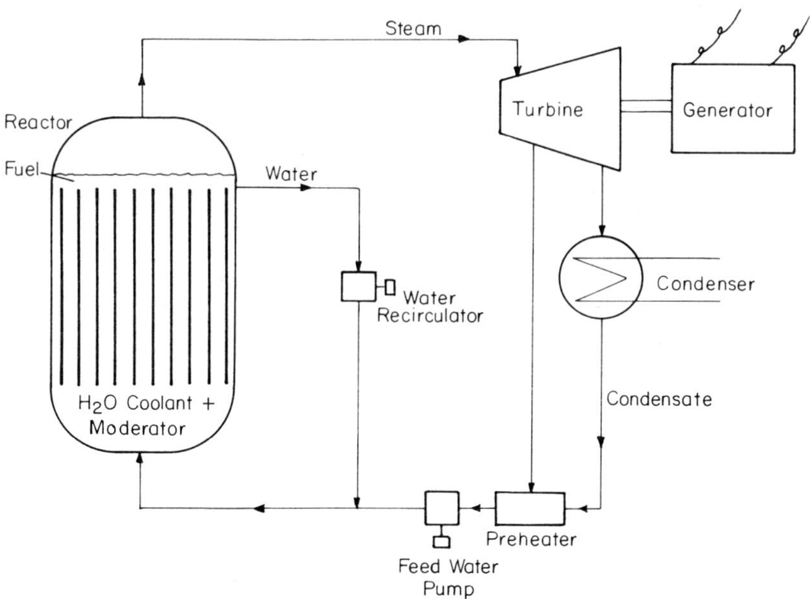

Figure 1.9 Schematic of boiling-water nuclear power plant.

reactors; helium gas has also been proposed. Fast reactors need a higher ratio of fissile to fertile isotopes than thermal reactors to support a chain reaction; a mixture of 20 percent plutonium and 80 percent ^{238}U is typical for a fast-reactor fuel. Mixed dioxides or mixed monocarbides are possible fuel materials. Although natural boron, which contains around 20 percent of the strong neutron-absorbing isotope ^{10}B, is satisfactory for control material in thermal reactors, concentrated ^{10}B is preferred for some fast reactors.

The molten-salt reactor differs from all reactors thus far described in that it uses a liquid solution of uranium as fuel and removes heat from the reactor by circulating hot fuel to an external heat exchanger. No reactor coolant is employed other than the fuel itself. The molten-salt breeder reactor (MSBR) uses as fuel a solution of UF_4 in a solvent salt consisting of mixture of BeF_2, ^7LiF, and ThF_4. Separated ^7Li is required instead of natural lithium because the 7.5 percent of ^6Li in natural lithium would absorb so many neutrons as to make breeding impossible. The MSBR is a thermal reactor that breeds ^{233}U from thorium; neutrons are thermalized by means of graphite moderator blocks, fixed in the reactor, containing channels through which the molten salt flows.

Table 1.3 summarizes the materials used for the principal services in pressurized-water and boiling-water reactors, the high-temperature gas-cooled reactor, fast reactors, and the molten-salt reactor, and indicates which materials are fixed in each reactor and which flow through it.

5 FUEL PROCESSING FLOW SHEETS

5.1 Uranium Fuel

The fuel processing operations to be used in conjunction with a nuclear power reactor and the amount of nuclear fuel that must be provided depend on the type of reactor and on the extent to which fissile and fertile constituents in spent fuel discharged from the reactor are to be recovered for reuse. Figures 1.10 and 1.11 outline representative fuel processing flow sheets for uranium-fueled thermal reactors generating 1000 MW of electricity, at a capacity factor of 80 percent.

Table 1.3 Materials for light-water, fast-breeder, and molten-salt reactors

	Pressurized-water reactor	Boiling-water reactor	Liquid-metal fast-breeder reactor	Gas-cooled fast-breeder reactor	Molten-salt breeder reactor†
Fuel	UO_2, 3.3% ^{235}U	UO_2, 2.6% ^{235}U	20% PuO_2–80% $^{238}UO_2$	20% PuO_2–80% $^{238}UO_2$	71.7 m/o ^7LiF 16 m/o BeF_2 12 m/o ThF_4 0.3 m/o $^{233}UF_4$
Cladding	Zircaloy	Zircaloy	Stainless	Stainless	None
Moderator	H_2O	H_2O	None	None	Graphite
Coolant	H_2O	H_2O	Na	He	Fuel
Control material	Hf or Ag-In-Cd	B_4C	B_4C or $^{10}B_4C$	B_4C or $^{10}B_4C$	B_4C
Fixed in reactor	Fuel	Fuel	Fuel	Fuel	Moderator
Circulating	Coolant and moderator		Coolant	Coolant	Fuel
Reference	[C3]	[C3]	[A1]	[E1]	[B1]

†m/o = mole percent.

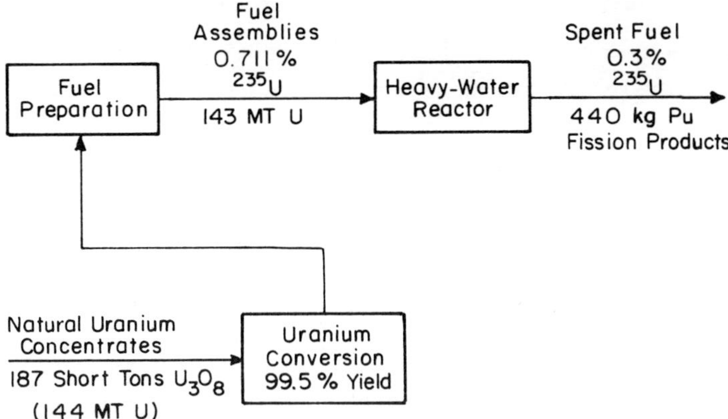

Figure 1.10 Fuel processing flow sheet for 1000-MWe heavy-water reactor. Basis: 1 year, 80 percent capacity factor.

The simplest flow sheet, Fig. 1.10, is applicable to heavy-water reactors fueled with natural uranium containing 0.711 w/o ^{235}U.† Feed preparation for this type of reactor consists of purifying natural uranium concentrates, converting the uranium to UO_2, and fabricating the UO_2 into fuel elements. In this type of heavy-water reactor, fission of ^{235}U initially present in the feed and fission of plutonium formed from ^{238}U will produce about 6800 MWd of heat per metric ton (1 MT = 1000 kg) of fuel before the fuel is so depleted in fissile material and so loaded with neutron-absorbing fission products that the reactor is no longer critical. Since the heat of fission is 0.95 MWd/g, complete utilization of 1 MT of fuel would generate 950,000 MWd of heat. In this type of thermal reactor, thus, 6800/950,000 = 0.0072 fraction of the natural uranium, about 0.7 percent, is converted to heat.

As the efficiency of conversion of heat to electricity in a heavy-water nuclear power plant is about 30 percent, the rate at which a 1000-MW plant would have to be supplied with natural uranium is

$$\frac{(0.8)(1000 \text{ MW})(1000 \text{ kg/MT})}{(6800 \text{ MWd/MT})(0.3)} = 392 \text{ kg/day} \tag{1.4}$$

or 143 MT of uranium per year.

In commercial transactions uranium concentrates are measured in short tons (2000 lb) of U_3O_8. In this unit, the annual uranium consumption of this reactor would be

$$\frac{(143 \text{ MT U})(1.1023 \text{ short tons/MT})(842 \text{ MT } U_3O_8/714 \text{ MT } U_3)}{0.995} = 187 \text{ short tons } U_3O_8 \tag{1.4a}$$

assuming 99.5 percent uranium recovery in conversion.

Spent fuel discharged from this reactor contains about 0.2 w/o plutonium and about 0.3 w/o ^{235}U. This content of fissile material is so low that its recovery is hardly economical, so that no recovery step has been shown.

Figure 1.11 shows three possible fuel processing flow sheets for reactors cooled and moderated by light water. The specific example shown is for a pressurized-water reactor. Fuel for this type of reactor consists of UO_2 enriched to around 3.3 w/o in ^{235}U. The expected performance of this type of reactor is described in some detail in Chap. 3, Sec. 7. After

†w/o = weight percent.

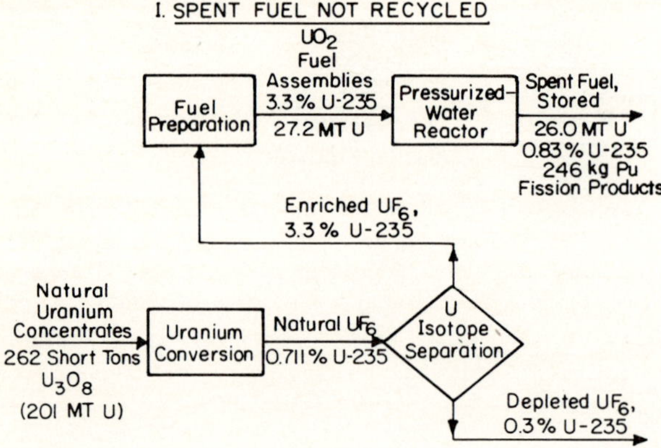

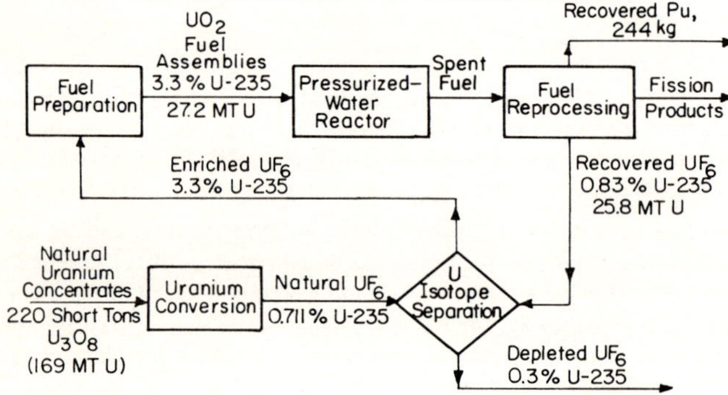

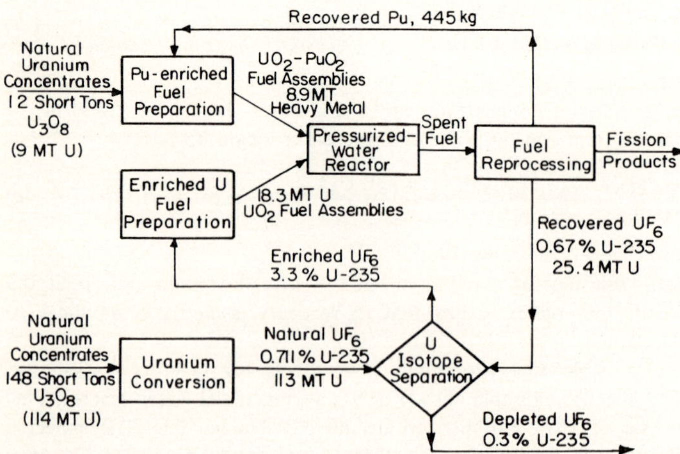

Figure 1.11 Fuel processing flow sheets for 1000-MWe pressurized-water reactor. Basis: 1 year, 80 percent capacity factor.

producing 33,000 MWd of heat per metric ton, the fuel ceases to support the fission chain reaction and must be discharged from the reactor. This spent fuel still contains around 0.83 w/o ^{235}U and about 0.6 w/o fissile plutonium. In part I of Fig. 1.11 this spent fuel is stored without reprocessing, as in the heavy-water reactor example of Fig. 1.10. The annual consumption of U_3O_8 for the light-water reactor, without reprocessing, is 262 short tons U_3O_8, substantially greater than for the heavy-water reactor.

Under some conditions it is economically attractive or environmentally preferable to reprocess spent fuel in order to (1) recover uranium to be recycled to provide part of the enriched uranium used in subsequent lots of fuel, (2) recover plutonium, and (3) reduce radioactive wastes to more compact form. In part II of Fig. 1.11 the recovered 0.83 percent enriched uranium is recycled and the 244 kg of plutonium recovered per year is stored for later use in either a light-water reactor or a fast-breeder reactor. This recycle of uranium to the isotope separation plant reduces the annual U_3O_8 feed rate to 220 short tons, still appreciably greater than for the heavy-water reactor.

In part III of Fig. 1.11, the recovered uranium is recycled and reenriched and the recovered plutonium is recycled to provide part of the fissile material in the reactor fuel assemblies. Two kinds of fuel assemblies are used. One kind is the same as used in cases I and II, which consist of UO_2 enriched to 3.3 w/o ^{235}U. The annual feed rate of these assemblies is 18.3 MT of enriched uranium. The other kind consists of mixed uranium and plutonium dioxides, in which the uranium is in the form of natural UO_2. Their annual feed rate is 8.9 MT of heavy metal (uranium plus plutonium), including 445 kg of recycle plutonium. The total annual U_3O_8 feed rate is 160 short tons, which is less than for the heavy-water reactor of Fig. 1.10.

In part III of Fig. 1.11, the 160 short tons of U_3O_8 consumed per year corresponds to a daily feed rate of 341 kg natural uranium. As this pressurized-water nuclear power plant has a thermal efficiency of 32.5 percent, the fraction of the natural uranium feed converted to energy is

$$\frac{(0.8)(1000\ \text{MW}/0.325)}{(341\ \text{kg/day})(950\ \text{MWd/kg})} = 0.0076 \tag{1.5}$$

Even with plutonium recycle, thus, this thermal reactor converts less than 1 percent of natural uranium to energy. This low uranium utilization results from the fact that the conversion ratio of ^{238}U to plutonium in a thermal reactor is less than unity.

In a fast reactor, on the other hand, the conversion ratio can be greater than unity, and almost all of the uranium can be converted to energy, in principle. Figure 1.12 shows the fuel processing operations associated with a fast-reactor power plant breeding plutonium from ^{238}U. Because of the low absorption cross section of plutonium for fast neutrons, it is necessary to use a mixture of about 20 percent plutonium and 80 percent ^{238}U in the core of such a reactor and to surround the core with a blanket of natural or depleted uranium to absorb neutrons leaking from the core and convert them to plutonium. Two types of fuel elements must be prepared for a fast-breeder reactor, then, blanket elements fabricated from natural or depleted uranium, and core elements containing around 20 w/o plutonium. Most fast reactors under development propose use of mixed PuO_2-UO_2 for core elements; mixed PuC-UC is also being considered. The core elements of a fast reactor are expected to generate from about 65,000 to 100,000 MWd of heat per metric ton before discharge; as they still contain nearly their original plutonium content, reprocessing is required. The blanket elements also must be reprocessed for plutonium recovery. Some savings can be effected by reprocessing both types of elements together, as shown in Fig. 1.12. Uranium recovered in the reprocessing plant can be recycled to provide most of the uranium used to prepare core and blanket elements. Plutonium recovered in the reprocessing plant provides all the enrichment needed for core elements, plus the net production of plutonium from the plant. With good conservation of neutrons in the reactor and efficient recovery of plutonium in reprocessing and core fabrication, a 1000-MWe fast-reactor

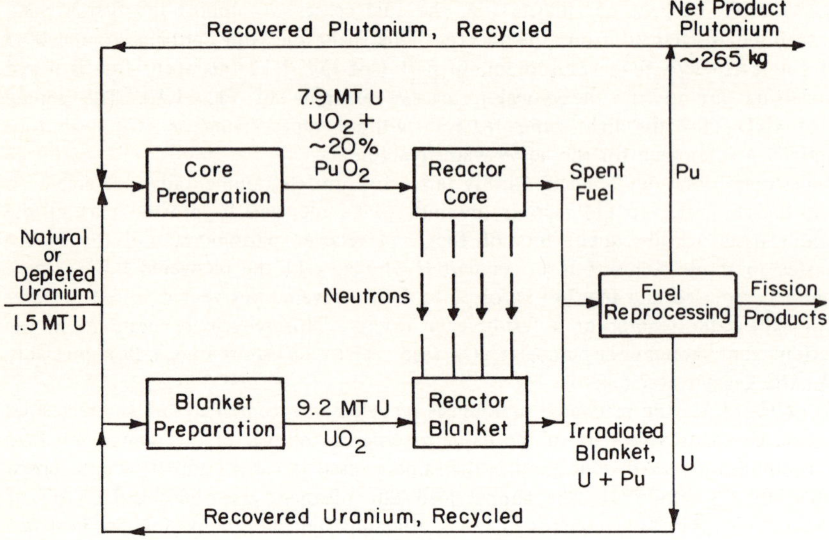

Figure 1.12 Fuel processing flow sheet for 1000-MWe fast-breeder reactor. Basis: 1 year, 80 percent capacity factor.

power plant is expected to breed about 265 kg/year of net plutonium product. A fast-reactor power plant cooled with sodium or helium is expected to have a thermal efficiency of 40 percent. If it could convert 100 percent of its uranium feed to heat, a 1000-MWe plant would consume only

$$\frac{1000 \text{ MW}/0.4}{950 \text{ MWd/kg}} = 2.6 \text{ kg/day} \tag{1.6}$$

of uranium. Because of reprocessing losses and conversion of some uranium to nonfissile isotopes, the uranium consumption of a practical fast-breeder system is expected to be somewhat greater, perhaps 4 kg/day, or 1.5 MT/uranium/year. This is much less than for a thermal reactor, and could be in the form of the depleted uranium tailings from the isotope separation plant of Fig. 1.11.

5.2 Thorium Fuel

Figure 1.13 shows fuel processing arrangements needed for the two types of thorium-fueled reactors mentioned in Sec. 4. As the conversion ratio of the high-temperature gas-cooled reactor (HTGR) is slightly less than unity, feed for this reactor consists of thorium plus some highly enriched ^{235}U from a uranium isotope separation plant. In the fuel preparation operation thorium, enriched UF_6, and uranium recovered from spent fuel and recycled are formed into fuel elements consisting of the carbides ThC_2 and UC_2 or the oxides ThO_2 and UO_2 clad with graphite. Fuel processing after irradiation consists of burning the carbon out of the fuel, followed by separation of the mixed oxides by solvent extraction into uranium to be recycled and radioactive fission products and thorium to be stored. The recycled uranium is a mixture of isotopes, mostly ^{233}U formed by absorption of neutrons in thorium. More detail is given in Chap. 3.

Fuel processing operations for the molten-salt breeder reactor are simpler in principle than for the HTGR. As the conversion ratio is expected to be above unity, no fissile feed is needed after the reactor and its fuel cycle are in steady state. As the reactor uses fluid fuel, no fuel fabrication is required. Net feed for the reactor consists merely of ThF_4, to replace thorium converted to ^{233}U, and BeF_2 and ^7LiF, to replace solvent salt withdrawn from the reactor to purge certain fission products. Fuel reprocessing for this reactor is conducted by high-temperature, nonaqueous methods. These methods remove fission products and net bred uranium and return the fissile uranium to solution in the molten salt, so that no reenrichment or fabrication of the recycle uranium is required.

6 FUEL-CYCLE OPERATIONS

Individual operations making up the nuclear fuel cycle for light-water power reactors of the type developed in the United States are shown in the pictorial flow sheet, Fig. 1.14. This follows case II of Fig. 1.11.

The first step is mining of uranium ore, which typically contains only a few pounds of uranium per ton. Uranium values in the ore are concentrated in a uranium mill, which is located near the mine, in order to reduce subsequent shipping charges. Concentration processes frequently used include leaching, precipitation, solvent extraction, and ion exchange. The principles of solvent extraction are described in Chap. 4; applications of solvent extraction and ion exchange to uranium ore processing are taken up in Chap. 5. Uranium concentrates are

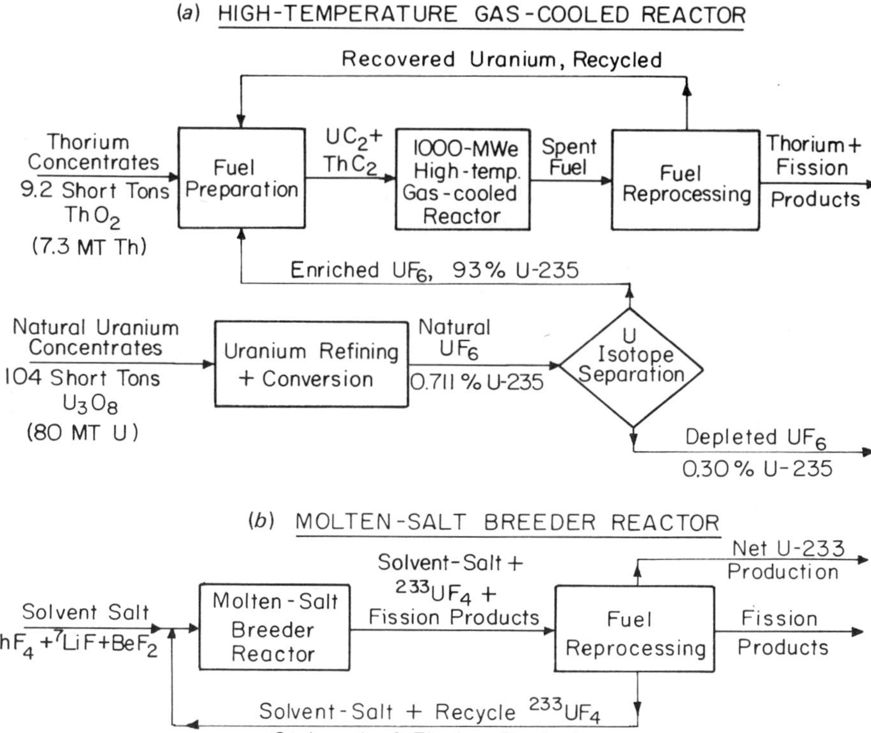

Figure 1.13 Fuel processing flow sheets for reactors using thorium as fertile material. Basis: 1 year, 80 percent capacity factor.

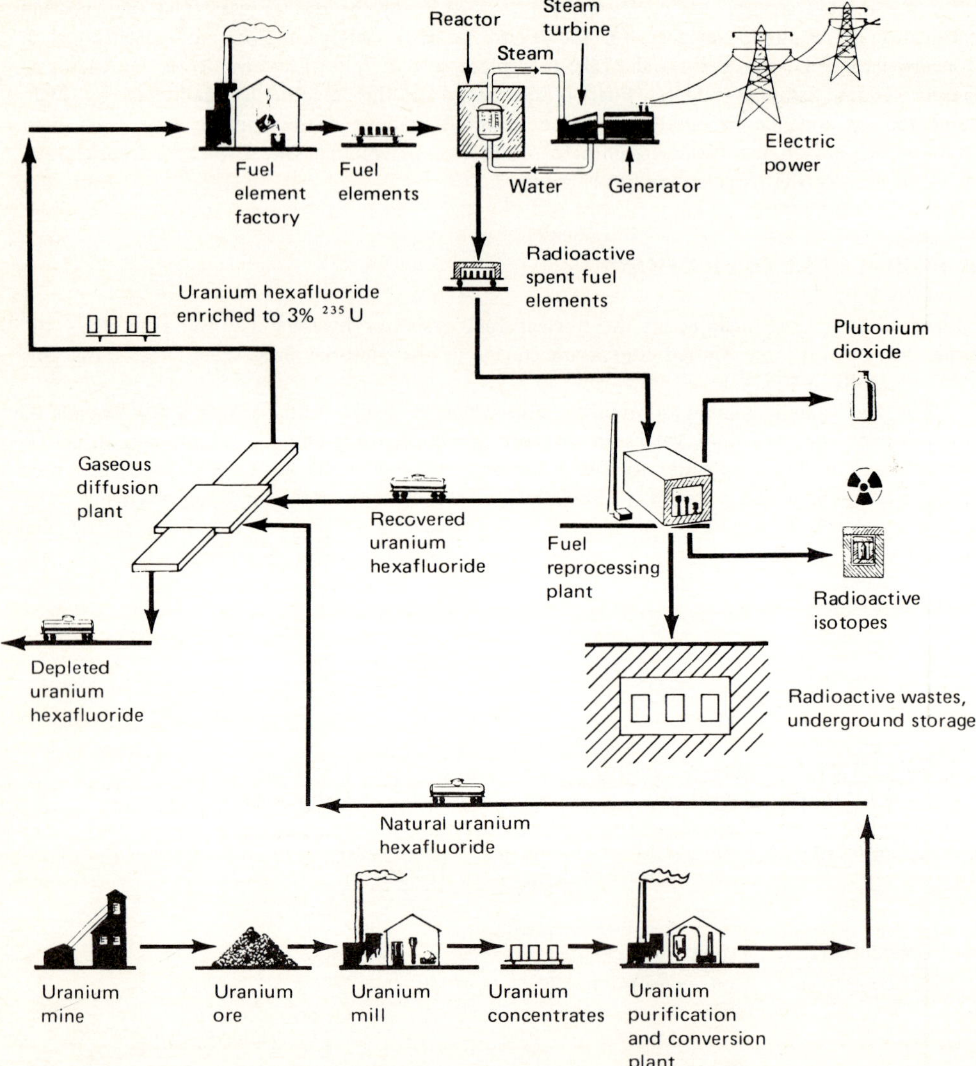

Figure 1.14 Fuel-cycle operations for light-water reactor.

known commercially as "yellow cake," because the sodium diuranate or ammonium diuranate commonly produced by uranium mills is a bright yellow solid. Figure 1.15 is a photograph of the uranium mill of Union Carbide Corporation.

Concentrates are shipped from the uranium mill to a uranium refinery or conversion plant. Here chemical impurities are removed and the purified uranium is converted into the chemical form needed for the next step in the fuel cycle. Figure 1.14 shows concentrates being converted into uranium hexafluoride (UF_6), the form used as process gas in the gaseous diffusion process for enriching ^{235}U. Other possible products of a uranium refinery used in other fuel cycles are uranium metal, uranium dioxide, or uranium carbide. Uranium purification and conversion processes are also described in Chap. 5.

Light-water reactors must be supplied with uranium having a higher content of fissile material than the 0.711 w/o ^{235}U present in natural uranium. This can be done by enriching ^{235}U in an isotope separation plant as depicted in Fig. 1.14, by adding plutonium to natural uranium, or by some combination. The gaseous diffusion process is the principal process that has been used thus far for enrichment of uranium on a commercial scale. As working fluid it uses UF_6, the only stable compound of uranium that is volatile at room temperature. UF_6 melts at 64°C, at which its vapor pressure is 1.5 atm. Natural UF_6 is shipped in large steel cylinders. As UF_6 reacts readily with water and organic materials, it must be handled in clean equipment, out of contact with moist air.

A gaseous diffusion plant consists of many gaseous diffusion stages connected in series. Each stage contains many porous tubes made of membranes with very fine holes, termed diffusion barriers. UF_6 gas at a relatively high pressure flows along the inner wall of these tubes, whose outer wall is maintained at a relatively low pressure. The UF_6 gas flowing through the tube wall is slightly enriched in ^{235}U relative to the gas remaining on the high-pressure side. Since one gaseous diffusion stage can increase the ratio of ^{235}U to ^{238}U by no more than a factor of 1.0043, it is necessary to repeat the process in hundreds of stages to obtain a useful

Figure 1.15 Uranium mill of Union Carbide Corporation, Uravan, Colorado. (*Courtesy of Union Carbide Corporation.*)

degree of separation, recompressing the UF_6 between stages. Large quantities of UF_6 must be recycled, and the power consumption is enormous. To produce 1 kg of uranium enriched to 3 percent ^{235}U while stripping natural uranium to 0.2 percent requires about 13,000 kWh of electric energy. The U.S. Atomic Energy Commission built three large gaseous diffusion plants at a cost of $2.3 billion. When operated at capacity they consume 6000 MW of electric power. Figure 1.16 is a photograph of the plant at Oak Ridge, Tennessee. The large number of stages is suggested by the repetition of the basic building structure. These plants and the gaseous diffusion process are described in more detail in Chap. 14.

Enriched UF_6 is shipped to the plant for fabricating reactor fuel elements in monel cylinders whose size is determined from the ^{235}U content, so as to prevent accumulation of a critical mass. At the fuel fabrication plant UF_6 is converted to UO_2 or other chemical form used in reactor fuel. For light-water reactors the UO_2 is pressed into pellets, which are sintered, ground to size, and loaded into zircaloy tubing, which is filled with helium and closed with welded zircaloy end plugs. These individual fuel rods are assembled into bundles, constituting the fuel elements shipped to the reactor. Conversion of UF_6 to UO_2 is described in Chap. 5. Extraction of zirconium from its ores and separation of zirconium from its companion element hafnium is described in Chap. 7.

The length of time that fuel can be used in a reactor before it must be discharged depends on the characteristics of the reactor, the initial composition of the fuel, the neutron flux to which it is exposed, and the way in which fuel is managed in the reactor, as described in more detail in Chap. 3. Factors that eventually require fuel to be discharged include deterioration of cladding as a result of fuel swelling, thermal stresses or corrosion, and loss of nuclear reactivity

Figure 1.16 Gaseous diffusion plant of U.S. Department of Energy, Oak Ridge, Tennessee. (*Courtesy of U.S. Atomic Energy Commission.*)

Figure 1.17 Purex plant of U.S. Department of Energy, Hanford, Washington. (*Courtesy of Atlantic Richfield Hanford Company.*)

as a result of depletion of fissile material and buildup of neutron-absorbing fission products. A typical fuel lifetime is 3 years.

When spent fuel is discharged from the reactor, it contains substantial amounts of fissile and fertile material, which, in the case of light-water reactors, are valuable enough to offset part or all of the cost of reclamation. Because of the fission products, spent fuel is intensely radioactive, with activities of 10 Ci/g† being common. Spent fuel is usually held in cooled storage basins at the reactor site for 150 days or more to allow some of the radioactivity to decay. If to be reprocessed, spent fuel would be shipped in cooled, heavily shielded casks, strong enough to remain intact in a shipping accident.

In the fuel reprocessing plant, fuel cladding is removed chemically or mechanically, the fuel material is dissolved in acid, and fissile and fertile materials are separated from fission products and from each other. The Purex process, commonly used in reprocessing plants, is described at somewhat greater length in Sec. 7, below, and in more detail in Chap. 10. Figure 1.17 is a photograph of the Purex plant of the U.S. Department of Energy at Hanford, Washington. The massive, windowless, concrete building is characteristic of these radiochemical fuel reprocessing plants. In the case of light-water reactor fuel, the most valuable products of the fuel reprocessing plant are plutonium, usually in the form of a concentrated aqueous solution of plutonium nitrate, and uranium, most conveniently in the form of UF_6. Some individual fission products such as ^{137}Cs, a valuable gamma-emitting radioisotope, may be separated for industrial or medical use. The remaining radioactive fission products are held at the reprocessing site for additional decay, then converted to solid form, packaged, and shipped to storage vaults where they

†Curies per gram.

must be kept out of human contact for thousands of years. Procedures for handling radioactive wastes are described in Chap. 11.

Plutonium nitrate from the reprocessing plant is converted to metal, oxide, or carbide and used in fuel for fast reactors or recycled to thermal reactors. UF_6 from the reprocessing plant is recycled to the gaseous diffusion plant to be reenriched in ^{235}U.

7 FUEL REPROCESSING

Because of the importance of reactor fuel reprocessing in nuclear power technology, some further discussion of this topic is warranted in this introductory chapter.

In addition to fissionable isotopes (^{235}U, ^{233}U, or plutonium) and fertile isotopes (^{238}U or thorium), spent fuel from a reactor contains a large number of fission product isotopes, in which all elements of the periodic table from zinc to gadolinium are represented. Some of these fission product isotopes are short-lived and decay rapidly, but a dozen or more need to be considered when designing processes for separation of reactor products. The most important neutron-absorbing and long-lived fission products in irradiated uranium are listed in Table 1.4.

Processing of spent reactor fuels is made especially difficult by their intense radioactivity. The process equipment must be surrounded by massive shielding, provision must be made to remove the substantial amounts of heat that are associated with this radioactivity, and in some instances damage to solvents and construction materials from the radiations emitted by the materials being processed is a problem. Another difficulty is the critical-mass hazard, which is present whenever fissionable material is handled at substantial concentrations. This often requires a limitation in the size of batches being processed or in the dimensions of individual pieces of equipment. A third difficulty is the high degree of recovery that is usually required because of the great value of the fissionable materials being processed. A fourth is the high degree of separation specified for the removal of radioactive fission products; in present

Table 1.4 Important isotopes in irradiated uranium

Heavy elements		Long-lived radioactive fission products	
Uranium	235, 236, 238	Krypton	85
Plutonium	239, 240, 241	Strontium	89, 90
		Yttrium	90, 91
Neutron-absorbing fission products		Zirconium	95
		Niobium	95
Technetium	99	Molybdenum	99
Rhodium	103	Technetium	99
Xenon	131, 133, 135	Ruthenium	103, 106
Neodymium	143, 145	Rhodium	106
Samarium	149, 151	Tellurium	129
Europium	155	Iodine	129, 131
Gadolinium	155	Xenon	133
		Cesium	137
		Barium	140
		Lanthanum	140
		Cerium	141, 144
		Praseodymium	143, 144
		Neodymium	147
		Promethium	147

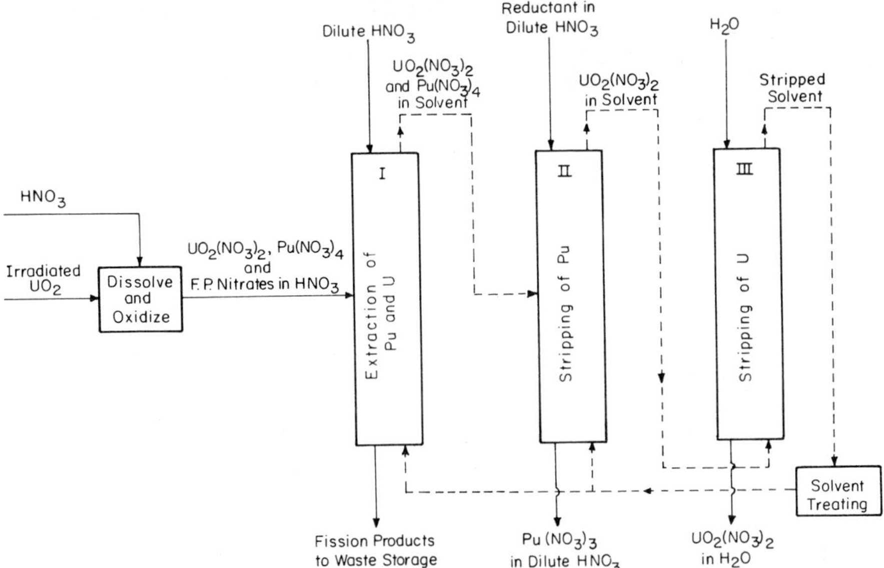

Figure 1.18 Principle of Purex process.

processes it is necessary to reduce the concentration of some of these elements by a factor of 10 million. Another difficulty is the large number of components present, with elements of such diverse properties as the alkali cesium and the manufactured elements technetium (resembling manganese) and promethium (one of the rare earths). A final difficulty, and one that was not originally anticipated, is the chemical similarity between uranium and plutonium.

The principle of the Purex process, now commonly used for processing irradiated uranium by solvent extraction, is illustrated in Fig. 1.18. The solvent used in this process is a solution of tributyl phosphate (TBP) in a high-boiling hydrocarbon, frequently *n*-dodecane or a mixture of similar hydrocarbons. TBP forms complexes with uranyl nitrate $[UO_2(NO_3)_2]$ and tetravalent plutonium nitrate $[Pu(NO_3)_4]$ whose concentration in the hydrocarbon phase is higher than in an aqueous solution of nitric acid in equilibrium with the hydrocarbon phase. On the other hand, TBP complexes of most fission products and trivalent plutonium nitrate have lower concentrations in the hydrocarbon phase than in the aqueous phase in equilibrium.

In the Purex process, irradiated UO_2 is dissolved in nitric acid under such conditions that uranium is oxidized to uranyl nitrate and plutonium to $Pu(NO_3)_4$. The resulting aqueous solution of uranyl, plutonium, and fission-product nitrates is fed to the center of countercurrent solvent extraction contactor I, which may be either a pulse column or a battery of mixer-settlers. This contactor is refluxed at one end by clean solvent and at the other by a dilute nitric acid scrub solution. The solvent extracts all the uranium and plutonium from the aqueous phase and some of the fission products. The fission products are removed from the solvent by the nitric acid scrub solution. Fission products leave contactor I in solution in aqueous nitric acid.

Solvent from contactor I containing uranyl nitrate and $Pu(NO_3)_4$ is fed to the center of contactor II. This is refluxed at one end by clean solvent and at the other by a dilute nitric acid solution of a reducing agent strong enough to reduce plutonium to the trivalent form, but not so strong as to reduce uranium from the hexavalent form. Ferrous sulfamate is frequently used. In contactor II plutonium is transferred to the aqueous phase, while uranium remains in the solvent. Solvent from contactor II is fed to one end of contactor III, which is stripped at the other end by water, which transfers the uranium to the aqueous phase leaving the contactor.

After chemical treatment to remove degradation products, the solvent leaving contactor III is reused in contactors I and II.

This brief discussion of the Purex process is expanded in Chap. 10, which discusses other processes for treating irradiated fuel and which deals with novel aspects of processing highly radioactive and fissile materials.

8 ISOTOPE SEPARATION

Although the isotopes of an element have very similar chemical properties, they behave as completely different substances in nuclear reactions. Consequently, the separation of isotopes of certain elements, notably ^{235}U from ^{238}U and deuterium from hydrogen, is of great importance in nuclear technology. Table 1.5 lists isotopes important in nuclear power applications, together with their natural abundance and processes that have been used or proposed for their separation. In addition to applications mentioned earlier in this chapter, Table 1.5 includes the use of ^{2}D and ^{6}Li as fuel for fusion power, a topic treated briefly in Sec. 9, following.

The fact that isotopes of an element have very similar chemical and gross physical properties makes their separation particularly difficult and has necessitated the development of concepts and processes especially adapted for this purpose. In almost all isotope separation processes the degree of separation obtainable in a single stage is very small, so that many identical stages must be used for practical, useful separation. An example of this is the use of more than 4000 stages in the Oak Ridge gaseous diffusion plant. Chapter 12 describes principles that have been developed for dealing with separation processes that consist of a large number of similar stages, and hence are applicable to all methods of isotope separation.

Table 1.5 indicates that for isotopes of the light elements hydrogen, lithium, and boron, separation methods used or proposed include distillation, electrolysis, and chemical exchange. These methods for separating isotopes of light elements are described at length in Chap. 13, with principal application to deuterium. Mention is also made of methods for concentrating ^{13}C, ^{15}N, ^{17}O, and ^{18}O. These are isotopes of elements important in living systems that are used extensively as stable tracers in biological and medical research.

None of the conventional separation processes, such as distillation, ion exchange, or solvent

Table 1.5 Isotopes in nuclear technology

Isotope	Atom percent in natural element	Use	Separation methods
^{2}D	0.015	Moderator, fuel for fusion	Distillation, electrolysis, chemical exchange
^{6}Li	7.5	Fuel for fusion	Distillation, electrolysis, chemical exchange (both ^{6}Li and ^{7}Li)
^{7}Li	92.5	Water conditioner	
^{10}B	20	Control material	Distillation, chemical exchange, ion exchange
^{235}U	0.711	Fissile material	Gaseous diffusion, laser isotope separation, gas centrifugation, aerodynamic methods (both ^{235}U and ^{238}U)
^{238}U	99.28	Fertile material	

extraction, has been used for large-scale separation for isotopes of uranium or other heavy elements. To separate isotopes of uranium or other heavy elements that exist in gaseous form at convenient temperatures, it has been necessary to use gaseous diffusion, gas centrifugation, or one of the other novel processes described in Chap. 14. Gases to which these processes are applicable include xenon, MoF_6, WF_6, and UF_6.

Another process that can be used to separate isotopes of all elements on a small scale, but that is too costly for large-scale production, is the electromagnetic method, which is based on the principle of the mass spectrometer. The electromagnetic method separated the microgram amounts of ^{235}U used to show [N1] that this was the fissile isotope of uranium and was later employed by the Manhattan District to produce the first kilogram quantities of ^{235}U. The cost was so high, however, that the electromagnetic method was replaced by gaseous diffusion. The electromagnetic method is now used [K1] to produce research quantities of separated isotopes of nearly all naturally occurring mixed elements. As the electromagnetic method is a physical rather than a chemical engineering process, it is not described further in this text.

9 NUCLEAR FUSION

When nuclei of certain light elements have speeds corresponding to temperatures of the order of tens of millions of degrees, they occasionally fuse together to form heavier elements with the concurrent release of large amounts of energy. These are the reactions from which the energy radiated by the sun and the stars is derived. The intense gravitational attraction in the sun and stars holds the reacting atoms together despite their high speed.

If fusion reactions are to be a practical method of generating energy on earth, other means than gravitational attraction must be found to confine the reacting atoms. The confinement principle on which most work is being done depends on the fact that atoms heated to the extremely high temperatures required for fusion are fully dissociated into positively charged ions and negatively charged electrons. Such a reacting mixture of positive and negative ions is called a thermonuclear plasma. By placing a plasma in a strong magnetic field, its positively and negatively charged particles are constrained to travel in helical paths around the magnetic lines of force. By proper shaping of the magnetic field, the charged particles can be confined for substantial periods of time, long enough to permit some fusion reactions to take place.

The fusion reaction easiest to bring about is between a deuterium ion (hydrogen of mass 2) and a tritium ion (hydrogen of mass 3), to produce a helium ion of mass 4 and a neutron:

$$\underset{\text{Deuterium}}{^{2}\text{D}} + \underset{\text{Tritium}}{^{3}\text{T}} \rightarrow \underset{\text{Helium}}{^{4}\text{He}} + \underset{\text{Neutron}}{^{1}n}$$

This reaction is favored because it occurs at an appreciable rate at a lower temperature (20,000,000 K) than other possible fusion reactions. Tritium is a radioactive isotope of hydrogen, with a half-life of 12 years, which does not occur significantly in nature. For use in this fusion reaction tritium must be made by reaction of the lithium isotope of mass 6 with a neutron:

$$\underset{\text{Lithium}}{^{6}\text{Li}} + \underset{\text{Neutron}}{^{1}n} \rightarrow \underset{\text{Helium}}{^{4}\text{He}} + \underset{\text{Tritium}}{^{3}\text{T}}$$

Natural lithium contains 7.5 percent ^{6}Li.

The energy released in these two reactions may be calculated from the decrease in mass between the reactants and the products:

Fusion reaction

Reactants, amu		Products, amu		Difference, amu
^{2}D	2.014102	^{4}He	4.002603	
^{3}T	3.016050	^{1}n	1.008665	
Total	5.030152		5.011268	0.018884

With the conversion factor 931.480 MeV/amu, this fusion reaction releases 17.6 MeV per pair of atoms fused.

Tritium production

Reactants, amu		Products, amu		Difference, amu
^{6}Li	6.015125	^{4}He	4.002603	
^{1}n	1.008665	^{3}T	3.016050	
Total	7.023790		7.018653	0.005137

Absorption of the neutron in ^{6}Li thus releases

$$(0.005137)(931.480) = 4.8 \text{ MeV} \tag{1.7}$$

The overall reaction is

$$\begin{array}{ccccc} ^{6}\text{Li} & + & ^{2}\text{D} & \rightarrow & 2\,^{4}\text{He} \\ 6.015125 & & 2.014102 & & 8.005206 \text{ amu} \end{array}$$

The fractional decrease in mass is

$$\frac{6.015125 + 2.014102 - 8.005206}{6.015125 + 2.014102} = 0.002992 \tag{1.8}$$

From the Einstein relation, Eq. (1.1), the energy released in this fusion reaction is

$$(0.002992)(2.997925 \times 10^{8})^{2} = 2.69 \times 10^{14} \text{ J} \tag{1.9}$$

per kilogram of ^{6}Li and deuterium reacting, or 3.11 MWd/g, or 115×10^{10} Btu/lb. This is about three times the heat of fission.

As the oceans of the world contain about 10^{17} kg of deuterium and resources of lithium minerals are of comparable magnitude, it is clear that if this fusion reaction could be utilized in a practical nuclear reactor, the world's energy resources would be enormously increased. Although intensive research is being conducted on confinement of thermonuclear plasmas, it is not yet clear whether a practical and economic fusion reactor can be developed. If fusion does become practical, isotope separation processes for extracting deuterium from natural water and for concentrating ^{6}Li from natural lithium will become of importance comparable to the separation of ^{235}U from natural uranium.

REFERENCES

A1. Argonne National Laboratory: *Proceedings of the International Conference on Sodium Technology and Large Fast Reactor Design,* Report ANL-7520, 1968, especially pp. 291–388.

B1. Bettis, E. S., and R. C. Robertson: "The Design and Performance Features of a Single-Fluid Molten-Salt Breeder Reactor," *Nucl. Appl. Tech.* **8**:190 (1970).
C1. "CANDU–Douglas Point Nuclear Power Station," *Nucl. Eng.* **9**:289 (1964).
C2. Central Electricity Generation Board, London: "Dungeness B AGR Nuclear Power Station," Report NF-15473, 1965.
C3. "Current Status and Future Technical and Economic Potential of Light Water Reactors," Report WASH-1082, Mar. 1968.
E1. "An Evaluation of Gas-Cooled Fast Reactors," Report WASH-1089, 1969.
E2. "An Evaluation of Heavy-Water-Moderated Organic-Cooled Reactors," Report WASH-1083, Mar. 1968.
H1. Hubbert, M. K.: "Energy Resources," in *Resources and Man,* National Academy of Sciences-National Research Council, NAS Publication No. 1703, 1969, chap. 8.
K1. See, for instance, Kistemaker, J., J. Bigeleisen, and A. O. C. Nier: *Proceedings of the International Symposium on Isotope Separation,* Interscience, New York, 1958, pp. 581-667.
L1. Lankton, C. S.: "Gas Cooled Reactors," in *Reactor Handbook,* vol. IV: *Engineering,* 2d ed., Interscience, New York, 1964, pp. 682-721.
N1. Nier, A. O., et al.: *Phys. Rev.* **57**: 546, 748 (1940).
S1. Stewart, H. B., and S. Jaye: "Economic and Technical Aspects of the HTGR," Report GA-7642, Jan. 1967.

PROBLEMS

1.1 In one mode of fission of ^{235}U by a slow neutron the end products are ^{98}Mo, ^{136}Xe, and two neutrons. The masses of ^{98}Mo and ^{136}Xe are 97.90541 and 135.9072 amu, respectively. How many megawatt-days of energy are released per kilogram of ^{235}U fissioned in this reaction?

1.2 Suppose that a fusion power system capable of generating electricity with a thermal efficiency of 40 percent could be developed. To supply a 1000-MWe power plant, how many kilograms of heavy water and natural lithium would be required per year?

1.3 The reaction

$$^{2}D + {}^{2}D \rightarrow {}^{3}He + {}^{1}n$$

has also been considered for a fusion power system. How many megawatt-days of heat could be obtained by fusion of the deuterium in 1 kg of natural water? The atomic mass of ^{3}He is 3.01603 amu.

CHAPTER

TWO

NUCLEAR REACTIONS

1 NUCLIDES

This chapter summarizes those aspects of nuclear physics and radiochemistry that are essential to an understanding of the chemical technology associated with nuclear reactors. No attempt is made to treat these subjects completely. A selected list of texts on nuclear physics and radiochemistry is given at the end of this chapter.

1.1 Make-up of Nuclides

A neutral atom consists of a small, dense central nucleus, about 10^{-13} cm in diameter, surrounded by a diffuse cloud of electrons whose outside diameter is around 10^{-8} cm. The nucleus contains most of the mass of the atom and carries a positive electric charge that equals a whole number times the electronic charge, 1.602101×10^{-19} C.† This whole number is called the *atomic number* Z of the atom. It is identical with the serial number of the element in the periodic table. Each nucleus is made up of Z protons and a definite number N of neutrons. The total number of particles in the nucleus, $N + Z$, is called the *mass number* and is denoted by A. The mass number turns out to be the whole number nearest to the atomic weight of the nuclide.

All neutral atoms having a given atomic number and given mass number are members of the same *nuclide* species. All atoms of a nuclide in a given energy state have the same nuclear properties, just as all atoms of an element have the same chemical properties. Nuclides having the same mass number A but different atomic numbers Z are called *isobars.* Nuclides having the same atomic number Z but different mass numbers A are called *isotopes*. Although isotopes have very similar chemical properties, their nuclear properties may be very different, e.g., ^{235}U and ^{238}U. It is customary to represent a nuclide by writing the mass number after the written

†A table of basic nuclear and physical constants is given in App. A.

chemical name or as a superscript preceding the chemical symbol; thus, the heaviest isotope of uranium would be represented as uranium-238, or ^{238}U.

The complete notation for a nuclide is $^{A}_{Z}$(element symbol). For example, $^{85}_{36}Kr$ is the isotope of krypton of mass number 85 ($Z = 36$, $A = 85$). Use of the atomic number in this symbol is redundant because all isotopes of an element have the same Z, but it is convenient in balancing equations for nuclear reactions.

In the published charts of the nuclides and in the compilation in App. C, the atomic masses are listed in physical mass units (amu), in which one atom of ^{12}C has a mass of 12.0000000.

Some nuclei with a given A and Z can exist temporarily in metastable states having more energy than the ground state, corresponding to that of A and Z. Nuclei with the same A and the same Z, but different energies, are called *isomers*. The higher-energy ones are represented by placing an m or * after the mass number, for example, $^{85m}_{36}Kr$.

1.2 Balancing Nuclear Reactions

It is characteristic of nuclear reactions of the type occurring in nuclear reactors that the sum of the number of neutrons and protons in the reactants equals the sum in the products. The same is true of the charge of the reactants and products. Consequently, in balancing nuclear reactions, the sum of the A's of the reactants must equal the sum of the A's of the products; and the sum of the Z's of the reactants must equal the sum of the Z's of the products. As an example of a balanced equation for a nuclear reaction, we may consider one of the fission reactions that occurs when ^{235}U absorbs a neutron:

$$^{235}_{92}U + ^{1}_{0}n \rightarrow ^{144}_{54}Xe + ^{89}_{38}Sr + 3^{1}_{0}n$$

The neutron is represented by $^{1}_{0}n$, a nuclide with nuclear charge 0 and mass number 1.

2 RADIOACTIVITY

2.1 Types

Radioactive nuclides break down spontaneously in six principal ways, illustrated by the following examples:

1. Alpha decay:

$$^{239}_{94}Pu \rightarrow ^{235}_{92}U + ^{4}_{2}He \qquad \text{(alpha particle)}$$

2. Beta decay:

$$^{89}_{38}Sr \rightarrow ^{89}_{39}Y\ ^{0}_{-1}e \qquad \text{(electron)}$$

3. Gamma emission:

$$^{85m}_{36}Kr \rightarrow ^{85}_{36}Kr + ^{0}_{0}\gamma \qquad \text{(gamma photon)}$$

4. Positron emission:

$$^{83}_{38}Sr \rightarrow ^{83}_{37}Rb + ^{0}_{1}e \qquad \text{(positron)}$$

$$^{0}_{1}e + ^{0}_{-1}e \rightarrow 2^{0}_{0}\gamma \qquad \text{(0.51 MeV photons)}$$

5. Electron capture:

$$^{83}_{38}Sr + ^{0}_{-1}e \rightarrow ^{83}_{37}Rb + \text{x-rays}$$

6. Spontaneous fission:

$$^{252}_{98}\text{Cf} \rightarrow \text{fission products} + \text{neutrons}$$

Some nuclides may decay alternatively in more than one way. For example, 14 percent of ^{85m}Kr decays by emission of a gamma ray, according to the above equation, and 86 percent decays by emission of a beta particle to form ^{85}Rb.

2.2 Rate of Radioactive Decay

The probability that a radioactive nucleus will decay in a given time is a constant, independent of temperature, pressure, or the decay of other neighboring nuclei. The disintegrations of individual nuclei are statistically independent events and are subject to random fluctuations. In a large number of nuclei, however, the fluctuations average out, and the fraction that decays in unit time is a constant and is numerically equal to the probability that a single nuclei will decay in that time. This rate of radioactive decay is known as the *decay constant* λ, with dimensions of reciprocal time.

Because the number of nuclei that decay in unit time is proportional to the number present, radioactive decay is a first-order reaction. If N is the number of nuclei present at time t, and if N changes with time only because of radioactive decay, then

$$\frac{dN}{dt} = -\lambda N \tag{2.1}$$

This integrates to

$$N = N^0 e^{-\lambda t} \tag{2.2}$$

where N^0 is the number of nuclei present at time zero. Thus, of N^0 nuclei originally present, $N^0 e^{-\lambda t}$ remain at time t. The number with lives between t and $t + dt$ is

$$-dN = \lambda N^0 e^{-\lambda t}\, dt \tag{2.3}$$

The mean life τ is the reciprocal of the decay constant, as may be seen from

$$\tau = \frac{1}{N^0} \int_0^{N^0} t\, dN = \int_0^{\infty} \lambda t e^{-\lambda t}\, dt = \frac{1}{\lambda} \tag{2.4}$$

It is customary to describe the specific rate of radioactive decay by the half-life $t_{1/2}$, which is the length of time required for half of the nuclei originally present to decay. The relation between the half-life and the decay constant is found from

$$\frac{N^0}{2} = N^0 e^{-\lambda t_{1/2}} \tag{2.5}$$

or

$$t_{1/2} = \frac{\ln 2}{\lambda} = \frac{0.693}{\lambda} \tag{2.6}$$

The *curie* (Ci) is a unit frequently used as a measure of the amount of radioactive material. It is defined as the amount of radioactive material that will produce 3.7×10^{10} disintegrations/s. This is approximately the number of disintegrations per second in 1 g of radium. A more up-to-date unit is the *Becquerel,* which is the amount of radioactive material that produces one disintegration per second.

Because the number of disintegrations per second in 1 g-atom is λN, where N is Avogadro's number, 6.02252×10^{23} atoms/g-atom,† the number of curies per gram of a nuclide of atomic weight M and decay constant λ is

$$\frac{\lambda N}{3.7 \times 10^{10} M} = 1.13 \times \frac{10^{13}}{t_{1/2}(\mathrm{s}) M} \tag{2.7}$$

2.3 Alpha Radioactivity

The alpha particle emitted in this type of radioactivity is a doubly charged ion of helium, $^4He^{2+}$. All alpha particles emitted by a given nuclide either have the same energy or have at most a few different energy values. Energies are in the range of 2 to 8 million electron volts (MeV), with higher energies associated with nuclides of shorter half-life.

In passing through matter, alpha particles give up their energy and become neutral helium atoms. Their range in solids and liquids is very short; an ordinary sheet of paper will stop alpha particles; the range-energy curve for air at standard conditions is shown in Fig. 2.1. Because of their short range, alpha particles do not constitute an external hazard to human beings. They are absorbed in the outer layers of the skin before they cause injury. On the other hand, if alpha-emitting elements are taken internally, they are very toxic, because of the large amount of energy released in a short distance within living tissue. For example, 1×10^{-7} g of radium is the maximum amount that may safely be allowed to accumulate in the human body.

Alpha radioactivity is found principally among elements beyond bismuth in the periodic table. All the nuclides important as fissionable or fertile material are alpha emitters, with half-lives and decay energies given in Table 2.1. These half-lives are so long that depletion of these fuel species by radioactive decay is not important, but all these nuclides are toxic, especially plutonium, which is even more toxic than radium.

†*Gram-atom* is that quantity of material whose mass in grams is equal to its atomic mass. Similarly, the mass in grams of 1 *gram-mole* of material is numerically equal to the molecular weight, and Avogadro's constant is also the number of molecules per gram-mole.

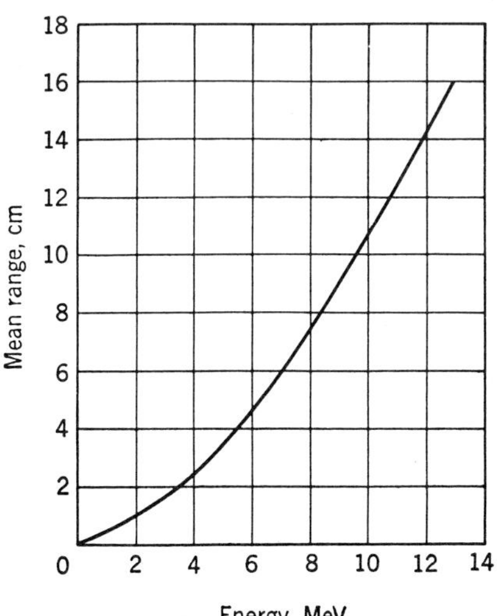

Figure 2.1 Range of alpha particles in air at 0°C, 760 Torr.

Table 2.1 Alpha energies and half-lives

Nuclide	Alpha energy, MeV	Half-life, yr
^{232}Th	3.95, 4.01, 4.08	1.41×10^{10}
^{233}U	4.78, 4.82	1.62×10^{5}
^{235}U	4.37, 4.40, 4.58	7.1×10^{8}
^{238}U	4.15, 4.20	4.51×10^{9}
^{239}Pu	5.11, 5.16	2.44×10^{4}

2.4 Beta Radioactivity

All the beta-radioactive nuclides important in nuclear reactors decay by emitting negative electrons. The daughter nuclide then has an atomic number one higher than the parent, as in the example of $^{89}_{38}Sr$ given in Sec. 2.1. Beta emission differs from alpha emission in that beta particles from a particular nuclide undergoing decay have all energies between zero and a maximum energy characteristic of that nuclide. Figure 2.2 is an example of how beta-particle energies are distributed. The average energy is usually around one-third the maximum. This distribution of energy is explained by postulating that a second particle, the neutrino, is emitted along with the electron and that the sum of the energies carried by the electron and the neutrino equals the maximum observed beta energy. The average neutrino energy is thus about twice the average electron energy.

Neutrinos carry no charge, have little if any mass, and have practically no observable effects. Their range in matter is so great that their energy cannot be utilized. They have no present practical importance.

Beta-radioactive isotopes are known for every element. The half-lives and maximum energies of a few of the most important are listed in Table 2.2, together with their source.

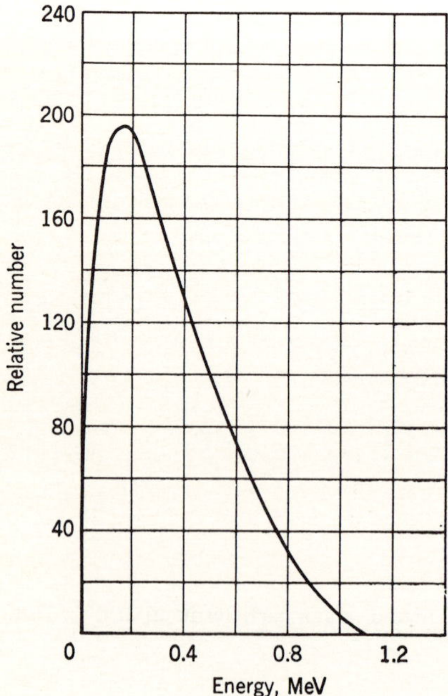

Figure 2.2 Energy distribution of beta rays from $^{210}_{83}Bi$.

Table 2.2 Beta-emitting radioactive nuclides

Nuclide	Maximum energy, MeV	Half-life	Source
$^{3}_{1}H$	0.0186	12.3 yr	$^{1}_{0}n + ^{6}_{3}Li \rightarrow ^{3}_{1}H + ^{4}_{2}He$
$^{14}_{6}C$	0.156	5730 yr	$^{1}_{0}n + ^{14}_{7}N \rightarrow ^{14}_{6}C + ^{1}_{1}H$
$^{32}_{15}P$	1.710	14.3 days	$^{1}_{0}n + ^{31}_{15}P$
$^{35}_{16}S$	0.167	88 days	$^{1}_{0}n + ^{35}_{17}Cl \rightarrow ^{35}_{16}S + ^{1}_{1}H$
$^{40}_{19}K$	1.314	1.26×10^{9} yr	Occurs in nature
$^{85m}_{36}Kr$	77% 0.82	4.4 h	Fission product
$^{85}_{36}Kr$	0.67	10.76 yr	Fission product
$^{89}_{38}Sr$	1.463	50.5 days	Fission product
$^{90}_{38}Sr$	0.546	28.1 yr	Fission product
$^{131}_{53}I$	0.7% 0.81 87.2% 0.608 9.3% 0.33 2.8% 0.25	8.05 days	Fission product
$^{233}_{90}Th$	1.23	22.2 min	$^{232}_{90}Th + ^{1}_{0}n$
$^{233}_{91}Pa$	5% 0.568 58% 0.257 37% 0.15	27.0 days	Decay of $^{233}_{90}Th$
$^{237}_{92}U$	0.248	6.75 days	$^{236}_{92}U + ^{1}_{0}n$
$^{239}_{92}U$	20% 1.29 80% 1.21	23.5 min	$^{238}_{92}U + ^{1}_{0}n$
$^{239}_{93}Np$	7% 0.713 4% 0.654 48% 0.437 13% 0.393 28% 0.332	2.35 days	Decay of ^{239}U

Maximum energies range from 10,000 eV to about 4 MeV. Half-lives range from microseconds to billions of years, with large half-lives tending to correlate with lower energies.

The dependence of range of beta particles in aluminum on energy is shown in Fig. 2.3. Although beta particles have a range greater than alpha particles, they can be stopped by relatively thin layers of water, glass, or metal. The range of beta particles in tissue is great enough, however, to cause burns when the skin is exposed. Beta-active isotopes that may become fixed in the body are very toxic. ^{90}Sr, which becomes fixed in bone, is an example. Those, like ^{85}Kr or ^{14}C, that are turned over quickly by the body, are much less toxic.

2.5 Gamma Radioactivity

Gamma rays are photons—electromagnetic radiation—given off when a nucleus undergoes transition from a state of higher energy to one of lower energy. The wavelength λ of the radiation is related to the energy change ΔE of the nucleus emitting this quantum of radiation (or photon) by the equation

$$\lambda = \frac{hc}{\Delta E} \tag{2.8}$$

where h is Planck's constant, 6.62559×10^{-34} J·s, and c is the velocity of light,

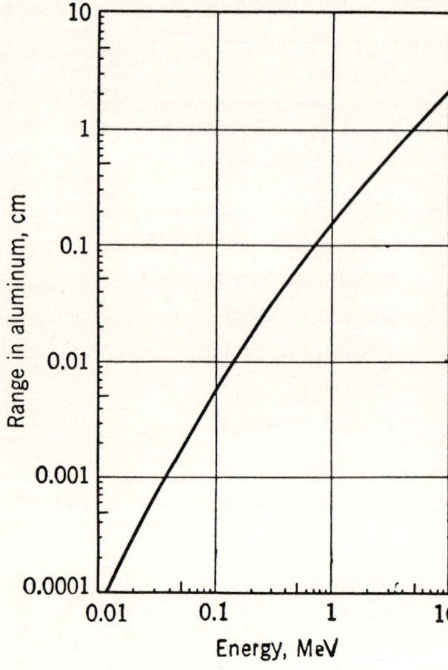

Figure 2.3 Range of beta particles in aluminum. For other materials, a useful approximation is that the range is inversely proportional to the density of electrons.

2.997925×10^8 m/s. Because energy changes of 0.1 MeV or more are common, gamma rays have wavelengths less than 1.2×10^{-9} cm. This is much shorter than the wavelength of visible light, around 10^{-5} cm. Gamma rays are in fact hard, or high-frequency, x-rays. They penetrate relatively great thicknesses of matter before being absorbed. Instead of having a well-defined range, like alpha or beta particles, a beam of gamma rays loses a certain fraction of its intensity per unit distance traveled through matter. The thickness of air, water, concrete, and lead required to dissipate one-half the intensity of a beam of gamma rays is plotted against energy per photon in Fig. 2.4.

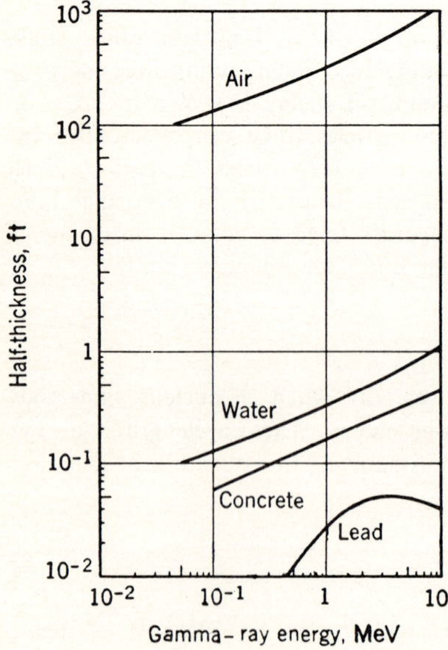

Figure 2.4 Thickness required to reduce the intensity of a beam of gamma radiation by a factor of 2.

Table 2.3 Long-lived gamma-emitting radioactive nuclides

Nuclide	Energy, MeV	Half-life	Source of nuclide	
$^{60m}_{27}Co$	0.059	10.5 min	$^{59}Co + ^{1}_{0}n$	
$^{80m}_{35}Br$	0.037, 0.049	4.38 h	$^{79}Br + ^{1}_{0}n$	
$^{91m}_{39}Y$	0.551	50 min	Decay of $^{91}_{38}Sr$	Fission products
$^{99m}_{43}Tc$	0.140	6.0 h	Decay of $^{99}_{42}Mo$	Fission products
$^{127m}_{52}Te$	0.059, 0.089, 0.67	109 days	Decay of $^{127}_{51}Sb$	Fission products
$^{135m}_{54}Xe$	0.527	15.6 min	Decay of $^{135}_{53}I$	Fission products
$^{137m}_{56}Ba$	0.662	2.55 min	Decay of $^{137}_{55}Cs$	Fission products

Because of the penetrating nature of gamma radiation, overexposure of the body to it results in deep-seated organic damage. Of the three types of radiation from radioactive substances, gamma radiation is by far the most serious external hazard and is the one that requires heavy shielding and remotely controlled operations.

Because a photon has neither charge nor mass, the parent and daughter nuclides in a gamma-radioactive transformation are nuclear isomers. A few gamma-active nuclides have half-lives long enough to be isolated and studied. Some of these are listed in Table 2.3. Many gamma-emitting nuclides resulting as products of alpha- or beta-radioactive decay have such short lives that the gamma ray appears to occur simultaneously with the alpha or beta emission that produced the gamma-active isomer. Data on gamma rays are customarily given with data on the parent alpha or beta emitter even though the gamma ray comes from the daughter nuclide. Frequently a number of gamma rays are emitted in cascade, as the unstable nuclide rapidly moves through several intermediate energy states before reaching its ground state. An example of this in the decay of $^{140}_{56}Ba$ is shown in Fig. 2.5.

2.6 Positron Emission

The transition involving the emission of a positron, i.e., a positively charged electron, is, in fact, another form of beta decay. Within the nucleus a proton is converted to a neutron. The positron is continuously distributed in energy up to some characteristic maximum energy, similar to the distributions of Fig. 2.2, accompanied by a corresponding distribution of neutrino energy. The emitted positively charged electron, as it passes through the field of atomic electrons in the surrounding matter, undergoes strong electrostatic attraction to these atomic electrons. The positron and negative electron then annihilate each other in a single reaction, and the resulting energy appears as two photons moving in opposite directions, each with an energy of 0.511 MeV. Further examples of positron-emitting nuclides are listed in Table 2.4.

2.7 Electron Capture

Some nuclei undergo radioactive decay by capturing an electron from the K or L shell of the atomic electron orbits. This results in the transformation of a proton to a neutron, the ejection of an unobservable neutrino of definite energy, and the emission of an x-ray where the electron vacancy of the K or L shell is filled by an atomic electron from an outer orbit. Because the net change in the radionuclide species is from atomic number Z to $Z-1$, similar to the nuclide change from positron emission, electron capture generally competes with all cases of positron beta decay.

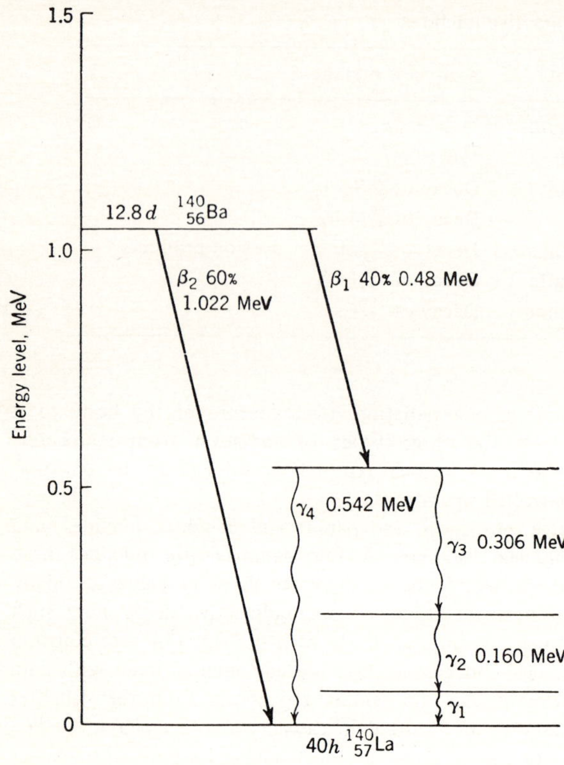

Figure 2.5 Decay scheme for $^{140}_{56}$Ba.

2.8 Spontaneous Fission

Many of the nuclides in the actinide family–U, Np, Pu, etc.–fission spontaneously as one of the modes of radioactive decay. Usually, for a nuclide with multiple modes of radioactive decay, the half-life of the nuclide is determined from the total decay rate, representing all the decay processes for that nuclide. However, in the case of spontaneous fission, a separate half-life for that process alone is used. Examples of nuclides that undergo spontaneous fission are given in Table 2.5.

The neutrons from spontaneous fission are emitted with average energies of a few million electron volts. Because the neutron carries no electrical charge, these fission neutrons penetrate quite readily through solids and liquids. They are stopped or slowed down only when they

Table 2.4 Examples of positron emitters

Nuclide	Maximum positron energy, MeV	Half-life	Fraction of decay, %
$^{11}_{6}$C	0.97	20.3 min	99+
$^{13}_{7}$N	1.19	10.0 min	100
$^{15}_{8}$O	1.72	124 s	100
$^{18}_{9}$F	0.635	109.7 min	97
$^{64}_{19}$Cu	0.657	12.8 h	19

Table 2.5 Examples of nuclides undergoing spontaneous fission

Nuclide	Half-life for spontaneous fission, yr
$^{235}_{92}U$	1.9×10^{17}
$^{238}_{92}U$	10^{16}
$^{239}_{94}Pu$	5.5×10^{15}
$^{240}_{94}Pu$	1.4×10^{11}
$^{242}_{94}Pu$	7×10^{10}
$^{244}_{96}Cm$	1.3×10^{7}
$^{252}_{98}Cf$	85

collide with nuclei of the material through which they are traveling. A neutron loses the greatest amount of energy per collision when it collides with a hydrogen nucleus, whose mass is almost identical with the neutron mass. Consequently, hydrogenous materials are used to degrade, or "moderate," energies of fission neutrons to energies in the few electron volt or kiloelectron volt range, where they are more easily absorbed by nuclear reactions. When energetic neutrons pass through animal tissue, the protons (hydrogen nuclei) recoiling from neutron collisions cause ionization within the tissue and can result in biological damage. Radionuclides with appreciable spontaneous fission, e.g., ^{252}Cf, must be shielded with mixtures of hydrogenous materials and neutron absorbers (e.g., boron) to protect against external hazards.

3 DECAY CHAINS

3.1 Batch Decay

Batch decay is concerned with the radioactive decay of a given amount of initially pure parent material. The decay products will build up and, if radioactive, will later die away as time progresses. An example is the decay chain resulting from the radioactive disintegration of ^{211}Pb, which is itself a member of the radioactive decay scheme of ^{235}U. Starting with ^{211}Pb, the decay chain is

Nuclide:	$^{211}_{82}Pb \xrightarrow{\beta^-}$	$^{211}_{83}Bi \xrightarrow{\alpha}$	$^{207}_{81}Tl \xrightarrow{\beta^-}$	$^{207}_{82}Pb$
Half-life:	36.1 min	2.15 min	4.79 min	stable
Denote by subscript:	1	2	3	4

Suppose that N_1^0 atoms of ^{211}Pb are freshly purified at time zero and there are no sources of ^{211}Pb present. The net rate of change of the number of ^{211}Pb atoms is

$$\frac{dN_1}{dt} = -\lambda_1 N_1 \tag{2.9}$$

The net rate of change of the number of ^{211}Bi atoms is

$$\frac{dN_2}{dt} = \lambda_1 N_1 - \lambda_2 N_2 \tag{2.10}$$

and the corresponding equations for ^{207}Tl and ^{207}Pb are

$$\frac{dN_3}{dt} = \lambda_2 N_2 - \lambda_3 N_3 \tag{2.11}$$

and

$$\frac{dN_4}{dt} = \lambda_3 N_3 \tag{2.12}$$

The solution to Eq. (2.9), subject to $N_1 = N_1^0$ at $t = 0$, is

$$N_1 = N_1^0 e^{-\lambda_1 t} \tag{2.13}$$

The solution to Eq. (2.10), subject to $N_2 = 0$ at $t = 0$, is

$$N_2 = \frac{\lambda_1 N_1^0}{\lambda_2 - \lambda_1}(e^{-\lambda_1 t} - e^{-\lambda_2 t}) \tag{2.14}$$

Likewise, with N_3 equal to zero at time $t = 0$, Eq. (2.11) integrates to

$$N_3 = \lambda_1 \lambda_2 N_1^0 \left[\frac{e^{-\lambda_1 t}}{(\lambda_2 - \lambda_1)(\lambda_3 - \lambda_1)} + \frac{e^{-\lambda_2 t}}{(\lambda_1 - \lambda_2)(\lambda_3 - \lambda_2)} + \frac{e^{-\lambda_3 t}}{(\lambda_1 - \lambda_3)(\lambda_2 - \lambda_3)}\right] \tag{2.15}$$

The amount of the stable fourth member of the chain is obtained directly from a material balance, as

$$N_4 = N_1^0(1 - e^{-\lambda_1 t}) - (N_2 + N_3) \tag{2.16}$$

Figure 2.6 shows the change with time of the number of atoms of each nuclide in the ^{211}Pb decay chain, per initial atom N_1^0 of ^{211}Pb. Figure 2.7 shows the variation with time of the activity, or disintegration rate λN, of each nuclide and the total activity of the mixture, relative to the initial activity $\lambda_1 N_1^0$ of ^{211}Pb.

In the general case of a radioactive decay chain

$$N_1 \to N_2 \to N_3 \to \cdots \to N_j \to \cdots \to N_i \to \cdots$$

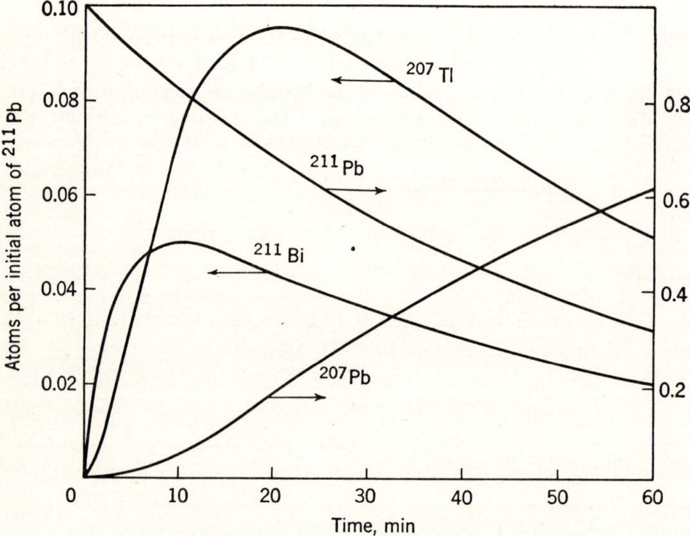

Figure 2.6 Concentration of nuclides in ^{211}Pb decay chain with pure ^{211}Pb initially.

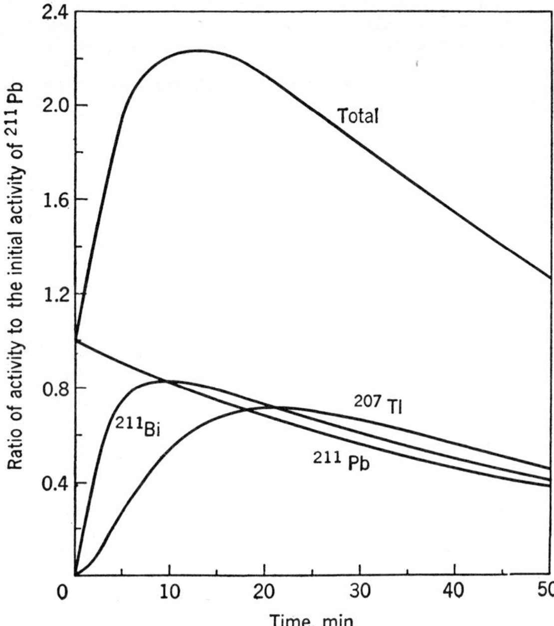

Figure 2.7 ^{211}Pb decay-chain activities.

in which the parent material is present in an amount N_1^0 at time zero, if none of the other members of the decay chain is initially present, and if there are no other sources of the parent material, the amount N_i of any nuclide present at time t can be written by analogy to Eq. (2.15):

$$N_i = N_1^0 \lambda_1 \lambda_2 \cdots \lambda_{i-1} \sum_{j=1}^{i} \frac{e^{-\lambda_j t}}{\prod_{\substack{k=1 \\ k \neq j}}^{i} (\lambda_k - \lambda_j)} \qquad (i > 1) \tag{2.17}$$

Equation (2.17) is known as the Bateman [B1] equation. It is derived in Sec. 7.

By superposition, the batch-decay equation can be further generalized for the case of arbitrary initial amounts N_l^0 of any of the radionuclides in the chain:

$$N_i = \sum_{l=1}^{i-1} \left[N_l^0 \lambda_l \lambda_{l+1} \cdots \lambda_{i-1} \sum_{j=l}^{i} \frac{e^{-\lambda_j t}}{\prod_{\substack{k=l \\ k \neq j}}^{i} (\lambda_k - \lambda_j)} \right] + N_i^0 e^{-\lambda_i t} \tag{2.18}$$

When a radionuclide decays to a daughter of half-life much shorter than that of its parent, the daughter builds up to an amount that remains in constant ratio to the amount of the parent, and the amount of the daughter then decreases at a rate controlled by the half-life of the parent. In this case, the daughter is said to be in equilibrium with the parent, even though the amount of the parent radionuclide may be changing with time. For example, for the batch decay scheme that led to Eq. (2.14), suppose that $\lambda_2 > \lambda_1$, and assume that for times of interest $\lambda_2 t \gg 1$. Equation (2.14), written in terms of decay rates, then reduces to

$$N_2\lambda_2 \approx \frac{N_1^0\lambda_1\lambda_2}{\lambda_2 - \lambda_1} e^{-\lambda_1 t} \tag{2.19a}$$

In the limit of $\lambda_2 \gg \lambda_1$ the daughter builds up to a concentration such that its decay rate is identical to that of the parent. This is the condition of ***transient equilibrium***, i.e., from Eq. (2.19*a*):

$$N_2\lambda_2 \approx N_1\lambda_1 \tag{2.19b}$$

Transient equilibrium is reached by ^{211}Bi from the batch decay of ^{211}Pb, as illustrated in Fig. 2.7. The time to reach this transient equilibrium is a few times the half-life of ^{211}Bi. The activities of ^{211}Bi and ^{211}Pb would approach *secular equilibrium*, i.e., equal activities, if the ratio of the half-life of ^{211}Pb to that of ^{211}Bi were even greater. The second daughter, ^{207}Tl, can also be said to be in transient equilibrium with ^{211}Pb, at times much greater than $1/(\lambda_2 + \lambda_3)$, because both its half-life and that of its immediate precursor are both short compared with the ^{211}Pb half-life.

3.2 Continuous Production

Consider a process, such as nuclear fission, that operates in a way so that P atoms of the first member of the chain are formed per unit time. The production and decay scheme, and assumed initial conditions for the radioactive chain, are

$$\begin{array}{lcccc} & P & & & \\ & \downarrow & & & \\ & N_1 & \xrightarrow{\lambda_1} N_2 & \xrightarrow{\lambda_2} \cdots \xrightarrow{\lambda_{i-1}} & N_i \xrightarrow{\lambda_i} \\ \text{Initial amounts at } t=0 & 0 & 0 & & 0 \end{array}$$

The net rate of change of the number N_1 of atoms of species 1 is

$$\frac{dN_1}{dt} = P - \lambda_1 N_1 \tag{2.20}$$

The net rate of change for species 2 is

$$\frac{dN_2}{dt} = \lambda_1 N_1 - \lambda_2 N_2 \tag{2.21}$$

The net rate of change for species i is

$$\frac{dN_i}{dt} = \lambda_{i-1} N_{i-1} - \lambda_i N_i \tag{2.22}$$

These amounts N will eventually reach a steady-state or "saturation" level N^* such that the rate of production equals the rate of decay, so that $dN/dt = 0$. Applying this condition to the above equations, we obtain

$$N_1^* = \frac{P}{\lambda_1} \tag{2.23}$$

$$N_2^* = \frac{\lambda_1 N_1^*}{\lambda_2} \tag{2.24}$$

or, substituting Eq. (2.23) for N_1^*,

$$N_2^* = \frac{P}{\lambda_2} \tag{2.25}$$

Similarly,

$$N_i^* = \frac{P}{\lambda_i} \tag{2.26}$$

For the transient case, with zero initial amount, the time-dependent solution of Eq. (2.20) is

$$N_1 = \frac{P}{\lambda_1}(1 - e^{-\lambda_1 t}) \tag{2.27}$$

and the solution of Eq. (2.21) is

$$N_2 = \lambda_1 P \left[\frac{1 - e^{-\lambda_1 t}}{\lambda_1(\lambda_2 - \lambda_1)} + \frac{1 - e^{-\lambda_2 t}}{\lambda_2(\lambda_1 - \lambda_2)}\right] \tag{2.28}$$

To obtain the amount $N_i(t)$, we consider a time interval from t' to $t' + dt'$, where $t' < t$. During this interval dt', the amount of species 1 produced is $P\,dt'$. The ultimate decay of this amount $P\,dt'$ of species 1 over the interval $t - t'$ results in a net amount $dN_i(t', t)$ obtained by applying the Bateman equation (2.17):

$$dN_i(t', t) = P\,dt'\,\lambda_1\lambda_2 \cdots \lambda_{i-1} \sum_{j=1}^{i} \frac{e^{-\lambda_j(t-t')}}{\prod\limits_{\substack{k=1\\k\neq j}}^{i} (\lambda_k - \lambda_j)} \qquad (i > 1) \tag{2.29}$$

Then, to determine $N_i(t)$ due to production of species 1 over all time t' from 0 to t, we integrate over t':

$$N_i(t) = \int_0^t dN_i(t', t) = \int_0^t \lambda_1\lambda_2 \cdots \lambda_{i-1} \sum_{j=1}^{i} \frac{e^{-\lambda_j(t-t')}}{\prod\limits_{\substack{k=1\\k\neq j}}^{i} (\lambda_k - \lambda_j)} P\,dt'$$

or

$$N_i(t) = P\lambda_1\lambda_2 \cdots \lambda_{i-1} \sum_{j=1}^{i} \frac{1 - e^{-\lambda_j t}}{\lambda_j \prod\limits_{\substack{k=1\\k\neq j}}^{i} (\lambda_k - \lambda_j)} \qquad (i > 1) \tag{2.30}$$

In terms of the saturation amount N_i^*, as given by Eq. (2.26),

$$\frac{N_i(t)}{N_i^*} = \lambda_1\lambda_2 \cdots \lambda_i \sum_{j=1}^{i} \frac{1 - e^{-\lambda_j t}}{\lambda_j \prod\limits_{\substack{k=1\\k\neq j}}^{i} (\lambda_k - \lambda_j)} \qquad (i > 1) \tag{2.31}$$

To illustrate, consider the decay chain of mass number 92:

$$^{92}_{36}\mathrm{Kr} \xrightarrow{\beta^-} {}^{92}_{37}\mathrm{Rb} \xrightarrow{\beta^-} {}^{92}_{38}\mathrm{Sr} \xrightarrow{\beta^-} {}^{92}_{39}\mathrm{Y} \xrightarrow{\beta^-} {}^{92}_{40}\mathrm{Zr}$$

Half-life:	3.0 s	5.3 s	2.71 h	3.53 h	stable

^{92}Kr is formed in fission at 0.063 times the rate of fission of ^{235}U. Let us assume that an experiment is conducted to fission ^{235}U at a constant rate for a period of 20 h. The half-lives of ^{92}Kr and ^{92}Rb are so short compared to the half-lives of ^{92}Sr and ^{92}Y that, for time scales

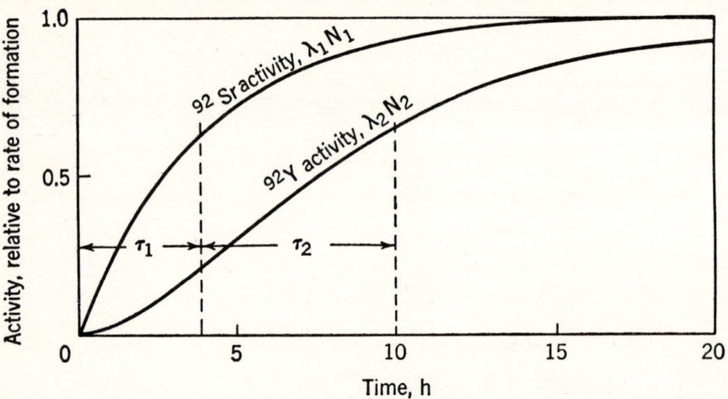

Figure 2.8 Buildup of activity in fission-product decay chain, mass 92.

of several hours, the decay of these first two nuclides in the chain may be assumed to be instantaneous, so ^{92}Sr will be treated as the first member of the decay chain.

Figure 2.8 shows the change with time of the activities (λN) of ^{92}Sr and ^{92}Y for unit rate of production of the first member of the chain, calculated by applying Eqs. (2.27) and (2.28). Each activity approaches a steady value equal to the rate of formation of the first member of the chain. The time to reach within $1/e$ of the steady activity is approximately equal to the sum of the mean lifetimes of all nuclides in the chain up to and including the nuclide in question. The amount of ^{92}Zr could be obtained by applying Eq. (2.30) for the third member of a decay chain, but since it is a stable nuclide ($\lambda_3 = 0$) its amount can be obtained simply by applying a material-balance equation

$$N_3 = Pt - (N_1 + N_2) \tag{2.32}$$

In many instances it is also necessary to consider sources that directly form intermediate nuclides in the decay chain, as in the case of some fission-product chains that have important direct fission yields of more than one nuclide in the chain. Defining P_l as the constant rate of formation of the lth nuclide in the chain, and for $N_i = 0$ at $t = 0$, we obtain from Eq. (2.30) by superposition:

$$N_i(t) = \sum_{l=1}^{i-1} \left[P_l \lambda_l \lambda_{l+1} \cdots \lambda_i \sum_{j=l}^{i} \frac{1 - e^{-\lambda_j t}}{\lambda_j \prod\limits_{\substack{k=1 \\ k \neq j}}^{i} (\lambda_k - \lambda_j)} \right] + P_i \frac{1 - e^{-\lambda_i t}}{\lambda_i} \tag{2.33}$$

At saturation the activity N_i^* is given by

$$\lambda_i N_i^* = \sum_{l=1}^{i} P_l \tag{2.34}$$

3.3 Continuous Production and Shutdown

Another case of practical interest in nuclear engineering is the buildup and decay of fission products formed in a nuclear reactor operating at a steady fission rate for a time T and that have been removed from the reactor and allowed to undergo radioactive decay for an additional time. The schematic diagram for continuous production of the first member of the chain at rate P is

$$\begin{array}{c} P \\ \downarrow \\ N_1 \end{array} \xrightarrow{\lambda_1} N_2 \xrightarrow{\lambda_2} N_3 \longrightarrow \cdots \longrightarrow N_j \longrightarrow \cdots \xrightarrow{\lambda_{i-1}} N_i \xrightarrow{\lambda_i} \cdots$$

	N_1	N_2	N_3	N_j	N_i
Initial amounts at $t = 0$	0	0	0	0	0

Time	*Production rate of first member of chain, atoms/s*
$0 - T$	P
$> T$	0

A general equation for the amount of any nuclide present at a time t after removal from the reactor can be derived by using the Bateman equation (2.17). Consider a time variable t', such that when $0 \leqslant t' \leqslant T$ the number of atoms of the first member of the chain produced during any interval dt' is $P\,dt'$. The relevant time scale is shown below:

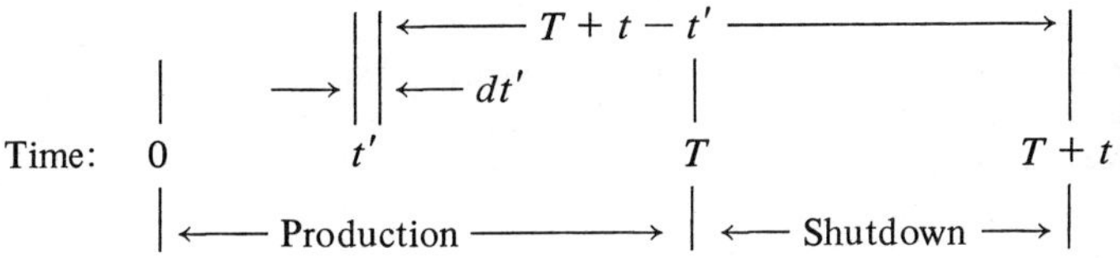

The number $dN_i(t', T+t)$ of atoms of the ith member of the chain at time $T+t$ resulting from the decay of the $P\,dt'$ atoms is obtained by applying Eq. (2.17) in the same manner as in developing Eq. (2.30):

$$dN_i(t', T+t) = \lambda_1 \lambda_2 \cdots \lambda_{i-1} \sum_{j=1}^{i} \frac{e^{-\lambda_j(T+t-t')}}{\prod_{\substack{k=1 \\ k \neq j}}^{i} (\lambda_k - \lambda_j)} P\,dt' \qquad (i > 1) \tag{2.35}$$

The total number of atoms of species i at time t is obtained by integrating Eq. (2.35) over the time interval $0 \leqslant t' \leqslant T$ during which P is finite:

$$N_i(T, t) = \int_0^T dN_i(t', T+t) = \int_0^T \lambda_1 \lambda_2 \cdots \lambda_{i-1} \sum_{j=1}^{i} \frac{e^{-\lambda_j(T+t-t')}}{\prod_{k \neq j}^{i} (\lambda_k - \lambda_j)} P\,dt'$$

or

$$N_i(T, t) = P \lambda_1 \lambda_2 \cdots \lambda_{i-1} \sum_{j=1}^{i} \frac{(1 - e^{-\lambda_j T}) e^{-\lambda_j t}}{\lambda_j \prod_{\substack{k=1 \\ k \neq j}}^{i} (\lambda_k - \lambda_j)} \qquad (i > 1) \tag{2.36}$$

For the first member of the chain ($i = 1$), the solution is

$$N_1(T, t) = \frac{P}{\lambda_1} (1 - e^{-\lambda_1 T}) e^{-\lambda_1 t} \tag{2.37}$$

For the second member of the chain, Eq. (2.36) yields

$$N_2(T, t) = \lambda_1 P \left[\frac{(1 - e^{-\lambda_1 T}) e^{-\lambda_1 t}}{\lambda_1(\lambda_2 - \lambda_1)} + \frac{(1 - e^{-\lambda_2 T}) e^{-\lambda_2 t}}{\lambda_2(\lambda_1 - \lambda_2)} \right] \tag{2.38}$$

To allow for the possibility of finite direct formation of any lth member of the chain during the production period from 0 to T, we obtain from Eq. (2.38) by superposition:

$$N_i(T, t) = \sum_{l=1}^{i-1} \left[P_l \lambda_1 \lambda_2 \cdots \lambda_{i-1} \sum_{j=l}^{i} \frac{(1 - e^{-\lambda_j T}) e^{-\lambda_j t}}{\lambda_j \prod\limits_{\substack{k=l \\ k \neq j}}^{i} (\lambda_k - \lambda_j)} \right] + P_i \frac{(1 - e^{-\lambda_i T}) e^{-\lambda_i t}}{\lambda_i} \tag{2.39}$$

Use of these equations is illustrated for the fission-product decay chain of mass number 92 considered in Sec. 3.2. Assume production of ^{92}Sr, the first nuclide of the chain, at a constant rate $P = 1$/h for a period of 3 h ($T = 3$ h), followed by several hours of radioactive decay with $P = 0$. The amounts of ^{92}Sr and ^{92}Y, calculated by applying Eqs. (2.37) and (2.38), respectively, are shown in Fig. 2.9. The amount of stable ^{92}Zr during the period of $P = 0$ is obtained from the material-balance equation:

$$N_3 = PT - (N_1 + N_2) \tag{2.40}$$

Figure 2.9 illustrates that when the parent nuclide ^{92}Sr has not reached equilibrium and when its radioactive daughter ^{92}Y has not reached transient equilibrium, the amount of the daughter nuclide continues to increase for a time period after the production of the initial member of the chain is discontinued.

4 NEUTRON REACTIONS

4.1 Capture Reactions

In fission reactors the transmutation reactions of principal importance involving neutrons are *capture* and *fission*. All nuclides (except ^{4}He) take part in the radiative capture reaction (n, γ), an example of which is

$$^{235}_{92}\text{U} + {}^{1}_{0}n \rightarrow {}^{236}_{92}\text{U} + {}^{0}_{0}\gamma$$

This reaction produces an isotope of the reacting nuclide with mass number increased by unity and one or more gamma rays, which carry off most of the energy of the reaction. Other capture reactions, possible for a few nuclides (mostly those of low mass number), result in emission of an alpha particle (n, α):

$$^{10}_{5}\text{B} + {}^{1}_{0}n \rightarrow {}^{7}_{3}\text{Li} + {}^{4}_{2}\text{He}$$

or a proton (n, p):

$$^{16}_{8}\text{O} + {}^{1}_{0}n \rightarrow {}^{16}_{7}\text{N} + {}^{1}_{1}\text{H}$$

4.2 Fission Reactions

The fission reaction is responsible for the sustained production of neutrons in a nuclear reactor and for most of the energy released. In this reaction, one neutron is absorbed by a heavy nuclide, which then splits into two nuclides each in the middle third of the periodic table, and several neutrons, which are available for initiating additional fissions. All elements beyond lead undergo fission with neutrons of sufficiently high energy; the only readily available long-lived nuclides that undergo fission with thermal neutrons are ^{233}U, ^{235}U, ^{239}Pu, and ^{241}Pu.

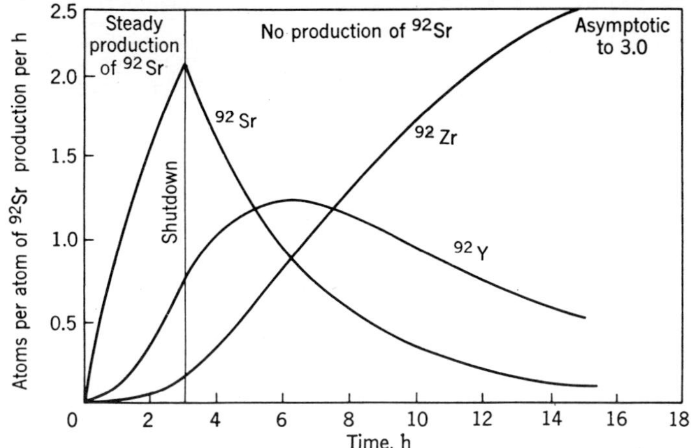

Figure 2.9 Concentration changes of fission products during steady production and after shutdown.

An example of the fission of ^{235}U into ^{144}Xe and ^{89}Sr has already been given. The fission reaction may take place in a number of alternative ways. Light fragments have been observed to have mass numbers from 72 to 118, heavy fragments from 118 to 162. At a given mass number fragments have also been observed with atomic numbers varying over a range of three or more. For example, ^{133}Te, ^{133}I, and ^{133}Xe have all been observed as primary fission fragments. Finally, the number of neutrons produced in an individual fission event may be anywhere from zero to four or more. As a result, a large number of alternative fission reactions take place, of the general form

$$^{235}_{92}\text{U} + ^{1}_{0}n \rightarrow ^{A_1}_{Z_1}\text{L} + ^{A_2}_{Z_2}\text{H} + X^{1}_{0}n$$

where L and H denote the light and heavy fission fragments, respectively, Z_1, A_1, and X all vary between limits, and Z_2 and A_2 are determined by the conditions

$$Z_1 + Z_2 = 92$$

$$A_1 + A_2 + X = 236$$

4.3 Reaction Rates

The number of nuclei reacting in a specified way with neutrons in unit time is proportional to the number of nuclei present and to the concentration of neutrons. In the language of chemical kinetics, neutron reactions are first-order with respect to concentration of nuclei and neutrons, and it is because neutron reactions are simple first-order irreversible processes that a very detailed quantitative treatment of the rate processes in a nuclear reactor can be given.

The expression for the rate of reaction of neutrons with reacting nuclei N is

$$\frac{\text{Reactions}}{\text{Volume} \times \text{time}} = K_R nN \tag{2.41}$$

where n is the concentration of neutrons, in number per unit volume, and K_R is the specific rate constant. It has become customary to express K_R as the product of another constant σ, called the *cross section*, and the neutron speed v, so that Eq. (2.41) becomes

$$\frac{\text{Reactions}}{\text{Volume} \times \text{time}} = \sigma v n N \tag{2.42}$$

The product vn is termed the *neutron flux* ϕ and is the measure most commonly used to describe the neutron intensity in a reactor.

For a given neutron density n and speed v, the product $\sigma\phi$ is the first-order rate constant and is the fraction of the reacting nuclei consumed by the reaction per unit time. It plays the same role in rate equations as the radioactive decay constant λ.

4.4 Cross Sections

The cross section σ has dimensions of length squared (cm^2) as is required to make Eq. (2.42) dimensionally consistent. Fundamentally, it is the fraction of the reacting nuclei consumed by the nuclear reaction per unit time per unit flux.

Cross sections for reactions with neutrons vary from a lower detectable limit of around 1×10^{-28} cm^2 to a maximum of 2.65×10^{-18} cm^2, which has been observed for ^{135}Xe. To avoid using such large negative exponents, cross sections are usually expressed in units of 10^{-24} cm^2, called *barns* (b). For instance, the xenon cross section is 2.65×10^6 b. The millibarn (mb) is 10^{-27} cm^2.

There is a different cross section for every different reaction of a nuclide with neutrons. Examples of cross sections for low-energy neutrons moving at a speed of 2200 m/s are given in Table 2.6.

The sum of the cross sections for all reactions in which a neutron is absorbed is called the *absorption cross section,* denoted by σ_a. In the examples of Table 2.6,

$$\sigma_a{}^{235}\mathrm{U} = 680.8 \text{ b}$$

$$\sigma_a{}^{14}\mathrm{N} = 1.88 \text{ b}$$

The neutron speed, or kinetic energy, is specified in the listing of neutron cross sections in Table 2.6 because the cross section generally varies with neutron speed, in many cases very strongly. Curves for the variation in capture or absorption cross sections with neutron energy for many nuclides are given in BNL-325 [M1]. A table of the published values of cross sections for neutron-absorption reactions, for 2200 m/s neutrons, is given in App. C. For most of the nuclides the absorption cross sections for low-energy neutrons vary nearly as the reciprocal of the neutron speed v.

4.5 Neutron Speeds in Reactors

Neutrons in a nuclear reactor have velocities, and energies, distributed over a wide range. Neutrons are born from the fission reaction at an average energy of about 2 MeV

Table 2.6 Examples of neutron reaction cross sections

Reaction	Example	Cross-section notation	Cross section for 2200 m/s neutrons, b
Fission	$^{235}_{92}\mathrm{U} + ^{1}_{0}n \rightarrow$ fission	σ_f	582.2
Neutron capture:			
Gamma emission	$^{235}_{92}\mathrm{U} + ^{1}_{0}n \rightarrow ^{236}_{92}\mathrm{U} + \gamma$	$\sigma(n,\gamma)$ or σ_c	98.6
	$^{14}_{7}\mathrm{N} + ^{1}_{0}n \rightarrow ^{15}_{7}\mathrm{N} + \gamma$	$\sigma(n,\gamma)$	0.075
Proton emission	$^{14}_{7}\mathrm{N} + ^{1}_{0}n \rightarrow ^{14}_{6}\mathrm{C} + ^{1}_{1}\mathrm{H}$	$\sigma(n,p)$	1.81
Alpha emission	$^{6}_{3}\mathrm{Li} + ^{1}_{0}n \rightarrow ^{3}_{1}\mathrm{H} + ^{4}_{2}\mathrm{He}$	$\sigma(n,\alpha)$	940

($v = 1.955 \times 10^7$ m/s). To maintain a steady-state nuclear chain reaction it is necessary that the rate constant for the neutron fission reaction be sufficiently high so that neutron production will compete favorably with processes that consume neutrons. In addition to neutron absorption, neutrons are consumed by diffusing to the outer surface of the reactor and escaping to the surroundings. The diffusion of neutrons through matter is similar to the diffusion of gas molecules, and the average rate of loss of neutrons of speed v from a volume element in a reactor due to diffusion, or "leakage," can be expressed as

$$\frac{\text{Average loss of neutrons by leakage}}{\text{cm}^3 \cdot \text{s}} = K_L n v \tag{2.43}$$

where n is the average concentration of neutrons of speed v throughout the reactor. The rate constant K_L varies as the surface-volume ratio of the reactor and is usually affected but little by neutron speed. From Eq. (2.43) it follows that neutron consumption by leakage increases with neutron speed. On the other hand, the cross sections for fission decrease markedly as neutron speed increases. Unless fissionable fuel in a highly concentrated form is available, it is then necessary to reduce the neutron speed to obtain the proper balance between neutron production and consumption. This is done by designing the reactor to contain sufficient atoms of low atomic weight, such as hydrogen, deuterium, beryllium, or carbon. The fast neutrons from fission undergo elastic collisions with these light nuclei, called *moderators*, and soon reach thermal equilibrium with the surrounding medium. In a *thermal* reactor enough moderator material is present so that the neutrons will be quickly degraded to thermal energies, and most of the fissions occur with the thermal neutrons.

A *fast* reactor is one in which no moderator is present and most of the fissions occur with neutrons of energies near the energies at which they were born. To overcome the high probability of neutron consumption by leakage in fast reactors, a high concentration of fissionable material is required, as may be obtained by fueling the reactor with plutonium or with uranium highly enriched in ^{235}U.

4.6 Neutron Flux

The neutron flux is the product of the number of neutrons per unit volume and the neutron speed. It has the physical significance of being the total distance traveled in unit time by all the neutrons present in unit volume. It seems reasonable that the rate of reaction of neutrons should be proportional to the distance they travel in unit time. The flux has the dimensions of neutrons per square centimeter per second. Typical values of the flux in nuclear reactors range from around 10^{11} to 10^{14} $n/(\text{cm}^2 \cdot \text{s})$.

To specify completely the neutron activity and to choose the proper cross sections for calculating the reaction rate constant, it is necessary to know the distribution of neutron concentration, or neutron flux, with respect to energy. In a thermal reactor the distribution of neutrons in thermal equilibrium with nuclei at an absolute temperature T is similar to the distribution of gas molecules in thermal equilibrium and can be approximated by the Maxwell-Boltzmann distribution

$$n_M(v)\, dv = n_M \left(\frac{2}{\pi}\right)^{1/2} \left(\frac{m}{kT}\right)^{3/2} v^2 e^{-mv^2/2kT}\, dv \tag{2.44}$$

where $n_M(v)\, dv$ = number of thermal neutrons per unit volume with speeds between v and $v + dv$

n_M = total number of thermal neutrons per unit volume

m = mass of neutron

k = Boltzmann's constant, 1.38054×10^{-23} J/K

The most probable speed v_0 is that for which $n_M(v)$ is a maximum, or

$$\text{Most probable speed} = v_0 = \sqrt{\frac{2kT}{m}} \tag{2.45}$$

For neutrons in thermal equilibrium at 20°C, the most probable speed from Eq. (2.45) is 2200 m/s.

The neutron kinetic energy E is related to the neutron speed by

$$E = \frac{mv^2}{2} \tag{2.46}$$

From Eqs. (2.45) and (2.46) the energy E_0 at the most probable speed is

$$E_0 = kT \tag{2.47}$$

and for thermal neutrons at 20°C, E_0 has the value of 0.0253 eV.

By means of Eq. (2.46), the speed distribution, Eq. (2.44), can be transformed into an energy distribution,

$$n_M(E)\,dE = n_M\left(\frac{4}{\pi}\right)^{1/2}\left(\frac{1}{kT}\right)^{3/2} E^{1/2}e^{-E/kT}\,dE \tag{2.48}$$

where $n_M(E)\,dE$ is the number of neutrons per unit volume with energies between E and $E + dE$.

The distributions presented in Eqs. (2.44) and (2.48) can be written in dimensionless form in terms of the most probable speed v_0 and the energy E_0 at the most probable speed as follows:

$$\frac{n_M(v/v_0)}{n_M} = \frac{4}{\sqrt{\pi}}\left(\frac{v}{v_0}\right)^2 e^{-(v/v_0)^2} \tag{2.49}$$

The left side of Eq. (2.48) is the fraction of the total thermal neutrons that have a speed ratio v/v_0, per unit increment in speed ratio v/v_0. Similarly,

$$\frac{n_M(E/E_0)}{n_M} = \frac{2}{\sqrt{\pi}}\left(\frac{E}{E_0}\right)^{1/2} e^{-E/E_0} \tag{2.50}$$

Dimensionless flux distributions may be obtained by multiplying the neutron density distributions by the neutron speed ratio v/v_0:

$$\frac{\phi_M(v/v_0)}{\phi_M} = 2\left(\frac{v}{v_0}\right)^3 e^{-(v/v_0)^2} \tag{2.51}$$

and

$$\frac{\phi_M(E/E_0)}{\phi_M} = \frac{E}{E_0}\,e^{-E/E_0} \tag{2.52}$$

where ϕ_M is the total flux of neutrons in thermal equilibrium, i.e.,

$$\phi_M = \int_0^\infty \phi_M\left(\frac{E}{E_0}\right)\frac{dE}{E_0} \tag{2.53}$$

The dimensionless neutron density and flux distributions, Eqs. (2.49) to (2.52), are plotted in Figs. 2.10 and 2.11.

4.7 Effective Cross Sections

If the energy dependence of a cross section is known, the total rate at which neutrons react with a nuclide is obtained by integrating the flux, cross-section product over all possible energies:

$$\text{Total reactions with neutrons per unit volume per unit time} = \int_0^\infty \phi(E)\sigma(E)\,dE \tag{2.54}$$

It is convenient to determine an effective cross section $\bar{\sigma}$ for the nuclide, so that when $\bar{\sigma}$ is multiplied by the total thermal flux ϕ_M the proper reaction rate is obtained:

$$\phi_M \bar{\sigma} \equiv \int_0^\infty \phi(E)\sigma(E)\,dE \tag{2.55}$$

If the cross section is one that varies inversely with the neutron speed, as in the case with many of the absorption cross sections, then

$$\sigma(E) = \sigma(E_0)\left(\frac{E_0}{E}\right)^{1/2} \qquad \text{for } 1/v \text{ absorbers} \tag{2.56}$$

where $\sigma(E_0)$ is arbitrarily chosen to be the cross section at the energy $E_0 = kT$ corresponding to the most probable neutron speed. We shall first assume that all neutrons are in a Maxwell-Boltzmann thermal equilibrium, so that $\phi(E) = \phi_M(E)$. The integral in Eq. (2.55) is then transformed to the variable E/E_0, and Eqs. (2.52) and (2.56) are substituted to yield

$$\bar{\sigma} = \sigma(E_0) \int_0^\infty \left(\frac{E}{E_0}\right)^{1/2} e^{-E/E_0}\, d\left(\frac{E}{E_0}\right) \tag{2.57}$$

The integral can be evaluated in terms of the gamma function, which in this case has a value of $\sqrt{\pi}/2$:

$$\bar{\sigma} = \frac{\sigma(E_0)\sqrt{\pi}}{2} \qquad \text{for } 1/v \text{ absorbers in a Maxwell-Boltzmann distribution} \tag{2.58}$$

Tables (see App. C) usually list values of the thermal absorption cross sections for monoenergetic neutrons of speed 2200 m/s. Because this happens to be the most probable speed for neutrons in thermal equilibrium at 293.2 K, the effective cross section at temperature T (K) can be obtained from

$$\bar{\sigma} = \sigma_{2200}\left(\frac{293.2}{T}\right)^{1/2}\frac{\sqrt{\pi}}{2} \qquad \text{for } 1/v \text{ absorbers in a Maxwell-Boltzmann distribution} \tag{2.59}$$

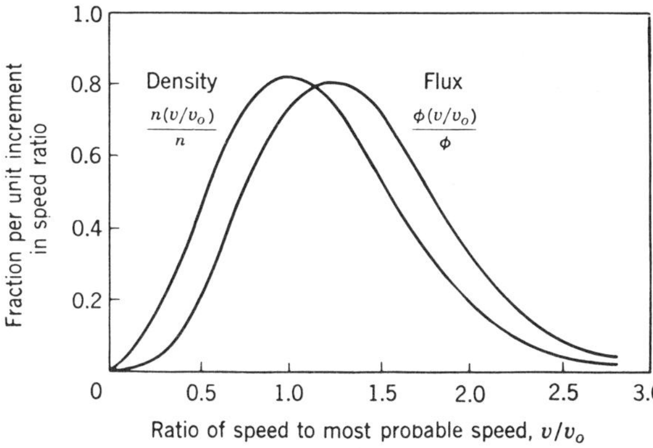

Figure 2.10 Neutron density and flux distributions with respect to speed ratio.

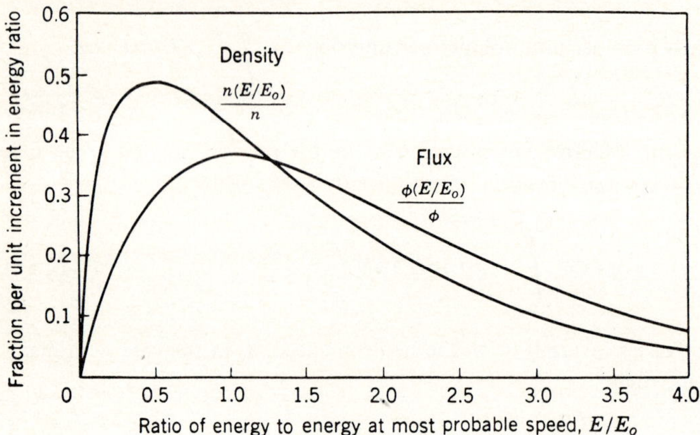

Figure 2.11 Neutron density and flux distributions with respect to energy ratio.

The effective cross section obtained from Eq. (2.59), when multiplied by the total flux of thermal neutrons, will give the proper value of reaction rate with thermal neutrons for a $1/v$ absorber. However, many of the most important nuclides entering into reactor calculations (e.g., the fissile nuclides) are not $1/v$ absorbers, and the integration of Eq. (2.54) must consider dependence of the neutron spectrum and the cross section on neutron energy (or speed). In refined calculations this integration is done stepwise by dividing the energy scale encountered in reactors (0 to ~12 MeV) into energy groups. An effective cross section is determined for each group and is multiplied by the flux of neutrons in that group to determine the group reaction rate. Digital computers are normally employed.

For simplified reactor calculations a "one-group" approximation can be employed. Westcott [W4] has developed a convention such that the total reaction rate with N_i atoms of nuclide i is given

$$\text{Total reactions per unit volume per unit time} = N_i\hat{\phi}\hat{\sigma} \tag{2.60}$$

where $\hat{\phi}$ is defined in terms of some arbitrary reference speed $\hat{v}$ as

$$\hat{\phi} \equiv \hat{v}\int_0^\infty n(E)\,dE = \hat{v}n \tag{2.61}$$

where n is the total density of neutrons in the reactor.

The reference speed $\hat{v}$ is arbitrarily chosen as 2200 m/s, which is the most probable speed for a Maxwell-Boltzmann distribution at temperature $\hat{T} = 293.2$ K. The cross section $\hat{\sigma}$ is now the specially defined effective cross section that, when multiplied by the "2200 m/s flux" $\hat{\phi}$, gives the proper reaction rate constant.

From Eqs. (2.54), (2.60), and (2.61),

$$\hat{\sigma} = \frac{\int_0^\infty \phi(E)\sigma(E)\,dE}{\hat{v}n} \tag{2.62}$$

In the Westcott formulation the energy distribution $\phi(E)$ is treated as a Maxwell-Boltzmann energy distribution $\phi_M(E)$ of thermalized neutrons on which is superimposed an epithermal distribution $\phi_E(E)$ of nonthermalized neutrons, so that

$$\phi(E)\,dE = [\phi_M(E) + \phi_E(E)]\,dE \tag{2.63}$$

From Eqs. (2.47) and (2.52):

$$\phi_M(E)\,dE = \phi_M \frac{E}{(kT)^2} e^{-E/kT}\,dE \tag{2.64}$$

where

$$\phi_M = \int_0^\infty \phi_M(E)\,dE \tag{2.65}$$

The epithermal flux distribution $\phi_E(E)$ can be approximated by a $1/E$ energy dependence above some lower cutoff energy of μkT, and it can be normalized to the integrated thermal flux ϕ_M by a factor β. Then

$$\phi_E(E)\,dE = \phi_M \frac{\beta\Delta}{E}\,dE \tag{2.66}$$

where Δ is the unit step function at μkT energy. A typical value of μ for a well-moderated reactor is 5.

By substituting Eqs. (2.64) and (2.66) into (2.63),

$$\phi(E)\,dE = \phi_M \left(\frac{E}{(kT)^2} e^{-E/kT} + \frac{\beta\Delta}{E}\right) dE \tag{2.67}$$

which will be used in solving the integral of Eq. (2.62).

To solve Eq. (2.62) we also need to formulate the total neutron density n as the sum of the densities of Maxwell-Boltzmann neutrons n_M and epithermal neutrons n_E:

$$n = n_M + n_E \tag{2.68}$$

To obtain n from a flux distribution,

$$n = \int_0^\infty n(E)\,dE = \int_0^\infty \phi(E)\,\frac{dE}{v} = \int_0^\infty \phi(E)\,\frac{dE}{\sqrt{2E/m}} \tag{2.69}$$

where Eq. (2.46) was used to change from v to E.

Using Eq. (2.64) in (2.69) to obtain n_M:

$$n_M = \int_0^\infty \phi_M \frac{E}{(kT)^2} e^{-E/kT} \frac{dE}{\sqrt{2E/m}} = \phi_M \sqrt{\frac{\pi m}{8kT}} \tag{2.70}$$

and using Eq. (2.66) in (2.69), we obtain n_E:

$$n_E = \int_{\mu kT}^\infty \phi_M \frac{\beta}{E} \frac{dE}{\sqrt{2E/m}} = \phi_M \beta \sqrt{\frac{2m}{\mu kT}} \tag{2.71}$$

From Eqs. (2.70) and (2.71),

$$\frac{n_E}{n_M} = \frac{4\beta}{\sqrt{\pi\mu}} \tag{2.72}$$

We now substitute Eqs. (2.67), (2.68), and (2.72) into (2.62) and perform the integration. The results can be written in the form

$$\hat{\sigma} = \sigma_{2200}(g + rs) \tag{2.73}$$

where

$$g = \frac{1}{\sigma_{2200}} \left(\frac{4T}{\pi\hat{T}}\right)^{1/2} \int_0^\infty \sigma(E) \frac{E}{(kT)^2} e^{-E/kT} \, dE \tag{2.74}$$

$$r = \frac{\sqrt{\pi\mu}}{4} \frac{n_E}{n} \tag{2.75}$$

$$s = \frac{1}{\sigma_{2200}} \left(\frac{4T}{\pi\hat{T}}\right)^{1/2} \int_{\mu kT}^\infty \sigma(E) \frac{dE}{E} - \frac{4}{\sqrt{\pi\mu}} g \tag{2.76}$$

and

$$\hat{T} = 293.2 \text{ K}$$

The fraction n_E/n in Eq. (2.75) is a parameter specified by the reactor designer. For a purely thermal spectrum $n_E = 0$, so that $r = 0$ and $\hat{\sigma} = \sigma_{2200}g$. For the pressurized-water reactor considered in Sec. 6.4 of Chap. 3, the epithermal ratio r is estimated to equal 0.222. When σ varies inversely with v, $\sigma(E) = \sigma_{2200}\sqrt{k\hat{T}/E}$, so that $g = 1$ and $s = 0$.

The factor g is called the "non-$1/v$ correction factor." It becomes greater than unity for a cross section that decreases with increasing neutron speed less rapidly then $1/v$, and it becomes less than unity when the cross section decreases more rapidly than $1/v$.

Values of g and s for ^{233}U, ^{235}U, and ^{239}Pu, as a function of the thermalization temperature T, are listed in Table 2.7. More detailed compilations are available in published reports [C1, W1, W4, W5].

The Westcott g and s factors can also be used to determine the effective thermal cross section $\bar{\sigma}$, such that when multiplied by the integrated Maxwell-Boltzmann thermal flux ϕ_M the proper reaction rate with a nuclide is obtained, as already defined by Eq. (2.55). From Eqs. (2.55) and (2.62), $\bar{\sigma}$ is related to $\hat{\sigma}$ by

$$\bar{\sigma} = \hat{\sigma} \frac{n\hat{v}}{\phi_M} \tag{2.77}$$

By substituting Eqs. (2.45), (2.68), and (2.70) into (2.77),

$$\bar{\sigma} = \hat{\sigma} \left(\frac{\pi\hat{T}}{4T}\right)^{1/2} \left(\frac{1}{1 - n_E/n}\right) \tag{2.78}$$

or, using Eq. (2.75) to introduce the spectrum parameter r:

$$\bar{\sigma} = \hat{\sigma} \left(\frac{\pi\hat{T}}{4T}\right)^{1/2} \left(\frac{1}{1 - 4r/\sqrt{\pi\mu}}\right) \tag{2.79}$$

$\bar{\sigma}$ is the effective cross section defined by Eq. (2.79), which is used later in this text (cf. Sec. 6 and Chap. 3).

The Westcott formulation for the effective cross sections $\hat{\sigma}$ and $\bar{\sigma}$ is useful only for well-moderated thermal reactors, where the approximations of the neutron spectra are more reasonable. Even in such reactors, more detailed calculations of actual neutron spectra and effective cross sections are necessary for precise reactor design. The Westcott cross sections are not applicable to fast-spectrum reactors, where neutron moderation and thermalization are suppressed.

Table 2.7 Westcott parameters for ^{233}U, ^{235}U, and ^{239}Pu†

T, °C	g (abs)	s (abs)	g (fiss)	s (fiss)
		^{233}U‡		
20	0.9983	1.286	1.0003	1.216
40	0.9979	1.330	1.0005	1.256
60	0.9976	1.372	1.0007	1.295
80	0.9973	1.412	1.0009	1.333
100	0.9972	1.452	1.0011	1.370
120	0.9971	1.490	1.0014	1.406
140	0.9971	1.527	1.0016	1.440
160	0.9971	1.562	1.0019	1.474
180	0.9972	1.597	1.0022	1.507
200	0.9973	1.631	1.0025	1.539
220	0.9975	1.664	1.0029	1.570
240	0.9978	1.697	1.0032	1.600
260	0.9980	1.728	1.0036	1.630
280	0.9984	1.759	1.0040	1.659
300	0.9987	1.789	1.0044	1.688
330	0.9993	1.833	1.0051	1.730
360	1.0000	1.876	1.0058	1.770
390	1.0007	1.918	1.0065	1.809
420	1.0015	1.958	1.0073	1.847
450	1.0024	1.998	1.0081	1.885
480	1.0033	2.036	1.0090	1.921
510	1.0042	2.074	1.0099	1.956
540	1.0052	2.111	1.0108	1.991
570	1.0062	2.147	1.0118	2.025
600	1.0072	2.182	1.0128	2.058
		^{235}U§		
20	0.9771	0.1457	0.9781	−0.0263
40	0.9723	0.1595	0.9735	−0.0178
60	0.9678	0.1729	0.9692	−0.0096
80	0.9636	0.1856	0.9650	−0.0017
100	0.9597	0.1977	0.9611	0.0058
120	0.9560	0.2092	0.9573	0.0131
140	0.9526	0.2201	0.9538	0.0197
160	0.9494	0.2302	0.9505	0.0260
180	0.9465	0.2396	0.9474	0.0317
200	0.9438	0.2484	0.9445	0.0368
220	0.9413	0.2565	0.9418	0.0416
240	0.9391	0.2640	0.9392	0.0459
260	0.9370	0.2711	0.9369	0.0496
280	0.9351	0.2774	0.9347	0.0530
300	0.9334	0.2833	0.9327	0.0559

(*See footnotes on page 52.*)

Table 2.7 Westcott parameters for ^{233}U, ^{235}U, and ^{239}Pu (*Continued*)

T, °C	g (abs)	s (abs)	g (fiss)	s (fiss)
330	0.9312	0.2913	0.9299	0.0597
360	0.9292	0.2987	0.9274	0.0629
390	0.9275	0.3054	0.9252	0.0655
420	0.9261	0.3117	0.9232	0.0680
450	0.9248	0.3180	0.9214	0.0703
480	0.9237	0.3242	0.9197	0.0727
510	0.9228	0.3304	0.9182	0.0750
540	0.9219	0.3370	0.9169	0.0776
570	0.9211	0.3439	0.9156	0.0805
600	0.9204	0.3510	0.9143	0.0837
		^{239}Pu§		
20	1.0723	2.338	1.0487	1.794
40	1.0909	2.369	1.0623	1.820
60	1.1117	2.389	1.0777	1.835
80	1.1350	2.396	1.0952	1.840
100	1.1611	2.390	1.1150	1.836
120	1.1903	2.373	1.1373	1.822
140	1.2227	2.343	1.1623	1.797
160	1.2582	2.298	1.1898	1.761
180	1.2970	2.239	1.2200	1.713
200	1.3388	2.166	1.2528	1.653
220	1.3836	2.077	1.2880	1.581
240	1.4313	1.974	1.3255	1.497
260	1.4817	1.857	1.3653	1.402
280	1.5345	1.727	1.4071	1.297
300	1.5895	1.586	1.4507	1.182
330	1.6758	1.356	1.5193	0.996
360	1.7658	1.110	1.5910	0.796
390	1.8588	0.854	1.6651	0.588
420	1.9539	0.594	1.7410	0.376
450	2.0505	0.334	1.8182	0.166
480	2.1477	0.081	1.8959	−0.040
510	2.2451	−0.163	1.9738	−0.238
540	2.3419	−0.395	2.0514	−0.426
570	2.4377	−0.614	2.1281	−0.604
600	2.5321	−0.817	2.2037	−0.770

† s is chosen here as the s_2 parameter in the Westcott formulation, consistent with the use of a cutoff energy for epithermal neutrons, as in Eq. (2.66).

‡ From Westcott [W4].

§ From Critoph [C1].

4.8 Half-life for Neutron Reactions

The change in number of atoms of neutron-absorbing nuclide N with time due to neutron reactions alone, and in the absence of a source of this nuclide, is

$$\frac{dN}{dt} = -\sigma_a \phi N \tag{2.80}$$

where the product $\sigma_a \phi$ represents the sum of the effective $\sigma\phi$ products defined in Sec. 4.7.

For time-independent effective cross sections and neutron flux, Eq. (2.80) integrates to

$$N = N^0 e^{-\sigma_a \phi t} \tag{2.81}$$

where N^0 is the number of atoms at $t = 0$. $\sigma_a \phi$ is sometimes referred to as the "burnout constant," and $\ln 2/\sigma_a \phi$ is the half-life for burnout. For example, in a flux of 10^{14} $n/(\text{cm}^2 \cdot \text{s})$, the half-life for burnout of a nuclide with an absorption cross section of 100 b is

$$\frac{0.693}{100 \times 10^{-24} \times 10^{14}} = 6.93 \times 10^7 \text{ s} = 2.20 \text{ yr} \tag{2.82}$$

If neutron absorption in species 1 results in a single nuclear reaction with a nonradioactive product, the number of product atoms N_2 formed is

$$N_2 = N^0 (1 - e^{-\sigma_a \phi t}) \tag{2.83}$$

If two or more competing reactions take place, the number of stable product atoms formed is

$$N_2 = N^0 \frac{\sigma_c}{\sigma_a} (1 - e^{-\sigma_a \phi t}) \tag{2.84}$$

where σ_c is the capture cross section for the reaction producing the product nuclide in question. Modification of these equations for simultaneous neutron reaction and radioactive decay will be treated in Sec. 6.

5 THE FISSION PROCESS

5.1 Fissile Materials

Table 2.8 lists capture and fission cross sections for the four nuclides fissile with thermal neutrons and gives the average number of neutrons produced per nuclide fissioned (ν) and per

Table 2.8 Properties of fissile nuclides for 2200 m/s neutrons†

	^{233}U	^{235}U	^{239}Pu	^{241}Pu
Cross sections, b				
Fission σ_f	531.1	582.2	742.5	1009
Capture σ_c	47.7	98.6	268.8	368
Absorption σ_a	578.8	680.8	1011.3	1377
$\alpha = \sigma_c/\sigma_f$	0.0898	0.169	0.362	0.3647
Neutrons produced				
Per fission ν	2.492	2.418	2.871	2.927
Per neutron absorbed η	2.287	2.068	2.108	2.145

†From App. C.

neutron absorbed (η). These properties are needed to calculate reactor neutron balances, evaluate fuel reactivity, and work out fuel cycles.

5.2 Fission Products

More than 300 different nuclides have been observed as the primary products of fission. The term *fission products* usually refers to the primary fission products, i.e., the fission fragments and their daughters resulting from radioactive decay and neutron absorption. Only a few of the primary fission products are stable, the rest being beta-emitting radionuclides. As a fission-product radionuclide undergoes beta decay, its atomic number increases whereas its mass number remains constant. The *direct yield* of a fission-product nuclide is the fraction of the total fissions that yield this nuclide, essentially as a direct-fission fragment. The cumulative yield of a given nuclide is the fraction of fissions that directly yield that nuclide and its radioactive decay precursors in the constant-mass fission-product chain; i.e., it is the sum of the direct yields of that nuclide and its decay precursors. Many of the fission products have such short half-lives that no accurate measure of their direct yields as primary fission products is available. However, reasonably reliable data have been secured on the cumulative yields of many of the long-lived radionuclides and on the cumulative yields of all the nuclides in a fission-product chain of given mass number [B3, W1]. The cumulative yields by mass number in the fission of ^{233}U, ^{235}U, and ^{239}Pu by slow neutrons and in the fission of ^{235}U, ^{239}Pu, ^{232}Th, and ^{238}U by fast neutrons are listed in Table 2.9 and are shown as the familiar double-hump mass-yield curves in Figs. 2.12 and 2.13.

This situation with regard to yield and radioactive decay at each mass number is illustrated for mass number 90 in Fig. 2.14. For accurate estimation of the amount of any nuclide produced at a given time, the differential equations appropriate to such a system of yields and decays must be set up and solved. This is illustrated in Secs. 6.3 through 6.5 for selected fission-product nuclides of mass 135 and masses 147, 149, 151, and 152, which are important neutron-absorbing poisons in thermal reactors.

5.3 Energy Release in Fission

In the steady state, when atoms undergoing fission are in equilibrium with their radioactive fission products, the energy released per fission is distributed approximately as in Table 2.10.

In a short burst of nuclear energy, such as in a fission bomb or in a rapid rise in reactor power, the total energy released is the sum of the first four terms, 182 to 191 MeV. When a reactor is shut down after reaching steady state, or when fuel from such a reactor is discharged, the energy of beta and gamma decay of the fission products, 13 MeV in all, is released gradually over a long period of time. The neutrino energy is not available. An average of 200 MeV of recoverable energy per fission is used in this text.

The rate of heat release and the intensity of radiation from the fuel are important factors in the design of emergency cooling systems for reactors, casks for shipping discharge fuel, fuel reprocessing plants, and facilities for storing fission-product wastes. These depend on the rate of fission of the fuel when it was in the reactor, the length of time the fuel was in the reactor, and the length of time the fuel was allowed to "cool" before shipping and processing. The exact calculation of these relationships is very tedious because of the large number of nuclides contributing to heat and radiation release, and large digital computers are required [B2]. An approximate statistical correlation by Way and Wigner [W2] provides simple equations suitable for quick approximations.

At a time t in days after fission, the products of a single fission undergo beta decay at a rate $\beta(t)$ given by

$$\beta(t) = 5.2 \times 10^{-6} t^{-1.2} \text{ disintegrations/s} \tag{2.85}$$

Table 2.9 Percent fission yield by mass number†

Mass number	Fission by slow neutrons ^{233}U	^{235}U	^{239}Pu	Fission by fast neutrons‡ ^{235}U	^{239}Pu	^{232}Th	^{238}U
3	2×10^{-4}	1.3×10^{-4}	2.3×10^{-4}	1.2×10^{-4}	2.5×10^{-4}	8.00×10^{-5}	1.4×10^{-4}
72	0.000200	0.000016	0.000120	0.00152	0.00120	0.000330	0.000100
73	0.000600	0.000110	0.000200	0.000190		0.000450	0.000200
74	0.00100	0.000350	0.000800	0.0332		0.00250	0.000700
75	0.00301	0.000804	0.000804	0.0758		0.00502	0.00100
76	0.00500	0.00250	0.00300	0.0190		0.0130	0.00200
77	0.0210	0.00830	0.0100	0.0883		0.0200	0.00380
78	0.0600	0.0200	0.0250	0.190		0.100	0.0160
79	0.100	0.0560	0.0400	0.379		0.180	0.0300
80	0.200	0.100	0.0700	0.152		0.337	0.0700
81	0.424	0.140	0.117	0.253		0.596	0.117
82	0.691	0.320	0.200	0.000072		1.30	0.220
83	1.17	0.544	0.290	0.910	0.580	1.99	0.445
84	1.95	1.00	0.468	1.90	0.940	3.65	0.848
85	2.64	1.30	0.539	1.42	0.539	3.80	0.736
86	3.27	2.02	0.769	1.92	0.760	6.00	1.38
87	4.56	2.49	0.920	2.56	0.920	6.50	1.80
88	5.37	3.57	1.42	3.51	1.42	6.70	2.50
89	5.86	4.79	1.71	4.55	1.71	6.70	2.90
90	6.43	5.77	2.21	5.59	2.25	6.80	3.20
91	6.43	5.84	2.61	5.41	2.36	7.23	4.04
92	6.64	6.03	3.14	5.79	3.14	7.20	4.50
93	6.98	6.45	3.97	6.16	3.97	7.08	4.99
94	6.68	6.40	4.48	6.16	4.48	6.99	5.31
95	6.11	6.27	5.03	6.07	5.80	6.90	5.70
96	5.59	6.33	5.17	6.08	6.16	6.61	5.91
97	5.37	6.09	5.65	5.87	7.33	5.20	6.00
98	5.15	5.78	5.89	5.49	5.88	3.60	6.20
99	4.80	6.06	6.10	5.98	6.10	2.70	6.30
100	4.41	6.30	7.10	5.98	7.10	1.11	6.40
101	2.91	5.00	5.91	4.74	5.90	0.550	6.50
102	2.22	4.19	5.99	3.98	5.99	0.220	6.60
103	1.80	3.00	5.67	2.85	5.66	0.160	6.60
104	0.940	1.80	5.93	1.71	5.93	0.0900	5.00
105	0.480	0.900	5.30	1.71	3.90	0.0700	3.30
106	0.240	0.380	4.57	0.901	4.57	0.0420	2.70
107	0.160	0.190	3.50	0.758	3.60	0.0600	2.00
108	0.0700	0.0650	2.50	0.304	2.10	0.0590	0.600
109	0.0440	0.0300	1.40	0.106	2.80	0.0550	0.320
110	0.0300	0.0200	0.500	0.0759	0.0	0.0550	0.150
111	0.0242	0.0192	0.232	0.0721	0.460	0.0525	0.0768
112	0.0160	0.0100	0.120	0.0417	0.240	0.0570	0.0460
113	0.0180	0.0314	0.0700	0.0417	0.0200	0.0353	0.0345
114	0.0190	0.0120	0.0520	0.0379	0.0200	0.0550	0.0400
115	0.0210	0.0104	0.0410	0.0398	0.00820	0.0750	0.0370
116	0.0180	0.0105	0.0380	0.0493	0.0	0.0550	0.0380
117	0.0170	0.0110	0.0390	0.0417	0.0220	0.0540	0.0400
118	0.0170	0.0110	0.0390	0.0382	0.00200	0.0550	0.0400
119	0.0170	0.0120	0.0400	0.0382	0.00800	0.0560	0.0400
120	0.0180	0.0130	0.0400	0.0382	0.00193	0.0570	0.0410

(See footnotes on page 56.)

Table 2.9 Percent fission yield by mass number (*Continued*)

Mass number	Fission by slow neutrons			Fission by fast neutrons‡			
	^{233}U	^{235}U	^{239}Pu	^{235}U	^{239}Pu	^{232}Th	^{238}U
121	0.0180	0.0150	0.0440	0.0591	0.0873	0.0590	0.0420
122	0.0300	0.0160	0.0450	0.0496	0.00193	0.0610	0.0450
123	0.0500	0.0173	0.0550	0.0580	0.00193	0.0660	0.0455
124	0.0700	0.0220	0.0700	0.0763	0.0	0.0670	0.0550
125	0.0840	0.0210	0.115	0.0878	0.139	0.0730	0.0650
126	0.200	0.0440	0.200	0.239	0.385	0.0800	0.0800
127	0.600	0.130	0.390	0.597	0.770	0.120	0.120
128	1.21	0.409	1.21	1.19	0.963	0.198	0.385
129	2.00	0.800	2.00	1.91	1.93	0.400	1.30
130	2.60	2.00	2.60	1.91	1.94	0.800	2.00
131	3.39	2.93	3.78	2.96	3.04	1.62	3.20
132	4.54	4.38	5.26	4.20	5.08	2.87	4.70
133	5.78	6.61	6.53	6.21	6.65	4.20	5.50
134	5.94	8.06	7.46	7.25	7.22	5.37	6.60
135	6.16	6.41	7.17	6.20	7.00	5.50	6.00
136	6.75	6.47	6.74	6.18	6.48	5.75	6.00
137	6.58	6.15	6.03	5.92	6.38	6.29	6.20
138	6.31	5.74	6.31	5.54	6.07	6.60	6.00
139	6.44	6.55	5.87	5.74	5.85	6.90	5.83
140	6.47	6.44	5.64	6.02	5.39	7.29	5.77
141	6.49	6.40	5.09	5.74	5.49	9.00	5.90
142	6.83	6.01	5.01	5.63	4.82	7.43	5.69
143	5.99	5.73	4.56	5.92	5.10	7.30	5.10
144	4.61	5.62	3.93	5.83	3.78	7.10	4.50
145	3.47	3.98	3.13	4.01	3.01	5.00	4.80
146	2.63	3.07	2.60	3.15	2.50	4.00	4.20
147	1.98	2.36	2.07	2.48	2.12	2.80	3.50
148	1.34	1.71	1.73	1.63	1.67	0.900	2.50
149	0.760	1.13	1.32	1.24	1.27	0.500	1.80
150	0.560	0.670	1.01	0.706	0.973	0.260	1.50
151	0.335	0.440	0.800	0.477	0.770	0.170	1.20
152	0.220	0.281	0.620	0.286	0.598	0.0550	0.850
153	0.130	0.169	0.417	0.143	0.356	0.0200	0.407
154	0.0450	0.0770	0.290	0.0858	0.280	0.0100	0.250
155	0.0230	0.0330	0.230	0.0592	0.443	0.00450	0.130
156	0.0110	0.0140	0.110	0.0248	0.212	0.00200	0.0710
157	0.00450	0.00780	0.0800	0.0141	0.143	0.000750	0.0350
158	0.00150	0.00200	0.0400	0.0191	0.385	0.000250	0.0130
159	0.000800	0.00107	0.0210	0.0105	0.202	0.000130	0.00840
160	0.000200	0.000390	0.00980	0.0258	0.0156	0.000030	0.00390
161	0.000060	0.000180	0.00300	0.0763	0.0376	0.000010	0.00160
162	0.000027	0.000060	0.00200		0.0173	0.000007	0.000800
163	0.000012		0.000900		0.00770		0.000360
164			0.000300		0.00289		0.000120
165			0.000130		0.00116		0.000050
166			0.000068		0.000855		0.000027
Sum	201	200	201	200	202	200	206

†Data from [B3, D1, G1, K1, W1].

‡^{235}U and ^{239}Pu yields are for a fast-reactor neutron spectrum; ^{232}U and ^{238}U yields are for fission-spectrum neutrons.

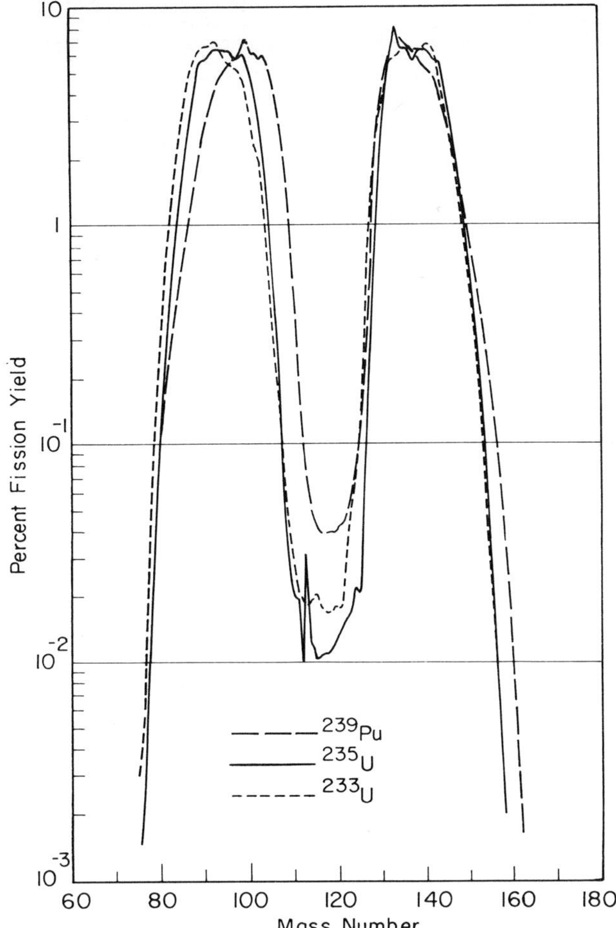

Figure 2.12 Fission yields for slow-neutron fission of ^{233}U, ^{235}U, and ^{239}Pu.

and release energy in the form of beta particles, gamma rays, and neutrinos at a rate $E(t)$ given by

$$E(t) = 3.9t^{-1.2} + 11.7t^{-1.4} \text{ eV/s} \tag{2.86}$$

The above equations apply after about 1 min after fission has taken place. Approximately one-fourth of this energy is due to gamma radiation and one-fourth to beta.

In a case of practical interest, a fuel sample will have been in a reactor liberating heat at some constant rate for T days, and will then have been cooled for t days. The rate of disintegration of fission products in the fuel sample in curies per watt of reactor power will be

$$\frac{\int_t^{T+t} [\beta(t) \text{ disintegrations/(s}\cdot\text{fission)}]\ (86{,}400 \text{ s/day})\ (dt \text{ days})}{(200 \text{ MeV/fission})\ [1.60 \times 10^{-13} \text{ (W}\cdot\text{s)/MeV}]\ [3.7 \times 10^{10} \text{ disintegrations/(s}\cdot\text{Ci)}]}$$

or

$$\frac{\text{Ci}}{\text{W}} = 1.9[t^{-0.2} - (T + t)^{-0.2}] \tag{2.87}$$

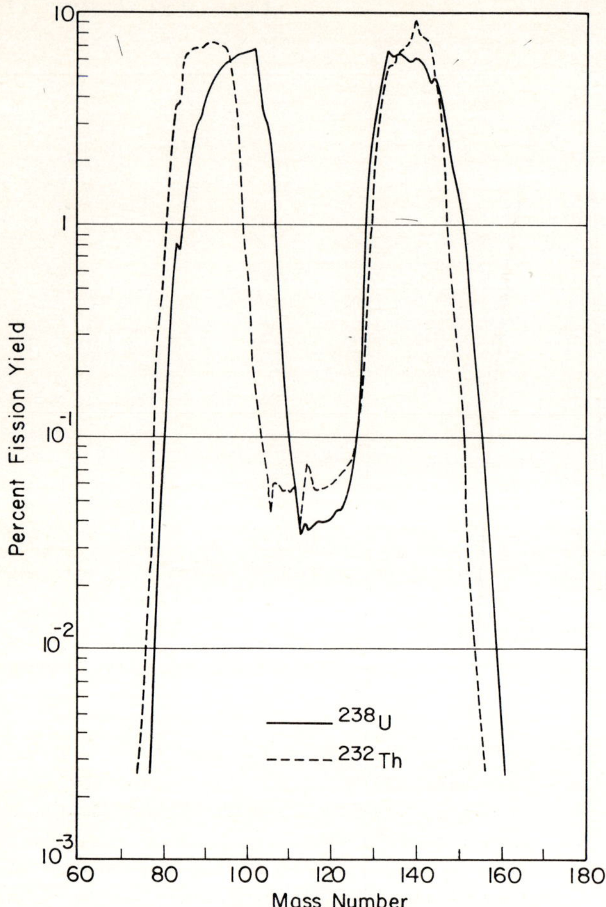

Figure 2.13 Fission yields for fast-neutron fission of ^{232}Th and ^{238}U.

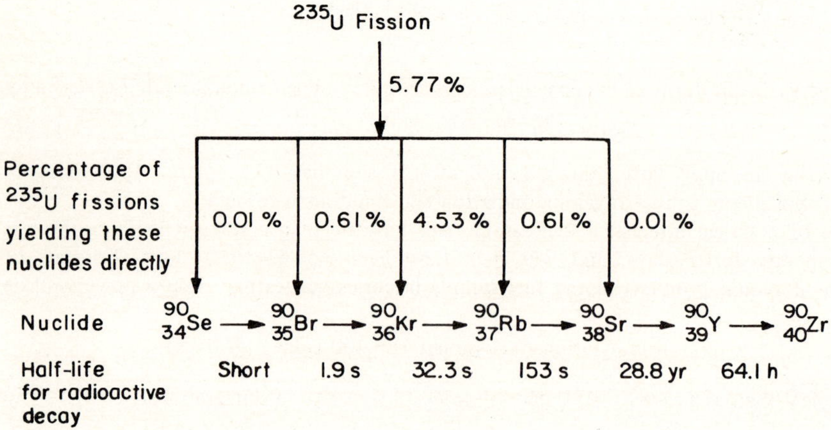

Figure 2.14 Fission-product decay chain for mass 90.

Table 2.10 Energy distribution in fission

	MeV per fission	
Kinetic energy of fission fragments	167	
Kinetic energy of neutrons	5	
Energy of instantaneous gamma rays	7	
Energy from absorption of excess neutrons†	3–12	
Subtotal		182–191
Energy from fission-product gamma rays	6	
Beta rays	8	
Neutrinos	(12)	
Subtotal (recoverable energy)		14
Total (recoverable energy)		196–205

†Dependent on how many excess neutrons are absorbed and how they are absorbed.

Similarly, the ratio of the rate of beta- and gamma-energy release $P_d(T, t)$ to the rate of heat release in fission P_f is

$$\frac{P_d(T, t)}{P_f} = 0.0042[t^{-0.2} - (T + t)^{-0.2}] + 0.0063[t^{-0.4} - (T + t)^{-0.4}] \tag{2.88}$$

Equation (2.88) can also be written as

$$\frac{P_d(T, t)}{P_f} = \frac{P_d(\infty, t)}{P_f} - \frac{P_d(\infty, T + t)}{P_f} \tag{2.89}$$

where the two quantities on the right-hand side are calculated from Eq. (2.88) for infinite irradiation time and for cooling times of t and $T + t$, respectively.

A more accurate estimate of the decay energy from fission products can be obtained from the ANS Standard [A2]. The data are presented here as the decay-heat rate $F(\infty, t)$ at cooling time t from fission products produced at a constant fission rate of unity, over an infinitely long operating period and without neutron absorption in the fission products. Values of $F(\infty, t)$ for the fission of ^{235}U by thermal neutrons are presented in Table 2.11. Data for the fission-product decay-heat rate from thermal fission of ^{239}Pu and from the fast fission of ^{238}U are also given in the ANS Standard [A2]. These data are applicable to light-water reactors containing ^{235}U as a major fissile material and ^{238}U as the fertile material. The time domain of the official ANS Standard extends from cooling times of 1 to 10^4 s.

The fission-product decay-heat rate $F(T, t)$ per unit fission rate for finite irradiation time T can be synthesized from

$$F(T, t) = \sum_{i=1}^{23} \frac{\gamma_i}{\lambda_i}(1 - e^{-\lambda_i T})e^{-\lambda_i t} \tag{2.90}$$

where γ_i and λ_i are empirical constants. Values of γ_i and λ_i for ^{235}U thermal fission are listed in Table 2.12. The data in Table 2.11 for infinite irradiation time can be constructed from Eq. (2.90) by choosing $T = 10^{13}$ s.

Alternatively, $F(T, t)$ can be obtained from the data in Table 2.11 by

$$F(T, t) = F(\infty, t) - F(\infty, T + t) \tag{2.91}$$

Data in Table 2.11 for cooling times greater than 10^4 s can be used in Eq. (2.91) to synthesize values of $F(T, t)$ within the time domain (1 to 10^4 s) of the ANS Standard.

Table 2.11 Decay-heat power from fission products from thermal fission of ^{235}U and for near-infinite reactor operating time†

Time after reactor shutdown, s	Decay-heat power $F(\infty, t)$, (MeV/s)/(fission/s)	Percent uncertainty
1	1.231×10^1	3.3
1.5	1.198×10^1	2.7
2.0	1.169×10^1	2.4
4.0	1.083×10^1	2.2
6.0	1.026×10^1	2.1
8.0	9.830	2.0
1.0×10^1	9.494	2.0
1.5×10^1	8.882	1.9
2.0×10^1	8.455	1.9
4.0×10^1	7.459	1.8
6.0×10^1	6.888	1.8
8.0×10^1	6.493	1.8
1.0×10^2	6.198	1.8
1.5×10^2	5.696	1.8
2.0×10^2	5.369	1.8
4.0×10^2	4.667	1.8
6.0×10^2	4.282	1.8
8.0×10^2	4.009	1.8
1.0×10^3	3.796	1.8
1.5×10^3	3.408	1.8
2.0×10^3	3.137	1.8
4.0×10^3	2.534	1.8
6.0×10^3	2.234	1.7
8.0×10^3	2.044	1.7
1.0×10^4	1.908	1.7
1.5×10^4	1.685	1.8
2.0×10^4	1.545	1.8
4.0×10^4	1.258	1.9
6.0×10^4	1.117	1.9
8.0×10^4	1.030	2.0
1.0×10^5	9.691×10^{-1}	2.0
1.5×10^5	8.734×10^{-1}	2.0
2.0×10^5	8.154×10^{-1}	2.0
4.0×10^5	6.975×10^{-1}	2.0
6.0×10^5	6.331×10^{-1}	2.0
8.0×10^5	5.868×10^{-1}	2.0
1.0×10^6	5.509×10^{-1}	2.0
1.5×10^6	4.866×10^{-1}	2.0
2.0×10^6	4.425×10^{-1}	2.0
4.0×10^6	3.457×10^{-1}	2.0

(*See footnotes on page 61.*)

Table 2.11 Decay-heat power from fission products from thermal fission of ^{235}U and for near-infinite reactor operating time (*Continued*)

Time after reactor shutdown, s	Decay-heat power $F(\infty, t)$, (MeV/s)/(fissions/s)	Percent uncertainty
6.0×10^6	2.983×10^{-1}	2.0
8.0×10^6	2.680×10^{-1}	2.0
1.0×10^7	2.457×10^{-1}	2.0
1.5×10^7	2.078×10^{-1}	2.0
2.0×10^7	1.846×10^{-1}	2.0
4.0×10^7	1.457×10^{-1}	2.0
6.0×10^7	1.308×10^{-1}	2.0
8.0×10^7	1.222×10^{-1}	2.0
1.0×10^8	1.165×10^{-1}	2.0
1.5×10^8	1.082×10^{-1}	2.0
2.0×10^8	1.032×10^{-1}	2.0
4.0×10^8	8.836×10^{-2}	2.0
6.0×10^8	7.613×10^{-2}	2.0
8.0×10^8	6.570×10^{-2}	2.0
1.0×10^9	5.678×10^{-2}	2.0

†For irradiation time of 10^{13} s. Calculated for no neutron absorption in fission products.

Source: American Nuclear Society Standards Committee Working Group ANS-5.1, "American National Standard for Decay Heat Power in Light Water Reactors," Standard ANSI/ANS-5.1, American Nuclear Society, La Grange Park, Ill., 1979. With permission of the publisher, the American Nuclear Society.

The total decay-heat power $P'_d(T, t)$ for fission products from a reactor operating at constant total thermal power P_f, and neglecting neutron absorption in fission products, is given by the following simplified method, from the ANS Standard:

$$P'_d(T, t) = 1.02 \frac{P_f F(T, t)}{Q} \tag{2.92}$$

where $F(T, t)$ is evaluated from ^{235}U data, using Eq. (2.90) or (2.91), and Q is the thermal energy per fission. The factor 1.02 corrects for the greater heat generation per fission from ^{238}U fission products during the period of about 100 s after reactor shutdown. The ratio $P'_d/P_f Q$ of fission-product decay heat rate at cooling time t to reactor power prior to shutdown is plotted as a function of T and t in Fig. 2.15.

Neutron absorption in fission products has a small effect on decay-heat power for $t \leqslant 10^4$ s and is treated by a correction factor G. The corrected total decay-heat power is given by the ANS Standard, in terms of thermal-neutron flux (in neutrons/cm$^2\cdot$s), reactor operating time T (in s), and cooling time t (in s) as

$$P(T, t) = P'(T, t)G \tag{2.93}$$

where

$$G = 1.0 + (3.24 \times 10^{-6} + 5.23 \times 10^{-10}\, t)T^{0.4}\psi \tag{2.94}$$

The parameter ψ is the total number of fissions after irradiation time T per initial fissile atom, calculated by techniques described in Chap. 3. Equation (2.94) applies for operating times $T < 1.2614 \times 10^8$ s (4 years), shutdown times $t < 10^4$ s, and $\psi < 3.0$. A more detailed technique for calculating fission-product decay-heat power from an arbitrary time-dependent fission power, including contributions from the fission of ^{235}U, ^{238}U, and ^{239}Pu, is given in the ANS Standard [A2].

To predict the decay-heat rate from fission products after cooling times of several years, additional corrections must be made for absorption of neutrons in long-lived fission products, particularly the absorption of neutrons in stable ^{133}Cs to form 2.05-year ^{134}Cs. Computer codes such as ORIGEN [B2] and CINDER [E1] are particularly useful for this purpose.

Estimated maximum values of the ratio G of fission-product decay-heat rate, with neutron absorption in fission products considered, to the decay-heat rate in the absence of neutron absorption in fission products are given in Table 2.13 [A2]. The data are calculated for ^{235}U-^{238}U fuel irradiated for 4 years in a light-water reactor. For cooling times of $< 10^4$ s, the

Table 2.12 Decay-heat parameters for fission products from thermal fission of ^{235}U

Group i	γ_i, MeV/(s·fission)	λ_i, s^{-1}
1	6.5057×10^{-1}	2.2138×10^{1}
2	5.1264×10^{-1}	5.1587×10^{-1}
3	2.4384×10^{-1}	1.9594×10^{-1}
4	1.3850×10^{-1}	1.0314×10^{-1}
5	5.5440×10^{-2}	3.3656×10^{-2}
6	2.2225×10^{-2}	1.1681×10^{-2}
7	3.3088×10^{-3}	3.5870×10^{-3}
8	9.3015×10^{-4}	1.3930×10^{-3}
9	8.0943×10^{-4}	6.2630×10^{-4}
10	1.9567×10^{-4}	1.8906×10^{-4}
11	3.2535×10^{-5}	5.4988×10^{-5}
12	7.5595×10^{-6}	2.0958×10^{-5}
13	2.5232×10^{-6}	1.0010×10^{-6}
14	4.9948×10^{-7}	2.5438×10^{-6}
15	1.8531×10^{-7}	6.6361×10^{-7}
16	2.6608×10^{-8}	1.2290×10^{-7}
17	2.2398×10^{-9}	2.7213×10^{-8}
18	8.1641×10^{-12}	4.3714×10^{-9}
19	8.7797×10^{-11}	7.5780×10^{-10}
20	2.5131×10^{-14}	2.4786×10^{-10}
21	3.2176×10^{-16}	2.2384×10^{-13}
22	4.5038×10^{-17}	2.4600×10^{-14}
23	7.4791×10^{-17}	1.5699×10^{-14}

Source: American Nuclear Society Standards Committee Working Group ANS-5.1, "American National Standard for Decay Heat Power in Light Water Reactors," Standard ANSI/ANS-5.1, American Nuclear Society, La Grange Park, Ill., 1979. With permission of the publisher, the American Nuclear Society.

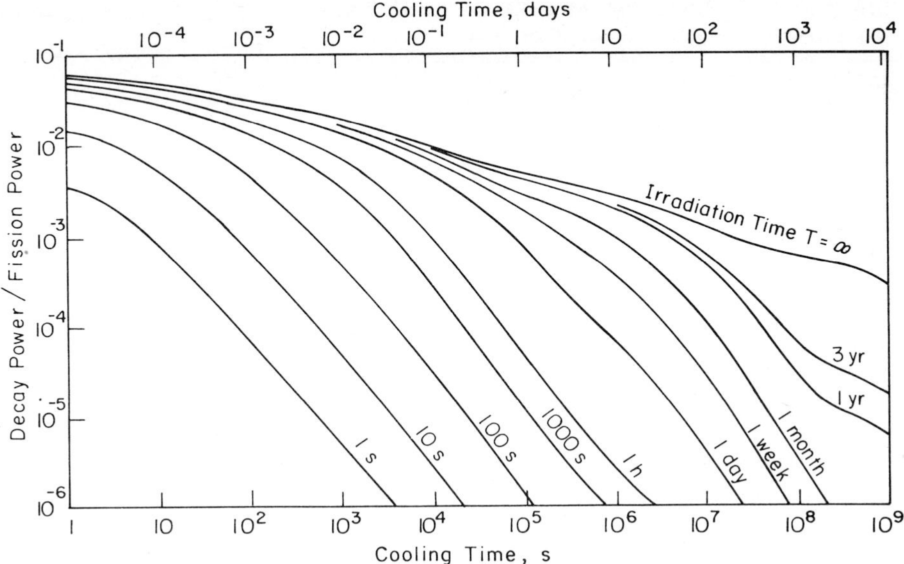

Figure 2.15 Decay power of fission products from ^{235}U.

correction is less than a 6 percent increase. For cooling times of about 3 years, neutron absorption causes the fission-product decay-heat rate to increase by about 60 percent.

Decay of the actinides formed by neutron capture is another source of decay heat, although during cooling times of less than a few hundred years it contributes much less decay heat than do the fission products. The actinide nuclides that contribute appreciably during the first few days after reactor shutdown are 23.5-min ^{239}U and 2.35-day ^{239}Np. The quantities of these actinides at the time of reactor shutdown can be calculated using the techniques described in Chap. 3, and their rate of decay after shutdown can be predicted from Eqs. (2.13) and 2.14). The decay-heat rate due to these two species can then be estimated as a function of T and t by multiplying the decay rates by the average thermal energy released per decay [A2]:

$$U = 0.474 \text{ MeV/decay}$$

$$Np = 0.419 \text{ MeV/decay}$$

For longer cooling times additional decay heat will be liberated by longer-lived actinides formed by neutron capture in the fuel material, e.g., ^{237}U, ^{238}Pu, ^{239}Pu, ^{240}Pu, ^{241}Pu, ^{241}Am, ^{242}Cm, ^{244}Cm, etc., and by radionuclides formed by neutron reactions with fuel structural material, such as metal cladding. Methods and illustrative data that can be used in estimating the concentrations of such radionuclides and their contributions to decay heat are discussed in Chaps. 3 and 8.

6 GROWTH AND DECAY OF NUCLIDES WITH SIMULTANEOUS RADIOACTIVE DECAY, NEUTRON ABSORPTION, AND CONTINUOUS PROCESSING

6.1 Batch Operation

We consider here the growth and decay of individual radionuclides in a chain in which individual radionuclides may be destroyed or removed by radioactive decay, neutron absorption,

Table 2.13 Ratio of fission-product decay-heat rate with neutron absorption to decay-heat rate without absorption†

Time after reactor shutdown, s	G_{max}	Time after reactor shutdown, s	G_{max}
1.0	1.020	6.0×10^4	1.111
1.5	1.020	8.0×10^4	1.119
2.0	1.020	1.0×10^5	1.124
4.0	1.021	1.5×10^5	1.130
6.0	1.022	2.0×10^5	1.131
8.0	1.022	4.0×10^5	1.126
1.0×10^1	1.022	6.0×10^5	1.124
1.5×10^1	1.022	8.0×10^5	1.123
2.0×10^1	1.022	1.0×10^6	1.124
4.0×10^1	1.022	1.5×10^6	1.125
6.0×10^1	1.022	2.0×10^6	1.127
8.0×10^1	1.022	4.0×10^6	1.134
1.0×10^2	1.023	6.0×10^6	1.146
1.5×10^2	1.024	8.0×10^6	1.162
2.0×10^2	1.025	1.0×10^7	1.181
4.0×10^2	1.028	1.5×10^7	1.233
6.0×10^2	1.030	2.0×10^7	1.284
8.0×10^2	1.032	4.0×10^7	1.444
1.0×10^3	1.033	6.0×10^7	1.535
1.5×10^3	1.037	8.0×10^7	1.586
2.0×10^3	1.039	1.0×10^8	1.598
4.0×10^3	1.048	1.5×10^8	1.498
6.0×10^3	1.054	2.0×10^8	1.343
8.0×10^3	1.060	4.0×10^8	1.065
1.0×10^4	1.064	6.0×10^8	1.021
1.5×10^4	1.074	8.0×10^8	1.012
2.0×10^4	1.081	1.0×10^9	1.007
4.0×10^4	1.098		

†Ratio based on ^{235}U thermal fission for 4 years, no depletion, typical spectrum for light-water reactor.

Source: American Nuclear Society Standards Committee Working Group ANS-5.1, "American National Standard for Decay Heat Power in Light Water Reactors," Standard ANSI/ANS-5.1, American Nuclear Society, La Grange Park, Ill., 1979. With permission of the publisher, the American Nuclear Society.

and by continuous processing. Examples of continuous-processing removal are the vaporization of one or more gaseous elements from a solid or liquid at high temperature or the continuous separation of one or more chemical elements from a well-stirred fluid mixture. Nuclides within the chain under consideration are linked by radioactive decay or neutron reactions. In the present analysis for batch operation we assume that there is a finite initial amount of only the first member of the chain, that there is no source for continuous formation of this first member, and that there is no source of any other member of the chain other than its precursor in the chain itself.

First we assume a chain in which adjacent members are linked by radioactive decay. The neutron flux and reaction cross sections are assumed to be time-independent. The chain and the removal-rate constants are indicated schematically below:

Chain:	$N_1 \xrightarrow{\lambda_1}$ $\downarrow \phi\sigma_1 + f_1$	$N_2 \xrightarrow{\lambda_2}$ $\downarrow \phi\sigma_2 + f_2$	$N_3 \xrightarrow{\lambda_3} \cdots$ $\downarrow \phi\sigma_3 + f_3$	$N_j \xrightarrow{\lambda_j} \cdots \xrightarrow{\lambda_{i-1}}$ $\downarrow \phi\sigma_j + f_j$	$N_i \xrightarrow{\lambda_i}$ $\downarrow \phi\sigma_i + f_i$
Initial amount at $t = 0$	N_1^0	0	0	0	0
Removal-rate constants:					
Radioactive decay	λ_1	λ_2	λ_3	λ_j	λ_i
Neutron absorption	$\phi\sigma_1$	$\phi\sigma_2$	$\phi\sigma_3$	$\phi\sigma_j$	$\phi\sigma_i$
Continuous processing	f_1	f_2	f_3	f_j	f_i
Total	μ_1	μ_2	μ_3	μ_j	μ_i

The processing removal constant f is interpreted as the fraction removed per unit time by continuous reprocessing. The total removal-rate constant μ is defined as

$$\mu = \lambda + \phi\sigma + f \tag{2.95}$$

In the absence of sources, the time-dependent equation for the number of atoms N_1 of the first member of the chain is

$$\frac{dN_1}{dt} = -(\lambda_1 + \phi\sigma_1 + f_1)N_1 \tag{2.96}$$

or, from Eq. (2.95),

$$\frac{dN_1}{dt} = -\mu_1 N_1 \tag{2.97}$$

For time-independent μ, Eq. (2.97) integrates to

$$N_1 = N_1^0 e^{-\mu_1 t} \tag{2.98}$$

Similarly, for the second member,

$$\frac{dN_2}{dt} = \lambda_1 N_1 - \mu_2 N_2 \tag{2.99}$$

which integrates to

$$N_2 = N_1^0 \lambda_1 \left(\frac{e^{-\mu_1 t}}{\mu_2 - \mu_1} + \frac{e^{-\mu_2 t}}{\mu_1 - \mu_2} \right) \tag{2.100}$$

Similarly, using the technique described in Sec. 7, the general batch equation is obtained:

$$N_1 = N_1^0 e^{-\mu_1 t} \tag{2.101a}$$

$$N_i = N_1^0 \lambda_1 \lambda_2 \cdots \lambda_{i-1} \sum_{j=1}^{i} \frac{e^{-\mu_j t}}{\prod\limits_{\substack{k=1 \\ k \neq j}}^{i} (\mu_k - \mu_j)} \qquad (i > 1) \tag{2.101b}$$

Equation (2.101*b*) was derived for the formation of second and subsequent members of the chain by radioactive decay of the precursors, and the product series $\lambda_1 \lambda_2 \cdots \lambda_{i-1}$ represents the product of the chain-linking decay constants.

However, we may wish to calculate nuclide amounts in a chain wherein some members may be formed by neutron reactions with their individual precursors. We define here a *linear chain* as one in which each nuclide other than the first is formed directly only from a single precursor, illustrated as follows:

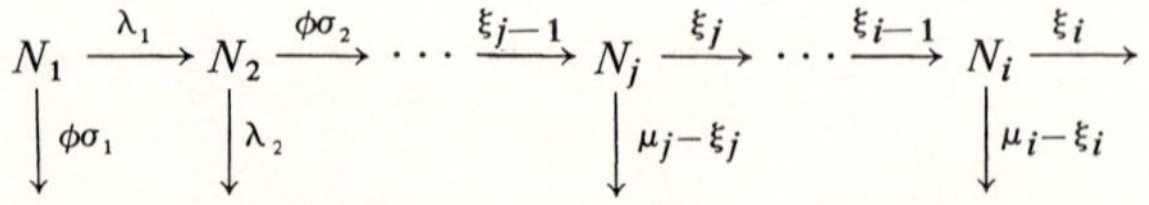

$$\begin{array}{ccccccccc} N_1 & \xrightarrow{\lambda_1} & N_2 & \xrightarrow{\phi\sigma_2} \cdots \xrightarrow{\xi_{j-1}} & N_j & \xrightarrow{\xi_j} \cdots \xrightarrow{\xi_{i-1}} & N_i & \xrightarrow{\xi_i} \\ \downarrow \phi\sigma_1 & & \downarrow \lambda_2 & & \downarrow \mu_j - \xi_j & & \downarrow \mu_i - \xi_i & \end{array}$$

where ξ represents the chain-linking rate constant, e.g., in the above sketch $\xi_1 = \lambda_1$ and $\xi_2 = \phi\sigma_2$.

Subject to the same initial conditions that led to Eq. (2.101), the amount N_i at time t is given by

$$N_1 = N_1^0 e^{-\mu_1 t} \tag{2.102a}$$

$$N_i = N_1^0 \, \xi_1 \xi_2 \cdots \xi_{i-1} \sum_{j=1}^{i} \frac{e^{-\mu_j t}}{\prod_{\substack{k=1 \\ k \neq j}}^{i} (\mu_k - \mu_j)} \qquad (i > 1) \tag{2.102b}$$

Consider now a branched nuclide chain that converges as follows:

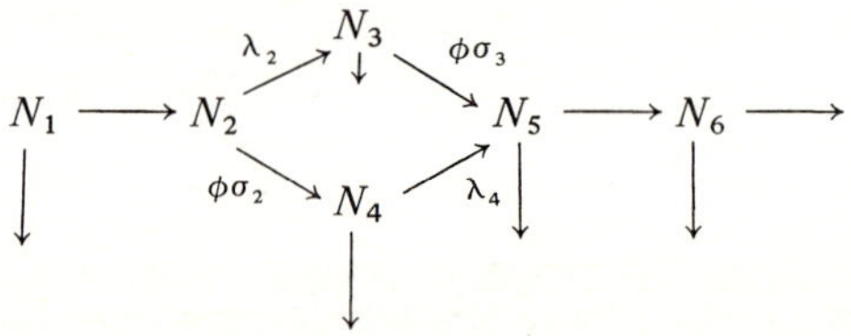

$$\begin{array}{ccccccc} & & & N_3 \downarrow & & & \\ & & \overset{\lambda_2}{\nearrow} & & \overset{\phi\sigma_3}{\searrow} & & \\ N_1 & \longrightarrow & N_2 & & & N_5 \longrightarrow N_6 \longrightarrow & \\ \downarrow & & & \underset{\phi\sigma_2}{\searrow} & & \underset{\lambda_4}{\nearrow} \downarrow & \downarrow \\ & & & N_4 \downarrow & & & \end{array}$$

N_3 and N_4 are each members of separate linear chains. Although the chain is now linear for N_5 and for subsequent members, the amount of each of these nuclides is obtained by the contribution from two linear chains:

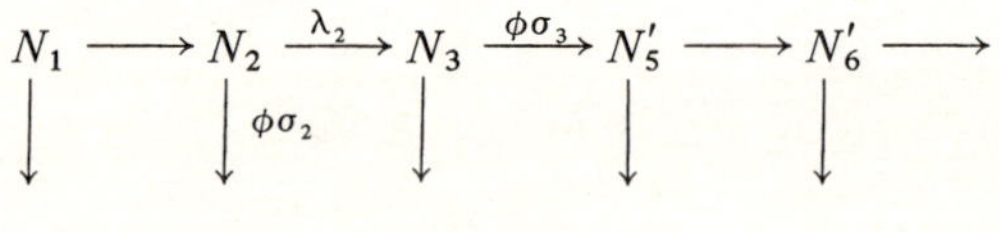

$$\begin{array}{cccccccccc} N_1 & \longrightarrow & N_2 & \xrightarrow{\lambda_2} & N_3 & \xrightarrow{\phi\sigma_3} & N_5' & \longrightarrow & N_6' & \longrightarrow \\ \downarrow & & \downarrow \phi\sigma_2 & & \downarrow & & \downarrow & & \downarrow & \end{array}$$

and

$$\begin{array}{cccccccccc} N_1 & \longrightarrow & N_2 & \xrightarrow{\phi\sigma_2} & N_4 & \xrightarrow{\lambda_4} & N_5'' & \longrightarrow & N_6'' & \longrightarrow \\ \downarrow & & \downarrow \lambda_2 & & \downarrow & & \downarrow & & \downarrow & \end{array}$$

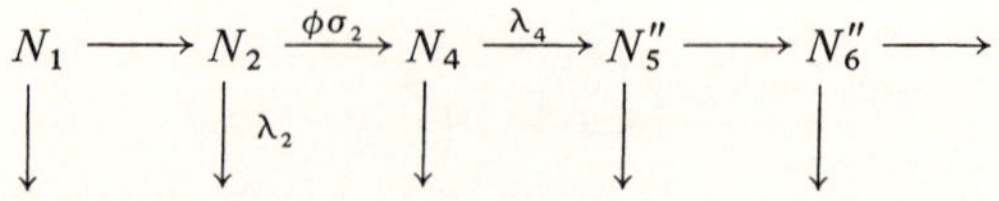

from which the amount N_i $(i \geqslant 5)$ can be obtained from

$$N_i = N_i' + N_i'' \tag{2.103}$$

where N_i' and N_i'' can be solved individually by applying Eq. (2.102).

Finally, consider a linear chain of nuclides with arbitrary initial amounts N_i^0 of any of the radionuclides in the chain, and with no sources of any of these nuclides other than by reactions within the chain. Each finite N_i^0 initiates a linear chain from which the contribution to the amounts of this and subsequent nuclides can be calculated by applying Eq. (2.102). By superposition, the total amount N_i is then

$$N_i = \sum_{l=1}^{i-1} \left[N_l^0 \xi_l \xi_{l+1} \cdots \xi_{i-1} \sum_{j=l}^{i} \frac{e^{-\mu_j t}}{\prod_{\substack{k=l \\ k \neq j}}^{i} (\mu_k - \mu_j)} \right] + N_i^0 e^{-\mu_i t} \qquad (2.104)$$

6.2 Continuous Production

Consider a process, such as nuclear fission, that operates in a way such that P atoms of the first member of the chain are formed per unit time. The production and removal processes and assumed initial conditions for a chain linked by rate constants ξ are shown below:

$$\begin{array}{c}P\\ \downarrow\\ N_1 \xrightarrow{\xi_1} N_2 \xrightarrow{\xi_2} \cdots \xrightarrow{\xi_{j-1}} N_2 \xrightarrow{\xi_j} \cdots \xrightarrow{\xi_{i-1}} N_i \xrightarrow{\xi_i}\end{array}$$

$$\downarrow \mu_1 - \xi_1 \qquad \downarrow \mu_2 - \xi_2 \qquad \downarrow \mu_j - \xi_j \qquad \downarrow \mu_i - \xi_i$$

Initial amount at $t = 0$	0	0	0	0
Removal-rate constants:				
Radioactive decay	λ_1	λ_2	λ_j	λ_i
Neutron absorption	$\phi\sigma_1$	$\phi\sigma_2$	$\phi\sigma_j$	$\phi\sigma_i$
Continuous processing	f_1	f_2	f_j	f_i
Total	μ_1	μ_2	μ_j	μ_i

Similar to the development of Eq. (2.30) in Sec. 3.2, the amount of $dN_i(t', t)$ resulting from decay of $P\,dt'$ atoms of species 1 formed in time interval dt' is obtained by applying Eq. (2.102):

$$dN_i(t', t) = P\,dt' \xi_1 \xi_2 \cdots \xi_{i-1} \sum_{j=1}^{i} \frac{e^{-\mu_j (t-t')}}{\prod_{\substack{k=1 \\ k \neq j}}^{i} (\mu_k - \mu_j)} \qquad (i > 1) \qquad (2.105)$$

Then, to determine $N_i(t)$ resulting from continuous production of species 1 over the time interval from 0 to t, we integrate over t':

$$N_i(t) = \int_0^t dN_i(t', t) = \int_0^t \xi_1 \xi_2 \cdots \xi_{i-1} \sum_{j=1}^{i} \frac{e^{-\mu_j (t-t')}}{\prod_{\substack{k=1 \\ k \neq j}}^{i} (\mu_k - \mu_j)} P\,dt'$$

or

$$N_i(t) = P\xi_1\xi_2 \cdots \xi_{i-1} \sum_{j=1}^{i} \frac{1 - e^{-\mu_j t}}{\mu_j \prod_{\substack{k=1 \\ k \neq j}}^{i} (\mu_k - \mu_j)} \qquad (i > 1) \tag{2.106}$$

The equilibrium amounts of nuclides in the chain are obtained from the differential equations:

$$\frac{dN_1}{dt} = P - \mu_1 N_1 \tag{2.107}$$

$$\frac{dN_2}{dt} = \xi_1 N_1 - \mu_2 N_2 \tag{2.108}$$

$$\frac{dN_i}{dt} = \xi_{i-1} N_{i-1} - \mu_i N_i \tag{2.109}$$

At equilibrium, $dN_i/dt = 0$, and from Eqs. (2.107), (2.108), and (2.109), the equilibrium amounts N^* are

$$N_1^* = \frac{P}{\mu_1} \tag{2.110}$$

$$N_2^* = \frac{P\xi_1}{\mu_1 \mu_2} \tag{2.111}$$

$$N_i^* = \frac{P \prod_{k=1}^{i-1} \xi_k}{\prod_{k=1}^{i} \mu_k} \qquad (i > 1) \tag{2.112}$$

Equations (2.111) and (2.112) also result from Eq. (2.106) as the time t approaches infinity. The time required for N_i to grow to $N_i^*\ (1 - e^{-1})$ is approximately $\Sigma_{j=1}^{i}\ (1/\mu_j)$ and is shorter than when radioactive decay is the only means of removal. Thus, in a chain linked by radioactive decay, the effect of removal by neutron absorption and continuous processing is to reduce the steady-state concentration of a nuclide and shorten the length of time required to reach steady state.

In many instances it is necessary to consider sources that directly form intermediate nuclides in the chain, as in some fission-product chains that have important direct-fission yields of more than one nuclide in the chain. Defining P_l as the constant rate of formation of the lth nuclide in the chain, and for $N_i = 0$ at $t = 0$, we obtain from Eq. (2.106) by superposition:

$$N_i(t) = \sum_{l=1}^{i-1} \left[P_l \xi_l \xi_{l+1} \cdots \xi_{i-1} \sum_{j=l}^{i} \frac{1 - e^{-\mu_j t}}{\mu_j \prod_{\substack{k=l \\ k \neq j}}^{i} (\mu_k - \mu_j)} \right] + P_i \frac{1 - e^{-\mu_i t}}{\mu_i} \tag{2.113}$$

and the steady-state amount is

$$N_i^* = \sum_{l=1}^{i-1} P_l \frac{\prod_{k=l}^{i-1} \xi_k}{\prod_{k=l}^{i} \mu_k} + \frac{P_i}{\mu_i} \tag{2.114}$$

6.3 ^{135}Xe Fission-Product Poisoning

The fission product ^{135}Xe has the largest absorption cross section of all the nuclides in a thermal-neutron flux, and its buildup is especially important in affecting the neutron balance in a thermal reactor. The fission-product decay chain involving the production and decay of ^{135}Xe is

$$\text{Fission} \longrightarrow {}^{135}_{52}\text{Te} \xrightarrow{\beta^-} {}^{135}_{53}\text{I} \xrightarrow{\beta^-} {}^{135}_{54}\text{Xe} \xrightarrow{\beta^-} {}^{135}_{55}\text{Cs} \xrightarrow{\beta^-} {}^{135}_{56}\text{Ba}$$

At mass 135, Table 2.9 shows that the total yield is 0.0641 from ^{235}U fission. Actually, this is broken down into 0.0609 for the yield of ^{135}Te from fission and 0.0032 for the direct yield of ^{135}Xe from fission. A summary of the nuclear properties of the above nuclides is given in Table 2.14.

Because the half-life of ^{135}Te is so short compared to the half-lives of the other members of the chain, ^{135}Te buildup may be ignored in calculating time variations in the amount of ^{135}Xe, and the chain is assumed to originate with ^{135}I, such that $y_\text{I} = 0.0609$. The production rate P_I of ^{135}I, which is now the first member of a fission-product decay chain, is

$$P_\text{I} = N_f \sigma_f \phi y_\text{I} \tag{2.115}$$

where N_f is the number of fissile atoms of effective fission cross section σ_f.

The number N_I^* of ^{135}I atoms at steady state is obtained by applying Eq. (2.110):

$$N_\text{I}^* = \frac{N_f \sigma_f \phi y_\text{I}}{\lambda_\text{I} + \phi\sigma_\text{I} + f_\text{I}} \tag{2.116}$$

For the neutron fluxes occurring in practical reactors, the $\phi\sigma_\text{I}$ term is very small relative to λ_I and may be neglected, so the above equation reduces to

$$N_\text{I}^* = \frac{N_f \sigma_f \phi y_\text{I}}{\lambda_\text{I} + f_\text{I}} \tag{2.117}$$

To use Eq. (2.114) to calculate the number N_Xe^* of ^{135}Xe atoms at steady state, the steady-state amount resulting from decay of ^{135}I (with $\xi_\text{I} = \lambda_\text{I}$) is added to the steady-state amount resulting from direct-fission yield:

$$N_\text{Xe}^* = \frac{\lambda_\text{I} N_f \sigma_f \phi y_\text{I}}{(\lambda_\text{I} + f_\text{I})(\lambda_\text{Xe} + \phi\sigma_\text{Xe} + f_\text{Xe})} + \frac{N_f \sigma_f \phi_f y_\text{Xe}}{\lambda_\text{Xe} + \phi\sigma_\text{Xe} + f_\text{Xe}}$$

or

$$N_\text{Xe}^* = \frac{N_f \sigma_f \phi}{\lambda_\text{Xe} + \phi\sigma_\text{Xe} + f_\text{Xe}} \left(\frac{\lambda_\text{I} y_\text{I}}{\lambda_\text{I} + f_\text{I}} + y_\text{Xe} \right) \tag{2.118}$$

Table 2.14 Nuclear properties of fission products of mass 135

		Radioactive decay constant		Absorption cross section, b		Direct yield from ^{235}U fission,‡
Nuclide	Half-life	s^{-1}	h^{-1}	2200 m/s	Effective†	y atoms per atom fissioned
$^{135}_{52}$Te	29 s	0.0239	86.0			0.0609
$^{135}_{53}$I	6.7 h	2.87×10^{-5}	0.1034			
$^{135}_{54}$Xe	9.2 h	2.09×10^{-5}	0.0753	2.65×10^{6}	2.64×10^{6}	0.0032
$^{135}_{55}$Cs	3×10^{6} yr	7.3×10^{-15}	2.6×10^{-11}	8.7	17.2	
$^{135}_{56}$Ba	Stable	0	0	5.8		

†Calculated for the neutron spectrum of a typical pressurized-water reactor.
‡Bennett [B3].

In reactor control problems and in reactor neutron balances, the quantity of interest is the *poisoning ratio* r, which is the ratio of neutrons absorbed by the poison to neutrons absorbed in fission. Assuming for simplicity that the neutron flux is constant throughout the reactor,† the xenon poisoning ratio at steady state is

$$r^*_{\mathrm{Xe}} = \frac{N_{\mathrm{Xe}}\sigma_{\mathrm{Xe}}\phi}{N_f\sigma_f\phi} = \frac{\sigma_{\mathrm{Xe}}\phi}{\lambda_{\mathrm{Xe}} + \phi\sigma_{\mathrm{Xe}} + f_{\mathrm{Xe}}}\left(\frac{\lambda_{\mathrm{I}}y_{\mathrm{I}}}{\lambda_{\mathrm{I}} + f_{\mathrm{I}}} + y_{\mathrm{Xe}}\right) \tag{2.119}$$

The ratio r^*_{Xe} has the maximum value 0.0641 for the case of no continuous processing (f_{I} and $f_{\mathrm{Xe}} = 0$) and very high flux ($\phi\sigma_{\mathrm{Xe}} \gg \lambda_{\mathrm{Xe}}$), where the ratio r^*_{Xe} becomes equal to the sum of the two yields ($y_{\mathrm{I}} + y_{\mathrm{Xe}}$). For no processing and a typical average thermal flux, $\phi = 3.5 \times 10^{13}$ $n/(\mathrm{cm}^2 \cdot \mathrm{s})$, the ^{135}Xe poisoning ratio is

$$r^*_{\mathrm{Xe}} = \frac{(2.64 \times 10^{-18})(3.5 \times 10^{13})0.0641}{2.09 \times 10^{-5} + (2.64 \times 10^{-18})(3.5 \times 10^{13})} = 0.052$$

For this case the time required for the neutron loss to ^{135}Xe to reach $1 - e^{-1}$ of the steady-state value is approximately

$$\frac{1}{0.1034} + \frac{1}{0.0753 + (2.64 \times 10^{-18})(3.5 \times 10^{13})\,3600} = 12.1 \text{ h}$$

If the xenon processing is to have any appreciable effect on the steady-state poisoning ratio, then the processing rate f_{Xe} must be sufficiently large to increase the value of the group ($\lambda_{\mathrm{Xe}} + \phi\sigma_{\mathrm{Xe}} + f_{\mathrm{Xe}}$) in Eq. (2.119). For example, for a flux of 3.5×10^{13} $n/(\mathrm{cm}^2 \cdot \mathrm{s})$, the processing rate required to halve the xenon poisoning ratio is 9.8/day, which means that the required processing is equivalent to a complete removal of xenon from the entire reactor contents 9.8 times per day.

From the above, it can be seen that for continuous processing to be effective in reducing the concentration of a particular fission-product nuclide, the fraction f_i removed by processing per unit time must be at least of the same order of magnitude as the sum of the other removal-rate constants λ_i and $\phi\sigma_i$. Hence, for nuclides with long half-lives and low cross sections, a very low processing rate f is sufficient to maintain their steady-state concentration at a low value. Greater processing rates are required as the half-lives and/or cross sections become large.

6.4 ^{135}Xe Transient after Reactor Shutdown

The equations of Sec. 6.2 give the number of atoms of each fission product after a reactor has been run at stated conditions for a specified time. If the reactor is then shut down, the fission products build up and decay in accordance with the laws of simple radioactive decay, which were outlined in Sec. 3. If the nuclides in the decay chain are removed only by radioactive decay during reactor operations, the equations of Sec. 3 describe the changes with time of the number of atoms of any nuclide in the decay chain. If a member of a fission-product decay chain or its precursors in the decay chain are removed by neutron absorption, equations for the amount of each nuclide present at time t after shutdown may be obtained by applying the equations of radioactive decay to the amount present at shutdown.

We may illustrate by calculating the number of atoms of ^{135}I and ^{135}Xe present in a reactor that had been operated at a flux ϕ long enough to build up a steady-state content of

†In reality, the neutron flux varies spatially throughout the reactor. The method of calculating effective xenon poisoning for spatially varying flux is developed in texts on reactor theory, such as Weinberg and Wigner [W3].

^{135}I and ^{135}Xe, and then shut down for a time t. No removal by processing is assumed, so that f_I and f_{Xe} are equal to zero. The steady-state contents of ^{135}I and ^{135}Xe have already been obtained as Eqs. (2.117) and (2.118), respectively. The number of N_I of ^{135}I atoms remaining at time t is obtained by applying Eq. (2.13), with $N_I^0 = N_I^*$:

$$N_I = N_I^* e^{-\lambda_I t} \tag{2.120}$$

Similarly, the number of N_{Xe} of ^{135}Xe atoms present at time t is obtained by applying Eq. (2.18), with $N_I^0 = N_I^*$ and $N_{Xe}^0 = N_{Xe}^*$:

$$N_{Xe} = N_I^* \lambda_I \frac{e^{-\lambda_I t} - e^{-\lambda_{Xe} t}}{\lambda_{Xe} - \lambda_I} + N_{Xe}^* e^{-\lambda_{Xe} t} \tag{2.121}$$

Substituting Eqs. (2.117) and (2.118) into (2.121):

$$N_{Xe} = N_f \sigma_f \phi \left[y_I \frac{e^{-\lambda_I t} - e^{-\lambda_{Xe} t}}{\lambda_{Xe} - \lambda_I} + \frac{(y_I + y_{Xe})}{\lambda_{Xe} + \phi \sigma_{Xe}} e^{-\lambda_{Xe} t} \right] \tag{2.122}$$

where ϕ is the neutron flux that existed prior to shutdown.

The transient poisoning ratio, which is the ratio of the neutron absorption in ^{135}Xe to fission absorption if the reactor is to be started up again after shutdown time t, is obtained from Eq. (2.122):

$$r_{Xe} = \frac{N_{Xe}\sigma_{Xe}}{N_f \sigma_f} = y_I \phi \sigma_{Xe} \frac{e^{-\lambda_I t} - e^{-\lambda_{Xe} t}}{\lambda_{Xe} - \lambda_I} + \frac{y_I + y_{Xe}}{1 + \lambda_{Xe}/\phi\sigma_{Xe}} e^{-\lambda_{Xe} t} \tag{2.123}$$

Figure 2.16 shows the growth of xenon to its steady-state value during reactor operation and its subsequent decay after the reactor is shut down. The quantity plotted is the xenon poison ratio, which is the ratio of the rate of absorption of neutrons by xenon to the rate of absorption of neutrons in fission of ^{235}U, $N_{Xe}\sigma_{Xe}/N_f\sigma_f$. Curves are given for fluxes of 1×10^{13}, 1×10^{14}, and 3×10^{14} $n/(\text{cm}^2 \cdot \text{s})$. Note that the steady-state poison ratio is higher the higher the flux. Note also that the poison ratio *increases* after the reactor is shut down and that the increase becomes very large for fluxes of 10^{14} $n/(\text{cm}^2 \cdot \text{s})$.

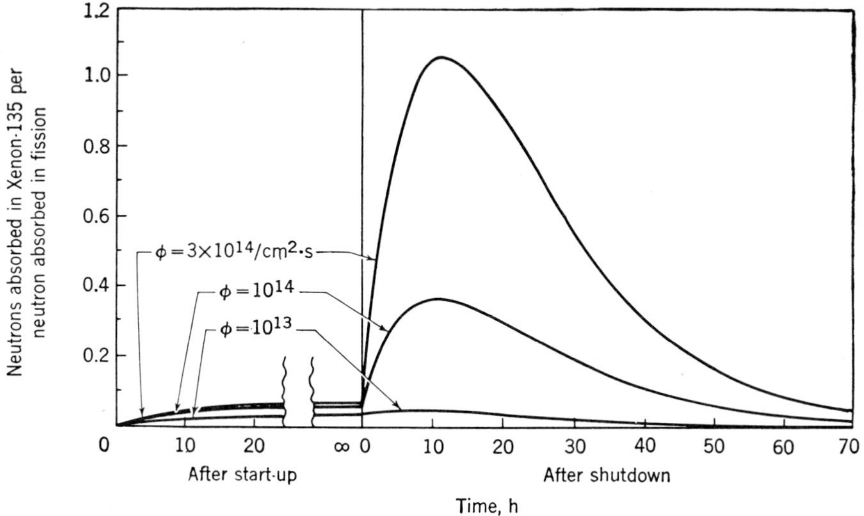

Figure 2.16 Xenon poison ratio during reactor operation at constant flux and after shutdown.

The increase is caused by the sudden reduction in the overall removal rate constant for xenon when the reactor is shut down, whereas the rate of production of xenon from its main source, the decay of ^{135}I, decreases only slowly with time as the iodine decays. For low neutron fluxes ($\phi < 10^{13}$) prior to shutdown the xenon buildup after shutdown is less important because the xenon burnout by neutron capture is then small relative to xenon removal by radioactive decay.

6.5 ^{149}Sm Chain

After ^{135}Xe, the fission product with highest cross section and appreciable yield is ^{149}Sm, whose cross section for 2200 m/s neutrons is 41,000 b and whose effective cross section in a typical water-cooled reactor is over 70,000 b. In addition, many of the fission-product nuclides that produce ^{149}Sm by neutron capture or radioactive decay and several of the nuclides produced from ^{149}Sm by successive neutron captures have high cross sections. Figure 2.17 illustrates the generic relationship between ^{149}Sm and the principal nuclides that lead to it or are produced from it.

Table 2.15 gives direct fission yields y [B3], effective thermal-neutron absorption cross sections σ and half-lives (cf. App. C) for radioactive decay that are used below to evaluate the poisoning ratio for this chain. Effective cross sections were calculated from cross sections for 2200 m/s neutrons and for neutrons of higher energy from cross-section data given by Bennett [B3], applied to the neutron spectrum of a typical pressurized-water reactor.

The set of 11 differential equations that describe the rate of change of each of the 11 nuclides in the ^{149}Sm fission-product chain, assuming no processing removal, are

$^{147}_{60}Nd$: $$\frac{dN_1}{dt} = y_1 N_f \sigma_f \phi - (\sigma_1 \phi + \lambda_1) N_1 \tag{2.124}$$

$^{147}_{61}Pm$: $$\frac{dN_2}{dt} = y_2 N_f \sigma_f \phi + \lambda_1 N_1 - (\sigma_2 \phi + \lambda_2) N_2 \tag{2.125}$$

$^{147}_{62}Sm$: $$\frac{dN_3}{dt} = y_3 N_f \sigma_f \phi + \lambda_2 N_2 - (\sigma_3 \phi) N_3 \tag{2.126}$$

$^{148m}_{61}Pm$: $$\frac{dN_4}{dt} = \sigma_{24} \phi N_2 - (\sigma_4 \phi + \lambda_4) N_4 \tag{2.127}$$

$^{148}_{61}Pm$: $$\frac{dN_5}{dt} = \lambda_{45} N_4 + \sigma_{25} \phi N_2 - (\sigma_5 \phi + \lambda_5) N_5 \tag{2.128}$$

$^{149}_{61}Pm$: $$\frac{dN_6}{dt} = y_6 N_f \sigma_f \phi + \sigma_4 \phi N_4 + \sigma_5 \phi N_5 - (\sigma_6 \phi + \lambda_6) N_6 \tag{2.129}$$

$^{149}_{62}Sm$: $$\frac{dN_7}{dt} = y_7 N_f \sigma_f \phi + \lambda_6 N_6 - \sigma_7 \phi N_7 \tag{2.130}$$

$^{150}_{61}Pm$: $$\frac{dN_8}{dt} = \sigma_6 \phi N_6 - (\sigma_8 \phi + \lambda_8) N_8 \tag{2.131}$$

$^{150}_{62}Sm$: $$\frac{dN_9}{dt} = \sigma_7 \phi N_7 + \lambda_8 N_8 - \sigma_9 \phi N_9 \tag{2.132}$$

$^{151}_{62}Sm$: $$\frac{dN_{10}}{dt} = y_{10} N_f \sigma_f \phi + \sigma_9 \phi N_9 - (\sigma_{10} \phi + \lambda_{10}) N_{10} \tag{2.133}$$

$^{152}_{62}Sm$: $$\frac{dN_{11}}{dt} = y_{11} N_f \sigma_f \phi + \sigma_{10} \phi N_{10} - \sigma_{11} \phi N_{11} \tag{2.134}$$

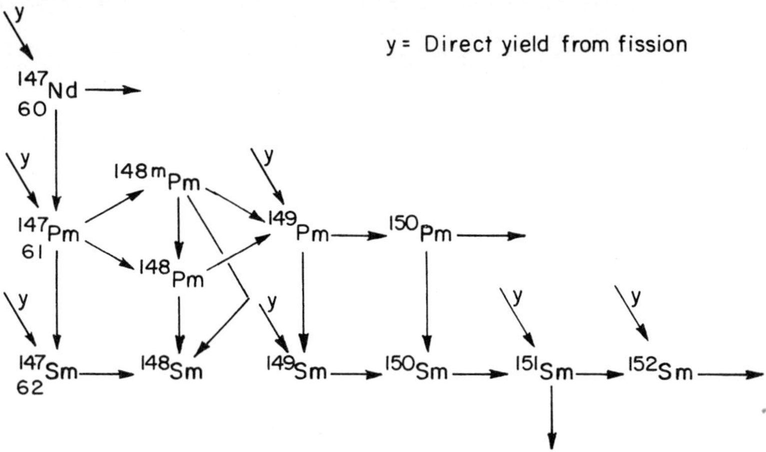

Figure 2.17 The fission-product chain leading to ^{149}Sm.

The poisoning ratio for this set of nuclides is

$$r = \sum_{i=1}^{11} \frac{N_i \sigma_i \phi}{N_f \sigma_f \phi} \tag{2.135}$$

The solution of this set of equations, with zero initial amount of each nuclide, can be written directly by applying Eq. (2.113). To do so, the nuclide chains of Fig. 2.17 are reformulated into an equivalent set of linear chains with constant formation rate of the first

Table 2.15 Nuclear properties for promethium-samarium decay chains

Nuclide	Nuclide designation	Half-life		Absorption cross section, b: 2200 m/s	Absorption cross section, b: Effective†	Direct yield from ^{235}U fission,‡ y atoms per atom fissioned
^{147}Nd	N_1	11.1 days			0	0.0236
^{147}Pm	N_2	2.62 yr		85	845.72 to ^{148m}Pm	0
				181	448.23 to ^{148}Pm	
^{147}Sm	N_3	∞		64	274.2	0
^{148m}Pm	N_4	42 days	(7% to ^{148}Pm; 93% to ^{148}Sm)	22,000	31,964	0
^{148}Pm	N_5	5.4 days		3,000	13,858	0
^{149}Pm	N_6	53.1 h		1,400	1,105.6	0.0113
^{149}Sm	N_7	∞		41,000	73,635	0
^{150}Pm	N_8	2.7 h		0	0	0
^{150}Sm	N_9	∞		102	158.38	0
^{151}Sm	N_{10}	87 yr		15,000	9,734.5	0.0044
^{152}Sm	N_{11}	∞		206	813.01	0.00281

†Calculated for the neutron spectrum of a typical pressurized-water reactor.
‡Bennett [B3].

member of each chain and with each subsequent member of a chain formed only by decay or neutron reaction of its single precursor within the chain. From the data in Table 2.15, only four of the nuclides, ^{147}Nd, ^{149}Pm, ^{151}Sm, and ^{152}Sm, have finite direct yields from fission. Each of these four nuclides is the first member of a chain formed at a rate

$$P_l = y_l N_f \sigma_f \phi \tag{2.136}$$

Two of these chains, i.e., those originating from the direct yields of ^{147}Nd and ^{149}Pm, involve chain branching. For the purpose of calculating the amount of the nuclide at which the branched chain converges, and to calculate the amount of the daughters of this nuclide in the chain, the branched chain must be subdivided into a subset of linear chains as illustrated in Sec. 6.1. For example, for the purpose of calculating the amounts of ^{150}Sm, ^{151}Sm, and ^{152}Sm formed from the chain initiated by the direct yield of ^{149}Pm, this chain is expressed as two subchains:

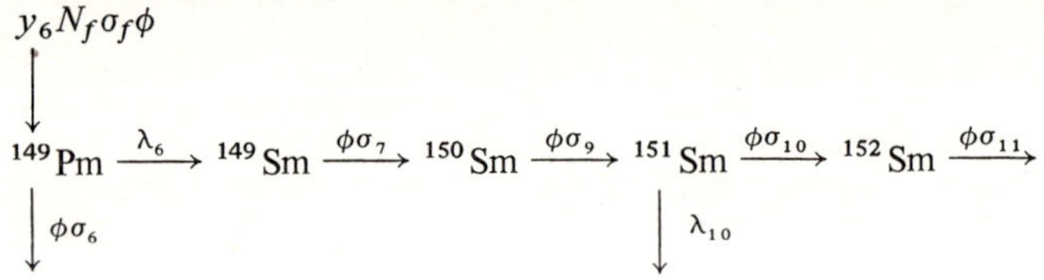

$$\begin{array}{lllllll} y_6 N_f \sigma_f \phi & & & & & & \\ \downarrow & & & & & & \\ {}^{149}\text{Pm} & \xrightarrow{\lambda_6} {}^{149}\text{Sm} & \xrightarrow{\phi\sigma_7} {}^{150}\text{Sm} & \xrightarrow{\phi\sigma_9} & {}^{151}\text{Sm} & \xrightarrow{\phi\sigma_{10}} {}^{152}\text{Sm} & \xrightarrow{\phi\sigma_{11}} \\ \downarrow \phi\sigma_6 & & & & \downarrow \lambda_{10} & & \end{array}$$

and

$$\begin{array}{lllllll} y_6 N_f \sigma_f \phi & & & & & & \\ \downarrow & & & & & & \\ {}^{149}\text{Pm} & \xrightarrow{\phi\sigma_6} {}^{150}\text{Pm} & \xrightarrow{\lambda_8} {}^{150}\text{Sm} & \xrightarrow{\phi\sigma_9} & {}^{151}\text{Sm} & \xrightarrow{\phi\sigma_{10}} {}^{152}\text{Sm} & \xrightarrow{\phi\sigma_{11}} \\ \downarrow \lambda_6 & \downarrow \phi\sigma_8 & & & \downarrow \lambda_{10} & & \end{array}$$

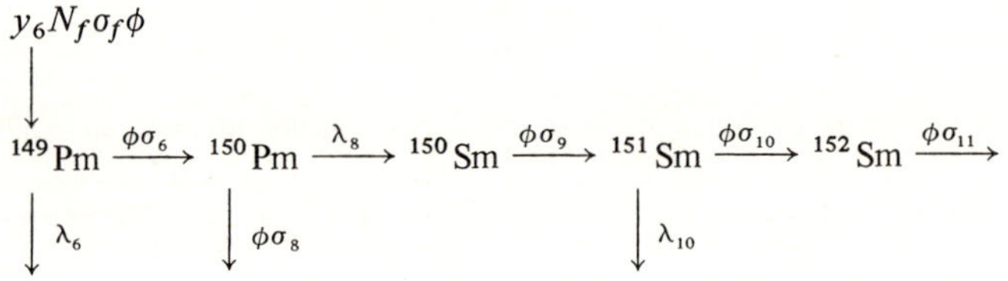

Similarly, the chain originating with the direct yield of ^{147}Nd branches at ^{147}Pm, ^{148m}Pm, and ^{149}Pm. It is subdivided into two linear chains to calculate the contribution to N_5, three to calculate the contribution to N_6, and six to calculate the contributions to N_9, N_{10}, and N_{11}. In this way the summation Eq. (2.113) can be used to write the solution for this series of chains.

To calculate the growth and decay of these nuclides after reactor shutdown, the assumed equilibrium amounts at the time T of shutdown are calculated as above, using Eq. (2.114). These become the initial amounts N_l^0 for application of the batch decay, Eq. (2.18) for time t after shutdown. During shutdown the branching and convergence involving neutron reactions disappear, and we have only four simple linear chains to solve by applying Eq. (2.18).

Alternatively, the differential equations may be solved directly by numerical methods with a digital computer [C2]. Results obtained from the latter approach are shown in Figs. 2.18 to 2.20. Calculations were made for a thermal-neutron flux of 3.5×10^{13} $n/(\text{cm}^2\cdot\text{s})$, considered representative of a 1060-MWe pressurized-water reactor similar to one manufactured by Westinghouse for the Donald C. Cook Nuclear Plant [A1].

Figure 2.18 shows the contribution of individual nuclides to the poisoning ratio as a function of time, starting with fresh, unirradiated fuel at time zero. The poisoning ratio of ^{149}Sm builds up very quickly to 0.0113, the fission yield at mass 149, and then increases more gradually because of additional ^{149}Sm production by neutron capture in nuclides of mass 147 and 148. Other nuclides of this chain that make appreciable contributions to the poisoning ratio include ^{147}Pm, ^{148m}Pm, ^{150}Sm, ^{151}Sm, and ^{152}Sm. The overall poisoning ratio, the sum of the contributions of individual nuclides, is shown in Fig. 2.19.

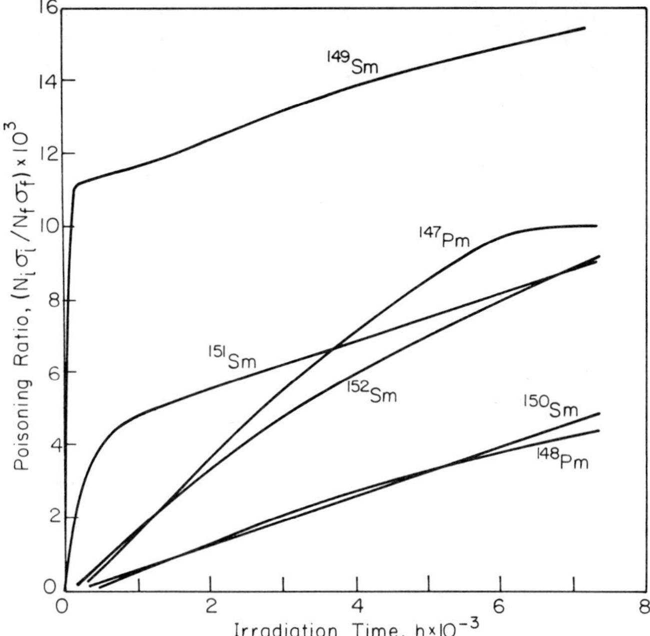

Figure 2.18 Individual nuclide contribution to total poisoning ratio of ^{149}Sm decay chain.

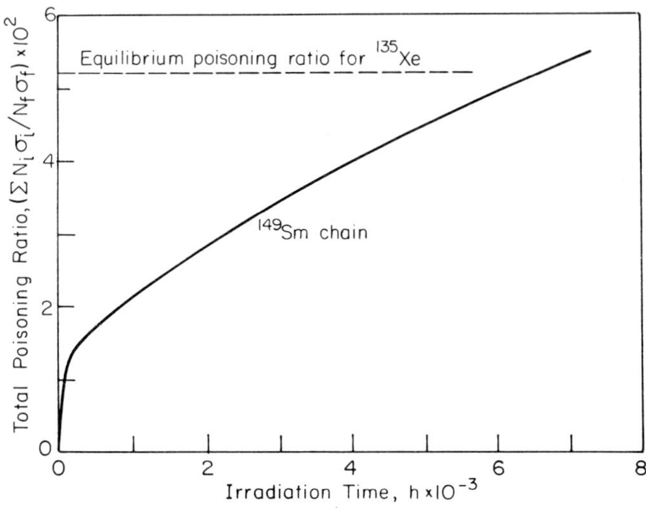

Figure 2.19 Buildup of poisoning ratio of ^{149}Sm chain in fresh pressurized-water reactor fuel containing 3.2 w/o ^{235}U.

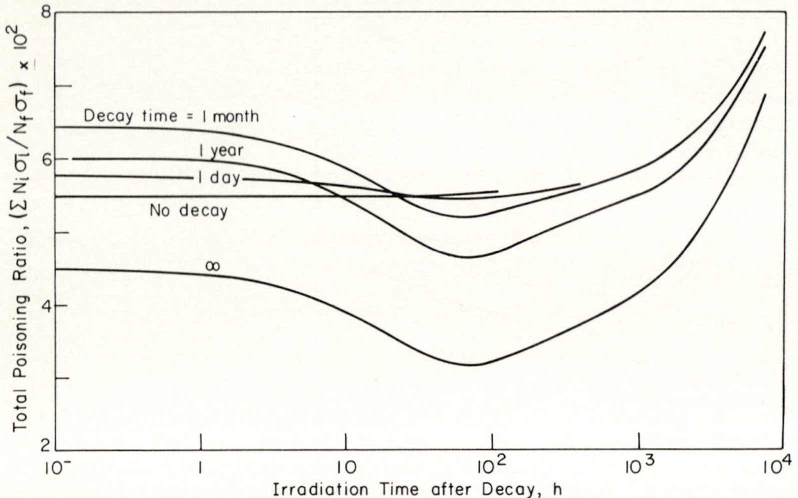

Figure 2.20 Buildup of poisoning ratio of ^{149}Sm chain after various decay times in fuel that has been previously irradiated for 7300 h.

Figure 2.20 shows how the poisoning ratio of this chain varies if the reactor is shut down after initial operation for 7300 h for various periods of time T' and then operated at a flux of 3.496×10^{13} $n/(\text{cm}^2 \cdot \text{s})$ for additional time T. The behavior shown in this figure is considered representative of this reactor after it has been refueled several times with one-third of the oldest fuel replaced by fresh fuel.

7 DERIVATION OF THE BATEMAN EQUATION (2.17) BY LAPLACE TRANSFORMS

7.1 Properties of Laplace Transforms

The *Laplace transform* $\bar{f}(t)$ of a function $f(t)$ is defined as

$$\bar{f}(t) \equiv \int_0^\infty e^{-st} f(t) = L(s) \tag{2.137}$$

It is a function of the transform variable s. The Laplace transform of a derivative is obtained by integration by parts:

$$\overline{\frac{df}{dt}} = \int_0^\infty e^{-st} \frac{df}{dt}\, dt = e^{-st} f \Big|_0^\infty + s \int_0^\infty e^{-st} f\, dt = -f(0) + s\bar{f} \tag{2.138}$$

The Laplace transform of the exponential function $e^{-\lambda t}$ is

$$\overline{e^{-\lambda t}} = \int_0^\infty e^{-st} e^{-\lambda t}\, dt = \frac{1}{s + \lambda} \tag{2.139}$$

The *inverse transform* of a function $L(s)$ of s is a function of the variable of which $L(s)$ is the Laplace transform. For example, functions of the variable t have been transformed in Eqs. (2.137), (2.138), and (2.139). It can be seen that the inverse transform of $1/s$ is e^{-0}, or unity.

These simple properties of the Laplace transform make it a very convenient tool for solving systems of first-order linear differential equations, such as the equations for growth and decay of nuclides in radioactive disintegrations and neutron irradiation. They permit these differential equations to be treated as if they were systems of simple transformed linear equations without derivatives.

7.2 Derivation of the Bateman Equation (2.17)

Consider the general radioactive decay chain

$$N_1 \longrightarrow N_2 \longrightarrow N_3 \longrightarrow \cdots \longrightarrow N_j \longrightarrow \cdots \longrightarrow N_i \longrightarrow \cdots$$

with N_1^0 atoms of the first member at time zero and none of the other members present at that time. The differential equations are

$$\frac{dN_1}{dt} = -\lambda_1 N_1 \tag{2.140a}$$

$$\frac{dN_2}{dt} = \lambda_1 N_1 - \lambda_2 N_2 \tag{2.140b}$$

$$\frac{dN_i}{dt} = \lambda_{i-1} N_{i-1} - \lambda_i N_i \tag{2.140i}$$

The boundary conditions at $t = 0$ are

$$N_1 = N_1^0 \tag{2.141a}$$

$$N_2 = N_3 = \cdots = N_i = \cdots = 0 \tag{2.141b}$$

The system of differential equations (2.140) may be transformed to a system of linear equations by taking the Laplace transform and using Eq. (2.138) for the Laplace transform of the first derivatives:

$$-N_1^0 + s\bar{N}_1 = -\lambda_1 \bar{N}_1 \tag{2.142a}$$

$$s\bar{N}_2 = \lambda_1 \bar{N}_1 - \lambda_2 \bar{N}_2 \tag{2.142b}$$

$$s\bar{N}_i = \lambda_{i-1} \bar{N}_{i-1} - \lambda_i \bar{N}_i \tag{2.142i}$$

where $\bar{N}$ is the transform of N. These equations may be solved successively for the $\bar{N}$'s:

$$\bar{N}_1 = \frac{N_1^0}{\lambda_1 + s} \tag{2.143a}$$

$$\bar{N}_2 = \frac{\lambda_1 \bar{N}_1}{\lambda_2 + s} = \frac{\lambda_1 N_1^0}{(\lambda_1 + s)(\lambda_2 + s)} \tag{2.143b}$$

$$\bar{N}_i = \frac{\lambda_{i-1} \bar{N}_{i-1}}{\lambda_i + s} = \frac{\prod_{k=1}^{i-1} \lambda_k N_1^0}{\prod_{k=1}^{i} (\lambda_k + s)} \tag{2.143i}$$

N_1 may be found by taking the inverse transform of Eq. (2.143*a*):

$$N_1 = N_1^0 e^{-\lambda_1 t} \tag{2.144}$$

To find the inverse transform of Eqs. (2.143*b*) to (2.143*i*) it is necessary to express the denominator as a sum of partial fractions. For Eq. (2.143*i*) this would be

$$\frac{1}{\prod_{k=1}^{i} (\lambda_k + s)} = \sum_{k=1}^{i} \frac{A_k}{(\lambda_k + s)} \tag{2.145}$$

To find a specific coefficient A_j, multiply each side of Eq. (2.145) by $(\lambda_j + s)$:

$$\frac{1}{\prod_{k \neq j}^{i} (\lambda_k + s)} = A_j + (\lambda_j + s) \sum_{k \neq j}^{i} \frac{A_k}{(\lambda_k + s)} \tag{2.146}$$

and let s approach $-\lambda_j$. When $s = -\lambda_j$,

$$\frac{1}{\prod_{k \neq j}^{i} (\lambda_k - \lambda_j)} = A_j \tag{2.147}$$

Hence, Eq. (2.143*i*) is equivalent to

$$\bar{N}_i = N_1^0 \lambda_1 \lambda_2 \cdots \lambda_{i-1} \sum_{j=1}^{i} \frac{1}{\prod_{\substack{k=1 \\ k \neq j}}^{i} (\lambda_k - \lambda_j)} \left(\frac{1}{\lambda_j + s} \right) \tag{2.148}$$

Because the inverse transform of $1/(\lambda_j + s)$ is $e^{-\lambda_j t}$,

$$N_i = N_1^0 \lambda_1 \lambda_2 \cdots \lambda_{i-1} \sum_{j=1}^{i} \frac{e^{-\lambda_j t}}{\prod_{\substack{k=1 \\ k \neq j}}^{i} (\lambda_k - \lambda_j)} \qquad (i > 1) \tag{2.149}$$

which is the Bateman equation (2.17). The product term $\Pi_{k \neq j}^{i} (\lambda_k - \lambda_j)$ has no meaning when the i species is the initial member of the chain, so Eq. (2.14) necessarily applies only to the daughter species, i.e., $i > 1$.

NOMENCLATURE

A	mass number
c	velocity of light
E	energy
E_0	kinetic energy of neutrons at most probable speed (Sec. 4.6)
$E(t)$	rate of heat release from fission-product decay per fission event
F	ratio of heat-generation rate from fission-product decay to fission rate, (MeV/s)/(fissions/s). F' denotes decay-heat value in the absence of neutron absorption in fission products.

f	fraction of material removed per unit time by processing
G	empirical correction factor, Eqs. (2.93) and (2.94)
g	non-$1/v$ correction factor for thermal-neutron cross section
h	Planck's constant
k	Boltzmann's constant
K_L	leakage-rate constant
K_R	reaction-rate constant
m	mass of neutron
M	atomic weight
n	neutron concentration
$n(E)$	distribution of neutron concentration with respect to neutron energy
$n(E/E_0)$	distribution of neutron concentration with respect to neutron energy ratio E/E_0
$n(v)$	distribution of neutron concentration with respect to neutron speed
$n(v/v_0)$	distribution of neutron concentration with respect to neutron speed ratio v/v_0
N	number of neutrons in nucleus (Sec. 1.1); number of nuclei
N', N''	number of nuclei contributed by a linearized chain, Eq. (2.103)
$\bar{N}$	Laplace transform of N
N_f	number of fissionable nuclei
P	production rate of initial member of chain
P_d	heat-generation rate due to radioactive decay of fission products (Sec. 5.3)
P_d'	fission-product decay-heat rate neglecting neutron absorption in fission products (Sec. 5.3)
P_f	heat-generation rate due to fission (Sec. 5.3)
Q	thermal energy per fission
r	neutron-spectrum index [Eq. (2.75)]; poisoning ratio, neutrons absorbed in poison per neutron absorbed in fission (Secs. 6.3 to 6.5)
s	epithermal-absorption correction factor for thermal-neutron cross section
t	time
$t_{1/2}$	half-life
T	operating time (Secs. 3.3 and 5.3); absolute temperature (Secs. 4.6 and 4.7)
$\hat{T}$	arbitrary reference temperature corresponding to $\hat{v}$
v	neutron speed
$\hat{v}$	arbitrary reference speed of a neutron
v_0	most probable neutron speed
y	fission yield, atoms of fission product per atom fissioned
α	ratio of capture cross section to fission cross section for fissionable nuclides
β	normalization of epithermal flux distribution to thermal flux, Eq. (2.66); number of beta disintegrations per second at time t after fission, Eq. (2.85)
γ	empirical constant in Eq. (2.90), MeV/s
Δ	unit step function, Eq. (2.66)
η	number of fission neutrons produced per neutron absorbed in a fissionable nuclide
λ	radioactive decay constant
μ	cutoff energy factor (Sec. 4.7); first-order removal-rate constant [cf. Eq. (2.95)]
ν	number of neutrons produced per fission
ξ	rate constant for formation of a nuclide from its precursor
σ	cross section
$\bar{\sigma}$	effective reaction cross section defined by Eq. (2.55)
$\hat{\sigma}$	effective reaction cross section defined by Eq. (2.62)
σ_a	cross section for neutron absorption
σ_c	cross section for nonfission capture
σ_f	cross section for fission

τ	mean life
ϕ	neutron flux
$\phi(v/v_0)$	distribution of neutron flux with respect to neutron speed ratio v/v_0
$\phi(E)$	distribution of neutron flux with respect to neutron energy
$\phi(E/E_0)$	distribution of neutron flux with respect to neutron energy ratio E/E_0
ψ	total fissions per initial fissile atom

Superscripts

A	mass number
*	steady state
0	amount at time zero

Subscripts

1	first member of a chain
2	second member of a chain
3	third member of a chain
4	fourth member of a chain
d	fission-product decay
D	nonthermalized epithermal neutrons
f	fission, fissionable species
I	^{135}I
i, j, k, l	a member of a nuclide chain
M	Maxwell-Boltzmann distribution of thermal neutrons
Xe	^{135}Xe
Z	atomic number

REFERENCES

A1. American Electric Power Co.: Donald C. Cook Nuclear Plant, Preliminary Safety Analysis Report, 1968.

A2. American Nuclear Society Standards Committee Working Group ANS-5.1: "American National Standard for Decay Heat Power in Light Water Reactors," Standard ANSI/ANS-5.1, American Nuclear Society, 1979.

B1. Bateman, H.: *Proc. Cambridge Phil. Soc.* **15**: 423 (1910).

B2. Bell, M.: "ORIGEN–The ORNL Isotope Generation and Depletion Code," Report ORNL-4628, May 1973.

B3. Bennett, L. L.: "Recommended Fission Product Chains for Use in Reactor Evaluation Studies," Report ORNL-TM-1658, Sept. 1966.

C1. Critoph, E.: "Effective Cross Sections for U-235 and Pu-239," Report CRRP-1191, Mar. 1964.

C2. Croff, A. G.: "Calculation of the Poisoning Ratio of the Nuclide Chains Associated with ^{135}Xe and ^{149}Sm," Communication to M. Benedict, 1973.

D1. Dudey, N. D.: "Review of Low-Mass Atom Production in Fast Reactors," Report ANL-7434, 1968.

E1. England, T. R., W. B. Wilson, and M. G. Stamatelatos: "Fission Product Data for Thermal Reactors, Part 1: A Data Set for EPRI-CINDER Using ENDF/B-IV," Report LA-6745-MS, Dec. 1976, and "Fission Product Data for Thermal Reactors, Part 2: Users Manual for EPRI-CINDER Code and Data," Report LA-6746-MS, Dec. 1976.

G1. Goode, J. H.: "Hot Cell Evaluation of the Release of Tritium and ^{85}Kr during Processing," Report ORNL-3956, June 1966.

K1. Katcoff, S.: *Nucleonics* **18**(11) (Nov. 1960).
L1. Lederer, C. M., J. M. Hollander, and I. Perlman: *Table of the Isotopes,* Wiley, New York, 1967.
M1. Mughabghab, S. F., and D. I. Garber: *Neutron Cross Sections,* vol. 1: *Resonance Parameters,* Report BNL-325, 3d ed., vol. 1, June 1973.
S1. Shure, K.: "Fission Product Decay Energy," Report WAPD-BT-24, Westinghouse Atomic Power Division, 1961.
W1. Walker, W. H.: "Fission Product Data for Thermal Reactors," Report AECL-3037, pt. I, 1973; pt. II, 1973.
W2. Way, K., and E. P. Wigner: "Rate of Decay of Fission Products," paper 43 in *National Nuclear Energy Series,* div. IV, vol. 9, McGraw-Hill, New York, 1951.
W3. Weinberg, A. M., and E. P. Wigner: *The Physical Theory of Neutron Chain Reactors,* University of Chicago Press, Chicago, 1958.
W4. Westcott, C. H.: "Effective Cross Section Values for Well-Moderated Thermal Reactor Spectra," Report AECL-1101, 1960 (corrected and reprinted Dec. 1964).
W5. Westcott, C. H.: "A Study of the Accuracy of *g*-Factors for Room-Temperature Maxwellian Spectra for U and Pu Isotopes," Report AECL-3255, 1969.

PROBLEMS†

2.1. In the fission of ^{235}U by a neutron, one of the fission fragments is identified as ^{94}Kr. What is the other fission fragment? Assume that three neutrons are released in this mode of fission. Write the complete equation for this reaction.

2.2. In one mode of fission of ^{239}Pu, three neutrons are observed, and ^{135}Xe is one of the fission products. What nuclide is the other fission product?

2.3. How many grams are there per curie of ^{14}C? Of ^{32}P?

2.4. How many curies are there per gram of ^{89}Sr; ^{210}Po; ^{226}Ra; ^{238}U?

2.5. The half-life of ^{137}Cs is 30.0 years. In 6.5 percent of the disintegrations a beta ray of 1.176 MeV maximum energy is emitted. In 93.5 percent of the disintegrations a beta ray of 0.514 MeV maximum energy is emitted to form 2.55-min ^{137m}Ba, which decays by isomeric transition to stable ^{137}Ba.

(*a*) What is the energy of the gamma ray emitted in decay of ^{137m}Ba?

(*b*) A 1-kg sample of ^{137}Cs is stored in a 1-mm-thick aluminum container, surrounded by a 30-cm-thick lead shield. At what rate is heat being liberated in the aluminum and sample? In the lead?

(*c*) What is the activity of the ^{137}Cs?

2.6. A sample of 1×10^{-10} g of beta-emitting radium E ($^{210}_{83}$Bi) is freed from other radioactive isotopes at time $t = 0$. As it decays, the activity of its daughter, alpha-emitting ^{210}Po, builds up and then decays. Sketch on semilog paper a plot of alpha and beta activity in disintegrations per second versus time. What is the time at which alpha activity reaches a maximum? What is the weight of ^{210}Po at that time?

2.7. An important fission-product chain is

$$^{140}_{56}\text{Ba} \xrightarrow{\lambda_1} {}^{140}_{57}\text{La} \xrightarrow{\lambda_2} {}^{140}_{58}\text{Ce} \qquad \text{(stable)}$$

A sample of pure ^{140}Ba is isolated at time zero. The activity of its daughter, ^{140}La, increases at first and then decreases. Derive a general expression for the time at which the daughter's

†Supplementary nuclear data needed for these problems will be found in App. C.

activity is a maximum and the ratio of daughter activity at that time to initial parent activity. From the half-lives for ^{140}Ba and ^{140}La, find the ratio of maximum ^{140}La activity to initial ^{140}Ba activity.

2.8. A sample of ^{101}Mo, initially pure at time zero, undergoes radioactive decay according to the scheme

$$^{101}_{42}\text{Mo} \xrightarrow{\beta^-} {}^{101}_{43}\text{Tc} \xrightarrow{\beta^-} {}^{101}_{44}\text{Ru} \qquad \text{(stable)}$$

The half-lives of ^{101}Mo and ^{101}Tc are nearly the same and for the purpose of this problem will be assumed equal. After a decay period of one half-life, how many atoms of ^{101}Tc are present per initial atom of ^{101}Mo? How many atoms of ^{101}Ru are present per initial atom of ^{101}Mo?

2.9. In 1941, Nier used a mass spectrograph to measure the relative abundances of lead isotopes in samples of lead and uranium ores. He found that in the uranium ores, which also contained thorium, there were higher isotopic concentrations of ^{206}Pb, ^{207}Pb, and ^{208}Pb than were found in lead ores not associated with uranium. The results of an analysis of Parry Sound uraninite are listed below. Each lead isotope is reported in terms of the amount in excess of that which would be expected from the natural lead content of the ore.

Atom ratios:	U/Th	^{206}Pb/^{238}U	^{208}Pb/Th	^{207}Pb/^{235}U
	23.4	0.166	0.0483	1.70

The present isotopic content of ^{235}U in natural uranium is 0.71 percent.

(*a*) From the above data, calculate three possible values of the age of the earth.

(*b*) Estimate the ^{235}U/^{238}U ratio at the time of the origin of the earth.

2.10. The fact that ^{232}Th and its decay products are found frequently in uranium deposits has led to the belief that ^{232}Th was not present as such in these deposits at the time of the origin of the earth, but is formed from the alpha decay of ^{236}U, an isotope no longer present in natural uranium. Thorium decays to stable ^{208}Pb with the overall reaction

$$^{232}_{90}\text{Th} \longrightarrow {}^{208}_{82}\text{Pb} + 6{}^{4}_{2}\text{He} + 14{}^{\;0}_{-1}e$$

The half-life of ^{232}Th is 1.41×10^{10} years. None of the nuclides intermediate between ^{232}Th and ^{208}Pb has a half-life greater than a few years. The ratio of ^{208}Pb to ^{232}Th in a typical uranium deposit is 0.0483. The estimated age of the earth is 10^9 years. What is the half-life of ^{236}U? Assume that no ^{232}Th and ^{208}Pb were present in this deposit when the earth was formed.

2.11. Radioactive cobalt (^{60}Co) is produced by exposing samples to neutrons in a reactor. What is the maximum number of curies that can be obtained from 1 g of cobalt exposed to a thermal-neutron flux of 10^{13} $n/(\text{cm}^2\cdot\text{s})$? How long must the cobalt be exposed to obtain an activity of 1 Ci/g?

1. Assume that the neutrons are in thermal equilibrium at 20°C.
2. The neutron-absorption cross section for ^{59}Co for 2200 m/s neutrons is 37.2 b.
3. Neglect neutron absorption by ^{60}Co.

2.12. Each fuel element of a reactor contains 150 g of ^{235}U. One such element has been irradiated for 30 days at a thermal-neutron flux of 10^{13} $n/(\text{cm}^2\cdot\text{s})$ and cooled for 5 days. What is the activity of fission products at that time, in curies per gram? At what rate is decay energy being released?

2.13. A fresh fuel element containing 1 g-atom of ^{235}U is exposed at time zero to a neutron flux of 10^{14} $n/(\text{cm}^2\cdot\text{s})$.

(*a*) What is the rate at which heat is generated at time zero? Express the answer in megawatts of heat.

(*b*) How long will it take for half the ^{235}U to be consumed by fission and neutron capture if

(1) The neutron flux is held constant?

(2) The heat-generation rate is held constant?

(*c*) At the end of the irradiations considered in part (*b*), how many gram-atoms of ^{236}U will have been produced? How many gram-atoms of fission products will have been produced?

The effective cross sections to be used for ^{235}U are as follows: fission = 539 b; nonfission capture = 99.5 b.

CHAPTER

THREE

FUEL CYCLES FOR NUCLEAR REACTORS

Section 1 of this chapter lists the principal fuels used in nuclear reactors, and Sec. 2 describes the effects of reactor irradiation on them, with emphasis on changes in fuel composition and reactivity. Section 3 describes methods of managing fuel and neutron-absorbing poisons aimed at increasing energy production, while reducing costs and controlling deterioration of fuel. Section 4 goes into some detail regarding fuel management in a pressurized-water reactor (PWR) and gives the results of computer calculations of fuel-cycle performance. Section 5 develops a procedure for calculating fuel-cycle costs and applies it to this PWR example, using cost bases anticipated for the year 1980. Section 6 develops an approximate method for calculating the fuel-cycle performance of a PWR suitable for hand calculation and compares the results with more precise ones obtained from a computer code. Section 7 presents fuel-cycle flow sheets for a PWR whose fuel is enriched with ^{235}U or plutonium, a high-temperature gas-cooled reactor (HTGR), and a liquid-metal fast-breeder reactor (LMFBR).

The principal objective of this chapter is to develop an appreciation of the demands made by the reactor on the steps in the nuclear fuel cycle that provide fuel for the reactor and reclaim fuel from it.

1 NUCLEAR FUELS

Nuclear fuels consist of fissile materials, which produce a net increase in neutrons when they absorb neutrons, and fertile materials, which produce fissile material when they absorb neutrons. The principal fissile materials are ^{235}U, ^{239}Pu, and ^{233}U; ^{241}Pu is also of some importance. The principal fertile materials are ^{238}U and ^{232}Th; ^{240}Pu and ^{234}U also play a role as fertile materials.

^{235}U is the only fissile material that occurs in nature in significant quantity. Natural uranium consists of 0 711 weight percent (w/o) ^{235}U, 99.283 w/o ^{238}U, and 0.0055 w/o ^{234}U as a negligible trace constituent. Until now, most power reactors have been fueled with either natural uranium or slightly enriched uranium containing from 2 to 5 w/o ^{235}U, produced from natural uranium in a gaseous diffusion plant.

The principal nuclear reactions that take place when mixtures of ^{235}U and ^{238}U are used as fuel in a reactor are illustrated in Fig. 3.1. Fissile materials are double underlined, and their fission cross sections for 2200 m/s neutrons are given on upward-slanting arrows. Fertile materials are single underlined, and their capture cross sections for 2200 m/s neutrons are given on horizontal arrows. Beta-decay reactions with short enough half-lives to be important are shown by vertical arrows.

When fissile ^{235}U absorbs a neutron, the principal reaction is fission, but some capture takes place to produce nonfissile ^{236}U. This ^{236}U is merely a poison, which can absorb another neutron to produce short-lived ^{237}U, which decays to nonfissile ^{237}Np.

Neutrons produced from ^{235}U fission are absorbed in ^{238}U to produce short-lived ^{239}U, which decays successively to ^{239}Np and fissile ^{239}Pu. In most fuel-cycle analyses, it is permissible to assume that neutron absorption by ^{238}U results in immediate formation of ^{239}Pu.

When ^{239}Pu absorbs a neutron, the more probable reaction is fission, but some atoms capture a neutron to produce fertile ^{240}Pu. Upon further irradiation this captures another neutron to produce fissile ^{241}Pu. When ^{241}Pu absorbs a neutron, either fission may take place or ^{242}Pu may be formed. ^{241}Pu also decays with a half-life of 13.2 years to nonfissile ^{241}Am. ^{242}Pu is neither fissile nor fertile and, like ^{236}U, is a poison. When it absorbs a neutron, ^{243}Pu is formed, which decays with a half-life of 5 h to nonfissile ^{243}Am.

Some nuclear reactors are fueled with a mixture of fissile ^{235}U and fertile thorium. Figure 3.2 is a similar diagram showing the principal nuclear reactions that take place in such fuel. The effect of irradiation on ^{235}U is the same as in Fig. 3.1. However, neutrons produced in fission of ^{235}U are now absorbed in ^{232}Th to produce short-lived ^{233}Th, which decays to 27-day ^{233}Pa. Most of this decays to fissile ^{233}U, but in reactors with a thermal-neutron flux above 5×10^{13} an appreciable fraction absorbs a neutron to make ^{234}Pa, which then decays to ^{234}U. In a

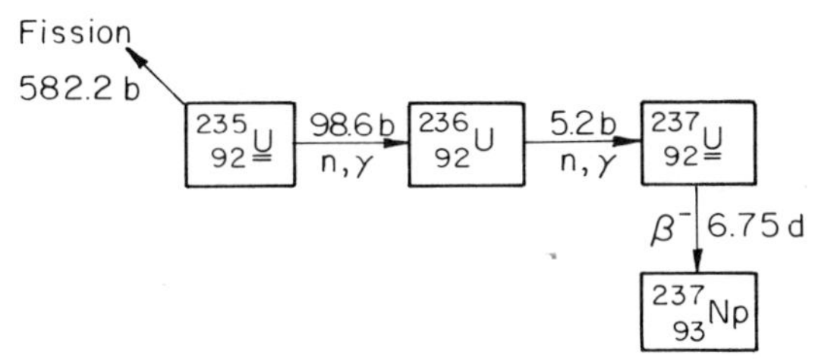

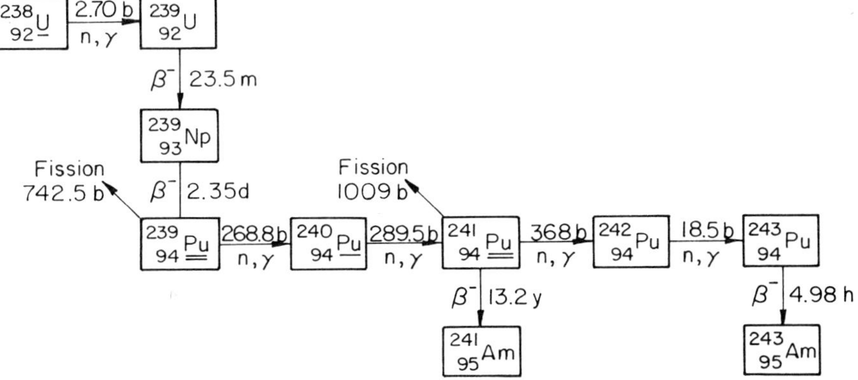

Figure 3.1 Principal nuclear reactions in uranium-fueled reactors.

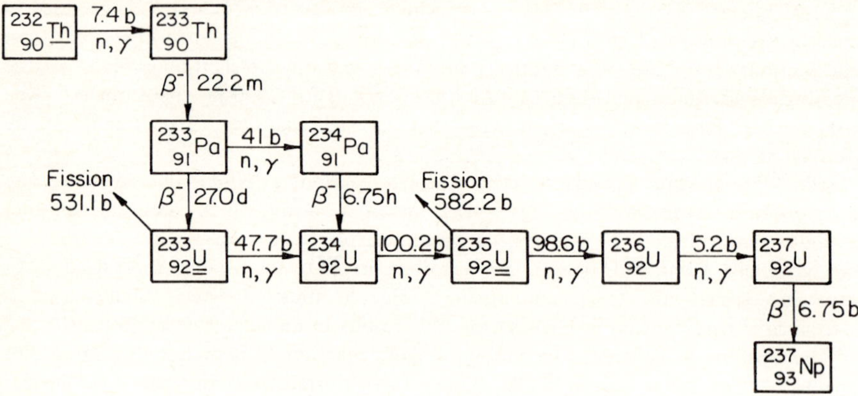

Figure 3.2 Principal nuclear reactions in thorium-fueled reactors.

thorium-fueled reactor, ^{233}U plays a role analogous to ^{239}Pu in a ^{238}U-fueled reactor. When ^{233}U absorbs a neutron the more probable reaction is fission, but some atoms capture a neutron to produce fertile ^{234}U. Upon further irradiation, this captures another neutron to produce fissile ^{235}U. Like ^{241}Pu, ^{235}U may either undergo fission or capture another neutron to produce the poison ^{236}U, analogous to ^{242}Pu.

Other less usual combinations of fissile and fertile materials may also be used for nuclear fuels, such as ^{238}U and ^{233}U or ^{232}Th and ^{239}Pu. This chapter, however, is concerned only with the ^{238}U, ^{235}U, plutonium system diagrammed in Fig. 3.1 and the thorium, ^{235}U system diagrammed in Fig. 3.2.

Effective neutron cross sections in a uranium-fueled PWR are given in Sec. 6. Because of resonance absorption, effective absorption cross sections for ^{236}U, ^{238}U, ^{240}Pu, and ^{242}Pu are much higher than the cross sections for 2200 m/s neutrons given in Fig. 3.1.

Nuclear properties of the three principal fissile nuclides are summarized in Table 3.1. The property η given in Table 3.1 is of interest in relation to the possibility of using these fissile nuclides in a breeder reactor. If a reactor is designed carefully for neutron economy, it is possible under certain conditions to generate fissile material at a rate equal to or greater than the consumption rate of fissile material. Such a reactor can be operated as a true breeder if the newly formed fissile material is returned to the reactor. The minimum requirements of a fuel to

Table 3.1 Properties of fissile nuclides

Property	^{235}U	^{239}Pu	^{233}U
Cross section, 2200 m/s			
Absorption, σ_a	680.8	1011.3	578.8
Fission, σ_f	582.2	742.5	531.1
Neutrons produced			
Per fission, ν	2.418	2.871	2.492
Per neutron absorbed, η			
2200 m/s	2.068	2.108	2.287
Typical light-water reactor†	1.96	1.86	2.2

†All entries except this row from BNL-325, 3d ed., vol. 1, June 1973.

maintain a breeder reaction can be expressed in terms of η, the number of fission neutrons produced per neutron absorbed in the fuel. To maintain a chain reaction, one of these neutrons must be absorbed in another fissile atom, and for breeding there must be still another neutron available for absorption in fertile material. Hence, the minimum requirement for breeding is that η be equal to or greater than 2. Referring to values of η in Table 3.1 we see that the η of ^{233}U is most favorable for breeding in a thermal reactor. Although the η listed for ^{235}U and ^{239}Pu for 2200 m/s neutrons is greater than 2, the effective value of η obtained by taking into account the neutron energy spectrum in a typical light-water reactor (LWR) is less than 2, and breeding in LWRs with ^{235}U or ^{239}Pu is not practical.

Nuclear properties of ^{239}Pu become more favorable for breeding if fission is carried out with fast neutrons, with kinetic energies of the order of 2×10^5 eV. In such a fast reactor the η for plutonium is around 2.3, and breeding with ^{238}U (or natural uranium) is possible.

2 EFFECTS OF IRRADIATION ON NUCLEAR FUELS

As the fuel in a nuclear reactor is irradiated, it undergoes nuclear transmutations that cause its composition to change in the following ways:

1. Fissionable material is consumed.
2. Neutron-absorbing fission products are formed.
3. Heavy nuclides, mainly isotopes of uranium and plutonium, are formed.

These changes in composition bring about changes in reactivity of the fuel, which eventually decreases to such an extent that the reactor will no longer be critical unless the spent fuel is replaced with fresh fuel.

The changes in fuel composition to be discussed in this chapter take place over a much longer period of time than the buildup of ^{135}Xe and ^{149}Sm to steady-state concentrations, because the cross sections of the nuclides involved are much smaller, being less than 2200 b for the most part. These changes continue to take place during the entire lifetime of the fuel charge, which may be as great as a year or more. The changes in reactivity caused by changes in composition of all nuclides except ^{135}Xe and ^{149}Sm are called long-term reactivity changes.

One of the principal objectives of fuel-cycle analysis is to follow quantitatively the changes in concentration of fissile and fertile nuclides and fission products during neutron irradiation.

Another important objective is to follow the changes in reactivity that take place as fissile nuclides are depleted or formed from fertile nuclides, and as neutron poisons are formed through buildup of fission products or burned out through reaction with neutrons.

A third important objective is to follow the shifts in flux and power density distribution that take place in a reactor as a result of spatially nonuniform changes in fuel composition and reactivity. Calculation of these shifts in flux and power density, however, requires very detailed attention to local changes in composition. These calculations cannot be readily carried out by simple analytic or graphic procedures and must be done with a high-speed computer. Consequently, this chapter is concerned primarily with changes in fuel composition and reactivity and discusses only briefly changes in flux and power density distribution. Primary emphasis is placed on determining the fraction of fuel that can be made to undergo fission before the reactor ceases to be critical, as this determines the amount of heat that can be produced from the fuel, and the composition of spent fuel discharged from the reactor, as this is related to its value if processed for reuse.

Figure 3.3 is an example of the change in composition of fuel in a PWR during irradiation, calculated by the computer code CELL [B2]. In this example fuel charged to the reactor contained 3.2 w/o ^{235}U in total uranium. The extent of irradiation, plotted along the x axis, is

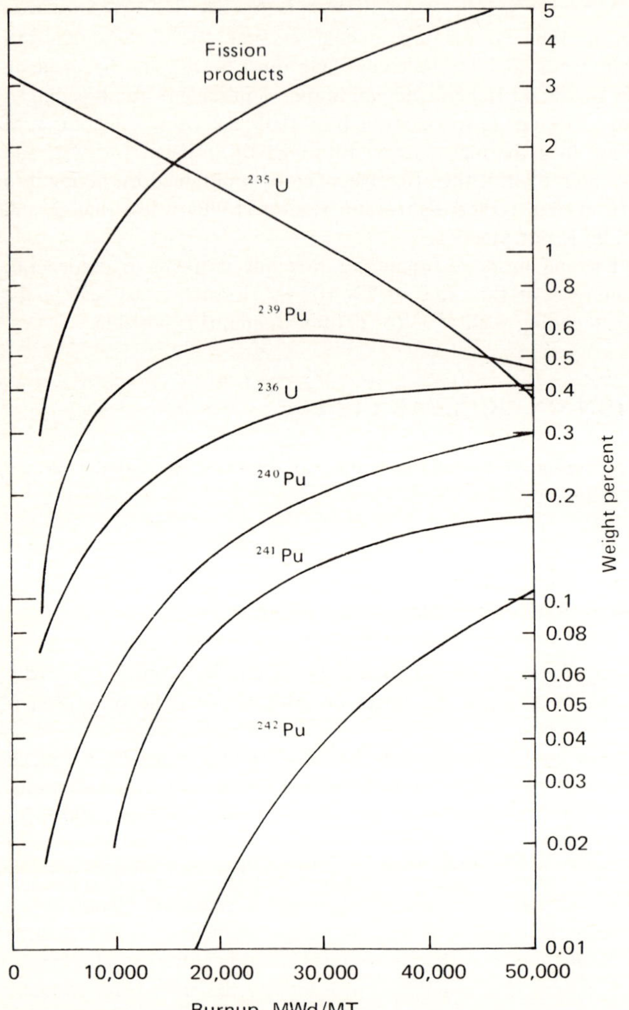

Figure 3.3 Change of nuclide concentration with burnup, 1060-MWe PWR.

expressed in terms of the "burnup," in megawatt-days per metric ton (MWd/MT), which is the same as kilowatt-days per kilogram. This is the amount of heat liberated by the fuel through fission and other nuclear reactions. Because complete fission of 1 g of ^{235}U produces 0.948 MWd of heat, burnup of 10,000 MWd/MT (1 Mg) corresponds to fission of around

$$\frac{(10{,}000)(100)}{(1{,}000{,}000)(0.948)} = 1.05 \text{ percent of the fuel}$$

This figure shows that ^{235}U concentration decreases almost exponentially with burnup. ^{236}U, a neutron-absorbing isotope of uranium, builds up to a concentration of around 0.4 percent of total fuel. ^{239}Pu, a fissionable isotope, builds up to a concentration of around 0.6 percent. ^{240}Pu builds up more slowly to around 0.3 percent. When ^{240}Pu absorbs a neutron, ^{241}Pu, another fissionable isotope, is formed. When this absorbs still another neutron, ^{242}Pu, a neutron absorber, results. The net effect is that ^{239}Pu and ^{241}Pu are desirable isotopes, which increase the reactivity of fuel, and ^{240}Pu is not detrimental because it makes a fissionable isotope. ^{242}Pu, however, like ^{236}U, is a deleterious, neutron-absorbing end product.

The changes in fuel composition just described cause the reactivity of the fuel to decrease with increasing burnup. The reactivity is defined as the difference between the rate of neutron production by fuel and the rate of neutron consumption, divided by the rate of neutron production. If the reactivity is zero, the reactor will be just critical without insertion of control poisons; if the reactivity is negative, the reactor power will die out; if the reactivity is positive, the reactor can be brought to a steady power level by insertion of sufficient neutron-absorbing control poison to reduce its reactivity to zero. Figure 3.4 shows how the reactivity of a PWR whose fuel composition is spatially uniform decreases with burnup. Lines are plotted for four different initial fuel compositions: 2.8, 3.2, 3.6, and 4.0 w/o ^{235}U. To a rough approximation, reactivity decreases linearly with burnup and increases linearly with w/o ^{235}U in fuel at the start of irradiation.

The reactivity of fuel in an actual reactor differs from Fig. 3.4 in two respects. First, Fig. 3.4 refers to a very large reactor, so large that neutron leakage to the outside has negligible effect on reactivity. A finite-sized reactor would have somewhat less reactivity than plotted here; the reactivity of a 1060-MWe PWR would be about 0.05 units less. Second, the composition of fuel in an actual reactor is nonuniform spatially, both because fuel of different composition may be placed in different positions in the reactor and because the composition of

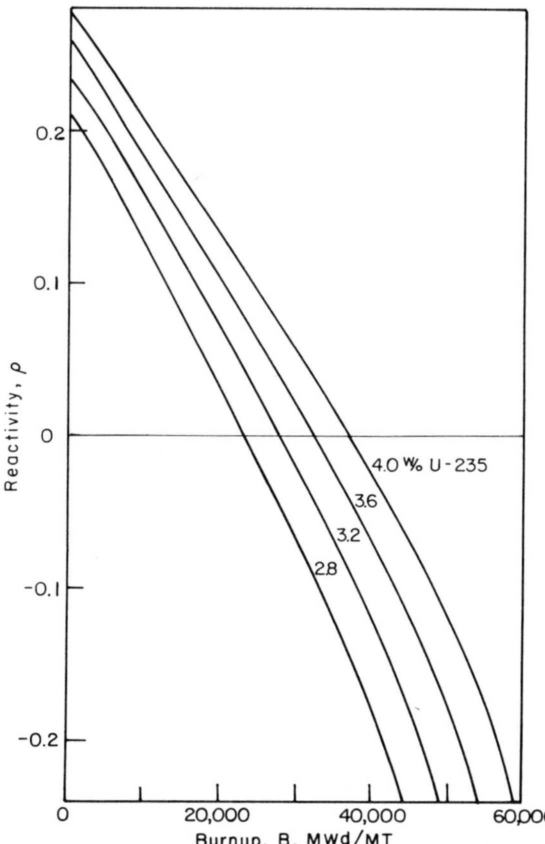

Figure 3.4 Change of reactivity with burnup for uniformly fueled infinite PWR.

fuel in different locations changes at different rates since the neutron flux is nonuniform. For these two reasons, Fig. 3.4 can give only general trends; determination of the change of reactivity with burnup in a practical reactor and the amount of energy that a given change of fuel can produce before the reactor ceases to be critical requires very detailed analysis of changes in composition and reactivity taking place at many different locations throughout the reactor.

Reactivity decreases with increasing burnup because the increase in ^{239}Pu and ^{241}Pu content is not sufficient to compensate for the decrease in ^{235}U content, and because ^{236}U, ^{240}Pu, ^{242}Pu, and fission products, whose content increases, are neutron-absorbing poisons.

Another very important effect of irradiation on fuel, which is noted here but not discussed further, is the change in physical properties that takes place. Fuels often change dimensions and swell, blister, or crack. Fission-product gases may be released and build up appreciable pressure inside cladding. Physical properties important in limiting fuel performance, such as thermal conductivity, may be changed. In many cases impairment of physical properties or intolerable dimensional changes limit the amount of heat that can be obtained from fuel rather than loss of reactivity. In a well-designed reactor, however, physical properties should remain satisfactory until fuel ceases to be critical. Currently, UO_2 fuel for LWRs is being designed to remain intact until about 3.5 percent fissions, corresponding to generation of around 35,000 MWd of heat per metric ton of fuel (35,000 MWd/MT). For fast reactors whose fuel is more expensive to fabricate, burnups of 100,000 MWd/MT are thought to be desirable for maximum economy.

3 FUEL AND POISON MANAGEMENT

When the time comes to replace fuel in a reactor, either because of loss of reactivity or because of changes in its physical properties, the reactor operator is faced with a number of alternative choices. The operator must decide whether to remove all or part of the fuel in the reactor, and whether to move some of the fuel remaining in the reactor from one location to another, and he or she must choose the composition of new fuel to replace the fuel removed.

The reactor operator may also elect to add neutron-absorbing poisons to the fuel when charged, and may change control-poison concentration or move poison from place to place in the reactor during fuel life. Procedures used in charging, discharging, or moving fuel and control poison are known collectively as fuel and poison management.

3.1 Objectives

The principal objectives of fuel and poison management are as follows:

1. To keep the reactor critical during long-term changes in fuel composition and reactivity
2. To shape power density distribution to maximize power output
3. To maximize heat production from fuel
4. To obtain uniform irradiation of fuel
5. To maximize productive use of neutrons

Not all these objectives can be achieved simultaneously in a given reactor, and some compromises among them are usually necessary. Each objective will be described briefly in turn.

Maintenance of criticality. As each fuel element in a reactor is irradiated, its composition changes, as does its contribution to overall reactivity. To maintain criticality in the face of these composition changes, it is necessary either to move control poison or change its concentration or to move fuel or change its concentration. Because reactivity changes caused by changes in fuel composition occur at low rates, seldom greater than a tenth of a percent per week, movement of fuel or poison to compensate for fuel composition changes may be very slow, in contrast to the rapid movement that may be required to compensate for load changes, operating disturbances, or emergencies.

Shaping power density distribution. A nuclear power reactor and its fuel are so costly that it is very desirable, economically, to obtain the maximum amount of power from a given charge of fuel and a given size of reactor, or conversely, to design a reactor in which a desired power output can be obtained from the minimum size of reactor and the minimum investment in fuel. The optimum use is made of fuel when each element is operating at the maximum allowable condition, i.e., at the maximum allowable cladding temperature, maximum allowable thermal stress, maximum allowable heat flux, and/or maximum allowable linear power density. A uniformly fueled and poisoned reactor is far from this ideal condition because of the wide variation of neutron flux and power density from point to point. In a cylindrical reactor whose fuel and poison distribution is spatially uniform, the neutron flux and power density vary with radius r and axial distance from midplane z as $J_0(2.405r/R) \cos(\pi z/H)$, where R is the effective radius and H the effective height of the fuel-bearing core of the reactor. The power density at the center is more than three times the average and the power density at the outer radius, top and bottom, is nearly zero. In all power reactors designed with economical performance in mind, fuel and/or poison is so managed that the power density distribution is more uniform than this cos J_0 distribution. The optimum power density distribution will depend on what factors limit power output, whether it be temperature, thermal stress, heat flux, or linear power, and usually is quite specific to a particular reactor.

Maximum heat production. Before fuel can be charged to a reactor, it is usually necessary to bring it into a closely specified chemical and physical condition and to seal it in pressure-tight cladding fabricated to narrowly specified dimensions. After fuel is discharged from a reactor, it usually still contains enough fissile material to justify its recovery through chemical reprocessing. These operations of fuel preparation and reprocessing often cost $200,000/ton of fuel or more. It is therefore economically desirable to obtain the maximum possible amount of heat from each fuel element before it is discharged from the reactor. Even at the burnup of 30,000 MWd/MT, now obtainable from oxide fuel before physical damage necessitates fuel replacement, fabrication and reprocessing contribute $6.7/MWd or more to the cost of heat, or 0.9 mills/kWh to the cost of electricity in a power plant that is 30 percent efficient. It is thus of considerable economic importance to strive for maximum burnup until limited either by physical damage or by offsetting economic factors such as the higher cost of the richer fuel needed for higher burnup. The economic optimum burnup will be discussed later in this chapter.

Uniform burnup. Because of the high cost of fuel fabrication and reprocessing, it is also important to manage fuel so that every element at discharge has been irradiated to nearly the same burnup. If this is not done, some of the fuel would have generated much less heat than elements that had received the maximum permissible irradiation, and the unit cost of heat from these underirradiated elements would be undesirably high.

Productive use of neutrons. In thermal reactors, the number of neutrons produced per neutron absorbed in fissile material (η) is of the order of 2.0. One of these neutrons is needed to keep the fission reaction going, but the second neutron, in theory, is available to produce valuable by-products of nuclear power. In practice, of course, some of these extra neutrons are necessarily lost through leakage and absorption in reactor materials, but around 0.6 neutron is available in water-moderated reactors for productive use. Examples of productive uses of neutrons are making plutonium from ^{238}U, ^{233}U from thorium, or ^{60}Co from natural cobalt. To maximize production of such by-products, it is desirable to use methods of fuel and poison management that minimize leakage of neutrons and their nonproductive absorption in control materials that upon neutron absorption produce relatively valueless materials. For example, it would be better to use ^{238}U or thorium to absorb extra neutrons than boron control poison, because plutonium from ^{238}U or ^{233}U from thorium may be worth as much as \$20/g as nuclear fuels, whereas boron produces only valueless helium and lithium. We shall see that some methods of fuel management conveniently permit the ^{238}U remaining in uranium fuel after ^{235}U is depleted to absorb the extra neutrons produced from fresh fuel of higher ^{235}U content. Such a method of fuel management is clearly more desirable economically than one that uses boron control poison to absorb extra neutrons produced in fresh fuel.

3.2 Drawbacks of Batch Irradiation of Uniform Fuel and Poison

To point to the importance of using improved methods of fuel and poison management, we shall discuss qualitatively the multiple drawbacks of the simplest method, which is batch irradiation of fuel initially uniform in composition, with spatially uniform distribution of boron control poison and with complete replacement of fuel at the end of its operating life. An example of this would be a PWR charged with fuel of uniform enrichment containing 4 percent ^{235}U and 96 percent ^{238}U and controlled by adjusting the concentration of boric acid dissolved in the water coolant to keep the reactor just critical at the desired power level. When this reactor starts operation, the compositions of fuel and poison are uniform throughout the core, and the flux and power density distribution are very nonuniform.

Figure 3.5 illustrates the spatial variation of power density in one-quarter of the core of a 1060-MWe PWR when the enrichment of ^{235}U and the concentration of boron control poison are uniform throughout the core. The lines plotted are lines of constant power density expressed as kilowatts of heat per liter of reactor volume, and also as kilowatts of heat per foot of fuel rod. The maximum permissible value of the latter is around 16 kW/ft, to ensure against overheating the fuel or cladding.

This figure illustrates immediately one of the disadvantages of batch fuel management. The power density, which is proportional to the product of the neutron flux and the fissile material concentration, is just as nonuniform as the neutron flux. If the local power density must be kept below some safe upper limit, to keep from overheating the fuel or cladding, only the fuel at the center of the reactor can be allowed to reach this power density, and fuel at all other points will be operating at much lower output. In a typical uniformly fueled and poisoned water-moderated reactor, the ratio of peak to average power density is over 3, so that the reactor puts out only one-third as much heat as it could if the power density were uniform.

The nonuniform flux is responsible for a second drawback of this method of fuel and poison management, the nonuniform change that takes place in fuel composition. In the center of the reactor, where the flux is highest, fuel composition changes more rapidly than at points nearer the outside of the reactor, where the flux is lower. As times goes on, therefore, the ^{235}U content at the center of the reactor becomes much lower and the burnup of the fuel much higher than toward the outside of the reactor. When the end of fuel life is reached, either because fuel at the center has reached the maximum burnup permitted because of radiation damage, or because the reactor has ceased to be critical with all boron removed, the outer fuel will have produced much less heat than the central fuel. If all fuel is discharged at end of life, the unit cost of heat from the outer fuel will be much higher than the central fuel. Figure 3.6 shows the final burnup distribution in a quarter of the core of a 1060-MWe PWR if charged initially with fuel of uniform composition.

A third drawback of this method of fueling is the large change in reactivity that takes place

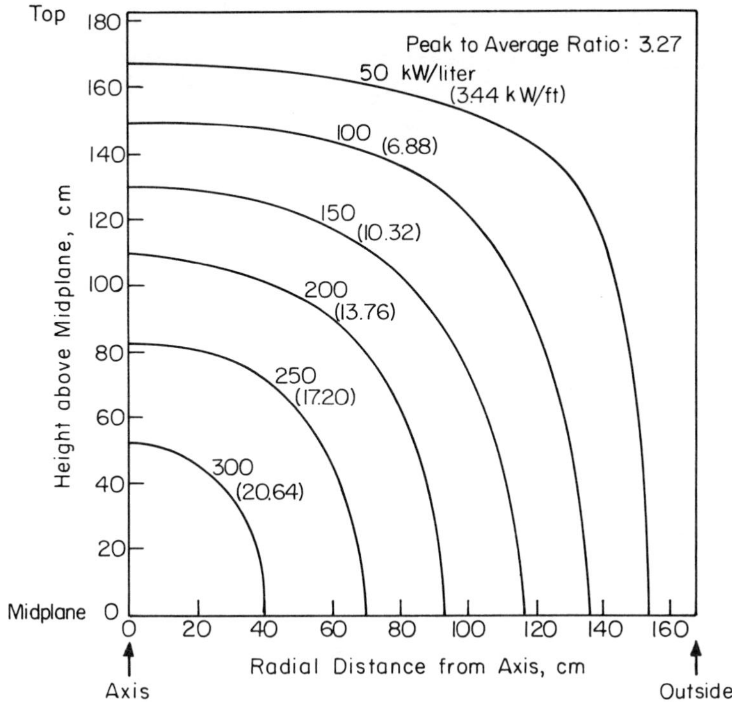

Figure 3.5 Power density distribution in 1060-MWe PWR at beginning of period, with uniform poison, moderator, and fuel containing 3.2 w/o ^{235}U.

between the beginning and end of fuel life. The reactivity of enriched uranium decreases steadily during irradiation. To compensate for this in simple batch irradiation, it is necessary to have a relatively large amount of control poison present at the beginning of fuel life and to withdraw this as irradiation progresses until at the end of life, ideally, all poison has been removed. When soluble poison such as boric acid is used, this means a high concentration at the beginning of life, with possible adverse effects on coolant corrosion and other chemical properties, and a large system for processing coolant to remove boron. When movable control rods are used, this means a large number of rods, which adds to cost; in some reactors the burnup obtainable is limited by the amount of room available for control rod insertion.

A fourth drawback of this simple batch irradiation is the waste of neutrons through absorption by boron at the beginning of the cycle. To give a rough example, to obtain an average burnup of 20,000 MWd/MT in a PWR with simple batch irradiation, it is necessary to absorb around 16 percent as many neutrons in boron at the beginning of life as are absorbed by ^{235}U at that time. In some of the more sophisticated methods of fuel management, these neutrons would be absorbed in ^{238}U to make plutonium. As the heat of fission is around 1 MWd/g and as about 0.8 g ^{235}U is fissioned per gram of ^{235}U consumed, $(0.16)(1/0.8) = 0.2$ g plutonium/MWd of heat could have been made with ^{238}U that are not made with boron. As plutonium has a value of around \$20/g, production of plutonium with these excess neutrons would be worth \$4/MWd of heat, or 0.5 mills/kWh of electricity in a nuclear power plant that is 30 percent efficient. At the end of fuel life this loss drops to zero, so that over fuel life the average loss due to absorbing neutrons in boron is about 0.25 mills/kWh. In a 1000-MW plant operating 7000 h/year, this is a loss of almost \$2 million/year, enough to make more sophisticated methods of fuel management well worth using.

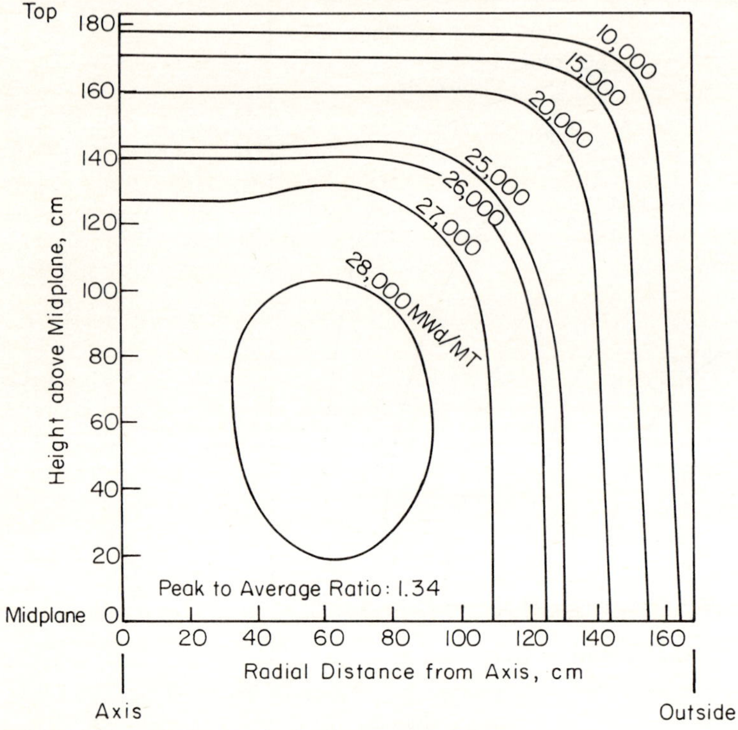

Figure 3.6 Burnup distribution in 1060-MWe PWR at end of period after batch irradiation of initially uniform fuel containing 3.2 w/o ^{235}U.

3.3 Idealized Methods of Fuel and Poison Management

Zoned loading. By charging fuel of different enrichments to different zones in the reactor, or by using a different concentration of poison in different parts of the reactor, it is possible to change the power density distribution from the undesirably nonuniform cos J_0 distribution to a distribution in which more of the reactor operates at the maximum permissible power density. One general type of zoned loading, which is close to optimum for a reactor in which the fuel linear power limits thermal output, is a reactor designed to have uniform power density throughout a substantial fraction of its core. This may be done by providing fuel in a central region, in which the flux is made uniform, of lower enrichment than in the peripheral regions of the reactor, the so-called buckled zones. A similar result may be obtained by poisoning fuel more heavily in the flattened central region than in the peripheral buckled zones. In addition to its advantage of providing more uniform power density, zoned loading also has the advantage of providing uniform burnup for at least the fuel in the part of the reactor where the flux is uniform. A disadvantage of zoned loading is the need to use in the buckled, unflattened zone fuel of higher enrichment, and hence greater cost, than would be necessary with uniform loading. The burnup of fuel in the buckled zone is also very nonuniform.

Partial batch replacement. Another method of fuel management, designed to deal with the nonuniform burnup of fuel, which is a second disadvantage of simple batch irradiation, is partial replacement of the fuel at the end of life instead of complete replacement. In this method, at the end of life only the most highly burned fuel is replaced by fresh fuel, and the rest of the charge is left in the reactor until the next time fuel has to be replaced. An example of how this might be done is shown in Fig. 3.7, which represents a cross section of a reactor core containing 320 square fuel assemblies, such as might be used in a large boiling- or pressurized-water reactor. Fuel assemblies are divided into groups containing equal numbers, each in a roughly annular zone. In the example of Fig. 3.7, five zones, each containing 64 assemblies, are shown, with zone 1 farthest from the center and zone 5 at the center. In the method of partial batch replacement, all zones initially are charged with fuel of the same composition. As irradiation proceeds, fuel in the central zone 5 is burned at a higher rate than fuel in the outer zones, because the flux is highest at the center of a uniformly fueled reactor. When it becomes necessary to replace fuel, either because fuel in zone 5 has reached the maximum permissible burnup, or because the reactor is no longer critical, only the most highly burned fuel, in zone 5, is replaced by fresh fuel, and irradiation is continued. When it again becomes necessary to refuel, the fuel then most highly burned, which will now probably be in zone 4, is replaced by fresh fuel, and so on. The advantage of this method of fuel management, of course, is that the fuel discharged each time has fairly uniform composition, because it comes from parts of the reactor where the flux has been fairly uniform. Disadvantages are (1) the need to open the reactor more frequently for refueling than when all the fuel is replaced at the same time, and (2) the peaking in flux and power density that occurs whenever fresh fuel is charged to the center of the reactor with partially depleted fuel elsewhere in the reactor, as in the first refueling of the foregoing example.

Scatter refueling. Flux peaking can be reduced by a different method of partial batch replacement, called scatter refueling, which is illustrated by Fig. 3.8. In this method, fuel is

						1	1	1	1	1	1								
				1	1	2	2	2	2	2	2	1	1						
		1	1	1	2	2	2	3	3	3	3	2	2	2	1	1	1		
		1	1	2	2	3	3	3	4	4	3	3	3	2	2	1	1		
		1	2	3	3	3	4	4	4	4	4	4	3	3	3	2	1		
	1	2	2	3	3	4	4	4	4	4	4	4	4	3	3	2	2	1	
	1	2	3	3	4	5	5	5	5	5	5	5	5	4	3	3	2	1	
1	2	2	3	4	4	5	5	5	5	5	5	5	5	4	4	3	2	2	1
1	2	3	3	4	4	5	5	5	5	5	5	5	5	4	4	3	3	2	1
1	2	3	4	4	4	5	5	5	5	5	5	5	5	4	4	4	3	2	1
1	2	3	4	4	4	5	5	5	5	5	5	5	5	4	4	4	3	2	1
1	2	3	3	4	4	5	5	5	5	5	5	5	5	4	4	3	3	2	1
1	2	2	3	4	4	5	5	5	5	5	5	5	5	4	4	3	2	2	1
	1	2	3	3	4	5	5	5	5	5	5	5	5	4	3	3	2	1	
	1	2	2	3	3	4	4	4	4	4	4	4	4	3	3	2	2	1	
		1	2	3	3	3	4	4	4	4	4	4	3	3	3	2	1		
		1	1	2	2	3	3	3	4	4	3	3	3	2	2	1	1		
		1	1	1	2	2	2	3	3	3	3	2	2	2	1	1	1		
				1	1	2	2	2	2	2	2	1	1						
						1	1	1	1	1	1								

Figure 3.7 Fuel zones for partial batch replacement or out-in fueling.

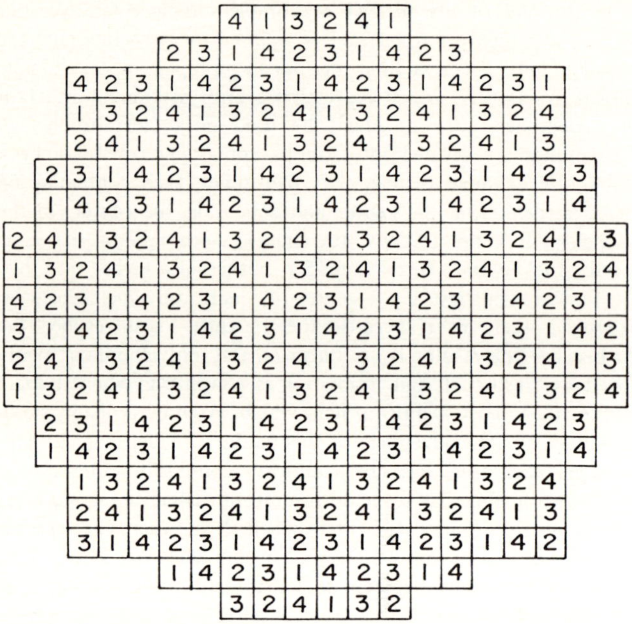

4 1 3 2 4 1
2 3 1 4 2 3 1 4 2 3
4 2 3 1 4 2 3 1 4 2 3 1 4 2 3 1
1 3 2 4 1 3 2 4 1 3 2 4 1 3 2 4
2 4 1 3 2 4 1 3 2 4 1 3 2 4 1 3
2 3 1 4 2 3 1 4 2 3 1 4 2 3 1 4 2 3
1 4 2 3 1 4 2 3 1 4 2 3 1 4 2 3 1 4
2 4 1 3 2 4 1 3 2 4 1 3 2 4 1 3 2 4 1 3
1 3 2 4 1 3 2 4 1 3 2 4 1 3 2 4 1 3 2 4
4 2 3 1 4 2 3 1 4 2 3 1 4 2 3 1 4 2 3 1
3 1 4 2 3 1 4 2 3 1 4 2 3 1 4 2 3 1 4 2
2 4 1 3 2 4 1 3 2 4 1 3 2 4 1 3 2 4 1 3
1 3 2 4 1 3 2 4 1 3 2 4 1 3 2 4 1 3 2 4
2 3 1 4 2 3 1 4 2 3 1 4 2 3 1 4 2 3
1 4 2 3 1 4 2 3 1 4 2 3 1 4 2 3 1 4
1 3 2 4 1 3 2 4 1 3 2 4 1 3 2 4
2 4 1 3 2 4 1 3 2 4 1 3 2 4 1 3
3 1 4 2 3 1 4 2 3 1 4 2 3 1 4 2
1 4 2 3 1 4 2 3 1 4
3 2 4 1 3 2

Figure 3.8 Fuel pattern in scatter refueling.

divided locally into groups containing the same number of assemblies, in this case into 80 groups each containing four assemblies. At the first refueling, an assembly in position 1 from each group is replaced by fresh fuel. At the second refueling, the assembly in position 2 from each group is replaced, at the third refueling the assembly from position 3 is replaced, and at the fourth refueling the assembly from position 4 is replaced. At the fifth refueling, each assembly from position 1 is replaced for the second time, and so on. After this stage is reached, at the beginning of every fueling cycle, each group of four assemblies will contain one fresh assembly, a second assembly that has been irradiated for one fueling cycle, a third that has been irradiated for two cycles, and a fourth that has been irradiated for three cycles. At the end of the fueling cycle, each group of four assemblies will contain one assembly that has been irradiated for one cycle, a second that has been irradiated for two, a third that has been irradiated for three, and a fourth that has been irradiated for four cycles and is then discharged and replaced by fresh fuel. The life of each assembly extends over four fueling cycles. When the individual assemblies are small, the neutron flux in each of the four assemblies of a group is nearly the same and flux peaking in the freshest, most reactive fuel is largely prevented. The overall flux distribution is flatter than in a uniformly fueled reactor, because the fuel in the center is more highly burned and less reactive than the fuel at the outside. Some power density peaking still occurs, however, because even though the flux is nearly uniform in a group of four assemblies, the freshest assembly has a higher fissile content than ones that have been in the reactor longer.

Scatter refueling also has two important advantages over simple batch irradiation: (1) Fuel of a given composition can be irradiated to a higher burnup before reactivity is lost in scatter refueling than in batch irradiation, and (2) less control poison is needed in scatter refueling than in simple batch irradiation. Both of these advantages of scatter refueling are a consequence of the fact that each part of the reactor contains some relatively fresh fuel and some fuel nearing the end of life. The fresh fuel maintains reactivity, while the older fuel is giving up more heat than it could in simple batch irradiation without ceasing to be critical. Furthermore, in scatter refueling, the more depleted fuel that is present at all times acts as a control poison to absorb excess neutrons from the more reactive fresh fuel. Moreover, many of the neutrons absorbed by depleted fuel are used productively to make plutonium. These advantages of scatter refueling are a feature of all methods of partial fuel replacement.

These advantages of scatter refueling may be expressed somewhat more quantitatively by considering how the reactivity ρ of fuel changes with burnup B. To a fair approximation, reactivity decreases linearly with burnup:

$$\rho = \rho_0 - aB \tag{3.1}$$

where ρ_0 is the reactivity of fresh fuel. In simple batch irradiation, the burnup of fuel at the end of life, B_1 when $\rho = 0$, is

$$B_1 = \frac{\rho_0}{a} \tag{3.2}$$

The amount of reactivity to be held down by control poison at the beginning of life, ρ_1 when $B = 0$, is

$$\rho_1 = \rho_0 \tag{3.3}$$

To find the reactivity-limited burnup of fuel in n-zone scatter refueling, B_n, note that at the end of life, the freshest nth fraction of fuel will have had burnup of approximately B_n/n, the next older nth fraction $2B_n/n$, etc., and the oldest nth fraction, ready for discharge, will have reached B_n burnup. The reactivity of this mixture of fuel is

$$\rho = \frac{1}{n}\sum_{i=1}^{n}\left(\rho_0 - \frac{aiB_n}{n}\right) = \rho_0 - \frac{a(n+1)B_n}{2n} \tag{3.4}$$

But $\rho = 0$ at the end of life, so that

$$B_n = \frac{2n\rho_0}{a(n+1)} \tag{3.5}$$

The ratio of the burnup obtainable in n-zone scatter refueling to that obtainable in simple batch irradiation is found by dividing Eq. (3.5) by (3.2):

$$\frac{B_n}{B_1} = \frac{2n}{n+1} \tag{3.6}$$

The reactivity of fuel in n-zone scatter refueling at the beginning of a cycle is

$$\rho_n = \frac{1}{n}\sum_{i=1}^{n}\left[\rho_0 - \frac{a(i-1)B_n}{n}\right] = \rho_0 - \frac{a(n-1)}{2n}B_n \tag{3.7}$$

By using (3.5),

$$\rho_n = \rho_0\left(1 - \frac{n-1}{n+1}\right) = \frac{2\rho_0}{n+1} \tag{3.8}$$

The ratio of the reactivity change per cycle in n-zone scatter refueling to the amount in simple batch irradiation is

$$\frac{\rho_n}{\rho_1} = \frac{2}{n+1} \tag{3.9}$$

Values of these ratios for several values of n, the number of zones of assemblies, are tabulated below.

Number of zones of assemblies, n	1	2	3	4	5	∞
Burnup ratio, scatter refueling/batch	1.00	1.33	1.50	1.60	1.67	2.00
Reactivity change, scatter/batch	1.00	0.67	0.50	0.40	0.33	0.00
Cycle time, scatter/batch	1.00	0.67	0.50	0.40	0.33	0.00

Thus, four-zone refueling permits attainment of 60 percent more burnup than simple batch refueling, with only 40 percent as much poison needed to control reactivity changes. The time between successive fuel replacements is only 40 percent as long in four-zone refueling as in batch, however.

Graded refueling. These equations and table show that increasing the number of zones continues to improve burnup and reduce reactivity changes, until the burnup approaches twice that obtainable from batch irradiation, and the reactivity change approaches zero. It is not feasible to approach these limits in water-moderated reactors, because fuel assemblies are relatively large, and even the biggest reactors contain only a few hundred assemblies at most, so for the reactor to contain a reasonable number of groups, six assemblies per group is practically an upper limit. Moreover, these reactors have to be shut down and opened to replace fuel, and fueling interruptions would occur too frequently with much more than six assemblies per group.

Graphite-moderated, gas-cooled reactors, on the other hand, make use of thousands of fuel assemblies and are equipped with fueling machines that permit replacement of individual assemblies without interrupting reactor operation. In these reactors it is possible, therefore, to have a large number of assemblies per group and to refuel continuously during operation. Under these conditions, fuel within the reactor is graded almost continuously in composition from fresh unburned fuel to fully burned fuel ready for replacement. The limiting continuous case of scatter refueling with n very large is sometimes called *graded refueling.* In graded fueling, when it is not necessary to shut down the reactor to refuel, it is possible to keep each assembly in the reactor until it has received the same burnup; whereas in scatter refueling, with a fixed fraction of fuel replaced at the same time, fuel removed from the center of the reactor is more heavily burned than fuel removed from the outside. Because the average composition and reactivity are constant in time, and all fuel discharged has the same composition, graded fueling is easier to treat analytically than scatter refueling with a finite number of assemblies per group, because of the changes in average composition and reactivity that then take place in each cycle.

Out-in refueling. Graded and scatter refueling have the disadvantage that the flux is higher in the center of the reactor than at the outside, although the nonuniformity is not so great as in simple batch irradiation because some highly burned fuel is always present at the center in graded and scatter refueling. An alternative method of fueling designed to depress the flux and power density further at the center of a reactor is out-in fueling. In this method, fuel is divided into annular zones of equal volume, such as those shown in Fig. 3.7. At the end of the first fueling cycle, fuel from central zone 5, the most heavily burned, is removed from the reactor; fuel from zone 4 is moved into zone 5; fuel from zone 3, to zone 4; fuel from zone 2, to zone 3; fuel from zone 1, to zone 2; and fresh fuel is charged to zone 1. At the end of each subsequent fueling cycle, this sequence of fuel movements is repeated. All cycles after the first few are similar, with the same cycle time, the same average burnup of discharged fuel, and the same change in reactivity. As fuel in the center of the reactor is most heavily depleted and least reactive, the flux and power density are depressed there relative to a uniformly fueled reactor. The upper half of Fig. 3.9 shows the power density distribution calculated by Westinghouse [D1] for three-zone out-in fueling of a 260-MWe PWR, with a core 1.25 m in radius operated at a burnup of 15,000 MWd/MT. The ratio of radial peak to average power density is 1.3, compared with about 1.5 for simple batch irradiation in this same reactor. The ratio of burnup with three-zone out-in fueling to burnup in batch fueling is about 1.5, as predicted by Eq. (3.6), which is approximately valid for this case also. Thus, out-in fueling has many advantages for a reactor of this size.

For larger reactors with high burnup, however, out-in fueling leads to too great a depression in the flux and power density at the center of the reactor. This may be seen from the lower half of Fig. 3.9, which shows the power density calculated by Westinghouse [D1] for three-zone out-in fueling of a 1000-MWe PWR, with a core 6.5 ft in radius, operated at a burnup of 24,000 MWd/MT. At the beginning of a cycle, the flux peaks heavily in the outside zone, and the peak-to-average radial power density ratio is 2.0. The reason for this poor

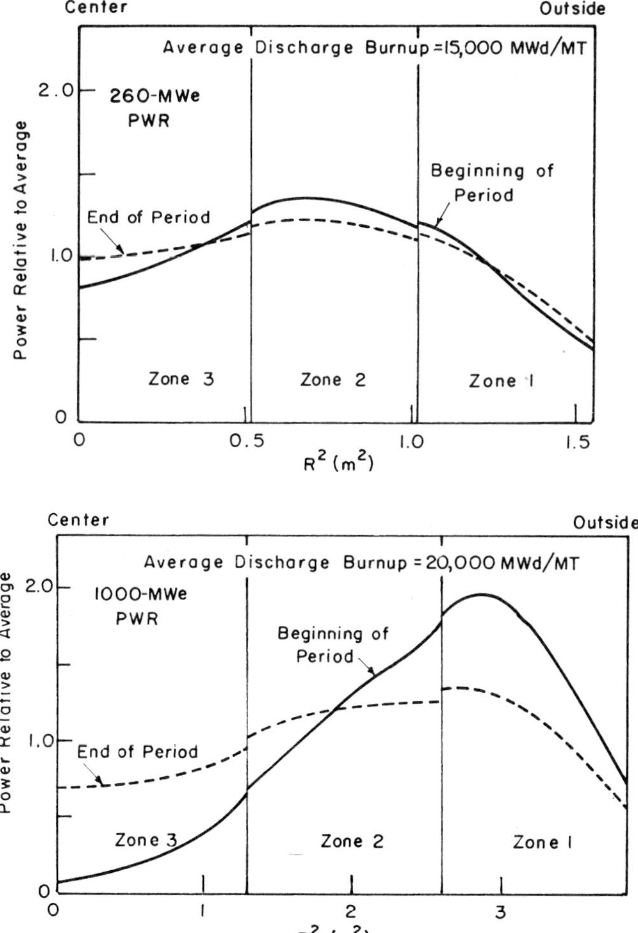

Figure 3.9 Radial power distribution with three-zone out-in fueling.

distribution is that the extra neutrons produced in the reactive outside zone 1, which are needed in the relatively unreactive central zone 3, must diffuse through a larger distance and hence require a greater flux difference than in a smaller core, with less reactivity difference.

Modified scatter refueling. For the largest reactors a combination of out-in and scatter refueling gives better results than either alone. Figure 3.10 shows how five-zone modified scatter refueling works. In this example for reactors with square fuel assemblies, fuel positions are divided into an outer zone 1 containing one-fifth of the fuel assemblies, and an inner zone containing the other four-fifths. Fuel in the inner zone is divided into groups of four for scatter refueling. At each refueling, the most heavily burned assembly in each group of four is removed from the inner zone and replaced by an assembly from the outer zone, which is moved in its entirety into the inner zone. Fresh fuel is then charged to the outer zone.

In this way the more depleted, less reactive inner zone is made to act rather like the flattened zone of zoned loading and the fresh fuel at the outside acts like the buckled zone. The peaking of power density at the center of a reactor using simple scatter refueling is reduced, without the overcompensation occurring with out-in fueling in a large reactor. The small reactivity change and high burnup obtainable with five-zone out-in or scatter refueling are realized.

3.4 Reactivity-limited Burnup in PWR with Modified Scatter Fueling

Watt [W2] has used the computer codes CELL and CORE to evaluate the reactivity-limited burnup of a 1060-MWe PWR operated with modified scatter refueling as a function of the

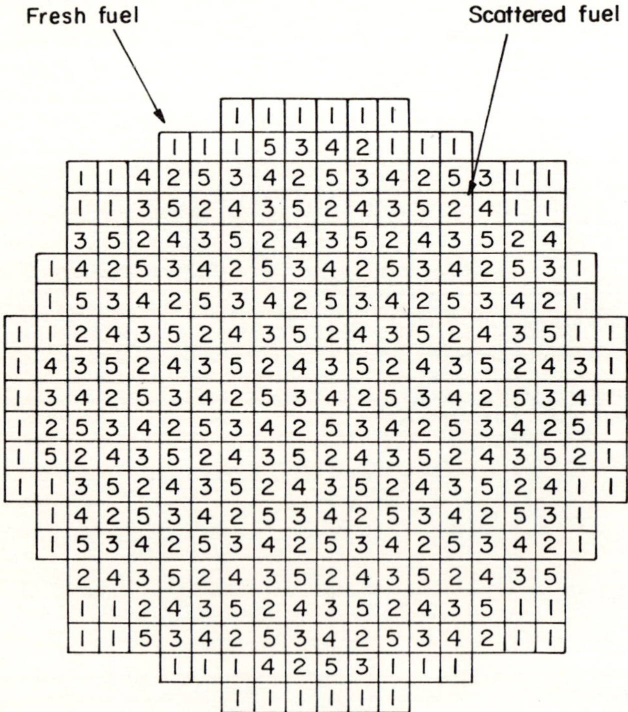

Figure 3.10 Modified scatter refueling.

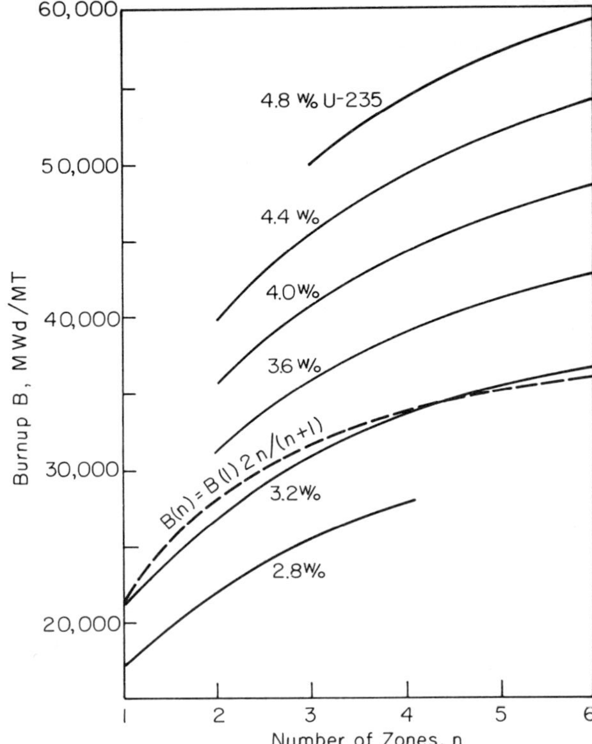

Figure 3.11 Reactivity-limited burnup versus number of fuel zones for various feed enrichments, 1060-MWe PWR, modified scatter refueling, steady state.

enrichment (w/o ^{235}U) of feed and number of fuel zones (n), with results shown in Fig. 3.11. The burnup increases roughly linearly with enrichment. The dashed line shows that the variation of reactivity-limited burnup of 3.2 w/o enriched fuel with the number of fuel zones predicted by the simple Eq. (3.6) is a fair representation of the more accurate computer result.

Since the burnup determines the amount of energy a lot of fuel produces during irradiation, it is an essential variable in determining the length of time the fuel spends in the reactor and the unit cost of that energy.

The number of megawatt-hours of electricity E generated by one lot of fuel during its entire stay in a reactor is

$$E = 24\eta B \left(\frac{U}{n}\right) \tag{3.10}$$

Here η is the thermal efficiency of the power plant (ratio of electricity generated to heat produced), U is the total number of metric tons of uranium in the reactor, n is the number of fuel zones, and U/n is the mass of uranium in one lot of fuel. With the dependence of burnup on enrichment and batch fraction given in Fig. 3.11, this equation permits evaluation of the electric energy that can be generated by a fuel batch of enrichment ϵ w/o ^{235}U, making up f $(= 1/n)$ fraction of the reactor. Figure 3.12 shows this relationship for the 1060-MWe PWR, with a thermal efficiency $\eta = 0.326$.

The number of megawatt-hours of electricity generated by the reactor during one fueling cycle is

$$E = 8766L'KT' \tag{3.11}$$

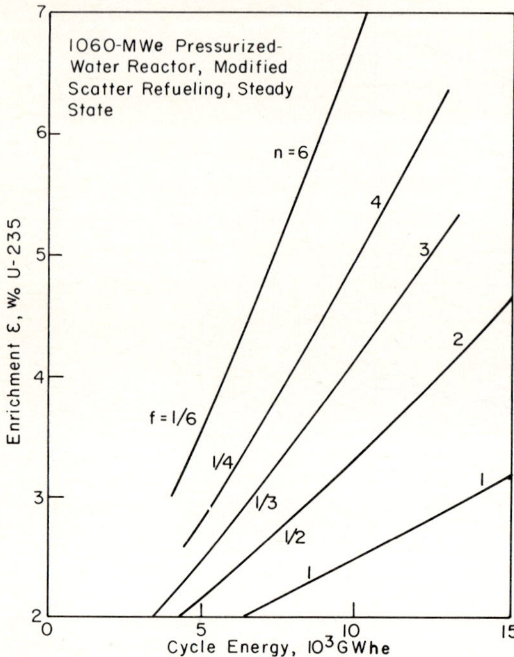

Figure 3.12 Enrichment versus cycle energy and batch fraction f.

Here 8766 is the average number of hours per year, K is the rated capacity of the power plant in electric megawatts, and T' is the number of years between the start of an irradiation cycle and the time the reactor is shut down for refueling. L' is the availability-based capacity factor, defined as the ratio of the amount of electricity generated by the power plant when not shut down for refueling to the amount it could have generated if operated at capacity for the same time.

When the reactor is operating in the steady state, with conditions in successive cycles repeating identically, these two energy amounts must be equal, so that

$$8766KL'T' = 24\eta B\left(\frac{U}{n}\right) \tag{3.12}$$

In addition to the time T' in which the plant is available for operation, light-water nuclear reactors must be shut down for refueling for a length of time ΔT between successive operating periods. Hence the duration of a full cycle is $T = T' + \Delta T$. For this chapter it will be assumed that the refueling downtime is $\Delta T = 0.125$ year, so that $T = T' + 0.125$. With this assumption, the relation between burnup B and steady-state cycle duration T becomes

$$T = 0.125 + \frac{B\eta(U/n)}{365.25KL'} \tag{3.13}$$

With the dependence of burnup on enrichment and batch fraction shown in Fig. 3.11, it is possible to express the cycle duration T as a function of these variables.

Figure 3.13 shows this relationship for a specific value, 0.9, for the availability-based capacity factor L'. This figure also shows the amount of electric energy produced per cycle. Because peak electric demands occur at intervals of 6 months or 1 year, Fig. 3.13 may be used to select combinations of number of fuel zones and enrichment that permit these desirable refueling intervals. Three-zone fueling with enrichment of 3.2 percent is one such combination.

3.5 Steady-State Fuel-Cycle Costs

Figure 3.14 shows the total steady-state fuel-cycle cost for an interval of 1.0 year between refuelings as a function of feed enrichment for batch fractions, f, of $\frac{1}{2}$, $\frac{1}{3}$, $\frac{1}{4}$, and $\frac{1}{6}$. The batch fraction is defined as $1/n$, where n is the number of fuel zones. Also plotted in this figure are levels of constant energy production (E) or capacity factor (L') and lines of constant burnup (B). The unit costs of fuel-cycle materials and services are those anticipated for the year 1980, to be described in more detail in Sec. 5.

To illustrate use of Fig. 3.14, the example of the line $L' = 0.9$ will be discussed. Suppose that this 1060-MWe reactor is expected to operate at an availability-based capacity factor $L' = 0.9$ with a 1-year interval between refuelings. The minimum fuel-cycle cost of $41 million will occur at a batch fraction $f = \frac{1}{4}$ and a feed enrichment of 3.75 w/o ^{235}U. This will require fuel to sustain an average burnup B of slightly over 40,000 MWd/MT. If average burnup should be limited for mechanical reasons to slightly over 30,000 MWd/MT, the minimum fuel cycle cost of $42 million will occur at $f = \frac{1}{3}$ and a feed enrichment of 3.2 w/o, the combination suggested by the manufacturer for this reactor.

Figure 3.15 shows the unit fuel-cycle cost in mills per kilowatt-hour as a function of the same variables. This unit cost is obtained by dividing the total cost in dollars by the electric energy in megawatt-hours. For example, the unit cost at $L' = 0.9$ and $f = \frac{1}{4}$ is $41,000,000/7317 $\times$ 10^3 MWh = 5.6 $/MWh or 5.6 mills/kWh. Because of the overlap of lines,

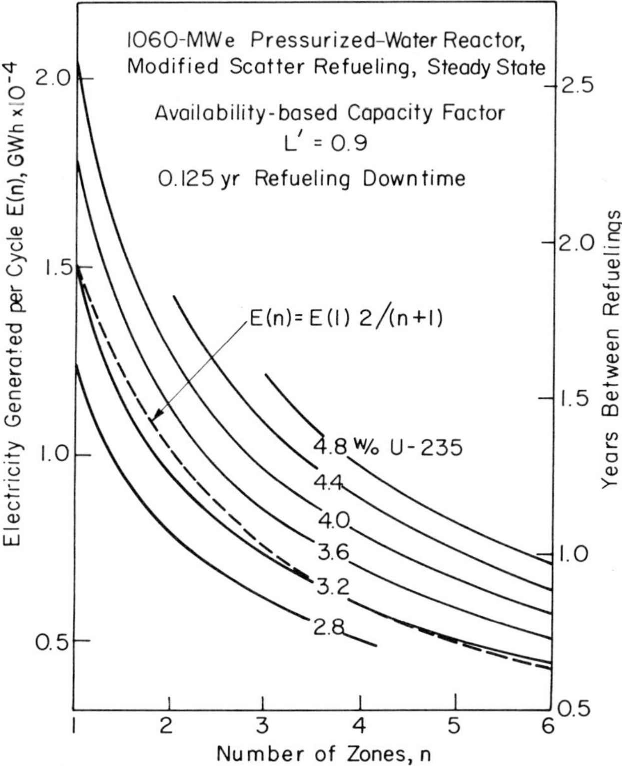

Figure 3.13 Energy produced per cycle and time between refuelings versus number of fuel zones for various feed enrichments.

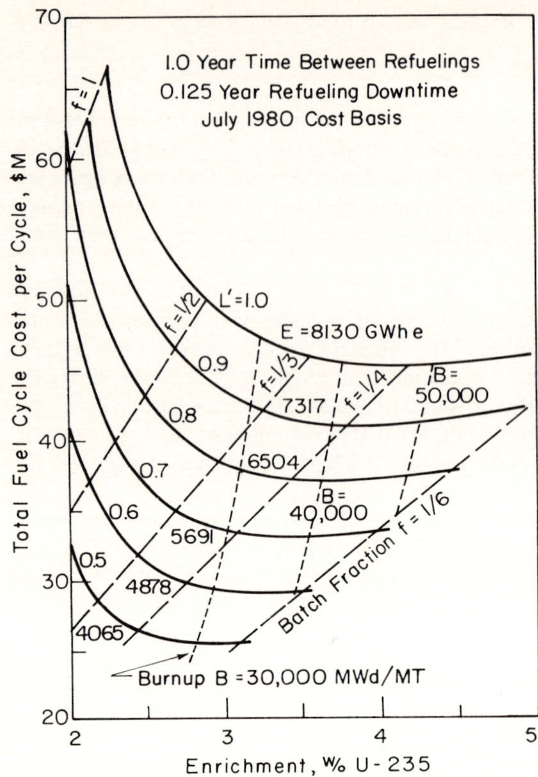

Figure 3.14 Effect of enrichment and batch fraction on total fuel cycle cost per steady-state cycle, electric energy per cycle (E), availability-based capacity factor (L'), and burnup (B).

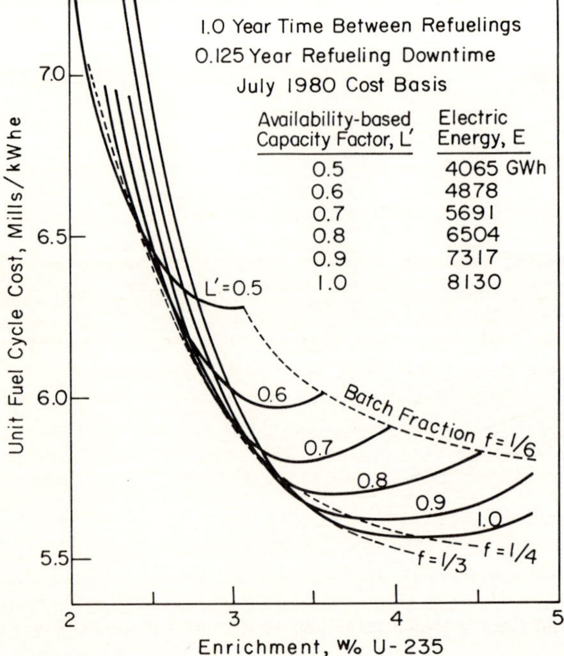

Availability-based Capacity Factor, L′	Electric Energy, E
0.5	4065 GWh
0.6	4878
0.7	5691
0.8	6504
0.9	7317
1.0	8130

Figure 3.15 Effect of enrichment and batch fraction on steady-state unit fuel-cycle cost.

representation of unit costs in Fig. 3.15 does not bring out the effect of the several variables on costs as well as representation of total costs in Fig. 3.14.

4 FUEL MANAGEMENT IN A LARGE PRESSURIZED-WATER REACTOR

Sections 3.4 and 3.5 have dealt with an idealized situation in which a PWR is operating in the steady state with an exact fraction (e.g., one-third) of the fuel replaced at each refueling. A real reactor seldom reaches a steady-state condition and may have a number of fuel assemblies that cannot be divided evenly into fuel zones containing equal numbers of assemblies. The purpose of this section is to describe briefly a real reactor and the results of a computer-based analysis of the fuel-cycle performance of this reactor through a succession of cycles.

4.1 Reactor Construction

The reactor to be discussed is the large PWR manufactured by the Westinghouse Electric Company, which has been built for the Diablo Canyon station of the Pacific Gas & Electric Company, the Donald C. Cook station of American Electric Power Corporation, and the Zion station of Commonwealth Edison Company. Rated capacities of 3250 MW (thermal) and 1060 MW (electric) have been used. The following brief description of this reactor was abstracted from the Safety Analysis Report of the Donald C. Cook station [A1].

Figure 3.16 is a cutaway view of this reactor. The reactor vessel is a cylinder 13 ft in diameter with an ellipsoidal bottom. The top of the vessel is closed with a flanged and bolted ellipsoidal head, which is removed for refueling. When in operation the reactor is filled with water at a pressure of 155 bar (15.5 MPa). The water enters the inlet nozzle at the left at a temperature of 282°C and leaves the outlet nozzle at the right at 317°C. The effective average temperature of the water is 301.6°C, which will be taken as the temperature of the Maxwell-Boltzmann component of the neutron flux.

There are 193 fuel assemblies held between the upper and the lower core plates. Figure 3.17 is a horizontal cross section through the portion of the reactor containing the assemblies. Inlet water flows down in the two annular spaces between the reactor vessel and the core barrel, turns at the bottom of the vessel, and flows upward through the fuel assemblies inside the core baffle.

Figure 3.18 is a dimensioned horizontal cross section of one fuel assembly. The assembly consists of a 15 × 15 square array of zircaloy-4 tubes set on 0.563-in square pitch. Two-hundred four of these tubes are filled with UO_2 pellets, pressurized with helium and closed with welded zircaloy end plugs. The zircaloy cladding for these fuel tubes is 0.422 in outside diameter, with a 0.0243-in wall. The overall length of tubing filled with UO_2 is 12 ft. At 20 points in the fuel assembly, zircaloy-4 guide tubes are provided for control rods. During normal operation these tubes are filled with water, burnable poison rods, or movable control rods. The central position in the fuel assembly is occupied by a zircaloy thimble for in-core instrumentation. It is sealed off from the water that surrounds the fuel assembly. The 225 zircaloy tubes of the assembly are held in place over their length by nine evenly spaced spring clip grids made of Inconel-718.

The mass of zircaloy in guide tubes and instrument thimble is 9.5 kg/assembly, and the mass of Inconel is 8.6 kg.

The reactor core consists of 193 fuel assemblies mounted on 8.466-in-square pitch. The initial loading of fuel and control poison in the core of this reactor is shown in Fig. 3.19. Fuel assemblies marked M are provided with movable control rods that can be inserted or withdrawn by control rod drives that enter through the head of the reactor vessel (Fig. 3.16). The numbers

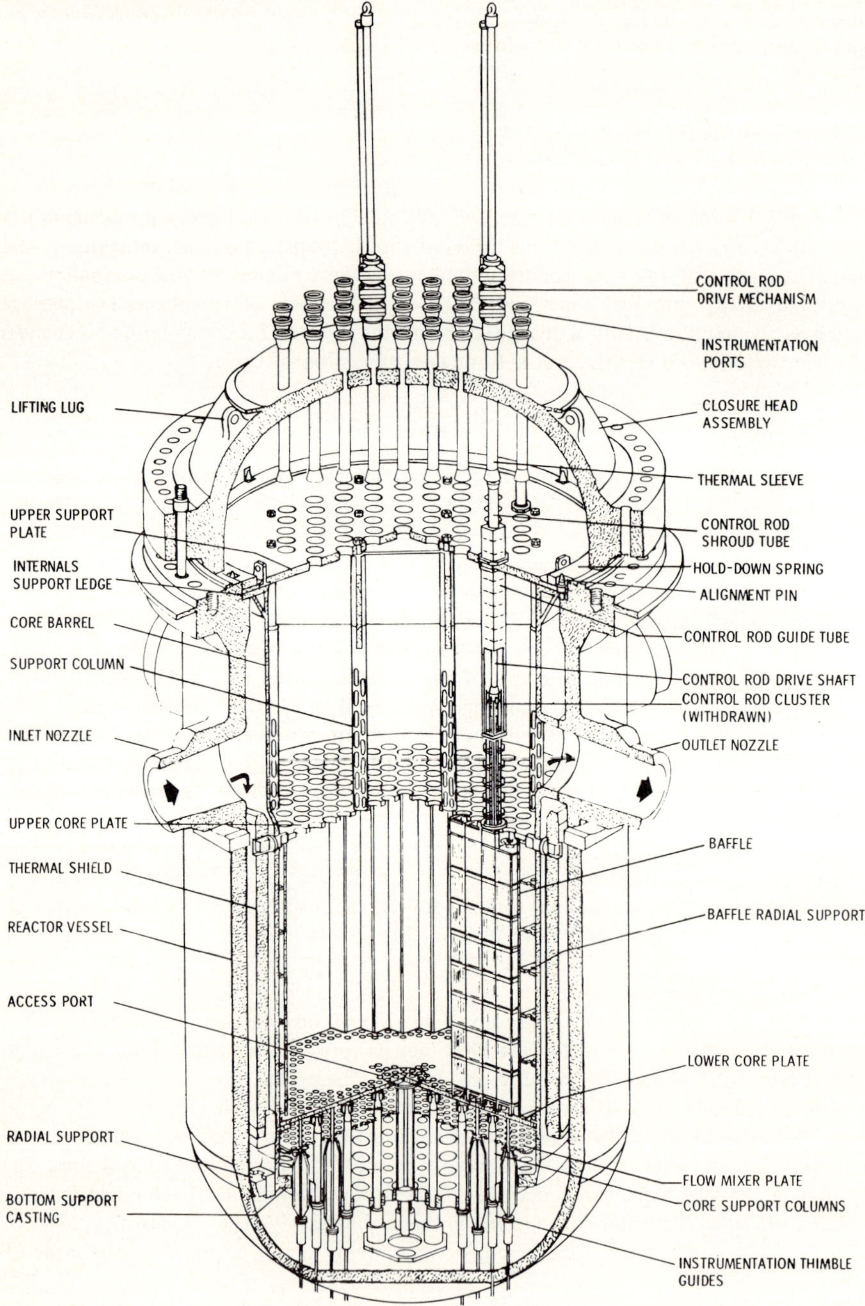

Figure 3.16 Cutaway view of large PWR.

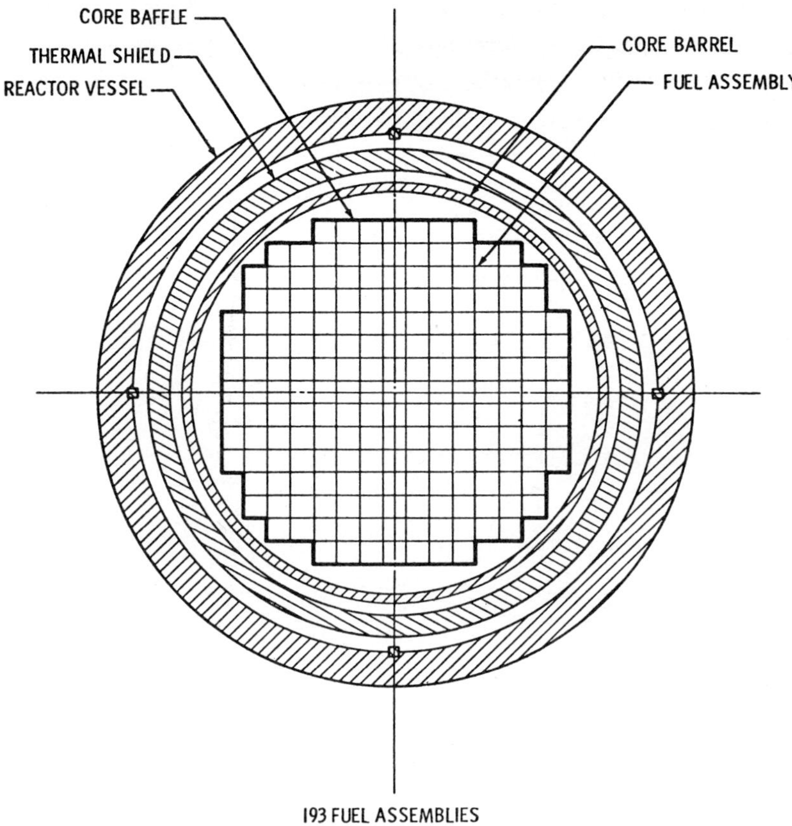

Figure 3.17 Core cross section of PWR.

(8, 9, 12, 16, or 20) placed at other fuel positions give the number of fixed burnable poison rods containing boron carbide placed in the indicated assembly during the first fuel cycle. During normal operation at full power, the movable control rods are fully withdrawn. Long-term reactivity changes are controlled by depletion of the burnable poison and by adjusting the concentration of boric acid dissolved in the cooling water.

4.2 Reactor Performance, Cycle 1

Rieck [R1] has used the computer codes LEOPARD [B1] and SIMULATE [F1] to predict the power distribution in the fuel and poison arrangement shown in Fig. 3.19 for the first fuel cycle for this reactor, and the amount of thermal energy produced by each assembly up to the time when the reactor ceases to be critical with all soluble boron removed from the cooling water. Figure 3.20 is a horizontal cross section of one-quarter of the core of this reactor. Each square represents one fuel assembly. The core arrangement has 90° rotational symmetry, about the central assembly 1AA at the upper left of the figure.

The first row of symbols in each square is the serial number of the assembly. The first symbol is the fuel lot number: lot 1 contains 2.25 w/o ^{235}U; lot 2 contains 2.8 w/o ^{235}U and boron burnable poison; and lot 3 contains 3.3 w/o ^{235}U and burnable poison. The second

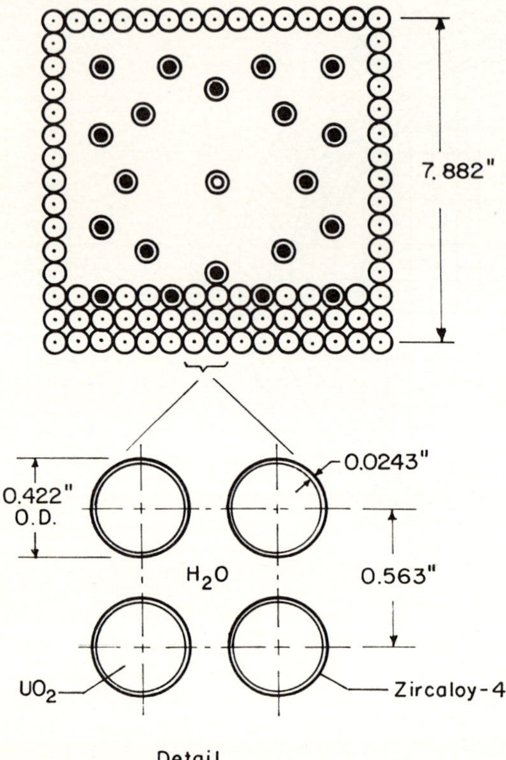

Figure 3.18 Section of fuel assembly.

symbol is the letter designating the row in which the assembly is placed when initially charged to the reactor. The third symbol is the letter designating the column in which the assembly is placed. The second row of symbols, here a dash (–), gives the burnup of the assembly at the start of the cycle, here zero. The third row gives the burnup at the end of the cycle when the reactivity has dropped to zero. The fourth row gives the power of the assembly relative to the core average. It is a requirement of fuel management in this reactor that the power of every assembly relative to the core average be kept below 1.58, to prevent the water leaving each assembly from reaching the boiling point at 155 bar. In this first cycle assembly power is controlled by the use of burnable poison and the placement of individual assemblies in the modified scatter pattern shown in Figs. 3.19 and 3.20.

The total thermal energy produced in the first cycle is evaluated by multiplying the burnup increment of each assembly in megawatt-days per metric ton by the mass of uranium in that assembly in metric tons and summing over all assemblies in the reactor, taking into account the total number of assemblies in positions equivalent to those shown in Fig. 3.20. For example, there are four assemblies in the BB position, four in BC, two in AB, and one in AA. The total thermal energy produced in the first cycle thus evaluated is 1341.1 GWd, or $32{,}188 \times 10^6$ kWh.

Table 3.2 gives the local power at 12 axial positions 1 ft apart in six selected assemblies relative to the average reactor power, at the beginning and end of the first cycle. Another requirement of fuel management in this reactor is that the ratio of local to average power at all points not exceed 2.33 to keep the linear power below 16 kW/ft. Table 3.2 shows that the maximum relative power of 1.68 at the beginning of the cycle (in EE) is well below this limit, and that the maximum relative power at the end of the cycle is even lower.

4.3 Reactor Performance, Cycle 2

At the end of cycle 1, 64 of the 65 assemblies of lot 1 (called lot 1A) are removed. One of the 1GC assemblies (called lot 1B) that had the lowest burnup of the lot 1 group is moved to the central AA position. Residual burnable poison is removed from the remaining lot 2 and lot 3 assemblies, which are shifted to the new positions shown in Fig. 3.21. Sixty-four new assemblies (called lot 4), enriched to 3.2 w/o ^{235}U and containing no burnable poison, are placed in the positions with heavy borders in this figure. This placement of assemblies was

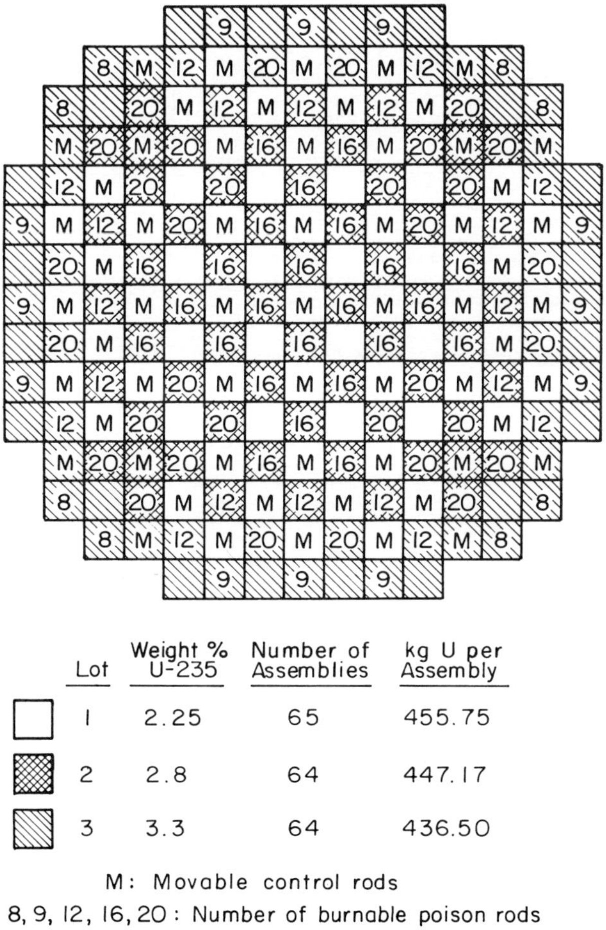

	Lot	Weight % U-235	Number of Assemblies	kg U per Assembly
(blank)	1	2.25	65	455.75
(cross-hatched)	2	2.8	64	447.17
(hatched)	3	3.3	64	436.50

M: Movable control rods

8, 9, 12, 16, 20 : Number of burnable poison rods

Figure 3.19 Arrangement of fuel and poison in initial loading of PWR.

Center-line	A	B	C	D	E	F	G	H
A	1AA 0 17302 1.04	2AB 0 16578 0.88	1AC 0 17063 1.04	2AD 0 16666 0.94	1AE 0 17442 1.19	2AF 0 17184 1.12	1AG 0 16220 1.16	3AH 0 12775 0.88
B	2BA 0 16578 0.88	1BB 0 17049 1.02	2BC 0 15844 0.82	1BD 0 17210 1.10	2BE 0 16716 0.99	1BF 0 17296 1.23	3BG 0 16041 1.00	3BH 0 13596 0.99
C	1CA 0 17063 1.04	2CB 0 15844 0.82	1CC 0 17050 1.05	2CD 0 16600 0.94	1CE 0 17260 1.16	2CF 0 16951 1.09	1CG 0 15950 1.15	3CH 0 11985 0.82
D	2DA 0 16666 0.94	1DB 0 17210 1.10	2DC 0 16600 0.94	1DD 0 17099 1.10	2DE 0 15981 0.90	1DF 0 16698 1.16	3DG 0 15945 1.06	3DH 0 10194 0.72
E	1EA 0 17442 1.19	2EB 0 16716 0.99	1EC 0 17260 1.16	2ED 0 15981 0.90	2EE 0 18303 1.24*	2EF 0 14934 0.87	3EG 0 14014 0.99	
F	2FA 0 17184 1.12	1FB 0 17296 1.23	2FC 0 16951 1.09	1FD 0 16698 1.16	2FE 0 14934 0.87	3FF 0 16484 1.16	3FG 0 9449 0.62	
G	1GA 0 16220 1.16	3GB 0 16041 1.00	1GC 0 15950 1.15	3GD 0 15945 1.06	3GE 0 14014 0.99	3GF 0 9449 0.62		
H	3HA 0 12775 0.88	3HB 0 13596 0.99	3HC 0 11985 0.82	3HD 0 10194 0.72				

* = Maximum Relative Power

Fuel Lot 1 2.25 w/o U-235
Fuel Lot 2 2.80 w/o U-235
Fuel Lot 3 3.30 w/o U-235

Cycle Average Burnup = 15,535 MWd/MT
Cycle Thermal Energy = 1341.1 GWd

Key

1AA	Assembly Number
0	BOC Burnup, MWd/MT
17302	EOC Burnup, MWd/MT
1.04	BOC Relative Power (Assembly/Average)

Figure 3.20 PWR, assembly power and burnup distribution, cycle 1.

found by Rieck [R1] to lead to an acceptably low maximum peak-to-average assembly power ratio of 1.34 in assemblies 4DG and 4GD. The burnup at the beginning of cycle 2 is shown in the second row of each square, and the burnup at the end in the third row. The total thermal energy produced in the second cycle is 835.2 GWd, or 20,044 × 10^6 kWh. At the end of cycle 2, assembly 1GC (called sublot 1B) and all assemblies from lot 2 except 2FE (called sublot 2A) are removed.

4.4 Reactor Performance, Cycle 3

Figure 3.22 shows the placement of assemblies at the start of cycle 3, with new assemblies containing 3.2 w/o ^{235}U placed in positions near the edge of the reactor, with heavy borders.

This refueling pattern is somewhat similar to modified scatter refueling. The maximum relative power, at 5DG and 5GD, is now 1.36. The average burnup in cycle 3 is 9894 MWd/MT. The total thermal energy is 866.5 GWd or 20,796 × 10^6 kWh.

4.5 Approach to Steady State

If this refueling pattern with 3.2 w/o fresh fuel is repeated through additional cycles, the fuel-cycle performance in cycles 7 and 8 will be as shown in Figs. 3.23 and 3.24. The relative power and burnup per cycle found in each location in cycle 7 and cycle 8 are almost identical, and the average burnup per cycle is exactly the same, 10,081 MWd/MT. This is evidence that a steady-state condition has been reached.

Table 3.3 summarizes the fuel-cycle performance of this reactor through the first eight cycles. The maximum value of the relative power, in the next-to-last column, levels off at a

Table 3.2 PWR, cycle 1: Axial and radial distribution of power relative to reactor average

Assembly location	AA	BB	CC	DD	EE	FF	
w/o ^{235}U	2.25	2.25	2.25	2.25	2.8	3.3	
Axial position			Relative power				
			Beginning of cycle				
12 (top)	0.35	0.34	0.35	0.36	0.39	0.36	
11	0.72	0.71	0.72	0.75	0.81	0.74	
10	0.99	0.97	1.00	1.03	1.12	1.03	
9	1.18	1.15	1.19	1.23	1.34	1.24	
8	1.30	1.27	1.31	1.36	1.50	1.39	
7 Mid-	1.38	1.35	1.39	1.45	1.61	1.50	Mid-
6 plane	1.41	1.37	1.42	1.49	1.67	1.57	plane
5	1.39	1.36	1.41	1.48	1.68†	1.59	
4	1.32	1.28	1.34	1.41	1.61	1.54	
3	1.16	1.13	1.18	1.25	1.44	1.38	
2	0.88	0.86	0.90	0.95	1.11	1.07	
1 (bottom)	0.43	0.42	0.44	0.47	0.55	0.53	
Average	1.04	1.02	1.05	1.10	1.24‡	1.16	
			End of cycle				
12 (top)	0.92	0.91	0.91	0.90	0.94	0.83	
11	1.16	1.17	1.16	1.16	1.25	1.13	
10	1.12	1.13	1.12	1.13	1.23	1.13	
9	1.05	1.05	1.05	1.06	1.16	1.08	
8	1.00	1.01	1.00	1.01	1.10	1.03	
7 Mid-	0.98	0.98	0.98	0.99	1.08	1.01	Mid-
6 plane	0.98	0.98	0.98	0.98	1.07	1.01	plane
5	0.99	1.00	0.99	0.99	1.09	1.02	
4	1.04	1.04	1.03	1.03	1.13	1.06	
3	1.12	1.12	1.12	1.12	1.22	1.14	
2	1.22	1.23	1.22	1.22	1.34†	1.23	
1 (bottom)	1.06	1.06	1.05	1.05	1.13	1.00	
Average	1.06	1.06	1.05	1.05	1.14‡	1.06	

†Maximum local power.

‡Maximum assembly power.

value of 1.37. The average burnup of fuel levels off at a value of 30,400 MWd/MT. Table 3.3 shows that when a reactor is refueled for a sufficient number of cycles in identical fashion, its performance in each cycle reaches a repetitive, steady-state behavior. When this occurs, the sum of the burnups of all assemblies discharged [(64)(30,400) = 1,945,600] approaches the burnup increment of all assemblies in the reactor [(193)(10,081) = 1,945,633].

Table 3.4 gives the relative power at the beginning of steady-state cycle 8 and the burnup of each assembly at the end of this cycle. The maximum relative power is 1.70 (acceptable), the maximum local burnup is 38,052 MWd/MT, and the maximum assembly burnup is 35,991 MWd/MT. Fuel can probably tolerate this much burnup without excessive mechanical deterioration.

Center-line	A (Center-line)	B	C	D	E	F	G	H
A	1GC 15950 23160 0.67	2AD 16666 25416 0.84	2AF 17184 25717 0.83	4AD 0 12776 1.33	2AB 16578 25877 0.95	3AH 12775 23416 1.14	3FF 16484 26896 1.15	4AH 0 10802 1.24
B	2DA 16666 25416 0.84	4BB 0 12191 1.22	2EF 14934 23655 0.85	3DG 15945 25682 0.97	3BG 16041 25728 0.98	3BH 13596 24194 1.12	3FG 9449 20735 1.25	4BH 0 10661 1.21
C	2FA 17184 15717 0.83	2FE 14934 23655 0.85	2EE 18302 25982 0.73	2BE 16716 25343 0.85	3EG 14014 23680 0.97	3CH 11985 22218 1.06	2CD 16600 26965 1.01	4CH 0 9950 1.10
D	4DA 0 12776 1.33	3GD 15945 25682 0.97	2EB 16716 25343 0.85	4DD 0 12242 1.25	2BC 15844 24770 0.88	2CF 16951 25607 0.88	4DG 0 12327 1.34*	4DH 0 7981 0.86
E	2BA 16578 25877 0.95	3GB 16041 25728 0.98	3GE 14014 23680 0.97	2CB 15844 24770 0.88	3DH 10194 19885 0.95	2DE 15980 24155 0.80	4EG 0 9757 1.01	
F	3HA 12775 23416 1.14	3HB 13596 24194 1.12	3HC 11985 22218 1.06	2FC 16951 25607 0.88	2ED 15980 24155 0.80	3HD 10194 18747 0.82	4FG 0 6828 0.66	
G	3FF 16484 26896 1.15	3GF 9449 20735 1.25	2DC 16600 25965 1.01	4GD 0 12327 1.34*	4GE 0 9757 1.01	4GF 0 6828 0.66		
H	4HA 0 10802 1.24	4HB 0 10661 1.21	4HC 0 9950 1.10	4HD 0 7981 0.86				

* = Maximum Relative Power

Fuel Lot 1 Initially 2.25 w/o U-235
Fuel Lot 2 Initially 2.80 w/o U-235
Fuel Lot 3 Initially 3.30 w/o U-235
Fuel Lot 4 Initially 3.20 w/o U-235

Cycle Average Burnup = 9,652 MWd/MT
Cycle Thermal Energy = 835.2 GWd

Key

Key	
1AA	Assembly Number
0	BOC Burnup, MWd/MT
17302	EOC Burnup, MWd/MT
1.04	BOC Relative Power (Assembly/Average)

Figure 3.21 PWR, assembly power and burnup distribution, cycle 2.

	A (Centerline)	B	C	D	E	F	G	H
A (Centerline)	2FE 23655 31695 0.75	4AH 10802 21262 1.01	3AH 23416 32433 0.85	4FG 6828 17879 1.08	4AD 12776 23317 1.04	3HD 18747 27348 0.83	3FF 26376 35427 0.83	5AH 0 10264 1.09
B	4HA 10802 21262 1.01	4BB 12191 22999 1.05	4CH 9950 20609 1.03	3CH 22218 31186 0.85	4DH 7981 18960 1.09	3BH 24194 32632 0.82	5BG 0 12387 1.31	5BH 0 10638 1.14
C	3HA 23416 32433 0.85	4HC 9950 20609 1.03	4DD 12242 22385 0.97	3FG 20738 29503 0.83	4EG 9757 20409 1.06	3EG 23680 31924 0.80	3BG 25728 34315 0.87	5CH 0 9686 1.02
D	4FG 6828 17879 1.08	3HC 22218 31186 0.85	3GF 20735 29503 0.83	3DH 19885 28920 0.87	4BH 10661 21655 1.12	4DG 12327 22787 1.08	5DG 0 12600 1.36*	5DH 0 7979 0.83
E	4DA 12776 23317 1.04	4HD 7981 18960 1.09	4GE 9757 20409 1.06	4HB 10661 21655 1.12	4GF 6828 18301 1.19	3DG 25682 34444 0.89	5EG 0 10268 1.09	
F	3HD 18747 27348 0.83	3HB 24194 32632 0.82	3GE 23680 31924 0.80	4GD 12327 22787 1.08	3GD 25682 34444 0.89	5FF 0 11628 1.22	5FG 0 7851 0.81	
G	3FF 26896 35427 0.86	5GB 0 12387 1.31	3BG 25728 34315 0.87	5GD 0 12600 1.36*	5GE 0 10268 1.09	5GF 0 7851 0.81		
H	5HA 0 10264 1.09	5HB 0 10638 1.14	5HC 0 9686 1.02	5HD 0 7979 0.83				

* = Maximum Relative Power

Fuel Lot 2 Initially 2.80 w/o U-235
Fuel Lot 3 Initially 3.30 w/o U-235
Fuel Lots 4, 5 Initially 3.20 w/o U-235

Cycle Average Burnup = 9894 MWd/MT
Cycle Thermal Energy = 866.5 GWd

Key

1AA	Assembly Number
0	BOC Burnup, MWd/MT
17302	EOC Burnup, MWd/MT
1.04	BOC Relative Power (Assembly/Average)

Figure 3.22 PWR, assembly power and burnup distribution, cycle 3.

5 FUEL-CYCLE COSTS

5.1 Procedure for Calculating Fuel-Cycle Costs

To calculate fuel-cycle costs, it is necessary to focus attention on individual fuel sublots and determine:

1. The amount and composition of each sublot when charged to the reactor
2. The amount of electricity generated by each sublot in each period in which electricity is paid for

3. The amount and composition of each sublot when discharged from the reactor
4. The cost incurred in each step for preparing fuel before it is charged to the reactor
5. The cost or credit incurred in each step for recovering fuel after it is discharged from the reactor
6. The time at which each cost is paid or each credit is received, and the time at which revenue is received for each increment of electricity generated by each lot of fuel

A somewhat simplified, approximate procedure for calculating fuel-cycle costs will be illustrated by the example of sublot 2A of the PWR whose fuel management was described in

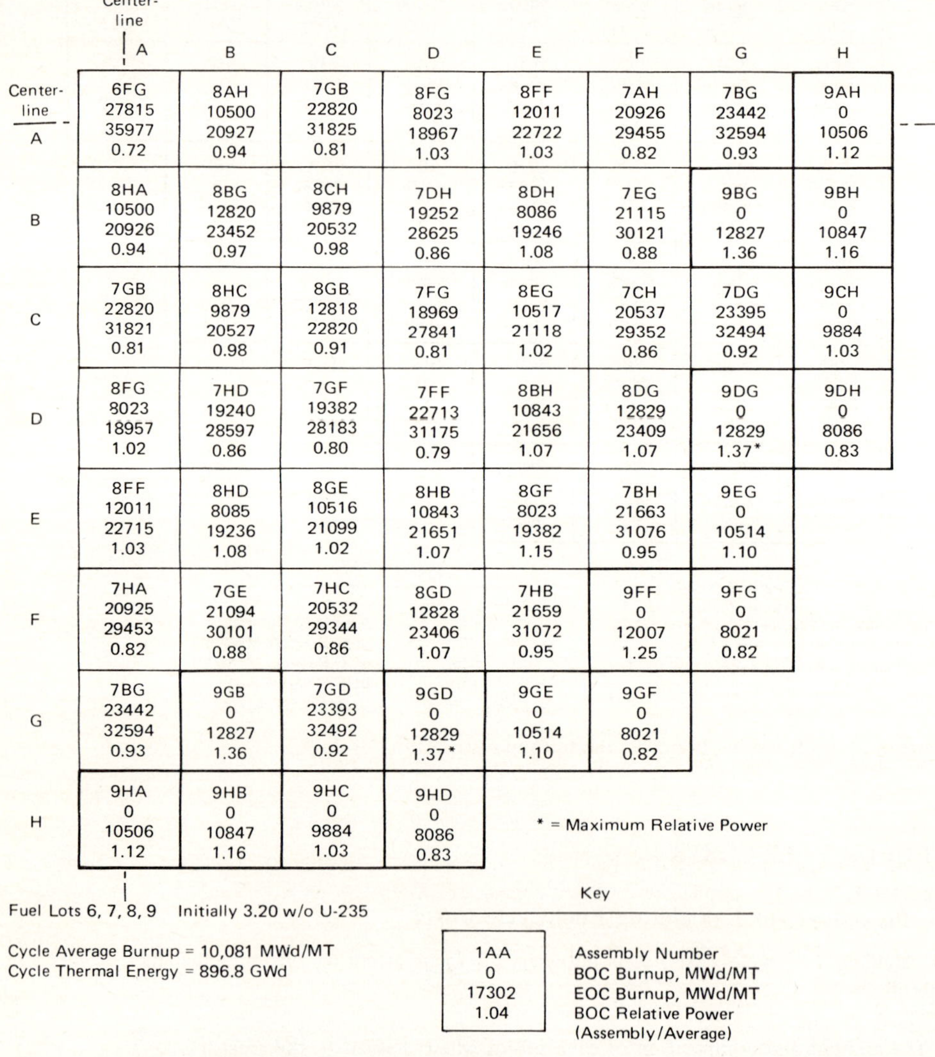

Figure 3.23 PWR, assembly power and burnup distribution, cycle 7.

Center-line	A	B	C	D	E	F	G	H
Center-line A	7FG 27841 35997 0.72	9AH 10506 20929 0.94	8GB 22820 31824 0.81	9FG 8021 18967 1.03	9FF 12007 22721 1.03	8AH 20927 29456 0.82	8BG 23452 32601 0.93	10AH 0 10505 1.12
B	9HA 10506 20928 0.94	9BG 12827 23454 0.97	9CH 9884 20535 0.98	8DH 19247 28621 0.86	9DH 8086 19249 1.08	8EG 21118 30125 0.88	10BG 0 12826 1.36	10BH 0 10847 1.16
C	8GB 22820 31820 0.81	9HC 9884 20530 0.98	9GB 12827 22827 0.91	8FG 18968 27840 0.81	9EG 10514 21117 1.02	8CH 20532 29348 0.86	8DG 23409 32506 0.92	10CH 0 9883 1.03
D	9FG 8021 18956 1.02	8HD 19236 28594 0.86	8GF 19382 28183 0.80	8FF 22722 31183 0.79	9BH 10847 21660 1.07	9DG 12829 23410 1.07	10DG 0 12830 1.37*	10DH 0 8086 0.83
E	9FF 12007 22714 1.03	9HD 8086 19238 1.08	9GE 10514 21098 1.02	9HB 10847 21656 1.07	9GF 8021 19382 1.15	8BH 21656 31071 0.95	10EG 0 10516 1.10	
F	8HA 20926 29454 0.82	8GE 21099 30105 0.88	8HC 20527 29340 0.86	9GD 12829 23408 1.07	8HB 21651 31067 0.95	10FF 0 12010 1.25	10FG 0 8022 0.82	
G	8BG 23452 32601 0.93	10GB 0 12825 1.36	8GD 23407 32504 0.92	10GD 0 12829 1.37*	10GE 0 10516 1.10	10GF 0 8022 0.82		
H	10HA 0 10505 1.12	10HB 0 10847 1.16	10HC 0 9883 1.03	10HD 0 8086 0.83				

* = Maximum Relative Power

Fuel Lots 7, 8, 9, 10 Initially 3.20 w/o U-235

Cycle Average Burnup = 10,081 MWd/MT
Cycle Thermal Energy = 896.8 GWd

Key

1AA	Assembly Number
0	BOC Burnup, MWd/MT
17302	EOC Burnup, MWd/MT
1.04	BOC Relative Power (Assembly/Average)

Figure 3.24 PWR, assembly power and burnup distribution, cycle 8.

Sec. 4. Information on material quantities and energy production from sublots 1A through 4B in cycles 1, 2, and 3, applicable to this example calculation, is given in Table 3.5.

The 63 assemblies of sublot 2A contain 28,171.8 kg of uranium enriched to 2.8 w/o ^{235}U. The average burnup experienced by this sublot is 16,448 MWd/MT in cycle 1 and 8700 MWd/MT in cycle 2, for a total burnup B_{2A} of 25,148 MWd/MT. The total thermal energy produced by this sublot is

$$(25{,}148 \text{ MWd/MT})(28.1718 \text{ MT})(24{,}000 \text{ kWh/MWd}) = 17{,}003 \times 10^6 \text{ kWh}$$

Calculations by computer code LEOPARD made by Rieck [R1] predict that this fuel when

Table 3.3 Fuel-cycle performance of PWR in successive cycles

Cycle number	Fuel lot number: Charged	Fuel lot number: Discharged	w/o ^{235}U charged	Burnup, MWd/MT: Cycle average	Burnup, MWd/MT: Fuel discharged	Peak radial power ratio: Max. value	Peak radial power ratio: Position
1	1	1	2.25	15,535	16,943	1.24	EE
	2†		2.8				
	3†		3.3				
2	4	2‡	3.2	9,652	25,115	1.34	DG
3	5	3‡	3.2	9,894	32,076	1.36	DG
4	6	4‡	3.2	10,284	30,306	1.34	DG
5	7	5‡	3.2	10,038	30,401	1.38	DG
6	8	6‡	3.2	10,084	30,419	1.37	DG
7	9	7‡	3.2	10,081	30,399	1.37	DG
8	10	8‡	3.2	10,081	30,400	1.37	DG

†Contains burnable poison.
‡Sixty-three assemblies from this lot and one from previous lot.

exposed to burnup of 25,148 MWd/MT will contain 27,991 kg of uranium enriched to 0.920 w/o ^{235}U and 161.330 kg of the fissile isotopes of plutonium ^{239}Pu and ^{241}Pu.

To obtain the durations of cycles 1 and 2, and from them the times at which payments are made, credits received, and revenue obtained from the sale of electricity, the total thermal energy per cycle H_i is required and is calculated from burnup increments in the next to the last row of the table. The duration of irradiation during the ith cycle, $t_i'' - t_i'$, in years is obtained from H_i in the last row of the table, using a rated thermal output of 3.250 GW.

Figure 3.25 is a schematic flow sheet for lot 2A fuel cycle. This shows the fuel-cycle steps to be considered and defines notation for the material and service quantities involved in each step, the total cost or credit associated with each step, and the timing of all transactions (dashed arrows).

Table 3.6 gives numerical values for material quantities, unit costs or credits, and total direct costs or credits involved in each fuel-cycle step and calculates the overall net direct cost for lot 2A as \$26.4 million. A total of \$27.8 million is paid out for UF_6 and fabrication in transactions 1 and 2 before any revenue is received from the sale of electricity. Because of this delay in receiving revenue, the total fuel-cycle cost includes also charges for carrying the \$27.8 million advanced several years before it is recovered through revenue from the sale of electricity. Similarly, there is a financing charge on the net credit of \$1.4 million in steps 3 through 6, delayed until after revenue is received from the sale of electricity.

The assumptions going into the calculation of direct costs in Table 3.6 will be described first. Then the procedure for calculating financing charges will be described, and finally a value will be given for the complete fuel-cycle cost.

Direct costs. The unit costs used in the examples of this chapter are those anticipated in 1975 for the year 1980. Because of changes since 1975, readers are cautioned to regard these costs more as examples than as firm numbers.

The unit cost of enriched uranium in the form of UF_6 depends on the ^{235}U content of the uranium, the price paid for the natural uranium from which the uranium was enriched, the cost of the separative work expended in enriching the uranium, and the composition of the tails stream containing depleted uranium leaving the uranium enrichment plant. The procedure for calculating the cost of enriched uranium is described in Chap. 12. The unit costs

$c_{U'}$ = \$848.66/kg U for uranium enriched to 2.8 w/o ^{235}U fed to fabrication and $c_{U''}$ = \$152.83/kg U for uranium containing 0.920 w/o ^{235}U recovered from reprocessing are based on the following assumptions:

Price of natural uranium ore concentrates, \$31.55/lb U_3O_8
Price of natural UF_6, \$89.11/kg U
Cost of separative work, \$100/separative work unit
^{235}U content of enrichment plant tails, 0.3 w/o

Table 3.4 Zion reactor, cycle 8: Relative power and burnup

Assembly location	DA	DB	DC	DD	DE	DF	DG	DH
Number of previous cycles	1	2	2	2	1	1	0	0
Axial position	Relative power at beginning							
12 (top)	0.72	0.64	0.60	0.57	0.68	0.62	0.61	0.35
11	1.03	0.87	0.81	0.78	1.00	0.94	1.07	0.64
10	1.13	0.95	0.89	0.86	1.13	1.10	1.32	0.81
9	1.17	0.98	0.92	0.89	1.20	1.18	1.48	0.91
8	1.18	0.99	0.92	0.90	1.23	1.23	1.58	0.97
7 Mid-	1.17	0.98	0.92	0.90	1.24	1.26	1.64	1.01
6 plane	1.15	0.96	0.90	0.89	1.23	1.26	1.68	1.04
5	1.11	0.93	0.87	0.86	1.21	1.25	1.70†	1.04
4	1.07	0.89	0.83	0.82	1.16	1.21	1.67	1.02
3	1.00	0.83	0.78	0.77	1.09	1.13	1.57	0.96
2	0.91	0.76	0.70	0.69	0.97	0.99	1.32	0.79
1 (bottom)	0.64	0.56	0.52	0.51	0.68	0.66	0.77	0.44
Average	1.02	0.86	0.80	0.79	1.07	1.07	1.37‡	0.83

Assembly location	AA	AB	AC	AD	AE	AF	AG	AH
Number of previous cycles	4	2	3	2	2	3	3	1
Axial position	Burnup, MWd/MT, at end of cycle							
12	27,179	14,426	23,474	13,087	15,589	21,108	22,962	6,100
11	35,525	19,961	31,028	18,154	21,530	28,394	31,095	9,347
10	37,482	21,536	32,909	19,575	23,259	30,387	33,421	10,543
9	37,850	21,992	33,348	19,972	23,794	30,932	34,124	11,018
8	37,893	22,174	33,478	20,120	24,028	31,129	34,421	11,267
7	37,897	22,303	33,564	20,221	24,199	31,256	34,631	11,454
6	37,913	22,430	33,653	20,320	24,364	31,378	34,833	11,629
5	37,952	22,568	33,762	20,431	24,537	31,513	35,042	11,803
4	38,028	22,710	33,894	20,546	24,710	31,653	35,247	11,956
3	38,052†	22,740	33,952	20,566	24,760	31,661	35,281	11,957
2	37,078	21,905	33,049	19,808	23,889	30,614	34,090	11,236
1	29,045	16,317	25,703	14,715	17,904	23,379	25,999	7,671
Average	35,991‡	20,922	31,818	18,959	22,714	29,450	32,595	10,498

†Maximum local value.
‡Maximum assembly average.

Table 3.5 Example of material quantities and energy production by lot and cycle

		Fuel charged		Cycle i					Fuel discharged		
				1	2	3					
				Average burnup increment of sublot k in cycle i, MWd/MT, ΔB_{ik}							
Fuel sublot number k	Number of assemblies	w/o ^{235}U	kg U, U'_k				Average burnup per lot, MWd/MT, $B_k = \Sigma_i \Delta B_{ik}$	Thermal energy per lot, GWh, $H_k = 24 \times 10^{-6} U'_k B_k$	w/o ^{235}U	kg U, U''_k	kg fissile Pu, P_k
1A	64	2.25	29,168.2	16,943	–	–	16,943	11,861	0.944	28,457.6	141.736
1B	1 (#1GC)	2.25	455.8	15,950	7,210	–	23,160	253.35	0.663	441.1	2.497
2A	63	2.8	28,171.8	16,448	8,700	–	25,148	17,003	0.920	27,199.1	161.330
2B	1 (#2FE)	2.8	447.2	14,934	8,721	8,040	31,695	340.17	0.653	428.2	2.719
3A	63	3.3	27,499.5	13,280	10,132	8,671	32,083	21,175	0.921	26,320.8	171.258
3B	1 (#3HD)	3.3	436.5	10,194	8,553	8,601	†	†	†	†	†
4A	63	3.2	29,039.1	–	10,242	10,743	†	†	†	†	†
4B	1 (#4FG)	3.2	460.9	–	6,828	11,051	‡	‡	‡	‡	‡
Thermal energy per cycle, GWh $H_i = 24 \times 10^{-6} \Sigma_k U'_k \Delta B_{ik}$				32,188	20,044	§					
Duration of irrad., yr $\dfrac{H_i}{8766 \times 3.250 \times 0.9}$				1.2553	0.7817	§					

† Requires information from cycle 4.
‡ Requires information from cycle 5.
§ Requires information from lot 5.

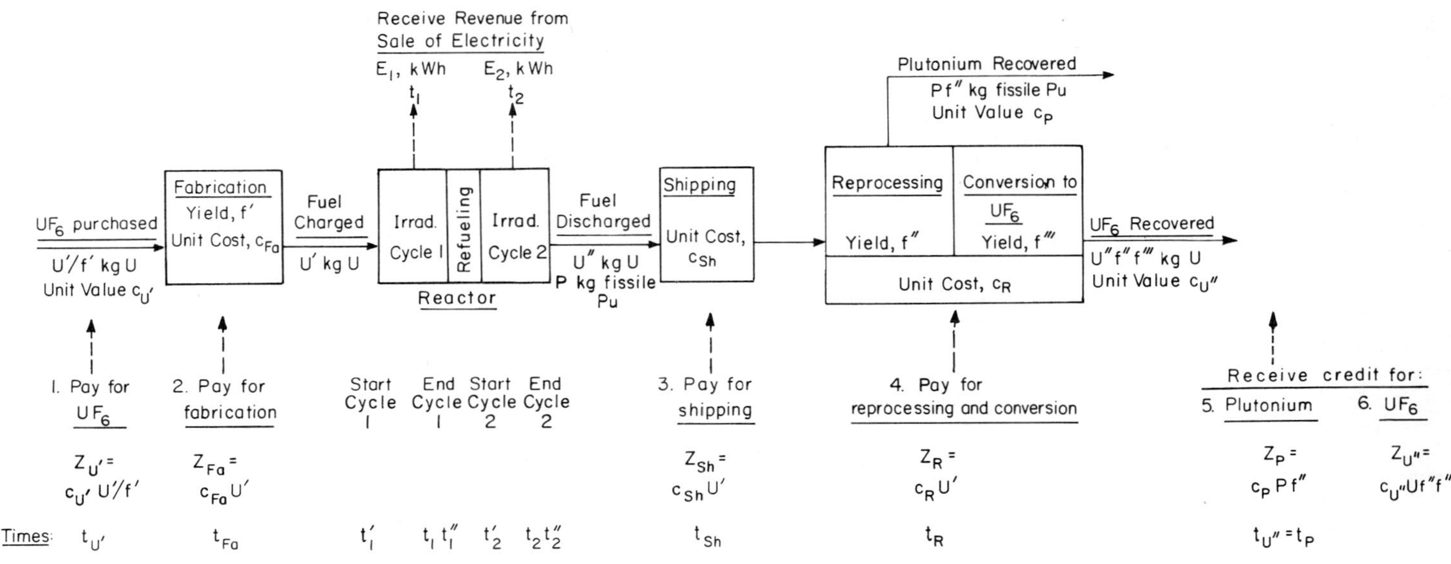

Figure 3.25 Schematic flow sheet for lot 2A fuel cycle, showing material and service quantities and timing of transactions.

Table 3.6 Direct fuel-cycle costs or credits, lot 2A

Transaction	Quantity	Unit cost or credit, \$/kg	Direct cost or credit, \$	
1. Pay for UF_6	$U'/f' = 28{,}171.8/0.99 = 28{,}456.364$ kg U	$c_{U'} = 848.66$	$Z_{U'} = (28{,}456.364)(848.66)$	$= 24{,}149{,}778$
2. Pay for fabrication	$U' = 28{,}171.8$	$c_{Fa} = 130$	$Z_{Fa} = (28{,}171.8)(130)$	$= 3{,}662{,}334$
3. Pay for shipping	$U' = 28{,}171.8$	$c_{Sh} = 30$	$Z_{Sh} = (28{,}171.8)(30)$	$= 845{,}154$
4. Pay for reprocessing and conversion	$U' = 28{,}171.8$	$c_R = 180$	$Z_R = (28{,}171.8)(180)$	$= 5{,}070{,}924$
5. Receive credit for Pu	$Pf'' = (161.330)(0.99) = 159.7167$ kg Pu	$c_P = 20{,}000$	$-Z_P = (-159.7167)(20{,}000)$	$= -3{,}194{,}334$
6. Receive credit for UF_6	$U''f''f''' = (27{,}199.1)(0.99)(0.995) = 26{,}792.473$	$c_{U''} = 152.83$	$-Z_{U''} = -(26{,}792.473)(152.83)$	$= -4{,}094{,}694$
			Overall net direct cost:	26,439,162

It is assumed that $f' = 0.99$ fraction of uranium charged in the form of UF_6 will be recovered as fabricated fuel. Hence, to provide $U' = 28{,}171.8$ kg of fabricated uranium, $28{,}171.8/0.99 = 28{,}456.364$ kg uranium in the form of UF_6 must be purchased. The direct cost of this UF_6 is $28{,}456.364 \times 848.66 = \$24{,}149{,}778$.

A similar procedure is used to calculate the other components of the direct cost shown in Table 3.6. Other assumptions are as follows:

Fraction of uranium and plutonium recovered in reprocessing, $f'' = 0.99$.
Fraction of recovered uranium converted to UF_6, $f''' = 0.995$.
The fabrication unit cost of \$130/kg includes cost of converting UF_6 to UO_2 and packaging UO_2 in fuel assemblies.
The shipping cost of \$30/kg includes storage charges at the reactor for around 150 days to permit fuel radioactivity to decrease.
The reprocessing and conversion cost of \$180/kg includes charges by the government for perpetual storage of radioactive wastes.

Financing charges. A company generating electricity that pays out Z dollars for fuel-cycle costs t years before it receives revenue from generation of electricity from that fuel must pay to the bondholders and stockholders who advanced the funds for the fuel the return they require on their investment, and must also pay income taxes to the government on the profits from which the stockholders' return is obtained. It is possible to represent all of these financing charges as the product yZt, where y is known as the annual cost of money before income taxes. For a privately owned U.S. electric company, a value of $y = 0.151$ per year is representative.

To find the total fuel financing charge, it is necessary to find the amount of money advanced for fuel as a function of time. Figure 3.26 is a schematic plot of the amount of

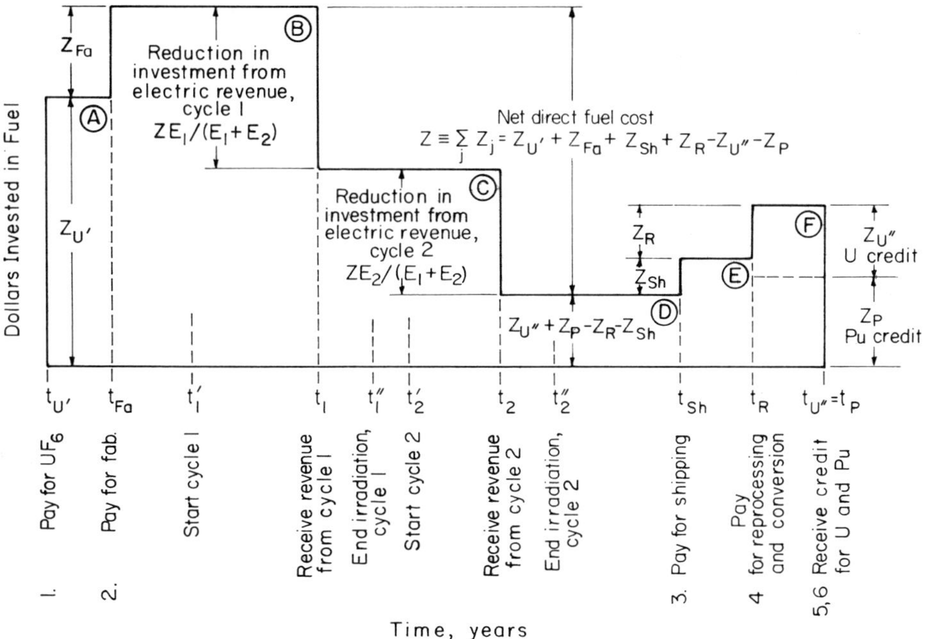

Figure 3.26 Plot of dollars invested in fuel versus time.

money invested in fuel cycle as a function of time for an example like sublot 2A, in which revenue is assumed received from the sale of electricity at two times during irradiation, from sale of E_1 kWh at time t_1 in cycle 1 and from sale of E_2 kWh at time t_2 in cycle 2. The generalization to a more realistic case, in which revenue is received at more times, should be clear.

At time $t_{U'}$, $Z_{U'}$ dollars are invested in enriched uranium. This amount of money is invested for $t_{Fa} - t_{U'}$ years, until t_{Fa}, when Z_{Fa} more dollars are paid out to fabricate fuel. The financing cost of carrying the initial investment of $Z_{U'}$ dollars for $t_{Fa} - t_{U'}$ years is the product of the cost of money before income taxes, y, and the area of region A, $Z_{U'}(t_{Fa} - t_{U'})$.

Between t_{Fa}, when fabrication is paid for, and t_1, when revenue is received from production of E_1 kWh of electricity, the dollars invested in fuel is $Z_{U'} + Z_{Fa}$, and financing costs are the product of y and area B, $(Z_{U'} + Z_{Fa})(t_1 - t_{Fa})$.

If the total electricity production of the lot of fuel is $\Sigma_m E_m$ ($E_1 + E_2$ for sublot 2A) and the net direct fuel-cycle cost is $\Sigma_j Z_j \equiv Z$ (where Z_j is the direct cost of the jth fuel-cycle step and $\Sigma_j Z_j \equiv Z = \$26{,}439{,}162$ for sublot 2A), the amount of money invested in fuel after t_1 should be reduced by $ZE_1/(E_1 + E_2)$. Between t_1 and t_2, when the second (and in this case, the last) revenue is received from production of E_2 kWh of electricity, the dollars invested in fuel is $Z_{U'} + Z_{Fa} - ZE_1/(E_1 + E_2)$, and financing costs are the product of y and area C, $[Z_{U'} + Z_{Fa} - ZE_1/(E_1 + E_2)]\ (t_2 - t_1)$.

The amount of electricity generated by sublot 2A in cycle 1, E_1, is obtained from the thermal efficiency of the power plant

$$\eta = \frac{1060}{3250} = 0.32615 \tag{3.14}$$

the average burnup increment of sublot 2A in cycle 1, $\Delta B_{1,2A} = 16{,}448$ MWd/MT from Table 3.5, and the mass of uranium, $U'_{2A} = 28{,}171.8$ kg. Hence

$$E_1 = 24\eta\ \Delta B_{1,2A} U'_{2A} = (24)(0.32615)(16{,}448)(28{,}171.8) = 3.6271 \times 10^9 \text{ kWhe} \tag{3.15}$$

Similarly, the amount of electricity generated by sublot 2A in cycle 2 is

$$E_2 = (24)(0.32615)(8700)(28{,}171.8) = 1.9185 \times 10^9 \text{ kWhe} \tag{3.16}$$

The total electric generation is

$$E = E_1 + E_2 = (3.6271 + 1.9185) \times 10^9 = 5.5456 \times 10^9 \text{ kWhe} \tag{3.17}$$

Between t_2 and t_{Sh} when payment is made for shipping fuel, the dollars invested in fuel is the difference between the initial outlay $Z_{U'} + Z_{Fa}$ and the direct fuel-cycle cost Z, which is equivalent to

$$Z_{U'} + Z_{Fa} - Z = Z_{U''} + Z_P - Z_{Sh} - Z_R \tag{3.18}$$

The financing cost for this time interval is the product of y and area D, $(Z_{U''} + Z_P - Z_R - Z_{Sh})(t_{Sh} - t_2)$.

Between t_{Sh} and t_R, when payment is made for reprocessing fuel and converting uranium to UF_6, the dollars invested in fuel is $Z_{U''} + Z_P - Z_R$. The financing cost for this time interval is the product of y and area E, $(Z_{U''} + Z_P - Z_R)(t_R - t_{Sh})$.

If credit for plutonium and uranium is received at a time $t_{U''}$ later than t_R, an additional financing charge is incurred on the value of this uranium and plutonium, $Z_{U''} + Z_P$, for the time interval $t_{U''} - t_R$, equal to the product of y and the area F, $(Z_{U''} + Z_P)(t_{U''} - t_R)$.

The sum of areas A, B, C, D, E, and F is

$$Z_{U'}(t_{Fa} - t_{U'}) + (Z_{Fa} + Z_{U'})(t_1 - t_{Fa}) + \left[Z_{U'} + Z_{Fa} - \frac{(Z_{U'} + Z_{Fa} + Z_{Sh} + Z_R - Z_{U''} - Z_P)E_1}{E_1 + E_2}\right](t_2 - t_1) \tag{3.19}$$

(*Cont. on p. 123*)

$$+ (Z_{U''} + Z_P - Z_R - Z_{Sh})(t_{Sh} - t_2) + (Z_{U''} + Z_P - Z_R)(t_R - t_{Sh})$$

$$+ (Z_{U''} + Z_P)(t_{U''} - t_R) = Z_{U'} \left(\frac{E_1 t_1 + E_2 t_2}{E_1 + E_2} - t_{U'}\right)$$

$$+ Z_{Fa} \left(\frac{E_1 t_1 + E_2 t_2}{E_1 + E_2} - t_{Fa}\right) + Z_{Sh} \left(\frac{E_1 t_1 + E_2 t_2}{E_1 + E_2} - t_{Sh}\right)$$

$$+ (Z_R)\left(\frac{E_1 t_1 + E_2 t_2}{E_1 + E_2} - t_R\right) - (Z_{U''} + Z_P)\left(\frac{E_1 t_1 + E_2 t_2}{E_1 + E_2} - t_{U''}\right) \quad \text{(3.19) (Cont.)}$$

Thus, the area under the curve is the sum of the product of each expenditure or credit times the difference between the mean time for receipt of revenue $\bar{t}$

$$\bar{t} \equiv \frac{E_1 t_1 + E_2 t_2}{E_1 + E_2} \quad (3.20)$$

and the time when the expenditure or credit is paid. This result generalizes to an expression for the area of

$$\text{Area} = \sum_j Z_j \left(\frac{\sum_m E_m t_m}{\sum_m E_m} - t_j\right) \quad (3.21)$$

where Z_j is the outlay for fuel-cycle step j (negative if a credit) at t_j and t_m is the time at which revenue is received for production of E_m kWh of electricity. The fuel-cycle cost e in mills per kilowatt-hour then is

$$e = \frac{10^3 \sum_j Z_j \left[1 + y(\Sigma_m E_m t_m / \Sigma_m E_m - t_j)\right]}{\sum_m E_m} \quad (3.22)$$

The assumptions of Table 3.7 are made to obtain the times needed to calculate the carrying-charge term for sublot 2A:

$$\sum_j Z_j y \left(\frac{E_1 t_1 + E_2 t_2}{E_1 + E_2} - t_j\right) \quad \text{in Eq. (3.22)}$$

A time basis of zero is taken for the start of cycle 1 ($t_1' = 0$).

The mean time for receipt of revenue from sale of electricity, $\bar{t}$, is

$$\bar{t} \equiv \frac{E_1 t_1 + E_2 t_2}{E_1 + E_2} = \frac{(3.6271)(0.7920) + (1.9185)(1.9355)}{3.6271 + 1.9185} = 1.1876 \text{ yr} \quad (3.23)$$

From the foregoing transaction times and the direct fuel-cycle costs or credits of Table 3.6, the carrying charges and total fuel-cycle costs may be calculated, as shown in Table 3.8 for sublot 2A. Division of the total fuel-cycle cost of \$33,173,168 by the electricity generated by sublot 2A, 5.5456×10^9 kWhe, and conversion to mills per kilowatt-hour of electricity by Eq. (3.22) gives 5.9819 mills/kWhe for the unit fuel-cycle cost of lot 2A.

Table 3.8 shows that more than 20 percent of the fuel-cycle cost arises from carrying charges.

5.2 Steady-State Fuel-Cycle Costs

The procedure for calculating fuel-cycle costs and the assumptions regarding unit costs described in Sec. 5.1, and the assumptions regarding the time displacement between financial

Table 3.7 Fuel-cycle times for sublot 2A

Activity	Time, years	Assumption or source
1. Pay for UF_6	$t_{U'} = -0.3474$	127 days before t_1'
2. Pay for fabrication	$t_{Fa} = -0.2656$	97 days before t_1'
Start cycle 1	$t_1' = 0$	Time basis
End irradiation, cycle 1	$t_1'' = 1.2553$	Table 3.5
Receive revenue, cycle 1	$t_1 = 0.7920$	60 days after $(t_1' + t_1'')/2$
Start cycle 2	$t_2' = 1.3803$	0.125 yr after t_1''
End irradiation, cycle 2	$t_2'' = 2.1620$	$t_2' + 0.7817$ (Table 3.5)
Receive revenue, cycle 2	$t_2 = 1.9355$	60 days after $(t_2' + t_2'')/2$
3. Pay for shipping	$t_{Sh} = 2.6603$	182 days after t_2''
4. Pay for reprocessing	$t_R = 2.7424$	212 days after t_2''
5. Receive credit for U and Pu	$t_{U''} = t_P = 2.7424$	212 days after t_2''

transactions and steps in the fuel cycle listed in Table 3.7, were used to obtain the total fuel-cycle costs of Fig. 3.14 and the unit fuel-cycle costs of Fig. 3.15. It will be recalled that these figures were for an idealized situation in which the reactor was treated as if operating under steady-state fuel-cycle conditions with an exact fraction (e.g., one-third) of the fuel replaced at each refueling.

Figure 3.27 is another example of steady-state fuel-cycle costs for this same large PWR with the same assumptions regarding transaction times and unit fuel-cycle costs. In Fig. 3.27 it has been assumed that the reactor uses three-zone, steady-state fueling, operates at a capacity factor of 90 percent when not down for refueling, and is down for 0.125 year at each refueling. Figure 3.27 shows how the total unit fuel-cycle cost and the principal components of this cost vary with the ^{235}U content of the fuel fed. Also plotted as independent variable, at the top, is the average burnup experienced by the fuel. A third independent variable, plotted at the bottom, is the time in years elapsing between the start of irradiation in the first cycle when the fuel is charged to the reactor and the end of irradiation in the third cycle when the fuel is discharged.

Table 3.8 Calculation of fuel-cycle costs for sublot 2A from schedule of payments and receipts

		Time, years		Cost, $		
Item	j	t_j	$\bar{t} - t_j$	Direct, Z_j	$0.151 Z_j (\bar{t} - t_j)$	Total
1. Payment for UF_6	U′	−0.3474	1.5350	24,149,778	5,597,556	29,747,334
2. Payment for fabrication	*Fa*	−0.2656	1.4532	3,662,334	803,638	4,465,972
3. Payment for shipping	*Sh*	2.6603	−1.4727	845,154	−187,943	657,211
4. Payment for reprocessing	*R*	2.7424	−1.5548	5,070,924	−1,190,525	3,880,399
5. Credit for plutonium	P	2.7424	−1.5548	−3,194,334	749,949	−2,444,385
6. Credit for UF_6	U″	2.7424	−1.5548	−4,094,694	961,331	−3,133,363
Total				26,439,162	6,734,006	33,173,168

Mean time for receipt of revenue, $\bar{t} = 1.1876$ yr

Unit fuel-cycle cost, from Eq. (3.22): $e = (10^3 \text{ mills/\$})(\$33{,}173{,}168)/5.5456 \times 10^9 \text{ kWhe} = 5.9819 \text{ mills/kWhe}$

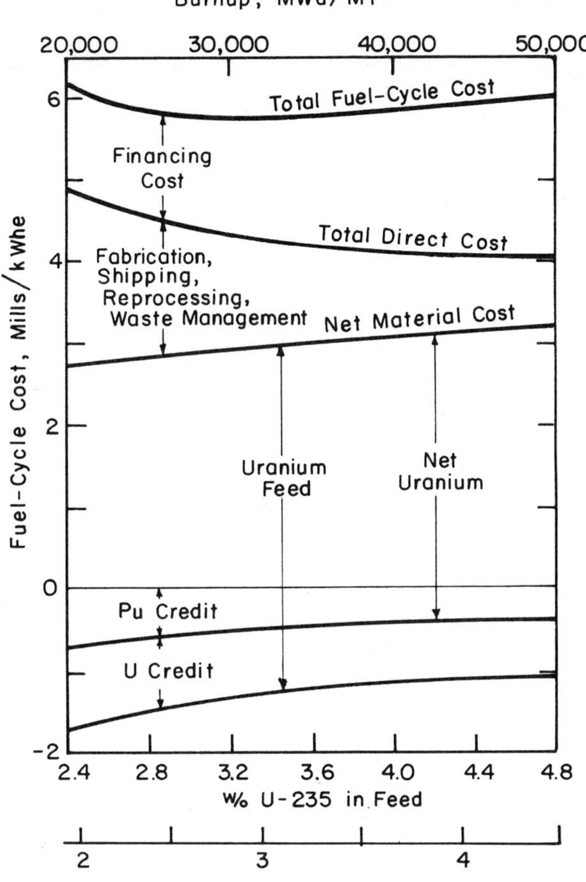

Figure 3.27 Components of steady-state unit fuel-cycle cost, PWR, three-zone fueling, 90 percent availability-based capacity factor, 0.125-year refueling downtime.

The minimum fuel-cycle cost of around 5.9 mills/kWhe results from use of feed containing around 3.3 w/o ^{235}U, which permits burnup of around 33,000 MWd/MT and an elapsed time of around 3 years between start and end of irradiation.

Existence of this optimum feed composition and burnup is a result of several conflicting factors, indicated by the components of the fuel-cycle cost shown in the figure. Credit for uranium and for plutonium in spent fuel per unit of electricity generated is smaller the higher the burnup, because the ^{235}U and fissile plutonium content of spent fuel is about the same for this range of feed compositions. Since the uranium feed cost per unit of electricity generated is nearly constant, the net material cost allowing for spent fuel credits increases with ^{235}U content of feed. Fabrication, reprocessing, and waste management costs per unit of electricity generated are inversely proportional to burnup, and decrease with increasing ^{235}U content of feed. Financing costs per unit of electricity generated increase with increasing ^{235}U content of feed because the feed costs more per kilogram and the time between purchase of feed and generation of electricity is longer—two effects that together more than offset the increased energy production from the richer feed. The net result of these conflicting trends is the flat minimum in the total fuel-cycle cost curve around 3.3 w/o ^{235}U and 33,000 MWd/MT.

6 HAND CALCULATION OF FUEL-CYCLE PERFORMANCE

The preceding example shows that accurate representation of fuel-cycle performance requires such detailed examination of the time-varying changes in power distribution and fuel composition at many points in a nuclear reactor as to necessitate use of a digital computer. The purpose of this section is to develop an approximate procedure for calculating fuel-cycle performance that, although complicated, can be carried out by hand calculation. The PWR described in Sec. 4.1 will be used as example, except that its rated electric output is taken as 1054 MWe instead of 1060 MWe.

6.1 Neutron Energy Cycle

The neutron energy cycle of the one-group reactor-physics model to be used is shown in Fig. 3.28.

Consider a unit volume containing N_m atoms of fissile material (^{233}U, ^{235}U, ^{239}Pu, or ^{241}Pu) of thermal absorption cross section σ_m, and N_g atoms of fertile material (^{238}U or ^{232}Th) of thermal absorption cross section σ_g. For this model we shall develop expressions for the number of neutrons produced or absorbed at any point in the neutron cycle per unit volume per unit time. Assume that the fissionable material absorbs only thermal neutrons. The rate of

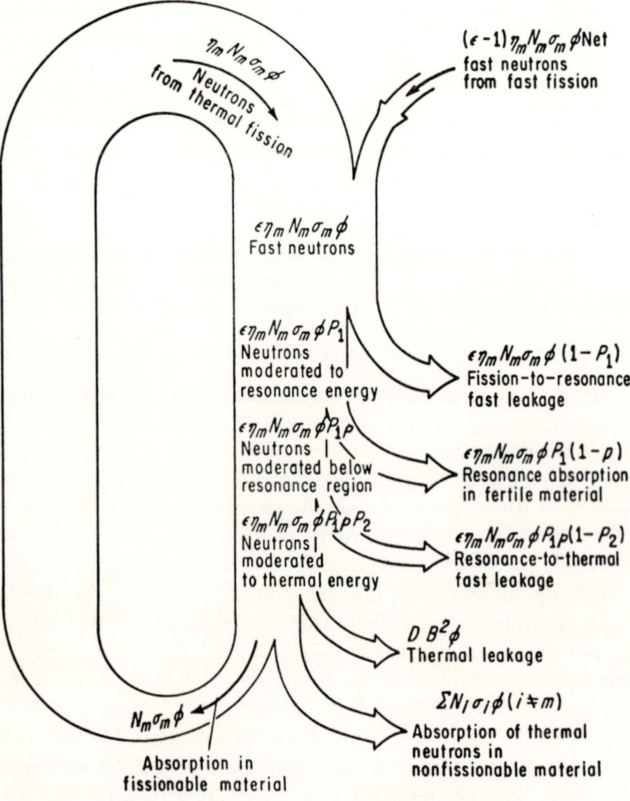

Figure 3.28 Neutron energy cycle in a thermal reactor.

absorption of neutrons by fissionable material is $N_m \sigma_m \phi$, where ϕ is the thermal-neutron flux. The resulting fissions produce fast neutrons at a rate $\eta_m N_m \sigma_m \phi$.

Those fast neutrons that have energies greater than about 1 MeV may cause a limited amount of fission of fertile material. To account for this, the reactor designer usually specifies a quantity ϵ, called the *fast-fission factor*, which is defined as the ratio of the net rate of production of fast neutrons to the rate of production of fast neutrons by thermal fission. The fraction $\epsilon - 1$ of the fast neutrons comes from fission of fertile material with fast neutrons; $\epsilon - 1$ may be of the order of a few hundredths in a thermal power reactor. The net production rate of fast neutrons from fission is $\epsilon \eta_m N_m \sigma_m \phi$.

As the fast neutrons undergo scattering collisions, they are degraded in energy and also tend to diffuse toward the outer surface of the reactor where they may escape. When those remaining in the reactor are degraded to energies of a few kilovolts, they have a good chance of being absorbed in the fertile material, which has large resonances in its absorption cross section in the kilovolt range. This resonance absorption is important in producing new fissionable material, i.e., ^{239}Pu from ^{238}U and ^{233}U from ^{232}Th.

The fraction of the fast neutrons that do *not* escape from the reactor as they degrade from fission to resonance energy depends on the size and moderating properties of the reactor. This fraction is denoted as P_1, the *fission-to-resonance nonleakage probability*. Hence, the rate at which fast neutrons degrade into the resonance region is $\epsilon \eta_m N_m \sigma_m \phi P_1$.

These resonance neutrons may be captured in fertile material or may escape resonance absorption by undergoing elastic collisions with the moderator, which degrades them to energies below resonance. A quantity p is defined as the fraction of the resonance neutrons that are not captured but are degraded to lower energies and is called the *resonance escape probability*. The fraction p is a function of the relative proportions and physical arrangement of the moderator and fertile material. Hence, $\epsilon \eta_m N_m \sigma_m \phi P_1 (1-p)$ neutrons undergo resonance absorption per unit volume per unit time, and $\epsilon \eta_m N_m \sigma_m \phi P_1 p$ are degraded to lower energies.

Of the latter, some diffuse to outer surfaces and escape, but the fraction P_2 remains in the reactor as thermal neutrons; P_2 is called the *resonance-to-thermal nonleakage probability*. Finally, the neutrons complete an energy cycle as $\epsilon \eta_m N_m \sigma_m \phi P_1 p P_2$ neutrons reach thermal energy per unit volume per unit time. The product $P_1 P_2$ is the *fission-to-thermal nonleakage probability*, which we shall denote as P_{th}.

Thermal neutrons are consumed by (1) absorption in fissionable material at a rate $N_m \sigma_m \phi$; (2) absorption in nonfissionable material at a rate $\Sigma_i N_i \sigma_i \phi$ $(i \neq m)$; and (3) leakage at a rate $DB^2 \phi$. The absorption in fissionable material leads to regeneration of fission neutrons, as shown in Fig. 3.28.

6.2 Neutron Balance for Reference Design

For this fuel-cycle analysis, it is convenient to specify a reactor design that satisfies a neutron balance and contains a charge of fresh fuel distributed uniformly throughout the reactor core.

Consider a portion of the reactor large enough to contain a representative sample of the entire contents of the reactor core. In a homogeneous reactor this might be any small amount of core volume; in a heterogeneous reactor it will be a region large enough to contain at least one set of repeating elements of core lattice structure.

We shall assume that a representative unit volume of this design contains N_M^* atoms of a single fissile species (e.g., ^{235}U) with absorption cross section σ_M^* and N_g^* atoms of a single fertile material (e.g., ^{238}U) with absorption cross section σ_g^*. The unit volume is assumed also to contain steady-state amounts of ^{135}Xe, ^{149}Sm, and other fission products with cross sections above 10,000 b, which build up to equilibrium concentration in a few days at the neutron fluxes typical of power reactors. It is assumed that no other fission products are present to an extent sufficient to affect the neutron balance. The items that affect the thermal-neutron balance for this reference design condition are listed in Table 3.9. The cross sections are effective values averaged over the energy distribution of neutron flux.

The thermal-neutron balance equation for the reference design condition states that the rate of production of thermal neutrons equals the rate of consumption:

$$\eta_M^* \epsilon P_{\text{th}} p N_M^* \sigma_M^* \phi = DB^{2*}\phi + N_M^* \sigma_M^* \phi + N_g^* \sigma_g^* \phi + \sum_P N_P^* \sigma_P^* \phi$$
$$+ N_{\text{Xe}}^* \sigma_{\text{Xe}}^* \phi + \sum_S N_S^* \sigma_S^* \phi + N_E^* \sigma_E^* \phi \tag{3.24}$$

The amount of control absorber present in the reactor N_E^* is set by the reactor operator so as to keep the reactor just critical at the desired power level.

The *reactivity* of the fuel in the reference design reactor ρ^* is defined as the ratio of the rate of absorption of thermal neutrons in control absorbers to the rate of production of thermal neutrons:

$$\rho^* = \frac{N_E^* \sigma_E^* \phi}{\eta_M^* \epsilon P_{\text{th}} p N_M^* \sigma_M^* \phi}$$
$$= \frac{(\eta_M^* \epsilon P_{\text{th}} p - 1) N_M^* \sigma_M - N_g^* \sigma_g^* - \sum_P N_P^* \sigma_P^* - N_{\text{Xe}}^* \sigma_{\text{Xe}}^* - \sum_S N_S^* \sigma_S^* - DB^{2*}}{\eta_M^* \epsilon P_{\text{th}} p N_M^* \sigma_M^*} \tag{3.25}$$

6.3 Neutron Balance for Operating Reactor

For an operating reactor, the terms of the neutron balance will differ from Table 3.9 and the neutron balance equation will differ from Eq. (3.25) because the fuel charged to the reactor may differ from the reference fuel and because the composition of fuel will change as a result of reaction with neutrons. The terms in the neutron balance of an operating reactor are listed in Table 3.10.

Table 3.9 Thermal-neutron balance for reference design condition

Item	Number of atoms per unit volume	Effective thermal absorption cross section	Neutron production or consumption rate, neutrons per unit volume per unit time
Production	N_M^*	σ_M^*	$\eta_M^* \epsilon P_{\text{th}} p N_M^* \sigma^* \phi$
Consumption:			
Thermal leakage	—	—	$DB^{2\,*}\phi$
Absorption by fissionable material	N_M^*	σ_M^*	$N_M^* \sigma_M^* \phi$
Absorption by fertile material	N_g^*	σ_g^*	$N_g^* \sigma_g^* \phi$
Absorption by coolant, moderator, structural material, etc.	N_P^*	σ_P^*	$\Sigma_P N_P^* \sigma_P^* \phi$
Absorption by ^{135}Xe	N_{Xe}^*	σ_{Xe}^*	$N_{\text{Xe}}^* \sigma_{\text{Xe}}^* \phi$
Absorption by ^{149}Sm, etc.	N_S^*	σ_S^*	$\Sigma_S N_S^* \sigma_S^* \phi$
Absorption by control absorbers	N_E^*	σ_E^*	$N_E^* \sigma_E^* \phi$

Table 3.10 Thermal-neutron balance for operating reactor

Item	Number of atoms per unit volume	Effective thermal absorption cross section	Neutron production or consumption rate, neutrons per unit volume per unit time
Production	N_m	σ_m	$\Sigma_m \eta_m \epsilon P_{\text{th}} p N_m \sigma_m \phi$
Consumption			
Thermal leakage			$DB^2 \phi$
Absorption by			
Fissionable material	N_m	σ_m	$\Sigma_m N_m \sigma_m \phi$
Fertile material	N_g	σ_g	$N_g \sigma_g \phi$
Higher isotopes	N_h	σ_h	$\Sigma_h N_h \sigma_h \phi$
Coolant, moderator, structural material, etc.	N_P	σ_P	$\Sigma_P N_P \sigma_P \phi$
^{135}Xe	N_{Xe}	σ_{Xe}	$N_{\text{Xe}} \sigma_{\text{Xe}} \phi$
^{149}Sm, etc.	N_S	σ_S	$\Sigma_S N_S \sigma_S \phi$
Fission-product pairs with low cross section	N_F	σ_F	$\Sigma_F N_F \sigma_F \phi$
Control absorbers	N_E	σ_E	$N_E \sigma_E \phi$

The neutron balance equation for the operating reactor is

$$\sum_m \eta_m \epsilon P_{\text{th}} p N_m \sigma_m \phi = DB^2 \phi + \sum_m N_m \sigma_m \phi + N_g \sigma_g \phi + \sum_h N_h \sigma_h \phi + \sum_P N_P \sigma_P \phi$$

$$+ N_{\text{Xe}} \sigma_{\text{Xe}} \phi + \sum_S N_S \sigma_S \phi + \sum_F N_F \sigma_F \phi + N_E \sigma_E \phi \qquad (3.26)$$

This differs from (3.24) in the following respects:

1. The production term on the left is now the sum of production rates from several fissionable species (such as ^{235}U, ^{239}Pu, and ^{241}Pu) instead of the single fissionable species of the reference design.
2. The second term on the right is also a sum of the consumption rate of neutrons by several fissionable species.
3. The fourth term on the right, $\Sigma_h N_h \sigma_h \phi$, is new, and represents additional consumption of neutrons by nonfissionable higher isotopes such as ^{236}U, ^{240}Pu, and ^{242}Pu, which were not present in the reference design.
4. The next to the last term on the right, $\Sigma_F N_F \sigma_F \phi$, is new, and represents additional consumption of neutrons by fission products of low cross section that were not present in significant amount in the reference design. It has become conventional to express N_F as *pairs* of fission products; for example, when n fissions take place, $N_F = n$.

To simplify the analysis of fuel cycles to be given in this chapter, the following assumptions are made about terms in Eqs. (3.24) through (3.26) that remain invariant between the reference design condition and the operating reactor:

1. The effective cross sections, σ, remain unchanged.
2. The reactor parameters ϵ, P_{th}, P_1, p, and DB^2 remain unchanged.

3. Absorption of thermal neutrons by fertile material, $N_g \sigma_g \phi$, remains unchanged.
4. Absorption of thermal neutrons by coolant, moderator, and structural material, $\Sigma_P N_P \sigma_P \phi$, remains unchanged.
5. The rate of absorption of thermal neutrons by ^{135}Xe, ^{149}Sm, and other fission products with high cross section is evaluated from the assumption that the poisoning ratios q for these nuclides remain unchanged between the reference design and the operating reactor. In the reference design the rate of absorption of neutrons by ^{135}Xe, ^{149}Sm, and other fission products with high cross sections produced by fission of ^{235}U is given by

$$N_{\mathrm{Xe}}^* \sigma_{\mathrm{Xe}}^* \phi + \sum_S N_S^* \sigma_S^* \phi = q_M^* N_M^* \sigma_M^* \phi \tag{3.27}$$

where q_M^* is the total poisoning ratio for all high-cross-section fission products from ^{235}U in the reference design reactor

$$q_M^* = q_{M,\mathrm{Xe}}^* + \sum_S q_{M,S}^* \tag{3.28}$$

defined as the ratio of absorption by these fission products to absorption by ^{235}U. In addition, a small amount of high-cross-section fission products comes from fast fissions in ^{238}U. If the poisoning ratio for high-cross-section fission products from ^{238}U in the reference design reactor

$$q_{28}^* = q_{28,\mathrm{Xe}}^* + \sum_S q_{28,S}^* \tag{3.29}$$

is defined as the ratio of absorption of thermal neutrons by these fission products to absorption of fast neutrons by ^{238}U, the total poisoning ratio for all high-cross-section fission products from ^{235}U and ^{238}U in the reference design reactor is

$$(q_M')^* = q_M^* + q_{28}^* \frac{\text{abs. of fast neutrons by } ^{238}\text{U}}{\text{abs. of thermal neutrons by } ^{235}\text{U}} \tag{3.30}$$

Table 3.14, in Sec. 6.4, shows that the ratio of absorptions in the above equation is

$$\frac{\alpha_{28}\eta_{25}(\epsilon - 1)}{\nu_{28} - 1 - \alpha_{28}} + \frac{\eta_{25}(\epsilon - 1)}{\nu_{28} - 1 - \alpha_{28}} = \frac{\eta_{25}(\epsilon - 1)}{\eta_{28} - 1} \tag{3.30a}$$

Hence

$$(q_M')^* = q_M^* + q_{28}^* \frac{\eta_{25}(\epsilon - 1)}{\eta_{28} - 1} \tag{3.30b}$$

In the operating reactor, it is assumed that the total poisoning ratio for all high-cross-section fission products from ^{235}U and the ^{238}U caused to undergo fission by fast neutrons from ^{235}U, q_M', has the same value as in the reference design:

$$q_M' = (q_M')^* \tag{3.31}$$

In addition, fission products come from other fissile species such as ^{239}Pu and ^{241}Pu, each of which has its own total poisoning ratio q_m' for high-cross-section fission products formed from that species and from ^{238}U caused to undergo fission by fast neutrons from it:

$$q_m' = q_m + \frac{\eta_m(\epsilon - 1)}{\eta_{28} - 1} q_{28} \tag{3.32}$$

Hence, the total absorption by ^{135}Xe, ^{149}Sm, and other fission products with high cross sections is

$$N_{\mathrm{Xe}}\sigma_{\mathrm{Xe}}\phi + \sum_S N_S\sigma_S\phi = \sum_m q'_m N_m\sigma_m\phi \tag{3.32a}$$

6. The cross section of low-cross-section fission products, σ_F, has a single constant value independent of the fissile nuclide from which the fission products were produced and independent of the flux time to which the fission products were exposed. This assumption is an oversimplification, because the yield of individual fission products is different from each fissile nuclide, and individual fission products with higher cross sections tend to be converted to those of lower cross section as irradiation progresses. Walker [W1] has given tables from which may be determined the effective cross sections of fission products from ^{235}U, ^{238}U, ^{239}Pu, and ^{241}Pu as a function of the flux and flux time to which the fission products have been exposed.

Assumptions 1 through 5 are slightly in error also for a number of reasons: Effective cross sections of nuclides with strong resonance absorption, such as the plutonium isotopes, will change as the ratio of resonance to thermal flux changes. This ratio increases as ^{235}U is burned out and ^{239}Pu builds up. Also, Eq. (3.26) neglects energy and spatial self-shielding of resonance neutrons, which becomes important as resonance absorbers such as ^{240}Pu build up to appreciable concentration. The effective thermal cross section is lower when self-shielding is taken into account. Absorption by fertile material will decrease as this material is used up; however, in most fuel cycles, fertile material is depleted by only a few percent. Also, the neutron flux level, which will vary during irradiation, will affect the concentration of ^{135}Xe. None of these effects introduces major error, and to take them into account would be impossible with hand calculation. They are neglected in the following treatment.

Since the first, third, and fifth terms on the right side of Eq. (3.26) are to be treated as unchanged during the operation of the reactor, it is convenient to replace them by a constant term $KN_M^*\sigma_M^*\phi$:

$$KN_M^*\sigma_M^*\phi = DB^2\phi + N_g\sigma_g\phi + \sum_P N_P\sigma_P\phi \tag{3.33}$$

The reactivity of the operating reactor ρ is defined as the ratio of the rate of absorption of thermal neutrons by control absorbers to the rate of production of thermal neutrons:

$$\rho = \frac{N_E\sigma_E\phi}{\sum_m \eta_m \epsilon P_{\mathrm{th}} p N_m\sigma_m\phi} \tag{3.34}$$

By substituting (3.32) and (3.33) into (3.26), solving for $N_E\sigma_E\phi$, and substituting into (3.34), we obtain Eq. (3.35) for the reactivity:

$$\rho = \frac{\sum_m (\eta_m\epsilon P_{\mathrm{th}} p - 1 - q'_m)N_m\sigma_m - KN_M^*\sigma_M^* - \sum_h N_h\sigma_h - \sum_F N_F\sigma_F}{\sum_m \eta_m\epsilon P_{\mathrm{th}} p N_m\sigma_m} \tag{3.35}$$

All of the quantities in this equation can be obtained from the properties of the reference design reactor except the nuclide concentrations N_m, N_h, and N_F. These can be obtained from equations for the rate of reaction of fuel with neutrons to be derived in Sec. 6.5. Thus, Eq. (3.35) permits evaluation of the changes in reactivity of nuclear fuel during irradiation.

One important use to be made of Eq. (3.35) is to find the length of time t that fuel may be irradiated before its reactivity drops to zero. The condition for this is that

$$0 = \frac{\sum_m (\eta_m \epsilon P_{\text{th}} p - 1 - q'_m) N_m(t) \sigma_m}{N_M^* \sigma_M^*} - \frac{\sum_h N_h(t) \sigma_h}{N_M^* \sigma_M^*} - \frac{\sum_F N_F(t) \sigma_F}{N_M^* \sigma_M^*} - K \tag{3.36}$$

6.4 Reactor Example

To illustrate the characteristics of different methods of fuel management, the example of the 1054-MWe PWR designed by the Westinghouse Electric Company for the Donald C. Cook Nuclear Plant of the American Electric Power System [A1] has been used.

We shall be interested in estimating the fuel-cycle performance of this reactor during a steady-state cycle, in which fresh fuel has a ^{235}U enrichment of 3.2 w/o ^{235}U and spent fuel has a ^{235}U enrichment of around 1.0 w/o and also contains around 0.6 w/o fissile plutonium. The reference design condition used to evaluate effective neutron cross sections and other reactor physics parameters during irradiation is taken to be 2.7 w/o ^{235}U. This value, slightly higher than the arithmetic mean of the fissile content of fuel at the beginning and end of irradiation, is intended to reflect the higher cross section of fissile plutonium compared with ^{235}U.

Table 3.11 gives the properties of each region of the core lattice for the reference design of this reactor. Information to be used directly in neutron balances and other fuel-cycle calculations are the volume fraction of each region v, the concentration N of each molecular or nuclear species in that region, and the ratio of the thermal flux in that region to the thermal flux in the fuel region ψ. The rate of reaction of thermal neutrons with species i in region j per unit lattice volume is $N_{ij} \sigma_i v_j \psi_j \phi$, where

N_{ij} is the number of atoms or molecules of species i per unit volume of region j
σ_i is the effective cross section of species i for thermal neutrons
v_j is the volume fraction of region j in the lattice
ψ_j is the ratio of thermal-neutron flux in region j to thermal flux in the fuel
ϕ is the thermal-neutron flux in the fuel region

Characteristics assumed for this reactor in the reference design condition are listed in Table 3.12.

Table 3.13 gives effective cross sections for thermal neutrons and other nuclear properties of the materials in the core of this reactor. These effective cross sections have been calculated by the procedure recommended by Westcott, which has been outlined in Chap. 2, from data provided by Westcott [W3] and Critoph [C1]. To obtain appropriate nuclear reaction rates, these effective cross sections are to be multiplied by the thermal-neutron flux, $n_{\text{MB}}\bar{v}$, where n_{MB} is the density of neutrons in the Maxwell-Boltzmann part of the spectrum and $\bar{v}$ is the average speed of the Maxwell-Boltzmann neutrons.

$$\bar{v} = 2200 \sqrt{\frac{4T}{\pi \hat{T}}} \quad \text{m/s} \tag{3.37}$$

The value of 80 b given for fission-product pairs is an approximate, constant value to be used independent of the fuel from which the fission products are formed and independent of the flux time to which the fission products are exposed after formation. For the present PWR, the effect of these variables on the cross sections of fission-product pairs, evaluated by extrapolation of Walker's [W1] tables, is as follows:

Thermal flux time, n/kb	Effective cross section σ_F, in barns, of fission-product pairs from			
	^{235}U	^{238}U	^{239}Pu	^{241}Pu
0	105	144	133	142
1	81	114	108	108
2	66	94	89	89
3	52	74	70	70

As irradiation progresses, the effective cross section of fission products from each fuel nuclide decreases. This decrease is partially offset by greater production of fission products from plutonium, which have a higher cross section than those from ^{235}U. The constant value of $\sigma_F = 80$ given in Table 3.13 takes these two effects approximately into account.

Table 3.14 gives the neutron balance for this reactor in the reference design condition, charged with UO_2 fuel containing 2.7 w/o ^{235}U uniformly distributed, operating at full power of 3250 MW, and with equilibrium xenon and samarium. Items 1 through 5 deal with events experienced by neutrons with energies high enough to cause fission in ^{238}U, above 1 MeV. Items 6 through 10 deal with events experienced by neutrons while being slowed down from fission to thermal energies. Items 11 through 23 give the production and consumption of thermal neutrons.

One significant feature of Table 3.14 is that the production of thermal neutrons, item 11, equals the total consumption, item 23. A second point to be noted is that the reactivity is given by the ratio of item 22 to item 23.

$$\rho = \frac{0.2112}{1.5655} = 0.1349 \tag{3.38}$$

A third point is that the constant term K in Eq. (3.33) is the sum of items 12, 15, 16, 17, 18, 19, and 20.

$$K = 0.0008 + 0.1431 + 0.0390 + 0.0043 + 0.0216 + 0.0724 + 0.0143 = 0.2955 \tag{3.39}$$

Finally, the initial conversion ratio ICR, the ratio of the atoms of ^{239}Pu produced per atom of ^{235}U consumed, is the sum of items 2, 8, and 15.

$$\text{ICR} = 0.0111 + 0.4619 + 0.1431 = 0.6161 \tag{3.40}$$

These terms determine the rate at which plutonium builds up during irradiation. One reason for giving the neutron balance in such detail is to be able to evaluate these terms.

6.5 Change of Composition with Flux Time

We are now in position to derive equations that will give the degree of burnup nuclear fuel can experience before it ceases to be critical. First, we must determine how the concentration of each nuclide that affects the neutron balance changes with time. We consider fuel that at time zero contains N_{25}^0 atoms of ^{235}U per cubic centimeter, N_{28}^0 atoms of ^{238}U, and no other uranium isotopes, plutonium, or fission products. This fuel is then exposed to a thermal-neutron flux $\phi(t)$, which may be a function of time. The variation in concentration of each nuclide in this fuel with time is obtained as follows.

Table 3.11 Volumes, masses, densities, and relative fluxes: PWR

Region	Material	Volume, cm^3	Volume fraction v	Mass m, kg	Molecular weight M	Density ρ, g/cm^3	Conc. N, molecules/cm^3 ($\times 10^{-24}$)	Relative thermal flux ψ
Fuel	UO_2 †	10,173,948	0.307095	99,123	269.969	9.7449	0.0217391	1.0000
	^{235}U			2,359.2	235.044		0.00059426	1.0000
	^{238}U			85,017.8	238.051		0.0211448	1.0000
Cladding	Zircaloy-4	2,820,738	0.085143	18,475.8	91.34	6.55	0.0432	1.0424
Guides + thimbles	Zircaloy-4	280,671	0.008472	1,838.4	91.34	6.55	0.0432	1.16
Thimble interior	Void	49,872	0.001505	0	–	0	0	1.16
Water in lattice	H_2O	16,454,134	0.496660	11,863.4	18.016	0.721	0.0241	1.1271
Extra water‡	H_2O	3,148,980	0.095050	2,270.4	18.016	0.721	0.0241	1.16
Spacers	Inconel-718	201,260	0.006075	1,654.6	57.941	8.221	0.0854	1.1271
		33,129,603	1.000000					

† 0.027 weight fraction ^{235}U in U.

‡ In spaces between assemblies.

Table 3.12 Characteristics of PWR in reference design condition

Power, thermal	3250 MW
Power, net electric	1054 MW
Core dimensions	
Equivalent radius R	169.80 cm
Height Z	365.76 cm
Effective core dimensions	
Radius R'	177.30 cm
Height Z'	380.76 cm
Temperature of Maxwell-Boltzmann neutrons T	574.75 K
Volume average Maxwell-Boltzmann flux in UO_2	3.496×10^{13} $n/(\text{cm}^2 \cdot \text{s})$
Fast fission factor ϵ	1.0476
Nonleakage probabilities	
Fission to ^{238}U resonance P_1	0.9889
^{238}U resonance to thermal P_2	0.9980
Resonance escape probability p	0.7725
Westcott epithermal flux factor r†	0.22166
Thermal leakage factor DB^2	0.0000819 cm^{-1}
Geometric buckling $B^2 = (2.405/R')^2 + (\pi/Z')^2$	0.000252 cm^{-2}
Fermi age τ	52.2 cm^2

†To be used with Westcott [W3, C1] s factor s_2.

Table 3.13 Effective properties of nuclides for thermal neutrons in PWR

Nuclide	Subscript†	Absorption cross section, σ_a, b	Neutrons produced: Per fission, ν	Neutrons produced: Per neutron absorbed, η	Ratio of capture to fission cross section, α	Poisoning ratio of high-cross section fission products, q
^{235}U	25	555.57	2.43‡	1.9600	0.2398	0.0541
^{236}U	26	123.9				
^{238}U	28	2.2342	2.79§	2.3432§	0.1907§	0.0683
^{239}Pu	49	1618.2	2.87‡	1.8600	0.5430	0.0549
^{240}Pu	40	2616.8				
^{241}Pu	41	1567.3	3.06‡	2.2230	0.3765	0.0547
^{242}Pu	42	381.0				
F.P.P.¶	F	80				
Zircaloy-4	Z	1.030				
Inconel-718	I	3.749				
Water	W	0.544				

†To designate isotopes of uranium, plutonium, and other actinide elements, it has become conventional to use two-digit subscripts, such as 49 for ^{239}Pu, in which the first digit is the atomic number minus 90 and the second digit is the last digit of the mass number.

‡These values of ν are from [C1]. They are used here because effective cross sections are from [C1] also. These values of ν differ slightly from App. C.

§In fast fission.

¶Pairs of fission products with cross sections less than 10,000 b.

Table 3.14 Neutron balance for reference design of PWR†

Item	Process		Number of neutrons
1.	Production of fast neutrons from fission of ^{235}U	η_{25}	1.9600
2.	Capture of fast neutrons by ^{238}U to produce ^{239}Pu	$\dfrac{\alpha_{28}\eta_{25}(\epsilon-1)}{\nu_{28}-1-\alpha_{28}}$	0.0111
3.	Fission of ^{238}U by fast neutrons	$\dfrac{\eta_{25}(\epsilon-1)}{\nu_{28}-1-\alpha_{28}}$	0.0583
4.	Production of fast neutrons from fission of ^{238}U	$\nu_{28}(3)=\dfrac{\nu_{28}\eta_{25}(\epsilon-1)}{\nu_{28}-1-\alpha_{28}}$	0.1627
5.	Net production of fast neutrons	$(1)-(2)-(3)+(4)=\eta_{25}\epsilon$	2.0533
6.	Neutron leakage during moderation from fission to ^{238}U resonance energy	$\eta_{25}\epsilon(1-P_1)$	0.0228
7.	Neutrons moderated to ^{238}U resonance energy	$(5)-(6)=\eta_{25}\epsilon P_1$	2.0305
8.	Neutrons captured in ^{238}U resonance	$\eta_{25}\epsilon P_1(1-p)$	0.4619
9.	Neutrons escaping ^{238}U resonance capture	$(7)-(8)=\eta_{25}\epsilon P_1 p$	1.5686
10.	Neutron leakage during moderation from ^{238}U resonance to thermal energy	$\eta_{25}\epsilon P_1 p(1-P_2)$	0.0031
11.	Production of thermal neutrons	$(9)-(10)=\eta_{25}\epsilon P_1 pP_2$	1.5655 ←
	Consumption of thermal neutrons by		
12.	Thermal leakage	$\dfrac{DB^2}{N_{25}^*\sigma_{25}v_U}$	0.0008
13.	^{235}U fission	$\dfrac{1}{1+\alpha_{25}}$	0.8066
14.	^{235}U capture	$\dfrac{\alpha_{25}}{1+\alpha_{25}}$	0.1934
15.	Absorption by ^{238}U	$\dfrac{N_{28}^*\sigma_{28}}{N_{25}^*\sigma_{25}}$	0.1431
16.	Absorption by zircaloy cladding	$\dfrac{N_Z\sigma_Z(\psi v)_{\text{clad}}}{N_{25}^*\sigma_{25}v_U}$	0.0390
17.	Absorption by zircaloy guides	$\dfrac{N_Z\sigma_Z(\psi v)_{\text{guides}}}{N_{25}^*\sigma_{25}v_U}$	0.0043
18.	Absorption by Inconel	$\dfrac{N_I\sigma_I(\psi v)_I}{N_{25}^*\sigma_{25}v_U}$	0.0216
19.	Absorption by water in lattice	$\dfrac{N_W\sigma_W(\psi v)_{\text{lat. water}}}{N_{25}^*\sigma_{25}v_U}$	0.0724
20.	Absorption by other water	$\dfrac{N_W\sigma_W(\psi v)_{\text{oth. water}}}{N_{25}^*\sigma_{25}v_U}$	0.0143
21.	Absorption by ^{135}Xe and Sm	$q_{25}'=q_{25}+q_{28}\dfrac{\eta_{25}(\epsilon-1)}{\eta_{28}-1}$	0.0588
22.	Absorption by control poison	$\dfrac{N_E\sigma_E}{N_{25}^*\sigma_{25}}$	0.2112
23.	Total consumption of thermal neutrons		1.5655 ←

†Basis: One thermal neutron absorbed by ^{235}U.

^{235}U. The rate at which ^{235}U is consumed is

$$\frac{dN_{25}}{dt} = -N_{25}\sigma_{25}\phi(t) \tag{3.41}$$

where N_{25} = number of ^{235}U atoms per unit volume
σ_{25} = effective absorption cross section of ^{235}U for thermal neutrons
$\phi(t)$ = thermal-neutron flux, which may vary with time t

The solution of (3.32) for the number of ^{235}U atoms at time t, N_{25}, is

$$N_{25} = N_{25}^0 \exp\left(-\sigma_{25}\int_0^t \phi\, dt'\right) \tag{3.42}$$

The flux time θ is defined by

$$\theta \equiv \int_0^t \phi(t')\, dt' \tag{3.43}$$

so that

$$N_{25} = N_{25}^0\, e^{-\sigma_{25}\theta} \tag{3.44}$$

The flux time is the fundamental variable used to express the extent of exposure to irradiation. Even when the flux varies with time, Eq. (3.44) is strictly correct. The flux time thus defined is in units of neutrons per square centimeter. Expressed in these units its magnitude is around 10^{21} in a typical reactor, as when fuel has been irradiated in a flux of 10^{14} $n/(\text{cm}^2\cdot\text{s})$ for 10^7 s. Consequently, it has become customary to express flux time in units of 10^{21} n/cm^2, termed neutrons per kilobarn and written n/kb. The flux time is often called the fluence.

^{236}U. ^{236}U is produced by capture of neutrons in ^{235}U at the rate per unit volume of $N_{25}\sigma_{25}\phi\alpha_{25}/(1+\alpha_{25})$, where α_{25} is the ratio of neutrons captured by ^{235}U to neutrons producing fission in ^{235}U. The rate of consumption of ^{236}U by neutron capture is $N_{26}\sigma_{26}\phi$. Hence the net rate of change of ^{236}U concentration is

$$\frac{dN_{26}}{dt} = \frac{N_{25}\sigma_{25}\alpha_{25}\phi}{1+\alpha_{25}} - N_{26}\sigma_{26}\phi \tag{3.45}$$

The solution of Eq. (3.45) subject to $N_{26} = 0$ at $t = 0$ is

$$N_{26} = \frac{N_{25}^0\sigma_{25}\alpha_{25}}{(\sigma_{25}-\sigma_{26})(1+\alpha_{25})}\left(e^{-\sigma_{26}\theta} - e^{-\sigma_{25}\theta}\right) \tag{3.46}$$

Plutonium isotopes. The net rate of formation of ^{239}Pu is

$$\frac{dN_{49}}{dt} = \underbrace{N_{28}^0\sigma_{28}\phi}_{\substack{\text{Absorption of thermal}\\ \text{neutrons in } ^{238}\text{U}}} + \underbrace{\eta_{25}\epsilon P_1(1-p)N_{25}\sigma_{25}\phi}_{\substack{\text{Absorption of resonance}\\ \text{neutrons from } ^{235}\text{U}\\ \text{fission in } ^{238}\text{U}}}$$

$$+ \underbrace{\eta_{49}\epsilon P_1(1-p)N_{49}\sigma_{49}\phi}_{\substack{\text{Absorption of resonance}\\ \text{neutrons from } ^{239}\text{Pu}\\ \text{fission in } ^{238}\text{Pu}}} + \underbrace{\eta_{41}\epsilon P_1(1-p)N_{41}\sigma_{41}\phi}_{\substack{\text{Absorption of resonance}\\ \text{neutrons from } ^{241}\text{Pu}\\ \text{fission in } ^{238}\text{U}}} \tag{3.47}$$

(*Cont. on p. 138*)

$$- N_{49}\sigma_{49}\phi$$

Absorption of thermal neutrons in ^{239}Pu

$$+ \frac{\alpha_{28}}{1+\alpha_{28}} \frac{\epsilon - 1}{\eta_{28} - 1} (\eta_{25}N_{25}\sigma_{25} + \eta_{49}N_{49}\sigma_{49} + \eta_{41}N_{41}\sigma_{41})\phi \qquad (3.47) \ (Cont.)$$

Absorption of fast neutrons from ^{235}U, ^{239}Pu, and ^{241}Pu in ^{238}U

This equation may be written as

$$\frac{dN_{49}}{d\theta} = N_{28}^0\sigma_{28} + \kappa_{25}N_{25}\sigma_{25} - \gamma N_{49}\sigma_{49} + \kappa_{41}N_{41}\sigma_{41} \qquad (3.48)$$

where

$$\kappa_m = \eta_m \epsilon P_1(1-p) + \eta_m \frac{\alpha_{28}}{1+\alpha_{28}} \frac{\epsilon - 1}{\eta_{28} - 1}$$

$$\gamma = 1 - \kappa_{49}$$

To solve this equation exactly, the dependence of ^{241}Pu concentration, N_{41}, on flux time must be derived. This is obtained by considering the rate of change of concentration of ^{240}Pu,

$$\frac{dN_{40}}{d\theta} = \frac{\alpha_{49}N_{49}\sigma_{49}}{1+\alpha_{49}} - N_{40}\sigma_{40} \qquad (3.49)$$

and of ^{241}Pu,†

$$\frac{dN_{41}}{d\theta} = N_{40}\sigma_{40} - N_{41}\sigma_{41} \qquad (3.50)$$

To obtain an exact solution, Eqs. (3.48), (3.49), and (3.50) should be solved simultaneously, with the dependence of N_{25} on flux time given by (3.44). The solution is of the form

$$N_{49} = N_{28}^0(A_{49} + B_{49}e^{-S_1\theta} + C_{49}e^{-S_2\theta} + D_{49}e^{-S_3\theta}) + N_{25}^0(E_{49}e^{-\sigma_{25}\theta} + F_{49}e^{-S_1\theta} + G_{49}e^{-S_2\theta} + H_{49}e^{-S_3\theta}) \qquad (3.51)$$

where S_1, S_2, and S_3 are the three roots of the cubic equation

$$\begin{vmatrix} S - \gamma\sigma_{49} & 0 & \kappa_{41}\sigma_{41} \\ \dfrac{\alpha_{49}\sigma_{49}}{1+\alpha_{49}} & S - \sigma_{40} & 0 \\ 0 & \sigma_{40} & S - \sigma_{41} \end{vmatrix} = 0 \qquad (3.52)$$

In this section, an approximate solution is derived that does not require finding the roots of the cubic equation (3.52).

†Equation (3.50) neglects the term $-\lambda_{41}N_{41}/\phi$ representing the decay of 13.2-yr ^{241}Pu to ^{241}Am. To have included this term would have made it necessary to use the time t as independent variable rather than the flux time θ, which would have greatly increased the complexity of integrating these equations. The magnitude of λ_{41}/ϕ in barns for ^{241}Pu (half-life 13.2 year) in a typical PWR ($\phi \simeq 5 \times 10^{13}$ n/(cm$^2 \cdot$s) is

$$10^{24} \times \frac{0.693}{(13.2)(3.15 \times 10^7)(5 \times 10^{13})} = 30 \text{ b}$$

This is small compared with the neutron-absorption cross section for ^{241}Pu, which is $\sigma_{41} = 1377$ b.

The approximation consists in neglecting the formation of ^{239}Pu by absorption of resonance neutrons from ^{241}Pu, a procedure justified as long as $\kappa_{41}N_{41}\sigma_{41} \ll \gamma N_{49}\sigma_{49}$. With this approximation Eq. (3.48) reduces to

$$\frac{dN_{49}}{d\theta} = N_{28}^0\sigma_{28} + \kappa_{25}N_{25}\sigma_{25} - \gamma N_{49}\sigma_{49} \tag{3.53}$$

With N_{25} given by (3.44), the solution of this equation, subject to $N_{49} = 0$ at $t = 0$, is

$$N_{49} = C_1 + C_2 e^{-\sigma_{25}\theta} - (C_1 + C_2)e^{-\sigma_{49}\gamma\theta} \tag{3.54}$$

where

$$C_1 \equiv \frac{N_{28}^0\sigma_{28}}{\sigma_{49}\gamma} \tag{3.55}$$

and

$$C_2 \equiv \frac{\kappa_{25}N_{25}^0\sigma_{25}}{\sigma_{49}\gamma - \sigma_{25}} \tag{3.56}$$

With N_{49} given by (3.54), the solution of Eq. (3.49), subject to $N_{40} = 0$ at $t = 0$, is

$$N_{40} = C_3 + C_4 e^{-\sigma_{25}\theta} + C_5 e^{-\sigma_{49}\gamma\theta} - (C_3 + C_4 + C_5)e^{-\sigma_{40}\theta} \tag{3.57}$$

where

$$C_3 \equiv \frac{N_{28}^0\sigma_{28}\alpha_{49}}{\sigma_{40}\gamma(1 + \alpha_{49})} \tag{3.58}$$

$$C_4 \equiv \frac{C_2\sigma_{49}\alpha_{49}}{(\sigma_{40} - \sigma_{25})(1 + \alpha_{49})} \tag{3.59}$$

and

$$C_5 \equiv \frac{C_3\sigma_{40}}{\sigma_{49}\gamma - \sigma_{40}} + \frac{C_4(\sigma_{40} - \sigma_{25})}{\sigma_{49}\gamma - \sigma_{40}} \tag{3.60}$$

With N_{40} given by (3.57), the solution of Eq. (3.50), subject to $N_{41} = 0$ at $t = 0$, is

$$N_{41} = C_6 + C_7 e^{-\sigma_{25}\theta} + C_8 e^{-\sigma_{49}\gamma\theta} + C_9 e^{-\sigma_{40}\theta} - (C_6 + C_7 + C_8 + C_9)e^{-\sigma_{41}\theta} \tag{3.61}$$

where

$$C_6 \equiv \frac{C_3\sigma_{40}}{\sigma_{41}} \tag{3.62}$$

$$C_7 \equiv \frac{C_4\sigma_{40}}{\sigma_{41} - \sigma_{25}} \tag{3.63}$$

$$C_8 \equiv \frac{C_5\sigma_{40}}{\sigma_{41} - \sigma_{49}\gamma} \tag{3.64}$$

and

$$C_9 \equiv \frac{(C_3 + C_4 + C_5)\sigma_{40}}{\sigma_{40} - \sigma_{41}} \tag{3.65}$$

^{242}Pu. Consumption of ^{242}Pu by absorption of neutrons will be neglected. The rate of accumulation of ^{242}Pu then is given by

$$\frac{dN_{42}}{dt} = \frac{\alpha_{41}N_{41}\sigma_{41}\phi}{1 + \alpha_{41}} \tag{3.66}$$

With N_{41} given by (3.61), the solution of Eq. (3.57), subject to $N_{42} = 0$ at $t = 0$, is

$$N_{42} = \frac{\alpha_{41}\sigma_{41}}{1 + \alpha_{41}}\left[C_6\theta + \frac{C_7(1 - e^{-\sigma_{25}\theta})}{\sigma_{25}} + \frac{C_8(1 - e^{-\sigma_{49}\gamma\theta})}{\sigma_{49}\gamma}\right.$$
$$\left. + \frac{C_9(1 - e^{-\sigma_{40}\theta})}{\sigma_{40}} - \frac{(C_6 + C_7 + C_8 + C_9)(1 - e^{-\sigma_{41}\theta})}{\sigma_{41}}\right] \tag{3.67}$$

Fission products. Burnout of fission products by neutron absorption will be neglected. The rate of formation of fission-product pairs from ^{235}U is

$$\frac{dN_F(25)}{dt} = \frac{N_{25}\sigma_{25}\phi}{1+\alpha_{25}} \tag{3.68}$$

With N_{25} given by (3.44), the solution of Eq. (3.68), subject to $N_F(25) = 0$ at $t = 0$, is

$$N_F(25) = \frac{N_{25}^0(1 - e^{-\sigma_{25}\theta})}{1+\alpha_{25}} \tag{3.69}$$

The rate of formation of fission-product pairs from ^{239}Pu is

$$\frac{dN_F(49)}{dt} = \frac{N_{49}\sigma_{49}\phi}{1+\alpha_{49}} \tag{3.70}$$

With N_{49} given by (3.54), the solution of this equation, subject to $N_F(49) = 0$ at $t = 0$, is

$$N_F(49) = \frac{\sigma_{49}}{1+\alpha_{49}}\left[C_1\theta + C_2\frac{1-e^{-\sigma_{25}\theta}}{\sigma_{25}} - (C_1 + C_2)\frac{1-e^{-\sigma_{49}\gamma\theta}}{\sigma_{49}\gamma}\right] \tag{3.71}$$

The rate of formation of fission-product pairs from ^{241}Pu is

$$\frac{dN_F(41)}{dt} = \frac{N_{41}\sigma_{41}\phi}{1+\alpha_{49}} \tag{3.72}$$

By comparing this equation with Eq. (3.66) for ^{242}Pu, we see that

$$\frac{dN_F(41)}{dt} = \frac{1}{\alpha_{41}}\frac{dN_{42}}{dt} \tag{3.73}$$

so that

$$N_F(41) = \frac{N_{42}}{\alpha_{41}} \tag{3.74}$$

with N_{42} given by Eq. (3.67).

The concentration of fission products formed in fast fission of ^{238}U is given by

$$N_F(28) = \frac{\epsilon - 1}{\eta_{28} - 1}\,\frac{\nu_{25}N_F(25) + \nu_{49}N_F(49) + \nu_{41}N_F(41)}{1+\alpha_{28}} \tag{3.75}$$

^{238}U. An equation for the decrease in ^{238}U concentration is obtained by considering the processes by which ^{238}U is used up:

$$N_{28}^0 - N_{28} = N_{28}^0\sigma_{28}\theta \qquad + \kappa_{25}(1+\alpha_{25})N_F(25)$$

Absorption of thermal neutrons — Absorption of resonance and fast neutrons from fission of ^{235}U

$$+ \kappa_{49}(1+\alpha_{49})N_F(49) \quad + N_F(28) \tag{3.76}$$

Capture of resonance and fast neutrons from fission of ^{239}Pu — Fast fission of ^{238}U

The term for capture of resonance and fast neutrons from fission of ^{241}Pu has been omitted to be consistent with the approximation used in Eq. (3.53).

Checks of numerical work. Two equations useful in checking numerical calculation of nuclide concentrations are

$$\alpha_{49}N_F(49) = N_{40} + N_{41} + N_{42} + N_F(41) \tag{3.77}$$

and $$N_{28}^0 - N_{28} = N_F(28) + N_{49} + N_F(49) + N_{40} + N_{41} + N_{42} + N_F(41) \tag{3.78}$$

Burnup. For this chapter we shall assume that the heat of fission is 0.95 MWd/g fissioned, or 9.5×10^5 MWd/MT. This corresponds roughly to a heat of fission of 200 MeV for ^{235}U. Fuel burnup B, in megawatt-days per metric ton, is thus related to the weight fraction w of fuel fissioned by

$$B = 9.5 \times 10^5 w \tag{3.79}$$

The weight fraction fissioned is

$$w = \frac{235N_F(25) + 238N_F(28) + 239N_F(49) + 241N_F(41)}{235N_{25}^0 + 238N_{28}^0} \tag{3.80}$$

Burnup can be measured experimentally, either from the amount of heat liberated in a particular fuel element or from the amount of fission products found in it. Flux time, on the other hand, is much less readily determined experimentally and is subject to more ambiguity because of the various ways in which neutron flux may be defined. Whereas flux time is the natural independent variable to characterize fuel exposure in calculations, it is preferable to use burnup in describing practical situations.

6.6 Composition Changes in PWR

Equations (3.44) through (3.80) have been used to calculate the effect of irradiation in this PWR on the composition of fuel initially containing 3.2 w/o ^{235}U, with results given in Table 3.15. Nuclide concentrations N_i from these equations have been converted to weight fractions w_i by

$$w_i = \frac{N_i M_i}{N_{25}^0 M_{25} + N_{28}^0 M_{28}^0} \tag{3.81}$$

where M_i is the atomic weight of nuclide i, and N_{25}^0 and N_{28}^0 are the atom concentrations of ^{235}U and ^{238}U, respectively, in fresh fuel.

Weight percents from this table are plotted against burnup as the points in Fig. 3.29. These points are to be compared with the lines, taken from Fig. 3.3, representing weight percents calculated by the more accurate point-depletion computer code CELL [B2]. Agreement is excellent for ^{235}U and ^{236}U, fair for ^{239}Pu, but poor for the higher plutonium isotopes at high burnup. This is because the effective absorption cross section of ^{240}Pu decreases as its

Table 3.15 Effect of irradiation in a PWR on composition, burnup, and reactivity of fuel containing initially 3.2 w/o ^{235}U

Flux time, n/kb	0	0.5	1.0	1.5	2.0	2.5
Weight percent						
^{235}U	3.2	2.4238	1.8360	1.3907	1.0534	0.7979
^{236}U	0.0	0.1459	0.2480	0.3166	0.3611	0.3874
^{239}Pu	0.0	0.3858	0.5605	0.6150	0.6126	0.5780
^{240}Pu	0.0	0.0408	0.0914	0.1207	0.1312	0.1300
^{241}Pu	0.0	0.0171	0.0723	0.1330	0.1772	0.2004
^{242}Pu	0.0	0.0011	0.0103	0.0325	0.0663	0.1073
Fission products	0.0	0.7977	1.6097	2.3945	3.1398	3.8187
Burnup, MWd/MT	0	7,578	15,292	22,748	29,828	36,277
Reactivity ρ	0.1643	0.1085	0.0435	−0.0134	−0.0675	−0.1272

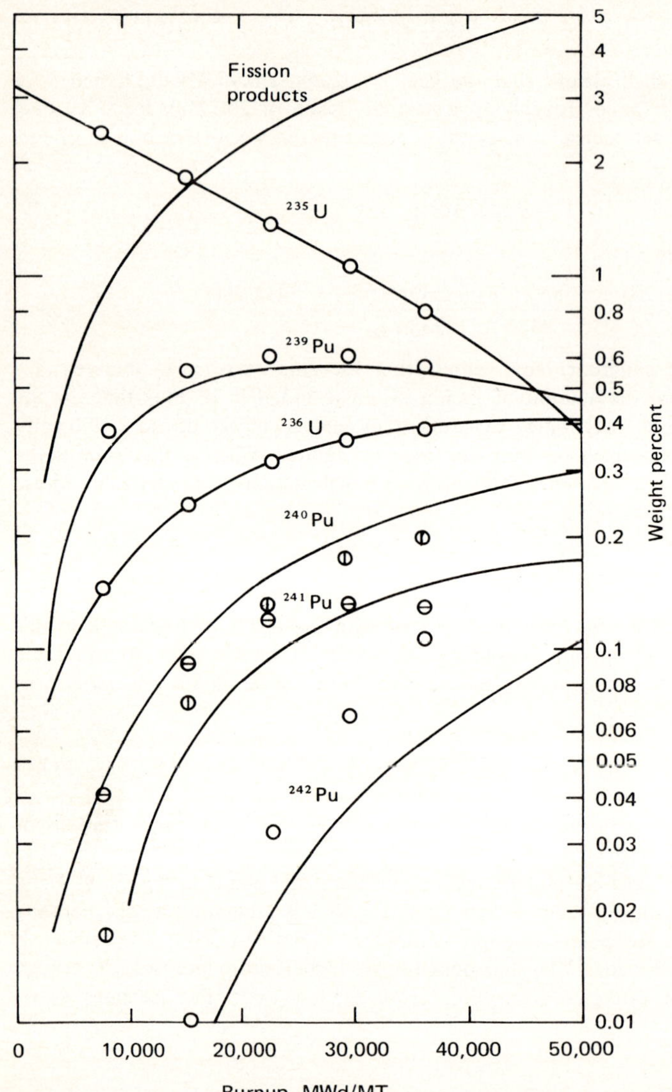

Figure 3.29 Change of nuclide concentrations in PWR with burnup. (○) Equations of this chapter; (⊖) ^{240}Pu, equations of this chapter; (⦶) ^{241}Pu, equations of this chapter; (—) computer code CELL.

concentration increases, as a result of its very high absorption cross section at resonance energies, an effect that the equations of this chapter cannot take into account.

6.7 Reactivity Changes in PWR

Despite the inability of these equations to represent accurately the concentration of higher plutonium isotopes, the reactivity-limited burnup attainable from fuel initially containing 3.2

w/o ^{235}U can be predicted quite satisfactorily from the results of this table. Reactivity ρ, calculated from these concentrations by Eq. (3.35), is listed in the last row of Table 3.15 and is plotted against burnup B at the circled points in Fig. 3.30. The points are represented quite accurately by the straight line

$$\rho = 0.17 - 8.16 \times 10^{-6} B \tag{3.82}$$

The reactivity equals zero at a burnup of 20,833 MWd/MT. This is in excellent agreement with the reactivity-limited burnup of 21,085 MWd/MT for batch irradiation of 3.2 w/o fuel in this reactor obtained by Watt [W2] using the computer codes CELL [B2] and CORE [K1].

6.8 Effect of Fuel Management Method on Burnup

Because the reactivity follows closely a linear relation with burnup, it is possible to predict the burnup B_n obtainable in n-zone fueling with 3.2 w/o ^{235}U by the approximate equation (3.6), which for this case becomes

$$B_n = 20{,}833 \frac{2n}{n+1} \tag{3.83}$$

Burnups for n-zone fueling predicted by this equation are compared below with those obtainable by Watt [W2] using computer codes CELL [B2] and CORE [K1].

Number of zones n	1	2	3	4	6
Burnup B_n, MWd/MT					
Eq. (3.83)	20,833	27,777	31,250	33,333	35,714
Watt	21,085	26,708	30,771	33,718	36,907

The agreement is remarkably good, considering that the computer codes take into account the changes in cross sections and reactor parameters that occur as fuel composition changes, and also follow spatial nonuniformities in flux and fuel composition. All these factors have been neglected in this section.

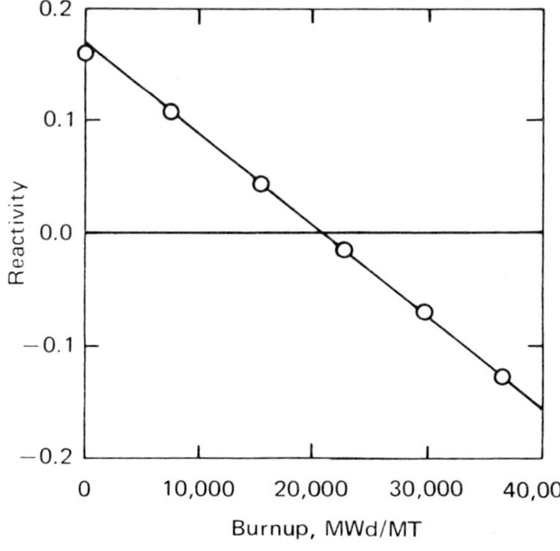

Figure 3.30 Change of reactivity in PWR with burnup. (○) Equation (3.35) and Table 3.15; (—) Eq. (3.82).

7 FUEL-CYCLE MATERIAL FLOW SHEETS

This section presents flow sheets giving the amount of nuclear materials consumed and fuel-cycle services required per year for four reactor systems: light-water reactor (LWR) fueled with slightly enriched uranium (Sec. 7.1); light-water reactor fueled with plutonium and natural uranium (Sec. 7.2); high-temperature gas-cooled reactor (HTGR) fueled with thorium, fully enriched ^{235}U, and recycle ^{233}U (Sec. 7.3); and liquid-metal fast-breeder reactor (LMFBR) (Sec. 7.4). These flow sheets have been adapted from studies by Pigford [P1, P2], which used consistent assumptions regarding cross sections and nuclide quantities transferred between different reactor types. As the bases for these studies were slightly different from those used for light-water reactors earlier in this chapter, the material quantities in Sec. 7.1 differ slightly from those in Secs. 5 and 6.

7.1 LWR Fueled with Slightly Enriched Uranium

Figure 3.31 shows the quantities of materials to be supplied or processed per year for a 1000-MWe LWR operated with recycle of uranium in spent fuel, after steady-state fuel-cycle conditions have been established. To generate electricity at this rate with a power-cycle thermal efficiency of 32.5 percent at a capacity factor of 80 percent requires that the reactor be supplied with 27,271 kg/year of uranium enriched to 3.3 w/o ^{235}U. This fuel is expected to support a burnup of 33,000 MWd/MT with three-zone fueling. Of the other material quantities shown in the flow sheet, the most significant are the following: 25,674 kg uranium containing 0.83 w/o ^{235}U is recovered in the form of UF_6 and recycled to isotope separation. Uranium isotope separation is assumed to reject tails containing 0.3 w/o ^{235}U. Under these conditions the uranium isotope separation facility needs 168,090 kg/year of natural uranium feed in the form of UF_6 and produces 106,974 separative work units per year (defined in Chap. 13). To provide UF_6 at this rate requires an annual supply of 439,199 lb U_3O_8 in uranium ore concentrates. The fuel reprocessing plant produces 243.5 kg/year of plutonium of the isotopic composition given at the right of Fig. 3.31.

7.2 LWR Fueled with Plutonium and Natural Uranium

Figure 3.32 is a material flow sheet on the same basis as Fig. 3.31 for a LWR fueled with natural uranium, recycle plutonium, and enough plutonium from the uranium-fueled LWR of Fig. 3.31 to provide the same burnup, 33,000 MWd/MT with three-zone fueling. The salient points to notice are the following: The 980 kg of plutonium recycled per year contains a much higher proportion of the higher isotopes 241 and 242 than the plutonium makeup from the uranium-fueled reactor. The depleted uranium recovered in reprocessing contains only 0.327 w/o ^{235}U, and has too little value to justify conversion to UF_6 and recycle to isotope separation. The amount of U_3O_8 to be supplied is 73,322 lb, only one-sixth that needed for Fig. 3.31.

It must be noted, however, that operation of the flow sheet of Fig. 3.32 requires 504.96 kg of plutonium from the flow sheet of Fig. 3.31, which would have to be provided by (1000 MWe)(504.96/243.5) = 2074 MWe of uranium-fueled LWR capacity, a total of 3074 MWe. Considering this self-contained reactor system, the specific annual consumption of U_3O_8 would be

$$\frac{(73{,}322)(1000) + (439{,}199)(2074)}{3074} = 320.2 \text{ lb } U_3O_8/(\text{MWe}\cdot\text{yr}) \tag{3.84}$$

Because Fig. 3.31 without plutonium recycle shows a specific uranium consumption of 439.2 lb U_3O_8/(MWe·year), plutonium recycle reduces U_3O_8 demand by 27 percent.

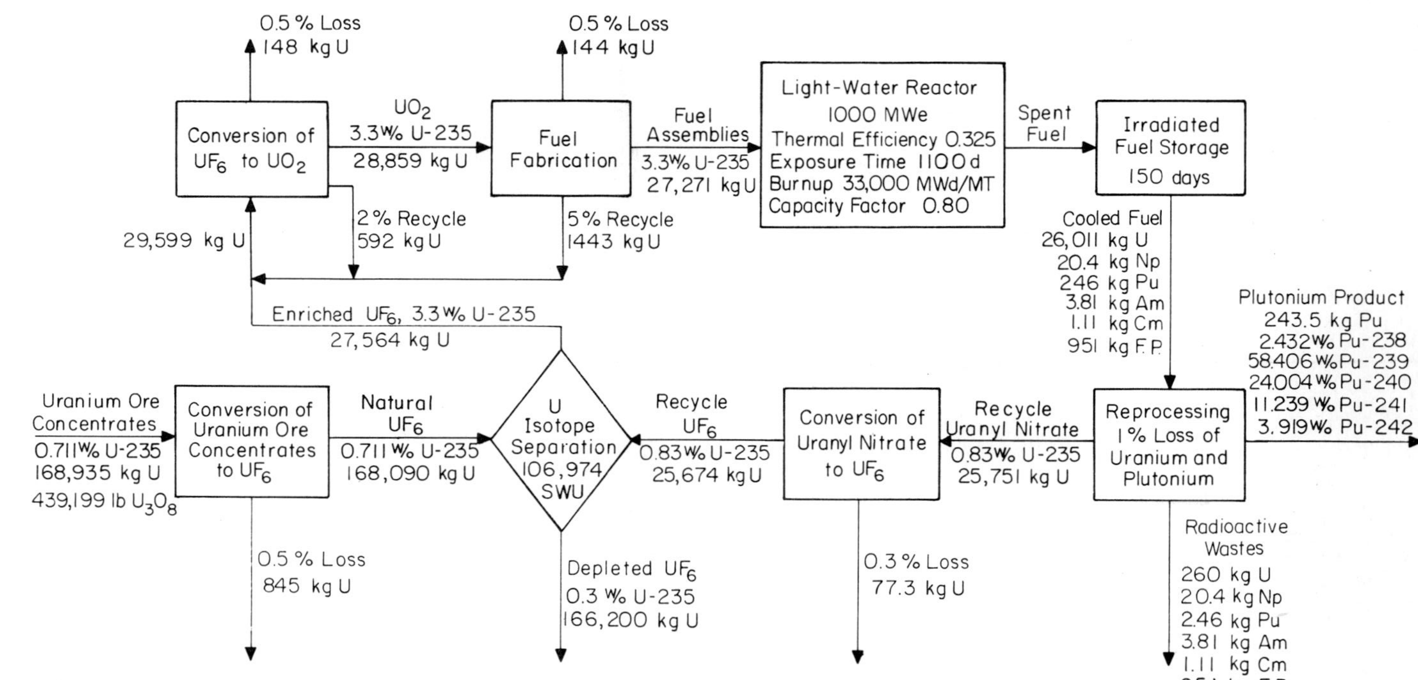

Figure 3.31 Fuel-cycle flow sheet for 1000-MWe LWR. Basis: 1 year, 80 percent capacity factor, enriched uranium fuel.

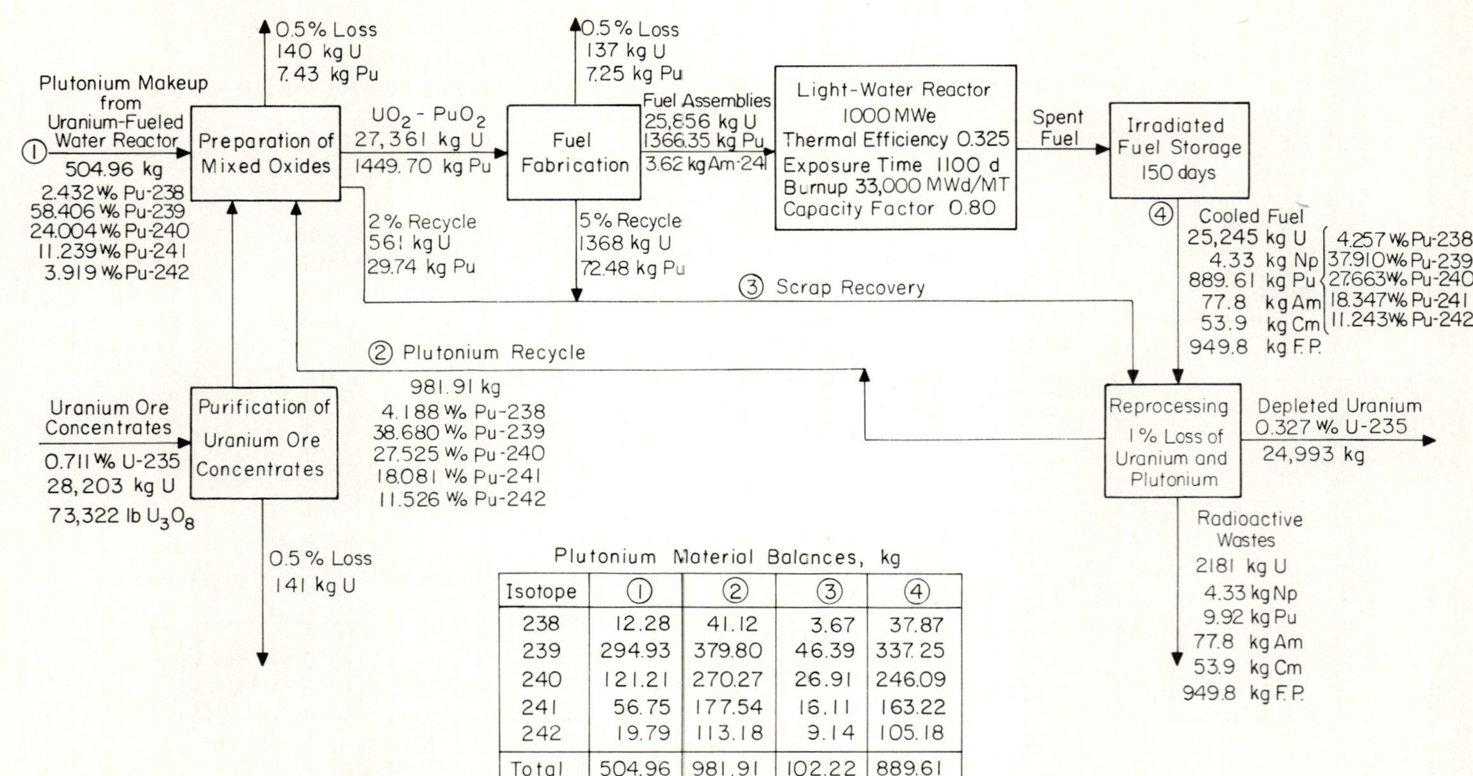

Isotope	①	②	③	④
238	12.28	41.12	3.67	37.87
239	294.93	379.80	46.39	337.25
240	121.21	270.27	26.91	246.09
241	56.75	177.54	16.11	163.22
242	19.79	113.18	9.14	105.18
Total	504.96	981.91	102.22	889.61

Figure 3.32 Fuel-cycle flow sheet for 1000-MWe LWR fueled with natural uranium, recycle plutonium, and plutonium recovered from reactor fueled with enriched uranium. Basis: 1 year, 80 percent capacity factor.

Because plutonium recycle makes possible the generation of 3074 MWe of electricity with only 2074 MWe requiring enriched uranium, the reduction in separative work demand made possible by plutonium recycle is $(100)(3074 - 2074)/3074 = 32.5$ percent, and the specific separative work demand is $(106{,}974)(2.074)/3074 = 72.1$ SWU/(MWe·year). This is to be compared with 106.974 SWU/(MWe·year) without plutonium recycle.

7.3 The HTGR

Although the HTGR is not being widely used commercially, its high thermal efficiency, its ability to produce process heat at high temperatures, and its relatively efficient use of natural uranium resources when fed with thorium as fertile material give it potential future importance.

The core of the HTGR consists of hexagonal blocks of graphite pierced with two sets of longitudinal holes. One set of holes permits flow of helium coolant, whose outlet temperature may reach 1500°F, thus making possible high thermal efficiency. The other set of holes is filled with rods in which microspheres of nuclear fuel are imbedded in a graphite matrix. When the reactor first goes into operation, before a supply of ^{233}U has been accumulated, two kinds of microspheres are used.

In one kind, microspheres of fully enriched (93.5 w/o ^{235}U) uranium carbide (UC_2), about 200 μm in diameter, are coated with three concentric layers, called a TRISO coating. The inner coating consists of porous graphite, to accommodate fuel swelling and fission-product gases. The intermediate coating, about 500 μm in outside diameter, consists of silicon carbide, to provide mechanical integrity for the spheres when the fuel is processed after discharge. The outer coating consists of impervious pyrolytic graphite, to retain fission products.

In the second kind, microspheres of thorium dioxide (ThO_2), about 500 μm in diameter, are coated with two concentric layers, called a BISO coating: an inner layer of porous graphite, to accommodate swelling and fission-product gases, and an outer layer of impervious pyrolytic graphite, to retain fission products.

During irradiation, about three-fourths of the TRISO-coated ^{235}U is consumed, leaving a residue of fission products and uranium whose isotopic content is around 20 percent ^{235}U, 25 percent ^{238}U, and 55 percent ^{236}U, formed by nonfission neutron capture in ^{235}U. At the same time, about 8 percent of BISO-coated thorium is converted to ^{233}U, some of which then undergoes fission.

When the first charge of fuel ceases to support a chain reaction, one-fourth of the fuel assemblies that have reacted most fully are replaced with fresh fuel. The spent assemblies are stored ("cooled") for 150 days to permit some fission products to decay, 6.75-day ^{237}U to change to ^{237}Np, and ^{233}Th and 27-day ^{233}Pa formed by neutron capture in ^{233}Th to change to ^{233}U.

In processing the fuel, the first steps are to crush the graphite blocks and then burn them. The BISO-coated particles lose their graphite coating and become spheres of mixed uranium oxide consisting mostly of the ^{233}U isotope, and fission-product oxides. The TRISO-coated spheres lose their outer graphite coating, but the silicon carbide and inner graphite coating remain intact. The product then is a mixture of dense oxide spheres from BISO particles and less dense silicon carbide-coated graphite and UO_2 spheres from TRISO particles, both about 500 μm in diameter. The two kinds of particles are separated by elutriation with CO_2 gas, so that they can be processed separately with minimal mixing.

The residue of BISO particles is dissolved in mixed HNO_3 and HF and then separated by the Thorex solvent extraction process (Chap. 10) into a decontaminated ^{233}U-rich uranium fraction, a thorium fraction containing 1.9-year radioactive ^{228}Th, and fission-product wastes.

The residue of TRISO particles is crushed to expose the remaining uranium and fission products. These are then dissolved in nitric acid and separated by a simplified version of the Purex process (Chap. 10) into a decontaminated uranium fraction containing around 20 percent ^{235}U and fission-product wastes.

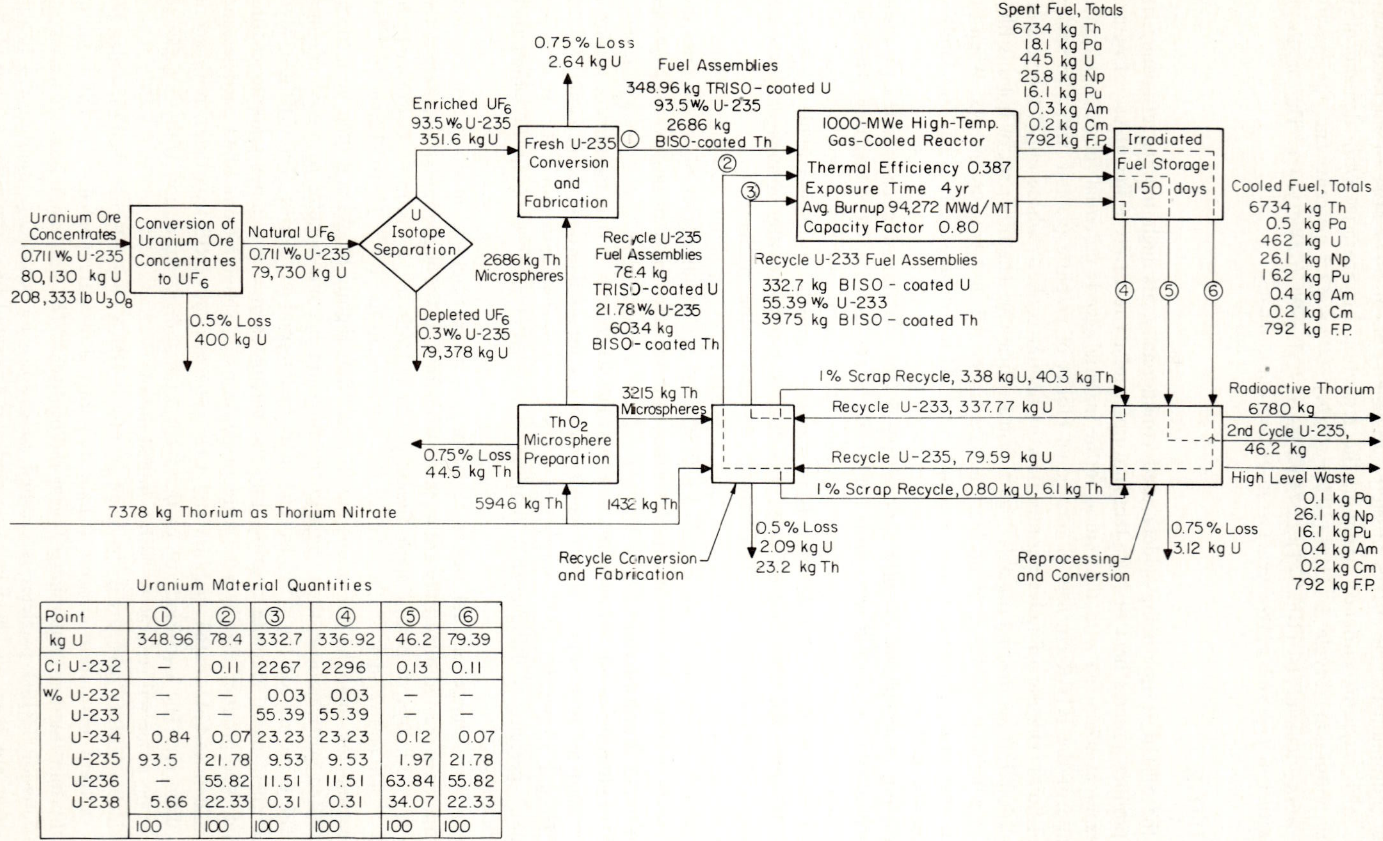

Point	①	②	③	④	⑤	⑥
kg U	348.96	78.4	332.7	336.92	46.2	79.39
Ci U-232	—	0.11	2267	2296	0.13	0.11
w/o U-232	—	—	0.03	0.03	—	—
U-233	—	—	55.39	55.39	—	—
U-234	0.84	0.07	23.23	23.23	0.12	0.07
U-235	93.5	21.78	9.53	9.53	1.97	21.78
U-236	—	55.82	11.51	11.51	63.84	55.82
U-238	5.66	22.33	0.31	0.31	34.07	22.33
	100	100	100	100	100	100

Figure 3.33 Fuel-cycle flow sheet for 1000-MWe HTGR fueled with thorium, enriched ^{235}U, once-recycled ^{235}U, and fully recycled ^{233}U. Basis: 1 year, 80 percent capacity factor.

In one possible fuel-cycle flow sheet for the HTGR, shown in Fig. 3.33, ^{233}U-rich uranium from the BISO particles and the 20 percent ^{235}U from the TRISO particles are recycled as part of the fissile fuel for a later HTGR fuel cycle. When this is done, some of the graphite fuel blocks are charged with TRISO-coated fully enriched ^{235}U and BISO-coated thorium (point 1, Fig. 3.33), others are charged with TRISO-coated 20 percent ^{235}U recycle uranium and BISO-coated thorium (point 2), and the rest are charged with BISO-coated ^{233}U-rich uranium and BISO-coated thorium (point 3). At the end of the cycle the spent fuel from the TRISO-coated fully enriched ^{235}U (point 4) is processed to recover uranium containing 20 percent ^{235}U to be recycled (point 2). The spent fuel from the TRISO-coated second-cycle ^{235}U (point 5) contains only 2 percent ^{235}U and is so highly contaminated by ^{236}U and fission products as to be discarded without reprocessing. The spent fuel from the BISO-coated recycle ^{235}U (point 6) and the BISO-coated thorium is processed to recover ^{233}U, to be recycled (point 3), and radioactive thorium, which is stored until 1.9-year ^{228}Th has decayed.

The quantities of uranium and thorium shown in the flow sheet Fig. 3.33 have been adapted from a study by Pigford [P2] of the fuel-cycle performance of the HTGR after a sufficient number of cycles have been operated to permit buildup of steady-state amounts of recycle ^{233}U and diluent ^{236}U. In earlier cycles the amounts of these two isotopes are lower.

The specific consumption of U_3O_8 and separative work in this HTGR cycle with ^{233}U recycle is compared in Table 3.16 with corresponding quantities for the LWR without or with plutonium recycle.

The HTGR with ^{233}U recycle thus consumes about the same amount of separative work as the LWR with plutonium recycle, but uses only 65 percent as much natural uranium.

7.4 The LMFBR

Figure 3.34 shows the quantities of materials to be supplied or processed for a 1000-MWe LMFBR having the characteristics of the Atomics International "follow-on" design [B3]. Because the LMFBR can be designed for a wide range of fuel-cycle conditions, the particular ones shown in this figure are only one example of how it is thought this type of reactor might perform when the first 1000-MWe units go into operation. Most of the quantities for this flow sheet were adapted from Pigford [P1], supplemented by data for the blankets from ERDA-1535 [E1].

The reactor contains three different fuel-bearing regions: core, axial blanket, and radial blanket. Fuel rods for the core and axial blanket consist of stainless steel tubing about $\frac{1}{4}$ in outside diameter, mounted vertically. The middle 4 ft of each tube is filled with core material, consisting of a mixture of 17 w/o PuO_2 and 83 w/o UO_2 made from depleted uranium

Table 3.16 Specific consumption of U_3O_8 and separative work, 80 percent capacity factor

	Reactor system		
	Light water	Light water	HTGR
Fissile feed	^{235}U	^{235}U and recycle Pu	^{235}U and recycle ^{233}U
Fertile feed	^{238}U	^{238}U	Th
Flow sheet figures	3.31	3.31 & 3.32	3.33
Annual consumption per MWe			
lb U_3O_8	439.2	320.2	208.3
kg SWU	107.0	72.1	70.8

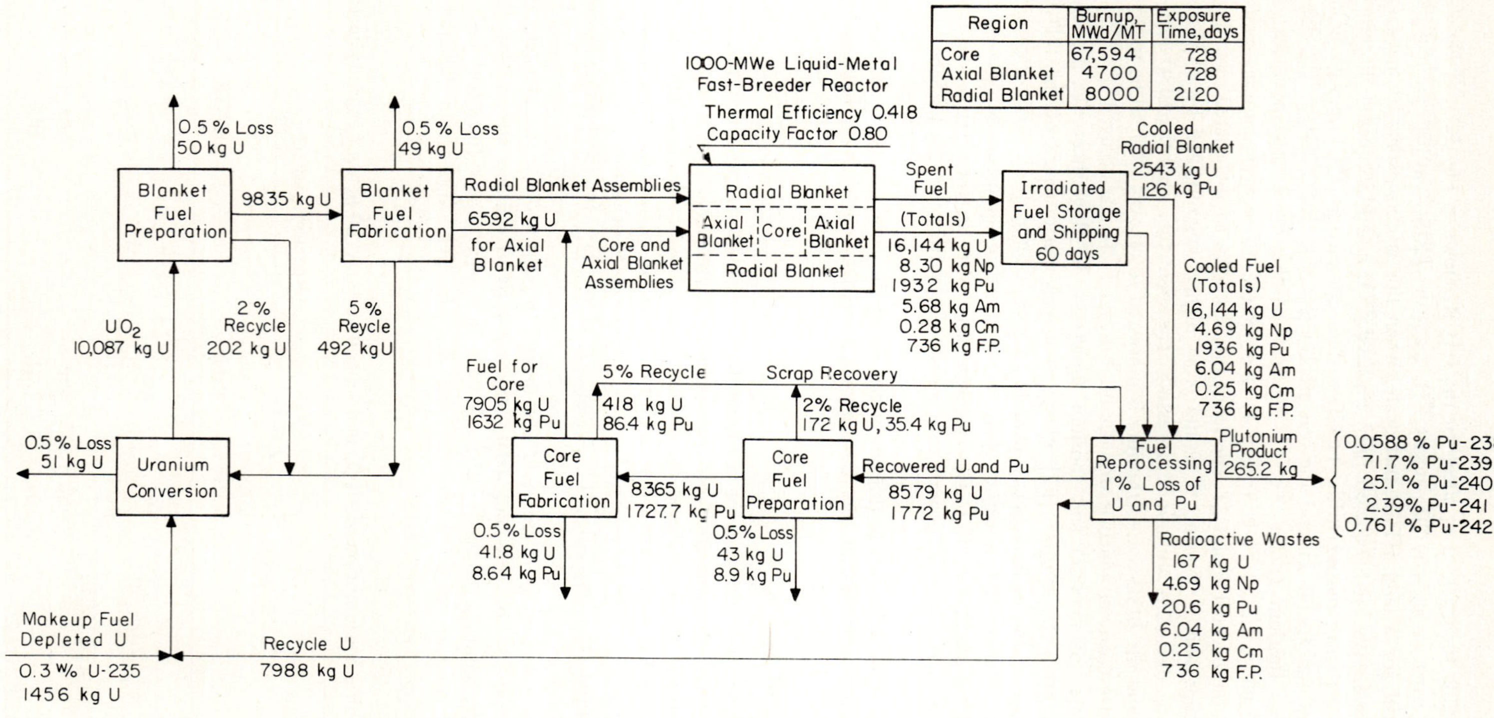

Figure 3.34 Fuel-cycle flow sheet for 1000-MWe LMFBR. Basis: 1 year, 80 percent capacity factor.

containing less than 0.3 w/o ^{235}U. Each tube, above and below the core material, contains axial blanket material, consisting of UO_2 also made from depleted uranium. Additional depleted uranium is contained in radial blanket rods that surround the rods containing core and axial blanket material.

Core and axial blanket rods remain in the reactor for an average exposure time of 728 days when the reactor is operated at 80 percent capacity factor. Radial blanket rods remain in the reactor for an average of 2120 days and contain a much lower Pu/U ratio when discharged. Although it would be possible to reprocess the radial blanket material separately from the core and axial blanket material, it is proposed that these be mixed in the proportions shown in Fig. 3.34, and reprocessed together. This makes the fission product and plutonium concentration in reprocessing plant feed lower than if core and axial blanket material were processed by itself, and reduces somewhat difficulties from high concentrations of radioactivity and fissile material. Nevertheless, the radioactivity and plutonium content of reprocessing plant feed is much higher than in the LWR flow sheets, Figs. 3.31 and 3.32. The high radioactivity is due partly to the higher burnup and higher specific power to which LMFBR fuel is exposed and partly to the shorter time assumed to elapse between discharge and reprocessing of LMFBR fuel compared with LWR fuel. For the LMFBR fuel it is assumed that fuel will be stored at the reactor for only 30 days to allow radioactivity to decay and then shipped, with only 60 days elapsing between discharge and reprocessing. The corresponding time assumed for the LWR is 150 days.

The shorter time allowed for LMFBR fuel to cool is desirable for economic reasons to reduce the amount of plutonium inventory outside of the reactor. The rate at which plutonium is discharged from the LMFBR is eight times as high as from the uranium-fueled LWR (Fig. 3.31) and twice as high as from the LWR with plutonium recycle (Fig. 3.32).

The net plutonium production of this LMFBR is 265.2 kg/year. This plutonium contains 71.7 w/o ^{239}Pu and much less of the higher isotopes ^{241}Pu and ^{242}Pu than plutonium from the LWR flow sheets Figs. 3.31 and 3.32. After the steady-state conditions of Fig. 3.34 have been attained, the LMFBR needs no external feed of fissile material. The only external feed is 1456 kg/year of depleted uranium, needed to supply the 1055 kg undergoing reaction in the core and blanket and the 401 kg assumed lost in the several fuel-cycle operations. This consumption of depleted uranium in the LMFBR is much lower than the consumption of natural uranium in the LWR or HTGR. This greatly reduced consumption of uranium is what makes development of the LMFBR so important in conserving uranium resources.

NOMENCLATURE

a	rate of change of reactivity with burnup
B	burnup, MWd/MT
B^2	geometric buckling, cm^{-2}
c	unit cost, \$/kg
C_1, etc.	constants defined by Eqs. (3.55) through (3.65)
D	neutron diffusion constant, cm
e	fuel-cycle cost, mills/kWhe
E	electric energy
f	fraction of fuel replaced
f'	fractional yield in fuel-cycle step
H	thermal energy
H	height of core
K	nuclear plant rated capacity, MWe
K	constant defined by Eq. (3.33)
L'	availability-based capacity factor

m	mass, kg
M	molecular weight
n	number of fuel zones
n	neutron density, per cm^3
p	resonance escape probability
P	nonleakage probability
q	poisoning ratio
q'	poisoning ratio including fission products from ^{238}U
r	radial distance, cm
r	Westcott epithermal flux factor
R	outer radius of core, cm
t	time
T	absolute temperature, K
T	years between refuelings
T'	years between start and end of irradiation
$\hat{T}$	298.2 K
U	mass of uranium, MT
v	volume fraction
$\bar{v}$	average speed of Maxwell-Boltzmann neutrons
w	weight fraction fissioned
y	cost of money before taxes, per year
z	distance from reactor midplane, cm
Z	height of core, cm
Z	fuel-cycle cost, \$/lot
α	ratio of capture to fission cross section
γ	$1 - \kappa_{49}$
ϵ	fuel enrichment, weight percent ^{235}U
ϵ	fast fission factor
η	thermal efficiency
η	number of neutrons produced per neutron absorbed
θ	flux time (fluence), cm^{-2}
κ	constant, defined following Eq. (3.48)
λ	radioactive decay constant, s^{-1}
ν	number of neutrons produced per fission
ρ	reactivity
ρ	density, g/cm^3
σ	neutron absorption cross section
τ	Fermi age, cm^2
ϕ	thermal-neutron flux, $n/(cm^2 \cdot s)$
ψ	ratio of thermal-neutron flux to thermal-neutron flux in fuel

Superscripts

$'$	beginning of irradiation
$'$	fabrication
$''$	end of irradiation
$''$	reprocessing
$'''$	conversion
*	reference design condition
0	fuel at the beginning of irradiation

Subscripts

a	absorption
E	control absorbers
F	fission products with low cross section
Fa	fabrication
g	fertile nuclide
h	higher isotope, ^{236}U, ^{240}Pu, or ^{242}Pu
i	nuclide
i	cycle number
I	Inconel
j	fuel-cycle step
j	lattice region
k	sublot number
m	fissile nuclide
M	single fissile nuclide in fuel charged
n	number of cycle at end of irradiation
P	nonfuel neutron absorber
P	plutonium
R	reprocessing
S	samarium and other high-cross-section fission products
Sh	shipping
th	thermal energy
U	uranium
U′	uranium charged
U″	uranium discharged
W	water
Xe	^{135}Xe
Z	zircaloy

REFERENCES

A1. American Electric Power Co.: Donald C. Cook Nuclear Power Plant, Preliminary Safety Analysis Report, New York, 1968.

B1. Barry, R. F.: LEOPARD–A Spectrum Dependent Non-Spatial Depletion Code for the IBM-7094, Westinghouse Electric Corp., Report WCAP-3269-26, Sep. 1963.

B2. Beaudreau, J. J.: "Development and Evaluation of the Computer Code CELL," thesis submitted to the Department of Nuclear Engineering, Massachusetts Institute of Technology, Cambridge, Mass., in partial fulfillment of requirements for the M.S. degree, 1967.

B3. Blomeke, J. O., C. W. Kee, and J. P. Nichols: "Projections of Radioactive Wastes to be Generated by the U.S. Nuclear Power Industry," Report ORNL-TM-3965, Feb. 1974.

C1. Critoph, E.: "Effective Cross Sections for U-235 and Pu-239," Report CRRP-1191, Mar. 1964.

D1. Dollard, W. J., and L. E. Strawbridge: "Nuclear Performance of Power Reactor Cores," Report TID-7672, 1963, p. 328.

E1. Energy Research and Development Administration: "Final Environmental Statement, Liquid Metal Fast Breeder Reactor Program," Report ERDA-1535, Dec. 1975.

F1. Ferguson, D. R., and D. M. ver Planck: "SIMULATE, Reactor Simulator Code," internal document, Yankee Atomic Electric Co., Westboro, Mass., 1972.

K1. Kearney, J. P.: "Simulation and Optimization Techniques for Nuclear In-Core Fuel Management Decisions," thesis submitted to the Department of Nuclear Engineering, Massachusetts Institute of Technology, Cambridge, Mass., in partial fulfillment of requirements for the Ph.D. degree, 1973.
P1. Pigford, T. H., and K. P. Ang: *Health Phys.* **29**: 451 (1975).
P2. Pigford, T. H., R. T. Cantrell, K. P. Ang, and B. J. Mann: "Fuel Cycle for 1000-MW High Temperature Gas Cooled Reactor," Report EPA 68-01-0561, Mar. 1975.
R1. Rieck, T. A.: "The Effect of Refueling Decisions and Engineering Constraints on the Fuel Management for a Pressurized Water Reactor," thesis submitted to the Department of Nuclear Engineering, Massachusetts Institute of Technology, Cambridge, Mass., in partial fulfillment of requirements for the Ph.D. degree, 1974.
W1. Walker, W. H.: "The Effect of New Data on Reactor Poisoning by Non-Saturating Fission Products," Report AECL-2111, Nov. 1964.
W2. Watt, H. Y.: Personal communication to M. Benedict, July 1973.
W3. Westcott, C. H.: "Effective Cross Section Values for Well-Moderated Thermal Reactor Spectra," 3d ed. (corr.), Report AECL-1101, Dec. 1964.

PROBLEMS

3.1 The 1060-MWe PWR discussed in Sec. 3.4 is to be operated with steady, four-zone modified scatter refueling, with 1.0 year between successive refuelings. The availability-based capacity factor is 0.8 and the refueling downtime is 0.15 year. The reactor fuel inventory is 88.961 MT heavy metal.

(*a*) What burnup should the fuel experience?

(*b*) What weight percent ^{235}U should reload fuel have to permit this reactivity-limited burnup and operating cycle?

3.2 Evaluate the fuel-cycle cost in mills per kilowatt-hour of electricity for lot 1B of fuel charged to the 1060-MWe PWR power plant described in Sec. 4, using the material quantities, burnup increments, unit costs, and transaction times given in that section. The unit costs of uranium charged and recovered are as follows:

Uranium	Charged	Recovered
% ^{235}U	2.25	0.663
Unit cost, \$/kg U	634.06	75.63

3.3 Dimensions, material content, and nuclear parameters of the Douglas Point CANDU heavy-water, natural uranium reactor are summarized in Table 3.17. Effective properties of nuclides in this reactor are listed in Table 3.18. The neutron balance for this reactor when charged with natural uranium fuel is given in Table 3.19.

(*a*) Assuming that the fuel in this reactor is irradiated in a uniform neutron flux, show that the reactivity drops to zero at a flux time of 1.4 n/kb.

(*b*) Find the burnup of the fuel in megawatt-days per metric ton at the flux time of 1.4 n/kb.

Table 3.17 Properties of heavy-water reactor

Power	
Thermal	728 MW
Net electric	200 MW
Fuel	Natural UO_2
Enrichment	0.7206 atom percent ^{235}U
Fuel element	Zircaloy-2 tube, 0.038 cm thick, 96% filled with UO_2 1.397 cm in diam. by 500 cm long.
Fuel channel	Bundle of 19 fuel elements centered in zircaloy-2 pressure tube 8.255 cm ID 0.414 cm thick.
Coolant	D_2O, mean temp. 273.1°C, mean pressure 139 bar
Moderator	D_2O, mean temp. 80°C, mean pressure 1 bar
Average neutron temp.	85.4°C
Lattice spacing	23.4 cm square pitch
Number of fuel channels	308
Core dimensions	Radius, $R = 231.7$ cm; length, $Z = 500$ cm
Effective core dimensions	Radius, $R' = 288.6$ cm; length, $Z' = 504.2$ cm

Region	Material	Volume, cm^3/cm	Density, g/cm^3	Relative thermal flux
Fuel	UO_2	27.957	10.2	1.000
Fuel sheath	Zircaloy-2	4.079	6.5	1.003
Pressure tube	Zircaloy-2	11.275	6.5	1.265
Insulating gap	Treat as vac.	16.263	0	1.31
Calandria tube	Zircaloy-2	4.596	6.5	1.357
Coolant	D_2O	21.484	0.845	1.1
Moderator	D_2O	461.89	1.076	1.823

Inventory	^{235}U, 275.42 kg Uranium, 38,710 kg
Fast-fission factor, ϵ	1.0173
Nonleakage probability	
Fission-to-resonance, P_1	0.98917
^{238}U resonance-to-thermal, P_2	0.99537
Thermal leakage factor, DB^2	1.085×10^{-4} cm^{-1}
Resonance escape probability, p	0.8954

Table 3.18 Effective properties of nuclides for thermal neutrons in heavy-water reactor

Nuclide	Subscript	Absorption cross section σ_a, b	Neutrons produced per fission ν	Neutrons produced per neutron absorbed η	Ratio of capture to fission cross section α	Poisoning ratio of high-cross-section fission products q
^{235}U	25	567.0	2.43†	2.0646	0.1769	0.0541
^{236}U	26	19.32				
^{238}U	28	2.3078	2.79‡	2.3432‡	0.1907‡	0.0683
^{239}Pu	49	1113.8	2.87†	1.9845	0.4638	0.0549
^{240}Pu	40	771.1				
^{241}Pu	41	1338.7	3.06†	2.2230	0.3765	0.0547
^{242}Pu	42	82.88				
F.P.P.§	F	55.3				
Zircaloy-2	Z	0.1782				
D_2O¶	D	0.00191				

†These values of ν are from [C1]. They are used here because effective cross sections are from [C1] also. These values of ν differ slightly from App. C.

‡In fast fission.

§Pairs of fission products with cross sections less than 10,000 b. Assumed yield, one pair per fission.

¶0.2% H_2O.

Table 3.19 Neutron balance for reference design of heavy-water reactor†

Item Process		Number of neutrons
1. Production of fast neutrons from fission of ^{235}U	η_{25}	2.0646
2. Capture of fast neutrons by ^{238}U to produce ^{239}Pu	$\dfrac{\alpha_{28}\eta_{25}(\epsilon-1)}{\nu_{28}-1-\alpha_{28}}$	0.0043
3. Fission of ^{238}U by fast neutrons	$\dfrac{\eta_{25}(\epsilon-1)}{\nu_{28}-1-\alpha_{28}}$	0.0223
4. Production of fast neutrons from fission of ^{238}U	$\nu_{28}(3)=\dfrac{\nu_{28}\eta_{25}(\epsilon-1)}{\nu_{28}-1-\alpha_{28}}$	0.0623
5. Net production of fast neutrons	$(1)-(2)-(3)+(4)=\eta_{25}\epsilon$	2.1003
6. Neutron leakage during moderation from fission to ^{238}U resonance energy	$\eta_{25}\epsilon(1-P_1)$	0.0227
7. Neutrons moderated to ^{238}U resonance energy	$(5)-(6)=\eta_{25}\epsilon P_1$	2.0776
8. Neutrons captured in ^{238}U resonance	$\eta_{25}\epsilon P_1(1-p)$	0.2173
9. Neutrons escaping ^{238}U resonance capture	$(7)-(8)=\eta_{25}\epsilon P_1 p$	1.8603
10. Neutron leakage during moderation from ^{238}U resonance to thermal energy	$\eta_{25}\epsilon P_1 p(1-P_2)$	0.0086
11. Production of thermal neutrons	$(9)-(10)=\eta_{25}\epsilon P_1 p P_2$	1.8517 ←
Consumption of thermal neutrons by		
12. Thermal leakage	$\dfrac{DB^2}{N_{25}^*\sigma_{25}v_U}$	0.0392
13. ^{235}U fission	$\dfrac{1}{1+\alpha_{25}}$	0.8497
14. ^{235}U capture	$\dfrac{\alpha_{25}}{1+\alpha_{25}}$	0.1503
15. Absorption by ^{238}U	$\dfrac{N_{28}^*\sigma_{28}}{N_{25}^*\sigma_{25}}$	0.5608
16. Absorption by zircaloy	$\dfrac{N_Z\sigma_Z(\psi v)_Z}{N_{25}^*\sigma_{25}v_U}$	0.0724
17. Absorption by heavy water	$\dfrac{N_D\sigma_D(\psi v)_D}{N_{25}^*\sigma_{25}v_U}$	0.0204
18. Absorption by ^{135}Xe and Sm	$q'_{25}=q_{25}+q_{28}\dfrac{\eta_{25}(\epsilon-1)}{\eta_{28}-1}$	0.0559
19. Absorption by control poison	$\dfrac{N_E\sigma_E}{N_{25}^*\sigma_{25}}$	0.1030
20. Total consumption of thermal neutrons		1.8517 ←

†Basis: One thermal neutron absorbed by ^{235}U.

CHAPTER

FOUR

SOLVENT EXTRACTION OF METALS

1 APPLICATIONS

Solvent extraction is the industrial-scale method preferred for purifying natural uranium, separating zirconium from hafnium [H3, H4, P1, S2], separating natural thorium from rare earths, and separating fissionable material, fertile material, and fission products in spent reactor fuel [H1]. For example, solvent extraction processes have been developed for separating ^{235}U from fission products [C8, C9, U1], and for separating uranium, plutonium, and fission products [G3]. The successful application of solvent extraction to these separations of importance in nuclear technology has stimulated its application to other metals, for example, to separating rare earths from one another [I1], cobalt from nickel [G5, R3], and tantalum from niobium [W3]. Table 4.1 summarizes some of the applications of solvent extraction to separating metals on an industrial scale and gives the solvents used and agents added to the aqueous phase to promote extraction or separation.

The value of solvent extraction arises from the ease with which it lends itself to multistage operation without increased consumption of heat or chemicals. This makes solvent extraction particularly useful when either extreme purification is necessary or when the metals are so similar in their properties that a single precipitation or crystallization would not give the requisite degree of separation.

Ion exchange is another method that lends itself to multistage operation. Generally speaking, solvent extraction is preferable when large amounts of metals are to be separated, and ion exchange is preferable for small quantities or low concentrations or for separating the alkalies or alkaline earths, to which solvent extraction is not readily applicable.

2 EXTRACTABLE METAL–ORGANIC COMPLEXES

For a metallic element to be extractable by an organic solvent immiscible with water, it appears to be necessary that the element be capable of forming an organic-soluble, electrically neutral complex compound with the solvent or with an added complexing agent. The compound

Table 4.1 Examples of solvent extraction of metals

Metals separated	Organic phase	Agent added to aqueous phase
U from ores	Trioctyl and tridecyl amines in isodecanol and kerosene	H_2SO_4
	Di(2-ethylhexyl) phosphoric acid in kerosene	H_2SO_4
	Tributyl phosphate in kerosene	HNO_3
Th from rare earths	Tributyl phosphate in kerosene	HNO_3
Zr from Hf	HCNS in methyl isobutyl ketone	NH_4CNS
	Tributyl phosphate in kerosene	HNO_3,$NaNO_3$
^{235}U from fission products	Methyl isobutyl ketone	$Al(NO_3)_3$
U, Pu, fission products	Tributyl phosphate in kerosene	HNO_3,$Al(NO_3)_3$
Ni,Co	Methyl isobutyl ketone	NH_4CNS
Nb,Ta	Methyl isobutyl ketone	HF, HCl

between uranyl nitrate and tri-*n*-butyl phosphate (TBP) is an example of the first kind, the compound between thorium nitrate and salicylic acid, of the second.

Formation of these extractable complexes involves coordination bonds with the metal cation, i.e., the sharing of electrons from the complexing agent to complete previously unfilled orbits of the cation. The alkalies and alkaline earths are not easily capable of forming such compounds because they have no empty electron orbits, and hence cannot be readily extracted with organic solvents immiscible with water. On the other hand, elements of the transition groups, such as the rare earths, uranium and the other actinides, iron, nickel, and cobalt, form coordination compounds with ease and are readily extracted by organic solvents immiscible with water.

The complex compounds formed by the metal cations in solvent extraction systems are illustrated first by the formation of complex compounds between cations and neutral molecules in aqueous solution. An example of an equilibrium reaction involving such formation is [P2]

$$Ag^+ + 2NH_3 \rightleftharpoons Ag(NH_3)_2^+$$

The complex compound $Ag(NH_3)_2^+$ is easily destroyed, i.e., the reaction is reversed, by adding acid to remove dissolved NH_3 or by adding a halide ion to precipitate silver halide.

Examples of stronger complexes formed with anions are [P2]

$$Fe^{2+} + 6CN^- \rightleftharpoons Fe(CN)_6^{4-}$$

$$Pt^{2+} + 4Cl^- \rightleftharpoons PtCl_4^{2-}$$

Cations that form such complexes are characterized by a *coordination number*, i.e., the number of complexing groups that are attached to the cation. Cations in the above reactions exhibit coordination numbers of 2 for Ag^+, 6 for Fe^{2+}, and 4 for Pt^{2+}.

Compounds with less than full coordination are also formed, such as

$$Th^{4+} + NO_3^- \rightleftharpoons ThNO_3^{3+}$$

and $$Th^{4+} + 2NO_3^- \rightleftharpoons Th(NO_3)_2^{2+}$$

with the amount of $Th(NO_3)_2^{2+}$ complex increasing with NO_3^- concentration. In a cation-anion complex with less than full coordination bonding from the anion, the cation may become fully coordinated by adding water molecules to the complex. Complexes formed by anions of weak acids are usually more stable than complexes formed by anions of strong acids.

When an organic ion or molecule is able to form coordination bonds with a metallic cation in more than one place in the organic molecule, an especially stable complex called a *chelate* compound is formed. Among the effective chelating agents are molecules that contain two ketone structures, such as the 1,3-diketones with the structure

```
–C–CH₂–C–
 ‖      ‖
 O      O
```

A molecule containing the diketone group forms a heterocyclic ring coordination compound with a metal cation by losing a proton and attaching itself to the cation through both oxygen atoms. The number of chelating molecules added is thus half the coordination number of the cation in the complex. A chelating agent of common use in laboratory extraction is the 1,3-diketone thenoyltrifluoracetone (TTA), with the enol form

```
HC–CH
‖   ‖
HC  C–C–CH=C–CF₃
 \ /  ‖     |
  S   O     OH
```

Representing the organic chelating compound by HK, the overall reaction involved in the chelate extraction of a metal in the ionic form M^{n+} is

$$M^{n+}(aq) + n\text{HK}(o) \rightleftharpoons \text{MK}(o) + n\text{H}^+(aq)$$

with an equilibrium constant K given by

$$K = \frac{[\text{MK}(o)]\,[\text{H}^+(aq)]^n}{[\text{M}^{n+}(aq)]\,[\text{HK}(o)]^n} \tag{4.1}$$

In these equations aq denotes the aqueous phase and o the organic. Quantities in brackets are the activities at equilibrium, i.e., the molar concentration of the indicated species times its activity coefficient. For example:

$$Th^{4+} + 4HK \rightleftharpoons ThK_4 + 4H^+$$

and $$UO_2^{2+} + 2HK \rightleftharpoons UO_2K_2 + 2H^+$$

As indicated by Eq. (4.1), the relative concentration of the metal chelate in the organic phase at equilibrium decreases with increasing concentration of hydrogen ion in the aqueous phase. When an aqueous solution containing extractable metal ions is contacted with an immiscible organic carrier, such as benzene, containing dissolved chelating agent, the chelating compound must dissolve in the aqueous phase, ionize, and react with the metal ion, and the metal chelate then dissolves in the organic phase. The low solubility of the chelates and their slow rates of formation limit the industrial-scale application of chelate separation [C8, S4].

More rapid extracting reactions result from the formation of relatively loose nonchelating complexes with organic molecules. A widely used organic complexing agent for the extraction of the actinide elements thorium, uranium, neptunium, and plutonium is TBP, which probably forms bonds by the electron from the phosphoryl oxygen atom in the structure [S4]

$$(BuO)_3P^+\text{–}O^-$$

where BuO denotes the butoxy group. Examples of overall extracting reactions involving covalent bonds with TBP are

$$UO_2^{2+}(aq) + 2NO_3^-(aq) + 2TBP(o) \rightleftharpoons UO_2(NO_3)_2 \cdot 2TBP(o)$$

$$Pu^{4+}(aq) + 4NO_3^-(aq) + 2TBP(o) \rightleftharpoons Pu(NO_3)_4 \cdot 2TBP(o)$$

The above reactions are shifted to the right, thereby increasing the relative amount of metal cation in the organic phase, by increasing the concentration of uncombined TBP in the organic phase and by increasing the concentration of aqueous nitrate ion. The latter is accomplished by adding a *salting* agent such as HNO_3 or $Al(NO_3)_3$. Nitric acid also forms a hydrogen bonded complex [S2] with TBP and extracts according to the overall reaction

$$H^+(aq) + NO_3^-(aq) + TBP(o) \rightleftharpoons HNO_3 \cdot TBP(o)$$

Similarly, TBP promotes solubility of water in the organic phase.

Subsequent chapters include discussion of TBP solvent extraction to purify uranium (Chap. 5) and thorium (Chap. 6) and to separate and recover actinides in irradiated reactor fuel (Chap. 10). Also discussed in Chap. 5 is a third type of organic extractant, consisting of an organic-soluble acid or base, of moderately high molecular weight, which extracts metals as simple or complex organic-soluble salts [P2], appropriately characterized as a "liquid ion exchanger." Examples are di(2-ethylhexyl) phosphoric acid and trioctylamine, both used in extracting uranium from ore leach liquors (cf. Table 5.14).

3 SOLVENT EXTRACTION PRINCIPLES

When an aqueous solution of an extractable component is brought into equilibrium with an immiscible solvent for the component and the two phases are then separated, the component will be found distributed between the two phases. This distribution may be characterized by the distribution coefficient D, defined as

$$D = \frac{\text{concentration of component in organic phase}}{\text{concentration of component in aqueous phase}}$$

for the two phases leaving the equilibrium contactor. The distribution coefficient is a function of the nature of the solvent, the temperature, and the equilibrium compositions of the aqueous and organic phases, but is independent of the amount of either phase.

The fraction of component initially present in the aqueous feed that is extracted in one stage of equilibrium contacting depends on the relative volumes of aqueous and solvent phases. Nomenclature for deriving an equation for the fraction extracted is given in Fig. 4.1.

The material balance on the extractable component may be expressed as

$$Fz = Fx + Ey \tag{4.2}$$

The definition of the distribution coefficient D is

$$D = \frac{y}{x} \tag{4.3}$$

Therefore

$$y = \frac{Dz}{1 + ED/F} \tag{4.4}$$

and the fraction extracted is

$$\rho = \frac{Ey}{Fz} = \frac{ED/F}{1 + ED/F} \tag{4.5}$$

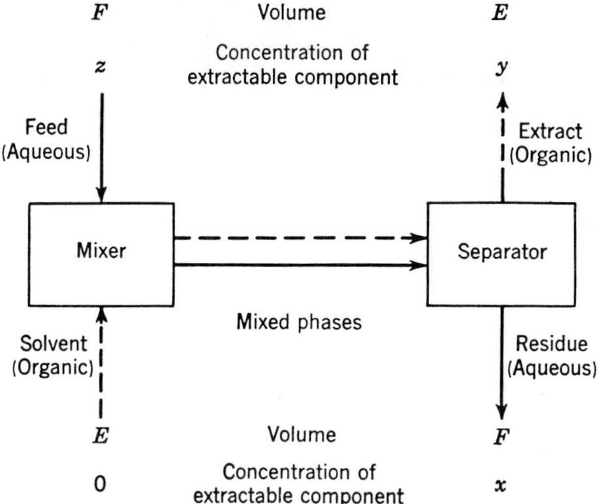

Figure 4.1 Single solvent extraction stage.

The ratio of solvent to feed needed for a given fraction extracted is

$$\frac{E}{F} = \frac{\rho}{D(1-\rho)} \tag{4.6}$$

The fraction extracted therefore becomes greater, the greater the ratio E/F of solvent to feed, but an infinite amount of solvent is needed for complete extraction in a single contact.

Solvent leaving this equilibrium contactor is capable of extracting more of the metallic component from additional aqueous feed, because the feed concentration z is greater than the concentration x in the aqueous phase with which this solvent is in equilibrium. It is therefore possible to reduce the amount of solvent needed for a given fraction extracted by using multiple contact between solvents and aqueous phases in a countercurrent cascade, as illustrated in Fig. 4.2. As the number of stages is increased indefinitely, the organic extract approaches equilibrium with the aqueous feed, so that in the limit

$$y_{\max} = Dz \tag{4.7}$$

where D is the distribution coefficient for the feed stage.

For a given ratio E/F, the $y_{\max}$ from an infinite number of contacting stages results in the maximum recovery $\rho_{\max}$, i.e.,

$$\rho_{\max} = \frac{Ey_{\max}}{Fz} = \frac{ED}{F} \tag{4.8}$$

Alternatively, for a specified recovery ρ, the minimum ratio $(E/F)_{\min}$ to achieve a specified recovery ρ occurs for an infinite number of stages and is given by

$$\left(\frac{E}{F}\right)_{\min} = \frac{\rho z}{y_{\max}} = \frac{\rho}{D} \tag{4.9}$$

Thus with a large number of stages it becomes possible to approach complete extraction, i.e., $\rho = 1$, with a finite amount of solvent. With a finite number of stages, the relative amount E/F of solvent for a given fraction extracted lies between the minimum, given by Eq. (4.9), and the single-stage maximum value given by Eq. (4.6).

Because the solvent (Fig. 4.2) ordinarily is valuable, it is desirable to wash the extracted component out of it and to recycle the solvent for reuse. A flow sheet of this type is shown in

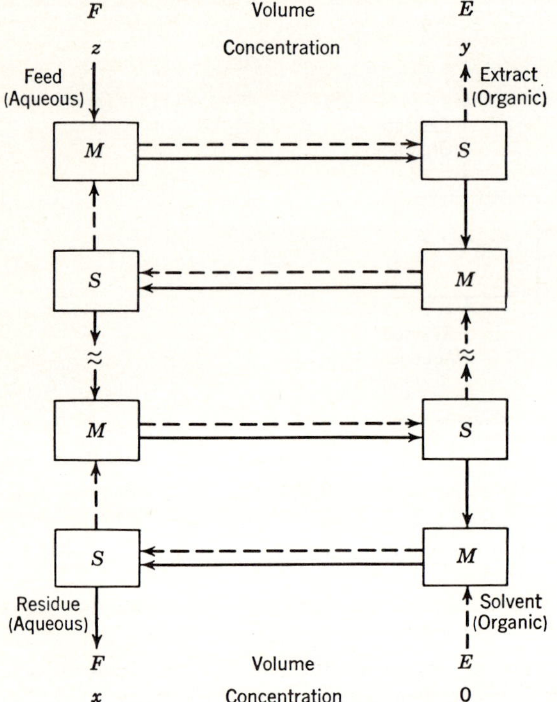

Figure 4.2 Multistage countercurrent solvent extraction. M, mixer; S, settler.

Fig. 4.3. Here the group of solvent extraction stages used to extract the desired component from feed has been designated as the extracting section, and the group used to wash this component back into the aqueous phase as the stripping section. In practice, each section may be either batteries of mixers and separators as shown in Fig. 4.2, in which phases are alternatively mixed and separated, or countercurrent columns, in which the two phases flow past one another in continuous contact and continuously exchange material.

A flow sheet similar to Fig. 4.3 could be used to extract uranium from sulfuric acid-leach solutions with organic amines, as illustrated in Fig. 5.9.

When more than one component of the feed is extractable, the flow sheet of Fig. 4.3 is not capable of producing any one component in pure form, because the organic phase leaving the extracting section will carry some of every component with it. To separate the most extractable component in relatively pure form, it is necessary to add an additional scrubbing section, as shown at the top of Fig. 4.4. The purpose of this scrubbing section is to scrub all but the most extractable component from the organic phase leaving the extracting section. This section functions somewhat like the enriching section of a fractional distillation column and provides partial reflux to wash back the components not wanted in the product. An operation of this kind, in which two or more extractable components are separated by distribution between two counterflowing solvents, is called fractional extraction.

Separation of two components by fractional extraction is possible when the distribution coefficient of one component D_i differs from that of the other component D_j.

The concentrations of component i in organic phase y_i and aqueous phase x_i leaving a stage are related by

$$y_i = D_i x_i \tag{4.10}$$

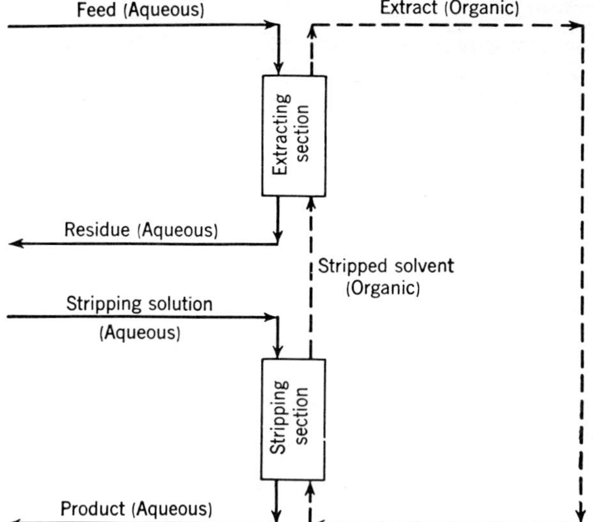

Figure 4.3 Solvent extraction flow sheet with recycle of solvent.

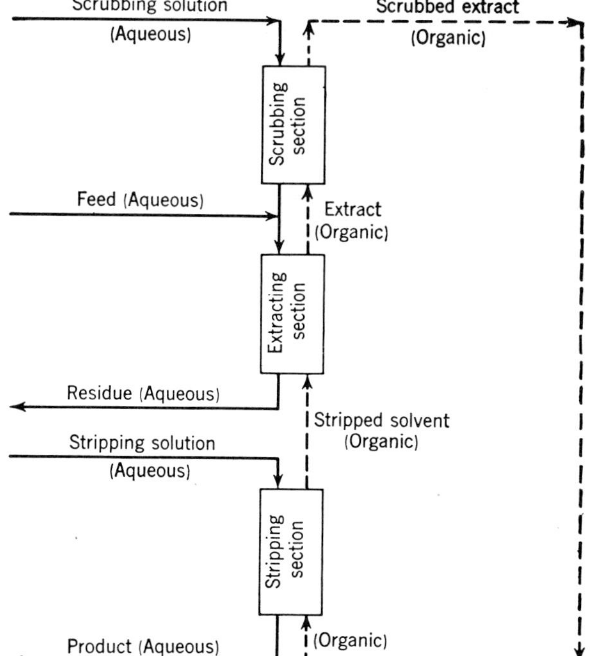

Figure 4.4 Flow sheet for fractional extraction of two extractable components.

and a similar equation holds for component j:

$$y_j = D_j x_j \tag{4.11}$$

The ratio of concentrations in the organic phase is related to the ratio of concentrations in the aqueous phase by

$$\frac{y_i}{y_j} = \frac{D_i x_i}{D_j x_j} \tag{4.12}$$

Thus, separation is possible when $D_i/D_j \neq 1$. The ratio of distribution coefficients is a measure of the ease or difficulty of a separation by fractional extraction and is known as the separation factor α:

$$\alpha = \frac{y_i/x_i}{y_j/x_j} = \frac{D_i}{D_j} \tag{4.13}$$

A flow sheet like Fig. 4.4 has been used to separate uranium from neutron-absorbing impurities (Chap. 5), and zirconium from hafnium (Chap. 7), by fractional extraction of an aqueous nitrate solution with an organic solution of TBP in kerosene.

For every additional extractable component to be separated in pure form, two additional sections are required, one for scrubbing and the other for stripping. As an example, Fig. 4.5 shows the flow sheet used in the Purex process to extract pure uranium and pure plutonium from fission products and to separate them from each other by fractional extraction between aqueous phases and TBP in kerosene.

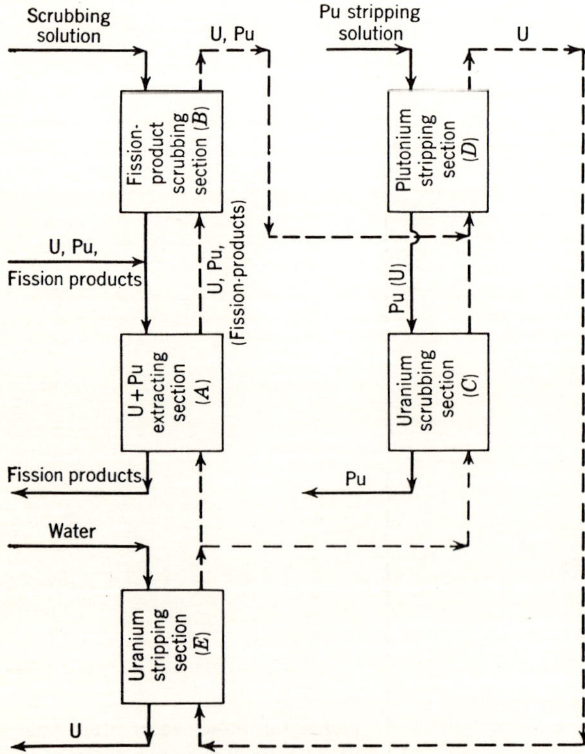

Figure 4.5 Fractional extraction of mixture of plutonium, uranium, and fission products. Solid line, aqueous; broken line, organic.

Here, sections C and D have been added to the flow sheet of Fig. 4.4 in order to separate the uranium and plutonium present in the extract leaving section B by fractional extraction.

4 DISTRIBUTION COEFFICIENTS

Control of values of distribution coefficients is one of the most important factors in achieving successful separations by solvent extraction. In simple extraction, without fractionation (as in Figs. 4.1 and 4.2), a high value of the distribution coefficient is desirable because the volume of solvent required is then small. In simple stripping, without fractionation (as in the stripping section of Fig. 4.3), a low value of the distribution coefficient is desired because the volume of stripping solution is then small and the product concentration high. In fractional extraction, when a separation is to be made between two extractable components, the ratio of their distribution coefficients should differ from unity by as much as possible. In addition, in this case, it is desirable that the geometric mean of the distribution coefficients not depart greatly from unity, because the optimum flow ratio of aqueous to organic phases is given approximately by

$$\left(\frac{S}{E}\right)_{\text{opt}} = \sqrt{D_1 D_2} \tag{4.14}$$

and most fractional extraction contactors operate best when this flow ratio is near unity.

The principal factors that affect the numerical value of the distribution coefficient are

1. Element being extracted
2. Oxidation-reduction potential of aqueous phase
3. Nature of solvent
4. Concentration of complexing agent
5. Concentration of salting agent
6. Hydrogen ion concentration in aqueous phase

4.1 Element Being Extracted

One of the reasons that solvent extraction is so successful in separating and purifying certain elements is that the distribution coefficients of different elements between certain solvents and aqueous solutions differ enormously. Table 4.2 lists distribution coefficients observed by Furman [F2] between diethyl ether and aqueous nitrate solutions. The much higher distribution coefficient for uranium is the reason for the successful use of diethyl ether in purifying uranium.

4.2 Oxidation-Reduction Potential

Ions of different valences of a metal behave like different elements with respect to extractability. The difference between Ce^{3+} and Ce^{4+} in Table 4.2 is one example. Another is afforded by Pu^{4+} and $Pu^{VI}O_2{}^{2+}$, which are readily extracted by TBP in kerosene, whereas Pu^{3+} has a very low distribution coefficient [G3]. Consequently, by adjusting the oxidation-reduction potential of the aqueous phase to control the proportion of an element in different valence states, it is possible to vary its distribution coefficient between wide limits. This is the means by which plutonium is stripped from aqueous solutions containing plutonium and uranium in sections C and D of Fig. 4.5 illustrating the Purex process. Addition of a reducing

Table 4.2 Distribution coefficients of various ions between aqueous and diethyl ether solutions[†]

	Distribution coefficient	
Element	10 N NH_4NO_3, 0.8 N HNO_3 ‡	12.2 N $Ca(NO_3)_2$, 0.8 N HNO_3 ‡
Al	(0.001)	(0.001)
As	0.007	0.048
Ba	(0.0005)	(0.0005)
Bi	0.0003	0.007
B	0.01	0.035
Cd	(0.00001)	(0.00001)
Ca	0.0005	0.0005
Ce^{3+}	(0.00005)	(0.003)
Ce^{4+}	–	27
Cr	(0.0001)	(0.0001)
Co	(0.0001)	(0.0001)
Cu	0.0002	0.0004
Dy	0.0003	0.002
Gd	0.00001	0.00076
In	(0.0004)	0.0003
Fe	0.0005	0.001
Pb	(0.0002)	(0.0002)
Li	0.0001	0.0002
Mg	0.0001	0.0002
Mn	(0.0001)	(0.0001)
Hg	(0.0001)	(0.0001)
Ni	(0.0001)	(0.0001)
K	0.0002	0.002
Ra	(0.00025)	(0.00025)
Re	(0.015)	(0.015)
Na	(0.0001)	(0.0001)
Sr	(0.0008)	(0.0008)
Tl	(0.0005)	(0.0005)
Th	0.001	
V^{5+}	0.0019	0.040
U	1.31	165
Zn	(0.0005)	(0.0005)
Zr	0.001	

[†]Values in parentheses are upper limits set by the sensitivity of analytical methods.

‡Salting agents.

agent such as a ferrous salt to the aqueous stripping solution entering section C reduces plutonium to Pu^{3+} and renders it readily back extracted into water. Uranium remains as UO_2^{2+} in the organic phase.

4.3 Nature of Solvent

Solvents differ greatly in their ability to extract compounds of metals from aqueous solution. For example, uranyl nitrate is strongly extracted by diethyl ether, hexone, TBP, and many other oxygenated organic solvents, but it is not extracted at all by benzene, kerosene, or other hydrocarbons, in the absence of complexing agents. Extractability of a salt by an organic solvent apparently requires that an uncharged coordination complex be formed between the solvent and the salt. As discussed in Sec. 4.2, this is possible when the solvent contains oxygen, nitrogen, or other electron-donor elements and when the metal of the salt is one of the transition elements with unfilled inner electron orbits capable of sharing electrons with the solvent molecules. Such compounds are not formed with saturated hydrocarbon solvents.

4.4 Complexing Agents

Complexing agents may be added to an extraction system either to increase or to decrease the distribution coefficient of a metallic component between aqueous solution and kerosene. For example, $Zr(NO_3)_4$ is not extracted at all from aqueous solution by kerosene. However, when TBP is added to the kerosene, a compound of probable composition [H4] $Zr(NO_3)_4 \cdot 2TBP$ is formed that is readily extracted. As an example of a complexing agent that reduces extraction, the effect of adding fluoride ion to the above system may be considered. The more stable, inextractable complex ion ZrF_6^{2-} is then formed according to the reaction

$$Zr(NO_3)_4 \cdot 2TBP + 6F^- \rightleftharpoons ZrF_6^{2-} + 4NO_3^- + 2TBP$$

and in the competition between the fluoride ion and TBP for the zirconium, the distribution coefficient of the zirconium is reduced.

4.5 Concentration of Salting Agent

In many solvent extraction systems, addition of solutes to the aqueous phase increases the distribution coefficient of extractable components. Data in Table 4.2 and Fig. 4.6 show how addition of nitrates to an aqueous solution of uranyl nitrate increases the distribution coefficient of uranyl nitrate between the aqueous phase and diethyl ether [F2]. The increase in distribution coefficients with increased nitrate concentration is explained as follows: Analysis of

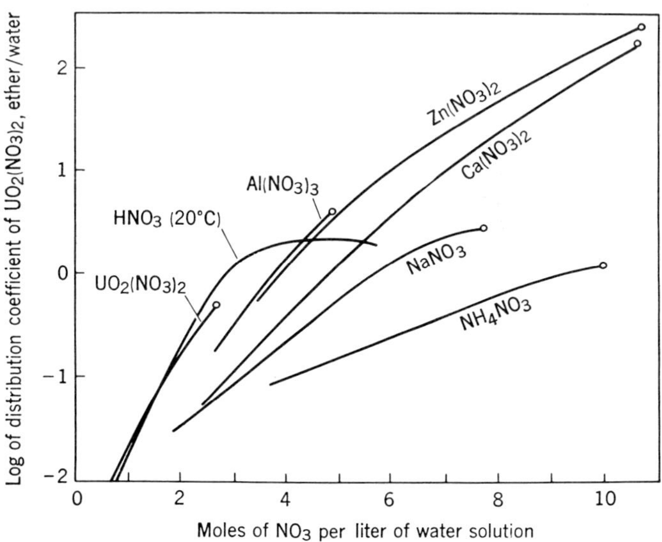

Figure 4.6 Effect of nitrates on distribution of $UO_2(NO_3)_2$ between diethyl ether and water. ○, saturated solution; temperature 25°C. (*From Furman et al. [F2].*)

the ether phase shows that uranium is extracted in the form of un-ionized uranyl nitrate. Addition of nitrate ion tends to increase the concentration of un-ionized uranyl nitrate by shifting the equilibrium to the right, and thus

$$UO_2{}^{2+} + 2NO_3{}^- \rightleftharpoons UO_2(NO_3)_2$$

converts more of the uranium to an extractable form. It may be noted that uranyl nitrate acts as a self-salting agent, probably also by displacement of this equilibrium. In addition, readily hydrated cations, such as Ca^{2+} and Al^{3+}, tie up much of the water in the aqueous phase, and thus increase the effective concentration of uranyl nitrate.

When the organic complexing agent in the solvent is nearly all combined as extracted complexes, further increase in concentration of the complex-forming metal ions in the aqueous phase will cause the distribution coefficient for metal extraction to decrease. This phenomenon has been observed for uranyl nitrate [G3, M1, M2] and for zirconium and hafnium nitrates [H4] when extracted by TBP in kerosene.

Table 4.3 gives distribution coefficients for uranyl nitrate between aqueous nitric acid and 40 percent TBP in kerosene observed by Goldschmidt et al. [G3]. At each nitric acid concentration, the uranium distribution coefficient decreases with increasing uranium concentration. This can be attributed to the following overall reaction equilibria [M2]:

$$UO_2{}^{2+}(aq) + 2NO_3{}^-(aq) + 2TBP(o) \rightleftharpoons UO_2(NO_3)_2 \cdot 2TBP(o)$$

and

$$H^+(aq) + NO_3{}^-(aq) + TBP(o) \rightleftharpoons HNO_3 \cdot TBP(o)$$

with the following equilibrium constants:

$$K_U = \frac{[UO_2(NO_3)_2 \cdot 2TBP(o)]}{[UO_2{}^{2+}(aq)]\,[NO_3{}^-(aq)]^2\,[TBP(o)]^2} \tag{4.15}$$

Table 4.3 Distribution coefficients for uranyl nitrate between aqueous nitric acid and 40 v/o TBP (1.464 *M*) in kerosene†

Moles per liter in aqueous phase		Distribution coefficient	
$UO_2(NO_3)_2$	HNO_3	Observed [G3]	Calculated‡
0.042	0.6	3.3	3.3
0.210	0.6	2.1	1.98
1.68	0.6	0.38	0.41
0.042	1.5	4.3	6.4
0.210	1.5	2.4	2.2
1.68	1.5	0.39	0.40
0.042	2.0	5.7	7.0
0.210	2.0	2.6	2.4
1.68	2.0	0.39	0.40
0.042	3.0	7.2	7.1
0.210	3.0	2.7	2.5

† v/o = volume percent.

‡ Calculated from Eq. (4.22) using $K_H = 0.145$ and $K_U = 5.5$

Table 4.4 Distribution coefficients for $Zr(NO_3)_4$ between water and TBP in kerosene

Aqueous phase: 3.0 *M* HNO_3
3.5 *M* $NaNO_3$
Organic phase: 60 v/o TBP (2.19 *M*)

Moles Zr per liter		
Aqueous	Organic	Distribution coefficient
0.012	0.042	3.5
0.039	0.083	2.1
0.074	0.114	1.54
0.104	0.135	1.30
0.123	0.147	1.20

and

$$K_H = \frac{[HNO_3 \cdot TBP(o)]}{[H^+(aq)]\,[NO_3{}^-(aq)]\,[TBP(o)]} \tag{4.16}$$

For the purposes of this chapter, activity coefficients of unity are assumed, so that the bracketed quantities in Eqs. (4.15) and (4.16) become identical with molar concentrations.

Assuming that at equilibrium all aqueous uranium is in the form of uranyl ion and all organic uranium is in the form of the $UO_2(NO_3)_2 \cdot 2TBP$ complex, the uranium distribution coefficient is

$$D_U = \frac{[UO_2(NO_3)_2 \cdot 2TBP(o)]}{[UO_2{}^{2+}(aq)]} \tag{4.17}$$

and combining Eqs. (4.15) and (4.17),

$$D_U = K_U [NO_3{}^-(aq)]^2 [TBP(o)]^2 \tag{4.18}$$

The TBP concentration appearing in this equation is that of uncombined TBP. For a given total amount of TBP in the organic phase, the uncombined TBP is lower the higher the concentration of uranium, and the uranium distribution coefficient should decrease as the uranium concentration increases. This is confirmed by the experimental data in Table 4.3.

Similar saturation effects are apparent from the data for the extraction of $Zr(NO_3)_4$ with TBP in kerosene [H4], as shown in Table 4.4.

4.6 Correlation of Equilibrium Extraction Data

Assuming that at equilibrium all aqueous HNO_3 is fully ionized and all organic HNO_3 is in the form of $HNO_3 \cdot TBP$, the distribution coefficient of nitric acid is

$$D_H = \frac{[HNO_3 \cdot TBP(o)]}{[H^+(aq)]} \tag{4.19}$$

and combining with Eq. (4.16),

$$D_H = K_H [NO_3{}^-(aq)]\,[TBP(o)] \tag{4.20}$$

For a uranium-nitric acid system, the concentration of uncombined TBP in terms of concentrations of uranium and acid in the organic phase is

$$[\mathrm{TBP}(o)] = C - [\mathrm{HNO_3 \cdot TBP}(o)] - 2[\mathrm{UO_2(NO_3)_2 \cdot 2TBP}(o)] \tag{4.21}$$

where C is the concentration of total TBP, both combined and uncombined, in the organic phase. The distribution coefficient of uranium in terms of total TBP concentration and aqueous concentrations is obtained by combining Eqs. (4.16), (4.18), and (4.21):

$$\frac{D_\mathrm{U}}{(C - 2D_\mathrm{U}[\mathrm{UO_2}^{2+}(aq)])^2} = \frac{K_\mathrm{U}[\mathrm{NO_3}^-(aq)]^2}{(1 + K_\mathrm{H}[\mathrm{H}^+(aq)][\mathrm{NO_3}^-(aq)])^2} \tag{4.22}$$

and solving for D_U,

$$D_\mathrm{U} = \frac{C + A}{2[\mathrm{UO_2}^{2+}(aq)]}\left[1 - \sqrt{1 - \left(\frac{C}{C + A}\right)^2}\,\right] \tag{4.23}$$

where

$$A = \frac{(1 + K_\mathrm{H}[\mathrm{H}^+(aq)][\mathrm{NO_3}^-(aq)])^2}{4K_\mathrm{U}[\mathrm{UO_2}^{2+}(aq)][\mathrm{NO_3}^-(aq)]^2} \tag{4.24}$$

Only the negative sign of the square root is used in Eq. (4.23); the positive value of the root yields a trivial solution that implies negative concentrations of uncombined TBP in the organic phase.

The extraction of nitric acid with TBP from aqueous solutions free of other extractable species has been studied by several investigators [G7, M2], and the equilibrium data lead to an average value of 0.145 for the acid equilibrium constant, assuming activity coefficients of unity. The correlation of nitric acid on the basis of Eq. (4.16) is poor when the acid concentration in the aqueous phase is greater than about 7 M, possibly because of the formation of the dinitrato and trinitrato complexes $2\mathrm{HNO_3 \cdot TBP}$ and $3\mathrm{HNO_3 \cdot TBP}$ or possibly because of solution of nitric acid in the organic without complexing [S4].

The uranium distribution data in Table 4.3 can be correlated reasonably well by using the following equilibrium constants in Eq. (4.23) and assuming activity coefficients of unity:

$$K_H = 0.145$$

$$K_\mathrm{U} = 5.5$$

The concentration C of total TBP is obtained from the volume percent (v/o) TBP in the organic phase by

$$C = \frac{(\mathrm{v/o})(0.972 \times 10^3)}{266.3}\ \mathrm{mol/liter} \tag{4.25}$$

where 0.972 is the density of pure TBP [G4, M1, S4] and 266.3 is the molecular weight of TBP, which has the chemical formula $(\mathrm{C_4H_9})_3\mathrm{PO_4}$. Equation (4.25) neglects the small volume change resulting from water solubility in the organic phase and from extraction of the uranium and acid complexes, and it neglects the solubility of TBP in the aqueous phase [S4]. At 40 percent TBP by volume, the total TBP molar concentration C is 1.464 mol/liter.

Uranium distribution coefficients calculated from Eq. (4.23) and from the above data are listed in Table 4.3.

The close agreement between observed and calculated distribution coefficients in Table 4.3 is surprising in view of the wide range of aqueous concentration and the assumption of unity for the activity coefficients. Distribution coefficients for an even greater range of parameters in the TBP extraction system are given in Chap. 10.

A number of attempts have been made to establish correlations of distribution coefficients in the $\mathrm{UO_2(NO_3)_2}$-$\mathrm{HNO_3}$-TBP system on a more fundamental thermodynamic basis. One approach [B1, H2, R4] has been to correlate on the basis of ionic strength in the aqueous phase, with assumed constant activity coefficients in the organic phase. A more advanced approach by Goldberg et al. [G2a] involved the correlation of distribution data based on the ratio of the activity of the $UO_2(NO_3)_2 \cdot 2TBP$ complex to the activity of free TBP in the organic phase, using activity coefficients derived from experimental data for the partial pressure of HNO_3 over aqueous solutions of HNO_3 and $UO_2(NO_3)_2$. With this approach the data of Codding et al. [C3] for the distribution coefficients of $UO_2(NO_3)_2$ in TBP extraction were correlated over a wide range of concentrations, with an average deviation of 5.8 percent.

4.7 Presence of Other Extractable Species

Uranium extraction by TBP may in some cases become poorer in the presence of other extractable components because of depletion of free TBP by components other than uranium. Such behavior is illustrated by the $UO_2(NO_3)_2$-HNO_3-TBP system analyzed above, as shown by the data [M1] in Fig. 4.7 for the distribution coefficient of uranium as affected by nitric acid concentration. For acid concentration less than about 5 *M*, the uranium distribution coefficient is greater the higher the acid concentration, because of the salting effect of nitrate ion from the acid. At acid concentrations greater than about 5 *M*, increasing acid concentration inhibits uranium extraction, because enough nitric acid has been extracted so that less free TBP is available to form the extractable complex with uranium. Similar effects have been observed in the extraction of other elements with TBP [M1].

In designing multistage extraction systems for extractive separations by TBP, or by other extractants that can change appreciably in noncomplexed concentration as a result of extraction, it is necessary to perform analyses similar to Eq. (4.15) through Eq. (4.24) for each of the extractable species present in other than trace quantities to determine the distribution coefficients for each of the species in each of the contacting stages [G6, H2, L3]. Such design procedures are illustrated in Sec. 6.6 for the separation of hafnium from zirconium by TBP extraction from a nitric acid solution.

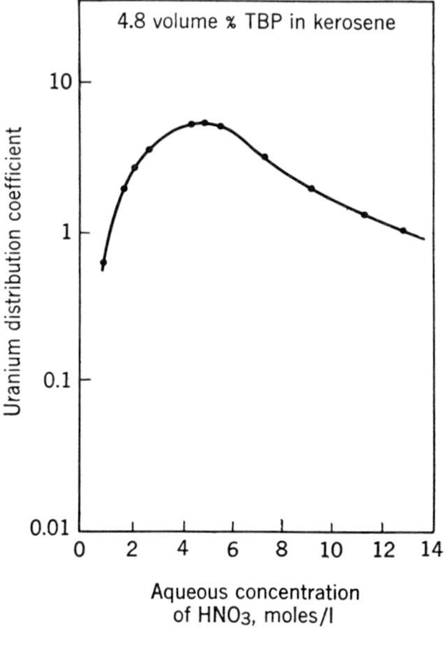

Figure 4.7 Effect of nitric acid concentration on extraction of uranyl nitrate with TBP. (Data of McKay [M1].)

4.8 Hydrogen Ion Concentration

When the distribution equilibrium reaction involves hydrogen ions, changing the hydrogen ion concentration will have a strong effect on the distribution coefficient. An example of this is the extraction of metal complexes of acetyl acetone (HAa) and other weakly acid complexing agents by benzene. The equilibrium reaction for extraction of thorium by this reagent is

$$Th^{4+}(aq) + 4HAa(o) \rightleftharpoons Th(Aa)_4(o) + 4H^+(aq)$$

Hence, the thorium distribution coefficient should be

$$D_{Th} = \frac{[Th(Aa)_4(o)]}{[Th^{4+}(aq)]} = \frac{K_{Th}[HAa(o)]^4}{[H^+(o)]^4} \tag{4.26}$$

where K_{Th} is the equilibrium constant for the above reaction. An inverse fourth-power dependence on hydrogen ion concentration is in fact observed for this distribution coefficient.

When an extractable cation, such as Zr^{4+}, is readily hydrolyzed, reduction of hydrogen ion concentration will reduce the distribution coefficient by increasing the proportion of the element in the form of partially hydrolyzed, nonextractable ions such as ZrO^{2+}. This principle was used in the Redox process [B2, C7, C8] for the hexone extraction of plutonium from irradiated uranium, wherein the aqueous phase was made slightly acid-deficient with ammonium hydroxide, to reduce the extraction of zirconium and rare-earth fission products.

5 SOLVENT REQUIREMENTS

Generally speaking, a solvent suitable for separation of metals by fractional extraction should meet the following requirements:

1. It should be selective; i.e., the ratio of distribution coefficients should be high.
2. It should have good capacity for extraction; i.e., distribution coefficients in the extracting section should be of the order of unity or higher.
3. It should be readily stripped; i.e., distribution coefficients in the stripping section should be no greater than unity.
4. It should be relatively immiscible with water, to reduce solubility losses.
5. Its density should be appreciably different from water, and it should have low viscosity and fairly high interfacial tension. These physical properties are important in promoting separation of phases following contact.
6. For safety reasons it should be relatively nonvolatile, nonflammable, and nontoxic.
7. It should be readily purified, preferably by fractional distillation.
8. It should be stable in the presence of chemical agents used in the process, such as nitric acid. Solvents used for radioactive materials should also have good radiation stability.

TBP meets most of these requirements except those of low viscosity and a density different from water. These deficiencies are corrected by diluting TBP with a light, saturated hydrocarbon, such as an aromatic-free kerosene. This solvent is the one most commonly used at present in fractional extraction of metals. The physical properties of TBP are summarized in Table 4.5 [F1, S4].

Although TBP is a relatively stable organic compound, it does undergo slow hydrolysis to form di-*n*-butyl phosphate (DBP). Although the presence of DBP increases the distribution coefficients of uranium, plutonium, and other actinides, it interferes with the separation of plutonium from uranium, and it makes complete stripping of these elements difficult. DBP forms an insoluble compound with thorium. DBP formation is appreciable only when the

Table 4.5 Physical properties of TBP†

Chemical formula	$(C_4H_9)_3PO_4$
Molecular weight	266
Color	Water white
Odor	Mildly sweet
Refractive index at 20°C	1.4223
Viscosity at	
25°C	3.32 cP
85°C	0.8 cP
Boiling point at	
760 Torr	289°C
15 Torr	173°C
1 Torr	121°C
Density at 25°C	0.9724
Freezing point	−80°C
Flash point, Cleveland open cup	145°C
Dielectric constant at 30°C	7.97
Solubility in water at 25°C	0.39 g/liter
Solubility of water in TBP at 25°C	64 g/liter

†Data from J. R. Flanary [F1] and T. H. Siddall, III [S4].

solvent is held for long periods at temperatures as high as 50 to 60°C, but it can be removed by periodically scrubbing the solvent with a basic solution [S4].

TBP can be decomposed explosively when heated to above 120°C in the presence of extractable nitrates [S4].

6 THEORY OF COUNTERCURRENT EQUILIBRIUM EXTRACTION

The calculation of the concentration of extractable components in a countercurrent cascade of equilibrium solvent extraction stages is first developed for the simple countercurrent extraction section of Fig. 4.3. The theory is then extended to the extracting-scrubbing system of Fig. 4.4 for fractional extraction and is illustrated by a numerical calculation for the separation of zirconium from hafnium, using TBP in kerosene as solvent.

6.1 Extracting Cascade

Here we consider an extracting cascade in which a feed solution containing one or more extractable components is contacted countercurrently with an organic solvent. Nomenclature for flow rates, concentrations, and stage numbers is shown in Fig. 4.8. It will be assumed that equilibrium is reached between the aqueous and organic phases leaving each stage. Changes in the volume flow rates of the aqueous and organic phases will be neglected. Consider the portion of the cascade below stage n. A material balance on one of the extractable components is

$$Ey_0 + Fx_n = Ey_{n-1} + Fx_1$$

or

$$y_{n-1} - y_0 = \frac{F}{E}(x_n - x_1) \tag{4.27}$$

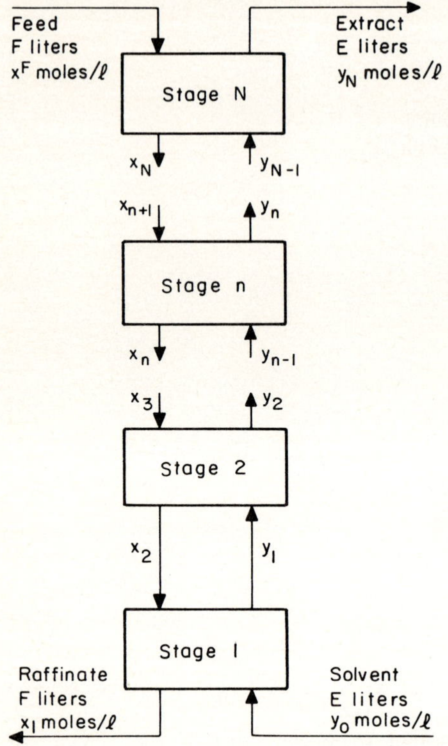

Figure 4.8 Nomenclature for cascade of solvent extraction stages.

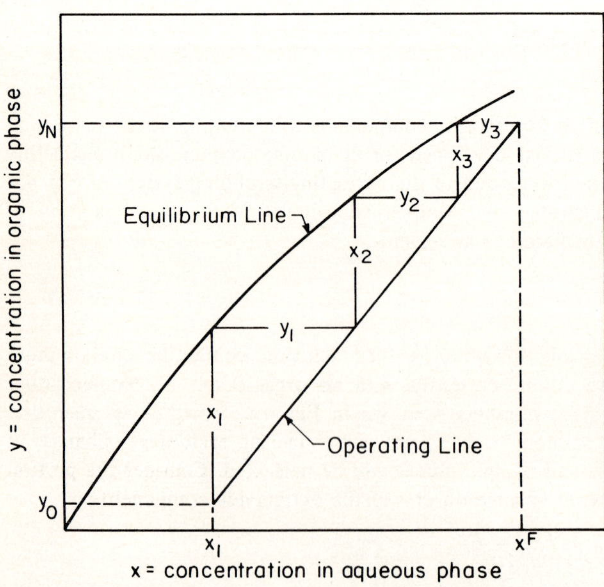

Figure 4.9 Stage concentration diagram for solvent extraction cascade.

Concentrations in the organic and aqueous phases leaving a stage are related by the equilibrium relation

$$y_n = D_n x_n \tag{4.28}$$

where D_n is the distribution coefficient at the conditions of the nth stage.

The meaning of Eqs. (4.27) and (4.28) may be visualized on a plot of y versus x, in the McCabe-Thiele diagram, Fig. 4.9. The material-balance relation (4.27) is represented by the operating line that passes through the point (x_1^E, y_0^E) and has the slope F/E. The equilibrium relation (4.28) is represented by the equilibrium line. When D is constant, the equilibrium line is a straight line, as would occur for the extraction of trace quantities of solutes in the presence of nonextractable salting agents, with constant concentration of uncombined complexing agent. More generally, as has been demonstrated in Sec. 4, D_n varies from stage to stage, resulting in a curved equilibrium line. Figure 4.9 illustrates the equilibrium line typical for the extraction of a single component in the presence of a nonextractable salting agent.

The McCabe-Thiele diagram is useful for constructing a graphic solution for the stagewise compositions. The operating line is the locus of points x_n, y_{n-1} of adjacent interstage flows. The vertical projection of any such point intersects the equilibrium line at x_n, y_n, thereby defining the compositions of the aqueous and organic phases leaving the equilibrium stage n.

Assume that the cascade is to reduce the concentration of the extractable component from x^F to x_1 by extraction with organic of relative volume E/F. The point x_1, y_0 is thereby specified. Beginning at x_1, y_0 and projecting upward in vertical and horizontal steps, the compositions for all of the other equilibrium stages are determined. The number of vertical projections between the operating line and equilibrium line necessary to step from x_1 to x^F gives the required number of equilibrium stages.

Given this number of stages, construction of a similar McCabe-Thiele diagram for other components in the feed, such as impurities, allows the calculation of the extent to which these impurities extract into the organic phase. If two or more extractable components are each in sufficient concentration to affect the distribution coefficient of the other species, e.g., TBP extraction of $UO_2(NO_3)_2$ and HNO_3, the equilibrium lines for the two components cannot be specified in advance but must be calculated by an iterative procedure, similar to that to be illustrated in Sec. 6.6 for the zirconium-hafnium separation.

From Eq. (4.27) or from the construction of Fig. 4.9 it is apparent that the ratio of organic flow rate to aqueous flow rate is given by

$$\frac{E}{F} = \frac{x^F - x_1}{y_N - y_0} \tag{4.29}$$

because x^F is the virtual aqueous effluent concentration from stage $N+1$. If the overall fractional recovery ρ of the extractable component is specified as

$$\rho = \frac{Ey_N}{Fx^F} \tag{4.30}$$

and, for the simple extraction cascade,

$$1 - \rho = \frac{x_1}{x^F} \tag{4.31}$$

then Eqs. (4.29) and (4.31) combine to yield

$$\frac{E}{F} = \frac{\rho x^F}{y_N - y_0} \tag{4.32}$$

For given compositions x^F, x_1, and y_0, or for given x^F, y_0, and ρ, reducing the relative amount of organic flow brings the operating line nearer to the equilibrium line and increases

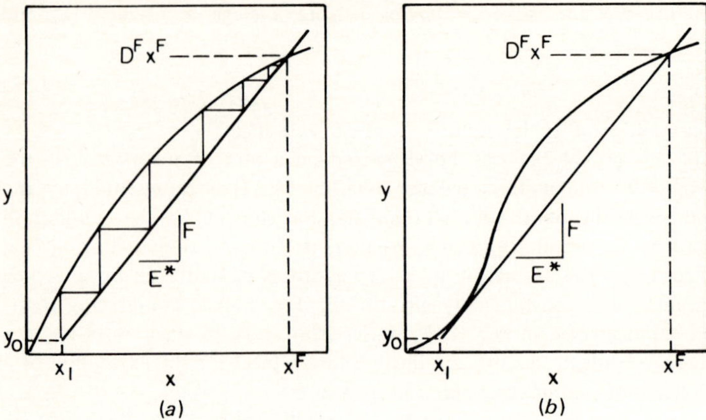

Figure 4.10 Limiting flow ratios for solvent extraction cascade.

the required number of stages. The minimum flow ratio $(E/F)_{\min}$ occurs when the operating line intersects the equilibrium line at x^F, requiring an infinite number of stages as illustrated in Fig. 4.10*a*. In this limiting condition, denoted by asterisks, the concentration y_N^* in the organic stream leaving the cascade is

$$y_N^* = D_N x^F \tag{4.33}$$

so that (4.29) becomes

$$\left(\frac{E}{F}\right)_{\min} = \frac{E^*}{F} = \frac{x^F - x_1}{D_N x^F - y_0} \tag{4.34}$$

and (4.32) becomes

$$\frac{E^*}{F} = \frac{\rho}{D_N(1 - y_0/x^F)} \tag{4.35}$$

If the equilibrium line is locally concave upward, as is possible in the extraction of a self-salting component with excess extracting agent, with sufficiently low x_1 the operating line may intersect the equilibrium at $x < x^F$. In this event Eqs. (4.33) through (4.35) are invalid unless x_1 is increased to allow intersection at x^F, as illustrated in Fig. 4.10*b*.

To carry out a specified separation in an actual extracting cascade with a finite number of stages, the flow ratio E/F must be greater than the minimum ratio E^*/F given in Eq. (4.34) or (4.35). The application of these equations for the case of constant distribution coefficients will be illustrated in Sec. 6.2.

6.2 Extracting Cascade with Constant Distribution Coefficients

When the distribution coefficients are independent of stage number, an equation can be derived for analytical calculation of the number of stages.

For any extractable component with a constant distribution coefficient, Eqs. (4.27) and (4.28) can be rewritten in terms of the constant *extraction factor* β:

$$y_n = \beta(y_{n-1} - y_0) + Dx_1 \tag{4.36}$$

where

$$\beta \equiv \frac{DE}{F} \tag{4.37}$$

Equation (4.28) is again written as

$$y_n = Dx_n \tag{4.38}$$

When $n = 1$, Eq. (4.36) becomes

$$y_1 = Dx_1 \tag{4.39}$$

When $n = 2$, Eq. (4.36) becomes

$$y_2 = \beta(y_1 - y_0) + Dx_1 \tag{4.40}$$

which, with (4.39) becomes

$$y_2 = (\beta + 1)Dx_1 - \beta y_0 \tag{4.41}$$

When $n = 3$, Eq. (4. 36) becomes

$$y_3 = \beta(y_2 - y_0) + Dx_1 \tag{4.42}$$

which, with (4.41) becomes

$$y_3 = (1 + \beta + \beta^2)Dx_1 - (\beta + \beta^2)y_0 \tag{4.43}$$

Proceeding in this way to stage N, we obtain

$$y_N = (1 + \beta + \cdots + \beta^{N-1})D_1x_1 - (\beta + \cdots + \beta^{N-1})y_0 \tag{4.44}$$

which is identical with

$$y_N = \frac{\beta^N - 1}{\beta - 1}(Dx_1 - y_0) + y_0 \tag{4.45}$$

Equation (4.45) is a form of the Kremser equation, originally derived for countercurrent gas absorption [S3].

The raffinate concentration x_1 may be eliminated by an overall material balance,

$$F(x^F - x_1) = E(y_N - y_0) \tag{4.46}$$

which combines with (4.45) to yield

$$\frac{y_N - y_0}{Dx^F - y_0} = \frac{\beta^N - 1}{\beta^{N+1} - 1} \tag{4.47}$$

Figure 4.11 provides a graphic solution of Eq. (4.47).

The overall recovery ρ of the extractable component can be expressed in terms of these variables by combining Eqs. (4.30), (4.37), and (4.47) to eliminate y_N:

$$\rho = \beta\left(\frac{\beta^N - 1}{\beta^{N+1} - 1}\right) + \beta^{N+1}\left(\frac{\beta - 1}{\beta^{N+1} - 1}\right)\frac{y_0}{Dx^F} \tag{4.48}$$

Thus, by specifying β for the cascade and the ratio y_0/Dx^F and recovery ρ for any one of the extractable components, the required number of equilibrium stages N can be calculated from (4.48).

If we have extractable components A and B in the feed to be separated in a simple extraction cascade, the constant distribution coefficients D_A and D_B result in extraction factors β_A and β_B, and the overall decontamination factor f_{AB} is obtained by applying Eq. (4.48) to each of the components, with

$$f_{AB} \equiv \frac{\rho_A}{\rho_B} \tag{4.49}$$

For the special case of $y_0 = 0$, Eqs. (4.48) and (4.49) combine to yield

$$f_{AB} = \frac{\beta_A}{\beta_B}\left(\frac{\beta_A^N - 1}{\beta_B^N - 1}\right)\left(\frac{\beta_B^{N+1} - 1}{\beta_A^{N+1} - 1}\right) \tag{4.50}$$

For a specified number of stages N and specified E/F, the decontamination factor for any two extractable components can be calculated from (4.48) and (4.49) or, for $y_0 = 0$, from (4.50).

To illustrate the use of these equations, consider the extraction of zirconium from an aqueous solution of zirconium and hafnium nitrates, as shown in Fig. 4.12. Although the distribution coefficients do, in fact, depend on the concentration of zirconium and hafnium, as shown later in Sec. 6.6, constant distribution coefficients are assumed here for the purpose of this illustration. The specified feed composition and the specified recovery to be obtained are listed in Table 4.6.

The distribution coefficients assumed above are those observed by Huré and Saint James

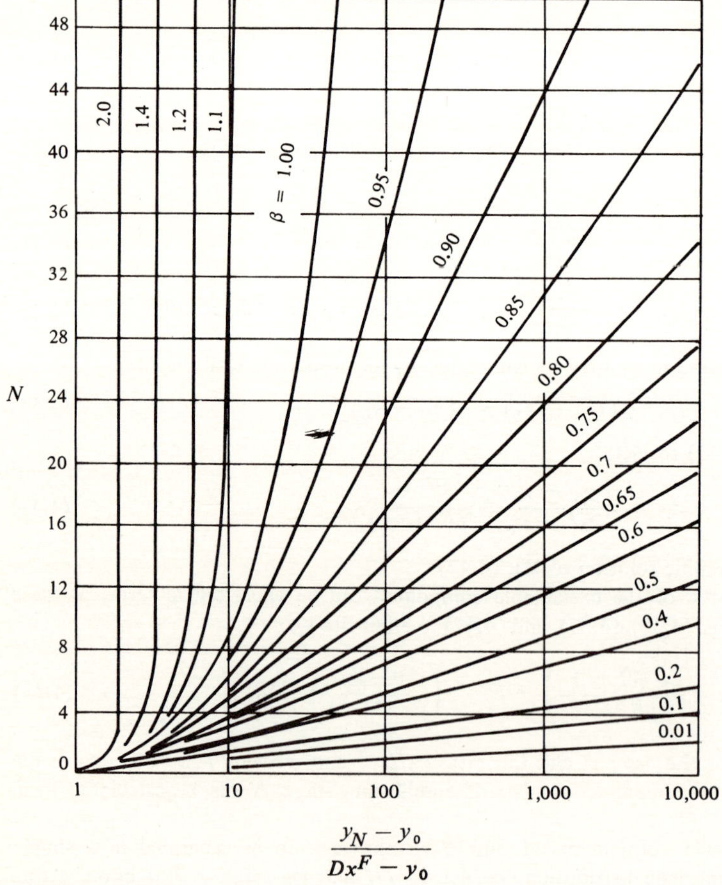

Figure 4.11 Number of equilibrium stages in an extraction cascade. (*Adapted from Sherwood et al. [S3].*)

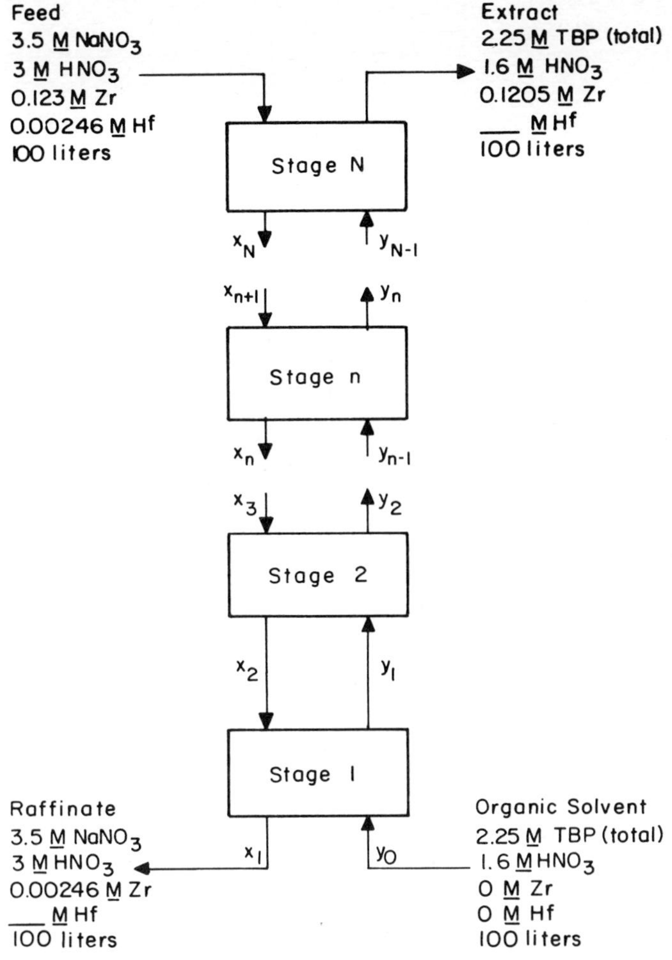

Figure 4.12 Flow sheet for zirconium-hafnium extraction example.

[H4] for an aqueous solution of the feed of concentration 3.5 N in $NaNO_3$ and 3.0 N in HNO_3, in contact with 60 percent TBP in kerosene. Distribution coefficients will be higher at the bottom of the cascade, where the aqueous zirconium concentration is lower; this will be neglected in the present treatment, but will be taken into account in Sec. 6.6.

By applying Eqs. (4.37) and (4.48) to zirconium,

$$\beta_{Zr} = 1.2$$

and
$$0.98 = \frac{1.2^{N+1} - 1.2}{1.2^{N+1} - 1}$$

from which $N = 12.2$.

The zirconium-hafnium decontamination factor is obtained from (4.48) and (4.49) with $\beta_{Hf} = 0.12$:

$$f_{\text{Zr-Hf}} = 0.98 \frac{0.12^{13.2} - 1}{0.12^{13.2} - 0.12} = 8.17$$

The McCabe-Thiele diagrams for this example are shown in Fig. 4.13.

It is interesting to compare the decontamination obtainable for $\rho_{Zr} = 0.98$ and $E/F = 1.0$ with that obtainable with an infinite number of stages, corresponding to operation at the same zirconium recovery but at $(E/F)_{\text{min}}$. From Eq. (4.35):

$$\frac{E}{F_{\text{min}}} = \left(\frac{E^*}{F}\right) = \frac{0.98}{1.20} = 0.817$$

and with $N = \infty$, Eq. (4.50) yields

$$\lim_{N \to \infty} f_{\text{Zr-Hf}} = \frac{1}{\beta_{\text{Hf}}} = 8.33$$

A more effective way to use an increased number of stages in a simple extracting cascade would be to increase the zirconium recovery. This would occur by allowing the slope of the operating line to approach D_{Zr}. In the limit of $N \to \infty$, $x_{Zr,1} \to 0$ and $\rho_{Zr} \to 1$. Because $D_{Hf} \ll D_{Zr}$, the operating line for hafnium can intersect the hafnium equilibrium line only at $x \geqslant x_{Hf}^F$ and not at $x_{Hf} = 0$. In this limit of

$$\frac{F}{E} \to D_{Zr}$$

$$N \to \infty$$

$$\rho_{Zr} \to 1$$

Table 4.6 Specifications for zirconium-hafnium separation example in an extracting cascade

Given	
Aqueous feed concentration	
Zirconium	$x_{Zr}^F = 0.123$ mol/liter
Hafnium	$x_{Hf}^F = 0.00246$ mol/liter
Solvent feed concentration	
Zirconium	$y_{Zr,0} = 0$
Hafnium	$y_{Hf,0} = 0$
Zirconium recovery	$\rho_{Zr} = 0.98$
Distribution coefficients, assumed to be constant for all stages	
Zirconium	$D_{Zr} = 1.20$
Hafnium	$D_{Hf} = 0.12$
Flow ratio	$E/F = 1.0$
Required	
Number of stages N	
Zr-Hf decontamination factor f	

Source: Adapted from J. Huré and R. Saint James, "Process for Separation of Zirconium and Hafnium," *Proceedings of the International Conference on the Peaceful Uses of Atomic Energy,* vol. 8, United Nations, New York, 1956, p. 551.

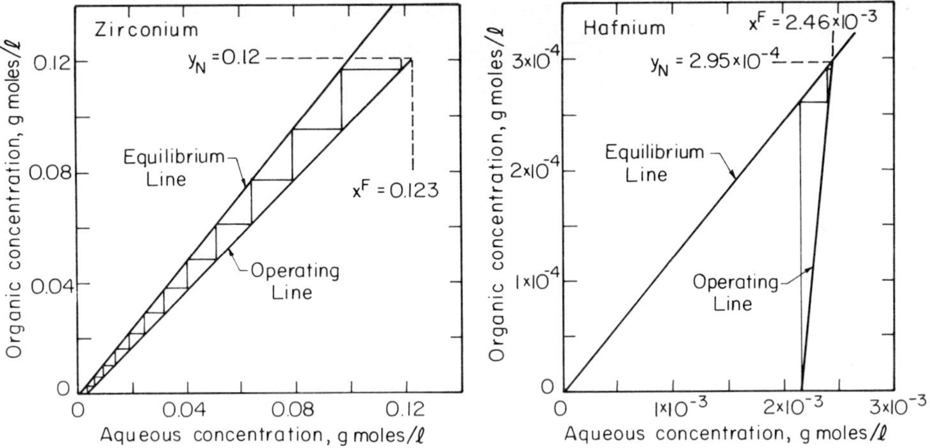

Figure 4.13 Stage concentration diagram for zirconium-hafnium extraction example.

we find from (4.50) that

$$\lim_{N \to \infty} f_{\text{Zr-Hf}} = \frac{D_{\text{Zr}}}{D_{\text{Hf}}} = 10$$

To obtain better decontamination a scrubbing section is added to the cascade, as illustrated in Fig. 4.4 and analyzed in the following section.

6.3 The Extracting-scrubbing Cascade

For more efficient fractional extraction of two or more extractable components, the extracting-scrubbing cascade of Fig. 4.4 is employed. Nomenclature for flow rates, concentration, and stage number is shown in Fig. 4.14. With the same assumptions and approach as in Sec. 6.1, a material balance for any one of the components in the portion of the cascade below stage n in the extracting section is

$$Ey_0^E + (S+F)x_n^E = Ey_{n-1}^E + (S+F)x_1^E$$

$$y_{n-1}^E - y_0^E = \frac{S+F}{E}(x_n^E - x_1^E) \tag{4.51}$$

The extracting-section operating line shown in the McCabe-Thiele diagram of Fig. 4.15 passes through the point (y_0^E, x_1^E) and has the slope $(S+F)/E$.

A material balance around the portion of the plant above stage m of the scrubbing section is

$$Sx_0^S + Ey_m^S = Sx_{m-1}^S + Ey_1^S$$

or

$$y_m^S - y_1^S = \frac{S}{E}(x_{m-1}^S - x_0^S) \tag{4.52}$$

In Fig. 4.15 this is represented by the operating line for the scrubbing section, which passes through (x_0^S, y_1^S) and has the slope S/E.

As illustrated in Fig. 4.15, different equilibrium lines can exist for the extracting and scrubbing sections, as might occur if the scrub solution contains a different salting agent

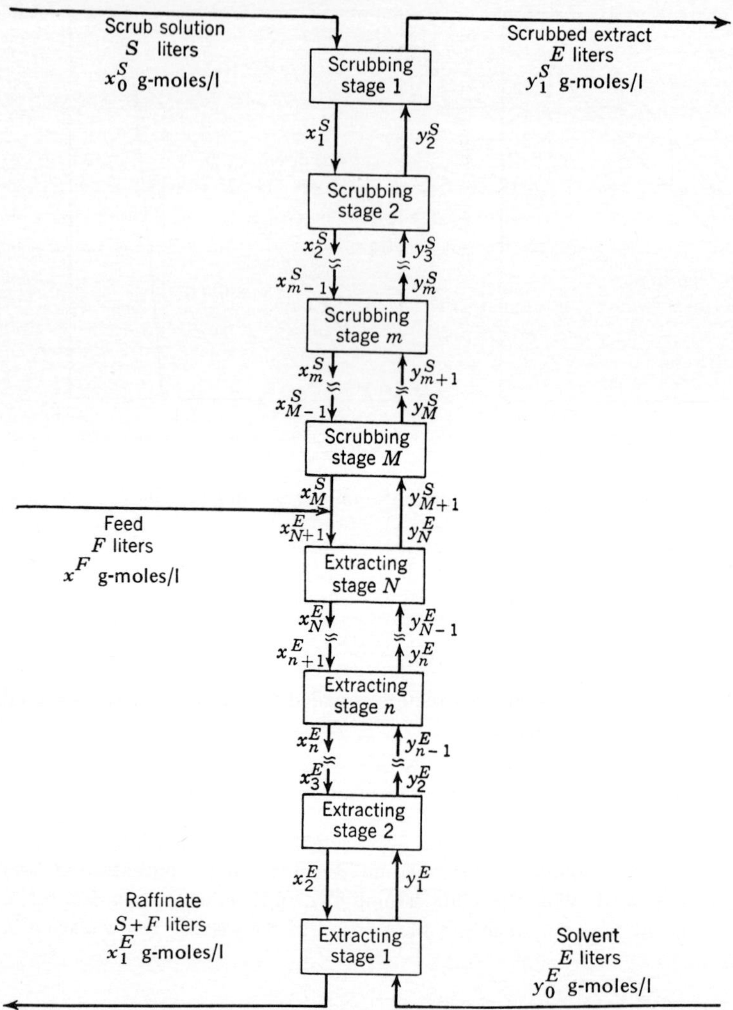

Figure 4.14 Nomenclature for cascade of extracting-scrubbing stages.

concentration than the feed solution, or from the effect of one extractable component on the distribution coefficients of other extractable components (cf. Sec. 6.6).

In Fig. 4.15 a graphic solution is illustrated for a specified number of stages N and M in the extracting and scrubbing sections, respectively. Proceeding upward by vertical and horizontal steps from the point x_1^E, y_0^E for N vertical steps between the extracting-section operating and equilibrium lines, the concentration in the solvent leaving the extracting section is identical with that entering the scrubbing section; i.e.,

$$y_N^E = y_{M+1}^S \tag{4.53}$$

where y_N^E is found at (x_N^E, y_N^E) on the equilibrium line for the extracting section. The horizontal projection of y_N^E onto the scrubbing section operating line yields (x_M^S, y_{M+1}^S), and this point is then projected downward to the equilibrium line for the scrubbing section.

The number M of equilibrium stages in the scrubbing section specifies the number of vertical steps between the operating line and the equilibrium line, starting at y_{M+1}^S and ending at y_1^S.

By thus determining the extract concentration for one of the extractable components in the feed, a similar graphic or numerical calculation is made for each of the other extractable components so that the composition of the organic product and aqueous raffinate can be determined. When two or more extractable components are each present in sufficient concentration to affect distribution coefficients of the other species, equilibrium lines must be calculated by an iterative procedure similar to that illustrated in Sec. 6.6 for TBP extraction in the Zr-Hf-HNO_3 system.

The operating lines for a given component in the extracting and scrubbing sections intersect at the feed composition x^F. This can be demonstrated by defining x_{mn}, y_{mn} as the intersection point, such that at the intersection

$$x_{mn} = x_n^E = x_{m-1}^S \tag{4.54}$$

and

$$y_{mn} = y_{n-1}^E = y_m^S \tag{4.55}$$

Substituting Eqs. (4.54) and (4.55) into (4.51) and (4.52) and solving, we find that

$$Fx_{mn} + Ey_0^E + Sx_0^S = (S + F)x_1^E + Ey_1^S \tag{4.56}$$

However, an overall material balance for the extractable component written for the entire separation cascade of Fig. 4.14 is

$$Fx^F + Ey_0^E + Sx_0^S = (S + F)x_1^E + Ey_1^S \tag{4.57}$$

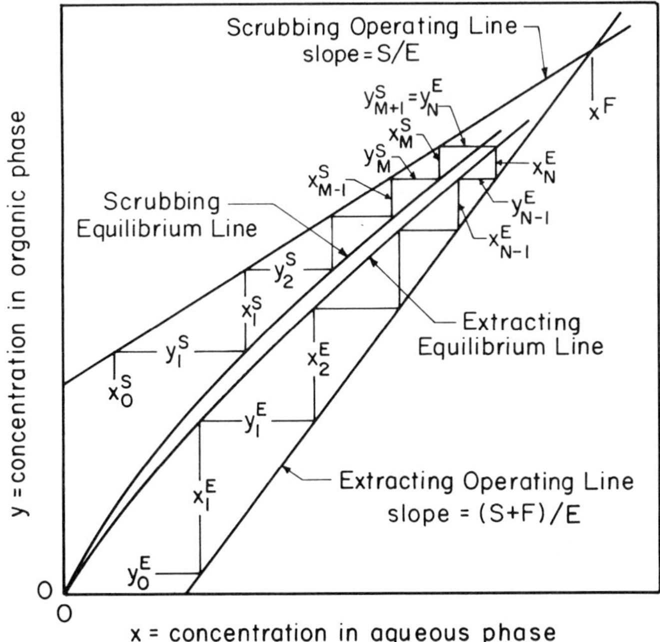

Figure 4.15 Stage concentration diagram for extracting-scrubbing cascade.

Comparison of Eqs. (4.56) and (4.57) shows that the operating lines intersect at

$$x_{mn} = x^F \tag{4.58}$$

In an extracting-scrubbing cascade with a finite number of stages, none of the aqueous streams entering or leaving a stage is at a concentration as high as x^F. The intersection of the two operating lines represents only an extrapolated point that is useful in graphic construction of the operating lines.

6.4 Limiting Flow Ratios for the Extracting-scrubbing Cascade

On a cascade operating with specified concentrations, x_1^E in the raffinate and y_0^E in the entering organic, and for a specified flow ratio $(F + S)/E$, an increase in the number of extracting stages results in an increase in the concentration x_{N+1}^E of the solute in the aqueous stream entering the extracting section. For an infinite number of extracting stages, x_{N+1}^E occurs at the intersection of the extracting equilibrium and operating lines, as illustrated in Fig. 4.16*a*. Similarly, for an infinite number of scrubbing stages, x_M^S occurs at the intersection of the scrubbing equilibrium and operating lines.

If there are infinite numbers of both extracting and scrubbing stages, the equilibrium lines intersect the operating lines at the operating-line intersection x^F, as illustrated in Fig. 4.16*b*; i.e.,

$$x_M^* = x_{N+1}^* = x^F \tag{4.59}$$

where the asterisks denote the limiting conditions of Fig. 4.16*b*.

The slopes of the operating lines in Fig. 4.16*b* are

$$\frac{S^*}{E^*} = \frac{D^F x^F - y_1^S}{x^F} \tag{4.60}$$

and

$$\frac{S^* + F}{E^*} = \frac{D^F x^F - y_0^E}{x^F - x_1^E} \tag{4.61}$$

When the limiting conditions of Fig. 4.16*b* and Eqs. (4.59), (4.60), and (4.61) are satisfied for

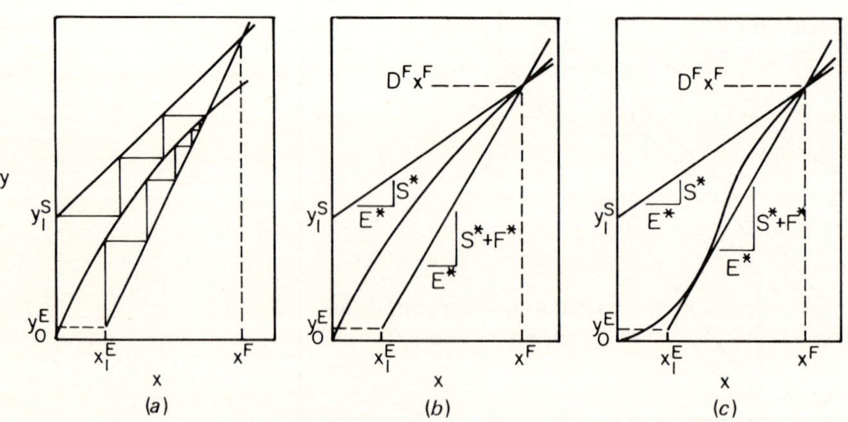

Figure 4.16 Limiting flow ratios for extracting-scrubbing cascade. (*a*) Infinite number of stages in extracting section only; (*b* and *c*) infinite number of stages in extracting and scrubbing sections; (*c*) double intersection in extracting section.

any one of the extractable components, they will be satisfied for all of the extractable components. For specified feed concentrations and for specified terminal concentrations or overall recoveries of two of the extractable components, these limiting conditions result in the minimum flow rates E^*/F and S^*/F of organic and scrub solution, relative to the feed rate, to achieve those recoveries.

For the separation of feed components A and B, and assuming zero concentration of these components in the entering organic ($y_0^E = 0$), Eq. (4.60) written for component A is

$$\frac{S^*}{E^*} = D_A^F - \frac{y_{A,1}^S}{x_A^F} \tag{4.62}$$

By writing Eq. (4.62) for components A and B and eliminating the flow ratios:

$$D_A^F - \frac{y_{A,1}^S}{x_A^F} = D_B^F - \frac{y_{B,1}^S}{x_B^F} \tag{4.63}$$

The performance of an extracting-scrubbing cascade may be defined in terms of the fractional recovery ρ of one of the desired components and the decontamination factor f of the organic product relative to the feed. These quantities are defined in terms of feed and product flow rates and composition as follows:

$$\rho \equiv \frac{Ey_1^S}{Fx^F} \tag{4.64}$$

$$f \equiv \frac{\rho_A}{\rho_B} = \frac{y_{A,1}^S/y_{B,1}^S}{x_A^F/x_B^F} \tag{4.65}$$

Writing Eq. (4.64) for the limiting conditions of Fig. 4.13*b* and combining with Eq. (4.62), we obtain

$$\frac{S^*}{E^*} = D_A^F - \frac{\rho_A F}{E^*} \tag{4.66}$$

Similarly, Eqs. (4.63), (4.64), and (4.65) combine to yield

$$D_A^F - \frac{\rho_A F}{E^*} = D_B^F - \frac{\rho_A F}{E^* f} \tag{4.67}$$

Equation (4.67) can be rewritten as

$$\frac{F}{E^*} = \frac{D_A^F - D_B^F}{\rho_A(1 - 1/f)} \tag{4.68}$$

Substituting Eq. (4.68) in (4.66):

$$\frac{S^*}{E^*} = \frac{fD_B^F - D_A^F}{f - 1} \tag{4.69}$$

From (4.68) and (4.69):

$$\frac{E^*}{F + S^*} = \frac{\rho_A(f - 1)}{D_A^F(f - \rho_A) - fD_A^F(1 - \rho_A)} \tag{4.70}$$

Equations (4.69) and (4.70) are the limiting flow ratios in the scrubbing and extracting sections, respectively. Equation (4.69) shows that, for the limiting conditions of an infinite number of extracting stages, no scrubbing is required, that is, $S^* = 0$, when

$$f \leqslant \frac{D_A^F}{D_B^F} \tag{4.71}$$

The limiting condition of infinite numbers of extracting and scrubbing stages can achieve complete recovery, that is, $\rho_A = 1$, provided that the extracting section flow ratio given in Eq. (4.70) is adjusted to

$$\frac{E^*}{F + S^*} = \frac{1}{D_A^F} \tag{4.72}$$

The above derivations are limited to the unique case of Fig. 4.16*b*. The solutions in terms of the limiting (asterisked) quantities are valid whenever the equilibrium curve intersects the operating lines only at $x = x_F$. If the equilibrium line is sufficiently curved, as illustrated in Fig. 4.16*c*, one of the operating lines may first intersect its equilibrium line at some point other than x_F, thereby making invalid the above limiting-condition equations for the triple-point intersection.

As a numerical example of these equations, consider an aqueous feed solution containing

	Mol/liter
$Hf(NO_3)_4$	0.00246
$Zr(NO_3)_4$	0.123
HNO_3	3
$NaNO_3$	3.5

to be extracted with 60 v/o TBP in kerosene, as considered in the simple extraction example of Sec. 6.2 (cf. Table 4.6).

From Table 4.4, the distribution coefficient for zirconium is

$$D_{Zr}^F = 1.20$$

Huré and Saint James [H4] have found that the zirconium-hafnium separation factor for this mixture is 10, so that

$$D_{Hf}^F = 0.12$$

Suppose that we wish to recover 98 percent of the zirconium and to obtain a zirconium-hafnium decontamination factor of 200. The limiting ratio of scrub to solvent, from (4.69), is

$$\frac{S^*}{E^*} = \frac{(200)(0.12) - 1.20}{199} = 0.1146$$

The limiting ratio of solvent to scrub plus feed, from (4.70), is

$$\frac{E^*}{F + S^*} = \frac{(0.98)(199)}{(1.2)(199) - (200)(1.2)(0.02)} = 0.833$$

6.5 Extracting-scrubbing Cascade with Constant Distribution Coefficients

To carry out a specified separation in an actual cascade with a finite number of stages, S/E must be greater than the minimum ratio given by (4.69) and $(F + S)/E$ must be less than the maximum ratio given by (4.70). For such an actual cascade it is important to know the number of extracting stages N and stripping stages M needed to effect a specified separation with given values of these flow ratios. When the distribution coefficients of each of the species to be separated are not affected by the concentration of the other extractable species, the number of stages can be calculated either by the graphic procedure illustrated in Fig. 4.15 or, for constant distribution coefficients, by means of the equations to be derived in this section. When the distribution coefficients of each of the extractable species are affected by the concentrations of the other extractable species, it is usually preferable to use the numerical, stage-to-stage calculation procedure to be described in Sec. 6.6. The graphic procedure is usually not useful under these conditions because the location of the equilibrium lines cannot be set in advance of the calculation of the stagewise concentrations of extractable species throughout the cascade.

When D is constant, equations for the number of stages can be outlined by applying the Kremser equation (4.45) derived in Sec. 6.2. For the extracting section the appropriate flow ratio is $E/(F+S)$, and the extraction factor β_E is

$$\beta_E = \frac{DE}{F+S} \tag{4.73}$$

so that, from (4.45), the concentration in the organic leaving the extracting section is

$$y_N^E = \frac{\beta_E^N - 1}{\beta_E - 1}(Dx_1^E - y_0^E) + y_0^E \tag{4.74}$$

The extraction factor β_S in the scrubbing section is

$$\beta_S = \frac{DE}{S} \tag{4.75}$$

Although derived for an extracting section, the Kremser equation (4.45) can be applied to the scrubbing section by the following transformation:

$$\left.\begin{aligned} x_1' &= x_M^S \\ y_0' &= y_{M+1}^S \\ y_1' &= y_1^S \end{aligned}\right\} \tag{4.76}$$

where the primed quantities are to be substituted in Eq. (4.45). This substitution results, with some rearrangement, in

$$y_1^S = \frac{\beta_S^M - 1}{\beta_S - 1} Dx_M^S - \frac{\beta_S^M - \beta_S}{\beta_S - 1} y_{M+1}^S \tag{4.77}$$

To eliminate x_M^S, a material balance is written around the scrubbing section:

$$S(x_M^S - x_0^S) = E(y_{M+1}^S - y_1^S) \tag{4.78}$$

Here we assume that $x_0^S = 0$ and combine (4.78) with (4.75) to obtain

$$y_{M+1}^S = \frac{\beta_S^{M+1} - 1}{\beta_S^{M+1} - \beta_S^M} y_1^S \tag{4.79}$$

By applying the continuity Eq. (4.53) for a cascade with aqueous feed, (4.74) and (4.79) are equated to result in

$$\frac{\beta_S^{M+1} - 1}{\beta_S^{M+1} - \beta_S^M} y_1^S = \frac{\beta_E^N - 1}{\beta_E - 1}(Dx_1^E - y_0^E) + y_0^E \tag{4.80}$$

There is one equation of the form of (4.80) for each of the extractable components. When flow ratios S/E and E/F have been specified, and when any three of the four terminal concentrations x^F, x_1^E, y_0^E, y_1^S, have been specified for two components, values of β_E and β_S for the two components can be calculated. Two equations of the form of (4.80) can be written, one for each component. These can then be solved for the number of stages N and M in the extracting and scrubbing sections, respectively.

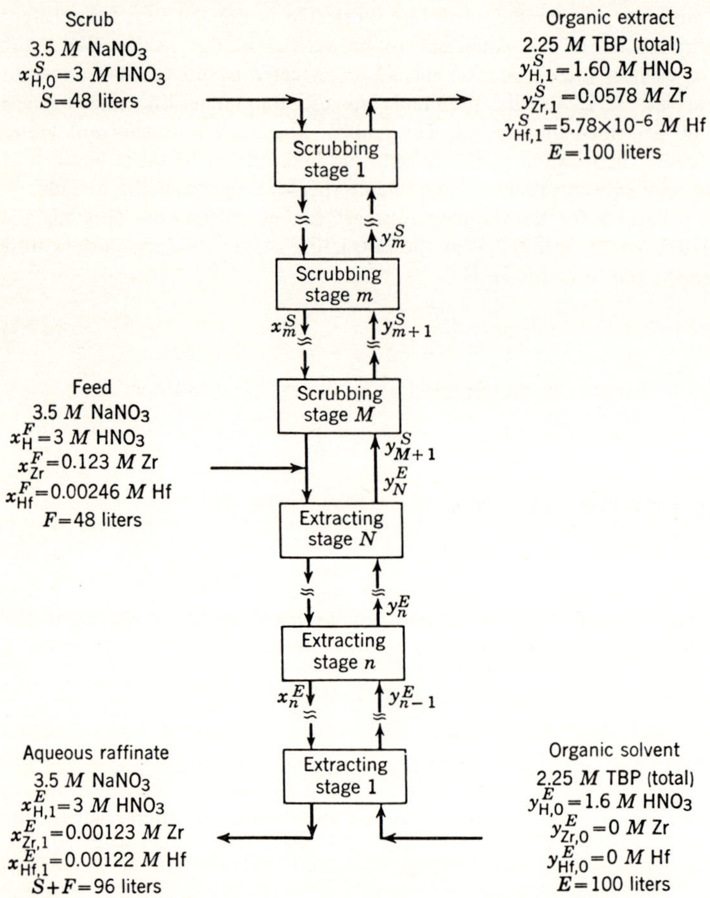

Figure 4.17 Flow sheet for zirconium-hafnium extracting-scrubbing example.

The above equations can be transposed to relatively simple equations for the recovery ρ of an extractable component and the decontamination factor f, as defined by Eqs. (4.64) and (4.65), respectively. To eliminate x^F from (4.64), an overall material balance is written:

$$Fx^F = (S + F)x_1^E + E(y_1^S - y_0^E)$$

or

$$\frac{Fx^F}{E} = \frac{Dx_1^E}{\beta_E} + y_1^S - y_0^E \tag{4.81}$$

Combining (4.64) and (4.81):

$$\rho = \frac{y_1^S}{(Dx_1^E/\beta_E) + y_1^S - y_0^E} \tag{4.82}$$

Now assume, for simplicity, that $y_0^E = 0$, so that Eq. (4.80) becomes

$$y_1^S = \left(\frac{\beta_S^{M+1} - \beta_S^M}{\beta_S^{M+1} - 1}\right)\left(\frac{\beta_E^N - 1}{\beta_E - 1}\right) Dx_1^E \tag{4.83}$$

Substituting (4.83) in (4.82):

$$\rho = \frac{\beta_E[(\beta_E^N - 1)/(\beta_E - 1)]}{(1/\beta_S^M)[(\beta_S^{M+1} - 1)/(\beta_S - 1)] + \beta_E[(\beta_E^N - 1)/(\beta_E - 1)]} \tag{4.84}$$

There is one equation of the form of (4.84) for each extractable component. The values of β for each component are determined by the value of the distribution coefficient for that component and the flow ratios. If the recovery ρ_A of one component is given and the overall decontamination f_{AB} for two extractable components is specified, the recovery ρ_B of the second component is obtained from (4.65). Thus, for known ρ_A and ρ_B and known β's Eq. (4.84) can be solved twice to obtain the necessary numbers M and N of equilibrium stages.

As an example of the use of these equations for an extracting-scrubbing cascade, consider the addition of a scrubbing section to the hafnium-zirconium separation example, which was first analyzed in Sec. 6.2 as a simple extraction problem. The modified flow sheet is shown in Fig. 4.17, and desired recoveries and decontamination are given in Table 4.7.

As noted in Sec. 6.2, and as will be demonstrated in Sec. 6.6, actual distribution coefficients will depend on concentration. Constant distribution coefficients are assumed for the purpose of this illustration, the same as used in the minimum-flow-ratio example of Sec. 6.4. Top and bottom concentrations are as follows.

Organic extract:

Zirconium: $$y_{\mathrm{Zr},1}^S = \frac{\rho F}{E} x_{\mathrm{Zr}}^F = (0.98)(0.48)(0.123) = 0.0578$$

Hafnium: $$y_{\mathrm{Hf},1}^S = \frac{y_{\mathrm{Zr},1}^S/f}{x_{\mathrm{Zr}}^F/x_{\mathrm{Hf}}^F} = \frac{0.0578/200}{0.123/0.00246} = 0.00000578$$

Aqueous residue (by material balance):

Zirconium: $$X_{\mathrm{Zr},1}^E = \frac{(1-\rho)Fx_{\mathrm{Zr}}^F}{S+F} = (0.02)(0.5)(0.123) = 0.00123$$

Hafnium: $$x_{\mathrm{Hf},1}^E = \frac{Fx_{\mathrm{Hf}}^F - Ey_{\mathrm{Hf},1}^S}{S+F} = (0.5)(0.00246) - (1.04)(0.00000578) = 0.00122$$

With the above quantities the extraction factors are as follows.

Zirconium:

Extracting: $$\beta_{E,\mathrm{Zr}} = \frac{D_{\mathrm{Zr}}E}{S+F} = (1.20)(1.04) = 1.248$$

Scrubbing: $$\beta_{S,\mathrm{Zr}} = \frac{D_{\mathrm{Zr}}E}{S} = \frac{1.20}{0.48} = 2.50$$

Hafnium:

Extracting: $$\beta_{E,\mathrm{Hf}} = \frac{D_{\mathrm{Hf}}E}{S+F} = (0.12)(1.04) = 0.1248$$

Scrubbing: $$\beta_{S,\mathrm{Hf}} = \frac{D_{\mathrm{Hf}}E}{S} = \frac{0.12}{0.48} = 0.250$$

Equation (4.80) written for zirconium is

$$\frac{2.50^{M+1} - 1}{2.50^{M+1} - 2.50^M} 0.0578 = \frac{1.248^N - 1}{1.248 - 1}(1.20)(0.00123)$$

or

$$16.2[1 - (0.40)^{M+1}] = (1.248)^N - 1$$

Equation (4.80) written for hafnium is

$$\frac{0.25^{M+1} - 1}{0.25^{M+1} - 0.25^M} 0.00000578 = \frac{(0.1248)^N - 1}{0.1248 - 1} (0.12)(0.00122)$$

or

$$(4.0)^{M+1} - 1 = 86.8[1 - (0.1248)^N]$$

Since $(0.40)^{M+1}$ and $(0.1248)^N$ are small relative to unity, approximate solutions of these equations are

$$N \approx \frac{\ln 17.2}{\ln 1.248} = \frac{2.84}{0.221} = 12.9$$

and

$$M + 1 \approx \frac{\ln 87.8}{\ln 4} = \frac{4.48}{1.396} = 3.21$$

Although slightly more accurate values could be obtained by using these results for a second approximate solution, it is evident that terms that have been neglected would not change the result appreciably.

The McCabe-Thiele diagrams of Fig. 4.18 for the zirconium-hafnium separation example illustrate the principles of the extracting-scrubbing separation system. In the extracting section the flow ratios are such that there are large changes in concentration of the more extractable component (Zr), but relatively little change in the aqueous concentration of the less extractable component (Hf). In the scrubbing section the flow ratios are such that the larger changes in concentration occur for the less extractable component (Hf).

Table 4.7 Specifications for zirconium-hafnium separation example in an extracting-scrubbing cascade

Given†	
Aqueous feed concentration	
Zirconium	$x^F_{Zr} = 0.123$ mol/liter
Hafnium	$x^F_{Hf} = 0.00246$ mol/liter
Solvent feed concentration	
Zirconium	$y^E_{Zr,0} = 0$
Hafnium	$y^E_{Hf,0} = 0$
Zirconium recovery	$\rho = \frac{Ey^S_{Zr,1}}{Fx^E_{Zr}} = 0.98$
Hafnium decontamination factor	$f = \frac{y^S_{Zr,1}/y^S_{Hf,1}}{x^F_{Zr}/x^F_{Hf}} = 200$
Flow ratios	$S/E = 0.48$ $E/(F+S) = 1.04$ $F/E = 0.48$
Distribution coefficients, assumed to be constant for all stages	$D_{Zr} = 1.20$ $D_{Hf} = 0.12$
Required	
Number of stages	
Extracting, N	
Scrubbing, M	

†As per J. Huré and R. Saint James [H4].

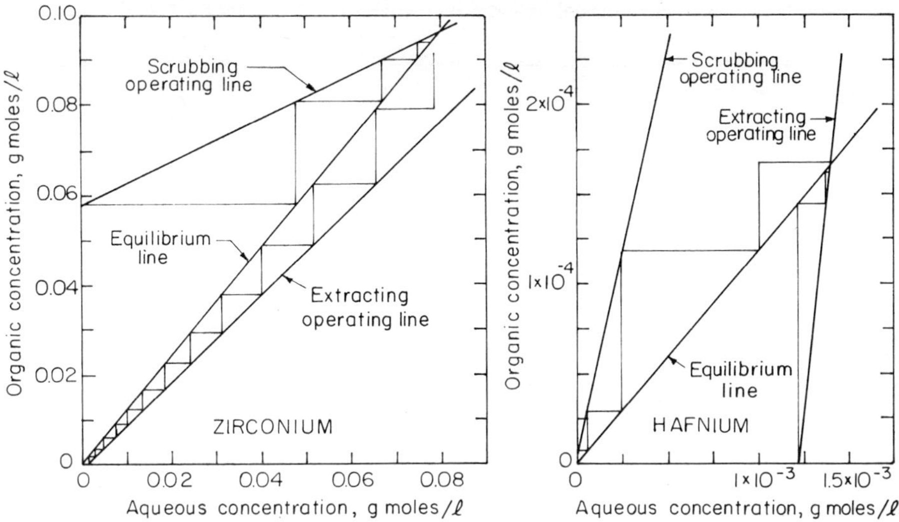

Figure 4.18 Stage concentration diagram for zirconium-hafnium extracting-scrubbing example, with constant distribution coefficients.

6.6 Extracting-scrubbing Cascade: Numerical Procedure for Use with Variable Distribution Coefficients

The algebraic procedure described in Sec. 6.5 is convenient for obtaining a rough estimate of the number of stages needed in fractional extraction, but is seldom accurate enough for design purposes, because distribution coefficients usually change as concentrations change from stage to stage. This section describes a numerical procedure that is generally applicable whenever distribution coefficients are known and the phases leaving an extraction stage are in equilibrium.

The procedure will be illustrated by the zirconium-hafnium separation example treated in Sec. 6.5. The material-balance quantities used for the present illustration are listed in Table 4.8. The relative flow rates of solvent, scrub, and feed streams are those recommended by Huré and Saint James [H4], as are the concentrations of HNO_3 and $NaNO_3$ in solvent, scrub, and feed streams. Changes in volume of the aqueous and organic streams within the scrubbing section and within the extracting section are to be neglected. The concentrations of total TBP in the organic stream is specified as 2.25 mol/liter, and the slight solubility of TBP in water is to be neglected. The concentration of zirconium and hafnium in the aqueous feed, the required zirconium recovery, and the required hafnium decontamination factor are the same as in the example of Sec. 6.5; the concentrations of $Zr(NO_3)_4$ and $Hf(NO_3)_4$ in the raffinate and extract streams are thereby specified.

After the above variables have been set, the cascade is fully specified and only one set of values for the number of stages in extracting and scrubbing sections will perform the specified separation. All other extractable components will be distributed in a determinate manner between aqueous-residue and organic-extract streams.

In the present example, nitric acid is an extractable component, whose split between extract and raffinate streams cannot be specified in advance. However, in the calculational procedure to be described, it is necessary to start with specified concentrations of all

Table 4.8 Material balance for zirconium-hafnium separation by fractional extraction with TBP†

	In				Out		
Stream:	Feed	Scrub	Solvent	Total	Residue	Extract	Total
Phase:	Aqueous	Aqueous	Organic		Aqueous	Organic	
Gram-moles/ liter:							
TBP	–	–	2.25	–	–	2.25	
$NaNO_3$	3.5	3.5	–	–	3.5		
HNO_3	3.0	3.0	1.6	–	3.03‡	1.576‡	
$Zr(NO_3)_4$	0.123	0.000	0.000	–	0.00123	0.0578	
$Hf(NO_3)_4$	0.00246	0.000	0.000	–	0.00122	5.78×10^{-6}	
Liters	48	48	100	–	96	100	
Gram-moles:							
TBP	–	–	225	225	–	225	225
$NaNO_3$	168	168	–	336	336	–	336
HNO_3	144	144	160	448	291	158	449
$Zr(NO_3)_4$	5.90	0.000	0.000	5.90	0.118	5.78	5.90
$Hf(NO_3)_4)$	0.118	0.000	0.000	0.118	0.117	0.006	0.118

†Basis: 100 liters of solvent.

‡These concentrations cannot be specified in advance and must be confirmed by calculation or experiment.

components in terminal streams from the cascade, so that provisional values must be assumed for nitric acid concentration in the extract streams, and the corresponding concentration of nitric acid in the raffinate is obtained from a material balance. If subsequent calculation fails to confirm the correctness of these provisional values, new values must be assumed for acid concentration in the extract stream and the calculation repeated. The particular values of exit acid concentrations listed in Table 4.8 were arrived at by several iterative calculations and give a consistent analytical solution to the separation problem.

To obtain distribution coefficients as a function of concentration, it will be assumed that equilibria are established in the three following reactions:

$$H^+(aq) + NO_3^-(aq) + TBP(o) \rightleftharpoons HNO_3 \cdot TBP(o) \qquad K_H = 0.145$$

$$Zr^{4+}(aq) + 4NO_3^-(aq) + 2TBP(o) \rightleftharpoons Zr(NO_3)_4 \cdot 2TBP(o) \qquad K_{Zr} = 0.0032$$

$$Hf^{4+}(aq) + 4NO_3^-(aq) + 2TBP(o) \rightleftharpoons Hf(NO_3)_4 \cdot 2TBP(o) \qquad K_{Hf} = 0.00032$$

The value of 0.145 for the equilibrium constant of the nitric acid complex is an average value derived from the equilibrium data of Moore [M2], Alcock et al. [A1], and Gruverman [G7]. The value of 0.0032 for the zirconium equilibrium is the average value derived from the equilibrium data in Table 4.4. The value of 0.00032 for the hafnium equilibrium is derived from the separation factor of 10 measured for zirconium-hafnium mixtures by Huré and Saint James [H4]. Distribution coefficients are then given by the following equations.

$$HNO_3: \qquad D_H = \frac{y_H}{x_H} = 0.145 y_{TBP} x_{NO_3^-} \tag{4.85}$$

$$Zr(NO_3)_4: \qquad D_{Zr} = \frac{y_{Zr}}{x_{Zr}} = 0.0032 (y_{TBP})^2 (x_{NO_3^-})^4 \tag{4.86}$$

$$\text{Hf(NO}_3)_4: \qquad D_{\text{Hf}} = \frac{y_{\text{Hf}}}{x_{\text{Hf}}} = 0.00032(y_{\text{TBP}})^2(x_{\text{NO}_3^-})^4 \tag{4.87}$$

where y_{TBP} is the concentration of *uncombined* TBP in the organic phase and $x_{\text{NO}_3^-}$ is the *total* nitrate concentration in the aqueous phase. For the concentrations assumed for Table 4.8, y_{TBP} is obtained from

$$y_{\text{TBP}} = 2.25 - [y_{\text{H}} + 2(y_{\text{Zr}} + y_{\text{Hf}})] \text{ mol/liter} \tag{4.88}$$

where 2.25 is the concentration of TBP in all forms in the organic phase. The aqueous nitrate concentration is obtained from

$$x_{\text{NO}_3^-} = 3.5 + x_{\text{H}} + 4(x_{\text{Zr}} + x_{\text{Hf}}) \text{ mol/liter} \tag{4.89}$$

where 3.5 is the constant concentration of sodium nitrate in the aqueous phase.

To find the required number of stages, a calculation is made of the concentrations of zirconium nitrate, hafnium nitrate, and nitric acid in the organic phase as a function of stage number in the scrubbing and extracting sections. A plot similar to Fig. 4.19 is then made of the concentration of zirconium versus the concentration of hafnium in each section. The point of intersection of the curves gives the concentrations of these components in the organic phase flowing between the two sections at the feed point, at which

$$\text{Zr:} \qquad y^S_{\text{Zr},M+1} = y^E_{\text{Zr},N} \tag{4.90}$$

$$\text{Hf:} \qquad y^S_{\text{Hf},M+1} = y^E_{\text{Hf},N} \tag{4.91}$$

The number of stages in each section (M and N) at which these concentrations are equal is the result of the first trial calculation.

A check is then made to determine if the nitric acid concentration in the organic phase leaving the extracting section equals that in the organic phase entering the scrubbing section. If

$$y^E_{\text{H},N} = y^S_{\text{H},M+1} \tag{4.92}$$

the provisional values assumed for nitric acid concentrations in residue and extract streams are correct. If these concentrations are not equal, new values must be assumed for nitric acid concentrations in the residue and extract streams and the entire calculational procedure repeated.

Steps in the calculation of the concentrations of zirconium nitrate, hafnium nitrate, nitric acid, and uncombined TBP as a function of stage number n in the extracting section are given in Table 4.9.

In order to calculate equilibrium concentrations y^E_1 in the organic stream leaving stage 1, $x^E_{\text{NO}_3^-,1}$ is calculated from Eq. (4.89), a trial value of $y^E_{\text{TBP},1}$ is assumed, and D^E_1 and y^E_1 are obtained from Eqs. (4.85) through (4.87). The trial value of $y^E_{\text{TBP},1}$ is checked by using the y^E_1 in Eq. (4.88), and when a trial value of $y^E_{\text{TBP},1}$ that satisfies this condition is found, a consistent set of concentrations in the streams leaving the first extracting stage has been obtained.

The next step is to apply the material-balance equation (4.51) to calculate the concentrations in the aqueous phase leaving stage 2. The calculation then proceeds up through the extracting section by a repetition of these steps.

Steps in the calculation of concentrations as a function of stage number m in the scrubbing section are given in Table 4.10. The calculation is started with the specified concentrations y^S_1 in the organic extract stream leaving stage 1, obtained from Table 4.8. Because the concentrations y^S_1 have been chosen, the value of $y^S_{\text{TBP},1}$ can be calculated from Eq. (4.88), and a value of $x^S_{\text{NO}_3^-,1}$ must be assumed to calculate values of D^S_1 and x^S_1 from Eqs. (4.85) through (4.87). When the assumed value of $x^S_{\text{NO}_3^-,1}$ agrees with that obtained by substituting the corresponding values of x^S_1 in Eq. (4.89), a consistent set of concentrations in the streams leaving the first scrubbing stage has been obtained.

Table 4.9 Concentrations in extracting section, zirconium-hafnium separation example

Stage no. n	Aqueous concentration, mol/liter $x_n^E = x_1^E + \frac{E}{S+F}(y_{n-1}^E - y_0^E)$ HNO$_3$ $x_{H,n}^E$	Zr(NO$_3$)$_4$ $x_{Zr,n}^E$	Hf(NO$_3$)$_4$ $x_{Hf,n}^E$	Total NO_3^- [a] $x_{NO_3^-,n}^E$	Assumed free TBP in organic phase $y_{TBP,n}^E$	Distribution coefficients HNO$_3$ D_H [b]	Zr(NO$_3$)$_4$ D_{Zr} [c]	Hf(NO$_3$)$_4$ D_{Hf} [d]	Organic concentration, mol/liter $y_n^E = Dx_n^E$ HNO$_3$ $y_{H,n}^E$	Zr(NO$_3$)$_4$ $y_{Zr,n}^E$	Hf(NO$_3$)$_4$ $y_{Hf,n}^E$	Free TBP [e] $y_{TBP,n}^E$
1	3.03[f]	0.001230[f]	0.001224[f]	6.53	0.570	0.540	1.897	0.1897	1.635	0.00233	0.000232	0.610
					0.580	0.550	1.967	0.1967	1.664	0.00242	0.000241	0.580
2	3.09	0.00375	0.001475	6.61	0.565	0.542	1.954	0.1954	1.678	0.00732	0.000288	0.577
					0.563	0.540	1.944	0.1944	1.671	0.00729	0.000287	0.563
3	3.10	0.00882	0.001523	6.64	0.555	0.534	1.918	0.1918	1.657	0.01692	0.000292	0.558
					0.556	0.535	1.924	0.1924	1.660	0.01696	0.000293	0.556
4	3.09	0.01900	0.001529	6.70	0.550	0.532	1.916	0.1916	1.643	0.0362	0.000293	0.534
					0.546	0.528	1.890	0.1890	1.632	0.0357	0.000289	0.546
5	3.06	0.0384	0.001525	6.72	0.537	0.523	1.880	0.1880	1.600	0.0723	0.000287	0.504
					0.530	0.516	1.830	0.1830	1.579	0.0703	0.000279	0.530
6	3.00	0.0745	0.001515	6.81	0.504	0.497	1.745	0.1745	1.494	0.1299	0.000264	0.496
					0.502	0.496	1.733	0.1733	1.489	0.1291	0.000263	0.502

$$\frac{E}{S+F} = 1.041 \qquad y_{H,0}^E = 1.60 \qquad y_{Zr,0}^E = 0 \qquad y_{Hf,0}^E = 0$$

[a] Calculated from Eq. (4.89).
[b] Calculated from Eq. (4.85).
[c] Calculated from Eq. (4.86).
[d] Calculated from Eq. (4.87).
[e] Calculated from Eq. (4.88).
[f] Specified concentrations.

Table 4.10 Concentrations in scrubbing section, zirconium-hafnium separation example

	Organic concentration, mol/liter											
	$y_m = y_1^S + \frac{S}{E}(x_{m-1}^S - x_0^S)$				Assumed total NO_3^- in aqueous phase	Distribution coefficients			Aqueous concentration, mol/liter $x_m^S = \frac{y_m^S}{D}$			Total NO_3^- concentration in aqueous phase[e]
Stage no. m	HNO_3 $y_{H,m}^S$	$Zr(NO_3)_4$ $y_{Zr,m}^S$	$Hf(NO_3)_4$ $y_{Hf,m}^S$	Free TBP[a] $y_{TBP,m}^S$	$x_{NO_3^-,m}^S$	HNO_3 D_H[b]	$Zr(NO_3)_4$ D_{Zr}[c]	$Hf(NO_3)_4$ D_{Hf}[d]	HNO_3 $x_{H,m}^S$	$Zr(NO_3)_4$ $x_{Zr,m}^S$	$Hf(NO_3)_4$ $x_{Hf,m}^S$	$x_{NO_3^-,m}^S$
1	1.576[f]	0.0578[f]	0.00000578[f]	0.558	6.57	0.532	1.859	0.1859	2.96	0.0311	31.1×10^{-6}	6.58
					6.58	0.533	1.872	0.1872	2.96	0.0309	30.9×10^{-6}	6.58
2	1.553	0.0727	0.0000206	0.549	6.60	0.526	1.837	0.1837	2.95	0.0396	0.0001122	6.61
					6.61	0.526	1.845	0.1845	2.95	0.0394	0.0001118	6.61
3	1.553	0.0768	0.0000594	0.543	6.64	0.523	1.834	0.1834	2.97	0.0418	0.000324	6.64
4	1.562	0.0780	0.0001613	0.532	6.68	0.516	1.809	0.1809	3.03	0.0431	0.000891	6.70
					6.70	0.517	1.822	0.1822	3.02	0.0428	0.000885	6.70
5	1.586	0.0784	0.000431	0.506								

$$\frac{S}{E} = 0.480 \qquad x_{H,0}^S = 3.0 \qquad x_{Zr,0}^S = 0 \qquad x_{Hf,0}^S = 0$$

[a] Calculated from Eq. (4.88).
[b] Calculated from Eq. (4.85).
[c] Calculated from Eq. (4.86).
[d] Calculated from Eq. (4.87).
[e] Calculated from Eq. (4.89).
[f] Specified concentrations.

The next step is to apply the material-balance equation (4.52) to calculate the concentrations in the organic stream leaving stage 2. The calculation then proceeds down through the scrubbing section by a repetition of these steps.

The large variation in distribution coefficients is noteworthy. This is caused principally by changes in the concentration of uncombined TBP. This large variation makes necessary the use of a numerical calculation method.

Figure 4.19 is a plot of zirconium concentration versus hafnium concentration in the organic phase, with points for the extracting section from Table 4.9 and points for the scrubbing section from Table 4.10. The point of intersection occurs at

$$\text{Extracting:} \qquad 5 < N < 6$$

$$\text{Scrubbing:} \qquad 4 < (M + 1) < 5$$

By interpolation, the point of intersection of the curves of Fig. 4.18 occurs at

$$\text{Extracting:} \qquad N = 5.2$$

$$\text{Scrubbing:} \qquad M + 1 = 4.6$$

Thus, six theoretical stages in the extracting section and four in the scrubbing section would result in higher values of zirconium recovery and hafnium decontamination than those specified.

Finally, a check must be made to determine if organic concentrations of nitric acid in the extracting and scrubbing sections match at the feed point. That this condition is satisfied can be seen from Fig. 4.20, a plot of organic concentrations leaving a stage versus stage number. If nitric acid does not match at the feed point, a new value of nitric acid concentration in the organic extract must be assumed and the entire calculational procedure repeated.

Figure 4.20 and the McCabe-Thiele diagrams of Fig. 4.21 illustrate the principle of separation of two metals by solvent extraction with complexing agents. The largest change in organic concentration of the more extractable component, zirconium, occurs in the extracting section. After the first extracting stage the hafnium soon reaches a concentration in the organic that is almost in equilibrium with the hafnium in the aqueous feed entering the succeeding stages. The hafnium concentration in organic is actually reduced somewhat as the feed stage is

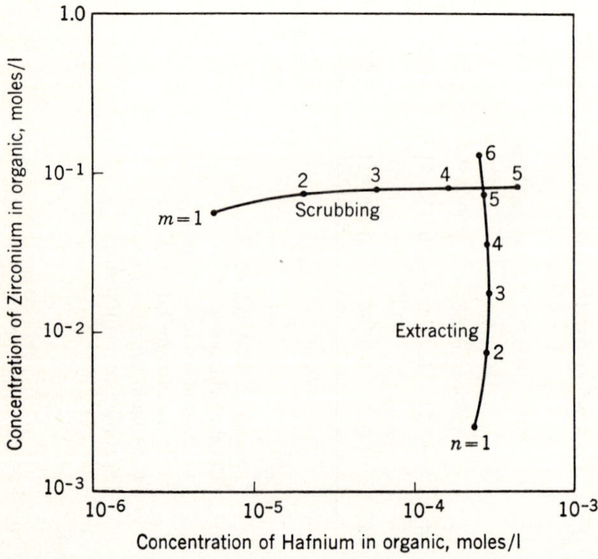

Figure 4.19 Zirconium and hafnium concentrations in organic phase, to determine feed point in zirconium-hafnium separation example.

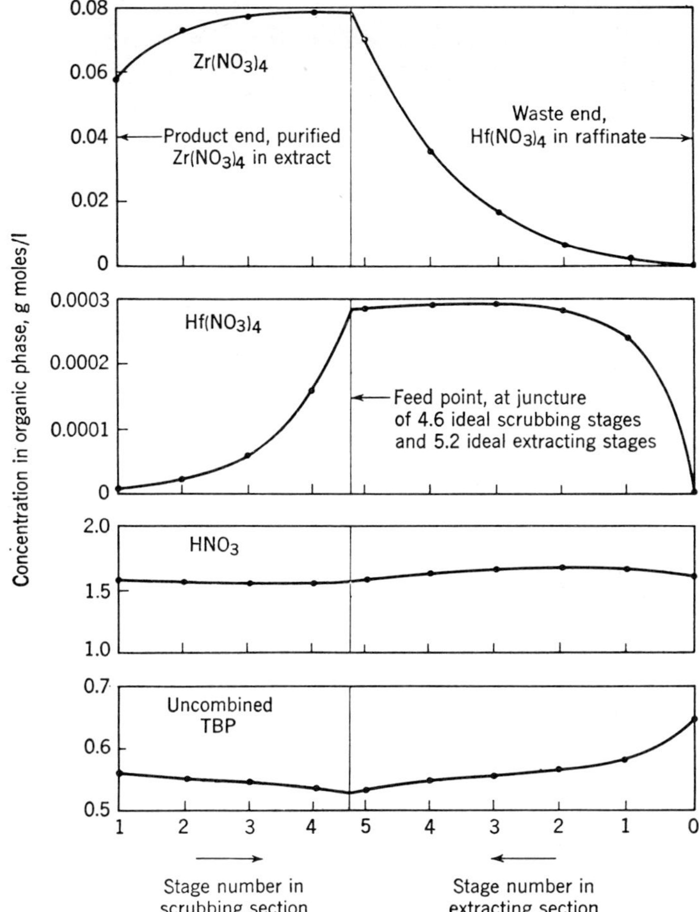

Figure 4.20 Concentrations in organic phase leaving a stage as a function of stage number for zirconium-hafnium separation.

approached, because the decreasing concentration of uncombined TBP reduces the hafnium distribution coefficient. In the scrubbing section the lower flow ratio of aqueous to organic allows a relatively large decrease in the concentration of hafnium in the organic. Because some zirconium is removed from the organic in the scrubbing section, the maximum zirconium concentration occurs near the feed point, resulting in the lowest concentration of uncombined TBP at this point. The equilibrium lines in the extracting and scrubbing sections are essentially identical for zirconium, but not for hafnium.

The feed conditions have been chosen so that nitric acid in the organic feed is nearly in equilibrium with the nitric acid in the aqueous feed and scrub solution, so that little change in nitric acid concentration occurs through the cascade. The small changes in nitric acid concentration are due principally to variations in amount of TBP complexed by zirconium.

The SEPHIS [G6, H2, W1] and SOLVEX [S1] computer codes are applications of the above techniques for calculating the number of equilibrium stages for the TBP extraction of uranium and plutonium in nitric acid solution. These codes include correlations of the

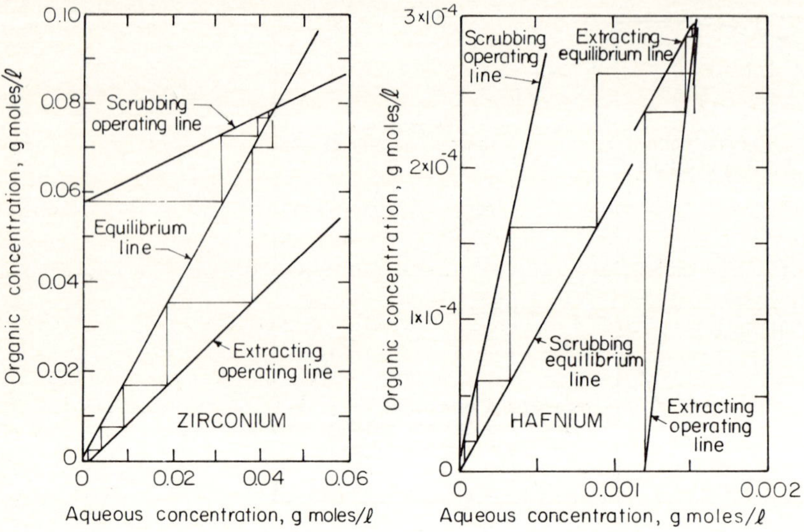

Figure 4.21 Stage concentration diagram for zirconium-hafnium extracting-scrubbing example.

extraction equilibrium constants and activity coefficients as affected by total ionic strength. Corrections for incomplete ionization of aqueous plutonium may also be important.

7 SOLVENT EXTRACTION EQUIPMENT

7.1 Requirements

The principal functional requirements of a solvent extraction contactor are as follows:

1. To develop sufficient interfacial area to promote transfer of extractable components between phases
2. To facilitate countercurrent flow of the two phases, without excessive entrainment

Additional considerations in selecting contacting equipment are as follows:

3. It should have flexibility to operate under varied conditions of flow ratios and concentrations.
4. It should be mechanically dependable and easy to operate and maintain.
5. It should be compact and have low holdup of process materials.
6. Initial cost and operating cost should be low.

The importance of these different factors varies with the application. Although reliability is important for any application, it is particularly important when processing highly radioactive materials, as in reprocessing discharged reactor fuel, where the intense radioactivity makes normal methods of maintenance difficult or impossible. Equipment handling such highly radioactive solutions must be enclosed in massive shielding and operated, and perhaps maintained, by remote control. If concrete is used as shielding material, thicknesses of 2 to 3 m are often required. To control the cost of such massive shielding, it is important to reduce its bulk, and this means using compact solvent extraction contactors. If vertical column contactors are used, their height should be kept to a minimum. If a horizontal array of mixer-settler contactors is used, the layout of equipment and piping should be compact.

Those applications that require remote operation and maintenance dictate the use of simple, rugged equipment, with a minimum of moving parts and with little tendency to foul, rust, clog, or corrode.

The intense radioactivity associated with the reprocessing of discharge fuel, and the degradation and decomposition of organic solvent when exposed to ionizing radiation, require that the amount of solvent exposed to radiation be kept to a minimum. Also, the length of time that the organic is exposed to radiation should be kept small. This places a premium on compact contacting equipment, with high throughput per unit volume.

Nuclear criticality places special constraints on the size of contacting equipment in fuel reprocessing. This is particularly important when reprocessing highly enriched uranium fuel or for the stripping-scrubbing contactors that separate plutonium from low-enriched uranium, as illustrated in Fig. 4.5. Both ^{235}U and plutonium fission. As shown by the criticality data in Table 4.11, as little as 760 g of ^{235}U or 510 g of plutonium can form a critical mass when dispersed at the optimum concentration through a hydrogenous medium, such as a nitric acid solution or organic, with relatively little fission products or nonfissile uranium [A2, T1]. The sizes of the contactors and other process equipment must be kept small enough to promote neutron leakage and make criticality impossible [C4]. Limiting dimensions may be as small as 14 cm in diameter for a cylindrical column contactor or 4.6 cm in height for an array of mixer-settlers in horizontal slab geometry [A2, T1]. Larger equipment sizes are acceptable for process operations that do not involve solutions of relatively pure fissile material, such as the extracting-scrubbing contactors that separate the fission products from low-enrichment uranium fuel. The allowable dimensions and throughput of criticality-limited process equipment can be increased by incorporating fixed neutron absorbers, i.e., "poisons," such as boron or gadolinium, without the equipment.

Contactors with low inventory of process solutions are also important when the material processed is valuable, such as the plutonium recovered from irradiated fuel. Low inventory is also important in maintaining a close accountability of the total inventory of fissionable material processed.

Table 4.11 Nuclear criticality limits for uniform aqueous solution reflected by an effectively infinite thickness of water[†]

Parameter	Subcritical limit[‡] for ^{235}U	^{233}U	^{239}Pu[§]
Mass of fissile nuclide, kg	0.76	0.55	0.51
Solution cylinder diameter, cm	13.9	11.5	15.7
Solution slab thickness, cm	4.6	3.0	5.8
Solution volume, liters	5.8	3.5	7.7
Concentration of fissile nuclide, g/liter	11.5	10.8	7.0
Areal density of fissile nuclide, g/cm^2	0.40	0.35	0.25

[†] Data from American National Standard for Nuclear Criticality Safety in Operations with Fissionable Materials Outside Reactors [A2] and J. T. Thomas [T1].

[‡] The fissile material is subcritical if any one of the listed conditions is met, with no other fissile species present.

[§] Provided the nitrogen-to-plutonium atom ratio is equal to or greater than 4.0.

7.2 Types of Equipment

Types of solvent extraction contactors in general commercial use for nonnuclear applications include [L1, M3, T2]:

1. Spray columns
2. Baffle-plate columns
3. Perforated-plate columns
4. Columns packed with Raschig rings or Berl saddles
5. Mixer-settlers

Some of this conventional equipment has been applied to the solvent extraction purification of natural uranium and thorium. All of these conventional gravity column contactors are less compact than is desirable for reprocessing irradiated reactor fuel. The height of a vertical-column gravity contactor equivalent to a single equilibrium stage of contacting is about 60 to 120 cm, so very tall columns would be needed for the 10 or more theoretical stages needed for some of the separations in fuel reprocessing.

The need for compact contactors in reprocessing irradiated reactor fuel, as well as the need for many stages and small inventory in purifying special organic materials and pharmaceutical products, has stimulated the development of solvent extraction contacting equipment with reduced holdup, reduced stage height, or both. Most of these devices have in common the input of mechanical energy to promote contacting of phases, separation of phases, and/or countercurrent flow. Contactors that have been used in reprocessing irradiated reactor fuel are listed in Table 4.12, adapted from a compilation by Davis and Jennings [D1]. Additional high-performance contactors that have been used in some nuclear applications and in the pharmaceutical industry include the Fenske-Long extractor, a vertical stack of mixer-settlers, and the Podbielniak centrifugal extractor [T2].

The contactors used in nuclear applications are described in the following sections.

7.3 Mixer-Settlers

One of the most compact and efficient of the mixer-settlers is the pump-mix mixer-settler, developed by Coplan et al. [C5, C6] specifically for radiochemical separations. One stage of this device is shown schematically in Fig. 4.22. A countercurrent cascade of mixer-settlers is shown in Fig. 4.23. The stage consists of a mixing chamber at the left of Fig. 4.22 and a settling chamber at the right. The rotating impeller in the mixing chamber serves to promote equilibrium contact between light and heavy phases, to pump phases between adjacent stages, and to control liquid levels. The mixing chamber is divided into an upper and a lower compartment by a horizontal baffle, pierced with a hole somewhat larger than the impeller shaft. This shaft, which is hollow, passes through the upper compartment of the mixing chamber and dips into the lower. The impeller draws liquid from the bottom compartment through the hollow shaft and discharges it through holes between the blades into the upper compartment. This pumping action of the impeller maintains an interface between mixed phases and the heavy phase at the bottom of the shaft. A free surface between mixed phases and air is maintained in the upper part of the mixing chamber. This free surface is progressively lower in adjacent stages in the direction of light-phase flow.

Mixed phases flow by gravity from the mixing chamber past a baffle into the settling chamber (closed at top) where the two phases separate. The heavy phase flows through a trapped outlet near the bottom (not shown) into the bottom compartment of the mixing chamber of the adjacent stage, where the interface controlled by the stirrer shaft is at a lower level. The light phase flows through an outlet near the top of the settling chamber directly into the mixing chamber of the adjacent stage in the opposite direction.

Table 4.12 Solvent extraction contactors for reprocessing irradiated fuels

	Relative capacity for processing low-enrichment uranium fuels, per unit of equipment	Relative capacity for processing enriched fuels in critically safe design	Amount of shielding per unit capacity	Flexibility	Reliability in plant service
Mixer-Settlers					
Pump mix	Medium	Medium[a]	Medium	Excellent[d]	Excellent [J1]
Centrifugal	Large	Large[b]	Small	–	–
Air ejector or air pulsed	Medium-small	Medium-small[a]	Medium	Good	–
Columns					
Pulsed sieve plate	Medium	Large[c]	Medium	Good	Excellent [G2]
Rotary extractor	Small	Medium-small	Large	Excellent[e]	–
Pulsed packed	Medium	Large[c]	Medium	Good	Good
Packed	Small	Medium[c]	Large	Fair	Good [I2]

[a]Height of mixer-settler limited to 7.6 cm. Not amenable to efficient neutron poisoning for criticality control.

[b]Built of stainless steel containing a neutron poison such as gadolinium or boron.

[c]Sieve plates or packing constructed of "poisoned" stainless steel, thus allowing large-diameter columns.

[d]With variable-speed drive and replaceable impellers.

[e]With variable-speed drive.

Source: Adapted from M. W. Davis and A. S. Jennings, "Equipment for Processing by Solvent Extraction," in *Chemical Processing of Reactor Fuels,* J. F. Flagg (ed.), Academic, New York, 1961, by permission.

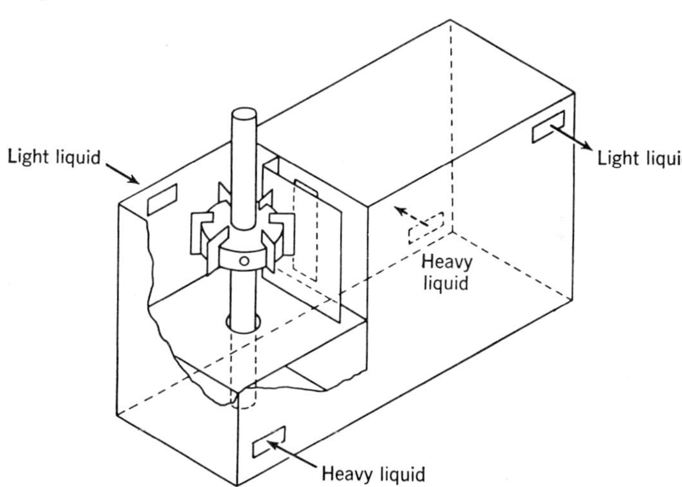

Figure 4.22 Pump-mix mixer-settler (schematic).

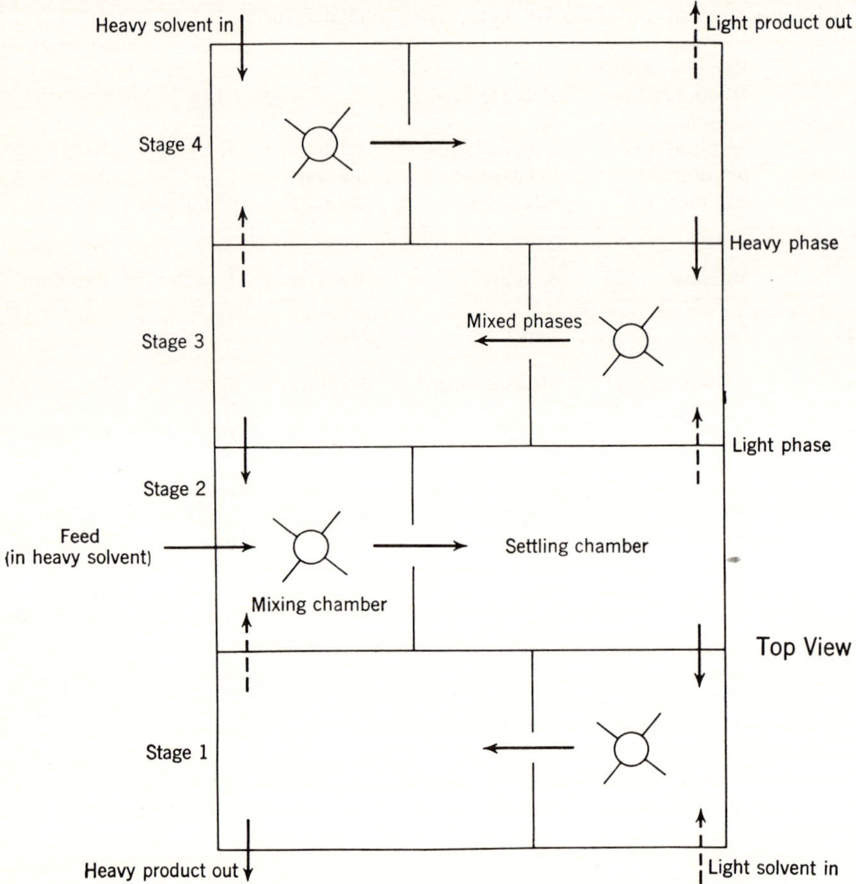

Figure 4.23 Flow through cascade of mixer-settlers.

Flow of heavy phase in one direction is induced by the pumping action of the stirrer, coupled with the trap between settling and mixing sections; flow of light phase in the opposite direction is induced by the progressively lower level of the free surface at the top of the mixing sections.

For an improved version, shown in Fig. 4.24, the impeller is a volute-vaned pump which recirculates an emulsion of the mixed phases within the mixing section [D1]. Interface control weirs provide flexibility for adjusting impeller speed and mixing intensity without upsetting the net interstage flow rate. Solvent from the previous stage flows into the vortex above the impeller and also into the interface weir section, resulting in pumping action to establish hydraulic gradients for the interstage flow of both solvent and aqueous streams.

Properties of two improved pump-mix mixer-settlers, one designed for a total interstage flow of 7.6 liters/min and the other for 380 liters/min, are summarized in Table 4.13 [D1]. The holdup times per contactor are 1 and 2 min, respectively. The small unit was designed for reprocessing highly enriched uranium fuel and, by limiting its height to 7.6 cm, is critically safe up to ^{235}U concentrations of 400 g/liter, provided it is not located near any dense material that can reflect neutrons. The large unit is suitable only for process solutions of low fissile enrichment. Aqueous-to-solvent flow ratios of 0.1 to 1.5 can be attained without excessive

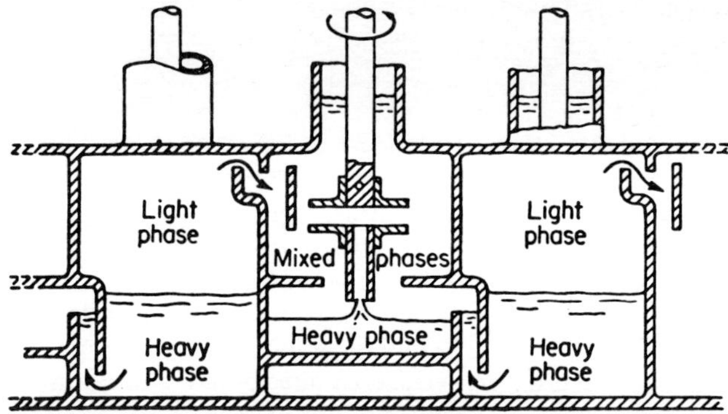

Figure 4.24 Mixing section of improved mixer-settler. (*From Davis and Jennings [D1], by permission.*)

entrainment in the interstage flow streams. Stage efficiencies, i.e., the actual interphase transfer per mixer-settler stage relative to that predicted for an equilibrium stage, of greater than 80 percent have been obtained in the TBP extraction of uranium and plutonium [D1].

The pump-mix mixer-settler is readily scaled over a wide range of throughputs, and because individual stages can be relied on to perform at high and reproducible efficiency, there is less risk in designing a production-scale separation plant than with some of the other types of solvent extraction contactors. A production plant can be designed with assurance on the basis of single-stage equilibrium data, data from a small-scale mixer-settler cascade, and hydraulic tests on a small section of a full-scale mixer-settler cascade. The horizontal arrangement of a mixer-settler cascade permits interruption of steady-state operation and shutdown for several hours without losing the concentration gradient of the cascade, so that the cascade can be restarted relatively easily.

Mixer-settler contactors of much larger scale are used in the solvent extraction operations associated with production of natural uranium (cf. Chap. 5), wherein nuclear criticality is not

Table 4.13 Description of improved pump-mix mixer-settlers

Specifications	Large	Small
Volume of settling section, liters	723	6.17
Volume of mixing section, liters	68	0.68
Volume of aqueous inlet section, liters	34	0.31
Total volume per stage, liters	825	8.85
Impeller		
Vane diameter, cm	23	8.9
Vane thickness, cm	3.8	1.6
Suction nozzle diameter, cm	5	1.6
Aqueous recirculation hole diameter, cm	23	2.5
Capacity, total flow of both phases, liter/min	380	7.6
Holdup time, min	2.2	1.2

Source: M. W. Davis and A. S. Jennings, "Equipment for Processing by Solvent Extraction," in *Chemical Processing of Reactor Fuels,* J. F. Flagg (ed.), Academic, New York, 1961, by permission.

an issue. In the Kerr-McGee uranium extraction plant at Shiprock, New Mexico, where uranium-bearing leach liquor is contacted with alkyl phosphate in kerosene, there are four stages of mixer-settlers [T2]. Each stage consists of a wood-stave settling tank 4.9 m in diameter and 2.1 m high in which a 1.2-m-diameter stainless steel mixing vessel with a 0.46-m-diameter turbine is placed. The aqueous flow of 6.3 liter/s is contacted with 1.3 liter/s of organic. The latter is pumped from one stage to the next by air-lift pumps. Interstage aqueous flow is by gravity, with elevation differences of 0.3 m between successive stages. The estimated holdup time per stage is 50 min.

The mixer-settler used at the Vitro uranium recovery operation near Salt Lake City, Utah, is shown schematically in Fig. 4.25. The contactor is a rubber-lined tank 6.1 m in diameter and 2.4 m straight height, capable of contacting 28 liter/s of aqueous solution [T2]. Organic and aqueous streams from adjacent stages in the cascade are introduced directly into a turbo-mixer, with a 46-cm-diameter impeller, mounted at the top of the tank. The mixer phases emerge into the tank, which acts as a large settling chamber. For an aqueous-to-organic flow ratio of 6, the holdup time per stage is estimated to be 46 min.

7.4 Centrifugal Contactor

The centrifugal contactor utilizes centrifugal force to achieve more rapid phase separation after mixing, thereby reducing the equipment size and holdup of process solutions below that required for the mixer-settler described in the previous section. A centrifugal contactor developed by Webster et al. [C2, D1, W2] for use in reprocessing irradiated fuel is illustrated schematically in Fig. 4.26. Organic and aqueous phases to be contacted enter at the centerline of the pump-mix chamber, which creates dispersion for interphase mass transfer and also supplies pumping action for interstage flow. The mixture flows upward through a perforated rotating plate into the rotating settling chamber, which contains radial vanes to keep the liquid

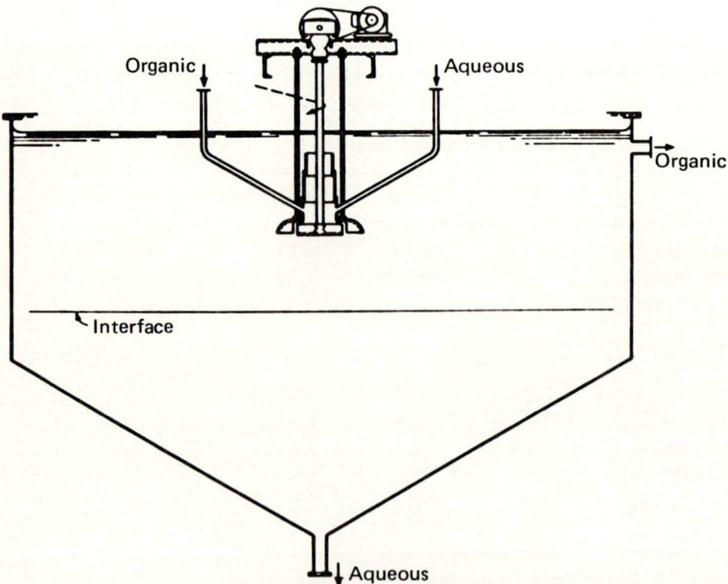

Figure 4.25 Vitro uranium extractor. (*From Treybal [T2], by permission.*)

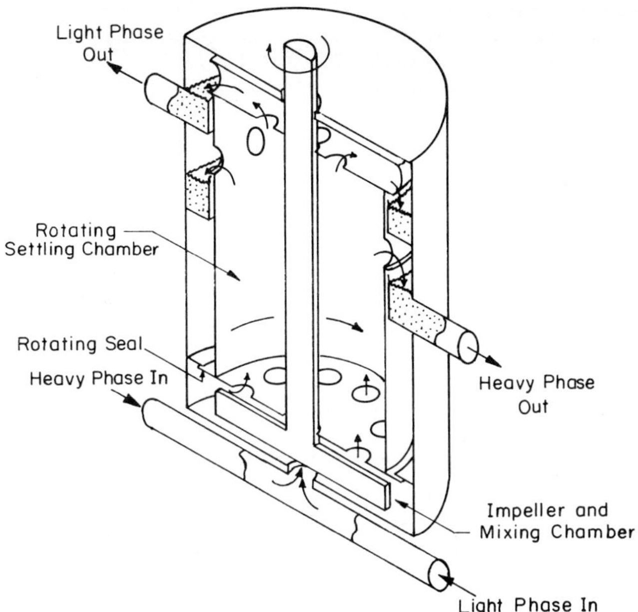

Figure 4.26 Centrifugal contactor (schematic).

rotating at the same speed as the bowl. Under the influence of the centrifugal force, the heavier aqueous phase collects at the outer periphery of the bowl and flows through holes in the bowl into a collecting trough located between the bowl and the stationary outer housing. The lighter organic phase collects near the rotating shaft, flows upward through holes in a top baffle, and out through holes into a collecting trough. The contactor requires no submerged bearings or external seals, but a seal at the bottom of the rotating bowl is necessary to prevent liquid in the bowl from leaking into the annular space between the rotating bowl and the stationary housing.

Dimensions and operating properties of critically safe centrifugal contactors are given in Table 4.14, adapted from Davis and Jennings [D1]. The holdup times for these centrifugal contactors are more than 20-fold less than the holdup times for the pump-mix mixer-settlers listed in Table 4.13. Contacting efficiencies approaching 75 percent or greater of the mass transfer obtainable with a single equilibrium stage have been reported [D1].

7.5 Rotary Annular Contactor

In the rotary annular contactor [D2, L2, T2] shown schematically in Fig. 4.27 [T2], the organic and aqueous phases flow countercurrently by gravity in the annular space between a rotating inner cylinder and a stationary outer cylinder. Taylor-instability vortices generated in the annulus promote dispersion and interfacial area. This is one of the simplest of the mechanically agitated contactors, and it has been developed for possible application to fuel reprocessing. In laboratory extractions of uranium from nitric acid with TBP in kerosene, Davis [D2] obtained values as low as 7.5 cm for the column height equivalent to a theoretical stage. The rotor speed varied from 1200 to 2000 r/min, with annular widths of 0.1 to 0.35 cm and a stator diameter of 2.2 cm. The residence time per theoretical stage was 10 s or less.

Table 4.14 Dimensions of critically safe centrifugal contactors†

Construction material	Unpoisoned stainless steel	Poisoned stainless steel‡	Poisoned stainless steel‡
Speed, r/min	1800	1800	1800
Rotating bowl diameter, cm	12	20	23
length, cm	25	61	39
Mixing chamber volume, liters	0.38	1.63	3.18
Settling chamber volume, liters	1.25	1.41	9.96
Maximum flow rate of both phases, liters/min	30	250§	250§
Holdup time, min	0.054	0.063	0.052

†Maximum critically safe size for highly enriched ^{235}U.

‡Stainless steel contains a "poison," such as gadolinium or boron, sufficient to absorb all neutrons entering the steel.

§Estimated for a Purex TBP solvent, with a maximum density difference between phases of 0.25 g/cm^3.

Source: Adapted from M. W. Davis and A. S. Jennings, "Equipment for Processing by Solvent Extraction," in *Chemical Processing of Reactor Fuels,* J. F. Flagg (ed.), Academic, New York, 1961, by permission.

7.6 Rotating Disk Contactor

Another type of gravity-flow, vertical contactor with a rotating axial shaft is the rotating disk contactor developed by the Shell Development Company [R1, R2], shown schematically in Fig. 4.28. It consists of alternate annular stator disks attached to the outer shell and circular rotor disks attached to the rotating shaft. Rotation of the central shaft, at peripheral speeds up to 6 m/s, provides controlled dispersion of the two phases and sets up a toroidal flow pattern within each stator compartment. There are no settling chambers, and the two phases drift past each other in countercurrent flow.

Dimensions of contactors for which test data are available [R2] are given in Table 4.15.

When the 20-cm-diameter column was tested in hexone-acetic acid-water, a height equivalent to a theoretical stage as low as 10 cm was observed, with a combined flow rate of both phases of 1.0 cm^3/(s·cm^2) of column cross-sectional area. The holdup time per equivalent theoretical stage is only 10 cm ÷ 1 cm/s = 10 s.

The rotating disk contactor has had extensive application in petroleum refining and organic chemical separations. A modified version, with holes in the horizontal stator disks to promote countercurrent flow, has been used in the recovery of uranium from solutions used in cleaning process equipment [D4, L2].

7.7 Spray Column

The spray column, shown schematically in Fig. 4.29, is the simplest of the contactors. The heavy aqueous phase enters the top of the vertical cylinder through a distributor and flows downward under gravity, usually as the continuous phase. A distributor at the bottom of the column disperses the entering organic phase into small drops, which rise through the continuous heavy phase and collect in a layer at the top. Coalescence of the dispersed phase drops and axial circulation and mixing of the continuous phase result in relatively low efficiency of contacting. Very tall columns may be required to obtain only a few theoretical stages.

Because of absence of internal structure, spray columns are sometimes selected for

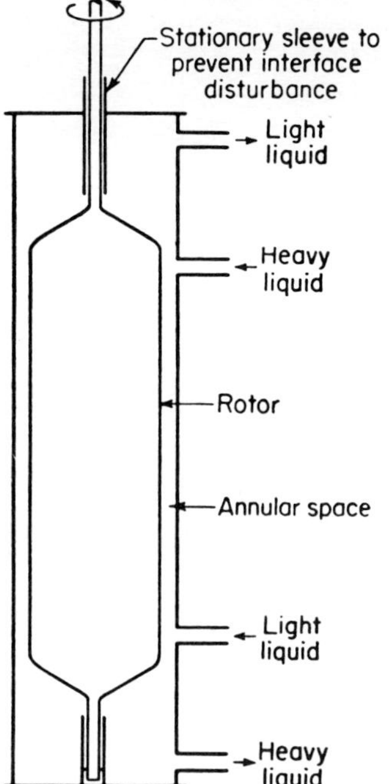

Figure 4.27 Rotary annular contactor. (*From Treybal [T2], by permission.*)

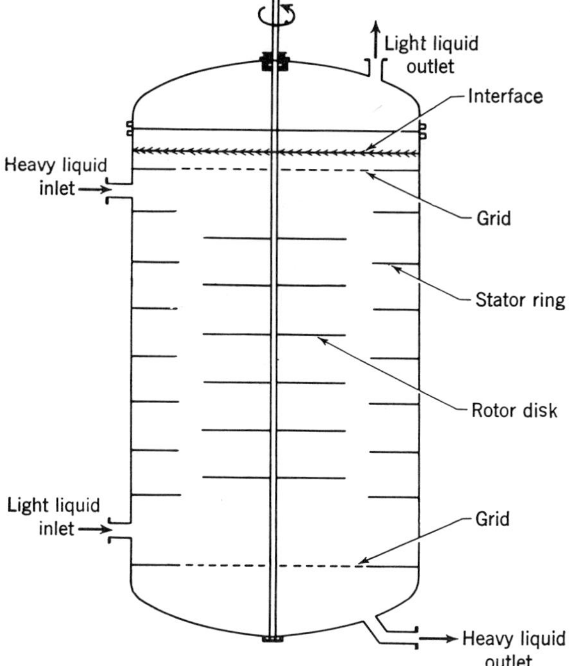

Figure 4.28 Rotating disk contactor (schematic).

Table 4.15 Rotating disk contactors

Tower diameter, cm	20	41	64	198
Stator opening, cm	12	30	41	135
Rotator diameter, cm	7.9	20	30	102
Stator spacing, cm	4.1	6.1	12	25
Number of compartments	21	20	30	20
Effective height, m	0.86	1.2	3.6	5.1
Peripheral speed, m/s	3.0	6.1	0.96	1.3

Source: G. H. Reman, U.S. Patent 2,619,280, Nov. 25, 1952.

liquid-liquid separation when suspended solids are present, as in the extraction of zirconium from hafnium in acidic thiocyanate solutions, wherein solid thiocyanate polymers tend to form (Chap. 7).

7.8 Packed Columns

A simple way to maintain interfacial area and dispersion in a vertical gravity-flow column, and to reduce axial mixing, is to fill the column with loose packing to provide tortuous flow paths. Typical packing consists of ceramic rings or saddle shapes, dumped in random arrangement. The

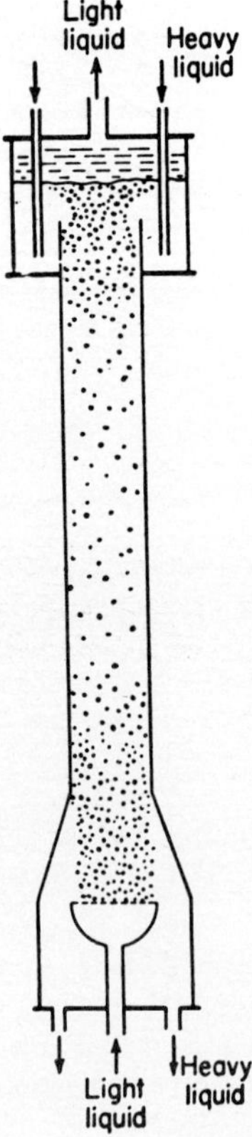

Figure 4.29 The spray column of Elgin [E1]. (*From Treybal [T2], by permission.*)

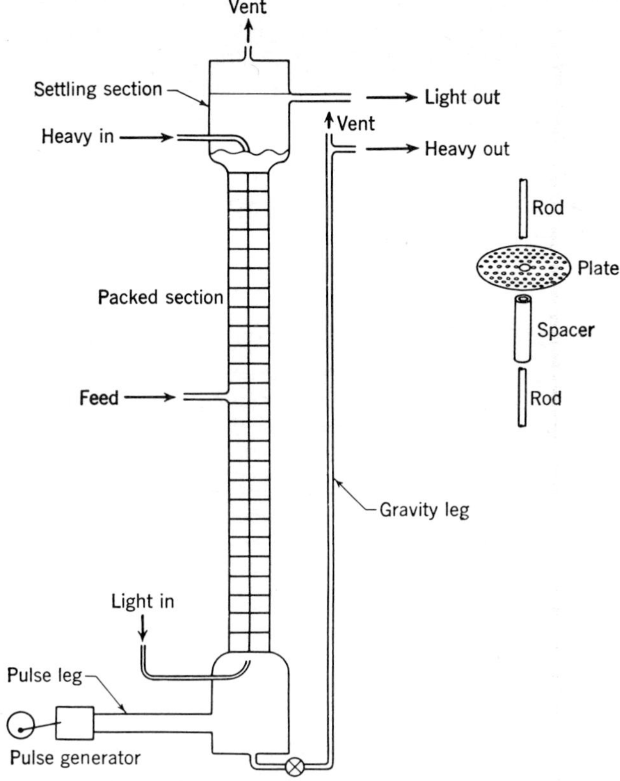

Figure 4.30 Schematic of pulse column.

packing reduces the available space for liquid flow and also introduces frictional drag, so the liquid throughout per unit of cross-sectional area is less than for spray columns. Neutron poisons can be incorporated into the packing to increase the criticality safe diameter.

Spray column contactors were used in the first large-scale solvent extraction plants at Hanford, Washington, for recovering plutonium from irradiated natural uranium and in the first chemical processing plant at Idaho for recovering enriched ^{235}U [L2].

Although packed columns are simple and have no moving parts, their large space requirements have resulted in the replacement of packed columns by pulsed columns, or by other more compact contactors, in more recent installations for reprocessing irradiated reactor fuel.

7.9 Pulse Columns

The pulse column is an outgrowth of a contactor patented by van Dijk [V1] in which perforated plates extending across the column were oscillated up and down to disperse the phases. Van Dijk also suggested that dispersion might be effected by leaving the plates stationary and pulsating the liquid contents of the column. Pulsating flow of liquid through a column bridged by perforated plates is the principle of the pulse columns in use today.

Figure 4.30 illustrates how a pulse column works. The column consists of a vertical tube packed for the major part of its length with perforated plates. At each end are calming, or disengaging, sections that may be enlarged in diameter, to prevent the carryover of entrainment. The principal interface may be carried at the top of the column, at the bottom, or anywhere in between by adjusting the back pressure applied by the gravity leg, as is done in any extraction column. The pulsating action is supplied by leading a pulse leg from the bottom of the column to the pulse generator. This may be simply a reciprocating pump with check valves removed, or it may be a bellows actuated by a cam. In a small installation, air pressure can be used to actuate the bellows. Larger columns require either hydraulic or mechanical operation.

There are at least two distinct modes of operation of a pulse column. One way is to pulse very gently. This action forces the discontinuous phase through the holes in the plate, forming bubbles. They rise (or fall) to the next plate, where they coalesce to await the next pulse. This is referred to as mixer-settler operation. If the pulsing is more vigorous, the bubbles never coalesce, but are repeatedly forced through holes in various plates as they work their way up and down the column. The passage of a bubble through a hole in a plate deforms it considerably, and the internal agitation thus produced improves the extraction. Also, passage through a hole tends to strip off any stagnant film of the continuous phase. Most pulse columns give optimum performance in this latter region, wherein the bubbles do not coalesce completely between pulses.

Since each cycle of pulsing pushes light phase upward through the holes and then pushes heavy phase downward, the net throughput through the column depends on the pulsing action. In the limit of low frequency and amplitude of pulsing, such that mixer-settler operation occurs between plates on each pulse cycle, the total volumetric flow rate of the light and heavy phases is equal to the product of the volumetric displacement of the pulse generator and the pulse frequency. Thus, throughput increases linearly with pulse frequency until, at higher frequencies, incomplete phase separation occurs between pulses and some back flow of an individual phase occurs in each pulse stroke. Finally, at a higher pulse frequency sufficient emulsification persists throughout the pulse cycle to cause flooding. Further increases in pulse frequency decrease the throughput. Pulse columns usually operate with pulse frequencies in the range of 30 to 120 cycles/min and amplitudes of 1.3 to 5 cm [L2].

Pulse columns provide more efficient phase dispersion and mass transfer than do packed columns, and pulse columns provide more uniform distribution of individual phases across the column cross section, with less tendency toward flow channeling. Pulse columns with phase-contacting sections as large as 0.86 m in diameter [I3, L2] and with heights as great as 10 m[G1, L2] have been used in recovering plutonium from irradiated uranium. A comprehensive study of the use of pulse columns 4 cm in diameter for the extraction of uranium by TBP has been reported by Durandet et al. [D3]. Operating conditions that resulted in the lowest values of the column height equivalent to a theoretical stage (HETS) of equilibrium contacting are listed in Table 4.16.

Table 4.16 Pulse column properties for uranium extracting and scrubbing[†]

Operation	Pulse frequency, cycles/min	Pulse amplitude, cm	HETS,[‡] cm	Total throughput, liters/min	Theoretical stage holdup time, min
Extracting	63	4.0	33	0.30	1.4
Scrubbing	165	1.1	15	0.30[§]	0.62

[†]From data of Durandet et al. [D3]. Column diameter = 4 cm, plate spacing = 5 cm, organic-to-aqueous flow ratio = 0.5, free area of sieve plates = 23%, sieve-plate hole diameter = 3 mm.

[‡]HETS = height equivalent to a theoretical stage.

[§]Near the flooding limit.

Packed columns can also be pulsed to improve their mass-transfer performance, with efficiency increases of as much as 300 percent reported. The flooding limits are lower than for a packed column without pulsing. The pulse amplitude and frequency must not be large enough to float the packing during a pulse, otherwise the packing will lose its random orientation, resulting in flow channeling and loss in mass-transfer efficiency [D1].

A disadvantage of pulsed columns is the need for a pulsing pump. However, similar pumps are needed in any case for transfer of liquids. In fact, the pulsing can also be used for transfer by installing check valves in the piping between the columns, although with some sacrifice in operating flexibility.

The great improvement in performance of pulse columns over other column-type contactors, and the simple and reliable equipment involved, have led to the widespread use of pulse columns in many solvent extraction operations separating and purifying nuclear materials. In addition to their use in some fuel reprocessing operations, as mentioned above, pulse columns have been used in uranium purification plants at Fernald, Ohio [C1], and Gore, Oklahoma (cf. Chap. 5).

NOMENCLATURE

A	defined by Eq. (4.24), mol/liter
C	molar concentration of total TBP in the organic phase, mol/liter
D	distribution coefficient, equilibrium ratio of concentration in organic phase to concentration in aqueous phase
E	volumetric flow rate of organic phase
F	volumetric flow rate of aqueous feed
K	equilibrium constant
M	number of stages in scrubbing section
N	number of stages in extracting section
S	volumetric flow rate of scrub solution
x	molar concentration in aqueous phase
y	molar concentration in organic phase
z	molar concentration in aqueous stream entering a stage
α	separation factor
β	extraction factor

Subscripts and Superscripts

E	extracting section
F	feed stream to cascade
H	nitric acid
Hf	hafnium
i,j	components to be separated
$0, 1, \ldots, m, \ldots$	stage in scrubbing section
M	stage in scrubbing section nearest feed point
$0, 1, \ldots, n, \ldots$	stage in extracting section
N	stage in extracting section nearest feed point
NO_3^-	total nitrate in aqueous phase
S	scrubbing section
TBP	uncombined TBP in organic phase
Th	thorium
U	uranium
Zr	zirconium
*	refers to minimum flow ratios

REFERENCES

A1. Alcock, K., S. S. Grimley, T. V. Healy, J. Kennedy, H. A. C. McKay: *Trans. Faraday Soc.* **52**: 39 (1959).

A2. American National Standard for Nuclear Criticality Safety in Operations with Fissionable Materials Outside Reactors, N 16.1-1975, American Nuclear Society, LaGrange Park, Ill., 1975.

B1. Baumgaertel, G., et al.: Reports KFK-671 (EUR-3703d) and KFK-680 (EUR-3707d), 1967.

B2. Bruce, F. R.: "Solvent Extraction Chemistry of the Fission Products," *PICG(1)*† 7: 100 (1956).

C1. *Chem. Eng.,* Oct. 1955, p. 112.

C2. Clark, A. T., Jr.: Performance of a 10-inch Centrifugal Contactor, USAEC Report DP-752, E. I. du Pont de Nemours & Co., Inc., Sept. 1962.

C3. Codding, J. W., W. O. Haas, and F. K. Heumann: "Equilibrium Data for Purex Systems," Report KAPL-602, Nov. 26, 1951; *Ind. Eng. Chem.* **50**: 145 (1958).

C4. Colven, T. J.: "Critically Safe Equipment for Aqueous Separations Processes," *PICG(2)* **17**: 555 (1958).

C5. Coplan, B. V., J. K. Davidson, and E. L. Zebroski: U.S. Patent 2,646,346, July 21, 1953.

C6. Coplan, B. V., J. K. Davidson, and E. L. Zebroski: *Chem. Eng. Progr.* **50**: 403 (1954).

C7. Culler, F. L.: *Chem. Eng. Progr.* **51**: 450 (1955).

C8. Culler, F. L.: "Reprocessing of Reactor Fuel and Blanket Materials by Solvent Extraction," *PICG(1)* **9**: 464 (1956).

C9. Culler, F. L., Jr., and F. R. Bruce: "The Processing of Uranium-Aluminum Reactor Fuel Elements," *PICG(1)* **9**: 484 (1956).

D1. Davis, M. W., and A. S. Jennings: "Equipment for Processing by Solvent Extraction," in *Chemical Processing of Reactor Fuels,* J. F. Flagg (ed.), Academic, New York, 1961.

D2. Davis, M. W., and E. J. Weber: "Liquid-Liquid Extraction between Rotating Concentric Cylinders," *Ind. Eng. Chem.* **52**: 929–934 (1960).

D3. Durandet, J., D. Defives, B. Choffe, and Y. L. Gladel: "A Study of Pulsed Columns in Solvent Extraction," *PICG(2)* **17**: 180 (1958).

D4. Dykstra, J., B. H. Thompson, and R. J. Clouse: "Solvent Extraction System for Enriched Uranium," *Ind. Eng. Chem.* **50**: 161–165 (1958).

E1. Elgin, J. C.: U.S. Patent 2,364,892, 1944.

F1. Flanary, J. R.: "Solvent Separation of Uranium and Plutonium from Fission Products by Means of Tributyl Phosphate," *PICG(1)* **9**: 528 (1956).

F2. Furman, N. F., R. J. Mundy, and G. H. Morrison: "The Distribution of Uranyl Nitrate from Aqueous Solutions to Diethyl Ether," Report AECD-2938, 1950.

G1. Geier, R. G.: "Application of the Pulse Column to the Purex Process," USAEC Report TID-7534 (bk.1), 1957, pp. 110–119.

G2. Geier, R. G.: "Improved Pulsed Extraction Column Cartridges," *PICG(2)* **17**: 192 (1958).

G2a. Goldberg, S. M., M. Benedict, and H. W. Levi: *Nucl. Sci. Eng.* **47**: 169–186 (1972).

G3. Goldschmidt, B., P. Regnaut, and I. Prevot: "Solvent Extraction of Plutonium from Uranium Irradiated in Atomic Piles," *PICG(1)* **9**: 492 (1956).

G4. Gresky, A. T.: "Solvent Extraction Separation of U^{233} and Thorium from Fission Products by Means of Tributyl Phosphate," *PICG(1)* **9**: 505 (1956).

G5. Griffith, W. L., G. R. Jasney, and H. T. Tupper: "The Extraction of Cobalt from Nickel in a Pulse Column," S.M. thesis in chemical engineering, Massachusetts Institute of Technology, Cambridge, Mass., 1952.

G6. Groenier, W. S.: "Calculation of the Transient Behavior of a Dilute-Purex Solvent

† *PICG(1), Proceedings of the International Conference on the Peaceful Uses of Atomic Energy, Geneva,* sponsored by the United Nations. The number in parentheses refer to the number of the conference.

Extraction Process Having Application to the Reprocessing of LMFBR Fuels," Report ORNL-4746, 1972.

G7. Gruverman, I. J.: "Extraction of Nitric Acid Solutions with Tributyl Phosphate," S.M. thesis in chemical engineering, Massachusetts Institute of Technology, Cambridge, Mass., 1955.

H1. Haas, Walter O., Jr.: "Solvent Extraction: General Principles," in *Chemical Processing of Reactor Fuels,* J. F. Flagg (ed.), Academic, New York, 1961.

H2. Horner, D. E.: "A Mathematical Model and a Computer Program for Estimating Distribution Coefficients for Plutonium, Uranium and Nitric Acid with Tri-n-Butyl Phosphate," Report ORNL-TM-2711, Feb., 1971.

H3. Hudswell, F., and J. M. Hutcheson: "Methods of Separating Zirconium from Hafnium and Their Technological Implications," *PICG(1)* 8: 563 (1956).

H4. Huré, J., and R. Saint James: "Process for Separation of Zirconium and Hafnium," *PICG(1)* 8: 551 (1956).

I1. *Ind. Eng. Chem.* **45**(6): 18A (1953).

I2. Irish, E. R.: Report TID-7534, bk. 1, 1957, pp. 69–82.

I3. Irish, E. R.: Description of Purex Plant Process, USAEC Report HW-60116, Hanford Atomic Products Operation, May 1959.

J1. Joyce, A. W., Jr., L. C. Peery, and E. B. Sheldon: *Chem. Eng. Progr. Sym. Ser. 28* **56**: 21 (1960).

L1. Logsdail, D. H., and L. Lowes: "Industrial Contacting Equipment," in *Recent Advances in Liquid-Liquid Extraction,* C. Hanson (ed.), Pergamon, Oxford, 1971.

L2. Long, J. R.: *Engineering for Nuclear Fuel Reprocessing,* Gordon and Breach, New York, 1967.

L3. Lowe, J. T.: "Calculation of the Transient Behavior of Solvent Extraction Processes," *Ind. Eng. Chem., Process Design Develop.* 7: 362–366 (1968).

M1. McKay, H. A. C.: "Tri-n-butyl Phosphate as an Extracting Agent for the Nitrates of the Actinide Elements," *PICG(1)* 7: 314 (1956).

M2. Moore, R. L.: "The Mechanism of Extraction of Uranium by Tributyl Phosphate," Report AECD-3196, 1951.

M3. Morello, V. S., and N. Poffenberger: *Ind. Eng. Chem.* **42**: 1021 (1950).

P1. Peterson, H. C., and G. H. Beyer: *J.A.I.Ch.E.* **2**: 38 (1956).

P2. Peterson, S., and R. G. Wymer: *Chemistry in Nuclear Technology,* Addison-Wesley, Reading, Mass., 1963.

P3. Pigford, R. L.: Private communication, July 1978.

R1. Reman, G. H.: U.S. Patent 2,619,280, Nov. 25, 1952.

R2. Reman, G. H., and R. B. Olney: *Chem. Eng. Progr.* **51**: 141 (1955).

R3. Rigiamonte, R., and E. Spaccamela: *Chim. e Ind. (Milan)* **35**: 787 (1954).

R4. Rozen, A. M.: *Atomic Energy Rev.* **6**: 2, 59–135 (1968).

S1. Scotten, W. C.: "SOLVEX: A Computer Program for Simulation of Solvent Extraction Processes," Report DP-1391, 1975.

S2. Shelton, S. M., E. D. Dilling, and J. H. McClain: "Zirconium Metal Production," *PICG(1)* 8: 505 (1956).

S3. Sherwood, T. K., R. L. Pigford, and C. R. Wilke: *Mass Transfer,* McGraw-Hill, New York, 1975, pp. 407–408.

S4. Siddall, T. H., III: "Solvent Extraction Processes Based on TBP," in *Chemical Processing of Reactor Fuels,* J. F. Flagg (ed.), Academic, New York, 1961.

T1. Thomas, J. T. (ed.): "Nuclear Safety Guide," Report NUREG/CR-0095, 1978.

T2. Treybal, R. E.: *Liquid Extraction,* 2d ed., McGraw-Hill, New York, 1963.

U1. U.S. Atomic Energy Commission: *Chemical Processing and Equipment,* U.S. Government Printing Office, Washington, D.C., 1955, pp. 1–44.

V1. van Dijk, W. J. D.: U.S. Patent 2,011,186, 1935.

W1. Watson, S. B., and R. H. Rainey: "Modification of the SEPHIS Computer Code for Calculating the Purex Solvent Extraction System," Report ORNL-TM-5123, 1975.

W2. Webster, D. S., J. F. Ward, Jr., and C. L. Williamson: "Hydraulic Performance of a 5-inch Centrifugal Contactor," USAEC Report DP-370, E. I. du Pont de Nemours & Co., Inc., Aug., 1962.

W3. Werning, J. R., et al.: *Ind. Eng. Chem.* **46**: 644 (1954).

PROBLEMS

4.1 For each point of Table 4.4, find the equilibrium constant K_{Zr} for the reaction

$$Zr^{4+} + 4NO_3^- + 2TBP \rightleftharpoons Zr(NO_3)_4 \cdot 2TBP$$

Assume that $K_H = 0.145$.

4.2 In a simple solvent extraction cascade without a scrubbing section, a feed stream containing an extractable component at concentration x^F is contacted countercurrently by an organic stream containing a concentration y_0 of extractable component. The distribution coefficient D and extraction factor β are constant.

(*a*) Show that the number of equilibrium stages required to reduce the aqueous concentration to x_1 is given by

$$\frac{x^F - x_1}{x^F - y_0/D} = \frac{\beta^{N+1} - \beta}{\beta^{N+1} - 1}$$

(*b*) Show that when $\beta = 1$, the above equation reduces to

$$\frac{x^F - x_1}{x^F - y_0/D} = \frac{N}{N+1}$$

(*c*) Show that when β is less than unity, complete extraction of the solute is impossible even if an infinite number of equilibrium stages is available.

4.3 Equilibrium constants for the formation of complexes of $Th(NO_3)_4$, $UO_2(NO_3)_2$, and HNO_3 with TBP are as follows:

$$Th^{4+}(aq) + 4NO_3^-(aq) + 2TBP(o) \rightleftharpoons Th(NO_3)_4 \cdot 2TBP(o) \qquad K = 0.6$$

$$UO_2^{2+}(aq) + 2NO_3^-(aq) + 2TBP(o) \rightleftharpoons UO_2(NO_3)_2 \cdot 2TBP(o) \qquad K = 5.5$$

$$H^+(aq) + NO_3^-(aq) + TBP(o) \rightleftharpoons HNO_3 \cdot TBP(o) \qquad K = 0.145$$

It is proposed that $^{233}UO_2(NO_3)_2$ be separated from $Th(NO_3)_4$ by fractional extraction between 3 *N* aqueous nitric acid and a 1 *M* solution of TBP (free and combined) in kerosene. The feed is 1 *M* in thorium and 0.2 *M* in uranium. What is the minimum volume ratio of 3 *N* nitric acid scrub solution to feed and the minimum volume ratio of TBP-kerosene to feed at which complete extraction of uranium uncontaminated by thorium is possible?

4.4 Uranyl nitrate containing 1 mol of boron per 100 mol of uranium is to be purified by fractional extraction with diethyl ether from a 10 *N* solution of ammonium nitrate. The extract is to contain no more than 1 mol of boron per million moles of uranium and is to contain 95 percent of the uranium in the feed. What are the minimum volumes of 10 *N* ammonium nitrate scrub solution and diethyl ether solvent needed per unit volume of feed?

4.5 In the uranium-boron separation example described in Prob. 4.4, an ether-feed ratio of 1.0 and a scrub-ether ratio of 0.1 are to be used. How many scrubbing stages are needed? How many extracting stages?

4.6 An aqueous solution containing initially 3 *N* nitric acid, 1 *M* thorium, and 0.2 *M* uranium is contacted with an equal volume of initially pure organic extractant consisting of 1 M TBP (free and combined) in kerosene. The two phases are brought to equilibrium. Using the equilibrium data of Prob. 4.3, calculate the resulting uranium-thorium separation factor α for this single stage of contacting.

4.7 Consider an extracting-scrubbing cascade separating components A and B, with constant distribution coefficients D_A and D_B. The concentrations x_A^F, x_B^F and the recovery ρ of component A are specified. Components A and B are not present in the entering organic and in the entering scrub solution.

(*a*) Show that there are an infinite number of flow ratios $e = E/F$ and $s = S/F$ that will satisfy these conditions. Show that one or the other of the allowable flow ratios can be less than the corresponding minimum ratios defined in Sec. 6.3.

(*b*) Show that the requirement of zero or finite number of stages in the scrubbing section results in [P3]

$$\frac{s+1}{s-1-D_A e} \geqslant \frac{1}{1-\rho_A}$$

How many extracting stages correspond to the equality limit?

(*c*) The decontamination factor f is also specified. Show that the requirement of zero or finite number of stages in the extracting section results in [P3]

$$\frac{1+D_B e}{D_B e - s} \geqslant \frac{f}{\rho_A}$$

How many scrubbing stages correspond to the equality limit?

(*d*) Show that the simultaneous application of the equality limits of (*b*) and (*c*) must be equivalent to Eqs. (4.69) and (4.70) for the limiting flow ratios.

4.8 A mixed vanadium-uranium concentrate contains 10 mole percent (m/o) uranium and 90 percent vanadium. It is desired to recover uranium containing less than 0.001 percent vanadium in 99 percent yield from this material by extracting with diethyl ether using ammonium nitrate as salting and scrubbing agent. What is the minimum ratio of solvent to feed and scrub solution to feed at which this would be possible? How many absorbing and scrubbing stages would be required at a solvent flow ratio 1.5 times the minimum and a scrubbing ratio of 0.1?

4.9 A waste solution containing nitrate salts and traces of uranium is to be purified by contacting it with initially pure ether in a countercurrent extractor. The flow rates of aqueous feed and ether are adjusted such that their ratio F/E is equal to the distribution coefficient D for uranium.

(*a*) How many equilibrium stages are required to recover 99 percent of the uranium from the aqueous waste?

(*b*) For the same number of stages, what changes would you make in the operating conditions to recover appreciable amounts of strontium from the waste solution? The distribution coefficient for uranium is about 1650 times greater than that for strontium.

4.10 Consider the zirconium-hafnium separation example given in Table 4.7. Suppose that input quantities remain unchanged but that output quantities are changed to correspond to a zirconium recovery of 95 percent and a hafnium decontamination factor of 400. Carry out a stage-to-stage calculation of concentrations in extracting and scrubbing sections similar to Tables 4.9 and 4.10, and determine how many theoretical stages are needed in each of these sections.

CHAPTER

FIVE

URANIUM

1 URANIUM ISOTOPES

Table 5.1 lists the isotopes of uranium that are important in nuclear technology and their most important nuclear properties.

1.1 Natural Uranium

The nominal isotopic content of natural uranium is 99.274 atom percent (hereafter a/o) ^{238}U, 0.7205 a/o ^{235}U, and 0.0054 a/o ^{234}U. Slight variations in ^{235}U content from this nominal value are discussed in Chap. 14. The ^{234}U:^{238}U ratio equals the ratio of their life-lives, as ^{234}U, a decay product of ^{238}U, is in secular equilibrium with its parent. ^{235}U is the only naturally occurring fissile nuclide.

1.2 ^{232}U and ^{233}U

^{233}U is a synthetic fissile nuclide produced by neutron capture in natural thorium, followed by two successive beta decays, as described in Fig. 3.2. ^{232}U is a short-lived (72 years) alpha-emitting contaminant that is always present in ^{233}U from the fast-neutron reactions in thorium and ^{233}U described in Chap. 8. The hard gamma rays emitted by daughters of ^{232}U makes nuclear fuel containing ^{233}U more difficult to handle than fuel enriched with ^{235}U.

1.3 ^{236}U and ^{237}U

^{236}U and ^{237}U are produced by successive neutron captures in fuel containing ^{235}U. Both isotopes are detrimental contaminants. Long-lived ^{236}U is a neutron absorber that reduces the fuel's reactivity. It has an atomic mass between ^{235}U and ^{238}U, which makes subsequent isotopic reenrichment more difficult, as described in Sec. 15 of Chap. 12. The 6.7-day half-life of ^{237}U necessitates storage of irradiated uranium for around 150 days if its radioactivity is to be no higher than that of natural uranium, as explained in Chap. 8. ^{237}U decays to 2.14-million-year ^{237}Np, the longest-lived member of the $4n + 1$ radioactive decay series.

Table 5.1 Isotopes of uranium

Mass, amu	Atom percent in natural uranium	Half-life	Radioactive decay: Type	Radioactive decay: Effective MeV	Reaction with 2200 m/s neutrons: Cross section, b: (n, γ)	Reaction with 2200 m/s neutrons: Cross section, b: Fission	Reaction with 2200 m/s neutrons: Neutrons per fission
232.037168	–	72 yr	α	5.414	73.1	75.2	3.13
233.039522	–	1.62E5 yr	α	4.909	47.7	531.1	2.492
234.040904	0.0056	2.47E5 yr	α	4.856	100.2	–	–
235.043915	0.7205	7.1E8 yr	α	4.681	98.6	582.2	2.418
236.045637	–	2.39E7 yr	α	4.573	5.2	–	–
237.048608	–	6.75 days	β	0.112	411	–	–
238.05077	99.274	4.51E9 yr	α	4.268	2.70	–	–
239.05430	–	23.5 min	β	0.400	22	14	–

1.4 ^{239}U

^{239}U is produced by neutron capture in fuel containing ^{238}U. It decays to ^{239}Pu through two successive beta emissions, as described in Fig. 3.1. Because of its short, 23.5-min half-life, it is not present after irradiated fuel has been stored. It is, however, a significant contributor to decay heat production immediately after reactor shutdown.

2 URANIUM RADIOACTIVE DECAY SERIES

Each of the uranium isotopes is a member of one of the four possible radioactive decay series involving successive alpha and beta decay reactions. ^{238}U is the longest-lived member and the parent of the $4n + 2$ series, which includes ^{234}U as a member. ^{235}U is the longest-lived member and the natural parent of the $4n + 3$ series. ^{236}U decays by alpha emission to ^{232}Th, the longest-lived member and natural parent of the $4n$ series, to be described in Chaps. 6 and 8. ^{232}U decays by alpha emission to ^{228}Th, also a member of the $4n$ series. Problems arising from the radioactivity of ^{232}U and its daughters are discussed in Chap. 8. ^{237}U decays by beta emission to ^{237}Np, the longest-lived member of the $4n + 1$ series, the only one not of natural occurrence. ^{233}U is an intermediate member of this series.

2.1 ^{238}U Decay Series

Figure 5.1 shows the nuclear reactions that occur successively as ^{238}U decays into its stable end product ^{206}Pb. As is conventional in such decay diagrams, each nuclide is plotted on a grid, with the mass number A vertical and the atomic number Z horizontal. Table 5.2 gives the half-lives of these radioactive species and their principal decay radiations. The last column of Table 5.2 gives the ratio of the number of atoms of each nuclide to the number of uranium atoms in natural uranium, assuming that the uranium in the ore has been undisturbed long enough to be in decay equilibrium with all its decay products. At equilibrium, the activity of each of these nuclides is the same. Per megagram of contained uranium, the activity of ^{238}U and each of its daughters is

$$\frac{(0.9927\ ^{238}\text{U/U})(10^6\ \text{g U/Mg})(6.0225 \times 10^{23}/\text{g-atom})(0.693)}{(238\ \text{g U/g-atom})(4.51 \times 10^9\ \text{yr})(3.154 \times 10^7\ \text{s/yr})[3.7 \times 10^{10}/(\text{Ci}\cdot\text{s})]} = 0.33\ \text{Ci/Mg U}, \tag{5.1}$$

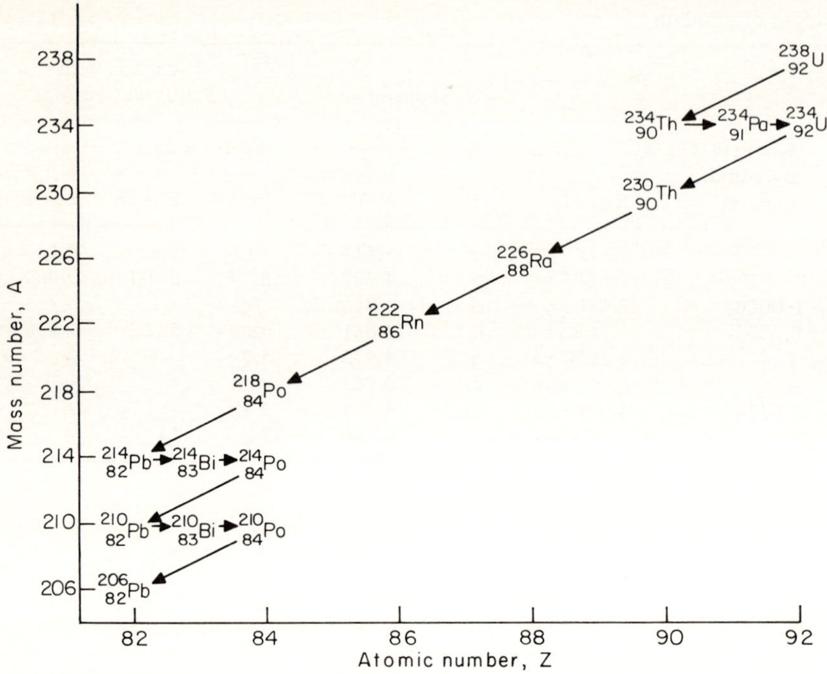

Figure 5.1 Radioactive decay of ^{238}U.

Table 5.2 Principal radioactive decay products of ^{238}U

Nuclide	Historical name	Half-life	Radiation	Atom ratio, ppb in natural uranium
$^{238}_{92}U$	Uranium I	4.51E9 yr	$\alpha(\gamma)$	9.927E8
$^{234}_{90}Th$	UX_1	24.1 days	$\beta(\gamma)$	0.0145
$^{234}_{91}Pa$	UX_2	1.17 min	β, γ	4.9E-7
$^{234}_{92}U$	Uranium II	2.47E5 yr	$\alpha(\gamma)$	5.44E4
$^{230}_{90}Th$	Ionium	8.0E4 yr	$\alpha(\gamma)$	1.76E4
$^{226}_{88}Ra$	Radium	1602 yr	$\alpha(\gamma)$	353
$^{222}_{86}Rn$	Radon	3.821 days	α	2.30E-3
$^{218}_{84}Po$	Radium A	3.05 min	α	1.28E-6
$^{214}_{82}Pb$	Radium B	26.8 min	β, γ	1.12E-5
$^{214}_{83}Bi$	Radium C	19.7 min	β, γ	8.25E-6
$^{214}_{84}Po$	Radium C′	164 μs	α	1.14E-12
$^{210}_{82}Pb$	Radium D	21 yr	$\beta(\gamma)$	4.62
$^{210}_{83}Bi$	Radium E	5.01 days	β	3.02E-3
$^{210}_{84}Po$	Polonium	138.4 days	α	0.0835
$^{206}_{82}Pb$	Radium G	Stable		

The daughters of ^{238}U of principal radiological concern in uranium mills and refineries are the long-lived nuclides ^{230}Th and ^{226}Ra (radium) and gaseous ^{222}Rn (radon). The amount of these nuclides in uranium mills and tailings piles is discussed in Sec. 8.9; their occurrence in uranium refineries is discussed in Secs. 9.2 and 9.7.

2.2 ^{235}U Decay Series

Figure 5.2 shows the nuclear reactions that occur successively as ^{235}U decays into its stable end product ^{207}Pb. Table 5.3 gives the half-lives of these radioactive species and their principal decay radiations. The last column of Table 5.3 gives the ratio of the number of atoms of each nuclide to the number of uranium atoms in natural uranium, again assuming that decay equilibrium has been established. At equilibrium, the activity of each of these nuclides is the same. Per megagram of natural uranium contained in the ore, the activity of ^{235}U and each of its daughters† is

$$\frac{(0.007205\ ^{235}\text{U/U})(10^6\ \text{g U/Mg})(6.0225 \times 10^{23}/\text{g-atom})(0.693)}{(238\ \text{g nat U/g-atom})(7.1 \times 10^8\ \text{yr})(3.154 \times 10^7\ \text{s/yr})[3.7 \times 10^{10}/(\text{Ci}\cdot\text{s})]} = 0.015\ \text{Ci/Mg U} \tag{5.2}$$

This is only one-twenty-second of the activity of ^{238}U and each of its daughters. This small contribution to the activity of natural uranium will be disregarded in the remainder of this chapter.

†Except ^{227}Th and ^{223}Fr.

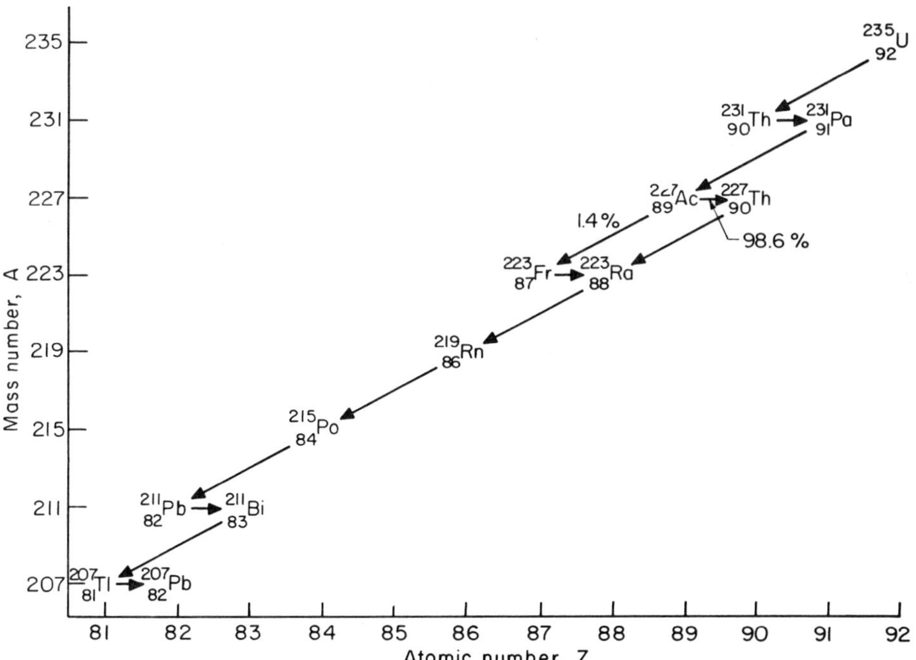

Figure 5.2 Radioactive decay of ^{235}U.

Table 5.3 Principal radioactive decay products of ^{235}U

Nuclide	Historical name	Half-life	Radiation	Atom ratio, ppb in natural uranium
$^{235}_{92}$U	Actinouranium	7.1E8 yr	α, γ	7.205E6
$^{231}_{90}$Th	Uranium Y	25.5 h	$\beta(\gamma)$	2.95E-5
$^{231}_{91}$Pa	Protactinium	3.25E4 yr	α, γ	330
$^{227}_{89}$Ac	Actinium†	21.6 yr	$\alpha, \beta(\gamma)$	0.219
$^{227}_{90}$Th	Radioactinium	18.2 days	α, γ	4.99E-4
$^{223}_{87}$Fr	Actinium K	22 min	α, γ	5.9E-9
$^{223}_{88}$Ra	Actinium X	11.43 days	α, γ	3.18E-4
$^{219}_{86}$Rn	Actinon	4.0 s	α, γ	1.29E-9
$^{215}_{84}$Po	Actinium A	1.78 ms	α	5.73E-13
$^{211}_{82}$Pb	Actinium B	36.1 min	β, γ	6.97E-7
$^{211}_{83}$Bi	Actinium C	2.15 min	$\alpha, (\beta), \gamma$	4.15E-8
$^{207}_{81}$Tl	Actinium C″	4.79 min	β, γ	9.25E-8
$^{207}_{82}$Pb	Actinium D	Stable		

†1.4% of decays of ^{227}Ac go to ^{223}Fr, 98.6% to ^{227}Th.

2.3 Radioactive Decay of ^{237}U, ^{237}Np, and ^{233}U

The synthetic isotopes ^{237}U and ^{233}U occur in the $4n + 1$ radioactive series, of which ^{237}Np is the longest-lived member. Figure 5.3 shows the nuclear reactions that occur successively as these nuclides decay into their nearly stable end product ^{209}Bi. Table 5.4 gives the half-lives of these radioactive species and their principal decay radiations. Radiations from ^{229}Th and its short-lived daughters will be the most important contributors to the remaining toxicity of high-level wastes from irradiated reactor fuels containing ^{237}Np, after such wastes have been in storage for several hundred thousand years, when these daughters will be in secular equilibrium with 162,000-year ^{233}U.

2.4 Radioactivity in Uranium Mines and Refineries

Because ^{238}U undergoes 14 successive reactions while decaying into stable ^{206}Pb, the activity of pure ^{238}U is only one-fourteenth as great as that of undisturbed uranium ore. Freshly extracted uranium consists of a mixture of ^{238}U and an equal activity of ^{234}U, and has one-seventh the activity of the ore from which it was extracted. Fresh uranium emits mainly alpha activity. As uranium ages, beta and gamma activity develop, owing to the growth of the first two decay products of ^{238}U:

$$24.1\text{-day } ^{234}\text{Th (UX-1)}$$

and

$$1.17\text{-min } ^{234}\text{Pa (UX-2)}$$

After a month these approach saturation activity, so that uranium older than this has four-fourteenths the activity of the uranium ore from which it came. The activity of uranium then remains constant for hundreds of years because of the long half-life of ^{230}Th, the first daughter of ^{234}U.

Removal of the elements responsible for the other ten-fourteenths of the activity of uranium ores is one of the important aspects of uranium concentration and purification.

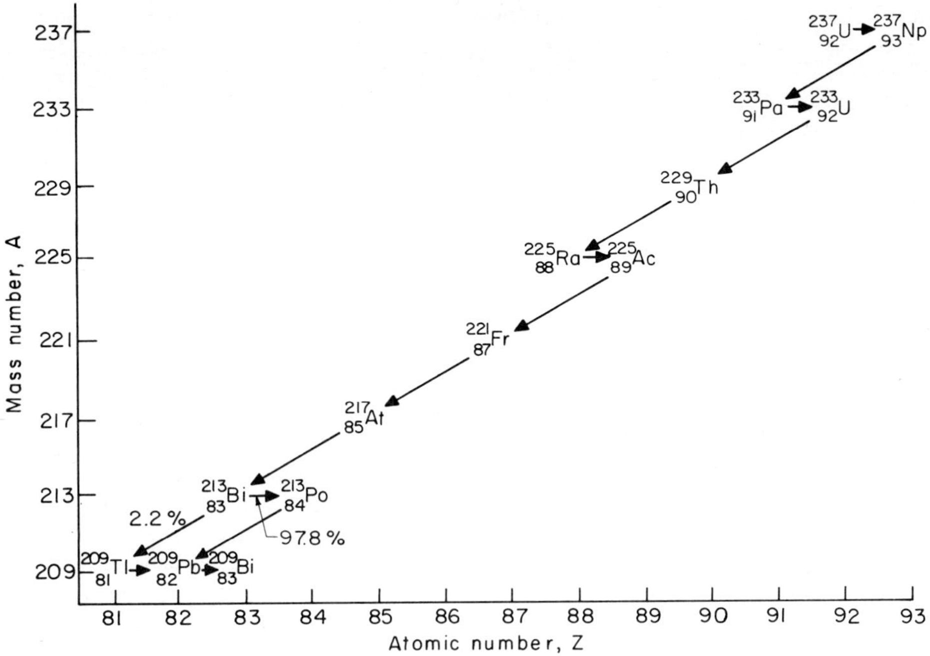

Figure 5.3 Radioactive decay of ^{237}U and ^{233}U.

Table 5.4 Principal radioactive decay products of ^{237}U, ^{237}Np, and ^{233}U

Nuclide	Half-life	Radiation
$^{237}_{92}$U	6.75 days	β
$^{237}_{93}$Np	2.14E6 yr	α, γ
$^{233}_{91}$Pa	27.0 days	β, γ
$^{233}_{92}$U	1.62E5 yr	$\alpha(\gamma)$
$^{229}_{90}$Th	7340 yr	α
$^{225}_{88}$Ra	14.8 days	$\beta(\gamma)$
$^{225}_{89}$Ac	10.0 days	α, γ
$^{221}_{87}$Fr	4.8 min	α, γ
$^{217}_{85}$At	0.032 s	α
$^{213}_{83}$Bi†	47 min	α, β, γ
$^{213}_{84}$Po	4.2 μs	α
$^{209}_{81}$Tl	2.2 min	β, γ
$^{209}_{82}$Pb	3.30 h	β
$^{209}_{83}$Bi	2E18 yr	α

†2.2% of decays of ^{213}Bi go to ^{209}Tl, 97.8% to ^{213}Po.

Purified uranium, freed from the decay products of ^{234}U, is much less toxic than uranium ores. For the same reason, the tailings of uranium mills are much more toxic than purified uranium. The most dangerous impurities among the decay products of ^{234}U are ^{230}Th and radium, relatively long-lived alpha emitters; radium's gaseous daughter radon, which disperses radioactivity in uranium mines and near mill tailing piles; and ^{210}Po, a very toxic alpha emitter. These, more than uranium itself, are responsible for the radioactive hazards of uranium mines and concentrating plants.

3 METALLIC URANIUM

3.1 Uses

As the uranium atom density is higher in uranium metal than in any uranium compound, metal is the preferred form of uranium for applications where the highest nuclear reactivity or highest density is wanted. For example, uranium metal was used by Fermi to create the world's first nuclear chain reaction with the limited amount of uranium then available. Uranium metal is still the fuel form preferred for nuclear reasons in reactors fueled with natural uranium and moderated by graphite, such as the Magnox nuclear power reactors used extensively in Great Britain. Because of its high density, uranium metal is used for compact shielding of x-rays or gamma rays and as counterweights in machinery. In such applications, uranium depleted in ^{235}U is preferred because nuclear reactivity is not needed.

3.2 Phases of Uranium

The phases of metallic uranium and their transition temperatures are listed in Table 5.5.

3.3 Density and Thermal Expansion

The density of uranium metal changes markedly with temperature. Table 5.6 summarizes densities inferred from x-ray diffraction data. The large density change at 662°C accompanying transition to the beta phase makes it undesirable to operate uranium metal reactor fuel above this temperature. Even at the lower temperatures at which the alpha phase is stable, its large density change with temperature, and its unequal temperature coefficient of thermal expansion along the three crystal axes (+19, −0.8, and $+20 \times 10^{-6}/°C$ at 25°C) cause severe distortion and elongation of fuel assemblies during temperature cycling unless the fuel is given special treatment prior to irradiation.

Table 5.5 Phases of uranium metal

Transition temperature, °C	Phase	Crystal system
	Solid α	Orthorhombic
662		
	Solid β	Tetragonal
772		
	Solid γ	Cubic
1133		
	Liquid	
~3900		
	Vapor	

Table 5.6 Density of uranium metal

Temperature, °C	Phase	Density, g/cm^3
25	α	19.070
662	α	18.369
662	β	18.17
772	β	18.07
772	γ	17.94
1100	γ	17.56

3.4 Chemical Reactivity

Uranium metal is very reactive. It tarnishes in air, with the oxide film preventing further oxidation of massive metal at room temperature. However, finely divided uranium ignites spontaneously at room temperature, and massive uranium burns steadily at 700°C, forming U_3O_8.

Water attacks massive uranium slowly at room temperature and rapidly at higher temperatures. UO_2 and UH_3 are formed, heat is evolved, and the metal swells and disintegrates. In water-cooled reactors uranium metal must be clad or canned in nonreacting metals such as aluminum, stainless steel, or zirconium. Nitric acid dissolves uranium readily.

4 URANIUM COMPOUNDS

4.1 Uranium Valence States

Uranium compounds have been prepared with positive valence states of 3, 4, 5, and 6. In addition, UO_3 is a weak acid, like MoO_3 and WO_3, and forms uranates such as Na_2UO_4 and diuranates such as $Na_2U_2O_7$.

Trivalent uranium ion reduces water to hydrogen. Hence, stable aqueous solutions of trivalent uranium compounds cannot be prepared. Compounds of tetravalent uranium are generally similar to those of zirconium or thorium, except that some uranium compounds can be oxidized to the hexavalent form. Compounds of pentavalent uranium are of little importance because they disproportionate readily into tetravalent and hexavalent forms. The properties of hexavalent uranium are generally similar to those of hexavalent molybdenum or tungsten. In aqueous solution hexavalent uranium forms the uranyl ion UO_2^{2+}

4.2 Uranium Oxides

Data on the most important oxides of uranium are given in Table 5.7.

Uranium dioxide UO_2 is the form in which uranium is most commonly used as a reactor fuel for light-water, heavy-water, and fast-breeder reactors. It is a stable ceramic that can be heated almost to its melting point, around 2760°C, without serious mechanical deterioration. It does not react with water, so that it is not affected by leakage of cladding in water-cooled reactors. Its principal disadvantages compared with uranium metal are its lower uranium atom density and lower thermal conductivity. At 100°C, thermal conductivities are metal, 0.25; UO_2, 0.09 W/(cm·°C).

U_3O_8 occurs naturally as the mineral pitchblende. It can be made by oxidizing UO_2 or heating UO_3.

Table 5.7 Uranium oxides

Oxide	Color	Melting point, °C	Density, g/cm^3	Method of formation
UO_2	Brown	2760	10.97	Reduction of UO_3 by H_2
U_3O_8	Black	Decomposes	8.38	Oxidation of UO_2
UO_3	Orange	Decomposes	7	Ignition of $UO_2(NO_3)_2$
$UO_4 \cdot 2H_2O$	Yellow	Decomposes	4.66	Precipitation by H_2O_2 from solutions of UO_2^{2+}

Uranium trioxide UO_3 is made by igniting uranyl nitrate, $UO_2(NO_3)_2 \cdot 6H_2O$, or $UO_4 \cdot 2H_2O$, the two principal forms in which uranium purified in aqueous solution is prepared. It is an intermediate in the preparation of UO_2 or UF_6.

Uranium peroxide $UO_4 \cdot 2H_2O$ is prepared by precipitation of an aqueous solution of uranyl nitrate at 70 to 80°C and a pH of 3 to 4 with H_2O_2. Because few other cations form precipitates under these conditions, this is an effective way of purifying uranium.

4.3 Uranium Carbides

Table 5.8 lists the carbides of uranium. The monocarbide UC is considered a preferred fuel for advanced fast-breeder reactors, because it has a higher uranium density, higher thermal conductivity, and lower moderating power than UO_2. The dicarbide UC_2 is specified as fuel for the high-temperature gas-cooled reactor. The sesquicarbide U_2C_3 has a limited range of stability and is of little importance. UC can be prepared by heating a mixture of 1 mol of UO_2 and 3 mol of graphite to 1800°C, or from 1 mol of metal and 1 mol of graphite. UC_2 can be prepared by reacting UC, UO_2, or U with additional graphite at 2400°C. UC forms solid solutions with UO_2 and UN. The carbides are stable only in dry air at room temperature. They react with moist air and react rapidly with water or steam, so they cannot be used in water-cooled reactors.

4.4 Uranium Nitride

Uranium nitride UN has a theoretical density of 14.32 g/cm^3 and melts around 2630°C. Made with ^{15}N, UN has been suggested as an advanced fuel for fast reactors because of its high U atom density, low moderation, and high melting point. UN is made by reacting UH_3 with the correct proportions of nitrogen or ammonia. UN reacts rapidly with moist air or water.

Table 5.8 Uranium carbides

	UC	UC_2	U_2C_3
Melting point, °C	2525	2350–2475	Transforms to UC + UC_2 at 1750–1820°C
Boiling point, °C		4100–4370	
Thermal conductivity, W/(cm·°C)	0.188 at 119–236°C	0.343 at 50°C	
Theoretical density, g/cm^3 at 25°C	13.63	11.68	12.88

4.5 Uranium Hydride

Uranium hydride UH_3 is made by reacting uranium metal with hydrogen at temperatures above 250°C and pressure above the dissociation pressure, p_{H_2} :

$$\log_{10} p_{H_2} \text{ (Torr)} = -\frac{4480}{T \text{ (K)}} + 9.20 \tag{5.3}$$

A solid solution of uranium and zirconium hydrides is used as fuel in TRIGA reactors. Uranium hydride is often pyrophoric and must be handled with care. It has been used to prepare finely divided uranium metal by reacting massive metal with hydrogen, then crushing the brittle hydride and heating it in vacuum to drive off hydrogen.

4.6 Uranium Halides

Table 5.9 lists uranium halides together with some of their more significant properties.

UF_4 is an important intermediate in the production of UF_6 and uranium metal. It is made by reacting UO_2 with an excess of HF vapor,

$$UO_2 + 4HF \rightleftharpoons UF_4 + 2H_2O$$

as described in more detail in Sec. 9.5. Dissolved in a low-melting eutectic of ZrF_4, BeF_2, and ^7LiF, $^{235}UF_4$ was used as fuel in the Molten Salt Reactor Experiment [H3].

UF_6 is the only compound of uranium volatile at room temperature. It is used as working fluid in the gaseous diffusion, gas centrifuge, and aerodynamic processes for uranium enrichment discussed in Chap. 14. Its principal physical and chemical properties are summarized in Sec. 4.7.

Table 5.9 Properties of uranium halides

Compound	Color	Temperature, °C Melts	Temperature, °C Boils at 1 atm	X-ray crystal density at 25°C, g/cm^3
UF_3	Black	~1427		8.95
UF_4	Green	1036	1457	6.70
U_4F_{17}	Black	430†	Disp.	6.94
U_2F_9		390†	Disp.	7.06
UF_5	White	348	Disp.	6.45
UF_6	Colorless	64.05	56.54‡	5.06
UCl_3	Olive green	837	1657	5.51
UCl_4	Dark green	590	789	4.87
UCl_5	Red brown	327†	Disp.	3.81
UCl_6	Black	179	392§	3.59
UBr_3	Dark brown	730	Disp.	6.53
UBr_4	Brown	519	777	5.35
UI_3	Black	~680		6.76
UI_4	Black	506	757§	

†With disproportionation (disp.)
‡Sublimes at 1 atm.
§With dissociation.

UCl_4, which boils without decomposition at 791°C, was used as feed material for the Y-12 electromagnetic uranium enrichment plant. It is hygroscopic and hydrolyzes in moist air.

The other uranium halides have not had significant practical uses.

4.7 Uranium Hexafluoride

The properties of UF_6 summarized in this section have been taken primarily from a comprehensive report by DeWitt [D3].

Vapor pressure, triple point, and critical point. Table 5.10 gives the triple-point pressure and temperature of UF_6 measured by Brickwedde et al. [B6], the critical pressure and temperature reported by Oliver et al. [O1], and values of the vapor pressure at temperatures between −200°C and the critical point from the following sources:

1. Solid, below 0°C, Eq. (5.4) fitted by Llewellyn [L1] to his measurements between −15°C and the triple point:

$$\log_{10} p_{\text{Torr}} = \frac{-2751}{T} - 75.0 \exp\left(\frac{-2560}{T}\right) - 1.01 \log_{10} T + 13.797 \qquad (T = t, °\text{C} + 273) \tag{5.4}$$

2. Solid at 0°C, measured by Weinstock et al. [W2].
3. Solid, between 0°C and triple point, Eq. (5.5) fitted by Oliver et al. [O1] to their measurements between 0°C and the triple point:

$$\log_{10} p_{\text{Torr}} = 6.38353 + 0.0075377t - \frac{942.76}{t + 183.416} \tag{5.5}$$

4. Liquid, between triple point and critical point, values quoted by DeWitt [D3] from a sixth-order polynomial fitted by Brooks and Wood [B7] to data of Oliver et al. [O1]. These appear to be the most reliable and consistent data among the many cited by DeWitt. Values below −15°C are of order-of-magnitude reliability only.

Density. Values of the density of liquid UF_6 between the triple point and 160°C given in Table 5.10 are from measurements of Wechsler and Hoge [W1]. The critical density is the estimate of Oliver et al. [O1]. The solid density at 25°C is calculated from x-ray diffraction data. Solid UF_6 formed when the liquid freezes or vapor condenses is much less dense because of contraction on solidification.

UF_6 vapor is unassociated, but displays van der Waals-type departures from the ideal gas law. Weinstock et al. [W2] proposed Eq. (5.6) as an equation of state for UF_6 vapor:

$$\rho_G \text{ (g/cm}^3) = \frac{4.291p}{T(1 - 1{,}376{,}900p/T^3)} \tag{5.6}$$

where p is the pressure in atmospheres and T is in kelvins.

Thermodynamic properties. Table 5.11 gives thermodynamic properties of UF_6 under its vapor pressure, selected from the comprehensive review of DeWitt [D3]. The original data sources were as follows:

Heat capacity and enthalpy of solid and liquid: measurements by Brickwedde et al. [B6].
Molal heat of vaporization: from the Clapeyron equation

$$H_G - H_C = T\,\Delta V\,\frac{dp}{dT} \tag{5.7}$$

Change of vapor pressure with temperature: dp/dT from the measurements of Oliver et al. [O1] summarized in Table 5.10.
Molal volume change ΔV: from

$$\Delta V = \frac{352}{\rho_G} - \frac{352}{\rho_C} \tag{5.8}$$

with ρ_G from Weinstock's Eq. (5.6) and ρ_C from Table 5.10.
Enthalpy of vapor: enthalpy of condensed phase plus heat of vaporization.

Table 5.10 Vapor pressure and density of UF_6

Temperature, °C	Vapor pressure, Torr	Density ρ_C, g/cm^3
−200	1E-26	
−150	1E-11	
−100	4.3E-5	
−90	0.0003	
−80	0.0017	
−70	0.0082	
−60	0.034	
−50	0.122	
−40	0.369	
−30	1.16	
−20	3.11	
−10	7.73	
0	17.70	
10	38.43	
20	79.37	
25	111.82	5.06 (solid)
30	155.66	
40	291.94	
50	526.56	
60	917.82	
62.5		4.87
64.052†	1,142	3.624 (liquid)
70	1,370.5	3.595
80	1,839.3	3.532
90	2,422.6	3.470
100	3,135.3	3.404
110	3,997.0	
120	5,025.9	3.263
130	6,240.8	
140	7,663.7	3.111
150	9,313.5	
160	11,210	2.948
170	13,379	
180	15,840	
190	18,626	
200	21,768	
210	25,313	
220	29,314	
230.2‡	34,580	1.36

†Triple point [B6].
‡Critical point [O1].

Table 5.11 Thermodynamic properties of UF_6 at its vapor pressure

Temperature		Condensed UF_6		Molal heat of vaporization	Molal enthalpy of UF_6 vapor
°C	K	Heat capacity C_p, cal/(g-mol·°C)	Molal enthalpy $H_C - H_{C,298}$, cal/g-mol	$H_G - H_C$, cal/g-mol	$H_G - H_{C,298}$, cal/g-mol
	0	0.00	−7,545	12,965	5,420
	20	4.06	−7,517		
	40	9.64	−7,379		
	60	14.90	−7,133		
	80	19.20	−6,790		
	100	22.49	−6,371		
	150	28.30	−5,093		
	200	32.65	−3,565		
	250	36.43	−1,837		
0		38.08	−974	12,080	11,106
10		38.77	−589	11,965	11,376
20		39.48	−197	11,842	11,645
25	298.15	39.86	0		11,790
30		40.27	201	11,739	11,940
40		41.16	608	11,636	12,244
50		42.10	1,025	11,506	12,531
60		43.09	1,452	11,331	12,783
64.052 (*s*)		43.49	1,627		
64.052 (*l*)		45.59	6,215		
70		45.78	6,488	6,634	13,122
80		46.13	6,949	6,470	13,419
90		46.42	7,413	6,264	13,677
100		46.71	7,880	6,067	13,947
110				5,867	
120				5,661	
130				5,446	
140				5,219	

The value of 12,965 for the molal heat of vaporization at 0 K was found by Weinstock et al. [W2] to provide the best correlation of measurements of vapor pressure, density of each phase, and heat capacity of each phase through the Clapeyron equation.

Table 5.12 gives thermodynamic properties of UF_6 in the ideal gas state at 1 atm pressure, denoted by the superscript (°). Values of C_p°, the free-energy function $-(G^\circ - H_0^\circ)/T$ and the entropy S° are taken from DeWitt's [D3] citation of calculations by Bigeleisen et al. [B3] from spectroscopic data. The enthalpy of UF_6 in the ideal gas state relative to the solid at 298.15 K was evaluated from

$$H_G^\circ - H_{C,298} = (H_G^\circ - H_0^\circ) + (H_0^\circ - H_{C,298})$$
$$= -T\left[\frac{-(G^\circ - H^\circ)}{T} + S^\circ\right] + 5420 \tag{5.9}$$

because

$$H_G^\circ = G^\circ - TS^\circ \tag{5.10}$$

and

$$H_0^\circ - H_{C,298} \equiv H_G^\circ \text{ (at } T = 0) - H_{C,298} = 5420 \tag{5.11}$$

from Table 5.11.

Barin and Knacke [B1] give for the heat of formation of solid UF_6 at 25°C

$$\Delta H_{C,298} = -523{,}000 \text{ cal/g-mol} \tag{5.12}$$

Hence the heat of formation of UF_6 in the ideal gas state at 25°C is

$$\Delta H^\circ_{G,298} = -523{,}000 + 11{,}807 = -511{,}193 \tag{5.13}$$

The free energy of formation of solid UF_6 at 25°C from the elements, from the same source, is

$$\Delta G_{C,298} = -492{,}252 \text{ cal/g-mol} \tag{5.14}$$

5 URANIUM SOLUTION CHEMISTRY

5.1 Oxidation States of Uranium in Aqueous Solution

In aqueous solution, uranium may occur as trivalent U^{3+}, the tetravalent uranous ion U^{4+}, pentavalent $U^{V}O_2{}^{+}$, or the hexavalent uranyl ion $U^{VI}O_2{}^{2+}$. However, U^{3+} is unstable, reducing water with production of hydrogen, and $U^{V}O_2{}^{+}$ is unstable, disproportionating into U^{4+} and $U^{VI}O_2{}^{2+}$:

$$2U^{V}O_2{}^{+} + 4H^{+} \rightarrow U^{4+} + U^{VI}O_2{}^{2+} + 2H_2O$$

Thus, only the uranous and uranyl ions are of practical importance.

5.2 Uranium(IV) Solutions

Solutions of tetravalent uranium salts are usually prepared by reduction of the corresponding uranyl compounds. Reduction may be effected by metallic zinc or at the cathode of an electrolytic cell:

$$UO_2{}^{2+} + 4H^{+} + 2e^{-} \rightarrow U^{4+} + 2H_2O$$

Table 5.12 Thermodynamic properties of UF_6 in ideal gas state at 1 atm pressure

Temperature T, K	Heat capacity C°_p, cal/(g-mol·°C)	Free-energy function $-(G^\circ - H^\circ_0)/T$, cal/(g-mol·°C)	Entropy S°, cal/(g-mol·°C)	Enthalpy relative to solid at 298.15 K, $H^\circ_G - H_{C,298}$, cal/g-mol
0				5,420
100	18.93	50.91	62.99	6,628
150	23.45	56.41	71.58	7,695
200	26.74	61.14	78.81	8,954
273	30.13	67.08	87.65	11,139
298	31.00	68.92	90.34	11,807
323	31.75	70.69	92.89	12,594
348	32.38	72.33	95.22	13,389
373	32.93	73.93	97.47	14,204
400	33.46	75.65	99.86	15,104
500	34.80	81.25	107.45	18,520
750	36.35	92.99	122.37	27,455
1000	36.94	101.37	132.58	36,630

Pure uranous compounds may be prepared by precipitating $U(OH)_4$ from aqueous solution with ammonium hydroxide and dissolving the precipitate in the appropriate acid. Uranous sulfate, the most common salt, is soluble in water, as are the chloride, bromide, and iodide. Uranous nitrate is unstable, gradually undergoing oxidation to uranyl nitrate with liberation of oxides of nitrogen.

Tetravalent uranium can be precipitated from aqueous solution as the insoluble oxalate, fluoride, or phosphate. UF_4 precipitated from aqueous solution contains water of crystallization. When this compound is heated to drive off the water, it is partially hydrolyzed to an oxyfluoride. The phosphate $U_3(PO_4)_4$ is soluble in hot, concentrated phosphoric acid and appears in this form when uranium in phosphate rock, $Ca_3(PO_4)_2$, is dissolved in sulfuric acid.

5.3 Uranyl Solutions

Solutions of uranyl salts can be readily prepared by dissolving UO_3 in the appropriate acid. Uranyl nitrate, sulfate, acetate, fluoride, chloride, bromide, and iodide all are very soluble, the solutions having a characteristic yellow-green fluorescence. Uranyl nitrate can be made by dissolving uranium metal or any of the oxides in nitric acid. It crystallizes from solution as the well-formed yellow uranyl nitrate hexahydrate $UO_2(NO_3)_2 \cdot 6H_2O$, often called UNH.

Uranium in solution as uranyl nitrate can be purified by addition of hydrogen peroxide, which selectively precipitates pale yellow uranyl peroxide $UO_2(O_2) \cdot 2H_2O$. Addition of sodium hydroxide to uranyl nitrate solution precipitates sodium diuranate $Na_2U_2O_7$. Addition of ammonium hydroxide precipitates ammonium diuranate:

$$2UO_2(NO_3)_2 + 6NH_4OH \rightarrow (NH_4)_2U_2O_7 + 4NH_4NO_3 + 3H_2O$$

Uranyl ion forms complexes with many anions. Table 5.13 gives equilibrium constants for complex formation at 20°C and an ionic strength around 1, arranged in order of increasing complex stability. At high concentrations of fluoride, sulfate, or carbonate ion, the complex uranyl ions then present are much less extractable by organic solvents than uranyl nitrate. The very stable uranyl sulfate and uranyl carbonate complex anions are strongly held by anion-exchange resins and are often used to recover and purify uranium from solutions obtained in leaching uranium ores (Sec. 8 of this chapter). The complex uranyl carbonate anions are very soluble in aqueous solutions of sodium or ammonium carbonate. This property has been used to separate uranium from elements such as radium, iron, or lead, which form insoluble precipitates with carbonate ion.

5.4 Solvent Extraction of Uranyl Compounds

Uranyl nitrate has an unusual property, shared only by nitrates of a few other actinides, of being very soluble in a number of organic solvents. When such an organic solvent is immiscible with water, it can be used in a solvent extraction process to extract uranium from aqueous solutions and separate it from associated impurities. Such applications of solvent extraction are very important in extracting and purifying uranium from leach solution of uranium ores or from nitric acid solution of irradiated nuclear fuel. Examples of extractants that have been used for such separation processes are listed in Table 5.14.

The ability of diethyl ether to extract uranyl nitrate from aqueous solution has been known for a hundred years and was the method chosen by the Manhattan Project to purify the uranium used in the first nuclear chain reactors. This solvent has numerous disadvantages. It is very volatile, very flammable, and toxic, and it requires addition of sodium, aluminum, or calcium nitrate to the aqueous phase to enhance extractions. When solvent extraction was first applied to recovery of uranium and plutonium from irradiated fuel, other oxygenated solvents less volatile than diethyl ether that were first used were methyl isobutyl ketone, dibutyl

Table 5.13 Complex formation constants of $UO_2{}^{2+}$

Reaction	Equilibrium constant
$UO_2{}^{2+} + NO_3{}^- \rightleftharpoons UO_2NO_3{}^+$	0.5 [A1]
$UO_2{}^{2+} + Cl^- \rightleftharpoons UO_2Cl^+$	0.8 [A1]
$UO_2{}^{2+} + HF \rightleftharpoons UO_2F^+ + H^+$	27
$UO_2{}^{2+} + SO_4{}^{2-} \rightleftharpoons UO_2SO_4$	50 [A1]
$UO_2{}^{2+} + 2SO_4{}^{2-} \rightleftharpoons UO_2(SO_4)_2{}^{2-}$	350 [A1]
$UO_2{}^{2+} + 3SO_4{}^{2-} \rightleftharpoons UO_2(SO_4)_3{}^{4-}$	2500 [A1]
$UO_2{}^{2+} + 2CO_3{}^{2-} \rightleftharpoons UO_2(CO_3)_2{}^{2-}$	4×10^{14} [M2]
$UO_2{}^{2+} + 3CO_3{}^{2-} \rightleftharpoons UO_2(CO_3)_3{}^{4-}$	2×10^{18} [M2]

carbinol, and triglycoldichloride. They had the disadvantages either of reacting with nitric acid or of requiring addition of solid salting agents. For solvent extraction of irradiated fuel, these have all been superseded by tributyl phosphate (TBP), in the Purex process described in Chap. 10. TBP has the advantage of being able to extract uranium efficiently from nitric acid solution without addition of solid nitrates. TBP is also used to purify natural uranium (Sec. 9.2 of this chapter).

The five solvents just discussed extract uranium in the form of neutral complexes of uranyl nitrate. With TBP the complex-forming equilibrium is

$$UO_2(NO_3)_2 \cdot 6H_2O(aq) + 2TBP(o) \rightleftharpoons UO_2(NO_3)_2 \cdot 2TBP(o) + 6H_2O(aq)$$

where (aq) denotes the aqueous phase and (o) the organic.

The last two solvents in Table 5.14 are sometimes called liquid ion-exchangers because they react with water-soluble uranium-bearing ions to form organic-soluble compounds. Di(2-ethylhexyl) phosphoric acid is an example of a liquid cation exchanger, which acts through the equilibrium

$$2(C_8H_{17}O)_2\overset{\overset{\displaystyle O}{\|}}{P}OH(o) + UO_2{}^{2+}(aq) \rightleftharpoons [(C_8H_{17}O)_2\overset{\overset{\displaystyle O}{\|}}{P}O]_2UO_2(o) + 2H^+(aq)$$

Because of the long octyl group, both the acid and the uranyl salt are soluble in a hydrocarbon diluent and insoluble in water.

Table 5.14 Solvents used in separation of uranium by solvent extraction

Solvent	Formula	Application	Name of process
Diethyl ether	$(C_2H_5)_2O$	Uranium purification	
Methyl isobutyl ketone	$CH_3(CO)C_4H_9$	Irradiated fuel	Redox
Dibutyl ether of diethylene glycol	$(C_4H_9OC_2H_4)_2O$	Irradiated fuel	Butex
Triglycol dichloride	$(ClC_2H_4O)_2C_2H_4$	Irradiated fuel	Trigly
Tributyl phosphate	$(C_4H_9)_3PO_4$	Uranium purification and irradiated fuel	Purex
Di(2-ethylhexyl) phosphoric acid	$(C_8H_{17}O)_2PO(OH)$	Extract uranium from leach liquors	Dapex
Trioctylamine	$(C_8H_{17})_3N$	Extract uranium from leach liquors	Amex

Trioctylamine is an example of a liquid anion-exchanger, which acts through the equilibrium

$$2[(C_8H_{17})_3NH]_2SO_4(o) + UO_2(SO_4)_3^{4-}(aq) \rightleftharpoons [C_8H_{17}NH]_4UO_2(SO_4)_3(o) + 2SO_4^{2-}(aq)$$

Because of the long octyl group, both the amine and the uranyl sulfate complex are soluble in a hydrocarbon diluent and insoluble in water.

These liquid ion-exchangers have two advantages over TBP for extraction of uranium from leach liquors. Distribution coefficients are higher, so that uranium may be extracted at higher concentration from dilute leach liquors. The involvement of hydrogen or sulfate ion in the distribution equilibrium makes it possible to drive the reaction to favor either the organic or aqueous phase by adjusting the H_2SO_4 concentration of the aqueous phase. Clegg and Foley [C1], Merritt [M3], and Brown et al. [B8] describe many other long-chain amines and organophosphorus compounds that have been used to extract uranium from leach liquors. These may be used either for the hexavalent uranyl or the tetravalent uranous ion.

6 SOURCES OF URANIUM

6.1 Principal Uranium-containing Minerals

At current prices of uranium (\$30 to \$50/lb U_3O_8) ores containing around 0.1 percent U_3O_8 or more are being mined primarily for their uranium content. The principal uranium-bearing minerals found in such ores are listed in Table 5.15, classified according to the type of treatment needed to extract the uranium.

The first group, minerals containing high concentrations of uranium, mostly in the tetravalent state, can be concentrated by specific gravity methods when in massive form. Frequently, however, the particle size is so small that the uranium-bearing mineral must be dissolved in sulfuric acid or sodium carbonate leach liquors. In either case, an oxidant must be added to bring uranium to the soluble, hexavalent state.

Uraninite, or pitchblende as it is more commonly known, is the form in which uranium was first discovered, at Joachimsthal, Czechoslovakia. Later, very rich deposits of massive uraninite were discovered at the Shinkolobwe mine in the Belgian Congo (Zaire) and were the principal source of uranium for the Manhattan Project. Leaner ores containing finely divided

Table 5.15 Principal uranium minerals

1. Ores containing tetravalent uranium	
Uraninite (pitchblende)	U_3O_8
Uranothorite	$Th_{1-x}U_xSiO_4$
Coffinite	$U(SiO_4)_{1-x}(OH)_{4x}$
2. Hydrated ores containing hexavalent uranium	
Gummite	$UO_3 \cdot nH_2O$
Carnotite	$K_2O \cdot 2UO_3 \cdot V_2O_5 \cdot 3H_2O$
Tyuyamunite	$CaO \cdot 2UO_3 \cdot V_2O_5 \cdot 8H_2O$
Autunite	$CaO \cdot 2UO_3 \cdot P_2O_5 \cdot 8H_2O$
Torbernite	$CuO \cdot 2UO_3 \cdot P_2O_5 \cdot 8H_2O$
Uranophane	$CaO \cdot 2UO_3 \cdot 2SiO_2 \cdot 6H_2O$
3. Refractory minerals containing tetravalent uranium	
Davidite	$UFe_5Ti_8O_{24}$
Brannerite	$(U,Th,Ca_2,Fe_2)Ti_2O_6$
Pyrochlore	$(Na_4,Ca_2,U,Th)(Nb,Ta)_4O_{12}$

Table 5.16 Low-grade sources of uranium†

Material	Location	Concentration range, g U/MT
Commercial sources		
Lignite	South Dakota	10–4000
Shale	Sweden	250–325
Gold tailings	South Africa	60–300
Phosphate rock	Florida	100–200
Copper tailings	Western United States	2–50
Sources presently not economic		
Shale	Chattanooga, Tennessee	50–70
Bostonite	Colorado	33
Granite	Conway, New Hampshire	10–30
Earth's crust	Average	1.7
Seawater		0.003

†Most of this information is from [B2].

uraninite are now being mined in the Lake Athabasca district of Canada, in the White Canyon and Big Indian Wash districts of Utah, and the Jackpile mine of New Mexico.

Coffinite is a major mineral in the important Ambrosia Lake district of New Mexico. Uranothorite is mined in the Bancroft district of Ontario.

The second group of minerals in Table 5.15, hydrated minerals containing hexavalent uranium, are usually soft, finely divided, of relatively low density, and readily soluble in dilute sulfuric acid or sodium carbonate solution. Their uranium cannot be concentrated by specific gravity or flotation methods and must be recovered by leaching and concentration by selective precipitation, solvent extraction, or ion exchange. These ores are usually of secondary origin, precipitated from uranium-bearing groundwaters.

Carnotite is a major ore of the Colorado plateau. Tyuyamunite occurs near Grants, New Mexico, in Utah, and in the Soviet Union. Autunite is found near Marysvale, Utah, and in Washington and Wyoming. Torbernite is found in the White Canyon district of Utah and in the upper zones of ore bodies in Zaire.

Minerals of the third group of Table 5.15 contain relatively small proportions of tetravalent uranium combined with a refractory oxide of titanium, niobium, or tantalum. To free the uranium from these minerals, they must be leached with hot, concentrated sulfuric acid. Davidite is one of the principal ores at Radium Hill in South Australia. Brannerite is found in the Blind River district of Ontario. Pyrochlore occurs in the Lake Nipissing district of Ontario and in Nigeria.

6.2 Low-Grade Sources of Uranium

In addition to the well-characterized minerals containing uranium listed in Table 5.15, uranium occurs as a minor constituent of many other materials, some of which have been used as commercial sources. Table 5.16 lists low-grade sources of uranium and gives the range of their uranium content. Uranium from the five commercial sources is being produced as a by-product or co-product of other materials whose value helps pay for the cost of producing the uranium.

Some lignites contain sufficient uranium so that they can be mined for the uranium alone; most are so lean that they must be used as fuel as well as a source of uranium.

Swedish shales contain 300,000 MT† of recoverable uranium, together with organic matter from which fuel can be made and pyrites that can be converted to sulfuric acid. The combined value of these products makes the operation economic.

Tailings from many South African gold mines contain sufficient uranium to permit its recovery at competitive costs, because the cost of mining and crushing the ore has been paid for by the gold previously extracted. Most South African uranium is produced in this way.

In the wet process for converting phosphate rock to fertilizer, the rock is first converted to phosphoric acid. In this process around 95 percent of the uranium goes into solution, and around 85 percent could be recovered if a uranium extraction step were added. Some uranium is now being produced from this source. A total of 65,000 MT of uranium could be recovered in this way by the year 2000 in the United States.

Pilot-plant studies confirm the feasibility of recovering uranium from the sulfuric acid solutions used to extract copper from tailing piles in the Western United States. One plant producing around 250 MT/year went into operation at Twin Buttes, Nevada, in 1975. It is estimated that total U.S. production from copper tailings might reach 800 MT/year.

Uranium produced from the uneconomic sources listed in Table 5.16 would cost several hundred dollars per pound and is not economic at present. If the uranium-fueled fast-breeder reactor becomes economic, it would generate so much electricity per ton of natural uranium that Chattanooga shale and even Conway granite might be used as economic uranium sources. These sources are estimated to contain 5 million and 6 to 9 million MT of uranium, respectively. Environmental problems from the large amount of earth disturbed in mining these low-grade sources would be severe.

Although seawater contains only 3.34 μg of uranium/liter, the oceans of the world are so vast that their total uranium content is estimated to be around 4 billion MT [D1]. Extraction of uranium from seawater is discussed in Sec. 8.8.

7 URANIUM RESOURCE ESTIMATES

7.1 World Resources

Table 5.17 summarizes information on the uranium resources and the annual uranium production capabilities of the principal uranium-producing countries of the non-Communist world compiled by the Organization for Economic Cooperation and Development [O2] in December 1977. The production *costs* of \$30 and \$50/lb U_3O_8 do not include exploration costs. The selling *price* of U_3O_8 is highly variable, as it depends on when the sale was negotiated, when the uranium is to be delivered, and whether it is a price for one-time, spot delivery or long-term supply. Typical prices in 1978 were in the range \$30 to 50/lb U_3O_8. In order of decreasing resources, the four countries with the largest resources are the United States, Canada, South Africa, and Australia.

Reference [O2] states:

> *Reasonably Assured Resources* refers to uranium which occurs in known ore deposits of such grade, quantity and configuration that it could be recovered within the given production cost range with currently proven . . . technology. Estimates of tonnage and grade are based on specific sample data and measurements of the deposits. . . .
>
> *Estimated Additional Resources* refers to uranium surmised to occur in unexplored extensions of known deposits in known uranium districts, and which is expected to be discoverable and could be produced in the given cost range.

†MT (metric ton) = 1 megagram = 1.102 short tons.

Table 5.17 Uranium resources and production of non-Communist world

	Resources, thousand MT uranium[†]						Production capability, thousand MT uranium/yr	
	Production cost <\$30/lb U_3O_8			Production cost \$30–\$50/lb U_3O_8				
Country	Reasonably assured	Estimated additional	Total	Reasonably assured	Estimated additional	Total, cost <\$50/lb U_3O_8	1977	1985
Algeria	28	50	78	0	0	78	‡	‡
Argentina	17.8	0	17.8	24	0	41.8	‡	‡
Australia	289	44	333	7	5	345	0.4	11.8
Brazil	18.2	8.2	26.4	0	0	26.4	‡	‡
Canada	167	392	559	15	264	838	6.1	12.5
France	37	24.1	61.1	14.8	20	95.9	2.2	3.7
Gabon	20	5	25	0	5	30	0.8	1.2
India	29.8	23.7	53.5	0	0	53.5	‡	‡
Niger	160	53	213	0	0	213	1.6	9.0
South Africa	306	34	340	42	38	420	6.7	12.5
Sweden	1	3	4	300	0	304	‡	‡
United States	523	838	1361	120	215	1696	14.7	36.0
Others	53.2	35	88.2	17.2	43	148.4	0.5	5.3
Total	1650	1510	3160	540	590	4290	33.0	92.0

[†] 1 MT uranium = 1 tonne uranium = 1 Mg uranium = 1.102 short tons uranium = 1.300 short tons U_3O_8.

[‡] Included in "others."

Source: Organization for Economic Cooperation and Development, and International Atomic Energy Agency, "Uranium Resources Production and Demand," Paris, Dec. 1977.

Table 5.18 U.S. uranium resources

	Uranium, thousand MT†					
Cost $/lb U_3O_8	Reserves	Probable potential resources	Subtotal	Possible potential resources	Speculative potential resources	Total
$15	285	416	701	377	127	1205
$15–30	246	365	611	496	192	1299
<$30	531	781	1312	873	319	2504
$30–50	154	292	446	292	115	853
<$50	685	1073	1758	1165	434	3357

†1 MT uranium = 1 tonne uranium = 1 Mg uranium = 1.102 short tons uranium = 1.300 short tons U_3O_8.

7.2 United States

An estimate of uranium resources in the United States more detailed and more recent than that of Table 5.17 provided by the U.S. Department of Energy in May 1978 [U1] is summarized in Table 5.18. "Reserves" corresponds approximately with the "Reasonably assured resources" category of Table 5.17, and "Probable potential resources" corresponds with "Estimated additional resources." The subtotal at <$30 of 1312 thousand MT in Table 5.18 may be considered an update of the 1361 for the United States in Table 5.17, and the subtotal at <$50 of 1758 thousand MT in Table 5.18, an update of the 1696 in Table 5.17.

"Possible potential resources" are defined [U2] as "those estimated to occur in undiscovered or partly defined deposits in formations ... productive elsewhere in the same geologic province." "Speculative potential resources" are defined [U2] as "those estimated to occur in undiscovered or partly defined deposits: 1. in formations ... not previously productive within a productive geologic province, or 2. within a geologic province not previously productive."

To relate these resource estimates to nuclear electric generation, it may be noted that a 1000-MWe pressurized-water reactor operating at 80 percent capacity factor without recycle, on uranium enriched to 3.3 w/o (weight percent) ^{235}U in an enrichment plant stripping natural uranium to 0.3 w/o ^{235}U, consumes around 200 MT of uranium per year. Thus the U.S. resource estimate of 1758 thousand MT available at less than $50/lb U_3O_8 would keep a 300,000-MWe nuclear power industry in fuel for

$$\frac{1{,}758{,}000 \text{ MT}}{(200 \text{ MT}/1000 \text{ MWe}\cdot\text{yr})(300{,}000 \text{ MWe})} = 29.3 \text{ yr} \tag{5.15}$$

Inclusion of uranium in the "Possible" and "Speculative" resource categories would increase the total to 3357 thousand MT and extend the life of a 300,000-MWe nuclear power industry to 56 years.

8 CONCENTRATION OF URANIUM

8.1 Steps in Producing Refined Uranium Compounds

The steps in producing refined uranium compounds from crude uranium ores may be conveniently classified into concentration, purification, and conversion to the chemical form finally wanted. *Concentration* consists in separating uranium from most of the nonuraniferous diluents that accompany uranium in nature. It increases the uranium oxide content from a few tenths of a percent in the ore to 85 to 95 percent in the concentrate, while rejecting most of the diluents as tailings. Concentration is usually carried out in uranium mills within a few miles of where the uranium is mined, to avoid the high cost of transporting the nonuraniferious bulk of the ore.

Purification consists in removing from the impure uranium the rest of the nonuraniferous contaminants and producing a pure uranium compound. In most uranium refineries purification is carried out before *conversion* of uranium to the chemical form finally wanted. This sequence of refining operations will be described in Secs. 9.2 through 9.6. However, in the process used by the Allied Chemical Company for producing UF_6 from uranium concentrates, the sequence is reversed, with conversion to UF_6 preceding purification of the impure UF_6 by fractional distillation. This process will be described in Sec. 9.7.

8.2 Concentration Methods

Because of the great variety of natural sources of uranium, no one process is uniquely suited to concentration of uranium from all ores. General methods that have been found useful for certain types of ores will be discussed first and then a few examples will be given of specific processes used for specific ores. Detailed accounts of uranium-concentration processes developed in the United States have been given by Marvin et al. [M1], Clegg and Foley [C1], and Merritt [M3].

Gravity concentration. Only the richest pitchblende ores can be concentrated by the specific gravity methods commonly used for other metals. Pitchblende particles are much denser than accompanying nonuraniferous rock (called *gangue*). In rich ores pitchblende particles are large enough to be concentrated by gravity methods such as were used at Shinkolobwe. Gravity methods are seldom used elsewhere because few other uranium ores have uranium-bearing particles large enough for selective concentration by this method. Most ores contain secondary minerals such as autunite or carnotite, which are soft and form such small particles as to be unrecoverable by gravity methods.

Flotation. Flotation, which is an extremely selective method of concentrating ores of many common elements, is seldom applicable to uranium because few uranium minerals can be selectively floated.

Leaching. Because of the unsuitability of physical methods such as specific gravity or flotation separation, the extraction of uranium from its ores is effected almost exclusively by chemical leaching. The principal leaching reagents are sulfuric acid or alkali carbonate solution. The type of leaching agent depends on the nature of the uranium mineral and of the gangue. When the gangue is silica or some other material insoluble in acid, sulfuric acid leaching is preferred, because it costs less than carbonate and dissolves uranium minerals more rapidly. When the uranium is combined as a silicate or titanate, as in brannerite, it is necessary to leach for longer times, or at higher temperatures, or in stronger acid than when it is more readily dissolved, as in carnotite or autunite. Acid consumption for U.S. ores treated for uranium recovery only is from 40 to 120 lb/t† of ore. When vanadium is also extracted, acid consumption may reach 300 lb/t. Leaching of ores containing tetravalent uranium requires addition of an oxidant to convert uranium to the soluble hexavalent condition. Manganese dioxide or sodium chlorate are the principal oxidants. Iron must be present in solution to catalyze the oxidation.

†t = short ton, 2000 lb.

When the gangue is limestone or some other rock that consumes acid, leaching with sodium or ammonium carbonate is preferred, to reduce chemical consumption and produce a cleaner solution containing lower concentrations of impurities than when acid leaching is used. The reactions that occur in carbonate leaching are

$$UO_2 + \tfrac{1}{2}O_2 \rightarrow UO_3$$

$$UO_3 + Na_2CO_3 + 2NaHCO_3 \rightarrow Na_4UO_2(CO_3)_3 + H_2O$$

As carbonate leaching is slower than acid leaching and carbonate solutions do not attack and penetrate gangue particles, it is usually necessary to grind the ore finer and to leach for longer times and at higher temperatures when leaching with carbonate solutions than with sulfuric acid.

Recovery of uranium from leach liquors. Uranium may be recovered from leach liquors by precipitation, ion exchange, or solvent extraction. Precipitation with sodium hydroxide was the recovery method used in the first uranium mills. When used on sodium carbonate leach liquors, the uranium precipitate is fairly free of other metallic contaminants, because sodium carbonate dissolves few other metals beside uranium. However, when used in sulfuric acid leach liquors, the uranium precipitate contains other metals, such as iron dissolved from the ore by the acid, and is no longer commercially acceptable. Consequently, in the United States, uranium mills employing acid leaching now follow it with selective recovery by either solvent extraction or ion exchange. These processes are described in Secs. 8.5 and 8.6, respectively.

8.3 U.S. Uranium Mills

Table 5.19 lists the uranium mills operating in the United States in January 1977 and their capacity, as reported by the Grand Junction Office of the U.S. Energy Research and Development Administration [U2]. Table 5.19 summarizes the processes used in the mills, when such information is available from a May 1975 report from the Oak Ridge National Laboratory [S2] and other industry sources. As uranium milling is a dynamic industry, with frequent changes in process technology and mill capacity, Table 5.19 serves more to illustrate the diversity of milling processes than to provide an invariant listing.

Part 1 of Table 5.19 lists mills using carbonate leaching. There is no standard process for recovery of uranium from carbonate leach liquors. The trend is away from precipitation with NaOH as crude $Na_2U_2O_7$ toward further purification as by UO_4 precipitation at Rio Algom or by ion exchange followed by precipitation as $(NH_4)_2U_2O_7$ at George West.

Part 2 of Table 5.19 lists mills using acid leaching and solvent extraction. All mills for which process information is available use a long-chain tertiary amine, Alamine-336 or Adogen-364, as extractant, with general features illustrated in Sec. 8.6.

Part 3 of Table 5.19 lists mills using acid leaching and anion exchange. The three types of contactors and the different eluting and final treating processes are described in Sec. 8.7.

8.4 Uranium Concentration by Carbonate Leaching

As an example of uranium concentration by carbonate leaching, a brief description will be given of the uranium mill owned by United Nuclear-Homestake Partners at Grants, New Mexico [J1, M3]. Mill capacity in 1977 was 3500 short tons ore per day. The ores treated are primarily sandstones, whose silica or limestone grains are cemented by intergranular uranium-bearing minerals such as coffinite, uraninite, tyuyamunite, and carnotite. Average ore composition is 0.21 w/o U_3O_8, with small amounts of vanadium, molybdenum, and selenium.

In this mill, uranium is recovered from carbonate leach liquor by precipitation with sodium

Table 5.19 U.S. uranium mills operating in January 1977

1. Alkaline leach

Company	Location	Short tons ore per day	Special process features
Rio Algom	La Sal, Utah	700	Uranium precipitated as $Na_2U_2O_7$, redissolved in H_2SO_4, and reprecipitated as UO_4 by H_2O_2
United Nuclear-Homestake	Grants, New Mexico	3500	Uranium precipitated as $Na_2U_2O_7$, roasted to convert vanadium to water-soluble $NaVO_3$, leached from $Na_2U_2O_7$
U.S. Steel-Niagara Mohawk	George West, Texas	Solution mining	NH_4HCO_3 leach, uranium recovery by anion exchange

2. Acid leach–Solvent extraction

Company	Location	Short tons ore per day	Extractant	Strippant
Anaconda	Grants, New Mexico	3000	Amine	
Atlas Corp.	Moab, Utah	1100	Amine	
Conoco-Pioneer Nuclear	Falls City, Texas	1750	Alamine-336 + Isodecanol	NH_4Cl
Cotter Corp.	Canon City, Colorado	450	Amine	
Exxon	Douglas, Wyoming	3000	Alamine-336 + Isodecanol or Adogen-364	$(NH_4)_2SO_4$
Kerr-McGee	Grants, New Mexico	7000	Alamine-336 + Isodecanol	NaCl
Sohio-Reserve Oil	Cebolleta, New Mexico	1660	Amine	

3. Acid leach–Anion exchange

Company	Location	Short tons ore per day	Contactor	Eluant	Final treatment
Dawn Mining	Ford, Washington	400	Fixed bed	NH_4NO_3-H_2SO_4	NH_3 ppt.
Federal-American	Gas Hills, Wyoming	950	Continuous RIP†	$(NH_4)_2SO_4$-H_2SO_4	Eluex + NH_3
Lucky Mc Uranium	Gas Hills, Wyoming	1650	Moving bed	$(NH_4)_2SO_4$-H_2SO_4	Eluex + NH_3
Lucky Mc Uranium	Shirley Basin, Wyoming	1800	Fixed bed	NaCl-H_2SO_4	NH_3 ppt.
Union Carbide	Uravan, Colorado	1300	Fixed bed	NaCl-H_2SO_4	NH_3 ppt.
Union Carbide	Gas Hills, Wyoming	1200	Continuous RIP†	$(NH_4)_2SO_4$-H_2SO_4	Eluex + NH_3
Western Nuclear	Jeffrey City, Wyoming	1700	Continuous RIP†	$(NH_4)_2SO_4$-H_2SO_4	Eluex + NH_3

†RIP = resin in pulp.

or ammonium hydroxide. Another U.S. mill employing carbonate leaching recovers uranium by anion exchange (Table 5.19).

Crushing and grinding. Figure 5.4 is a schematic flow sheet of the principal process steps in this mill. Ore is crushed dry to particles smaller than 0.5 in. Crushed ore is ground in two conical ball mills with recycled filter washings containing sodium carbonate and bicarbonate and some recycled uranium. Ball mills are operated in closed circuit with a spiral classifier which returns oversize particles to the mill. Classifier overflow is 16 to 20 percent solids, with 95 percent finer than 48 mesh and 65 percent coarser than 200 mesh. This is finer grinding than used in the Kerr-McGee acid-leach mill (see Sec. 8.5).

Leaching. Ground slurry is divided between two similar leaching circuits. The slurry is thickened to 52 to 54 percent solids in 20-in (0.5-m) diameter cyclones and 100-ft (30-m)

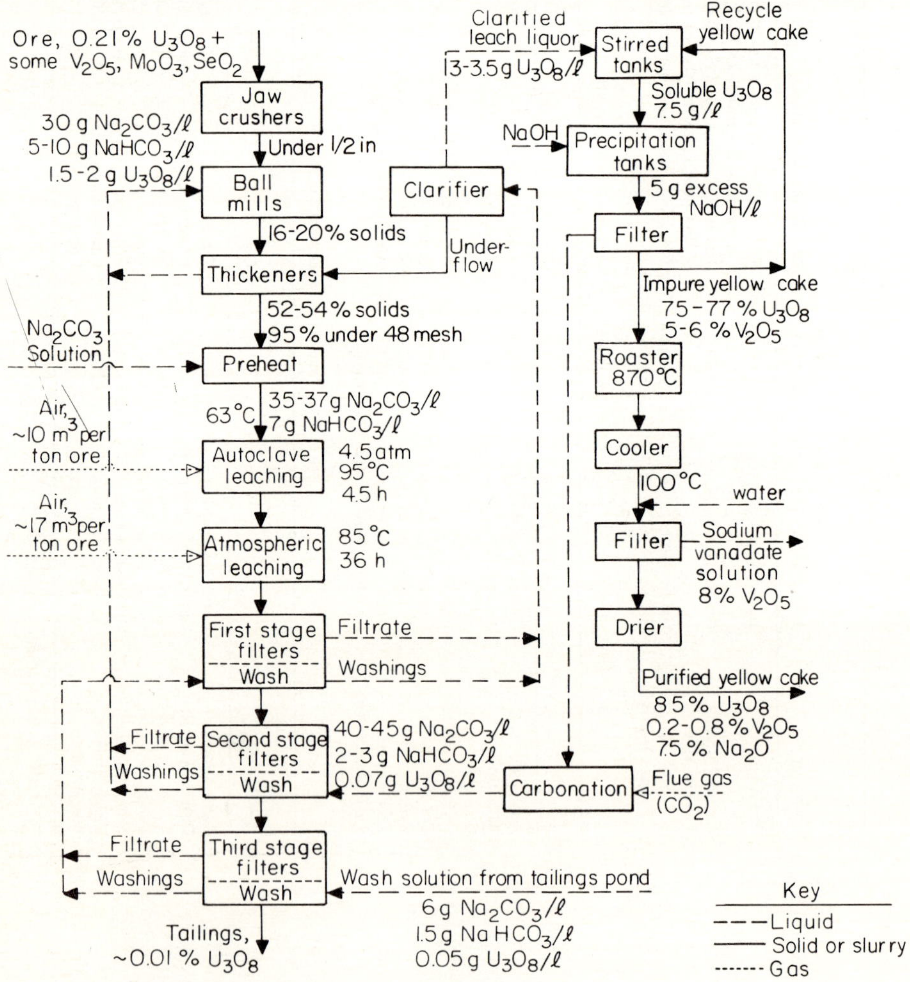

Figure 5.4 Schematic flow sheet, United Nuclear-Homestake Partners uranium mill.

diameter settling tanks, with overflow recycled to grinding. The slurry is preheated to 63°C, and sodium carbonate is added up to 35 to 37 g Na_2CO_3/liter, with about 7 g $NaHCO_3$/liter. Slurry is pumped at a pressure of about 4.5 atm in series through two lines of four stirred autoclaves 12 ft (3.7 m) in diameter and 16 ft (4.9 m) high, which are heated to about 95°C, with a residence time of about 4.5 h. Air is pumped through the autoclaves to oxidize tetravalent uranium.

Less soluble ores are leached for an additional 36 h at 85°C and atmospheric pressure in nine large tanks 19 ft (5.8 m) in diameter by 38 ft (11.6 m) deep, stirred, and supplied with additional air. Enough excess sodium carbonate is used to bring the pH up to 11.

Filtration. Uranium-bearing solution is separated from leached solids in three filter stages with countercurrent washing. Each stage consists of five 650-ft^2 (60-m^2) and two 570-ft^2 (53-m^2) rotary-drum vacuum filters, operated batchwise. Filter cake from the first stage is washed with filtrate from the third stage; filtrate and washings from the first stage constitute the leach liquor, which contains 3 to 3.5 g U_3O_8/liter. Filtrate and washings from the third stage are used in the first stage to wash the filter cake and then to reslurry it and move it to the second stage. Filter cake from the second stage is washed and reslurried with sodium carbonate solution; the filtrate and washings containing some uranium are returned to the ball mills. Filter cake from the third stage is washed with dilute carbonate solution from the tailings pond; filtrate and washings are used to wash and reslurry filter cake in the first stage. Washed filter cake from the third stage is reslurried and pumped to the tailings pond.

Leaching and washing reduce the U_3O_8 content of tailings to around 0.01 percent.

Precipitation. Leach liquor from the first stage filters is clarified in a thickener, mixed with about five times its content of recycled yellow cake,† heated to 74°C, and held about 5 h in stirred tanks. Recycle increases the soluble uranium content from about 3.5 to 7.5 g U_3O_8/liter and was found needed to improve the completeness of uranium precipitation in the next step.

Precipitation of uranium by the reaction

$$2Na_4UO_2(CO_3)_3 + 6NaOH \rightarrow Na_2U_2O_7 + 6Na_2CO_3 + 3H_2O$$

is accomplished by adding sufficient sodium hydroxide to leave an excess of 5 g NaOH/liter after reaction. The mixture flows through eight heated and stirred tanks in series for an 8-h residence time at 74°C. Reaction product is filtered to produce an impure solid $Na_2U_2O_7$ and a solution of NaOH and Na_2CO_3. The solution is recarbonated with flue gas and returned to the leaching circuit via the second-stage filter wash.

Purification. The impure $Na_2U_2O_7$ contains 5 to 6 percent V_2O_5, probably coprecipitated with the uranium. This impurity is converted to soluble sodium vanadate, $NaVO_3$, by roasting the impure yellow cake with about an equal mass of Na_2CO_3 at 860°C for a half-hour, cooling the mass, and leaching it with water. The leached product is filtered and washed, producing a solution from which vanadium is recovered, and a yellow cake that, after drying, contains about 85 percent U_3O_8, 0.2 to 0.8 percent V_2O_5, and up to 7.5 percent Na.

8.5 Acid Leaching of Uranium Ores

As an example of the acid-leaching, solvent extraction class of uranium-concentration processes, a description will be given of the process used in the large uranium mill of the Kerr-McGee Corporation at Grants, New Mexico. This has been condensed from a 1960 paper by Hazen

†Yellow cake is the name conventionally used for uranium ore concentrates.

[H4] and from Merritt's [M3] account of operations in 1971. At that time the mill's capacity was 5000 short tons ore (4500 Mg) per day. In 1978 its capacity was 6200 short tons per day. Figure 5.5 shows this mill.

Leaching operations in the Kerr-McGee mill are described in this section, with reference to Fig. 5.6. Recovery of uranium from leach liquor by solvent extraction with organic amines in the Amex process is to be described in Sec. 8.6.

The ore processed in the Kerr-McGee mill is primarily a sandstone, with uranium minerals in the material bonding the sand grains. The ore contains about 0.2 w/o U_3O_8, 0.01 to 0.03 w/o MoO_3, and 0.05 to nearly 0.20 w/o V_2O_5. Uranium and molybdenum are leached and recovered, with uranium recovery exceeding 97 percent. Some vanadium is also leached, but was not being recovered in 1971. The ore also contains acid-soluble calcium minerals, equivalent to from 2 to 5 w/o CaO, which are the principal consumers of sulfuric acid.

Crushing and grinding. The ore is first crushed dry to particles smaller than 1 in. Crushed ore is then processed in two parallel, identical systems, of which circuit A is shown in Fig. 5.6. In each circuit 2500 t per day of crushed ore is ground with heated water in rod mills until 97 to 98 percent passes 28 mesh, with 70 percent coarser than 150 mesh.

Leaching. Slurry from the rod mills flows in series by gravity through 14 rubber-lined steel leaching tanks 13 ft (4 m) in diameter and 14 ft (4.25 m) high, equipped with turbine agitators. The holding time in the 14 tanks is around 4.5 h. The rod-mill slurry fed to the first tank is mixed with recycle water containing slimes, and steam and sulfuric acid sufficient to bring the temperature in the first tank to 43 to 54°C and the pH to 0.6 to 0.7. Here the most readily dissolved minerals react, and gaseous reaction products such as CO_2, H_2, and H_2S are liberated.

Figure 5.5 Kerr-McGee Corporation uranium mill, near Grants, New Mexico. (*Courtesy of Kerr-McGee Corporation.*)

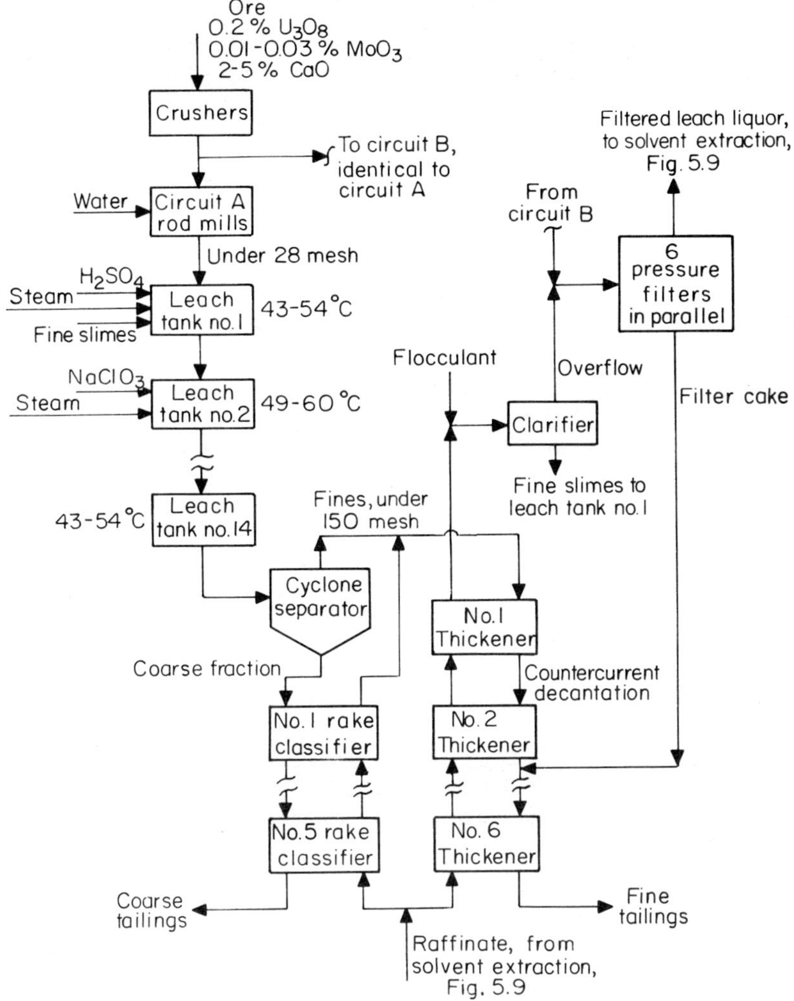

Figure 5.6 Principal steps in leaching uranium ore in Kerr-McGee uranium mill.

To leach the more acid-resistant minerals containing tetravalent uranium, steam is fed to the second tank to bring the temperature to 49 to 60°C, and sodium chlorate $NaClO_3$ is added to bring the oxidation-reduction potential e, measured relative to the calomel electrode, to from −0.47 to −0.51 V. At −0.51 V, the equilibrium ratio of ferric iron to ferrous iron in the solution is 0.52.† Ferric iron catalyzes the oxidation of insoluble tetravalent uranium to the soluble hexavalent uranyl form:

†Oxidation-reduction potentials and their effect on the valence state of materials being processed are explained in Chap. 9. Because the emf of the saturated KCl-calomel electrode relative to the standard hydrogen electrode is −0.244 eV at 25°C ([M3], p. 70), the relation between the emf e relative to the saturated calomel electrode and the emf E° relative to the standard hydrogen electrode used elsewhere in this text is $E^\circ = e - 0.244$. Thus, the emf in the second tank relative to the standard hydrogen electrode is from −0.714 to −0.754 V.

$$U^{4+} + 2Fe^{3+} + 2H_2O \rightarrow U^{VI}O_2{}^{2+} + 2Fe^{2+} + 4H^+$$

Addition of sodium chlorate to the second tank instead of to the first avoids consumption of this relatively expensive material by metallic iron introduced in grinding or by reducing gases, such as H_2 or H_2S, which are vented from the first tank.

As the hot, acid, oxidizing slurry flows through the remaining 13 tanks, dissolution of the more resistant uranium and molybdenum minerals is completed. After 4.5 h, when the slurry leaves the fourteenth tank, its temperature has dropped to 43 to 54°C, acid has been consumed with pH increased to 0.9 to 1.2, and sodium chlorate has been used up in oxidizing uranium, with the oxidation-reduction potential relative to the calomel electrode in the range −0.41 to −0.43 V. This is sufficiently negative to convert substantially all soluble uranium to the hexavalent state, while leaving only 2 percent of the iron oxidized to ferric, thus minimizing consumption of sodium chlorate.

In some other U.S. mills and in South Africa and Australia, manganese dioxide MnO_2, synthetic or in the form of the mineral pyrolusite, is used as oxidant. Typical oxidant consumption is 3 lb (1.5 kg) $NaClO_3$ or from 3 to 6 lb (1.5 to 3 kg) MnO_2 per short ton of ore.

Sulfuric acid consumption depends on the amount of reactive minerals present in the ore. In the United States, acid requirements range from 40 to 120 lb (20 to 60 kg) H_2SO_4 per short ton of ore.

Liquid-solid separation. Separation of the slurry leaving leach tank #14 into (1) tailings relatively free of uranium-containing liquid and (2) uranium-bearing leach liquor free of suspended solids is shown at the bottom and right of Fig. 5.6. A rough separation of slurry from leach tank #14 at about 150-mesh particle size is made in two 20-in (0.5-m) diameter cyclone separators in parallel. The coarse fraction from the cyclones passes through five rake classifiers in series, where the sand is washed countercurrently with acid-bearing aqueous raffinate from the solvent extraction system, Fig. 5.9. The fines fraction from the cyclone separators and the overflow from #1 rake classifier are combined and washed countercurrently with additional raffinate in six large countercurrent decantation thickeners 120 ft (36 m) in diameter and 17 ft (5 m) deep. Wash ratios are 2.5 to 3.0 in the classifiers and 3.0 to 4.0 in the thickeners. This recycle of raffinate returns H_2SO_4 from the solvent extraction system to the washing circuits and maintains the pH in them at 1.5 or less, thus preventing precipitation of uranium during washing. Tailings from #5 rake classifer are about 75 v/o (volume percent) solids; from #6 thickener, about 31 v/o. The thickeners handle about 25 percent of total solids. Soluble uranium losses in the washed tailings represent only about 0.2 percent of the uranium in mill feed.

Overflow from #1 thickener contains 150 to 200 ppm of solids. This is reduced to 50 to 75 ppm by addition of flocculant and settling in a 60-ft (18-m) diameter by 20-ft (6-m) deep reactor-clarifier. Fine slimes from the clarifier are recycled to leach tank #1.

Overflow from the clarifier is combined with a similar stream from leaching circuit B and passed through six 600-ft^2 (58-m^2) U.S. pressure filters in parallel, each operated at a feed rate of 300 to 500 gal/min (1.1 to 1.9 m^3/min). Filters are precoated with about 0.1 lb (0.05 kg) of precoat per short ton of ore, and 0.35 lb (0.17 kg) of filter aid per ton is added to the filter feed. The filter loading cycle lasts from 4 to 24 h, depending on the solids contents of clarifier overflow. Filter cake is returned to #2 or #3 thickener. Filtered leach liquor containing about 1 g U_3O_8/liter is the product of the leaching system and the feed to the solvent extraction system.

8.6 Solvent Extraction of Uranium from Leach Liquors

Processes for recovering uranium from acid leach liquors used in the United States include solvent extraction with organic amines, solvent extraction with organophosphorus compounds, and anion exchange. Amine extraction in the so-called Amex process is described in this section with specific reference to the Kerr-McGee mill. Solvent extraction with organophosphorus compounds in the so-called Dapex process was used in several U.S. mills, but is being phased out. It will be discussed briefly at the end of this section. Uranium recovery by anion exchange is to be discussed in Sec. 8.7.

Three papers from Oak Ridge National Laboratory provide a comprehensive summary of developments in solvent extraction of uranium from leach liquors. Coleman et al. [C2] describe studies of a number of possible amine extractants. Blake et al. [B4] describe work with organophosphorus compounds. Brown et al. [B8] describe processes based on both types of extractants.

Amine extraction. As explained in Sec. 5.4, long-chain organic amines act as liquid anion-exchange media for the uranyl sulfate complex anion through a reversible reaction such as

$$2(R_3NH)_2SO_4(o) + UO_2(SO_4)_3^{4-}(aq) \rightleftharpoons (R_3NH)_2UO_2(SO_4)_3(o) + 2SO_4^{2-}(aq)$$

The first reported use of high-molecular-weight amines to extract anions from aqueous solution was by the British workers Smith and Page [S3], who in 1948 used this method to separate strong acids from weak and suggested its use for recovering metals that form anionic complexes from aqueous solutions. Starting in 1953, Oak Ridge workers investigated a number of amines capable of extracting uranium as an anionic complex. Amines acting as liquid anion exchangers were found to be more selective in extracting uranium than organophosphorus compounds, which act as liquid cation-exchangers, because fewer of the metals associated with uranium in leach liquors form extractable anions than form extractable cations.

Of the numerous amines investigated [C2, B8], the one now used industrially in the Amex process is a mixture of straight-chain, saturated trioctyl and tridecyl amines, sold by General Mills Chemicals, Inc., under the trade name Alamine-336 and by Archer Daniels Midland Company under the name Adogen-364. Choice of this amine mixture has resulted partly from its commercial availability at an acceptable price of around $1.00/lb and partly from its having the necessary physical properties of good chemical stability, low aqueous solubility, high uranium distribution coefficient, and good selectivity for uranium.

To use this amine, it is dissolved in a high-boiling kerosene diluent, at a concentration of around 3 v/o, approximately 0.1 *M*. To this solution is added around 3 v/o (0.2 *M*) of a long-chain alcohol, such as isodecanol, to increase solubility in the kerosene diluent of the sulfate and acid sulfate salts of the amine and its complexes with compounds of uranium and molybdenum.

Extraction equilibria with amines. Figure 5.7 shows the distribution of uranium between aqueous and organic phases observed [C2] at conditions similar to those used in the Amex process. Reported distribution coefficients range from nearly 200 in very dilute solutions to 90 to 130 at organic uranium concentrations around 0.01 *M*, with still lower values at higher concentrations. The leveling off of organic uranium concentration as aqueous concentration increases is attributed to approaching saturation of the amine with uranium, which would occur at 0.025 *M* for a complex of four amine molecules per uranium atom, as in the foregoing reaction. Physicochemical studies [C2] suggest that the complex actually contains around 4.7 amine molecules per uranium atom.

Distribution coefficients for uranium and other metals in trioctylamine are compared in Table 5.20. Although the conditions are not exactly those used in the Amex process, they indicate that the only element normally present in leach liquors that extracts readily with uranium is molybdenum. Ferric iron, which is always present in leach liquors and extracts in the Dapex process, is not extracted in the Amex process. Vanadium, if pentavalent, can be extracted by raising the pH from 1 to 2.

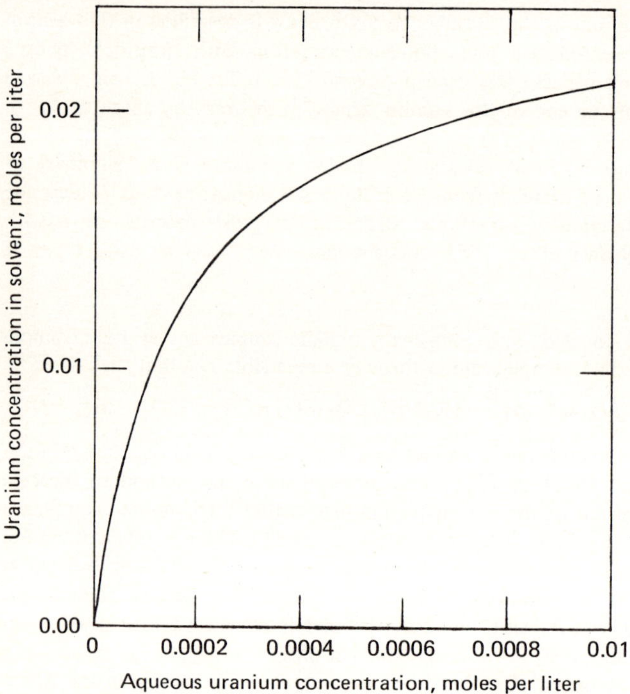

Figure 5.7 Uranium extraction by 0.1 *M* tri-*n*-octylamine in 98 percent kerosene-2 percent tridecanol. Aqueous phase: pH = 1; 0.5 *M* SO_4^{2-}

The uranium distribution coefficient decreases with increasing SO_4^{2-} concentration, as shown in Fig. 5.8, owing to reversal of the preceding equilibrium. This permits stripping uranium from the amine by aqueous sulfate solution, as practiced at the Exxon mill, Table 5.19. The Kerr-McGee mill strips with 1.5 *M* NaCl solution by the reaction

$$(R_3NH)_4UO_2(SO_4)_3(o) + 4NaCl(aq) \rightleftharpoons 4R_3NHCl(o) + 2Na_2SO_4(aq) + UO_2SO_4(aq)$$

Because the chloride salt of the amine is more stable than the sulfate salt or the uranyl sulfate complex, quite high aqueous uranium concentrations can be obtained with chloride stripping. Molybdenum is not stripped by sodium chloride and, if present, must be stripped by other means to prevent precipitation when its solubility of around 0.03 g/liter is exceeded.

Amine extraction in the Kerr-McGee mill. As a practical example of the use of organic amines to extract uranium from leach liquors, a description will be given of the solvent extraction section of the Kerr-McGee uranium mill, whose leaching section was described in Sec. 8.5 of this chapter [M3, H4]. The solvent extraction plant consists of two similar circuits; process conditions approximating those of one circuit are shown in Fig. 5.9.

Leach liquor containing about 1 g U_3O_8/liter at a pH around 1.0 is fed at the rate of 3800 liters/min to the first of four mixer-settler states in series, where the uranium is extracted by a solution containing 3 v/o Alamine-336 (mixed *n*-trioctyl- and *n*-tridecylamines) and 3 v/o isodecanol in a high-boiling kerosene diluent. These four stages reduce the uranium content of the aqueous stream from 1 g U_3O_8/liter to around 0.001, while increasing that of the solvent from 0.002 g/liter to 3.33.

Table 5.20 Distribution coefficients between aqueous sulfate solution and triisooctylamine†

Metal	Valence	Distribution coefficient
Uranium	6	90
Uranium	4	<1
Molybdenum	6	150
Zirconium	4	200
Vanadium	5	<1
Vanadium	5	~ 20 (pH = 2)
Vanadium	4	<0.01
Titanium	4	<0.1
Iron	2, 3	<0.01
Magnesium, calcium, manganese, copper, zinc	2	<0.01

†pH = 1; $SO_4^{2-} = 1\ M$; amine 0.1 *M* in aromatic hydrocarbon diluent; 1 g metal per liter in aqueous feed; organic/aqueous volume ratio, 1:1.

Source: C. F. Coleman et al., "Amine Salts as Solvent Extraction Reagents for Uranium and Other Metals," *PICG(2)* **28**: 278 (1958).

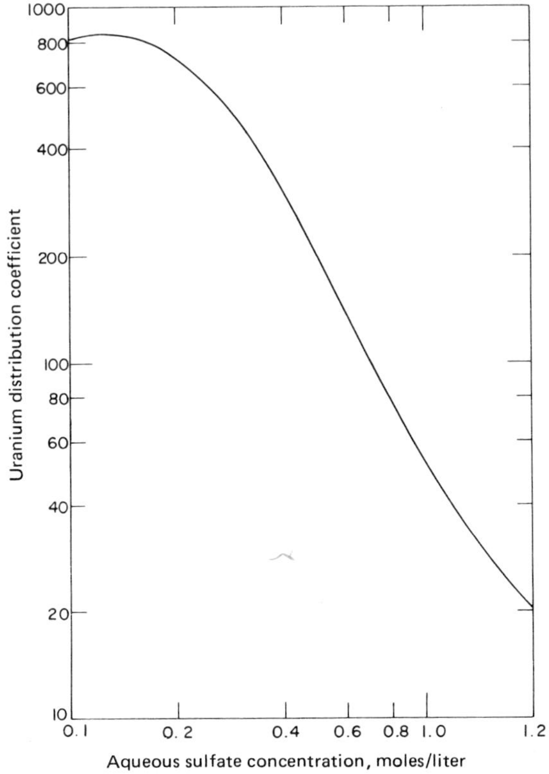

Figure 5.8 Variation of uranium distribution coefficient in 0.1 *M* tri-*n*-octylamine with aqueous sulfate concentration. pH = 1; 0.01 mol uranium/liter in solvent.

Each stage consists of a central steel mixer tank 8 ft (2.5 m) in diameter and 9.5 ft (2.9 m) deep set in the center of a wooden settler tank 40 ft (12 m) in diameter. Mixing is by a turbine-type impeller. Mixed aqueous and solvent phases from the central tank flow through holes in its lateral wall to the settler annulus, where the phases separate. Aqueous phase leaves through holes at the outside of the settler in the bottom and flows down to the next stage, which is set 1 ft (0.3 m) lower. Solvent phase is collected in a circular launder surrounding the top of the outside of the settler and is pumped up to the preceding stage. A portion of the solvent phase is recycled from the settler to the mixer to permit the latter to operate with solvent phase continuous, a condition that reduces solvent losses by entrainment in aqueous effluent.

Uranium in rich solvent leaving the extraction section is transferred to the uranium product solution by back extraction into 1.5 *M* sodium chloride solution flowing at the rate of 114 liters/min in four uranium-stripping mixer-settler stages. Each of these consists of a separate wood mixer tank 8 ft (2.4 m) in diameter by 9 ft (2.7 m) high and a wood settler tank 22 ft (6.7 m) in diameter by 8 ft (2.4 m) high. In this section, solvent phase flows up by gravity and aqueous phase is pumped at such a rate as to control the interphase level in each settler. Solvent recycle is unnecessary, because solvent is the continuous phase, owing to its flow rate being higher than the aqueous.

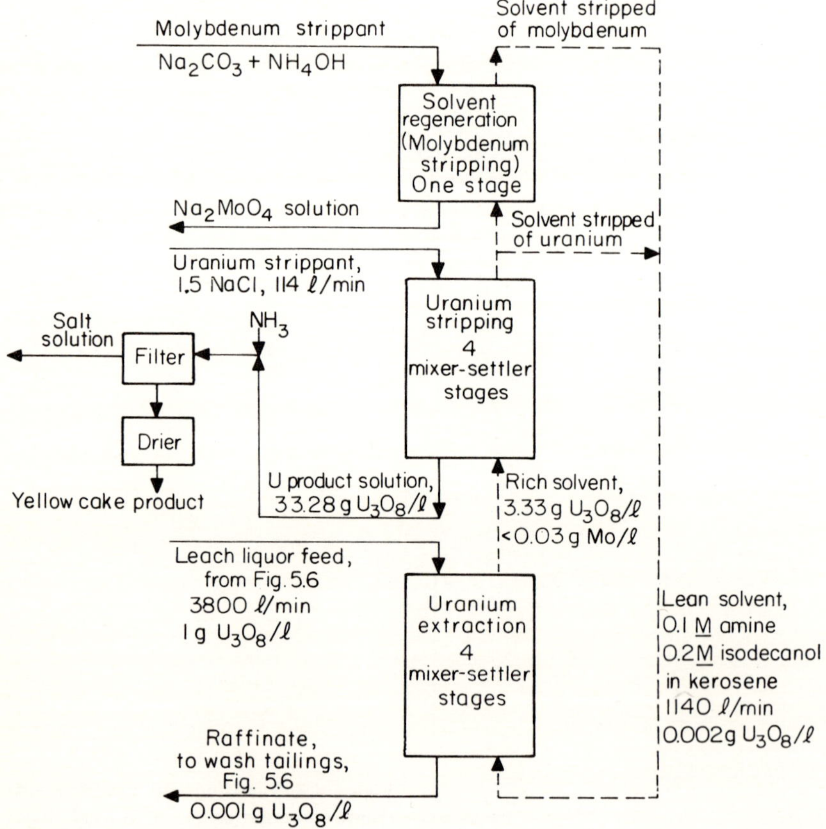

Figure 5.9 Amex process for recovering uranium from leach liquor. Conditions approximately those of one circuit of Kerr-McGee mill.

Product solutions leaving the two solvent extraction circuits at a concentration around 30 g U_3O_8/liter are combined and flow through four stirred precipitation tanks 8 ft (2.4 m) in diameter by 12 ft (3.7 m) high in series. Steam is added to the first tank to heat its contents to 60°C. A mixture of two to four volumes of air and one volume of ammonia is added to the last three tanks to raise the pH to 7.0. This precipitates uranium as mixed sodium and ammonium diuranate.

The diuranate precipitate is separated from the mixed NaCl and Na_2SO_4 salt solution by a system of thickeners and filters. Filter cake from the first filter is washed with water, reslurried with water, filtered a second time, and washed again to reduce its content of NaCl and Na_2SO_4. When it is necessary to reduce the amount of sodium diuranate, a third stage of filtration is used, and the filter cake is reslurried with ammonium sulfate instead of water to replace most of the sodium with ammonium ion. Washed filter cake is dried by heating to 160 to 180°C.

Salt solution leaving the filtration system contains about 0.01 g U_3O_8 and 15 to 30 g Cl^-/liter. Most of this is recycled to make up part of the stripping solution, but some is bled to tailings to keep sulfate ion from building up.

Molybdenum is not stripped from the amine solvent by sodium chloride. If not kept below around 0.03 g Mo/liter, it precipitates as a sludge and interferes with uranium extraction. To control molybdenum concentration, a portion of the solvent leaving the uranium stripping section is contacted in a single mixer-settler with an aqueous solution of Na_2CO_3 and NH_4OH. This converts the molybdenum to sodium molybdate, Na_2MoO_4, and transfers it to the aqueous phase, from which molybdenum is recovered as a by-product.

The solvent makeup requirement reported by Hazen [H4] was only 0.21 volumes per 1000 volumes leach liquor treated.

Solvent extraction of uranium with organophosphorus compounds. The first reported use of organophosphorus compounds for solvent extraction of uranium from minerals was recovery of uranium from commercial phosphoric acid using as extractant the reaction product of phosphorus pentoxide and octyl alcohol [L2]. This led to research on many organophosphorus compounds for extraction of uranium from sulfuric acid uranium leach liquors by Dow Chemical Company and Oak Ridge National Laboratory, among others. The numerous compounds investigated have been described by Merritt [M3], Blake et al. [B4], and Brown et al. [B8]. The compound selected for use in three U.S. uranium mills [M3] in the late 1960s was di(2-ethylhexyl) phosphoric acid (EHPA) dissolved in kerosene, in the Dapex process, so-named by its developers at Oak Ridge National Laboratory.

Because of the long hydrocarbon chain, EHPA and its salts are insoluble in water, but are soluble in hydrocarbons such as kerosene. The reaction by which EHPA reacts with the uranyl cation in the aqueous phase and transfers it to the organic phase may be represented by

$$2\mathrm{HO}\overset{\overset{\mathrm{O}}{\|}}{\mathrm{P}}(\mathrm{OR})_2(o) + \mathrm{UO_2}^{2+}(aq) \rightleftharpoons \mathrm{UO_2}[\mathrm{O}\overset{\overset{\mathrm{O}}{\|}}{\mathrm{P}}(\mathrm{OR})_2]_2(o) + 2\mathrm{H}^+(aq)$$

although the actual reaction is more complex [B4]. Thus, EHPA acts as a liquid cation-exchanger.

At the conditions typical of the Dapex process (0.1 M EHPA, 0.5 M SO_4^{2-}, pH = 1) distribution coefficients at very low concentration are [B4]

UO_2^{2+}	200
Fe^{3+}	135
Al^{3+}	0.03
Th^{4+}	20,000
V^{4+} (0.01 M)	1,000

To maintain solvent capacity for uranium and to prevent contamination of extracted uranium by iron, it is necessary to reduce iron to the unextractable ferrous condition before solvent extraction. This is done by contacting the leach liquor with scrap iron, SO_2, or sodium sulfide. Because the iron content of leach liquor is high, reduction is costly, and the Amex process, in which ferric iron does not extract, is preferred for sulfuric acid leach liquors. The high distribution coefficient of other polyvalent cations such as Th^{4+} and V^{4+} in EHPA makes the Dapex process less selective for uranium than the Amex process.

In the Dapex process, uranium in the organic phase is usually stripped by contact with an aqueous solution of sodium carbonate, through the reaction

$$UO_2[\overset{\overset{\displaystyle O}{\|}}{O}P(OR)_2]_2(o) + 3Na_2CO_3(aq) \rightleftharpoons 2NaO\overset{\overset{\displaystyle O}{\|}}{P}(OR_2)(o) + Na_4UO_2(CO_3)_3(aq)$$

The sodium uranyl carbonate is very soluble in the aqueous phase, but the sodium salt of EHPA has only limited solubility in the organic phase and tends to form a third liquid phase containing the salt, some diluent, and water. To prevent this, the organic phase is also made 0.1 *M* in TBP, in which the sodium salt of EHPA is soluble.

Although the Dapex process is no longer being widely used to extract uranium from sulfuric acid leach liquors, organic phosphoric acids are favored for extracting by-product uranium from commercial phosphoric acid. Organic amines are impractical for this application because they are too fully saturated by the strong acid.

8.7 Uranium Concentration by Anion Exchange

Because uranium is present in sulfuric acid leach liquors as a mixture of the species ${UO_2}^{2+}$, UO_2SO_4, $UO_2(SO_4)_2{}^{2-}$, and $UO_2(SO_4)_3{}^{4-}$, it is possible to extract uranium either on cation-exchange resins or anion-exchange resins. However, the concentrations of other metal cations such as Fe^{2+}, Fe^{3+}, Ca^{2+}, Mg^{2+}, Al^{3+}, and Na^+ are so much higher than that of ${UO_2}^{2+}$ that uranium extracted by a cation-exchange resin is heavily contaminated by other metals. On the other hand, only a few metals other than uranium, such as Fe(III), V(V), and Mo(VI), form anions in sulfuric acid, and their concentration can be made small by control of oxidation-reduction potential. Consequently, anion exchange has been recognized as a selective means for recovering uranium from sulfuric acid leach liquors since pioneering research sponsored by the U.S. Atomic Energy Commission in 1948 to 1952 [C1]. The first reported commercial use of anion exchange in the processing of uranium ore was that of the West Rand Consolidated Mines, Ltd., mill in the Republic of South Africa, in 1952 [M4]. Anion exchange is also effective and selective in extracting uranium from carbonate leach liquor, where uranium is present mainly as the anion $UO_2(CO_3)_3{}^{4-}$.

For a general discussion of ion exchange, reference may be made to standard texts, such as Helfferich [H5].

Anion-exchange resins. The type of anion-exchange resin first used in uranium mills and still most commonly used is the quaternary ammonium, strong base type, in which the active nitrogen atom is attached to four carbon atoms. Figure 5.10 shows the molecular structure of a typical quaternary ammonium resin. This is made by copolymerizing styrene and divinyl benzene, chloromethylating the polymer, and reacting the chlorinated polymer with trimethylamine. The principal reaction by which the resin adsorbs uranium in the form of a complex anion from solution is

$$4RCH_2(CH_3)_3NX + UO_2(SO_4)_3{}^{4-} \rightleftharpoons [RCH_2(CH_3)_3N]_4UO_2(SO_4)_3 + 4X^-$$

Here R is the styrene-divinyl benzene radical, and X^- is an inorganic anion such as Cl^-, ${NO_3}^-$,

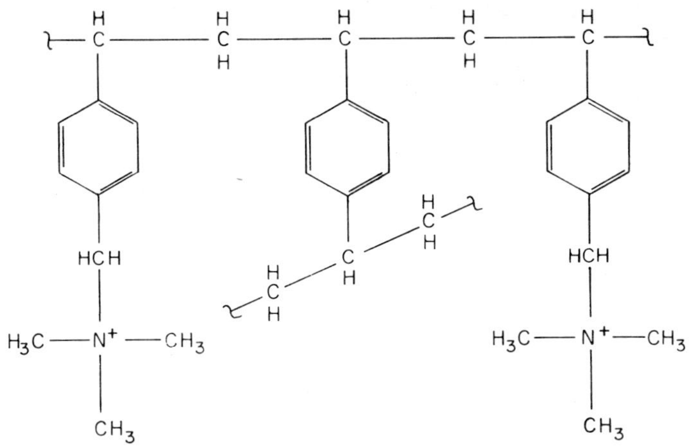

Figure 5.10 Quaternary ammonium anion-exchange resin.

or $\frac{1}{2}SO_4^{2-}$. Some uranium in the form of the $UO_2(SO_4)_2^{2-}$ ion is also adsorbed, together with SO_4^{2-} and HSO_4^- ions and other anions possibly present in the solution.

Resins are used in the form of small spherical beads. For fixed-bed operation on clear solutions, the standard particle size range is −16 to +50 mesh (1.0 to 0.6 mm). For operation on unclarified solutions, in the resin-in-pulp type of operation to be described later, larger particles in the size range −14 to +20 mesh (1.1 to 0.9 mm) are preferred.

The density of the material of the beads is around 1.15 g/ml, so they sink in leach liquors. They are light enough, however, to be readily transported by upflowing liquids, as required in moving-bed operations.

Resin consumption and cost. Resin deteriorates in use. Volume changes during adsorption and elution and mechanical wear cause breakage and attrition. Chemical poisoning and fouling cause gradual loss of adsorptive capacity. In three South African plants, useful resin life was estimated ([M3], p. 143) at from 19,000 to 27,000 volumes of solution treated per volume of resin. This represents about 3 years' usage in normal service.

In 1978, resins used in uranium mills cost \$95 to \$110/ft³. For leach liquors containing 0.5 g U_3O_8/liter or 0.031 lb/ft³, the replacement cost of resin at the highest price and lowest life would be \$110/(19,000)(0.031), or \$0.19/lb U_3O_8.

The anion-adsorption capacity of some commercially available anion-exchange resins is given in Table 5.21. Merritt [M3] uses 1.25 meq/ml resin as a representative number. For uranium adsorbed as quadruply charged $UO_2(SO_4)_3^{4-}$, this is equivalent to

$$(1.25\ \text{eq/liter})\left(\frac{842\ \text{g}\ U_3O_8}{\text{g-mol}\ U_3O_8}\right)\left(\frac{1\ \text{g-mol}\ U_3O_8}{3 \times 4\ \text{eq}}\right) = 87.7\ \text{g}\ U_3O_8/\text{liter}$$

or 5.5 lb U_3O_8/ft³. Partial saturation of the resin with SO_4^{2-}, HSO_4^-, and other anions decreases resin capacity for uranium; some adsorption of $UO_2(SO_4)_2^{2-}$ increases it.

Adsorption equilibria. The upper curve of Fig. 5.11 shows how the U_3O_8 concentration of Amberlite IRA-400 resin measured in laboratory experiments [P1] varies with U_3O_8 content of a UO_2SO_4 solution containing 5 g free H_2SO_4/liter and enough $MgSO_4$ to bring total SO_4^{2-} to 30 g/liter. This approximates conditions in a typical leach liquor. Because of the presence of other anions and deterioration of resin in service, the actual uranium content of resin in

Table 5.21 Exchange capacity of anion exchange resins

Resin	Manufacturer	Capacity, meq/ml wet resin
Amberlite IRA-400	Rohm & Haas Co.	1.2
Amberlite IRA-430	Rohm & Haas Co.	1.10–1.25†
Dowex 1	Dow Chemical Co.	1.33
Dowex 11	Dow Chemical Co.	1.24
Dowex 21K	Dow Chemical Co.	1.25
Duolite A-101D	Diamond Alkali Co.	1.4
Ionac A-580	Ionac Chemical Co.	1.30 (minimum)
Ionac A-590	Ionac Chemical Co.	1.30 (minimum)

†Manufacturer's pamphlet.

Source: R. C. Merritt, *The Extractive Metallurgy of Uranium*, Colorado School of Mines Research Institute, Boulder, Colo., 1971.

uranium mills may be lower. In the Dawn Mining Company's uranium mill to be described later, a resin concentration of 3 lb U_3O_8/ft^3 was reported [H1] for a leach liquor concentration of 0.5 g U_3O_8/liter. Values at other mills range from 2 to 5 lb U_3O_8/ft^3, depending on the uranium content of the solution and other conditions. The lower dashed curve of Fig. 5.11 will be used in subsequent predictions of ion-exchanger performance.

Ion-exchange equipment. Three types of ion-exchange equipment used in U.S. uranium mills are the fixed-bed type, the moving-bed type, and the continuous resin-in-pulp (RIP) type. These will be described in turn.

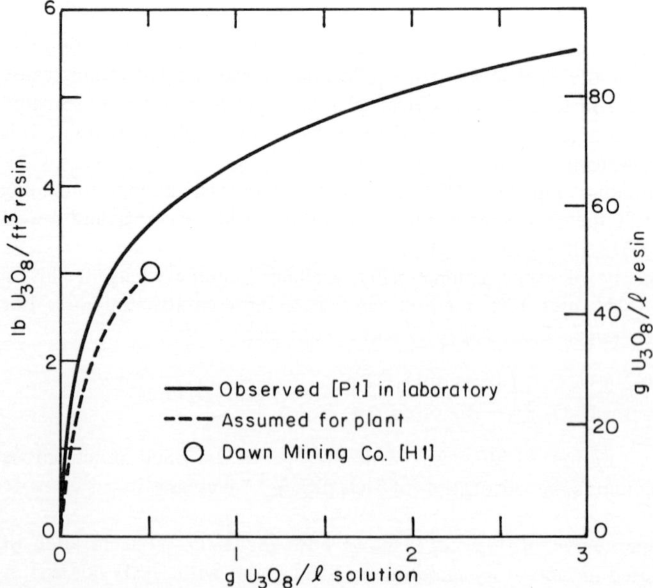

Figure 5.11 Equilibrium distribution of UO_2^{2+} between anion-exchange resin Amberlite IRA-400 and aqueous solution containing 5 g H_2SO_4/liter and enough $MgSO_4$ to bring total SO_4^{2-} to 30 g/liter.

Adsorption in fixed-bed ion exchange. A fixed-bed ion-exchanger is a column of ion-exchange resin through which the solution whose ions are to be adsorbed flows downward. Ions collect in the bed starting at the top and work downward as the resin becomes saturated. When the height of resin remaining unsaturated becomes too short to remove the ions completely, they begin to appear in the effluent. This is known as "break-through." With further flow, the ion concentration of the effluent steadily increases and reaches the feed concentration when the resin bed is fully saturated. Curve A of Fig. 5.12 shows a typical curve of uranium effluent concentration versus effluent volume for a condition in which the resin contained no uranium when flow began. Effluent volume is expressed as the ratio of effluent volume to volume of the resin bed. At point B, the break-through volume, uranium begins to appear in the effluent. Operation between points B and S deposits additional uranium in the bed, because the effluent concentration is less than the feed, but uranium is being lost in the effluent. At point S, effluent concentration has reached feed concentration, and the bed is saturated with uranium.

To prevent loss of uranium while still saturating the bed with uranium, it is customary to resort to cyclic operation of two or more beds in series. Figure 5.13 is an example of cyclic operation of four beds, with three beds in series adsorbing uranium while the fourth bed is having its uranium eluted. In cycle 1, feed solution is charged to bed 1, which has been on line for two previous cycles. Solution then flows through bed 2, which has been on line for one previous cycle. Solution finally flows through bed 3, which was freshly eluted and free of uranium at the beginning of the cycle. At the end of cycle 1, the feed point is moved to bed 2, freshly eluted bed 4 is put in series after bed 3, and bed 1 is taken out of adsorption and put on elution. This progression continues through four cycles, after which the sequence is repeated. This is sometimes called "merry-go-round" operation.

Figure 5.14 shows how the uranium concentration in the resin beds of Fig. 5.13 might be distributed along the beds at different times. To make the illustration concrete, it has been assumed that the beds and flow rates have approximately the characteristics of the ion-exchange beds of the Dawn Mining Company's uranium mill [H1]. Specifically, the beds have a cross

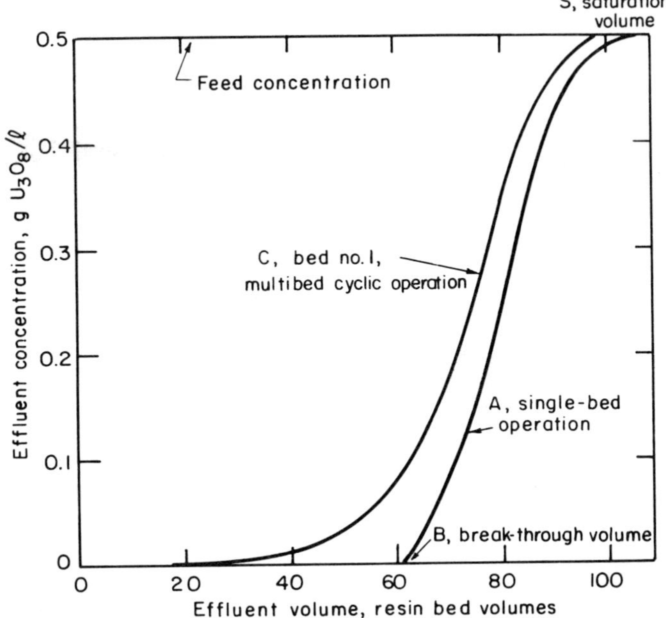

Figure 5.12 Effluent concentration versus bed volumes passed through ion-exchange column.

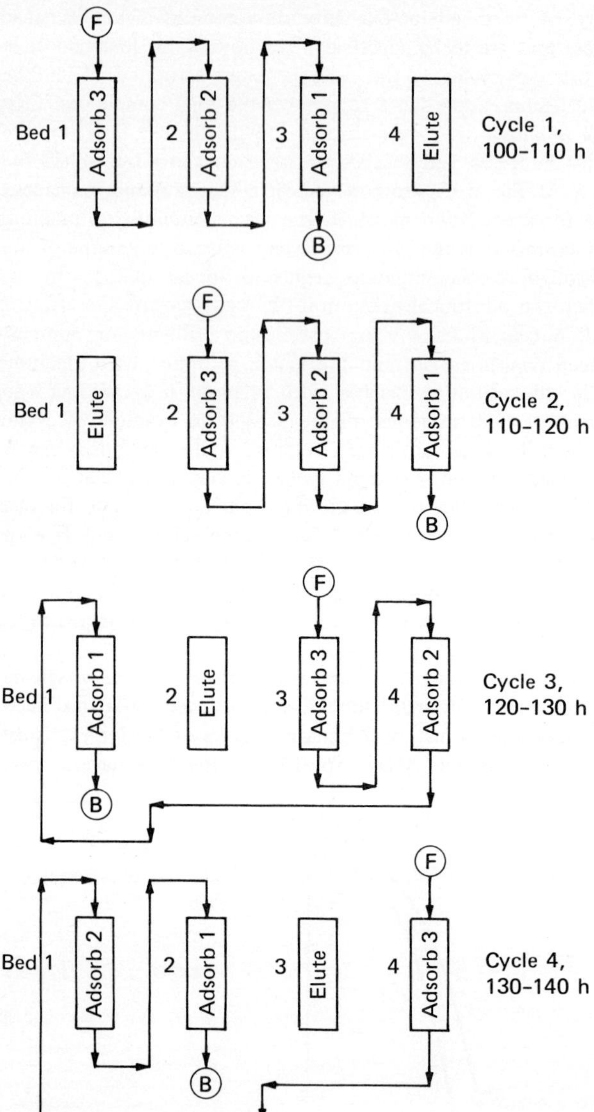

Figure 5.13 Four cycles of fixed-bed ion-exchange operation. F = feed; B = barren solution.

section of 50 ft^2 (4.64 m^2), a depth of 5.4 ft (1.65 m), and a volume of 270 ft^3 (7646 liters). To load a bed fully with uranium at 3 lb U_3O_8/ft^3 (48.04 g U_3O_8/liter) with a solution containing 0.5 g U_3O_8/liter would involve feeding 48.04/0.5 = 96.1 bed volumes of solution. To provide a cycle time of 10 h for this example, the feed flow rate should be 9.61 bed volumes/h, or 9.61 × 7646 = 73478 liters/h, or 323 gal/min. The flow rate reported [H1] for the Dawn plant was 300 gal/min.

At the start of cycle 1, at 100 h, bed 1 has been partially loaded with uranium, distributed over the top two-thirds of the bed. Beds 2 and 3 contain no uranium. At 102 h, uranium distribution in bed 1 has increased and extends over the entire bed, so that some is being deposited at the top of bed 2. This continues until at the end of the cycle, at 110 h, bed 1 is fully loaded with uranium and the distribution in bed 2 has increased until it is the same as in bed 1 at the beginning of the cycle. Concentration distribution in bed 3 tracks that of bed 2, 10 h later.

The change in effluent solution concentration with time is determined from the uranium concentration at the bottom of the bed at 5.4 ft, the equilibrium relation between bed and solution concentrations (Fig. 5.11), and mass transfer rates in ion exchange. Curve C of Fig. 5.12 is an estimate of how the effluent concentration from the first bed would change with time in cyclic operation.

Elution. In the elution step, an aqueous solution about 1 *M* in nitrate, chloride, or sulfate ion is passed through the bed to reverse the reaction given previously, transfer the uranium back from the resin to an aqueous solution, and regenerate the resin so that it can adsorb more uranium in a later cycle. The feed solution is called the eluant; the uranium-bearing product solution, the eluate. The eluant should be at least 0.1 *N* in free acid to prevent precipitation of uranium.

Figure 5.15 shows how the uranium concentration of the eluate changes with bed volumes of eluate for three commonly used eluants, as reported by Greer et al. [G1]. The area under each curve represents the original, uniform uranium concentration of the bed, apparently around 80 g U_3O_8/liter in this example.

Disadvantages of sulfuric acid are (1) the low uranium concentration of eluate, (2) the high volume of eluant needed for complete elution, and (3) the high acid concentration to be neutralized if uranium is to be recovered from eluant by precipitation. The advantages of sulfuric acid are that it leaves the resin in the sulfate form, which adsorbs uranium more readily than the nitrate or chloride form, and it introduces no extraneous anions that must later be purged from the system. To facilitate use of sulfuric acid, the Eluex process, to be described later, was developed; in this process uranium is removed from sulfuric acid eluate by solvent extraction rather than by neutralization and precipitation, and the acid is recycled to a subsequent elution cycle.

Sodium chloride has the advantage of lowest reagent cost. Nitrate eluant has the advantage of providing the highest uranium concentration in the eluate. However, chloride and nitrate eluants require conversion of resin back to sulfate form with sulfuric acid to improve uranium recovery in the next adsorption cycle.

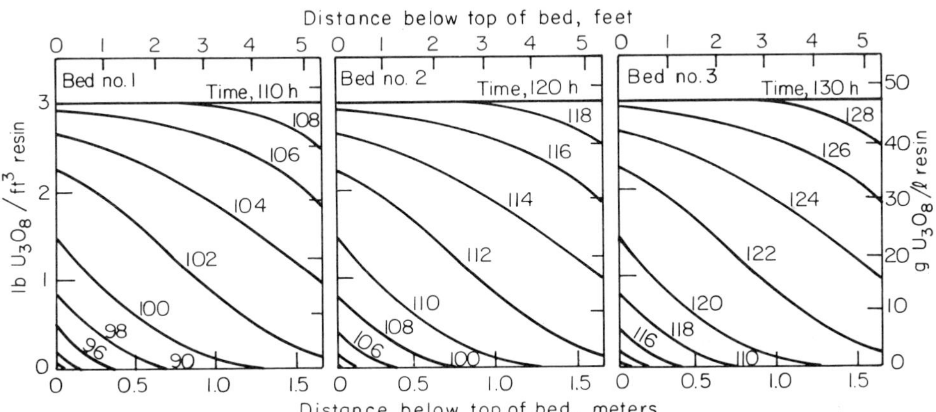

Figure 5.14 Variation of uranium concentration with time and position in fixed-bed ion-exchange columns. Feed concentration, 0.5 g U_3O_8/liter; feed rate, 9.61 bed volumes/h.

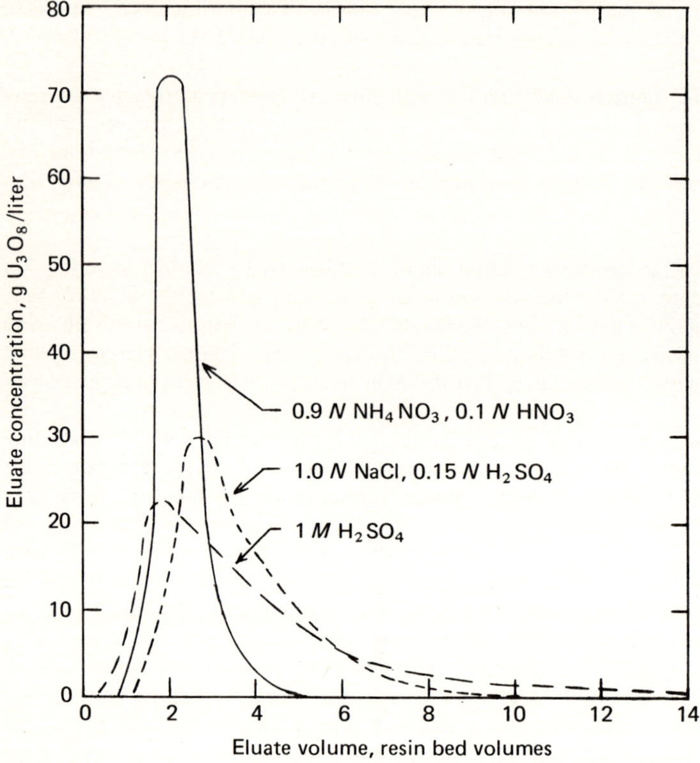

Figure 5.15 Elution of uranium with various eluants. (*From Greer et al. [G1].*)

Elution in fixed-bed ion exchange. As an example of elution of uranium with nitrate ion from a fixed-bed ion-exchange system, a brief description will be given of the elution cycle in the Dawn Mining Company's uranium mill [H1]. Figure 5.16 shows the eight steps of the elution cycle. Table 5.22 gives the composition and flow rate of inflow in each step. Figure 5.17 shows the change with time of the effluent uranium, nitrate ion and total acid concentration throughout the elution cycle.

In step 1 leach liquor remaining in the column at the end of an adsorption cycle is flushed from the column by fresh water into the next column on adsorption.

In step 2 the column is flushed with water from the bottom to the top to remove fines that settle at the top of the bed during adsorption from leach liquor. Washings, which may contain traces of uranium, are used to wash filters in the leaching section of the mill.

In step 3 approximately one bed volume of recycle eluate displaces the solution remaining in the bed into the feed tank, for recovery of any uranium that may have started to appear in the effluent. Recycle eluate contains approximately 30 g NO_3^-/liter and 25 g free acid/liter, expressed as H_2SO_4, and a low concentration of U_3O_8.

In step 4 approximately three volumes of recycle eluate removes most of the uranium from the bed, and transfers it to the product precipitation tank. At the start of this step, high concentrations of uranium and acid appear in the effluent while the bed is becoming converted to nitrate; toward the end of the step the nitrate concentration in the effluent approaches feed concentration and the uranium concentration declines.

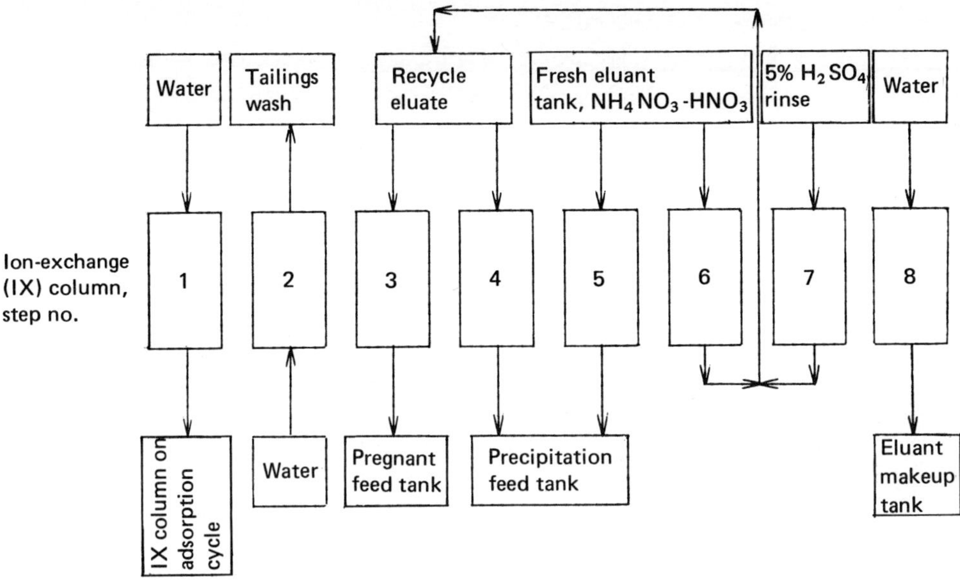

Figure 5.16 Steps in elution of uranium from fixed-bed ion-exchange column. (*From Hargrove [H1].*)

In step 5 one volume of fresh eluant is used to flush more of the uranium from the bed into the product precipitation tank. Fresh eluant contains about 50 g NO_3^- and 25 g acid as H_2SO_4 per liter.

In step 6 fresh eluant continues to purge uranium, now into the recycle eluate tank.

In step 7 the resin is converted from nitrate form to sulfate with 5% H_2SO_4. This effluent, with high concentrations of nitrate ion and acid, is added to recycle eluate.

In step 8 the acid remaining in the bed is flushed with water into the eluant makeup tank. The bed is now in the sulfate form, the liquid in the bed is water, and the bed is ready to be used as the third bed in series in the adsorption cycle. The eight elution steps take 462 min, which is less than the 10-h duration of an adsorption cycle.

Reagent consumption in the elution section of the Dawn mill was 10 lb H_2SO_4 and 4 lb

Table 5.22 Elution cycle in Dawn uranium mill

Step	Duration, min	Inflow	Total inflow, bed volumes	Disposition of effluent
1	15	Water	2.22	To next column on adsorption cycle
2	60	Water backwash	5.93	To wash tailings in leaching section
3	36	Recycle eluate	1.02	To pregnant feed (uranium-rich IX feed)
4	105	Recycle eluate	2.96	To product precipitation
5	35	Fresh eluant	0.99	To product precipitation
6	121	Fresh eluant	3.42	To recycle eluate tank
7	30	5% H_2SO_4	0.58	To recycle eluate tank
8	60	Water	0.92	To eluant makeup tank
	462			

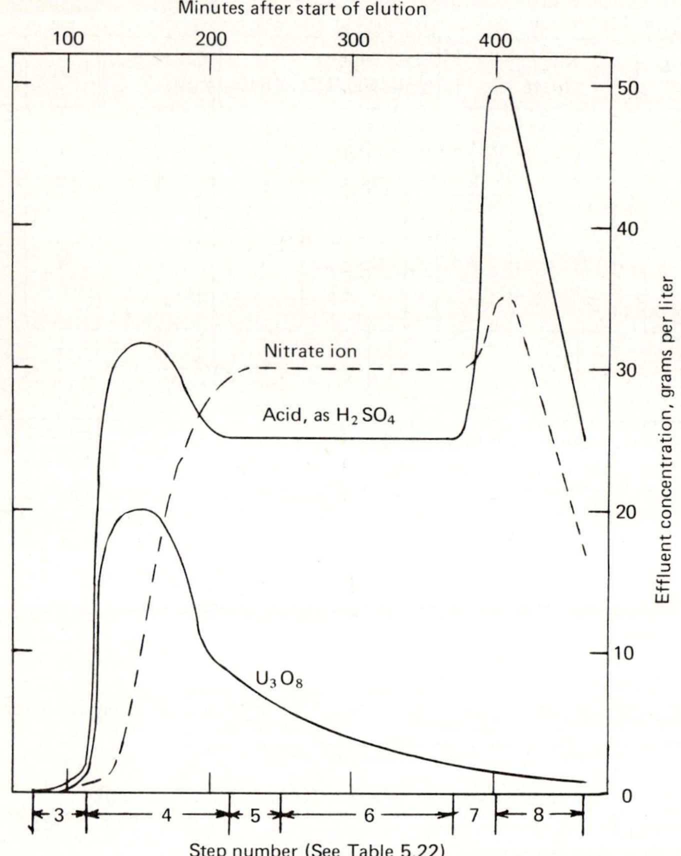

Figure 5.17 Change in column effluent concentration with time during elution cycle of Table 5.22.

HNO_3 per short ton of ore. For ore containing 0.2 w/o U_3O_8 this would represent 2.5 lb H_2SO_4 and 1 lb HNO_3 per pound U_3O_8.

Precipitation of uranium from eluate. In the Dawn mill, uranium was recovered from eluate by two-stage precipitation with alkali. In the first stage, pH is increased to 3.3 to 3.6 by addition of CaO slurry. This precipitates most of the ferric iron that may have been adsorbed with the uranium and removes most of the sulfate as $CaSO_4$. Because the filter cake contains 1 to 2 percent uranium, it is returned to the leaching circuit.

After filtration, in the second stage, ammonia is added to bring the pH to 7.0, which precipitates the uranium as ammonium diuranate. The precipitate is filtered, washed with water, and dried at 160°C. Filtrate is added to eluant makeup, to conserve nitrate.

Moving-bed ion exchange. Ion-exchange resin particles are small spherical beads. These can be readily moved out of an ion-exchange column with up-flowing liquid and pumped with the liquid to another tank or column. Transport of resin is used in two types of uranium extraction processes, (1) the moving-bed type and (2) the continuous RIP type.

Among U.S. mills, the moving-bed system is used in the mill of the Lucky Mc Uranium Company at Gas Hills, Wyoming [D2a]. The moving-bed process is a modified batch-operated, fixed-bed ion-exchange system. Adsorption of uranium from leach liquor is carried out in two parallel sets of three columns in series. Each set is operated cyclically as in the fixed-bed system described earlier, with the feed point moved progressively around the cycle as the last bed in flow sequence becomes saturated. The novelty of the moving-bed system is in the physical transfer of loaded resin from an adsorption column to one of three elution columns, also operated cyclically in series. After elution is complete, the stripped resin is transferred back to one of the two adsorption sets, where it is placed last in flow sequence.

Physical transfer in the moving-bed system involves the following operations. The resin is first washed with down-flowing water in its original column. It is then transported with up-flowing water to a resin transfer and backwash tank. There it is given another wash and then transferred by up-flowing water to the column where it is next to be used.

Advantages of this system are simplification of fluid valving and piping: The adsorption columns need be connected only to leach liquor supply, to barren solution disposal, and to resin transfer tank; and the elution columns need be connected only to eluant and wash supply, to eluate and recycle storage, and to resin transfer tank. This reduces the chance of cross contamination of solutions between the adsorption and elution systems.

Continuous RIP ion exchange. As an illustration of the continuous RIP ion-exchange process used in several U.S. uranium mills, a brief summary will be given of Merritt's [M3] description of the principal steps in the uranium mill of Federal-American Partners at Gas Hills, Wyoming, with reference to Fig. 5.18.

Ore containing around 0.15 percent U_3O_8 is crushed dry to particles smaller than 1 in, then ground wet with heated water and dilute sulfuric acid recycled from subsequent ion-exchange operations at B. Grinding is done in closed circuit with classifiers. Classifier overflow is a slurry, or pulp, 55 percent solids, 95 percent finer than 28 mesh. In a leaching period of 12 to 13 h, the pulp passes through six leaching tanks in series, to which are added enough sulfuric acid and sodium chlorate to bring the effluent solution to a free sulfuric acid concentration of 10 g/liter and the oxidation-reduction potential relative to the calomel electrode to −390 to −400 mV. The pulp passes through a system of cyclones and several classifiers. There it is separated into a sand fraction coarser than 325 mesh, which is washed with water and sent to tailings, and a fine slurry, or pulp, fraction, which contains 11 to 12 percent of slimes (particles finer than 325 mesh) and about 0.8 g dissolved U_3O_8/liter. The pH of pulp is raised to 1.65 by addition of ammonia to increase adsorption of $UO_2(SO_4)_3^{4-}$ while decreasing that of SO_4^{2-} and HSO_4^-.

The pulp flows through seven continuous ion-exchange stages countercurrent to ion-exchange resin beads in the size range 20 to 50 mesh. In these seven stages, dissolved uranium is transferred from the pulp to the resin leaving stage #1, while the slime tailings leaving stage #7 are substantially free of dissolved uranium.

Each adsorption stage consists of a mechanically stirred tank 11 ft (3.3 m) in diameter by 10 ft (3.0 m) deep, from which the mixture of resin and pulp flows to an airlift, which in turn discharges to a set of screens S1. These separate the coarser resin particles from the finer slimes. The resin drops by gravity into the next lower numbered tank, toward the feed end of the cascade, and the slurry flows to the next higher numbered tank, toward the tailings end. This counterflow is made possible by the absence of particles coarser than 325 mesh in the pulp and the absence of particles finer than 50 mesh in the resin. Residence time in each adsorbing stage is 18 to 20 min.

Resin leaving adsorbing stage #1 contains 2 to 3 lb U_3O_8/ft^3 (32 to 48 g/liter). This resin is washed partially free of entrained slurry, with the washings and some dissolved uranium returned to the grinding circuit. Washed resin then passes countercurrent to eluant in 12 eluting

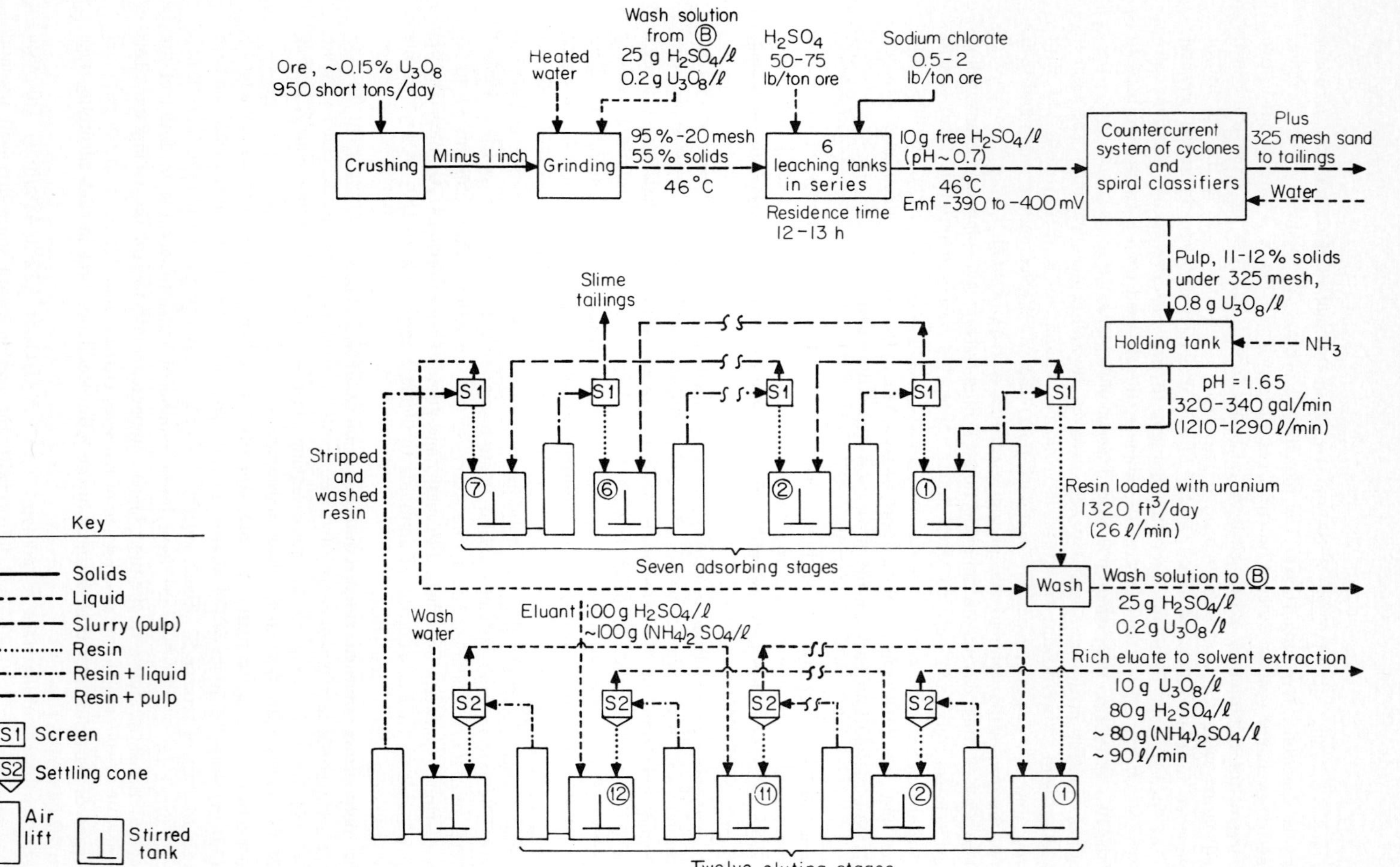

Figure 5.18 Principal steps in leaching and ion-exchange sections of Federal-American Partners' uranium mill.

stages. Each stage consists of a mechanically stirred tank 6 ft (1.8 m) in diameter by 6 ft (1.8 m) deep. Interstage flow is driven by airlifts, with separation of resin from eluate in settling cones S2. This simpler separator is feasible because the slime content of the liquid in the eluting stages is low. Residence time in each eluting stage is 25 to 30 min.

Eluant containing 100 g H_2SO_4/liter and about the same content of $(NH_4)_2SO_4$ is charged to stage #12. In its flow through the 12 eluting stages it picks up uranium and water from the counterflowing resin and leaves with a concentration of around 10 g U_3O_8/liter and 80 g H_2SO_4/liter. This rich eluate is sent to solvent extraction, Fig. 5.19, for recovery of uranium.

Stripped, barren resin leaving eluting stage #12 is washed with water in a thirteenth stage and charged to adsorbing stage #7. Barren resin washing is used to wash loading resin and then returned to grinding at B.

The total resin inventory is 825 ft^3; the resin circulation rate is 1320 ft^3/day or 26 liters/min. Despite the continuous flow of resin, attrition rate has been very low, averaging only 0.076 percent per day. Absence of abrasive sand particles coarser than 325 mesh is important in keeping the attrition rate low.

Eluex process. Although it would be possible to recover uranium from the eluate leaving Fig. 5.18 by neutralization with ammonia, this would be costly because of its high H_2SO_4 content. The Eluex process was developed to concentrate the uranium by solvent extraction before precipitation, thus reducing the ammonia requirement and simultaneously purifying the uranium further.

Figure 5.19 is a schematic flow sheet showing the Eluex process as used at the Federal-American Partners' uranium mill, and the final steps for precipitating and calcining its uranium product. The Eluex process is a variant of the amine extraction (Amex) process described in Sec. 8.6.

Eluate containing 10 g U_3O_8/liter flowing at the rate of 90 liters/min is extracted in four countercurrent mixer-settler stages with 150 liters/min of a solution of 6 v/o tertiary amine and 3 v/o isodecanol in kerosene. Organic extract contains about 6 g U_3O_8/liter. Uranium in aqueous raffinate is reduced to under 0.01 g U_3O_8/liter, and H_2SO_4 is reduced from 80 to 65 g/liter. Raffinate passes through a settling tank to recover entrained solvent, is reacidified, and is returned to serve as ion-exchange eluant.

Organic extract is stripped of uranium with 23 liters/min of an aqueous ammonium sulfate solution in four additional countercurrent mixer-settler stages, to produce an aqueous product containing about 39 g U_3O_8/liter. To drive uranium into the aqueous product, pH in the first mixer stage is brought to 4.1 to 4.3 by addition of ammonia gas to react with most of the 9 g H_2SO_4/liter carried by the extract.

Uranium in the aqueous product is precipitated by enough additional ammonia gas to bring the pH to 7.0. The precipitated $(NH_4)_2U_2O_7$ is separated from the ammonium sulfate solution in a thickening tank and centrifuge and is washed with water. Ammonium sulfate filtrate is returned to the stripping section, with excess bled off to the IX circuit. $(NH_4)_2U_2O_7$ precipitate is dried and converted to U_3O_8 in a roaster at 600°C. Product contains 95 to 96 percent U_3O_8 and 3 to 4 percent SO_4.

8.8 Uranium from Seawater

Despite the very low concentration of uranium in seawater, 3.34 mg/m^3, the large total amount in the world's oceans, around 4×10^9 MT of uranium [D1], has provided incentive for study of means for extracting uranium from this ubiquitous source. To produce 1 MT (1 Mg) of uranium requires processing $10^6/3.34 \times 10^{-3} = 300$ million m^3 of seawater. This enormous volume gives rise to the principal problems in extracting uranium from seawater. These problems are (1) providing at low cost a continual supply of feed water undiluted by depleted

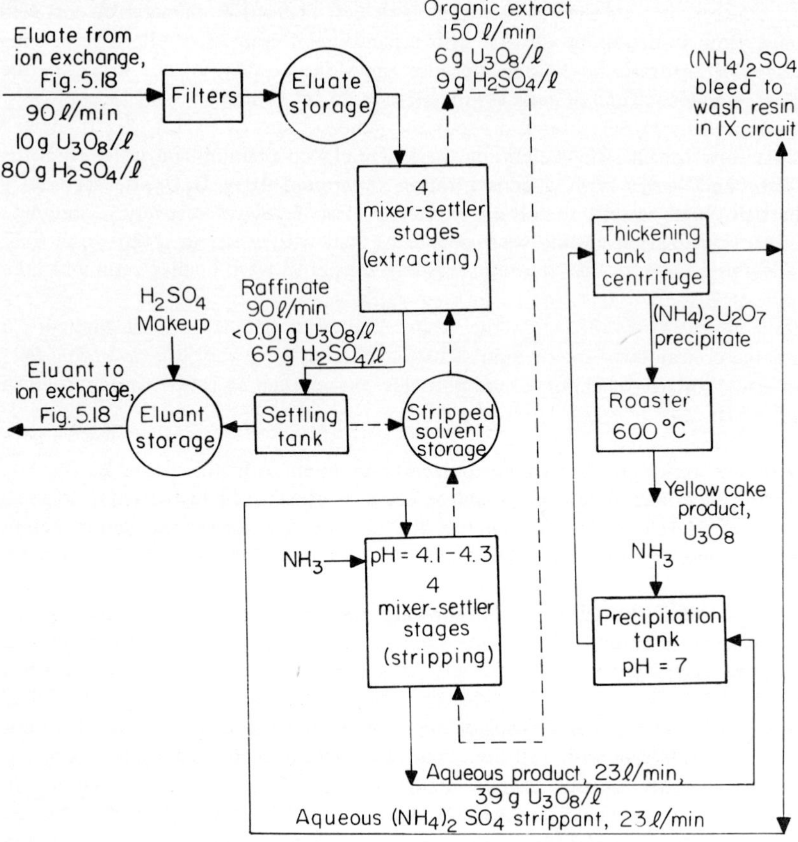

Figure 5.19 Eluex solvent extraction and yellow cake precipitation at Federal-American Partners' uranium mill.

water from which uranium has been extracted, (2) preventing fouling of equipment by contaminants dissolved or suspended in seawater, (3) minimizing energy input, (4) preventing loss of reagents through solution or entrainment in reject water or reaction with it, and (5) recovering uranium selectively in the presence of millions of times greater concentrations of other substances.

In a December 1974 report, Battelle Pacific Northwest Laboratory [B2] summarized the principal processes that had been proposed for extracting uranium from seawater and gave references to more detailed descriptions of these processes. That report concluded that the most promising process was the selective adsorption of uranium from seawater on hydrous titanium oxide (titania).

Uranium is present in seawater as the very stable uranyl tricarbonate anion, which is thought to react with titania as follows:

$$UO_2(CO_3)_3{}^{4-}(aq) + TiO(OH)_2(s) \rightarrow TiOUO_4(s) + 2HCO_3{}^-(aq) + CO_3{}^{2-}(aq)$$

Three important advantages of this reaction are the following: (1) It proceeds nearly to completion at the hydrogen ion concentration of seawater (pH = 8), so that chemical preconditioning of the large volumes of water is not required; (2) titania has very low solubility in seawater, so that solution losses of this relatively expensive reagent (around \$3/kg Ti) should be low; (3) it has a fairly high absorptive capacity for uranium from seawater; a value of 240 mg U/kg·Ti is considered representative [B2, D1].

This process was first publicized by Dr. R. Spence of the United Kingdom Atomic Energy Authority (UKAEA) at the 1964 United Nations Geneva Conference on the Peaceful Uses of Atomic Energy and was described briefly in reference [D1]. More details of the proposed process were given by British workers in references [K2] and [D2]. These workers made preliminary civil and chemical engineering designs of a plant located on the Menai Straits in the west of England where the mean tide amplitude of 5 m and the local coast configuration is favorable to construction of the 20 km of dams and sea gates needed to provide the desired once-through tidal flow of seawater through the absorption beds. The initial conclusion of these workers was that 840 MT of uranium/year could be recovered in this plant for a total cost of from \$11 to \$22/lb U_3O_8, assuming 80 percent recovery of uranium put through the plant. Later British experimental data indicated that the recovery in the titania absorbers originally proposed would be only 46 percent on the first cycle and would drop to 23 percent after eight cycles, with a cost increase to \$26 to \$42/lb U_3O_8 [K1]. A critique of the original design of this plant (assuming 80 percent uranium recovery) and more complete process engineering and cost estimation by Oak Ridge National Laboratory [H2] indicated that the cost in 1966 dollars of an optimized plant at the Menai Straits site producing uranium at the rate of 430 MT/year would be around \$1.5 billion, and that the cost of uranium produced in it would be around \$260/lb U_3O_8. When the reduced capacity found in reference [K1] and cost inflation since 1966 are taken into account, it seems likely that extraction of uranium from seawater would cost of the order of \$500/lb U_3O_8.

The principal steps in the process proposed by the UKAEA for recovery of uranium from seawater are shown in Fig. 5.20. The titania recovery system consists of 60 beds, 1.3 ft (0.4 m) deep, each with a flow area of 188,000 ft^2 (17,500 m^2), filled with hydrous titanium oxide supported on an inert carrier. The inventory of the entire system is 71 million lb (32.2 million kg) of Ti, valued at \$71 million in 1966.

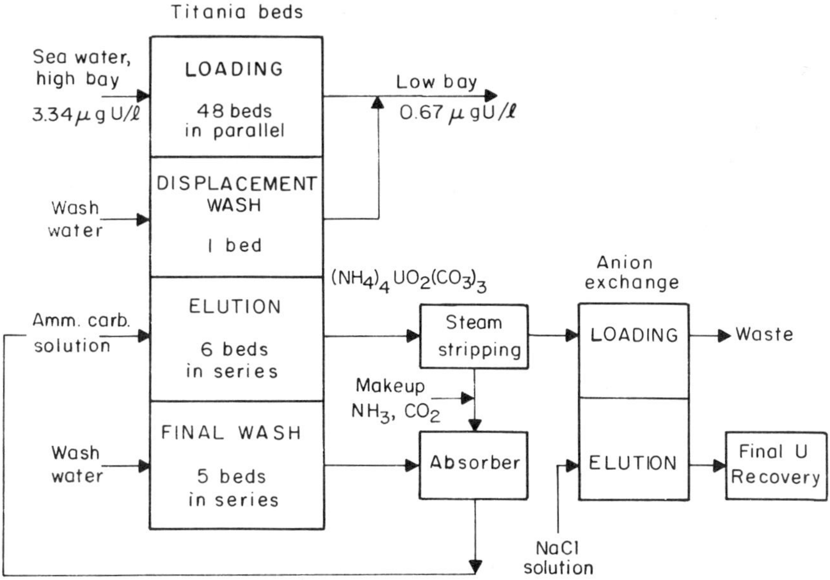

Figure 5.20 Steps in recovering uranium from seawater by adsorption on titania.

Table 5.23 Operating cycle for 1 of 60 beds for recovery of uranium from seawater

Operation	Feed	Flow configuration	Flow rate, $m^3/(bed \cdot h)$	Duration, h
Uranium loading	Seawater	48 beds in parallel	466,000	96
Displacement wash	Fresh water	Single bed	10,400	2
Uranium elution	Ammonium carbonate solution	Six beds in series	13,900	12
Final wash	Fresh water	Five beds in series	13,900	10
				120

Table 5.23 summarizes the 120-h operating cycle for 1 of the 60 beds. The phase of each bed is displaced by 2 h from its preceding neighbor in flow sequence. At the start of a cycle, a bed will have been stripped of uranium and filled with fresh water.

In the first 48 phases of a cycle, seawater flows through the bed for 96 h and deposits 80 percent of its uranium on the titania. Seawater flow is then terminated.

In the forty-ninth phase, a displacement wash of fresh water flows for 2 h, to flush seawater from the bed.

In the next six elution phases, the bed is connected in series with the five other beds that preceded it in flow sequence and fed with ammonium carbonate solution that has flowed through and picked up uranium from the other five beds through the reaction

$$TiOUO_4(s) + (NH_4)_2CO_3(aq) + 2NH_4HCO_3(aq) \rightarrow TiO(OH)_2(s) + UO_2(CO_3)_3^{4-} + 4NH_4^+$$

Every 2 h the ammonium carbonate feed point is changed so that the bed is moved in flow sequence toward the ammonium carbonate feed point. After six phase changes the bed is being fed with fresh ammonium carbonate solution and has been stripped of uranium.

In the final five wash phases the bed is connected in series with the four other beds that preceded it in flow sequence and washed with water in countercurrent flow sequence to recover ammonium carbonate and prepare the bed to receive seawater at the start of the next cycle.

To provide steady flow of seawater through the recovery beds during the twice-daily change of tides, an elaborate system of tidal basins, dikes, and sea gates is required, responsible for two-thirds of the plant's high cost.

Ammonium carbonate in the eluate is removed by steam stripping and then recycled. Uranium in the steam-stripped eluate is concentrated further by conventional anion exchange.

8.9 Radioactive Effluents from Uranium Mills

The principal effluents carrying radioactive material from a uranium mill are the following:

1. Airborne effluents, carrying radon gas (^{222}Rn) and radioactive dust particles
2. Liquid effluents, carrying water-soluble radionuclides
3. Solid effluents, in mill tailings

In present mills, radioactive liquid effluents are held in storage ponds with mill tailings and eventually evaporate to a solid.

The amounts of these radioactive effluents have been estimated [S2] for two model uranium mills, each with a capacity of 2000 short tons of ore containing 0.2 percent U_3O_8/day. One mill uses a carbonate leach, sodium hydroxide precipitation flow sheet such as that described in Sec. 8.4. The other mill uses an acid-leach, amine extraction flow sheet such

Table 5.24 Airborne radioactive effluents from model uranium mill and 20 years' tailings storage†

	Process	
Nuclide	Acid leach, amine extraction, Ci/yr	Carbonate leach, NaOH precipitation, Ci/yr
^{234}U, ^{238}U	0.090	0.090
^{234}Th	0.0096	0.0048
^{230}Th	0.014	0.0087
^{226}Ra	0.0090	0.010
^{222}Rn	3700	5800
^{210}Pb, ^{210}Bi, ^{210}Po	0.0087	0.0088

†Capacity 2000 t 0.2% ore per day; wind speed 7 mi/h.

as described in Secs. 8.5 and 8.6. Each mill is assumed to be associated with a storage pond and tailings pile in which 20 years of mill effluents have accumulated. Two alternative sites were studied for each mill type, one in New Mexico in an arid region with average wind speed of 7 mi/h and the other in Wyoming in a region with more vegetation and an average wind speed of 10 mi/h.

For each model mill a number of cases were examined, with progressively better retention of airborne radioactive effluents. The results to be summarized here are for case 1, at the New Mexico site. Case 1 represents 1975 practice, with least complete removal of airborne dust and no holdup of gaseous effluents to permit 3.8-day ^{222}Rn to decay.

Airborne effluents. Table 5.24 lists the annual emission rate of airborne radionuclides from the two types of model uranium mill, each after 20 years' accumulation of tailings.

Yellow cake. Table 5.25 gives the percent of the uranium, thorium, and radium in the ore assumed [S2] to be recovered in the yellow cake uranium mill product, and the activity of the yellow cake due to the thorium and radium.

Table 5.25 Radioactive impurities in yellow cake concentrate

	Process	
	Acid leach, amine extraction	Carbonate leach, NaOH precipitation
Percentage of nuclide recovered in yellow cake		
Uranium	91	93
^{230}Th	5	0
^{226}Ra	0.2	1.8
Activity in yellow cake, $\mu Ci/g\ U_3O_8$, from		
^{230}Th	0.014	0
^{226}Ra	0.00055	0.0055

Tailings. The first two columns of Table 5.26 give the percent of the uranium, ^{230}Th, and ^{226}Ra and its daughters assumed [S2] to be recovered in the tailings sand and the tailings slime and liquid effluents for the acid-leach and carbonate-leach processes, and the calculated [S2] concentration of the principal radionuclides in the two classes of tailings. The third column gives the resulting nuclide concentration in the composite tailings. The fourth column gives the calculated total curies of each radionuclide in the tailings after 20 years of mill operation. The potential hazard from insecurely impounded tailings is suggested by the large amount of radioactivity.

Differences among uranium mills and the ores they process will cause the amounts of radioactivity in individual mills to vary considerably from the estimates given above.

9 URANIUM REFINING

9.1 Uranium Refineries

Table 5.27 lists the principal uranium refineries of the Western world and their feed and products. In all these refineries except Allied Chemical's, the sequence of operations follows some or all of the steps shown in Fig. 5.21, in which uranium ore concentrates are first *purified* by solvent extraction and then *converted* to the materials of principal practical importance, uranium dioxide, uranium metal, or uranium hexafluoride. The steps in these refining operations will be described in process sequence in Secs. 9.2 through 9.6.

In Allied Chemical's uranium refinery the sequence of process operations is reversed, with *conversion* to UF_6 preceding *purification*, and with UF_6 as the sole purified product. The Allied Chemical process will be described briefly in Sec. 9.7.

9.2 Purification of Uranium Concentrates

As received by the uranium refinery, uranium ore concentrates now usually consist of uranium oxide or sodium, magnesium, or ammonium diuranate. These concentrates still contain appreciable amounts of elements other than uranium and some of uranium's radioactive decay products present in the original uranium ore, such as radium and radon.

The first step in the conventional process for refining uranium is dissolution in nitric acid. When the concentrates have been produced by chemical leaching and are in the form of diuranates, dissolution proceeds rapidly and leaves little solid residue. When the concentrates have been separated mechanically and are in the form of the original uranium mineral, dissolution may require more concentrated acid, higher temperatures, longer times, and addition of oxidants such as MnO_2. Also, filtration to remove undissolved residues is usually required. In either case, dissolution produces an aqueous solution of uranyl nitrate hexahydrate $UO_2(NO_3)_2 \cdot 6H_2O$, containing excess nitric acid and variable amounts of nitrates of metallic impurities present in the concentrates.

The next step in purification is separation of uranyl nitrate from the other metallic impurities in the dissolver solution by solvent extraction. Practically all uranium refineries now use as solvent tributyl phosphate (TBP) dissolved in an inert hydrocarbon diluent. The first U.S. refinery used diethyl ether as solvent and later refineries have used methyl isobutyl ketone or organic amines, but practically all have now adopted TBP. It is nonvolatile, chemically stable, selective for uranium, and has a uranium distribution coefficient greater than unity when the aqueous phase contains nitric acid or inorganic nitrates.

Although uranium refineries use widely different types of solvent extraction contactors, their basic process flow sheets are similar, along the lines of Fig. 5.22, which illustrates the

Table 5.26 Radionuclides in tailings from model uranium mills processing 2000 t ore/day containing 0.2 w/o U_3O_8

	Sand, over 200 mesh	Slime, under 200 mesh, plus evaporated liquid waste	Composite	Curies after 20 years
	A. Acid leach, amine extraction			
Percent of tailings	70	30		
Percent of uranium	1.4	7.6		
Percent of ^{230}Th	7.5	92.5		
Percent of radium	15	85		
	Picocuries per gram solids			
Natural uranium	10	150	52	688
^{234}Th	10	150	52	688
^{230}Th	60	1750	567	7510
^{226}Ra	120	1610	567	7510
^{210}Pb, ^{210}Bi, ^{210}Po	120	1610	567	7510
	B. Carbonate leach, NaOH precipitation			
Percent of tailings	50	50		
Percent of uranium	1	6		
Percent of ^{230}Th	15	85		
Percent of radium	15	83		
	Picocuries per gram solids			
Natural uranium	10	70	40	530
^{234}Th	10	70	40	530
^{230}Th	170	960	565	7483
^{226}Ra	170	950	560	7417
^{210}Pb, ^{210}Bi, ^{210}Po	170	960	565	7483

Source: M. B. Sears et al., Report ORNL/TM-4903, vol. 1, May 1975, p. 174.

specific process developed by the Mallinckrodt Chemical Company for the Weldon Springs refinery. A similar flow sheet is used in the Kerr-McGee plant.

Uranium ore concentrates are digested with hot 40% nitric acid. The resulting mixture is about 1 N in nitric acid and contains about 400 g uranium/liter and some suspended solids. The aqueous mixture is fed to a series of pumper-decanter mixer-settlers, where the uranyl nitrate is extracted by countercurrent flow of 30 v/o TBP in normal hexane. The flow ratio of organic to aqueous is about 13:1. Uranium concentration in the organic extract leaving the first stage is about 95 g uranium/liter and in the aqueous raffinate leaving the last stage is under 0.1 g uranium/liter. The raffinate is neutralized with lime. It contains most of the radioactive impurities in the ore concentrates, principally ^{230}Th and ^{226}Ra.

In the scrubbing section all nonuranium metallic impurities and some uranium are removed from the organic phase by counterflowing dilute nitric acid, which is returned to the extracting section. In the stripping section purified uranium in the organic phase leaving the scrubbing

Table 5.27 Principal uranium refineries

Owner	Location	Feed	Products
U.S. Dept. of Energy	Weldon Springs, Missouri†	Ore concentrates	UO_3, UF_4, uranium metal
U.S. Dept. of Energy	Fernald, Ohio	Ore concentrates, uranium-bearing scrap	UO_3, UF_4, uranium metal
U.S. Dept. of Energy	Paducah, Kentucky	UO_3, UF_4	UF_6
U.S. Dept. of Energy	Portsmouth, Ohio	Enriched uranium scrap	UF_6
Kerr-McGee Corp.	Gore, Oklahoma	Ore concentrates	Natural UO_2, UF_4, UF_6
Allied Chemical Corp.	Metropolis, Illinois	Ore concentrates	Natural UF_6
Eldorado Nuclear Ltd.	Port Hope, Ontario	Ore concentrates	Natural UO_2, UF_4, UF_6
British Nuclear Fuels Ltd.	Springfields, United Kingdom	Ore concentrates	Natural uranium metal, UF_6
Comurhex	Malvesi, France	Ore concentrates	Natural uranium metal, UF_4
		Depleted UNH	Depleted uranium metal, UF_4
Comurhex	Pierrelatte, France	UF_4	UF_6

† Decommissioned in 1966.

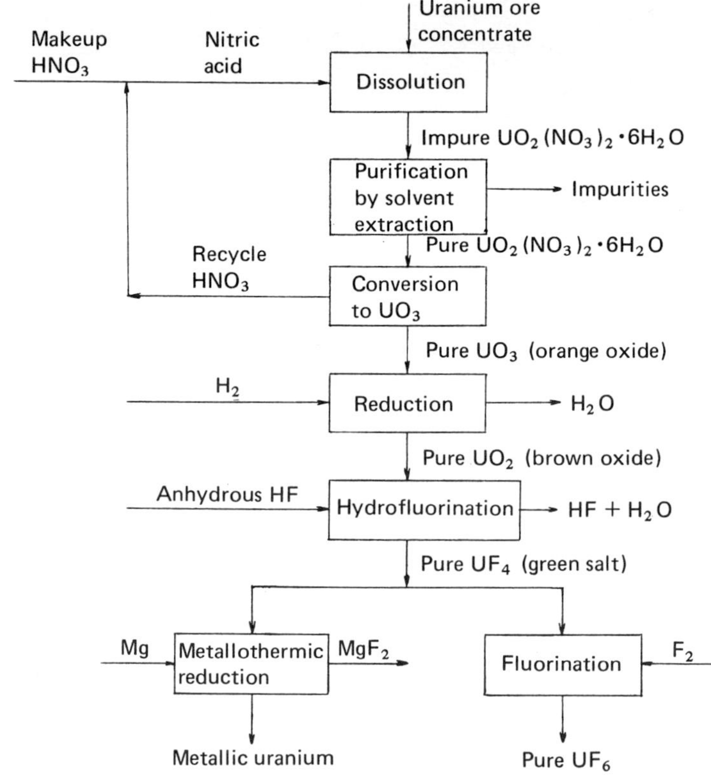

Figure 5.21 Steps in conventional uranium refining processes.

section is transferred to an aqueous phase by back-extraction with 0.01 normal nitric acid. Pulse columns are used for the scrubbing and stripping sections.

A portion of the aqueous stream leaving the stripping section is withdrawn, washed with hexane to remove dissolved and entrained TBP, and leaves the TBP-removal column as product uranyl nitrate solution (UNH).

All TBP is washed with an aqueous solution of sodium carbonate in a spray column to remove any hydrolyzed TBP and impurities that might accumulate in the TBP if it were not cleaned in this way. Sodium hydroxide is added to the aqueous sodium carbonate stream leaving the spray column to precipitate any uranium that might have been carried to this point. This impure uranium is recycled to the dissolver.

Variants of this basic process are used in other plants. For example, the Comurhex plant at Malvesi [B5] filters the output from the dissolver, uses pulse columns in the extracting section, and dilutes TBP with *n*-dodecane instead of *n*-hexane.

9.3 Conversion of UNH to UO_3

In U.S. plants the aqueous solution of UNH is converted to UO_3 in two steps, concentration and denitration. In the concentration step the UNH solution is evaporated in a boil-down tank to a syrupy liquid with the approximate composition of the hexahydrate. Three types of equipment have been used for denitration: a heated pot; a fluidized bed; and a stirred, heated

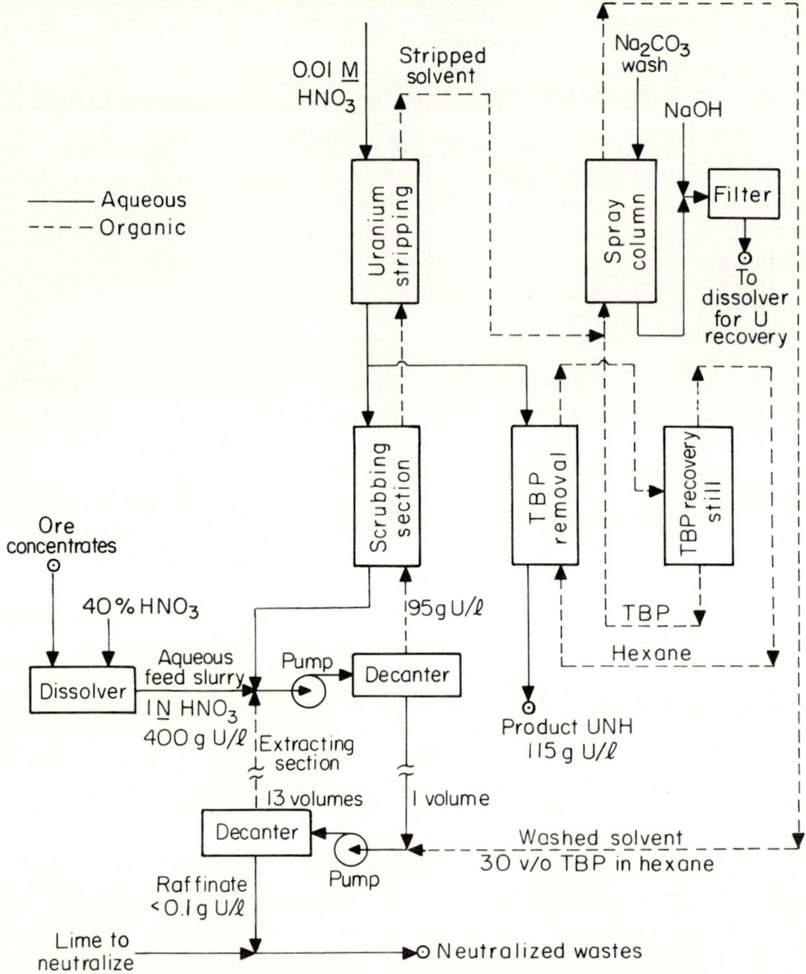

Figure 5.22 Purification of uranium ore concentrates by solvent extraction with TBP.

trough. The last was used at Weldon Springs. This consists of an enclosed, heated trough in which a horizontal, rotating agitator keeps the bed thoroughly mixed. The UO_3 product overflows an adjustable weir and is cooled and collected for subsequent grinding. Water, nitric acid, and oxides of nitrogen given off in these steps are recovered and recycled to the dissolver.

In European refineries a different process is used to convert UNH to UO_3. The uranyl nitrate solution from solvent extraction is neutralized with gaseous ammonia to precipitate $(NH_4)_2U_2O_7$. This ammonium diuranate is filtered off, dried, and calcined to drive off ammonia and form UO_3.

9.4 Reduction of UO_3 to UO_2

UO_3 is converted to UO_2 by reduction with cracked ammonia gas ($3H_2$:$1N_2$) around 590°C in two fluidized reactors through which solids and reducing gases flow countercurrently. Exhaust gases are filtered to remove entrained dust, cooled to condense steam formed in the reaction

$$UO_3 + H_2 \rightarrow UO_2 + H_2O$$

and the unreacted hydrogen is burned. Conditions must be carefully controlled to prevent sintering of the oxides, to obtain a product that will react satisfactorily with HF in the next hydrofluorination step. If the UO_2 is to be used directly as reactor fuel, as in CANDU reactors, reduction is carried out at a higher temperature to make a denser oxide.

9.5 Hydrofluorination of UO_2 to UF_4

The hydrofluorination reaction to convert UO_2 to UF_4,

$$UO_2 + 4HF \rightleftharpoons UF_4 + 2H_2O$$

is exothermic. It proceeds rapidly at 500°C, but the equilibrium mixture of H_2O and HF contains around 35 percent HF. At 300°C, nearly complete utilization of HF can be obtained, but the reaction rate is slow. Problem 5.3 illustrates calculation of the HF content of the equilibrium mixture of HF and H_2O at these two temperatures from free-energy data.

In U.S. plants hydrofluorination is carried out in two stirred fluidized-bed reactors in series, with counterflow of solids and gases. The bed to which UO_2 is fed and from which exhaust gases are discharged runs at 300°C, partially converts UO_2 to UF_4, and reduces the HF content of the effluent gases to around 15 percent. The bed to which anhydrous HF and the partially converted UO_2 are fed runs at 500°C and converts more than 95 percent of the UO_2 to UF_4. To prevent caking of the fluidized beds, it has been found necessary to provide each reactor with a vertical-shaft, slow-speed stirrer to scrape the reactor walls. Production rates around 700 to 900 kg/h are obtained in 0.75-m-diameter reactors. Effluent gases are filtered to remove entrained solids, cooled to condense aqueous HF, and scrubbed to remove the last traces of HF.

In the Comurhex plant at Malvesi [B5], reduction of UO_3 and conversion of UO_2 to UF_4 are carried out in a single L-shaped, moving-bed reactor. Reduction takes place in a vertical section and hydrofluorination in a horizontal section. Practically complete utilization of HF is obtained.

9.6 Fluorination of UF_4 to UF_6

In the U.S. Department of Energy (DOE) plant at Paducah and the Comurhex plant at Pierrelatte [B5], UF_4 is converted to UF_6 by reaction with fluorine in a tower reactor. Solid UF_4 and a slight excess of fluorine gas are fed at the top of a monel tower with walls cooled to around 500°C. Most of the UF_4 reacts almost instantaneously with a flame temperature of around 1600°C. Small amounts of unreacted UF_4 and uranium oxides are removed from the bottom of the tower and recycled to the hydrofluorination step.

The effluent gases containing UF_6, fluorine, and diluent gases such as oxygen and nitrogen are cooled to around 150°C and passed through filters to remove entrained solids. Most of the UF_6 is condensed as solid in cold traps cooled to −10°C. Residual fluorine in the gases leaving the cold trap is removed by reaction with additional UF_4 in a fluid-bed reactor which forms additional UF_6 and nonvolatile intermediate fluorides such as UF_5. Solids from this bed are fed to the primary fluorination reactor.

Exhaust gases from the second reactor go to a second cold trap at −50°C, which condenses most of the UF_6. The last traces of UF_6 are removed by a second UF_4 fluid-bed reactor, which reduces the UF_6 content of exhaust gases to less than 10 ppm.

UF_6 produced in this way is exceptionally pure. The UF_6 content is above 99.97 percent, and the overall process yield exceeds 99.5 percent. Table 5.28 summarizes U.S. DOE specifications that UF_6 must meet to be fed to U.S. gaseous diffusion plants.

Table 5.28 Specifications for UF_6 delivered to U.S. DOE

Minimum w/o UF_6	99.5
Maximum m/o hydrocarbons and halocarbons	0.01
Maximum ppm of elements forming volatile fluorides, in total uranium	
Antimony	1
Bromine	5
Chlorine	100
Niobium	1
Phosphorus	50
Ruthenium	1
Silicon	100
Tantalum	1
Titanium	1
Maximum ppm of nonvolatile fluorides	300
Maximum ppm in ^{235}U	
Chromium	1500
Molybdenum	200
Tungsten	200
Vanadium	200
^{233}U	500
^{232}U	0.110
Maximum thermal-neutron absorption, equivalent ppm boron in total uranium	8
Maximum gamma activity of ^{237}U and fission products, expressed as percent of gamma activity of aged natural uranium	20
Maximum beta activity of fission products, same basis	10
Maximum alpha activity of transuranics	1500 disintegrations/(min·g U)

Source: *Federal Register,* July 15, 1971, p. 286a.

9.7 Allied Chemical Process for Converting Uranium Concentrates to UF_6

In the Allied Chemical Company's plant for converting uranium ore concentrates to UF_6 at Metropolis, Illinois, ore concentrates are first converted to impure UF_6, which is then purified by fractional distillation, through the steps shown in Fig. 5.23. The plant and processes have been described by Ruch et al. [R1] and Sutton et al. [S5].

To minimize HF consumption and avoid formation of low-melting compounds of NaF and UF_4, this plant prefers feed consisting of $(NH_4)_2U_2O_7$ or uranium oxide. Feed containing high concentrations of sodium or magnesium is converted to $(NH_4)_2U_2O_7$ by reaction with hot $(NH_4)_2SO_4$ solution, at an extra charge. $(NH_4)_2U_2O_7$ is converted to UO_3 by heating to 450°C. The UO_3 is reduced to UO_2 in a fluidized bed by reaction at 540 to 620°C with 1.5 times the stoichiometric amount of hydrogen, made by cracking ammonia gas. Although the reaction is exothermic, heat must be added to bring the reactants to the required temperature and to compensate for heat losses. Careful temperature control is necessary; too low a temperature leaves UO_3 unreduced, and too high a temperature causes sintering and loss of reactivity in the hydrofluorination step.

Conversion of UO_2 to UF_4 by reaction with HF gas is carried out in a two-stage countercurrent fluidized-bed system as described in Sec. 9.5. This step removes as volatile fluorides any silicon, sulfur, and boron present in the feed, and some of the molybdenum and vanadium. Off-gases are scrubbed first with water and then with aqueous KOH to remove these effluents and the excess HF used to complete conversion of UO_2 to UF_4. The water scrub solution is neutralized with lime.

UF_4 is converted to UF_6 by reaction with fluorine at 425 to 535°C in an air-cooled, monel fluid-bed reactor charged with CaF_2 diluent to improve heat transfer. Because of nonvolatile fluorides present in the crude UF_4 feed, a small amount of CaF_2 is continuously removed from the bed and processed for uranium recovery by reaction with fluorine in an ash cleanup reactor. Product gases from the primary reactor are passed through a cold trap to condense most of the UF_6. Unreacted fluorine in off-gas from the cold trap is removed by reaction with UF_4 in a fluorine cleanup reactor. Effluent passes through a filter, additional cold traps, and a KOH scrubber.

The UF_6 condensed in the cold traps contains as possible impurities fluorides with the following normal boiling points, among others:

HF	19.5°C
TeF_6	35.5°C
MoF_6	35.6°C
RuF_6	75°C
VF_5	111.2°C
MoF_4	180°C
VOF_3	480°C

Because UF_6 has a vapor pressure of 1 atm at 56°C, it is necessary to use two columns to purify it, one to remove low-boiling impurities and the other, high-boiling. Because the triple point of UF_6 is 64°C and 1140 Torr, to prevent freeze-up it is necessary to operate both columns at a pressure over 1140 Torr with condenser cooling at a temperature over 64°C.

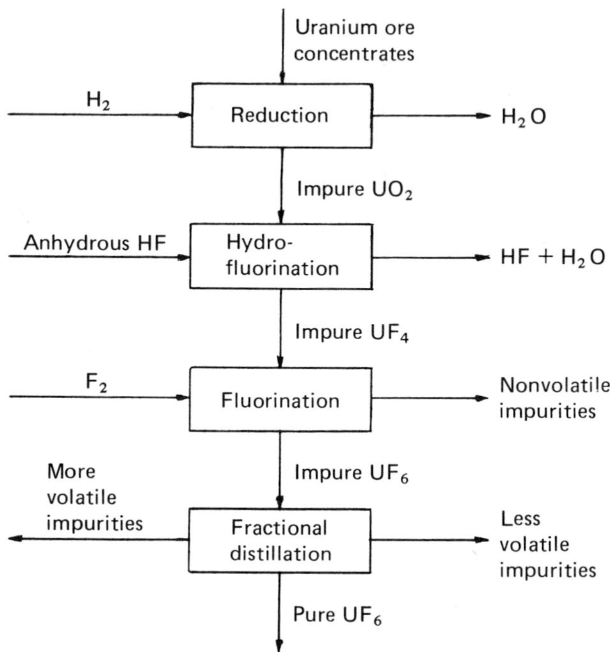

Figure 5.23 Allied Chemical UF_6 process.

Most of the radioactive contaminants in the yellow cake fed to the plant leave in the residue from the fluorination reactor, where the nonvolatile fluorides of ^{226}Ra and thorium are concentrated. In addition, ^{234}Th and ^{234}Pa, which form in UF_6 from decay of ^{238}U, build up in bottoms of the UF_6 distillation column (Prob. 5.4).

Uranium recovery at the Allied Chemical plant has exceeded 99.5 percent. UF_6 product meets DOE specifications. The charge for converting concentrates to UF_6 was around \$3.50/kg uranium in 1977.

The advantage of the Allied Chemical dry process over the conventional process described in Secs. 9.2 through 9.6 is the smaller number of steps in the dry process, which makes its costs lower for certain kinds of concentrates, e.g., those not containing large amounts of Na, Ca, Mg, or Fe. Disadvantages of the process are the difficulty of handling large amounts of these impurities and its inability to produce pure UO_2 or pure UF_4.

9.8 Conversion of UF_6 to UF_4 and UO_2

Enriched or depleted uranium is usually produced in the form of UF_6, but is used as metallic uranium or UO_2. This requires conversion of UF_6 to UF_4 or UO_2. UF_6 is converted to UF_4 by vapor-phase reduction with hydrogen. Because the heat of reaction is small, the mixture must be heated. In small reactors used for converting highly enriched uranium, heat is provided internally by reacting fluorine with hydrogen.

Three processes have been used for converting UF_6 to UO_2. In one, UF_6 is reduced to UF_4, which is then hydrolyzed by steam,

$$UF_4 + 2H_2O \rightarrow UO_2 + 4HF$$

the reverse of the reaction used to make UF_4. In a second process, UF_6 is hydrolyzed to UO_2F_2 by solution in water, after which ammonia is added to precipitate ammonium diuranate,

$$2UO_2F_2 + 6NH_4OH \rightarrow 4NH_4F + (NH_4)_2U_2O_7 + 3H_2O$$

The diuranate is then reduced to UO_2 with hydrogen at 820°C. In the third, AUC (ammonium uranyl carbonate) process, developed in West Germany by Nukem, streams of gaseous UF_6, CO_2, and NH_3 are fed batchwise into demineralized water, whereby $(NH_4)_4UO_2(CO_3)_3$ is precipitated. The AUC is converted batchwise to UO_2 by contacting it with steam and hydrogen at 500°C in a fluidized bed, with recovery of CO_2 and NH_3. Subsequently, steam at 650°C is supplied to the fluidized bed to reduce fluorine content to 50 to 60 ppm by pyrohydrolysis.

10 PRODUCTION OF URANIUM METAL

10.1 Difficulties

Production of uranium metal sufficiently pure for use in nuclear reactors is difficult. Uranium forms very stable compounds with oxygen, nitrogen, and carbon, and it reduces the oxides of many common refractories. Methods that yield uranium metal at temperatures below its melting point result in a fine powder that oxidizes rapidly in air and is difficult to consolidate into massive metal. Uranium cannot be deposited electrolytically from aqueous solution. It is not practical to purify uranium by distillation because of its very high boiling point, 3900°C. Any nonvolatile impurities introduced into uranium during production will remain in it during subsequent operations and contaminate the final product.

10.2 Alternative Methods

Four methods that have been used to produce uranium metal are

1. Electrolysis of fused salts
2. Reduction of UO_2
3. Reduction of UF_4
4. Reduction of UCl_4

Electrolysis. Electrolysis of KUF_5 or UF_4 dissolved in a molten mixture of 80 percent $CaCl_2$ and 20 percent NaCl was the method used by the Westinghouse Electric Company to produce the first pure uranium metal for the Manhattan Project [S4]. Because electrolysis was carried out below the melting point of uranium metal, 1130°C, the crude metal contained salt that had to be leached with water and had to be remelted before acceptably pure metal could be obtained. By 1943 this method was superseded by the less costly reduction of UF_4 by magnesium, to be described later.

Possible reductants. Elements that might be considered for reducing UO_2, UF_4, or UCl_4 to metallic uranium are hydrogen, sodium, magnesium, or calcium. Carbon is impractical because of formation of uranium carbide, and aluminum is undesirable because it forms an intermetallic compound with uranium. Sodium, magnesium, and calcium do not do this.

To show which combinations of uranium compound and reductant are thermodynamically favorable, the free-energy change in reducing UO_2, UF_4, or UCl_4 by hydrogen, sodium, calcium, or magnesium has been evaluated in Table 5.29. These data are for a temperature of

Table 5.29 Free-energy change in production of uranium metal

Part 1. Free energies of formation from elements at 1500 K [N1]

Element	Atoms to reduce one atom of uranium	Reference	Free energy of formation at 1500 K, calories for number of gram-moles equivalent to 1 g-mol uranium: Oxide	Fluoride	Chloride
U	–	[B1]	−197,874	−351,544	−173,190
H	4	[N1]	−78,594	−267,640	−99,252
Na	4	[N1]	−80,674	−383,776	−249,920
Mg	2	[N1]	−202,406	−407,219	−187,950
Ca	2	[N1]	−228,140 [N2]	−465,266	−278,042

Part 2. Free-energy change per gram-mole uranium in reduction of uranium compounds at 1500 K

Reductant	Free-energy change, cal/g-mol, for uranium compound: UO_2	UF_4	UCl_4
H	+119,280	+83,904	+73,938
Na	+117,200	−32,232	−76,730
Mg	−4,532	−55,675	−14,760
Ca	−30,266	−113,722	−104,852

1500 K (1227°C), which is high enough for rapid reaction and is above the melting point of uranium, 1130°C, so that the uranium product would be consolidated rather than a fine powder. Part 1 of Table 5.29 lists the free energy of formation of 1 mol of each of the uranium compounds from its elements and the free energy of formation of the number of gram-moles of the oxide, fluoride, or chloride of the four possible reductants needed to produce 1 g-mol of uranium. Part 2 of Table 5.29 lists the free-energy change in reaction of each possible combination of uranium compound and reductant. For example, the free-energy change in the reaction

$$UF_4 + 2Mg \rightarrow U + 2MgF_2$$

is

$$-407{,}219 - (-351{,}544) = -55{,}675 \text{ cal/g-mol uranium} \tag{5.16}$$

For reduction of the uranium compound to be complete without requiring a large excess of reductant, the free-energy change at 1500 K should be more negative than 10,000 cal/g-mol. Table 5.29 shows that hydrogen is completely impractical and that the only feasible reductant for UO_2 is calcium. For UF_4 or UCl_4, sodium, magnesium, or calcium meet the free-energy criterion.

Reduction of UO_2. Production of metallic uranium by reacting UO_2 with calcium metal is thermodynamically possible and was practiced in Germany in 1942 [S1]. However, the melting point of CaO is so high, 2615°C, and the heat of reaction is so small, 44 kcal/g-mol uranium, that it is impossible to melt the lime to make a clean separation between it and uranium metal. The result is that the uranium product is in the form of small particles and the recovery of clean uranium is usually no more than 35 to 40 percent.

Reduction of halides. Table 5.30 lists pertinent properties of substances that might take part in the reduction of the uranium halides UF_4 or UCl_4. As this table shows, the heat available per mole of uranium produced is much higher than in the reduction of UO_2. In addition, the melting point of the halide by-product is much lower than that of CaO and is near that of uranium metal. Consequently, the reaction temperature can be raised enough to melt both the uranium metal and the halide by-product, so that a clean separation between metal and slag can be obtained.

Production of reactive metals by reduction of the tetrachlorides is the basis of the well-known Kroll process for titanium or zirconium (Chap. 7). Although metallothermic reduction of uranium tetrachloride is thermodynamically possible for uranium also, its use in practice is made difficult by the hygroscopic character of UCl_4. This salt picks up water from moist air, which would contaminate uranium metal with uranium oxide after reduction. Moreover, the low boiling point of UCl_4 (789°C) relative to the higher melting point of uranium (1130°C) means that the UCl_4 would have to be fed into the reaction zone as vapor, a complication avoided with UF_4, whose boiling point is higher (1457°C). For these reasons, UF_4 is generally used.

Of possible reductants for UF_4, sodium is less desirable than magnesium or calcium because, like UCl_4, its boiling point is far below that of uranium metal. Use of sodium would require either feed of sodium vapor from an external source or operation of the reactor at very high pressure. Choice of reductant for UF_4 thus is essentially limited to calcium or magnesium. In practice magnesium is used for production of large batches of uranium, with calcium being used for smaller quantities, as when criticality considerations limit batch size. For example, at the French refinery at Malvesi [B5], calcium is used to produce uranium in small batches, under 100 kg, with magnesium used for larger batches. The same general practice is followed in the United States. Magnesium reduction has been used in the United States since 1943, when F. H. Spedding and co-workers at Iowa State College developed the process for the Manhattan Project [S4] to supersede fused-salt electrolysis.

Table 5.30 Thermodynamic data for metallothermic reduction of UF_4 and UCl_4

	Metal			
	U	Na	Mg	Ca
Melting point, K	1406	371	922	1112
Boiling point, K	~4200	1156	1378	1767
	Fluoride			
	UF_4	NaF	MgF_2	CaF_2
Melting point, K	1309	996	1536	1691
Boiling point, K	1730	1787	2499	2806
Heat of formation at 298 K, kcal/g-mol	−453.7	−137.52	−268.7	−293.0
Available heat† per gram-mole of U, kcal at 298 K		96.38	83.7	132.3
	Chloride			
	UCl_4	NaCl	$MgCl_2$	$CaCl_2$
Melting point, K	863	1074	987	1045
Boiling point, K	1062	1738	1710	2209
Heat of formation at 298 K, kcal/g-mol	−251.3	−98.26	−153.35	−190.2
Available heat† per gram-mole of U, kcal at 298 K		141.7	55.7	129.1
Reference	[B1]	[N1]	[N1]	[N1]

†Example of calculation of available heat: For $UF_4 + 4Na \rightarrow U + 4NaF$, available heat = $-453.7 - 4(-137.52) = 96.38$.

The advantages of calcium include the following:

1. The reaction can be carried out at atmospheric pressure, because the melting point of calcium fluoride is below the boiling point of calcium metal.
2. The heat of reaction is sufficient to melt both uranium metal and CaF_2 slag, with reactants initially at room temperature, so that preheating is unnecessary.

The advantages of magnesium include the following:

1. Magnesium costs much less than calcium, and only 60 percent as much mass of reductant is needed.
2. It is easier to obtain magnesium of the requisite purity than calcium, and magnesium does not pick up oxygen from air or moisture.

The problems in using magnesium are these:

1. The reaction must be carried out in a sealed reactor to contain the superatmospheric vapor

pressure of magnesium developed when the reactants are heated to the melting point of MgF_2.

2. It is necessary to preheat the charge because the heat of reaction is too small to melt the reaction products when the reactants are initially at room temperature.

10.3 Heat Balances in Uranium Metal Production

The enthalpy data of Table 5.31 may be used to show that the reaction of UF_4 and calcium metal initially at 25°C (298 K) provides sufficient heat to melt the reaction products (uranium metal and CaF_2) at the melting point of CaF_2 (1691 K), with an excess of 35.3 kcal/g-mol uranium available to offset heat losses. However, with magnesium metal, the same table shows that its heat of reaction with UF_4 initially at 25°C is insufficient to melt the reaction products completely at the melting point of MgF_2 (1536 K), and that 6.8 kcal of heat must be supplied per gram-mole uranium produced.

The lowest temperature to which the stoichiometric mixture of UF_4 and magnesium must be preheated to just melt the reaction products without heat loss may be found by reference to Fig. 5.24, in which the enthalpies of 1 mol of UF_4, 2 mol of magnesium, and their mixture are plotted against temperature. The temperature at which the enthalpy change of the mixture from 25°C is 6.8 kcal, or 192°C, is the desired minimum preheat temperature. In practice, it is necessary to preheat the charge to a higher temperature to compensate for heat losses and because a slight excess of magnesium is used, as illustrated by Prob. 5.6.

10.4 Production of Uranium Metal by Reduction of UF_4 with Magnesium

The process for producing uranium metal by reaction of UF_4 and magnesium developed at Iowa State College in the United States was described by Wilhelm [W3] at the first Geneva Conference. In this process, the additional heat needed to melt the products was provided by preheating the reactor and its charge of UF_4 and magnesium according to a schedule designed to heat the reactor and its entire contents by several hundred degrees before any part reached a high enough temperature to set off the reaction. The reactor used to contain the high pressure produced momentarily when the reaction mixture reached the melting point of MgF_2 consisted of a steel tube lined with an insulating layer of specially purified dolomitic lime. Lime is one of the few oxides that does not react with and contaminate molten uranium.

The largest reactors used at Iowa State College were 35 cm in diameter and 120 cm high, made of standard-wall seamless steel pipe. They were closed with a welded steel plate at the bottom and a cover bolted to a flange at the top. The liner was made of electrically fused dolomitic lime which was finely powdered and packed into the steel tube to form a dense layer 2.5 cm thick on the bottom and 1.25 cm thick on the sides.

A mixture of finely ground UF_4 and a slight excess of granular, oxide-free magnesium was packed tightly in the reactor to within 8 cm of the top. The charge was covered with a tightly fitting graphite plug, and the space above the graphite was filled with powdered lime. Closure was by a steel cap bolted to the reactor flange.

The reactor was placed in a preheating furnace, whose optimum temperature was in the range 550 to 700°C. With the 35-cm reactor, a furnace temperature of 565°C initiated the reaction after about 3.5 h. The reaction was over in about 1 min, and the uranium remained molten for about 10 min. The reactor was then cooled, first by air and then by water, opened, and the contents removed. An average of 108 kg of massive metal was obtained per batch, in 98.3 percent yield based on UF_4 charged.

The process used in France at Malvesi [B5] is generally similar, except that the reactor is lined with MgF_2. A preheating period of 10 h is used to produce a 210-kg ingot, and a 20-h period for 450 kg.

Table 5.31 Heat to be supplied in reduction of UF_4

	Reducing agent M			
	Magnesium		Calcium	
Melting point of MF_2, K (T_2)	1536		1691	
Temperature of reactants, K (T_1)	298		298	
	Kilocalories per gram-mole U			
Products				
2 X heat of formation of $MF_2(s)$ at 298 K [N1]	−537.4		−586.0	
2 X enthalpy of $MF_2(l)$ at T_2 above $MF_2(s)$ at 298 K [N1]	74.168		78.904	
Enthalpy of $U(l)$ at T_2 above $U(s)$ at 298 K [B1]	16.354		18.129	
Enthalpy of products (l) at T_2 above elements		−446.9		−489.0
Reactants				
Heat of formation of $UF_4(s)$ at 298 K [B1]	−453.7		−453.7	
Enthalpy of UF_4 at T_1 above $UF_4(s)$ at 298 K	0		0	
2 X enthalpy of M at T_1 above M(s) at 298 K	0		0	
Enthalpy of reactants at T_1 above elements		−453.7		−453.7
Heat to be supplied (differences)		+6.8		−35.3

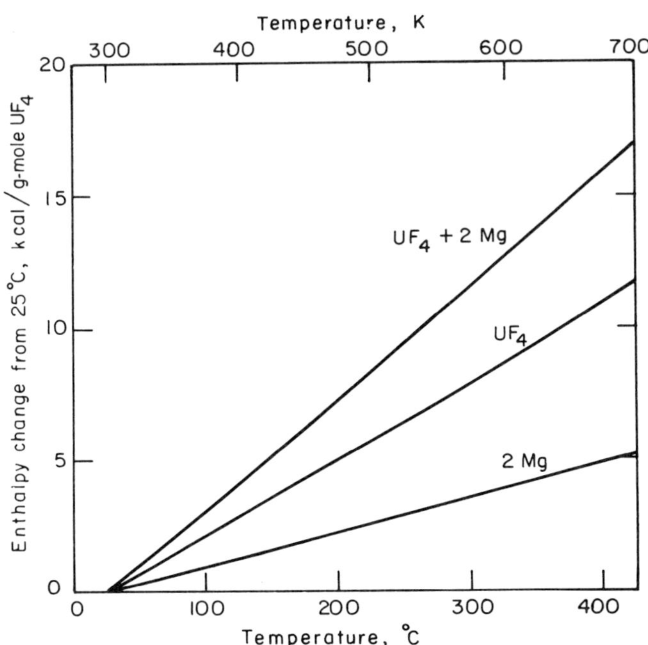

Figure 5.24 Enthalpy change of reactants in reduction of UF_4 by magnesium.

REFERENCES

A1. Ahrland, S.: *Acta Chem. Scand.* **5**: 1151 (1951).

B1. Barin, I., and O. Knacke: *Thermochemical Properties of Inorganic Substances,* Springer-Verlag, New York, 1973.

B2. Battelle Pacific Northwest Laboratories: "Assessment of Uranium and Thorium Resources in the United States and the Effect of Policy Alternatives," Richland, Wash., Dec. 1974.

B3. Bigeleisen, J., et al.: *J. Chem. Phys.* **16**: 442 (1948).

B4. Blake, C. A., Jr., et al.: "Solvent Extraction of Uranium and Other Metals by Acidic and Neutral Organophosphorus Compounds," *PICG(2)* **28**: 289 (1958).

B5. Bodu, R. L.: "From Uranium Ore to Metal and UF_6. Description and Recent Developments of French Methods," Amer. Nuclear Soc. Conf., Miami Beach, Fla., Oct. 17–21, 1971, CONF-71109. See also *Trans. Amer. Nuclear Soc.* **14**(2): 481 (1971).

B6. Brickwedde, F. G., H. J. Hoge, and R. B. Scott: *J. Chem. Phys.* **16**: 429 (1948).

B7. Brooks, A. A., and P. Wood: "Vapor Pressure Tables for Liquid Uranium Hexafluoride," Report K-722, Nov. 21, 1957.

B8. Brown, K. B., et al.: "Solvent Extraction Processing of Uranium and Thorium Ores," *PICG(2)* **3**: 472 (1958).

C1. Clegg, J. W., and D. D. Foley: *Uranium Ore Processing,* Addison-Wesley, Reading, Mass., 1958.

C2. Coleman, C. F., et al.: "Amine Salts as Solvent Extraction Reagents for Uranium and Other Metals," *PICG(2)* **28**: 278 (1958).

D1. Davies, R. V., et al.: *Nature* **203**: 1110 (1964).

D2. Davies, R. V., et al.: Report AERE-R-5024, 1965.

D2a. Dayton, S. H.: *Mining World* **21**(2): 42 (Feb. 1959).

D3. DeWitt, R.: "Uranium Hexafluoride: A Summary of the Physico-Chemical Properties," Report GAT-280, Aug. 12, 1960.

G1. Greer, A. H., A. B. Mindler, and J. P. Termini: *Ind. Eng. Chem.* **50**(2): 167 (1958).

H1. Hargrove, D.: *Mining World,* Feb. 1958, p. 34.

H2. Harrington, F. E., et al.: Report ORNL-TM-4757, Nov. 1974.

H3. Haubenreich, P. N., and J. R. Engel: *Nucl. Appl. Tech.* **8**: 118 (1970).

H4. Hazen, W. C.: *Mining Congress Journal,* July 1960, pp. 56–59.

H5. Helfferich, F.: *Ion Exchange,* McGraw-Hill, New York, 1962.

J1. Jones, J. Q., and L. C. DeJong: in *Milling Methods in the Americas,* N. Arbiter (ed.), Gordon and Breach, New York, 1964, pp. 283–312.

K1. Keen, N. J.: *J. Brit. Nucl. Energy Soc.* **7**: 168 (1968).

K2. Kennedy, J.: Report AERE-R-5023, 1965.

L1. Llewellyn, D. R.: *J. Chem. Soc.,* p. 28 (1953).

L2. Long, R. S., D. A. Ellis, and R. H. Bailes: "Recovery of Uranium from Phosphates by Solvent Extraction," *PICG(1)* **8**: 77 (1956).

M1. Marvin, G. G., et al.: "Recovery of Uranium from Its Ores," *PICG(1)* **8**: 3 (1956).

M2. McClaine, L. A., E. P. Bollwinkel, and J. C. Huggins: *PICG(1)* **8**: 26 (1956).

M3. Merritt, R. C.: *The Extractive Metallurgy of Uranium,* Colorado School of Mines Research Institute, Boulder, Colo., 1971.

M4. Mindler, A. B., and J. P. Termini: *Eng. Mining J.* **157**(9): 100 (1956).

N1. National Bureau of Standards: *JANAF Thermochemical Tables,* 2d ed., U.S. Government Printing Office, Washington, D.C., June 1971.

N2. National Bureau of Standards, *JANAF Thermochemical Tables,* 2d suppl., U.S. Government Printing Office, Washington, D.C., 1975.

O1. Oliver, G. D., H. T. Milton, and J. W. Grisard: "The Vapor Pressure and Critical Constants of Uranium Hexafluoride," *J. Amer. Chem. Soc.* **75**: 2827 (1953).

O2. Organization for Economic Cooperation and Development, and International Atomic Energy Agency: "Uranium Resources, Production and Demand," Paris, Dec. 1977.
P1. Preuss, A. F., and R. Kunin: "Uranium Recovery by Ion Exchange," in *Uranium Ore Processing,* J. W. Clegg and D. D. Foley (eds.), Addison-Wesley, Reading, Mass., 1958, p. 199.
R1. Ruch, W. C., et al.: *Chem. Eng. Progr. Symp. Ser. 28,* **56**: 35 (1960).
S1. Schulenberg, W.: "Fiat Report on Non-Ferrous Metallurgy," His Majesty's Stationery Office, London, 1946, p. 35.
S2. Sears, M. B., et al.: Report ORNL/TM-4903, vol. 1, May 1975.
S3. Smith, E. L., and J. E. Page: *J. Soc. Chem. Ind.* **67**: 48 (1948).
S4. Smyth, H. D.: *Atomic Energy for Military Purposes,* Princeton University Press, Princeton, N.J., 1945.
S5. Sutton, A. H., et al.: *Chem. Eng. Progr. Symp. Ser. 65,* **62**: 20 (1966).
U1. U.S. Department of Energy, Office of Public Affairs, Press Release, May 25, 1978.
U2. U.S. Energy Research and Development Administration: "Statistical Data of the Uranium Industry," Report GJO-100(77), Jan. 1, 1977.
W1. Wechsler, M. T., and H. J. Hoge: *J. Chem. Phys.* **17**: 617 (1949).
W2. Weinstock, B., E. E. Weaver, and J. G. Malm: *J. Inorg. Nuclear Chem.* **11**: 104 (1959).
W3. Wilhelm, H. A.: "The Preparation of Uranium Metal by the Reduction of Uranium Tetrafluoride by Magnesium," *PICG(1)* **8**: 162 (1956).

PROBLEMS

5.1 The principal chemicals that were consumed in the Dawn Mining Company's acid-leach, ion-exchange uranium mill [H1] and their present approximate unit prices are as follows:

Chemical	Amount consumed	Per	Unit price
H_2SO_4 (leaching)	100 lb	Ton (2000 lb) ore	\$0.02/lb
Amberlite-400 resin	1 ft^3	120,000 gal leach liquor	\$100/ft^3
H_2SO_4 (elution)	10 lb	Ton ore	\$0.02/lb
HNO_3 (elution)	4 lb	Ton ore	\$0.05/lb
CaO (precipitation)	5 lb	Ton ore	\$0.01/lb
NH_3 (precipitation)	2.5 lb	Ton ore	\$0.10/lb

Ore contains 0.1 w/o U_3O_8. Uranium recovery is 95 percent. Leach liquor contains 0.5 g U_3O_8/liter.

What is the cost of chemicals per pound U_3O_8 recovered?

5.2 It has been proposed that some of the natural uranium needed to fuel a pressurized-water nuclear power plant be obtained by extracting uranium from seawater used to cool the plant. If the seawater temperature rise is 10°C and the reactor and fuel-cycle conditions are as given in Fig. 3.31, how many kilograms of uranium per year could be recovered at 80 percent yield from cooling water? What fraction is this of the annual fuel requirement of the reactor?

5.3 The reaction to convert UO_2 to UF_4 is

$$UO_2(s) + 4HF(g) \rightleftharpoons UF_4(s) + 2H_2O(g)$$

The free energies of formation of the products and reactants from the elements at 300 and 500°C are as follows:

	Substance			
	UO_2	HF	UF_4	H_2O
Source of data	[B1]	[N1]	[B1]	[N1]
Free energy of formation, cal/g-mol				
300°C	−235,263	−66,060	−412,603	−51,481
500°C	−227,253	−66,310	−398,931	−48,989

(*a*) What is the equilibrium constant for the above reaction at 500°C?

(*b*) What is the mole fraction of HF in its equilibrium mixture with H_2O at 1 atm and 500°C?

(*c*) Repeat for 300°C.

5.4 After being in storage for 1 year, UF_6 is fed continuously through a still at the rate of 1000 kg/day. Still bottoms are allowed to accumulate in the equipment. What is their steady-state activity in curies?

5.5 UF_6 is made by reacting UF_4 with fluorine in a highly exothermic reaction. If UF_4 and fluorine are charged to the reactor at 25°C and UF_6 vapor is discharged at 750 K, how much heat must be removed per gram-mole UF_6 produced?

5.6 In the production of uranium metal by reaction of UF_4 with Mg, a 5 percent excess of magnesium is added to ensure complete reaction of UF_4. What is the minimum temperature to which the reactants should be preheated to ensure complete melting of the products at 1536 K? The enthalpy change of magnesium between 298 and 1536 K is 41.57 kcal/g-mol.

5.7 Show that the beta radioactivity of aged natural uranium is 0.68 μCi/g uranium.

CHAPTER

SIX

THORIUM

1 USES OF THORIUM

Thorium is important in nuclear technology as the naturally occurring fertile nuclide from which neutron capture produces fissile ^{233}U by the succession of reactions

$$^{232}_{90}\text{Th} + ^{1}_{0}n \xrightarrow[7.40 \text{ b}]{} ^{233}_{90}\text{Th} \xrightarrow[22.2 \text{ min}]{\beta^-} ^{233}_{91}\text{Pa} \xrightarrow[27.0 \text{ days}]{\beta^-} ^{233}_{92}\text{U}$$

In thermal-neutron reactors ^{233}U has an important advantage over ^{235}U or ^{239}Pu in that the number of neutrons produced per thermal neutron absorbed, η, is higher for ^{233}U than for the other fissile nuclides. Table 6.1 compares the 2200 m/s cross sections and neutron yields in fission of these three nuclides. Thorium has not heretofore been extensively used in nuclear reactors because of the ready availability of the ^{235}U in natural or slightly enriched uranium. As natural uranium becomes scarcer and the conservation of neutrons and fissile material becomes more important, it is anticipated that production of ^{233}U from thorium will become of greater significance.

Compared with ^{239}Pu, the other synthetic fissile nuclide, ^{233}U, has the advantage that it can be "denatured," made less available for use as a nuclear explosive, by isotopic dilution with ^{238}U in a mixture containing less than 12 percent ^{233}U. Production of a nuclear explosive from such a mixture would require costly and difficult isotope separation (Chap. 14). No similar means exists for denaturing ^{239}Pu, which can be more readily separated from ^{238}U by chemical reprocessing (Chap. 10).

In addition to its potential use in nuclear power systems, thorium has had minor industrial use in Welsbach mantles for incandescent gas lamps, in magnesium alloys to increase strength and creep resistance at high temperatures, and in refractories.

2 THORIUM ISOTOPES

Table 6.2 lists the most important isotopes of thorium, together with their properties of greatest significance in nuclear technology.

Table 6.1 Nuclear properties of ^{233}U, ^{235}U, and ^{239}Pu

	2200 m/s cross sections		Neutron yield	
Nuclide	Fission, σ_f	Absorption, σ_a	Per fission, ν	Per absorption, $\eta = \nu\sigma_f/\sigma_a$
^{233}U	531.1	578.8	2.492	2.287
^{235}U	582.2	680.8	2.418	2.068
^{239}Pu	742.5	1011.3	2.871	2.108

2.1 Naturally Occurring Thorium Isotopes

^{232}Th and ^{228}Th. Natural thorium consists almost entirely of ^{232}Th, plus 1.35×10^{-8} percent of ^{228}Th (radiothorium), and small, but variable, amounts of ^{234}Th, ^{230}Th, ^{231}Th, and ^{227}Th.

The ^{238}Th/^{232}Th ratio in natural thorium equals the ratio of their half-lives, as ^{228}Th, a decay product of ^{232}Th, is in secular equilibrium with its parent. In irradiated thorium, however, the ^{228}Th/^{232}Th ratio may be much higher because of formation of additional ^{228}Th by alpha decay of ^{232}U, as explained in Chap. 8.

^{234}Th and ^{230}Th. The members of the ^{238}U decay chain, ^{234}Th (UX-1) and ^{230}Th (ionium), Table 5.2, are found in natural thorium when uranium is present in thorium ores. Because of the short half-life of ^{234}Th, its concentration in thorium is inconsequential, and it soon decays from separated thorium. Eighty-thousand-year ^{230}Th, on the other hand, is a significant constituent of thorium from ores containing uranium. The ^{230}Th/^{232}Th atom ratio is given by

$$r_{0,2} = \frac{(0.9927)(8 \times 10^4 r_{\text{U,Th}})}{4.51 \times 10^9} = 1.76 \times 10^{-5} r_{\text{U,Th}} \tag{6.1}$$

Here 0.9927 is the atom fraction of ^{238}U in natural uranium, $r_{\text{U,Th}}$ is the atomic ratio of U to Th in the ore, and 4.51×10^9 years is the half-life of ^{238}U. The ratio of the activity of ^{230}Th to ^{232}Th is

$$\frac{A_{230}}{A_{232}} = \frac{\tau_{232} r_{0,2}}{\tau_{230}} = \frac{1.41 \times 10^{10}}{8 \times 10^4} 1.76 \times 10^{-5} r_{\text{U,Th}} = 3.1 r_{\text{U,Th}} \tag{6.2}$$

Table 6.2 Isotopes of thorium

			Radioactive decay		Cross section for reaction with 2200 m/s neutrons	
Mass, amu	Atom percent in natural thorium	Half-life	Type	Effective MeV	(n, γ)	Fission
227.027706	Very small	18.2 days	α	6.145		200
228.02875	1.35E-8	1.910 yr	α	5.521	123	
229.031652	–	7340 yr	α	5.167	54	30.5
230.033087	Small and variable, see Sec. 2.1	8E4 yr	α	4.767	23.2	
231.036291	Very small	25.5 h	β	0.21		
232.038124	100	1.41E10 yr	α	4.08	7.40	3.9E-5
233.041469	–	22.2 min	β	0.427	1500	15
234.043583	Very small	24.1 days	β	0.060	1.8	

where τ is the half-life of the designated isotope.

Thorium has been recovered as a by-product of uranium production from ores of the Blind River district in Ontario in which the uranium:thorium ratio is 6:1 [C5]. In such thorium the ^{230}Th activity is $3.1 \times 6 = 18.6$ times the activity of the ^{232}Th.

^{231}Th and ^{227}Th. The members of the ^{235}U decay chain, ^{231}Th and ^{227}Th, Table 5.3, are trace constituents of thorium from ores containing uranium. Because of their short half-lives and the small proportion of ^{235}U in natural uranium, they are of no significance in thorium technology.

2.2 Synthetic Thorium Isotopes

^{233}Th. The isotope ^{233}Th, produced when ^{232}Th captures a neutron, is an important short-lived intermediate in production of ^{233}U, as explained in Sec. 1.

^{229}Th. The isotope ^{229}Th, with a half-life of 7340 years, is the longest-lived radioactive daughter of ^{233}U (Table 5.4). It is present at very low concentration in aged irradiated thorium.

3 THORIUM RADIOACTIVITY

^{232}Th is the parent of the $4n$ radioactive decay series shown in Fig. 6.1 and listed in Table 6.3. The last column of Table 6.3 gives the ratio of the number of atoms of each decay product of natural thorium to ^{232}Th, assuming that the thorium has been undisturbed long enough, around 40 years, for its decay products to reach equilibrium. At equilibrium, the activities of all these radioactive nuclides are equal, except for ^{212}Po and ^{208}Tl, which are alternative decay products of ^{212}Bi.

Freshly purified natural thorium contains significant amounts of three radioactive nuclides: long-lived ^{232}Th, an equal activity of its daughter ^{228}Th, and an amount of ^{230}Th whose relative

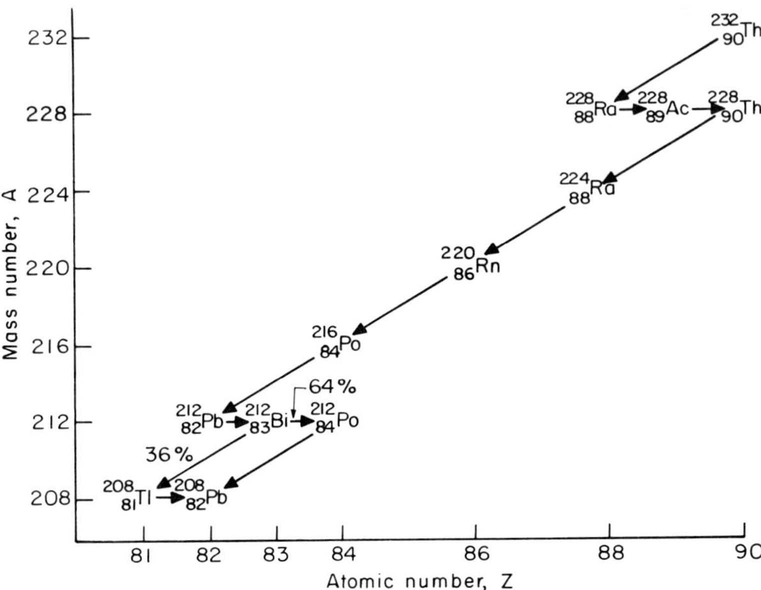

Figure 6.1 Radioactive decay of ^{232}Th.

Table 6.3 Principal radioactive decay products of ^{232}Th

Nuclide	Historical name	Half-life	Radiation		Atom ratio, ppb in natural thorium
			Type	Max γ MeV	
$^{232}_{90}$Th	Thorium	1.41E10 yr	α		10^9
$^{228}_{88}$Ra	Mesothorium 1	6.7 yr	β		0.48
$^{228}_{89}$Ac	Mesothorium 2	6.13 h	β, γ	1.84	5.0E-5
$^{228}_{90}$Th	Radiothorium	1.910 yr	α		0.135
$^{224}_{88}$Ra	Thorium X	3.64 days	α(γ)		7.1E-4
$^{220}_{86}$Rn	Thoron	55s	α		1.24E-7
$^{216}_{84}$Po	Thorium A	0.15 s	α		3.4E-10
$^{212}_{82}$Pb	Thorium B	10.64 h	β, γ	0.42	8.6E-5
$^{212}_{83}$Bi	Thorium C†	60.6 min	α, β, γ	1.81	8.2E-6
$^{212}_{84}$Po	Thorium C′	304 ns	α		4.4E-16
$^{208}_{81}$Tl	Thorium C″	3.10 min	β, γ	2.61	1.50E-7
$^{208}_{82}$Pb	Thorium D	Stable			

† Sixty-four percent of decays of $^{212}_{83}$Bi go to $^{212}_{84}$Po, 36 percent to $^{208}_{81}$Tl.

activity in natural thorium depends on the uranium:thorium ratio in the ore, Eq. (6.1). Because of the long, 80,000-year half-life of ^{230}Th, its activity remains practically constant. Because of the 1600-year half-life of its first decay product, ^{226}Ra, the daughters of ^{230}Th contribute little to the radioactivity of separated thorium for several decades. On the other hand, the short-lived daughters of 1.910-year ^{228}Th build up quickly and give rise to hazards with purified thorium that are not met with purified uranium. The most significant hazards from thorium

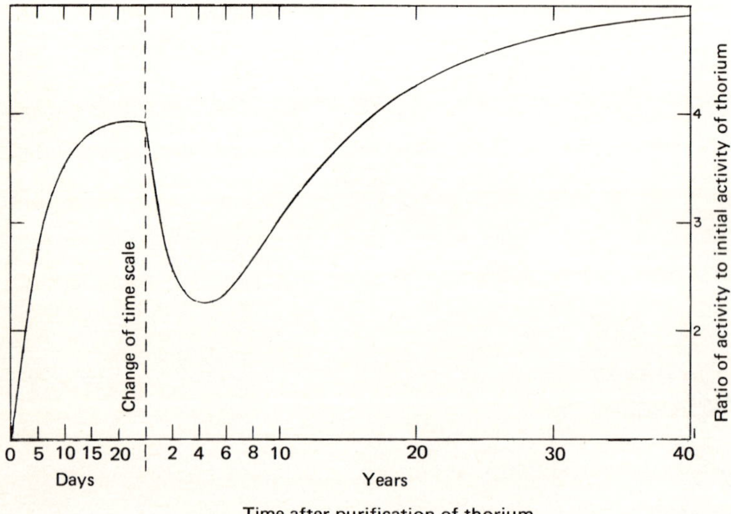

Figure 6.2 Change in radioactivity with time of natural ^{232}Th plus ^{228}Th after separation from their daughters at time zero.

decay products are caused by gaseous ^{220}Rn and high-energy gamma rays from ^{212}Bi and ^{208}Tl. These reach steady-state concentrations within a month after thorium is purified, on a time scale controlled by the 3.64-day half-life of $^{224}_{88}Ra$, the parent of ^{220}Rn.

Figure 6.2 shows how the radioactivity of freshly separated thorium containing no ^{230}Th changes with time. During the first month, the activity increases by a factor of nearly four because of buildup to steady state of the six short-lived decay products of ^{228}Th. The activity then decays with a half-life of 1.91 years as the original ^{228}Th in the purified thorium at time zero decays. After about 4 years the activity again increases, as 6.7-year ^{228}Ra produces new ^{228}Th. After about 40 years, the activity levels off at five times that of the original ^{232}Th plus ^{228}Th, as all decay products of ^{232}Th reach asymptotic levels.

Thorium that has been irradiated in a nuclear reactor will contain much higher concentrations of ^{228}Th and its daughters than natural thorium because of the sequence of reactions

$$\left.\begin{array}{l} ^{230}_{90}\mathrm{Th} \xrightarrow[23\ b]{(n,\gamma)} \\ ^{232}_{90}\mathrm{Th} \xrightarrow[\sim 0.01\ b]{(n,2n)} \end{array}\right\} {}^{231}_{90}\mathrm{Th} \xrightarrow[25.5\ \mathrm{h}]{\beta^-} {}^{231}_{91}\mathrm{Pa} \xrightarrow[210\ \mathrm{b}]{(n,\gamma)} {}^{232}_{91}\mathrm{Pa} \xrightarrow[1.31\ \mathrm{days}]{\beta^-} {}^{232}_{92}\mathrm{U} \xrightarrow[72\ \mathrm{yr}]{\alpha} {}^{228}_{90}\mathrm{Th}$$

This makes irradiated thorium much more toxic than natural thorium until it is separated from ^{232}U and stored sufficiently long (10 to 20 years) for the excess 1.91-year ^{228}Th to decay (Prob. 6.1).

4 METALLIC THORIUM

4.1 Uses

Because the thorium atom density is higher in thorium metal than in any thorium compound, metal is the preferred form of thorium where the highest nuclear reactivity or highest density is wanted. One likely nuclear application is in a sodium-cooled fast reactor where thorium would capture a neutron and be converted to ^{233}U.

4.2 Phases

The phases of thorium metal and their transition temperatures are listed in Table 6.4. Equations for the vapor pressure of thorium metal are [I1]

$$\text{Solid:} \qquad \log_{10} p(\text{atm}) = 6.648 - \frac{30{,}970}{T(\text{K})} \tag{6.3}$$

$$\text{Liquid:} \qquad \log_{10} p(\text{atm}) = 5.955 - \frac{29{,}620}{T(\text{K})} \tag{6.4}$$

4.3 Density and Thermal Expansion

The theoretical density of thorium at 25°C, from x-ray crystal measurements, is 11.72 g/cm^3. The density of cast metal is between 11.5 and 11.6 g/cm^3.

The mean thermal expansion coefficient is given in Table 6.5. Because thorium crystallizes in the cubic system, it expands equally in all directions and is not subject to as much distortion on thermal cycling as uranium. For this reason, and because the α-β transition temperature in thorium is much higher than in uranium, thorium metal reactor fuel has much better limensional stability than uranium metal.

Table 6.4 Phases of thorium metal

Transition temperature, °C	Phase	Crystal system
	Solid α	Face-centered cubic
1360 ± 10	Solid β	Body-centered cubic
1750 ± 10	Liquid	
4702 (1 atm)	Vapor	

Source: International Atomic Energy Agency, "Thorium: Physicochemical Properites of Its Compounds and Alloys," *Atomic Energy Rev.*, Special Issue No. 5, 1975.

4.4 Thermodynamic Properties

The heat capacities of the two solid phases of thorium and the liquid metal and the heats of transformation and fission are given in Table 6.6.

4.5 Thermal and Electrical Conductivity

The thermal conductivity of thorium metal is given in Table 6.7. The electrical conductivity of thorium metal is very dependent on its impurity content. Chiotti [C3] found that at room temperature the resistivity of thorium metal containing 0.2 w/o (weight percent) carbon was 37×10^{-6} Ω·cm, and that of metal containing 0.03 w/o carbon was 18×10^{-6} Ω·cm. An extrapolated value for carbon-free thorium metal is 13 to 15×10^{-6} Ω·cm. The temperature coefficient of resistivity is 3.6 to 4.0×10^{-3} per °C.

4.6 Chemical Reactivity

Thorium metal is slowly tarnished by air at room temperature, but further attack is prevented by an adherent oxide film. At temperatures above 200°C, however, progressive attack takes place. Weight gains of 0.03, 0.43, and 8.7 g/(cm^2·h) have been reported at 300, 400, and 500°C, respectively [W2]. The product is primarily ThO_2. Finely divided thorium is pyrophoric.

Thorium reacts with hydrogen at temperatures above 250°C to form ThH_2 and Th_4H_{15}. Thorium reacts with nitrogen at temperatures above 670°C to form ThN. For these reasons, melting of pure thorium metal must be carried out in vacuum, helium, or argon.

At temperatures below 100°C, thorium metal is only slowly corroded by water because of formation of a protective oxide film. At temperatures above 178°C, the film spalls off and

Table 6.5 Mean thermal expansion coefficient of thorium

Temperature range, °C	Mean linear expansion coefficient, per °C
20–200	11.2×10^{-6}
200–675	12.3×10^{-6}
675–1000	13.8×10^{-6}
25–1000	12.6×10^{-6}

Source: J. R. Murray, "The Preparation, Properties and Alloying Behavior of Thorium," Report AERE-M/TN-12, 1952.

Table 6.6 Thermodynamic properties of thorium metal

Phase	Temperature range T, K		Heat capacity, cal/(g-mol·°C) $C_p = a + bT + c/T^2$			Heat of transformation or fusion, cal/g-mol
	From	To	a	$10^3\ b$	$10^{-5}\ c$	
Solid α	298	1633	6.09	1.87	−0.122	860 ± 30
Solid β	1633	2023	3.75	2.86		3330 ± 300
Liquid	2023	5000	11.0			

Source: International Atomic Energy Agency, "Thorium: Physicochemical Properties of Its Compounds and Alloys," *Atomic Energy Rev.*, Special Issue No. 5, 1975.

oxidation is rapid. At 315°C, the rate of weight loss is around 56 mg/(cm^2·h) [W2]. For this reason, thorium metal is not considered a suitable fuel material for water-cooled power reactors.

Because thorium metal is unattacked by sodium at temperatures up to 500°C, it is compatible with the coolant in sodium-cooled reactors.

Thorium metal is slowly attacked by dilute hydrochloric, hydrofluoric, sulfuric, or nitric acids. It dissolves readily in concentrated hydrochloric acid or aqua regia. It is passive in concentrated nitric acid, but dissolves readily if 0.05 *M* of fluoride ion is added. This would be the preferred reagent to prepare feed for reprocessing irradiated thorium metal by the Thorex process.

5 THORIUM COMPOUNDS

5.1 Thorium Valence States

The tetravalent state, exemplified by compounds such as ThO_2, is the only valence state of thorium of practical importance.

5.2 Thorium Dioxide

Thorium dioxide ThO_2 is the form in which thorium is proposed for use as reactor fuel for light-water, heavy-water, and liquid-metal fast-breeder reactors. It is a stable ceramic that can be

Table 6.7 Thermal conductivity of thorium metal

Temperature, °C	Thermal conductivity, W/(cm·°C)
100	0.377
200	0.389
300	0.402
400	0.419
500	0.427
600	0.444
650	0.452

Source: J. R. Murray, "The Preparation, Properties and Alloying Behavior of Thorium," Report AERE-M/TN-12, 1952.

Table 6.8 Physical properties of ThO_2

Color	White
Crystal system	Face-centered cubic
Density (x-ray, 25°C)	10.00 g/cm^3
Linear expansion from 25°C	9×10^{-3} at 1000°C; 20×10^{-3} at 2000°C
Thermal conductivity, W/(cm·°C)	0.10 at 100°C; 0.04 at 600°C
Melting point	3370 ± 30°C
Vapor pressure	$\log_{10} p_{atm}$† $= 8.00 - 35{,}170/T$; $2180 < T < 2865$ K
Heat capacity, cal/g-mol	$C_p = 16.56 + 2.232 \times 10^{-3} T - 2.195 \times 10^5 T^{-2}$; $298 < T < 3000$ K
Entropy at 25°C	15.59 cal/(K·g-mol)
Heat of formation from elements at 25°C	−293.2 kcal/g-mol
Free energy of formation from elements at 25°C	−279.43 kcal/g-mol
Approximate equation for free energy of formation,	$\Delta G = -291.93 + 0.04350\, T$ kcal/g-mol; $298 < T < 2023$ K

†ThO_2 dissociates partially into ThO and O; at low oxygen pressures, vapor pressure is somewhat higher.

heated almost to its melting point of 3370°C without serious deterioration. Physical properties of ThO_2, from reference [I1], are summarized in Table 6.8.

ThO_2 forms solid solutions with UO_2 or PuO_2 over the entire composition range from 0 to 100 percent ThO_2.

ThO_2, either as the mineral thorianite or as synthetic thoria produced by heating thorium nitrate, oxalate, or hydroxide, reacts only slowly with mineral acids. It can be dissolved in hot, concentrated sulfuric acid or in hot nitric acid containing 0.05 *M* HF.

5.3 Thorium Carbides

Reference [I1] summarizes the somewhat conflicting data on the system thorium-carbon. From maxima in the melting point curve, it is concluded that two compounds exist:

Compound	Formula	Crystal system	Density, g/cm^3	Melting point, °C
Monocarbide	$ThC_{0.97}$	Face-centered cubic		2500 ± 35
Dicarbide	$ThC_{1.90}$	Monoclinic	9.6	2640 ± 30
	>1410°C	Tetragonal	8.76	

The dicarbide, either by itself, mixed with uranium dicarbide, or in solid solution with uranium dicarbide, is used as fuel material in some versions of high-temperature gas-cooled reactors. Like uranium carbides, the thorium carbides react rapidly with water or moist air and must be protected from moisture in storage and fuel fabrication.

ThC_2 and $(Th,U)C_2$ particles are made by granulating oxides and graphite flour in the proper proportions for the reaction

$$ThO_2 + 4C \rightarrow ThC_2 + 2CO$$

reacting the granules at high temperature, and then melting to consolidate and spheroidize the particles.

5.4 Thorium Nitrides

Thorium forms two nitrides, ThN and Th_3N_4. Th_3N_4 loses nitrogen at temperatures above 1500°C. At low pressures and temperatures above 2200 K, ThN also dissociates, into thorium and nitrogen. The N_2 pressure over solid ThN in equilibrium with liquid Th-N alloy is [I1]

$$\log_{10} p(\text{atm}) = 8.086 - \frac{33{,}224}{T} + 0.958 \times 10^{-17} T^5 \qquad 2689 < T < 3063 \text{ K} \qquad (6.5)$$

At a nitrogen partial pressure around 2 atm, ThN melts at 2820°C without dissociation [B4]. The density of ThN is 11.9 g/cm^3.

Th_3N_4 is made by reacting thorium hydride with nitrogen at temperatures increasing from 200 to 900°C. ThN is made by pressing Th_3N_4 in vacuo at 1500°C. The nitrides react rapidly with water or moist air.

5.5 Thorium Hydrides [M2]

Thorium forms two hydrides: ThH_2, density 9.50 g/cm^3; and Th_4H_{15}, density 8.28 g/cm^3. The limited information on temperature-composition relations for condensed phases in the system thorium-hydrogen is shown in Fig. 6.3. Approximate values for the equilibrium pressure of hydrogen in the system αTh-ThH_{2-x} obtained from measurements of Mallett and Campbell [M1] made with impure metal are given in Table 6.9.

Like other metal hydrides, thorium hydride is pyrophoric and must be handled with care.

5.6 Thorium Halides

The tetrahalides are the thorium halides of greatest practical importance. The tetrafluoride ThF_4 is the preferred starting material for large-scale production of thorium metal (Sec. 10.4). ThF_4 has been proposed as fertile material in the fuel mixture of the molten-salt reactor. The tetraiodide has been used as feed material in the iodide process for making very pure thorium metal (Sec. 10.4).

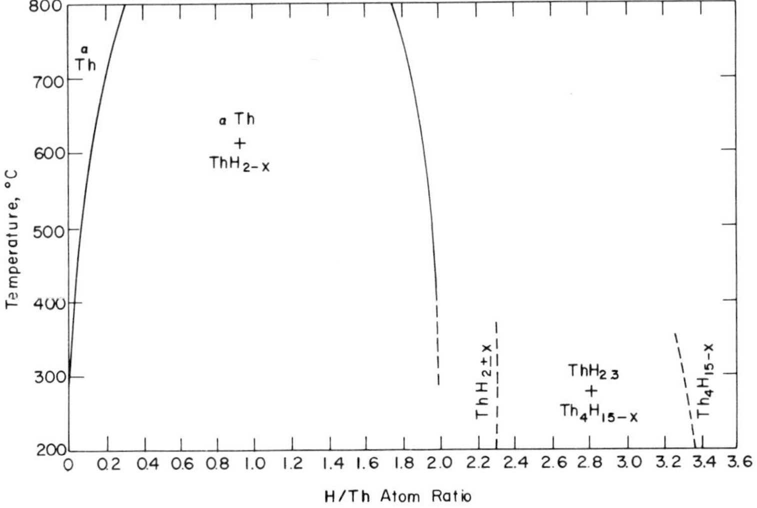

Figure 6.3 Thorium-hydrogen phase diagram [M2]. *(Reprinted with permission from the copyright holder, Academic Press, Inc., New York, and Dr. G. G. Libowitz.)*

Table 6.9 Hydrogen pressure in system thorium-thorium hydride, Th/H = 1.6

Temperature, °C	651	676	701	726	751	776	801	826	851	876
Hydrogen pressure, Torr	20	35	55	90	130	195	280	400	580	800

The more important properties of the tetrahalides, from reference [I1], are listed in Table 6.10. Many of these properties, especially for $ThCl_4$, $ThBr_4$, and ThI_4, are known only semiquantitatively.

Anhydrous ThF_4 is made by passing an excess of HF vapor over ThO_2 or $ThOF_2$ at temperatures between 550 and 600°C. The anhydrous double fluoride $KThF_5$ is precipitated from aqueous solutions of thorium nitrate by addition of an excess of KF. It has been used for electrolytic production of thorium metal.

Table 6.10 Thorium Tetrahalides

Property	ThF_4	$ThCl_4$	$ThBr_4$	ThI_4
Color	White	White	White	Yellow
Density at 25°C, g/cm³	6.12	4.62	5.72	6.00
Crystal system				
Low-temperature form	Monoclinic	Tetragonal	Orthorhombic	Monoclinic
High-temperature form	–	Orthorhombic	Tetragonal	Orthorhombic
Transition temperature, °C	–	(406)	~420	?
Melting temperature, °C	1110	770	679	570
Normal boiling point, °C	1782	942	905	853
Vapor pressure equation, $\log_{10} p(\text{atm}) = A - B/T$				
Solid, A	9.345	9.426	9.498	9.747
B, K	17,089	10,630	10,151	9,894
Liquid, A	6.395	5.229	5.260	5.714
B, K	13,080	6,346	6,187	6,425
Heat capacity equation, $C_p = A + 10^{-3} BT - 10^5 C/T^2$				
Solid, A, cal/(g-mol·K)	26.75	28.75	30.5	31.0
B, cal/(g-mol·K²)	5.854	5.561	3.6	3.1
C, (cal·K)/g-mol	1.805	1.470	1.47	1.47
Liquid, cal/(g-mol·K)	36.5	40.0	41.0	42.0
Heat of transition, cal/g-mol	–	1,200	1,000	?
Heat of fusion, cal/g-mol	10,510	14,700	(13,000)	11,500
Heat of vaporization at normal boiling point, cal/g-mol	55,700	?	26,500	27,500
Heat of formation at 25°C, cal/g-mol	–504,600	–283,600	–230,800	–158,800
Free energy of formation, $\Delta G = A + BT$				
Solid, A, cal/g-mol	–502,140	–281,100	–243,450	–187,040
Solid, B, cal/(g-mol·K)	70.11	67.70	70.09	69.46
Liquid, A, cal/g-mol	–485,080	–261,120	–225,190	–171,000
B, cal/(g-mol·K)	57.38	48.09	50.55	50.25

$ThCl_4$ can be made by reacting ThO_2 with chlorine mixed with CCl_4 or $COCl_2$.

All tetrahalides react with water to form oxyhalides:

$$ThX_4 + H_2O \rightarrow ThOX_2 + 2HX$$

For this reason, ThF_4, precipitated from aqueous solution, cannot be dried without contamination by oxygen. When $ThCl_4$ is dissolved in water, soluble $ThOCl_2$ is formed and crystallizes out on evaporation. The oxyhalides are stable against disproportionation into oxide and tetrahalide at pressures near atmospheric and temperatures under 2000 K, as may be seen from the positive free-energy change ΔG_{disp} in the reaction

$$2ThOX_2 \rightarrow ThO_2 + ThX_4$$

The free-energy change ΔG_{disp} may be calculated from the enthalpy change ΔH_{disp} and entropy change ΔS_{disp} for the disproportionation reaction given in Table 6.11 by Eq. (6.6):

$$\Delta G_{disp} = \Delta H_{disp} - T\,\Delta S_{disp} \tag{6.6}$$

Thorium di- and triiodides have been prepared by Scaife and Wylie [S1] and are of some practical significance in the iodide process for making thorium metal (Sec. 10.4). Other lower halides have only limited stability and are not well known.

6 THORIUM SOLUTION CHEMISTRY

6.1 Solubility of Thorium Compounds in Aqueous Solution

Thorium forms relatively stable tetravalent salts of many of the oxyacids. These can be prepared by reacting thorium hydroxide or basic carbonate with the appropriate acid.

Thorium nitrate is very soluble in water, to the extent of 65.6 g $Th(NO_3)_4$/100 g solution at 20°C. It can be crystallized from solution as the nominal tetrahydrate. Thorium nitrate solutions are used for purifying thorium by solvent extraction, Sec. 9.

Although anhydrous thorium sulfate dissolves in water at 0°C to the extent of 20 w/o, the solution is metastable and deposits hydrates on standing. Stable solutions at higher temperature require the presence of free sulfuric acid, as in solutions used to leach thorium minerals, Sec. 8.5.

A solution of $ThOCl_2$ is produced when $ThCl_4$ reacts with water. Evaporation to dryness produces a succession of ill-defined hydrates that can be converted to anhydrous $ThOCl_2$ by heating to 250°C.

Hydrated thorium fluoride is precipitated when a soluble fluoride is added to a solution of thorium nitrate. Precipitation can be prevented by addition of aluminum nitrate to complex the fluoride ion, an expedient used in the Thorex process (Chap. 10, Sec. 5).

Table 6.11 Enthalpy and entropy changes in disproportionation of thorium oxyhalides: $2ThOX_2 \rightarrow ThO_2 + ThX_4$

Property	F	Cl	Br	I
ΔC_p, assumed	0	0	0	0
ΔH_{disp}, cal/g-mol	4,200	12,800	16,000	22,400
ΔS_{disp}, cal/(g-mol·K)	1.14	5.09	7.49	9.89

Source: International Atomic Energy Agency, "Thorium: Physico-chemical Properties of Its Compounds and Alloys," *Atomic Energy Rev.*, Special Issue No. 5, 1975.

Thorium hydroxide $Th(OH)_4$ is precipitated from solutions of thorium salts by addition of alkali hydroxides.

Thirty percent hydrogen peroxide precipitates thorium peroxide Th_2O_7 from solutions of thorium salts. As few cations other than thorium and uranium precipitate under these conditions, this method has been used to purify thorium.

Thorium orthophosphate $Th_3(PO_4)_4$ is precipitated by phosphate ion from neutral or slightly acid solutions of thorium nitrate or sulfate. It is soluble in concentrated phosphoric or sulfuric acid, such as is present when monazite is dissolved in sulfuric acid.

Addition of an alkali carbonate to an aqueous solution of a thorium salt first precipitates a basic thorium carbonate of variable composition. Like uranyl carbonate, thorium carbonate dissolves in an excess of alkali carbonate, in this case forming the complex ion $[Th(CO_3)_4(OH)_2]^{6-}$.

Thorium oxalate is precipitated when a solution of oxalic acid is added to a solution of a thorium salt. This is a commonly used intermediate step in producing thorium dioxide from thorium nitrate solution (Sec. 10.1). Quantitative precipitation of thorium from solutions up to 1.8 N in nitric acid can be obtained by use of five times the stoichiometric amount of oxalic acid [A2]. Thorium oxalate can be dissolved by concentrated nitric acid or by sodium oxalate solution, with which it forms a double oxalate [B7].

6.2 Solvent Extraction of Thorium Compounds

Thorium compounds can be extracted from aqueous solution by many of the immiscible organic solvents that have been used for extraction of uranium (Table 5.14). As with uranium, tributyl phosphate (TBP) is now the universal choice for extracting thorium from aqueous nitric acid solutions and for purifying thorium compounds by solvent extraction. However, in sulfuric acid solutions of thorium compounds or in solutions containing phosphoric acid such as are obtained from acid leaching of monazite, thorium is too strongly complexed to be readily extracted by TBP. To extract thorium from such solutions, processes using other organophosphorus compounds or organic amines have been developed, just as they have for uranium. Audsley and co-workers [A1] conducted pilot-plant experiments on extraction of thorium from solutions simulating the composition of sulfuric acid leach liquors from Canadian uranium ores after removal of uranium. They found that di(2-ethylhexyl)phosphoric acid, the solvent used in the Dapex process (Chap. 5, Sec. 8.6), was a satisfactory extractant, provided that ferric iron in the leach liquor (which would extract with thorium) was first reduced to ferrous by reaction with iron filings.

Because of the complications introduced by ferric iron, Audsley et al. concluded that a solvent that would be selective for thorium in the presence of ferric iron and that would not be inhibited by phosphate would be preferable to the Dapex solvent. They concurred with the conclusion of Crouse and co-workers at Oak Ridge National Laboratory [C5], that long-chain primary amines selectively extract thorium in the presence of uranyl, ferric, and phosphate ions. Compounds of this type are now the preferred extractant for thorium in such systems. Application of this so-called Amex process to thorium extraction from monazite is described in Sec. 8.6.

7 THORIUM RESOURCES

7.1 Principal Thorium-containing Minerals

Heretofore, most of the world's thorium has come from monazite in beach sands where coproduction of rare earths, titanium, and zirconium has defrayed much of the cost of extracting thorium. Recent increased interest in thorium as an alternative feed material for nuclear power systems has led to more extensive search for thorium deposits and to interest in other thorium minerals that could be produced if the demand for thorium (and its price) increased. Table 6.12 lists the principal thorium-containing minerals and gives their nominal composition and examples of where they have been found.

7.2 World Thorium Resources

Table 6.13 gives the thorium resources of the non-Communist world as estimated by the Organization for Economic Cooperation and Development [O1] in December 1977. The definitions of the two resource categories are the same as for Table 5.17. The production cost of these thorium resources was not stated in [O1], but was probably \$15/lb ThO_2, from similar statistics cited by Nininger and Bowie [N3].

Since 1975, renewed interest in thorium as source material for production of ^{233}U has led to extensive prospecting for thorium, discovery of numerous potentially commercial deposits and substantial increase in U.S. resource estimates over those listed in Table 6.13.

7.3 U.S. Thorium Resources

Until recently, most U.S. thorium production has been as a by-product of monazite processing from placer deposits in Florida, Georgia, and South Carolina. Some bastnaesite has been mined at Mountain Pass, San Bernardino, California.

The U.S. Geological Survey was scheduled to publish a revised study of U.S. thorium resources in August 1979. Partial results of this study, which cover most of these resources but do not include the beach placers of Florida, Georgia, and the Carolinas, were presented orally by Staatz [S5] of the U.S. Geological Survey in 1978. Table 6.14 lists the types of deposit, the principal districts in which potentially economic thorium-bearing deposits have been found, the principal thorium minerals, and estimates of thorium reserves and resources. Thorium from the vein deposits, the first type, could be produced for less than \$30/lb. Thorium is the principal salable product in these deposits. Thorium could be coproduced with other elements from disseminated deposits, massive carbonatites, and placers; the amount of thorium that might be produced from them, and its cost, depends on the marketability of the other minerals that occur with the thorium.

Table 6.12 Principal thorium-containing minerals

Mineral	Nominal composition	Examples of where found
Monazite	$(La,Ce,Th)PO_4$	Brazil, India, Sri Lanka, Australia, South Africa, United States
Brockite	$(Ca,Ce,Th)[(PO_4)\cdot(CO_3)]\cdot H_2O$	United States
Thorianite	ThO_2	Sri Lanka, Canada
Uranothorianite	$(U,Th)O_2$	Malagasay Republic
Thorogummite	$ThO_2\cdot UO_3$	Brazil
Thorite	$ThSiO_4$	Idaho and Montana
Aueralite	$ThSiO_4\cdot YPO_4$	Idaho and Montana
Uranothorite	$(U,Th)SiO_4$	Blind River, Ontario
Brannerite	$(U,Th,Ca_2,Fe_2)Ti_2O_6$	Blind River, Ontario
Bastnaesite	$(La,Ce,Th)FCO_3$	California
Pyrochlore	$(Na_4,Ca_2,U,Th)(Nb,Ta)_4O_{12}$	Colorado
Allanite	$(Ca,Ce,Th)_2(Al,Fe,Mn,Mg)_3(SiO_4)_3OH$	Idaho and Montana

Table 6.13 Thorium resources of non-Communist world

	Thousand metric tons† thorium		
	Reasonably assured resources	Estimated additional resources	Total
Australia	18.5	0	18.5
Brazil	58.2	3	61.2
Canada	0	250	250
Denmark	15	0	15
Egypt	14.7	280	294.7
India	320	0	320
Iran	0	30	30
Liberia	0.5	0	0.5
South Africa	11	0	11
Turkey	0.5	0	0.5
United States	52	270	322
Total	490.4	833	1323.4

†One metric ton = 1 tonne = 1 Mg.

Source: Organization for Economic Cooperation and Development, and International Atomic Energy Agency, "Uranium Resources, Production and Demand," Paris, Dec. 1977.

7.4 Thorium Production

Table 6.15 summarizes two sources of information on the annual rate of thorium production, by country. The first three columns give the production rate of monazite concentrates for the more recent years of 1976, 1977, and 1978 [E1]. We have estimated total thorium production from a typical monazite thorium content of 6 weight percent (w/o). These columns do not include monazite production in the United States or Soviet Union, nor the small production of other thorium minerals. The last two columns give the U.S. Bureau of Mines figures [U1] for total thorium production in 1973 and an estimate of total thorium production capacity in 1980, if demand were such as to support it.

U.S. consumption has averaged about 70 MT thorium/year for all uses, nonnuclear included.

7.5 Thorium Requirements

Thorium makeup requirements for one reactor system, the HTGR (high-temperature gas-cooled reactor), may be estimated from Fig. 3.33. A 1000-MWe HTGR requires 7.4 MT of thorium as feed per year. Reprocessing recovers 6.8 MT, which can be recycled after storage for 20 to 30 years to permit excess ^{228}Th to decay. The net thorium consumption of a 1000-MWe reactor then is 0.6 MT/year. Thus, the 441,000 MT of U.S. ThO_2 thorium reserves listed in Table 6.14 would provide thorium fuel for

$$\frac{441{,}000 \text{ MT}}{0.6 \text{ MT}/1000 \text{ MW-yr}} = 7.35 \times 10^8 \text{ MW-year} \tag{6.7}$$

of HTGR operation. The point to be made is that U.S. thorium resources are more than ample for any likely use of thorium to supplement uranium as a nuclear fuel.

It should be noted, however, that Fig. 3.33 indicates that more than 80 MT of natural uranium are consumed in an HTGR for every 0.6 MT of net thorium consumption. But in

Table 6.14 Principal U.S. thorium resources, metric tons thorium[†]

Type of deposit	District	Principal thorium minerals	Other elements	Reserves	Probable potential resources
Vein	Lemhi Pass, Idaho-Montana	Thorite	RE[‡]	54,000	117,100
	Wet Mountains, Colorado	Monazite	↓	53,200	126,000
	Powderhorn, Colorado	Allanite, brockite	↓	1,500	6,800
	Hall Mountain, Idaho	↓	↓	4,150	23,800
	Diamond Creek, Idaho	↓	↓	200	12,700
	Bear Lodge Mountains, Wyoming	↓	↓	44	220
	Mountain Pass, California	↓	↓	223	925
			Subtotal	113,317	287,545
Disseminated deposits	Bear Lodge Mountains, Wyoming	Monazite, thorite, brockite	RE	269,000	1,020,000
	Powderhorn, Colorado	Monazite, brockite	RE	4,000	7,200
	Other	Thorite, brockite	Nb,Ti,RE,Be,Ba	17,500	640,000
Massive carbonatites	Powderhorn, Colorado	Pyrochlore	Nb,RE,U	24,800	91,600
	Mountain Pass, California	Bastnaesite	RE,Ba	7,770	7,770
Placers	Carolinas	Monazite	RE,U	4,200	29,300
			Rounded total	441,000	2,083,000

[†] Converted from [S5] by factor 0.79722 MT (Mg) thorium/short ton ThO_2.
[‡] RE, rare earths.

Table 6.15 Thorium production rate

	Monazite concentrates,† MT/yr,‡ for			Thorium,† MT/yr,‡ for	
	1976	1977	1978 (est.)	1973	1980 capacity
United States	NA§	NA	NA	¶	544
Canada	0	0	0	0	64
Brazil	1,628	1,814	1,814	82	179
Soviet Union	NA	NA	NA	¶	138
Zaire	¶	¶	¶	12	28
Other African	¶	¶	¶	6	28
India	3,027	3,027	3,118	266	367
Malaysia	1,899	2,018	2,201	138	183
Thailand	¶	¶	¶	14	46
Australia	4,601	8,847	8,621	275	459
Other	262	120	92	138	–
Total	11,417	15,846	15,867	930	2,041
Tons thorium at 6%	657	951	952		

†Monazite concentrates from [E1]; thorium from [U1].

‡Metric tons converted from short tons in original sources by factor 0.91718 MT/short ton. One metric ton = 1 tonne = 1 Mg.

§NA, not available.

¶Included in "other."

other reactor systems not treated in this text, such as the Canadian thorium-heavy water reactor, or thorium-fueled liquid-metal fast-breeder reactors, natural uranium consumption would be much lower and could be reduced to zero when and if self-sustaining breeding is reached.

8 CONCENTRATION AND EXTRACTION OF THORIUM

The principal steps in producing refined thorium compounds from thorium-bearing ores are concentration of thorium minerals, extraction of thorium from concentrates, purification or refining of thorium, and conversion to metal or the thorium compound finally wanted. This section describes the concentration of monazite, the principal source of thorium in the past; the extraction of thorium from monazite; and the recovery of thorium from leach liquors by solvent extraction. Purification of thorium is described in Sec. 9 and conversion in Sec. 10.

8.1 Concentration of Monazite

Monazite is usually a minor constituent of deposits of other minerals, all of which must be separated and processed for a profitable venture. As an example, the mineral constituents of beach sands in Travancore, India, which are dredged for their zirconium, titanium, thorium, and rare-earth content, are as follows:

Monazite, $(RE,Th,U)PO_4$, 0.5 to 1.0%	Rutile, TiO_2, 3 to 6%
Ilmenite, $FeTiO_3$, 65 to 80%	Zircon, $ZrSiO_4$, 4 to 6%
Garnet, $(Fe,Mg,Ca)_3Al_2SiO_4$, 1 to 5%	Sillimanite, Al_2SiO_5, 2 to 5%

Beach sands are usually first treated by crude specific gravity methods to separate the denser minerals from silica and produce concentrate that, at Travancore, has approximately the above composition. Then the heavier minerals are separated in a series of strong electromagnets of progressively increasing intensity. The low-intensity magnet removes the most magnetic constituent, magnetite. The first pole of the high-intensity magnet separates ilmenite; the second pole, garnet; the third pole, coarse grains of monazite; and the strongest pole, fine grains of monazite. The rare earth elements in monazite make it paramagnetic. The nonmagnetic residue is treated by flotation and other means to recover rutile, zircon, and some gold. A further specific gravity treatment of the monazite fraction produces a monazite concentrate 98% percent pure.

8.2 Composition of Monazite

Although the principal constituents of monazite are rare earth and thorium phosphates, its composition varies widely within a given deposit and from place to place. Table 6.16 gives the composition of monazite concentrates from different locations.

8.3 Processes for Opening Up Monazite

Monazite is chemically very inert. Two general methods that are used for opening up monazite and making its constituents sufficiently reactive to permit extraction and separation of thorium, uranium, and rare earths are

1. Reaction with hot, concentrated caustic soda solution
2. Dissolution in hot, concentrated sulfuric acid

The caustic soda process has been used on a large scale in Brazil [B6] and India [D1] and has been investigated on a pilot-plant scale at Battelle Memorial Institute [B2, C1] in the United States. A brief description is given in Sec. 8.4.

Sulfuric acid has been used to dissolve monazite in Europe, Australia, and the United States. The numerous processes used to separate thorium from the acid leach liquors are listed

Table 6.16 Composition of monazite concentrates

	Weight percent				
Constituent	India	Brazil	Florida beach sand†	South Africa Monazite Rock	Malagasay Republic
ThO_2	8.88	6.5	3.1	5.9	8.75
U_3O_8	0.35	0.17	0.47	0.12	0.41
$(RE)_2O_3$ ‡	59.37	59.2	40.7	46.41	46.2
Ce_2O_3	(28.46)	(26.8)	–	(24.9)	(23.2)
P_2O_5	27.03	26.0	19.3	27.0	20.0
Fe_2O_3	0.32	0.51	4.47	4.5	–
TiO_2	0.36	1.75	–	0.42	2.2
SiO_2	1.00	2.2	8.3	3.3	6.7

†Florida beach sand contains about 70% monazite.
‡Rare-earth oxides, including Ce_2O_3.
Source: R. K. Garg et al., "Status of Thorium Technology," in *Nuclear Power and Its Fuel Cycle*, vol. 2, International Atomic Energy Agency, Vienna, 1977, p. 457.

in Sec. 8.5, with a more detailed description of one process, developed at Iowa State College [B1].

Solvent extraction processes for recovering thorium from monazite sulfuric acid leach liquor are described in Sec. 8.6.

8.4 Caustic Soda Process

A flow sheet of the caustic soda process described by Bearse et al. [B2] is shown in Fig. 6.4. The sand is ground with water in a ball mill until 96.5 percent passes 325 mesh. A wet classifier recycles coarse particles and delivers a slurry of fine particles to a stainless steel reactor. A liquid caustic solution containing 73% NaOH is fed to the reactor. At the beginning of the reaction the slurry contains 1.5 kg of NaOH and 1.7 kg of water per kilogram of sand. The mixture is heated to 140°C, and after 3 h at this temperature the sand is completely reacted. The mixture is then diluted with the wash solution of caustic and trisodium phosphate from a later step and digested at 105°C for 1 h to facilitate later filtration. The resulting hot slurry contains practically all the original phosphorus in solution as trisodium phosphate, and thorium, cerium, and rare earths are present as solid hydrous metal oxides. The trisodium phosphate and unreacted caustic are removed by filtering the slurry through Monel wire cloth, and the metal oxide cake is washed with water. The filtration is carried out at 80°C to keep caustic and trisodium phosphate in solution. The filtrate, which contains about two-thirds of the original caustic soda charged, is evaporated in an open steel kettle until the NaOH

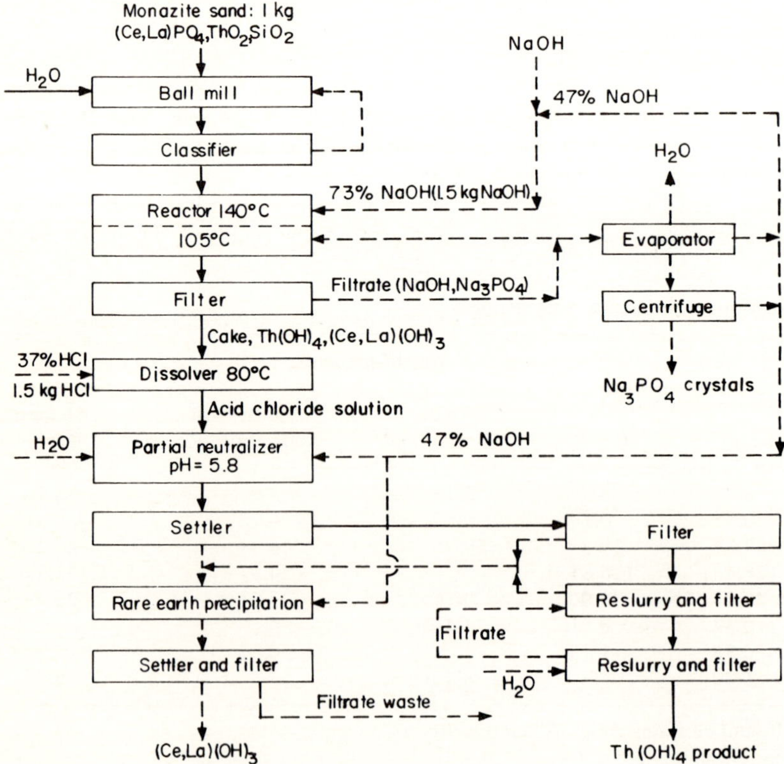

Figure 6.4 Caustic soda process for the recovery of thorium from monazite.

Table 6.17 Recoveries and compositions in thorium concentrate

Constituent	Recovery, %	Composition, w/o
Thorium	99.7	36.4
Rare earths	2.3	7.45
Uranium	96.2	0.74
Iron	–	2.21
Titanium	–	6.73
Silicon	–	4.47
Phosphorus	0.3	0.44
Chlorine	–	0.36
Acid insolubles	100	23

concentration is 47%, corresponding to a boiling point of 137°C. The concentrated solution is cooled to room temperature. More than 95 percent of the sodium phosphate crystallizes out of solution and is removed by filtration. The caustic soda liquor is recycled for sand digestion and for later neutralization steps.

The hydrous oxide cake is brought into solution by dissolving in 37% hydrochloric acid (1.5 kg acid/kg sand) at 80°C for 1 h in a glass-lined vessel. Hydrochloric rather than sulfuric acid is recommended because of more selective precipitation from chloride solutions in the later step. About 2 percent of the weight of the original sand is left as residue, which contains undissolved monazite and rutile (TiO_2), an impurity in the sand.

The acid solution and undissolved material are transferred to a neutralizer vessel and diluted with water.

Thorium is separated from the rare earths by selective precipitation of thorium hydroxide at a pH of 5.8. This is effected by neutralizing the diluted chloride solution with caustic recovered from the evaporator. The wet cake is reslurried in water solution, filtered, and again reslurried and filtered to effect a high degree of separation of the thorium precipitate from any occluded rare earth solution.

The percentage of thorium and other constituents of the monazite sand recovered in the precipitate is given in the first column of Table 6.17. The composition of the precipitate is given in the second column.

Rare earths are recovered from the combined decantates and filtrates by further neutralization with NaOH. The hydroxide precipitate is removed by filtration.

The process used in Brazil [B6] is generally similar.

8.5 Sulfuric Acid Processes

Dissolution of monazite. The first step in all of the sulfuric acid processes is dissolution of monazite. The procedure recommended by workers at Iowa State [B1] is as follows. Monazite ground to minus 65 mesh is digested with 93% sulfuric acid for 4 h at 210°C in a stirred reactor. The mass ratio of acid to sand, based on 100% H_2SO_4, is 1.56. The temperature must be kept below 230°C to prevent formation of water-insoluble ThP_2O_7. The monazite is converted into a thick paste soluble in cold water. The reaction mass is cooled to 70°C and diluted with about 10 kg cold water/kg monazite. Most of the thorium, rare earths, and uranium go into solution, leaving a sludge of silica, rutile, zircon, and some unreacted monazite. Most of the solution is decanted from the silica sludge and unreacted monazite. The denser monazite is separated from the sludge and recycled. The sludge is filtered and washed to recover additional solution.

Radium in the monazite may be removed with the sludge by adding barium carbonate before decantation. This forms barium sulfate, which removes radium as insoluble radium sulfate.

This process produces a solution of thorium, rare earths, and uranium cations with sulfate and phosphate anions.

Thorium recovery processes. Because of the many elements in the solution, their chemical similarity, and the presence of phosphoric acid, separation of thorium from this acid solution has proved to be difficult. Wylie [W5] has reviewed the numerous separation processes that have been developed. Figure 6.5 shows the principal steps in seven of these processes and gives references for more details. Processes 4 and 6 appear to be the most economic when thorium, rare earths, and uranium all are to be recovered. Process 4, involving separation of thorium and rare earths from phosphate and uranium by precipitation with oxalic acid, is described next. Process 6, involving separation by solvent extraction with organic amines, is described in Sec. 8.6.

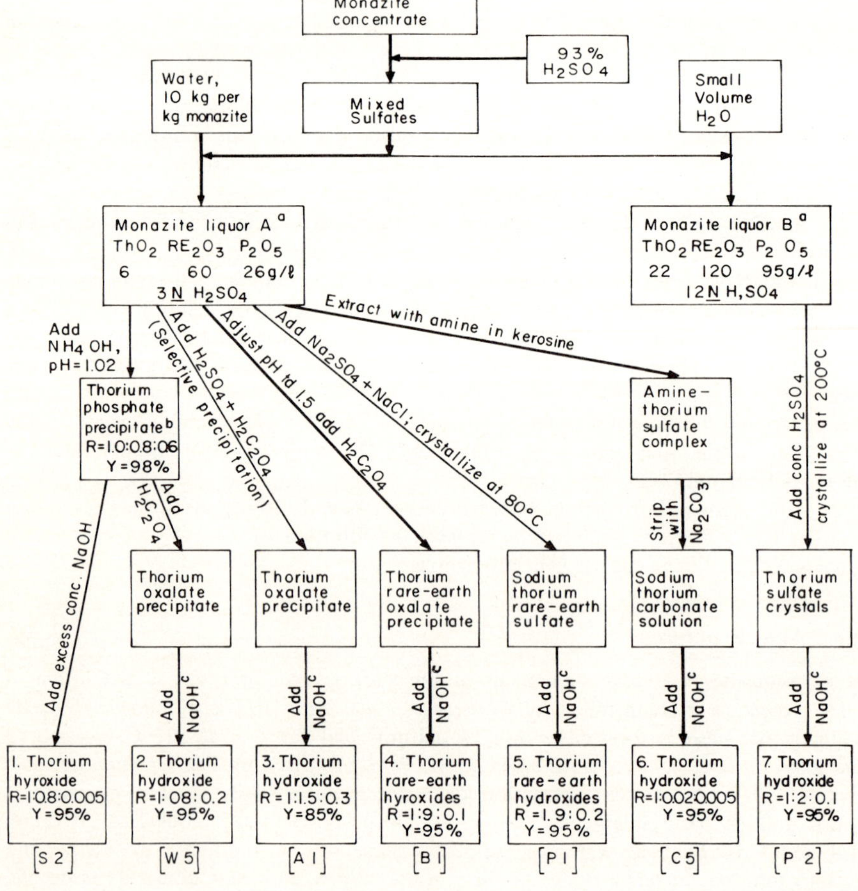

Figure 6.5 Principal processes for extracting thorium from monazite acid leach liquor. R = mass ratio $ThO_2:RE_2O_3:P_2O_5$; Y = approximate overall ThO_2 yield in concentrate. a, filtered; b, washed; c, 10 percent excess.

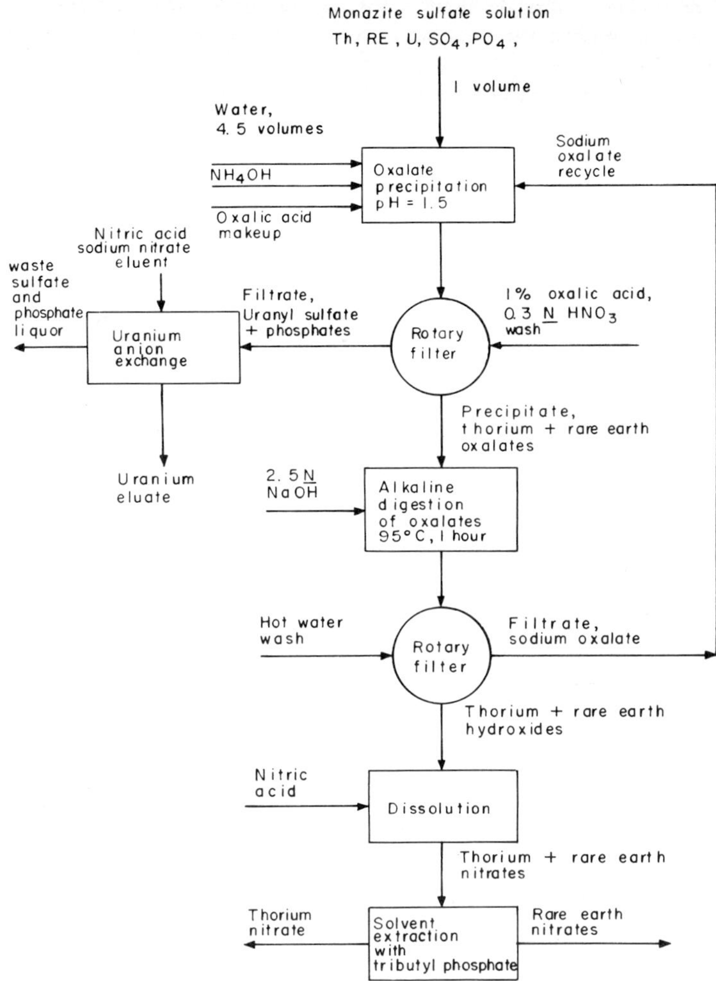

Figure 6.6 Iowa process for separating thorium, rare earths, and uranium from monazite sulfate solutions.

Precipitation with oxalic acid. Figure 6.6 shows the principal steps in the process for separating the sulfuric acid solution of monazite into a thorium concentrate, a rare earth concentrate, and a uranium concentrate developed at the Ames, Iowa, Laboratory of the U.S. Atomic Energy Commission [B1].

The solution of monazite in sulfuric acid containing about 50 to 60 g of thorium and rare earths per liter is diluted with about 4.5 volumes of water and brought to a pH of 1.5 by addition of NH_4OH. Oxalate ion is added in the form of recycle sodium oxalate, plus sufficient oxalic acid in 10% aqueous solution to provide 110% of the oxalate ion needed to precipitate thorium and rare earth oxalates. The precipitate is filtered and washed with 1 % oxalic acid in 0.3 N nitric acid. A clean separation of uranium from rare earths plus thorium is claimed.

Because of the comparatively high cost of oxalic acid, economics requires recovery of oxalate ion. This is effected by digesting the thorium and rare-earth oxalates with a stoichiometric equivalent of sodium hydroxide at 95°C for 1 h, to convert the precipitate to hydroxide, which is filtered and washed with hot water. Oxalate ion is recovered as sodium oxalate, of which 95% is recycled.

Uranium is recovered from the sulfate and phosphate filtrate by anion exchange (Chap. 5).

Thorium and rare earths in the hydroxide precipitate are dissolved in nitric acid and separated by solvent extraction with TBP (Sec. 8.7).

8.6 Separation of Thorium, Rare Earths, and Uranium from Monazite by Solvent Extraction

Attempts to separate thorium and uranium from sulfuric acid solution of monazite by solvent extraction with TBP were unsuccessful because distribution coefficients of uranium and thorium from monazite solutions were too low, as these elements are complexed by phosphate ion. Development of extractants with higher distribution coefficients for these metals has made solvent extraction a practical process for recovering uranium and thorium from monazite sulfate solutions and from sulfuric acid solutions of other thorium ores. This section describes processes tested on a pilot-plant scale by Oak Ridge National Laboratory [C5].

Table 6.18 summarizes distribution coefficients of hexavalent uranium, thorium, and trivalent cerium (representative of rare earths) for four different types of long-chain amines, in sulfate solution with phosphate ion absent. Primary amines have the highest coefficient for thorium and the lowest for uranium, with the converse true of tertiary amines such as were cited for uranium extraction in Chap. 5. Secondary amines extract both metals, with thorium extraction favored by branching distant from the nitrogen. Either primary or secondary amines provide good separation of thorium from cerium.

Table 6.18 Distribution coefficients for uranium, thorium, and cerium between organic amines and aqueous sulfate solution[a]

		Distribution coefficient		
Amine type	Examples of amines	U(VI)	Th	Ce(III)
Branched primary	Primary JM[b] and 1-(3-ethylpentyl)-4-ethyloctylamine	5–30	>20,000	10–20
Secondary with alkyl branching distant from the nitrogen	Di(tridecyl)amine[c]	80	>500	<0.1
Secondary with alkyl branching on the first C	Amberlite LA-1[d] and bis(1-isobutyl-3,5-dimethylhexyl)amine	80–120	5–15	<0.05
Tertiary with no branching or branching no closer than the third C	Alamine 336[e,f] and triisooctyl-amine[f,g]	140	<0.03	<0.01

[a] 1 M SO_4; pH = 1; ~1 g metal/liter; 0.1 M amine in kerosine; 1:1 phase ratio.
[b] Trialkylmethylamine, homologous mixture, 18–24 carbons.
[c] Mixed C_{13} alkyls from tetrapropylene by oxo process.
[d] Dodecenyl-trialkylmethylamine, homologous mixture, 24–27 carbons.
[e] Trialkylamine with mixed n-octyl and n-decyl radicals.
[f] Kerosine diluent modified with 3 v/o (volume percent) tridecanol.
[g] Mixed C_8 alkyls from oxo process.

Source: D. J. Crouse and K. B. Brown, "Recovery of Thorium, Uranium, and Rare Earths from Monazite Sulfate Leach Liquors by the Amine Extraction (Amex) Process," Report ORNL-2720, July 16, 1959.

Table 6.19 Distribution coefficients for uranium between organic amines and aqueous monazite sulfate solution after extraction of thorium†

Amine	Phase ratio, aqueous/organic	Uranium distribution coefficient
N-benzyl-1-(3-ethylpentyl)-4-ethyloctyl	3	50
N-(1-nonyldecyl)benzyl	3	25
N-(1-undecyldodecyl)benzyl	3	25
Amberlite LA-1	0.5	3
	3	2
Triisooctyl	0.5	4
	3	2
Alamine 336‡	0.5	5
Primene JM‡	0.5	2

†0.18 g uranium/liter; pH = 0.1; 0.05 *M* amine sulfate in kerosine.
‡In 97% kerosine, 3% tridecanol.

Because of the high acidity and high sulfate and phosphate content of sulfuric acid monazite leach solutions, distribution coefficients with primary and secondary amines are lower than in Table 6.18. In monazite sulfate solutions, thorium distribution coefficients with the primary amines of Table 6.18 are still greater than 500, however. The coefficient with di(tridecyl)amine is 4.6. These are still high enough for practical processes [C5].

Table 6.19 lists distribution coefficients for amines considered [C5] for extracting uranium from monazite sulfate solutions after removal of thorium. Except for Primene JM, all coefficients were judged [C5] to be large enough and sufficiently greater than those of the rare earths to provide efficient solvent extraction separation of uranium.

Crouse and Brown [C5] give a number of alternative flow sheets for separating quite pure thorium, uranium, and rare earth products from monazite sulfate solution by solvent extraction, using several alternative organic amines and with different orders of separation. They named this type of process using amine extractants the Amex process. Figure 6.7 is a composite of several of their flow sheets showing one possible arrangement for separating these three components by solvent extraction.

Monazite sulfate solution containing 5.9 g thorium/liter, 0.2 g uranium/liter, and 34 g rare-earth oxides/liter and about 3 *N* in sulfuric acid is extracted with a 0.1 *M* solution of the primary amine Primene JM at an organic/aqueous flow ratio of 80:59. The solvent is 97% kerosine, 3% tridecanol. The flow ratio and solvent composition are so chosen that the solvent leaving the extracting section is effectively saturated with thorium (3 g/liter), to minimize extraction of uranium and rare earths. To ensure high thorium loading of solvent, the solvent-to-feed ratio is set below that which would extract all thorium in the feed. To prevent thorium loss, 25% of the aqueous stream is withdrawn from the second stage of the extracting contactor and recycled to feed. This permits reduction of thorium content of the raffinate leaving the fourth stage to <0.01 g/liter.

Rich solvent leaving the extracting section is scrubbed with 0.2 *M* H_2SO_4 in four scrubbing stages to remove traces of uranium and rare earths. Finally, the thorium is stripped into the aqueous phase by 0.75 *M* Na_2CO_3.

A similar sequence of operations in the uranium separation section separates uranium from rare earths. The difference here is use of triisoocytylamine as solvent because of its high selectivity for uranium.

Finally, rare earths are extracted from the raffinate leaving the uranium separation section

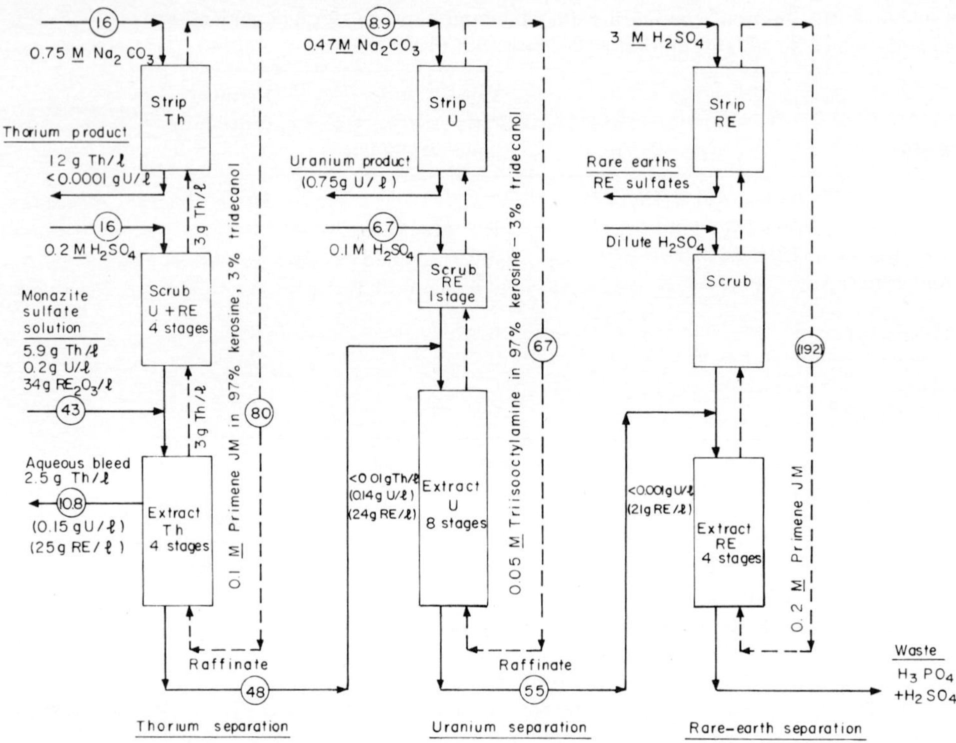

Figure 6.7 Separation of thorium, uranium, and rare earths from monazite by solvent extraction in Amex process. Circles, relative flow; ( ), estimated. Other data from Oak Ridge Laboratory runs [C5].

by another solvent extraction with Primene JM. Because the rare-earth distribution coefficient is lower than thorium's and the rare-earth concentration higher, the Primene concentration and the organic-to-aqueous flow ratio are higher than in the thorium extraction section. Rare earths may be stripped from this solvent by 3 *M* sulfuric acid. Evaporation of the sulfuric acid strip solution to 8 *M* H_2SO_4 precipitates the rare earths as sulfates and, after filtration, provides acid for recycle (not shown).

An alternative method for separating rare earths from the raffinate from uranium extraction is salting out the sodium double sulfate with sodium chloride or sodium sulfate [C4].

Although there has been no reported use of solvent extraction for commercial processing of monazite sulfate solutions, it seems likely that this efficient method would be used if the demand for thorium increased sufficiently to require construction of new extraction plants.

8.7 Separation of Thorium from Other Minerals by Solvent Extraction

Solvent extraction has been used commercially for recovery of thorium from minerals other than monazite, in which complexing by phosphate is not a problem. Braun et al. [B5] describe the combined extraction of thorium and uranium from nitric acid solution of uranothorianite ore, $(U,Th)O_2$, by 33 v/o (volume percent) TBP at the Le Bouchet plant of the French Commissariat à l'Energie Atomique.

Williams [W3] reported that from 1959 to 1968 thorium was recovered as a by-product of Rio Algom's uranium mill at Elliot Lake, Ontario. Thorium-bearing uraninite ore was dissolved in sulfuric acid. Uranium was first recovered by anion exchange. Then thorium was recovered from the acid solution by solvent extraction with "alkyl phosphoric acid," probably di(2-ethylhexyl)phosphoric acid.

Crouse and Brown [C5] describe pilot-plant studies on recovery of uranium and thorium from Canadian uraninite by sulfuric acid leaching followed by solvent extraction in a two-cycle amine extraction process using trioctylamine to extract uranium and di(tridecyl)amine to extract thorium.

9 PURIFICATION OF THORIUM

Thorium concentrate produced by the processes described in Sec. 8 is too impure to be used as nuclear fuel. Especially objectionable impurities, which frequently are associated with thorium in its ores, are neutron-absorbing rare earths and uranium, the latter because it would dilute isotopically ^{233}U formed in thorium during subsequent neutron irradiation. The objective of thorium purification is removal of these and other impurities to concentrations below a few parts per million.

Solvent extraction with TBP has become the standard procedure for purifying thorium, just as for uranium. Processes used in different countries differ, however, in details such as the solvent used to dilute TBP, its concentration, and the means used to strip thorium and coextracted uranium from TBP. Table 6.20 summarizes the main features of processes used for purification of thorium on an industrial scale in the principal thorium-producing countries. Wylie [W5] gives more detail on early pilot-plant thorium-purification runs. Most of the published U.S. work on thorium purification on an industrial scale deals with irradiated thorium rather than natural; this will be described under the Thorex process, in Sec. 5 of Chap. 10.

Here, a summary will be given of Callow's [C2] description of a process used in England for purifying thorium concentrate and separating it from associated uranium. Figure 6.8 shows relative flow rates and nitric acid and thorium concentrations in this process. Feed is a nitric acid solution of thorium concentrates containing about 200 g ThO_2/liter of nitrate, a smaller concentration of uranyl nitrate, and considerable amounts of nitrates of other metals, such as iron and rare earths (RE). In the first contacting unit, consisting of five extracting stages and five scrubbing stages, one volume of feed is extracted with four volumes of recycle solvent, 40 v/o TBP in kerosine. At the 4 *N* nitric acid concentration of the feed, this solvent extracts effectively all of the uranium and thorium in the feed and a little of the associated impurities. Counterflow of 0.8 volume of 4 *N* HNO_3 in the scrubbing section removes these impurities from the solvent. Rare-earth content of extracted thorium is less than 5 ppm if the rare earth-to-thorium ratio of feed is less than 1:4.

In the second contacting unit, thorium is stripped from the rich solvent by 0.1 *N* HNO_3; uranium is scrubbed from the thorium product by additional solvent.

Uranium in solvent leaving the second contacting unit is stripped into an aqueous phase by 5% sodium carbonate solution. Stripped solvent is washed and reacidified with 4 *N* nitric acid for recycle to the process.

At the high thorium and nitric acid concentrations used in this flow sheet, two solvent phases may form, one rich in thorium and TBP and the other, lean. Callow [C2] states that formation of the two solvent phases does not interfere with operation of a mixer-settler cascade, whereas difficulty would be experienced with a pulse column. Conditions at which a

Table 6.20 Examples of purification of thorium on an industrial scale by solvent extraction with TBP

Reference	Braun et al. [B5]	Jamrack [J1]	Callow [C2]	Dar et al. [D1]	Ross[†] [R2]	Bril & Krumholz [B6]
Country	France	United Kingdom	United Kingdom	India	United States	Brazil
v/o TBP						
To extract uranium	–	5	40	10	–	46
To extract thorium	33	40	40	40	30	46
Diluent	Kerosene	Xylene	Kerosene	Kerosene	Solvesso 100	Varsol
Uranium strippant	–	0.02 N HNO_3	5% Na_2CO_3	Water	–	Na_2CO_3
Thorium strippant	Oxalic acid, to ppt. $Th(C_2O_4)_2$	0.02 N HNO_3	0.1 N HNO_3	Water	Water	4 N H_2SO_4, to ppt. $Th(SO_4)_2$

[†] Pilot-plant studies for this operation were described by Ewing et al. [E2].

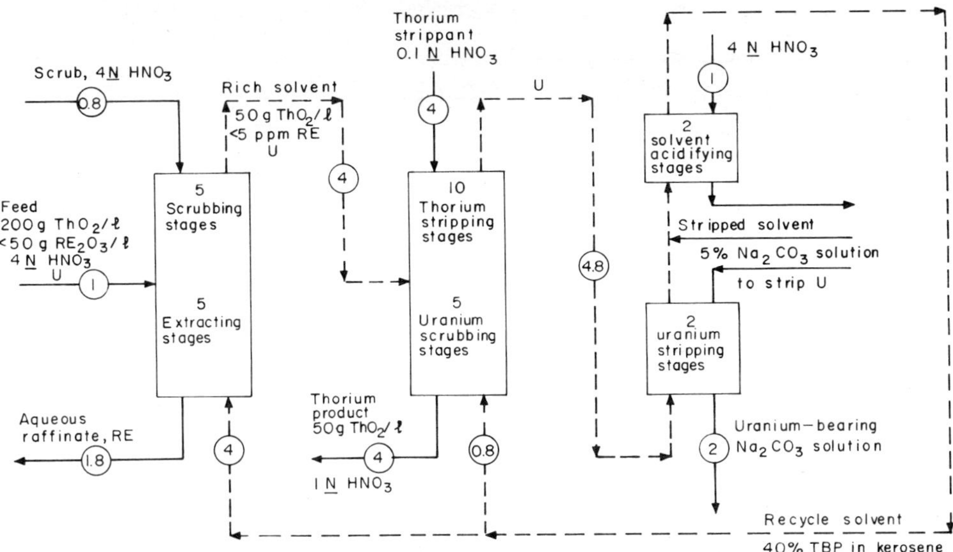

Figure 6.8 Thorium purification by solvent extraction with TBP. Circles, relative flow; —— aqueous; ––– 40 v/o TBP in kerosene. (*From Callow [C2].*)

second solvent phase forms are sketched in Sec. 5 of Chap. 10. Distribution equilibrium data for thorium, nitric acid, uranium, and impurities are also given there.

10 CONVERSION OF THORIUM NITRATE TO OXIDE, FLUORIDE, CHLORIDE, OR METAL

Purified thorium is usually produced in the form of an aqueous solution of thorium nitrate or crystals of hydrated thorium nitrate. The principal forms in which thorium is used in nuclear systems are the oxide ThO_2, the carbide ThC_2, the fluoride ThF_4, the chloride $ThCl_4$, or the metal. Conversion to oxide, fluoride, chloride, and metal are discussed in this section; production of thorium carbide was discussed in Sec. 5.3.

10.1 Conversion of Thorium Nitrate to ThO_2

Three methods that have been used to convert thorium nitrate to ThO_2 are as follows:

1. Thermal denitration,

$$Th(NO_3)_4 \cdot 4H_2O \xrightarrow{\text{heat}} ThO_2 + 4HNO_3 + 2H_2O$$

2. Precipitation of thorium hydroxide from aqueous solution with NH_3,

$$Th(NO_3)_4 + 4NH_3 + 4H_2O \rightarrow Th(OH)_4 + 4NH_4NO_3$$

followed by ignition of the hydroxide,

$$Th(OH)_4 \xrightarrow{\text{heat}} ThO_2 + H_2O$$

3. Precipitation of thorium oxalate with oxalic acid,

$$Th(NO_3)_4 + 2H_2C_2O_4 \cdot 2H_2O \rightarrow Th(C_2O_4)_2 \cdot 2H_2O + 4HNO_3 + 2H_2O$$

followed by ignition of the oxalate in air,

$$Th(C_2O_4)_2 \cdot 2H_2O + O_2 \xrightarrow{\text{heat}} ThO_2 + 4CO_2 + 2H_2O$$

Precipitation with NH_3 has been used to prepare a colloidal sol of $Th(OH)_4$ for formation into small spheres of $Th(OH)_4$ gel, in the so-called sol-gel process, followed by ignition to small ThO_2 spheres [Z1] to be incorporated in fuel elements for an HTGR (Chap. 3, Sec. 7.3).

Precipitation with oxalic acid followed by ignition to oxide has the advantage of separating thorium from several impurities (uranium, iron, and titanium) that remain in nitric acid solution. Following is a brief description of the process developed at Iowa State College [W2], pp. 70–75, for the U.S. Atomic Energy Commission (AEC) and used on a production scale by the National Lead Company at Fernald, Ohio [C6], pp. 150–152. To an aqueous solution of $Th(NO_3)_4$ containing about 200 g thorium/liter and 0.5 N in HNO_3 at 60°C is added about 105% of the oxalic acid, $H_2C_2O_4 \cdot 2H_2O$, needed to convert all thorium to the oxalate. The solution is stirred for about 5 min to complete precipitation. This results in an easily filtered crystalline precipitate of $Th(C_2O_4) \cdot 2H_2O$. This is filtered on a vacuum filter and washed with about half the feed solution volume of distilled water at 35°C. The precipitate is dried in a twin-screw drier with a jacket temperature of 120°C and a screw temperature of 154°C to a water content of 10 w/o.

The dried oxalate is converted to oxide in an externally fired rotary kiln, with counterflow of air. The exit gas temperature is controlled at 820°C. This produces a reactive, free-flowing oxide containing less than 0.5% carbon and about 0.5% moisture.

10.2 Production of ThF_4

The principal uses of ThF_4 are as intermediate in the production of thorium metal or, potentially, as a compound in the fuel mixture of the molten-salt breeder reactor. For both applications anhydrous, oxide-free ThF_4 is required.

Such ThF_4 cannot be made by precipitation from aqueous solution, as the precipitate contains water that, during heating or evaporation, hydrolyzes some ThF_4 to ThO_2 or $ThOF_2$. Instead, ThF_4 is produced by gas-phase hydrofluorination of ThO_2 with anhydrous HF:

$$ThO_2 + 4HF \rightleftharpoons ThF_4 + 2H_2O$$

This reaction is exothermic and proceeds rapidly at 566°C, but the equilibrium gas mixture contains some unreacted HF (Prob. 6.3). At lower temperatures nearly complete utilization of HF can be obtained, but the reaction is slow. The process was developed at Iowa State College

Table 6.21 Principal processes for producing metallic thorium

Electrolysis of fused salts
Electrolysis of $KThF_5$ in NaCl
Electrolysis of ThF_4 in NaCl/KCl
Electrolysis of $ThCl_4$ in NaCl/KCl
Reduction with Reactive Metals
Reduction of ThO_2 with Ca
Reduction of $ThCl_4$ with Mg
Reduction of ThF_4 with Ca
Thermal Dissociation of ThI_4

and used industrially for the U.S. AEC by the National Lead Company at Fernald, Ohio. A summary of the process, described in detail by Cuthbert [C6], pp. 152–154, follows.

Equipment consists of four externally heated, screw-fed, horizontal reactors positioned vertically one above another. The reactors are made of 309 Nb stainless steel, and the screw of Inconel and Illium R. Solids flow through the four reactors in series. In the first reactor, counterflowing air at 650 to 675°C removes residual H_2O and CO_2. Anhydrous HF vapor enters the fourth reactor and flows counter to the solids through the fourth reactor held at 566°C, the third at 370°C, and the second at 260°C. In this way, ThO_2 can be converted completely to ThF_4, the highly exothermic reaction can be controlled and most of the HF can be reacted. However, the process was usually operated to produce 70 w/o aqueous hydrofluoric acid, which was sold as a by-product.

10.3 Production of $ThCl_4$

Several of the processes for producing thorium metal start with anhydrous $ThCl_4$. As with ThF_4, anhydrous $ThCl_4$ cannot be prepared from aqueous solution, but must be made by gas-phase chlorination. A process used in England [B3] involved chlorinating a mixture of ThO_2 and carbon at or above 600°C:

$$ThO_2 + 2C + 2Cl_2 \rightarrow ThCl_4 + 2CO$$

The chloride must be purified by distillation to free it from unreacted solids and from impurities in the carbon. This is difficult because of the hygroscopicity of $ThCl_4$ and its high boiling point, 942°C. An alternative process [C6] reacts thorium oxalate with an excess of carbon tetrachloride and a small amount of chlorine as catalyst,

$$Th(C_2O_4)_2 + CCl_4 \rightarrow ThCl_4 + 2CO + 3CO_2$$

batchwise in a vertical graphite reactor at 600°C. This uses more expensive materials but produces pure solid $ThCl_4$ in a single step.

10.4 Production of Thorium Metal

Production of pure thorium metal is beset by all the difficulties cited for uranium metal in Chap. 5, Sec. 10.1, complicated further by the higher melting point of thorium, 1750°C. Table 6.21 lists the principal processes that have been used on a semiindustrial scale to produce thorium metal.

Electrolysis of fused salts. The first electrolytic processes used thorium fluorides, as these are less hygroscopic than thorium chloride. In a process developed for the U.S. AEC [S3], a solution of dried 15 to 20% $KThF_5$ in molten sodium chloride was electrolyzed at 800 to 900°C in a graphite anodic cell with a molybdenum cathode under an inert argon atmosphere. As only chlorine was produced at the anode, fluorides accumulated in the electrolyte and required its periodic renewal. A somewhat similar process, involving electrolysis of ThF_4 in molten KCl/NaCl has been used in the Soviet Union [K1].

To permit continuous operation, workers in both the United States [F1] and the United Kingdom [G2] have developed processes for electrolysis of $ThCl_4$ in molten NaCl [R1] or KCl/NaCl [W4]. By rigorous exclusion of moisture and oxygen, coarsely crystalline thorium of high purity was produced through electrolysis at 800°C in an argon atmosphere.

Reduction with reactive metals. As with uranium, processes for producing thorium by reduction with reactive metals have been developed starting with thorium oxide, chloride, or fluoride. To show which combinations of thorium compound and reactive metal are thermodynamically favorable, the free-energy changes in reducing ThO_2, $ThCl_4$, or ThF_4 by sodium, calcium, or magnesium have been evaluated in Table 6.22 at the temperature used in practice for the respective thorium compound. This table shows that calcium is the only metal capable of reducing ThO_2 or ThF_4, but that any one of sodium, calcium, or magnesium could reduce $ThCl_4$. Magnesium has been preferred because it costs less and can be handled in air without picking up oxygen.

Reduction of ThO_2. Because the heat of reaction of ThO_2 with calcium is small, it is incapable of melting the thorium product. However, finely divided thorium metal powder has been prepared by reacting ThO_2 and calcium metal in an argon atmosphere. In a U.S. process [F2], calcium chloride was added to promote thorium particle growth. A mixture of ThO_2, $CaCl_2$, and calcium chips in weight ratio 1:0.4:0.45 was heated to 950°C to start the reaction and held there for 2 to 5 h. In a U.K. process [B8], no $CaCl_2$ flux was used. The reaction was carried out in a CaO-lined, argon-purged metal cylinder at 1200°C. In both cases, the reaction products were cooled and the calcium compounds were leached from the thorium metal by dilute acid.

Reduction of $ThCl_4$. Of possible reductants for $ThCl_4$, magnesium is the most convenient. Magnesium alloys with thorium metal and reduces its melting point so that it is possible to produce massive metal instead of a fine, reactive powder. After reaction, magnesium and adherent $MgCl_2$ are removed from the alloy by vacuum distillation. This process is a variant of the Kroll process, which is used commercially for production of titanium and zirconium, with the latter application described in Chap. 7. Its use on a pilot-plant scale for production of thorium at the Albany Station of the U.S. Bureau of Mines has been described by Cuthbert [C6], pp. 182-184. Because of the comparatively small free-energy change with magnesium, it was necessary to use 100% percent excess, and yields of acceptably pure thorium metal were less than 50% of the $ThCl_4$ fed.

Reduction of ThF_4. Reduction of ThF_4 by calcium is the process used to produce most of the nuclear-grade thorium metal in the United States. The process was developed by workers at the

Table 6.22 Free-energy change in reduction of thorium compounds to metallic thorium†

	Thorium compound					
	Oxide		Fluoride		Chloride	
Reaction temperature T, K	1223		2023		1150	
Free energy of formation from elements at T, kcal/g-mol	$ThO_2(s)$	−238.45	$ThF_4(l)$	−369.16	$ThCl_4(l)$	−205.82
	$2Na_2O(s)$	−116.49	$4NaF(l)$	−311.02	$4NaCl(l)$	−291.31
	$2MgO(s)$	−223.17	$2MgF_2(l)$	−351.58	$2MgCl_2(l)$	−222.57
	$2CaO(s)$	−242.58	$2CaF_2(l)$	−423.02	$2CaCl_2(l)$	−296.44
Free-energy change in reaction, kcal/g-mol Th reduced with						
4 mol Na		+121.96		+58.14		−85.49
2 mol Mg		+15.28		+17.58		−16.75
2 mol Ca		−4.13		−53.86		−90.62

†Sources of data: Thorium [I1]; CaO [N2]; all others [N1].

Ames Laboratory of the U.S. AEC at Iowa State College under the direction of F. H. Spedding, H. A. Wilhelm, and W. H. Keller [S4]. Details of the process have been described by Wilhelm [W2], pp. 78–103, and Cuthbert [C6], pp. 175–180.

Because of the high melting point of thorium and the high heat of formation of ThF_4, the metallothermic reduction process used for making massive uranium metal (Chap. 5, Sec. 10.4) will not liberate enough heat to melt thorium, even when calcium is used as reductant. To get around this difficulty, the Iowa workers added $ZnCl_2$ and additional calcium to the charge, to act as a "booster." The reaction

$$ZnCl_2 + Ca \rightarrow Zn + CaCl_2$$

is more exothermic than reduction of ThF_4 by calcium, and the extra heat brings the reactants to higher temperature. Use of the booster has two other advantages:

1. Zinc alloys with thorium, reducing its melting point.
2. Calcium chloride reduces the melting point of the CaF_2 slag.

The reactor was a flanged steel cylinder lined with dolomitic lime similar to the one used for producing metallic uranium (Chap. 5, Sec. 10.4). It was charged with a mixture of 75.3 kg ThF_4, 27.22 kg granular calcium metal, and 7.26 kg anhydrous $ZnCl_2$. This amount of $ZnCl_2$ produced an alloy containing 18 m/o (mole percent) Zn. This amount of calcium was 25 percent more than needed for stoichiometric reduction of the ThF_4 and $ZnCl_2$. The excess was needed to drive the reduction of ThF_4 to completion.

The charge was covered with a graphite disk and a layer of lime, and a steel cover was bolted to the steel flange. The reactor was placed in a furnace preheated to 660°C. After about 40 min when the charge reached an average temperature around 475°C, the highly exothermic reactions between calcium and $ZnCl_2$ and ThF_4 took place, the molten zinc-thorium alloy settled to the bottom, and the molten CaF_2-$CaCl_2$ slag rose to the top.

After the reactor was cooled to room temperature, it was opened and the mass of metal was mechanically freed of frozen slag. Ninety percent of the zinc in the alloy was removed by distillation in a retort heated to 1150°C at a vacuum lower than 0.2 Torr. The retort was then filled with argon or helium to prevent oxidation of the spongy thorium and cooled to room temperature. The thorium was transferred to a beryllia crucible in an induction-heated vacuum furnace for melting, evaporation of the residual zinc, and casting into a graphite mold. Thorium metal yield was 94 to 96 percent.

Heat balances for Iowa process. To show the need for addition of the "booster" $ZnCl_2$ to the charge, Table 6.23 shows that when 1 mol ThF_4 and 2 mol $CaCl_2$ at 475°C (748 K) react to produce 1 mol liquid Th and 2 mol liquid CaF_2 at the melting point of thorium (1750°C or 2023 K), there is an enthalpy deficiency of 12.84 kcal/g-mol thorium. It would thus be impossible to melt the thorium product and obtain massive thorium metal free of CaF_2 from these reactants preheated to 475°C.

Table 6.24 shows that 49.57 kcal of heat is available when 1 mol $ZnCl_2$ and calcium at 475°C (748 K) react to form liquid zinc and $CaCl_2$ at 1750°C (2023 K). Thus, simultaneous reduction of $12.84/49.57 = 0.26$ mol $ZnCl_2$ and 1 mol ThF_4 initially at 475°C with stoichiometric amounts of calcium would bring the products to 1750°C and melt all the thorium.

In the actual Iowa process it was sufficient to heat the products only to around 1360°C (1633 K) because the thorium-zinc alloy produced has a lower melting point than pure thorium. This permitted use of only 0.218 mol $ZnCl_2$/mol ThF_4 and provided enough heat to melt the alloy and slag, bring the 25 percent excess calcium to reaction temperature, and still allowed for heat losses.

Table 6.23 Heat to be supplied in reaction 2Ca (*s*, 748 K) + ThF_4 (*s*, 748 K) → Th (*l*, 2023 K) + $2CaF_2$ (*l*, 2023 K)

	kcal/g-mol thorium	
Products at 2023 K		
2 × heat of formation of $CaF_2(l)$ at 298 K [N1]	−566.95	
2 × enthalpy change of $CaF_2(l)$ from 298 to 2023 K [N1]	+75.71	
2 × enthalpy change of Th(*l*) at 2023 K from Th(*s*) at 298 K [I1]	+18.17	
Enthalpy of products at 2023 K above elements at 298 K		−473.07
Reactants at 748 K		
Heat of formation of $ThF_4(s)$ at 298 K [I1]	−504.6	
Enthalpy of $ThF_4(s)$ at 748 K above $ThF_4(s)$ at 298 K [I1]	+12.57	
2 × enthalpy of Ca(*s*) at 748 K above Ca(*s*) at 298 K [N1]	+6.12	
		−485.91
Heat to be supplied (difference), kcal/g-mol thorium		+12.84

Table 6.24 Heat provided in reduction of $ZnCl_2$ (*l*) by Ca(*s*)

	kcal/g-mol zinc	
Products at 2023 K		
Heat of formation of $CaCl_2(l)$ at 298 K [N1]	−185.01	
Enthalpy of $CaCl_2(l)$ at 2023 K above $CaCl_2(l)$ at 298 K [N1]	+39.80	
Enthalpy of Zn(*l*) at 2023 K above Zn(*s*) at 298 K [W1]	+14.32	
Enthalpy of products at 2023 K above elements at 298 K		−130.89
Reactants at 748 K		
Heat of formation of $ZnCl_2(s)$ at 298 K [W1]	−99.60	
Enthalpy of $ZnCl_2(l)$ at 748 K above $ZnCl_2(s)$ at 298 K [W1]	+15.22	
Enthalpy of Ca(*s*) at 748 K above Ca(*s*) at 298 K [N1]	+3.06	
		−81.32
Heat provided (130.89–81.32), kcal/g-mol zinc		+49.57

Table 6.25 Material balance for production of 1 mol thorium metal by reduction of ThF_4 with excess calcium, with heat supplied by simultaneous reduction of 0.218 mol $ZnCl_2$

Charge at 748 K	Products at 1633 K
1 mol $ThF_4(s)$	1 mol Th(*l*)
0.218 mol $ZnCl_2(l)$	0.218 mol Zn(*l*)
2.778 mol Ca(*s*)	2 mol $CaF_2(l)$
	0.218 mol $CaCl_2(l)$
	0.56 mol Ca(*l*)

Table 6.26 Heat balance for production of thorium metal under conditions of Table 6.24

	Number of moles per mole thorium	Enthalpy,† kcal		
		Per mole substance	Per mole thorium	
Products at 1633 K				
Th(*l*), enthalpy change from Th(*s*) at 298 K	1	~+14.7	+14.7	
Zn(*l*), enthalpy change from Zn(*s*) at 298 K	0.218	~+11.4	+2.5	
CaF_2(*l*), enthalpy change from CaF_2(*l*) at 298 K	2	+28.54	+57.08	
CaF_2(*l*), enthalpy of formation at 298 K	2	−283.48	−566.96	
$CaCl_2$(*l*), enthalpy change from $CaCl_2$(*l*) at 298 K	0.218	+30.24	+6.59	
$CaCl_2$(*l*), enthalpy of formation at 298 K	0.218	−185.01	−40.33	
Ca(*l*), enthalpy change from Ca(*s*) at 298 K	0.56	+12.22	+6.84	
Total				−519.6
Reactants at 748 K				
ThF_4(*s*), enthalpy change from ThF_4(*s*) at 298 K	1	+12.57	+12.57	
ThF_4(*s*), enthalpy of formation at 298 K	1	−504.6	−504.6	
$ZnCl_2$(*l*), enthalpy change from $ZnCl_2$(*s*) at 298 K	0.218	+15.22	+3.32	
$ZnCl_2$(*s*), enthalpy of formation at 298 K	0.218	−99.60	−21.71	
Ca(*s*), enthalpy change from Ca(*s*) at 298 K	2.778	+3.06	+8.50	
Total				−501.9
Difference				−17.7

†References for enthalpies: thorium [I1]; zinc [W1]; calcium [N1].

Table 6.25 gives the material balance for the Iowa process as described by Wilhelm [W2]. Table 6.26 gives the heat balance for this process, with reactants at 475°C (748 K) and products molten at 1360°C (1633 K). An excess of 17.7 kcal/mol thorium product was available to compensate for heat losses.

Thermal dissociation of ThI_4. Veigel et al. [V1] have prepared massive thorium metal of high purity in lots of several hundred grams each by the Van Arkel-de Boer "hot-wire" process, which has been used for semicommercial production of zirconium as described in Sec. 8.4 of Chap. 7. The process is less suitable for thorium because the thorium metal product is less coherent, so that batch sizes are small. In this process, ThI_4 is evaporated at 455 to 480°C in an evacuated vessel containing a metal filament heated to 900 to 1700°C. The iodide dissociates at the higher temperature,

$$ThI_4 \rightarrow Th + 2I_2$$

and deposits thorium metal on the heated wire. Temperatures in the range 500 to 900°C must be avoided to prevent formation of nonvolatile ThI_2 and ThI_3, which are stable in this range [S1].

REFERENCES

A1. Audsley, A., et al.: "Recently Developed Processes for Extraction and Purification of Thorium," *PICG(2)* **3**: 216 (1958).

A2. Ayers, A. S.: "Precipitation of Thorium Oxalate from Nitric Acid Solutions," Paper 7.9 of Report TID-5223, 1952.

B1. Barghusen, J., Jr., and M. Smutz: *Ind. Eng. Chem.* **50**: 1754 (1958).

B2. Bearse, A. E., et al.: *Chem. Eng. Progr.* **50**: 235 (1954).
B3. Bellamy, R. G., and N. A. Hill: *Extraction and Metallurgy of Uranium, Thorium and Beryllium,* Macmillan, New York, 1963.
B4. Benz, R., C. G. Hoffman, and G. N. Rupert: *J. Amer. Chem. Soc.* **89**: 191 (1967).
B5. Braun, C., et al.: "The Manufacture of Pure Thorium Nitrate at Le Bouchet Plant," *PICG(2)* **4**: 202 (1958).
B6. Bril, K. J., and P. Krumholz: "Developments in Thorium Production Technology," *PICG(3)* **12**: 167 (1964).
B7. Britton, H. T. S., and M. E. D. Jarrett: *J. Chem. Soc. (London),* 1936, p. 1494.
B8. Buddery, J. H.: "The Extraction Metallurgy of Thorium," in *Progress in Nuclear Energy Series V,* vol. 2: *Metallurgy and Fuels,* H. M. Finniston and J. P. Howe (eds.), Pergamon, London, 1959, p. 35.
C1. Calkins, G. D., et al.: "Recovery of Thorium and Uranium from Monazite Sand," Reports BMI-243 and BMI-243A, 1950.
C2. Callow, R. J.: *The Industrial Chemistry of the Lanthanons, Yttrium, Thorium and Uranium,* Pergamon, New York, 1967, pp. 115–119.
C3. Chiotti, P.: "Thorium-Carbon System," Report AECD-3072, June 5, 1950.
C4. Crouse, D. J., and K. B. Brown: "Recovery of Thorium, Uranium, and Rare Earths from Monazite Sulfate Leach Liquors by the Amine Extraction (Amex) Process," Report ORNL-2720, July 16, 1959.
C5. Crouse, D. J., Jr., and K. B. Brown: *Ind. Eng. Chem.* **51**: 1461 (1959).
C6. Cuthbert, F. L.: *Thorium Production Technology,* Addison-Wesley, Reading, Mass., 1958.
D1. Dar, K. K., et al.: "Uranium and Thorium Resources and Development of Technology for their Extraction in India," *PICG(4)* **8**: 99 (1972).
E1. *Engineering and Mining Journal,* March 1979, p. 164.
E2. Ewing, R. A., S. J. Kiehl, Jr., and A. E. Bearse: "Refining of Thorium by Solvent Extraction," Report BMI-955, Oct. 19, 1954.
F1. Fisher, C. E., et al. (Horizons, Inc.): "Research and Development in the Field of Thorium Chemistry and Metallurgy," Reports SRO-11, 12, and 13, June 30, 1956.
F2. Fuhrman, N., et al.: "The Production of Thorium Powder by Calcium Reduction of Thorium Oxide," Report SCNC-185, 1957.
G1. Garg, R. K., et al.: "Status of Thorium Technology," in *Nuclear Power and Its Fuel Cycle,* vol. 2, International Atomic Energy Agency, Vienna, 1977, p. 457.
G2. Gibson, A. R., and J. R. Chalkley: *Trans. Inst. Min. Metall.* **69**: 281 (1960).
I1. International Atomic Energy Agency: "Thorium: Physicochemical Properties of Its Compounds and Alloys," *Atomic Energy Rev.,* Special Issue No. 5, 1975.
J1. Jamrack, W. D.: *Rare Metal Extraction by Chemical Engineering Techniques,* Macmillan, New York, 1963, pp. 176–178.
K1. Kaplan, G. E.: "Metallurgy of Thorium," *PICG(1)* **8**: 184 (1955).
M1. Mallett, M. W., and I. E. Campbell: *J. Amer. Chem. Soc.* **73**: 4850 (1951).
M2. Mueller, W. M., J. P. Blackledge, and G. G. Libowitz: *Metal Hydrides,* Academic, New York, 1968.
M3. Murray, J. R.: "The Preparation, Properties and Alloying Behavior of Thorium," Report AERE-M/TN-12, 1952.
N1. National Bureau of Standards: *JANAF Thermochemical Tables,* 2d ed., U.S. Government Printing Office, Washington, D.C., June 1971.
N2. National Bureau of Standards: *JANAF Thermochemical Tables,* 1974 suppl., U.S. Government Printing Office, Washington, D.C., 1974.
N3. Nininger, R. D., and S. H. U. Bowie: "Technological Status of Nuclear Fuel Resources," Paper given at 1976 Winter Meeting of American Nuclear Society, Washington, D.C., Nov. 1976.

O1. Organization for Economic Cooperation and Development, and International Atomic Energy Agency: *Uranium Resources, Production and Demand,* Paris, Dec. 1977.
P1. Pilkington, E. S., and A. W. Wylie: *J. Appl. Chem. (London)* **4**: 568 (1954).
P2. Powell, A. R.: In *Thorpe's Dictionary of Applied Chemistry,* 4th ed., vol. XI, Longmans Green, London, 1954, p. 600.
R1. Raynes, B. C., et al.: "Investigations for the Production of Thorium Metal," Report TID-5246, 1954.
R2. Ross, A. M.: Quoted by F. L. Cuthbert, *Thorium Production Technology,* Addison-Wesley, Reading, Mass., 1958, p. 128.
S1. Scaife, D. E., and A. W. Wylie: *J. Chem. Soc.*, 1964, p. 5450; *PICG(2)* **4**: 215 (1958).
S2. Shaw, K. G., M. Smutz, and G. L. Bridger: "A Process for Separating Thorium Compounds from Monazite Sands," Report ISC-407, 1954.
S3. Sibert, M. E., and M. A. Steinberg (Horizons, Inc.): "Investigations for the Production of Thorium Metal by Fused Salt Electrolysis," Report NYO-3725, 1952.
S4. Spedding, F. H.: "Progress Report in Metallurgy," Report ISC-6, 1947.
S5. Staatz, M. H.: "Update on Thorium Resources," Paper presented at Department of Energy Symposium, Grand Junction, Colorado, Oct. 1978.
U1. U.S. Department of the Interior, Bureau of Mines: *Mineral Facts and Problems, Bulletin 667,* 1975, p. 1115.
V1. Veigel, N. D., et al.: "The Preparation of High-Purity Thorium by the Iodide Process," Report AECD-3586, 1953.
W1. Wicks, C. E., and F. E. Block: *Thermodynamic Properties of 65 Elements–Their Oxides, Halides, Carbides and Nitrides,* Bulletin 605, Bureau of Mines, U.S. Government Printing Office, Washington, D.C., 1963.
W2. Wilhelm, H. A. (ed.): *The Metal Thorium,* American Society for Metals, Cleveland, Ohio, 1958.
W3. Williams, R. M.: "Uranium and Thorium in Canada," *PICG(4)* **8**: 37 (1972).
W4. Wyatt, J. L.: "The Preparation of Thorium Metal Powder by Fused Salt Electrochemical Techniques," in Report TID-7521, Part 1, 1955.
W5. Wylie, A. W.: *Rev. Pure Appl. Chem.* **9**: 169 (1959).
Z1. Zimmer, E., P. Naefe, and H. Ringel: "Continuous Working Process for the Production of ThO_2 and $(Th,U)O_2$ Fuel Kernels," *Proc. Eur. Nucl. Conf.*, vol. 7, Pergamon, New York, 1976, p. 1.

PROBLEMS

6.1 Suppose that freshly purified thorium separated from HTGR spent fuel contains 30 Ci ^{228}Th for every curie of ^{232}Th. How long must the fuel be stored for the ^{228}Th activity to decrease to 1.1 times the ^{232}Th activity? Consider only 6.7-year ^{228}Ra in the decay chain between ^{232}Th and ^{228}Th.

6.2 Suggest three methods that might be used to separate uranium from thorium.

6.3 The free energies of formation of HF and H_2O at 566°C are HF, −66,382; H_2O, −48,141 cal/g-mol. From the data of Tables 6.8 and 6.10, find the equilibrium constant for the reaction

$$ThO_2(s) + 4HF(g) \rightleftharpoons ThF_4(s) + 2H_2O(g)$$

at this temperature.

What is the composition of the equilibrium mixture of H_2O and HF?

What is the minimum number of moles of HF needed to produce 1 mol ThF_4 at this temperature?

CHAPTER
SEVEN

ZIRCONIUM AND HAFNIUM

1 USES OF ZIRCONIUM AND HAFNIUM

Zirconium and hafnium have very similar chemical properties, invariably occur together in nature, and are difficult to separate. Yet their absorption cross sections for thermal neutrons are very different:

	Absorption cross section for 2200 m/s neutrons
Zirconium, Zr	0.185
Hafnium, Hf	102

The thermal absorption cross section of zirconium is the lowest of all mechanically strong, high-melting, corrosion-resistant metals. For this reason, zirconium and zirconium-based alloys are the materials preferred for cladding and structural materials in water-cooled, thermal-neutron power reactors.

When this type of reactor was under development for the U.S. nuclear submarine program in the early 1950s, the good chemical and mechanical properties of zirconium were recognized, but its low neutron absorption was obscured in the zirconium then available commercially by the hafnium present in natural zirconium. This caused the neutron-absorption cross section reported for commercial zirconium to be high and variable. Workers at Oak Ridge National Laboratory deduced that the variability was due to the presence of small amounts of hafnium, with its high cross section. They devised processes for removing hafnium and showed that the cross section of pure zirconium was 0.18 b. Reactor-grade zirconium has less than 100 ppm hafnium by weight.

When separated from zirconium, hafnium also has valuable nuclear applications. The high cross section, good mechanical strength, and corrosion resistance of hafnium make it an excellent material for control elements in water-cooled reactors, where it can be used without cladding.

The amount of hafnium-free zirconium used in nuclear applications is much smaller than the ordinary commercial uses of zirconium metal and compounds, for which costly removal of hafnium is not required. Zirconium metal is used in corrosion-resistant equipment for chemical plants, refractory alloys, and photo flashbulbs. The mineral zircon is used extensively in foundry sands, abrasives, and ceramics. Zircon and zirconia are widely used as refractories.

Hafnium metal is used in refractory alloys and in photo flashbulbs where especially high light output is wanted.

2 NATURAL OCCURRENCE

Zirconium is the eleventh most abundant element in the earth's crust, which contains 0.028 percent of this element. It is more abundant than copper, lead, nickel, or zinc. Zirconium minerals always contain from 0.5 to 2 percent of chemically similar hafnium, which seldom occurs naturally by itself.

The principal natural sources of zirconium and hafnium are the minerals zircon $(Zr,Hf)SiO_4$, and baddleyite $(Zr,Hf)O_2$.

3 PRODUCTION AND PRICE

Table 7.1 gives the annual production of zirconium concentrate by the principal producing countries of the non-Communist world, excluding the United States.

U.S. production in these years was around 150,000 short tons. Thus, Australia and the United States are the principal zirconium-producing nations. Most of their production was from dredging of black sands on beaches and in stream beds, where zircon has been concentrated hydraulically along with other relatively dense minerals such as rutile (TiO_2), ilmenite ($FeTiO_3$), and monazite (Chap. 6).

Principal U.S. producers of zircon concentrates in these years were E. I. du Pont de Nemours and Company and Titanium Enterprises, Inc., with operations primarily in northern Florida and southern Georgia.

In the 1960s, hafnium-free zirconium had been produced by several U.S. companies

Table 7.1 Annual production of zirconium concentrates

	Short tons† concentrate per year		
Country	1972	1973	1974
Australia	393,187	393,336	406,648
Brazil	4,645	3,411	3,500
India	5,500	6,800	6,800
Korea	14	25	44
Malagasy Republic	15	0	0
Malaysia	1,820	3,463	3,035
South Africa	744	5,463	13,203
Sri Lanka	33	31	23
Thailand	403	443	2,207
Total	406,361	412,972	435,460

†One short ton = 0.917 MT = 0.917 Mg.

Source: S. G. Ampian, in *Minerals Yearbook, 1974*, vol. I, *Metals, Minerals and Fuels*, U.S. Government Printing Office, Washington, D.C., 1976.

including Amax, Inc., National Distillers and Chemicals Corporation, Columbia-National Corporation, and Wah Chang Corporation, but in 1978, the only U.S. producer of hafnium-free zirconium sponge was Teledyne Wah Chang Albany Corporation, with annual capacity of 7.5 million lb [N2]. Western Zirconium then announced plans to produce 3 to 4 million lb/year. In France, Pechiney Ugine Kuhlmann was increasing capacity to 4 million lb/year. Indian zirconium production capacity was around 0.1 million lb/year. Hafnium-free zirconium has also been produced in England, Canada, Japan, West Germany, and the Soviet Union.

Prices in 1974 were [A2]

Zircon concentrate	\$250/short ton
Zirconium, hafnium-free	
Sponge	\$5.50–7.00/lb
Sheets, strip, bars	\$12–17/lb
Hafnium	
Sponge	\$75/lb
Bar and plate	\$120/lb

4 ZIRCONIUM AND HAFNIUM METAL AND ALLOYS

4.1 Phases

The phases of metallic zirconium and hafnium and their transition temperatures are listed in Table 7.2.

Zirconium and hafnium form nearly ideal solid solutions, with melting and transition temperatures between those of the pure components.

Equations for the vapor pressure of zirconium are [I1]

$$\text{Solid:} \qquad \log_{10} p(\text{atm}) = 6.950 - \frac{30{,}810}{T(\text{K})} \tag{7.1}$$

$$\text{Liquid:} \qquad \log_{10} p(\text{atm}) = 6.541 - \frac{29{,}940}{T(\text{K})} \tag{7.2}$$

Table 7.2 Phases of metallic zirconium and hafnium

		Transition temperature, °C	
Phase	Crystal system	Zirconium	Hafnium
Solid α	Hexagonal		
		863	1740
Solid β	Body-centered cubic		
		1852	2227
Liquid			
		4304	4603
Gas, 1 atm			
Reference		[I1]	[H4]

Table 7.3 Thermodynamic properties of metallic zirconium

Phase	Temperature range T, K: From	To	Heat capacity $C_p = a + bT + c/T^2$, cal/(g-mol·°C): a	$10^3 b$	$10^{-5} c$	Heat of transformation or fusion, cal/g-mol
Solid α	298	1136	5.463	2.144	−0.166	
	1136					930
Solid β	1136	2125	5.137	1.5705	8.776	
	2125					4500
Liquid			8.00	–	–	

Source: International Atomic Energy Agency, "Zirconium: Physico-Chemical Properties of Its Compounds and Alloys," *Atomic Energy Rev.*, Special Issue No. 6, 1976.

4.2 Density and Thermal Expansion

Russell's [R4] measurements of the x-ray crystal density and coefficients of thermal expansion of alpha zirconium may be summarized as follows:

Density, 20°C: 6.490 g/cm^3
Expansion coefficients, per °C:
Linear, parallel to c axis, $(6.106 + 0.01398t) \times 10^{-6}$, t = temperature, °C
Linear, perpendicular to c axis, $(5.599 + 0.002241t) \times 10^{-6}$
Volume, $(17.304 + 0.018462t) \times 10^{-6}$

The density of hafnium is 13.09 g/cm^3.

4.3 Thermodynamic Properties

The heat capacities of the two solid phases and the liquid phase of zirconium and the heats of transformation and fusion are given in Table 7.3.

4.4 Thermal and Electrical Conductivity

Figure 7.1 shows the temperature dependence of the thermal conductivity of zirconium and zircaloy-2. Figure 7.2 shows the temperature dependence of the electrical resistivity of zirconium and a zirconium-1.65 percent tin alloy, which approximates zircaloy-2 and -4.

4.5 Chemical Reactivity

Massive zirconium is unaffected by air or oxygen at room temperature. However, finely divided zirconium in the form of powder or sponge is pyrophoric and may ignite spontaneously in air. Ignition of chips formed in mechanical decladding of zircaloy-clad fuel rods is a hazard to be guarded against. Zirconium and hafnium foil are used in photographers' flashbulbs.

At high temperatures zirconium reacts with nitrogen to form ZrN and with carbon to form ZrC. Thus, in melting and casting zirconium it is impossible to use a nitrogen atmosphere or graphite electrodes or molds. Vacuum, or an argon or helium atmosphere, is required, and water-cooled, copper molds.

Massive zirconium has excellent resistance to chemical attack by water, seawater, or steam

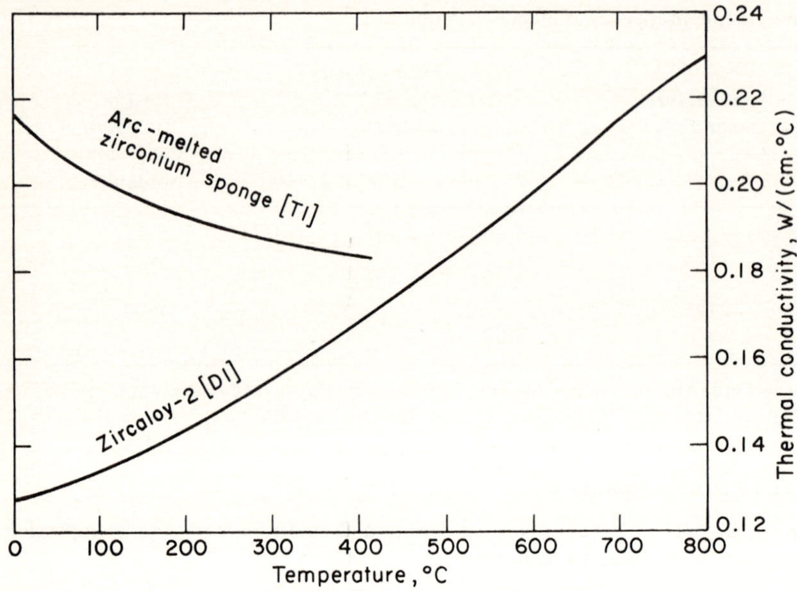

Figure 7.1 Thermal conductivity of zirconium and zircaloy-2.

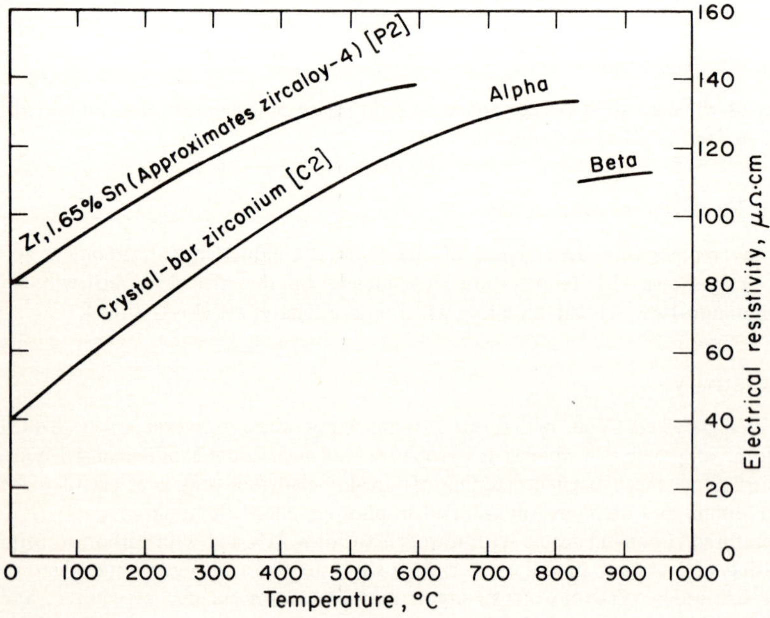

Figure 7.2 Electrical resistivity of zirconium and zircaloy.

below 150°C, owing to a protective surface oxide film. At higher temperatures pure zirconium is slowly attacked by water owing to gradual diffusion of oxygen through the metal, resulting in loss of ductility. Alloying with small amounts of tin and other metals, in the zircaloy series of alloys (Sec. 4.6), reduces corrosion by water to such an extent that zircaloy has excellent resistance to corrosion by water or steam at temperatures up to 350°C. Zircaloy thus is an excellent cladding and structural material for water-cooled reactors. Zircaloy is not recommended at the higher temperatures in a reactor producing supercritical steam, because zirconium oxide and hydride are slowly formed, making the metal less ductile. These reactions occur quickly in the even higher temperatures encountered in a loss-of-coolant accident.

Finely divided zirconium, such as the zirconium sponge produced in the Kroll process (Sec. 8.3), reacts sufficiently rapidly with water at ordinary temperatures to impair its mechanical properties. Thus it is not feasible to use water leaching to separate zirconium sponge from the magnesium chloride by-product of the Kroll process.

Massive zirconium and zircaloy are resistant to hot nitric acid. In fuel reprocessing (Chap. 10), uranium metal or uranium dioxide fuel can be dissolved by hot nitric acid while leaving the zircaloy cladding unattacked.

Chapter 11 of Lustman and Kerze [L1] and Chap. 2 of ASTM Special Technical Publication 639 [S2] describe the generally excellent corrosion resistance of zirconium to most aqueous solutions. It is corroded, however, by hot concentrated sulfuric or phosphoric acids and is attacked by fluoride ion at concentrations as low as 0.001 percent [S2].

Methods that have been proposed for chemical removal of zircaloy cladding in nonaqueous processing of spent fuel include conversion to gaseous $ZrCl_4$ by reaction with HCl above 350°C or solution in molten zinc, which dissolves 10 a/o (atom percent) zirconium at 900°C.

4.6 Zirconium Alloys

The grades of zirconium and zirconium-base alloys commercially available in the United States are described in ASTM Special Technical Publication 639 [S2]. The most important of these are zircaloy-2, zircaloy-4, and Zr-2.5 Nb. Table 7.4 gives the ASTM composition specifications for these three alloys and zirconium sponge.

The zircaloy series of alloys was developed by the U.S. Navy Nuclear Propulsion Program for service in the core of water-cooled nuclear reactors [R3]. Compared with pure zirconium, these alloys have greater strength and better resistance to corrosion by water or steam. Zircaloy-4 was developed later than zircaloy-2 and became the preferred material, because the nickel in zircaloy-2 promoted the absorption of hydrogen, leading to reduction in ductility.

Zr-2.5 Nb [zirconium alloyed with 2.5 w/o (weight percent) niobium] has better mechanical properties than zircaloy, but is corroded more rapidly by water containing oxygen, such as is found in boiling-water reactors. It was the material preferred in 1971 [E1] for pressure tubes in Canadian pressurized-water reactors.

References [S2], [R3], and [H3] give more detailed information on zirconium alloys and their history and corrosion behavior.

5 ZIRCONIUM AND HAFNIUM COMPOUNDS

5.1 Valence States

The principal valence of zirconium and hafnium is +4. Halides of valence +2 and +3 have been prepared, but they are of less practical importance as they disproportionate when heated and react with water in aqueous solution.

Table 7.4 Composition specifications for zirconium sponge and zirconium alloys

Form	Sponge	Ingots		
Common name	Zirconium sponge	Zircaloy-2	Zircaloy-4	Zr-2.5Nb
ASTM grade	R60001	R60802	R60804	R60901
ASTM specification	B349-73	B350-73	B350-73	B350-73
Maximum impurities, ppm by weight				
Aluminum	75	75	75	75
Boron	0.5	0.5	0.5	0.5
Cadmium	0.5	0.5	0.5	0.5
Carbon	250	270	270	270
Chlorine	1300	–	–	–
Chromium	200	See below		200
Cobalt	20	20	20	20
Copper	30	50	50	50
Hafnium	100	100	100	100
Hydrogen	–	25	25	25
Iron	1500	See below		1500
Manganese	50	50	50	50
Nickel	70	See below	70	70
Nitrogen	50	65	65	65
Oxygen	1400	To be specified in order		See below
Silicon	120	200	120	120
Titanium	50	50	50	50
Tungsten	50	100	100	100
Uranium	3.0	3.5	3.5	3.5
Alloying elements, w/o				
Tin		1.20–1.70	1.20–1.70	–
Iron		0.07–0.20	0.18–0.24	–
Chromium		0.05–0.15	0.07–0.13	–
Nickel		0.03–0.08	<0.007	–
Iron + chromium + nickel		0.18–0.38	0.28–0.37	–
Niobium		–	–	2.40–2.80
Oxygen		–	–	0.09–0.13

Source: American Society for Testing and Materials, *1976 Annual Book of ASTM Standards:* part 8, *Nonferrous Metals–Nickel, Lead and Tin Alloys, Precious Metals, Primary Metals; Reactive Metals*, ASTM, Philadelphia, 1976.

5.2 Zirconium Dioxide

Zirconium dioxide, zirconia, is the only oxide of zirconium stable chemically at temperatures below 2000 K. At higher temperatures some dissociation into ZrO and oxygen takes place. The phases of ZrO_2, their densities, and phase-transition temperatures are listed in Table 7.5. Zirconia stabilized in the high-temperature cubic phase by addition of 3 to 5 percent calcium oxide is used as a refractory at temperatures up to 2200°C. ZrO_2 has been used to dilute $^{235}UO_2$ in fuel elements.

Zirconium dioxide can be partially reduced to zirconium metal by reactive metals such as calcium or magnesium, but the product is contaminated by some unreduced oxide, owing to the solubility of oxide in the metal.

Zirconium dioxide reacts with carbon to form the carbide ZrC. Zirconium dioxide is inert to the halogens, but when mixed with carbon reacts to form tetrahalides at high temperatures. This is the basis of one process for extracting zirconium from zircon.

Zirconium dioxide is unattacked by all mineral acids except concentrated HF and H_2SO_4, which slowly dissolve it. ZrO_2 can be converted to ZrF_4 by reaction with gaseous HF at 550°C. Fusion with a fluosilicate converts ZrO_2 to fluozirconate:

$$K_2SiF_6 + ZrO_2 \rightarrow K_2ZrF_6 + SiO_2$$

Fusion with alkali hydroxides converts ZrO_2 to a zirconate:

$$2NaOH + ZrO_2 \rightarrow Na_2ZrO_3 + H_2O$$

As alkali fluozirconates are soluble in water, and alkali zirconates are soluble in strong acid, fusion with fluosilicates or with alkali hydroxides are useful steps for getting ZrO_2 into aqueous solution.

5.3 Zirconium Carbide

Zirconium carbide, ZrC, is made by reacting ZrO_2 or zircon with graphite in an electric furnace. It has a very high melting point, 3420°C. It reacts with chlorine at 500°C to produce $ZrCl_4$.

5.4 Zirconium Nitride

Zirconium nitride, ZrN, is formed when zirconium metal is heated with nitrogen above 1200°C.

5.5 Zirconium Hydrides

The phase diagram for the system zirconium-hydrogen is shown in Fig. 7.3. The maximum hydrogen content of the solid phase approaches that of ZrH_2, $66\frac{2}{3}$ a/o H.

The δ and ϵ phases have been used extensively as moderator in several types of nuclear reactor. The compact SNAP† reactors [D2] employed in power plants for space vehicles use as fuel and moderator a mixture of 7 to 10 w/o enriched (90 a/o) ^{235}U with zirconium hydride

†Systems for Nuclear Auxiliary Power.

Table 7.5 Phases of zirconium dioxide†

Phase	Crystal form	Density, g/cm^3	Transition temperature, K
Solid	Monoclinic	5.68	
			1447 ± 30
Solid	Tetragonal	6.10	
			2566 ± 8
Solid	Cubic	6.27	
			2953 ± 15
Liquid			
			~4300 (1 atm)
Gas			

†Data from [I1] and [B3].

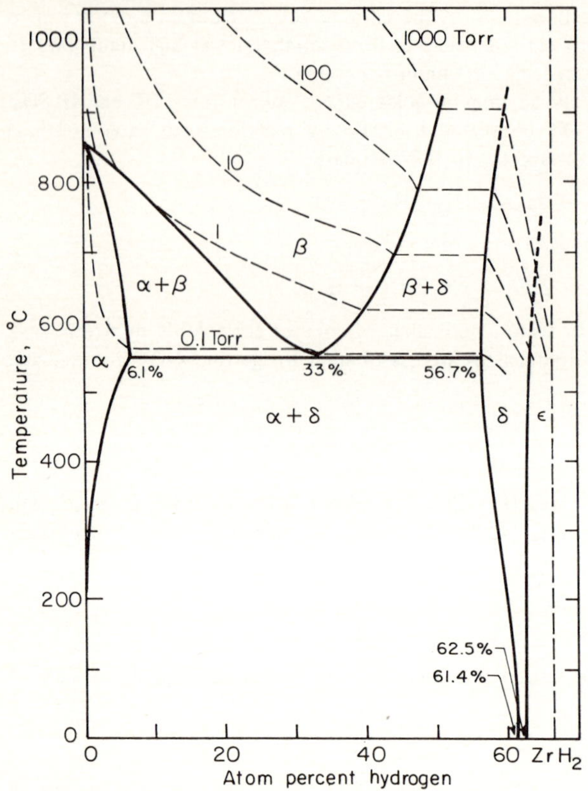

Figure 7.3 Zirconium-hydrogen phase diagram. *(Reprinted with permission from Dr. W. M. Mueller [M5] and the copyright holder, Academic Press, Inc., New York.)*

containing from 63 to 64.9 a/o hydrogen. The TRIGA research reactor developed by General Atomic [W1] uses a mixture of uranium, zirconium, and hydrogen in atomic proportions 0.03:1:1. The uranium contains 20 a/o ^{235}U. Zirconium hydride has been suggested as moderator for high-temperature power reactors producing superheated steam.

The desirable properties of zirconium hydride for these systems are (1) the low equilibrium pressure of hydrogen at temperatures up to 650°C, and (2) the high atomic density of hydrogen in them. Because the ϵ phase has a density of 5.62 g/cm³, the atomic density of hydrogen in zirconium hydride containing 64.9 a/o hydrogen is

$$\frac{(5.62 \text{ g/cm}^3)(6.02252 \times 10^{23} \text{ molecules/g-mol})[(0.649/0.351) \text{ atoms H/molecule}]}{[91.22 + (0.649/0.351)1.008] \text{ g/g-mol}}$$

$$= 6.7 \times 10^{22} \text{ atoms H/cm}^3 \tag{7.3}$$

This hydrogen density in ϵ zirconium hydride is as high as in water at room temperature and is appreciably higher than in water at the 300°C used in power reactors. Another advantage of the uranium-zirconium hydride fuel-and-moderator mixture is its high prompt negative temperature coefficient of reactivity, a consequence of the intimate thermal contact between ^{235}U and hydrogen atoms.

The left side of Fig. 7.3 shows that hydrogen is appreciably soluble in α zirconium, up to a maximum concentration of 6.1 a/o at 550°C. Dissolved hydrogen reduces the impact strength of zirconium [M4] and has been responsible for the failure of fuel cladding through hydrogen embrittlement.

Zirconium hydride is made by reacting zirconium metal with hydrogen.

Massive zirconium hydride is stable in air at temperatures below 600°C, but finely divided hydride will ignite at 430°C [H2] and should be kept out of contact with air.

For additional information on zirconium hydrides, see reference [B1].

5.6 Zirconium Halides

The principal halides of zirconium are listed in Table 7.6, together with the melting points, vapor-pressure equations, and temperatures at which the vapor pressure of each equals 760 Torr. The tetrahalides sublime without melting at atmospheric pressure, like UF_6. The lower halides disproportionate before melting.

The halides of greatest practical importance are ZrF_4, as a component of fluozirconates; $ZrCl_4$, as the feed material for production of zirconium in the Kroll process; and ZrI_4, as feed material in the hot-wire process. These will be discussed later in the chapter.

The heat capacity, enthalpy, heat of formation, and free energy of formation of $ZrCl_4$ are given in Table 7.7.

The tetrahalides are hydrolyzed to ZrO_2 by steam. They dissolve in water, forming oxyhalides such as $ZrOCl_2$.

5.7 Compounds of Hafnium

Compounds of hafnium have physical and chemical properties very similar to the corresponding compounds of zirconium, except for the much higher density of hafnium compounds. The melting and subliming temperatures of some hafnium compounds are compared with corresponding zirconium compounds in Table 7.8. As the vapor pressures of corresponding hafnium and zirconium compounds are so nearly equal, separation by fractional distillation is impractical. As corresponding compounds, such as $HfCl_4$ and $ZrCl_4$, form solid solutions miscible over the entire composition range and have nearly the same melting point, separation by fractional crystallization is also difficult.

The free energies of formation of corresponding hafnium and zirconium compounds are also nearly equal, so that separation by preferential reaction of one species is difficult, too. Table 7.9 compares the free energies of formation of hafnium and zirconium tetrahalides at 1000 K.

Successful processes for separating hafnium from zirconium take advantage of rather rare occurrences of substantial differences in solubilities of corresponding hafnium and zirconium compounds in water, organic solvents, fused salts, or liquid metals.

5.8 Aqueous Chemistry of Zirconium and Hafnium

Aqueous solutions of zirconium or hafnium compounds may be obtained by dissolving the corresponding hydrous dioxide in the appropriate strong acid. Because zirconium and hafnium oxides are such weak bases, these aqueous solutions tend to hydrolyze, with formation of zirconyl salts such as $ZrO(NO_3)_2$.

Crystals of zirconium nitrate, $Zr(NO_3)_4 \cdot 5H_2O$, can be obtained by evaporating a solution of ZrO_2 in strong nitric acid at a temperature not higher than 15°C. Evaporation at higher temperatures or low acid concentration yields zirconyl nitrate, $ZrO(NO_3)_2 \cdot 2H_2O$. Zirconyl nitrate is less readily extracted by tributyl phosphate than zirconium nitrate, a property made use of in separating thorium from fission-product zirconium, Chap. 10.

Evaporation of aqueous solutions of $ZrCl_4$ or of solutions of hydrous zirconia in hydrochloric acid yields the oxychloride, $ZrOCl_2$.

Sulfuric acid solutions of hydrous zirconia contain very little Zr^{4+} because of complex

Table 7.6 Phase relations of zirconium halides

Compound	Melting temperature, K	Vapor-pressure equations, $\log_{10} \Pi$ (Torr) $= A - B/T$ (K)			Temperature, K		
		Π	A	B (K)	From	To	Vapor pressure = 760 Torr
$ZrF_4(s)$	1205	p_{ZrF_4}	12.77	11,640	650	1200	1177
$ZrCl_2(s)$	Disp.†						
$ZrCl_3(s)$	Disp.	p_{ZrCl_4}‡	11.632	6,246	613	723	714
$ZrCl_4(s)$	710	p_{ZrCl_4}	11.78	5,409	400	710	608
$ZrCl_4(l)$	710	p_{ZrCl_4}	9.21	3,580	710	800	
$ZrBr_3(s)$	Disp.						
$ZrBr_4(s)$	723	p_{ZrBr_4}	12.04	5,800	500	650	633
$ZrI_3(s)$	Disp.	p_{ZrI_4}§	12.47	8,700	548	598	
$ZrI_4(s)$	772	p_{ZrI_4}	11.84	6,280	423	671	701

†Disp. = disproportionates.

‡$2ZrCl_3(s) \rightleftharpoons ZrCl_2(s) + ZrCl_4(g)$.

§$8ZrI_3(s) \rightleftharpoons 6ZrI_3 \cdot ZrI_2(s) + ZrI_4(g)$.

Source: International Atomic Energy Agency, "Zirconium: Physico-Chemical Properties of Its Compounds and Alloys," *Atomic Energy Rev.*, Special Issue No. 6, 1976.

Table 7.7 Thermodynamic properties of $ZrCl_4$

Temperature, K	Phase	Heat capacity, cal/(g-mol·K)	Enthalpy $H - H^\circ_{298}$, cal/g-mol	Heat of formation ΔH°_f, cal/g-mol	Free energy of formation ΔG°_f, cal/g-mol
298	Solid	28.630	0	−234,170	−212,545
300	Solid	28.660	53	−234,158	−212,410
400	Solid	29.970	2,991	−233,508	−205,259
500	Solid	30.760	6,030	−232,621	−198,276
600	Solid	31.340	9,136	−232,131	−191,432
700	Solid	31.820	12,294	−231,435	−184,704
600	Ideal gas	25.176	33,820	−207,477	−191,158
700	Ideal gas	25.345	36,347	−207,382	−188,450
800	Ideal gas	25.456	38,887	−207,347	−185,749
900	Ideal gas	25.534	41,437	−207,339	−183,049
1000	Ideal gas	25.590	43,993	−207,360	−180,351
1500	Ideal gas	25.724	56,828	−208,171	−166,545

Source: National Bureau of Standards, *JANAF Thermochemical Tables*, 2d ed., U.S. Government Printing Office, Washington, D.C., June 1971.

formation. Zirconium-containing species present include un-ionized $Zr(SO_4)_2$ and a number of complex sulfatozirconic anions, of which the best known is the disulfatozirconic acid anion $ZrO(SO_4)_2^{2-}$. Evaporation of such solutions produces crystals of disulfatozirconic acid trihydrate, $H_2ZrO(SO_4)_2 \cdot 3H_2O$, formerly regarded as zirconium sulfate tetrahydrate, $Zr(SO_4)_2 \cdot 4H_2O$. Dehydration of these crystals at 100°C produces the anhydrous acid $H_2ZrO(SO_4)_2$. Further heating at 380°C produces anhydrous $Zr(SO_4)_2$. A number of salts of this acid have been prepared.

ZrF_4 is partially hydrolyzed by water and is only slightly soluble in it. It is soluble in aqueous solutions of HF. From such solutions a fluozirconate, M_2ZrF_6, can be crystallized by

Table 7.8 Comparison of melting and subliming temperatures of compounds of hafnium and zirconium†

Compound‡	Temperature, K: Melting, Hafnium	Melting, Zirconium	Vapor pressure = 760 Torr, Hafnium	Vapor pressure = 760 Torr, Zirconium
XO_2	3063	2953		
XC	4110	3805		
XF_4	(1200)	1205	(1200)	1177
XCl_4	705	710	590	608
XBr_4	693	723	595	633
XI_4	(750)	772	(700)	701

†Sources of data: hafnium, Lustman and Kerze [L1]; zirconium, *JANAF Tables* [N1] and IAEA [I1]; ( ) = estimated.

‡X = hafnium or zirconium.

Table 7.9 Comparison of free energies of formation of halides of hafnium and zirconium at 1000 K†

	Free energy of formation from elements at 1000 K, kcal/g-mol	
Element	Hafnium	Zirconium
Tetrafluoride (*s*)	−363	−378
Tetrachloride (*g*)	−203	−180
Tetrabromide (*g*)	−172	−154
Tetraiodide (*g*)	−118	−104

†Sources of data: ZrF_4 and $ZrCl_4$ [N1]; others [L1].

addition of an alkali chloride, carbonate, or hydroxide. Potassium fluozirconate can be made by reacting zircon with potassium fluosilicate at 1000°C.

$$ZrSiO_4 + K_2SiF_6 \rightarrow K_2ZrF_6 + 2SiO_2$$

Its solubility in water at 20°C is 0.055 g-mol/liter. Fractional crystallization of a fluozirconate was one of the early methods used to separate hafnium from zirconium.

A much more complete description of the chemistry of zirconium compounds is given by Blumenthal [B3].

6 EXTRACTION OF ZIRCONIUM AND HAFNIUM FROM ZIRCON

6.1 Composition of Zircon

A chemical analysis of a typical Florida zircon is given in Table 7.10 [B2]. In addition to the compounds listed in this table, zircon often contains a few hundredths of a percent of uranium and thorium. These elements must be removed from zirconium in subsequent processing because they would form fission products if present in zirconium cladding.

6.2 Zircon Extraction Processes

Zircon is quite resistant to chemical attack. The three most important processes for breaking down zircon chemically and separating zirconium plus hafnium from silicon are

Table 7.10 Composition of Florida zircon, w/o

$(Zr,Hf)O_2$	65 min
Total SiO_2	34 max
Free SiO_2	1 max
Fe_2O_3	0.15 max
Al_2O_3	1 max
TiO_2	0.25 max

1. Chlorination in the presence of carbon
2. Fusion with alkalis, followed by solution in acid
3. Fusion with K_2SiF_6

The first process makes $ZrCl_4$; the second, a zirconium salt; and the third, K_2ZrF_6. The most suitable extraction process depends, in part, on the chemical form in which the zirconium is wanted and on the subsequent steps to be used in separating hafnium (Sec. 7) and in preparing zirconium metal (Sec. 8).

6.3 Chlorination of Zircon

Chlorination of zircon has been the process mainly used in the United States because it produces $ZrCl_4$, which is used in the Kroll process for making zirconium metal (Sec. 8.3), and because $ZrCl_4$ was the feed material for the first process developed for separating hafnium from zirconium, using thiocyanate extraction (Sec. 7.3).

In the early zircon chlorination plants such as used by W. J. Kroll [K3] at Albany, Oregon, zircon was first converted to zirconium carbide by reaction with graphite in a graphite-lined arc furnace at 1800°C:

$$ZrSiO_4 + 4C \rightarrow ZrC + SiO + 3CO$$

The silicon monoxide, volatile at 1800°C, distilled off. The ZrC was then converted to $ZrCl_4$ by chlorination at 500°C:

$$ZrC + 2Cl_2 \rightarrow ZrCl_4 + C$$

In the newer plants, such as the Wah Chang plant, a mixture of zircon and carbon is chlorinated at 1200°C, to produce $ZrCl_4$ in a single step:

$$ZrSiO_4 + 4C + 4Cl_2 \rightarrow ZrCl_4 + SiCl_4 + 4CO$$

This process has the advantages of operating at lower temperature and of converting silicon to a useful by-product ($SiCl_4$) instead of to a troublesome airborne contaminant (SiO). The reaction is endothermic and requires good thermal contact between carbon and zircon. A fluid-bed reactor is used, and energy is provided either chemically by addition of silicon carbide or physically by electrical resistance heating of the bed.

The principal steps in this direct chlorination process for converting zircon to $ZrCl_4$ are shown in Fig. 7.4. Gases from the chlorinating furnace are cooled to around 100°C to condense crude solid $ZrCl_4$ and $FeCl_3$, then cooled further to condense $SiCl_4$, $TiCl_4$, and $AlCl_3$. The crude $ZrCl_4$ is purified by sublimation with hydrogen in a stainless steel retort. Hydrogen reduces volatile $FeCl_3$ to nonvolatile $FeCl_2$, which remains in the retort with ZrO_2 and other nonvolatile impurities. This process removes most of the metals associated with zirconium in zircon except hafnium.

6.4 Alkali Fusion

The alkali-fusion process was developed by the Ames Laboratory of the U.S. Atomic Energy Commission [B2] to provide a method for producing zirconium salts that did not need the high temperature of an electric furnace. A flow sheet for this process is shown in Fig. 7.5. In this process, zircon sand is mixed with from 1.0 to 1.5 times its weight of sodium hydroxide, and the mixture is heated in a furnace at 565°C. The sodium hydroxide melts at 318°C, and as its temperature rises it reacts with the zircon:

$$4NaOH + ZrSiO_4 \rightarrow Na_2ZrO_3 + Na_2SiO_3 + 2H_2O$$

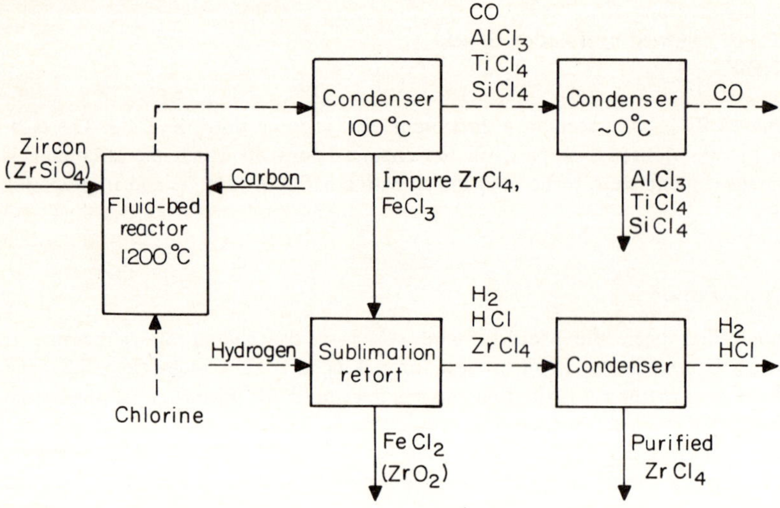

Figure 7.4 Production of $ZrCl_4$ from zircon.

Steam is evolved, the mix becomes viscous, and finally is converted to a fragile, porous solid ("frit") when the temperature reaches 530°C. After cooling, this solid is ground and leached with water, which extracts the Na_2SiO_3. The residue then is leached with acid, which dissolves the Na_2ZrO_3. The final residue consists of unreacted zircon, which may be recycled. Any desired zirconyl salt can be made by using the appropriate acid in the final leaching step.

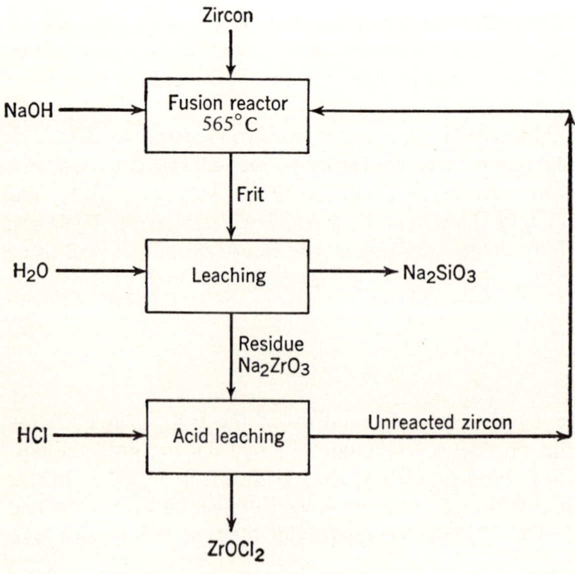

Figure 7.5 Alkali-fusion process.

This process would appear to be especially suitable for preparing feed for separating hafnium from zirconium by solvent extraction from an aqueous solution.

6.5 Fluosilicate Fusion

Fluosilicate fusion has been used in the Soviet Union [S1] to produce feed for separation of hafnium from zirconium by fractional crystallization of K_2MF_6. Zircon is ground to pass 200 mesh and mixed with potassium fluosilicate and potassium chloride (to act as promoter). The mixture is sintered in a rotary furnace at 650 to 700°C. The following reaction takes place:

$$ZrSiO_4 + K_2SiF_6 \rightarrow K_2ZrF_6 + 2SiO_2$$

The sinter is cooled, crushed to pass 100 mesh, and leached at 85°C with 1 percent HCl. The product is filtered at 80°C, then cooled, to crystallize $K_2ZrF_6(+K_2HfF_6)$, which are filtered off and washed with water.

7 SEPARATION OF ZIRCONIUM AND HAFNIUM

7.1 Methods

Lustman and Kerze [L1], pp. 115-129, list a number of processes that have been used on the laboratory scale for separation of zirconium and hafnium. Of these, three that have been used on an industrial scale are

1. Fractional crystallization of double fluorides
2. Solvent extraction of the thiocyanates by hexone
3. Solvent extraction of the nitrates by tributyl phosphate, TBP

These processes are described in Secs. 7.2 through 7.4. Another recently patented [M2, M3] process,

4. Selective reduction of the molten double fluorides by aluminum dissolved in molten zinc

seems promising and is described in Sec. 7.5.

7.2 Fractional Crystallization

Solubilities of corresponding salts of hafnium and zirconium are compared in Table 7.11.

Fractional crystallization of the double potassium fluorides was the method originally used to separate hafnium from zirconium. Because these salts form solid solutions and the ratio of solubilities is close to unity (1.54 at 20°C), multiple recrystallizations are necessary for the necessary completeness of separation. In the United States, Kawecki [K1] has found that 10 recrystallizations of the double potassium fluorides reduced the hafnium content of zirconium from 2.0 to 0.1 percent.

A similar process has been used in the Soviet Union [S1], where the operating temperatures and solubilities were as follows:

	Temperature, °C	Moles of K_2ZrF_6 per liter
Dissolver	100	0.88
Crystallizer	19	0.058

Table 7.11 Solubility of salts of hafnium and zirconium

			Solubility, g mol/liter		
Salt[†]	Solvent	Temperature, °C	Zirconium	Hafnium	Hafnium/zirconium ratio
$(NH_4)_2MF_6$	H_2O	0	0.611	0.890	1.46
$(NH_4)_3MF_7$	H_2O	0	0.360	0.425	1.18
K_2MF_6	0.125 N HF	20	0.0655	0.1008	1.54
$MOCl_2$	11.6 N HCl	20	0.33	0.15	0.46

[†]M = zirconium or hafnium.

After 16 to 18 recrystallizations, the hafnium content of zirconium was reduced to 0.003 percent. The yield of zirconium was about 80 percent.

Because of the large number of independent steps, this fractional crystallization process has been superseded by the solvent extraction processes next to be described.

7.3 Solvent Extraction of Thiocyanates

History. In 1947, Fischer and co-workers [F1, F2] described a solvent extraction method for separating hafnium from zirconium in which an aqueous solution of sulfates containing ammonium thiocyanate was extracted with diethyl ether containing thiocyanic acid. Hafnium concentrates preferentially in the organic phase; in one reported experiment zirconium in the aqueous phase contained 0.35 percent hafnium, while the organic phase zirconium contained more than 5 percent. Six to eight batch laboratory separations concentrated hafnium from 0.5 percent in zirconium to 70 to 90 percent.

In a study of methods for separating hafnium from zirconium, the Oak Ridge National Laboratory [O2] concluded that solvent extraction of the thiocyanate was the method best suited for commercial use. The aqueous phase recommended was an HCl solution of the oxychlorides $(Zr,Hf)OCl_2$, the solution obtained when the tetrachlorides are dissolved in water. The organic solvent recommended was methylisobutyl ketone (hexone) containing around 2.3 mol thiocyanic acid (HCNS) per liter. Hexone was preferred over diethyl ether because it is less volatile and less flammable. A hafnium-zirconium separation factor of about 5 is obtained in this system, with hafnium concentrating in the organic phase. A plant to separate 24 kg of natural zirconium per hour into hafnium and reactor-grade zirconium was built at the Y-12 plant of the Union Carbide Corporation at Oak Ridge in 1951 [R1]; its performance has been analyzed by Googin [G1]. In 1952 operation was transferred to a similar plant built at the Albany, Oregon, station of the U.S. Bureau of Mines; its construction and performance have been described by McClain and Shelton [M1]. Another similar plant was used by the Carborundum Metals Corporation at Akron, New York. Both of these plants have since been shut down. They have been superseded by a larger plant of Teledyne Wah Chang Albany Corporation at Albany, Oregon, with a capacity of around 400 kg hafnium-free zirconium per hour. This hexone-thiocyanate separation process has also been used in France and England [J1] and was studied by Fischer et al. [F3] in Germany.

U.S. Bureau of Mines plant. Figure 7.6 is a process flow sheet for the zirconium-hafnium separation portion of the U.S. Bureau of Mines zirconium plant at Albany, Oregon [M1]. Commercial-grade zirconium tetrachloride containing about 2 w/o hafnium was dissolved in water together with ammonium thiocyanate (NH_4CNS) and NH_4OH, to make a feed solution containing around 120 g zirconium + hafnium per liter that was 1.0 to 1.1 *M* in HCl and 2.7 to 2.9 *M* in NH_4CNS. The zirconium and hafnium were in the form of thiocyanate complexes that could be extracted from the aqueous solution by a solution of thiocyanic acid HCNS in hexone. This aqueous solution was fed at a rate of 189 liters/h to a solvent extraction system consisting of spray columns made of sections of Pyrex glass pipe 10.2 cm in diameter. At the feed point the feed joined the aqueous stream flowing at the rate of 76 liters/h from the scrubbing section B and entered the extracting section C, 62.8 m in total length, made up of four shorter columns. Here countercurrent extraction by a solution of HCNS in hexone reduced the hafnium content of zirconium to 0.004 w/o.

Hexone flowing from the extracting section to the scrubbing section B contained almost all the hafnium in the feed and about 30 percent of the zirconium. The scrubbing section consisted of three columns of 10.2 cm Pyrex pipe 45.4 m in total length. Countercurrent flow

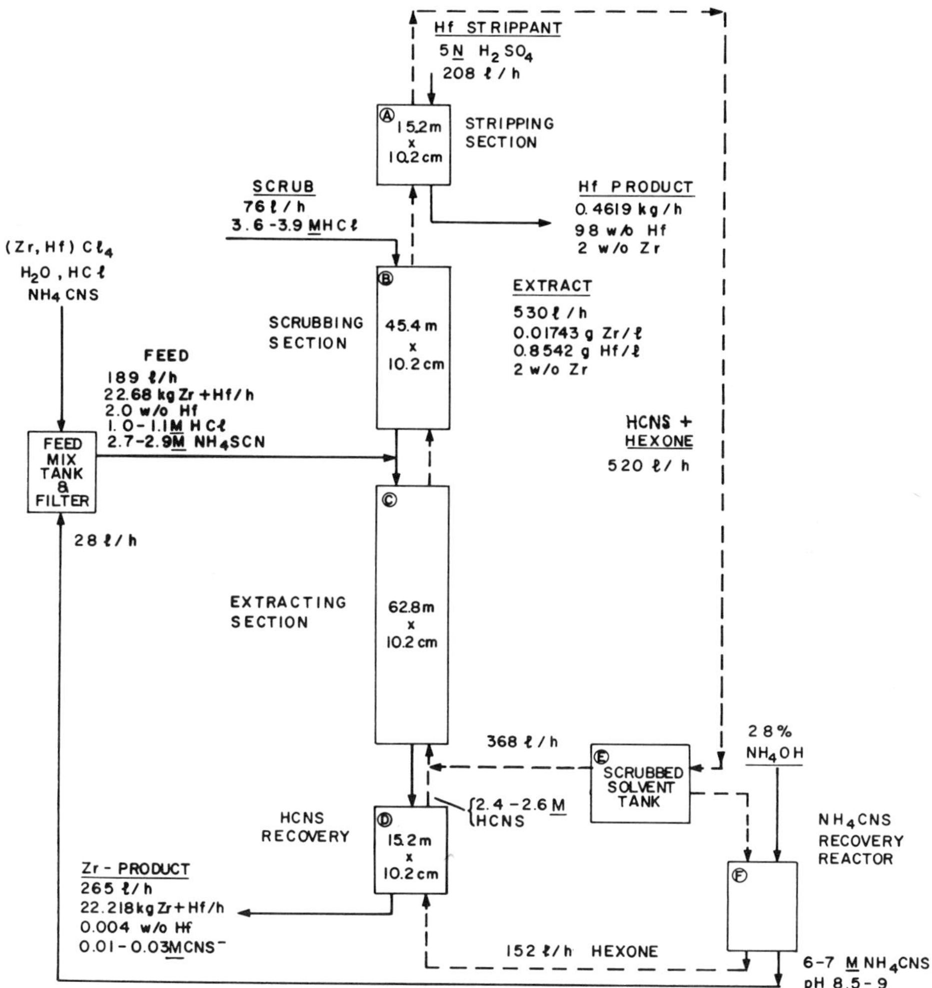

Figure 7.6 Process flow sheet for zirconium-hafnium separation portion of the U.S. Bureau of Mines plant at Albany, Oregon.

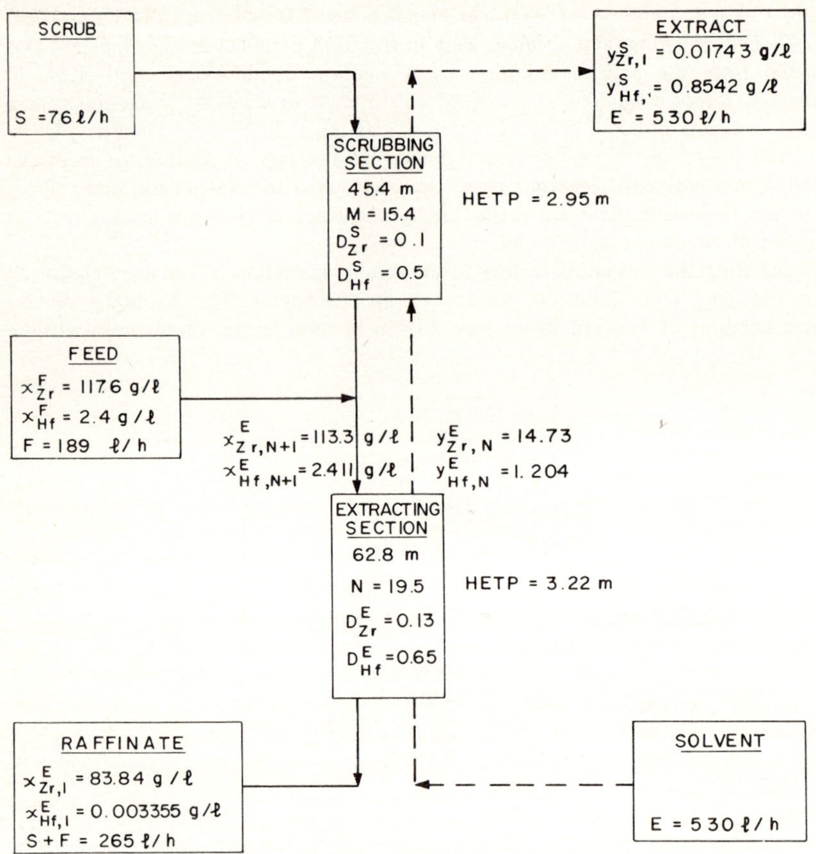

Figure 7.7 Calculated separation performance of Bureau of Mines thiocyanate solvent extraction columns for separation of zirconium and hafnium.

of 76 liters/h of 3.6 to 3.9 *M* HCl scrub solution transferred most of the zirconium from the hexone back to the aqueous phase and reduced the zirconium content of the extract to 2 w/o.

Hafnium was removed from the extract by stripping with 5 *N* H_2SO_4 in the stripping column A, which was 15.2 m long.

HCNS was recovered from the aqueous product stream by extraction with pure hexone in the HCNS recovery column D. The pure hexone was prepared by reacting a hexone solution of HCNS from the scrubbed solvent tank E with 28% NH_4OH in a cooled NH_4CNS recovery reactor F. The 6 to 7 *M* NH_4CNS thus recovered was recycled to feed.

A rough estimate of the number of theoretical plates in the scrubbing and extracting sections can be obtained from the material flow sheet for these sections, Fig. 7.7. The distribution coefficients for zirconium and hafnium assigned to the two sections were obtained from the following conditions:

1. The hafnium/zirconium separation factor for this system of 5 implies that $D_{Hf}/D_{Zr} = 5$.
2. Tests on the columns of the Y-12 plant reported by Googin [G1] showed that distribution coefficients were in the following ranges:

Scrubbing, Hf	0.5
Zr	0.012–0.10
Extracting, Hf	0.5–1.0
Zr	0.13–0.17

3. The height of an equivalent theoretical plate (HETP) in each section should be about the same.

With these assumptions the distribution coeffients and number of theoretical plates in the two sections are as given in Fig. 7.7 (Prob. 7.1). Assignment of constant distribution coefficients to each section is an oversimplification, but does permit semiquantitative representation of the more complex actual system.

Two significant points may be noted. (1) The concentrations of zirconium and hafnium in the feed are approximately equal to the corresponding concentrations in the aqueous stream entering the extracting section, a condition that minimizes loss of separation at this point. (2) The fraction of feed zirconium recycled through the scrubbing section is (14.73 g/liter)(530 liters)/(117.6 g/liter)(189 liters) = 0.35; McClain and Shelton state that about 30 percent was recycled.

Materials of construction. It is necessary to use materials of construction for this separation that resist corrosion at the high concentrations of HCl used in the process. The Bureau of Mines plant [M1] used glass columns, glass- or rubber-lined equipment, and rubber tubing connectors. The British plant [J1] used polythene mixer-settlers.

Zirconium purification and conversion to zirconium dioxide. The zirconium-product stream leaving the HCNS recovery column D of Fig. 7.6 contained most of the metal impurities in the $ZrCl_4$ feed other than hafnium. Purified zirconium was obtained by precipitating $Zr(OH)_4$ at a pH low enough to prevent precipitation of other metal hydroxides. The precipitation procedure used by the Bureau of Mines was as follows. Zirconium content of the raffinate was diluted to 19 g/liter. To every cubic meter of diluted raffinate were added 5.7 liters of concentrated 33 *N* sulfuric acid, followed by sufficient 28% ammonium hydroxide to bring the pH to 1.2 to 1.6 at a temperature controlled at 88°C. The basic zirconium sulfate precipitate was filtered off. The precipitate was twice reslurried with 28% ammonium hydroxide and filtered off to complete conversion to zirconium hydroxide. The hydroxide was dried in a rotary stainless steel drier at 350 to 400°C and converted to ZrO_2 in a rotary Thermalloy-40 retort at 700°C.

7.4 Solvent Extraction with TBP

Huré and Saint-James [H5] of the French Atomic Energy Commission have shown that TBP diluted with kerosene is a selective solvent for the fractional extraction of zirconium from hafnium. These workers recommend using an organic phase consisting of 60 v/o (volume percent) TBP and 40 percent refined kerosene, and an aqueous phase 3 *N* in nitric acid and 3.5 *N* in sodium nitrate, containing no more than 30 g zirconium/liter. Under these conditions, the distribution coefficient of zirconium is around 1.5, favoring the organic phase, and that of hafnium is only one-tenth as great, so that the separation factor is 10. Unlike thiocyanate extraction, zirconium concentrates in the organic phase with TBP.

This process was demonstrated in a pilot plant built by the French Atomic Energy Commission, which used the flow sheet shown in Fig. 7.8. The zirconium-extracting section consisted of six mixer-settler stages, and the hafnium-scrubbing section consisted of three stages. Each was 75 percent efficient. A single contact was used for the zirconium-stripping section. The plant produced 24 kg zirconium/day, containing less than 0.02 percent hafnium, from feed

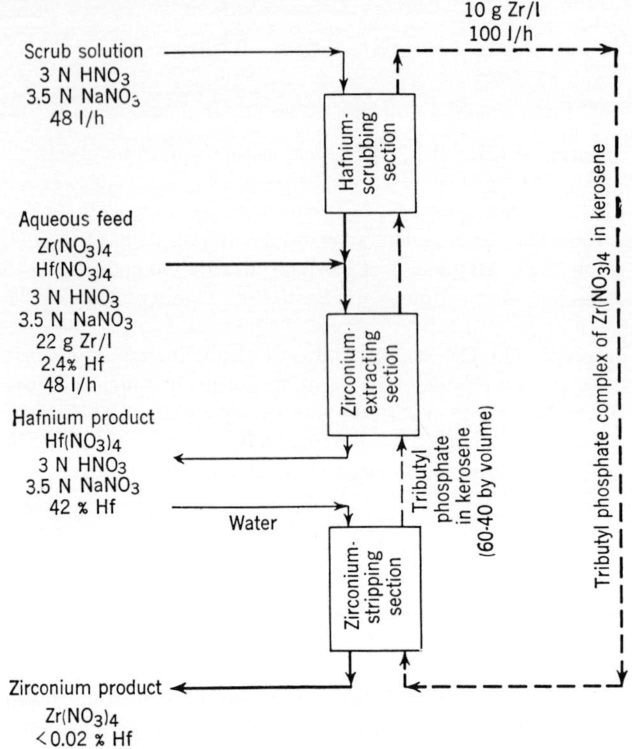

Figure 7.8 Pilot plant of French Atomic Energy Commission for separation of zirconium from hafnium by solvent extraction with TBP. Solid line, aqueous; broken line, organic.

containing 2.4 percent, and made a hafnium concentrate containing 42 percent hafnium. Solvent losses by hydrolysis and solution in water were around 2 percent per cycle. Separation performance in this plant was analyzed in the examples given in Chap. 4.

Pilot-plant work on a similar process was conducted by Cox et al. [C3] at the Ames, Iowa, Laboratory of the U.S. Atomic Energy Commission. A 14-stage mixer-settler cascade was used, with 10 extracting and 4 scrubbing stages. Table 7.12 gives reported flow rates and compositions. This process differs from the one developed by Huré and Saint-James in using only HNO_3 as salting agent, without $NaNO_3$. As HNO_3 is easier to recover and recycle (by distillation) than $NaNO_3$, HNO_3 alone is preferable in a commercial process. Reported separation factors ranged between 2.5 and 36. This pilot-plant work provided the design data for separation of hafnium from zirconium in the Columbia National Corporation plant [K2].

7.5 Selective Reduction of Double Fluorides by Aluminum

Two recent patents [M2, M3] by J. A. Megy describe a process in which zirconium metal is reduced from a salt and separated from hafnium in the same step, thus shortening the long series of steps in present processes for producing reactor-grade zirconium from natural zircon.

The Megy process has evolved from the finding of Petenev and Ivanovskii [P1] that when a mixture of K_2ZrF_6 and K_2HfF_6 dissolved in molten alkali chlorides was reduced electrolytically at a molten zinc cathode, the metal phase was enriched in zirconium relative to the residual salt. Megy found that when a mixture of Na_2ZrF_6 and Na_2HfF_6 is reduced by aluminum dissolved in liquid zinc, a very high separation factor between hafnium and zirconium is obtained, with very little contamination of the zirconium by aluminum. Examples given in Megy's patent [M2] indicate that the ratio of zirconium to hafnium in the metal phase may be as high as 328 times the ratio of these elements in the residual salt phase. This high separation factor, $\alpha = 328$, makes possible production of reactor-grade zirconium containing less than 0.01 w/o hafnium from typical natural zirconium containing 2 w/o hafnium in two stages of salt-metal contact, such as shown in Fig. 7.9.

Material quantities in Fig. 7.9 are based on production of 1.000 mol zirconium at point 6. Feed to this process (point 1) consists of 1.058 mol Na_2ZrF_6 and 0.011 mol Na_2HfF_6, corresponding to 2 w/o hafnium in zirconium + hafnium. This feed is combined with 0.0558 mol Na_2ZrF_6 and Na_2HfF_6 of the same composition (point 5) recycled from a previous batch. In step A the salt feed is reacted in a graphite-lined container at 900°C with a metallic solution of 4 w/o aluminum in molten zinc containing 1.4078 mol aluminum. Reactions taking place are

$$3Na_2ZrF_6 + 4Al(Zn) \rightarrow 4(NaF)_{1.5}AlF_3 + 3Zr(Zn)$$

and

$$3Na_2HfF_6 + 4Al(Zn) \rightarrow 4(NaF)_{1.5}AlF_3 + 3Hf(Zn)$$

At the reaction temperature the products, $(NaF)_{1.5}AlF_3$ and the solution of zirconium and hafnium in zinc, are two immiscible liquids. The lighter salt phase, enriched in hafnium to 27 w/o, is drawn off at point 4, leaving a heavier metallic zinc solution of zirconium containing 0.1126 w/o hafnium at point 3. These compositions are consistent with the separation factor of 328:

$$\frac{(27)(100-0.1126)}{(100-27)(0.1126)} = 328 \tag{7.4}$$

To reduce the hafnium content below the 0.01 w/o specified for reactor-grade zirconium, the solution of zirconium and hafnium in molten zinc at point 3 is contacted in step B with a liquid mixture of 0.1116 mol ZnF_2† and 0.1116 mol NaF. Reactions

$$Zr + 2ZnF_2 + 2NaF \rightarrow Na_2ZrF_6 + 2Zn$$

†Alternatively, an equivalent amount of natural sodium fluozirconate could be used, with only slight increase in hafnium content of product zirconium.

Table 7.12 Flow rates and compositions in TBP solvent extraction pilot plant for hafnium-zirconium separation

Stream	Solvent feed, 60 v/o TBP in heptane	Aqueous: Feed	Raffinate	Scrub	Strip	Product
Flow rate, liters/h	5.0	1.0	2.2	1.18	1.62	1.56
Mol HNO_3/liter	0.65	5.1	2.65	5.4	0	3.96
g $(Zr,Hf)O_2$/liter	0	127	3.5	0	0	76.4
w/o hafnium in zirconium + hafnium	–	2.2	43.8†	–	–	<0.01

†This reported composition does not satisfy a hafnium material balance, probably because of unsteady cascade conditions; 30.9 w/o hafnium would balance.

Source: R. P. Cox et al., *Ind. Eng. Chem.* **50**: 141 (1958).

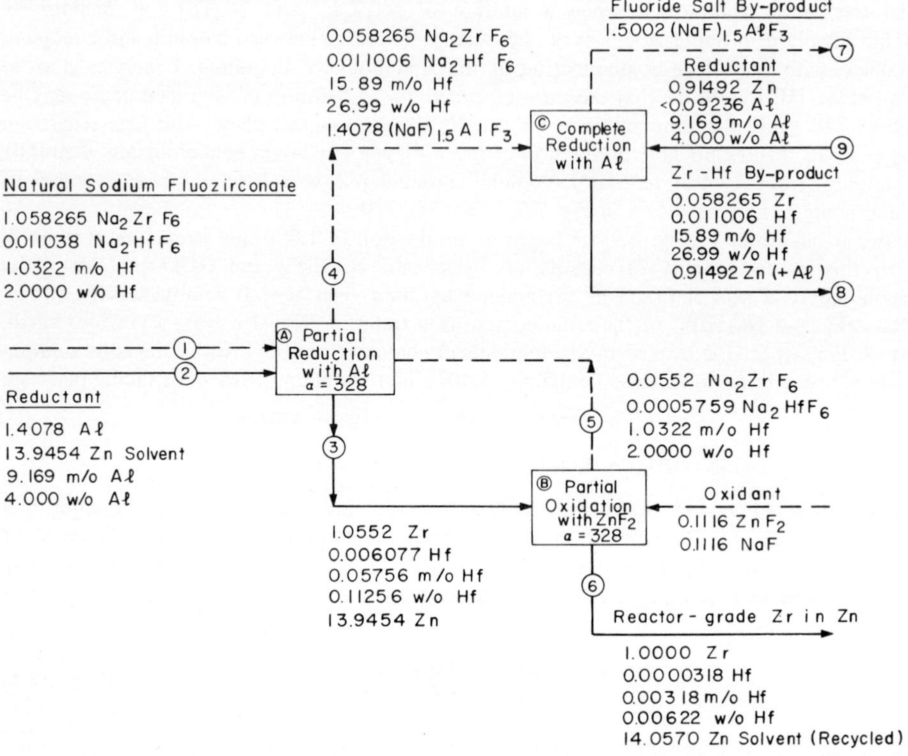

Figure 7.9 Megy process for producing reactor-grade zirconium from natural sodium fluozirconate. Material quantities in moles. Basis, 1 mol zirconium product. ––– salt; —— metal.

and $$Hf + 2ZnF_2 + 2NaF \rightarrow Na_2HfF_6 + 2Zn$$

take place, with most of the hafnium and a small fraction of the zirconium reacting. The hafnium content of the residual zirconium metal at point 6 is 0.00622 w/o, thus meeting the 0.01 w/o specification of reactor-grade zirconium.

The hafnium content of the salt at point 5 is at the feed level of 2 w/o, again satisfying the separation factor condition:

$$\frac{(2)(100-0.00622)}{(100-2)(0.00622)} = 328 \tag{7.5}$$

The salt at point 5 is recycled to step A of a later batch. The zinc solvent at point 6 is distilled from the zirconium product and recycled to a later batch at point 2.

The mixture of $(NaF)_{1.5}AlF_3$, Na_2ZrF_6, and Na_2HfF_6 is converted to more useful by-products by reduction with 4 w/o aluminum in zinc in step C. This produces a zirconium and hafnium-free fluoride salt by-product 7, $(NaF)_{1.5}AlF_3$, and a solution of 27 percent Hf, 73 percent Zr in zinc, 8. The $(NaF)_{1.5}AlF_3$ can be sold as a substitute for cryolite Na_3AlF_6, in electrolytic production of aluminum. The zinc can be distilled from the 27 percent Hf, 73 percent Zr and recycled to point 9, and the hafnium-zirconium alloy can be sold for metallurgical applications in which the high cross section of hafnium is not harmful.

Figure 7.10 shows how the Megy selective reduction process can be combined with K_2SiF_6

fusion to produce reactor-grade zirconium from zircon ore. The ore is fused with K_2SiF_6 in a graphite-lined arc furnace A at 1000°C to convert zirconium to K_2ZrF_6 and K_2HfF_6:

$$K_2SiF_6 + (Zr,Hf)SiO_4 \rightarrow K_2(Zr,Hf)F_6 + 2SiO_2$$

Potassium is preferred to sodium because the potassium complex fluorides are more stable at this temperature. The $K_2(Zr,Hf)F_6$ is dissolved in water and filtered from insoluble SiO_2 at B and crystallized at C. To recover the relatively expensive potassium, the $K_2(Zr,Hf)F_6$ crystals are dissolved in heated NaCl brine and cooled to precipitate the less soluble $Na_2(Zr,Hf)F_6$ at D:

$$K_2(Zr,Hf)F_6 + 2NaCl \rightarrow Na_2(Zr,Hf)F_6 + 2KCl$$

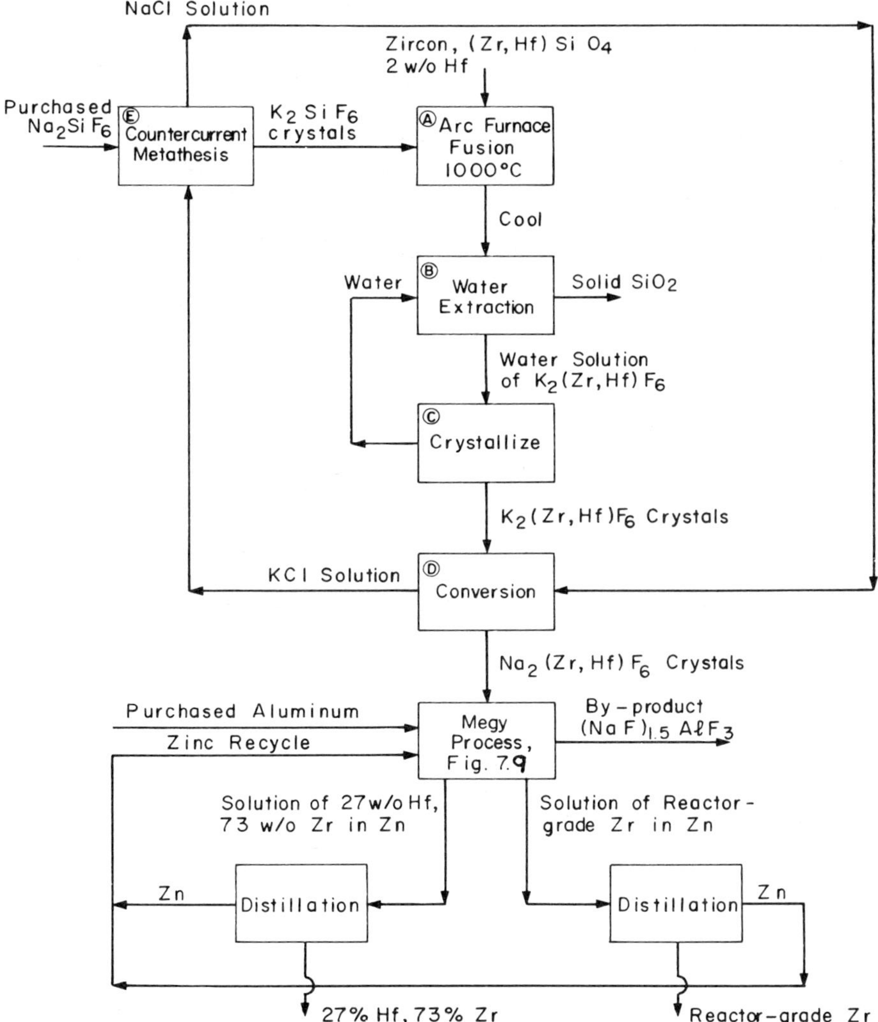

Figure 7.10 Production of reactor-grade zirconium from zircon by combination of K_2SiF_6 fusion and Megy process.

The $Na_2(Zr,Hf)F_6$ is converted to zirconium metal by the Megy process described earlier. The KCl is recycled and converted to K_2SiF_6 by countercurrent metathesis with purchased Na_2SiF_6 at E.

$$2KCl + Na_2SiF_6 \rightarrow K_2SiF_6 + 2NaCl$$

In this way zircon, Na_2SiF_6, and aluminum are converted to zirconium metal, hafnium-rich zirconium, and by-product $(NaF)_{1.5}AlF_3$ and SiO_2.

8 PRODUCTION OF METALLIC ZIRCONIUM AND HAFNIUM

8.1 Difficulties

The high melting point of these metals and their reactivity make their production in pure form very difficult. They form oxides, hydrides, nitrides, and carbides that are soluble in the metal, diffuse through it, and make it hard and brittle even at concentrations of a few tenths of a percent. The oxides are especially stable; once the metal has been contaminated by oxygen, no reducing agent can remove it completely. The metals react with air or nitrogen at temperatures above 300°C and, when finely divided, react with water even at room temperature. Consequently, they must be protected by helium, argon, or a vacuum during high-temperature reduction operations, casting, or hot forming, and finely divided metal cannot be cleaned by washing with water or aqueous solutions. The molten metal reacts with all known refractories, even graphite or lime, which can be used for uranium.

All these difficulties require that chemicals for producing these metals be exhaustively purified, especially of oxygen, water, and nitrogen, and limit the number of processes that can be used.

8.2 Available Processes

The principal processes that have been used for producing zirconium and hafnium metal of the requisite purity are as follows:

1. The Kroll process, involving reduction of tetrachloride vapor by molten magnesium
2. The hot-wire process, involving thermal decomposition of the iodide
3. Electrolysis of the double potassium fluoride dissolved in fused salts

8.3 Kroll Process

In the United States, practically all zirconium metal is now being made by the Kroll process. This process was an adaptation to zirconium of a similar process for titanium developed by W. J. Kroll. The work of Kroll and metallurgists of the Albany, Oregon, station of the Bureau of Mines culminated in a plant to produce 135,000 kg zirconium/year at the station. A similar plant was operated by the Carborundum Metals Corporation, at Akron, New York. These have been superseded by the plant of the Teledyne Wah Chang Albany Company, at Albany, Oregon, with a capacity in 1978 of 3.4 million kg/year.

The form of the Kroll process used in this plant is believed to be generally similar to the process used at the Bureau of Mines plant, which has been described in detail by Shelton et al. [S3]. The principal steps in the Kroll process as practiced in the Bureau of Mines plant are shown in Fig. 7.11.

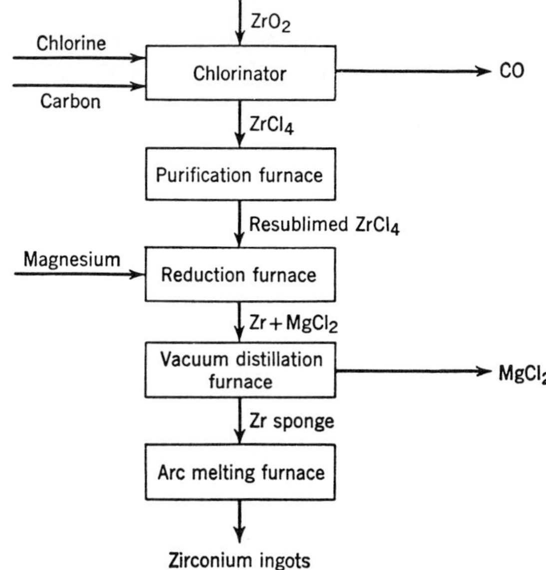

Figure 7.11 Kroll zirconium process as practiced at Albany, Oregon.

Production of $ZrCl_4$. Zirconium oxide from the hafnium-separation step was mixed with carbon black, dextrin, and water in proportions 142 ZrO_2, 142 C, 8 dextrin, and 8 water. The mixture was pressed into small briquettes (3.8 × 2.5 × 1.9 cm) and dried at 120°C in a tray drier. The oxide briquettes were charged to the reaction zone of a vertical-shaft chlorinator lined with silica brick. The charge was first heated by carbon resistance strips until it became conductive. During production, the bed temperature was maintained at 600 to 800°C by an electric current passed directly through the bed. After steady conditions were reached, a reactor 66 cm in diameter produced about 25 kg $ZrCl_4$/h. The $ZrCl_4$ was condensed from the reaction products in two cyclone-shaped aftercondensers in series, and the chlorine off-gas was removed in a water scrubbing tower.

The Wah Chang plant is believed to use an electrically heated fluidized-bed chlorination reactor.

Reduction of $ZrCl_4$. The furnace used for reducing $ZrCl_4$ to zirconium metal is shown in Fig. 7.12. The outer shell is a stainless steel cylinder 178 cm high and 70 cm inside diameter. An annular trough 5 cm wide and 23 cm deep is welded to the top lip of the cylinder. The top lid of the furnace carries a cylindrical ring that dips into this trough. The trough is filled with a low-melting lead alloy. This arrangement facilitates opening and closing the furnace.

The top lid also carries stainless steel cooling coils, through which air may be circulated, to control top temperatures and prevent loss of $ZrCl_4$ vapor in gases discharged from the furnace. The furnace is provided with three external electric resistance heating elements to provide the heat of sublimation of $ZrCl_4$ and control temperature distribution.

A stainless steel reduction crucible rests on the bottom of the furnace. Before a run is started, this is charged with 55 kg of distilled magnesium.

Resting on the top of the crucible is an Inconel can charged with raw $ZrCl_4$. The total amount of $ZrCl_4$ charged here and possibly present on the air-cooling coils from a previous run is 236 kg, an amount that provides a magnesium excess of 10 or 15 percent for the reduction reaction

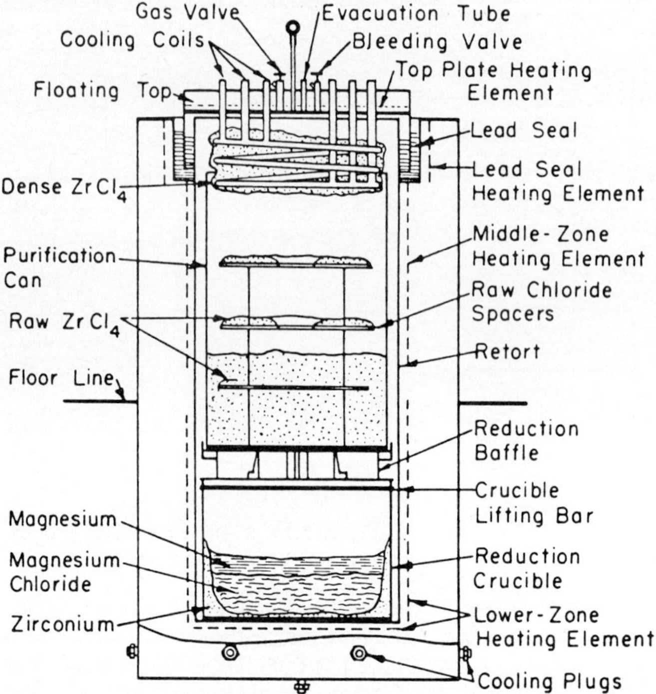

Figure 7.12 Zirconium-reduction furnace, Kroll process.

$$ZrCl_4 + 2Mg \rightarrow Zr + 2MgCl_2$$

Inconel rods extending into the $ZrCl_4$ improve heat transfer.

The detailed operating cycle has been described by Shelton et al. [S3]. A brief summary of the procedure follows. With the reactor first at 300°C and cooling air flowing through the top coils, the furnace is evacuated and flushed with helium three times to remove gases originally present. Temperatures are then raised to 450°C while the internal pressure is kept near atmospheric by bleeding helium. This purges additional gas occluded in the $ZrCl_4$ charge. Top temperatures are then reduced to lower the $ZrCl_4$ vapor pressure, while additional helium is fed to hold pressure near atmospheric. The temperature of the reduction crucible containing the magnesium is next raised to 825°C, at which reaction with $ZrCl_4$ vapor commences. The $ZrCl_4$ transport rate is controlled by the rate at which heat is supplied to the middle-zone heater. The rate is kept as high as possible without raising the temperature in the magnesium reaction zone over 875°C. Completion of reaction is indicated by a fall in pressure, which is countered by supplying additional helium. Heaters are then turned off and the vessel is cooled to 150°C by air blown over the outer reactor surface.

The total cycle time is around 40 h, of which 18 h is for the reduction reaction itself. The product of the reaction is a lower layer of spongy zirconium metal mixed with $MgCl_2$, covered by a layer of frozen $MgCl_2$.

Vacuum distillation of $MgCl_2$. To remove $MgCl_2$ from the zirconium sponge it is necessary to resort to vacuum distillation. Water leaching cannot be used because the finely divided zirconium sponge would become contaminated by oxide corrosion product.

The crucible containing the zirconium sponge and $MgCl_2$ is transferred to the vacuum distillation retort, shown in Fig. 7.13, where it is supported, upside down, over a perforated, stainless steel funnel. The air in the retort is evacuated, and the crucible is heated to 900 to 920°C to melt the $MgCl_2$, which partially drains off the sponge. Salt still wetting the sponge is distilled at this temperature to a water-cooled jacket inside the retort.

Arc melting. As a last step, the salt-free sponge is fed into an arc-melting furnace and cast into the desired shapes in a water-cooled, copper mold. The furnace atmosphere is helium.

Details of the Kroll process are given in papers by Kroll and his co-workers [K3, K4], Shelton [S3], and Lustman and Kerze [L1].

Reduction with sodium. A modified process in which $ZrCl_4$ was reduced by sodium was used by National Distillers and Chemicals Corporation in Ashtabula, Ohio, during the late 1950s [C1].

Production of hafnium metal. Hafnium metal has been produced from $HfOCl_2$ by the same methods used in making zirconium. Up to the vacuum distillation step, separate equipment was used than for zirconium, to avoid contaminating the zirconium.

8.4 The Hot-Wire Process

The hot-wire process was developed by Van Arkel and de Boer [V2], who used it to produce the first pure, massive specimens of many refractory metals, notably titanium, zirconium, hafnium, and thorium. An interesting account of early uses of this process is given in

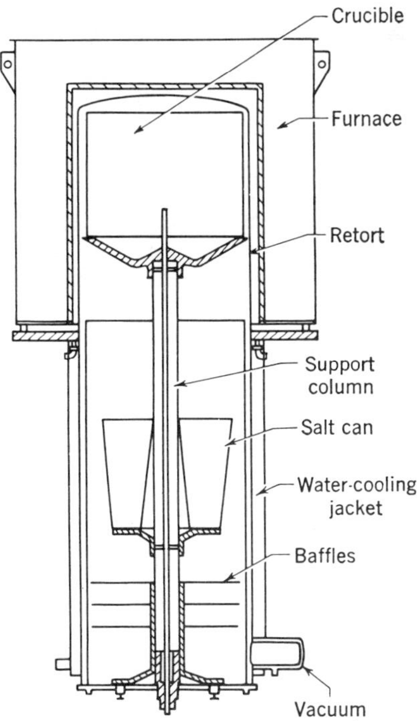

Figure 7.13 Vacuum-distillation furnace, Kroll process.

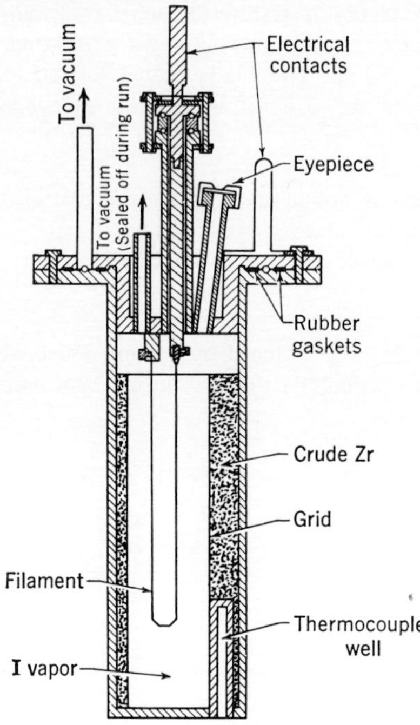

Figure 7.14 Hot-wire reactor for zirconium production.

Van Arkel's book *Reine Metalle* [V1]. This method was used by the Foote Mineral Company and the Westinghouse Electric Corporation to produce the first zirconium pure enough for nuclear reactors.

The apparatus used by Foote is shown in Fig. 7.14. It consists of an Inconel tube through which are led insulated tungsten leads capable of carrying a heavy electric current. Inside the tube, the leads are connected to a thin tungsten wire. The tube is charged with crude zirconium and evacuated, and a few grams of iodine are distilled into it. The tube is heated to a temperature at which iodine reacts with the zirconium and at which the iodide produced has a vapor pressure of several torrs. The tungsten wire is heated electrically to a temperature high enough to dissociate the iodide, but below that at which the metal melts or has a substantial vapor pressure. Tetraiodide, formed from the crude metal, diffuses through the iodine vapor and deposits pure metal on the tungsten wire. As the latter increases in cross section, the electric current through it is increased to keep it above the dissociation temperature of the iodide. The run is concluded when the tungsten leads are carrying the maximum possible current.

In this way rods, or "crystal bars," of compact, ductile zirconium or hafnium have been prepared. The usual crystal bar is 0.25 to 0.4 in in diameter in lengths up to 2 ft, but Westinghouse and Battelle Memorial Institute have produced zirconium bars as large as 1.7 in in diameter and 50 ft overall length [H1].

The hot-wire process eliminates oxygen, nitrogen, and carbon, the impurities most difficult to keep out of zirconium in other processes, but other metals that form volatile iodides are not removed completely. The main disadvantage of the process is its low capacity, the rate of production being limited by the rate of diffusion of iodide vapor to the small wire. Temperatures used for producing metals of the IVA group by the hot-wire process are listed in Table 7.13.

Eighty-five percent of the zirconium used in the first land-based prototype of a submarine reactor was made by the hot-wire process. In 1952, the hot-wire process for zirconium was superseded by the lower-cost Kroll process. However, the hot-wire process is still used to produce hafnium for control rods in U.S. naval reactors.

8.5 Electrolysis of Fused Salts

Processes for making ductile zirconium by electrolysis of K_2ZrF_6 dissolved in molten chlorides have been described by Steinberg et al. in the United States [S4, R2] and by Ogarev et al. [O1] in the Soviet Union. An advantage over the Kroll process is that a coarsely crystalline product is obtained from which coproduced halides can be removed by leaching with acidified water without undue contamination of zirconium by oxygen. The washed crystals are then vacuum dried and consolidated by arc melting.

Steinberg et al. [S4], of Horizons, Inc., developed a process for making ductile zirconium by electrolysis of K_2ZrF_6 dissolved in fused sodium chloride. The essential features of the process were as follows:

1. A gastight cell filled with a purified argon atmosphere
2. A carbon crucible to hold the fused-salt bath and act as anode
3. Electrolyte consisting of high-purity K_2ZrF_6 and NaCl, free from oxygen and water
4. Steel cathode
5. 850°C bath temperature
6. Initial concentration, 30 to 35 w/o K_2ZrF_6 in NaCl
7. 3.5 to 4.0 V
8. Current density, 250 to 400 A/dm^2

The current efficiency was about 60 percent and the current yield around 0.5 g zirconium/A·h.

The largest cell developed [R2] produced zirconium at a rate of from 4 to 6 lb/h, and 30 to 40 lb of zirconium metal per run. A cell could be used for a minimum of six runs before the operation was terminated by buildup of NaF and KF formed in the overall reaction

$$K_2ZrF_6 + 4NaCl \rightarrow Zr + 2Cl_2 + 2KF + 4NaF$$

A somewhat similar process has been described by Ogarev et al. [O1], the principal differences being the use of KCl-K_2ZrF_6 as electrolyte and a steel cell with frozen salt wall to contain it.

Table 7.13 Temperatures for production of group IVA metals by the hot-wire process

	Temperature, °C	
Metal	Crude metal	Filament
Titanium	200	1300
Zirconium	340	1300
Hafnium	600	1600
Thorium	480	1700

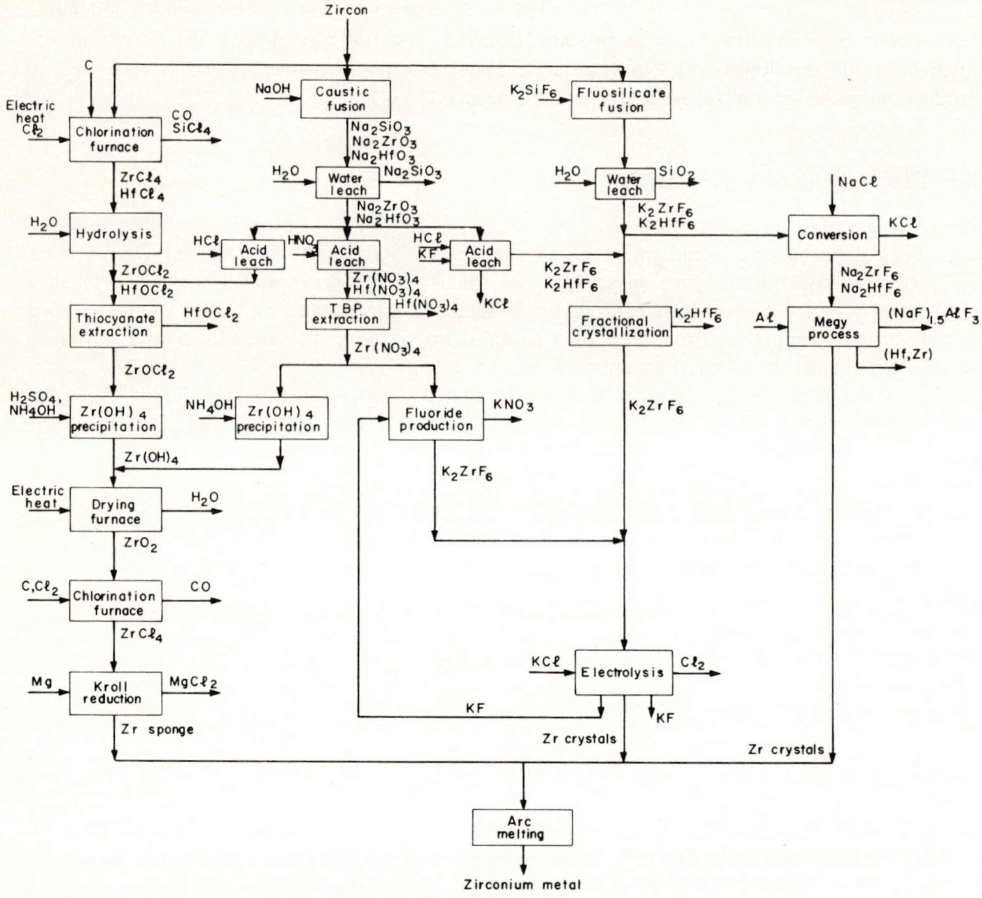

Figure 7.15 Routes from zircon to zirconium metal.

9 ALTERNATIVES FOR PRODUCING HAFNIUM-FREE ZIRCONIUM FROM ZIRCON

The foregoing processes for extracting zirconium from zircon, separating zirconium from hafnium, and reducing zirconium compounds to metal can be combined in a number of ways, some of which are shown in Fig. 7.15. The combination of processes used in the United States in 1978 is at the left of this figure.

REFERENCES

A1. American Society for Testing and Materials: *1976 Annual Book of ASTM Standards:* part 8, *Nonferrous Metals–Nickel, Lead and Tin Alloys, Precious Metals, Primary Metals; Reactive Metals*, American Society for Testing and Materials, Philadelphia, 1976.

A2. Ampian, S. G.: In *Minerals Yearbook, 1974*, vol. I, *Metals, Minerals and Fuels*, U.S. Government Printing Office, Washington, D.C., 1976, pp. 1397–1415.

B1. Beck, R. L., and W. M. Mueller: "Zirconium Hydrides and Hafnium Hydrides," in *Metal Hydrides*, W. M. Mueller, J. P. Blackledge, and G. G. Libowitz (eds.), Academic Press, New York, 1968, chap. 7.
B2. Beyer, G. H., et al.: *Chem. Eng. Progr. Symposium Ser. 12* **50**: 67 (1954).
B3. Blumenthal, W. B.: *The Chemical Behavior of Zirconium*, Van Nostrand, Princeton, N.J., 1958.
C1. *Chem. Eng. News:* "How USI Will Make Zirconium," **35**: 92 (Mar. 18, 1957).
C2. Cook, L. A., L. S. Castleman, and W. E. Johnson: USAEC Report WAPD-25, 1950.
C3. Cox, R. P., H. C. Peterson, and G. H. Beyer: *Ind. Eng. Chem.* **50**: 141 (1958).
D1. Dayton, R. W., and C. R. Tipton: USAEC Report BMI-1273, 1958.
D2. Dieckamp, H., R. Balent, and J. R. Welch: *Nucleonics* **19**(4): (Apr. 1961).
E1. Evans, W., P. A. Ross-Ross, and J. E. LeSurf: "Metallurgical Properties of Zirconium-Alloy Pressure Tubes and Their Steel End-Fittings for CANDU Reactors," *PICG(4)* **10**: 513 (1972).
F1. Fischer, W., and W. Chalybaens: *Z. Anorg. Chem.* **255**: 79 (1947).
F2. Fischer, W., W. Chalybaens, and M. Zumbusch: *Z. Anorg. Chem.* **255**: 277 (1947).
F3. Fischer, W., et al.: *Angew. Chem. Internat. Edit.* **5**: 15 (1966).
G1. Googin, J. M.: "The Separation of Hafnium from Zirconium," in *Progress in Nuclear Energy, Process Chemistry*, series III, vol. 2, F. R. Bruce et al. (eds.), Pergamon, New York, 1958, p. 194.
H1. Hampel, C. A. (ed.): *Rare Metals Handbook*, Reinhold, New York, 1954, chap. 28.
H2. Hartmann, I.: USAEC Report NYO-1562, June 20, 1951.
H3. Hillner, Edward: "Corrosion of Zirconium-Base Alloys," in *Zirconium in the Nuclear Industry*, Special Technical Publication 633, A. L. Lowe and G. W. Parry (eds.), American Society for Testing and Materials, Philadelphia, 1977.
H4. Hultgren, R., et al.: *Selected Values of the Thermodynamic Properties of the Elements*, American Society for Metals, Metals Park, Ohio, 1973.
H5. Huré, J., and R. Saint-James: *PICG(1)* **8**: 551 (1956).
I1. International Atomic Energy Agency: "Zirconium: Physico-Chemical Properties of Its Compounds and Alloys," *Atomic Energy Rev.*, Special Issue No. 6, 1976.
J1. Jamrack, W. D.: *Rare Metal Extraction by Chemical Engineering Techniques*, Macmillan, New York, 1963, pp. 181–183.
K1. Kawecki, H. C.: Private communication to S. H. Shelton, in B. Lustman and F. Kerze, Jr., *The Metallurgy of Zirconium*, McGraw-Hill, New York, 1955, p. 117.
K2. Keller, W. H., and I. Zonis: *Chem. Eng. Progr. Symp. Ser. 27* **55**: 107 (1959).
K3. Kroll, W. J., et al.: *Trans. Electrochem. Soc.* **92**: 99 (1948).
K4. Kroll, W. J., and W. W. Stephens: *Ind. Eng. Chem.* **42**: 395 (1950).
L1. Lustman, B., and F. Kerze, Jr.: *The Metallurgy of Zirconium*, McGraw-Hill, New York, 1955.
M1. McClain, J. H., and S. M. Shelton: "Zirconium-Hafnium Separation," in *The Reactor Handbook*, 2d ed., vol. I, *Materials*, C. R. Tipton (ed.), Interscience, New York, 1960.
M2. Megy, J. A.: Method of Separating Hafnium from Zirconium, U.S. Patent 4,072,056, Feb. 7, 1978.
M3. Megy, J. A.: Method of Reducing Zirconium, U.S. Patent 4,127,409, Nov. 28, 1978.
M4. Mudge, W. L., Jr.: "Hydrogen Embrittlement of Zirconium," in *Zirconium and Zirconium Alloys*, American Society for Metals, Cleveland, Ohio, 1953, pp. 146–167.
M5. Mueller, W. M., J. P. Blackledge, and G. G. Libowitz: *Metal Hydrides*, Academic, New York, 1968.
N1. National Bureau of Standards: *JANAF Thermochemical Tables*, 2d ed., U.S. Government Printing Office, Washington, D.C., June 1971.
N2. Nielsen, R. H.: Personal communication to M. Benedict, Sept. 5, 1978.

O1. Ogarev, A. N., et al.: "Preparation of Ductile Zirconium by Fused Salt Electrolysis," *PICG(2)* **4**: 280 (1958).
O2. Overholser, L. G., C. J. Barton, and W. R. Grimes: "Separation of Hafnium from Zirconium by Extraction of Thiocyanate Complexes," Reports Y-421, June 28, 1949 and Y-477, Sept. 9, 1949.
P1. Petenev, O. S., and L. E. Ivanovskii: *Trans. Inst. Elektrochim., Akad. Nauk SSR, Ural. Filial* **11** (1968); **12**: 66 (1969); **14**: 106 (1970).
P2. Powell, R. W., and R. T. Tye: *J. Less-Common Metals* **3**(3): 202 (June 1961).
R1. Ramsey, J. W., and W. K. Whitson, Jr.: "Production of Zirconium at Y-12," Reports Y-817, Oct. 12, 1951 and Y-824, Nov. 15, 1951.
R2. Raynes, B. C., et al.: *J. Electrochem. Soc.* **102**: 137 (1955).
R3. Rickover, H. G., L. D. Geiger, and B. Lustman: "History of the Development of Zirconium Alloys for Use in Nuclear Reactors," USERDA Report TID-26740, Mar. 21, 1975.
R4. Russell, R. B.: *Trans. AIMME* **200**: 1045 (1954).
S1. Sajin, N. P., and E. A. Pepelyaeva: *PICG(1)* **8**: 559 (1956).
S2. Schemel, J. H.: *ASTM Manual on Zirconium and Hafnium*, Special Technical Publication 639, American Society for Testing and Materials, Philadelphia, 1977.
S3. Shelton, S. M., E. D. Dilling, and J. H. McClain: "Zirconium Metal Production," *PICG(1)* **8**: 505 (1956).
S4. Steinberg, M. A., M. E. Siebert, and E. Wainer: *J. Electrochem. Soc.* **101**: 63 (1954).
T1. Tipton, C. R. (ed.): *Reactor Handbook*, 2d ed., vol. 1, *Materials*, Interscience, New York, 1960, Fig. 7.1.
V1. Van Arkel, A. E.: *Reine Metalle*, Springer-Verlag, Berlin, 1939.
V2. Van Arkel, A. E., and J. H. de Boer: *Z. Anorg. Allgem. Chem.* **148**: 345 (1925).
W1. Wallace, W. P., M. T. Simnad, and B. Turovlin: USAEC Report GA-422, 1958.

PROBLEMS

7.1 Refer to Fig. 7.7 and show that the number of theoretical plates in the scrubbing and extracting sections of the zirconium-hafnium separation plant are 15.4 and 19.5, respectively, for the flow rates and distribution coefficients stated in the figure. Show that the zirconium and hafnium concentrations in the aqueous feed to the extracting section and the solvent phase leaving that section are as stated.

Equations needed for this problem have been derived in Chap. 4.

7.2 Zirconium metal is to be made from zircon by a combination of the following steps:

1. Extraction of zirconium and hafnium by caustic fusion
2. Separation of zirconium from hafnium by solvent extraction with TBP
3. Electrolysis of K_2ZrF_6 in NaCl
4. Arc melting

Draw up a detailed material flow sheet showing the quantities of electric energy and chemicals required to produce 1 lb of hafnium-free zirconium metal. Where information is lacking, make what you consider to be plausible assumptions regarding material losses and the extent to which recycling of materials would be necessary and practical. What is the cost of chemicals and electrical energy consumed in producing 1 lb of zirconium?

Enthalpy changes are

$$Zr(s,25°C) \rightarrow Zr(l,2400\ K) \qquad \Delta H = 23\ kcal/g\text{-}mol$$

$$4NaOH + ZrSiO_4(25°C) \rightarrow Na_2SiO_3 + Na_2ZrO_3 + 2H_2O(530°C) \qquad \Delta H = 50\ kcal/g\text{-}mol\ ZrSiO_4$$

$$NaCl(s,25°C) \rightarrow NaCl(l,875°C) \qquad \Delta H = 18\ kcal/g\text{-}mol$$

$$K_2ZrF_6(s,25°C) \rightarrow K_2ZrF_6(l,875°C) \qquad \Delta H = 67\ kcal/g\text{-}mol$$

Assume that electric energy for heating (AC) costs \$0.02/kWh, and for electrolysis (DC), costs \$0.03/kWh. Neglect heat losses. Use the following prices for purchased chemicals:

	\$/lb (100% basis)
HNO_3	0.10
KF	0.75
NaCl (USP)	0.05
NaF	0.25
$NaNO_3$	0.05
NaOH	0.10
NH_4OH (29%)	0.10
TBP (sp gr = 1)	1.00
Zircon	0.15

CHAPTER

EIGHT

PROPERTIES OF IRRADIATED FUEL AND OTHER REACTOR MATERIALS

Irradiated fuel discharged from nuclear power reactors can be stored or it can be reprocessed to recover and recycle the fissile and fertile nuclides. In the latter case the discharge fuel is stored at the reactor site for several months until the intense radioactivity has decayed to a level suitable for shipment and reprocessing. Criteria for such cooling times are considered in Sec. 4. The fission products are the dominant contributors to radioactivity in discharge fuel, and their intense radioactivity persists in stored fuel or in radioactive wastes from fuel reprocessing for periods of several hundred years after the fuel is discharged from the reactor. Some fission products, especially ^{129}I, contribute appreciable radioactivity for even longer periods, but the principal longer-term radioactive nuclides are the actinides and their decay daughters, as discussed in Sec. 3. The radioactivity of the actinides is also important in determining the necessary fission-product decontamination to be achieved in fuel reprocessing and in determining the shielding and containment required for fabricating recycled fuel. As discussed in Sec. 3, activation products in reactor structural materials and fuel cladding are important contributors to the radioactive wastes, and ^{14}C, formed by neutron activation of nitrogen impurities in the fuel, may be important as a potential environmental contaminant.

1 FISSION-PRODUCT RADIOACTIVITY

1.1 Activity in Irradiated Fuel

For irradiation in a constant neutron flux, the activity of any fission-product nuclide can be evaluated from the equations in Chap. 2. When fissions occur at a constant rate and when neutron-absorption reactions in the fission product and its precursors can be neglected, the activity of a nuclide with relatively short-lived precursors can be evaluated by applying Eq. (2.37):

$$N\lambda = Fy(1 - e^{-\lambda T_R})e^{-\lambda T_c} \tag{8.1}$$

where F = fission rate, fissions/s
N = atoms of long-lived fission product present after cooling for a time T_c
T_R = irradiation time, s
T_c = cooling time, s
y = cumulative fission yield, atoms/atom fissioned
λ = decay constant for the nuclide, s^{-1}

When the half-lives of a fission product and of its decay precursors are short compared to the irradiation time ($T_{1/2} \ll T_R$), the fission-product nuclide reaches saturation prior to the end of the irradiation. Its saturation activity per unit of reactor power is a constant, so that

$$\frac{N\lambda}{F} = ye^{-\lambda T_c} \qquad (\text{when } T_{1/2} \ll T_R) \tag{8.2}$$

The saturation activity is conveniently expressed in curies per watt of thermal power, or

$$\frac{\text{Curies}}{\text{Watt}} = ye^{-\lambda T_c}\left(\frac{\text{disintegrations/s}}{\text{fission/s}}\right)\left(\frac{\text{fission}}{200\text{ MeV}}\right)$$
$$\times\left(\frac{\text{MeV}}{1.6\times 10^{-13}\text{ W}\cdot\text{s}}\right)\left(\frac{\text{Ci}}{3.7\times 10^{10}\text{ disintegrations/s}}\right)$$

or

$$\frac{\text{Curies}}{\text{Watt}} = 0.845ye^{-\lambda T_c} \qquad (\text{when } T_{1/2} \ll T_R) \tag{8.3}$$

Practical irradiation periods for fuel in power reactors are in the range of about 1 to 4 years. Most of the fission-product nuclides reach saturation in this period. An example is 8.05-day ^{131}I, which is formed in 2.93 percent of ^{235}U fissions. Its saturation activity is 0.023 Ci/W.

Many radioactive fission-product nuclides have half-lives that are long compared to reactor irradiation periods, i.e., ^{3}H, ^{85}Kr, ^{90}Sr, ^{129}I, and ^{137}Cs. In these cases, Eq. (8.1) simplifies to

$$\frac{N\lambda}{F} \approx \frac{0.693yT_Re^{-\lambda T_c}}{T_{1/2}} \qquad (\text{when } T_{1/2} \gg T_R) \tag{8.4}$$

or

$$\frac{\text{Curies}}{\text{Watt}} = \frac{0.586yT_Re^{-\lambda T_c}}{T_{1/2}} \tag{8.5}$$

Because these long-lived nuclides do not reach saturation in the reactor fuel, their yearly production rate is important. This is obtained by dividing Eq. (8.5) by T_R and setting T_c equal to zero:

$$\frac{\text{Curies}}{\text{Watt}\times\text{time}} = \frac{0.586y}{T_{1/2}} \tag{8.6}$$

The short-lived daughter of a long-lived parent nuclide contributes significantly to the activity even after long cooling periods because it is constantly being formed from the parent (e.g., ^{90}Y from ^{90}Sr). If the half-life of the daughter is very small relative to that of its parent, the two are in secular equilibrium and the daughter activity is equal to that of the parent.

Only a few fission-product nuclides have half-lives too long for saturation but too short for the assumption of linear buildup that led to Eq. (8.4). Examples are ^{106}Ru, ^{144}Ce, and ^{147}Pm. A few radionuclides, such as ^{95}Nb, ^{140}La, and ^{147}Pm, have precursors that must be considered in the calculation of activity after a few months of postirradiation cooling.

In Table 8.1 are listed those fission-product nuclides that contribute appreciably to the activity of fission products formed after long irradiation and cooled for periods of a few months or more. Fission-product activities have been calculated for uranium fuel irradiated for 3 years in the 1000-MWe pressurized-water reactor (PWR) operating as shown in Fig. 3.31. Activities are listed for fuel at the time it is discharged from the reactor and after

Table 8.1 Long-lived radioactive fission products†

Radionuclide	Half-life	In discharge fuel 10^6 Ci/yr: At discharge‡	150-day decay	10-yr decay	Elemental boiling temperature, °C§
^3H	12.4 yr	1.93×10^{-2}	1.88×10^{-2}	1.09×10^{-2}	100 (as tritiated water)
^{79}Se	$\leqslant 6.5 \times 10^4$ yr	1.08×10^{-5}	1.08×10^{-5}	1.08×10^{-5}	
Total¶		10.0	1.08×10^{-5}	1.08×10^{-5}	657
^{85}Kr	10.76 yr	0.308	0.300	0.162	
Total		85.0	0.300	0.162	−153.4
^{86}Rb	18.66 days	1.34×10^{-2}	5.18×10^{-3}	0	
Total		1.34×10^{-2}	5.18×10^{-3}	0	705
^{89}Sr	52.7 days	19.6	2.65	0	
^{90}Sr	27.7 yr	2.11	2.09	1.65	
Total		1.38×10^2	4.74	1.65	1357
^{90}Y	64.0 h	2.20	2.09	1.65	
^{91}Y	58.8 days	25.5	4.39	0	
Total		2.08×10^2	6.48	1.65	3337
^{93}Zr	1.5×10^6 yr	5.15×10^{-5}	5.15×10^{-5}	5.15×10^{-5}	
^{95}Zr	65.5 days	37.3	7.54	0	
Total		96.2	7.54	5.15×10^{-5}	4325
^{93m}Nb	13.6 yr	3.95×10^{-6}	4.98×10^{-6}	2.3×10^{-5}	
^{95m}Nb	90 h	0.762	0.160	0	
^{95}Nb	35.0 days	37.6	14.2	0	
Total		2.30×10^2	14.4	2.3×10^{-5}	4842
^{99}Tc	2.12×10^5 yr	3.90×10^{-4}	3.90×10^{-4}	3.90×10^{-4}	
Total		29.7	3.90×10^{-4}	3.90×10^{-4}	3927
^{103}Ru	39.5 days	33.2	2.41	0	
^{106}Ru	368 days	14.8	11.2	1.50×10^{-2}	
Total		75.7	13.6	1.50×10^{-2}	4227
^{103m}Rh	57.5 min	33.2	2.41	0	
^{106}Rh	30 s	20.2	11.2	1.50×10^{-2}	
Total		1.17×10^2	13.6	1.50×10^{-2}	3667
^{107}Pd	7×10^6 yr	3.00×10^{-6}	3.00×10^{-6}	3.00×10^{-6}	
Total		9.10	3.00×10^{-6}	3.00×10^{-6}	3112
^{110m}Ag	255 days	0.100	6.64×10^{-2}	4.52×10^{-6}	
^{110}Ag	24.4 s	4.33	8.65×10^{-3}	5.88×10^{-7}	
^{111}Ag	7.5 days	1.08	1.03×10^{-6}	0	
Total		10.4	7.51×10^{-2}	5.11×10^{-6}	2163
^{113m}Cd	13.6 yr	2.86×10^{-4}	2.81×10^{-4}	1.74×10^{-4}	
^{115m}Cd	43 days	0.0150	1.34×10^{-3}	0	
Total		0.981	1.62×10^{-3}	1.74×10^{-4}	770
^{117m}Sn	14.0 days	1.62×10^{-3}	9.65×10^{-7}	0	
^{119m}Sn	250 days	4.47×10^{-4}	2.95×10^{-4}	1.79×10^{-8}	

(*See footnotes on page 356.*)

Table 8.1 Long-lived radioactive fission products† *(Continued)*

Radionuclide	Half-life	In discharge fuel 10^6 Ci/yr			Elemental boiling temperature, °C§
		At discharge‡	150-day decay	10-yr decay	
^{123}Sn	125 days	0.242	1.05	3.87×10^{-10}	
^{125}Sn	9.4 days	0.368	5.81×10^{-6}	0	
^{126}Sn	10^5 yr	1.49×10^{-5}	1.49×10^{-5}	1.49×10^{-5}	
Total		72.2	1.05	1.49×10^{-5}	2722
^{124}Sb	60.4 days	1.11×10^{-2}	1.95×10^{-3}	0	
^{125}Sb	2.71 yr	0.237	0.215	1.85×10^{-2}	
^{126m}Sb	19.0 min	6.13×10^{-4}	1.49×10^{-5}	1.49×10^{-5}	
^{126}Sb	12.5 days	1.55×10^{-3}	1.50×10^{-5}	1.47×10^{-5}	
Total		1.31×10^{2}	0.217	1.85×10^{-2}	1625
^{123m}Te	117 days	1.66×10^{-5}	6.82×10^{-6}	0	
^{125m}Te	58 days	8.47×10^{-2}	8.69×10^{-2}	7.66×10^{-3}	
^{127m}Te	109 days	0.420	0.167	0	
^{127}Te	9.4 h	1.96	0.62	0	
^{129m}Te	34.1 days	1.56	7.38×10^{-2}	0	
^{129}Te	68.7 min	9.18	3.87×10^{-2}	0	
Total		1.63×10^{2}	0.986	7.66×10^{-3}	1012
^{129}I	1.7×10^7 yr	1.01×10^{-6}	1.02×10^{-6}	1.02×10^{-6}	
^{131}I	8.05 days	23.5	5.94×10^{-5}	0	
Total		2.66×10^{2}	6.04×10^{-5}	1.02×10^{-6}	183
^{131m}Xe	11.8 days	0.174	8.50×10^{-5}	0	
^{133}Xe	5.270 days	43.9	1.46×10^{-7}	0	
Total		1.78×10^{2}	8.51×10^{-5}	0	−108.2
^{134}Cs	2.046 yr	6.70	5.83	0.228	
^{135}Cs	3.0×10^6 yr	7.79×10^{-6}	7.79×10^{-6}	7.79×10^{-6}	
^{136}Cs	13.7 days	1.66	5.42×10^{-4}	0	
^{137}Cs	30.0 yr	2.94	2.92	2.33	
Total		1.56×10^{2}	8.75	2.56	686
^{137m}Ba	2.554 min	2.75	2.72	2.18	
^{140}Ba	12.80 days	39.5	1.18×10^{-2}	0	
Total		1.51×10^{2}	2.73	2.18	1634
^{140}La	40.22 h	40.9	1.34×10^{-2}	0	
Total		1.49×10^{2}	1.34×10^{-2}	0	3370
^{141}Ce	32.5 days	37.9	1.53	0	
^{144}Ce	284 days	30.2	21.0	4.11×10^{-3}	
Total		1.48×10^{2}	22.5	4.11×10^{-3}	3470
^{143}Pr	13.59 days	32.7	1.85×10^{-2}	0	
^{144}Pr	17.27 min	30.5	21.0	4.11×10^{-3}	
Total		1.23×10^{2}	21.0	4.11×10^{-3}	3017
^{147}Nd	11.06 days	16.0	2.58×10^{-3}	0	
Total		24.9	2.58×10^{-3}	0	3111
^{147}Pm	4.4 yr	2.78	2.65	0.211	
^{148m}Pm	41.8 days	1.06	8.91×10^{-2}	0	

(See footnotes on page 356.)

Table 8.1 Long-lived radioactive fission products† (*Continued*)

		In discharge fuel 10^6 Ci/yr			
Radionuclide	Half-life	At discharge‡	150-day decay	10-yr decay	Elemental boiling temperature, °C§
^{148}Pm	5.4 days	5.42	7.08×10^{-3}	0	
Total		31.6	2.74	0.211	3200
^{151}Sm	≈87 yr	3.41×10^{-2}	3.41×10^{-2}	3.16×10^{-2}	
Total		11.5	3.41×10^{-2}	3.16×10^{-2}	1670
^{152}Eu	12.7 yr	3.41×10^{-4}	3.32×10^{-4}	1.92×10^{-4}	
^{154}Eu	16 yr	0.191	0.187	0.123	
^{155}Eu	1.811 yr	0.204	0.174	4.44×10^{-3}	
^{156}Eu	15.4 days	6.16	5.94×10^{-3}	0	
Total		6.56	0.367	0.127	1430
^{160}Tb	72.1 days	3.49×10^{-2}	8.24×10^{-3}	0	
Total		4.01×10^{-2}			2470
Total, all fission products		3.76×10^3	1.14×10^2	8.66	

†Uranium-fueled 1000-MWe PWR, 3-year fuel life.
‡Total elemental activities for fuel at discharge include short-lived radionuclides not listed here.
§G. V. Samsonov [S1].
¶Total activity of the element whose principal radionuclide(s) is (are) listed above.

postirradiation cooling periods of 150 days and 10 years. The variation of beta activity of the long-lived fission products with cooling time is shown in Fig. 8.1.

Gaseous fission products are important when possible releases of radioactive species to the air are to be considered. At the reactor site such releases can result when gaseous fission products diffuse from the fuel material and escape through defects in the fuel cladding. These radioactive nuclides are still confined within the coolant circuit of the reactor. However, coolant leaks and the need for occasional venting of insoluble and noncondensable gases from a liquid coolant system result in some handling of radioactive fission gases at the reactor site. Gaseous radioiodine is removed by adsorption in activated carbon. Radioactive noble gases are held for radioactive decay for periods of time varying from over a week to a month, after which the ^{85}Kr and possibly some remaining ^{133}Xe are discharged to the atmosphere. Atmospheric dilution brings the concentration of these radionuclides to levels well below tolerance. Alternatively, these vented gases may be treated by various means, such as absorption, adsorption, condensation, and/or compression into storage cylinders, for removal and long-term storage.

Most of the long-lived radioactive fission gases are still present in the fuel when it is processed to recover the uranium and plutonium. In many separation processes the first step involves mechanical chopping of the fuel rods, followed by acid dissolution of the fuel material. Gaseous and volatile fission products liberated in these steps must be disposed of safely. Of the noble fission gases, ^{85}Kr is the only radionuclide that is present in significant quantities after reprocessing cooling periods of a few months. At many reprocessing plants ^{85}Kr is discharged directly to the atmosphere through a tall release stack provided to ensure sufficient mixing with the air. Alternatively, krypton can be recovered from the off-gases by condensation, adsorption, or absorption, as discussed in Chaps. 10 and 11.

In addition to the fission-product tritium listed in Table 8.1, additional tritium is produced in the reactor by neutron reactions with boron control absorbers, with lithium contaminants, and with deuterium in the water coolant-moderator. A portion of the tritium that is produced in the coolant of light-water reactors (LWRs) is released to the environment as diluted tritiated water at the reactor site. Solid boron control absorbers containing tritium are ultimately stored as solid radioactive wastes. During fuel reprocessing a portion of the fission-product tritium is evolved as gaseous hydrogen and the remainder appears as tritiated water (HTO) or as zirconium tritide in the chopped fuel cladding. If not collected prior to fuel dissolution, the tritiated water follows the water carrier in fuel reprocessing and at present is released to the environment as tritiated water vapor or liquid.

Although most of the fission-product radioiodine will have decayed away during the preprocessing cooling period, the extremely low tolerance concentration of radioiodine requires that ^{131}I and ^{129}I be removed from reprocessing effluents. Also, radioactive iodine remaining in the dissolved-fuel solution extracts readily and reacts with the organic extracting solvents. Only about 1 Ci/year of ^{129}I is formed in a 1000-MWe reactor, but its long half-life and relatively high biological toxicity make ^{129}I an important long-term environmental hazard. Special processes for recovering and sequestering radioactive iodine from the off-gas in fuel reprocessing are discussed in Chaps. 10 and 11.

^{137}Cs and ^{90}Sr, elements of groups I and II of the periodic table, are important in determining the radioactivity of fission products after long decay periods. They are both easy

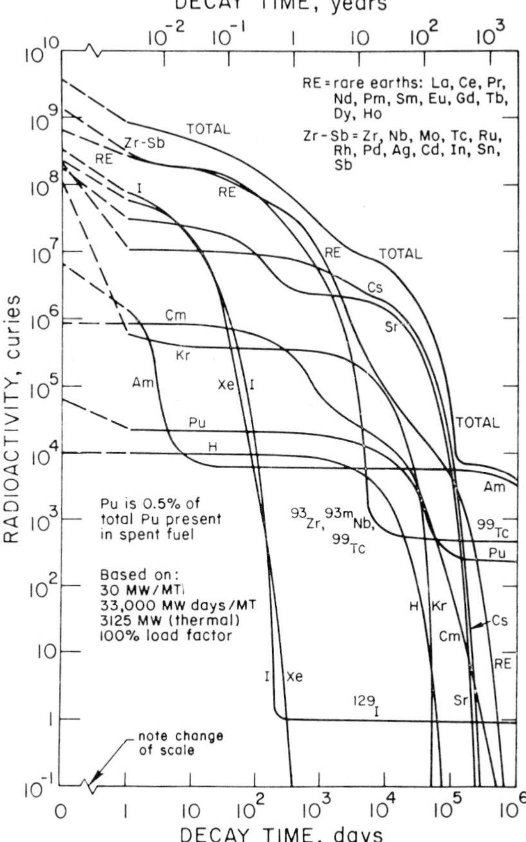

Figure 8.1 Radioactivity of fission products and actinides in high-level wastes produced in 1 year of operation of a uranium-fueled 1000-MWe PWR.

to remove from uranium in aqueous processing because of their very low solubility in organic solvents.

Yttrium and the lanthanides, which are grouped together under group IIIB, likewise are easily separable from uranium in aqueous processing, with the possible exception of cerium. The troublesome activity from cerium contamination is due to the beta and gamma decay of ^{144}Pr, the short-lived daughter of ^{144}Ce. ^{140}La emits penetrating gamma radiation and is one of the most important rare-earth fission products to be considered if the decay period is of the order of 30 days or less. ^{147}Nd is relatively short-lived, and its long-lived daughter ^{147}Pm emits no gammas; both are easily removed in aqueous processing.

Zirconium and niobium, of groups IVB and VB, are both amphoteric† in character, and their complex hydrolytic behavior makes zirconium and niobium two of the most difficult fission products to separate by aqueous processing. Group VI fission products have either very short or very long half-lives, and the most troublesome fission product in this group, ^{99}Mo, will be present in appreciable activity only for very short cooling periods. Its group VII decay daughter, 2.12×10^5 year ^{99}Tc, contributes to the long-term radioactivity of stored fission-product wastes. ^{99}Tc may be important to the long-term transport of fission products stored in geologic media.

^{106}Ru, of group VIII, is one of the most important fission-product contaminants in fuel reprocessing because of its multiple valence states and complex chemistry in aqueous solutions. In the presence of strong oxidizing agents ruthenium may appear in gaseous form as RuO_4.

1.2 Chemical Composition of Fission Products

The chemical composition of fission products in discharge fuel is controlled by the long-lived and stable species. The amounts of most of the fission-product chemical elements change but little for thousands of years after discharge. Those elements that do change significantly in amount over long decay periods include:

1. Cesium, which includes appreciable quantities of 30-year ^{137}Cs
2. Hydrogen, which consists entirely of 12.3-year ^{3}H
3. Niobium, which consists almost entirely of 35-day ^{95}Nb
4. Promethium, which consists entirely of 2.62-year ^{147}Pm
5. Strontium, which contains appreciable quantities of 27.7-year ^{90}Sr
6. Technetium, which consists entirely of 2.12×10^5-year ^{99}Tc

The elemental composition of the fission products in fuel discharged from the uranium-fueled PWR (Fig. 3.31) is listed in Table 8.2 and is plotted in Fig. 8.2. The composition expressed in elemental atoms per fission-product pair is the effective long-term elemental yield per fission, so the sum over all elements is equal to 2.

1.3 Neutron Absorption by Long-lived Fission Products

Also shown in Table 8.2 are the effective thermal cross sections for the individual nuclides, calculated for the neutron spectrum of a typical PWR and including the contributions from resonance absorption. The cross sections are multiplied by the atoms per fission-product pair to obtain the effective cross sections per fission-product pair listed in Table 8.2. Although the total effective cross section of 89.2 b/fission-product pair is calculated for the mixture of radionuclides existing 150 days after fuel discharge, it is a good approximation for the effective

† Acts both as an acid and as a base.

Table 8.2 Nuclide composition, elemental composition and neutron absorption of fission products in discharge uranium fuel†

Nuclide	Half-life (S = stable)	Atoms per fission-product pair‡	Effective thermal cross sections,§ b	Neutron absorption, barns per fission-product pair
^{3}H	12.3 yr	1.26×10^{-4}	–	–
^{73}Ge	S	1.38×10^{-6}	11.5	1.59×10^{-5}
^{74}Ge	S	4.94×10^{-6}	0.369	1.83×10^{-6}
^{76}Ge	S	2.61×10^{-5}	0.295	7.70×10^{-6}
Total¶		3.29×10^{-5}		2.54×10^{-5}
^{75}As	S	7.98×10^{-6}	14.5	1.16×10^{-4}
Total		7.98×10^{-6}		1.16×10^{-4}
^{77}Se	S	8.06×10^{-5}	42.7	3.44×10^{-3}
^{78}Se	S	2.16×10^{-4}	0.352	7.60×10^{-5}
^{79}Se	$\leqslant 6.5 \times 10^{4}$ yr	5.00×10^{-4}	3.74	1.87×10^{-3}
^{80}Se	S	9.05×10^{-4}	0.737	6.67×10^{-4}
^{82}Se	S	2.87×10^{-3}	1.638	4.70×10^{-3}
Total		4.58×10^{-3}		1.08×10^{-2}
^{81}Br	S	1.29×10^{-3}	20.0	2.58×10^{-2}
Total		1.29×10^{-3}		2.58×10^{-2}
^{82}Kr	S	2.75×10^{-5}	93.0	2.56×10^{-3}
^{83}Kr	S	3.51×10^{-3}	222	7.79×10^{-1}
^{84}Kr	S	9.73×10^{-3}	1.47	1.43×10^{-2}
^{85}Kr	10.76 yr	2.48×10^{-3}	9.89	2.45×10^{-2}
^{86}Kr	S	1.65×10^{-2}	0.065	1.07×10^{-3}
Total		3.22×10^{-2}		8.22×10^{-1}
^{85}Rb	S	8.14×10^{-3}	0.937	7.63×10^{-3}
^{87}Rb	4.7×10^{10} yr	2.03×10^{-2}	0.147	2.98×10^{-3}
Total		2.84×10^{-2}		1.06×10^{-2}
^{88}Sr	S	2.94×10^{-2}	0.005	1.47×10^{-4}
^{89}Sr	52 days	2.82×10^{-4}	0.466	1.31×10^{-4}
^{90}Sr	28.1 yr	4.43×10^{-2}	1.34	5.94×10^{-2}
Total		7.40×10^{-2}		5.96×10^{-2}
^{89}Y	S	3.82×10^{-2}	1.29	4.93×10^{-2}
^{90}Y	64 h	1.16×10^{-4}	3.27	3.79×10^{-4}
^{91}Y	58.8 days	1.06×10^{-3}	0.996	1.06×10^{-3}
Total		3.87×10^{-2}		5.07×10^{-2}
^{90}Zr	S	2.05×10^{-3}	0.093	1.91×10^{-4}
^{91}Zr	S	4.81×10^{-2}	3.81	1.83×10^{-1}
^{92}Zr	S	5.19×10^{-2}	0.363	1.88×10^{-2}
^{93}Zr	1.5×10^{6} yr	5.65×10^{-2}	8.93	5.05×10^{-1}
^{94}Zr	S	5.92×10^{-2}	0.118	6.99×10^{-3}
^{95}Zr	65 days	9.20×10^{-4}	~0	–
^{96}Zr	$> 3.6 \times 10^{17}$ yr	6.00×10^{-2}	0.063	3.78×10^{-3}
Total		2.78×10^{-1}		7.18×10^{-1}
^{95}Nb	35.0 days	9.28×10^{-4}	4.10	3.80×10^{-3}
Total		9.35×10^{-4}		3.80×10^{-3}

(*See footnotes on page 362.*)

Table 8.2 Nuclide composition, elemental composition and neutron absorption of fission products in discharge uranium fuel[†] (*Continued*)

Nuclide	Half-life (S = stable)	Atoms per fission-product pair[‡]	Effective thermal cross sections,[§] b	Neutron absorption, barns per fission-product pair
^{95}Mo	S	5.47×10^{-2}	40.8	2.23
^{96}Mo	S	2.50×10^{-3}	8.44	2.11×10^{-2}
^{97}Mo	S	5.93×10^{-2}	6.39	3.79×10^{-1}
^{98}Mo	S	5.88×10^{-2}	2.04	1.20×10^{-1}
^{100}Mo	$\geqslant 3 \times 10^{17}$ yr	6.52×10^{-2}	1.60	1.04×10^{-1}
Total		2.40×10^{-1}		2.86
^{99}Tc	2.12×10^{5} yr	5.77×10^{-2}	44.4	2.56
Total		5.77×10^{-2}		2.56
^{100}Ru	S	2.89×10^{-3}	10.9	3.15×10^{-2}
^{101}Ru	S	5.19×10^{-2}	25.1	1.30
^{102}Ru	S	4.90×10^{-2}	4.33	2.12×10^{-1}
^{103}Ru	39.6 days	1.66×10^{-4}	~0	–
^{104}Ru	S	3.10×10^{-2}	1.70	5.20×10^{-2}
^{106}Ru	367 days	6.28×10^{-3}	0.693	4.35×10^{-3}
Total		1.41×10^{-1}		1.60
^{103}Rh	S	2.36×10^{-2}	426	1.01×10^{1}
Total		2.36×10^{-2}		1.01×10^{1}
^{104}Pd	S	9.43×10^{-3}	10.4	9.81×10^{-2}
^{105}Pd	S	1.67×10^{-2}	30.8	5.14×10^{-1}
^{106}Pd	S	1.42×10^{-2}	1.95	2.77×10^{-2}
^{107}Pd	$\approx 7 \times 10^{6}$ yr	1.16×10^{-2}	19.6	2.27×10^{-1}
^{108}Pd	S	7.35×10^{-3}	54.2	3.98×10^{-1}
^{110}Pd	S	1.56×10^{-3}	3.06	4.77×10^{-3}
Total		6.71×10^{-2}		1.27
^{109}Ag	S	2.94×10^{-3}	487	1.43
Total		2.94×10^{-3}		1.43
^{110}Cd	S	1.14×10^{-3}	8.76	9.99×10^{-3}
^{111}Cd	S	8.06×10^{-4}	16.54	1.33×10^{-2}
^{112}Cd	S	4.30×10^{-4}	3.75	1.61×10^{-3}
^{113}Cd	S	9.35×10^{-6}	1.66×10^{4}	1.55×10^{-1}
^{114}Cd	S	6.50×10^{-4}	6.78	4.41×10^{-3}
^{116}Cd	S	1.95×10^{-4}	2.06	4.02×10^{-4}
Total		3.23×10^{-3}		1.85×10^{-1}
^{115}In	6×10^{14} yr	7.24×10^{-5}	1.14×10^{3}	8.25×10^{-2}
Total		7.24×10^{-5}		8.25×10^{-2}
^{116}Sn	S	1.06×10^{-4}	4.02	4.26×10^{-4}
^{117}Sn	S	2.02×10^{-4}	6.80	1.37×10^{-3}
^{118}Sn	S	2.05×10^{-4}	~0	–
^{119}Sn	S	2.11×10^{-4}	3.94	8.31×10^{-4}
^{120}Sn	S	2.21×10^{-4}	0.347	7.67×10^{-5}
^{122}Sn	S	2.56×10^{-4}	0.147	3.76×10^{-5}
^{124}Sn	S	3.69×10^{-4}	0.115	4.24×10^{-5}

(*See footnotes on page 362.*)

Table 8.2 Nuclide composition, elemental composition and neutron absorption of fission products in discharge uranium fuel† *(Continued)*

Nuclide	Half-life (S = stable)	Atoms per fission-product pair‡	Effective thermal cross sections,§ b	Neutron absorption, barns per fission-product pair
^{126}Sn	$\approx 10^5$ yr	4.71×10^{-4}	0.280	1.32×10^{-4}
Total		2.05×10^{-3}		2.92×10^{-3}
^{121}Sb	S	2.32×10^{-4}	46.3	1.07×10^{-2}
^{123}Sb	$> 1.3 \times 10^{16}$ yr	2.72×10^{-4}	54.6	1.49×10^{-2}
^{125}Sb	2.71 yr	3.36×10^{-4}	1.46	4.91×10^{-4}
Total		8.44×10^{-4}		2.61×10^{-2}
^{125m}Te	58 days	7.98×10^{-6}	–	–
^{125}Te	S	1.59×10^{-4}	8.16	1.30×10^{-3}
^{126}Te	S	4.50×10^{-4}	3.32	1.49×10^{-3}
^{127m}Te	109 days	2.98×10^{-5}	–	–
^{128}Te	S	6.21×10^{-3}	3.00	1.86×10^{-2}
^{129m}Te	34 days	1.03×10^{-5}	–	–
^{130}Te	8×10^{20} yr	2.16×10^{-2}	0.270	5.83×10^{-3}
Total		2.85×10^{-2}		2.73×10^{-2}
^{127}I	S	1.79×10^{-3}	55.8	9.99×10^{-2}
^{129}I	1.7×10^{7} yr	1.07×10^{-2}	37.4	4.00×10^{-1}
Total		1.25×10^{-2}		5.00×10^{-1}
^{130}Xe	S	3.95×10^{-4}	2.46	9.72×10^{-4}
^{131}Xe	S	2.18×10^{-2}	322	7.02
^{132}Xe	S	5.68×10^{-2}	0.869	4.94×10^{-2}
^{134}Xe	S	7.83×10^{-2}	0.689	5.39×10^{-2}
^{136}Xe	S	1.19×10^{-1}	0.230	2.74×10^{-2}
Total		2.76×10^{-1}		7.15
^{133}Cs	S	5.37×10^{-2}	158	8.48
^{134}Cs	2.046 yr	6.94×10^{-3}	129	8.95×10^{-1}
^{135}Cs	3.0×10^{6} yr	1.42×10^{-2}	30.2	4.29×10^{-1}
^{137}Cs	30.0 yr	6.02×10^{-2}	0.176	1.06×10^{-2}
Total		1.35×10^{-1}		9.82
^{134}Ba	S	3.91×10^{-3}	0.819	3.20×10^{-3}
^{136}Ba	S	9.20×10^{-4}	4.05	3.73×10^{-3}
^{137}Ba	S	2.37×10^{-3}	4.75	1.13×10^{-2}
^{138}Ba	S	5.91×10^{-2}	0.574	3.39×10^{-2}
Total		6.63×10^{-2}		5.21×10^{-2}
^{139}La	S	6.25×10^{-2}	9.87	6.17×10^{-1}
Total		6.25×10^{-2}		6.17×10^{-1}
^{140}Ce	S	6.37×10^{-2}	0.631	4.02×10^{-2}
^{141}Ce	33 days	9.66×10^{-5}	23.7	2.29×10^{-3}
^{142}Ce	$> 5 \times 10^{16}$ yr	5.73×10^{-2}	1.15	6.59×10^{-2}
^{144}Ce	284 days	1.16×10^{-2}	1.57	1.82×10^{-2}
Total		1.33×10^{-1}		1.27×10^{-1}

(*See footnotes on page 362.*)

Table 8.2 Nuclide composition, elemental composition and neutron absorption of fission products in discharge uranium fuel† (*Continued*)

Nuclide	Half-life (S = stable)	Atoms per fission-product pair‡	Effective thermal cross sections,§ b	Neutron absorption, barns per fission-product pair
^{141}Pr	$>2 \times 10^{16}$ yr	5.90×10^{-2}	6.40	3.78×10^{-1}
Total		5.90×10^{-2}		3.78×10^{-1}
^{142}Nd	S	8.75×10^{-4}	16.8	1.47×10^{-2}
^{143}Nd	S	3.69×10^{-2}	288	1.06×10^{1}
^{144}Nd	2.4×10^{15} yr	5.23×10^{-2}	7.54	3.94×10^{-1}
^{145}Nd	$>6 \times 10^{16}$ yr	3.43×10^{-2}	86.7	2.97
^{146}Nd	S	3.37×10^{-2}	15.4	5.19×10^{-1}
^{148}Nd	S	1.75×10^{-2}	7.74	1.35×10^{-1}
^{150}Nd	$>10^{16}$ yr	8.37×10^{-3}	6.47	5.42×10^{-2}
Total		1.84×10^{-1}		1.47×10^{1}
^{147}Pm	2.62 yr	5.70×10^{-3}	1.11×10^{3}	6.33
Total		5.70×10^{-3}		6.33
^{147}Sm	1.05×10^{11} yr	3.67×10^{-3}	274	1.01
^{148}Sm	$>2 \times 10^{14}$ yr	1.04×10^{-2}	21.7	2.26×10^{-1}
^{149}Sm	$>1 \times 10^{15}$ yr	2.19×10^{-4}	3.52×10^{4}	7.71
^{150}Sm	S	1.35×10^{-2}	149	2.01
^{151}Sm	≈ 87 yr	1.70×10^{-3}	2.17×10^{3}	3.88
^{152}Sm	S	4.46×10^{-3}	1.03×10^{3}	4.59
^{154}Sm	S	1.43×10^{-3}	11.7	1.67×10^{-2}
Total		3.54×10^{-2}		1.94×10^{1}
^{153}Eu	S	4.70×10^{-3}	629	2.96
^{154}Eu	16 yr	1.39×10^{-3}	1.32×10^{3}	1.83
^{155}Eu	1.811 yr	1.56×10^{-4}	1.22×10^{4}	1.90
Total		6.26×10^{-3}		6.69
^{155}Gd	S	2.84×10^{-5}	4.51×10^{4}	1.28
^{156}Gd	S	2.49×10^{-3}	16.0	3.98×10^{-2}
^{157}Gd	S	1.20×10^{-6}	2.08×10^{5}	2.50×10^{-1}
^{158}Gd	S	4.33×10^{-4}	11.18	4.84×10^{-3}
^{160}Gd	S	3.06×10^{-5}	0.655	2.00×10^{-5}
Total		3.06×10^{-3}		1.58
^{159}Tb	S	5.90×10^{-5}	218	1.28×10^{-2}
Total		5.90×10^{-5}		1.28×10^{-2}
^{160}Dy	S	1.06×10^{-5}	377	4.00×10^{-3}
^{161}Dy	S	6.96×10^{-6}	970	6.75×10^{-3}
^{162}Dy	S	6.01×10^{-6}	1.08×10^{3}	6.50×10^{-3}
^{163}Dy	S	4.92×10^{-6}	664	3.27×10^{-3}
^{164}Dy	S	1.16×10^{-6}	2.32×10^{3}	2.69×10^{-3}
Total		2.96×10^{-5}		2.32×10^{-2}
Total, all fission products		2.00		89.2

†One hundred fifty days after discharge from uranium-fueled PWR.

‡Some elemental totals include minor contributions for nuclides not shown in table.

§Effective thermal cross sections for a typical neutron spectrum of a PWR, including contributions from nonthermal resonance absorption.

¶Total yield of element whose principal radionuclides are listed above

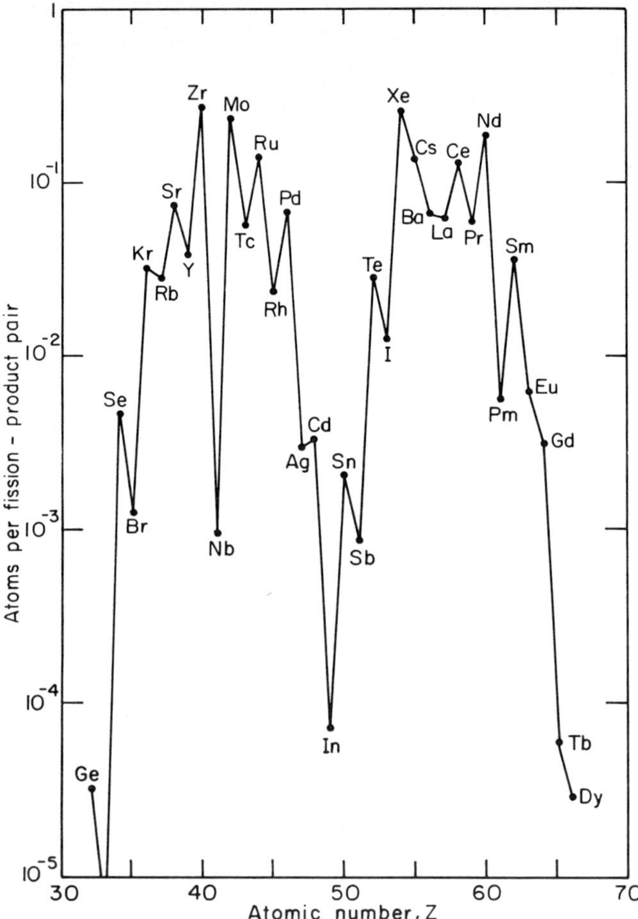

Figure 8.2 Chemical composition of fission products (for uranium-fueled PWR 150 days after discharge).

cross section for all fission products other than ^{135}Xe at the time of fuel discharge. Except for ^{135}Xe, the shorter-lived species that are also present at the time of discharge do not exist in sufficient concentration to contribute appreciably to neutron absorption. Neutron absorption in ^{135}Xe is usually treated separately, by the techniques discussed in Sec. 6.3 of Chap. 2.

The elemental contribution to neutron absorption by fission products tends to follow the effective fission yield of the elements, but with exceptions for several individual elements. The rare earths neodymium, promethium, samarium, europium, and gadolinium, as well as xenon and cesium, are the important neutron-absorbing elements resulting from the high-mass fission-yield peak, and rhodium and its near neighbors are the important neutron absorbers from the low-mass peak.

1.4 Toxicity of Inhaled or Ingested Fission Products

The rate of radioactive disintegration, e.g., curies, is only a crude measure of the importance of individual fission products in irradiated fuel and in radioactive wastes. A more meaningful measure of potential biological hazard must also include the sensitivity of humans to inhalation or ingestion of these radionuclides. For this purpose we use the radioactivity concentration limit C, which is the concentration of radioactivity (curies) of a given radionuclide in air or water such that an individual who obtains his or her total intake of air or water from this source will receive a radiation dose from this radionuclide at the rate of 0.5 rem/year.† Values of the public-exposure radioactivity concentration limit C for selected radionuclides are listed in App. D. A more complete listing appears in the Federal Regulations 10 CFR 20 [F2]. Assuming that the biological hazard to an individual exposed to low levels of radiation is proportional to the accumulated radiation dose, then the potential biological hazard from inhalation or ingestion of a mixture of radionuclides is proportional to the toxicity index, defined as

$$\text{Toxicity index} = \sum_i \frac{\lambda_i N_i}{C_{ik}} \tag{8.7}$$

where λ_i = radioactive decay constant for nuclide i

N_i = number of atoms of nuclide i

C_{ik} = radioactivity concentration limit for nuclide i in medium k (i.e., air or water)

The toxicity index is the volume of air or water with which the mixture of radionuclides must be diluted so that breathing the air or drinking the water will result in accumulation of radiation dose at a rate no greater than 0.5 rem/year. However, the toxicity index still does not measure ultimate hazards and risk, because it does not take into account the mechanisms by which the radionuclides could be released to air or water and transported to humans.

The inhalation-toxicity indices of the fission products in the fuel discharged yearly from the 1000-MWe uranium-fueled LWR are shown in Fig. 8.3 as a function of storage time. Ingestion toxicity indices for the same fission products are shown in Fig. 8.4. If Fig. 8.4 is compared with the activity plot of Fig. 8.1, it is apparent that the relatively high toxicity, i.e., low C, of bone-seeking ^{90}Sr makes this nuclide more important than any other fission product in terms of potential inhalation or ingestion toxicity during the first few hundred years after discharge from the reactor. Thereafter, the long-lived thyroid-seeking ^{129}I is potentially the most important of the fission products, even though only about 1 Ci of ^{129}I is produced yearly in a 1000-MWe reactor.

1.5 Effects of Fuel-Cycle Alternatives on Fission Products in Irradiated Fuel

Because the nuclides ^{232}Th, ^{233}U, ^{235}U, ^{238}U, ^{239}Pu, and ^{241}Pu yield different amounts of individual fission products, different fuel cycles such as uranium fueling without recycle, uranium-plutonium fueling, and thorium-uranium fueling will result in different amounts of fission products in the discharge fuel. Calculated yearly production and composition of some of the principal fission products for some of the alternative fuel cycles described in Chap. 3 are listed in Table 8.3.

2 RADIOACTIVITY OF THE ACTINIDES

2.1 Actinide Radioactivity in Uranium and Uranium-Plutonium Fuel

The important actinides in irradiated uranium fuel are uranium, neptunium, plutonium, americium, and curium, which are produced according to the reactions of Fig. 8.5. ^{236}U,

†The terminology "radioactivity concentration limit" is that used in the U.S. Federal Regulations. In the publications of the International Committee in Radiation Protection [I1], a similar concentration limit is referred to as the "maximum permissible concentration."

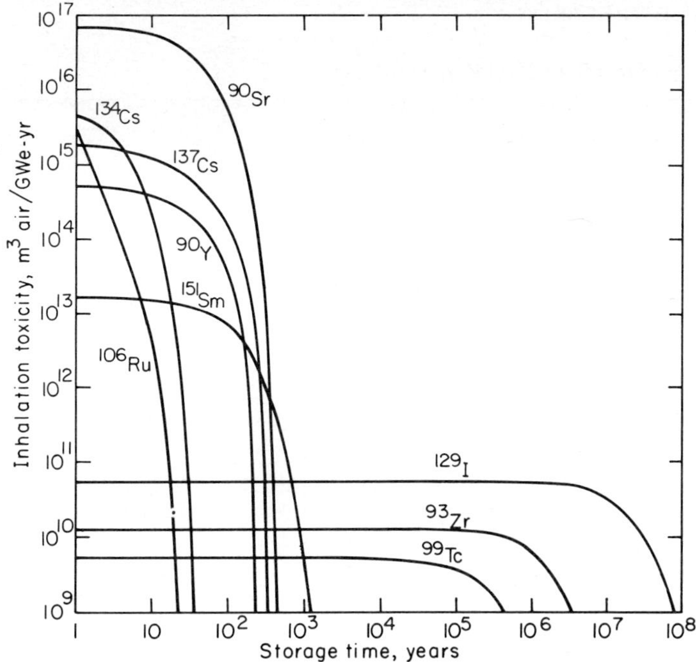

Figure 8.3 Inhalation toxicity of the fission products from a uranium-fueled LWR.

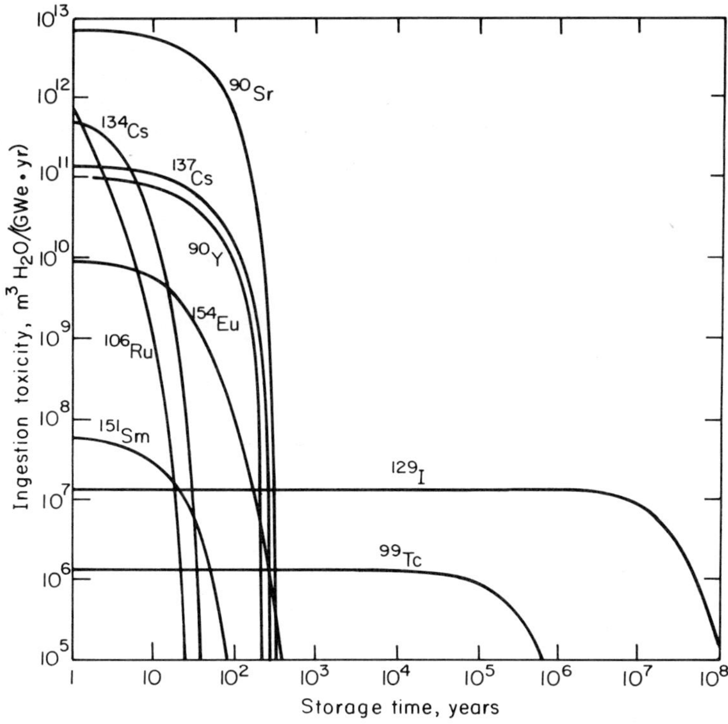

Figure 8.4 Ingestion toxicity of the fission products from a uranium-fueled LWR.

Table 8.3 Long-lived fission products from 1000-MWe power reactors

Reactor type†	PWR	PWR	HTGR	LMFBR
Fuel	Uranium (3.3% ^{235}U)	Uranium and recycled plutonium	^{235}U, thorium, and recycled uranium	Uranium and recycled plutonium
		Volatile fission products, Ci/yr		
^{3}H‡	1.88×10^4	2.47×10^4	1.03×10^4	1.98×10^4
^{85}Kr	3.00×10^5	1.87×10^5	4.90×10^5	1.59×10^5
^{129}I	1.02	1.31	1.00	0.742
		Nonvolatile fission products, Ci/yr		
^{89}Sr	2.65×10^6	1.84×10^6	3.18×10^6	2.16×10^6
^{90}Sr	2.09×10^6	1.24×10^6	2.32×10^6	8.93×10^5
^{91}Y	4.39×10^6	3.24×10^6	4.10×10^6	3.92×10^6
^{95}Zr	7.54×10^6	6.95×10^6	5.24×10^6	8.53×10^6
^{95}Nb	1.60×10^7	1.30×10^7	9.86×10^6	1.60×10^7
^{99}Tc	3.90×10^2	3.95×10^2	2.70×10^2	3.11×10^2
^{103}Ru§	2.41×10^6	2.70×10^6	7.02×10^5	3.39×10^6
^{106}Ru§	1.12×10^7	1.86×10^7	9.26×10^6	1.94×10^7
^{134}Cs	5.83×10^6	5.09×10^6	5.52×10^6	4.86×10^5
^{137}Cs	2.92×10^6	3.00×10^6	2.42×10^6	2.37×10^6
^{141}Ce	1.53×10^6	1.42×10^6	1.19×10^6	1.40×10^6
^{144}Ce	2.25×10^7	1.79×10^7	1.43×10^7	1.65×10^7
Rare earths	2.42×10^7	2.15×10^7	3.02×10^7	4.01×10^7
Total¶	1.14×10^8	1.24×10^8	1.02×10^8	1.26×10^8

†PWR, pressurized-water reactor; HTGR, high-temperature gas-cooled reactor; LMFBR, liquid-metal-cooled fast-breeder reactor. Data are calculated for 150 days after discharge. Calculated from data in [B2].

‡Additional ^{3}H produced by neutron activation is shown in Table 8.11.

§Ruthenium may also form volatile compounds.

¶Total includes radionuclides not listed here.

produced by (n, γ) reactions in ^{235}U, is important because of its neutron absorption. If uranium containing ^{236}U is recycled, a slightly greater fissile concentration in the fresh fuel to the reactor is required. Neutron capture in ^{236}U and the $(n, 2n)$ reaction in ^{238}U lead to 6.75-day ^{237}U, which dominates the uranium radioactivity during the first several months that irradiated fuel is stored after discharge. Because of its relatively short half-life, ^{239}U disappears rapidly after the fuel is discharged.

Decay of ^{237}U forms ^{237}Np, which is important because its (n, γ) and $(n, 2n)$ reactions lead to ^{238}Pu and ^{236}Pu. Also, ^{237}Np is an important long-term constituent of radioactive wastes, particularly because its transport through some geologic media is not as delayed as that of other actinides and because of the toxicity of radionuclides in its decay chain, especially ^{233}U, ^{229}Th, and ^{225}Ra.

Although only small quantities of ^{238}Pu are formed, its half-life of 86 years is long enough that ^{238}Pu persists in plutonium recovered for recycle and is short enough that ^{238}Pu is the greatest contributor to the alpha activity of plutonium in irradiated fuel. Although the quantities and activities of 2.85-year ^{236}Pu are relatively small, its decay daughter ^{232}U can build up when recovered plutonium is stored prior to fuel refabrication. As discussed in Sec.

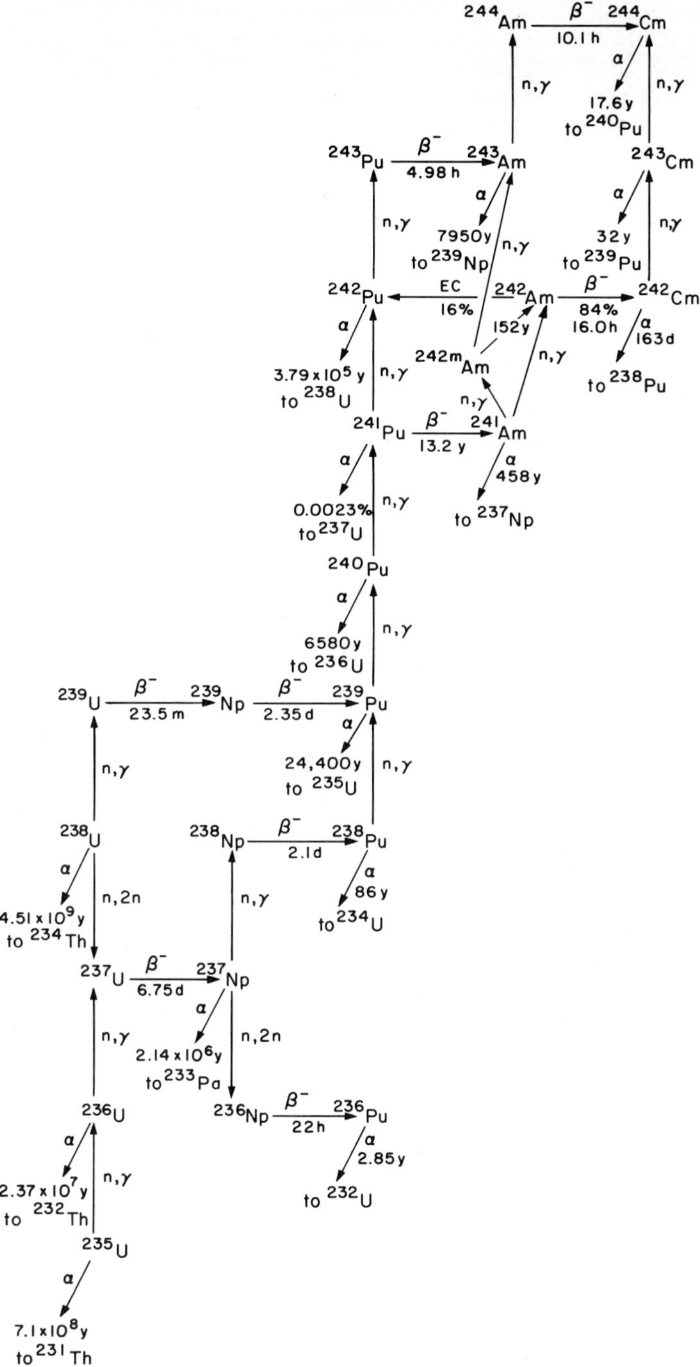

Figure 8.5 Nuclide chains producing plutonium, americium, and curium.

2.3, the ^{232}U decay daughters emit high-energy gammas and may contribute to the shielding requirements for handling recycled plutonium. The largest material quantities of plutonium are produced by neutron capture in ^{238}U, leading through short-lived ^{239}U and ^{239}Np to fissile ^{239}Pu, with a half-life of 24,400 years. Nonfission capture of neutrons in ^{239}Pu results in ^{240}Pu, and its neutron capture results in fissile ^{241}Pu. Because of its half-life of 6580 years, ^{240}Pu is a strong and persistent alpha source in reactor plutonium, and 13.2-year ^{241}Pu is an extremely intense beta source. Because of the long half-life of ^{242}Pu its radioactivity is not important compared to the other plutonium isotopes. Its neutron-capture daughter ^{243}Pu is short-lived and decays away within a few days after plutonium is removed from the neutron environment of the reactor.

Radioactive decay of ^{241}Pu and ^{243}Pu form ^{241}Am and ^{243}Am, which are also important and persistent sources of alpha radioactivity in discharge fuel. Another persistent americium radioisotope is 152-year ^{242m}Am, formed by neutron capture in ^{241}Am. Its isomeric decay and the beta decay of its short-lived daughter result in 163-day ^{242}Cm, which is the most intense source of alpha activity in discharged uranium fuel. Successive neutron captures lead to ^{243}Cm, ^{244}Cm, and ^{245}Cm. Higher-mass curium nuclides are usually not important in power reactor fuel.

^{242}Cm alpha decays to ^{238}Pu, ^{243}Cm to ^{239}Pu, ^{244}Cm to ^{240}Pu, and ^{245}Cm to ^{241}Pu. Also, the alpha decay of ^{243}Am results in ^{239}Np, which decays quickly to ^{239}Pu. The decay of ^{242}Cm prior to fuel reprocessing adds to the quantity of ^{238}Pu in recovered plutonium. Also, these decay reactions are the most significant sources of plutonium in the high-level wastes resulting from reprocessing uranium fuel. Although the ^{242}Cm decay daughter ^{238}Pu is not an important contributor to the alpha activity of high level wastes, the subsequent decay daughter ^{226}Ra is one of the most important contributors to the long-term ingestion toxicity of these wastes.

Material quantities and activities of the actinides in the discharge fuel can be calculated from the equations in Chap. 2. If the irradiation is at constant neutron flux, Eqs. (2.104) and (2.113) can be applied directly, as in the example of Sec. 6.5 of Chap. 2. However, power reactors usually operate at constant power, and because of the changing inventory and composition of the fissile material the neutron flux usually increases between refueling intervals. Equations (2.104) and (2.113) can still be applied to calculate the amount of a nuclide in an actinide chain by assuming constant neutron flux during a small but finite time increment, solving the nuclide equations for that time increment, recalculating the flux, and proceeding stepwise through subsequent time steps. This is the calculational method of the ORIGEN code [B2], which was used to calculate [P1] the quantities of actinides in discharge fuel for the pressurized-water and fast-breeder reactors. The results appear in Table 8.4.

The data in Table 8.4 show that curium is the strongest alpha source during fuel reprocessing, assuming that fuel is reprocessed 150 days after discharge from the reactor. The 246 kg of plutonium to be recovered yearly from the discharge fuel contains 1.2×10^5 Ci of alpha activity and 2.8×10^6 Ci of beta activity. The remaining actinide activity is associated with americium and curium, which will normally follow the high-level reprocessing wastes, along with the fission products.

The effect of plutonium recycle is to increase the production of higher-mass isotopes of plutonium and of americium and curium, because the recycled plutonium is exposed to neutrons throughout the entire irradiation cycle. The actinide quantities calculated [P1] for the same 1000-MWe reactor operating on an equilibrium fuel cycle with self-generated plutonium recycle are shown in Table 8.5. The alpha activity of the plutonium processed yearly is increased by a factor of 14 by plutonium recycle, the americium activity is increased by a factor of 5, and the curium activity by a factor of 7.

Also shown in Table 8.5 are the actinide quantities of a 1000-MWe fast-breeder reactor operating on an equilibrium fuel cycle with recycle of plutonium and uranium [P1]. The quantity of plutonium to be recovered and fabricated into recycled breeder fuel is greater than

Table 8.4 Actinides in discharge uranium fuel†

Radionuclide	Half-life	kg/yr	Ci/yr	Elemental boiling temperature, °C‡
^{234}U	2.47×10^5 yr	3.14	1.94×10^1	
^{235}U	7.1×10^8 yr	2.15×10^2	4.61×10^{-1}	
^{236}U	2.39×10^7 yr	1.14×10^2	7.22	
^{237}U	6.75 days	9.15×10^{-7}	7.47×10^1	
^{238}U	4.51×10^9 yr	2.57×10^4	8.56	
Total		2.60×10^4	$\alpha\ 3.56 \times 10^1$ $\beta\ 7.47 \times 10^1$	4135
^{237}Np	2.14×10^6 yr	2.04×10^1	1.44×10^1	
^{239}Np	2.35 days	2.05×10^{-6}	4.78×10^2	
Total		2.04×10^1	$\alpha\ 1.44 \times 10^1$ $\beta\ 4.78 \times 10^2$	–
^{236}Pu	2.85 yr	2.51×10^{-4}	1.34×10^2	
^{238}Pu	86 yr	5.99	1.01×10^5	
^{239}Pu	24,400 yr	1.44×10^2	8.82×10^3	
^{240}Pu	6,580 yr	5.91×10^1	1.30×10^4	
^{241}Pu	13.2 yr	2.77×10^1	2.81×10^6	
^{242}Pu	3.79×10^5 yr	9.65	3.76×10^1	
Total		2.46×10^2	$\alpha\ 1.23 \times 10^5$ $\beta\ 2.81 \times 10^6$	3508
^{241}Am	458 yr	1.32	4.53×10^3	
^{242m}Am	152 yr	1.19×10^{-2}	1.16×10^2	
^{243}Am	7,950 yr	2.48	4.77×10^2	
Total		3.81	$\alpha\ 5.01 \times 10^3$ $\beta\ 1.16 \times 10^2$	2880
^{242}Cm	163 days	1.33×10^{-1}	4.40×10^5	
^{243}Cm	32 yr	1.96×10^{-3}	9.03×10^1	
^{244}Cm	17.6 yr	9.11×10^{-1}	7.38×10^4	
^{245}Cm	9,300 yr	5.54×10^{-2}	9.79	
^{246}Cm	5,500 yr	6.23×10^{-3}	1.92	
Total		1.11	$\alpha\ 5.14 \times 10^5$	–
Total		2.63×10^4	$\alpha\ 6.42 \times 10^5$ $\beta\ 2.81 \times 10^6$	

†Uranium-fueled 1000-MWe PWR, 150 days after discharge.
‡G. V. Samsonov [S1].

for the LWR operating with plutonium recycle, because of the higher fissile concentration required for fast-breeder fuel. However, the breeder produces much less ^{238}Pu, so the total alpha activity in the breeder plutonium is almost 10-fold less than in the water-reactor plutonium. Also, the breeder does not build up such large concentrations of ^{241}Pu and ^{242}Pu, and the yearly production of americium and curium is less [P1].

2.2 Preprocessing Storage Time for Irradiated Uranium Fuel

There are several reasons why it is useful to store or "cool" irradiated uranium fuel for several months prior to shipment for reprocessing:

Table 8.5 Actinide quantities in discharge fuel with plutonium recycle†

	Pressurized-water reactor, self-generated Pu recycle‡		Liquid-metal fast-breeder reactor§		
Radionuclide	kg/yr	Ci/yr	Core, kg/yr	Blankets kg/yr	Core and blankets, Ci/yr
^{234}U	2.66	1.65×10^1	8.38×10^{-2}	7.91×10^{-2}	1.01
^{235}U	1.71×10^2	3.69×10^{-1}	1.68	1.06×10^1	1.45×10^{-2}
^{236}U	8.34×10^1	5.28	4.25	4.72	5.69×10^{-1}
^{237}U	2.24×10^{-6}	1.83×10^2	9.58×10^{-5}	1.54×10^{-4}	1.04×10^4
^{238}U	2.55×10^4	8.51	7.22×10^3	8.91×10^3	5.37
Total	2.58×10^4	$\alpha\ 3.07 \times 10^1$ $\beta\ 1.83 \times 10^2$	7.23×10^3	8.92×10^3	$\alpha\ 6.97$ $\beta\ 1.04 \times 10^4$
^{237}Np	1.51×10^1	1.07×10^1	3.07	1.62	3.31
^{239}Np	1.80×10^{-5}	4.19×10^1	1.64×10^{-6}	2.19×10^{-8}	3.87×10^2
Total	1.51×10^1	$\alpha\ 1.07 \times 10^1$ $\beta\ 4.19 \times 10^1$	3.07	1.62	$\alpha\ 3.31$ $\beta\ 3.87 \times 10^2$
^{236}Pu	2.77×10^{-4}	1.48×10^1	2.68×10^{-5}	3.03×10^{-6}	1.59×10^1
^{238}Pu	1.61×10^1	1.66×10^6	1.27	6.21×10^{-2}	2.25×10^4
^{239}Pu	2.05×10^2	1.23×10^4	1.09×10^3	2.99×10^2	8.51×10^4
^{240}Pu	1.20×10^2	2.64×10^4	4.71×10^2	1.49×10^1	1.07×10^5
^{241}Pu	7.27×10^1	7.40×10^6	4.56×10^1	4.15×10^{-1}	4.67×10^6
^{242}Pu	4.16×10^1	1.62×10^2	1.47×10^1	1.01×10^{-2}	5.75×10^1
Total	4.55×10^2	$\alpha\ 1.70 \times 10^6$ $\beta\ 7.40 \times 10^6$	1.62×10^3	3.14×10^2	$\alpha\ 2.15 \times 10^5$ $\beta\ 4.67 \times 10^6$
^{241}Am	6.00	2.06×10^4	4.02	2.98×10^{-2}	1.39×10^4
^{242m}Am	7.93×10^{-2}	7.68×10^2	7.11×10^{-2}	9.68×10^{-5}	6.92×10^2
^{243}Am	2.18×10^1	4.19×10^3	1.92	3.05×10^{-4}	3.69×10^2
Total	27.9	$\alpha\ 2.48 \times 10^4$ $\beta\ 7.68 \times 10^2$	6.01	3.02×10^{-2}	$\alpha\ 1.43 \times 10^4$ $\beta\ 6.92 \times 10^2$
^{242}Cm	7.14×10^{-1}	2.37×10^6	1.13×10^{-1}	1.14×10^{-4}	3.76×10^5
^{243}Cm	8.61×10^{-3}	3.96×10^2	6.25×10^{-3}	1.21×10^{-6}	2.87×10^2
^{244}Cm	1.56×10^1	1.27×10^6	1.27×10^{-1}	3.32×10^{-6}	1.03×10^4
^{245}Cm	1.74	3.07×10^2	3.56×10^{-3}	2.15×10^{-8}	6.29×10^{-1}
^{246}Cm	1.74×10^{-1}	5.27×10^1	9.49×10^{-5}	1.36×10^{-10}	2.93×10^{-2}
Total	1.82×10^1	$\alpha\ 3.64 \times 10^6$	2.50×10^{-1}	1.19×10^{-4}	$\alpha\ 3.87 \times 10^5$
Total	2.63×10^4	$\alpha\ 5.36 \times 10^6$ $\beta\ 7.40 \times 10^6$	8.86×10^3	9.23×10^3	$\alpha\ 6.16 \times 10^5$ $\beta\ 4.68 \times 10^6$

†1000-MWe reactor, 80% capacity factor.

‡33 MWd/kg, 32.5% thermal efficiency, calculated for 150 days after discharge, equilibrium fuel cycle.

§Core: 67.6 MWd/kg, 41.8% thermal efficiency, calculated for 60 days after discharge, equilibrium fuel cycle. Residence time of radial blanket = 2120 days.

1. The decay of 8.05-day ^{131}I avoids troublesome quantities of gaseous and dissolved radioiodine in fuel reprocessing.
2. The decay of 6.75-day ^{237}U eliminates the need for remote handling of the purified uranium recovered by fuel reprocessing. Also, presence of high activities of ^{237}U would interfere with monitoring for fission-product decontamination of the recovered uranium.

3. Decay of fission-product activity and heat generation simplifies fuel shipment, and the lower activity reduces radiation damage to the organic solvents used in fuel reprocessing.
4. The decay of 5.27-day ^{133}Xe leaves ^{85}Kr as the only radioactive noble gas liberated in fuel reprocessing.

Preprocessing cooling is useful for iodine decay until the ^{131}I activity has decayed to a level equal to the activity of ^{129}I. The time $T_{c,\text{I}}$ at which these two activities become equal can be calculated by applying Eq. (8.3) for ^{131}I and Eq. (8.6) for ^{129}I, with the simplification that for ^{129}I, $T_{1/2} \gg T_c$. Using the yield data for ^{235}U fission given in Table 2.9, we obtain

$$T_{c,\text{I}} = 11.6 \ln \frac{8.98 \times 10^7}{T_R} \text{ days} \tag{8.8}$$

where the fuel irradiation time T_R is in years. Assuming a typical T_R of 3 years,

$$T_{c,\text{I}} = 200 \text{ days}$$

This is the length of time such that further cooling produces no appreciable reduction in the iodine activity. Shorter cooling times are possible for aqueous reprocessing, because it is not necessary to reduce the ^{131}I activity to quite as low a level as the ^{129}I activity.

A common specification of the permissible activity remaining in separated and decontaminated uranium is that the specific beta activity not exceed that of natural uranium in equilibrium with its short-lived decay products ^{234}Th, $^{234\text{m}}$Pa, ^{234}Pa, and ^{231}Th. These activities are

β 0.68 μCi/g
α 0.69 μCi/g

These specific activities correspond to 1.5×10^6 beta disintegrations/(min·g uranium). This is rounded off to the specification of 10^6 disintegrations/(min·g uranium) as the allowable ^{237}U activity in uranium to be recovered and recycled to isotope separation. The actual allowable ^{237}U content must depend on the amount of material to be handled and the allowable dose rate to operating personnel. ^{237}U activity at this level of 10^6 disintegrations/(min·g) would result in a radiation dose on the surface of uranium metal at the rate of 2.6 mrem/h. This is less than 9 percent of the surface dose due to gammas in normal uranium and is a safe level for direct-contact handling of uranium.

The required cooling time $T_{c,\text{U}}$ for ^{237}U decay can be determined if the atoms of ^{237}U per atom of uranium $N_{27}(T_R)/N_\text{U}$ at the end of the irradiation period are known[†]:

$$\left[\frac{N_{27}(T_R)}{N_\text{U}} \lambda_{27} e^{-\lambda_{27} T_{c,\text{U}}} \text{ disintegrations/(s·atom U)}\right] \left(\frac{6.02 \times 10^{23} \text{ atoms U}}{A_\text{U} \text{ g U}}\right)$$
$$\times \text{ (60 s/min)} = 10^6 \text{ disintegrations/(min·g U)} \tag{8.9}$$

or

$$T_{c,\text{U}} = \frac{1}{\lambda_{27}} \ln \frac{3.61 \times 10^{19} N_{27}(T_R)\lambda_{27}}{N_\text{U} A_\text{U}} \tag{8.10}$$

where A_U is the average atomic weight of the isotopic mixture of uranium in the reactor product and λ_{27} is the decay constant of ^{237}U.

The concentration $N_{27}(T_R)/N_\text{U}$ depends on the ^{236}U concentration $N_{26}(T_R)/N_\text{U}$ at the end of the irradiation. Because of its relatively short half-life, ^{237}U will be in secular equilibrium with ^{236}U, and its concentration is obtained from

[†]The notation for nuclides is the same as that used in Chap. 3 and is defined under Nomenclature at the end of that chapter.

$$\frac{\lambda_{27}N_{27}(T_R)}{N_U} = \frac{N_{26}(T_R)\sigma_{26}\phi}{N_U} \tag{8.11}$$

where ϕ is the neutron flux at the end of the irradiation. Eliminating $N_{27}(T_R)/N_U$ from Eqs. (8.10) and (8.11), we obtain

$$T_{c,U} = 0.75 \ln \left[3.61 \times 10^{19} \frac{N_{26}(T_R)\sigma_{26}\phi}{N_U A_U} \right] \text{days} \tag{8.12}$$

The concentration $N_{26}(T_R)/N_U$ of ^{236}U can be obtained by applying the equations of Chap. 3. For the PWR example of Fig. 3.31:

$$\frac{N_{26}(T_R)}{N_U} = \frac{3.85}{957} = 0.00402$$

The neutron flux to which the fuel is exposed is $3.5 \times 10^{13} n/(\text{cm}^2 \cdot \text{s})$. The effective absorption cross section† for ^{236}U is 123.9 b for this reactor, and the average atomic weight of uranium in the reactor product is 238. Using the above data in Eq. (8.12), the required decay time is

$$T_{c,U} = 145 \text{ days}$$

If there were sufficient incentive to reduce the fuel-cycle inventory of plutonium, it would be possible to operate with shorter preprocessing cooling times and to take the remaining ^{237}U decay time after the plutonium-uranium separation. In the fast-breeder fuel cycle, where there is usually the greatest incentive to reduce fuel cycle fissile inventory and thereby to reduce the fissile doubling time, the ^{237}U content of the recovered uranium need not be as low as 10^6 disintegrations/(min·g), because the uranium is not to be recycled to isotope separation.

2.3 Radioactive Decay of Recycled Plutonium

If the plutonium recovered from discharge fuel by fuel reprocessing is stored for long periods, there is a loss of fuel value due to the radioactive decay of fissile ^{241}Pu. Even during storage periods as short as a few months, ^{241}Am, the beta-decay daughter of ^{241}Pu, builds up. Its decay is accompanied by gammas that increase the shielding required in the fabrication of fuel from recycled plutonium. Small quantities of ^{237}U, formed by the alpha decay of ^{241}Pu, also increase the gamma activity. The decay of 2.85-year ^{236}Pu forms ^{232}U, ^{228}Th, and short-lived decay daughters that also contribute to the shielding requirement. The growth of radioactive daughters in plutonium recovered from the fuel discharged each year by the uranium-fueled 1000-MWe LWR of Fig. 3.31 is shown in Fig. 8.6 [P1]. The radioactivity of the ^{228}Th daughters, which will be in secular equilibrium with ^{228}Th, is not included.

2.4 Long-term Radioactivity of Actinides from Uranium-Plutonium Fuel

The long-term radioactivities of neptunium, americium, and curium in the high-level reprocessing wastes from the uranium-fueled water reactor are shown in Fig. 8.7. Except for ^{241}Am and ^{237}Np, these curves are also applicable to unprocessed discharge fuel. The curves ^{241}Am and ^{237}Np have been calculated for 0.5 percent of the plutonium in discharge fuel to appear in the wastes, so that there is not sufficient ^{241}Pu to significantly increase the amounts of ^{241}Am and

†This effective cross section is greater than the cross section for thermal neutrons because of resonance absorption in ^{236}U.

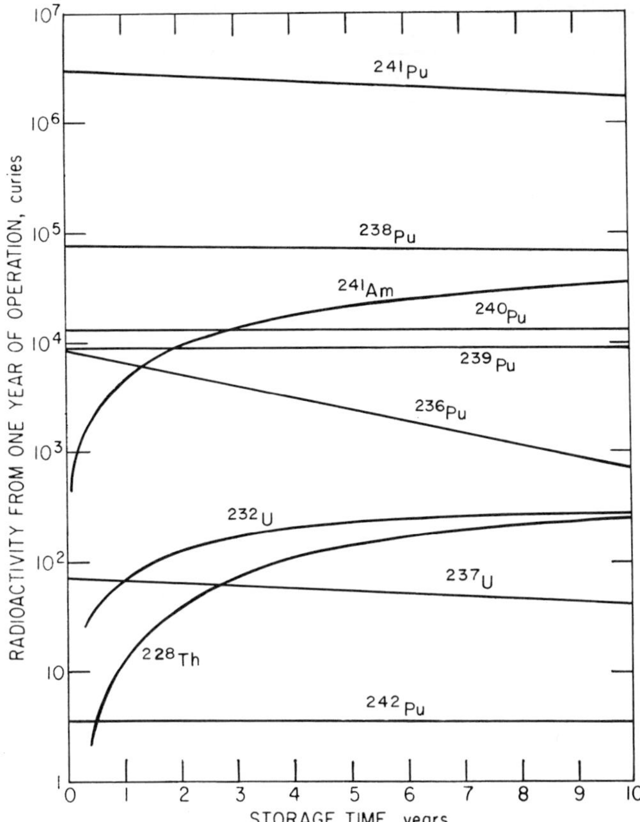

Figure 8.6 Radioactivity in separated plutonium as a function of storage time. (Amount in the plutonium recovered from the fuel discharged annually from a 1000-MWe uranium-fueled PWR.)

^{237}Np during the decay periods. The high activities of americium persist for thousands of years and are greater than the fission-product activity after a few hundred years of storage.

The radioactivities of the plutonium radionuclides in the high-level wastes from fuel reprocessing are shown as a function of storage time in Fig. 8.8 [P1]. Because the initial plutonium quantities are due only to the small fraction, e.g., 0.5 percent, of the plutonium that is lost to these wastes in reprocessing, larger quantities appear after a few years due to the decay of americium and curium. The ^{238}Pu increases with time because of the decay of ^{242m}Am and ^{242}Cm, ^{239}Pu increases from the decay of ^{243}Am and ^{243}Cm, and ^{240}Pu increases due to the decay of ^{244}Cm. Therefore, even though the total actinide activity in these wastes is dominated by plutonium after the americium has decayed, the plutonium in the wastes at this time is due mainly to the earlier decay of americium and curium and not to the small fraction of plutonium lost to the wastes in fuel reprocessing.

The ingestion toxicity indices of the actinides in the wastes are shown as a function of decay time in Fig. 8.9 [P2]. Because the actinides are nonvolatile and because the wastes are expected to be geologically isolated, ingestion toxicity is probably a more important measure than inhalation toxicity. During the first 600 years the total toxicity index is controlled by the fission products, mainly ^{90}Sr. It is thereafter controlled by ^{241}Am and ^{243}Am, followed by

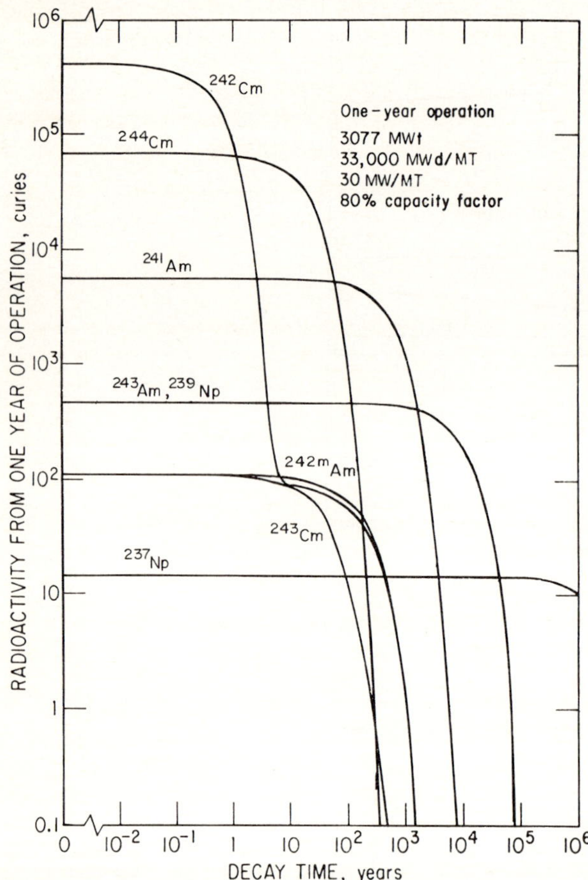

Figure 8.7 Radioactivity in curium, americium, and neptunium as a function of decay time. (Amount in the wastes produced annually by reprocessing fuel discharged from a 1000-MWe uranium-fueled PWR.)

^{239}Pu and ^{240}Pu. Subsequently, the most important radionuclide is ^{226}Ra, which is formed from the decay of ^{234}U, ^{238}Pu, ^{242m}Am, ^{242}Cm, and ^{238}U. The time for ^{226}Ra to build up is controlled by the half-life of its precursor ^{230}Th. The next important radionuclide is ^{225}Ra, which is the third decay daughter of ^{237}Np. The time for its buildup is controlled by the half-life of its precursor, 1.62×10^5 year ^{233}U. At about 10^6 years, the toxicity index is governed primarily by ^{129}I; finally, after approximately 10^8 years it reaches a level due to ^{226}Ra in secular equilibrium with the residual ^{238}U in the wastes.

The curve for ^{226}Ra in Fig. 8.9 was calculated for an assumed preprocessing cooling period of 150 days. When high-level waste is formed at this time after fuel discharge, the principal contributor to the long-term growth of ^{226}Ra is 163-day ^{242}Cm in the waste. However, if the discharge fuel is stored for longer periods prior to reprocessing, the long-term growth of ^{226}Ra will be considerably reduced. Longer storage allows more time for decay of ^{242}Cm, and the ^{238}Pu decay daughter is recovered in reprocessing. For example, increasing the preprocessing storage period from 150 days to 3 years reduces the long-term peak activity of ^{226}Ra in these high-level wastes by a factor of 2.9.

Although the true hazards of radioactive wastes are not measured by these toxicity indices, some perspective can be obtained by comparing the total ingestion toxicity index of the high-level wastes to the similar toxicity index for the ore used to fuel the reactor to generate these wastes. In Fig. 8.10 the toxicity indices are shown relative to the ingestion toxicity of the ore [P2]. The ore toxicity is due mainly to the ^{226}Ra, which is in secular equilibrium. Also shown are the relative toxicity indices for the uranium mill tailings, which contain ^{230}Th and ^{226}Ra separated from the uranium ore, and for the depleted uranium from isotope separation, neglecting the likely later use of this uranium as fuel for breeder reactors. Because the uranium ore ingestion toxicity is dominated by ^{226}Ra, all of this toxicity is transferred to the mill tailings and is preserved for over 100,000 years because of the long half-life of ^{230}Th. The tailings toxicity then dècays to a lower value due to the residual uranium, e.g., about 5 percent, which remains with the mill tailings.

The ingestion toxicity of the high-level waste decays to a level below that of the initial ore after the fission-product period of about 600 years, and it ultimately decays to a toxicity that is a fraction of a percent of the toxicity of the original ore consumed to generate these wastes.

Because in the LWR fuel cycle most of the uranium in the ore appears in the depleted uranium from isotope separation, this depleted uranium if not used as breeder fuel, will slowly build up its decay daughters and ^{226}Ra toxicity. Ultimately, a toxicity level within a few percent of that of the original ore will be reached.

The toxicity indices are not measures of hazards, in part because they take no account of the barriers that isolate these wastes from the biosphere or of the behavior of different radioactive elements with respect to these barriers. However, the long-term toxicities of the high-level reprocessing wastes are due to radium, which is the same element that controls the ore toxicity. The long-term radium toxicity of the reprocessing wastes is considerably less than the radium toxicity of the ore. It seems reasonable that high-level wastes can be geologically

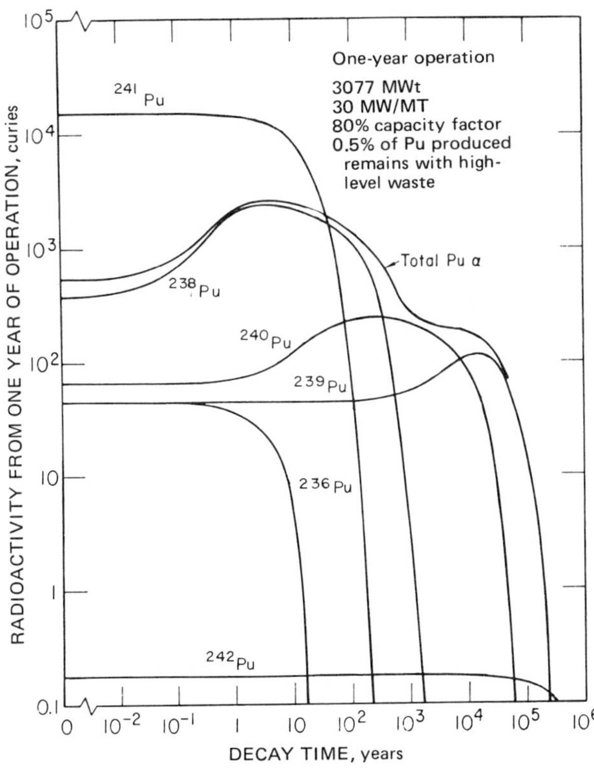

Figure 8.8 Radioactivity in plutonium in high-level wastes as a function of decay time (in wastes produces annually by reprocessing fuel discharged from a 1000-MWe uranium-fueled PWR).

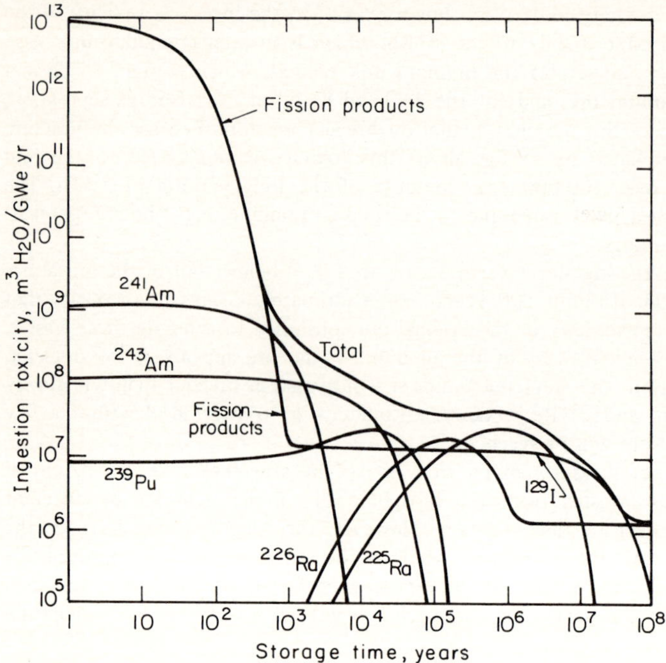

Figure 8.9 Principal contributions to the long-term ingestion toxicity of high-level waste from reprocessing uranium fuel (fuel from uranium-fueled PWR, 33 MWd/kg, 0.5 percent of uranium and plutonium appear in waste).

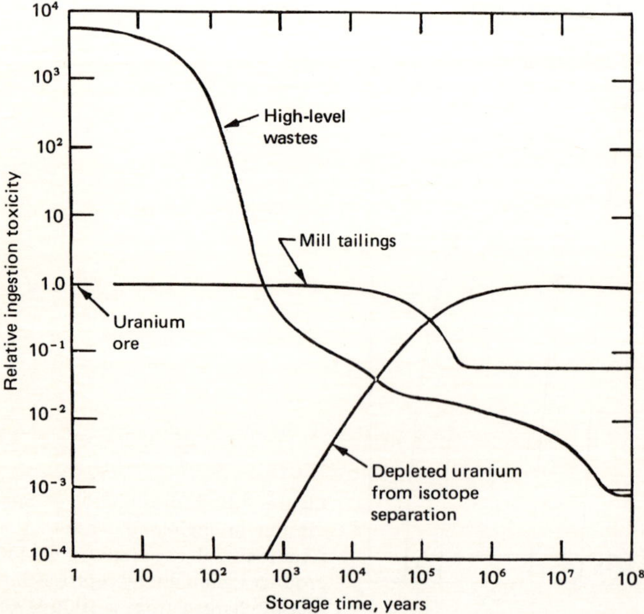

Figure 8.10 Relative ingestion toxicity of solid residuals from LWR fuel cycle (uranium fuel, 0.5 percent of uranium and plutonium in high-level wastes).

isolated so that the waste material has less access to the environment than the radium in the natural ore. Therefore, it is likely that the longer-term hazards from geologically isolated high-level wastes will be less than those already experienced due to the naturally occurring uranium minerals. The period of greatest importance in high-level waste management is probably the earlier, 600-year period of high fission-product toxicities.

2.5 Actinide Reactions in Thorium Fuel

The principal actinides involved in using thorium-uranium fuel are shown in the actinide chains of Fig. 8.11. The important reactions are the fission of ^{233}U and ^{235}U and the absorption of neutrons in ^{232}Th to form ^{233}U.

The relatively long 27.0-day half-life of ^{233}Pa, the precursor of ^{233}U, affects the time that irradiated fuel must be stored prior to reprocessing. If the discharged fuel is stored only for 150

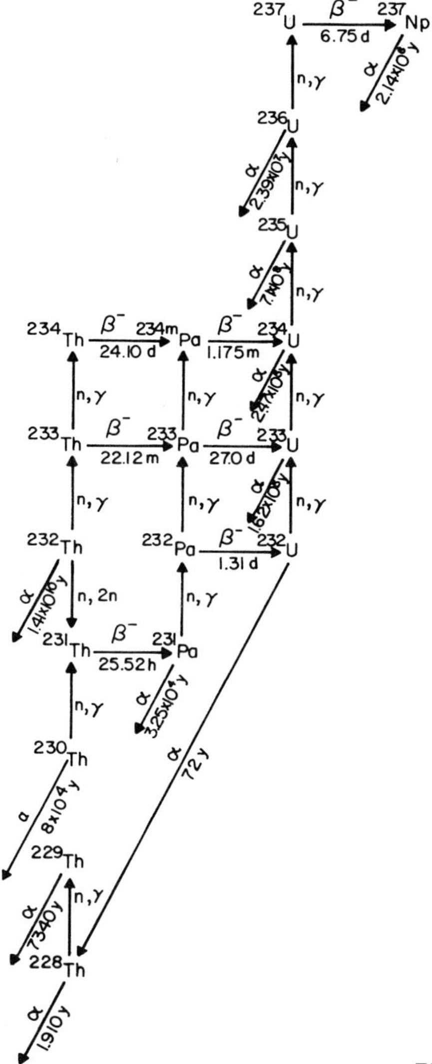

Figure 8.11 Actinide chains in thorium fuel.

days, as is frequently specified for sufficient decay of ^{131}I, some of the ^{233}Pa will remain during reprocessing. Protactinium is one of the more difficult elements to separate from uranium, and the high radioactivity of protactinium will contribute to the problem of decontaminating the uranium product after it is separated from the fission products and thorium. Also, if protactinium is not recovered, the loss of undecayed ^{233}Pa will represent some loss in the production of ^{233}U for recycle.

Another problem of the thorium fuel cycle results from the radioactivity of 72-year ^{232}U, and its daughters. ^{232}U is formed by $(n, 2n)$ reaction with ^{232}Th according to

$$^{232}\text{Th} \xrightarrow{n,2n} {}^{231}\text{Th} \xrightarrow[25.5\text{ h}]{\beta^-} {}^{231}\text{Pa} \xrightarrow{n,\gamma} {}^{232}\text{Pa} \xrightarrow[1.31\text{ days}]{\beta^-} {}^{232}\text{U} \tag{8.13}$$

and by

$$^{233}\text{U} \xrightarrow{n,2n} {}^{232}\text{U} \tag{8.14}$$

The threshold neutron energy for the ^{232}Th$(n, 2n)$ reaction is 6.37 MeV.

^{232}U is also formed by the chain initiating with ^{235}U:

$$^{235}\text{U} \xrightarrow{n,\gamma} {}^{236}\text{U} \xrightarrow{n,\gamma} {}^{237}\text{U} \xrightarrow[6.75\text{ days}]{\beta^-} {}^{237}\text{Np} \xrightarrow{n,2n} {}^{236}\text{Np} \xrightarrow[22\text{ h}]{\beta^-} {}^{236}\text{Pu} \xrightarrow[2.85\text{ yr}]{\alpha} {}^{232}\text{U} \tag{8.15}$$

Also, many thorium ores as well as thorium, which is obtained as a by-product of uranium mining, contain traces of ^{230}Th, a radionuclide in the decay chain of ^{238}U. Neutron absorption in ^{230}Th also results in the formation of ^{232}U:

$$^{230}\text{Th} \xrightarrow{n,\gamma} {}^{231}\text{Th} \xrightarrow[25.52\text{ h}]{\beta^-} {}^{231}\text{Pa} \xrightarrow{n,\gamma} {}^{232}\text{Pa} \xrightarrow[1.31\text{ days}]{\beta^-} {}^{232}\text{U} \tag{8.16}$$

Although significant alpha activity results from ^{232}U in the ^{233}U to be recovered and recycled, more of a problem results from the ^{232}U daughters. The ^{232}U decay daughter is 1.91-year ^{228}Th, a radionuclide that is also formed by the radioactive decay of ^{232}Th. As shown in Table 6.3, the decay daughters of ^{228}Th are all short-lived, so they reach secular equilibrium with ^{228}Th after a delay time of only a few days. The decays of ^{212}Bi and ^{208}Tl are accompanied by very energetic and penetrating gammas, so gamma shielding is required when fabricating fuel from recycled uranium containing ^{232}U.

Although chemical reprocessing yields essentially pure uranium, storage after separation and time elapsed in shipping to fabrication allow the buildup of ^{228}Th and its decay daughters. Consequently, the gamma activity in separated uranium containing ^{232}U increases continuously with storage time, until it reaches a maximum at about 10 years after separation. Once uranium has been separated from thorium, there is considerable incentive to complete the uranium purification and fuel fabrication quickly to avoid the increasing gamma radiation due to the buildup of ^{228}Th. Hydrogenous shielding is also necessary because of the high-energy neutrons from alpha decay in recycled uranium. The alphas from the decay of ^{233}U, ^{232}U, and ^{228}Th interact with light elements such as oxygen and carbon to form neutrons, so the neutron activity also increases with storage time.

The ^{228}Th and ^{234}Th appearing with irradiated thorium fuel results in appreciable radioactivity in the separated thorium. Consequently, as discussed in Sec. 2.9, it may not be practicable to recycle the recovered thorium until it has been stored for about 5 to 20 years.

When ^{235}U is used as fissile makeup in the thorium cycle, as in the reference high-temperature gas-cooled reactor (HTGR) fuel cycle, the high burnup and uranium recycle result in considerable production of ^{237}Np, according to the reactions shown in Fig. 8.11. The ^{237}Np then forms a relatively large activity of ^{238}Pu. These plutonium activities are important because of the problems of decontaminating uranium from plutonium when reprocessing the uranium. Also, even though fissile plutonium is formed by neutron absorption in the ^{238}U accompanying the highly enriched ^{235}U makeup, the high activities of ^{238}Pu may discourage the utilization of the fuel value of plutonium in the discharge fuel.

Relatively little ^{239}Pu, ^{240}Pu, ^{241}Pu, americium and curium are formed in the irradiation of thorium-uranium fuel with ^{235}U fissile makeup. However, when plutonium is used as fissile makeup for a thorium fuel cycle, considerable quantities of americium and curium are formed. As discussed in Sec. 2.4, these are the radionuclides that are the greatest contributors to radioactivity and ingestion toxicity after about 600 years of waste isolation, when the fission products have decayed.

Material quantities and activities of the actinides calculated [H1, P3] in the cooled discharge fuel from the uranium-thorium-fueled HTGR (cf. Fig. 3.33) are listed in Table 8.6. The natural thorium is assumed to contain 100 ppm ^{230}Th, so the quantities of ^{228}Th and ^{232}U in the discharge fuel are greater than would occur for thorium consisting of pure ^{232}Th. The strongest actinide beta source is ^{233}Pa, which contributes 7.58×10^6 Ci/year after 150 days of cooling. In the uranium, which is to be recovered and fabricated into recycle fuel, the main contributors to alpha activity are ^{232}U and ^{233}U. Both are important as potential environmental contaminants, but the activity of the ^{232}U daughters, which grow into separated uranium prior to fabrication, dictate the requirements for semiremote and remote fabrication. By comparison with the data in Table 8.5, the total alpha activity of 5.16×10^3 Ci/year in the uranium to be fabricated as recycle HTGR fuel is much less than the 1.70×10^6 Ci/year of alpha activity in the plutonium to be fabricated for recycle in a 1000-MWe LWR.

The total alpha activity in the plutonium in the HTGR discharge fuel is within 20 percent of the total alpha activity in plutonium from the uranium-fueled LWR (Table 8.4). In both cases the plutonium alpha activity is dominated by ^{238}Pu. However, the HTGR plutonium consists of 66 percent ^{238}Pu, and the high alpha activity, the high heat generation rate, and the low fissile content mitigate against the recycle of HTGR plutonium.

Because of the relatively small amount of high-mass plutonium nuclides produced in uranium-thorium fueling, the amounts of americium and curium produced are about two orders of magnitude less than in a uranium-fueled reactor with plutonium recycle.

2.6 Growth of ^{232}U in Irradiated Uranium-Thorium Fuel

When fresh thorium is irradiated, ^{231}Th builds up quickly to equilibrium because of its relatively short half-life of 25.5 h. After a time T_R of irradiation, the amount N_{11} of ^{231}Pa is obtained by applying Eq. (2.101). For simplicity, we shall assume an essentially constant amount N_{02} of ^{232}Th during the irradiation and will assume no ^{230}Th in the thorium:

$$N_{11} = N_{02} \frac{\sigma_{n,2n}}{\sigma_{11}} (1 - e^{-\sigma_{11}\phi T_R}) \tag{8.17}$$

where σ_{11} is the effective absorption cross section of ^{231}Pa and $\sigma_{n,2n}$ is the $(n, 2n)$ cross section for ^{232}Th. Even though the ^{232}Th$(n, 2n)$ reaction occurs for neutrons at energies above 6.37 MeV, we may define an effective $(n, 2n)$ cross section such that when multiplied by the thermal flux, the proper $(n, 2n)$ reaction rate is obtained. The effective $(n, 2n)$ cross section will depend, in part, on the reactor core composition.

Because of its relatively short half-life, 1.31-day ^{232}Pa will be in secular equilibrium with ^{231}Pa, so that the concentration $N_{22}(T_R)$ of ^{232}U as a function of irradiation time T_R is given by an extension of Eq. (2.101):

$$\frac{N_{22}(T_R)}{N_{02}} = \phi^2 \sigma_{n,2n}\sigma_{11} \left[\frac{1 - e^{-\mu_{11} T_R}}{\mu_{11}(\mu_{22} - \mu_{11})} + \frac{1 - e^{-\mu_{22} T_R}}{\mu_{22}(\mu_{11} - \mu_{22})} \right] + \frac{N_{22}^0}{N_{02}} e^{-\mu_{22} T_R} \tag{8.18}$$

Table 8.6 Actinides in discharge thorium fuel†

Radionuclide	Half-life	kg/yr	Ci/yr
^{228}Th‡	1.910 yr	2.54×10^{-3}	2.08×10^{3}
^{229}Th	7,340 yr	2.96×10^{-3}	6.29×10^{-1}
^{230}Th	8×10^{4} yr	2.71×10^{-1}	5.26
^{231}Th	25.5 h	1.35×10^{-7}	7.20×10^{1}
^{232}Th	1.41×10^{10} yr	6.75×10^{3}	7.37×10^{-1}
^{234}Th	24.1 days	1.39×10^{-5}	3.22×10^{2}
Total		6.75×10^{3}	α 2.16×10^{3}
			β 3.22×10^{2}
^{233}Pa	27.0 days	2.18×10^{-1}	4.52×10^{6}
^{234}Pa	6.75 h	2.28×10^{-7}	4.52×10^{2}
Total		2.18×10^{-1}	β 4.52×10^{6}
^{232}U§	72 yr	1.39×10^{-1}	2.97×10^{3}
^{233}U	1.62×10^{5} yr	1.89×10^{2}	1.79×10^{3}
^{234}U	2.47×10^{5} yr	6.20×10^{1}	3.83×10^{2}
^{235}U	7.1×10^{8} yr	4.90×10^{1}	1.05×10^{-1}
^{236}U	2.39×10^{7} yr	1.04	6.59×10^{-2}
^{237}U	6.75 days	6.69×10^{-9}	5.46×10^{2}
^{238}U	4.51×10^{9} yr	2.91×10^{1}	9.68×10^{-3}
Total		3.30×10^{2}	α 5.14×10^{3}
			β 5.46×10^{2}
^{237}Np	2.14×10^{6} yr	1.10×10^{1}	7.75
Total		1.10×10^{1}	7.75
^{236}Pu¶	2.85 yr	4.95×10^{-6}	2.62
^{238}Pu	86 yr	5.68	9.92×10^{4}
^{239}Pu	24,400 yr	1.20	7.35×10^{1}
^{240}Pu	6,580 yr	5.59×10^{-1}	1.26×10^{2}
^{241}Pu	13.2 yr	5.36×10^{-1}	6.02×10^{4}
^{242}Pu	3.79×10^{5} yr	5.45×10^{-1}	2.12
Total		8.52	α 9.94×10^{4}
			β 6.02×10^{4}
^{241}Am	458 yr	2.17×10^{-2}	7.02×10^{1}
^{242m}Am	152 yr	3.03×10^{-4}	2.94
^{243}Am	7,950 yr	1.56×10^{-1}	2.88×10^{1}
Total		1.78×10^{-1}	α 9.90×10^{1}
			β 2.94
^{242}Cm	163 days	4.35×10^{-3}	1.44×10^{4}
^{243}Cm	32 yr	1.31×10^{-4}	6.02
^{244}Cm	17.6 yr	7.04×10^{-2}	5.86×10^{3}
^{245}Cm	9,300 yr	2.90×10^{-4}	4.55×10^{-2}
Total		7.52×10^{-2}	α 2.03×10^{4}
Total		7.08×10^{3}	α 1.27×10^{5}
			β 4.58×10^{6}

†1000-MWe uranium-thorium-fueled HTGR. 95 MWd/kg heavy metal, 38.7% thermal efficiency, 80% capacity factor, 150-day cooling, equilibrium fuel cycle.

‡Natural thorium is assumed to contain 100 ppm ^{230}Th. Discharge thorium is not recycled.

§Includes 59.0 kg/year of second-cycle uranium, from initial makeup ^{235}U, which is not to be recycled. Composition of discharged second-cycle uranium: 0.8% ^{234}U, 3.6% ^{235}U, 75.5% ^{236}U, 20.1% ^{238}U.

¶Plutonium is not recycled.

where $\mu_{11} = \phi\sigma_{11}$
$\mu_{22} = \phi\sigma_{22} + \lambda_{22}$
N_{22}^0 = initial amount of ^{232}U, which may be finite due to recycled uranium

It has been assumed that no ^{231}Pa is recycled. Because the concentration of ^{233}U is much less than that of thorium in uranium-thorium thermal reactors, $(n, 2n)$ reactions in ^{233}U have been neglected.

For neutron fluxes in excess of $10^{13}/(\text{cm}^2 \cdot \text{s})$, the term λ_{22} in Eq. (8.18) is relatively unimportant and the concentration N_{22} becomes

$$\frac{N_{22}}{N_{02}} = \frac{\sigma_{n,2n}}{\sigma_{11} - \sigma_{22}} \left[\frac{\sigma_{11}}{\sigma_{22}} (1 - e^{-\sigma_{22}\theta}) - (1 - e^{-\sigma_{11}\theta}) \right] + \frac{N_{22}^0}{N_{02}} e^{-\sigma_{22}\theta} \tag{8.19}$$

where

$$\theta \equiv \int_0^{T_R} \phi(t)\, dt \tag{8.20}$$

The concentration of ^{232}U in initially pure ^{232}Th ($N_{22}^0 = 0$) as a function of irradiation time is shown in Fig. 8.12.

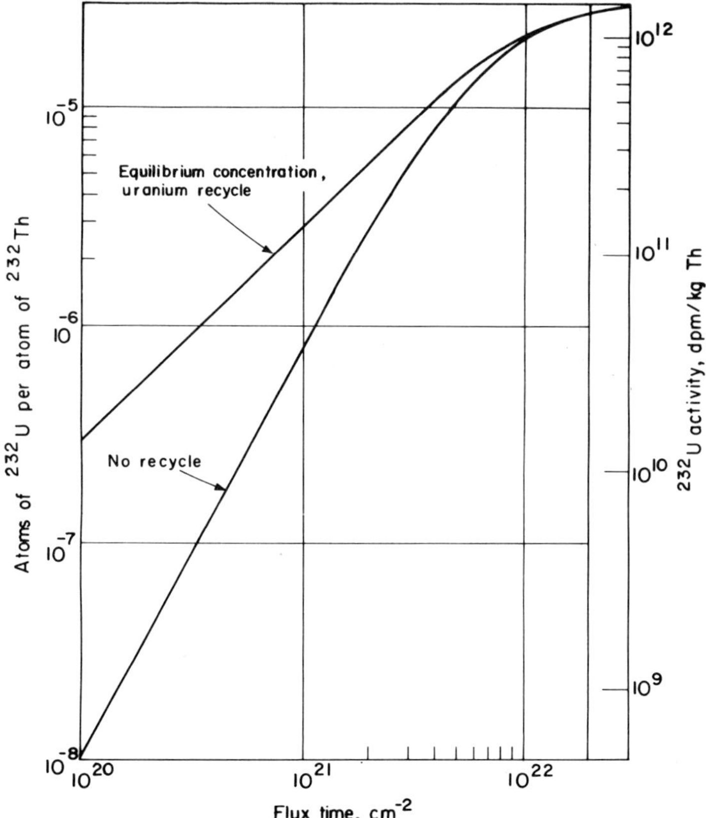

Figure 8.12 ^{232}U concentration in irradiated thorium. Basis: $\sigma_{02}(n, 2n) = 0.010$ b, $\sigma_{11} = 200$ b, $\sigma_{22} = 339$ b.

When uranium is recycled, the initial amount of ^{232}U for fuel generation n is related to the final concentration from generation $n-1$ by

$$N_{22,n}^{0} = N_{22,n-1}(T_R) \tag{8.21}$$

where process losses and decay of ^{232}U in the external fuel-cycle operation have been neglected.

For the equilibrium fuel cycle,

$$N_{22}^{0} = N_{22}(T_R) = N_{22}^{\infty} \tag{8.22}$$

and Eq. (8.19) becomes

$$\frac{N_{22}^{\infty}}{N_{02}} = \frac{\sigma_{n,2n}\sigma_{11}}{\sigma_{22}(\sigma_{11}-\sigma_{22})}\left[1-\left(\frac{1-e^{-\sigma_{11}\theta}}{1-e^{-\sigma_{22}\theta}}\right)\frac{\sigma_{22}}{\sigma_{11}}\right] \tag{8.23}$$

where θ is the flux time at the end of the irradiation.

For the first "generation" of thorium-uranium fuel, for which $N_{22}^{0}=0$, Eqs. (8.19) and (8.23) show that the ^{232}U content N_{22}^{1} at the end of the first cycle is related to the equilibrium content N_{22}^{∞} by

$$\frac{N_{22}^{1}}{N_{22}^{\infty}} = 1-e^{-\sigma_{22}\theta} \tag{8.24}$$

which assumes the same flux time for all cycles. Equation (8.24) is also valid if ^{230}Th is present as an additional source of ^{232}U.

In the case of equilibrium recycle, the concentration of ^{232}U in the discharged thorium is the same as that in the makeup thorium containing the recycled uranium. In Fig. 8.12 this concentration is shown as a function of the total flux time of the fuel irradiation. However, during irradiation the ^{232}U in the fuel decreases below its initial concentration and then recovers as ^{231}Pa is formed.

2.7 Growth of ^{228}Th and Gamma Activity in Separated Uranium

During chemical separation, the ^{232}U follows the uranium product and the ^{228}Th follows the thorium. The activity $\lambda_{08}N_{08}(t)$ of ^{228}Th that has again built up in the separated uranium during a time t after separation is obtained by applying Eq. (2.14):

$$\lambda_{08}N_{08}(t) = \frac{\lambda_{08}\lambda_{22}N_{22}^{0}}{\lambda_{08}-\lambda_{22}}\left(e^{-\lambda_{22}t}-e^{-\lambda_{08}t}\right) \tag{8.25}$$

where N_{22}^{0} is the amount of ^{232}U present after separation. For a time scale in years, the ^{228}Th daughters will be in secular equilibrium and the beta activity $(\lambda N)_\beta$ at time t is just twice the activity given by Eq. (8.25).

During the first few years, decay of ^{232}U is negligible, so that Eq. (8.25) becomes

$$\lambda_{08}N_{08}(t) = \lambda_{22}N_{22}^{0}\left(1-e^{-\lambda_{08}t}\right) \tag{8.26}$$

and for a time scale in days, the growth of beta activity due to ^{212}Pb, ^{212}Bi, and ^{208}Tl is given by

$$(\lambda N) = 2\lambda_{224}\lambda_{08}\lambda_{22}N_{22}^{0}\left[\frac{1-e^{-\lambda_{08}t}}{\lambda_{08}(\lambda_{224}-\lambda_{08})}+\frac{1-e^{-\lambda_{224}t}}{\lambda_{224}(\lambda_{08}-\lambda_{224})}\right] \tag{8.27}$$

where λ_{224} is the decay constant for ^{224}Ra. Buildup and decay of beta activity and gamma dose as a function of days after separation are illustrated in Fig. 8.13. It is important that the uranium product from thorium irradiation be carried rapidly through the refabrication operations soon after chemical separation.

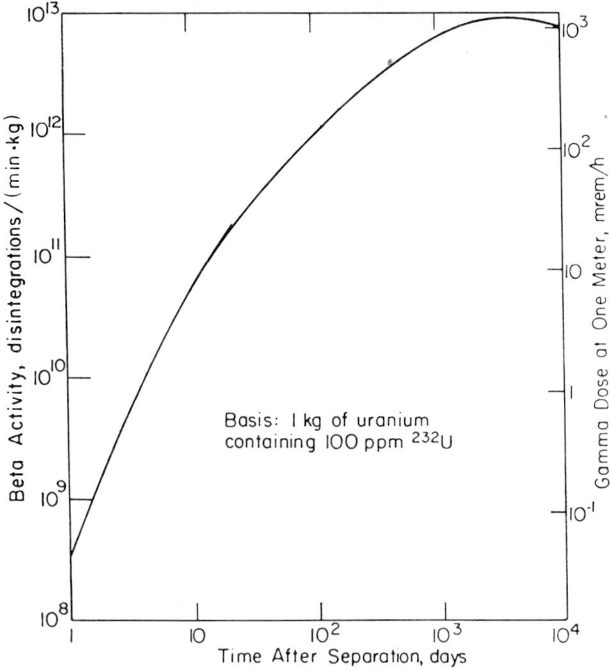

Figure 8.13 Growth of beta activity and gamma dose due to ^{232}U in uranium.

The high surface activities of the uranium require semiremote refabrication methods, whereby direct bodily contact with the material is avoided, but only distance or light shielding need be used to avoid above-tolerance radiation doses. To illustrate, assume that personnel performing the fabrication operations are separated from the work by an average distance of 1 m and that each person is exposed to 1 kg of separated uranium containing 100 ppm ^{232}U. From Fig. 8.13, we see that the refabrication must be completed within only 6 days after separation if the workers handling the last–and usually most delicate–stages of fabrication are to receive a typical tolerance dose of no more than 2.5 mr/h† at the end of the fabrication period. Alternatively, for an allowable dose of 2.5 mr/h, averaged over the entire period since separation, the allowable time to complete semiremote fabrication is 11 days.

A longer wait may necessitate remote fabrication, whereby all operations must be carried out behind heavy shields. For example, uranium containing 100 ppm ^{232}U and aged 35 days since final separation would yield an unshielded dose of 38 mr/h per kilogram. From the data in Fig. 2.4, it is estimated that the fabrication must be carried out behind about 7 cm of lead or 35 cm of concrete if the dose to operators is to be 2.5 mr/h or less.

As the ^{232}U concentration increases, the allowable time for semiremote fabrication decreases rapidly, and greater shielding thicknesses for remote fabrication are required.

2.8 ^{234}Th in Separated Thorium

The principal sources of activity in irradiated and chemically purified thorium are ^{234}Th and its short-lived daughter ^{234m}Pa, and ^{228}Th and its daughters. Beta and gamma activity from these

†Corresponding to a weekly total of 0.100 rem for continued exposure.

nuclides constitute the greatest danger in external exposure; neutrons from (α, n) reactions with light contaminants are relatively unimportant in this regard. Prediction of activities due to ^{234}Th is similar to the analyses of ^{237}U activity in Sec. 2.2.

Nuclides in the ^{234}Th chain reach equilibrium concentration during irradiation exposures of a few months or greater, with the concentrations given by

$$N_{03}\lambda_{03} = N_{02}\sigma_{02}\phi \tag{8.28}$$

and

$$N_{04}\lambda_{04} = N_{03}\sigma_{03}\phi \tag{8.29}$$

where σ_{02} is the equivalent thermal cross section for (n, γ) reactions in ^{232}Th, and is greater than the true thermal value to allow for absorption of resonance neutrons. By combining Eqs. (8.28) and (8.29), the equilibrium concentration of ^{234}Th is

$$\frac{N_{04}}{N_{02}} = \frac{\sigma_{02}\sigma_{03}\phi^2}{\lambda_{04}\lambda_{03}} \tag{8.30}$$

and the concentration at a time T_c after irradiation is

$$\frac{N_{04}(T_c)}{N_{02}} = \frac{\sigma_{02}\sigma_{03}\phi^2}{\lambda_{04}\lambda_{03}} e^{-\lambda_{04}T_c} \tag{8.31}$$

The beta activity $2(\lambda N)_{04}$ due to ^{234}Th and ^{234}Pa is

$$\frac{2(\lambda N)_{04}}{N_{02}} = \frac{2\sigma_{02}\sigma_{03}\phi^2}{\lambda_{03}} e^{-\lambda_{04}T_c} \tag{8.32}$$

If the ^{232}Th is irradiated in a neutron flux with a negligible component above 6.37 MeV so that no ^{232}U-^{228}Th are formed, postirradiation cooling can reduce the beta activity to a tolerable level. Even if ^{228}Th is present, preprocessing decay of ^{234}Th may be useful to aid beta decontamination of the separated thorium product. From Eq. (8.32) the time required for the ^{234}Th-^{234}Pa beta activity to reach the beta activity of natural thorium of 4.37×10^{-7} Ci/g is given by

$$T_c = 34.8 \ln (6.17 \times 10^{20} \sigma_{02}\sigma_{03}\phi^2) \text{ days} \tag{8.33}$$

where $\sigma\phi$ are expressed in reciprocal seconds.

For the uranium-thorium-fueled reactor of Fig. 3.33, $\sigma_{02} = 6.1$ b, $\sigma_{03} = 520$ b, and $\phi = 1.2 \times 10^{14}$ $n/(\text{cm}^2 \cdot \text{s})$, resulting in ^{234}Th-^{234}Pa beta activity at discharge of

$$\frac{2(N\lambda)_{04}}{N_{02}} = 1.2 \times 10^{-2} \text{ Ci/g } ^{232}\text{Th}$$

The time for this to decay to the equilibrium beta activity of the ^{232}Th daughters is

$$T_c = 356 \text{ days}$$

Cooling for this length of time will ensure that in chemical reprocessing thorium can undergo total beta decontamination to twice the level of natural ^{232}Th. The decontamination can be verified with total beta monitoring. For shorter cooling times beta discrimination techniques must be used to ensure that long-lived beta contaminants are not present in the separated thorium.

^{228}Th is also present in irradiated thorium and is accompanied by beta-emitting daughters in its decay chain. These daughters are removed from thorium in chemical reprocessing, but they appear again in the separated thorium, growing with a time constant of about 4 days. Thereafter, the beta activity in the separated thorium approaches the level in secular equilibrium with ^{228}Th. It is therefore important that monitoring for beta decontamination of thorium separated in fuel reprocessing be carried out promptly after the separation is performed.

2.9 ^{228}Th in Irradiated Thorium

Contrasted to benefits from reduction in ^{234}Th activity, preprocessing cooling increases the ^{228}Th content of irradiated thorium. The amount $N_{08}(T)$ of ^{228}Th present at the end of an irradiation period T_R, due to ^{232}U decay, is given by applying Eq. (2.106):

$$\frac{N_{08}}{N_{02}} = \phi\sigma_{02}(n, 2n)\mu_{11}\lambda_{22}\left[\frac{1-e^{-\mu_{11}T_R}}{\mu_{11}(\mu_{22}-\mu_{11})(\mu_{08}-\mu_{11})} + \frac{1-e^{-\mu_{22}T_R}}{\mu_{22}(\mu_{11}-\mu_{22})(\mu_{08}-\mu_{22})} + \frac{1-e^{-\mu_{08}T_R}}{\mu_{08}(\mu_{11}-\mu_{08})(\mu_{22}-\mu_{08})}\right] \tag{8.34}$$

where

$$\mu_{08} = \phi\sigma_{08} + \lambda_{08} \tag{8.35}$$

During the preprocessing cooling period, the atom ratio of ^{232}U to ^{232}Th remains essentially constant because of the long half-life of ^{232}U. An equation for the activity $\lambda_{08}N_{08}(T_c)$ of ^{228}Th present after a time T_c of preprocessing cooling is obtained by applying Eqs. (2.13) and (2.27):

$$\frac{\lambda_{08}N_{08}(T_c)}{N_{02}} = \frac{\lambda_{08}N_{08}(T_R)}{N_{02}}e^{-\lambda_{08}T_c} + \frac{\lambda_{22}N_{22}(T_R)}{N_{02}}(1-e^{-\lambda_{08}T_c}) \tag{8.36}$$

where the activities $\lambda_{08}N_{08}(T_R)$ and $\lambda_{22}N_{22}(T_R)$ at the end of the irradiation are obtained from Eqs. (8.34) and (8.19), respectively.

The growth of ^{228}Th activity during irradiation at various neutron fluxes is shown in Fig. 8.14. At a given flux time of irradiation, the ^{228}Th activity is lower at the higher flux levels. This is because the actual time since the beginning of irradiation is shorter at the higher fluxes and less of the ^{232}U formed has undergone radioactive decay.

Because ^{228}Th is usually not in secular equilibrium with ^{232}U, its activity continues to grow during preprocessing cooling. Although the total of the ^{228}Th and ^{234}Th activities decreases with time, the activity from ^{228}Th daughters is the most troublesome when chemically purified thorium is being refabricated. The highly energetic betas from both ^{228}Th and ^{234}Th chains give large skin doses on surface contact with separated thorium, but the hard (i.e., highly energetic) gammas (2.3 MeV) from the ^{228}Th chain can result in serious dose rates even with semiremote fabrication techniques.

When the separated thorium is eventually to be recycled and blended with low-activity uranium streams, such as makeup ^{235}U, the activity of ^{228}Th after a preprocessing cooling time T_c and a postprocessing storage time T_s is given by

$$(\lambda N)_{08} = [N_{22}(T_R)\lambda_{22}(1-e^{-\lambda_{08}T_c}) + N_{08}(T_R)\lambda_{08}e^{-\lambda_{08}T_c}]e^{-\lambda_{08}T_s} \tag{8.37}$$

where $N_{22}(T_R)$ = quantity of ^{232}U in discharge fuel
$N_{08}(T_R)$ = quantity of ^{228}Th in discharge fuel

Thorium can be recycled for fabrication with low-activity uranium if the ^{228}Th activity is no more than a factor ψ greater than the ^{228}Th activity in natural thorium,

$$(\lambda N)_{08} = \psi(\lambda N)_{02} \tag{8.38}$$

Arnold [A1] suggests a value of $\psi = 5$ for thorium to avoid the requirement of semiremote fabrication. Combining Eqs. (8.37) and (8.38), we obtain

$$T_s = \frac{1}{\lambda_{08}}\ln\left[\frac{1}{\psi}\frac{N_{22}(T_R)\lambda_{22}}{N_{02}\lambda_{02}}(1-e^{-\lambda_{08}T_c}) + \frac{N_{08}(T_R)\lambda_{08}}{N_{02}\lambda_{02}}e^{-\lambda_{08}T_c}\right] \tag{8.39}$$

For an HTGR [H1, P3] with discharge concentrations of $(\lambda N)_{22}/(\lambda N)_{02} = 4.04 \times 10^3$, $(\lambda N)_{08}/(\lambda N)_{02} = 2.54 \times 10^3$, $T_c = 150$ days, and $\psi = 5$, we obtain

$$T_s = 21.3 \text{ years}$$

for thorium to be used when fabricating fuel with makeup ^{235}U. In the HTGR about two-thirds of the thorium is used to fabricate fuel containing makeup or recycled uranium containing no ^{232}U, so about two-thirds of the separated thorium would be subjected to the storage time estimated above.

For that portion of the separated thorium that is eventually to be recycled and blended with the recycled bred uranium, less time for thorium storage is possible. A reasonable criterion is that the thorium be stored for a sufficient period such that its ^{228}Th activity is equal to the activity of ^{228}Th in the recycled uranium at the time of fabrication. Ignoring process losses, the recycled bred uranium contains all of the ^{232}U that was present in the discharge thorium. If this recovered uranium has been stored for a time T_F prior to fuel fabrication, the activity of ^{228}Th in the uranium is

$$(\lambda_{08} N_{08})^{U} = N_{22} \lambda_{22} (1 - e^{-\lambda_{08} T_F}) \tag{8.40}$$

Applying the above criterion, we equate the ^{228}Th activity in the bred uranium to the activity

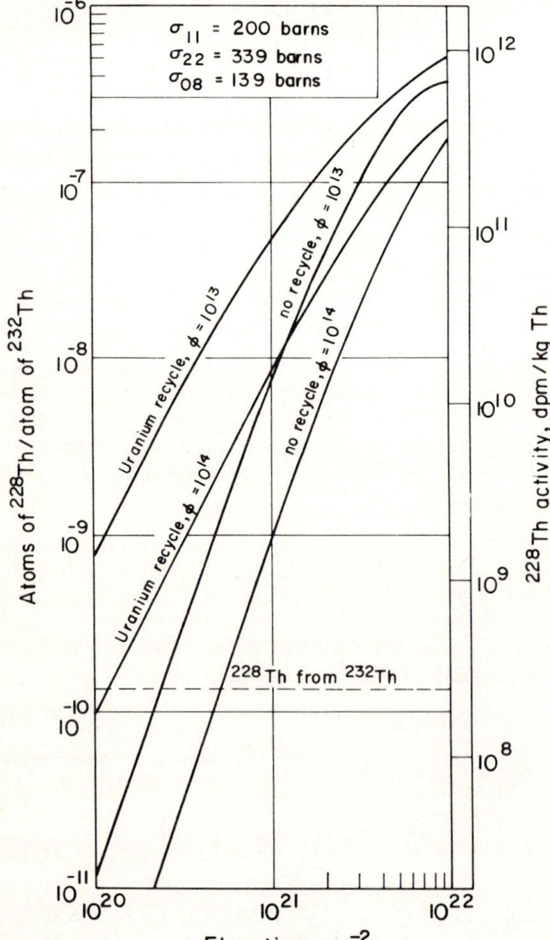

Figure 8.14 ^{228}Th concentration in irradiated thorium. Basis: $\sigma_{02}(n, 2n) = 0.010$ b.

of ^{228}Th in the fraction β of the recovered thorium that is eventually to be recycled for fabrication with the bred uranium, i.e.,

$$(\lambda_{08} N_{08})^{U} = \beta(\lambda_{08} N_{08})^{Th} \tag{8.41}$$

where $(\lambda_{08} N_{08})^{Th}$ is given by Eq. (8.34). Combining Eqs. (8.34), (8.36), (8.40), (8.41), and (8.37), we obtain

$$T_s = \frac{1}{\lambda_{08}} \ln \left[\beta \frac{1-(1-N_{08}\lambda_{08}/N_{22}\lambda_{22})e^{-\lambda_{08} T_c}}{1-e^{-\lambda_{08} T_F}} \right] \tag{8.42}$$

For the HTGR, $\beta = 0.36$. Assuming that $T_c = 150$ days and $T_F = 60$ days, we obtain

$$T_s = 4.2 \text{ years}$$

As the prefabrication time of uranium storage increases, less time is required for thorium storage. For the parameters listed above, if the recovered uranium is stored for 312 days before fabrication, the ^{228}Th activity in the uranium becomes equal to that in 36 percent of the separated thorium, so no thorium storage is then required to meet the ^{228}Th criterion.

3 EFFECT OF FUEL–CYCLE ALTERNATIVES ON PROPERTIES OF IRRADIATED FUEL

The calculated elemental composition, radioactivity, and decay-heat rate for discharge fuel are shown in Table 8.7 for the uranium-fueled PWR (cf. Fig. 3.31), in Table 8.8 for the liquid-metal fast-breeder reactor (LMFBR) (cf. Fig. 3.34), and in Table 8.9 for the uranium-thorium-fueled HTGR (cf. Fig. 3.33). These quantities, expressed per unit mass of discharge fuel, are useful in the design of reprocessing operations. For the purpose of comparison, all quantities are calculated for 150 days of postirradiation cooling.

When expressed in terms of radioactivity per unit amount of energy produced, as in Table 8.1, there is little variation in the fission-product radioactivity and toxicity due to the different fuel-cycle options. However, the long-term actinide activity is considerably affected. The greater quantities of americium and curium resulting from plutonium recycle increase the amounts of all of the actinides and ^{226}Ra, which control the ingestion toxicity of wastes after the fission products have decayed. The resulting total ingestion toxicity for the 1000-MWe LWR operating with self-generated plutonium recycle is compared with that for uranium fueling in Fig. 8.15 [P2]. The greatest long-term ingestion toxicity results if the discharge fuel is not reprocessed, because all of the plutonium and uranium in the discharge fuel then contribute to the long-term radioactivity. The toxicity for the radioactive wastes from the uranium-plutonium fast-breeder fuel cycle is similar to that for self-generated plutonium recycle in the LWR.

The toxicity of the high-level wastes from a uranium-thorium HTGR fuel cycle is initially smaller, after the fission-product decay period of 600 years, because of the relatively small quantities of americium, curium, ^{239}Pu, and ^{240}Pu formed in this thorium fuel cycle. However, after about 100,000 years of isolation the theoretical ingestion toxicity of the wastes is governed by ^{226}Ra, formed by

$$^{234}\text{U} \xrightarrow[2.47 \times 10^5 \text{ yr}]{\alpha} {}^{230}\text{Th} \xrightarrow[8.0 \times 10^4 \text{ yr}]{\alpha} {}^{226}\text{Ra} \xrightarrow[1602 \text{ yr}]{\alpha} \tag{8.43}$$

and

$$^{238}\text{Pu} \xrightarrow[86 \text{ yr}]{\alpha} {}^{234}\text{U} \xrightarrow[2.47 \times 10^5 \text{ yr}]{\alpha} \text{etc.} \tag{8.44}$$

Because ^{234}U is formed in the irradiation of recycled ^{233}U, fractional losses of uranium to the radioactive wastes result in considerable long-term production of ^{226}Ra. Also, the relatively large ^{238}Pu formation in thorium fueling is a further contributor to long-term ^{226}Ra. Therefore,

Table 8.7 Elemental constituents in uranium fuel discharged from a PWR†

	g/Mg	Ci/Mg	W/Mg
Actinides			
Uranium	9.54×10^5	4.05	4.18×10^{-2}
Neptunium	7.49×10^2	1.81×10^1	5.20×10^{-2}
Plutonium	9.03×10^3	1.08×10^5	1.52×10^2
Americium	1.40×10^2	1.88×10^2	6.11
Curium	4.70×10^1	1.89×10^4	6.90×10^2
Subtotal	9.64×10^5	1.27×10^5	8.48×10^2
Fission products			
Tritium	7.17×10^{-2}	6.90×10^2	2.45×10^{-2}
Selenium	4.87×10^1	3.96×10^{-1}	1.50×10^{-4}
Bromine	1.38×10^1	0	0
Krypton	3.60×10^2	1.10×10^4	6.85×10^1
Rubidium	3.23×10^2	1.90×10^2	0
Strontium	8.68×10^2	1.74×10^5	4.50×10^2
Yttrium	4.53×10^2	2.38×10^5	1.05×10^3
Zirconium	3.42×10^3	2.77×10^5	1.45×10^3
Niobium	1.16×10^1	5.21×10^5	2.50×10^3
Molybdenum	3.09×10^3	0	0
Technetium	7.52×10^2	1.43×10^1	9.67×10^{-3}
Ruthenium	1.90×10^3	4.99×10^5	3.13×10^2
Rhodium	3.19×10^2	4.99×10^5	3.99×10^3
Palladium	8.49×10^2	0	0
Silver	4.21×10^1	2.75×10^3	4.16×10^1
Cadmium	4.75×10^1	5.95×10^1	2.13×10^{-1}
Indium	1.09	3.57×10^{-1}	1.04×10^{-3}
Tin	3.28×10^1	3.85×10^4	1.56×10^2
Antimony	1.36×10^1	7.96×10^3	2.74×10^1
Tellurium	4.85×10^2	1.34×10^4	1.66×10^1
Iodine	2.12×10^2	2.22	8.98×10^{-3}
Xenon	4.87×10^3	3.12	3.04×10^{-3}
Cesium	2.40×10^3	3.21×10^5	2.42×10^3
Barium	1.20×10^3	1.00×10^5	3.93×10^2
Lanthanum	1.14×10^3	4.92×10^2	8.16
Cerium	2.47×10^3	8.27×10^5	7.87×10^2
Praseodymium	1.09×10^3	7.71×10^5	5.73×10^3
Neodymium	3.51×10^3	9.47×10^1	2.65×10^{-1}
Promethium	1.10×10^2	1.00×10^5	9.17×10^1
Samarium	6.96×10^2	1.25×10^3	2.18
Europium	1.26×10^2	1.35×10^4	7.19×10^1
Gadolinium	6.29×10^1	2.32×10^1	3.34×10^{-2}
Terbium	1.25	3.02×10^2	2.54
Dysprosium	6.28×10^{-1}	0	0
Subtotal	3.09×10^4	4.18×10^6	1.96×10^4
Total	9.95×10^5	4.31×10^6	2.04×10^4

†Quantities are expressed per metric ton of uranium in the fresh fuel charged to the reactor. Average fuel exposure = 33 MWd/kg. Average specific power = 30 MW/Mg. 150 days after discharge.

Table 8.8 Elemental constituents in fuel discharged from LMFBR†

	g/Mg	Ci/Mg	W/Mg
Actinides			
Uranium	8.56×10^{5}	4.25×10^{-1}	9.75×10^{-3}
Neptunium	2.49×10^{2}	2.07×10^{1}	0
Plutonium	1.03×10^{5}	2.57×10^{5}	3.69×10^{2}
Americium	3.53×10^{2}	9.39×10^{2}	2.89×10^{1}
Curium	1.11×10^{1}	1.42×10^{4}	5.21×10^{2}
Subtotal	9.60×10^{5}	2.72×10^{5}	9.19×10^{2}
Fission products			
Tritium	1.05×10^{-1}	1.05×10^{3}	3.73×10^{-2}
Selenium	7.36	5.95×10^{-1}	2.26×10^{-4}
Bromine	2.50	0	0
Krypton	3.49×10^{2}	8.43×10^{3}	5.25×10^{1}
Rubidium	1.99×10^{2}	1.66×10^{1}	0
Strontium	5.91×10^{2}	1.62×10^{5}	4.75×10^{2}
Yttrium	2.85×10^{2}	2.55×10^{5}	1.06×10^{3}
Zirconium	3.09×10^{3}	4.53×10^{5}	2.37×10^{3}
Niobium	2.32×10^{1}	8.58×10^{5}	4.08×10^{3}
Molybdenum	3.96×10^{3}	0	0
Technetium	9.79×10^{2}	1.65×10^{1}	1.11×10^{-2}
Ruthenium	3.37×10^{3}	1.21×10^{6}	6.49×10^{2}
Rhodium	9.41×10^{2}	1.21×10^{6}	9.99×10^{3}
Palladium	1.95×10^{3}	2.68×10^{-1}	2.22×10^{-5}
Silver	4.08×10^{2}	8.01×10^{2}	1.21×10^{1}
Cadmium	1.41×10^{2}	3.23×10^{2}	1.04
Indium	2.29	4.81×10^{-1}	1.46×10^{-3}
Tin	8.31×10^{1}	8.29×10^{3}	3.29×10^{1}
Antimony	3.46×10^{1}	2.38×10^{4}	8.24×10^{1}
Tellurium	6.07×10^{2}	4.26×10^{4}	5.27×10^{1}
Iodine	5.00×10^{2}	3.55	1.44×10^{-2}
Xenon	4.77×10^{3}	5.27	5.12×10^{-3}
Cesium	4.30×10^{3}	1.52×10^{5}	4.75×10^{2}
Barium	1.46×10^{3}	1.18×10^{5}	4.65×10^{2}
Lanthanum	1.28×10^{3}	7.43×10^{2}	1.23×10^{1}
Cerium	2.91×10^{3}	8.76×10^{5}	7.67×10^{2}
Praseodymium	1.23×10^{3}	8.76×10^{5}	6.51×10^{3}
Neodymium	3.88×10^{3}	7.84×10^{1}	2.19×10^{-1}
Promethium	3.92×10^{3}	3.21×10^{5}	2.25×10^{2}
Samarium	9.45×10^{2}	5.66×10^{3}	9.86
Europium	1.54×10^{2}	4.90×10^{4}	5.48×10^{1}
Gadolinium	2.06×10^{2}	6.05×10^{-1}	8.71×10^{-4}
Terbium	4.27×10^{1}	7.13×10^{2}	6.00
Dysprosium	1.68×10^{1}	0	0
Subtotal	3.91×10^{4}	6.71×10^{6}	2.71×10^{4}
Total	9.99×10^{5}	6.98×10^{6}	2.80×10^{4}

†Quantities are expressed per metric ton of uranium and plutonium in the combined fuel charged to the reactor core and blanket. Overall average fuel exposure = 37 MWd/kg. Overall average specific power = 49.3 MW/Mg. 150 days after discharge.

Table 8.9 Elemental constituents in fuel discharged from HTGR†

	g/Mg	Ci/Mg	W/Mg
Actinides			
Thorium	8.49×10^5	3.12×10^2	7.89
Protactinium	4.59×10^1	9.54×10^5	2.42×10^3
Uranium	5.44×10^4	6.49×10^2	2.00×10^1
Neptunium	1.37×10^3	9.67×10^{-1}	0
Plutonium	1.06×10^3	1.99×10^4	4.10×10^2
Americium	2.20×10^1	1.18×10^1	3.78×10^{-1}
Curium	9.54	2.77×10^3	1.02×10^2
Subtotal	9.06×10^5	9.78×10^5	2.96×10^3
Fission products			
Tritium	1.13×10^{-1}	1.09×10^3	3.88×10^{-2}
Selenium	2.76×10^2	1.83	6.95×10^{-4}
Bromine	9.62×10^1	0	0
Krypton	1.98×10^3	6.08×10^4	9.87×10^1
Rubidium	1.86×10^3	2.16×10^1	1.02×10^{-1}
Strontium	3.73×10^3	6.84×10^5	1.80×10^3
Yttrium	1.99×10^3	7.99×10^5	3.64×10^3
Zirconium	1.25×10^4	6.55×10^5	3.43×10^3
Niobium	3.13×10^1	1.24×10^6	5.92×10^3
Molybdenum	9.17×10^3	1.83×10^{-10}	7.47×10^{-13}
Technetium	1.99×10^3	3.40×10^1	2.30×10^{-2}
Ruthenium	3.90×10^3	2.45×10^5	3.04×10^2
Rhodium	4.22×10^2	2.45×10^5	1.68×10^3
Palladium	1.26×10^3	4.85×10^{-2}	4.02×10^{-6}
Silver	1.59×10^1	1.01×10^3	1.60×10^1
Cadmium	6.63×10^1	8.27×10^1	2.71×10^{-1}
Indium	1.48	7.67×10^{-1}	2.20×10^{-3}
Tin	1.12×10^2	8.89×10^3	3.14×10^1
Antimony	4.24×10^1	2.00×10^4	8.37×10^1
Tellurium	1.79×10^3	6.40×10^4	9.23×10^1
Iodine	9.47×10^2	4.07	1.39×10^{-2}
Xenon	1.50×10^4	5.93	1.15×10^{-2}
Cesium	7.15×10^3	9.98×10^5	7.85×10^3
Barium	4.20×10^3	2.85×10^5	1.12×10^3
Lanthanum	3.69×10^3	1.01×10^3	1.79×10^1
Cerium	9.07×10^3	1.93×10^6	1.75×10^3
Praseodymium	3.85×10^3	1.79×10^6	1.38×10^4
Neodymium	1.16×10^4	1.11×10^2	3.59×10^{-1}
Promethium	1.85×10^2	1.76×10^5	1.43×10^2
Samarium	1.77×10^3	7.10×10^2	1.24
Europium	3.35×10^2	2.62×10^4	1.30×10^2
Gadolinium	5.23×10^2	0	0
Terbium	8.30×10^{-1}	1.82×10^2	1.55
Dysprosium	4.85×10^{-1}	3.23×10^{-13}	2.81×10^{-16}
Subtotal	9.95×10^4	9.23×10^6	4.19×10^4
Total	1.00×10^6	1.02×10^7	4.49×10^4

†Quantities are expressed per metric ton of uranium and thorium in the combined fuel charged to the reactor. Average fuel exposure = 95 MWd/kg. Overall average specific power = 64.6 MW/Mg. 150 days after discharge.

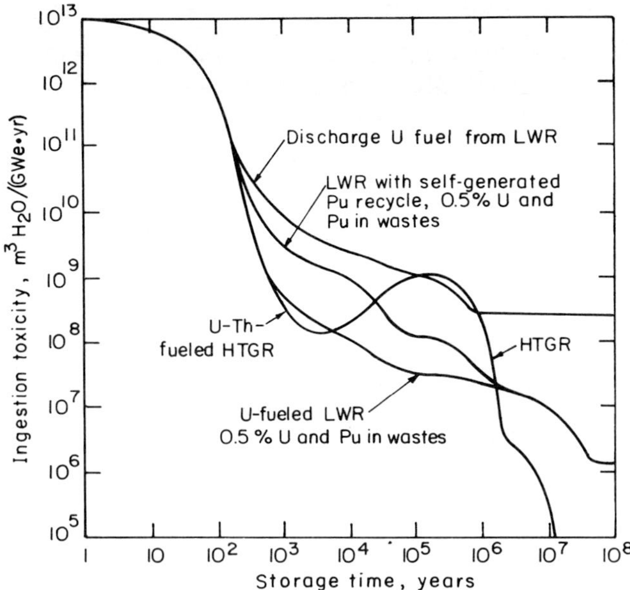

Figure 8.15 Ingestion toxicity of high-level wastes from LWR, with and without reprocessing, and from HTGR.

these actinide reactions in uranium-thorium fuel result in a relatively large growth in the theoretical toxicity of the radioactive wastes after storage periods of a few hundred thousand years.

Following the long-term buildup and decay of ^{226}Ra, which peaks at about 200,000 years, the main contributor to the waste ingestion toxicity is ^{225}Ra, a daughter from the decay of ^{233}U:

$$^{233}\text{U} \xrightarrow[1.62\times 10^5\ \text{yr}]{\alpha} {}^{229}\text{Th} \xrightarrow[7340\,\text{yr}]{\alpha} {}^{225}\text{Ra} \xrightarrow[14.8\,\text{days}]{\alpha} \tag{8.45}$$

Although much of the ^{225}Ra results from the decay of ^{233}U lost directly to the wastes in reprocessing and fabrication, more results from the formation and decay of ^{233}U formed in the wastes by the decay of ^{237}Np:

$$^{237}\text{Np} \xrightarrow[2.14\times 10^6\ \text{yr}]{\alpha} {}^{233}\text{Pa} \xrightarrow[27.0\ \text{days}]{\beta} {}^{233}\text{U} \xrightarrow[1.62\times 10^5\ \text{yr}]{} \text{etc.} \tag{8.46}$$

Consequently, the actinide content and theoretical ingestion toxicity of the radioactive wastes from uranium-thorium fuel are relatively small for waste disposal times of 1000 to 100,000 years but are relatively large for waste disposal times of 100,000 to 1 million years.

4 RADIOACTIVITY FROM NEUTRON ACTIVATION

4.1 Tritium from Neutron Activation

In addition to tritium produced by ternary fission, as shown in Table 8.1, tritium is also produced in reactors by neutron reactions with lithium, boron, and deuterium. Reactors can be designed to produce tritium by irradiating lithium targets with thermal neutrons, resulting in the (n, α) reaction:

$$^{6}_{3}\text{Li} + ^{1}_{0}n \rightarrow ^{4}_{2}\text{He} + ^{3}_{1}\text{H} \tag{8.47}$$

with a 2200 m/s cross section of 940 b. Lithium contaminants in reactor fuel, structure, or coolant will produce tritium by reaction (8.47). Also, the more predominant natural isotope ^{7}Li reacts with high-energy neutrons according to

	Fast-neutron cross section [S3]	
$^{7}_{3}\text{Li} + ^{1}_{0}n \rightarrow ^{5}_{2}\text{He} + ^{3}_{1}\text{H}$	55 mb	(8.48)
$^{7}_{3}\text{Li} + ^{1}_{0}n \rightarrow ^{1}_{0}n + ^{4}_{2}\text{He} + ^{3}_{1}\text{H}$	330 mb	(8.49)

Although relatively little tritium is produced from natural lithium contaminant in thermal reactors by reactions (8.48) and (8.49), the ^{7}Li source of tritium is also produced by the (n, α) reaction with boron used for reactivity control:

$$^{10}_{5}\text{B} + ^{1}_{0}n \rightarrow ^{7}_{3}\text{Li} + ^{4}_{2}\text{He} \tag{8.50}$$

The cross section for reaction (8.50) is 3837 b for 2200 m/s neutrons. Boron also reacts with high-energy neutrons in reactors to produce tritium by the reactions:

	Fast-neutron cross section [S3]	
$^{10}_{5}\text{B} + ^{1}_{0}n \rightarrow 2^{4}_{2}\text{He} + ^{3}_{1}\text{H}$	42 mb	(8.51)
$^{11}_{5}\text{B} + ^{1}_{0}n \rightarrow ^{9}_{4}\text{Be} + ^{3}_{1}\text{H}$	15 mb	(8.52)

The cross section for reaction (8.51) can be interpreted as the spectrum-averaged value for neutrons of energy greater than 1 MeV. The threshold neutron energy for reaction (8.52) is 10.4 MeV. The flux of neutrons with energies above this threshold is negligible in fission reactors, so tritium production from reaction (8.52) is negligible.

Neutron absorption in deuterium in water coolant-moderator produces tritium by the (n, γ) reaction

$$^{2}_{1}\text{H} + ^{1}_{0}n \rightarrow ^{3}_{1}\text{H} \tag{8.53}$$

for which the 2200 m/s cross section is 0.53 mb. This reaction is most important as a tritium source in reactors cooled and/or moderated by heavy water, but it is negligible in LWRs.

The activity $(N\lambda)_T$ of tritium produced in a reactor can be estimated by assuming irradiation in a constant neutron flux for a period T_R and applying Eq. (2.27). For these tritium-producing reactions it is sometimes a good approximation to assume that the parent material is present in nearly constant amount during the irradiation period. The high (n, α) cross section for ^{10}B might suggest that this nuclide would decrease considerably in amount if exposed to the full reactor flux over a period of even 1 year, which is the typical time interval for reactivity adjustment between refueling intervals. However, in boiling-water reactors (BWRs), which use solid control absorbers for long-term reactivity control, the effect of the large thermal cross section of boron is to self-shield all but the surface of these absorbers from thermal neutrons, so that very little of the boron is actually consumed during a refueling interval or even during the period T_R of fuel irradiation. The boron cross section for fast neutrons is relatively small, so fast neutrons are not self-shielded and essentially homogenous exposure of all the boron to the average fast-neutron flux in the reactor can be assumed. In PWRs boron is dissolved in the coolant for long-term control of reactivity, with the boron concentration controlled by chemical means during the irradiation period between refueling intervals. Because this concentration change occurs over a time period short compared to the half-life of tritium, and because the boron concentrations are repeated from one refueling cycle to another, a constant average concentration of boron in the coolant can be assumed for the purpose of estimating tritium production. Therefore, for those tritium sources in which the parent nuclide can be assumed to be of constant amount, Eq. (2.27) takes the form

$$(N\lambda)_T = \sum_i N_i \sigma_i \phi (1 - e^{-\lambda_T T_R}) \tag{8.54}$$

where N_i = number of atoms of species i producing tritium by neutron reactions
σ_i = cross section for species i to produce tritium
λ_T = radioactive decay constant for tritium
T_R = time of constant-flux irradiation

For an irradiation period T_R much smaller than the tritium half-life of 12.3 years, $\lambda_T T_R \ll 1$, and Eq. (8.54) simplifies to

$$(N\lambda)_T \cong \lambda_T T_R \sum_i N_i \sigma_i \phi \tag{8.55}$$

To illustrate, we shall consider a 1000-MWe PWR with the same core composition and power density as the reactor described in Chap. 3. The in-core inventory of water is approximately 13,400 kg. The tritium produced by ^{2}H(n, γ) during one calendar year in an average thermal-neutron flux of 3.5×10^{13} $n/(\text{cm}^2 \cdot \text{s})$ with an effective ^{2}H(n, γ) cross section of 3.35×10^{-4} b is

$$(N\lambda)_T = \left(\frac{\ln 2}{12.3 \text{ yr}}\right)(0.8 \text{ yr})(1.34 \times 10^7 \text{ g})\left(\frac{2 \times 6.02 \times 10^{23} \text{ atoms}}{18.02 \text{ g}}\right)(1.5 \times 10^{-4} \text{ atoms } ^2\text{H/atom H})$$
$$\times (3.35 \times 10^{-28} \text{ cm}^2)[3.5 \times 10^{13} (\text{cm}^2 \cdot \text{s})^{-1}]\left(\frac{\text{Ci}}{3.7 \times 10^{10} \text{ disintegrations/s}}\right) = 1.9 \text{ Ci}$$

The actual irradiation time T_R is 0.8 years because of the assumed 0.8 capacity factor of the power plant.

Assuming an average dissolved boron concentration of 600 ppm in the coolant, the tritium produced from reaction (8.51) in an average fast-neutron flux of 7.2×10^{13} $n/(\text{cm}^2 \cdot \text{s})$ is similarly obtained by applying Eq. (8.55), resulting in an estimated yearly production of 360 Ci.

In a water-cooled reactor the coolant is processed continuously for control and removal of chemical and radioactive contaminants. In a PWR the lithium formed by (n, α) reactions in dissolved boron will add to whatever natural lithium is present as a contaminant and for corrosion control, but the continued processing will hold it at some steady concentration. For the purpose of this estimate we shall assume a concentration of 1.0 ppm of lithium in the coolant and will neglect the additional ^{7}Li produced by reaction (8.50). However, after the coolant lithium has been exposed to thermal neutrons for a few years it will become depleted in the ^{6}Li, because of the high absorption cross section of ^{6}Li. A typical isotopic composition of lithium in the coolant of a PWR is 99.9 percent ^{7}Li [S2]. Applying Eq. (8.55) for tritium produced by ^{6}Li (n, α) yields the yearly production of 34 Ci listed in Table 8.10. The yearly production of the tritium from ^{7}Li reactions is estimated at 4 Ci [S2].

The total yearly production of neutron-activation tritium in the PWR coolant is 400 Ci, as shown in Table 8.10. Another source of tritium in the coolant is fission-product tritium that diffuses through the fuel cladding and escapes through pin-hold penetrations through the cladding. Estimates of the amount of fission-product tritium reaching the coolant in LWRs with zircaloy fuel range from 0.2 to 1 percent of the total fission-production tritium produced within the fuel.

Table 8.10 Estimated tritium production in the coolant of a 1000-MWe PWR

Source	Tritium production, Ci/yr
$^{2}H(n, \gamma)$	2
$^{10}B(n, {}^{8}Be)$	360
$^{6}Li(n, \alpha)$	34
$^{7}Li(n, n\alpha)$	4
Total from activation reactions	400
Fission-product tritium†	149
Total	549

†Assumes fission-product tritium diffusing through fuel cladding or escaping through pin-hole cladding failures is equivalent to release of fission-product tritium from 0.5% of the fuel. Calculated as average over irradiation cycle.

In the HTGR the principal nonfission sources of tritium are from lithium and boron contaminants in the graphite fuel elements. Typical contaminant concentrations assumed in the HTGR designs [H1] are

$$\frac{\text{Li}}{\text{C}} = 1.2 \times 10^{-6}$$

$$\frac{\text{B}}{\text{C}} = 1.36 \times 10^{-4}$$

At such low concentrations the lithium and boron are exposed homogenously to the neutron flux. Because of the large thermal-neutron cross sections for ^{6}Li and ^{10}B, these isotopes are depleted significantly during the typical fuel irradiation time of 4 years. Therefore, to calculate the tritium activity $(N\lambda)_T$ in a fuel element after an irradiation time T_R, we rewrite Eq. (2.100), recognizing that the chain-linking term here is $\phi\sigma$ instead of λ. For the ^{6}Li reaction of Eq. (8.47),

$$(N\lambda)_T = \frac{\lambda_T N_6^0 \phi\sigma_6}{\lambda_T - \phi\sigma_6}\left(e^{-\phi\sigma_6 T_R} - e^{-\lambda_T T_R}\right) \tag{8.56}$$

where N_6^0 = initial number of atoms of ^{6}Li

σ_6 = (n, α) cross section for ^{6}Li

For an effective ^{6}Li cross section of 294 b and an average thermal-neutron flux of $1.2 \times 10^{14}\, n/(\text{cm}^2 \cdot \text{s})$, the tritium in discharge fuel due to $^{6}Li(n, \alpha)$ is calculated to be

$$(N\lambda)_T = 308 \text{ Ci/Mg of graphite}$$

The tritium from fast-neutron reactions with ^{10}B is estimated to be about 0.6 Ci/Mg of graphite, and tritium from ^{7}Li and other sources is even less.

The fuel discharged yearly from the 1000-MWe HTGR of Fig. 3.33 contains 90.5 Mg of graphite [P3]. The yearly production of tritium from neutron activation of lithium impurities is then estimated to be

$$(308)(90.5) = 27{,}900 \text{ Ci/year}$$

This compares with 9.59×10^3 Ci/year of fission-product tritium calculated to be present in the discharge fuel from a 1000-MWe HTGR [G1].

Tritium is also produced in the HTGR helium coolant by neutron reactions with small amounts (1.7×10^{-5} percent) of ^{3}He present in underground sources of natural helium:

$$^3_2\text{He} + ^1_0n \rightarrow ^1_1\text{H} + ^3_1\text{H} \tag{8.57}$$

with a 2200 m/s cross section of 5327 b. For an inventory of natural helium of 618 kg in the core of a 1000-MWe HTGR [B1], ^{3}H is initially formed at the rate of about 8,020 Ci/year and is trapped by forming tritides with hot titanium in the coolant cleanup system. However, because of its large cross section, ^{3}He is rapidly depleted by neutron absorption. It is replaced by fresh helium introduced to make up for coolant leakage. If a fraction f_{He} of the coolant leaks from the coolant system per unit time, the steady-state concentration $X_{^3\text{He}}$ of ^{3}He within the reactor coolant can be calculated by

$$N_{\text{He}}^T f_{\text{He}} X_{^3\text{He}}^0 = N_{\text{He}}^R X_{^3\text{He}} \phi \sigma_{^3\text{He}} + N_{\text{He}}^T X_{^3\text{He}} f_{\text{He}} \tag{8.58}$$

where N_{He}^T = total inventory of helium in the coolant system
N_{He}^R = total inventory of helium within the reactor core
$X_{^3\text{He}}^0$ = atom fraction of ^{3}He in natural helium (1.7×10^{-7})

Solving for $X_{^3\text{He}}$, we obtain

$$X_{^3\text{He}} = X_{^3\text{He}}^0 \frac{1}{1 + N_{\text{He}}^R \phi \sigma_{^3\text{He}} / N_{\text{He}}^T f_{\text{He}}} \tag{8.59}$$

From HTGR design data, it is estimated [B1] that

$$\frac{N_{\text{He}}^R}{N_{\text{He}}^T} = 0.09$$

$$f_{\text{He}} = 0.015/\text{yr}$$

For an effective $\bar{\sigma}_{^3\text{He}} = 2800$ b, and for $\phi = 1.2 \times 10^{14} n/(\text{cm}^2 \cdot \text{s})$, we obtain

$$X_{^3\text{He}} = 2.63 \times 10^{-9}$$

The resulting steady-state rate of production of tritium in the coolant from ^{3}He(n, p) is 124 Ci/year.

In the CANDU heavy-water reactor the dominant source of tritium is the deuterium activation reaction of Eq. (8.53). The data given in Prob. 3.3 for the Douglas Point Nuclear Power Station provide a basis for estimating the rate of production of tritium in the heavy-water moderator and coolant:

Electrical power = 203 MWe
Inventory of D_2O coolant in reactor core = 2.82×10^6 g
Average thermal-neutron flux in coolant = $6.10 \times 10^{13} n/(\text{cm}^2 \cdot \text{s})$
Inventory of D_2O moderator in reactor core = 7.72×10^7 g
Average thermal-neutron flux in moderator = $1.01 \times 10^{14} n/(\text{cm}^2 \cdot \text{s})$
Average ^{2}H(n, γ) cross section = 4.45×10^{-4} b

The rate of production of ^{3}H in the moderator is then

$$\{[(2.82 \times 10^6)(6.10 \times 10^{13}) + (7.72 \times 10^7)(1.01 \times 10^{14})] \text{ g/(cm}^2 \cdot \text{s)}\}(4.45 \times 10^{-28} \text{ cm}^2)$$
$$\times \left(\frac{6.02 \times 10^{23} \times 2 \text{ atoms } ^2\text{H}}{20.02 \text{ g } D_2O}\right)\left(\frac{\ln 2}{12.3 \text{ yr}}\right)\left(\frac{\text{Ci}}{3.7 \times 10^{10}/\text{s}}\right)(0.8) = 2.60 \times 10^5 \text{ Ci/yr}$$

For a 1000-MWe CANDU power plant with the same reactor lattice and with the same ratio of D_2O in-core inventory to uranium inventory as in the Douglas Point Reactor, the yearly production of tritium in the heavy water is then

$$\left(\frac{1000}{203}\right)(2.60 \times 10^5) = 1.28 \times 10^6 \text{ Ci/yr}$$

Because of this large rate of tritium generation, it is necessary to operate a small isotope-separation unit to prevent the buildup of large concentrations of tritium in the heavy water. The losses of heavy water are kept small enough so that only a very small fraction of the tritium is released to the environment. The yearly release of tritium reported for the Douglas Point Station is typically about 4000 Ci/year, which is about 0.2 percent of the allowable release [D1].

4.2 ^{14}C

^{14}C is an activation product of potential environmental importance in the nuclear fuel cycle because of its long half-life of 5730 years and because it easily appears in volatile form, such as CO_2. Most of the ^{14}C formed in reactors results from the (n, p) reaction with ^{14}N:

$$^{14}_{7}N + ^{1}_{0}n \rightarrow ^{14}_{6}C + ^{1}_{1}H \tag{8.60}$$

The ^{14}N, which constitutes 99.6 percent of natural nitrogen, is present as residual nitrogen impurity in oxide fuel of water reactors and fast-breeder reactors, as air dissolved in the coolant of water-cooled reactors, and as residual nitrogen in the graphite of HTGRs. The ^{14}N activation cross section for 2200 m/s neutrons is 1.85 b.

^{14}C also results from the (n, α) reaction on ^{17}O, which is present as 0.03 percent of natural oxygen, with a 2200 m/s cross section of 0.235 b:

$$^{17}_{8}O + ^{1}_{0}n \rightarrow ^{14}_{6}C + ^{4}_{2}He \tag{8.61}$$

In graphite-moderated reactors another source of ^{14}C is the (n, γ) reaction with ^{13}C, which is present as 1.108 percent of the natural carbon in graphite:

$$^{13}_{6}C + ^{1}_{0}n \rightarrow ^{14}_{6}C + ^{0}_{0}\gamma \tag{8.62}$$

However, the 2200 m/s cross section is only about 0.9 mb. Additional but less important reactions are

$$^{15}_{7}N + ^{1}_{0}n \rightarrow ^{14}_{6}C + ^{2}_{1}H \tag{8.63}$$

with a 2200 m/s cross section of 2.4×10^{-7} b, and

$$^{16}_{8}O + ^{1}_{0}n \rightarrow ^{14}_{6}C + ^{3}_{2}He \tag{8.64}$$

The activity $(N\lambda)_C$ of ^{14}C produced in a reactor can be estimated by assuming irradiation in a constant-neutron flux for a period T_R and applying Eq. (2.27). Because of the long half-life of ^{14}C, the approximation $\lambda_C T_R \ll 1$ leads, as in the case of Eq. (8.55), to

$$(N\lambda)_C = \lambda_C T_R \sum_i N_i \sigma_i \phi \tag{8.65}$$

where N_i = number of atoms of species i producing ^{14}C by neutron reactions
σ_i = cross section for species i to produce ^{14}C
λ_C = radioactive decay constant for ^{14}C

^{14}C produced in water coolant is important because of its possible environmental release at the reactor site. If ^{14}C forms carbon dioxide or a hydrocarbon such as CH_4, and if no processes are provided to recover the gaseous ^{14}C, the coolant-produced ^{14}C will be discharged along with the noncondensable gases removed by the main condenser air ejector in a BWR and through the gaseous waste disposal system for a PWR.

We consider here the production of ^{14}C by reactions (8.60) and (8.61) in the reactor coolant, which requires estimates of the inventories of ^{17}O and dissolved nitrogen in the coolant within the reactor core. For the 1000-MWe PWR with an in-core water inventory of 13,400 kg, an effective $^{17}O(n, \alpha)$ thermal cross section of 0.149 b, and an average thermal-neutron flux of 3.5×10^{13} $n/(cm^2 \cdot s)$, the ^{14}C production from ^{17}O is estimated to be 2.2 Ci/year.

To obtain the ^{14}C from dissolved nitrogen in the coolant, a dissolved nitrogen concentration of 1 ppm (by weight) is assumed, with an effective $^{14}N(n, p)$ cross section of 1.17 b, resulting in a yearly production of 0.061 Ci of ^{14}C. The total yearly production of ^{14}C in the PWR coolant is then about 2.3 Ci/year, which is the source term for possible environmental release at the reactor site. A 1000-MWe BWR would contain about 33,000 kg of water in the core under operating conditions. Assuming the same values of neutron flux and cross sections, the yearly production of 5.6 Ci of ^{14}C in the BWR coolant is estimated.

The ^{14}C produced by $^{17}O(n, \alpha)$ in UO_2 fuel, calculated as the yearly production per metric ton (Mg) of uranium originally in the makeup fuel, is again obtained by applying Eq. (8.65):

$$\left(\frac{10^6 \times 6.02 \times 10^{23}}{238} \frac{\text{atoms U}}{\text{Mg U}}\right)\left[\frac{(2)(3.74 \times 10^{-4} \text{ atoms } ^{17}\text{O})}{\text{atom U}}\right](6.47 \times 10^{-25} \text{ cm}^2)$$

$$\times [3.5 \times 10^{13} (\text{cm}^2 \cdot \text{s})^{-1}] \left(\frac{\text{Ci}}{3.7 \times 10^{10} \text{ disintegrations/s}}\right)\left(\frac{\ln 2}{5730 \text{ yr}}\right)(0.8)$$

$$= 2.54 \times 10^{-2} \text{ Ci/(yr} \cdot \text{Mg U)}$$

For the ^{14}N source in the fuel, it is assumed that the nitrogen impurity is present in UO_2 at a weight ratio of 25 ppm, although nitrogen contents from 1 to 100 ppm have been reported [K1]. The yearly production per metric ton of uranium is

$$\left(\frac{10^6 \text{ g U}}{\text{Mg U}}\right)\left(\frac{270 \text{ g UO}_2}{238 \text{ g U}}\right)\left(\frac{25 \times 10^{-6} \text{ g N}}{\text{g UO}_2}\right)\left(\frac{6.02 \times 10^{23} \text{ atoms}}{14 \text{ g}}\right)\left(\frac{0.996 \text{ atoms } ^{14}\text{N}}{\text{atom N}}\right)$$

$$\times (1.17 \times 10^{-24} \text{ cm}^2)[3.5 \times 10^{13} (\text{cm}^2 \cdot \text{s})^{-1}] \left(\frac{\text{Ci}}{3.7 \times 10^{10} \text{ disintegrations/s}}\right)\left(\frac{\ln 2}{5730 \text{ yr}}\right)(0.8)$$

$$= 0.130 \text{ Ci/(yr} \cdot \text{Mg U)}$$

The total amount of ^{14}C produced yearly in the fuel is then 0.155 Ci/Mg of uranium.

To obtain the ^{14}C in the discharge fuel, we use the fuel life of 3 calendar years, as calculated in Chap. 3 for the reference PWR. Because there is negligible decay of the ^{14}C during this 3-year period, the concentration in the discharge fuel is

$$3 \times 0.155 = 0.465 \text{ Ci/Mg}$$

The quantity of ^{14}C in the total fuel discharged yearly, which initially contained 27.2 Mg of uranium, is

$$0.465 \times 27.2 = 12.7 \text{ Ci/yr}$$

In a PWR operating with plutonium recycle the thermal-neutron flux is lower than for uranium fueling because of the higher fission cross section for plutonium. As a result, less ^{14}C is produced by thermal-neutron activation within the fuel, as shown in Table 8.11.

Fast-breeder oxide fuel is also assumed to contain 25 ppm of residual nitrogen [K1]. Typical average fast-spectrum cross sections are 0.135 mb for $^{17}O(n, \gamma)$ and 14 mb for $^{14}N(n, p)$

Table 8.11 Volatile radionuclides in discharge fuel from neutron activation†

	Activated radionuclides, Ci/yr			
	PWR‡	PWR	HTGR‡	LMFBR‡
Radio-nuclide	Uranium ($3.3\%\ ^{235}U$)	Uranium and recycled uranium + plutonium	^{235}U, thorium, and recycled uranium	Uranium and recycled plutonium
^3H (tritium)	–	–	2.79×10^4	–
^{14}C	1.27×10^1	6.67	1.20×10^2	3.3
^{35}S	–	–	2.15×10^2	–
^{33}P	–	–	1.1	–
^{36}Cl	–	–	1.02	–

†1000-MWe reactors, 80% capacity factor.

‡PWR, pressurized-water reactor; HTGR, high-temperature gas-cooled reactor; LMFBR, liquid-metal-cooled fast-breeder reactor. Data are calculated for 150 days after discharge for PWR and HTGR, 60 days after discharge for LMFBR.

within the reactor core [C1]. For an average fast-spectrum core flux [C1] of $3.8 \times 10^{15}\ n/(cm^2 \cdot s)$, and for the breeder parameters of Fig. 3.34, the estimated yearly production of ^{14}C for a 1000-MWe fast breeder is estimated to be 3.3 Ci/year. Relatively little ^{14}C is produced in the blanket fuel because of the lower neutron flux there.

The fuel of the HTGR consists of uranium and thorium particles, as oxides and carbides, distributed through a graphite matrix. The important ^{14}C-producing reactions in this fuel are $^{14}N(n, p)$ and $^{13}C(n, \gamma)$. Residual nitrogen is assumed to be present in graphite at a weight ratio of 30 ppm [B4]. In the thermal-neutron energy spectrum of an HTGR the effective activation cross sections [B4] are 0.683 b for ^{14}N and 3.3×10^{-4} b for ^{13}C. For an average thermal-neutron flux of $1.2 \times 10^{14}\ n/(cm^2 \cdot s)$ and a 4-year fuel life, the estimated concentration of ^{14}C in the discharged graphite fuel is calculated from Eq. (8.65), with the result:

Source	Ci ^{14}C/kg of graphite in discharge fuel
$^{14}N(n, p)$, 30 ppm N	1.10×10^{-3}
$^{13}C(n, \gamma)$	2.29×10^{-4}
Total	1.33×10^{-3}

The fuel discharged yearly from the 1000-MWe HTGR of Fig. 3.33 contains 7.95 Mg of heavy metal and 90.5 Mg of graphite. The yearly production of ^{14}C by this reactor is then estimated to be

$$(1.33 \times 10^{-3})(90{,}500) = 120 \text{ Ci/yr}$$

In another HTGR calculation 1 ppm of N_2 in the graphite is assumed [H1], resulting in an estimated yearly production of 24 Ci/year for a 1000-MWe plant.

When HTGR fuel is reprocessed the graphite matrix is to be incinerated in oxygen, exposing the fuel particles for dissolution. The combustion gas, which contains the ^{14}C and all of the normal carbon from the graphite, is to be recovered to avoid release of ^{14}C to the environment.

4.3 ^{35}S, ^{33}P, and ^{36}Cl in HTGR Fuel

The graphite fuel blocks of the HTGR contain sulfur contaminant, which originates from the pitch used to form the fuel-rod matrix material. Neutron activation of the 4.22 percent ^{34}S in natural sulfur results in 88-day ^{35}S, according to the reaction

$$^{34}_{16}S + ^{1}_{0}n \rightarrow ^{35}_{16}S + ^{0}_{0}\gamma \tag{8.66}$$

for which the 2200 m/s cross section is 0.24 b. Assuming that sulfur is present at 193 ppm in the HTGR fuel [H1], it is estimated that 215 Ci of ^{35}S are present in the fuel discharged yearly from a 1000-MWe HTGR, after 150 days of storage. In HTGR fuel reprocessing the stable and radioactive sulfur will volatilize to follow the carbon dioxide from graphite incineration. The radioactive sulfur is a potential environmental contaminant that must be recovered. The amount of ^{35}S activity is greater than that of ^{14}C, and the radioactivity concentration limit for inhalation is more than an order of magnitude lower for ^{35}S. The stable sulfur may interfere chemically with some of the recovery processes in the off-gas system.

Natural sulfur also contains 0.76 percent ^{33}S, which undergoes (n, p) reactions to form 25-day ^{33}P according to

$$^{33}_{16}S + ^{1}_{0}n \rightarrow ^{33}_{15}P + ^{1}_{1}H \tag{8.67}$$

with a 2200 m/s cross section of 0.14 b. The estimated activity of ^{33}P in the fuel discharged annually from a 1000-MWe HTGR, after 150 days of storage, is 1.1 Ci.

Another volatile radionuclide formed in HTGR fuel is 3.1×10^5 year ^{36}Cl, formed by neutron activation of chlorine contaminant in the fuel, according to the reaction

$$^{35}_{17}Cl + ^{1}_{0}n \rightarrow ^{36}_{17}Cl + ^{0}_{0}\gamma \tag{8.68}$$

Natural chlorine contains 75.77 percent ^{35}Cl, for which the 2200 m/s activation cross section is 43 b. Assuming 3 ppm chlorine in the fabricated HTGR fuel [H1], the estimated yearly production of ^{36}Cl from a 1000-MWe reactor is 1.02 Ci.

These additional radionuclides volatilized in HTGR fuel reprocessing are summarized in Table 8.11.

4.4 Nonvolatile Radionuclides Activated in Fuel-Element Structure

Fuel elements discharged from PWRs also contain radionuclides formed by neutron activation in the zircaloy cladding, stainless steel end fittings, and Inconel spacers. A typical 3-year irradiation of the metallic structure produces the radionuclides listed in Table 8.12, calculated for fuel elements discharged from a LWR and stored for 150 days [B3]. Neutron capture in stable ^{94}Zr forms 65-day ^{95}Zr and its decay daughter, 35-day ^{95}Nb. The radioactivity produced is large, but it is still smaller than the radioactivity of these two nuclides formed as fission products (cf. Table 8.1). Other large contributors to the cladding radioactivity are ^{60}Co, resulting from neutron capture in stable ^{59}Co, and ^{51}Cr, ^{55}Fe, ^{58}Co, and ^{68}Ni.

After 10 years of decay there is still appreciable radioactivity remaining, so irradiated cladding must be treated as a long-lived radioactive waste. The only species that persist after about 1000 years of decay are 1.5×10^6 year ^{93}Zr and 2.12×10^5 year ^{99}Tc. The activity of ^{93}Zr in irradiated cladding is about the same as the activity of fission-product ^{93}Zr (cf. Table 8.1), but the activity of ^{99}Tc in cladding is about 1000 times less than the activity of fission-product ^{99}Tc.

The fast-breeder fuel cladding and structure, typically of 316 stainless steel, result in the radionuclides listed in Table 8.12 [B3]. Because the structure is entirely an austenitic alloy, the most radioactive nuclides are ^{54}Mn, ^{55}Fe, and ^{60}Co.

Fuel cladding hulls will also contain uranium, plutonium, and other transuranic radio-

Table 8.12 Nonvolatile radionuclides in discharge fuel from neutron activation†

		Activity in discharge fuel, Ci/yr		
		PWR‡	HTGR‡	LMFBR‡
Radio-nuclide	Half-life	Uranium (3.3% ^{235}U)	^{235}U, thorium, and recycled uranium	Uranium and recycled plutonium
^{10}Be	2.5×10^6 yr		1.20×10^{-1}	
^{22}Na	2.60 yr			5.16
^{32}P	14.3 days			23.7
^{33}P	25 days			3.16
^{45}Ca	165 days	4.61×10^{-2}		
^{46}Sc	83.9 days	3.37×10^1		
^{49}V	330 days			7.04×10^{-1}
^{51}Cr	27.8 days	1.91×10^4		2.03×10^4
^{54}Mn	303 days	4.79×10^3		1.74×10^6
^{55}Fe	2.6 yr	4.89×10^4		1.30×10^6
^{59}Fe	45 days	6.17×10^2		1.47×10^4
^{58}Co	71.3 days	5.92×10^4	1.05	2.24×10^6
^{60}Co	5.26 yr	1.66×10^5	4.71×10^{-1}	3.22×10^4
^{59}Ni	8×10^4 yr	1.05×10^2	1.72	7.46×10^1
^{63}Ni	92 yr	1.56×10^4	2.28×10^2	2.37×10^3
^{89}Sr	52 days	1.41×10^2		
^{91}Y	58.8 days	4.69×10^2		
^{93}Zr	1.5×10^6 yr	2.81		
^{95}Zr	65 days	1.59×10^5		
^{92m}Nb	10.16 days			2.09×10^{-1}
^{93m}Nb	13.6 yr	2.90×10^{-1}		4.86
^{95}Nb	35 days	2.96×10^5		4.88×10^1
^{93}Mo	>100 yr	5.45×10^{-1}		7.46×10^1
^{99}Tc	2.12×10^5 yr	3.81×10^{-1}		7.25
^{117m}Sn	14.0 days	1.96×10^2		
^{119m}Sn	250 days	4.31×10^2		
^{121m}Sn	76 yr	9.16		
^{123}Sn	125 days	5.30		
^{124}Sb	60 days	2.28×10^1		
^{125}Sb	2.7 yr	1.10×10^3		
^{125m}Te	58 days	4.97×10^2		
Total		7.72×10^5	2.31×10^2	5.33×10^6

†1000-MWe reactors, 80% capacity factor.

‡PWR, pressurized-water reactor; HTGR, high-temperature gas-cooled reactor; LMFBR, liquid-metal-cooled fast-breeder reactor. Data are calculated for 150 days after discharge for PWR and HTGR, 60 days after discharge for LMFBR.

nuclides as contaminants on the inner surfaces of the cladding. These transuranics can be removed by chemical treatment of the cladding-hull surfaces, or the cladding hulls can be classified as transuranic wastes.

The HTGR fuel contains no metallic structure, but impurities in the graphite fuel blocks result in the production of relatively small amounts of radioactive cobalt and nickel, as listed in Table 8.12 [H1, P3]. The total activity from metallic contaminants in HTGR fuel is considerably lower than that in the fuels from light-water and breeder reactors.

5 NEUTRON ACTIVITY IN RECYCLED FUEL

5.1 Light-Element (α, n) Reactions

Additional biological hazard in the handling of plutonium recovered from irradiated uranium or of uranium from irradiated thorium arises from fast neutrons produced by (α, n) reaction. Alpha particles from actinide decay react with light elements–lithium, beryllium, carbon, oxygen, etc.–to produce energetic neutrons such as

$$^{9}_{4}\text{Be} + ^{4}_{2}\text{He} \rightarrow ^{12}_{6}\text{C} + ^{1}_{0}n \tag{8.69}$$

The fast neutrons are very penetrating and may require some hydrogenous shielding for protection of operating personnel. Also, techniques to ensure low concentration of light-element contaminants in the recycled actinide material may be required.

The allowable concentration of light elements in recycled fuel depends on the alpha-decay rate in the material, the energy of the alpha particle, the probability of an (α, n) reaction, the energy and relative biological effectiveness of the neutron produced, and the allowable surface dose rate of these (α, n) neutrons. The average energies of neutrons from (α, n) reactions in light elements are listed in Table 8.13 along with the tolerance flux for these neutrons. Also listed in Table 8.13 is the neutron emission rate per gram of uranium or plutonium metal that would result in a dose of 1 rem per 40-h exposure at the surface of a kilogram of this metal. This dose rate is about 30 percent less than the official tolerance for radiation exposure localized to the hands and forearms of radiation workers.

The rate of neutron generation from (α, n) reactions in a fuel containing alpha-emitting actinides and various light elements is predicted from

$$\dot{n} = \sum_i \left(\chi_i \sum_j a_{ij} A_j \right) \tag{8.70}$$

where $\dot{n}$ = the neutron production rate

χ_i = the concentration of the light element i

Table 8.13 Energies and tolerances for neutrons from (α, n) reactions

Element	Average energy of emitted neutron,† MeV	Neutron flux required to give 100 mrem in 40 h,† $n/(\text{cm}^2 \cdot \text{s})$	Neutron emission rate due to contaminant, to give 1000 mrem/40-h wk exposure at surface of 1-kg sphere of uranium or plutonium metal,‡ n/min per gram of metal
Lithium	2.34	20	45
Beryllium	>5	<18	<40
Boron	5.47	18	40
Carbon	~0.1	~80	~180
Nitrogen	1.7	18	40
Oxygen	~0.1	~80	~180
Fluorine	>5	<18	<40
Sodium	3.7	19	43
Calcium	0.8	21	47

†From *Federal Register* [F1].

‡Based on data supplied by Arnold [A2].

Table 8.14 Reaction constants for (α, n) reactions†, ‡

Contaminant element	Neutrons per 10^{10} alpha disintegrations/ppm of contaminant element				
	^{228}Th§	^{232}U	^{233}U	^{238}Pu	^{239}Pu
Lithium	1.9	2.37×10^{-1}	1.74×10^{-1}	2.82×10^{-1}	2.14×10^{-1}
Beryllium	7.03×10^{1}	3.89	2.22	4.95	3.16
Boron	7.35	1.13	5.07×10^{-1}	1.65	9.21×10^{-1}
Carbon	3.95×10^{-1}	6.9×10^{-3}	3.1×10^{-3}	9.95×10^{-3}	5.35×10^{-3}
Nitrogen	$< 1.1 \times 10^{-2}$	$< 4 \times 10^{-4}$	$< 5 \times 10^{-7}$	1.42×10^{-3}	4.83×10^{-5}
Oxygen	5.4×10^{-2}	2.76×10^{-3}	1.33×10^{-3}	4.01×10^{-3}	2.12×10^{-3}
Fluorine	9.20	3.52×10^{-1}	1.50×10^{-1}	5.20×10^{-1}	2.55×10^{-1}
Sodium	2.17	2.58×10^{-2}	1.27×10^{-1}	3.86×10^{-2}	1.94×10^{-2}
Calcium	4.73×10^{-1}	0	0	4.73×10^{-3}	6.9×10^{-4}

†Based on data by Arnold [A1, A2].
‡Reaction constant, a.
§Based on 10^{10} alphas directly from ^{228}Th, but includes effect of ^{228}Th daughters.

a_{ij} = the number of (α, n) neutrons per alpha disintegration per unit concentration of the light element, with j identifying the energy of the alpha particle

A_j = the alpha-disintegration rate of actinide j

Values of a for various light elements, calculated from the data of Arnold [A2], are listed in Table 8.14. The values of a for ^{239}Pu and ^{240}Pu are assumed equal for a given light element because of the nearly equal energies of the alphas from these two isotopes. Similarly, the a for ^{242}Pu should be very nearly the same as that for ^{233}U.

For a given mass and isotopic composition of plutonium and contaminant concentration, neutron production rate and allowable concentration of a given contaminant can be estimated from data in Tables 8.13 and 8.14. Estimates for ^{239}Pu containing 1000 ppm ^{238}Pu and for ^{233}U containing 100 ppm ^{232}U are given in Table 8.15. By comparison, plutonium undergoing

Table 8.15 (α, n) surface dose from light elements in recycled plutonium and uranium

Element	Light-element concentration (ppm)† to give 1 rem/40-h wk at surface of 1-kg sphere	
	^{239}Pu + 1000 ppm ^{238}Pu	^{233}U + 100 ppm ^{232}U‡
Lithium	10	80
Beryllium	0.6	4
Boron	2	21
Carbon	1,600	980
Nitrogen	6,500	60,000
Oxygen	4,000	28,000
Fluorine	7	46
Sodium	100	320
Calcium	1,700	2,300

†Each light-element concentration is calculated on the basis of no other light elements present.
‡Includes ^{232}U daughters present after 100 days of postprocessing storage.

Table 8.16 Typical contaminants in plutonium product following oxalate precipitation [A1]

Element	Concentration in plutonium, ppm
Lithium	0.2–1
Beryllium	0.2–0.4
Boron	35–300†
Sodium	20–500†
Magnesium	40–600†
Calcium	1,000–10,000†
Aluminum	5–70
Potassium	—
Silicon	1–5
Nickel	3–40
Chromium	5–10
Iron	35–600

†Contaminants that are borderline or much above the allowable concentrations.

final purification by the typical technique of oxalate precipitation may contain the contaminant concentrations listed in Table 8.16. Boron, sodium, and calcium are easily found above the allowable concentration. Either neutron shielding or special precautions to maintain low contaminant concentrations are necessary.

A special problem arises when chemical compounds of plutonium with light elements are handled as massive solids. For example, PuO_2 fuel will produce above-tolerance fluxes of (α, n) neutrons at surface contact. Even greater neutron production occurs with PuF_4, which is an intermediate in the conversion of plutonium compounds to plutonium metal. Approximately 12 neutrons are produced per 10^6 alphas in $^{239}PuF_4$, and some thickness of neutron-shield material may be required.

5.2 Neutrons from Spontaneous Fission

Another radiation problem arises from fast neutrons produced in spontaneous fission of the even-mass plutonium isotopes. Half-lives and specific activities for spontaneous fission of the plutonium isotopes are listed in Table 8.17.

In determining the biological hazard from spontaneous fission, it is assumed that each fission produces an average of three neutrons, each with an energy of 2 MeV. Estimated surface fluxes and dose rates are given in Table 8.18.

Table 8.17 Half-lives for spontaneous fission

	Spontaneous fission half-life, yr
^{238}Pu	4.9×10^{10}
^{240}Pu	1.4×10^{11}
^{242}Pu	7.1×10^{10}

Table 8.18 Dose rates from spontaneous fission of plutonium†

	Spontaneous fission neutrons per min	Spontaneous fission dose rate, mrem/h		Surface flux of spontaneous fission neutrons, $n/(\text{cm}^2 \cdot \text{s})$
		Surface	1 m	
^{239}Pu	1.82×10^3	8.6×10^{-2}	2.4×10^{-5}	0.67
^{239}Pu plus 1000 ppm ^{238}Pu	2.1×10^5	9.8	2.7×10^{-3}	76

†Basis: 1-kg sphere of plutonium; 3 n/fission, 2 MeV/n; 58 $n/(\text{cm}^2 \cdot \text{s})$ = 7.5 mrem/h.

NOMENCLATURE

a	number of (α, n) neutrons per alpha particle
A	alpha disintegration rate
f	fraction of coolant lost by leakage per unit time
F	fission rate
$\dot{n}$	neutron production rate
N	number of atoms
C	radioactivity concentration limit (Sec. 1.4)
t	time
T	time
y	fission yield, atoms of fission product per atom fissioned
β	fraction of thorium recycled with recycled uranium
ψ	ratio of ^{228}Th activity in recycled thorium to ^{228}Th activity in natural thorium
λ	radioactive decay constant
μ	removal-rate constant, Eq. (8.18)
ϕ	neutron flux, $n/(\text{cm}^2 \cdot \text{s})$
χ	concentration of light element, atoms per atom of nuclide undergoing alpha disintegration
σ	microscopic cross section, cm^2
θ	flux time, Eq. (8.20)

Subscripts

c	cooling time
C	^{14}C
F	storage time between reprocessing and fabrication
He	helium
^{3}He	^{3}He
i	identifies nuclear species; identifies light element in Eq. (8.70)
I	iodine
j	identifies nuclide undergoing alpha disintegration, Eq. (8.70)
k	identifies medium (air or water), Eq. (8.7)
n	identifies generation of fuel in reactor, Eq. (8.21)
$n, 2n$	$(n, 2n)$ neutron reaction
R	reactor irradiation time
s	storage time after reprocessing
T	tritium
U	uranium
1/2	half-life

6 ^{6}Li
02 ^{232}Th
03 ^{233}Th
04 ^{234}Th
08 ^{228}Th
11 ^{231}Pa
22 ^{232}U
26 ^{236}U
27 ^{237}U

Superscripts

R within the reactor, Eq. (8.58)
T total coolant system, Eq. (8.58)
Th in recycled thorium
U in recycled uranium
0 initial quantity
1 quantity at end of first-generation irradiation
∞ quantity at end of irradiation for equilibrium fuel cycle

REFERENCES

A1. Arnold, E. D.: *PICG(2)* **13**:248 (1958).
A2. Arnold, E. D.: Private communication, 1961.
B1. Baxter, A. (General Atomic): Private communication, Feb. 1978.
B2. Bell, M. J.: "The ORNL Isotope Generation and Depletion Code (ORIGEN)," Report ORNL-4628, 1973.
B3. Blomeke, J. O., C. W. Kee, and J. P. Nichols: "Projections of Radioactive Wastes to Be Generated by the U.S. Nuclear Power Industry," Report ORNL-TM-3965, Feb. 1974.
B4. Brooks, L. H., C. A. Heath, B. Kirstein, and D. G. Robert: "Carbon-14 in the HTGR Fuel Cycle," Report GA-A13174, Nov. 1974.
C1. Croff, A. G. (Oak Ridge National Laboratory): Private communication, Aug. 1977.
D1. Drolet, T. S., E. C. Choi, and J. A. Sovka: *Trans. Amer. Nuc. Soc.* **22**:354 (Nov. 1975).
F1. *Federal Register* **25**, no. 174: 8597 (Sept. 1960).
F2. Federal Regulations 10 CFR 20, app. B, Table II.
F3. Fowler, T. W., R. L. Clark, J. M. Gruhlbe, and J. L. Russel: "Public Health Considerations of Carbon-14 Discharges from the Light-Water-Cooled Nuclear Power Reactor Industry," Report ORP/TAD-76-3, July 1976.
G1. Gainey, B. W.: "A Review of Tritium Behavior in HTGR Systems," Report GA-A 13461, Apr. 1976.
H1. Hamilton, C. J., N. D. Holder, V. H. Pierce, and M. W. Robertson: "HTGR Spent Fuel Composition and Fuel Element Block Flow," Report GA-A13886, vol. 1, July 1976.
I1. International Committee on Radiation Protection: "Report of the Committee II on Permissible Dose for Internal Radiation," ICRP Publication 2, 1959.
K1. Kee, C. W., A. G. Croff, and J. O. Blomeke: "Updated Projections of Radioactive Wastes to Be Generated by the U.S. Nuclear Power Industry," Report ORNL/TM-5427, Dec. 1976.
P1. Pigford, T. H., and K. P. Ang: *Health Phys.* **29**:451 (1975).
P2. Pigford, T. H., and J. S. Choi: "Effect of Fuel Cycle Alternatives on Nuclear Waste Management," *Proc. Symp. Waste Management*, CONF-761020, Oct. 1976.
P3. Pigford, T. H., and C. S. Yang: "Thorium Fuel-Cycle Alternatives," Report EPA 68-01-1962, UCB-NE 3227, 1978.

S1. Samsonov, G. V.: *Handbook of the Physicochemical Properties of the Elements*, IFI/Plenum, New York, 1968.

S2. Shapiro, N. L. (Combustion Engineering, Inc.): Private communication, Sept. 1977.

S3. Smith, J. M., and R. S. Gilbert: "Tritium Experience in Boiling Water Reactors," *Proc. EPA Tritium Symp.* Aug. 1971.

PROBLEMS

8.1 Assume that a reactor is operated for a very long time so that all fission products reach saturation activity. On the average there are three radioactive decays in each fission-product decay chain. Calculate the fission-product radioactivity at saturation in units of curies per watt of thermal power from fission.

8.2 A small percentage of the fuel elements in a water-cooled reactor release gaseous fission products to the coolant. The insoluble noble gases are collected and stored for radioactive decay prior to their release to the atmosphere. Calculate the required storage time such that the radioactivity levels of ^{133}Xe and ^{85}Kr in the released gas are equal. Assume fissions at constant power only in ^{235}U, an average irradiation time of 2 years, and assume that these noble gas radionuclides are released to the coolant in the same proportion as they exist within the fuel. Obtain mass yields from Table 2.9. Twenty-three percent of the fissions at mass 85 produces ^{85}Kr.

8.3 Calculate the time required for preprocessing cooling of irradiated fuel such that the inhalation toxicity of the contained ^{131}I will have decayed to that of ^{129}I. Assume ^{235}U fission at constant power and an irradiation time of 3 years. Obtain mass-yield data from Table 2.9.

8.4 Extended preprocessing cooling can reduce the amount of those radionuclides in the high-level radioactive wastes that decay eventually to form ^{226}Ra in the wastes. By providing more time for the decay of ^{242}Cm prior to reprocessing, the decay daughter ^{238}Pu is recovered and recycled with the plutonium product, rather than appearing in the high-level wastes where its subsequent decay would eventually lead to ^{226}Ra. The ^{226}Ra in the wastes reaches its peak concentration after about 200,000 years. Using the data in Table 8.5, calculate the percentage reduction in the peak ^{226}Ra concentration if the preprocessing decay period is extended from 150 days to 3 years. Assume that 0.5 percent of the uranium and plutonium in the reprocessed fuel appears in the high-level wastes.

CHAPTER

NINE

PLUTONIUM AND OTHER ACTINIDE ELEMENTS

1 GENERAL CHEMICAL PROPERTIES OF THE ACTINIDES

1.1 Electronic Configurations

Many of the elements of importance in fuel reprocessing are found within the sixth and seventh periods of the periodic system. In the sixth period the rare-earth fission products, lanthanum to dysprosium, are difficult to separate from each other by chemical means. Their close similarity in chemical properties is explained on the basis of their electronic configurations [S1] as shown in Table 9.1.

The chemical properties of elements are determined by the behavior of electrons in the outermost shells. The chemical properties of cesium, an alkali metal, and barium, an alkaline earth, differ appreciably because of the different number of electrons in the 6*d* shell. The 6*s* shell is filled for barium, and for lanthanum the next electron is apparently added to the previously empty 5*d* shell. Additional electrons for most of the elements of higher atomic numbers, extending through lutetium, are added to the 4*f* shell, which is so deep within the atoms as to have little influence on the chemical properties of these elements. The 4*f* shell is filled for lutetium, and succeeding elements, hafnium, tantalum, tungsten, etc., add electrons to the 5*d* shell. The 5*d* electrons are not strongly bound, and each member of the sixth-period transition series, hafnium to tungsten, shows chemical properties quite distinct from those of its neighbors.

The series of 15 elements, lanthanum to lutetium, is known as the *lanthanide series.* These elements all form trivalent ions in solution; quadrivalent oxidation states of cerium, praseodymium, and terbium, and bivalent states of samarium and europium are also obtained.

The seventh period of the periodic table is occupied by a similar series called the *actinide series.* Beginning with actinium the 5*f* electron shell is populated in a manner analogous to filling the 4*f* electron shell in the lanthanide series. A suggested electronic configuration [K2, M6], is shown in Table 9.2. After the alkaline earth radium, additional electrons are added to the 6*d* and 5*f* shells, beginning the actinide series. At the beginning of the actinide series electrons are added

Table 9.1 Suggested electronic configuration of elements in sixth period

		Number of electrons								
			Shell 4				Shell 5			Shell 6
Element	Atomic number	Shells 1, 2, 3	4s	4p	4d	4f	5s	5p	5d	6s
Cesium	55	28	2	6	10	–	2	6	–	1
Barium	56	28	2	6	10	–	2	6	–	2
Lanthanum	57	28	2	6	10	–	2	6	1	2
Cerium	58	28	2	6	10	2	2	6	–	2
Praseodymium	59	28	2	6	10	3	2	6	–	2
Neodymium	60	28	2	6	10	4	2	6	–	2
Promethium	61	28	2	6	10	5	2	6	–	2
Samarium	62	28	2	6	10	6	2	6	–	2
Europium	63	28	2	6	10	7	2	6	–	2
Gadolinium	64	28	2	6	10	7	2	6	1	2
Terbium	65	28	2	6	10	9	2	6	–	2
Dysprosium	66	28	2	6	10	10	2	6	–	2
Holmium	67	28	2	6	10	11	2	6	–	2
Erbium	68	28	2	6	10	12	2	6	–	2
Thulium	69	28	2	6	10	13	2	6	–	2
Ytterbium	70	28	2	6	10	14	2	6	–	2
Lutetium	71	28	2	6	10	14	2	6	1	2

Table 9.2 Suggested electronic configuration of elements in seventh period

		Number of electrons								
			Shell 5				Shell 6			Shell 7
Element	Atomic number	Shells 1,2,3,4	5s	5p	5d	5f	6s	6p	6d	7s
Francium	87	60	2	6	10	–	2	6	–	1
Radium	88	60	2	6	10	–	2	6	–	2
Actinium	89	60	2	6	10	–	2	6	1	2
Thorium	90	60	2	6	10	–	2	6	2	2
Protactinium	91	60	2	6	10	2	2	6	1	2
Uranium	92	60	2	6	10	3	2	6	1	2
Neptunium	93	60	2	6	10	4	2	6	1	2
Plutonium	94	60	2	6	10	6	2	6	–	2
Americium	95	60	2	6	10	7	2	6	–	2
Curium	96	60	2	6	10	7	2	6	1	2
Berkelium	97	60	2	6	10	8	2	6	1	2
Californium	98	60	2	6	10	10	2	6	–	2
Einsteinium	99	60	2	6	10	11	2	6	–	2
Fermium	100	60	2	6	10	12	2	6	–	2
Mendelevium	101	60	2	6	10	13	2	6	–	2
Nobelium	102	60	2	6	10	14	2	6	–	2
Lawrencium	103	60	2	6	10	14	2	6	1	2

to the $6d$ shell rather than to the $5f$, so actinium and thorium behave chemically as homologs of lanthanum and hafnium, respectively. For gaseous elements the $5f$ shell is preferred beginning with protactinium. However, for metallic crystals at room temperature the $6d$ shell is preferred as far as uranium and possibly neptunium, which is consistent with the initial contraction of the metallic radii with increasing atomic number as shown in Table 9.3 [A1].

The chemical properties of the actinides are much less similar to each other than those of the lanthanides, because the additional electrons added to the $5f$ and $6d$ are bound less tightly than those of the $4f$ and $5d$ shells of the lanthanides. As shown in Table 9.4, the lanthanides in aqueous solutions exist principally in a single, trivalent oxidation state, whereas four or more oxidation states are observed in the aqueous chemistry of uranium, neptunium, and plutonium. The actinide ions normally formed in solution by the oxidation states II through VI are M^{2+}, M^{3+}, M^{4+}, MO_2^+, MO_2^{2+}, respectively.

The oxidation states of the lanthanide and actinide elements are summarized [K2] in Table 9.4. The most stable oxidation states in the actinide series are italicized. Numbers in parentheses indicate unstable or unusual states of oxidation. Actinides of the same oxidation state are similar in chemical properties, but different oxidation states show appreciably different chemical properties.

1.2 Hydrolytic Behavior

Hydrolysis of actinide ions is one of their most important reactions in aqueous solution. Cations with high positive charge generally tend to undergo hydrolysis, such as

$$Pu^{4+} + H_2O \rightleftharpoons Pu(OH)^{3+} + H^+ \tag{9.1}$$

and

$$UF_6 + 2H_2O \rightleftharpoons UO_2^{2+} + 2F^- + 4HF \tag{9.2}$$

which accounts for the acidity of water solutions of these cations. The tendency to displace the hydrogen ion from a water molecule by hydrolysis is usually greater the greater the charge of the cation and the smaller the physical size of the hydrolyzing cation. For example, the "acidity" of the trivalent ions of uranium, neptunium, and plutonium should increase in the order

$$U^{3+} < Np^{3+} < Pu^{3+}$$

which is the order of decreasing ionic radii shown in Table 9.3.

Table 9.3 Metallic and ionic radii of the actinides and the interatomic distances in the actinyl (V and VI) ions (Å)

Element	Atomic number	M^0	M^{3+}	M^{4+}	M^{5+}	M^{6+}	V M-O	VI M-O
Actinium	89	1.88	1.076					
Thorium	90	1.80		0.984				
Protactinium	91	1.63		0.944	0.90			
Uranium	92	1.56	1.005	0.929	0.88	0.83		1.71
Neptunium	93	1.55	0.986	0.913	0.87	0.82	1.98	
Plutonium	94	1.60	0.974	0.896	0.87	0.81	1.94	
Americium	95	1.74	0.962	0.888	0.86	0.80	1.92	
Curium	96	1.75	0.946	0.886				
Berkelium	97		0.935	0.870				

Source: S. Ahrland et al., "Solution Chemistry," in *Comprehensive Inorganic Chemistry*, vol. 5, J. C. Bailar, Jr., et al. (eds.), Pergamon, Oxford, 1973.

Table 9.4 Oxidation states of lanthanide and actinide elements†, ‡

Lanthanides															
Atomic number	57	58	59	60	61	62	63	64	65	66	67	68	69	70	71
Element	La	Ce	Pr	Nd	Pm	Sm	Eu	Gd	Tb	Dy	Ho	Er	Tm	Yb	Lu
Oxidation states				(2)		2	2						(2)	2	
	3	*3*	*3*	*3*	*3*	*3*	*3*	*3*	*3*	*3*	*3*	*3*	*3*	*3*	*3*
		4	(4)						(4)						

Actinides (+ transactinides)																	
Atomic number	89	90	91	92	93	94	95	96	97	98	99	100	101	102	103	104	105
Element	Ac	Th	Pa	U	Np	Pu	Am	Cm	Bk	Cf	Es	Fm	Md	No	Lr	Ku (Rf)	Ha
Oxidation states							(2)		(2)				2	*2*			
	3	(3)	(3)	3	3	3	*3*	*3*	*3*	*3*	*3*	*3*	*3*	3	3		
		4	4	4	4	*4*	4	4	4							4	
			5	5	*5*	5	5										(5)
				6	6	6	6										
					7	7											

†The most stable oxidation states are italicized. Those not known in solution are within parentheses.
‡Data through atomic number 103 are from Ahrland et al. [A1]. Data for atomic numbers 104 and 105 are from Keller [K2].

For the same reason, the tendency of tetrapositive ions to hydrolyze should increase in the order

$$Th^{4+} < U^{4+} < Np^{4+} < Ce^{4+} < Hf^{4+} < Zr^{4+}$$

The actinyl MO_2^+ ions of the pentapositive oxidation states and the actinyl MO_2^{2+} ions of the hexapositive states can also undergo hydrolysis, because the electrostatic attraction to form a hydrolysis complex, such as $MO_2(OH)^+$ from MO_2^{2+}, is influenced both by the net charge of the ion and the charge on the M atom. Furthermore, the ionic radii vary between oxidation states. For example, as shown in Table 9.3, Pu^{4+} ion is smaller than Pu^{3+}. The tendency toward hydrolysis of a given M atom of various oxidation states, and within the chemical series considered here, increases according to

$$MO_2{}^+ < M^{3+} < MO_2{}^{2+} < M^{4+}$$

The degree of hydrolysis depends on acidity of the solution; low acidity promotes hydrolysis to the extent that basic salts containing hydroxyl groups such as MOH^{3+}, $M(OH)_2^{2+}$, $M(OH)_3^+$, or $M(OH)_4$ precipitate from solution. A common example is the precipitation of hexavalent uranium as the diuranate with sodium or ammonium hydroxide to form $Na_2U_2O_7$ and $(NH_4)_2U_2O_7$ by the reactions

$$UO_2{}^{2+} + 2OH^- \rightarrow UO_2(OH)_2(s) \tag{9.3}$$

$$2UO_2(OH)_2(s) + 2NaOH \rightarrow Na_2U_2O_7(s) + 3H_2O \tag{9.4}$$

Hydrolyzed precipitates frequently involve polymers of higher molecular weight, especially in the cases of Th(IV) and Pu(IV). For example, the early stages of hydrolysis of Th(IV) proceeds with the dimerization reaction [A1]

$$2Th(OH)^{2+} \rightarrow Th_2(OH)_2{}^{4+} \tag{9.5}$$

The hydrolysis of such metal ions can be suppressed by high acidity, i.e., low OH^- concentration, or at low acidity by strong complexing agents.

1.3 Complex Formation

The tendency of positive ions to form stable complexes with anions, e.g., ThF^{3+}, $ThF_2{}^{2+}$, $PuNO_3{}^{3+}$, $PuF_6{}^{2-}$ in aqueous solution, is analogous to the tendency toward hydrolytic behavior.

In an aqueous solution of uranyl nitrate there are present not only hydrated uranyl ions $UO_2{}^{2+}$, but also a series of nitrate complexes $UO_2NO_3{}^+$, $UO_2(NO_3)_2$, and $UO_2(NO_3)_3{}^-$. At room temperature there is rapid equilibrium between these complexed species. The formation of the high-nitrate complexes is promoted by the presence of nitrate ions; thus $UO_2NO_3{}^+$ may predominate in solutions 4 *M* in nitrate concentrations, and $UO_2(NO_3)_2$ in more concentrated nitrate solutions [F3].

The relative tendencies toward hydrolysis mentioned in Sec. 1.2 above apply generally to complex formation, except for the reversal of the order of M^{3+} and $MO_2{}^{2+}$, acording to [K2]

$$MO_2{}^+ < MO_2{}^{2+} < M^{3+} < M^{4+}$$

For example, the tripositive ions such as La^{3+} or Pu^{3+} show relatively little tendency to complex, but stable complexes are formed with the tetrapositive ions M^{4+} and with hydrated ions of hexapositive metals $MO_2{}^{2+}$. The stability of the complex depends also on the properties of the complexing anion; strongest complexes usually occur with anions of weak acids, or small atomic radius, and with large negative charge. The tendency of anions to form complexes increases approximately in the order

$$ClO_4^- < Cl^- < NO_3^- < SO_4^- < PO_4^{3-} < F^-$$

The presence of strongly complexing anions in solution, such as F^-, inhibits hydrolysis.

Cations that complex easily generally form stable complexes with oxygenated organic compounds, such as diethyl ether, methyl isobutyl ketone, and tributyl phosphate (TBP). The purification of uranium by solvent extraction of hexavalent uranium from nitrate solutions, with TBP forming $UO_2(NO_3)_2 \cdot 2TBP$, was described in Chap. 5. These metals are extracted most easily from aqueous solutions free of the more highly complexing anions, such as F^-, PO_4^{3-}, or SO_4^{2-}.

From the above discussion it follows that tetravalent and hexavalent thorium, uranium, and plutonium can be separated from the trivalent rare-earth fission products by taking advantage of differences in complexing properties. More highly charged cation fission products, such as tetravalent cerium and the fifth-period transition elements zirconium, niobium, molybdenum, technetium, and ruthenium, complex more easily than the trivalent rare-earths and are more difficult to separate from uranium and plutonium by processes involving complex formation.

The tendency toward hydrolysis of some of these elements can be used to advantage in separation processes. For example, in the Redox process for separating uranium and plutonium from fission products, the aqueous feed to the separation plant is made acid-deficient to promote hydrolysis of zirconium to a less extractable species, probably a colloidal hydrate [B5].

Separation by ion exchange also involves tendency toward hydrolytic behavior and complex formation. A cationic resin in the acid form will exchange hydrogen ions for n-valent ions, M^{n+}, in aqueous solution according to

$$n\mathrm{HR}_{(\mathrm{resin})} + \mathrm{M}^{n+}_{(\mathrm{soln})} \rightleftharpoons \mathrm{MR}_{n\,(\mathrm{resin})} + n\mathrm{H}^{+}_{(\mathrm{soln})} \tag{9.6}$$

where R represents the insoluble resin group.

For an anionic resin in the hydroxyl form the "adsorption" of n-valent anions A^{n-} from aqueous solution is given by

$$n\mathrm{ROH}_{(\mathrm{resin})} + \mathrm{A}^{n-}_{(\mathrm{soln})} \rightleftharpoons \mathrm{R}_n\mathrm{A}_{(\mathrm{resin})} + n(\mathrm{OH})^{-}_{(\mathrm{soln})} \tag{9.7}$$

For the rare-earths and actinides the relative tendencies toward adsorption on the exchange resin are roughly in the same order as the relative tendencies toward hydrolysis and complex formation, as described above. For example, for the trivalent lanthanides the atomic radius increases with atomic number, and the smaller-radii elements like lanthanum can be adsorbed preferentially from the trivalent elements of high atomic number in this series. The same applies to the separability of trivalent actinide elements, where trivalent elements of higher atomic number adsorb less easily. The adsorbed cations are easily removed from the resin by adjusting the solution acidity and/or eluting with a solution containing a strongly complexing anion, such as citric acid.

Tetravalent plutonium can be adsorbed preferentially from most fission products on cation-exchange resins. Eluting with a solution more acidic than that from which the plutonium was adsorbed removes plutonium from the resin because the higher hydrogen ion concentration displaces the equilibrium of reaction (9.6) and the higher anion concentration, such as NO_3^-, tends to complex plutonium in solution. This provides a means of concentrating plutonium in solution as well as decontamination. Separation can occur with elution by proper choice of the complexing nature of the eluent. For example, uranium can be removed preferentially from a cation-exchange resin on which Pu^{4+} and UO_2^{2+} have been sorbed by eluting with dilute sulfuric acid solution.

The negatively charged complexes that result with many of the fission products and the actinides in solutions of high concentrations of complexing anion are easily adsorbed on anion-exchange resins. Such adsorption occurs readily with tetravalent and hexavalent uranium, with tetravalent plutonium, and with the fission products that complex easily, such as zirconium, niobium, and ruthenium. Because the extent to which negative complexes are formed for a given metal in solution depends on the concentration of the complexing anion, control of the anion concentration provides a sensitive means of controlling the exchange equilibria for a given metal and for affecting partition between two or more metals in solution. Extensive studies of the adsorption of a wide variety of elements by means of anion exchange have been reported by Kraus and Nelson [K4].

1.4 Oxidation-Reduction Reactions in Aqueous Solutions

As a result of the wide range of oxidation states for some of the actinide ions, as shown in Table 9.4, the control of the oxidation state may play an important role in chemical separations, particularly if plutonium is involved. The oxidation-reduction chemistry is also important to the behavior of actinide elements and compounds if released to the environment.

The selective extraction of plutonium from uranium or fission products depends on proper adjustment of the valence state of plutonium relative to the other ions from which it is to be separated. For instance, in decontaminating plutonium by extraction with TBP, plutonium must be oxidized to the tetravalent state, without bringing cerium into the tetravalent, ceric state. Again, to separate plutonium from uranium and the fission products in the tributyl phosphate extraction process, plutonium must be trivalent and uranium hexavalent.

A typical oxidation-reduction reaction of the type met in processing plutonium is the reduction of Pu^{4+} to Pu^{3+} by ferrous ion:

$$Fe^{2+} + Pu^{4+} \rightarrow Fe^{3+} + Pu^{3+} \tag{9.8}$$

In dealing with groups of such equilibria it is convenient to break them up into two half-reactions, or oxidation-reduction couples, which indicate the mechanism by which electrons are transferred from the reducing agent to the oxidizing agent.

The two couples for the above reaction are

$$Pu^{3+} \rightarrow Pu^{4+} + e^- \tag{9.9}$$

and

$$Fe^{2+} \rightarrow Fe^{3+} + e^- \tag{9.10}$$

The electrons from each couple may be thought of as exerting an electromotive force, which is the oxidation-reduction potential of the couple. When equilibrium is reached in a solution containing a number of oxidation-reduction couples, the potentials of all couples must be equal, otherwise electrons would be transferred from one couple to another and further reaction would take place. Thus, if we can evaluate the dependence of oxidation-reduction potentials on concentration, we can determine equilibrium concentrations in solution of ions of mixed valence.

The oxidation-reduction potential of a couple E, in volts, is related to the change in free energy ΔG when 1 g-equiv of electrons is produced:

$$\Delta G = -\mathfrak{F} E \tag{9.11}$$

where $\mathfrak{F}$ is Faraday's constant, 96,487 J/(Volt·g-equiv). The minus sign is used because of the negative charge on the electron. When the components of a couple are in their standard states of unit activity and electrons are in their standard state, the free-energy change is the standard free-energy change ΔG°, and the potential is the standard oxidation potential E°. These are related by

$$\Delta G^\circ = -\mathfrak{F} E^\circ \tag{9.12}$$

The standard state for electrons is customarily taken to be that corresponding to equilibrium in the reaction

$$\tfrac{1}{2}H_2(g,\ 1\ \text{atm}) \rightarrow H^+(\text{unit activity}) + e^- \tag{9.13}$$

whose standard oxidation potential by definition is zero.

The free-energy change in the couple (9.9) is

$$\Delta G_P = \Delta G_P^\circ + RT \ln \frac{[Pu^{4+}]}{[Pu^{3+}]} \tag{9.14}$$

where the quantities in brackets denote activities. By using (9.11) and (9.12) this becomes

$$-E_P = -E_P^\circ + \frac{RT}{\mathfrak{F}} \ln \frac{[Pu^{4+}]}{[Pu^{3+}]} \tag{9.15}$$

Similarly, for the iron couple,

$$-E_F = -E_F^\circ + \frac{RT}{\mathfrak{F}} \ln \frac{[Fe^{3+}]}{[Fe^{2+}]} \tag{9.16}$$

Because these two potentials must be equal at equilibrium,

$$E_F^\circ - E_P^\circ = \frac{RT}{\mathfrak{F}} \ln \frac{[Pu^{3+}]\,[Fe^{3+}]}{[Pu^{4+}]\,[Fe^{2+}]} = \frac{RT}{\mathfrak{F}} \ln K_{FP} \tag{9.17}$$

where K_{FP} is the equilibrium constant for the overall iron-plutonium reaction (9.8).

At 25°C, we have for K:

$$K_{FP} = e^{38.93(E_F^\circ - E_P^\circ)} \tag{9.18}$$

For the plutonium-iron case, the values are [A1]

$$Pu^{3+} \rightarrow Pu^{4+} + e^- \qquad E_P^\circ = -0.9819 \text{ V}$$

$$Fe^{2+} \rightarrow Fe^{3+} + e^- \qquad E_F^\circ = -0.7701 \text{ V}$$

The equilibrium constant for the reaction (9.8) is

$$K_{FP} = e^{38.93(-0.7701 + 0.9819)} = 3810 \tag{9.19}$$

This shows that reduction of plutonium from tetravalent to trivalent can be made substantially complete with only a slight excess of ferrous ion.

A large negative oxidation-reduction potential means that the first member is a strong oxidizing agent. A positive potential means that the first member is a strong reducing agent (and could, in fact, reduce water) and the second member is a very weak oxidizing agent.

The general equation relating the equilibrium constant for a reaction involving the transfer of n electrons and standard oxidation-reduction potentials is

$$K_{AB} = e^{38.93(E_A^\circ - E_B^\circ)n} \tag{9.20}$$

For example, in the reaction

$$2Pu^{4+} + Sn^{2+} \rightarrow 2Pu^{3+} + Sn^{4+} \tag{9.21}$$

the individual potentials are [A1]

$$Pu^{3+} \rightarrow Pu^{4+} + e^- \qquad E_P^\circ = -0.9819 \text{ V} \tag{9.22}$$

$$Sn^{2+} \rightarrow Sn^{4+} + 2e^- \qquad E_S^\circ = -0.154 \text{ V} \tag{9.23}$$

Equation (9.21), as written, involves the transfer of 2 equivalents of electrons from tin to plutonium, so that the equilibrium constant is

$$K_{SP} = \frac{[Sn^{4+}]\,[Pu^{3+}]^2}{[Sn^{2+}]\,[Pu^{4+}]^2} = e^{38.93(-0.154 + 0.9819)2} = 9.88 \times 10^{27} \tag{9.24}$$

Oxidation-reduction potentials for plutonium. Oxidation reduction potentials [A1] for plutonium ions are given in Table 9.5.

Other potentials may be obtained as linear combinations of these. For instance,

$$Pu^{4+} + 2H_2O \rightarrow PuO_2{}^{2+} + 4H^+ + 2e^-$$

$$E^\circ_{46} = \frac{E^\circ_{45} + E^\circ_{56}}{2} = -1.0433 \text{ V} \tag{9.29}$$

Equilibria for the reduction of $PuO_2{}^+$ or $PuO_2{}^{2+}$ to Pu^{3+} or Pu^{4+} depend on the fourth power of the hydrogen ion concentration. Increasing acidity displaces equilibrium toward the reduced states.

If the salts and acids in a given system were completely dissociated into simple ions with no complex formation and if the activity coefficients were always the same, oxidation-reduction potentials would be independent of the anionic or cationic species present. However, as we saw in Sec. 1.1, these conditions are rarely obtained with the higher-valent actinides. Therefore, oxidation-reduction potentials for the actinides are usually measured in perchlorate solutions, where relatively little complexing and a high degree of dissociation occur [A1, S1]. These are referred to as "formal potentials," which are more accurate for practical calculations. To obtain standard potentials these formal potentials, or equivalent thermodynamic data, must be extrapolated to zero ionic strength. Formal potentials for other solutions, such as 1 M HNO_3 or 1M NaOH, are also reported in the literature [A1]. As pointed out in Sec. 1.3, in a practical system such as nitric acid solutions the equilibria estimated from oxidation-reduction potentials must be corrected for complexing, as well as incomplete dissociation. For example, the oxidation potential for the couple Pu(III)-Pu(IV) is -0.9819 V in 1 M $HClO_4$ but -0.92 V in 1 M HNO_3 [S1].

Oxidation-reduction potentials of the actinides. The formal potentials for transition between the valence states of the actinides are listed in Table 9.6.

The stability of an intermediate oxidation number against disproportionation can be obtained as follows. Consider the disproportionation of U(V) according to the following reaction:

$$U^VO_2{}^+ \rightarrow U^{VI}O_2{}^{2+} + e^- \quad E^\circ_{5,6} = -0.063 \text{ V} \tag{9.30}$$

$$U^{4+} + 2H_2O \rightarrow U^VO_2{}^+ + 4H^+ + e^- \quad E^\circ_{4,5} = -0.613 \text{ V} \tag{9.31}$$

$$2U^VO_2{}^+ + 4H^+ \rightarrow U^{VI}O_2{}^{2+} + U^{4+} + 2H_2O \quad \Delta E = -0.063 - (-0.613) = 0.550 \text{ V} \tag{9.32}$$

The equilibrium constant for the overall reaction (9.32) is

$$K = e^{38.93(-0.063+0.613)} = 1.99 \times 10^9 \tag{9.33}$$

Hence, pentavalent uranium is unstable in aqueous solution at $[H^+] \geqslant 1$.

Disproportionation of valence n to valences $n + 1$ and $n - 1$ will proceed spontaneously at pH = 0 if the potential for the oxidation from n to $n + 1$ is larger or more nearly positive than the potential for oxidation from $n - 1$ to n. Applying this criterion to the data of Table 9.6,

Table 9.5 Oxidation-reduction potentials for plutonium

$Pu \rightarrow Pu^{3+} + e^-$	$E^\circ_{03} = 2.08$ V	(9.25)
$Pu^{3+} \rightarrow Pu^{4+} + e^-$	$E^\circ_{34} = -0.9819$ V	(9.26)
$Pu^{4+} + 2H_2O \rightarrow Pu^VO_2{}^+ + 4H^+ + e^-$	$E^\circ_{45} = -1.1702$ V	(9.27)
$Pu^VO_2{}^+ \rightarrow Pu^{VI}O_2{}^{2+} + e^-$	$E^\circ_{56} = -0.9164$ V	(9.28)

Table 9.6 Formal oxidation-reduction potentials for actinides, V†

Element	0/III	II/III	III/IV	IV/V	V/VI	IV/VI	VI/VII
Actinium	2.62						
Thorium	1.8(0/IV)		2.4				
Protactinium	0.97(0/V)			0.29‡			
Uranium	1.85		0.631	−0.613	−0.063	−0.338	
Neptunium	1.83		−0.1551	−0.7391	−1.1364	−0.9377	<−2.07
Plutonium	2.08		−0.9819	−1.1702	−0.9164	−1.0433	−0.847
Americium	2.42		−2.34	−1.16	−1.60	−1.38	
Curium	2.31	5.0	−3.24				
Berkelium		3.4	−1.64				
Californium	2.32	1.9	<−1.60				
Einsteinium		1.60					
Fermium		1.3					
Mendelevium		0.15					
Nobelium		−1.45					

†In 1 *M* $HClO_4$.
‡In 6 *M* HCl.
Source: S. Ahrland et al., "Solution Chemistry," in *Comprehensive Inorganic Chemistry*, vol. 5, J. C. Bailar, Jr., et al. (eds.), Pergamon, Oxford, 1973.

we see that in the tetravalent state uranium, neptunium, and plutonium are stable. In the pentavalent state protactinium, neptunium, and americium are stable (cf. Table 9.4).

The positive potential for U(III)-U(IV) indicates that the unstable U(III) would be rapidly oxidized by water in aqueous solution. The relatively low negative potentials for the oxidation of U(III) through intermediate states to U(VI) indicate that the latter should be quite stable in aqueous solutions. The elements of higher atomic number become progressively more difficult to oxidize to the hexavalent state.

Solutions containing U(VI) and Pu(III) or Pu(IV), as used in aqueous separation processes, are stable against oxidation of plutonium by uranium because the potentials for the transitions U(IV) to U(VI) and U(V) to U(VI) are more nearly positive than the plutonium potentials. Plutonium may be reduced from Pu(IV) to Pu(III) without affecting uranium oxidation by choosing a reducing agent, such as Fe^{2+}, whose oxidation potential is less negative than the −0.9819 V required for Pu(IV) reduction and more negative than the −0.338 V that would reduce U(VI).

The data indicate that U(VI) should oxidize Np(III) to Np(IV).

Oxidation-reduction potentials for couples consisting of the actinides or the fission products in acid solution (1 *M* $HClO_4$) are listed in Table 9.7. Potentials for a selected group of oxidizing and reducing agents are listed in Table 9.8. The couples are listed in order of decreasing strength as reducing agents. In the cases where the molecular and ionic species involved in a given valence transition are different in acidic and basic solutions, the acid system (1 *M* $HClO_4$) has been chosen.

The oxidation-reduction schemes of the more important multivalent elements encountered in aqueous fuel reprocessing are summarized in Fig. 9.1.

Rate of oxidation-reduction reactions. Oxidation-reduction reactions that involve only the transfer of electrons from one uncomplexed ion to another in an ionizing solvent are reversible and, for all practical purposes, instantaneous. Equation (9.8) is an example. On the other hand, reactions involving molecular rearrangements, even though thermodynamically possible, may be

Table 9.7 Formal oxidation-reduction potentials for actinides and fission products in acid solutions†

Couple	E°, V
$Cm^{2+} \rightarrow Cm^{3+} + e^-$	5.0
$Bk^{2+} \rightarrow Bk^{3+} + e^-$	3.4
$Ac \rightarrow Ac^{3+} + 3e^-$	2.62
$La \rightarrow La^{3+} + 3e^-$	2.52
$Ce \rightarrow Ce^{3+} + 3e^-$	2.48
$Nd \rightarrow Nd^{3+} + 3e^-$	2.44
$Am \rightarrow Am^{3+} + 3e^-$	2.42
$Sm \rightarrow Sm^{3+} + 3e^-$	2.41
$Gd \rightarrow Gd^{3+} + 3e^-$	2.40
$Th^{3+} \rightarrow Th^{4+} + e^-$	2.4
$Y \rightarrow Y^{3+} + 3e^-$	2.37
$Cf \rightarrow Cf^{3+} + 3e^-$	2.32
$Cm \rightarrow Cm^{3+} + 3e^-$	2.31
$Pu \rightarrow Pu^{3+} + 3e^-$	2.08
$Cf^{2+} \rightarrow Cf^{3+} + e^-$	1.9
$Th \rightarrow Th^{4+} + 4e^-$	1.8
$U \rightarrow U^{3+} + 3e^-$	1.85
$Np \rightarrow Np^{3+} + 3e^-$	1.83
$Es^{2+} \rightarrow Es^{3+} + e^-$	1.60
$Zr \rightarrow Zr^{4+} + 4e^-$	1.53
$Fm^{2+} \rightarrow Fm^{3+} + e^-$	1.3
$Nb \rightarrow Nb^{3+} + 3e^-$	~1.1
$Pa + 2H_2O \rightarrow PaO_2^+ + 4H^+ + 5e^-$	0.97
$2Nb + 5H_2O \rightarrow Nb_2O_5 + 10H^+ + 10e^-$	0.65
$U^{3+} \rightarrow U^{4+} + e^-$	0.631
$Ga \rightarrow Ga^{3+} + 3e^-$	0.53
$Eu^{2+} \rightarrow Eu^{3+} + e^-$	0.43
$Pa^{4+} + 2H_2O \rightarrow PaO_2^+ + 4H^+ + e^-$	0.29‡
$Mo \rightarrow Mo^{3+} + 3e^-$	~0.2
$Md^{2+} \rightarrow Md^{3+} + e^-$	0.15
$2Ru + 3H_2O \rightarrow Ru_2O_3 + 6H^+ + 6e^-$	0.1
$U^{3+} + 2H_2O \rightarrow UO_2^+ + 4H^+ + 2e^-$	0.009
$H_2(g) \rightarrow 2H^+ + 2e^-$	0
$U^{3+} + 2H_2O \rightarrow UO_2^{2+} + 4H^+ + 3e^-$	−0.015
$UO_2^+ \rightarrow UO_2^{2+} + e^-$	−0.063
$Ru_2O_3 + H_2O \rightarrow 2RuO_2 + 2H^+ + 2e^-$	−0.1
$Np^{3+} \rightarrow Np^{4+} + e^-$	−0.1551
$U^{4+} + 2H_2O \rightarrow UO_2^{2+} + 4H^+ + 2e^-$	−0.338
$RuO_2 + 2H_2O \rightarrow RuO_4^- + 4H^+ + 2e^-$	−0.4
$UO_2 \rightarrow UO_2^{2+} + 2e^-$	−0.427
$Np^{3+} + 2H_2O \rightarrow NpO_2^+ + 4H^+ + 2e^-$	−0.4471
$2I^- \rightarrow I_2 + 2e^-$	−0.5355
$3I^- \rightarrow I_3^- + 2e^-$	−0.6
$Te + 2H_2O \rightarrow TeOOH^+ + 3H^+ + 4e^-$	−0.559
$RuO_4^{2-} \rightarrow RuO_4^- + e^-$	−0.6
$U^{4+} + 2H_2O \rightarrow UO_2^+ + 4H^+ + e^-$	−0.613
$Np^{3+} + 2H_2O \rightarrow NpO_2^{2+} + 4H^+ + 3e^-$	−0.6769
$Np^{4+} + 2H_2O \rightarrow NpO_2^+ + 4H^+ + e^-$	−0.7391

(See footnotes on page 418.)

Table 9.7 Formal oxidation-reduction potentials for actinides and fission products in acid solutions† *(Continued)*

Couple	E°, V
$Rh \rightarrow Rh^{3+} + 3e^-$	~−0.8
$RuO_4^- \rightarrow RuO_4 + e^-$	−0.9
$PuO_2^+ \rightarrow PuO_2^{2+} + e^-$	−0.9164
$Np^{4+} + 2H_2O \rightarrow NpO_2^{2+} + 4H^+ + 2e^-$	−0.9377
$Pu^{3+} \rightarrow Pu^{4+} + e^-$	−0.9819
$Pd \rightarrow Pd^{2+} + 2e^-$	−0.987
$Pu^{3+} + 2H_2O \rightarrow PuO_2^{2+} + 4H^+ + 3e^-$	−1.0228
$Pu^{4+} + 2H_2O \rightarrow PuO_2^{2+} + 4H^+ + 2e^-$	−1.0433
$Pu^{3+} + 2H_2O \rightarrow PuO_2^+ + 4H^+ + 2e^-$	−1.0761
$NpO_2^+ \rightarrow NpO_2^{2+} + e^-$	−1.1364
$Am^{4+} + 2H_2O \rightarrow AmO_2^+ + 4H^+ + e^-$	−1.16
$Pu^{4+} + 2H_2O \rightarrow PuO_2^+ + 4H^+ + e^-$	−1.1702
$Am^{4+} + 2H_2O \rightarrow AmO_2^{2+} + 4H^+ + 2e^-$	−1.38
$No^{2+} \rightarrow No^{3+} + e^-$	−1.45
$AmO_2^+ \rightarrow AmO_2^{2+} + e^-$	−1.60
$Cf^{3+} \rightarrow Cf^{4+} + e^-$	<-1.60
$Ce^{3+} \rightarrow Ce^{4+} + e^-$	−1.61
$Bk^{3+} \rightarrow Bk^{4+} + e^-$	−1.64
$Am^{3+} + 2H_2O \rightarrow AmO_2^{2+} + 4H^+ + 3e^-$	−1.70
$Am^{3+} + 2H_2O \rightarrow AmO_2^+ + 4H^+ + 2e^-$	−1.75
$NpO_2^{2+} \rightarrow NpO_2^{3+} + e^-$	<-2.07
$Am^{3+} \rightarrow Am^{4+} + e^-$	−2.34
$Cm^{3+} \rightarrow Cm^{4+} + e^-$	−3.24

†Actinide potentials are from Ahrland et al. [A1]. In 1 *M* $HClO_4$.
‡In 6 *M* HCl.

very slow or may not go at all. A familiar example of a slow reaction is the gradual approach to the end point in titration of ferrous ion with permanganate ion in acid solution:

$$8H^+ + 5Fe^{2+} + MnO_4^- \rightarrow 5Fe^{3+} + Mn^{2+} + 4H_2O \quad (9.34)$$

The oxidation of an actinide ion from M^{3+} or M^{4+} to MO_2^+ or MO_2^{2+}, or its reduction from M(V or VI) to M(IV or III), is inconveniently slow, apparently because of the sluggish combination of M and O in the oxidation step or the slow breaking of the M—O bond in the reduction step. Of the three couples involving plutonium, Eq. (9.27) is very slow, whereas (9.26) and (9.28) are practically instantaneous. Further discussion of the rate of oxidation-reduction of plutonium solution appears in Sec. 4.6.

1.5 Summary

In summary, actinides in the oxidation stage M(III) form trivalent ions in aqueous solution with chemical properties similar to those of the trivalent rare earths, e.g., lanthanum. The M(IV) actinides form tetravalent ions with properties characteristic of Th^{4+}. The M(V) actinides form MO_2^+ ions, and the M(VI) form MO_2^{2+} ions whose properties are characteristic of $U^{VI}O_2^{2+}$. The chemical properties of uranium and thorium are discussed in Chaps. 5 and 6, respectively. The following sections summarize the properties of protactinium, neptunium, plutonium, americium, and curium, which are the bred actinides important in reprocessing thorium and uranium fuels.

Table 9.8 Standard oxidation-reduction potentials for oxidizing and reducing agents in acid solutions[†]

Couple	$E°$, V
$Zn \rightarrow Zn^{2+} + 2e^-$	0.7628
$Fe \rightarrow Fe^{2+} + 2e^-$	0.440
$Cr^{2+} \rightarrow Cr^{3+} + e^-$	0.41
$Cd \rightarrow Cd^{2+} + 2e^-$	0.4025
$Ti^{2+} \rightarrow Ti^{3+} + e^-$	0.37
$Sn \rightarrow Sn^{2+} + 2e^-$	0.1406
$Ti^{3+} + H_2O \rightarrow Ti(OH)^{3+} + H^+ + e^-$	0.055
$H_2 \rightarrow 2H^+ + 2e^-$	0.00
$2I^- \rightarrow I_2(s) + 2e^-$	−0.0536
$Cu^+ \rightarrow Cu^{2+} + e^-$	−0.153
$Sn^{2+} \rightarrow Sn^{4+} + 2e^-$	−0.154
$H_2O + H_2SO_4 \rightarrow SO_4{}^{2-} + 4H^+ + 2e^-$	−0.17
$Cu \rightarrow Cu^{2+} + 2e^-$	−0.337
$Fe(CN)_6{}^{4-} \rightarrow Fe(CN)_6{}^{3-} + e^-$	−0.36
$2NH_3OH^+ \rightarrow H_2N_2O_2 + 6H^+ + 4e^-$	−0.496[‡]
$Cu \rightarrow Cu^+ + e^-$	−0.521
$MnO_4{}^{2-} \rightarrow MnO_4{}^- + e^-$	−0.564
$H_2O_2 \rightarrow O_2 + 2H^+ + 2e^-$	−0.682
$H_2N_2O_2 \rightarrow 2NO + 2H^+ + 2e^-$	−0.71
$Fe^{2+} \rightarrow Fe^{3+} + e^-$	−0.7701
$2Hg \rightarrow Hg_2{}^{2+} + 2e^-$	−0.789
$N_2O_4 + 2H_2O \rightarrow 2NO_3{}^- + 4H^+ + 2e^-$	−0.80
$H_2N_2O_2 + 2H_2O \rightarrow 2HNO_2 + 4H^+ + 4e^-$	−0.86
$Hg_2{}^{2+} \rightarrow 2Hg^{2+} + 2e^-$	−0.920
$HNO_2 + H_2O \rightarrow NO_3{}^- + 3H^+ + 2e^-$	−0.94
$NO + 2H_2O \rightarrow NO_3{}^- + 4H^+ + 3e^-$	−0.96
$NO + H_2O \rightarrow HNO_2 + H^+ + e^-$	−1.00
$VO^{2+} + 3H_2O \rightarrow V(OH)_4{}^+ + 2H^+ + e^-$	−1.000
$2NO + 2H_2O \rightarrow N_2O_4 + 4H^+ + 4e^-$	−1.03
$2Br^- \rightarrow Br_2(l) + 2e^-$	−1.0652
$2HNO_2 \rightarrow N_2O_4 + 2H^+ + 2e^-$	−1.07
$ClO_3{}^- + H_2O \rightarrow ClO_4{}^- + 2H^+ + 2e^-$	−1.19
$\frac{1}{2}I_2(s) + 3H_2O \rightarrow IO_3{}^- + 6H^+ + 5e^-$	−1.195
$HClO_2 + H_2O \rightarrow ClO_3{}^- + 3H^+ + 2e^-$	−1.21
$2H_2O \rightarrow O_2 + 4H^+ + 4e^-$	−1.229
$Mn^{2+} + 2H_2O \rightarrow MnO_2 + 4H^+ + 2e^-$	−1.23
$2Cr^{3+} + 7H_2O \rightarrow Cr_2O_7{}^{2-} + 14H^+ + 6e^-$	−1.33
$NH_4{}^+ + H_2O \rightarrow NH_3OH^+ + 2H^+ + 2e^-$	−1.35
$Cl^- \rightarrow \frac{1}{2}Cl_2(g) + e^-$	−1.354
$\frac{1}{2}Cl_2(g) + 3H_2O \rightarrow ClO_3{}^- + 6H^+ + 5e^-$	−1.47
$Mn^{2+} \rightarrow Mn^{3+} + e^-$	−1.51
$Mn^{2+} + 4H_2O \rightarrow MnO_4{}^- + 8H^+ + 5e^-$	−1.51
$\frac{1}{2}Br_2(l) + 3H_2O \rightarrow BrO_3{}^- + 6H^+ + 5e^-$	−1.52
$Ce^{3+} \rightarrow Ce^{4+} + e^-$	−1.61
$\frac{1}{2}Cl_2 + H_2O \rightarrow HClO + H^+ + e^-$	−1.63
$HClO + H_2O \rightarrow HClO_2 + 2H^+ + 2e^-$	−1.64
$MnO_2 + 2H_2O \rightarrow MnO_4{}^- + 4H^+ + 3e^-$	−1.695
$2H_2O \rightarrow H_2O_2 + 2H^+ + 2e^-$	−1.77
$O_2 + H_2O \rightarrow O_3 + 2H^+ + 2e^-$	−2.07[§]
$2HF(aq) \rightarrow F_2 + 2H^+ + 2e^-$	−3.06

[†] From Ahrland et al. [A1] and Latimer [L1].
[‡] Forward reaction only.
[§] Reverse reaction only.

Additional information on the properties of these elements and their compounds may be found in texts by Seaborg and co-workers [S1, S2].

2 PROPERTIES OF PROTACTINIUM

2.1 Protactinium Isotopes

Table 9.9 lists the isotopes of protactinium important in nuclear technology and some of the important nuclear properties.

^{231}Pa. The only isotope of protactinium with a half-life longer than 1 month is ^{231}Pa. It is a member of the $4n + 3$ decay series of ^{235}U, occurring in secular equilibrium with natural uranium at a concentration about the same as that of radium. The activity of ^{231}Pa in natural uranium at secular equilibrium is 0.022 Ci/Mg of uranium.

^{231}Pa is contained in the waste sludges and mill tailings formed when processing uranium ore. Sludges from high-grade uranium ores have been processed at the Springfields (U.K.) plant. In a process developed by Goble et al. [G2] and applied by Nairn et al. to 56 Mg of sludge, about 130 g of ^{231}Pa was extracted. The sludge was leached with nitric acid and residual uranium was removed by TBP extraction. Protactinium was precipitated with $AlCl_3$, redissolved in NaOH, purified by dibutyl ketone extraction, and finally purified by ion exchange from HCl solution [N1].

^{231}Pa is also formed in the neutron irradiation of thorium. The $(n, 2n)$ reactions in ^{232}Th, and (n, γ) reactions in any ^{230}Th that may be present in the natural thorium, result in 25.5-h ^{231}Th, which decays to ^{231}Pa.

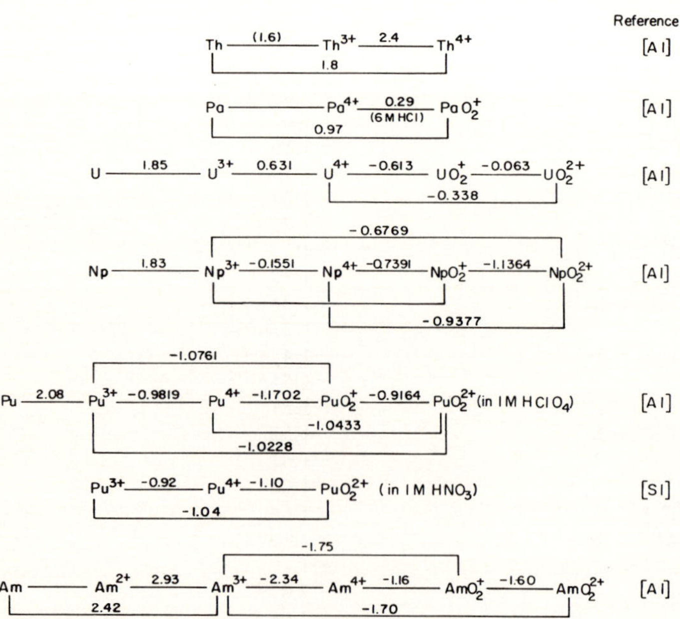

Figure 9.1 Oxidation-reduction diagrams. Formal and standard potentials in volts. Calculated or uncertain couples are listed in parentheses.

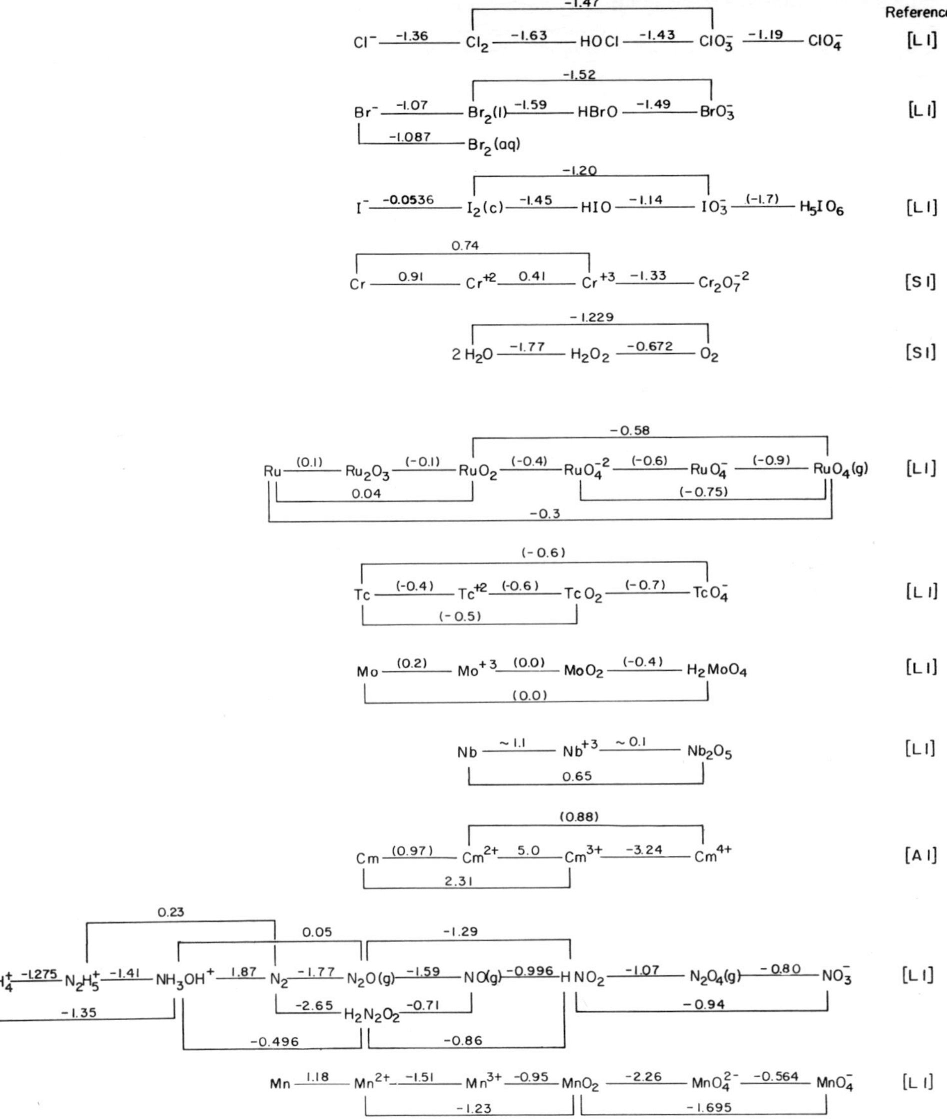

Figure 9.1 Oxidation-reduction diagrams. Formal and standard potentials in volts. Calculated or uncertain couples are listed in parentheses. (*Continued*)

^{232}Pa. The isotope ^{232}Pa is a beta emitter with a half-life of 1.31 days. It is formed in irradiated thorium by (n, γ) reactions in ^{231}Pa. Most of it undergoes beta decay into 72-year ^{232}U, which is an important radioactive contaminant in the ^{233}U recovered from irradiated thorium, as discussed in Chap. 8.

^{233}Pa. The isotope ^{233}Pa is an important intermediate nuclide in the formation of ^{233}U, in the chain originating by neutron capture in ^{232}Th. In this sense, ^{233}Pa is functionally analogous to

Table 9.9 Isotopes of protactinium

Mass, amu	Half-life	Radioactive decay: Type	Radioactive decay: Effective MeV	Reaction with 2200 m/s neutrons, Cross section, b: (n, γ)	Reaction with 2200 m/s neutrons, Cross section, b: Fission
231.035877	3.25×10^4 yr	α	5.148	210	
232.038612	1.31 days	β	1.289	760	700
233.040132	27.0 days	β	0.228	41 (n, a)	
234.043298	6.75 h	β	1.533		
234.______†	1.17 min	β	0.868		

† ^{234m}Pa.

^{239}Np in the chain leading to ^{239}Pu by neutron capture in ^{238}U. An important difference, however, is the relatively long 27.0-day half-life of ^{233}Pa. As a result, ^{233}Pa persists much longer in irradiated thorium fuel and may contribute significantly to actinide radioactivity during thorium fuel reprocessing.

Because of epithermal resonance absorption of neutrons in ^{233}Pa, its effective cross section in a thermal-neutron spectrum is much greater than the 2200 m/s cross section listed in Table 9.9. During thorium irradiation ^{233}Pa may exist in sufficient concentration that its destruction by chain-branching neutron absorption can reduce the rate of formation of ^{233}U. For this reason, thorium-uranium breeder reactors tend to optimize at lower neutron fluxes, and at lower specific power, than do uranium-plutonium breeders.

^{233}Pa can be recovered from irradiated thorium fuel as a separated actinide, with possible value because of its subsequent decay to form isotopically pure ^{233}U.

^{234}Pa. The isotope ^{234}Pa is formed in irradiated thorium from (n, γ) reactions in ^{233}Pa and, to a limited extent, from the decay of ^{234m}Pa, the daughter of ^{234}Th.

^{234m}Pa. The isotope ^{234m}Pa is formed by the beta decay of ^{234}Th, and by (n, γ) reactions in ^{233}Pa. A small fraction (0.13 percent) of the decays of ^{234m}Pa are isomeric transitions to ^{234}Pa, and the rest are beta transitions to ^{234}U.

2.2 Metallic Protactinium

Earlier laboratory processes used to prepare metallic protactinium include vacuum decomposition of the oxide by 35-keV electrons [G6] and thermal decomposition of the pentahalides [G2]. More recently, protactinium has been prepared by reducing the tetrafluoride by barium vapor [C5, S3, Z1], by calcium at 1250°C [M2] and by zinc-magnesium. The purest protactinium has been prepared by reduction in a barium-fluoride crucible at 1300°C [L2].

The two known crystalline phases of the metal, up to its melting point of 1575°C, are shown in Table 9.10 [K2].

2.3 Protactinium Compounds

Protactinium oxides. Protactinium forms the fcc dioxide PaO_2, which, like UO_2, adds additional oxygen to form the hyperstoichiometric PaO_{2+x}. The most stable oxide is the pentavalent Pa_2O_5, which exists in five different crystalline forms. The pentoxide results from heating any binary Pa(IV) or (V) compound in oxygen to temperatures above 650°C. Hydrogen reduction at 1550°C transforms Pa_2O_5 to black PaO_2. Like thoria, high-fired Pa_2O_5 is only slightly soluble in mineral acids [K2].

Protactinium halides. Presently known halides of protactinium are PaF_4, Pa_2F_9, PaF_5, $PaCl_4$, $PaCl_5$, $PaBr_4$, $PaBr_5$, PaI_3, PaI_4, and PaI_5. The pentafluoride is formed by the high-temperature reactions of fluorine with protactinium compounds. Hydrogen-HF mixtures stabilize the tetrafluoride PaF_4 [B4].

The volatilization of $PaCl_5$ has been used in the analytic separation of ^{231}Pa in wastes, which are heated in CCl_4 vapor in a closed vessel at 400 to 500°C [C7].

2.4 Protactinium Solution Chemistry

In aqueous solutions trivalent protactinium is unknown. Tetravalent protactinium is stable in the absence of air, but it is rapidly oxidized to Pa(V) by oxygen. It can be prepared from Pa(V) by using strong reducing agents such as zinc dust, amalgamated zinc, or Cr(II) salts, or by electrolytic reduction. In moderately acidic ($H^+ < 1$ *M*) aqueous solutions Pa(IV) appears to exist as the protactinyl ion, PaO^{2+} or $Pa(OH)_2^{2+}$, which forms moderately stable complexes [L2].

Pentavalent protactinium is the most stable oxidation state in aqueous solution. It shows a strong tendency toward irreversible hydrolysis in solution, but it differs from the other pentavalent actinides in that it does not hydrolyze to form an actinyl ion of the form MO_2^+. Instead, the postulated ionic species in noncomplexing solutions are $PaOOH^{2+}$ and $PaO(OH)_2^+$. With strong complexing agents, even in aqueous solutions, Pa(V) can form a nonoxygenated complex, such as PaF_6^- [A1, L6].

The distribution coefficients of Pa(V) between nitric acid solutions and solvents containing TBP are less than those of uranium [C6], and are less than those of thorium except at high concentrations of HNO_3 [H2]. In extraction measurements it is found that the fraction of Pa(V) that can be extracted decreases with time, due evidently to the slow polymerization of Pa(V) colloids. The more highly condensed forms cannot be depolymerized by acid treatment [K2].

Protactinium can be recovered from irradiated thorium, after fission-product decontamination, by exchange onto Dowex 1-X8 anion-exchange resin from a 9 *M* HCl solution. Thorium is then eluted with 9 *M* HCl, followed by elution of Pa(V) with 9 *M* HCl-0.25 *M* HF. Uranium is eluted with 0.25 *M* HF [K1, K4].

Protactinium can be recovered from an aqueous nitrate solution of fission products and protactinium by adding sodium chromate, which brings down protactinium on the aluminum chromate precipitate. After dissolution of the precipitate in acid, protactinium may be recovered by solvent extraction, or it may be allowed to decay to ^{233}U, which is more easily extracted [G4]. Protactinium can also be recovered by adsorption on powdered Vycor glass.

Table 9.10 Phases of protactinium metal

Transition temperature, °C	Phase	Crystal system	Density, g/cm^3
	Solid α	Body-centered tetragonal	15.37
$\leqslant \sim 1170$			
	Solid β	Body-centered cubic	13.87
1575			
	Liquid		

Specific extractants for Pa(V) from fairly concentrated (6 *M*) HCl solutions are branched-chain ketones and alcohols such as diisobutyl carbinol or diisobutyl ketone [H2, K4].

3 PROPERTIES OF NEPTUNIUM

3.1 Neptunium Isotopes

Table 9.11 lists the isotopes of neptunium important in nuclear technology and some of their important nuclear properties.

^{236}Np. The isotope ^{236}Np is formed in reactors by $(n, 2n)$ reactions in ^{237}Np. It undergoes beta decay, with a half-life of 22 h, to form ^{236}Pu.

^{237}Np. The isotope ^{237}Np is formed in considerable quantities in reactors, by the nuclide chains initiated by (n, γ) reactions in ^{235}U and by $(n, 2n)$ reactions in ^{238}U. Neutron capture by ^{237}Np leads through ^{238}Np to ^{238}Pu, which is the principal alpha-emitting constituent of plutonium in power reactors. To produce ^{238}Pu for use as a heat source for thermoelectric devices, neptunium has been recovered from irradiated uranium to form target elements for further irradiation in reactors. Commercial processes designed for this recovery are discussed in Chap. 10.

In normal reprocessing of irradiated uranium fuel neptunium appears in the high-level wastes. Because of its long half-life of 2.14×10^6 years, ^{237}Np persists in these wastes long after most of the fission products and other actinides have decayed. It undergoes alpha decay in the $2n + 1$ decay chain to form ^{233}Pa, which subsequently decays to ^{233}U, to ^{229}Th, and thence to ^{225}Ra and its decay daughters. Because of its half-life and the radiotoxicity of its daughters, ^{237}Np is the source of important long-term toxicity in high-level wastes. If the radionuclides in these wastes ever become dissolved in groundwater, the chemistry of neptunium is such that it may not be as effectively retarded by sorption in geologic media as are the other actinides in these wastes.

^{238}Np. The isotope ^{238}Np is the 2.1-day beta emitter formed by neutron capture in ^{237}Np. With the availability of separated ^{237}Np from fuel reprocessing, ^{238}Np is easily made by irradiation of the ^{237}Np target. It has displaced ^{239}Np as a tracer for chemical studies. The high capture-to-fission ratio for ^{237}Np results in only a relatively small contamination by fission products, which are easily removed chemically [K2].

Table 9.11 Isotopes of neptunium

		Radioactive decay		Reaction with 2200 m/s neutrons, Cross section, b	
Mass, amu	Half-life	Type	Effective MeV	(n, γ)	Fission
236.046624	22 h	β, EC	0.17		
237.048056	2.14×10^6 yr	α	4.956	169	0.019
238.050896	2.1 days	β	0.839		2070
239.052924	2.35 days	β	0.69	50(n, a)	

Table 9.12 Phases of neptunium metal

Transition temperature, °C	Phase	Crystal system	Density, g/cm^3
	Solid α	Orthorhombic	20.48
<280			
	Solid β	Tetragonal	19.40
577			
	Solid γ	Body-centered cubic	18.04
637			
	Liquid		

Source: C. Keller, *The Chemistry of the Transuranium Elements*, Verlag Chemie, Weinheim, 1971.

^{239}Np. The isotope ^{239}Np is formed by neutron capture in ^{238}U or by decay of ^{243}Am. The latter method is the easiest for laboratory preparation, if separated americium is available. Reactor-produced americium will not produce pure ^{239}Np, however, because of the presence of ^{241}Am, which decays to ^{237}Np.

3.2 Metallic Neptunium

The phases of metallic neptunium, and their densities and transition temperatures, are listed in Table 9.12.

Metallic neptunium is prepared by reducing NpF_4 with calcium. Neptunium yields of about 99 percent have been obtained from 100- to 400-g quantities of NpF_4, with 30 percent excess calcium and with 0.25 to 0.35 mol of iodine booster per mole of NpF_4. Metallic neptunium forms a protective oxide layer in air at room temperature, but it rapidly oxidizes at higher temperatures. It dissolves readily in HCl and H_2SO_4 [K2].

3.3 Neptunium Compounds

Neptunium oxides. Keller [K2] reports five binary oxides or oxide hydrates of neptunium: NpO_2, Np_2O_5, Np_3O_8, $NpO_3 \cdot 2H_2O$, and $NpO_3 \cdot H_2O$. Anhydrous Np(VI) oxide has not been prepared.

Neptunium dioxide, NpO_2, is the most stable of the neptunium oxides. It crystallizes with the fluorite structure of all the actinide dioxides, with a crystalline density of 11.14 g/cm^3. It can be formed from the thermal decomposition of other neptunium compounds, such as the hydroxide, the nitrate, or the oxalate, in the temperature range of 600 to 1000°C. High-fired NpO_2 can be dissolved in hot concentrated nitric acid containing small amounts of fluoride.

The mixed oxide Np_3O_8 is structurally analogous to U_3O_8. Above 500°C it decomposes to NpO_2.

Neptunium halides. Neptunium forms binary halides in the oxidation states of Np (III, IV, and VI). The trifluoride is prepared by hydrofluorination of the dioxide, hydroxide, carbonate, oxalate, or nitrate in the presence of hydrogen:

$$NpO_2 + \tfrac{1}{2}H_2 + 3HF = NpF_3 + 2H_2O \tag{9.35}$$

The trifluoride can be carried to the tetrafluoride by hydrofluorination in the presence of oxygen:

$$NpF_3 + \tfrac{1}{4}O_2 + HF = NpF_4 + \tfrac{1}{2}H_2O \tag{9.36}$$

The above hydrofluorination reactions are carried out at 500°C [F4].

Fluorination of NpO_2 or NpF_4 by F_2, BrF_3, or BrF_5 at 300 to 500°C results in NpF_6 [M2], which has a triple point of 55.759°C and 758 Torr [04, W2]. Neptunium hexafluoride, like PuF_6, is decomposed in the presence of light [K2]. The hexafluoride is readily hydrolyzed by moisture to form the Np(VI) oxyfluoride NpO_2F_2, which can be reduced by hydrogen to form the Np(V) oxyfluoride $NpOF_3$.

3.4 Neptunium Solution Chemistry

In aqueous solutions neptunium exists in the five oxidation states Np(III), Np(IV), Np(V), Np(VI), and Np(VII), although the heptavalent Np(VII) is stable only in alkaline solutions. In the absence of complexing agents the first four oxidation states exist as Np^{3+}, Np^{4+}, NpO_2^+, and NpO_2^{2+}, usually in the hydrated form, whereas in strong alkaline solutions the heptavalent state is NpO_5^{3-} [K2].

Pentavalent neptunium is the most stable state in solution. It hydrolyzes only in basic solutions, disproportionates only at high acidity, and forms no polynuclear complexes. As shown by the oxidation-reduction potentials of Table 9.6, hexavalent neptunium is much less stable in solution than is hexavalent plutonium; in fact, hexavalent neptunium is a strong oxidizing agent and is easily reduced in the presence of oxidizable substances, such as those present in ion-exchange and solvent extraction separations [K2].

The data in Table 9.6 demonstrate that all oxidation states of neptunium are stable to disproportionation in 1 *M* $HClO_4$, in the absence of complexing agents. Only in very strong acids (pH $\leqslant -1$) does NpO_2^+ disproportionate to Np^{4+} and NpO_2^{2+} [A1]. The disproportionation of Np(V) is promoted by addition of complexing agents, because Np^{4+} and NpO_2^{2+} form more stable complexes than does NpO_2^+.

Trivalent neptunium is stable only in the absence of oxygen, being oxidized to Np(IV) in aqueous solutions exposed to air [A1].

Tetravalent neptunium forms strong complexes with anions, but Np(V) forms only weak complexes, evidently a result of the low charge of the neptunyl ion NpO_2^+ and its small size [K2]. This may account for the relatively low distribution coefficients for Np(V) in solvent extraction and in ion exchange. To adsorb neptunium onto anion exchange resins it is necessary to reduce neptunium to Np(IV). Pentavalent neptunium is only weakly adsorbed, and Np(VI) is reduced by most exchange resins to Np(V). Because of the relatively weak organic complexes of NpO_2^+, the distribution coefficient of its TBP complex is much lower than the distribution coefficients of the TBP complexes of Np^{4+} and NpO_2^{2+} ions.

4 PROPERTIES OF PLUTONIUM

4.1 Plutonium Isotopes

Table 9.13 lists the isotopes of plutonium important in nuclear technology and some of their important nuclear properties. Plutonium isotopes are produced in reactors by the nuclide chains shown in Fig. 8.5. Typical quantities and isotopic compositions of plutonium in various reactor fuel cycles are listed in Tables 8.4, 8.5, 8.6, and 8.7. In reactors fueled with uranium and plutonium, ^{239}Pu is the principal isotopic constituent, but ^{238}Pu contributes the greatest amount of alpha activity. With ^{235}U-thorium fueling, ^{238}Pu is the principal isotopic constituent.

^{236}Pu. The isotope ^{236}Pu results from the $(n, 2n)$ reaction in ^{237}Np, as shown in Fig. 8.5. It undergoes alpha decay, with a half-life of 2.85 years, to form ^{232}U, which subsequently decays to ^{228}Th,

Table 9.13 Isotopes of plutonium

Mass, amu	Radioactive decay: Half-life	Type†	Effective MeV	Fraction of decays	Reaction with 2200 m/s neutrons: Cross section, b (n, γ)	Fission	Neutrons per fission
236.04607	2.85 yr	α	5.868			165	
		SF		8×10^{-10}			2.22
238.049511	86 yr	α	5.592		547	16.5	2.90
		SF		1.7×10^{-9}			2.33
239.052146	24,400 yr	α	5.243		268.8	742.5	2.871
		SF		4.4×10^{-12}			
240.053882	6,580 yr	α	5.255		289.5		
		SF		4.7×10^{-8}			2.143
241.056737	13.2 yr	α	0.007	2.3×10^{-5}	368	1009	2.927
		β					
242.058725	3.79×10^{5} yr	α	4.98		18.5	<0.2	
		SF		5×10^{-6}			2.15
243.061972	4.98 h	β	0.239		60	196	
244.0641	8×10^{7} yr	α	4.66		1.7		
		SF		3×10^{-3}			2.30

†SF, spontaneous fission.

a member of the $4n$ decay series of natural thorium discussed in Chaps. 6 and 8 (cf. Fig. 6.1). Problems arising from the energetic gammas accompanying the decay of ^{228}Th daughters are discussed in Chap. 8.

^{238}Pu. The isotope ^{238}Pu is produced by neutron capture in ^{237}Np and the subsequent beta decay of ^{238}Np. It alpha decays with a half-life of 86 years to form ^{234}U, a daughter in the $4n + 2$ series of ^{238}U, discussed in Chap. 5. Essentially pure ^{238}Pu, containing about 1 ppm ^{236}Pu, is prepared by irradiation of ^{237}Np recovered from discharged reactor fuel [L2]. ^{238}Pu of even greater isotopic purity can be prepared by irradiating isolated ^{241}Am to form ^{242m}Am and ^{242}Cm, followed by chemical recovery of the ^{238}Pu decay daughter (see Fig. 8.5). ^{238}Pu has been used as a heat source in the Apollo program and for thermoelectric power devices employed in communication satellites and in heart pacemakers. Isotopically pure ^{238}Pu is desirable to reduce the radiation dose in biological applications. When used as an energy source, ^{238}Pu is formed into refractory compounds. Radiation dose from (α, n) reactions with light anions, as discussed in Chap. 8, can be reduced by forming ^{238}Pu into nitrides or by forming oxides from oxygen depleted in ^{18}O.

^{239}Pu. The isotope ^{239}Pu results from neutron capture in ^{238}U followed by two beta decays. It is the principal isotopic constituent of plutonium formed by the irradiation of low-enrichment uranium. It is the principal fissile constituent in plutonium fuel used in thermal and fast reactors. ^{239}Pu alpha decays, with a half-life of 24,400 years, to form the ^{235}U parent of the $4n + 3$ decay series discussed in Chap. 5. Relatively pure ^{239}Pu can be made by the short-term low-exposure irradiation of natural uranium. Plutonium containing more than 99 percent ^{239}Pu results from the irradiation of uranium at fuel exposures of less than 0.7 MWd/kg [K2]. Because of the high radiological toxicity, laboratory work on reactor plutonium must be carried out in airtight glove boxes.

^{240}Pu. The isotope ^{240}Pu is produced by neutron capture in ^{239}Pu. It is not fissionable by thermal neutrons, but, like all other plutonium isotopes, it fissions with fast neutrons. ^{240}Pu is converted to a fissionable nuclide by neutron capture. Therefore, like ^{232}Th and ^{238}U, it is a fertile material. It undergoes alpha decay, with a half-life of 6580 years, to form ^{236}U, which then decays to ^{232}Th, the parent of the $4n$ decay series discussed in Chaps. 6 and 8. Like the other even-mass plutonium isotopes, ^{240}Pu produces neutrons by spontaneous fission. It is present in greater concentration in reactor plutonium than any of the other even-mass plutonium isotopes.

^{241}Pu. The isotope ^{241}Pu results from neutron capture in ^{240}Pu. It is fissionable with thermal neutrons and contributes significantly to the energy production in uranium irradiated to high exposure and in recycled plutonium. It undergoes beta decay, with a half-life of 13.2 years, to form ^{241}Am, which then decays to ^{237}Np in the $4n + 1$ decay series. The decay of ^{241}Pu results in only low-energy electrons and weak x-rays. Alpha particles are formed in only 2.3×10^{-3} percent of the decays. However, the beta-decay daughter ^{241}Am emits gamma radiation when decaying, thereby adding to shielding requirements when working with separated reactor-grade plutonium.

^{242}Pu. The isotope ^{242}Pu is formed by neutron capture in ^{241}Pu. With a half-life of 3.79×10^5 years, it is the longest-lived of all the plutonium isotopes present in any appreciable amount in reactor-produced plutonium. It alpha decays to ^{238}U in the $4n + 2$ decay series. Because ^{242}Pu has a small neutron-absorption cross section relative to ^{239}Pu, ^{240}Pu, and ^{241}Pu, and because its neutron-capture daughter ^{243}Pu is relatively short lived, ^{242}Pu of high isotopic purity can be produced by the long irradiation of separated reactor plutonium. After a neutron-exposure fluence of 1.6×10^{22} thermal neutrons/cm^2, about 60 g of ^{242}Pu of approximately 99 percent isotopic purity is produced per kilogram of original reactor plutonium [K2]. Because of its long half-life and correspondingly lower radiotoxicity, ^{242}Pu is useful for laboratory chemical research.

^{243}Pu. The isotope ^{243}Pu, formed by neutron capture in ^{242}Pu, undergoes beta decay to ^{243}Am with a half-life of 4.98 h. Because of its short half-life, ^{243}Pu is present only in very small concentration during reactor irradiation, and it disappears after irradiated fuel has been stored for a few weeks. The low concentration of ^{243}Pu results in negligible production of the long-lived ^{244}Pu in reactors.

^{244}Pu. The isotope ^{244}Pu is the longest-lived of the plutonium isotopes, with a half-life of 8×10^7 years. It can be produced by neutron absorption in ^{243}Pu, but because of the short half-life and low concentration of ^{243}Pu only minute quantities of ^{244}Pu, of the order of 10^{-10} percent, are present in reactor-produced plutonium [K2]. Small quantities of ^{244}Pu, as well as ^{245}Pu and ^{246}Pu, are present in the residues from nuclear explosions, resulting from the decay of the neutron-rich uranium isotopes ^{244}U, ^{245}U, and ^{246}U formed by multiple neutron capture in the high neutron flux at the initiation of the explosion.

4.2 Plutonium Radioactivity

The radioactive decay properties of the plutonium isotopes that appear in irradiated reactor fuel are listed in Table 9.14. All but ^{241}Pu and ^{243}Pu are alpha emitters. Because it penetrates matter only weakly, alpha radiation is stopped by the outer layer of dead skin and is not a hazard outside the body. However, plutonium is very effective biologically when deposited in or on living tissue, particularly if by inhalation or by contaminated injuries. ^{241}Pu is a relatively short-lived (13.2-year

Table 9.14 Alpha and beta decay properties of plutonium

Nuclide	Half-life	Ci/g	Decay	Energy, MeV†	Yield, %†
^{236}Pu	2.85 yr	532	α	5.77	69
				5.72	31
^{238}Pu	86 yr	17.5	α	5.49	72
				5.45	28
				5.35	0.13
^{239}Pu	24,400 yr	0.0613	α	5.147	73.3
				5.134	15.1
				5.095	11.5
^{240}Pu	6,580 yr	0.226	α	5.159	75
				5.115	25
^{241}Pu	13.2 yr	112	β	0.022	99.99+
			α	4.896	83% of 0.0023%
				4.853	12% of 0.0023%
				4.797–5.055	5% of 0.0023%
^{242}Pu	3.79×10^5 yr	3.90×10^{-3}	α	4.903	76
				4.863	24
^{243}Pu	4.98 hr	2.60×10^6	β	0.579	62
				0.490	38

†Data for ^{236}Pu through ^{242}Pu from Valentine [V1]; data for ^{243}Pu from Keller [K2]. The energy listed for beta decay is the maximum beta energy.

half-life) beta emitter and is of radiological significance because it is the parent of ^{241}Am, an alpha emitter that accumulates in tissues and constitutes a hazard comparable to that of plutonium [B2].

Personnel working with plutonium must be protected by light shielding. The external radiation to be shielded includes gammas from alpha and beta decay, internal conversion x-rays, gammas, and neutrons from spontaneous fission, and neutrons from (α, n) reactions in materials of low atomic number. Neutron yields for various types and forms of plutonium are listed in Table 9.15.

Kilogram quantities of plutonium are fabricated in shielded glove-box facilities [V1]. A

Table 9.15 Neutron yields for plutonium

	Neutron yield, $n/(\text{g Pu}\cdot\text{s})$	
Type of plutonium	Metal†	Oxide‡
Low-exposure plutonium§	51	60
High-exposure reactor plutonium	340	538
^{238}Pu heat source	2,150	13,500

†From spontaneous fission.

‡From spontaneous fission and from (α, n) reactions.

§Plutonium with a relatively low content of ^{240}Pu, resulting from irradiation of ^{238}U at low burnup.

Source: A. Valentine, "Capabilities for Control of Plutonium in Processing," Plutonium Information Meeting, Jan. 1974.

typical box consists of a $\frac{1}{4}$ in of lead sandwiched between $\frac{3}{16}$ in of stainless steel on the interior and $\frac{1}{16}$ in on the exterior. Windows consist of $\frac{1}{4}$ in of lead glass. For neutron shielding 4 in of water, paraffin, or Plexiglas is added. Exhaust ventilation from the glove boxes passes through several layers of high-efficiency particulate filters to remove plutonium aerosols and to provide essentially complete containment of the plutonium being processed.

4.3 Plutonium Electronic Structure

The electronic structures of the ions are simpler than those of the metals. In the case of plutonium, removal of the first two (7*s*) electrons increases the stability of the 5*f* level relative to the 6*d* level, and the electrons become firmly placed in the 5*f* shell. After depletion of the 7*s* electrons, the next four electrons are removed from the 5*f* shell. A summary of the electronic structures of plutonium (in addition to the Rn core) is given in Table 9.16.

4.4 Plutonium Metal

The phases of plutonium metal and their transition temperatures at atmospheric pressure are shown in Table 9.17. The delta-prime phase exists only in high-purity plutonium; as little as 0.15 w/o (weight percent) impurities results instead in a continuous delta phase [C4]. Because of the high densities of the metallic states of uranium and plutonium, there is considerable incentive to use metal fuel in fast-breeder reactors to obtain high breeding ratio. However, the solid-phase transformations in uranium and plutonium and the susceptibility of these metals to radiation damage have resulted in greater emphasis on nonmetallic forms for high-burnup breeder fuel. The large density changes of plutonium metal, particularly between the alpha and beta plutonium, and the large thermal expansion coefficients, as shown in Table 9.17, can result in serious distortion and deformation of the fuel elements when subjected to internal stresses from repeated thermal cycling and from radiation damage. Solutions of molten uranium and plutonium have been considered as a fluid fuel for breeder reactors, but the extensive corrosion of structural materials by this molten metallic fuel is a formidable problem.

Plutonium metal is prepared by calcium reduction of plutonium fluorides or oxides in induction-heated MgO crucibles, under an inert atmosphere of helium or argon. The thermodynamics of plutonium reduction are discussed later in this chapter.

Plutonium metal oxidizes readily in the presence of humid air at elevated temperatures. The massive metal is relatively inert to atmospheric oxidation at room temperature, although the presence of water vapor causes unalloyed plutonium metal to disintegrate over long periods even with relatively little oxidation [K2]. The finely divided metal is pyrophoric. Plutonium reacts with halogens at moderate temperatures to form the trihalides. The metal is readily soluble in

Table 9.16 Electronic structures of plutonium ions

Plutonium valence	Shell				
	6*s*	6*p*	5*f*	6*d*	7*s*
0	2	6	6	0	2
	2	6	5	1	2
+3	2	6	5	0	0
+4	2	6	4	0	0
+5	2	6	3	0	0
+6	2	6	2	0	0

Table 9.17 Properties of plutonium metal

Phase	Temperature range,† °C	Crystal structure‡	Density,‡ g/cm³	Linear expansion coefficient,§ per °C (X 10⁶)
α	25–122	Simple monoclinic	19.86	59
β	122–205	Body-centered monoclinic	17.70	30.3
γ	205–318	Face-centered orthorhombic	17.14	33.3
δ	318–452	Face-centered cubic	15.92	−8.8
δ'	452–476	Body-centered tetragonal	16.00	−63
ϵ	476–640	Body-centered cubic	16.51	25.6
Liquid				

†From Rand [R2].
‡From Miner and Schonfeld [M5].
§Mean value over the indicated temperature range (25 to 122°C for α phase) [C4].

HCl of all concentrations and dissolves also in 72% $HClO_4$, 85% H_3PO_4, and concentrated trichloroacetic acid. Nitric acid shows no visible attack on massive plutonium over a period of several hours. Plutonium reacts slowly with H_2SO_4 of moderate concentrations, but with concentrated H_2SO_4 it forms a protective coating that resists further attack.

4.5 Plutonium Compounds

Plutonium oxides. The phase diagram of the plutonium-oxygen system is shown in Fig. 9.2. The observed compounds are the stoichiometric Pu_2O_3 and PuO_2 and the nonstoichiometric $PuO_{1.52}$ and $Pu_{1.61}$. PuO has also been shown to exist, but only under extreme conditions. No oxide of higher oxidation state than PuO_2 has been formed.

Plutonium dioxide is the form of plutonium most commonly specified for fuel for power reactors. It has the same general features already described for pure UO_2 fuel, such as high melting point, irradiation stability, compatability with metals and with reactor coolants, and ease of preparation. In most designs of plutonium-fueled power reactors the fuel is a mixture of uranium and plutonium oxides.

PuO_2 is formed when plutonium or its compounds, except the phosphates, are ignited in air. The most common starting materials are the nitrate or oxalate. Heating Pu(III) or Pu(IV) oxalate at 1000°C in air results in pure crystalline PuO_2. The physical appearance of the dioxide depends on its origin, ranging from yellow-black to green and from powder to shiny particles. The PuO_2 crystalline density is 11.46 g/cm³. The melting point varies from 2280°C in helium to about 2400°C in air [C1].

The only other binary oxide of plutonium of practical importance is the peroxide, which is the basis of a process for the purification of plutonium and its conversion to the metal. Addition of H_2O_2 to an aqueous plutonium solution first converts plutonium ions to the tetravalent state.

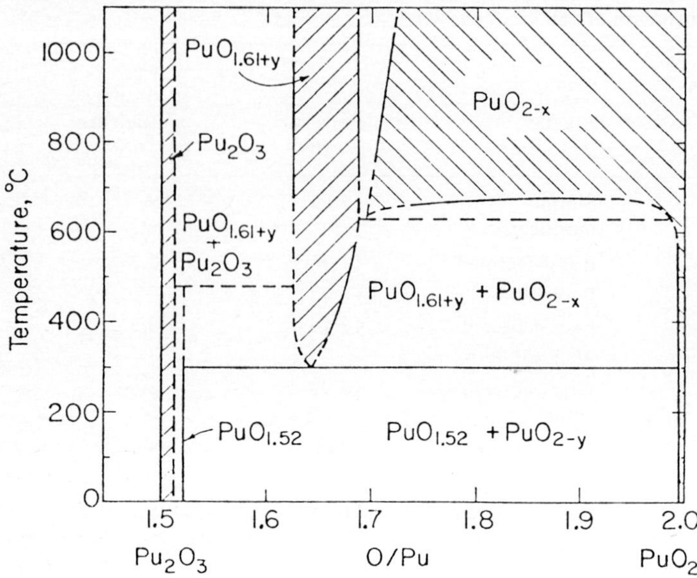

Figure 9.2 Phase diagram of the plutonium-oxygen system. (*From Mattys [M4] and Olander [O2], by permission.*)

Further addition of peroxide precipitates the plutonium peroxide complex, a nonstoichiometric compound whose composition and crystalline form depend on precipitating conditions. Manageable hexagonal precipitates are promoted by adding sulfate ions or by precipitating at acidities as high as 4.7 *M*. The dry peroxide is unstable, decomposing rapidly and sometimes explosively, especially when iron is present. Plutonium peroxide is a stable solid in acid of concentrations up to 5 *N*. Dry sulfate-free plutonium peroxide can be fluorinated directly at 600°C in HF containing small quantities of oxygen, yielding PuF_4 which can be readily reduced to the metal [C1, C2, M1].

Most of the designs for power-reactor fuel utilizing recycled plutonium involve the use of the mixed oxides of plutonium and natural or depleted uranium. The mixed-oxide fuel is formed either from mechanically mixed powders of the individual PuO_2 and UO_2 binary oxides or by calcining a coprecipitated uranium-plutonium compound. A portion of the phase diagram of the uranium-plutonium-oxygen system at 20°C, in the region of UO_2-U_3O_8-Pu_2O_3-PuO_2, is shown in Fig. 9.3 [I1, K2]. The phase boundaries deduced for the same region of the uranium-plutonium-oxygen system at 400, 600, and 800°C [I1] are shown in Fig. 9.4. The mixed uranium-plutonium oxides with the stoichiometric dioxide composition form a continuous solid solution from UO_2 to PuO_2, with the fcc fluorite structure, which is stable also at high temperature.

Oxidation of mixed oxides to overall oxygen-to-metal ratios greater than 2, and subsequent cooling to 20°C, results in a two-phase region $MO_{2+x} + M_4O_9$ up to the oxygen-to-metal ratio of 2.20 and up to a Pu/(U + Pu) ratio of 0.30. For overall oxygen-to-metal ratios of 2.20 to 2.27, a single phase M_4O_9 exists that is stable up to 1000°C. Oxidation of mixed uranium-plutonium oxides containing more than 39 percent plutonium results in the oxidation of uranium from U(IV) to U(V). An equimolar uranium-plutonium oxide forms a single phase of overall composition M_4O_9 when all uranium has been oxidized to U(V). At 1400°C a single fluorite phase exists for all plutonium concentrations and for oxygen-to-metal ratios up to 2.27.

The complete miscibility of the stoichiometric uranium-plutonium dioxide results in the simple liquidus-solidus melting-point curves of Fig. 9.5. The curves are consistent with ideal-

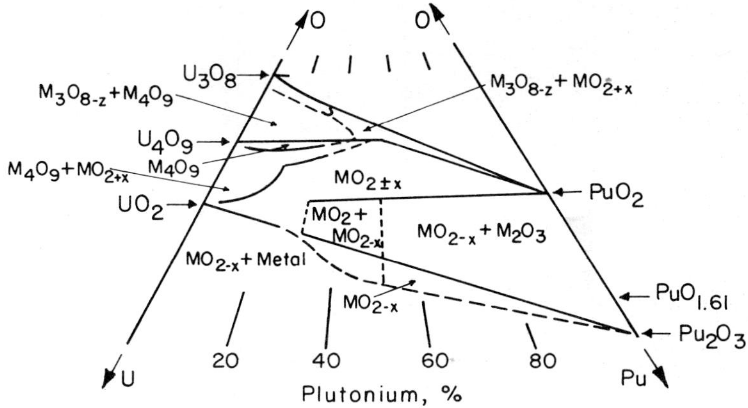

Figure 9.3 Phase diagram of the uranium-plutonium-oxygen system at 20°C [K2, I1].

solution theory for heats of fusion of 11.2 ± 1.3 kcal/mol for UO_2 and 16.8 ± 1.3 kcal/mol for PuO_2 [I1].

The preferred industrial process for manufacturing mixed-oxide uranium-plutonium fuel involves mechanical mixing of UO_2 and PuO_2 powder, followed by compaction and sintering above 1200°C. At the temperatures normally used in commercial sintering of the mixed oxides only a small portion of the sintered material contains the solid solution of UO_2-PuO_2. Even at temperatures as high as 1400 to 1750°C long sintering times are required for complete homogenization of the binary oxides.

It is important that the size of the remaining discrete particles of PuO_2 be small enough so that fission heat generated in the particles, particularly during rapid power transients, is not sufficient to locally overheat the PuO_2 particles. This requirement, which is most stringent for mixed-oxide fuel for fast-breeder reactors, is fulfilled by using PuO_2 powder with particle sizes of less than about 0.01 cm [B1].

Another important consideration is the problem of dissolving the mixed-oxide fuel for subsequent reprocessing and plutonium recovery after the irradiated mixed-oxide fuel has been discharged from the reactor. When plutonium dioxide is in solid solution with uranium dioxide at low concentrations, as in the case of plutonium created during the irradiation of uranium dioxide

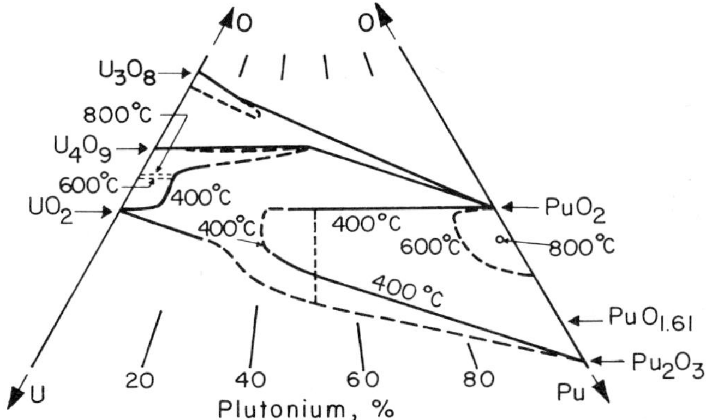

Figure 9.4 Phase diagram of the uranium-plutonium-oxygen system at 400, 600, and 800°C [I1].

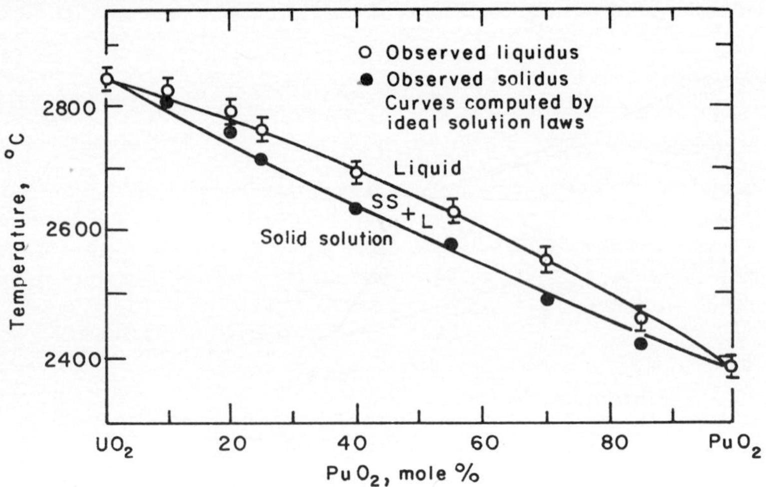

Figure 9.5 Solid-liquid phase diagram for the UO_2-PuO_2 system [I1].

fuel, the contained plutonium is soluble in the nitric acid normally used in fuel dissolution. However, pure PuO_2 is very difficult to dissolve in nitric acid, unless a fluoride catalyst is used, so problems with the dissolution of fuel containing crystallites of pure plutonia can be expected.

The fabrication and irradiation histories of uranium-plutonium mixed-oxide fuels strongly affect their solubility in nitric acid [L3, G3]. Fuel containing up to 28 percent PuO_2 can be dissolved in a few hours in boiling 6 to 10 M HNO_3, provided that all of the plutonium is in solid solution with the uranium. However, poorly fabricated fuels consist of very small islands of essentially pure plutonia in a matrix of UO_2-PuO_2 solid solution. Consequently, most of the fuel dissolves readily, leaving a refractory residue containing 1 to 10 percent of the plutonium. This residue is readily dissolved in 8 M HNO_3-0.05 M HF. Irradiation improves the solubility of poorly fabricated fuels, but it cannot be depended on to erase the solubility defect [O1, L3]. Fuels completely soluble in HNO_3 have been prepared successfully by methods of coprecipitation, sol-gel, and mechanical mixing of the separate oxides followed by sintering [G3].

Plutonium carbides. Carbides of plutonium and uranium are of interest as high-performance fuel for advanced breeder reactors. As compared with oxide fuel, the higher density of the monocarbide (U, Pu)C results in higher breeding ratio. Although the carbides cannot operate at as high a temperature as can the oxides, the much greater thermal conductivity of the carbides allows greater heat generation rates in the fuel. However, the technology for carbide fuel is not as far advanced as that for oxide fuel.

Plutonium carbides that have been prepared are PuC, PuC_2, and Pu_2C_3, formed by the reaction of graphite with metallic plutonium or PuH_3 at high temperature. The products are usually mixtures of PuC and Pu_2C_3. Plutonium oxide can also be reduced by carbon, but the stability of Pu(O, C) prevents the formation of plutonium carbide of high purity. PuC_2 exists only at temperatures above 1750°C. Plutonium monocarbide exists only as a substoichiometric compound, with a crystalline density of 13.58 g/cm^3, in the presence of excess carbon. It undergoes peritectic decomposition to the metallic liquid and Pu_2C_3 at 1654°C [K2].

For reactor fuel, the ternary uranium-plutonium-carbon monocarbide is prepared by reduction of (U, Pu)O_2 with graphite [F1], by melting a uranium-plutonium alloy with graphite, or by melting separately prepared individual carbides in an electric arc [K2]. Even though at low temperatures UC exists in the stoichiometric composition, the need for excess carbon for the existence of PuC limits the region of PuC-UC miscibility to a maximum of 35 a/o (atom percent) plutonium at room temperature. At higher plutonium concentrations the excess carbon is precipitated as Pu_2C_3.

The unirradiated mixed carbide (U, Pu)C readily hydrolyzes in water or acid, but neutron irradiation profoundly reduces the tendency toward hydrolysis.

Plutonium nitride. Unlike the corresponding uranium-nitrogen system, only the one plutonium nitride PuN exists. It is prepared by heating plutonium hydride in nitrogen at 250 to 400°C, by reacting plutonium metal with a hydrogen-ammonia mixture at 600°C, or by direction reaction of molten plutonium with nitrogen at 1000°C. Plutonium nitride forms solid solutions with UN. However, because of the appreciable volatility and dissociation of PuN at temperatures at about 1600°C and above, the ternary (U, Pu)N is less attractive as a nuclear fuel than pure UN [K2, S4].

Plutonium hydrides. Plutonium hydrides are made by reacting plutonium metal with hydrogen at atmospheric pressure and at temperatures between 50 and 300°C, forming a series of hydrides up to PuH_3. Plutonium hydride is a useful intermediary in the formation of other plutonium compounds.

Plutonium halides. Table 9.18 lists plutonium halides together with some of their more significant properties.

PuF_3 and PuF_4 are important intermediates in the production of plutonium metal. The trifluoride is made by reacting PuO_2 with a mixture of HF and H_2 at 600°C:

$$PuO_2 + 3HF + \tfrac{1}{2}H_2 \rightarrow PuF_3 + 2H_2O \tag{9.37}$$

The tetrafluoride is made by reacting oxide or oxalate in HF at 550°C, in the presence of oxygen to prevent reduction of tetravalent plutonium:

$$PuO_2 + 4HF \rightarrow PuF_4 + 2H_2O \tag{9.38}$$

The volatile plutonium hexafluoride can be prepared by fluorination of the tetrafluoride at 550°C:

$$PuF_4 + F_2 \rightarrow PuF_6 \tag{9.39}$$

Table 9.18 Properties of binary plutonium halides†

		Temperature, °C		
Compound	Color	Melts	Boils at 1 atm	X-ray crystal density at 25°C, g/cm^3
PuF_3	Purple-violet	1426		9.35
PuF_4	Light brown	1027		6.95
$PuF_4 \cdot 2.5H_2O$	Pink			4.87
PuF_6	Red-brown	51.59	62.16	4.97
$PuCl_3$	Green	767	1767	5.70
$PuCl_4$	(exists only as vapor)			
$PuBr_3$	Green	681	1463	6.75
$PuBr_3 \cdot 6H_2O$	Blue			3.47
PuI_3	Green	777		6.84

†Data from Cleveland [C1] and Rand [R2].

or by direct fluorination of the oxide. In contrast to UF_6, PuF_6 does not sublime when heated at atmospheric pressure. It first melts, at 51.59°C, and then boils at 62.16°C. Its triple point is 51.59°C and 533 Torr. The vapor pressure of PuF_6 above liquid PuF_6, in the range from 51.59 to 77.17°C, is given by [C1]

$$\log_{10} p(\text{Torr}) = -\frac{1807.5}{T(K)} - 1.5340 \log_{10} T + 12.14545 \tag{9.40}$$

Unlike the stable uranium hexafluoride, which has a negative free energy of formation, plutonium hexafluoride is thermodynamically unstable. It dissociates to F_2 and the relatively nonvolatile PuF_4, although the rate of thermal decomposition is very low at room temperature. If the specific alpha activity of plutonium is equivalent to that of ^{239}Pu, the rate of decomposition of solid PuF_6 at room temperature is controlled by radiolytic decomposition, amounting to 1.5 percent per day [W2].

Plutonium also forms the ternary oxyhalides PuOF, PuO_2F_2, PuOCl, PuOBr, and PuOI.

4.6 Plutonium Solution Chemistry

Oxidation states. In aqueous solution plutonium can exist in the five oxidation states Pu(III), Pu(IV), Pu(V), Pu(VI), and Pu(VII), which occur as hydrated ions Pu^{3+}, Pu^{4+}, PuO_2^{+}, PuO_2^{2+}, and PuO_2^{3+}. Although the most stable oxidation state in solution is usually considered to be Pu(IV), the differences in oxidation potentials of Pu(III) through Pu(VI) are small enough that all of the first four states can exist simultaneously in aqueous solution. Although Pu(V) is unstable to disproportionation, it is less so than pentavalent uranium. The Pu(IV) state also disproportionates to a measurable extent, in part due to radiolytic decomposition. Unless the oxidation potential is controlled, solutions that contain other than the pure Pu(III) or pure Pu(VI) will react to form stable mixtures containing appreciable concentrations of all four states through Pu(VI). Pu(IV) usually predominates in nitric acid solutions free of oxidizing and reducing substances. In Purex reprocessing nitrite ion is added to the nitric acid-plutonium solution to oxidize Pu(III) to Pu(IV) and to reduce Pu(VI) to Pu(IV).

The heptavalent state Pu(VII) is observed when Pu(VI) in a basic solution is oxidized by ozone, persulfate, or electrolytically. The resulting PuO_2^{3+} is highly unstable and readily reverts to Pu(VI) in a less oxidizing environment or in acidic solutions.

In the Purex aqueous separation process [C6] plutonium is maintained as Pu(IV) for decontamination from fission products and as Pu(III) for partition from uranium. The stability of these oxidation states in nitrate solutions depends on rate phenomena, because the potentials of the nitrate-reduction couples NO_3^--$NO(g)$ and NO_3^--$N_2O_4(g)$ are near those of the plutonium-oxidation couples Pu(III)-Pu(IV) and Pu(IV)-Pu(VI). The rate is in part determined by the concentration of the nitrate-reduction products $NO(g)$ and $N_2O_4(g)$ and their rate of removal from solution. Experimental data summarized by Seaborg [S1], show that Pu(IV) in nitric acid solutions at room temperature undergoes essentially no disproportionation or oxidation for several days. Addition of sulfuric acid to the solution further represses the oxidation. Data taken at temperatures of 75 to 100°C indicate no detectable oxidation of Pu(IV) in 14 M HNO_3, a half-life for Pu(IV) of about 4 h in 2 M HNO_3, and a half-life of only about 10 min for 0.25 M HNO_3. The high acid concentration represses the equilibrium of the hydrolytic oxidation of Pu(IV) to $Pu^{VI}O_2^{2+}$.

The disproportionation of Pu(IV) proceeds by means of two reactions, first the slow,

$$2Pu^{4+} + 2H_2O \rightarrow Pu^{V}O_2^{+} + Pu^{3+} + 4H^{+} \tag{9.41}$$

followed by rapid establishment of the equilibrium

$$Pu^{V}O_2^{+} + Pu^{4+} \rightleftharpoons Pu^{VI}O_2^{2+} + Pu^{3+} \tag{9.42}$$

The calculated equilibrium percentages of plutonium in the various oxidation states in 0.5 *M* HCl at 25°C, assuming an average oxidation state of IV and assuming no complexing, are listed in Table 9.19.

The rate of the Pu(IV) disproportionation reaction (9.41) is proportional to the square of the concentration of Pu(IV), with a rate constant [S1] of 0.75 liter/(mol·h) at 25°C in 0.481 *M* $HClO_4$. The rate is faster the lower the acid concentration and the higher the temperature.

The practical consequences of these findings is that plutonium may be stabilized in the Pu^{4+} state by keeping the solution at 25°C, by having a high HNO_3 concentration to retard reaction (9.41) and to complex the Pu^{4+} ion, and by keeping the solution dilute in plutonium.

Seaborg [S1] states that the rate of oxidation of Pu(III) to Pu(IV) is slow in dilute nitric acid at room temperature but proceeds rapidly in dilute nitric acid at 100°C and in concentrated nitric acid (16 *M*) at room temperature. The suggested mechanism is the oxidation of Pu(III) by nitrous acid HNO_2 formed by the reaction of dissolved NO with NO_3^- or with NO_2.

Another complication in plutonium solution is the gradual, spontaneous reduction of Pu(VI) to Pu(IV), and Pu(IV) to Pu(III), caused by ionization products of alpha particles emitted in radioactive decay [S1]. The rate of alpha reduction is slow, however. For example, the observed rate of reduction of Pu(VI) in 0.5 *M* HCl at 25°C is 0.0035 g-equiv/day per mole of plutonium, which corresponds to a half-life of 199 days for reduction of Pu(VI) to Pu(IV). From these rates and the known alpha-decay rate and decay energies of plutonium, it is estimated that approximately 80 eV of dissipated alpha energy in this solution brings about the addition of one electron in reducing plutonium ions. After several hundred days the plutonium reaches an average oxidation state intermediate between Pu(III) and Pu(IV).

For Pu(IV) in 0.481 *M* $HClO_4$ at 25°C, the rate of spontaneous reduction is 0.00106 g-equiv/day per mole of plutonium, corresponding to a half-life of 653 days for reduction of Pu^{4+} to Pu^{3+}. To keep plutonium in the hexavalent and tetravalent states over long periods of time, it is necessary to have an oxidant present to reoxidize lower valence states as fast as they form.

Trivalent plutonium. Solutions of trivalent plutonium salts are generally similar to the trivalent rare-earths. Like the rare-earths, the hydroxide, fluoride, oxalate, and phosphate are insoluble. Plutonium forms double sulfates with alkalis of the form $MPu(SO_4)_2 \cdot 4H_2O$, again like the rare-earths. In the absence of air, aqueous solutions of trivalent plutonium salts are stable against hydrolysis; they are readily oxidized by air to the tetravalent form.

Tetravalent plutonium. Solutions of tetravalent plutonium salts are generally similar to tetravalent cerium and uranium. The fluoride PuF_4, potassium complex fluoride K_2PuF_6, iodate $Pu(IO_3)_4$, and phosphate $Pu_3(PO_4)_4$ are insoluble. Excess soluble hydroxides precipitate $Pu(OH)_4$. The

Table 9.19 Relative amounts of plutonium oxidation states

	Percentage of total plutonium†
Pu(III)	27.2
Pu(IV)	58.4
Pu(V)	0.8
Pu(VI)	13.6
	100.0

†Average oxidation state = Pu(IV). In 0.5 *M* HCl at 25°C, no complexing.

hydroxide is easily converted to PuO_2 by heating. When hydrogen peroxide is added to acid solutions of Pu^{3+}, Pu^{4+}, $PuO_2{}^{+}$, and $PuO_2{}^{2+}$, a Pu(IV) peroxide is precipitated. The precipitate composition is variable, such as $PuO_{4-x}(NO_3)_{2x}\cdot 2\text{–}3H_2O$. Tetravalent plutonium forms chelate compounds with thenoyl trifluoracetone (TTA) or acetylacetone, which may be extracted from aqueous solution into benzene. Tetravelent plutonium nitrate is the form of plutonium most readily extractable by TBP.

Pentavalent plutonium. Pentavalent plutonium salts have only limited stability in aqueous solution. At pH above 1, they begin to hydrolyze, and at lower pHs they tend to disproportionate to $PuO_2{}^{2+}$ and Pu^{4+} or Pu^{3+}. All the common salts are soluble.

Hexavalent plutonium. $PuO_2{}^{2+}$ in acid solution is a much stronger oxidizing agent than $UO_2{}^{2+}$. The two ions also differ in that the solubilities of plutonium are greater than those of the corresponding uranium compounds. In most other respects the two ions are similar. Plutonyl nitrate is very soluble in water and is extracted by methyl isobutyl ketone and other oxygenated organic solvents. Soluble hydroxides precipitate plutonates, such as Na_2PuO_4. These dissolve in sodium carbonate solution as complex carbonates. Plutonyl phosphate, arsenate, and double sodium acetate, $NaPuO_2(CH_3CO_2)_3$, are relatively insoluble.

Plutonium complexes. Plutonium ions form complexes with many anions. The most important of the complexes are those that form with Pu^{4+}, some of which are listed in Table 9.20, in order of increasing stability.

At sufficiently high concentrations of HNO_3 or HCl, e.g., > 2.5 *M* HCl, plutonium forms anionic complexes that are strongly sorbed by anion-exchange resins. Because the complexing ability to form anions varies with the plutonium oxidation state, which can be preferentially adjusted with respect to the other actinides, anion exchange is useful in the separation of plutonium from other actinide elements and in the separation from cationic impurities that do not easily complex. Because of its high ionic potential, plutonium is also readily adsorbed onto cation-exchange resins. Elution of sorbed Pu(III) from such resins by means of dilute nitric acid, or of sorbed Pu(IV) by a complexing acid such as HCl, is a means of concentrating plutonium in solution.

Plutonium readily complexes with organic complexing agents, such as TBP, according to the overall reactions

$$Pu(NO_3)_4(aq) + 2TBP(o) \rightleftharpoons Pu(NO_3)_4\cdot 2TBP(o) \tag{9.43}$$

Table 9.20 Complex formation constants of Pu^{4+}

Reaction	Equilibrium constant	Ionic strength
$Pu^{4+} + 3Cl^- \rightleftharpoons PuCl_3{}^+$	2.10	1
$Pu^{4+} + NO_3{}^- \rightleftharpoons PuNO_3{}^{3+}$	2.9	>2
$Pu(SO_4)_2 + HSO_4{}^- \rightleftharpoons Pu(SO_4)_3{}^{2-} + H^+$	5.0	2.33
$Pu(C_2O_4)_2 + H_2C_2O_4 \rightleftharpoons Pu(C_2O_4)_3{}^{2-} + 2H^+$	2.51×10^1	>0.75
$Pu^{4+} + HSO_4{}^- \rightleftharpoons PuSO_4{}^{2+} + H^+$	8.5×10^2	2.33
$PuC_2O_4{}^{2+} + H_2C_2O_4 \rightleftharpoons Pu(C_2O_4)_2 + 2H^+$	9.65×10^2	>0.75
$PuSO_4{}^{2+} + HSO_4{}^- \rightleftharpoons Pu(SO_4)_2 + H^+$	1.1×10^3	2.33
$Pu^{4+} + HF \rightleftharpoons PuF^{3+} + H^+$	4.25×10^3	1

Source: S. Peterson and R. G. Wymer, *Chemistry in Nuclear Technology*, Addison-Wesley, Reading, Mass., with permission.

and $$PuO_2(NO_3)_2(aq) + 2TBP(o) \rightleftharpoons PuO_2(NO_3)_2 \cdot 2TBP(o) \tag{9.44}$$

Weaker complexes are formed by Pu(III) and Pu(V). The use of these extractable complexes in fuel reprocessing to separate uranium and plutonium from fission products and to partition plutonium from uranium is discussed in Chap. 10.

Hydrolysis. Hydrolysis is one of the most important reactions in the chemistry of plutonium in aqueous solutions. The tendency of plutonium ions to hydrolyze decreases in the order:

$$Pu^{4+} > PuO_2^{2+} > Pu^{3+} > PuO_2^{+}$$

As would be expected from the relative sizes of ions listed in Table 9.3, the plutonium ions in any oxidation state are more readily hydrolyzed than their larger neptunium and uranium analogues.

The hydrolysis of Pu^{3+} is not extensive, forming the hydrolyzed species $PuOH^{2+}$ according to the reaction

$$Pu^{3+} + H_2O \rightleftharpoons PuOH^{2+} + H^+ \tag{9.45}$$

In solutions of lower acidity $Pu(OH)_3$ precipitates, with a solubility product of about 2×10^{-20} [C3].

Tetravalent Pu^{4+} hydrolyzes more readily than any other plutonium species. In hydrogen ion concentrations of less than 0.3 *M*, the hydrolysis is initiated by the reversible reaction

$$Pu^{4+} + H_2O \rightleftharpoons PuOH^{3+} + H^+ \tag{9.46}$$

In Pu^{4+} solutions of low acidity the highly insoluble hydroxide $Pu(OH)_4$ precipitates, with an estimated solubility product of 7×10^{-56} [C3].

The hydrolysis of Pu(V) is very slight, because of the low charge on the $Pu^VO_2^+$ ion. The hydrolyzed Pu(V) rapidly disproportionates. Hydrolyzed monomers and polymers of Pu(VI) are also formed.

Polymers. In weakly acidic solutions the reversible hydrolysis may be followed by an irreversible formation of a colloidal product polymerized to a high molecular weight, quite similar to the hydrolytic behavior of Th^{4+} and U^{4+}. These polymers consist of very small, discrete, amorphous particles, which invariably convert to the crystalline form when aged [L4]. The rate of formation of plutonium polymers is greater at high temperatures. Once formed, these plutonium polymers do not readily disperse or dissolve in solutions of acidities sufficiently high to have prevented their formation. The hydrolysis and polymerization of Pu(IV) is suppressed in a sufficiently acid solution and in the presence of some complexing agents. The Pu(IV) polymer is destroyed only slowly by highly concentrated acid at room temperature, but it is destroyed rapidly at 90°C. Although the presence of fluoride ion appears to have little effect on the formation of Pu(IV) polymers, its presence does accelerate the depolymerization. The plutonium polymers cannot be extracted by organic complexing agents, such as TBP. The polymers can also form in the organic solvents.

The tendency toward Pu(IV) polymerization is of considerable practical importance in process operations involving plutonium solutions. Dilution of an acidic plutonium solution with water can result in polymerization in localized regions of low acidity, so plutonium solutions should be diluted instead with acid solutions. Polymerization can result from leaks of steam or water into plutonium solutions or by overheating during evaporation. Polymer formation can clog transfer lines, interfere with ion-exchange separations, cause emulsification in solvent extraction and excessive foaming in evaporation, and can result in localized accumulation of plutonium that may create a criticality hazard [C3].

The hydrolytic chemistry of Pu^{4+} is important in that it affects the behavior and mobility of plutonium in the environment [A2] and in geologically isolated radioactive wastes that may be subjected to slow leaching by ground water. The absorption spectra of the Pu(IV) polymer is similar to that of the plutonium hydroxide precipitate $Pu(OH)_4$ [L4]. Experimental data in Fig.

9.6 show that the solubility of $Pu(OH)_4$ as a function of acidity (pH) is quite similar to the plutonium concentrations at which Pu(IV) polymer formation has been detected [R1]. The comparison suggests that if the total plutonium concentration at any given pH falls above the line of $Pu(OH)_4$ solubility, the solution will be saturated with respect to $Pu(OH)_4$ or to the Pu(IV) polymer, so that these species can form. If the plutonium concentration falls below the $Pu(OH)_4$ solubility line, the Pu(IV) polymer or precipitated $Pu(OH)_4$ will be absent. In the environmental pH range of 4 to 8 the concentration of soluble and unpolymerized Pu(IV) is so low that the environmental chemistry may be governed by the possible oxidation or reduction to other valence states of plutonium, as well as by the presence of complexing agents.

The controlled formation of polymeric Pu(IV) is important to the sol-gel process for the production of spherical PuO_2 particles. Ammonia is added to a nitric acid solution of Pu(IV) to precipitate $Pu(OH)_4$, which is subsequently peptized at 50°C with dilute nitric acid to produce sols of 1 to 3 *M* Pu with a NO_3^-/Pu ratio of 0.1 to 0.3. The sols, which remain stable over periods of a few months, are dispersed in a dehydrating organic solvent to form a gel, which is ignited to form spherical particles of PuO_2 [K2].

4.7 Plutonium Conversion

Plutonium nitrate solution from fuel reprocessing is to be converted either to plutonium dioxide for fuel fabrication or it may be converted to intermediate compounds suitable for reduction to plutonium metal. Direct thermal decomposition of the nitrate solution to plutonium dioxide is possible, but it requires very pure feed solutions. Industrial-scale operations usually begin with the precipitation of plutonium peroxide or plutonium oxalate, which result in further decontamination of impurities. Routes to the production of plutonium metal may involve hydrofluorination of plutonium dioxide, peroxide, or oxalate to form the anhydrous tetrafluoride for metallothermic reduction. Alternatively, anhydrous plutonium halides may be formed by precipitation of PuF_3 or $CaF_2 \cdot PuF_4$, either of which can be reduced directly to the metal, or PuF_3 may be converted to PuF_4 or PuF_4-PuO_2 for subsequent reduction. Preparation of $PuCl_3$ from calcined PuO_2 is another alternative.

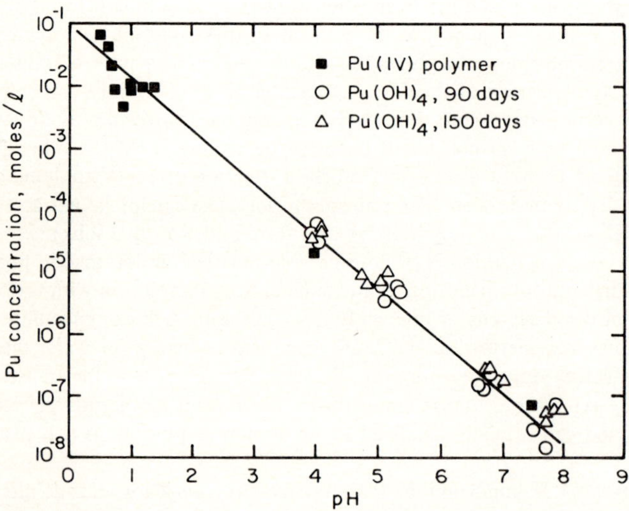

Figure 9.6 Solubility of $Pu(OH)_4$ and Pu(IV) polymer as a function of acidity. (*From Rai and Serne [R1].*)

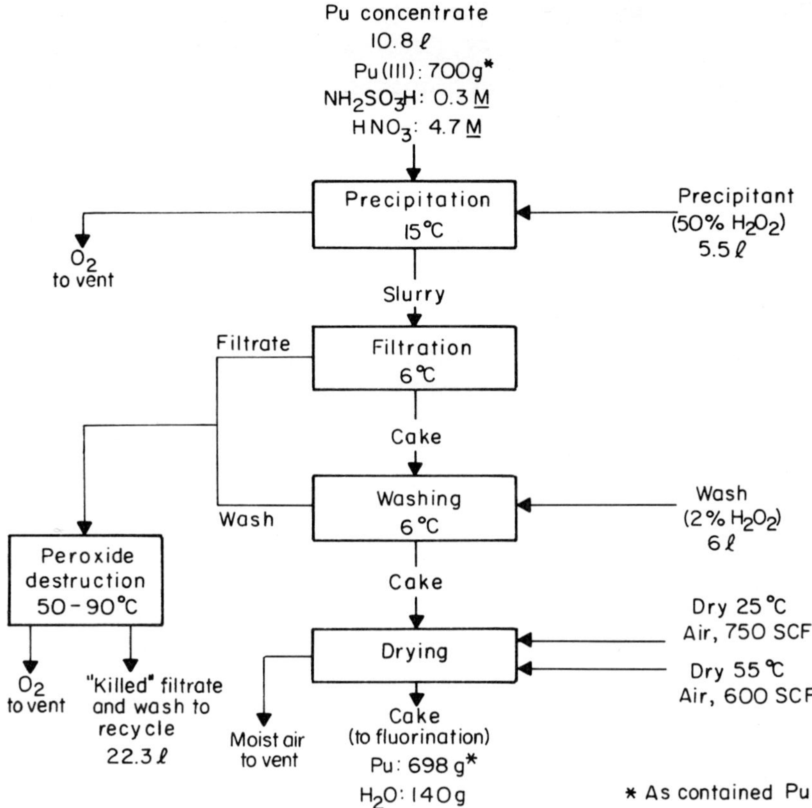

Figure 9.7 Flow sheet for the precipitation of plutonium peroxide. (*From Mainland et al. [M1] and Cleveland [C2], by permission.*)

The important industrial-scale conversion operations are described below, followed in Sec. 4.8 by the description of processes to produce plutonium metal.

Plutonium peroxide. Plutonium nitrate solutions containing plutonium of any oxidation state can be used as the feed solution for peroxide precipitation, because plutonium in aqueous solutions is converted to the tetravalent state by hydrogen peroxide. A flow sheet of the peroxide precipitation process is shown in Fig. 9.7 [M1, C2]. A solution of 30 to 50 percent H_2O_2 is added to the plutonium nitrate-nitric acid solution slowly to promote precipitation of large and easily filtered hexagonal crystals. About three peroxide oxygen atoms are required per plutonium atom. A low temperature, about 30°C or less, is desirable to reduce peroxide decomposition. Peroxide precipitation equipment used at the U.S. Savannah River plant is refrigerated so that precipitation can occur at 15°C, followed by digestion at 6°C, to minimize decomposition of H_2O_2 by impurities in the feed [M1]. The filtered plutonium peroxide cake can be calcined at 150°C to form plutonium oxide, although a final temperature of 900°C is required to produce stoichiometric PuO_2. Alternatively, the peroxide cake can be dried at 25 to 55°C and fluorinated with HF to form PuF_4, for subsequent reduction to the metal.

Peroxide precipitation achieves excellent decontamination from cationic impurities, because there are so few metals that form peroxide precipitates. However, special care must be taken in the preparation and handling of solid and dissolved peroxides, which can decompose explosively, especially in the presence of iron and other impurities that catalyze decomposition.

Plutonium(IV) oxalate. The tetravalent plutonium oxalate can be precipitated from a nitric acid solution of Pu(IV) nitrate according to the flow sheet of Fig. 9.8 [H3]. Hydrogen peroxide is added for valence adjustment to Pu(IV), either before or during the addition of oxalic acid. Best precipitations occur at a temperature of 50 to 60°C and for time periods of oxalic acid addition in the region of 10 to 60 min. Higher temperatures and more rapid addition of oxalic acid result in finely divided and gummy precipitates. The nitric acid concentration must be adjusted so that the final oxalate slurry is in a solution between 1.5 and 4.5 *M* HNO_3. At lower acid concentrations coprecipitation of impurities is favored and the precipitate is finely divided, whereas at higher acid concentrations the solubility of Pu(IV) oxalate is unsuitably high and the precipitate is thixotropic. The oxalate cake is washed and then calcined at 300°C, followed by restructuring at 900°C to form stoichiometric PuO_2. If the final product is to be plutonium metal, the oxalate can be fluorinated directly with HF and oxygen to form PuF_4 for subsequent reduction.

The Pu(IV) oxalate process achieves decontamination factors of about 3 to 6 for zirconium-niobium, 12 for ruthenium, 60 for uranium, and 100 for aluminum-chromium-nickel. As compared with peroxide precipitation, the oxalate process achieves less decontamination from impurities, but the solutions and solids are more stable and safer to handle. It is more suitable for processing solutions containing high concentrations of impurities that would catalyze peroxide decomposition.

Plutonium(III) oxalate. If the plutonium solution from fuel reprocessing is concentrated by sorption on a cation-exchange resin, the plutonium eluent will be a nitrate solution of stabilized trivalent plutonium. This solution may be a logical candidate for the relatively simple precipitation of Pu(III) oxalate. The Pu(III) oxalate $Pu_2(C_2O_4)_3 \cdot 9H_2O$ can be easily precipitated by adding oxalic acid, either as a solution or a solid. Precipitation conditions are not critical, and the Pu(III)

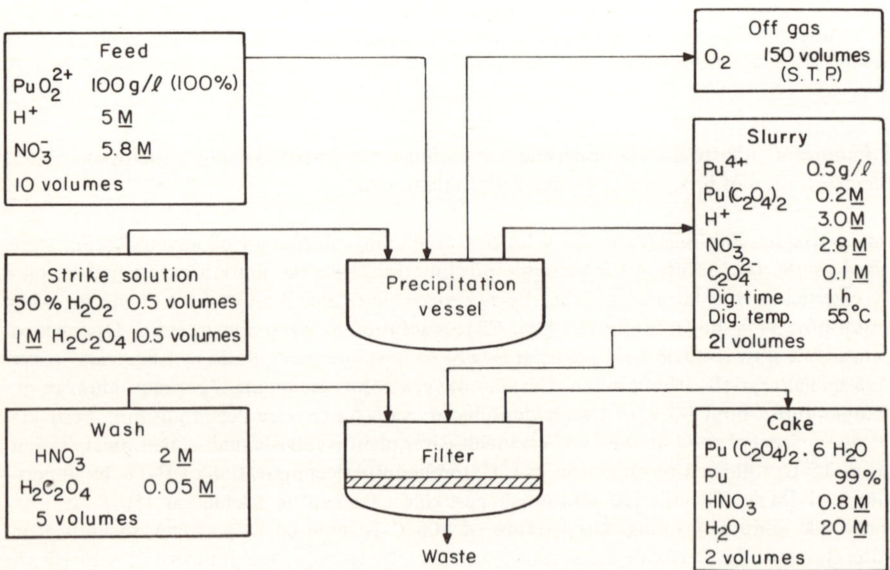

Figure 9.8 Flow sheet for the precipitation of Pu(IV) oxalate. (*From Cleveland [C2], by permission.*)

oxalate precipitate settles rapidly and is easy to filter. The oxalic acid can be added rapidly, with a digestion period of about a half hour.

If the Pu(III) nitrate is not stabilized against oxidation, hydriodic acid, hydroxylamine or ascorbic and sulfamic acid may be used as a reducing agent. Problems of handling and corrosion result with HI. If appreciable Pu(IV) were present, more stringent care would have to be taken to avoid unmanageable precipitates of Pu(IV) oxalate, as discussed above.

The Pu(III) oxalate is calcined to form the dioxide or hydrofluorinated with HF and oxygen to form PuF_4, as has been described above for the tetravalent oxalate.

When process solutions of Pu(III) are available, Pu(III) oxalate precipitation may be desirable if impurity levels are too high for the PuF_3-precipitation process described later.

PuO_2 from direct calcination of $Pu(NO_3)_4$. The precipitation steps of the above processes can be avoided by the direct calcination of the plutonium nitrate solution to PuO_2. Calcination has been carried out at 350°C in a liquid-phase screw calciner. Half a mole of ammonium sulfate per mole of plutonium is added to the feed solution to increase the production of reactive PuO_2. The calcination time and temperature must be low enough to minimize sintering, which would otherwise reduce the chemical reactivity of the oxide particles for subsequent conversion to a halide.

Direct calcination of $Pu(NO_3)_4$ involves no chemical separations that could remove impurities, so a highly pure plutonium nitrate feed solution is required. The plutonium dioxide product can be hydrofluorinated to PuF_4, or it can be used as a feed for the formation of $PuCl_3$. Direct calcination has received less industrial-scale application than the precipitation processes described above [C2].

Plutonium trifluoride. Plutonium trifluoride can be converted directly to plutonium metal, or it is an intermediate in the formation of PuF_4 or PuF_4-PuO_2 mixtures for thermochemical reduction, as described in Sec. 4.8. The stabilized Pu(III) solution, produced by cation exchange in one of the Purex process options for fuel reprocessing, is a natural feed for the formation of plutonium trifluoride, as is shown in the flow sheet of Fig. 9.9 [O3]. A typical eluent solution from cation exchange consists of 30 to 70 g plutonium/liter, 4 to 5 M nitric acid, 0.2 M sulfamic acid, and 0.3 M hydroxylamine nitrate. The sulfamic acid reacts rapidly with nitrous acid to reduce the rate of oxidation of Pu(III) to about 4 to 6 percent per day. Addition of ascorbic acid to the plutonium solution just before fluoride precipitation reduces Pu(IV) rapidly and completely to Pu(III).

Addition of 2.7 to 4 M hydrofluoric acid results in the precipitation reaction

$$Pu(NO_3)_3(aq) + 3HF(aq) \rightleftharpoons PuF_3(s) + 3HNO_3(aq) \tag{9.47}$$

with a solubility product for PuF_3 of $2.4 \pm 0.4 \times 10^{-16}$. Easily filterable precipitates result from controlled rate of addition of the reagents and by maintaining a HNO_3/HF ratio of at least 4. In contrast to PuF_4, the PuF_3 precipitate is crystalline and contains no water of crystallization, so that it is easily dried to the anhydrous salt desirable for metallothermic reduction to plutonium metal. The filtered trifluoride cake is washed with 0.8 M HF and dried by ambient air. Anhydrous PuF_3 is produced by further drying with warm air, followed by heating to 600°C in argon to remove remaining volatile impurities [B6].

The PuF_3 process does not attain the degree of decontamination from cationic impurities that can be achieved in peroxide or oxalate precipitation, but it is acceptable when the plutonium nitrate feed contains no more than a few hundred parts of uranium and aluminum per million of plutonium. This process has been in routine use at the U.S. plant at Savannah River [B6, O3].

Plutonium tetrafluoride. Precipitation of PuF_4 by adding hydrofluoric acid to a Pu(IV) nitrate solution is impractical because the hydrated precipitate $PuF_4 \cdot 2.5H_2O$ is amorphous and difficult to filter, and it is difficult to dehydrate to the anhydrous material necessary for subsequent reduc-

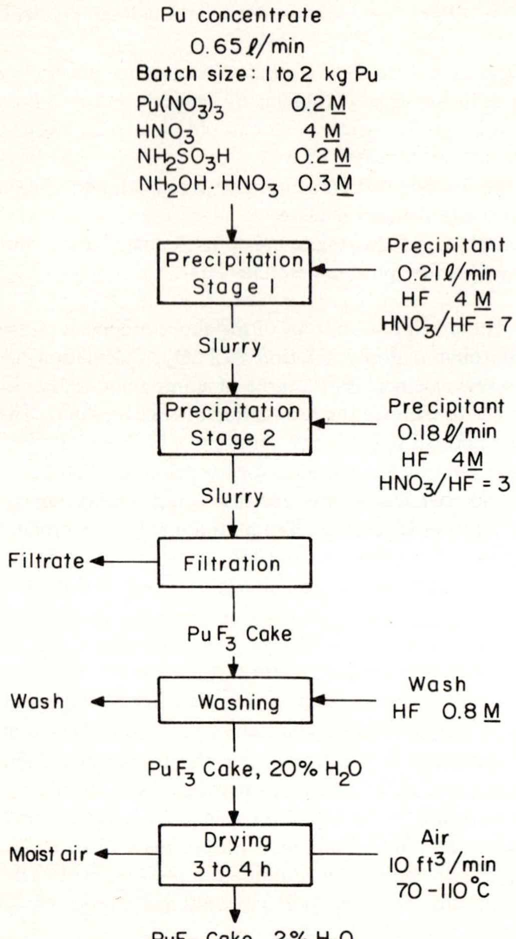

Figure 9.9 Flow sheet for the precipitation of PuF_3 [O3].

tion to the metal. Vacuum dehydration of the precipitate at 200°C yields $PuF_4 \cdot H_2O$, and further heating results in the trifluoride [C1]. When heated in a moist atmosphere above 300°C, PuF_4 hydrolyzes to PuO_2 [K2]. Therefore, the more difficult process of hydrofluorination of a solid is necessary to obtain anhydrous PuF_4.

If the plutonium to be fluorinated is the plutonium peroxide cake, as in one of the processes used at the U.S. Savannah River plant, the air-dried cake is reacted with HF gas at 600°C. The reaction time is quite sensitive to sulfate containment in the oxalate cake, which interferes with fluorination and requires a longer time for reaction of the oxalate with HF. The interfering sulfate is that present due to a sulfuric acid wash of the cation-exchange resin prior to peroxide precipitation.

Alternatively, PuF_4 can be formed from Pu(III) or Pu(IV) oxalate cake by drying the cake in air at 100 to 120°C and fluorinating in HF at 400 to 600°C, or the dried oxalate can be calcined at 130 to 300°C in air to form PuO_2, which is then hydrofluorinated in HF to form PuF_4. The hydrofluorination temperature must equal or exceed the calcination temperature, and the latter must be kept below 480°C to prevent formation of refractory oxide [M1]. Similarly, PuF_4 can be prepared by hydrofluorinating PuO_2 or PuF_3.

With the availability of anhydrous plutonium trifluoride, as discussed in the previous section, the equipment problems associated with the direct conversion of PuF_3 to PuF_4 with HF and oxygen can be avoided by roasting the trifluoride in oxygen to form a mixture of PuF_4 and PuO_2, according to the reaction

$$4PuF_3 + O_2 \rightarrow 3PuF_4 + PuO_2 \tag{9.48}$$

The PuF_4-PuO_2 mixture is suitable for metallothermic reduction, as discussed in Sec. 4.8.

Air-dried PuF_3 cake is roasted in an inert atmosphere at 150 to 200°C for 1.5 to 3 h, and then in an oxygen atmosphere at 400 to 600°C. This is one of the processes that has been employed at the U.S. plant at Savannah River [C2, O3].

The $CaF_2 \cdot PuF_4$ process. A process for forming plutonium tetrafluoride, without the attendant corrosion problems of dry hydrofluorination, involves the precipitation of the double salt $CaF_2 \cdot PuF_4$ from a solution of $Pu(NO_3)_4$. By contrast to the $PuF_4 \cdot 2.5H_2O$ precipitate, the $CaF_2 \cdot PuF_4$ precipitate is less soluble, is readily dried, and may be directly reduced to the metal.

A flow sheet for the precipitation of $CaF_2 \cdot PuF_4$ is shown in Fig. 9.10 [C2]. A solution of plutonium and calcium nitrates in 4 to 5 *M* HNO_3 is added to HF solution of 6 *M* or less. The precipitate is washed with dilute HNO_3-HF solution and dried at 300°C in argon or nitrogen to form the anhydrous $CaF_2 \cdot PuF_4$, which must be crushed to particles suitable for reduction to the metal.

The process is attractive in its simplicity. However, in the subsequent metallothermic reduction the CaF_2 diluent absorbs a portion of the heat of reaction otherwise needed for slag melting. Also, there is less decontamination from impurities than in the case of the other precipitation processes described earlier.

Plutonium trichloride. Although $PuCl_3$ is more hygroscopic than the plutonium fluorides, and although it generates less heat of reaction in subsequent metallothermic reduction to the metal,

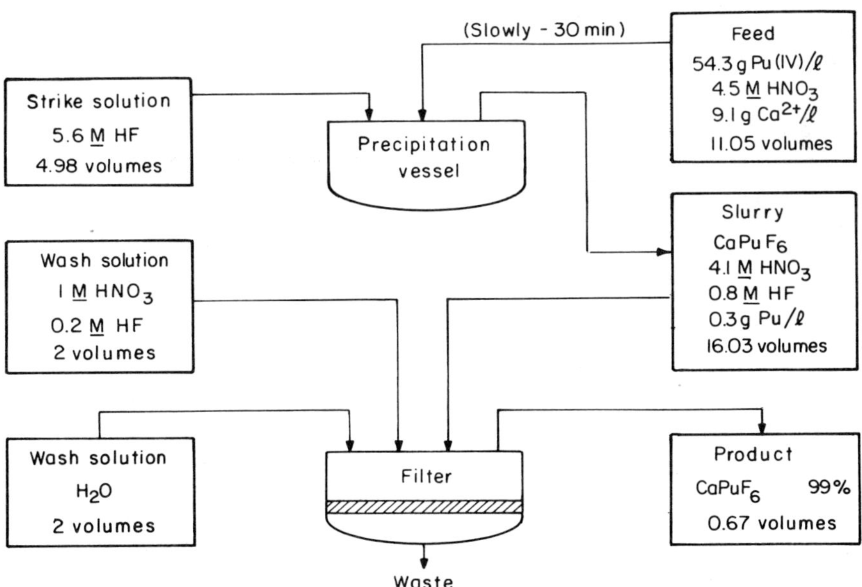

Figure 9.10 Flow sheet for the precipitation of $CaF_2 \cdot PuF_4$. (*From Cleveland [C2], by permission.*)

the production of $PuCl_3$ is motivated by the reduced shielding requirements. Because of the relatively weak reaction of plutonium alphas with chlorine to produce neutrons, the neutron emission of $PuCl_3$ is one-sixty-fourth that of PuF_4 [C2]. Also, the slag from $PuCl_3$ reduction melts at a much lower temperature than the fluoride slags (see Sec. 4.8).

Plutonium dioxide, prepared by direct calcination of the nitrate or calcination of the peroxide or oxalate precipitates, can be chlorinated to $PuCl_3$ by HCl-H_2, gaseous CCl_4, or phosgene ($COCl_2$), the latter resulting in the most rapid reaction.

Chlorination of nitrate-calcined oxide has been carried out in a fluidized bed at 500°C. Oxide from oxalate calcination has been chlorinated in a continuous screw calciner at 250 to 350°C. Because many impurities form volatile chlorides under these conditions, relatively good decontamination from impurities results. Consequently, this is a logical conversion step to follow the direct calcination of $Pu(NO_3)_4$.

It is essential that $PuCl_3$ be handled only in a very dry atmosphere, otherwise hygroscopic moisture accumulation can result in excessive pressures during subsequent reduction to the metal.

4.8 Production of Plutonium Metal

As in the case of uranium metal, production of pure plutonium metal presents many difficulties. It forms very stable compounds with oxygen and carbon, it oxidizes rapidly in air when in the form of powder, it cannot be deposited electrolytically from aqueous solution, and it boils at too high a temperature to be purified by distillation. Additionally, the extreme radiotoxicity of plutonium and neutron production from (α, n) reactions require that production operations be carried out in airtight and shielded enclosures. Nuclear criticality limits the amount of plutonium produced in any production operation to batch sizes of no more than a few kilograms.

Methods that have been used to produce plutonium metal are

1. Reduction of PuF_4
2. Reduction of PuF_3
3. Reduction of PuF_4-PuO_2
4. Reduction of PuO_2
5. Reduction of the double salt $CaF_2 \cdot PuF_4$
6. Electrolysis of fused salts
7. Thermal decomposition of plutonium hydride

Thermochemical reduction. Elements that might be considered for reducing PuO_2 or the plutonium halides are hydrogen, sodium, or calcium. Carbon is impractical because of carbide formation, and magnesium or aluminum are undesirable because they form intermetallic compounds with plutonium. Excess reductant must be used to ensure high yields, and it is important that the remaining reductant not dissolve in or react with the molten plutonium. Lithium and potassium could also be considered, although the volatility of potassium and the chemical reactivity of lithium would make handling difficult.

The free-energy changes in reducing PuO_2, PuF_3, PuF_4, or $PuCl_3$ by hydrogen, sodium, or calcium are shown in Table 9.21. Magnesium is included in the table because it can be vacuum distilled from the plutonium product, and its thermochemistry is relevant to the use of MgO as a crucible in the reduction operation. The free-energy data are evaluated at 1500 K (1227°C), which is high enough for rapid reactions, is above the melting point of plutonium 640°C, and is within the temperature range required to obtain molten slags formed by reduction of the plutonium halides. A molten slag is important in aiding the coalescence of the molten plutonium metal that is produced. The melting points of the slags produced are indicated by the melting temperatures of the reductant compounds shown in Table 9.22. For reduction to be complete without requiring a large excess in reductant, the free-energy change at 1500 K should be more negative

Table 9.21 Free-energy change in production of plutonium metal

	Plutonium compounds			
	PuO_2	PuF_3	PuF_4	$PuCl_3$
Free energy of formation at 1500 K, kcal/g-mol†	−190	−288	−319	−156
Free-energy change for plutonium-reduction reaction at 1500 K,‡ kcal/g-mol plutonium				
Reductant: H	+111	+87.3	+51.4	+81.6
Na	+10.9	+0.17	−64.8	−31.4
Mg	−12.4	−17.4	−88.2	+15.0
Ca	−38.1	−61.0	−146	−52.5

†From data of Brown [B4] and Rand [R2].

‡Free energies of formation of reductant compounds are listed in Table 5.29, from reference [N2].

than about 10 kcal/g-mol of plutonium. The positive free-energy changes for hydrogen reduction, shown in Table 9.21, indicate that hydrogen is impractical. Sodium results in a suitably negative free-energy change with PuF_4 or $PuCl_3$, and magnesium is thermodynamically suitable for the reduction of PuO_2, PuF_3, and PuF_4. Calcium is the only feasible reductant for all four plutonium compounds, including PuO_2 and PuF_3. Negative free-energy changes also result for the calcium reduction of $PuBr_3$ or PuI_3, but these plutonium salts are both hygroscopic and corrosive [C3].

Because of the small scale and correspondingly high heat loss in plutonium reduction, it is important that the reaction be sufficiently exothermic so that the reacting mixture will self-heat to the temperature range of 1500 K, after the reaction is initiated at lower temperature by an external heat source. Data on heats of formation and available heat of reaction are given in Table 9.22. The greatest heat of reaction, 161 kcal/g-mol of plutonium, results from the calcium reduction of PuF_4. PuF_4 is also easy to reduce because it is not hygroscopic, so it does not absorb moisture which could create excessive pressures during reduction operations.

Even the highly exothermic calcium reduction of PuF_4 may be insufficient to result in slag temperatures above the CaF_2 melting temperature of 1418°C, particularly in smaller-scale reductions because of the high heat loss. To furnish additional heat, elemental iodine is added to the reactants as a booster, to react with excess calcium. The heat of formation of the resultant CaI_2 at 298 K is −143.4 kcal/g-mol [B3]. No iodine booster is needed for larger-scale reductions of PuF_4, in the range of about 1 kg of plutonium.

The thermodynamically favorable reduction of PuO_2 with calcium has the disadvantage that the CaO coproduct is not molten, so that the resulting plutonium metal and unreacted calcium metal remain finely dispersed throughout the slag. However, the dispersed plutonium can be recovered as a massive metal by preferentially extracting the calcium oxide and unreacted calcium with molten calcium chloride at a temperature above the melting points of plutonium and calcium, leaving consolidated plutonium metal with yield efficiencies in excess of 99.9 percent [W1].

Electrolytic processes. Because of the positive oxidation potential for plutonium metal to displace hydrogen from aqueous solution, as shown in Table 9.7, nonaqueous solutions such as fused salts must be used for the electrodeposition of plutonium metal. One process involves the electrolysis of a molten equimolar mixture of LiCl-KCl containing 30 w/o $PuCl_3$. The melt is contained in a $MgO\text{-}TiO_2$ crucible heated to 950°C, with an anode through which chlorine gas can be introduced

Table 9.22 Thermodynamic data for metallothermic reduction of plutonium†

Metal		Pu	Na	Mg	Ca
Melting point, K		913	371	922	1112
Boiling point, K		3460	1156	1378	1767
Fluoride	PuF_3	PuF_4	NaF	MgF_2	CaF_2
Melting point, K	1699	1300	996	1536	1691
Boiling point, K	2300	1550	1787	2499	2806
Heat of formation at 298 K, kcal/g-mol	−371	−425	−137.52	−268.7	−293.0
Available heat at 298 K, kcal/g-mol of Pu:					
$PuF_3 + \frac{3}{x} M \rightarrow Pu + \frac{3}{x} MF_x$			41.6	32.0	68.5
$PuF_4 + \frac{4}{x} M \rightarrow Pu + \frac{4}{x} MF_x$			125	112	161
Chloride	$PuCl_3$	$PuCl_4$	NaCl	$MgCl_2$	$CaCl_2$
Melting point, K	1040	Exists only in gaseous state	1074	987	1045
Boiling point, K	2000		1738	1710	2209
Heat of formation at 298 K, kcal/g-mol	−229.8	−189.7‡	−98.26	−153.35	−190.2
Available heat at 298 K, kcal/g-mol of Pu:					
$PuCl_3 + \frac{3}{x} M \rightarrow Pu + \frac{3}{x} MCl_x$			65.0	0.23	55.5
Oxide		PuO_2	Na_2O	MgO	CaO
Melting point, K		2553	1558§	3073	2888
Heat of formation at 298 K, kcal/g-mol		−252.8	−62.0	−143.84	−151.7
Available heat at 298 K, kcal/g-mol of Pu:					
$PuO_2 + 2xM \rightarrow Pu + 2M_xO$			−128.8	34.9	50.6

†From references [B4, G1, N2, N3, R2].
‡Gaseous state.
§Sublimes.

to chlorinate any PuO_2 impurity. Fresh $PuCl_3$ containing as much as 15 percent PuO_2 is added continuously, with the molten plutonium product drawn off continuously at the cathode at demonstrated rates of 300 to 400 g/h. Another process demonstrated on a laboratory scale involves $PuCl_3$ in a mixture of 28 percent $BaCl_2$ and 42 percent KCl at 800°C, also in a $MgO\text{-}TiO_2$ crucible [C2].

Electrodeposition as a means of reducing plutonium to the metal has been limited by the corrosiveness of the chlorine and fused chloride environment. It also provides relatively little separation from impurities, except for those elements that form volatile chlorides [C2].

5 PROPERTIES OF AMERICIUM

5.1 Americium Isotopes

Table 9.23 lists the isotopes of americium important in nuclear technology and some of their important nuclear properties.

^{241}Am. The isotope ^{241}Am is formed by the decay of ^{241}Pu. It undergoes alpha decay, with a half-life of 458 years, to form ^{237}Np. Isotopically pure ^{241}Am can be extracted from aged reactor-grade plutonium. Irradiation of separated ^{241}Am is the basis of technology to produce gram quantities of ^{242}Cm. This is also one route to the production of the transcurium elements. However, the first neutron-capture products are ^{242}Am and ^{242m}Am, which have appreciable fission cross sections and which result in considerable heat evolution as contrasted to the production of transcurium isotopes from the irradiation of plutonium rich in the isotope ^{242}Pu.

^{241}Am will grow with time in plutonium recycled for fabricating reactor fuel, as discussed in Chap. 8. The gamma radiation accompanying the decay of ^{241}Am will contribute external radiation and may require personnel protection.

^{241}Am is a source of alpha particles in neutron sources for laboratory experiments and for reactor start-up. It is alloyed with beryllium to form $AmBe_{13}$, which produces high-energy neutrons by (α, n) reactions.

^{242}Am. The isotope ^{242}Am is the 16.0-h beta-EC emitter formed in 83.8 percent of the neutron captures in ^{241}Am.

^{242m}Am. The isotope ^{242m}Am is the long-lived (152-year) isomeric state that results from 16.2 percent of the neutron captures in ^{241}Am. It is one of the sources of the $4n + 2$ decay series in high-level wastes from reprocessing irradiated uranium or uranium-plutonium fuel.

^{243}Am. The isotope ^{243}Am is the 7950-year alpha emitter resulting from neutron capture in ^{242}Pu and, to a less extent, from neutron capture in ^{242m}Am. It is an intermediate in the production of

Table 9.23 Isotopes of americium

	Radioactive decay				Reaction with 2200 m/s neutrons		
					Cross section, b		Neutrons per fission
Mass, amu	Half-life	Type†	Effective MeV	Fraction of decays	(n, γ)	Fission	
241.05674	458 yr	α	5.640		832	3.15	3.219
		SF		2.2×10^{-12}			
242.059502	16.0 h	β	0.225	0.84	–	2900	–
		EC		0.16			
242.____‡	152 yr	γ	0.075		1400	6600	3.264
		SF		1.6×10^{-10}			
243.061367	7950 yr	α	5.439		79.3	–	–
		SF		2.3×10^{-10}			
244.064355	10.1 h	β	1.256		–	2300	–

†SF, spontaneous fission.
‡^{242m}Am.

transcurium elements by the long-term irradiation of plutonium. In the U.S. process ^{239}Pu is converted to ^{242}Pu, ^{243}Am, and ^{244}Cm by extensive irradiation in neutron fluxes of 3 to 5 $\times$ 10^{14} $n/(cm^2 \cdot s)$ for about 18 months, resulting in a yield of about 60 g ^{242}Pu, 30 g of ^{243}Am and ^{244}Cm, and 910 g of fission products per kilogram of initial ^{239}Pu. The remaining actinides are extracted from the fission products and are then further irradiated in the high-flux isotope reactor (HFIR), which was especially constructed at the Oak Ridge National Laboratory to produce transcurium elements. The overall production is about 0.2 g of ^{252}Cf per year, with a yield of about 0.1 to 0.3 percent of the original ^{239}Pu [K2].

^{243}Am is also important as a source of ^{239}Pu in the high-level wastes from fuel reprocessing.

^{244}Am. The isotope ^{244}Am is the 10.1-h beta emitter formed by neutron capture in ^{243}Am. It is an intermediate in the nuclide chain leading to ^{244}Cm and thence to the transcurium elements.

5.2 Metallic Americium

The properties of metallic americium are listed in Table 9.24.

Alternative processes for preparing metallic americium are the reduction of AmF_3 with barium vapor in high vacuum at about 1300°C, reduction of AmF_4 with calcium, and reduction of AmO_2 with lanthanum or thorium at about 1500°C in high vacuum. The vapor pressure of americium is much higher than that of lanthanum or thorium, so that pure americium is condensed in the colder parts of the apparatus [K2, L2]. Metallic americium dissolves readily in mineral acids.

5.3 Americium Compounds

Americium oxides. Keller [K2, K3] reports three stoichiometric binary oxides of americium: AmO, Am_2O_3, AmO_2. The dioxide AmO_2 is the most stable of the americium oxides. It crystallizes with the cubic fluorite structure of all the actinide dioxides. It can be formed as a dark brown powder, stable up to 1000°C, by heating trivalent americium nitrate, hydroxide, or oxalate in oxygen to 700 to 800°C. Americium dioxide is readily soluble in mineral acids. Hydrogen reduction of the dioxide yields Am_2O_3.

Americium monoxide AmO is observed as a surface layer on americium metal if oxygen is present during preparation.

Americium halides. Americium forms binary halides in the oxidation states of III and IV. The trifluoride is prepared by hydrofluorination of AmO_2 with HF at 400 to 500°C and by precipita-

Table 9.24 Phases of americium metal

Transition temperature, °C	Phase	Crystal system	Density, g/cm^3
	Solid	Hexagonal double close packed	13.671
1079			
	Solid	Face-centered cubic	
1176			
	Liquid		
2600†			
	Vapor		

†Extrapolated.

Source: C. Keller, *The Chemistry of the Transuranium Elements*, Verlag Chemie, Weinheim, 1971.

tion from aqueous solutions. Fluorination of AmF_3 or AmO_2 with F_2 at 400 to 500°C yields AmF_4. Similar compounds are formed with chlorine, bromine, and iodine.

5.4 Americium Solution Chemistry

In aqueous solution americium exists in the four oxidation states Am(III), Am(IV), Am(V), and Am(VI). In the absence of complexing agents trivalent, pentavalent, and hexavalent americium exist as Am^{3+}, AmO_2^{+}, and AmO_2^{2+}, usually in hydrated form. In aqueous solution tetravalent americium rapidly disproportionates, except in concentrated fluoride and phosphate solutions.

Trivalent americium is the most stable state in solution. As shown by the oxidation-reduction potentials of Table 9.6, the higher oxidation states of americium are strong oxidizing agents, so they exist only in solutions that contain no oxidizable species.

Trivalent americium forms relatively unstable complexes with Cl^- and NO_3^- and more stable complexes with the thiocyanate ion CNS^-. These americium complexes are more stable than those of the corresponding lanthanide compounds, so that americium can be separated from trivalent lanthanides by anion exchange with concentrated solutions of LiCl, $LiNO_3$, or NH_4CNS. Trivalent americium can be extracted with TBP from a concentrated nitrate solution. It can also be extracted with TBP from a molten $LiNO_3$-KNO_3 eutectic at 150°C, with much higher distribution coefficients than in extraction from aqueous solutions. Americium is more readily extracted by this process than is trivalent curium [K2].

6 PROPERTIES OF CURIUM

6.1 Curium Isotopes

Table 9.25 lists the isotopes of curium important in nuclear technology and some of their important nuclear properties.

^{242}Cm. The isotope ^{242}Cm is the largest contributor to the alpha activity of irradiated uranium fuel from power reactors. It is an important source of the $2n + 2$ decay chain in the high-level wastes from fuel reprocessing. The alpha activity of ^{242}Cm results in an internal heat-generation rate of 120 W/g of pure ^{242}Cm. Separated ^{242}Cm, prepared by the neutron irradiation of ^{241}Am, provides a useful alternative for a thermoelectric source and for radionuclide batteries when relatively high outputs are desired over short periods of the order of its half-life of 163 days. For example, a space power generator denoted as SNAP-11 contained 7.5 g of ^{242}Cm and produced 20 W of thermoelectric power. ^{242}Cm is also the decay source of ^{238}Pu, which is used as a longer-lived radioisotope heat source.

To optimize isotopic purity when producing ^{242}Cm it is desirable that the irradiation of the ^{241}Am target be carried out at low neutron flux. At higher fluxes the increased chain branching by fission of ^{242}Am and the increased neutron capture in ^{242}Cm result in greater contamination by ^{243}Cm per unit amount of ^{242}Cm produced. The actual production rate of ^{242}Cm optimizes at a neutron flux of about 8×10^{14} $n/(\text{cm}^2 \cdot \text{s})$. At higher fluxes the increased chain branching from ^{242}Am fission more than offsets the increased rate of neutron absorption in ^{241}Am [K2].

^{243}Cm. The isotope ^{243}Cm is an alpha emitter with a half-life of 32 years. About 0.3 percent of its decays occur by beta emission, and the accompanying gammas contribute to shielding problems when ^{242}Cm heat sources are contaminated with ^{243}Cm.

^{244}Cm. The isotope ^{244}Cm, an alpha emitter with a half-life of 17.6 years, is useful as a longer-lived decay-heat source and as a source for radionuclide batteries. Its specific heat-generation

Table 9.25 Isotopes of curium

Mass, amu	Radioactive decay: Half-life	Type†	Effective MeV	Fraction of decays	Reaction with 2200 m/s neutrons: Cross section, b: (n, γ)	Fission	Neutrons per fission
242.058788	163 days	α	6.217				
		SF		6×10^{-8}	16		
243.06137	32 yr	α	6.15		225	600	3.430
244.062821	17.6 yr	α	5.902		13.9	1.2	
		SF		1.3×10^{-6}			
245.065371	9.3×10^3 yr	α	5.624		345	2020	3.832
246.067202	5.5×10^3 yr	α	5.476		1.3	0.7	
		SF		3×10^{-4}			
247.07028	1.6×10^7 yr	α	5.3		60	90	
248.0722	350 days	α	21.41	0.89	4	0.34	
		SF		0.11			3.157
249.07581	64 min	β	0.3		1.6		

†SF, spontaneous fission.

rate due to alpha decay is 2.8 W/g ^{244}Cm. It is also an intermediate in the chain to produce transcurium elements by neutron irradiation beginning with ^{239}Pu, as has already been discussed in Sec. 5.1 under ^{243}Am. Repeated irradiation of the ^{242}Pu and ^{243}Am and extraction of ^{244}Cm yields as much as 80 g ^{244}Cm per kilogram of initial ^{239}Pu [F2]. Further irradiation of the ^{244}Cm yields the higher-mass curium isotopes and the transcurium elements.

^{244}Cm in high-level wastes from reprocessing irradiated fuel from power reactors is the principal long-term source of ^{240}Pu in these wastes.

^{245}Cm. The isotope ^{245}Cm, with a half-life of 9300 years, is the long-lived alpha emitter resulting from neutron capture in ^{244}Cm. It is an intermediate in the production of transcurium elements by neutron irradiation. It has a high cross section (2020 b) for fission by thermal neutrons, which accounts for most of the chain branching in this process. ^{245}Cm and its two higher-mass neighbors ^{246}Cm and ^{247}Cm are desirable for studies of the chemistry of curium because of their long half-lives. They must be isolated from other curium isotopes by mass separation, and consequently they are available only in small quantities.

^{246}Cm. The isotope ^{246}Cm, with a half-life of 5500 years, is a member of the long-lived group of alpha-emitting curium isotopes. It is produced by neutron capture in ^{245}Cm.

^{247}Cm. The isotope ^{247}Cm, with a half-life of 1.6×10^7 years, is the third member of the long-lived group of curium isotopes. It is produced by neutron capture in ^{246}Cm.

^{248}Cm. The isotope ^{248}Cm, with a half-life of 350 days, is the highest-mass curium isotope produced in appreciable quantities in the irradiation of ^{244}Cm. Very pure ^{248}Cm is now being produced by the alpha decay of ^{252}Cf, which is the principal transcurium isotope produced in the long-term neutron irradiation of plutonium, americium, curium, and berkelium. ^{252}Cf decays with a half-life of 2.65 years, 3 percent by spontaneous fission and 97 percent by alpha emission.

^{249}Cm. The isotope ^{249}Cm is the 64-min beta emitter that terminates the curium chain formed by neutron irradiation. It decays to ^{249}Bk.

6.2 Metallic Curium

The properties of metallic curium are listed in Table 9.27. Microgram quantities of metallic curium have been prepared by reducing curium trifluoride CmF_3 with barium vapor in a vacuum chamber at 1315 to 1375°C, using a tantalum crucible, followed by heating at 1235°C to volatilize excess barium and BaF_2 slag [C8]. Larger quantities have been produced by suspending CmO_2 in a $MgCl_2$-MgF_2 melt and adding magnesium, in the form of a magnesium-zinc alloy, to reduce the curium. Unreacted magnesium and zinc are removed by evaporation [E1].

6.3 Curium Compounds

Curium oxides. The stoichiometric binary oxides of curium are Cm_2O_3 and CmO_2. The dioxide CmO_2 is not stable above about 350°C, changing into nonstoichiometric phases with $O/Cm < 2$. The sesquioxide Cm_2O_3 is stable up to its melting point of 2002°C. A black product near the composition of CmO_2 is formed when the sesquioxide Cm_2O_3 or aqueous compounds such as $Cm(OH)_3$ or $Cm_2(C_2O_4)_3$ are heated in air or oxygen at 500 to 600°C. The sesquioxide Cm_2O_3 is formed by the decomposition of CmO_2 at 600 to 700°C in high vacuum, by reduction of CmO_2 with hydrogen, or by the decomposition of the dioxide in air at higher temperatures. The stable form of curium sesquioxide at room temperature is the hexagonal lattice.

Curium halides. Like americium, curium forms binary halides in the oxidation states of III and IV. The trifluoride can be precipitated by adding fluoride ions to a solution of Cm(III), drying in vacuum over P_2O_5, and reducing in HF gas at 600°C. The tetrafluoride CmF_4 is prepared by heating anhydrous CmF_3 in elementary fluorine at 400°C. The trichloride $CmCl_3$ is very hygroscopic [K2]. The tribromide and triiodide are also formed.

6.4 Curium Solution Chemistry

In aqueous solution curium exists in the oxidation states Cm(III) and Cm(IV). Solutions of Cm(IV) can be prepared only by dissolving CmF_4 in a solution containing a high concentration of fluoride to form the stabilized complexes CmF_5^- and/or CmF_6^{2-}. As shown by the oxidation-reduction potentials of Table 9.6, Cm(IV) is a strong oxidizing agent. In the absence of sufficient complexing for stability, it is rapidly reduced to Cm(III) under the influence of the alpha activity from curium decay. Trivalent curium forms complexes with Cl^-, NO_3^-, SO_4^{2-}, and $C_2O_4^{2-}$ which are less stable than those with americium, but the CNS^- complex of curium is more stable than that of americium. Organocomplexes and chelates are also formed [K2].

^{242}Cm is recovered from irradiated AmO_2/Al cermets by dissolution of the aluminum in hot NaOH, followed by dissolution in 6 *M* HCl. The curium and americium are separated from fission products by anion exchange from 11 *M* LiCl and elution with 12 *M* HCl, and the curium is then separated from the americium by selective elution from a cation-exchange column with lactic acid solution [H4].

^{244}Cm is recovered from irradiated Pu/Al alloys and $AmO_2(PuO_2)$/Al cermets by dissolution, extraction of plutonium with TBP in *n*-dodecane, extraction of americium and curium from the aqueous raffinate with 50 percent TBP in kerosene, purification of the americium and curium fraction by extraction with tertiary amines, and separation of americium by precipitation of the double carbonate $K_5AmO_2(CO_3)_3$ [G5]. A high-pressure ion-exchange system for the separation of curium, americium, and rare earths from feed solutions of dilute HNO_3 has been applied at the Savannah River Laboratory [H1, L5].

REFERENCES

A1. Ahrland, S., J. O. Liljenzin, and J. Rydberg: "Solution Chemistry," in *Comprehensive Inorganic Chemistry*, vol. 5, J. C. Bailar, Jr., H. J. Emeléus, R. Nyholm, and A. F. Trotman-Dickenson (eds.), Pergamon, Oxford, 1973, pp. 465–635.

A2. Andelman, J. B., and T. C. Rozzell: "Plutonium in the Water Environment," in *Radionuclides in the Environment*, Advances in Chemistry Series 93, American Chemical Society, Washington, D.C., 1970, pp. 118–137.

B1. Bailey, H. S., P. C. Vaughan, and G. L. Gyorey: *Trans. Amer. Nucl. Soc.* **12**, no. 2: 703 (1969).

B2. Bair, W. J., and R. C. Thompson: "Plutonium Biomedical Research," *Science* **183**: 715 (1974).

B3. Brewer, L., L. A. Bromley, P. W. Gilles, and N. F. Lofgren: "The Thermodynamic Properties of the Halides," in *The Chemistry and Metallurgy of Miscellaneous Materials*, L. Quill (ed.), National Nuclear Energy Series, div. IV, vol. 19B, McGraw-Hill, New York, 1950.

B4. Brown, D.: "Halides, Halates, Perhalates, Thiocyanates, Selenocyanates, Cyanates and Cyanides," in *Comprehensive Inorganic Chemistry*, vol. 5, J. C. Bailar, Jr., H. J. Emeléus, R. Nyholm, and A. F. Trotman-Dickenson (eds.), Pergamon, Oxford, 1973, p. 151.

B5. Bruce, F. R.: "Solvent Extraction Chemistry of the Fission Products," *PICG (1)* **9**: 459 (1956).

B6. Burney, G. A., and F. W. Tober: *I&EC Process Design and Development* **4**: 28 (1965).

C1. Cleveland, J. M.: "Compounds of Plutonium," in *Plutonium Handbook*, vol. 1, O. J. Wick (ed.), Gordon & Breach, New York, 1967.

C2. Cleveland, J. M.: "Plutonium Conversion Processes," in *Plutonium Handbook*, vol. 2, O. J. Wick (ed.), Gordon & Breach, New York, 1967.

C3. Cleveland, J. M.: "Solution Chemistry of Plutonium," in *Plutonium Handbook*, vol. 1, O. J. Wick (ed.), Gordon & Breach, New York, 1967, p. 403.

C4. Coffinberry, A. S., and W. F. Miner: *The Metal Plutonium*, University of Chicago Press, Chicago, 1961.

C5. Coffinberry, A. S., F. W. Schonfeld, J. T. Waber, L. R. Kehman, and C. R. Tipton, Jr.: "Plutonium and Its Alloys," in *Reactor Handbook*, 2d ed., vol. 1, Interscience, New York, 1960, p. 248.

C6. Culler, F. L.: "Reprocessing of Reactor Fuel and Blanket Materials by Solvent Extraction," *PICG (1)* **9**: 464 (1956).

C7. Cunningham, B. B.: *Physico-Chimie du Protactinium*, Centre National de la Recherche Scientifique, Paris, 1966, p. 45.

C8. Cunningham, B. B., and J. C. Wallmann: *J. Inorg. Nucl. Chem.* **26**: 271 (1964).

E1. Eubanks, I. D., and M. C. Thompson: *Inorg. Nucl. Chem. Lett.* **5**: 187 (1969).

F1. Federer, J. I., and V. J. Tennery: "Synthesis of (U, Pu)C by Carbothermic Reduction of Mixed Oxides and Evaluation of Sintering Behavior," Report ORNL/TM-6089, 1978.

F2. Ferguson, D. E., and J. E. Bigelow: *Actinides Rev.* **1**: 213 (1969).

F3. Fletcher, J. M.: "Chemical Principles in the Separation of Fission Products from Uranium and Plutonium by Solvent Extraction," *PICG (1)* **9**: 459 (1956).

F4. Fried, S., and N. Davidson: *J. Amer. Chem. Soc.* **70**: 3539 (1948).

G1. Glassner, A.: "The Thermochemical Properties of the Oxides, Fluorides, and Chlorides to 2500°K," Report ANL-5750, 1957.

G2. Goble, A., J. Golden, A. G. Maddock, and D. T. Toms: *Progress in Nuclear Energy*, Series 3, vol. II, F. R. Bruce, J. M. Fletcher, and H. H. Hyman (eds.), Pergamon, London, 1959, p. 86.

G3. Goode, J. H., C. L. Fitzgerald, and V. C. A. Vaughen: *Amer. Ceram. Soc. Bull.* **52**: 720 (1973).

G4. Gresky, A. T.: "Solvent Extraction Separation of U-233 and Thorium from Fission Products by Means of Tributyl Phosphate," *PICG (1)* **9**: 528 (1956).

G5. Groh, J. J., R. T. Huntoon, C. S. Schlea, J. A. Smith, and F. H. Springer: *Nucl. Appl.* **1**: 327 (1965).

G6. Grosse, A. V., *J. Amer. Chem. Soc.* **56**: 2200 (1934).

H1. Hale, W. H., and J. T. Lowe: *Inorg. Nucl. Chem. Lett.* **5**: 363 (1969).

H2. Hardy, C. T., D. Scargill, and J. M. Fletcher: *J. Inorg. Nuclear Chem.* **7**: 257 (1958).

H3. Harmon, K. M., B. F. Judson, W. L. Lyon, R. H. Pugh, and R. C. Smith: "Plutonium Reconversions," in *Reactor Handbook*, vol. 2, S. M. Stoller and R. B. Richards (eds.), Interscience, New York, 1961, p. 441.

H4. Higgins, G. H., and W. W. T. Crane: *PICG (2)* **17**: 245 (1958).

I1. International Atomic Energy Agency: "The Plutonium-Oxygen and Uranium-Plutonium-Oxygen Systems: A Thermochemical Assessment," Report of a Panel on Thermodynamic Properties of Plutonium Oxides, Vienna, Oct. 1966, Tech. Rept. Series No. 79, Vienna, 1967.

K1. Keller, C.: "Binary and Ternary Oxides, Hydroxides and Hydrous Oxides, Peroxides, Phosphates and Arsenates," in *Comprehensive Inorganic Chemistry*, vol. 5, J. C. Bailar, Jr., H. J. Emeléus, R. Nyholm, and A. F. Trotman-Dickenson (eds.), Pergamon, Oxford, 1973, p. 219.

K2. Keller, C.: *The Chemistry of the Transuranium Elements*, Verlag Chemie, Weinheim, 1971.

K3. Keller, C.: *Radiochim. Acta* **1**: 147 (1963).

K4. Kraus, K. A., and F. Nelson: "Anion Exchange Studies of the Fission Products," *PICG (1)* **7**: 113 (1958).

L1. Latimer, W. M.: *Oxidation Potentials*, 2d ed., Prentice Hall, Englewood Cliffs, N. J., 1952.

L2. Lee, J. A., and J. A. C. Marples: "The Elements," in *Comprehensive Inorganic Chemistry*, vol. 5, J. C. Bailar, Jr., H. J. Emeléus, R. Nyholm, and A. F. Trotman-Dickenson (eds.), Pergamon, Oxford, 1973.

L3. Lerch, R. E., and C. R. Cooley: *Trans. Amer. Nuc. Soc.* **15**: 86 (1972).

L4. Lloyd, M. H., and R. G. Haire, XXIVth IUPAC CONF 730927-2, 1973.

L5. Lowe, J. T.: USAEC Report DP-1194, 1969.

L6. Lundqvist, R.: *Acta Chemica Scand.* **A28**: 243 (1974).

M1. Mainland, E. W., D. A. Orth, E. L. Field, and J. H. Radke: *Ind. Eng. Chem.* **52**: 685 (1961).

M2. Malm, J. G., B. Weinstock, and E. E. Weaver: *J. Phys. Chem.* **62**: 1506 (1958).

M3. Marples, J. A. C., *Physico-Chimie du Protactinium*, Centre National de la Recherche Scientifique, Paris, 1966, p. 39.

M4. Mattys, H. M.: *Actinides Rev.* **1**: 165 (1968).

M5. Miner, W. N., and F. W. Schonfeld: "Physical Properties," in *Plutonium Handbook*, vol. 1, O. J. Wick (ed.), Gordon & Breach, New York, 1967.

M6. Moeller, Th.: *J. Chem. Ed.* **47**: 417 (1970).

N1. Nairn, J. S., D. A. Collins, H. A. McKay, and A. G. Maddock, *PICG (2)* **16**: 218 (1958).

N2. National Bureau of Standards, *JANAF Thermochemical Tables*, 2d ed., U.S. Government Printing Office, Washington, D.C., June 1971.

N3. National Bureau of Standards, *JANAF Thermochemical Tables*, 2d Suppl., U.S. Government Printing Office, Washington, D.C., 1975.

O1. Oak Ridge National Laboratory: "LMFBR Fuel Cycle Studies Progress Report for July 1971, No. 29," Report ORNL-TM-3534, 1971.

O2. Olander, D. R.: "Fundamental Aspects of Nuclear Reactor Fuel Elements," Report TID-26711-P1, NTIS, 1976.

O3. Orth, D. A.: *I&EC Process Design and Development* **2**: 121 (1963).

O4. Osborne, D. W., B. Weinstock, and J. H. Burns: *J. Chem. Phys.* **52**: 1803 (1970).

P1. Peterson, S., and R. G. Wymer: *Chemistry in Nuclear Technology*, Addison-Wesley, Reading, Mass., 1963.

R1. Rai, D., and R. J. Serne: "Solution Species of ^{239}Pu in Oxidizing Environments: I. Polymeric Pu(IV)," Report PNL-SA-6994, 1978.

R2. Rand, M. H.: "Thermochemical Properties," in *Atomic Energy Review*, vol. 4, *Plutonium: Physico-Chemical Properties of Its Compounds and Alloys*, Special Issue No. 1, International Atomic Energy Agency, Vienna, 1966.

S1. Seaborg, G. T., and J. J. Katz: *The Actinide Elements*, National Nuclear Energy Series, div. IV, vol. 14A, McGraw-Hill, New York, 1954.

S2. Seaborg, G. T., J. J. Katz, and W. M. Manning: *The Transuranium Elements Research Papers*, National Nuclear Energy Series, div. IV, vol. 14B, McGraw-Hill, New York, 1949.

S3. Sellers, P. A., S. Fried, R. E. Elson, and W. H. Zachariasen: *J. Amer. Chem. Soc.* **76**: 5935 (1954).

S4. Skavdahl, R. E., and T. D. Chikalla: "Plutonium Refractory Compounds," in *Plutonium Handbook*, vol. 1, O. J. Wick (ed.), Gordon & Breach, New York, 1967.

V1. Valentine, A.: "Capabilities for Control of Plutonium in Processing," Plutonium Information Meeting, Los Alamos, Jan. 1974.

W1. Wade, W. Z., and T. Wolf: *J. Nucl. Sci. Technol.* **6**: 402 (1969).

W2. Weinstock, B., E. E. Weaver, and J. G. Malm: *J. Inorg. Nucl. Chem.* **11**: 104 (1959).

Z1. Zachariasen, W. H.: *Acta. Cryst.* **5**: 19 (1952).

PROBLEMS

9.1 (*a*) Use App. C to find the rate, in joules per gram per hour, at which heat is produced by ^{238}Pu, ^{244}Cm, and ^{252}Cf.

(*b*) What is the neutron flux 1 m from 1 mg of ^{252}Cf?

9.2 What are the relative concentrations of trivalent, tetravalent, and hexavalent plutonium when a small amount of PuO_2 is dissolved in a large excess of 1 N HNO_3? Repeat for 3 N HNO_3. Assume that activities are proportional to molarities. For oxidation-reduction potentials use Fig. 9.1, sixth diagram from the top.

9.3 The solution from the dissolver of a reprocessing plant is 1 M in nitric acid, 0.01 M in $Pu(NO_3)_4$, and 0.001 M in $Np^{VI}O_2(NO_3)_2$. To 1 liter of solution is added 0.012 mol of ferrous sulfamate. After oxidation-reduction equilibria are established, what are the molarities of Pu^{3+}, Pu^{4+}, Np^{4+}, and $Np^{V}O_2^{+}$? To simplify the solution, assume that (1) no $Np^{VI}O_2^{2+}$ remains and (2) HNO_3 molarity remains at 1.0.

CHAPTER

TEN

FUEL REPROCESSING

1 OBJECTIVES OF REPROCESSING

As described in Chap. 3, fuel discharged from a nuclear reactor after irradiation to the end of its useful life still contains most of the fertile material (^{238}U or thorium) that was present in the fuel when charged, appreciable concentrations of valuable fissile nuclides (^{235}U, plutonium, and/or ^{233}U) and large amounts of radioactive, neutron-absorbing fission products. The principal objectives of reprocessing are (1) to recover uranium and plutonium, and thorium if present, for reuse as nuclear fuels; (2) to remove radioactive and neutron-absorbing fission products from them; and (3) to convert the radioactive constituents of spent fuel into forms suitable for safe, long-term storage. There may be some interest in recovering individual fission products such as ^{90}Sr or ^{137}Cs for use as radiation sources or in recovering by-product transuranic elements such as neptunium, americium, or curium.

2 COMPOSITION OF IRRADIATED FUEL

The composition of irradiated fuel to be fed to a reprocessing plant varies widely. It depends on the composition of the fresh fuel charged to the reactor, the neutron spectrum in which the fuel is irradiated, the specific power or rate of heat generation in the fuel, the duration of irradiation, and the length of time the fuel is "cooled"—the interval between end of irradiation and start of reprocessing.

Table 8.7 gave an example of the type of irradiated fuel from a commercial nuclear power plant that will predominate in the feed to reprocessing plants. That table listed the composition of fuel resulting from irradiation of fuel initially containing 3.3 w/o (weight percent) ^{235}U after irradiation in a pressurized-water reactor (PWR) at a specific power of 30 MW/Mg (metric ton, MT) to a burnup of 33 MWd/kg (33,000 MWd/MT), followed by cooling for 150 days. The table gave the grams of each element per megagram of fuel, the radioactivity of each element in curies, and the rate of energy production by each element in watts. Noteworthy are the large number of elements present; the intense radioactivity, totaling 4.3 million Ci/Mg; and the substantial decay power, more than 20 kW/Mg. Because almost 50 percent of the fission-product energy production is in the form of gamma rays, spent fuel and radioactive wastes separated from it require massive shielding.

The nominal design period for storing fuel after irradiation before reprocessing is 150 days.† This allows the radioactivity and decay power to decrease to the levels of of Table 8.7, high though they be. Other reasons for storing fuel this long are to permit all gaseous fission products except tritium, ^{85}Kr, and ^{129}I to decay to inconsequential levels, to reduce the activity of 8-day ^{131}I to manageable levels, to complete decay of 2.35-day ^{239}Np to ^{239}Pu, and to complete decay of 6.75-day ^{237}U. Some reprocessing plants require longer cooling periods; the proposed plant for oxide fuel at Windscale, England, requires that fuel be cooled for a year [B17] before reprocessing.

3 HISTORY OF REPROCESSING

3.1 Bismuth Phosphate Process

The first microgram quantities of plutonium were produced [S6] in 1942 by irradiation of natural uranium with deuterons in the cyclotron of Washington University in St. Louis. This plutonium was separated at the Chicago Metallurgical Laboratory of the Manhattan Project by Seaborg and his collaborators, who employed the method of carrier precipitation frequently used by radiochemists to extract small amounts of radioactive material present at low concentration. As wartime urgency required that a plutonium separation plant be designed and built before macro quantities of plutonium could be available for process development, it was decided to use the same carrier precipitation process that had successfully produced the first small quantities of this element.

Seaborg and associates [L1] had found that tetravalent plutonium [Pu(IV)] could be coprecipitated from aqueous solution in good yield with insoluble bismuth phosphate $BiPO_4$, made by adding bismuth nitrate and sodium phosphate to an aqueous solution of plutonium nitrate. The bismuth phosphate process was developed at the Metallurgical Laboratory, demonstrated at the X-10 pilot plant at Oak Ridge National Laboratory in 1944, and put into operation for large-scale recovery of plutonium from irradiated fuel at Hanford in early 1945.

The bismuth phosphate process consisted of a number of steps in which plutonium is made alternatively soluble and insoluble. Fuel elements containing plutonium, uranium, and fission products were first dissolved in nitric acid. Plutonium was reduced to the tetravalent state by addition of sodium nitrite. Plutonium phosphate $Pu_3{}^{IV}(PO_4)_4$ was coprecipitated with bismuth phosphate $BiPO_4$, by addition of bismuth nitrate and sodium phosphate. Coprecipitation of uranium was prevented by the presence of sufficient sulfate ion to form anionic $UO_2(SO_4)_2{}^{2-}$. The $BiPO_4$ precipitate was redissolved in nitric acid and subjected to two decontamination cycles to purify the plutonium. In each cycle the plutonium was oxidized to the soluble hexavalent state by $NaBiO_3$ or other strong oxidant. Next bismuth phosphate was again precipitated, to remove fission products while hexavalent plutonium remained in solution. Then plutonium was reduced to the tetravalent state and again coprecipitated with bismuth phosphate.

After the third precipitation with bismuth phosphate, the plutonium was put through a similar cycle in which lanthanum fluoride LaF_3 was used as carrier precipitate, to remove fission products not completely scavenged by bismuth phosphate in previous steps.

Despite the numerous steps, the overall recovery of plutonium exceeded 95 percent and the

†For some years most fuel will be stored much longer than 150 days, because of the large backlog of spent fuel awaiting reprocessing.

overall decontamination factor from fission products was 10^7. Serious disadvantages of the process were its batch operation, its inability to recover uranium, the large amount of process chemicals used, and the large volume of wastes.

3.2 Redox Process

As soon as sufficient plutonium was available for pilot-plant studies of more efficient processes, research was initiated on solvent extraction processes, which were already in large-scale use for recovery of uranium from ore leach liquors. Compared with the bismuth phosphate process, solvent extraction had the advantages that it could be operated continuously rather than batchwise, and that it could recover both uranium and plutonium in good yield and with high decontamination factors. The first solvent extraction process used in the United States for large-scale separation of uranium and plutonium from irradiated fuel was the Redox process. This process was developed by Argonne National Laboratory, tested in a pilot plant at Oak Ridge National Laboratory in 1948 to 1949, and installed by the General Electric Company at the Hanford plant of the U.S. Atomic Energy Commission (AEC) in 1951 [L1].

The solvent used in the Redox process was hexone, methyl isobutyl ketone, an extractant already in use for purifying uranium ore concentrates (Chap. 5). Hexone is immiscible with water and will extract uranyl nitrate and plutonyl nitrate selectively from fission-product nitrates if the aqueous solution has a sufficiently high nitrate ion concentration. In the Redox process, aluminum nitrate was used as salting agent because high concentrations of nitric acid would decompose the hexone solvent.

Figure 10.1 is a material flow sheet for the first cycle of one form of the Redox process [F3]. Plutonium in the feed was converted to hexavalent plutonyl nitrate $Pu^{VI}O_2(NO_3)_2$, by oxidation with dichromate ion $Cr_2O_7^{2-}$, as this is the plutonium valence with highest distribution coefficient into hexone. In the decontamination contactor, hexavalent uranium and plutonium nitrates were extracted into hexone solvent, and fission-product nitrates were removed from the solvent by a scrub solution containing aluminum nitrate, sodium nitrate, and sodium dichromate.

In the partition contactor, plutonium was converted to inextractable, trivalent $Pu^{III}(NO_3)_3$ by a reductant solution of ferrous sulfamate containing aluminum nitrate to keep uranium in the hexone phase. Plutonium was thus separated from uranium and transferred back to the aqueous phase along with the aluminum nitrate. Impure plutonium nitrate was purified by additional cycles of solvent extraction, not shown.

In the uranium stripping contactor, uranyl nitrate was transferred back to the aqueous phase by 0.1 *M* nitric acid strip solution. Fission products were separated from the impure uranyl nitrate by additional cycles of solvent extraction and by adsorption on silica gel, not shown.

A modification of the Redox process, the ^{235}U-hexone process, was used at the Idaho Chemical Processing Plant of the U.S. AEC, to recover highly enriched uranium from ^{235}U-Al alloy fuel elements irradiated in the Materials Testing Reactor. The aluminum nitrate needed as salting agent was provided when the fuel was dissolved in nitric acid. The plutonium content of the fuel was too low to warrant recovery. Plutonium was made trivalent and inextractable before solvent extraction and thus routed to the aqueous high-level waste.

Disadvantages of the Redox process were the volatility and flammability of the hexone solvent and the large amount of nonvolatile reagents such as $Al(NO_3)_3$ added to the radioactive wastes.

3.3 Trigly Process

At about the same time that the Redox process was being developed in the United States, a group of Canadian chemists [C1] at the Chalk River Laboratory were developing the Trigly

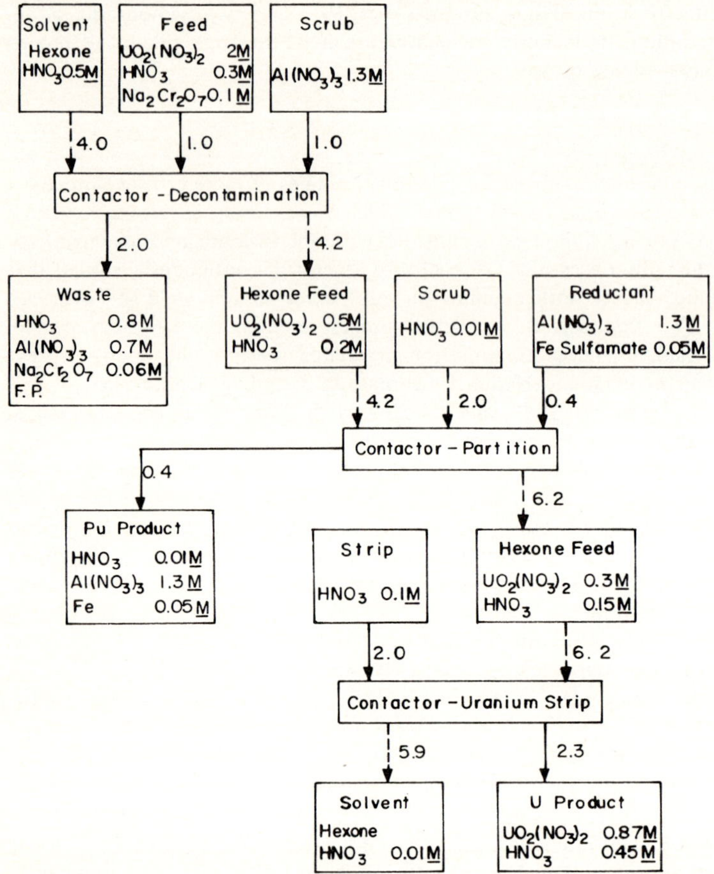

Figure 10.1 First cycle of acid Redox process. —— aqueous; – – – organic. Numbers give relative volumetric flow rates.

process, for recovering plutonium from natural uranium irradiated in the NRX reactor. This process used triglycol dichloride ($ClC_2H_4OC_2H_4OC_2H_4Cl$) as solvent and nitric acid and ammonium nitrate as salting agents. Hexavalent plutonium has a higher distribution coefficient than uranium in this solvent; seven batch extractions of the aqueous phase, each with one-fourth its volume of trigly, recovered 97 percent of the plutonium, 5 percent of the uranium, and only 0.01 percent of the fission products. Further purification was with hexone as in the Redox process.

3.4 Butex Process

Both the Redox and Trigly process had the disadvantage of adding large amounts of nitrate salts to the high-level wastes. The first process free from this disadvantage to be developed was the Butex process, which used dibutyl carbitol ($C_4H_9OC_2H_4OC_2H_4OC_4H_9$) as solvent and nitric acid as salting agent. The nitric acid was evaporated from the aqueous high-level wastes and reused. The Butex process was developed by a group of British chemists [N4] working at the Chalk River Laboratory in the late 1940s. As possible solvents they evaluated diethyl ether, hexone, triglycol dichloride, dibutyl cellosolve ($C_4H_9OC_2H_4OC_4H_9$), and dibutyl carbitol. They concluded that hexone and dibutyl carbitol were satisfactory solvents, but that dibutyl carbitol was preferable because of its greater stability toward nitric acid and lower vapor pressure.

Following successful pilot-plant tests at Chalk River [N4], the Butex process was adopted for large-scale separation of plutonium, uranium, and fission products from natural uranium irradiated to low burnup at the Windscale plant of the U.K. Atomic Energy Authority [H8]. Even after its use in this application was replaced by the Purex process, the Butex process remained in use at Windscale for primary decontamination of high-burnup fuel until the 1970s. Then an explosion, probably due to reaction of nitric acid with solvent, terminated its use.

3.5 Purex Process

The Purex process uses a mixture of tributyl phosphate (TBP) and a hydrocarbon diluent to extract uranyl nitrate and tetravalent plutonium nitrate from an aqueous solution containing nitric acid. The Purex process was suggested by the discovery of Warf [W2] in 1949 that tetravalent cerium nitrate could be separated from the nitrates of trivalent rare earths by solvent extraction with TBP. The Purex process was developed by the Knolls Atomic Power Laboratory of the General Electric Company and carried through the pilot-plant stage at Oak Ridge National Laboratory from 1950 to 1952. It was adopted by E. I. duPont de Nemours and Company for the Savannah River plutonium-production plant that company built for the U.S. AEC at Aiken, South Carolina, where the Purex process was put into operation in November 1954. Its success there led to replacement of the Redox process by the Purex process by the General Electric Company at Hanford in January 1956. The Purex process was used in a plant owned by Nuclear Fuel Services, Inc., which that company operated at West Valley, New York, from 1966 to 1972. The plant was designed to reprocess 1 MT/day of irradiated, slightly enriched uranium fuel. It also reprocessed irradiated thorium and irradiated plutonium, with appropriate flow-sheet modifications [R8]. This plant was noteworthy for being the only one to reprocess fuel from privately owned nuclear power plants in the United States.

Although the West Valley plant met all safety and environmental requirements in effect when it first went into operation, in the 1970s the plant was required to meet increasingly strict licensing requirements on permissible radioactive effluents and resistance to ground motion in an earthquake. It was also required to provide facilities for converting plutonium nitrate to oxide and for solidifying high-level acid wastes instead of neutralizing them and storing as liquid. Because of the high cost of retrofitting the plant to meet these later requirements and because its capacity was by then too small to permit it to compete with a larger plant under construction by Allied-General Nuclear Services (Sec. 4.14), the West Valley plant was permanently shut down in 1976 [N8].

A solvent extraction process similar to Purex using TBP was developed by the Commissariat à l'Energie Atomique [G1] for use in the French plutonium separation plant at Marcoule. Since then, the Purex process has replaced the Butex process at Windscale [W3], has been used in the Soviet Union [S11], India [S7], and Germany [S3], and by now is the universal choice for separation of uranium and plutonium from fission products in irradiated slightly enriched uranium. Fuel from the liquid-metal fast-breeder reactor (LMFBR) is also reprocessed by the Purex process, with modifications to accommodate the higher concentrations of plutonium and fission products.

The Purex process has four significant advantages over the Redox process: (1) Waste volumes can be made much lower, as the nitric acid used as salting agent can be removed by evaporation. (2) The solvent, TBP, is less volatile and less flammable than hexone. (3) TBP is more stable against attack by nitric acid. (4) Operating costs are lower.

The Purex process will be described in considerable detail in Sec. 4.

3.6 Thorex Process

The stability of TBP and its selectivity for tetravalent and hexavalent metal nitrates led to its consideration and selection for processing irradiated thorium to separate ^{233}U and thorium and decontaminate them from fission products. The form of the Thorex process first developed used aluminum nitrate as salting agent to enhance the distribution coefficients of uranyl and thorium nitrates. It was used in the early 1960s by E. I. duPont de Nemours and Company [R5] to process thorium irradiated in the U.S. AEC's Savannah River production reactors. Because the aluminum nitrate used as salting agent added undesirably to the nonvolatile material in the high-level wastes, it was replaced by nitric acid in the acid Thorex process developed by Oak Ridge National Laboratory in the early 1960s [R1]. More details of the history and applications of the acid Thorex process will be given in Sec. 5.

3.7 Other Aqueous Processes

Culler and Blanco [C18] have summarized other aqueous processes that have been studied for processing power reactor fuels not readily handled by the standard Purex or Thorex processes. Many of these require reagents other than nitric acid to dissolve either the cladding or the fuel, but finally use solvent extraction with TBP to separate and purify fissile materials. Details of these other processes are given in references cited by Culler and Blanco [C18].

3.8 Nonaqueous Processes

Early in the development of reprocessing, it was recognized that nonaqueous processes could have numerous advantages over aqueous processes. (1) There would be no radiation damage to water or solvent, so that fuel could be reprocessed soon after irradiation. This would be especially advantageous for fast reactors, for which the value of fissile material in spent fuel is high. (2) For some, fewer chemical steps would be involved, because dissolution of fuel as nitrates and conversion of nitrates back to metal or oxide would be eliminated. (3) Processing equipment would be more compact, and shielded volumes would be smaller. (4) Larger batches could be reprocessed because the critical mass is greater in the absence of moderating materials.

However, nonaqeuous processes were found to have a number of disadvantages, which have discouraged their widespread use. (1) Except for fluoride volatility processes, separations are incomplete, fuel is not completely decontaminated, and refabrication must be done remotely. (2) Operation at high temperature with corrosive or reactive reagents requires special, costly construction materials and makes maintenance difficult.

The three types of nonaqueous processes on which most development work has been done are (1) pyrometallurgical processes, involving high-temperature processing of metallic fuels; (2) pyrochemical processes, involving high-temperature processing of oxide or carbide fuels; and (3) fluoride volatility processes, in which elements in fuel are converted to fluorides, which are then separated by fractional distillation.

3.9 Pyrometallurgical Processes

In the 1950s and early 1960s, when metal fuel was still being considered for power reactors, many pyrometallurgical processes were proposed and some were carried through pilot-plant demonstration. Subsequently, when it was realized that the susceptibility of metal fuel to radiation damage limited its usefulness in power reactors, these processes became of less practical importance.

Table 10.1 lists the principal types of pyrometallurgical processes on which experimental work has been conducted. These have been grouped into physical separation processes, in which no chemical reactions take place, and chemical separation processes.

Table 10.1 Pyrometallurgical processes

Physical separations
- Volatilization or distillation
- Fractional crystallization
- Extraction with liquid metals

Chemical separations
- Liquid extraction with fused salts
  - From molten fuel
  - From liquid-metal solution of fuel
- Electrolysis through fused salts
- Partial oxidation

Volatilization. Many fission-product elements, including krypton, xenon, iodine, cesium (normal boiling point 705°C), strontium (1380°C), barium (1500°C), the rare earths (3200°C), and plutonium (3235°C), are more volatile than uranium (3813°C). Cubicciotti [C17], McKenzie [M5], and Motta [M8], in laboratory experiments, showed that around 99 percent of these more volatile elements could be separated from uranium by vacuum distillation at 1700°C. Because of the high temperature and severe materials problems, volatilization has not been used as a primary separation process, but does contribute to removal of the most volatile fission products in conventional reprocessing. In fractional crystallization or extraction with liquid metals, distillation is used to separate uranium and plutonium from more volatile solvent metals.

Fractional crystallization. Volatile metals with much lower boiling points than uranium, such as magnesium (1103°C), zinc (906°C), and cadmium (767°C), have been extensively studied as solvents for separating constituents of irradiated metal fuel by fractional crystallization, followed by evaporation of the solvent metal from the separated fractions. For example, in liquid magnesium, the solubility of plutonium or thorium is high, but uranium is very low. A process of this type was developed at Argonne National Laboratory [P6] for concentrating plutonium in the uranium metal blanket of a breeder reactor from 1 percent to 40 percent.

Liquid extraction with metals. By operating at temperatures above the melting point of uranium (1133°C), liquid metals partially miscible with uranium, such as magnesium [B3], silver [V3], (m.p. 960°C), and calcium [M4] (m.p. 1482°C), have been used to separate uranium, plutonium, and some fission products. These processes have not proved attractive. The vapor pressure of magnesium at the uranium melting point is too high. The high boiling point of silver (2212°C) makes its separation from uranium difficult. Calcium is so reactive that no suitable container material is available. Another disadvantage of liquid extraction is that metals less extractable than uranium remain in the uranium metal phase.

Liquid extraction with fused salts. Liquid extraction of metal fuels is made more flexible by use of fused salt extractants. The distribution coefficient of elements into the salt phase can be increased by adding to the salt a compound more readily reduced than the corresponding compound of the metal to be extracted. For example, addition of $ZnCl_2$ to $MgCl_2$ increases the distribution coefficient of uranium into the salt phase because the reaction

$$2U + 3ZnCl_2 \rightarrow 2UCl_3 + 3Zn$$

is more strongly displaced to the right than

$$2U + 3MgCl_2 \rightarrow 2UCl_3 + 3Mg$$

Distribution coefficients may be further modified and operating temperatures reduced by dissolving uranium fuel in a low-melting metal such as bismuth or zinc. Separation of uranium from fission products by liquid extraction between molten bismuth and fused chlorides was extensively studied at Brookhaven National Laboratory [D5] in connection with the liquid-metal fuel reactor (LMFR), which used a dilute solution of ^{235}U in bismuth as fuel. Extraction of fission products from molten plutonium by fused chlorides was studied at Los Alamos [L2] in connection with the LAMPRE reactor.

Workers at Argonne [B10] extended chloride extraction to higher-melting uranium alloys such as the 20 percent Pu-80 percent U alloy proposed for breeder-reactor cores. In a modified process [A9], this alloy was dissolved in molten zinc and contacted first with a low-melting LiCl-NaCl-$MgCl_2$ salt phase containing sufficient $ZnCl_2$ to transfer uranium, plutonium, and the more reactive fission products to the salt phase. The salt was then contacted with a cadmium-zinc alloy containing sufficient magnesium to return uranium and plutonium, but not the fission products, to the metal phase, from which cadmium, zinc, and magnesium were finally distilled.

Oak Ridge National Laboratory [R9] has studied liquid extraction with bismuth containing controlled amounts of lithium as a process for removing UF_3 and PaF_4 from LiF, BeF_2, ThF_4, and fission-product fluorides in fuel from the molten-salt reactor.

Electrolysis through fused salts. Electrodeposition of metal fuel through a fused-salt electrolyte to separate uranium and plutonium from fission products was studied at Knolls Atomic Power Laboratory [N6].

Partial oxidation. Spent fuel from the core of the EBR-II reactor was reprocessed [H7] by melting the ^{235}U fuel in a ZrO_2-lined crucible at 1400°C for from 1 to 3 h, after which the remaining metal was poured into Vycor glass molds for refabrication into fuel pins. Volatile fission products were vaporized from the fuel, and rare earths, strontium, and other metals more reactive than uranium, plus 5 to 10 percent of the uranium, were oxidized by the ZrO_2 liner and remained in the crucible. This process, called melt refining by Argonne, operated from September 1964 through February 1969 and recycled 2300 kg of irradiated metal whose burnup ranged from 1.0 to 1.3 percent. Most of the fission-product metals less reactive than uranium, such as zirconium, niobium, molybdenum, ruthenium, rhodium, and palladium, remained in the metal. Known collectively as fissium, these were found to improve the stability of the recycled alloy to radiation.

3.10 Pyrochemical Processes

Three examples of pyrochemical processes that have been developed for purifying uranium or plutonium oxides are listed in Table 10.2.

Skull-reclamation process. The skull-reclamation process was developed by Argonne National Laboratory and used at the Idaho EBR-II Fuel Cycle Facility in the 1960s to recover uranium

Table 10.2 Pyrochemical processes

Name	Developer	Reference
Skull reclamation	Argonne	[H7]
Salt transport	Argonne	[S22]
Salt cycle	Battelle	[H4]

from the crucible oxide residues, or skulls, remaining after the partial oxidation, melt-refining process described above. The process involved selective reduction and extraction of the oxides by magnesium-zinc alloys at controlled temperatures and reductant metal concentrations, followed by removal of the magnesium-zinc solvents by distillation.

Salt-transport process. The salt-transport process was studied by Argonne, with the objective of reprocessing short-cooled, high-burnup LMFBR fuel oxide with nonaqueous systems in which radiation damage of solvents would not be a problem. In this process, stainless steel cladding is removed from the fuel by solution in molten zinc at 850°C. The UO_2-PuO_2 fuel is then reduced by a copper-magnesium-calcium alloy with a $CaCl_2$-CaF_2 flux at 800°C. This produces a salt solution of the more stable fission-product oxides (Cs_2O, SrO, BaO, and some rare earth oxides), a copper-magnesium solution of plutonium, the rest of the rare earths and the more noble fission-product metals (ruthenium, molybdenum, palladium, etc.), and a solid phase consisting mostly of uranium metal. Plutonium in the liquid copper-magnesium phase is purified by countercurrent extraction with 50 w/o $MgCl_2$, 30 w/o NaCl, 20 w/o KCl, which extracts the rare earths selectively. Finally, the plutonium is separated from the noble fission-product metals by transport through a second 50 w/o $MgCl_2$ salt phase to a 95 w/o Zn-5 w/o Mg Pu-acceptor alloy. This last salt-transport step, suggested by Chiotti and Klepfer [C7], gave the process its name. Argonne [V2] tested individual steps of this process, but did not conduct a complete demonstration with full-burnup fuel.

Salt-cycle process. The salt-cycle process was developed by Battelle Northwest Laboratory [H4] with the following objectives:

To permit reprocessing short-cooled fuel at the reactor site
To handle UO_2 and UO_2-PuO_2 fuel without requiring conversion to other chemical forms
To recover 99 percent of the plutonium and remove at least 80 percent of the neutron-absorbing fission products
To permit control of the plutonium/uranium ratio in recovered fuel

In this process, oxide fuel is dissolved in a molten chloride salt mixture through which Cl_2-HCl gas is flowing. Dissolved uranium and plutonium are then recovered as oxides by cathodic electrodeposition at 500 to 700°C. The process was demonstrated with kilogram quantities of irradiated fuel, with production of dense, crystalline UO_2 or UO_2-PuO_2 reactor-grade material. Difficulties were experienced with process control, off-gas handling, electrolyte regeneration, and control of the plutonium/uranium ratio. Development has been discontinued.

3.11 Fluoride Volatility Processes

The unusual property of uranium, neptunium, and plutonium of forming volatile hexafluorides has led to extensive work on fluoride volatility processes for separating these elements from irradiated fuel and from each other. Major programs were carried out at Brookhaven, Argonne, Oak Ridge, and European laboratories. These programs have been summarized by Jonke [J2], Barghusen et al. [B1] and Schmets [S2].

Brookhaven made engineering-scale studies of a process in which uranium metal fuel was dissolved in a liquid interhalogen compound such as BrF_3. The reaction was difficult to control; work was terminated after an explosion [B18]. Brookhaven later developed the Nitrofluor process [B19], in which fuel was converted to UF_4 and PuF_3 by a liquid mixture of HF and oxides of nitrogen. After dissolution, UF_4 was converted to UF_6 by BrF_3 and distilled off. Finally, PuF_3 was converted to PuF_6 by fluorine and distilled off.

Gas-phase fluorination reactions were studied at Argonne [B1] and in Europe [C13]. Fuel was first oxidized to U_3O_8 and PuO_2. The crushed oxides were charged to a fluidized bed of alumina through which gases containing fluorinating agents, F_2, ClF_3, or BrF_5, were passed. Uranium was readily separated as volatile, stable UF_6. Separation of neptunium and plutonium was less satisfactory. Although these also form volatile hexafluorides, they are less stable than UF_6. Stronger fluorinating conditions are needed to form them, and PuF_6, in particular, is so unstable that it tended to decompose and deposit solid fluorides throughout the equipment.

Experience has shown that fluoride volatility processes are most useful when applied either to fuel containing little plutonium and neptunium or to fuel from which these elements have been largely removed by other processes. Oak Ridge National Laboratory has successfully separated and purified multikilogram amounts of irradiated, highly enriched uranium relatively free of plutonium from zirconium-^{235}U fuel used in submarine reactors [O3], from aluminum-^{235}U fuel used in research reactors [O4], and from the mixture of fused salts used in the aircraft reactor experiment [C2]. More recently, $^{235}UF_6$ and $^{233}UF_6$ were recovered from the BeF_2-^7LiF-UF_3 fuel melt used in the molten-salt reactor experiment [L3]. This work led to design of a process to separate ^{233}U and fission products from the BeF_2-^7LiF-UF_3-ThF_4 mixture proposed as fuel for the molten-salt breeder reactor nuclear power system [R9].

In the Aquafluor process [G4] developed by the General Electric Company, most of the plutonium and fission products in irradiated light-water reactor (LWR) fuel are separated from uranium by aqueous solvent extraction and anion exchange. Final uranium separation and purification is by conversion of impure uranyl nitrate to UF_6, followed by removal of small amounts of PuF_6, NpF_6, and other volatile fluorides by adsorption on beds of NaF and MgF_2 and a final fractional distillation. A plant to process 1 MT/day of irradiated low-enriched uranium fuel was built at Morris, Illinois, but was never used for irradiated fuel because of inability to maintain on-stream, continuous operation even in runs on unirradiated fuel. The difficulties at the Morris plant are considered more the fault of design details than inherent in the process. They are attributed to the attempt to carry out aqueous primary decontamination, denitration, fluorination, and distillation of intensely radioactive materials in a close-coupled, continuous process, without adequate surge capacity between the different steps and without sufficient spare, readily maintainable equipment [G5, R8].

4 THE PUREX PROCESS

4.1 Steps in Purex Process

The Purex process has become the process quite generally used for reprocessing slightly enriched uranium fuel from power reactors. For this reason, it will be described in more detail in this chapter than other fuel separation processes.

The principal steps in the Purex process as applied to fuel clad with stainless steel or zircaloy are shown schematically in Fig. 10.2. Each of these steps will be described in more detail later in Sec. 4.

In preparation for dissolution, step 1, cladding is opened to permit subsequent dissolution of the oxide fuel. For steel or zircaloy this is done by mechanical shearing or sawing. Off-gases from decladding contain up to 10 percent of the radiokrypton and xenon in the fuel and some of the $^{14}CO_2$,† tritium, and other volatile fission products. If voloxidation (Sec. 4.3) is used after decladding to remove tritium, more of the other volatile radionuclides will then be evolved also.

†^{14}C is produced primarily by the (n, p) reaction on ^{14}N present as a contaminant in the fuel.

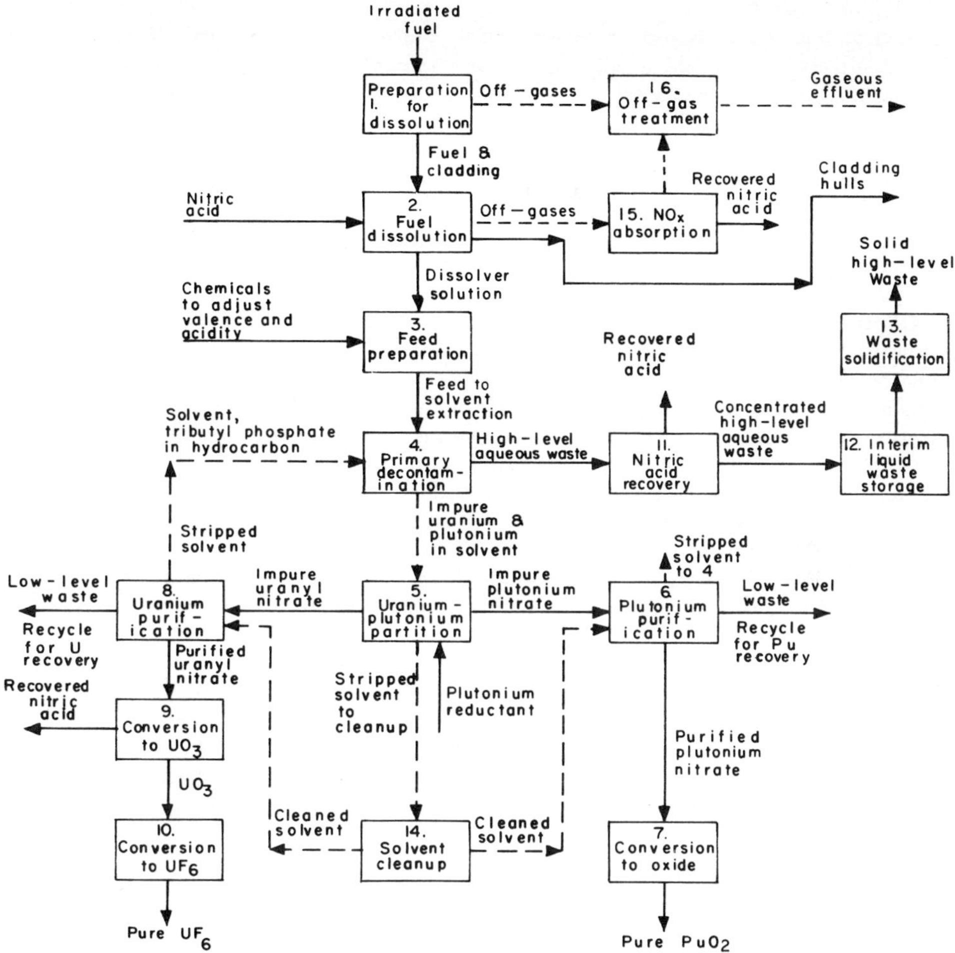

Figure 10.2 Principal steps in Purex process.

Fuel and cladding next are charged to a dissolver where, in step 2, they are reacted with hot nitric acid. This dissolves oxides of uranium and most other elements in the fuel while leaving cladding essentially unreacted. Dissolver off-gases are primarily a mixture of steam, air, and oxides of nitrogen, NO_x. Off-gases also contain all of the radiokrypton and xenon remaining in the fuel after voloxidation and any ^{14}C convertible to $^{14}CO_2$ or other volatile form. Practically all of the radioiodine can also be driven into the dissolver off-gases.

In NO_x absorption, step 15, the off-gases are cooled and scrubbed with water to recover oxides of nitrogen for conversion to nitric acid and recycle to the dissolver and to other process steps.

Gases from NO_x absorption are combined with off-gases from decladding and treated in step 16 for as complete removal of radioactivity as practical. Substantially complete retention of radioiodine and entrained liquids and solids is required; technology for retaining them is fully developed. Retention of radiokrypton has been mandated by the U.S. Environmental Protection Agency in reprocessing fuel irradiated after 1982. Separation and packaging of radiokrypton from dissolver off-gases has been demonstrated on a pilot-plant scale and is practiced at the Idaho Chemical Processing Plant. Retention of tritium and ^{14}C may also be required in the future, but the requisite technology is not yet completely worked out.

Cladding hulls from the dissolver are washed with water, monitored to show that most of the fission products and fuel have been removed, packaged, and shipped to waste storage.

In feed preparation, step 3, the acidity of the dissolver solution is adjusted to the desired value around a pH of 2.5, and plutonium is brought into its most extractable valence of 4, usually by controlled addition of N_2O_4 or, formerly, sodium nitrite.

In primary decontamination, step 4, uranium and plutonium are separated from more than 99 percent of the fission products by solvent extraction with 30 v/o (volume percent) TBP in a paraffinic hydrocarbon diluent.

In partition, step 5, plutonium is separated from uranium by reducing plutonium to the organic-insoluble, trivalent state with a reductant strong enough to act on plutonium but not so strong as also to reduce uranium. Ferrous sulfamate was used in early plants; tetravalent uranium, hydroxylamine, or controlled cathodic reduction is now preferred. In some plants, reduction of plutonium and its return to the aqueous phase is carried out in a single step. In others, plutonium and uranium are first returned to the aqueous phase, then plutonium is reduced, and finally uranium is separated by reextraction into the organic phase.

Impure plutonium nitrate is purified, step 6, by one or more additional solvent extraction cycles, plus ion exchange in some plants.

Purified plutonium nitrate is converted to the preferred product form, PuO_2, in step 7 either by evaporation to dryness and calcination or by precipitation as the oxalate or peroxide and calcination.

Impure uranyl nitrate from step 5 is purified in step 8 by one or more additional cycles of solvent extraction and, in some plants, by treatment with silica gel or hydroxamic acid.

Purified uranyl nitrate solution from step 8 is evaporated to dryness and calcined to UO_3 in step 9. Nitric acid vapors are condensed and recycled.

If the UO_3 product is to be reenriched, it is converted to UF_6 in step 10, either by one of the processes described in Chap. 5, Secs. 9.4 through 9.7, or by direct reaction with fluorine.

Low-level aqueous wastes from steps 6 and 8 are processed for further recovery of plutonium and uranium, then concentrated for recovery of water and nitric acid. High-level aqueous wastes from step 4 are concentrated by evaporation, with recovery of condensed nitric acid in step 11.

After a period of interim storage as liquid, step 12, to permit the rate of heat generation to decrease, high-level wastes are solidified in step 13. The duration of interim storage as liquid can be reduced if the spent fuel is stored as solid an equivalent time before reprocessing.

In a well-designed reprocessing plant, materials are recycled to the maximum extent practicable, to minimize the volume of effluents and reduce the cost of chemicals. Water and nitric acid evaporated from products and wastes are recycled. Nitrogen oxides are converted to nitric acid and reused. Solvent stripped of uranium and plutonium is cleaned of degradation products and contaminants in step 14 and reused.

The technology for the more important of these steps will be described in Secs. 4.3 through 4.13.

4.2 Principal Reprocessing Plants

U.S. plants. The principal U.S. reprocessing plants are listed in Table 10.3, together with their main process features. All use some form of the Purex process. In 1979, the only ones operating were the Savannah River and Idaho plants of the U.S. Department of Energy (DOE). The Hanford plant had been used primarily for recovery of plutonium and uranium from irradiated natural uranium, but was versatile and had been used, for example, for Thorex

Table 10.3 Principal U.S. reprocessing plants

Name				Idaho Chemical Processing Plant		Barnwell Nuclear Fuel Plant
Location	Hanford, Wash.	Savannah River, S.C.		Idaho National Engineering Lab.	West Valley, N.Y.	Barnwell, S.C.
Owner	U.S. AEC-DOE	U.S. AEC-DOE		U.S. AEC-DOE	Nuclear Fuel Services, Inc.	Allied General Nuclear Services
Year operational	1956†	1954		1953	1966	–
Status	Inactive since 1974	Operating		Operating	Shut down 1972	Waiting for license
		F area	H area			
Capacity, MT/day	–	9	0.05	–	1	5
Maximum % ^{235}U	1.8	Natural	100	100	93	5
Maximum burnup MWd/MT	–	1500	–	–	~30,000	40,000
Minimum cooling, days	135	200	150	120	150	160
Decladding	Chemical	Chemical	Codissolution	Chemical and electrolytic	Shear-leach	Shear-leach
Dissolver	Annular	Batch	Batch, or electrolytic	Batch	Annular	Semicontinuous
Solvent extraction						
Contactors	Pulse columns	Centrifugal + mixer-settlers	Mixer-settlers	Pulse and packed columns	Pulse columns	Centrifugal + pulse columns
v/o TBP	30	30	7.5	5	30	30
Number of decontamination cycles	1	1		1	1	1
Pu reductant	FeSAm‡	FeSAm		–	FeSAm	Electrolytic
Pu purification	1 cycle TBP	1 cycle TBP	None	None	1 cycle TBP + anion exchange	2 cycles TBP
U purification	1 cycle TBP	1 cycle TBP		2 cycles TBP	2 cycles TBP + silica gel	1 cycle TBP + silica gel
Waste treatment	Neutralize	Neutralize		Evaporate + calcine	Neutralize	Evaporate
Maintenance	Remote	Remote		Direct	Remote	Remote + direct
Reference	[G2]	[J4, P8]		[A1]	[R10]	[A3, B21, M10]

†Purex plant. Reprocessing was first done by the bismuth phosphate process, operational in 1944.
‡Ferrous sulfamate.

process runs (sec. 5.5). The F area at Savannah River is used primarily for irradiated natural uranium, but it, too, has been used for Thorex runs. The H area at Savannah River is a multipurpose facility used for processing highly enriched uranium from production and test reactors.

The Idaho Chemical Processing Plant is a versatile, multipurpose facility used for recovering highly enriched uranium from a variety of fuels in naval propulsion, research, and test reactors. Materials processed [A1] include aluminum-alloyed, zirconium-alloyed, stainless steel-based, and graphite-based fuels. The West Valley plant, although designed primarily for low-enriched uranium fuel from power reactors, also processed plutonium-enriched and thorium-based fuels. It is the only U.S. plant to have reprocessed fuel from commercial nuclear power plants.

The Barnwell Nuclear Fuel Plant is the newest U.S. reprocessing plant. In 1979, it was nearly complete, but standing unused because of U.S. government policy unfavorable to reprocessing fuel from power reactors. Its main process features are to be described in Sec. 4.14 as an example of a modern Purex plant.

Overseas plants. Table 10.4 lists the reprocessing plants outside of the United States and the Soviet Union with capacities greater than 100 kg heavy metal per day and gives their principal process features. In addition to these plants, smaller plants have been operated in Italy, India, and, probably, other countries.

The Cogema plant at Marcoule, France, designed originally for natural uranium fuel from plutonium-production reactors, has also been used to reprocess natural uranium fuel from Magnox power reactors. Since 1976 the Cogema plant at La Hague, France, has been operating head-end facilities that enable it to handle slightly enriched fuel from water reactors. The capacity at La Hague is being increased, first by expansion of the plant to 800 t/year and, later, by construction of a larger plant [C5]. Figure 10.3 is an aerial view of the plant at La Hague.

The British Magnox reprocessing plant at Windscale, designed originally for natural uranium fuel from plutonium-production reactors, is being used to reprocess slightly enriched, low-burnup fuel from British gas-cooled power reactors. From 1970 to 1973 this plant also operated a Butex head-end facility that enabled it to process higher-burnup oxide fuel from LWRs. A second plant at Windscale, termed THORP, using the Purex process to treat oxide fuel, is planned [B17] as the British participation in United Reprocessors GmbH, a joint Anglo-French-German company created to coordinate commercial fuel reprocessing in Europe.

The 0.17 MT/day WAK Purex pilot plant at Karlsruhe, West Germany, has operated since 1971 [S3]. A joint venture of Kernforschungszentrum Karlsruhe (KFK) and Gesellschaft zur Wiederaufarbeitung von Kernbrennstoffen mbH (GWK), this plant has provided operating experience to guide design of the full-scale Deutsche Gesellschaft für Wiederaufbereitung von Kernbrennstoffen (DKW) plant.

Additional European reprocessing experience was gained from the Eurochemic plant at Mol, Belgium [D1]. This joint undertaking of the Organization for Economic Cooperation and Development Nuclear Energy Agency operated a demonstration reprocessing plant from 1966 until the mid-1970s. This multipurpose plant could reprocess either 0.35 MT/day of slightly enriched uranium or 10 to 20 kg/day of 93 percent enriched ^{235}U.

The 0.7 MT/day plant at Tokai-Mura is a prototype of a larger plant that Japan expects to build.

4.3 Decladding

The method used for decladding depends on the composition of the cladding and fuel and the bond between them, if any. The two general decladding methods are chemical and mechanical.

Chemical decladding. In chemical decladding, the clad is removed and the fuel exposed by dissolving the clad, to leave the fuel as a separable solid. Chemical decladding has the disadvantage that the clad reaction products require more storage space than the original cladding. However, it sometimes is the more practical method. For example, fuel for the first U.S. production reactors consisted of a uranium metal slug bonded with aluminum-silicon alloy to an aluminum can. Because of the bond, mechanical decladding was impractical. Chemical decladding consisted in dissolving the aluminum can and the bond in hot, aqueous, 10 w/o sodium hydroxide solution containing about 20 w/o sodium nitrate to prevent evolution of hydrogen. The overall reaction was [B12]

$$Al + 0.85NaOH + 1.05NaNO_3 \rightarrow NaAlO_2 + 0.9NaNO_2 + 0.15NH_3 + 0.2H_2O$$

With this reactant, uranium metal fuel is relatively unattacked.

Another example of chemical decladding is afforded by the Zirflex process, which was proposed for zircaloy-clad UO_2 fuel before mechanical decladding was fully developed. In the Zirflex process [S17], zirconium or zircaloy cladding is dissolved as ammonium fluozirconate in a boiling solution of ammonium fluoride containing ammonium nitrate, the latter added to reduce hydrogen evolution. Overall reaction is approximately

$$Zr + 6NH_4F + 0.5NH_4NO_3 \rightarrow (NH_4)_2ZrF_6 + 5NH_3 + 1.5H_2O$$

Because of limited solubility of the ammonium fluozirconate product, there is an optimum NH_4F concentration, around 5.5 *M*, with an initial molar ratio of fluoride to zirconium between 6.5 and 7.0.

A third example of chemical decladding is afforded by the Sulfex process, which was proposed for stainless steel-clad UO_2 or ThO_2 fuel before mechanical decladding was fully developed. In the Sulfex process [F2], stainless steel cladding is dissolved in hot 4 to 6 *M* sulfuric acid. Disadvantages of the process are the slow and variable dissolution rate, passivation of the steel by nitrate ion unavoidably present if the same dissolver is used alternately to dissolve cladding and fuel, some attack of UO_2 by H_2SO_4, and evolution of hydrogen.

Electrolytic dissolution in nitric acid has been used at the Savannah River [B22] and Idaho Chemical Processing plants [A10, A11] to dissolve a wide variety of fuels and cladding materials, including uranium alloys, stainless steel, aluminum, zircaloy, and nichrome. The electrolytic dissolver developed by du Pont [B22], pictured in Fig. 10.4, uses niobium anodes and cathodes, with the former coated with 0.25 mm of platinum to prevent anodic corrosion. Metallic fuel to be dissolved is held in an alundum insulating frame supported by a niobium basket placed between anode and cathode and electrically insulated from them. Fuel surfaces facing the cathode undergo anodic dissolution in a reaction such as

$$Fe \rightarrow Fe^{3+} + 3e^-$$

At nitric acid concentrations above 2 *M*, the cathode reaction is

$$NO_3^- + 3H^+ + 2e^- \rightarrow HNO_2 + H_2O$$

so that hydrogen evolution is suppressed. The reverse reaction takes place at the platinum-coated anode, together with evolution of some oxygen in

$$2H_2O \rightarrow 4H^+ + O_2 + 4e^-$$

Anodic corrosion of the niobium basket that supports the fuel is inhibited by an electrically conducting oxide film, which forms on the niobium.

Advantages of this method are its wide applicability, the absence of anions other than the nitrate ion, and the fact that little hydrogen is evolved. A disadvantage is the presence of cladding metal nitrate in the dissolver solution and its eventual routing to the high-level wastes. When applied to zircaloy cladding, most of the zirconium is converted to hydrous ZrO_2, which can be filtered from the dissolver solution.

Table 10.4 Principal overseas reprocessing plants

Location	Marcoule, France	La Hague, France		Windscale, England		Karlsruhe, W. Germany	Hessen (WA 350) W. Germany	Mol, Belgium	Tokai-Mura, Japan
Owner	Cogema	Cogema		British Nuclear Fuels, Ltd.		KFK/GWK	DKW	Eurochemic	Power Reactor and Nucl. Fuel Devel. Corp.
Type of fuel	Magnox	Magnox	Oxide	Magnox	Oxide	Oxide	Oxide	Oxide or metal	Magnox or oxide
Year operational	1958	1967	1976	1964	–	1971	1992	1966	1975
Status	Operating	Operating	Operating	Operating	Planned	Operating	Planned	Shutdown	Operating
Capacity, MT/day	~2	~2	1 (to 5 in 1985)	5	4	0.17	2†	0.35‡	0.7
Maximum % ^{235}U	Natural	Natural	3.5	Natural to 1%	–	3	3.5	5‡	4
Maximum burn-up, MWd/MT	–	3,000	39,000	4,000	37,000	39,000	40,000	17,000	28,000
Minimum cooling, days	120	140	–	130	360	250	2500	190	180
Decladding	Mechanical	Mechanical or chemical	Shear-leach	Mechanical	Shear-leach	Shear-leach	Shear-leach	Chemical	Mechanical or shear-leach
Dissolver	Continuous or batch, fumeless	Continuous-fumeless		Continuous	Batch	Batch	Batch	Batch-fumeless	Batch
Feed clarifier	Filter	Filter	Centrifuge	Decanter	Centrifuge	Filter	Centrifuge	Centrifuge	–

Solvent extraction									
Contactors	Mixer-settlers + pulse columns	Mixer-settlers	Centrifugal + mixer-settlers	Mixer-settlers	Pulse columns	Mixer-settlers	Pulse columns + mixer-settlers	Pulse columns + mixer-settlers	Mixer-settlers
v/o TBP	30	30	30	20	30	30	30	30	30
Number of codecontamination cycles	1	2	3	2	1	1	1	1	2
Pu reductant[§]	U(IV)		U(IV)	FeSAm	U(IV)	U(IV)	U(IV)	U(IV)	U(IV) + N_2H_4
Pu purification	1 cycle TBP + oxalate ppt.		1 cycle TBP + oxalate ppt.	2 cycles TBP + oxalate ppt.	2 cycles TBP + oxalate ppt.	1 cycle TBP + anion exchange	2 cycles TBP	1 cycle TBP + oxalate ppt.	1 cycle TBP
U purification	1 cycle TBP		None	1 cycle TBP	2 cycles TBP (in M-S)	1 cycle TBP + SiO_2	2 cycles TBP	1 cycle TBP + SiO_2	1 cycle TBP
Waste treatment	CH_2O denitration		CH_2O denitration	Evaporation	Evaporation	Evaporation	Evaporation + denitration	Evaporation	HCOOH denitration
Maintenance	Direct		Direct	Direct	Direct	Direct	Direct	Direct	Remote and direct
References	[C14, C15, J3]		[C5, C12, C14, D2]	[W3]	[B17]	[S3]		[D3]	[I1, U1]

[†]Expected to operate 175 days/yr.
[‡]Also processed 93% ^{235}U at reduced rate and lower v/o TBP.
[§]U(IV), tetravalent uranium; FeSAm, ferrous sulfamate.

Figure 10.3 Cogema reprocessing plant at La Hague, France. (*With permission of Cogema.*)

Figure 10.4 Electrolytic dissolver at Savannah River Plant. (*Photo courtesy of E. I. duPont de Nemours & Company.*)

Mechanical decladding. The objective of mechanical decladding is to break or cut the cladding so as to expose the fuel to reaction with a dissolvent that does not attack the cladding. The best decladding method is one that requires minimum disassembly of fuel elements, minimizes production of fines, produces fuel fragments that can be readily and completely leached, and uses equipment with minimum maintenance requirements. Methods that have been used include transverse chopping with a shear, transverse cutting with a saw or abrasion wheel, longitudinal slitting, and longitudinal extrusion.

Transverse methods have the advantage of not requiring disassembly of a fuel element into individual rods before decladding. Transverse chopping with a specially designed shear is the method now generally favored for decladding fuel bundles from U.S. boiling- and pressurized-water reactors. This method, developed at Oak Ridge National Laboratory [W4] and Hanford [K3], uses a stepped blade such as shown in Fig. 10.5 moved horizontally past a circular or V-shaped anvil. Blade wear is minimized by rounding the cutting edge to a $\frac{1}{32}$-in (0.8-mm) radius before use. Blade life is from 10,000 to 50,000 cuts at a stroke rate of from 1 to 2 in/s (2.5 to 5 cm/s). For a 36-rod boiling-water reactor (BWR) fuel assembly, forces of from 45 to 80 t are required. Optimum length of segments is from $\frac{1}{2}$ to 2 in (1.25 to 5 cm), depending on degree of oxide fragmentation, dissolving conditions, blade life, and cost of blade replacement. Such shearing leaves the cut lengths open for subsequent leaching and produces little metal fines. With zircaloy cladding, the shear must be operated in an inert atmosphere to prevent zirconium fires. The first production use of such a shear, at the West Valley plant of Nuclear Fuel Services, was very satisfactory, with necessary maintenance carried out by remote means.

Fuel bundles have been declad by transverse sawing with a hacksaw blade operated under water to provide cooling and prevent zircaloy fires. The preferred saw consists of a hardened tool-steel cutting edge welded to a tough-steel blade [H2]. More fines are produced than in shearing.

Unbonded aluminum jackets are removed from uranium metal fuel rods in the British [C4] and French Magnox reactors by longitudinal extrusion through a hardened steel die. The hole in the die is large enough to admit the uranium rod but small enough to reject and peel off the jacket, which is scored lengthwise before meeting the die. The method is not applicable to oxide fuel because the fuel would crumble and jam the die.

A longitudinal cut with a remotely operated milling cutter was used to remove stainless steel jackets from uranium metal fuel rods in the first core of the Experimental Breeder Reactor-1 [C18].

Gas evolution in decladding. Fuel elements with unvented cladding contain fission-product gases under pressure. Some fuel elements also contain helium charged during fabrication to improve heat transfer in subsequent reactor operation. In decladding, this helium is evolved, together with around 10 percent of the fission-product krypton and xenon and a small fraction of the iodine, tritium, and ^{14}C. Gas evolved in decladding is routed to the off-gas treatment system for

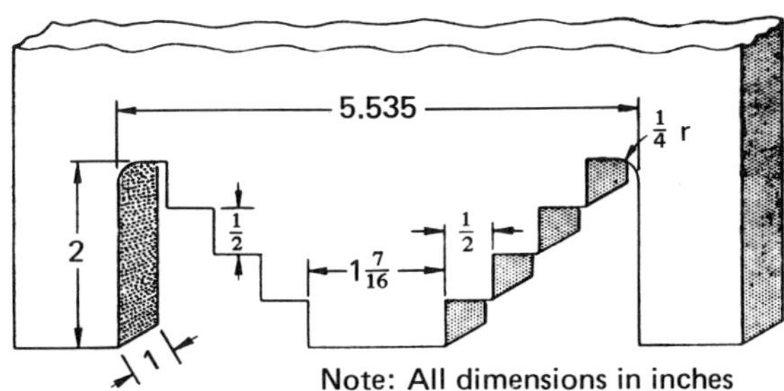

Figure 10.5 Stepped blade used for shearing metal-clad uranium oxide fuel bundles. (*From J. T. Long, Engineering for Nuclear Fuel Reprocessing, Gordon & Breach, New York, 1967, with permission.*)

retention of iodine and entrained solids. Processes have been developed for retention of krypton and xenon. Processes are being developed for tritium and ^{14}C, but are not yet in general use.

Voloxidation. If tritium is to be separated in the plant, it is highly desirable to do so before dissolution, when the tritium would be diluted isotopically with the large amount of hydrogen added as water and nitric acid in the dissolver. The voloxidation process has been developed by Oak Ridge National Laboratory for this purpose [F1, G8]. In this process, sheared or cut fuel is oxidized in a rotating kiln to convert UO_2 to U_3O_8. As U_3O_8 is less dense than UO_2, the fuel swells and is pulverized, thus exposing occluded tritium to oxidizing gases and converting it to tritiated water. More than 99 percent of the tritium and the remaining krypton and xenon escape from the fuel. The gases are filtered, passed over heated copper oxide to convert any unreacted hydrogen to water, and cooled, after which the tritiated water is absorbed by a molecular sieve or anhydrous $CaSO_4$.

In voloxidation, stainless steel-clad UO_2 is oxidized with flowing air or oxygen at 575 to 650°C. With zircaloy-clad fuel, these gases may be unsafe because of the danger of a zirconium fire. Less reactive N_2O_4 has been proposed as an oxidant for such fuel.

4.4 Dissolution

Objectives. The objectives of fuel dissolution are (1) to bring the uranium and plutonium in the fuel completely into aqueous solution; (2) to complete the separation of fuel from cladding; (3) to determine as accurately as possible the amounts of uranium and plutonium charged to reprocessing; and (4) to convert uranium, plutonium, and fission products into the chemical states most favorable for their subsequent separation.

Reactions. Because the Purex process requires that the elements to be separated be present in aqueous solution as nitrates, the dissolvent is always nitric acid. The principal reactions that take place are

$$3UO_2 + 8HNO_3 \rightarrow 3UO_2(NO_3)_2 + 2NO + 4H_2O$$

and

$$UO_2 + 4HNO_3 \rightarrow UO_2(NO_3)_2 + 2NO_2 + 2H_2O$$

Ordinarily, both reactions take place to some extent, with the first dominant at acid concentrations below 10 *M* and the second at higher concentrations [S8]. In principle, formation of gaseous reaction products could be avoided by addition of oxygen directly to the dissolver:

$$2UO_2 + 4HNO_3 + O_2 \rightarrow 2UO_2(NO_3)_2 + 2H_2O$$

This process is known as "fumeless dissolving" and is used in European plants. Practically, small amounts of nitrogen, nitrogen oxides, and gaseous fission products are also formed. Reference [O7] gives an example of fumeless dissolving.

Plutonium in oxide fuel dissolves as a mixture of tetravalent and hexavalent plutonyl nitrates, both of which are extractable with TBP. Neptunium dissolves as a mixture of inextractable pentavalent and extractable hexavalent nitrates.

Most of the fission products go into aqueous solution. However, at high burnups above 30,000 MWd/MT, some elements such as molybdenum, zirconium, ruthenium, rhodium, palladium, and niobium may exceed their solubility limits and be present as solids.

In the solution, americium, curium, and most of the fission products are in a single, relatively inextractable valence state. Iodine and ruthenium are important exceptions. Iodine may appear as inextractable iodide or iodate or as elemental iodine, which would be extracted by the solvent and react with it. Ruthenium may appear in any valence state between 0 (insoluble metal) and 8 (volatile ruthenium tetroxide) and, at valence 4, may form a number of nitrosyl ruthenium ($Ru^{IV}NO$) complexes of varying extractability.

An important objective of dissolution and the preconditioning of feed solution prior to extraction is to convert these fission-product elements into states that will not contaminate uranium, plutonium, or solvent in subsequent solvent extraction.

Dissolution rates. Uranium dioxide dissolves more rapidly than PuO_2 or ThO_2. The time required for dissolving more than 99.5 percent of the UO_2 from unirradiated stainless steel-clad UO_2 pellets was found to be 40, 70, and 110 min for $\frac{1}{2}$-, 1-, and 2-in chopped lengths, respectively [W4]. Irradiated fuel usually dissolves faster, probably because of cracking during irradiation. These tests were made with from 150 to 200 percent excess of 10 *M* HNO_3 at a temperature just below boiling. The instantaneous dissolution rate in 8 *M* HNO_3 is about one-half that in 10 *M* acid. Because extensive foaming results when fuel is added directly to boiling nitric acid, the preferred procedure in a batch dissolver is to add UO_2 to cold acid, then bring the solution to just below the boiling point, with adequate cooling available to deal with heat evolved from chemical reaction and radioactive decay.

The rate of dissolution of PuO_2 in nitric acid is slower than UO_2 and depends on the plutonium/uranium ratio, the methods used to fabricate fuel, and the conditions of irradiation. At one extreme, plutonium produced at low concentration in UO_2 by transmutation dissolves almost as rapidly as the associated UO_2. At the other extreme, plutonium present as PuO_2 mixed mechanically with UO_2 without proper sintering dissolves much more slowly and less completely than UO_2. Plutonium present as a solid solution $(U,Pu)O_2$ at the concentration of 20 to 25 percent used in breeder-reactor fuel dissolves at an intermediate rate.†

In all cases, however, dissolution of irradiated fuel in nitric acid leaves some plutonium associated with undissolved fission products. This plutonium can be leached from the residue with mixed nitric and hydrofluoric acids or with mixed nitric acid and ceric nitrate, $Ce(NO_3)_4$ [U2]. Residue from irradiated mixed UO_2-PuO_2 fuel was 99.94 percent dissolved in 4 h by treatment with 4 *M* HNO_3-0.5 *M* Ce(IV). Ceric nitrate is preferred to HF in the secondary dissolution step because cerium is already present as a fission product, and its addition does not complicate subsequent solvent extraction. Use of Ce(IV) in the primary dissolution step is undesirable because it would convert all plutonium to the less extractive hexavalent state and would volatilize much of the ruthenium as RuO_4.

Separation from cladding. After reaction of fuel with acid has been completed, the resulting solution and any suspended fine particles are drained from the coarser cladding fragments. The cladding is washed, first with dilute nitric acid and then with water. The cladding is checked by gamma spectroscopy to establish removal of adherent fuel and then discharged for packaging as radioactive waste. The fuel solution, possibly containing suspended particles, is clarified by centrifugation. Centrifuged solids are accumulated and periodically leached as described above for recovery of plutonium and uranium.

Accountability measurements. The dissolver solution and washings are collected in a calibrated accountability tank and mixed thoroughly. The volume and density of the solution are measured as accurately as possible, and samples are taken for determination of uranium and plutonium concentrations. This is the first point in reprocessing at which a quantitative measure of input amounts can be made. Even here, measurement is difficult because of the intense radioactivity.

†Dissolution of PuO_2-UO_2 fuel is also discussed in Sec. 6.8.

Conditioning of feed. Before solvent extraction, the concentrations of nitric acid and uranyl nitrate are brought to the desired values by addition of water and/or nitric acid, as required. Preferred concentrations are HNO_3, 2 to 2.5 *M*; $UO_2(NO_3)_2$, 1.2 to 1.4 *M*.

It is usually considered desirable to bring all plutonium to the most extractable, tetravalent state, although this step was not found necessary at West Valley. Sodium nitrite was formerly used for this purpose, but N_2O_4 or hydroxylamine is now favored because each adds no nonvolatile material to the aqueous phase. With N_2O_4, hexavalent plutonium is reduced:

$$Pu^{VI}O_2{}^{2+} + N_2O_4 + 2H^+ \rightarrow Pu^{4+} + 2HNO_3$$

Any trivalent plutonium that might be present would be oxidized:

$$4Pu^{3+} + N_2O_4 + 4H^+ \rightarrow 4Pu^{4+} + 2NO + 2H_2O$$

Prevention of criticality. In dissolving fuel obtained from irradiating material more enriched than natural uranium, precautions must be taken to prevent accumulation of a critical mass in the dissolver. Three general methods are (1) use of subcritical geometry, (2) control of fissile material concentration, or (3) addition of a soluble neutron absorber with the dissolver solvent. For subcritical geometry, dissolvers have been built as thin slabs or long cylinders of subcritical diameter. A good example of subcritical geometry combined with concentration control is the dissolver used in the West Valley plant of Nuclear Fuel Services, Inc., shown in horizontal cross section in Fig. 10.6. Fuel baskets were 7 ft high and 8 in or less in diameter. The basket diameter selected for a particular fuel was one that limited the concentration in the 3-in

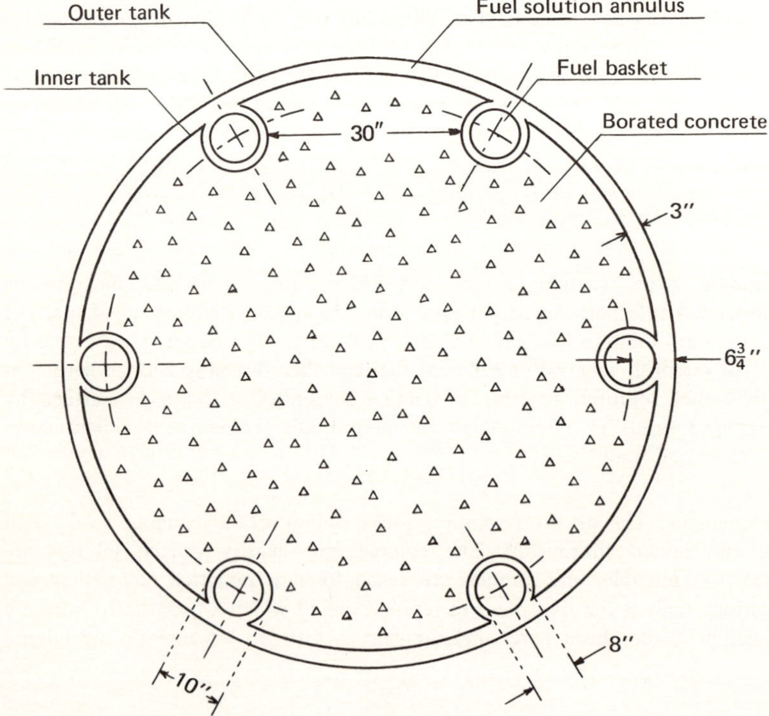

Figure 10.6 Horizontal section of annular dissolver used in Nuclear Fuel Services, Inc., plant.

annulus and 10-in cylinders after dissolving to 60 percent of the critical value. Nuclear interaction between the cylinders was prevented by addition of 0.5 w/o natural boron to the concrete which provided 30-in separation between cylinders.

The Barnwell plant of Allied-General provides an example of use of soluble poison. There it is proposed that 5.6 g natural gadolinium, as nitrate, per liter be added to the nitric acid solvent. At the design concentrations of plutonium and uranium in dissolver solution, this will prevent criticality even with fully enriched ^{235}U (Prob. 10.1).

Dissolution equipment. Dissolution equipment, termed *dissolvers,* must provide for (1) adding fuel and dissolvent; (2) removing the product solution, undissolved solids, and gaseous effluents; (3) maintaining proper contact between fuel and dissolvent; and (4) controlling the dissolution rate. Dissolvers may be characterized by the mode of fuel addition as either *batch* or *continuous*, or by their shape as *column, slab, annular,* or *pot*. The first three shapes are used for geometric control of criticality. Of the many types of dissolver that have been used, only a few examples can be described.

Batch pot dissolvers have been widely used, especially for low-enrichment fuel. The big advantage of batch operation is simplification of charging fuel and discharging residues. A disadvantage is the variable reaction rate, which is highest at the start when a large quantity of fuel is present, and which becomes much smaller toward the end when most of the fuel has been dissolved. The dissolution rate can be made more uniform by varying the concentration of dissolvent during the cycle. Dissolver product from a previous cycle, partially saturated with uranium, may be charged at the beginning of a cycle, to produce the most concentrated solution. When the reaction slows down, this solution may be replaced by fresh nitric acid, to produce a partially saturated solution to be used as solvent at the beginning of a subsequent cycle. Finally, at the end of a cycle, the residue in the dissolver would be washed with water to remove the remaining acid and fuel solution.

A batch dissolver typically is provided with heating coils to bring the solution to the desired temperature, cooling coils to remove the heat of reaction when it is most rapid, corrosion-resistant baskets or other containers to hold the fuel undergoing dissolution and retain cladding hulls at the end of the operation, and a cover to prevent escape of steam, nitric acid vapors, and volatile fission products and lead them to a condenser and fission-product traps. For ease of placement and removal, the cover may be sealed by a flanged ring that dips into a trough containing a sealing liquid. Recirculation of liquid through the dissolver is sometimes used to provide more uniform conditions and increase reaction rates. Product solution and undissolved sediment are withdrawn from the bottom. For dissolvers using nitric acid, heavy-gauge stainless steel has satisfactory corrosion resistance.

The volume of a batch of fuel to be charged to a batch dissolver may be evaluated from the product of the desired fuel reprocessing rate and the time required to complete the dissolving cycle. The cycle time may be estimated from small-scale experiments that simulate the geometry and time-temperature-concentration variations in the production dissolver, with a substantial allowance for inability to mock up accurately all relevant conditions of a production dissolver.

Continuous dissolution is especially advantageous when fuel and cladding are to be dissolved completely, as there is then no problem in removing undissolved solids from the dissolver. In such a case, fuel may be charged continuously at the top, dissolvent may be fed continuously, and dissolved solution removed continuously. The volume of undissolved fuel in the dissolver adjusts itself automatically so that the rate of solution balances the rate of addition. The big advantages over batch dissolution are smaller dissolver volume, more uniform product solution composition, steady gas evolution rate, and smaller and more efficient absorption system. It is estimated [B12] that the volume of a continuous dissolver may be from one-tenth to one-twentieth that of a batch dissolver of the same average dissolving rate.

This is especially advantageous for enriched fuels, where criticality limits dissolver dimensions.

The overall dissolution rate in a continuous dissolver is controlled by the metal feed rate, the temperature, the concentration of feed solution, and its flow rate. The metal composition of product solution at steady-state operation is just the ratio of the metal mass feed rate to the solution volume feed rate.

In a continuous dissolver with no solid residue, the solid necessarily flows downward. Liquid flow may be either down or up. With liquid downflow the dissolver is sometimes called a *trickle dissolver*. With liquid downflow it is preferable to remove off-gases at the bottom, to prevent flooding. With liquid upflow the dissolver is sometimes called a *flooded dissolver*. Off-gases separation from liquid is simpler with liquid upflow than with downflow.

When a solid residue such as cladding hulls remain, design of a continuous dissolver is much more complicated. Provision must be made for washing dissolver solution from the residue and for discharging the residue without escape of off-gases. A number of possible design concepts for continuous dissolvers have been tested by Oak Ridge National Laboratory [G15, O12].

4.5 NO_x Absorption

Dissolver off-gases are processed for recovery of oxides of nitrogen, known collectively as NO_x, in step 15. This is necessary to prevent corrosion of downstream equipment and acid contamination of the environment, and it reduces HNO_3 makeup. The preferred procedure is to pass the hot dissolver off-gases, which contain water vapor, nitric acid vapor, N_2O_4, NO_2, NO, N_2O, and added air or oxygen, successively through a downdraft condenser and a water scrubber.

In a downdraft condenser gases and condensed dilute nitric acid flow concurrently down the condenser wall. In this way the leanest gas is contacted with the maximum volume of coolest condensate, thus improving absorption. The following reactions take place:

$$2NO_2(g) \rightleftharpoons N_2O_4(g)$$

$$2NO_2 \ (\text{or } N_2O_4)\ (l) + H_2O(l) \rightleftharpoons HNO_3(l) + HNO_2(l)$$

$$3HNO_2(l) \rightleftharpoons H_2O(l) + HNO_3(l) + 2NO(g)$$

and

$$2NO(g) + O_2(g) \rightarrow 2NO_2(g)$$

The first three equilibrium reactions are rapid, but the fourth reaction, which eventually proceeds to completion, is slow even with excess oxygen and determines the extent of absorption.

Gases leaving the downdraft condenser are passed through a bubble-plate or packed water scrubber, where additional absorption of NO_x takes place. Laboratory studies [M2] indicate that the nitrogen oxide content can be reduced to from 0.1 to 0.5 percent with a residence time of 2 min in the condenser and water scrubber. The design of scrubbers for recovery of nitrogen oxides is described in standard texts, e.g., [P5].

The nitrogen oxide content of dissolver off-gases can be further reduced to 10 ppm by adding NH_3 to the gases leaving the absorber and passing the mixture over a hydrogen mordenite catalyst [P4], which reduces NO_x to N_2 and H_2O.

4.6 Off-gas Treatment

Off-gases from decladding, voloxidation if practiced, and dissolution are passed through high-efficiency particulate filters, processed for radioiodine absorption and, in some plants, for krypton and xenon retention before discharge through the plant stack. Gases vented from downstream process equipment are also passed through high-efficiency particulate filters and radioiodine absorbers. This section describes briefly processes that have been developed for absorbing radioiodine and removing and packaging krypton and xenon. Retention of tritium and ^{14}C may also be required in the future.

Radioiodine removal. Radioiodine removal is important because of its toxicity, the comparatively high iodine content of fission products (0.69 w/o, Table 8.7), and the high fission yields at the mass numbers of the two principal radioiodines, 1.7×10^7 year ^{129}I (1 percent) and 8.05-day ^{131}I (2.09 percent). Removal of radioiodine is complicated because of the numerous process streams in which iodine may appear and the variety of chemical forms it assumes. About 1 percent of the iodine is volatilized during decladding, some during voloxidation and a significant but incomplete amount during dissolution. If iodine is allowed to remain in the feed to solvent extraction, it reacts with solvent to form hard-to-remove compounds that eventually contaminate the entire system. It is thus important to remove as much of the iodine as possible before solvent contacting. Iodine may appear as I_2, HI, HIO, or organic iodides in off-gases or aqueous or organic phases, or as HIO_3 in concentrated nitric acid solutions.

The preferred procedure for removing iodine is to route the gases from decladding and voloxidation to iodine absorbers and to distill iodine from dissolver solution before solvent extraction. Experiments at Oak Ridge showed that 95 percent of the iodine could be removed by distilling 2 percent of the volume from 4 M HNO_3 and 99 percent by distilling 20 percent [O6]. Some of the remaining iodine is evolved with vent gases.

Of the numerous iodine-removal methods discussed by Goode and Clinton [G9], the most significant are characterized below.

Absorption by aqueous NaOH removes HI and I_2 but not organic iodides. No good procedure is available for disposal of spent solution.

The Iodox process under development by Oak Ridge National Laboratory uses absorption in boiling 21 to 23 M HNO_3 to convert iodine and its compounds to solid, nonvolatile I_2O_5.

An alternative process developed by Oak Ridge [O8] uses boiling 8 to 14 M HNO_3 containing 0.2 to 0.4 M $Hg(NO_3)_2$ to absorb all forms of iodine as HgI_2. The absorber solution is evaporated from vermiculite, which retains the iodine in stable form suitable for storage. The Barnwell plant proposes [A3] use of a similar process.

Unglazed Berl saddles coated with silver nitrate and operated at 135°C were used at Hanford [M1] to remove HI and I_2 from dissolver off-gases. In 1958, an explosion occurred, which was attributed to an unstable compound of silver and ammonia formed when the reactor was periodically cleaned by washing with ammonium sulfite. After this was replaced by sodium thiosulfate, the reactor operated for 14 years without incident.

A more effective way of using silver is to impregnate with silver a zeolite catalyst of the type used in hydrocarbon processing. With moist air at 150°C all volatile iodine species are absorbed as stable silver iodide in a form suitable for packaging and permanent storage. Silver zeolites for iodine absorption have been developed at Idaho Nuclear [P3] and Karlsruhe, Germany [W7]. Wilhelm et al. [W7] give data for fractional penetration of I_2 and CH_3I through an amorphous silicic acid zeolite impregnated with 0.06 to 0.08 g silver/g zeolite. More than 98 percent of the silver is available for reaction, permitting loadings of 0.1 g iodine/g zeolite. Fractional penetration of iodine is a function of many variables, as described in [W7]. Decontamination factors of from 10^2 to 10^4 have been reported. Long-term management of radioiodine as a radioactive waste is discussed in Chap. 11.

Krypton and xenon removal. The number of curies of krypton and xenon per megagram (metric ton) of spent fuel from pressurized-water, liquid-metal fast-breeder, and high-temperature gas-cooled reactors from Tables 8.7, 8.8, and 8.9 are listed in Table 10.5, together with the number of standard liters per megagram, assuming atomic weights of 85 and 133 for krypton and xenon.

Table 10.5 Curies and liters of krypton and xenon in spent fuel 150 days after discharge

	Reactor		
	Pressurized-water	Liquid-metal fast-breeder	High-temperature gas-cooled
Burnup, MWd/kg	33	37	95
Ci/Mg			
Krypton	11,000	8,430	60,800
Xenon	3.12	5.27	5.93
Std. liters/Mg			
Krypton	95	92	522
Xenon	821	804	2,528
Total	916	896	3,050

The volume of gas is appreciable; 80 percent or more is xenon. Practically all of the radioactivity is due to ^{85}Kr. One year after discharge the xenon activity would be negligible. This xenon could be a significant commercial source.

Processes that have been studied for krypton-xenon removal are listed in Table 10.6 together with comments on the process from reference [M6]. All have achieved 99 percent krypton removal.

Room-temperature adsorption is used for off-gases from nuclear power plants to delay escape of krypton and xenon long enough for all radionuclides except ^{85}Kr to decay to innocuous levels. Retention of ^{85}Kr would require very large bed volumes and a more complex system for bed regeneration. There is a fire hazard when treating reprocessing off-gases with charcoal, so that O_2 and NO_x must be removed from the feed.

In cryogenic adsorption, smaller bed volumes suffice, but the feed must be pretreated to remove condensibles. The fire hazard with charcoal remains and may be worse, because of the possibility of adsorption of ozone produced by radiolysis of oxygen.

Development of permselective membranes is only at the laboratory stage. For reprocessing

Table 10.6 Processes for removal of ^{85}Kr from reprocessing off-gas

Process	Development status	Comments
Adsorption on charcoal or molecular sieves at room temperature	Used for decay storage of xenon in off-gases from nuclear power plants	Simple operation, but very large beds. Charcoal can ignite.
Adsorption on charcoal or silica gel at low temperature	Pilot-plant test at reprocessing plant	Smaller bed volumes. Charcoal can ignite.
Separation by permselective membranes	Nonradioactive pilot-plant tests	Small equipment; no fire hazard.
Cryogenic distillation	Radioactive pilot plant at Idaho Chemical Processing Plant	Small equipment; ozone explosion hazard.
Absorption in chlorofluoromethane	Radioactive pilot-plant tests at Oak Ridge	Small equipment; no fire hazard.

off-gases, disadvantages are the serious consequence of mechanical failure and deterioration from radiation and exposure to ozone and NO_x.

The last two processes are the ones favored for reprocessing plants.

Cryogenic distillation has been extensively operated at Harwell [W8] and the Idaho Chemical Processing Plant [B9]. The principal concerns are (1) plugging of low-temperature equipment by condensed ice, solid CO_2 or Xe, or solid nitrogen oxides and (2) possible explosion from accumulation of solid hydrocarbons in the presence of condensed oxygen and ozone. To deal with them, feed gases must be pretreated for removal of impurities before condensing the krypton and xenon. At the Idaho plant [B9], NO_2 and CO_2 were removed from feed gas by scrubbing with sodium hydroxide solution. No attempt was made to package the CO_2 containing ^{14}C, but this could have been done by precipitation as $CaCO_3$ with lime. N_2O was dissociated into N_2 and O_2 by passage over a rhodium catalyst at 650°C. The hydrocarbon content of feed gas was low enough that hazardous accumulation in low-temperature equipment was prevented by warm-up once per shift. More generally applicable practice would be to oxidize hydrocarbons by passing feed gas over copper oxide at 600°C.

After purification the feed gas at Idaho was cooled to −160°C by passage through regenerators, precooled by outflowing cold gas, in which H_2O and remaining traces of CO_2 and nitrogen oxides were condensed and removed. Finally, the purified feed gas was washed with liquid nitrogen to condense krypton and xenon, which were then concentrated by fractional distillation. The concentrate was separated periodically by batch distillation into an oxygen fraction, which was recycled to prevent loss of small amounts of accompanying krypton, and a krypton fraction and a xenon fraction, which were bottled separately for storage.

Absorption in halogenated solvents, such as refrigerant R-12, CF_2Cl_2, has been extensively studied at Brookhaven [S21], Harwell [T3], and Oak Ridge [M6, V4]. The process has several advantages. Fire or explosion hazards are minimal, and gas purification prior to absorption is not required. The process is flexible and does not use extremely low temperatures. Disadvantages are operation at 8 to 10 bar pressure, a fairly complex flow sheet, and the need for an auxiliary system to separate krypton from xenon and CO_2.

Figure 10.7 shows one of the flow sheets for removing radioisotopes from reprocessing plant off-gases by refrigerated absorption tested by Oak Ridge National Laboratory [V4]. Contaminated feed gas, consisting of H_2O, CO_2, N_2O, Xe, Kr, Ar, N_2, and O_2, and possibly containing I_2, CH_3I, and NO_2 not previously removed, is compressed to 8 bar (100 psig) and cooled to −28°C in a cold trap. This removes most of the H_2O, NO_2, I_2, and iodine compounds. The gas is then fed to a 5-m absorber-fractionator column refluxed with refrigerant R-12 at −28°C at the top and reboiled at 31°C at the bottom. Decontaminated Ar, N_2, and O_2 are taken off the top, and a solution of Kr, Xe, N_2O, and CO_2 in R-12 is taken off the bottom. This solution also contains traces of Ar, N_2, O_2, NO_2, and water, and iodine compounds if present. In the stripper, Kr, Xe, N_2O, and CO_2 diluted with small amounts of Ar, N_2, and O_2 are taken off as overhead product, together with some R-12. Solvent from the bottom of the stripper is distilled to separate it from small amounts of water and other less volatile impurities prior to recycle to the absorber-fractionator.

Overhead product from the stripper, although greatly reduced in volume from feed gas, requires further treatment (not shown) to separate and package CO_2 containing ^{14}C, Kr, and Xe. One possible sequence of operations would be

1. Absorb CO_2 for permanent storage on solid soda lime.
2. Remove R-12 for recycle with a selective molecular sieve.
3. Decompose N_2O over a rhodium catalyst at 650°C.
4. Remove O_2 with copper at 600°C.
5. Condense Xe, Kr, and some Ar in a cold trap refrigerated with liquid nitrogen.
6. Separate the condensate by low-temperature distillation into (a) an Ar-Kr fraction to be bottled for permanent storage as radioactive waste and (b) an Xe fraction.

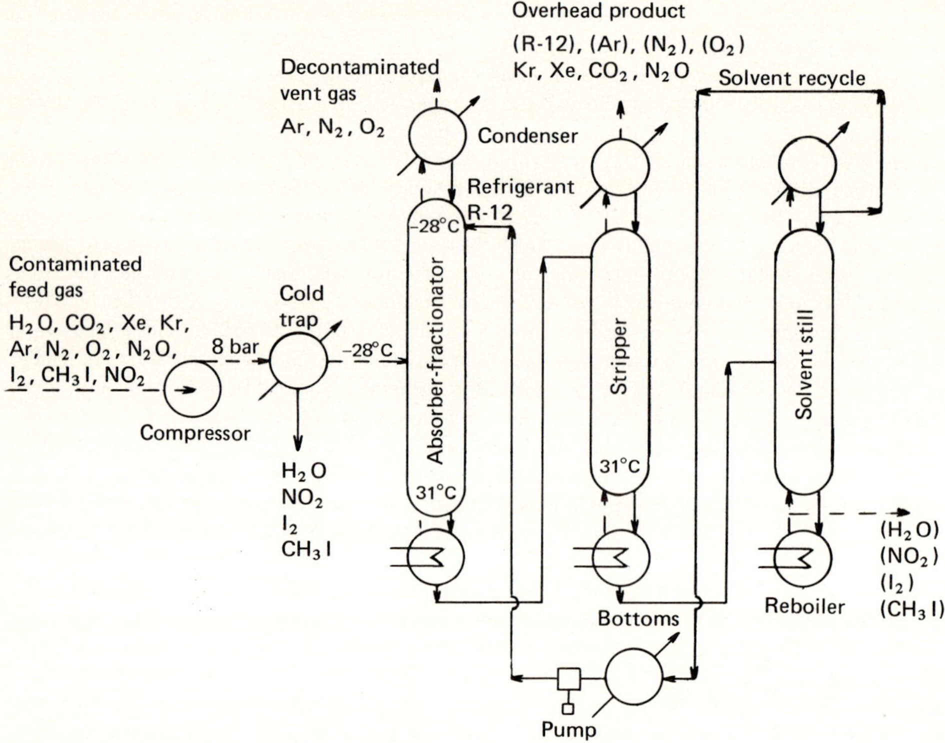

Figure 10.7 Refrigerated absorption system for removing krypton and xenon from reprocessing off-gases.

4.7 Primary Decontamination

Dissolution, described in Sec. 4.4, produces an aqueous solution of uranyl nitrate, plutonium(IV) nitrate, nitric acid, small concentrations of neptunium, americium, and curium nitrates, and almost all of the nonvolatile fission products in the fuel. With fuel cooled 150 days after burnup of 33,000 MWd/MT, the fission-product concentration is around 1700 Ci/liter. The first step in the solvent extraction portion of the Purex process is primary decontamination, in which from 99 to 99.9 percent of these fission products are separated from the uranium and plutonium. Early removal of the fission products reduces the amount of required shielding, simplifies maintenance, and facilitates later process operations by reducing solvent degradation from radiolysis.

The decontamination system consists of an extracting section, often designated HA, and a scrubbing section, HS. In the extracting section, uranium and plutonium are extracted from the dissolver solution by multistage countercurrent contacts with 30 v/o TBP in a normal-paraffin diluent. Fission products, which have much lower distribution coefficients than uranium and plutonium, largely remain in the aqueous phase and leave the extracting section in the aqueous raffinate. Americium and curium are predominantly trivalent, have low distribution coefficients like the rare-earth fission products, and also leave in the raffinate. Neptunium in the feed is partly in the extractable hexavalent state and partly in the inextractable pentavalent state and divides between aqueous raffinate and organic extract.†

†Neptunium behavior in Purex systems is discussed further in Sec. 7.

In the scrubbing section, most of the small amounts of fission products from the feed carried by the solvent leaving the extracting section are moved from the solvent and returned to the extracting section by countercurrent washing with aqueous nitric acid, about 3 *M*.

Contacting equipment used in the extracting section must have low holdup to minimize solvent degradation from the intense fission-product radioactivity. Here, centrifugal contactors or pulse columns are preferred to mixer-settlers. In the scrubbing section and in the balance of the solvent extraction plant, mixer-settlers are often used.

The extracting section is usually run at or near room temperature, to reduce solvent degradation and because the uranium distribution coefficient is higher the lower the temperature. It has been found advantageous to operate the scrubbing section at higher temperature, around 60°C, primarily because decontamination of ruthenium is more complete at higher temperature.

Figure 10.8 illustrates the effect of nitric acid concentration on distribution coefficients in the Purex process, for 80 percent saturation of the solvent with uranium, a condition that obtains near the feed point. Increasing acid concentration improves separation of ruthenium

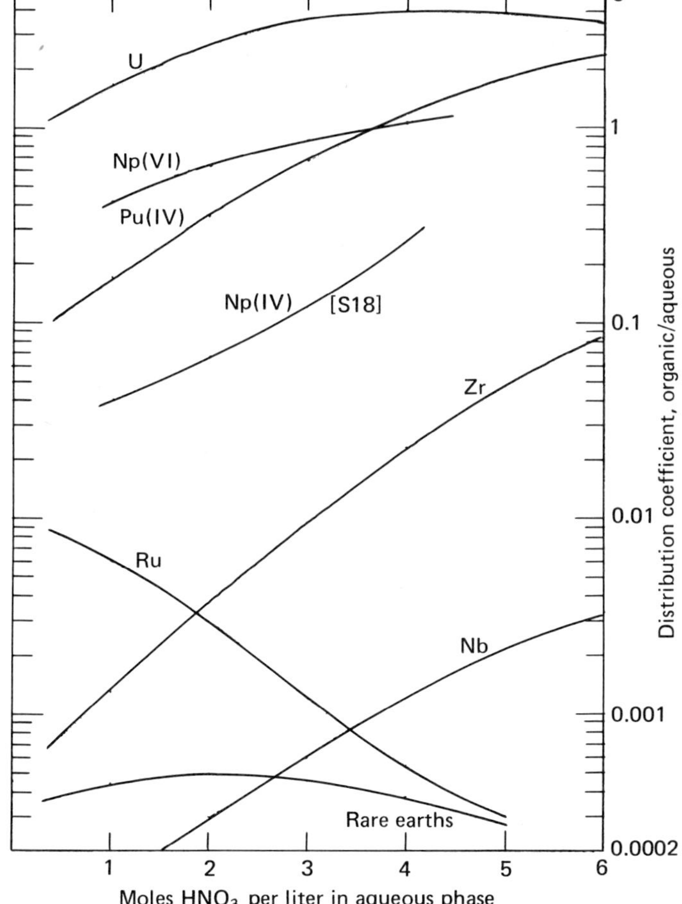

Figure 10.8 Effect of nitric acid concentration on distribution coefficients in 30 v/o TBP 80 percent saturated with uranium at 25°C. (*From [S23]*.)

from uranium and plutonium, but impairs slightly separation of zirconium. An acid concentration around 2.5 to 3.0 M is a practical optimum for these contaminants.

Figure 10.9 illustrates the effect of uranium saturation of solvent on distribution coefficients, at nitric acid concentrations approximately those in the extracting and scrubbing sections. The ratio of plutonium distribution coefficient to fission products is improved at high uranium loadings, a condition sought at the feed point.

More quantitative data on distribution coefficients for uranium, plutonium, and HNO_3 are given in Sec. 4.15.

4.8 Plutonium Partitioning

The next step in the Purex process after primary decontamination is separation of plutonium from uranium. This is done by reducing plutonium to the trivalent state, in which it is inextractable by TBP, while leaving the uranium in the extractable hexavalent condition. Reductants that have been used for this purpose include Fe^{2+}, U^{4+}, hydroxylamine, or cathodic reduction.

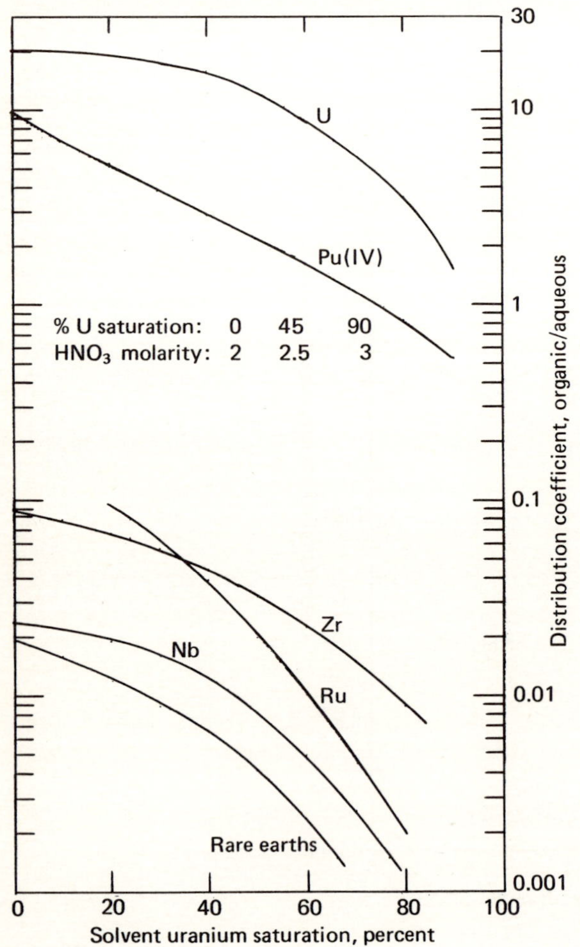

Figure 10.9 Effect of solvent saturation on distribution coefficients in 30 v/o TBP at 25°C. (*From [S23].*)

With ferrous ion or cathodic reduction, conversion of plutonium from Pu^{4+} to Pu^{3+} is so rapid that back extraction of plutonium to the aqueous phase and reduction there to Pu^{3+} can be carried out simultaneously in a single multistage contactor. With tetravalent uranium, reduction of plutonium is slower, so that additional contactor volume is desirable to complete back extraction. With hydroxylamine, reduction of plutonium is so much slower that it is preferable first to return both uranium and plutonium to the aqueous phase by stripping with dilute nitric acid and then to reduce the plutonium in equipment providing sufficient residence time for reduction to proceed to completion. Finally, the uranium is reextracted by TBP.

The kinetics of the reduction of Pu^{4+} to Pu^{3+} with U(IV) have been studied by Newton [N3]. Reduction kinetics with hydroxylamine have been studied by Barney [B2] and Koltunov et al. [K5].

Reduction with ferrous ion was the reaction used in the first Purex flow sheets, at Hanford and Savannah River. The specific reductant used was ferrous sulfamate $Fe(SO_3NH_2)_2$, a compound selected because it stabilized ferrous ion against oxidation in a nitric acid-nitrous acid system. The process was satisfactory in all respects except its addition of extraneous, nonvolatile components to the wastes.

The other three reductants are free of this disadvantage, but introduce process complications. Simultaneous cathodic reduction and partitioning has been patented by Allied-General [G13] and is proposed for use in the Barnwell Nuclear Fuel Plant (Sec. 4.14), but it has not yet been used commercially. It requires a novel, electrolytic reduction extraction contactor.

Tetravalent uranium has been used in French reprocessing plants and in the German WAK plant. It requires auxiliary equipment for reducing uranium to the tetravalent form.

Hydroxylamine nitrate (HAN) requires addition of hydrazine as a holding reductant, to prevent destruction of hydroxylamine by the nitrous acid present as a result of radiolysis, in the reaction

$$3NH_2OH + HNO_2 \rightarrow 2N_2 + 5H_2O$$

Hydrazine removes HNO_2 in the reaction

$$HNO_2 + N_2H_4 \rightarrow HN_3 + 2H_2O$$

Hydrazoic acid HN_3 is a volatile, potentially explosive compound, but it is extracted by Purex solvent and can be removed safely in the solvent wash system.

4.9 Uranium Purification

Uranium leaving the partitioning step in the organic phase is back extracted to the aqueous phase by 0.01 *M* HNO_3. It is then purified by one or more additional cycles of solvent extraction by TBP, while plutonium is kept in the inextractable trivalent state. To purify this uranium sufficiently to permit its use as feed to a UF_6 plant, the U.S. DOE requires that the total beta-gamma activity be less than twice that of aged natural uranium and that the alpha activity be less than 1500 disintegrations per minute per gram of uranium, corresponding roughly to a plutonium-uranium ratio less than 1×10^{-8}. To meet these strict specifications, a final cleanup step is usually needed. The first Purex plants passed the concentrated uranyl nitrate solution through a silica gel bed, which adsorbs fission products, primarily zirconium and niobium. A recent, more versatile process [B4, S1], developed in Italy, removes zirconium, niobium, and tetravalent neptunium and plutonium from aqueous nitrate solution by batch extraction with a 0.4 *M* solution of oleyl hydroxamic acid in 20 v/o octyl alcohol, 80 v/o *n*-dodecane. Distribution coefficients for these contaminants in this solvent are very high. When the solvent becomes too contaminated it is regenerated by washing with aqueous oxalic and nitric acids.

4.10 Plutonium Purification

Plutonium in the aqueous phase leaving partitioning contains around 1 percent of the feed uranium, an uncertain fraction of the feed neptunium, and fission products. This plutonium is usually purified by two additional cycles of solvent extraction. In each, plutonium is made tetravalent and, in the A contactor, is extracted by TBP, together with the uranium, thus separating it from most of the fission products and neptunium, here mostly pentavalent. In the B contactor, the plutonium is stripped selectively from uranium into the aqueous phase, either by use of 0.35 *M* HNO_3 or by reducing plutonium to the trivalent state prior to stripping. In some plants, final plutonium purification is by anion exchange from nitrate solution. Some others use precipitation as plutonium oxalate with oxalic acid. Optimum conditions are reported [S23, p. 449] to be as follows: HNO_3, 1.5 to 4.5 *M*; $H_2C_2O_4$, 0.05 to 0.15 *M*. A problem in plutonium purification systems is heating and radiolysis of solvent or resin owing to the high alpha activity.

Many reprocessing plants are required to convert purified nitrates containing plutonium to oxides before shipment. Conversion processes are described in Chap. 9, Sec. 4.7.

4.11 Solvent Reuse

A well-designed Purex plant aims for as complete recycle of solvent as possible, to minimize costs of solvent makeup and disposal. Solvent from the uranium purification section usually contains so few contaminants or degradation products that it can be reused a number of times without cleanup. On the other hand, solvent that has processed solutions containing high activity of fission products and plutonium carries traces of these contaminants, uranium, nitric acid, dibutyl phosphate, and other radiolytic degradation products of TBP and dodecane. Uranium and plutonium should be recovered because of their value. Fission products should be removed to prevent product contamination in later cycles. Dibutyl phosphate should be removed because it forms strong complexes with tetravalent zirconium and plutonium that would impair ability of the solvent to reject zirconium and separate plutonium from uranium.

A typical solvent cleanup process for a 5 MT/day oxide fuel-reprocessing plant would be as follows.

Solvent from the first extraction cycle is transferred to a decanter of 5-m^3 volume. Small amounts of water are separated and sent to waste evaporation. The solvent is then pumped to an interim storage tank of 50-m^3 volume, which serves as feed tank for the wash columns.

The solvent is expected to contain approximately

Uranium:	1 mg/liter
Plutonium:	0.2 mg/liter
Fission products:	20 mCi/liter
Nitric acid:	0.001 *M*

It is washed successively with 0.01 *M* HNO_3, 0.2 *M* Na_2CO_3, 0.2 *M* NaOH, and 0.02 *M* HNO_3. The operations are performed in mixer-settler batteries. The uranium content is reduced by a factor of 10, the plutonium content by a factor of 50, and the fission-product content by a factor of 2. Dibutyl phosphate is removed as the water-soluble sodium salt. If sodium ion in the wastes from solvent cleanup is objectionable, ammonium carbonate has been proposed as an alternative [G3].

The cleaned solvent is collected in a 20-m^3 tank and, via sintered stainless steel filters, transferred to a larger storage tank for reuse. As the metal filters may become very radioactive, provision is made for back washing and remote replacement. Final solvent polishing by adsorption on anion-exchange resin has been found advantageous at Hanford [S4].

Solvent cleanup methods at the principal reprocessing plants have been summarized by Naylor [N2]. Methods used at Savannah River have been described by Orth et al. [O13].

4.12 Aqueous Waste Processing

Characterization of aqueous wastes. Reprocessing plants generate many aqueous waste streams, which differ widely with respect to their content of radioactivity, solids, and nitric acid. Radioactivity is characterized as *low-level, intermediate* (or *medium*)*-level,* or *high-level*, but with no generally accepted quantitative definition for each category. The terms *low, medium,* or *high activity* are also used. Until around 1975 low-level liquid wastes were regarded as those that could be discharged directly to groundwaters or the ocean because after natural dilution their radionuclide concentrations were below the maximum permissible values for general population exposure. More recently, the requirement that the concentration and amount of radioactive effluents be made as low as practicable has led to a preference for discharging no liquid wastes to ground or surface waters and disposing of excess water by evaporation into plant off-gases. About the only universally accepted usage is characterization of the aqueous waste stream from the first, codecontamination cycle as *high-level,* or *high-activity* waste. This waste contains many curies per liter and must be cooled to prevent self-boiling.

Liquid wastes are sometimes characterized as low-salt or high-salt wastes. Low-salt wastes are those that can be greatly reduced in volume by evaporation without precipitation of solids. High-salt wastes are those that can be only moderately reduced in volume.

Low-acid wastes are those whose nitric acid content is too low to justify fractionating the distillate for acid recovery. If necessary to remove the little acid present, this is better done by neutralization or ion exchange. High-acid wastes are those whose nitric acid can advantageously be recovered by fractional distillation.

Steps in aqueous waste processing. Because of the great variety of aqueous waste streams and differences in process flow arrangements in different plants, there is no standard flow sheet for processing aqueous wastes from the Purex process. Figure 10.10 shows the principal steps in one possible scheme for concentrating the wastes and recovering water and nitric acid from them.

Low-level, low-acid, low-salt wastes are neutralized if necessary and concentrated in a simple flash or vapor-compression evaporator to produce low-level waste concentrates and water sufficiently decontaminated for return to process. With simple wire-mesh entrainment separators, decontamination factors of several thousand are easily obtained.

The intermediate-level waste concentrator handles the low-level waste concentrate, contaminated aqueous solutions from solvent washing, and many other streams with appreciable solids content. With more exhaustive entrainment removal, as by partial reflux of condensate through a bubble-plate or sieve-plate column, water sufficiently pure for return to process can be produced. If concentrator bottoms are concentrated to the point of incipient crystallization, they are routed to waste storage. If still unsaturated, they are routed to the high-level waste concentrator.

The principal feed to the high-level waste concentrator is the high-level waste stream (HAW) from the codecontamination solvent extraction cycle. This typically contains about 2.5 mol HNO_3, 3 to 9 g fission products, and 400 to 1200 Ci/liter and generates heat at the rate of 2 to 6 W/liter. Additional feed may be intermediate-level waste concentrate and nitric acid evaporator bottoms. The high-level waste concentrator is usually operated at subatmospheric pressure and made of a corrosion-resistant material such as titanium, to extend life and minimize maintenance. Wastes are concentrated as far as possible without appreciable solids formation. If solids other than fission products are absent, a concentration of about 90 g fission products per liter can be obtained. Products are contaminated nitric acid overhead, slightly under 2.5 M, and evaporator bottoms, about 7 M in HNO_3. Because evaporator bottoms self-heat at a rate up to 1°C/min, the evaporator and the bottoms storage tanks must be provided with reliable cooling.

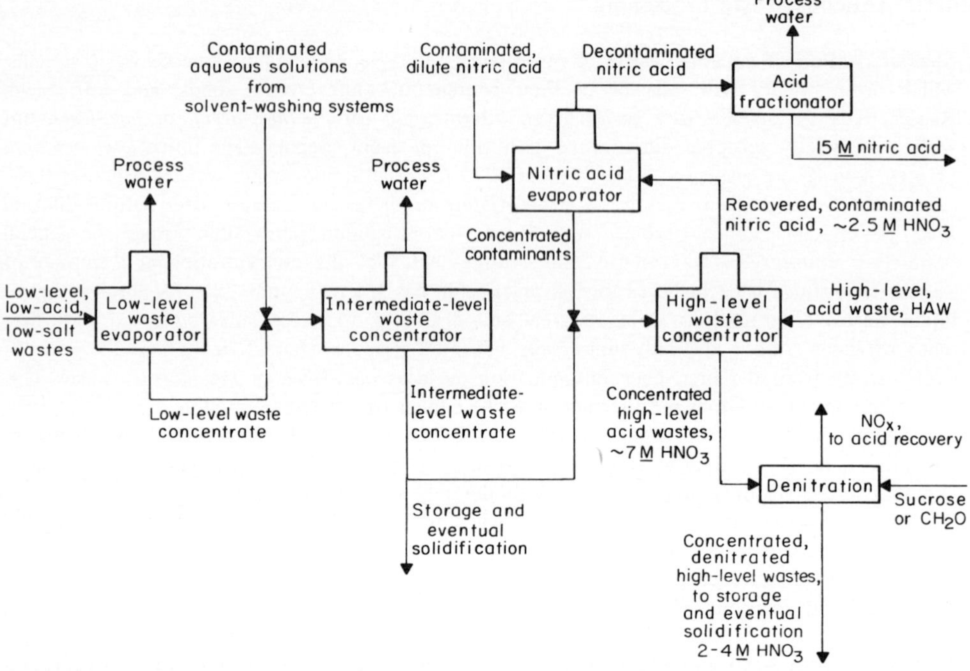

Figure 10.10 Steps in Purex waste processing and acid recovery.

Denitration of high-level wastes. To reduce corrosion during subsequent storage of concentrated high-level wastes, it is desirable to reduce their nitric acid content from around 7 to between 2 and 4 *M*, a value high enough to prevent major precipitation of hydrolyzed fission-product nitrates. To avoid loading the wastes with additional nonvolatile solids, nitric acid concentration can be reduced either by steam distillation or by reduction to gaseous nitrogen oxides by an organic reducing agent such as sucrose or formaldehyde.

Steam distillation is used in British Magnox plants. There, when the HAW stream has been evaporated to the desired solids content, water is substituted for radioactive feed for about 2 days and evaporation continued at constant volume. In this way the acidity of the concentrate is reduced to about 4 *M*.

Some of the reactions that occur with organic reducing agents are

With formaldehyde: $$HCHO + 4HNO_3 \rightarrow CO_2 + 3H_2O + 4NO_2$$

With sucrose: $$C_{12}H_{22}O_{11} + 24HNO_3 \rightarrow 12CO + 23H_2O + 24NO_2$$

Denitration with formaldehyde was first studied at Harwell [H5, H6]. Above 80°C the reaction proceeds smoothly, and the acidity can be reduced to 1 *M* in from 1 to 2 h. At lower temperature, if unreacted formaldehyde is allowed to accumulate, the reaction may become uncontrollable.

Denitration with sucrose was studied at Hanford [B16]. If the temperature is above 85°C the reaction proceeds smoothly through many intermediate stages, after an incubation period of several minutes. As with formaldehyde, sugar should not be added at lower temperature.

Treatment of process water. Water recovered in processing aqueous wastes usually contains so little radioactivity that it can be recycled to the reprocessing plant, but it must be treated further if it is to be discharged. Some water must be discharged because more water is fed to the plant in nitric acid and aqueous solutions than leaves it in aqueous wastes. This excess water is evaporated into gases and ventilating air discharged through the plant stack. Before excess water vapor is discharged in this way, it is treated to remove radioiodine and filtered to remove suspended solids.

4.13 Nitric Acid Recovery

The nitric acid evaporated from the high-level waste concentrator is too dilute and contains too much entrained radioactivity to be recycled without additional treatment. This acid, together with dilute acid waste streams from the uranium and plutonium purification solvent extraction systems, is decontaminated in the nitric acid evaporator. Entrainment can be suppressed by providing partial reflux through a few bubble-plate or perforated-plate trays, backed up by wire-mesh mist eliminators.

Decontaminated acid is separated in the acid fractionator into 15 *M* acid, water, and acid of intermediate concentration as needed in the reprocessing plant. The 15 *M* upper limit is the concentration of the nitric acid-water azeotrope. The acid evaporator and fractionator are made of stainless steel and usually run at an absolute pressure of 70 to 200 Torr to reduce corrosion and reaction of nitric acid with traces of TBP dissolved or entrained in the acid feed. At 200 Torr the azeotrope boils at 86.5°C.

4.14 Barnwell Nuclear Fuel Plant

As an example of a nuclear fuel reprocessing plant that makes use of some of the newest design concepts, a brief description will be given of the Barnwell Nuclear Fuel Plant, built and owned by Allied-General Nuclear Services (AGNS) at Barnwell, South Carolina. Construction of this plant, with the exception of its plutonium-conversion and waste solidification facilities, had been practically completed in 1977, when work was halted by President Carter's decision to suspend indefinitely commercial reprocessing in the United States. Allied-General is jointly owned by the Allied Chemical Corporation and the General Atomic Company, the latter jointly owned by the Gulf Oil Corporation and a U.S. affiliate of the Royal Dutch/Shell Group of Companies.

The Barnwell plant is designed to process fuel from commercial PWRs and BWRs. It will process routinely fuel having no more than 3.5 percent ^{235}U (or equivalent plutonium) prior to irradiation. On an annual basis, the average burnup is expected to be less than 35,000 MWd/MT and the average specific power less than 40 MW/MT. However, the design will permit the plant to process fuel containing up to 5 percent ^{235}U before irradiation, with burnups reaching 40,000 MWd/MT at a specific power of 50 MW/MT. Such high-enrichment batches will be handled with special techniques, including higher concentrations of soluble poison in the dissolver. Fuel will be cooled a minimum of 160 days before reprocessing.

Flow sheet and material quantities. Figure 10.11 is a schematic flow sheet showing the principal components of the reprocessing sections of the Barnwell plant [A3]. Table 10.7 [B21, M10] gives the flow rates and concentrations of the principal plant streams for the high-enrichment, high-burnup case. In Table 10.7 stream numbers in the first column correspond with stream numbers used in Fig. 10.11. The second column gives the stream designations used in AGNS reports [A2, A3]. Table 10.8 gives the fission-product content of the most important streams of Fig. 10.11. These material quantities were kindly provided by AGNS [B21, M10].

Performance specifications. The Barnwell plant is designed to recover at least 98.5 percent of the plutonium and 98.5 percent of the uranium in the plant feed [M10].

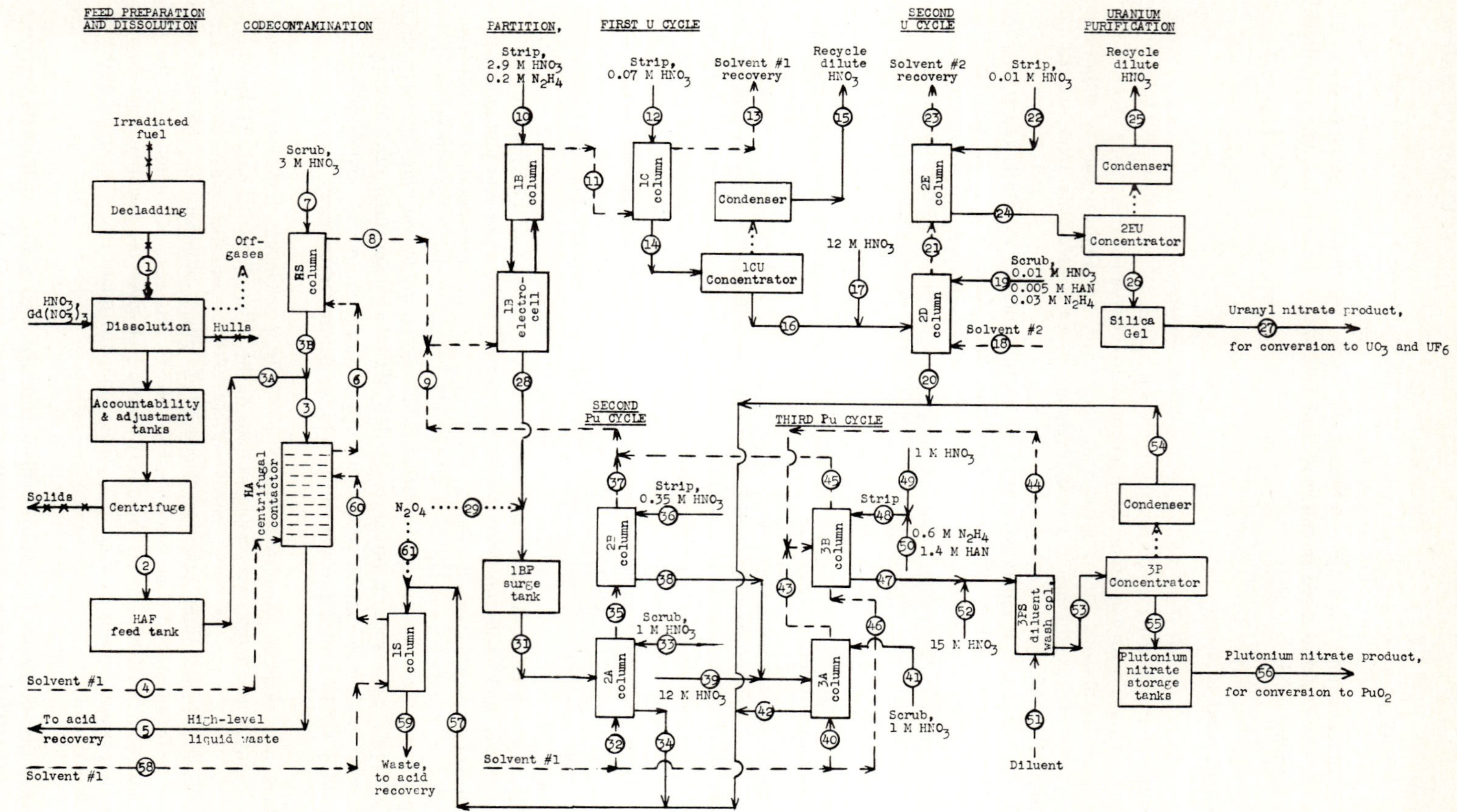

Fig. 10.11 Process flow diagram for solvent extraction section of Barnwell Nuclear Fuel Plant. x x x solids; —— aqueous; – – – organic; . . . . vapor.

Table 10.7 Flow rates and concentrations of principal streams in Barnwell Nuclear Plant[†,‡]

Stream number, Fig. 10.11	AGNS designation	Phase	Temperature, °C	Flow rate, liters/h	Moles uranyl nitrate per liter	Grams plutonium per liter	Moles nitric acid per liter	Concentration of other materials
1	Fuel	Solid	–	–	208 kg U/h	2.08 kg Pu/h	–	
2	HAF feed	A		723	1.21	(2.877)	2.44	(13.7 g FP/liter); 5.6 g Gd/liter
3A	HAF product	A	29	723	1.21	(2.877)	2.44	(13.7 g FP/liter); 5.6 g Gd/liter
3B	HSR	A	60	369	0.08	1.2	3.7	
3	HAF	A		1065	0.85	2.36	3.0	(9.3 g FP/liter); 4 g Gd/liter
4	HAX	O	31	1837	0	0	0	
5	HAW	A	39	994	0.002	0.02	2.5	10 g FP/liter; 4 g Gd/liter
6	HAP	O	40	2625	0.35	0.97	0.28	
7	HSS	A	35	350	0	0	3.0	
8	HSP	O	52	2632	0.34	0.80	0.15	
9	POR	O		257	0.065	(0.23)	0.04	
10	1BX	A	35	419	0	0	2.9	0.2 M N_2H_4
11	1BU	O	34	2864	0.31	0.0037	0.18	
12	1CX	A	60	3085	0	0	0.07	
13	1CW	O	60	2806	[0]	[0]	[0]	
14	1CU	A	47	3195	0.28	(0.0033)	0.17	
15	1CU Ohd	A		2540	[0]	[0]	[0]	
16	1UC	A		626	1.43	(0.017)	0.86	
17	2D makeup	A		107	0	0	12	
18	2DX	O	35	2206	0	0	0	
19	2DS	A	41	340	0	0	0.01	0.005 M HAN; 0.03 M N_2H_4
20	2DW	A	39	1000	0.02	(0.01)	1.8	0.002 M HAN; 0.01 M N_2H_4
21	2DU	O	41	2305	0.38	[0]		
22	2EX	A	60	2225	0	0	0.01	
23	2EW	O	60	2275	[0]	[0]	[0]	

(*See footnotes on page 495.*)

Table 10.7 Flow rates and concentrations of principal streams in Barnwell Nuclear Plant†,‡ (*Continued*)

Stream number, Fig. 10.11	AGNS designation	Phase	Temperature, °C	Flow rate, liters/h	Moles uranyl nitrate per liter	Grams plutonium per liter	Moles nitric acid per liter	Concentration of other materials
24	2E Prod	A		2330	0.38	[0]	0.01	
25	2UD	A		1708	0	0	0.002	
26	2UC	A		580	1.51	[0]	0.04	
27	U Product	A		557	1.57	3.7E-6	0.04	10 g Pu/billion g U
28	1BP	A	30	433	0.04	4.8	2.6	0.19 *M* N_2H_4
29	N_2O_4	V		–	–	–	–	N_2O_4
31 (no 30)	2AF	A	35	437	0.04	4.78	3.1	
32	2AX	O	31	146	0	0	0	
33	2AS	A	38	61	0	0	1.0	
34	2AW	A	34	498	[0]	0.04	2.85	
35	2AP	O	34	152	0.12	14	0.13	
36	2BX	A	35	132	0	0	0.35	
37	2BW	O	35	151	(0.059)	(0.44)	(0.05)	
38	2BP	A	35	133	0.07	15.55	0.44	
39	3S Butt	A		46	0	0	12	
40	3AX	O	31	83	0	0	0	
41	3AS	A	30	36	0	0	1	

42	3AW	A	34	213	[0]	0.08	2.95	
43	3AP	O	34	85	0.11	24	0.12	
44	3PSW	S		1.06	[0]	[0]	[0]	
45	3BW	O	35	105	0.09	[0]	0.054	
46	3BX	O	35	21.3	0	0	0	
47	3BP	A	35	35	<0.0058	58.5	0.58	0.56 *M* HAN; 0.29 *M* N_2H_4
48	3BX	A		34	0	0	0.2	0.7 *M* HAN; 0.3 *M* N_2H_4
49	3B acid	A		17	0	0	1.0	
50	3B red.	A		17	0	0	0	1.4 *M* HAN; 0.6 *M* N_2H_4
51	3PS scrub	S		1.06	0	0	0	
52		A		2	0	0	15	
53	3PSP	A		37	<0.0055	55.4	1.38	0.53 *M* HAN; 0.28 *M* N_2H_4
54	3PD	A		29	[0]	[0]	0.33	14.5 g-mol/h NO_x
55	3PC	A		8.4	<0.024	244	2.93	U/Pu < 1E-4
56	3PC	A		8.4	<0.024	244	2.93	U/Pu < 1E-4
57	1SF	A	38	1743	0.01	0.02	2.2	
58	1SX	O	31	620	0	0	0	
59	1SW	A	37	1730	[0]	[0]	2.2	
60	1SP	O	38	622	0.03	0.06	0.1	
61	N_2O_4	V		–	–	–	–	N_2O_4

†Data from Buckham [B21] and Murbach [M11].

‡Feed enrichment, 5 w/o ^{235}U; specific power, 50 MW/MT; burnup, 40,000 MWd/MT. A, aqueous; O, organic 30 v/o TBP in dodecane; S, solvent, dodecane; V, vapor; HAN, hydroxylamine nitrate; FP, fission products; [0] concentration not stated, but known to be small (assume zero); ( ), calculated by material balance.

Table 10.8 Fission-product content of streams in Barnwell Nuclear Fuel Plant, design estimate†

Stream number		Ruthenium-rhodium	Zirconium-niobium	Total fission products
Fig. 10.11	AGNS No.		Ci‡/kg uranium fed	
3A	HAF prod.	1650	1150	5980
3B	HSR	12.2	9.6	24.6
3	HAF	1660	1160	6010
5	HAW	1650	1150	5980
6	HAP	16.4	15.4	34.6
8	HSP	4.16	5.82	10.0
9	POR	0.0128	0.0175	0.0303
11	1BU	1.61	2.32	4.00
13	1CW	0.805	1.16	2.00
14	1CU	0.805	1.16	2.00
24	2E prod.	1.93E-4	1.06E-3	1.26E-3
27	U prod.	4.80E-5	3.37E-5	9.68E-5
34	2AW	2.53	3.47	6.00
35	2AP	0.0252	0.0348	0.0602
37	2BW	0.0127	0.0174	0.0301
38	2BP	0.0127	0.0174	0.0301
42	3AW	0.0126	0.0173	0.0299
43	3AP	1.26E-4	1.74E-4	3.01E-4
45	3BW	6.30E-5	8.70E-5	1.50E-4
47, 55	3BP, 3PC	6.30E-5	8.70E-5	1.50E-4
			μCi/g uranium product	
27	U prod.	0.048	0.034	0.097
			μCi/g plutonium	
35	2AP	2460	3400	5880
38	2BP	1280	1750	3030
43	3AP	12.8	17.7	30.7
47, 55	3BP, 3PC	6.4	8.8	15.2

†Data from Buckham [B21] and Murbach [M10].
‡To convert to hourly basis, multiply by 208 kg uranium fed/h.

Specifications for uranyl nitrate product call for a total beta-gamma activity less than 200 percent that of aged natural uranium if no less than 75 percent of the activity is due to ruthenium-rhodium; otherwise, a total beta-gamma activity less than 100 percent that of aged natural uranium, which is 0.68 μCi/g uranium (Chap. 8). Thus, uranium product betters this specification. The specified alpha activity of uranium product due to transuranium elements is less than 1500 disintegrations/min per g uranium. This is roughly equivalent to a plutonium content of 10 ppb (10^{-8}).

Specifications for plutonium product call for less than 100 ppm of uranium, a total gamma activity less than 40 μCi/g plutonium, and a zirconium-niobium activity less than 5 μCi/g plutonium. Plutonium nitrate product 3PC, stream 55, meets the total activity specification but does not quite meet the zirconium-niobium specification with the high burnup, 40,000 MWd/MT, feed used in this process example.

The plant is designed to discharge no liquid radioactive effluents to the environment [A2].

Gaseous effluents will be processed for the maximum practicable removal of radioactivity. Annual radiation exposure with the plant operating at capacity is expected to be 4.1 mrem maximum whole-body dose at the plant boundary and 116 man-rem to the entire population within 50 mi of the plant [A2]. These figures represent only 3 and 0.12 percent, respectively, of the average exposure from natural radiation.

Process building. The process steps shown in Fig. 10.11 are carried out in the main process building, a reinforced concrete structure with overall dimensions of 105 m by 87 m by 27.4 m high. Inside are the process cells, special heavily shielded enclosures for handling highly radioactive materials, and the surrounding work aisles and support facilities. The central control room is located on the second floor and is shielded and provided with special ventilation to permit occupancy during any radiation emergency.

Equipment within the high-radiation area that may experience mechanical or electrical failure or is subject to severe corrosion can be replaced remotely; the rest of this equipment is designed for direct maintenance after cleanout and decontamination.

Decladding, dissolving, and feed preparation. The plant will use the shear-leach method of feed preparation. Fuel elements up to 6 m long and 0.3 by 0.3 m in cross section can be sheared into 2.5- to 12.5-cm lengths.

Sheared fuel is fed to a receiving basket held in one compartment of a three-compartment semicontinuous dissolver. Nitric acid is fed continuously to the compartment and leach liquor is discharged continuously from it to the feed adjustment tank. When the basket is filled with fuel, the compartment holding it is rotated to a second position in which dissolution of the remaining oxide fuel in additional acid is completed. The basket and fuel hulls are finally rotated to a third position where water washes the residual leach liquor into the feed adjustment tank. The nitric acid contains 5.6 g of gadolinium as nitrate, to prevent criticality.

In the feed-adjustment tank plutonium is brought to the tetravalent state by addition of N_2O_4, if needed, and the uranium and nitric acid molarities are brought to about 1.2 M and 2.44 M, respectively, by addition of water and nitric acid as required. Samples are taken for input material accounting.

A centrifuge removes suspended solids from the dissolver solution ahead of the HAF feed tank.

Codecontamination. HAF product, stream 3A, is fed continuously to the codecontamination section of the solvent extraction system. This consists of the 10-stage HA centrifuge contactor, which serves as the extracting section, and the HS pulse column, which serves as the scrubbing section. In the HA contactor solvent HAX, stream 4, extracts more than 99 percent of the uranium and plutonium from the feed and less than 1 percent of the fission products. More than 99 percent of the fission products from the feed and the gadolinium leave the solvent extraction plant in high-level waste HAW, stream 5. In the scrubbing section HS, 3 M nitric acid scrub HSS, stream 7, removes about 70 percent of the residual fission products from HAP, the extract stream 6 leaving the extracting section.

The novel item in this part of the plant is the HA centrifugal contactor, named the Robatel, after its inventor. It was developed and built by the French firm Saint-Gobain Techniques Nouvelles. It consists of 10 centrifugal contacting stages, each similar in principle to the single stage described in Sec. 7.4 of Chap. 4, stacked vertically on a single rotating shaft. More information on this device has been given by Bebbington [B7] and Tarnero and Dollfus [T2].

Partitioning. Plutonium nitrate and uranyl nitrate in the extract HSP stream 8 leaving the HS column are separated in the 1B electrocell and pulse column. To this end, plutonium is reduced from the organic-soluble tetravalent state to the organic-insoluble trivalent state in the electrocell and returned to the aqueous phase 1BP, stream 28, by the strip solution 1BX, stream 10.

The novel feature of the partition section is the electrocell, an electrolytic pulse column patented by Gray, Schneider, Cermak, and Ayers [G13] of AGNS. One concept of the electrocell column is shown schematically in vertical section in Fig. 10.12. A porous, electrolytically conducting alundum tube D separates the outer annular anolyte compartment G from the inner cylindrical catholyte compartment C. Across the inner compartment are placed at regular intervals perforated plates F which enable this compartment to be operated as a pulse column.

The outer compartment G is provided with a cylindrical anode E made of corrosion-resistant metal wire mesh. The anode compartment is filled with aqueous nitric acid, around 3 *M*.

The inner compartment C is provided with a set of disk-shaped cathodes B made of similar mesh. These disks are spaced between the perforated plates. They are mounted on a cathode center post A, which also supports the perforated plates through insulators. The inner cathode compartment C contains a dispersion of the organic phase flowing up through a down-flowing continuous aqueous phase, as in a conventional pulse column.

The organic phase is a solution of uranyl nitrate and tetravalent plutonium nitrate in 30 v/o TBP. The aqueous phase is a solution of uranyl nitrate and tetravalent and trivalent

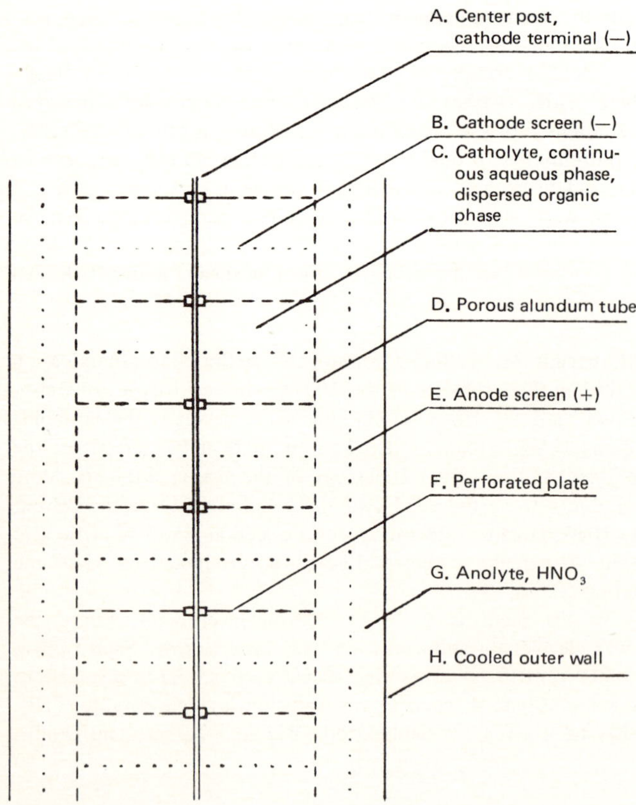

Figure 10.12 Vertical section of AGNS electrocell pulse column.

plutonium nitrates in nitric acid. When an electric current flows from anode to cathode, the following net reactions take place:

In the outer, anode compartment, $OH^- \rightarrow \frac{1}{2}H_2O + \frac{1}{4}O_2 + e^-$

In the inner, cathode compartment, $Pu^{4+} + e^- \rightarrow Pu^{3+}$

As the organic phase rises through the inner compartment, tetravalent plutonium is extracted from it by the counterflowing aqueous phase. There the tetravalent plutonium is reduced electrolytically to the organic-insoluble trivalent form. In this way, the organic phase leaving the top of the electrocell column becomes stripped of plutonium, and the aqueous phase 1BP, stream 28, leaving the bottom of the column carries practically all the plutonium in the feed. This aqueous effluent contains less than 2 percent of the uranium in the feed because of its relatively high, 2.6 *M*, nitric acid content and the high, 7:1, organic-to-aqueous flow ratio.

In the 1B column remaining traces of plutonium are stripped from the solvent by a strippant 1BX, stream 10, containing hydrazine as holding reductant. A decontamination factor of 200 for removal of plutonium from uranium is anticipated for the 1B columns. The big advantage of this partitioning system is that it adds no nonvolatile materials such as ferrous sulfamate to the system.

The uranium in the 1BU extract stream 11 is returned to the aqueous phase by 0.07 *M* HNO_3 strippant 1CX, stream 12, in the 1C column.

Uranium purification. Uranium is purified by a second solvent extraction cycle and by silica gel adsorption. To this end, the uranium-bearing aqueous stream 14, 1CU, leaving the 1C column is concentrated by evaporation, reacidified, and passed through the 2D column. There uranium is extracted by 30 v/o TBP in solvent 2DX, stream 18. Extract 2DU, stream 21, leaving this column is scrubbed with dilute nitric acid 2DS, stream 19, containing hydroxylamine and hydrazine. The scrub stream is intended to free the uranium of traces of plutonium and fission products, which leave column 2D in the aqueous raffinate 2DW, stream 20.

Uranium in the extract 2DU, stream 21, is returned to the aqueous 2E Prod, stream 24, by stripping with 0.01 *M* HNO_3, stream 22. This stream is concentrated by evaporation and passed through silica gel, which removes most of the remaining fission products from the uranyl nitrate product stream 27.

Plutonium purification. Plutonium in the aqueous 1BP stream 28 leaving the electrocell is purified by two additional cycles of solvent extraction. This plutonium is oxidized to the tetravalent state and reacidified by addition of N_2O_4, stream 29. In the 2A column extraction with solvent 2AX, stream 32, and scrubbing with 1 *M* HNO_3 2AS, stream 33, reduces the fission-product content of the extract 2AP, stream 35, to 1 percent that of the feed 2AF, stream 31. Plutonium and traces of uranium in the extract 2AP, stream 35, are returned to the aqueous phase in column 2B by stripping with 0.35 *M* HNO_3 2BX, stream 36. In streams containing plutonium at 35°C, the nitric acid concentration must not drop below 0.35 *M*; otherwise, insoluble plutonium polymers will form. These are inextractable by TBP and deposit in equipment, plug lines, and represent a criticality hazard. Conditions under which plutonium polymer forms are detailed at the end of this chapter.

In column 3A of the third plutonium cycle the fission-product content of the plutonium is reduced another factor of 100 by another extraction with TBP and scrubbing with 1 *M* HNO_3. The 3B column returns plutonium in the 3AP extract stream 43 to the aqueous phase by stripping with 1 *M* HNO_3 3BX, stream 48, containing hydrazine and hydroxylamine to reduce plutonium to the trivalent state. Uranium in the organic feed to 3B remains in the organic effluent 3BW, stream 45. Scrubbing with solvent 3BX, stream 46, reduces the uranium content of the aqueous plutonium product 3BP, stream 47, to less than 0.01 percent.

The dilute aqueous solution of plutonium nitrate, stream 47, is acidified with nitric acid and washed with dodecane diluent in the 3PS column to remove traces of TBP. After concentration by evaporation in the 3P concentrator, this becomes the concentrated plutonium nitrate product solution 56 of the solvent extraction portion of the plant.

Recycle streams. Uranium and plutonium remaining in solvent leaving columns 2B (stream 37) and 3B (stream 45) are recovered by recycling these as stream 9 (POR) to the 1B electrocell. Plutonium and uranium in aqueous streams leaving the 2D column (stream 20), the 3A column (42), the 2A column (34) and the 3P concentrator (54) are combined as stream 1SF (57). Plutonium in stream 57 is made tetravalent by N_2O_4 (61). This uranium and plutonium are extracted with 30 v/o TBP in the 1S column. The extract 1SP, stream 60, is returned to the third stage from the top of the HA centrifugal contactor.

Pulse columns. Dimensions of the pulse columns of the Barnwell plant are given in Table 10.9.

Solvent recovery. To prevent cross-contamination of products and to allow for the greater degradation of solvent by high concentrations of fission products and plutonium, two independent solvent recovery systems are provided. Solvent recovery system 1 processes solvent 1CW, stream 13, which has been used in the high-activity codecontamination, partitioning, and plutonium purification cycles. System 2 processes the low-activity solvent 2EW, stream 23, which has been used only for final uranium decontamination. Solvent in both systems is processed before recycle by a sodium carbonate wash, filtration and a nitric acid wash. System 1 also uses a second sodium carbonate wash.

Aqueous wastes. High-level aqueous wastes HAW (stream 5) and 1AW (stream 59) are concentrated by evaporation. The condensate, containing recovered nitric acid, is recycled. Processes for converting the concentrated, unneutralized liquid wastes to solid suitable for long-term storage have not been finalized.

Table 10.9 Pulse columns of Barnwell plant

	Inside diameter		Separating height	
Column number	in	cm	ft	m
HS	12	30	35	10.7
1B electrocell	22	56	11.5	3.5
1B	16	41	34.7	10.6
1C	20	51	21	6.4
2D	12	30	31.3	9.5
2E	20	51	21	6.4
2A	8	20	36.5	11.1
2B	8	20	23	7.0
3A	6	15	36.5	11.1
3B (upper)	7	18	31.7	9.7
3B (lower)	6	15	9.2	2.8
3PS	3	7.6	10	3.0
1S	12	30	25	7.6

Nitric acid recovery. Aqueous nitric acid is recovered from the high-level waste evaporator, from recycle streams 15 and 25, and from nitrogen oxides in dissolver off-gases as described in Secs. 4.11 and 4.13.

Effluent treatment. All gaseous effluents are processed for maximum practicable recovery of contained radioactivity. Most of this is in the fission-product gases vented from the shear and in the dissolver off-gases. These pass through a dust screen, condenser, mercuric nitrate iodine scrubber, and NO_x absorber. Gases leaving the NO_x absorber are combined with vessel off-gases from the solvent extraction system and pass through a second iodine scrubber, a silver zeolite iodine absorber, and a high-efficiency particle filter, before discharge with ventilating air through a 100-m stack.

Because of the extensive recycle of liquids, the net water feed is only 2 m^3/h. Excess water is vaporized into stack effluent. No liquid radioactive wastes are to be released to the environment.

The principal radionuclides to be discharged from the plant as originally designed are gaseous tritium and ^{85}Kr. The 100-m stack provides adequate dilution and dispersion. Equipment for removing ^{85}Kr will be added when fuel irradiated after 1982 is to be reprocessed.

4.15 Distribution Equilibria in Purex Systems

Sources of data. Knowledge of distribution equilibria in Purex systems is useful in designing the solvent extraction contactors for a Purex reprocessing plant and in predicting the change in performance of an existing plant when operating conditions are changed. The first experimental measurements were those of Codding et al. [C10] at the Knolls Atomic Power Laboratory. These were for an aqueous phase containing only water, uranyl nitrate, and nitric acid; an organic phase consisting of a 30 v/o solution of TBP in a commercial solvent (Gulf BT or Amsco 123-15); and for a temperature of 25°C. Distribution equilibria were represented graphically as plots of molarities of nitric acid and uranyl nitrate in the organic phase as functions of the corresponding molarities in the aqueous phase. After TBP became generally accepted as the preferred solvent for fuel reprocessing, many additional studies were made of distribution equilibria between aqueous nitric acid and TBP dissolved in a hydrocarbon diluent. These extended the early work to other hydrocarbon diluents, to temperatures other than 25°C, to TBP concentrations between 5 and 100 v/o, and to additional distributed components including plutonium, neptunium, and thorium.

It was found that distribution equilibria are not very sensitive to the composition of the hydrocarbon diluent, provided that it consists mostly of saturated (paraffinic or naphthenic) hydrocarbons containing about 12 carbon atoms per molecule. However, Purex plants now usually specify a synthetic, straight-chain, saturated hydrocarbon made by polymerizing and hydrogenating lower olefins, which contains an average of 12 carbon atoms and is mostly *n*-dodecane.

The SEPHIS[†] computer program was developed by Gronier [G16] for Purex equilibria in 15 v/o TBP. The program was adapted to the conventional 30 percent TBP Purex process by Richardson at Hanford [R7], and was further modified and generalized by Watson and Rainey [W5] at Oak Ridge. The SEPHIS code predicts the equilibrium distribution of uranium, tetravalent plutonium, nitric acid, and water between an aqueous phase containing these components and an organic phase containing TBP at any concentration between 2.5 and 100

[†]*S*olvent *E*xtracting *P*rocesses *H*aving *I*nteracting *S*olutes.

v/o, at temperatures between 0 and 70°C. The SEPHIS code may be used for uranium concentrations up to 2 *M*, Pu(IV) up to 0.2 *M*, and nitric acid up to about 6 *M*.

Scotten [S5] at Savannah River developed a similar program, SOLVEX.

Distribution coefficients. The equations used in the SEPHIS code to correlate distribution equilibria are too complex for hand calculation or for graphic representation in a few figures. To provide a semiquantitative basis for stage-to-stage calculation of the separation performance of Purex solvent extraction contactors described in Sec. 4.14, Figs. 10.13 through 10.16 have been plotted from computer printouts from the SEPHIS code kindly provided by Vaughen [V1]. These give distribution coefficients for nitric acid and uranyl nitrate at 40 and 55°C between an aqueous phase and 30 v/o TBP in normal dodecane.

Plutonium and fission-product and other inextractable nitrates are present in significant amounts in contactors HA and HS. Their effects on distribution coefficients of nitric acid and uranyl nitrate may be taken into account *approximately* by reading distribution coefficients of nitric acid or uranyl nitrate given in these figures at a value on the horizontal, x, axis equal to

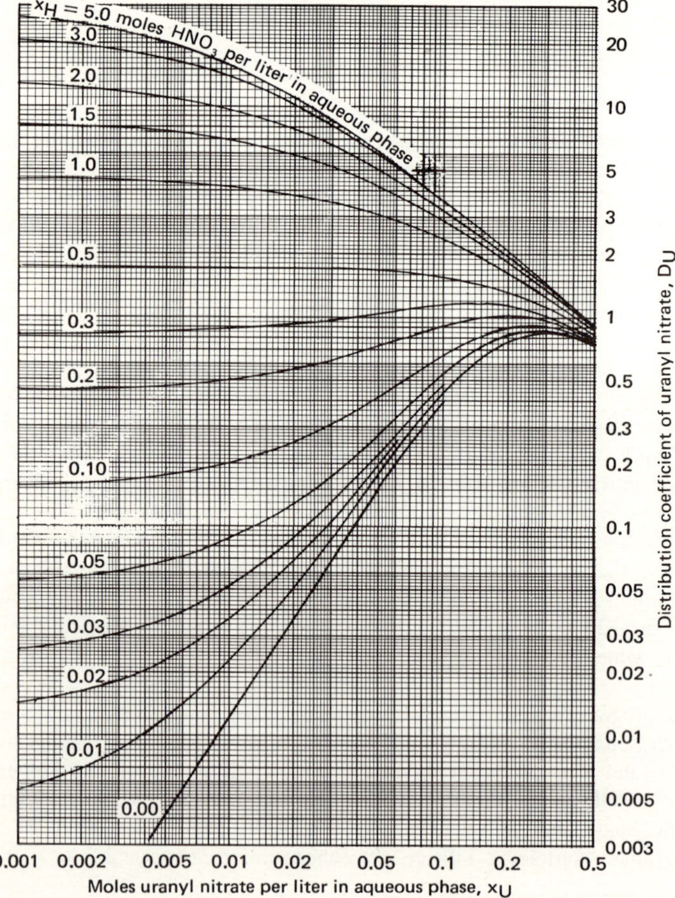

Figure 10.13 Distribution coefficient of uranyl nitrate between 30 v/o TBP in hydrocarbon diluent and aqueous nitric acid at 40°C, from SEPHIS code.

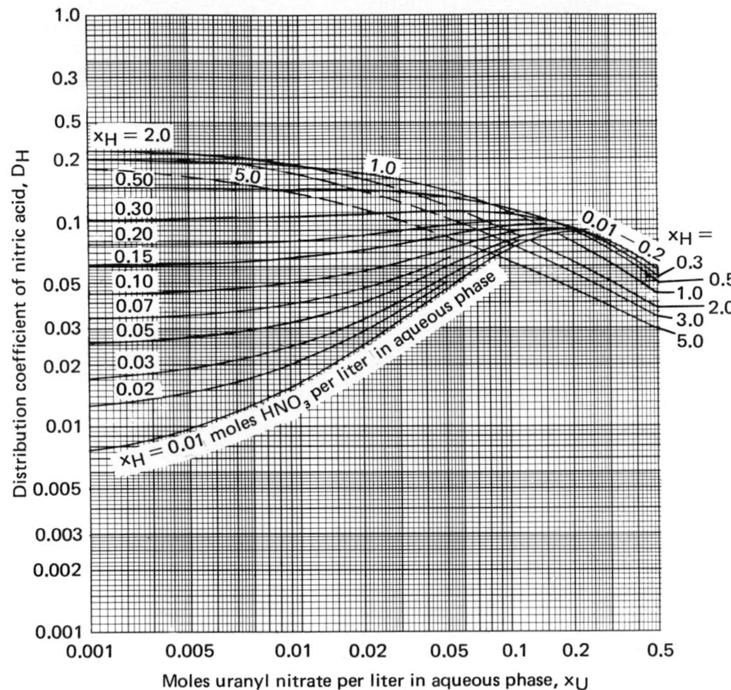

Figure 10.14 Distribution coefficient of nitric acid between 30 v/o TBP in hydrocarbon diluent and aqueous uranyl nitrate at 40°C, from SEPHIS code.

the *combined* molarities of $Pu(NO_3)_4$ and $UO_2(NO_3)_2$, from the curve whose designated nitric acid molarity equals the sum of the actual HNO_3 molarity *plus* the normality of inextractable nitrates. Problem 10.3 requires application of these adjustments.

Distribution coefficients for Pu(IV) may be obtained from Fig. 10.17. This plots the ratio of distribution coefficients of tetravalent plutonium to hexavalent uranium.

In the SEPHIS code [W5] this distribution coefficient ratio D_{Pu}/D_U is evaluated from

$$\frac{D_{Pu}}{D_U} = \left[0.20 + 0.55F^{1.25} + \frac{0.0074(x_{NO_3^-})^2}{(1.0 - 0.0724x_U - 0.13x_{Pu} - 0.0309x_H - 0.031x_S)^2}\right] \times \left[\exp 2700\left(\frac{1}{298} - \frac{1}{T}\right)\right] \tag{10.1}$$

where F = volume fraction of TBP in dry solvent

$x_{NO_3^-}$ = total nitrate molarity in aqueous phase

$$= 2x_U + 4x_{Pu} + x_H + x_S \tag{10.2}$$

x_U = uranyl ion molarity in aqueous phase

x_{Pu} = molarity of Pu^{4+} in aqueous phase

x_H = hydrogen ion molarity in aqueous phase

and x_S = molarity of NO_3^- associated with other nitrates in aqueous phase

$$= x_{NO_3^-} - 2x_U - 4x_{Pu} - x_H \tag{10.3}$$

The coefficients of x_U, x_{Pu}, x_H, and x_S are so nearly proportional to the charges of the respective cations that this equation may be simplified to

$$\frac{D_{\mathrm{Pu}}}{D_{\mathrm{U}}} = \left[0.3221 + \frac{0.0074(x_{\mathrm{NO_3^-}})^2}{(1 - 0.031x_{\mathrm{NO_3^-}})^2}\right]\left[\exp 2700\left(\frac{1}{298} - \frac{1}{T}\right)\right] \tag{10.4}$$

for $F = 0.30$ (30 v/o TBP). For uranium molarities under 0.5, the difference between Eq. (10.1) and Eq. (10.4) is less than 0.1 percent. Figure 10.17 is a plot of Eq. (10.4) for temperatures of 25, 40, and 55°C.

4.16 Example of Use of Purex Equilibrium Data

Use of these Purex equilibrium charts will be illustrated by calculating the number of equilibrium extracting and scrubbing stages needed in the uranium decontamination unit 2D of the Barnwell Nuclear Fuel Plant, data for which were given in Fig. 10.11 and Tables 10.7 and 10.8.

Table 10.10 gives flow rates and compositions for the streams to and from this unit. Process quantities taken directly from Tables 10.7 and 10.8, or calculated from them in

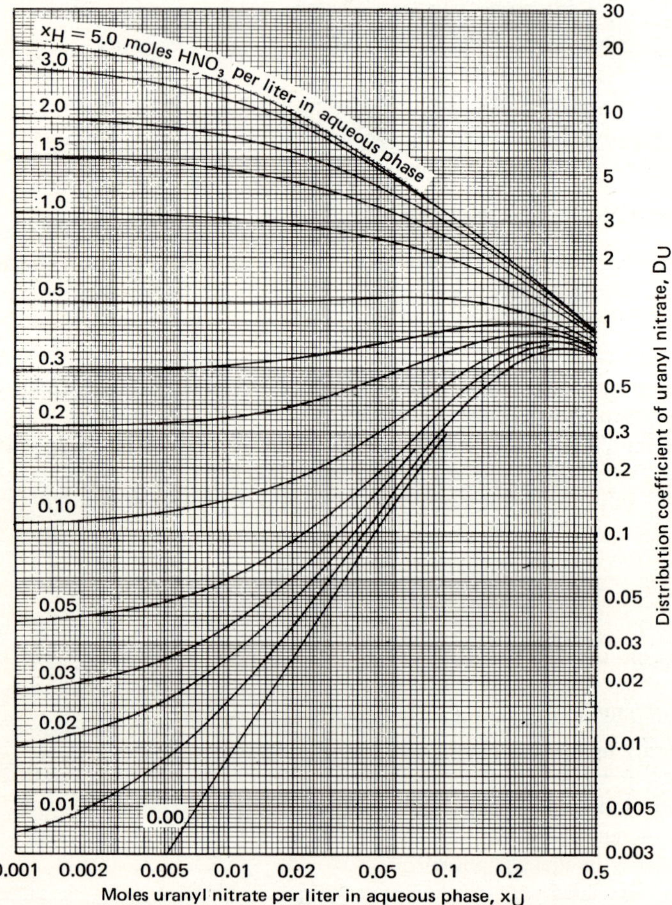

Figure 10.15 Distribution coefficient of uranyl nitrate between 30 v/o TBP in hydrocarbon diluent and aqueous nitric acid at 55°C, from SEPHIS code.

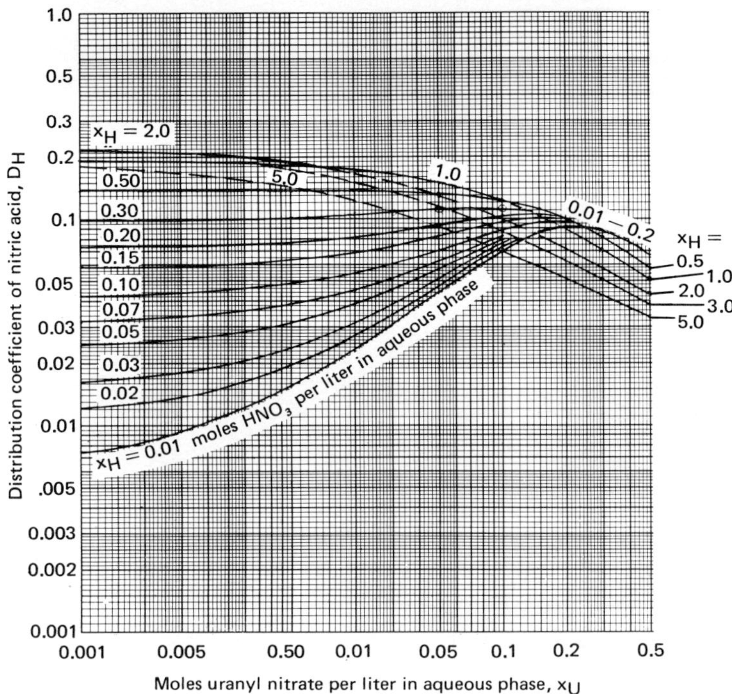

Figure 10.16 Distribution coefficient of nitric acid between 30 v/o TBP in hydrocarbon diluent and aqueous uranyl nitrate at 55°C, from SEPHIS code.

equivalent units, are in *italics*. The remaining quantities in Table 10.10 have been adjusted as stated in the footnotes for two reasons. (1) Component flow rates in some streams have been changed slightly to provide exact material balances. (2) Volume flow rates, which change by a few percent in Fig. 10.11, have been held constant to simplify calculation; compositions were adjusted where necessary to keep component flow rates unchanged.

Figure 10.18 shows the solvent extracting system to be analyzed and the nomenclature to be used. Input and output flow rates and concentrations are from Table 10.10. Three extracting and two scrubbing stages are shown, because the calculation next to be described indicates that between two and three theoretical extracting stages and between one and two scrubbing stages would be sufficient for the specified separation.

In this example, uranium and ruthenium are the key components whose compositions in feed, aqueous waste, and organic extract are specified. Nitric acid concentration in feed is specified, but its distribution between the two product streams must be found by trial, as in the zirconium-hafnium separation example in Sec. 6.5 of Chap. 4. The HNO_3 concentrations of 0.02 *M* in organic extract and 1.658 *M* in aqueous waste were found by trial to require the same number of scrubbing and extracting stages as the specified uranium and zirconium separation, as will now be shown, and hence represent the calculated distribution of nitric acid.

Table 10.11 gives steps in calculating concentrations of nitric acid, uranyl nitrate, and ruthenium as a function of stage number in the extracting section. Starting from the given aqueous concentrations x_1^E, marked with a †, the calculation proceeds through alternative distribution-equilibrium and material-balance calculations. No iterations are required, as the distribution coefficients of uranyl nitrate and nitric acid are available in Figs. 10.13 and 10.14 as functions of the first calculated *aqueous* concentrations.

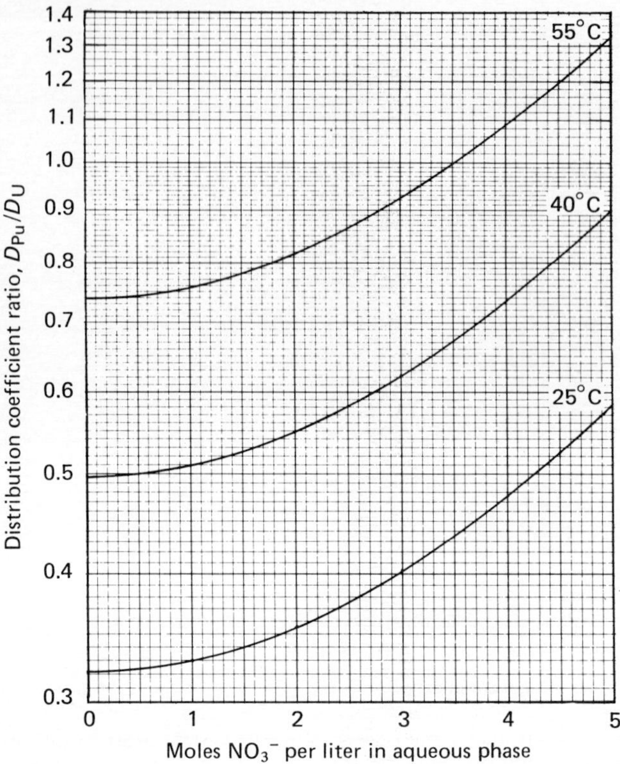

Figure 10.17 Distribution coefficient ratio, tetravalent plutonium to hexavalent uranium in 30 v/o TBP, from SEPHIS code.

Table 10.12 gives the steps in calculating concentrations of these three components in the scrubbing section, starting from the specified composition of the *organic* extract stream y_1^S, marked with a †. As the distribution coefficients of Figs. 10.13 and 10.14 are given as functions of the to-be-calculated aqueous composition, these must be found by the successive approximation procedure shown in the table. In this, distribution coefficients are assumed, trial aqueous concentrations are calculated, distribution coefficients are obtained from Figs. 10.13 and 10.14, and the process is repeated until calculated distribution coefficients are the same as assumed.

Figure 10.19 is a plot of the concentrations of ruthenium (bottom, circles) and nitric acid (top, squares) versus uranium concentration in the organic streams leaving the designated stages of the extracting section (filled symbols) and scrubbing section (open symbols). The intersection of the two bottom lines shows that the specified ruthenium-uranium separation would be obtained at a value of $n = 2.1$ ($N = 2.1$ theoretical extracting stages) and $m = 2.4$ ($M = 1.4$ theoretical scrubbing stages, because organic stream from stage m flows *into* scrubbing stage $m - 1$). The intersection of the two top lines shows that the assumed nitric acid-uranium separation would be obtained at the same values of $m = 2.4$ and $n = 2.1$, thus establishing that the assumed nitric acid concentrations in aqueous waste and organic extract streams are correct. Concentrations at the intersections of the two curves are the calculated values for the organic stream flowing from the extracting to the scrubbing section.

Table 10.10 Adjusted material balance for calculation of number of theoretical stages in uranium decontamination unit 2D of Barnwell Nuclear Fuel Plant[a]

Stream	In					Out		
	Feed	Acid	Scrub	Solvent	Total	Waste	Extract	Total
Number, Fig. 10.11	16	17	19	18		20	21	
Phase	Aqueous	Aqueous	Aqueous	Organic		Aqueous	Organic	
g-mol/liter								
HNO_3	*0.86*	*12*	*0.01*	0		1.658[b]	0.02[e]	
$UO_2(NO_3)_2$	1.431[b]	0	0	0		0.01864[d]	*0.38*	
Ci Ru/liter	*0.2675*	0	0	0		0.1560[b]	*0.0000174*	
Liters/h	*626*	*107*	*340*	2305[c]		1073[c]	*2305*	
g-mol/h								
HNO_3	*538*	*1284*	*3*	0	*1825*	1779[b]	46[e]	*1825*
$UO_2(NO_3)_2$	*896*	0	0	0	*896*	*20*	*876*	*896*
Ci Ru/h	*167.4*	0	0	0	*167.4*	167.36[b]	*0.0401*	*167.4*

[a] Quantities in italics evaluated from Tables 10.7 and 10.8.
[b] Adjusted to close material balance.
[c] Adjusted to keep constant volume flow rate.
[d] Adjusted to keep mass flow rate in residue the same as evaluated from Table 10.7.
[e] Nitric acid distribution cannot be specified in advance, but is confirmed by calculation of Tables 10.11 and 10.12.

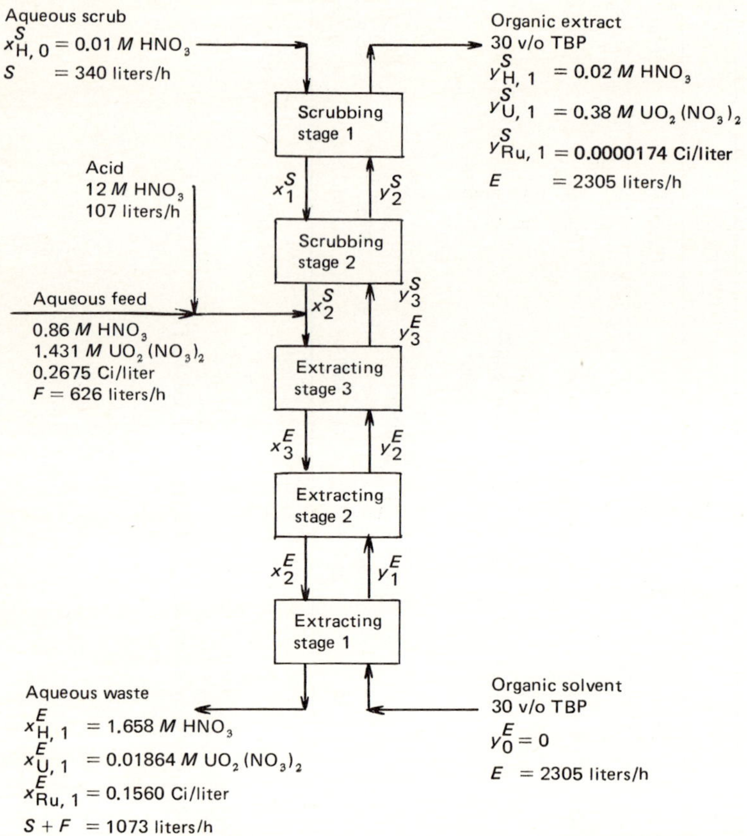

Figure 10.18 Nomenclature for uranium-ruthenium-nitric acid separation example, unit 2D of Barnwell Nuclear Fuel Plant.

4.17 Physical Properties of TBP and Its Mixtures with Hydrocarbons, Water, and Nitric Acid

The general use of TBP as extractant in reprocessing nuclear fuel is due to its selectivity for the actinides, its reasonably good stability against radiolysis and reaction with nitric acid, its nonflammability, and its ready availability at low cost. Because its density is close to water's and because of its high viscosity, for reprocessing TBP is diluted with a less dense, less viscous hydrocarbon. The diluents most inert to nitric acid and radiation are straight-chain paraffins. The diluent now usually chosen is a mixture of normal paraffins, mostly *n*-dodecane, because it is commercially available and provides a reasonable compromise between the desired low viscosity and high flash point. Maximum capacity of pulse columns is obtained at TBP concentrations between 20 and 30 v/o, the composition usually used in processing irradiated natural or slightly enriched uranium. TBP concentrations in the range of 2.5 to 7.5 v/o are used in processing fully enriched uranium or plutonium as one of the measures to avoid criticality.

Physical properties of pure TBP have been given in Table 4.5. Physical properties of *n*-dodecane and a 30 v/o solution of TBP in *n*-dodecane are summarized in Table 10.13.

When TBP and a hydrocarbon such as *n*-dodecane are mixed, a slight volume increase takes

Table 10.11 Concentrations in extracting section, uranium decontamination unit 2D

Stage number, n	1	2	3
Aqueous concentration, mol/liter			
$x_n^E = x_1^E + \dfrac{E}{S+F} y_{n-1}^E$ †			
HNO_3, $x_{H,n}^E$	1.658‡	2.264	1.891
$UO_2(NO_3)_2$, $x_{U,n}^E$	0.01864‡	0.2990	0.924
Ruthenium (Ci/liter), $x_{Ru,n}^E$	0.1560‡	0.1835	0.1570
Distribution coefficient			
HNO_3, $D_{H,n}$ (Fig. 10.14)	0.17	0.048	~0.026
$UO_2(NO_3)_2$, $D_{U,n}$ (Fig. 10.13)	7	1.41	~0.51
Organic concentration, mol/liter			
HNO_3, $y_{H,n}^E = D_{H,n} x_{H,n}^E$	0.282	0.1087	0.0492
$UO_2(NO_3)_2$, $y_{U,n}^E = D_{U,n} x_{U,n}^E$	0.1305	0.4216	0.47
Ruthenium			
Distribution coefficient, $D_{Ru,n}$			
From Fig. no.:	10.9	10.8	10.8
Value	0.082	0.0026	0.0031
Organic concentration, Ci/liter			
$y_{Ru,n}^E = D_{Ru,n} x_{Ru,n}^E$	0.0128	0.00048	0.00049

† $\dfrac{E}{S+F} = \dfrac{2305}{1073} = 2.148$

‡ Given concentrations.

Table 10.12 Concentrations in scrubbing section, uranium decontamination unit 2D

Stage number, m	1		2		3
Organic concentration, mol/liter					
$y_m^S = y_1^S + \dfrac{S}{E}(x_{m-1}^S - x_0^S)$ †					
HNO_3, $y_{H,m}$	0.02‡		0.0676		0.256
$UO_2(NO_3)_2$, $y_{U,m}^S$	0.38‡		0.442		0.447
Ruthenium (Ci/liter), $y_{Ru,m}^S$	0.0000174‡		0.000302		0.0112
Assumed distribution coefficient					
HNO_3, $D_{H,m}$	0.055	0.060	0.044	0.042	
$UO_2(NO_3)_2$, $D_{U,m}$	0.90	0.90	0.93	0.97	
Aqueous concentration, mol/liter					
HNO_3, $x_{H,m}^S = y_{H,m}^S / D_{H,m}$	0.364	0.333	1.536	1.609	
$UO_2(NO_3)_2$, $x_{U,m}^S = y_{U,m}^S / D_{U,m}$	0.422	0.422	0.475	0.456	
Distribution coefficient, from figures					
HNO_3, $D_{H,m}$ (Fig. 10.14)	0.058	0.060	0.042	0.042	
$UO_2(NO_3)_2$, $D_{U,m}$ (Fig. 10.13)	0.90	0.90	0.94	0.97	
Ruthenium, $D_{Ru,m}$ (Fig. 10.8)		0.0090		0.0040	
Aqueous ruthenium concentration, Ci/liter					
$x_{Ru,m}^S = y_{Ru,m}^S / D_{Ru,m}$		0.00193		0.076	

† $\dfrac{S}{E} = \dfrac{340}{2305} = 0.1475$; $x_{H,0}^S = 0.01$; $x_{U,0}^S = x_{Ru,0}^S = 0$.

‡ Given concentrations.

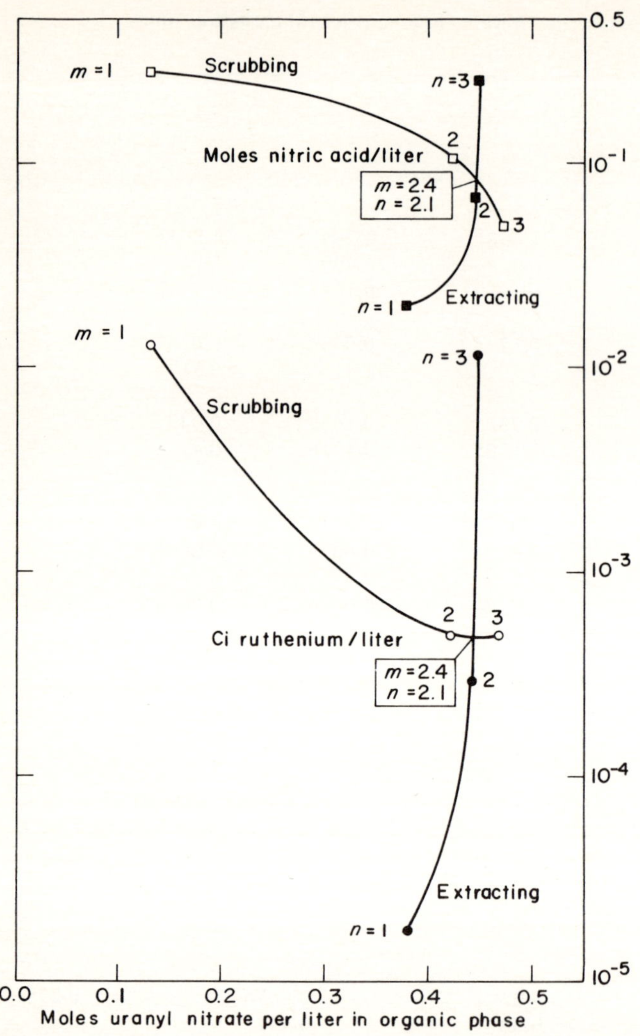

Figure 10.19 Ruthenium and nitric acid concentrations as function of uranium concentration in organic streams leaving stages of Fig. 10.18. Circles, ruthenium; squares, nitric acid; filled symbols, extracting section; open symbols, scrubbing section; top curves, moles nitric acid per liter in organic phase; bottom curves, curies ruthenium per liter in organic phase.

Table 10.13 Physical properties of dodecane and 30 v/o TBP in dodecane

	n-Dodecane	30 v/o TBP in dodecane
Molecular weight	170.34	
Density at 25°C, g/ml	0.749	0.814
Viscosity at 25°C, poise	0.0140	0.0173
Normal boiling point, °C	216	
Freezing point, °C	−9.6	
Refractive index at 25°C	1.4226	
Flash point, Tag closed cup, °C	74	78 (calc.)

place. For TBP contents between 15 and 45 v/o, the increase is about 0.2 percent of the volume of the separate constituents. In precise work it is thus necessary to specify whether v/o TBP is referred to the sum of the volumes of the separate constituents or to the mixture, as is done in this text. Then, the molarity of TBP is related to its volume percent by

$$\frac{\text{mol TBP}}{\text{liter}} = \frac{(972.4 \text{ g TBP/liter})(\text{v/o TBP}/100)}{266.32 \text{ g TBP/mol TBP}} = 0.03651 \text{ (v/o TBP)} \tag{10.5}$$

The viscosity η of mixtures of TBP and hydrocarbons is given within 20 percent by

$$\log \eta_{\text{mix}} = \upsilon \log \eta_{\text{TBP}} + (1 - \upsilon) \log \eta_{\text{HC}} \tag{10.6}$$

where υ is volume fraction TBP.

The mutual solubilities of water and TBP-dodecane mixtures are given in Table 10.14. The volume change when water dissolves in TBP-dodecane mixtures is negligible.

4.18 Degradation of TBP-Hydrocarbon Mixtures

Although TBP and the hydrocarbon diluent are comparatively stable compounds, they slowly react in Purex systems with formation of degradation products that impair separation performance. The principal deleterious reactions are reaction with radioiodine, hydrolysis, and radiolysis. These will be discussed in turn.

Reaction with radioiodine. Any iodine left in dissolver solutions slowly reacts with TBP and diluent to form iodine compounds that cannot be removed by subsequent alkaline washing. Thus, it is important to remove as much iodine as possible from the dissolver solution before solvent extraction and to use low-inventory contactors in the first, HA, extraction step.

Hydrolysis of TBP. Hydrolysis of TBP occurs stepwise via dibutyl and monobutyl phosphoric acid ($[C_4H_9O]_2PO_2H$ and $C_4H_9OPO_3H_2$) and leads eventually to phosphoric acid. Dibutyl phosphoric acid is the most abundant degradation product. Its rate of formation is influenced by temperature, the nitric acid concentration, the uranium content, and the presence of a diluent, beside the radiation dose. The acidic nature of these hydrolysis products allows in principle cleanup by an alkaline wash.

The effect of TBP degradation products, particularly of dibutyl phosphoric acid, is the formation of strong complexes with uranium(VI), plutonium(IV), zirconium, and niobium. The sequence of complexing strength is Zr > Pu(IV) > U(VI) > Nb.

The uranium and plutonium complexes are strong enough to remain in the organic phase during stripping. In reprocessing LWR fuel, uranium is mainly affected because of its great

Table 10.14 Mutual solubility of water and TBP-dodecane mixtures at 25°C

	g/liter	
v/o TBP	TBP in H_2O	H_2O in organic
10	0.18	1.2
20	0.24	3.5
30	0.27	7.2
40	0.285	11.5
60	0.31	23.7
100	0.42	64.6

excess. As a consequence, uranium is lost to the waste in the solvent wash. In LMFBR fuel, plutonium is taken up by the degradation products to a significant extent. It can be removed only by an alkaline wash with fluoride addition.

The other detrimental effect of TBP degradation is its complexing of zirconium. This increases the zirconium distribution coefficient and consequently decreases the decontamination coefficient. Moreover, solvent residual radioactivity is increased because of incomplete zirconium reextraction. Another and even more troublesome consequence of zirconium complexing is the formation of precipitates known as crud. This is a severe problem, particularly in mixer-settlers, and has led to a preference for pulsed columns or centrifugal contactors in the first extraction cycle when high-burnup fuel is to be processed.

Table 10.15 shows the effect of temperature and organic-phase nitric acid concentration on the rate of formation of dibutyl phosphate in 30 v/o TBP, as reported by Siddall [S15].

With the relative volumes of aqueous and organic phases usually present in Purex systems, the amount of DBP formed in the aqueous phase is much smaller than in the organic phase because of the low aqueous solubility of TBP. The practical consequence of these rates is that if the solvent is washed with water to remove HNO_3 and with aqueous sodium carbonate to remove DBP after less than 15 min contact with process solutions at temperatures under 70°C, the concentration of DBP in process contactors can be held so low that solvent separation performance is not degraded. The DBP concentration after 15 min at 70°C is approximately

$$\frac{(0.048 \text{ v/o/day})(15 \text{ min})}{1440 \text{ min/day}} = 0.0005 \text{ v/o} \tag{10.7}$$

Nitration and oxidation. Nitric acid does not react appreciably with TBP at temperatures up to 70°C. At sufficiently high temperatures, however, nitration and oxidation take place. In two instances reaction of TBP-hydrocarbon mixtures with hot, concentrated solutions of nitric acid and uranyl nitrate led to destructive explosions. At Savannah River in 1953 [C11], an evaporator was destroyed while concentrating a solution of nitric acid and uranyl nitrate that contained TBP and a kerosene diluent. At Oak Ridge in 1959 [A8], an explosion occurred in a radiochemical plant evaporator that was concentrating a nitric acid solution of plutonium nitrate possibly contaminated by TBP, diluent, and their radiation degradation products.

Because of these accidents laboratory studies were made at Hanford [W1] and Savannah River [C11, N5] to determine the conditions under which nitric acid solutions possibly containing TBP could be safely evaporated. Wagner [W1] reported that a "red oil" formed by extended refluxing of a concentrated aqueous solution of uranyl nitrate, nitric acid, and TBP decomposed autocatalytically when heated to 150°C. Nichols [N5] found that a mixture of 10.5 M HNO_3 and TBP enters into a runaway reaction when heated rapidly to 130°C, but not at 125°C. Protective measures recommended to prevent future explosive reactions were to (1) minimize the amount of TBP added to the evaporator; (2) permit TBP to steam distill during evaporation; (3) hold the temperature below 130°C until all the TBP has been distilled.

Table 10.15 Rate of formation of DBP in 30 v/o TBP

Moles HNO_3 per liter in solvent	Rate of DBP formation, v/o per day, at 25°C	40°C	70°C
0.2	0.0002	0.0010	0.033
0.4	0.0003	0.0017	0.043
0.6	0.0003	0.0015	0.048
0.8	0.0003	0.0015	0.051

Table 10.16 Effect of radiolytic energy density on decontamination in Purex process

Wh/liter	Effect
0.1	Process performance unimpaired
1.0	Noticeable but not too serious effects
10	Catastrophic loss in decontamination

Radiolysis. Radiation degrades both TBP and hydrocarbon diluent in Purex systems, with formation of molecular fragments, polymers, and nitration products. The main product, however, is the same as from hydrolysis, namely, DBP. The yield of DBP in radiolysis of TBP varies somewhat with the diluent used, water content, type of radiation, and dose rate. Baumgärtner and Ochsenfeld [B6] cite production of 20 to 30 mg DBP/liter in 30-min exposure of 30 v/o TBP to 0.2 Wh/liter of radiation in mixer-settlers processing fuel cooled 240 days after 33,000 MWd/MT burnup. Because the density of DBP is 1065 g/liter, the volume percent DBP was

$$(100)\left(\frac{0.02 \text{ to } 0.03}{1065}\right) = 0.0019 \text{ to } 0.0028 \text{ v/o} \tag{10.8}$$

Hence the radiation exposure that would produce the same amount of DBP as the 0.0005 v/o produced by hydrolysis for 15 min at 70°C, which has negligible effect on decontamination, is

$$\frac{(0.0005)(0.2 \text{ Wh/liter})}{0.0019 \text{ to } 0.0028} = 0.05 \text{ to } 0.04 \text{ Wh/liter} \tag{10.9}$$

In addition to DBP, ionizing radiation produces in TBP-hydrocarbon mixtures long-chain acid phosphate esters, nitrohydrocarbons, and nitrate esters that also complex uranium, plutonium, and zirconium, and that cannot be removed by simple alkaline washing. These must eventually be removed either by purging a fraction of the solvent or treating it with strong oxidants [B8].

Siddall [S15] summarizes the effects of increasing radiation exposure on decontamination in the first Purex extraction contactor as shown in Table 10.16. The power density, or dose rate, also has an effect on solvent performance. Baumgärtner [B5] cited experiments in which 1.2 Wh/liter, delivered to 20 v/o TBP in one pass through the HA and HS contactors, reduced the zirconium decontamination factor from 1000 to 10.

The principal causes of radiolysis in a Purex plant are beta and gamma radiation in the first extracting unit (HA) and alpha radiation from plutonium in the plutonium purification units. Accurate calculation of radiation absorption by solvent is difficult, because it depends on details of the dispersion of aqueous and organic phases and contactor geometry. Blake [B11] has given equations for estimating solvent radiation absorption when these details are known.

An upper bound for the exposure in the HA unit may be obtained by treating the solvent as uniformly dispersed as small droplets in the aqueous phase, assumed as containing all of the radiation sources. Then the radiation exposure is

$$R = vD[f + \theta(1 - f)]\, t \tag{10.10}$$

where R = exposure of organic phase in watt-hours per liter
t = residence time, hours, of organic phase in contact with aqueous
v = volume fraction of aqueous phase
D = power density in aqueous phase, watts per liter
f = fraction of radiation as beta radiation

$1 - f$ = fraction of radiation as gamma radiation

θ = fraction of gamma radiation absorbed in contactor

This equation will be applied to the HA contactor of the Barnwell plant. From Tables 10.7 and 10.8, the activity of the aqueous waste HAW stream is

$$\frac{(5980 \text{ Ci/kg U})(208 \text{ kg U/h})}{994 \text{ liters/h}} = 1251 \text{ Ci/liter} \tag{10.11}$$

The watts per curie in this stream may be obtained approximately from Table 8.7 as

$$\frac{1.96 \times 10^4 \text{ W/Mg}}{4.18 \times 10^6 \text{ Ci/Mg}} = 0.00469 \text{ W/Ci} \tag{10.12}$$

Hence

$$D = (1251 \text{ Ci/liter})(0.00469 \text{ W/Ci}) = 5.9 \text{ W/liter} \tag{10.13}$$

Blake [B11] gives approximate values for f (0.65) and θ (0.4). The volume fraction of aqueous phase v is lower with organic phase continuous than with aqueous. It will be assumed that the HA contactor will be run with organic phase continuous, with $v = 0.25$. Then

$$R = (0.25)(5.9 \text{ W/liter})[0.65 + (0.4)(0.35)]\,t = 1.17 t_h \text{ Wh/liter} \tag{10.14}$$

To keep organic exposure below 0.1 (Wh)/liter, the organic residence time should be below 0.1/1.17 h, or 5 min. Thus, short-residence-time contactors, like the centrifugal contactor specified for Barnwell, are desirable for this primary decontamination service.

5 REPROCESSING THORIUM-BASED FUELS

5.1 History

A mixture of thorium and fissile uranium (^{235}U and/or ^{233}U) has been used as fuel in the Indian Point 1 reactor of Consolidated Edison Company, the high-temperature gas-cooled reactor (HTGR) under development by General Atomic Company, and the light-water breeder reactor (LWBR) in the United States, and in the Arbeitsgemeinschaft Versuchsreaktor (AVR) reactor in Germany. This mixture also has been proposed as fuel for a heavy-water reactor and the German thorium high-temperature reactor (THTR).

At the end of irradiation in such reactors, fuel consists of a mixture of thorium, uranium containing fissile isotopes, and fission products. Figure 3.33 showed a fuel-cycle flow sheet for an HTGR. The Thorex process has been developed for recovering the uranium and thorium from such fuel cycles, freeing them from fission products and separating them from each other. The Thorex process will be described in this section. When the fuel being irradiated contains appreciable ^{238}U, the plutonium thus formed requires that a combination of the Thorex and Purex processes be used.

As in the Purex process, the Thorex process uses a solution of TBP in hydrocarbon diluent to extract the desired elements, uranium and thorium, from an aqueous solution of nitrates. Thorium nitrate however, has a much lower distribution coefficient between an aqueous solution and TBP than uranium or plutonium. To drive thorium into the TBP, the Thorex process as first developed at the Knolls Atomic Power Laboratory [H1] and the Oak Ridge National Laboratory [G14] added aluminum nitrate to the thorium nitrate in dissolved fuel. This had the disadvantage of increasing the bulk of the high-level wastes, which then contained almost as many moles of metallic elements as the original fuel. To reduce the metal content of the waste, the Oak Ridge National Laboratory in the late 1950s [R1, R2] developed the acid Thorex process, in which nitric acid is substituted for most of the aluminum nitrate in the first extraction section. The nitric acid is later evaporated from the wastes, as in the Purex process.

To simulate recovery of uranium and thorium from irradiated 6 percent uranium, 94 percent thorium fuel from the first loading of Consolidated Edison Company's Indian Point 1 nuclear power plant, Oak Ridge National Laboratory [R3] made small-scale experiments on application of the acid Thorex process to fuel containing the appropriate amounts of uranium and thorium, with tracer quantities of the principal fission products. Spent uranium-thorium fuel from the Indian Point 1 plant was subsequently processed by Nuclear Fuel Services, Inc., at West Valley, New York, for recovery of uranium, but without separation of thorium from fission products. No account of this separation has been published.

The other full-scale applications of the Thorex process have been to separation of ^{233}U from thorium irradiated at the U.S. Atomic Energy Commission's production reactors at Savannah River and Hanford. As the object of these irradiations was to produce ^{233}U of high isotopic purity for use in the first core of the LWBR, the burnup to which the fuel was exposed was low, and the concentrations of uranium and fission products in the irradiated thorium were much lower than will exist in power reactor fuel irradiated to full burnup. Nevertheless, the successful separation of uranium and thorium from each other and from fission products is significant confirmation of the workability of the Thorex process.

The Thorex separation campaign at Savannah River and the planned separation at Hanford have been described by Rathvon et al. [R5]. The Savannah River campaign used the older form of the Thorex process, with aluminum nitrate salting, and will not be described further.

The first separation campaign at Hanford, in 1966, was described in a classified report [I2]. The second campaign, in 1970, was described in detail by Jackson and Walser [J1] of the Atlantic Richfield Hanford Company. This used less aluminum nitrate than the Savannah River campaign and is closer to the acid Thorex process presently favored. A summary of the Hanford operation will be given in Sec. 5.5.

Irradiated fuel from the HTGR, AVR, and THTR differs from the fuel processed at West Valley, Savannah River, and Hanford in two important respects:

1. Uranium, thorium, and fission products are imbedded in a matrix of graphite instead of being in the form of oxides surrounded by metal cladding.
2. Design fuel burnup is between 60,000 and 100,000 MWd/MT, instead of a few hundred in the irradiated thorium processed at Savannah River and Hanford. This causes uranium and fission-product concentrations and fission-product radiation levels to be much higher than in the irradiated thorium processed at Savannah River and Hanford.

In the early 1970s Küchler and associates [K6, K7] of Farbwerke Hoechst adapted the acid Thorex process to fuel irradiated to burnups to 100,000 MWd/MT, such as are expected from the HTGR, AVR, and THTR. They made laboratory runs on spiked synthetic fuel simulating chemically high-burnup fuel. They also made hot-laboratory runs at the Kernforschungsanlage Jülich on 1 kg/day of fuel irradiated to burnups up to 54,000 MWd/MT. The flow sheet demonstrated in these hot-laboratory runs is described in Sec. 5.6.

Merz, Kaiser, and associates [M7, K1] are building the JUPITER (*JU*lich *PI*lot Plant for *T*horium *E*lement *R*eprocessing) at Jülich, planned for operation in the early 1980s. This plant is designed to reprocess 2 kg/day of heavy metal in mixed ThO_2-U_3O_8 fission products resulting from burning graphite-matrix fuel elements of the THTR reactor after irradiation to high burnup.

5.2 Decladding Thorium-based Fuels

The principal types of thorium-based fuel to which the Thorex process has been applied are

1. ThO_2-UO_2 fuel with stainless steel cladding
2. Irradiated ThO_2 with aluminum cladding

3. HTGR, AVR, or THTR fuel, consisting of particles of ThO_2 or ThC_2 and particles of UO_2 or UC_2 imbedded in a graphite matrix

Stainless steel-clad fuel. Mechanical decladding such as has been used extensively for UO_2 fuel is also the preferred method for decladding stainless steel-clad ThO_2-UO_2 fuel. Mechanical decladding by shearing fuel bundles was used successfully at the West Valley plant of Nuclear Fuel Services, Inc., for decladding this type of fuel from the first core of the Indian Point 1 reactor. Mechanical decladding could also be used for zircaloy-clad fuel, but with complications in subsequent dissolving because of reactivity of zircaloy with the HNO_3-HF reagent used in the dissolver.

Aluminum-alloy-clad fuel. With aluminum-alloy-clad fuel mechanical decladding, followed by *selective* dissolution of fissile and fertile material in nitric acid (such as was used for fuel clad with stainless steel), is not feasible because the aluminum alloy also is dissolved by nitric acid. Consequently, for aluminum alloy fuel, chemical decladding has been the preferred method. Two methods have been used for chemical removal of aluminum cladding from thoria-based fuel. In the Savannah River ^{233}U production campaign described by Rathvon et al. [R5], aluminum cladding was dissolved together with the ThO_2-UO_2 fuel in a mixture of nitric acid, hydrofluoric acid, and mercuric nitrate, the latter used to accelerate reaction of aluminum with nitric acid by intermediate formation of an aluminum amalgam. In the Hanford ^{233}U production campaign described by Jackson and Walser [J1], aluminum cladding was dissolved in a mixture of sodium hydroxide and sodium nitrate, which left undissolved the ThO_2-UO_2 fuel. After removal of most of the resulting solution of sodium aluminate, the solid ThO_2-UO_2 was dissolved in a mixture of nitric and hydrofluoric acids.

Fuel imbedded in graphite. Fuel for the HTGR under development in the United States was described briefly in Sec. 7.3 of Chap. 3. The preferred fuel [B20, R6] consists of hexagonal graphite blocks 79.3 cm high by 35.9 cm across flats, bearing longitudinal holes, some empty for coolant flow, others filled with fuel rods. The fuel rods consist of a graphite matrix in which are imbedded two kinds of microspheres. Thorium-bearing microspheres consist of ThO_2 with a two-layer "BISO" pyrolytic graphite coating. Uranium-bearing microspheres consist of uranium carbide UC_2, with a three-layer "TRISO" coating consisting of two layers of pyrolytic graphite separated by a mechanically strong, combustion-resistant layer of silicon carbide.

After irradiation this fuel is declad by crushing the graphite blocks to free the fuel-bearing particles. The mixture of graphite and fuel particles is then burned with oxygen at 875°C in a fluidized bed [N7, Y1]. This removes the graphite matrix and the BISO graphite coating of the ThO_2 particles. The TRISO-coated particles retain their silicon carbide coating. The denser ThO_2 residue from the BISO particles is separated from the lighter silicon carbide-coated residue from the TRISO particles by elutriation with CO_2 gas. The irradiated ThO_2 from the BISO particles is dissolved in thorex dissolvent, HNO_3-HF-$Al(NO_3)_3$ and reprocessed by the Thorex process.

The TRISO particles are crushed to break their silicon carbide coating. The UC_2 thus freed is then oxidized to U_3O_8, dissolved in nitric acid, and freed from fission products by the Purex process. More details of the two types of fuel particles and their treatment prior to reprocessing is given in Chap. 3, Sec. 7.3.

Other reactors of this type, such as the AVR and THTR reactors in Germany, use somewhat different fuel particles, such as mixed ThO_2-UO_2. However, the decladding procedure recommended still involves crushing the fuel, burning the graphite, and converting carbides to oxides.

A disadvantage of burning the graphite moderator is the larger volume of carbon dioxide, radioactive because of 5730-year ^{14}C, which must be removed completely from combustion products, probably by absorption as $CaCO_3$ with $Ca(OH)_2$ slurry [C16], and then stored as long-lived, low-level radioactive waste.

Attempts in the 1960s [B14] to avoid burning the graphite blocks were not successful. Even when the blocks were ground to under 200 mesh, the fuel was not completely leached from the graphite. And hydrocarbon reaction products of the carbides with nitric acid caused poor phase separation in later solvent extraction. Thus, all HTGR reprocessing systems now plan to convert fuel to UO_2 and ThO_2 before dissolution.

A reactor design in which a clean mechanical separation could be made of components bearing fuel nuclides and fission products from the bulk of the graphite moderator would obviate the need for burning the moderator and would reduce greatly the amount of CO_2 to be formed, absorbed, and stored as $CaCO_3$.

5.3 Dissolution of ThO_2-UO_2 Fuel

Unlike irradiated UO_2 fuel, which dissolves readily in hot nitric acid, irradiated ThO_2-UO_2 fuel dissolves only very slowly and incompletely in this reagent. After extensive research, Oak Ridge National Laboratory concluded [B14] that the best reagent for producing a nitrate solution from ThO_2-UO_2 fuel was a mixture of nitric and hydrofluoric acids. This reagent has two serious drawbacks:

1. Mixed nitric and hydrofluoric acids react also with stainless steel and zircaloy, so that both the cladding and the stainless steel dissolver itself are attacked
2. Dissolution of ThO_2 is much slower than that of UO_2 in nitric acid.

Oak Ridge found [E1] that corrosion of 304L and 309SCb stainless steel by the mixture of HNO_3 and HF could be reduced to an acceptable level by addition of aluminum nitrate, $Al(NO_3)_3$, to the mixture of HNO_3 and HF without decreasing the rate of solution of ThO_2 by more than 20 percent. The aluminum nitrate acts by complexing the fluoride ion. The composition recommended for the solvent is 13 *M* HNO_3, 0.05 *M* HF, 0.1 *M* $Al(NO_3)_3$. Reddick [R6] has summarized the extensive research that has been conducted on dissolution of ThO_2 in HNO_3-HF-$Al(NO_3)_3$ mixtures. This procedure still has drawbacks. Parts of a stainless steel dissolver exposed to hot HNO_3-HF vapors, which contain no complexing $Al(NO_3)_3$, are not protected and will corrode. Aluminum nitrate is only partially effective in preventing reaction of HF with zircaloy cladding. Aluminum nitrate increases the volume of nonvolatile solids in the waste.

The rate of dissolution of ThO_2-UO_2 fuel is higher the lower the density of the fuel and the smaller the particle size. For example, with 200 percent stoichiometric excess of reagent, from 25 to 40 h were required to dissolve completely pellets 0.66 cm in diameter having 90 to 95 percent of theoretical density, whereas in 5 h, 99 percent of such fuel dissolved when the fuel was first crushed to under 100 mesh [B13]. Fuel with only 60 percent of theoretical density dissolved almost 10 times as fast as fuel having 90 percent density.

5.4 Feed Pretreatment

The thorium nitrate solution from the dissolver will be about 9 *M* in nitric acid. To obtain satisfactory decontamination of thorium from fission-product protactinium, ruthenium, and zirconium-niobium, it was found necessary to remove all of the nitric acid from the solution *and* make the solution around 0.15 *M* acid-deficient in nitrate ion by converting a fraction of the $Al(NO_3)_3$ to a water-soluble basic nitrate. This also converts the readily hydrolyzed nitrates of these fission products to basic nitrates that are less extractable than the species present in the acid dissolver solution.

In the initial development of the Thorex process [S9], the feed was made acid-deficient by evaporation until the boiling point reached 155°C. Trouble was experienced with corrosion and with precipitation of solids. The procedure finally adopted [R2] is shown in Fig. 10.20. The dissolver solution is evaporated until its boiling point reaches 135°C, at which point about 70 percent of the original volume has been evaporated and the nitric acid concentration is down to 3 *M*. Further stripping at constant volume and a constant temperature of 135°C is carried out by adding water and boiling off aqueous nitric acid until the solution is 0.2 *M* acid-deficient. The solution is finally diluted with water to make it around 0.15 *M* acid-deficient.

Prior to solvent extraction, the solution is treated with 0.02 *M* $NaHSO_3$ at 55°C for 1 h to convert ruthenium to a less extractable form.

If the irradiated thorium to be processed still contains appreciable protactinium activity from 27-day ^{233}Pa, 97 percent of this can be removed and recovered by adsorption on unfired, porous Vycor glass [G10]. At a flow rate of 1.57 ml/(min·cm^2), columns containing 100-120 mesh Vycor, water-cooled to prevent boiling, could be loaded with 3.1 g protactinium per kg glass with only 3 percent break-through. Washing with eight volumes of 10 *M* HNO_3, 0.1 *M* $Al(NO_3)_3$ removed most of the uranium and thorium. Then, elution with 0.5 *M* oxalic acid recovered 98.5 percent of the protactinium at an average concentration of 1.46 g/liter.

5.5 Thorex Solvent Extraction at Hanford

Jackson and Walser [J1] have given a detailed description of the operations conducted in 1970 by the Atlantic Richfield Hanford Company to separate and decontaminate ^{233}U and its associated thorium from aluminum-clad thorium dioxide irradiated to low burnup in the Hanford reactors.

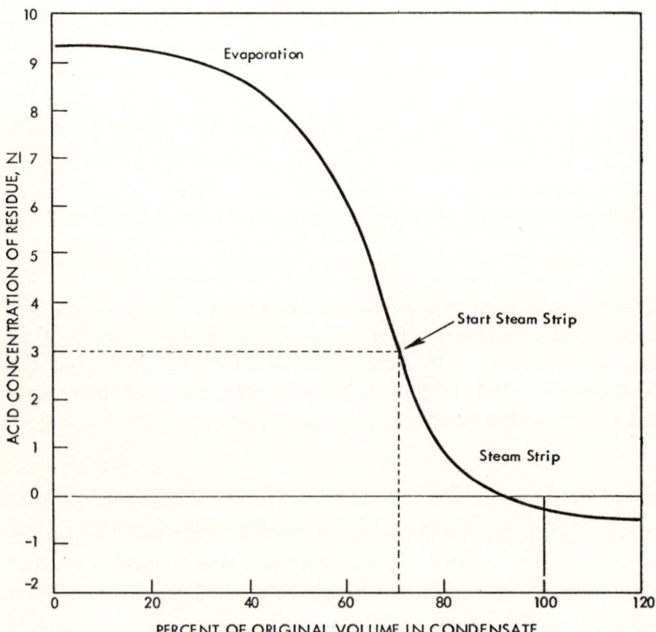

Figure 10.20 Evaporation and steam stripping of thorium nitrate solution for solvent extraction feed.

Table 10.17 Feed compositions in Hanford Thorex operations

	Dissolver solution	Solvent extraction feed
Molarity HNO_3	6.15	−0.10
$Th(NO_3)_4$	0.736	1.50
KF	0.039	0.071
$Al(NO_3)_3$	0.118	0.26
$NaNO_3$	0.20	0.41
KNO_3	–	0.0078
Grams ^{233}U per liter	0.30	0.61

Dissolution. The aluminum cladding was dissolved in a solution of mixed sodium nitrate and sodium hydroxide, and the undissolved uranium, thorium, and fission-product oxides were separated by filtration and centrifugation. The oxides, together with some adherent sodium hydroxide, sodium nitrate, and sodium aluminate, were dissolved in a mixture of boiling 13 *M* nitric acid, aluminum nitrate, and potassium fluoride. From 16 to 48 h were required for dissolution. After dissolution was complete, the composition of the solution was as given in the first column of Table 10.17.

Feed adjustment. Dissolver solution was concentrated by evaporation till the boiling temperature reached 135°C and the density 2.35 g/liter. This increased the thorium concentration to 3.0 *M*. One difficulty with this operation was volatilization of some of the ruthenium as RuO_4. The solution was next made 0.20 *M* acid-deficient by steam stripping at constant volume. After cooling to 70°C, water was added to bring the solution to the desired composition of 1.5 *M* thorium and 0.10 *M* nitric acid-deficient for solvent extraction feed.

Feed composition was as given in the second column of Table 10.17.

Solvent extraction. The aqueous feed solution of nitrates was separated into decontaminated uranium product, decontaminated thorium product, and high-level fission-product waste by solvent extraction at 30°C with 30 v/o TBP in normal paraffin hydrocarbon diluent ($nC_{10} - nC_{14}$), with controlled amounts of nitric acid used as salting agent. The flow sheet, described in detail by Jackson and Walser [J1], used a codecontamination and partition cycle to separate feed into partially decontaminated uranium, partially decontaminated thorium, and high-level waste, as described in the following section. The uranium was purified by two additional cycles of solvent extraction followed by cation exchange. The thorium was purified by one additional cycle of solvent extraction. Overall separation performance and fission-product decontamination in the final group of runs, batches 2-1 to 2-44, is summarized in Table 10.18. A total of 285.5 short tons of thorium containing 452.6 kg of uranium was processed.

Codecontamination and partition cycle. Because the codecontamination and partition cycle is the critical step in the acid Thorex process, it will be described in more detail. In this cycle, shown in Fig. 10.21, most of the fission products were separated from the uranium and thorium, which were then separated from each other. The four solvent extraction units, HA, 1BX, 1BS, and 1C, were pulse columns with dimensions given in the figure.

Feed HAF to the decontamination column HA had been adjusted to −0.1 *M* HNO_3 acid-deficient and 1.5 *M* $Th(NO_3)_4$. It contained 0.61 g uranium/liter and the amounts of KF, $Al(NO_3)_3$, $NaNO_3$, and KNO_3 shown in the figure. The KF had been added to catalyze

Table 10.18 Separation performance of batches 2-1 to 2-44 in Hanford 1979 Thorex campaign

	Thorium product	Uranium product
Product yield, based on reactor receipts	96.3%	96.4%
ppm uranium in thorium product	9.3	–
ppm thorium in uranium product	–	60
Alpha activity, counts/g-min	–	2.2×10^{10}
Gamma activity, μCi/g	0.45†	0.35
Decontamination factor based on feed		
Protactinium	180	3.5×10^6
Zirconium-niobium	5300	3.3×10^7
Ruthenium-rhodium	110	6.1×10^5

† Fission products.

dissolution of ThO_2. The $Al(NO_3)_3$ and $NaNO_3$ came from a residue of the Na_3AlO_3 produced by reaction of cladding with NaOH. The $Al(NO_3)_3$ was left in the dissolver to complex fluoride ion and reduce corrosion of stainless steel.

In the extracting section of the HA column, uranium and thorium in the aqueous feed were transferred to the organic phase by HAS solvent, 30 v/o TBP in dodecane, flowing at a ratio of 940 volumes to 100 volumes of aqueous feed HAF plus 130 volumes of aqueous scrub HAS. The solvent reduced concentrations in high-level waste HAW to 0.29 g thorium/liter and <0.0003 g uranium/liter. These represented losses to waste of 0.2 percent thorium and <0.12 percent uranium. To keep these losses this low, it was necessary to raise the distribution coefficients of thorium and uranium by increasing the HNO_3 concentration in the column above the acid-deficient condition of the feed. HNO_3 addition was divided between 130 volumes of 1.0 *M* acid added at the top at HAS and 18 volumes of 13 *M* acid added near the bottom at HAS. The HAS scrub added at the top provided enough nitrate ion to drive uranium into the organic phase, without greatly increasing the distribution coefficients of fission products. Thorium distribution coefficients above the HAF feed point were also high enough to drive thorium there into the organic phase. Below the feed point, however, in the extracting section, as the $Th(NO_3)_4$ concentration decreased, it was necessary to increase its distribution coefficient. This was done by adding the HAX stream, 13 *M* in HNO_3, near the bottom of the extracting column.

By splitting the HNO_3 addition in this way, with high HNO_3 concentrations only near the bottom of the column where the $Th(NO_3)_4$ concentration was low, formation of a second organic phase was avoided. Conditions for second-phase formation are shown in Fig. 10.27. This split feed of HNO_3 also reduced extraction of fission products.

In the scrubbing section of the HA column, fission products were scrubbed from the organic phase leaving the extraction section by the HAS scrub stream. It contained 0.01 *M* H_3PO_4 to complex protactinium and zirconium-niobium and reduce their extraction. It also contained 0.01 *M* ferrous sulfamate to reduce plutonium and chromium corrosion product to inextractable species.

The HA column also processed 223 volumes of recycle solvent HAO containing low concentrations of thorium and uranium.

In the 1BX column thorium in solvent HAP leaving the HA column was returned to the aqueous phase by stripping with 0.2 *M* HNO_3 thorium strippant 1BX. This reduced the thorium/uranium ratio in the solvent 1BU leaving the column to about 1:5.5.

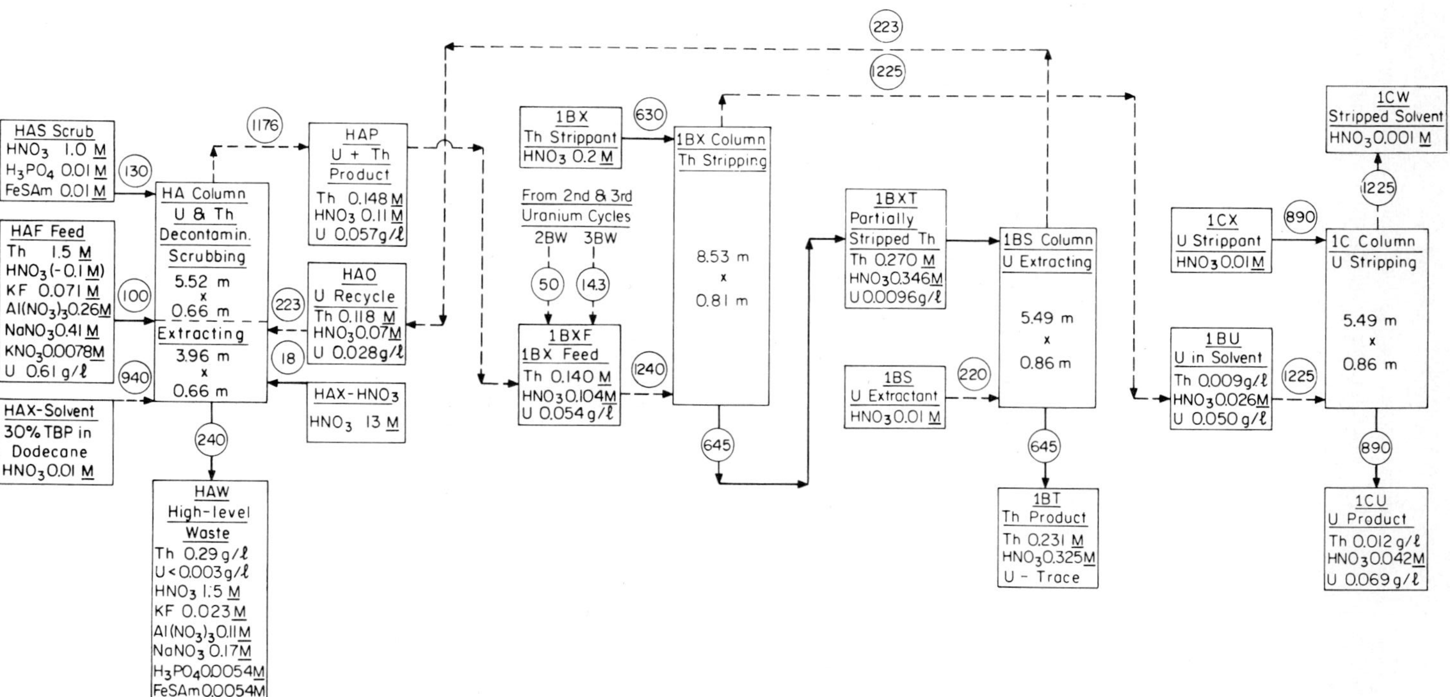

Figure 10.21 Codecontamination and partition cycle of 1970 Hanford acid Thorex operation. —— aqueous; – – – organic. Relative flow in numbered circles: 100 = 0.73 m^3/h.

Uranium in the 1BU solvent stream was transferred to the aqueous phase in the uranium-stripping column 1C by back-extraction with 0.01 *M* HNO_3 and left this section of the plant as crude uranium product 1CU. Thorium was removed from this crude uranium in the second and third uranium cycles (not shown) and was returned to 1BXF feed in the 2BW and 3BW streams.

Uranium in the partially stripped thorium stream 1BXT leaving the 1BX column was extracted from the thorium in the 1BS column by additional 30 v/o TBP in dodecane containing 0.01 *M* HNO_3. The aqueous stream leaving 1BX was the crude thorium product 1BT. The organic stream HAO leaving 1BS contained some uranium and thorium and was recycled to the HA column.

Decontamination factors in this codecontamination and partition cycle were as follows:

	Thorium 1BT	Uranium 1CU
Protactinium	85	50,000
Zirconium-niobium	180	32,000
Ruthenium-rhodium	105	115

Two additional uranium purification cycles (not shown) removed the thorium and small amounts of fission products remaining in the crude uranium stream ICU and returned the thorium to the first cycle in streams 2BW and 3BW.

5.6 Two-Stage Acid Thorex Process for High Burnup Fuel

Küchler and associates [K6, K7] of Farbwerke Hoechst have investigated the modifications necessary in the acid Thorex process to enable it to handle (1) the high concentration of fission products present in fuel with the burnups of up to 100,000 MWd/MT expected in fuel from the HTGR, AVR, and THTR, and (2) uranium concentrations of up to 20 percent in thorium, which may be used in these reactors when fissile uranium is diluted with ^{238}U to deter its use as a nuclear explosive. They found two difficulties with the acid Thorex process flow sheets previously used at Oak Ridge [B14] and Hanford [J1]:

1. A second organic phase formed when the thorium concentration in first-stage solvent extraction feed was as high as 1.5 *M*.
2. Hydrolysis products of fission products precipitated when the feed was made acid-deficient.

To avoid these difficulties they reduced the thorium content of solvent extraction feed to 1.15 *M* and developed a two-stage acid Thorex process. In this process thorium and uranium were coextracted from an acid feed to separate them from most of the fission products and then stripped back into the aqueous phase. By this means fission products were removed to such an extent that the Thorex process with acid-deficient feed could be used in the second stage without causing them to precipitate.

First stage. The flow sheet recommended by Küchler et al. [K7] for the first stage of this two-stage process is shown in Fig. 10.22. Adjusted feed is 1.15 *M* in thorium and is assumed to contain from 4 to 20 percent as much uranium. The nitric acid content of feed is made from 0.7 to 1.1 *M*, depending on its uranium content. One volume of feed is extracted with 9.5 volumes of 30 v/o TBP in unit 1A, with eight extracting stages and eight scrubbing stages. One volume of 0.1 *M* HNO_3 is used for aqueous scrub, and 0.22 volume of 13 *M* HNO_3 is added to

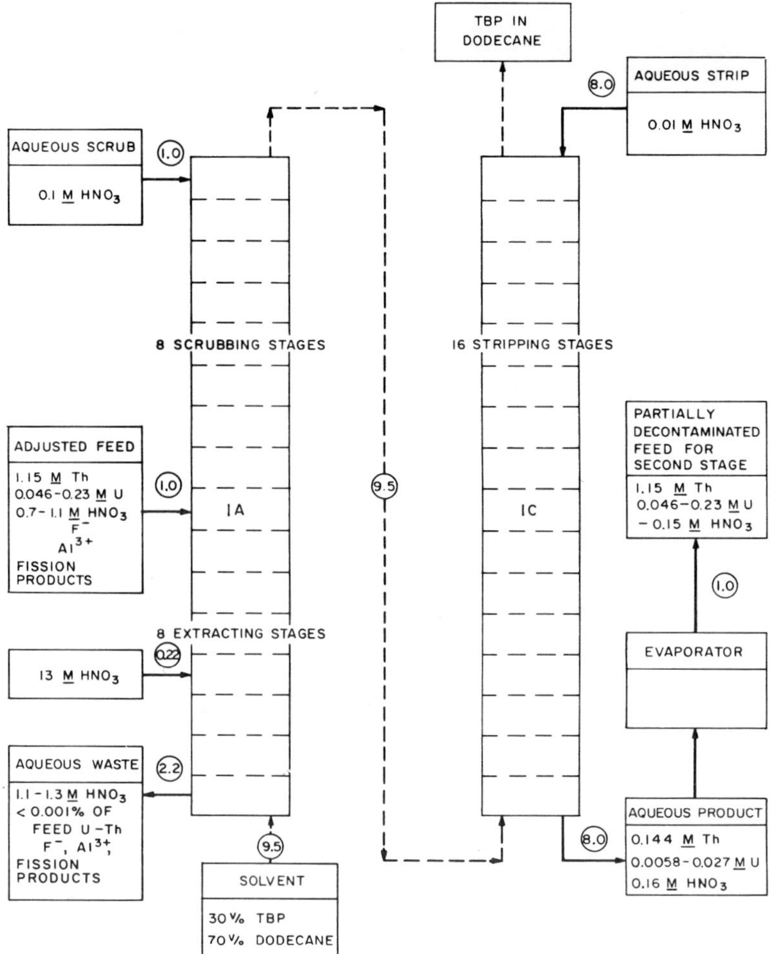

Figure 10.22 First stage of two-stage acid Thorex process for high-burnup fuel. (*From Küchler et al. [K7].*)

the third extracting stage to complete extraction of thorium, as in the Hanford flow sheet Fig. 10.21. Uranium and thorium are returned to the aqueous phase by eight volumes of 0.01 *M* HNO_3 in 16 stripping stages 1C. Aqueous product from 1C is concentrated and made 0.15 *M* acid-deficient in the evaporator and becomes partially decontaminated feed for the second stage.

Second stage. The second stage is shown in Fig. 10.23 with material quantities for the lower, 4 percent, uranium feed. In unit 2A, one volume of feed is extracted in eight stages with eight volumes of 30 v/o TBP and scrubbed in eight stages with one volume of 1 *M* HNO_3. The scrub contains 0.01 *M* H_3PO_4 to improve decontamination from protactinium and zirconium-niobium, as in the Hanford flow sheet Fig. 10.21. An additional scrub of 13 *M* HNO_3 is added to the third extracting stage to complete recovery of thorium.

In unit 2B, thorium is returned to the aqueous phase by stripping in eight stages with 4.8

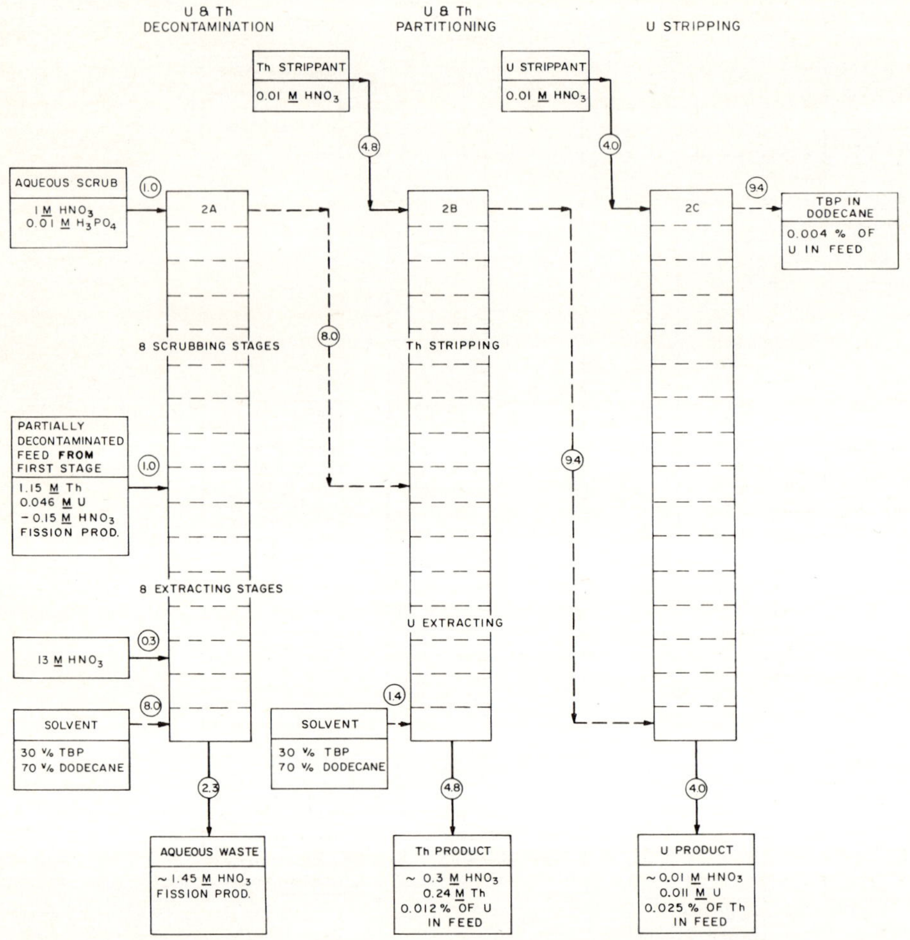

Figure 10.23 Second stage of two-stage acid Thorex process for high-burnup fuel. (*From Küchler et al. [K7].*)

volumes of 0.01 *M* HNO_3. Uranium is extracted from thorium product in eight stages by an additional 1.4 volumes of solvent.

In unit 2C uranium is returned to the aqueous phase by stripping in 16 stages with an additional 4.0 volumes of 0.01 *M* HNO_3.

Uranium product is further decontaminated by a third cycle of extraction with 5 v/o TBP in *n*-dodecane and stripping with 0.01 *M* HNO_3.

Process results. Decontamination factors observed by Küchler et al. [K7] in processing 54,000 MWd/MT fuel with thorium/uranium ratio of 5.9, cooled 346 days, are listed in Table 10.19. Uranium losses were 0.012 percent to thorium product, 0.004 percent to solvent from 2C, and 0.0018 percent to solvent from third uranium cycle. Thorium loss was 0.025 percent to uranium product.

In these experiments, no mention was made of the disposition of the plutonium that will be present in fuel containing uranium irradiated to high burnup. This plutonium could either be routed to high-level waste by adding ferrous sulfamate to the scrub solution for the second stage (as in Fig. 10.21) or could be made to accompany uranium into the third cycle. There, prior to extraction of uranium, plutonium could be reduced and made inextractable by addition of hydroxylamine.

5.7 JUPITER Pilot Plant

The 2-kg HM/day JUPITER pilot plant being built by Merz and associates [M7] at Jülich will test a flow sheet for reprocessing high-burnup graphite-clad fuel generally similar to the two-stage process just described. In addition, the pilot plant will demonstrate fluidized-bed combustion of crushed graphite-clad fuel assemblies and refabrication of fuel from recovered uranium and thorium. Radiokrypton in combustion off-gases will be recovered by cryogenic absorption in liquid CO_2, and tritiated water vapor by adsorption on molecular sieves.

5.8 Phase Equilibria in Thorex Systems

The equilibrium distribution of thorium nitrate and nitric acid between their solutions in water and in 30 v/o TBP in a hydrocarbon diluent at temperatures between 30 and 60°C has been reported by Siddall [S13] and Weinberger et al. [W6]. Siddall's diluent was Ultrasene, a mixture of normal, iso-, and cycloparaffins with an average molecular weight of 175. Weinberger et al. used practical-grade *n*-dodecane, molecular weight 170. Rainey and Watson [R4] modified the SEPHIS computer program to represent the distribution coefficients of nitric acid and thorium nitrate between an aqueous phase and 30 v/o TBP. Figures 10.24 and 10.25 display the distribution coefficients predicted by the 1978 version of the SEPHIS code [V1] for equilibria at 30°C. Distribution coefficients agree with measurements of Siddall [S13] and Weinberger et al. [W6] except for thorium at aqueous concentrations below 0.06 *M* within the dashed line, where the code predicts lower values than observed.

Adequate data on distribution coefficients of uranyl nitrate between 30 v/o TBP and aqueous solutions of thorium nitrate and nitric acid are not available. Examination of concentrations of coexistent phases in Thorex process mixer-settler runs reported in references [R11], [O1], and [O2] indicate that the distribution coefficient of uranium D_U when present at uranium concentrations below 0.02 *M* in Thorex systems at thorium concentrations above 0.1 *M* is given approximately by

$$D_U = 20D_{Th} \tag{10.15}$$

Table 10.19 Decontamination factors observed by Küchler in hot-cell run with two-stage Thorex process followed by third uranium cycle

Decontamination factor	Thorex two-stage process: Thorium	Thorex two-stage process: Uranium	Third uranium cycle
Total gamma	6E4	1E5	170
Total beta	3E5	2E6	2.7E3
^{144}Ce	6E6	3E7	2E5
^{106}Ru	5E4	5E4	174
^{95}Zr	6E5	>6E6	2E4
^{233}Pa	1E3	2E4	–

Source: L. Küchler, L. Schäfer, and B. Wojtech, "The Thorex Two-Stage Process for Reprocessing Thorium Reactor Fuel with High Burnup," *Kerntechnik* **13**: 319 (1971).

However, column separation performance in the Hanford Thorex campaign correlates better with a D_U/D_{Th} ratio of 14 (Prob. 10.5). Because the distribution coefficient of thorium is so much less than that of uranium, the Thorex process requires a much higher organic/aqueous flow ratio than the Purex process.

Figure 10.26 compares the low-concentration distribution coefficients of uranium, thorium, plutonium, protactinium, and the principal fission products. The spread between thorium and fission-product zirconium is greatest between 1 and 2 *M* HNO_3, the range used in the decontamination step of the acid Thorex process. Because the distribution coefficient of protactinium is close to that of thorium, it is necessary to remove protactinium or complex it with fluoride or phosphate ion to prevent its extraction with thorium.

A complication of the Thorex process is appearance of a second organic phase at high concentrations of thorium nitrate and nitric acid. To obtain reproducible separation, Thorex process systems are designed to stay below the thorium concentrations at which the second organic phase forms. Figure 10.27 shows these conditions for *n*-dodecane diluent and Ultrasene at 30°C. Siddall [S14] has pointed out that substitution of triamyl phosphate for TBP would essentially eliminate formation of a second organic phase with thorium.

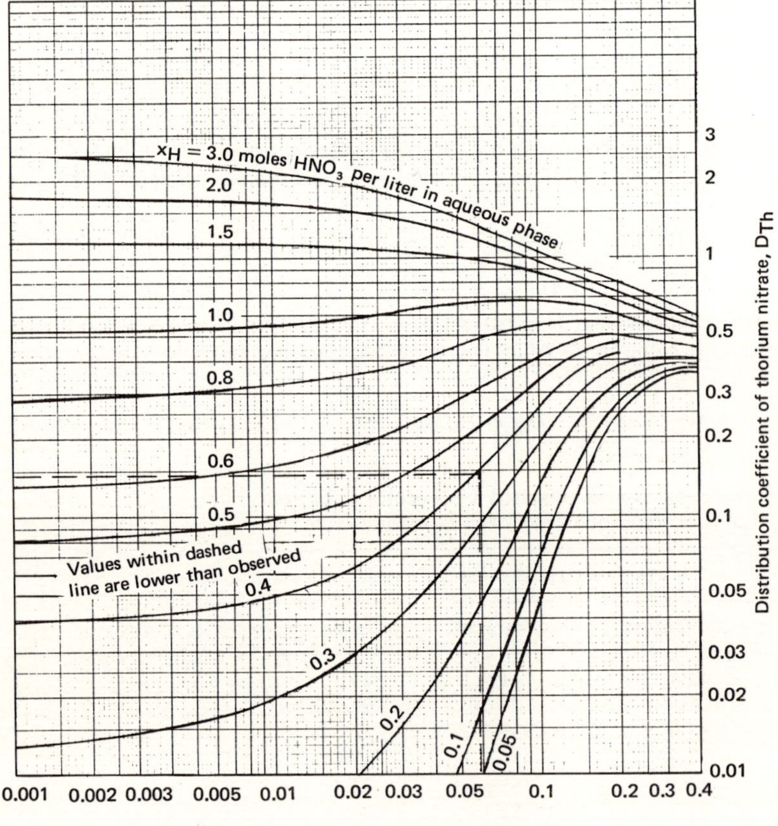

Figure 10.24 Distribution coefficient of thorium nitrate between 30 v/o TBP in hydrocarbon diluent and aqueous nitric acid at 30°C, from SEPHIS code.

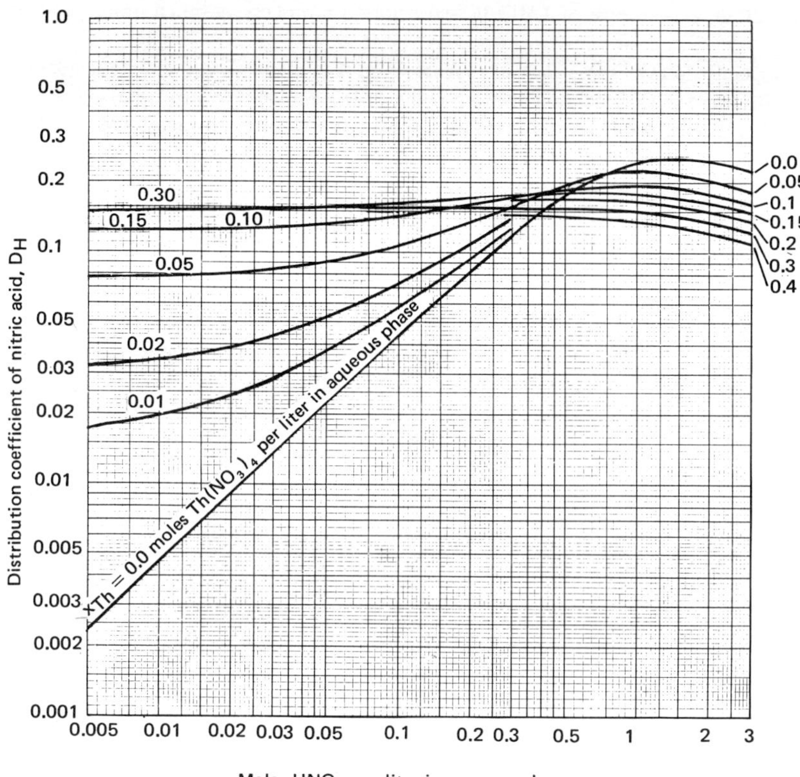

Figure 10.25 Distribution coefficient of nitric acid between 30 v/o TBP in hydrocarbon diluent and aqueous thorium nitrate at 30°C, from SEPHIS code.

6 REPROCESSING LMFBR FUELS

6.1 Differences from LWR Fuels

Table 10.20 lists the principal differences between irradiated fuel to be reprocessed from an LMFBR and an LWR. The data have been excerpted from Figs. 3.34 and 3.31 and Tables 8.8 and 8.7.

Because some of the sodium coolant used in the LMFBR fuel that may have adhered to the cladding or penetrated leaks in it would react vigorously with water or nitric acid, it is necessary to oxidize all sodium by exposing the fuel to an inert gas containing a controlled amount of water vapor before the dissolution step. LMFBR fuel may not be stored with water cooling till after all sodium has been removed.

Use of stainless steel cladding in the LMFBR instead of zircaloy has little effect on mechanical decladding or on dissolution. Stainless steel, like zircaloy, is not rapidly dissolved by nitric acid of the concentrations used in the Purex process.

Fuel in the core of the LMFBR is operated at a specific power over three times that of the LWR. During the cooling period, the specific power of LMFBR core fuel from radioactive decay remains about three times that of LWR fuel cooled for the same length of time. This makes shipping, handling, and storage of LMFBR fuel prior to reprocessing much more difficult than LWR fuel.

To reduce the specific power somewhat in reprocessing, it is planned to combine irradiated fuel from the LMFBR core with irradiated fuel from the LMFBR blankets in proportion to the rates at which they are discharged from the reactor. Even so, the specific power of LMFBR fuel cooled 150 days is 1.4 times that of LWR fuel cooled the same length of time.

The burnup of fuel in the LMFBR core is two or more times that of LWR fuel, leading to higher concentrations of fission products, gaseous and solid, and greater radiation effects on cladding and fuel. The average burnup of combined LMFBR core and blanket material is about 15 percent higher than that of LWR fuel.

The concentration of plutonium in combined core and blanket fuel from the LMFBR is more than 10 times that of LWR fuel. This is the most significant difference between the two fuels with respect to reprocessing. Other important differences are the greater amounts of tritium and ^{131}I, the 140 percent greater ruthenium activity, and the 60 percent greater overall specific activity of 150-day cooled LMFBR fuel.

Because of the high plutonium content of spent fuel from the LMFBR, there is strong

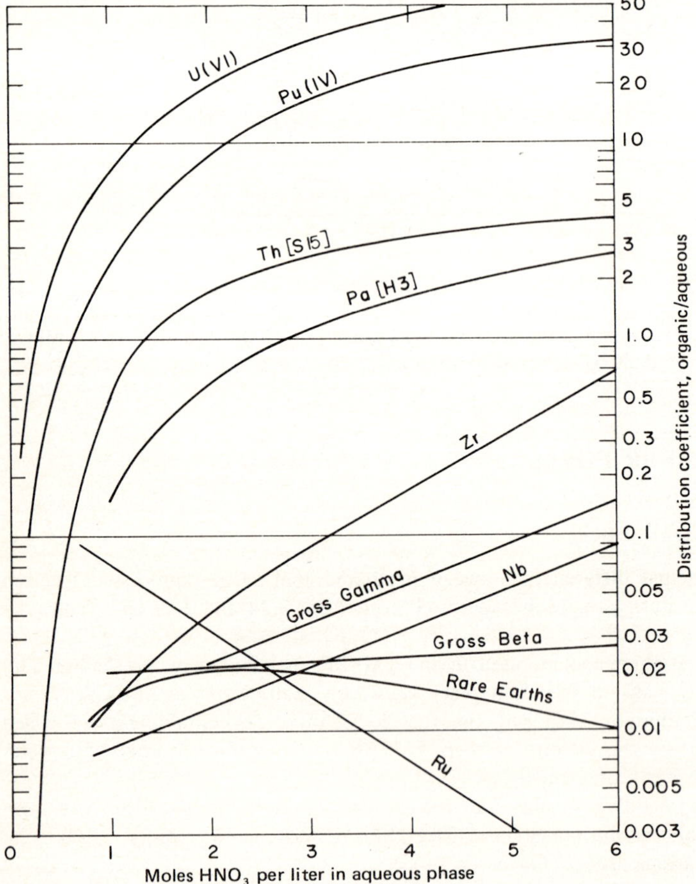

Figure 10.26 Distribution coefficients of principal metal nitrates in acid Thorex process at low concentration.

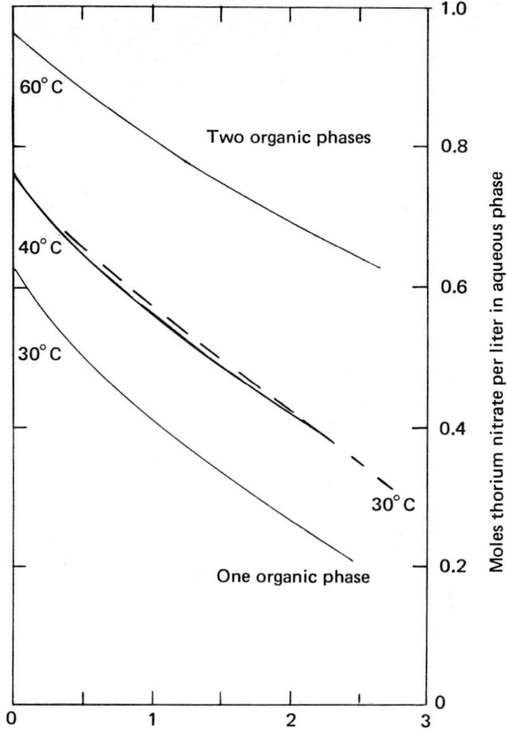

Figure 10.27 Aqueous phase concentration at which second organic phase forms. —— *n*-dodecane [W6]; – – – Ultrasene [S23].

economic incentive to return this plutonium to the reactor with minimum delay for cooling, reprocessing, and refabrication. Consequently, the foregoing comparison of LMFBR and LWR reprocessing conditions for equal cooling periods of 150 days does not tell the whole story. For example, if LMFBR fuel were reprocessed 90 days after discharge from the reactor instead of 150 days, the activity of 8.05-day ^{131}I would be

$$2^{(150-90)/8.05} = 175 \tag{10.16}$$

times greater, and the specific power of fuel from the core would be 1.5 times greater.

The following discussion of reprocessing LMFBR fuels outlines the principal process steps, lists the main problem areas, and discusses possible solutions. Since 1973, international dissemination of reprocessing information has been restricted. This discussion of reprocessing LMFBR fuel is thus less complete and less up to date than would be desired.

6.2 Principal Steps in Reprocessing LMFBR Fuel

Figure 10.28 shows the principal steps in reprocessing LMFBR fuel. Feed quantities are for a plant fed with 5 MT/day of irradiated heavy metal (uranium plus plutonium). Feed is combined core and blanket assemblies from LMFBRs operated under conditions nearly the same as those on which Fig. 3.34 and Tables 8.8 and 10.20 were based. The head-end steps 1 through 6 follow one alternative of several sketched in Report ORNL-4422 [O5].

Fuel assemblies for the core and axial blanket consist of long bundles of stainless steel tubes, each about 0.6 cm in diameter, in which the spent fuel and fission products are sealed.

Table 10.20 Principal differences between irradiated fuel from LMFBR and LWR

	Reactor	
	LMFBR	LWR
Coolant	Sodium	Water
Cladding material	Stainless steel	Zircaloy
Fuel rod diameter, cm	0.6–0.8	1.0–1.2
Reactor specific power, MW/Mg HM†		
Core	98	
Average, core and blankets	49.3	30
Burnup, MWd/MT		
Core	67,600	
Axial blanket	4,700	
Radial blanket	8,000	
Mixed core and blankets	37,000	33,000
Specific power of fuel cooled 150 days, kW/Mg HM		
Core	52	
Mixed core and blankets	28	20
Composition of mixed core and blanket cooled 150 days, w/o		
Uranium	85.6	95.4
Neptunium	0.025	0.075
Plutonium	10.3	0.90
Americium	0.035	0.014
Curium	0.0011	0.0047
Fission products	3.9	3.1
Specific activity of mixed core and blanket cooled 150 days, Ci/Mg HM		
Tritium	1,050	690
^{85}Kr	8,430	11,000
^{131}I	3.55	2.22
Strontium	162,500	174,000
Cesium	152,000	321,000
Ruthenium	1.21E6	0.50E6
Total	6.98E6	4.31E6

†Mg HM, megagrams (metric tons) heavy metal (uranium + plutonium) charged to reactor.

The lower end of each tube contains irradiated depleted UO_2, the middle portion irradiated mixed depleted UO_2 and PuO_2, an upper portion irradiated depleted UO_2, and the top a plenum to accommodate buildup of fission-product gases. The rod bundles are surrounded by a square or hexagonal stainless steel sheath to the top and bottom of which are attached end fittings to direct sodium flow in the reactor and to facilitate handling outside. Fuel assemblies for the radial blanket are of the same length but contain rods of larger diameter charged initially with depleted UO_2.

In Fig. 10.28 it is assumed that assemblies from the core and radial blanket are reprocessed in the proportion in which they are discharged from the reactor. The average composition of feed to the reprocessing plant then is 10 w/o plutonium, 3.56 w/o fission products, and 86.44 w/o uranium. The 5000 kg of fuel processed per day is associated with 6858 kg of stainless steel and an indeterminate amount of metallic sodium that coats exterior surfaces of the assembly and possibly has penetrated imperfections in some of the fuel rods. Sodium is used as coolant in the LMFBR and is a likely candidate for removing decay heat in shipping irradiated fuel from the reactor to reprocessing.

The first step in Fig. 10.28 is deactivation of sodium coating the outside the fuel rods, either by dissolving it off or converting it to a less reactive sodium compound. In the second step, as much of the stainless steel as possible is removed without permitting fission products to escape. End fittings are removed and fuel rod bundles are extracted from the enclosing sheath, if possible. In the third step, the plenum is sheared from fuel rod bundles, thus releasing some of the fission product gases to a retention system. The portion of the rod bundle containing

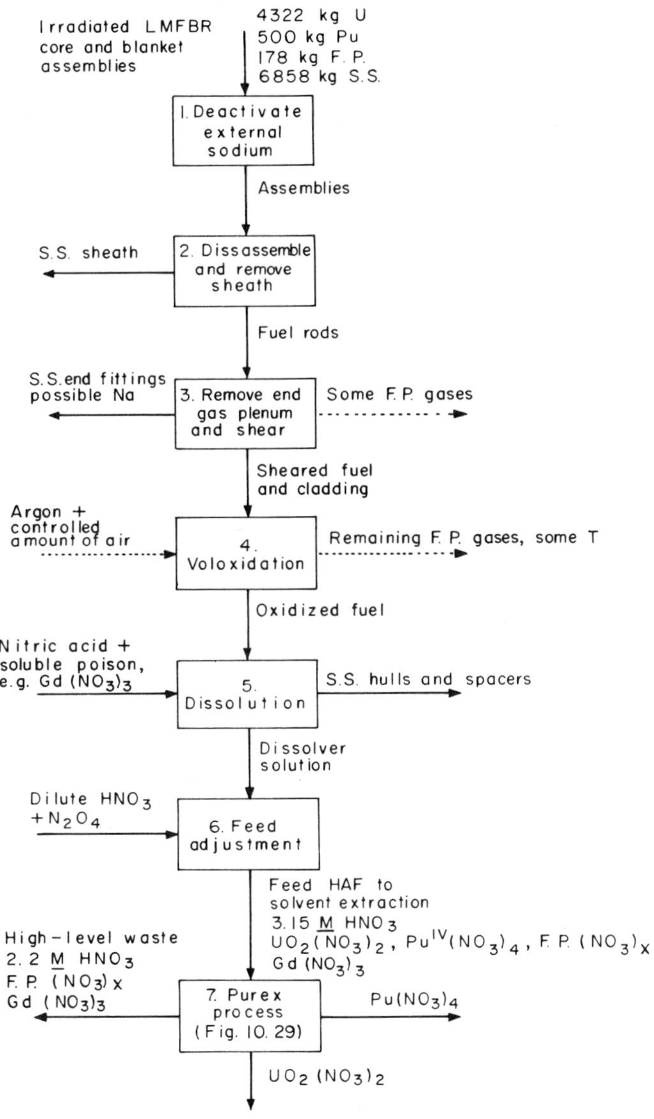

Figure 10.28 Principal head-end steps in preparing irradiated LMFBR core and blanket assemblies for Purex process. F.P. = fission products; S.S. = stainless steel.

fuel and blanket material is sheared into short lengths to facilitate subsequent processing. In step 4, voloxidation, the sheared fuel is heated to 550 to 600°C first in argon, to which is then added an increasing amount of air, to react with possible entrained sodium, convert UO_2 to U_3O_8, and release tritium. In step 5, the fuel is dissolved in 8 *M* nitric acid to which sufficient gadolinium nitrate, boric acid, or other soluble poison is added to control criticality. Undissolved residues rich in PuO_2 are treated with special reagents. In step 6, feed adjustment, nitric acid concentration of solvent extraction feed is brought to 3 *M* and plutonium is made tetravalent by addition of N_2O_4. In the Purex process, step 7, solvent extraction with 30 v/o TBP is used to separate dissolver solution into high-level waste, decontaminated uranyl nitrate, and decontaminated plutonium.

6.3 Problem Areas

Special problems in reprocessing LMFBR fuels compared with LWR fuels are as follows:

Removal of greater decay heat
Deactivation of sodium
Voloxidation of mixed UO_2-PuO_2 fuel of high PuO_2 content
More complete retention of iodine, if LMFBR fuel cooling time is reduced
More difficult dissolution
The higher plutonium concentration in the first Purex cycle

These problem areas are discussed in order in the following pages.

Another problem is the greater rate of solvent degradation caused by the higher specific activity of the fuel. This makes the use of short-contact-time contactors even more necessary than with LWR fuels. A final problem is control of criticality, which is vital throughout all fuel reprocessing and is discussed in general terms in Sec. 8.

In one respect, requirements for reprocessing fuel from an LMFBR is less demanding than from an LWR. It is not necessary to purify the uranium and plutonium products so completely. The uranium is not converted to UF_6 and does not have to meet the strict purity requirements of feed for a uranium enrichment plant. When the plutonium is recycled to an LMFBR it is diluted with uranium, so complete separation from uranium is not necessary.

6.4 Decay Heat Removal

The necessity for reliable and uninterrupted cooling of irradiated LMFBR fuel can be seen from its rate of adiabatic temperature rise owing to radioactive decay. From Table 10.20, the average specific power of core fuel cooled 150 days is 0.052 W/g HM.† At a nominal specific heat of 0.08 cal/(g·°C) for UO_2 or PuO_2, the adiabatic rate of temperature rise is

$$\frac{dT}{dt} = \frac{(0.052 \text{ W/g HM})(238 \text{ g HM}/270 \text{ g UO}_2)}{[0.08 \text{ cal/(g UO}_2 \cdot {}^\circ\text{C})]\,[4.187(\text{W}\cdot\text{s})/\text{cal}]} = 0.14^\circ\text{C/s} \tag{10.17}$$

or 500°C/h. The actual temperature rise would be less because of heat losses, but is still rapid and inexorable. When fuel is removed from the sodium in which it was shipped and stored in preparation for disassembly and reprocessing, it is necessary to provide reliable alternative cooling. Until the sodium adhering to the fuel element can be removed, the preferred cooling medium is an inert gas such as flowing argon.

†HM = heavy metal, uranium + plutonium.

6.5 Deactivation of Sodium

Goldberg [G7] has listed a number of procedures that have been used for removing or deactivating sodium adhering to LMFBR assemblies. Use of a relatively nonvolatile alcohol, such as the *n*-butyl ether of ethylene glycol, is reported [C8] to remove sodium metal and oxide completely in 24 h. A difficulty is subsequent complete removal of solvent. Reaction with water vapor carried by an inert gas such as argon has been used extensively to deactivate sodium adhering to fuel assemblies. The principal disadvantage is the residue of sodium hydroxide, which reacts with acid in subsequent dissolution. Amalgamation with mercury has been used in the United Kingdom and the United States. In one application, mercury removed sodium from a 40-fuel-pin batch in 0.5 h [B15]. In the Soviet Union [S12], molten lead at 400 to 500°C has been used to wash sodium from fuel assemblies and as a substitute for water in storage of fast-reactor fuel for extended periods. A disadvantage in reprocessing is the layer of lead that coats the fuel. Sodium was washed from fuel assemblies from the Enrico Fermi LMFBR [K2] by ultrasonic cleaning with a high-boiling hydrocarbon oil at a temperature above the melting point of sodium. A disadvantage is the need to remove the flammable oil before voloxidation.

Thus, all methods have disadvantages. Deactivation with moist argon seems the simplest.

6.6 Voloxidation

For voloxidation to remove tritium completely from $(U,Pu)O_2$ fuel, it has been found necessary that the oxygen content of the fuel be increased to that corresponding to mixed U_3O_8 and PuO_2. With LWR fuel, containing 1 percent or less plutonium, oxidation of UO_2 to U_3O_8 proceeds relatively rapidly and completely at temperatures around 600°C, with almost quantitative release of tritium. This favorable result is attributed to the swelling and disintegration of the fuel accompanying the phase change from cubic UO_2 to less dense orthorhombic U_3O_8.

With the $Pu/(U + Pu)$ ratio of 0.20 or 0.25 proposed for the core of an LMFBR, the voloxidation process is more complex, very dependent on method of fuel fabrication, and sometimes incomplete. Oak Ridge National Laboratory [O9] found that cubic dioxide fuel with a $Pu/(U + Pu)$ atom ratio of 0.2 was oxidized to the desired mixture of U_3O_8 and PuO_2 in a two-step process only within a narrow temperature range of 500 to 600°C. First oxidation takes place rapidly to tetragonal $U_{2.4}Pu_{0.6}O_7$. Little swelling or disintegration occurs in this step, and little tritium is released. The second, slower oxidation step results in conversion of the fuel to a less dense mixture of orthorhombic U_3O_8 and cubic PuO_2 (oxygen/metal ratio of 2.53), with crumbling of the fuel and release of 98 percent or more of the contained tritium [F1]. The rate at which this second phase change occurs is strongly dependent on how the uranium and plutonium were homogenized, how the fuel was made, or possibly some other factor. The second change takes place most rapidly with coprecipitated UO_2-PuO_2, next most rapidly with sol-gel fuel, and slowest with mechanically mixed fuel. Furthermore, Oak Ridge reported [O9, p. 17]: "Under some conditions as yet undetermined, even fuel with $Pu/(U + Pu) = 0.2$ will not readily oxidize to oxygen/metal ratios in excess of 2.4," and hence will not release tritium.

Dioxide fuel with a $Pu/(U + Pu)$ atom ratio of 0.25 oxidizes readily to a cubic phase of empirical formula $(U,Pu)_4O_{9.4}$, but with little swelling or tritium release. Further oxidation to a mixture of U_3O_8 and PuO_2, with the desired swelling and tritium release, proceeds only with great difficulty.

Another difficulty with voloxidation of mixed UO_2-PuO_2 fuel is conversion of some of the PuO_2 to a form insoluble in nitric acid. For these reasons the German workers Baumgärtner and Ochsenfeld concluded [B6] that "voloxidation is no longer considered as a treatment step preceding dissolution of LMFBR elements." However, work on voloxidation of mixed UO_2-PuO_2 fuel is continuing at Oak Ridge.

6.7 Retention of Iodine

The processes described in Sec. 4.6 are, in principle, applicable to off-gases from LMFBR reprocessing plants. The problem is the greater iodine activity per ton of fuel processed. This is due to the 60 percent higher specific activity of iodine for 150-day cooled LMFBR fuel compared with similar LWR fuel noted in Table 10.20 and the incentive to reprocess LMFBR fuel with shorter cooling. Oak Ridge National Laboratory has estimated that if LMFBR fuel were to be reprocessed only 30 days after irradiation, an iodine retention factor as high as 10^{10} would be required. This seems completely unattainable. However, some improvement over the retention factor of 10^2 (99 percent retention), feasible with the technology described in Sec. 4.6, would be possible if iodine could be stripped more completely from the dissolver solution. Retention factors of 10^4 or better have been reported for individual silver-zeolite absorbers.

6.8 Dissolution

Mixed UO_2-PuO_2 fuel dissolves more slowly in nitric acid than straight UO_2 fuel. Moreover, the rate of solution varies widely depending on plutonium content, method of fuel preparation, irradiation history, and voloxidation conditions. Reaction of irradiated, mechanically mixed UO_2-PuO_2 fuel without sintering with nitric acid always leaves an undissolved residue containing several percent of the original plutonium combined with such less soluble fission products as ruthenium, niobium, palladium, molybdenum, and zirconium. As an example, when mechanically blended, irradiated, and voloxidized 25 percent PuO_2, 75 percent UO_2 fuel was reacted with boiling 8 M HNO_3 for 8 h, 4 percent of the plutonium and 0.5 percent of the uranium remained undissolved [O9]. Under these conditions about 75 percent of the ruthenium, 50 percent of the niobium, 4 percent of the zirconium, and 3 percent of the cerium was also undissolved [O10]. Coprecipitated, solid-solution $(U,Pu)O_2$ dissolves more rapidly and completely, but still leaves some plutonium-enriched refractory residue. Two procedures have been studied for dissolving this residue: (1) leaching the residue in a secondary dissolver with boiling 8 M HNO_3 containing 0.05 M KF, or (2) leaching the residue with boiling 4 M HNO_3 containing 0.05 M tetravalent cerium.

In one test [O10], leaching the residue with the HNO_3-HF mixture dissolved all but 0.05 percent of the original plutonium and 0.01 percent of the original uranium. This addition of the fluoride ion has the disadvantages cited under the Thorex process of corroding stainless steel and complexing plutonium. Corrosion of the dissolver could be dealt with by making it of Inconel 625 or 690, but it would be necessary to protect downstream stainless steel equipment and reduce plutonium complexing by addition of aluminum nitrate, after dissolution, to complex fluoride ion. This would increase waste solids.

Boiling 4 M HNO_3 containing tetravalent cerium [U2] leaches plutonium from the residue but has the disadvantages of making plutonium hexavalent and converting some of the ruthenium in the residue to volatile RuO_4.

6.9 Purex Process for LMFBR Fuel

Figure 10.29 shows the principal steps in applying the Purex process to irradiated LMFBR fuel, step 7 of Fig. 10.28. The flow scheme and the compositions and locations of solvent, scrubbing, and stripping streams have been taken from the process flow sheet of a 1978 Oak Ridge report [O11] describing a planned experimental reprocessing facility designed for 0.5 MT of uranium-plutonium fuel or 0.2 MT of uranium-plutonium-thorium fuel per day. As that report gave process flow rates only for the uranium-plutonium-thorium fuel, Fig. 10.29 does not give flow rates for the uranium-plutonium fuel of present interest. This flow sheet shows the codecontamination step, in which fission products are separated from uranium and plutonium; the partitioning step, which produces an aqueous stream of partially decontaminated

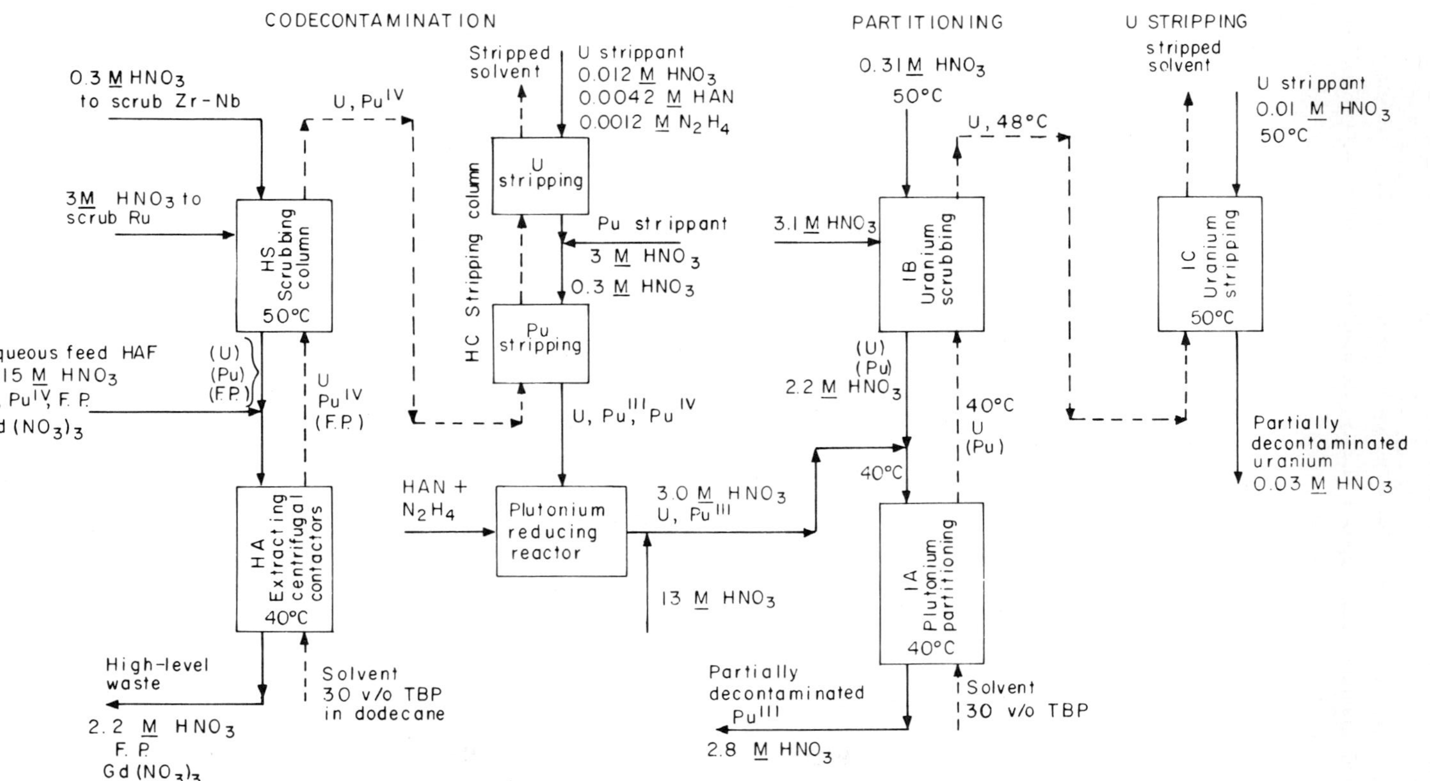

Figure 10.29 Principal steps in Purex process for LMFBR fuel. F.P. = fission products; HAN = hydroxylamine nitrate; —— aqueous; – – – organic.

plutonium; and the uranium stripping step, which produces an aqueous stream of partially decontaminated uranium. The proposed facility has additional solvent extraction cycles, not shown, for completing decontamination.

Codecontamination. The codecontamination section consists of the HA extraction section equipped with short-contact-time centrifugal contactors and the HS scrubbing section equipped with pulse columns. In the HA section, uranium and plutonium in the aqueous feed and reflux from the HS section are extracted into the organic stream containing 30 v/o TBP. In the HS section any ruthenium extracted by TBP is scrubbed into the aqueous phase with 3 M HNO_3. Then any zirconium-niobium in the TBP is scrubbed with 0.3 M HNO_3. Scrubbing is at 50°C to enhance decontamination of ruthenium.

Stripping. This flow sheet uses hydroxylamine† to reduce plutonium to inextractable Pu(III). Because the reduction of plutonium by hydroxylamine takes place almost entirely in the aqueous phase and requires many minutes for completion, before reduction it is necessary to return the uranium and plutonium in the organic phase leaving the HS contactor to the aqueous phase. This is done in two stages in the HC stripping column. In the bottom stage, plutonium is stripped with 0.3 M HNO_3. Lower acid concentration must be avoided because plutonium polymer would then form. In the top stage, solvent, now containing no plutonium, is stripped of uranium by 0.012 M HNO_3. Some hydroxylamine is added to this strippant to start reduction of plutonium in the HC unit.

Plutonium reduction. Reduction of plutonium to Pu(III) is completed by adding concentrated hydroxylamine (with hydrazine as holding reductant) to the aqueous raffinate leaving the HC column. The mixture must be held long enough, half an hour or more [B2], to complete the rather slow reduction to Pu(III). To hasten the reaction, the hydroxylamine concentration should be high and the nitric acid concentration as close to 0.3 M as possible without risking plutonium polymer formation.

Partitioning. Feed for partitioning is made 3 M in HNO_3 by addition of 13 M HNO_3 to reducing reactor effluent, to enhance extraction of uranium from inextractable Pu(III). In the plutonium partitioning pulse column 1A uranium is extracted from Pu(III) by 30 v/o TBP. In the uranium scrubbing section 1B, any Pu(III) that may have been extracted with uranium and traces of extracted fission products are scrubbed with two aqueous streams, 3.1 M HNO_3 to remove ruthenium and 0.31 M HNO_3 to remove zirconium-niobium.

Uranium stripping. Uranium in solvent leaving the 1B column is stripped into the aqueous phase by counterflowing 0.01 M HNO_3 in the 1C column. This is run at 50°C to reduce the uranium distribution coefficient.

Prevention of criticality. Because the plutonium content of feed to this LMFBR solvent extraction flow sheet is 10 times that of the Barnwell plant, Sec. 4.14, extra precautions must be taken to prevent criticality in the dissolver; the HA, HS, HC, and 1A contactors; and the plutonium reduction reactor. Addition of sufficient soluble poison to the feed will prevent criticality in the dissolver, feed adjustment tanks, and centrifugal HA contactors. The other sections of the plant processing plutonium must either have small enough dimensions to be

†Hydroxylamine is used for plutonium reduction instead of cathodic reduction as in the Barnwell flow sheet Fig. 10.11, because the plutonium/uranium ratio in this LMFBR fuel is 10 times that in LWR fuel and because electrolytic reduction has not been demonstrated for this high plutonium content.

subcritical (small-diameter columns or small-diameter or thin-slab tanks) or be provided with neutron-absorbing inserts such as boron steel plates or borosilicate glass Raschig rings. General procedures for guarding against criticality are discussed in Sec. 8.

7 NEPTUNIUM RECOVERY IN REPROCESSING

This section describes processes for recovering neptunium from irradiated uranium. Neptunium is an example of one of the numerous elements in irradiated fuel that could be recovered as by-products of extraction of uranium and plutonium in the Purex process.

7.1 Use of Neptunium

^{237}Np, a beta-emitting nuclide with a half-life of 2.14×10^6 years, is used as target material for production of ^{238}Pu by irradiation with thermal neutrons in the reactions

$$^{237}\text{Np}(n, \gamma) \xrightarrow{169\text{ b}} {}^{238}\text{Np} \xrightarrow[2.1\text{ day}]{\beta^-} {}^{238}\text{Pu}$$

^{238}Pu is an important alpha-emitting radioactive energy source, which has been used extensively in space missions and in cardiac pacemakers. Its advantages for these applications are its relatively high specific power of about 0.5 W/g, its rather long half-life of 89.6 years, and the absence of appreciable gamma radiation, making heavy shielding unnecessary.

7.2 Sources of Neptunium

In thermal reactors ^{237}Np is formed in the following reactions:

$$^{235}\text{U}(n, \gamma) \xrightarrow{98.6\text{ b}} {}^{236}\text{U}(n, \gamma) \xrightarrow{5.2\text{ b}} {}^{237}\text{U} \xleftarrow{0.013\text{ b}} {}^{238}\text{U}(n, 2n)$$

$$^{237}\text{U} \xrightarrow[6.75\text{ days}]{\beta^-} {}^{237}\text{Np}$$

In fast reactors, the ^{238}U$(n,2n)$ reaction predominates. Typical concentrations of ^{237}Np in irradiated fuel, from Chap. 8, are

Thermal reactors:	749 g/Mg HM
Fast reactors:	249 g/Mg HM

Neptunium concentration in fuel from thermal reactors could be increased by recycling uranium containing ^{236}U.

7.3 Oxidation-Reduction Equilibria in Neptunium Recovery

Figure 10.8 has shown that the neptunium ions Np^{4+} and $Np^{VI}O_2{}^{2+}$ have sufficiently high distribution coefficients to be extractable by 30 v/o TBP in the Purex process. On the other hand, the distribution coefficient of $Np^{V}O_2{}^{+}$ is of the order of 0.001, so that pentavalent neptunium is essentially inextractable. The distribution of neptunium between aqueous and organic phases in the Purex process is thus determined by oxidation-reduction equilibria among the three valences of neptunium in the presence of the oxidizing or reducing agents used in that process. A semiquantitative indication of neptunium distribution among the three valence states is afforded by the comparison of their standard oxidation-reduction potentials with those of plutonium and possible reductants and oxidants, given in Table 10.21.

Comparison of the potentials for the two neptunium couples with those for plutonium

Table 10.21 Oxidation-reduction potentials at 25°C in neptunium recovery processes (Chap. 8)

	$E°$, V
Neptunium	
$Np^{4+} + 2H_2O \rightarrow Np^{V}O_2^{+} + 4H^{+} + e^{-}$	−0.7391
$Np^{V}O_2^{+} \rightarrow Np^{VI}O_2^{2+} + e^{-}$	−1.1364
Plutonium	
$Pu^{3+} \rightarrow Pu^{4+} + e^{-}$	−0.9819
$Pu^{4+} + 2H_2O \rightarrow Pu^{VI}O_2^{2+} + 4H^{+} + 2e^{-}$	−1.0433
Reductants	
Uranium(IV)	
$U^{4+} + 2H_2O \rightarrow U^{VI}O_2^{2+} + 4H^{+} + 2e^{-}$	−0.338
Hydroxylamine	
$2NH_3OH^{+} \rightarrow H_2N_2O_2 + 6H^{+} + 4e^{-}$	−0.496
Ferrous iron	
$Fe^{2+} \rightarrow Fe^{3+} + e^{-}$	−0.7701
Oxidants	
Nitrate ion	
$HNO_2 + H_2O \rightarrow NO_3^{-} + 3H^{+} + 2e^{-}$	−0.94
Vanadium(V)	
$V^{IV}O^{2+} + 3H_2O \rightarrow V^{V}(OH)_4^{+} + 2H^{+} + e^{-}$	−1.000
Ceric ion	
$Ce^{3+} \rightarrow Ce^{4+} + e^{-}$	−1.61

indicate that conditions that bring plutonium into the most extractable, tetravalent state used in the conventional Purex process will oxidize neptunium above the tetravalent state and leave most of it in the inextractable pentavalent state.

Because the distribution coefficient in TBP of hexavalent neptunium is higher than tetravalent, the hexavalent form is preferred for the first extraction from fission products. The first part of Table 10.22 gives equations for the concentration ratio of hexavalent to pentavalent neptunium calculated for the three oxidants listed there, with the coefficient evaluated from $\exp(-38.93\ \Delta E°)$.

With nitrate ion or pentavalent vanadium, high neptunium concentration ratio requires a high ratio of oxidant to reductant and is favored by high hydrogen ion concentration. High oxidant ratio would tend to convert plutonium to the less extractable hexavalent state, but this tendency is inhibited by high hydrogen ion concentration and by the complexing of tetravalent plutonium with nitrate ion. At nitric acid concentrations of 2.5 or 3 M, 90 percent or more of the neptunium can be extracted as Np(VI) without converting more than a few percent of plutonium to the hexavalent state.

Strong oxidants, such as tetravalent cerium, would make neptunium almost completely hexavalent but would make plutonium hexavalent also and would volatilize substantial amounts of ruthenium as RuO_4. For these reasons, strong oxidants are not favored when extracting hexavalent neptunium.

Because of solution nonidealities, observed concentration ratios differ significantly from ratios calculated from the standard oxidation-reduction potentials in Table 10.22. Figure 10.30 shows equilibrium ratios K_{Np}

$$K_{Np} \equiv \frac{[Np(VI)]\,[HNO_2]^{1/2}}{[Np(V)]\,[H^{+}]^{3/2}\,[NO_3^{-}]^{1/2}} \tag{10.24}$$

for oxidation of Np(V) by nitric acid observed by Gourisse [G11, 12] at 25, 35, and 50°C and

Table 10.22 Neptunium valence ratios in oxidation and reduction reactions

Oxidant	ΔE°	1. Oxidation Equilibrium equation	
NO_3^-	0.1964	$\frac{[Np(VI)]}{[Np(V)]} = 0.00048 \frac{[H^+]^{3/2}[NO_3^-]^{1/2}}{[HNO_2]^{1/2}}$	(10.18)
$V^V(OH)_4^+$	0.1364	$\frac{[Np(VI)]}{[Np(V)]} = 0.0049 \frac{[H^+]^2[V(V)]}{[V(IV)]}$	(10.19)
Ce^{4+}	−0.4836	$\frac{[Np(VI)]}{[Np(V)]} = 1.5 \times 10^8 \frac{[Ce^{4+}]}{[Ce^{3+}]}$	(10.20)
Reductant	**ΔE°**	**2. Reduction Equilibrium equation**	
U(IV)	−0.4011	$\frac{[Np(IV)]}{[Np(V)]} = 6.05 \times 10^6 \frac{[U(IV)][H^+]^2}{[U(VI)]}$	(10.21)
NH_3OH^+	−0.2431	$\frac{[Np(IV)]}{[Np(V)]} = 12{,}900 \frac{[NH_3OH^+]^{1/2}[H^+]^{5/2}}{[H_2N_2O_2]^{1/4}}$	(10.22)
Fe^{2+}	+0.0310	$\frac{[Np(IV)]}{[Np(V)]} = 0.299 \frac{[Fe^{2+}][H^+]^4}{[Fe^{3+}]}$	(10.23)

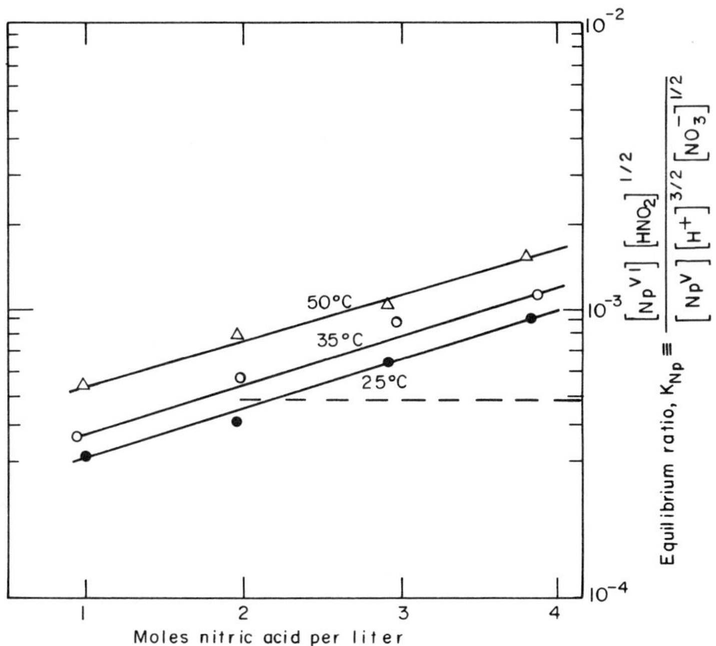

Figure 10.30 Equilibrium ratios for oxidation of Np(V) by nitric acid. —— observed by Gourisse [G12]; – – – calculated, 25°C, Eq. (10.18).

compares them with the coefficient 0.00048 in Eq. (10.18). The kinetics of this reaction are discussed in Sec. 7.5.

The second part of Table 10.22 gives equations for the concentration ratio of tetravalent to pentavalent neptunium calculated for the three reductants listed there. In the older Purex plants the ferrous sulfamate used to reduce plutonium to inextractable Pu^{3+} reduced neptunium partly to inextractable Np(V) and partly to extractable Np(IV). The reductants now preferred, tetravalent uranium (possibly made electrolytically) or hydroxylamine, are sufficiently strong, in sufficient time, to make neptunium almost completely tetravalent, but the reactions are much slower than reduction of tetravalent plutonium, because of slow deoxidation of the NpO_2^+ radical. Kinetics of these reductions are also discussed in Sec. 7.5.

7.4 Distribution Coefficients in Neptunium Recovery

Distribution coefficients of neptunium in 30 v/o TBP depend on neptunium valence, temperature, and concentrations of uranyl nitrate, nitric acid, and other nitrates. At the nitric acid concentrations below 4 *M* usually used in Purex processes, the distribution coefficient of hexavalent neptunium is higher than that of tetravalent neptunium at the same nitric acid and uranyl nitrate concentrations. Both are much higher than that of pentavalent neptunium. Both tetravalent and hexavalent neptunium are extracted as the complexes with two molecules of TBP, $Np^{IV}(NO_3)_4 \cdot 2TBP$ and $Np^{VI}O_2(NO_3)_2 \cdot 2TBP$.

Table 10.23 lists principal sources of information on distribution coefficients of neptunium between 30 v/o TBP and aqueous solutions of uranyl nitrate and nitric acid.

Distribution coefficients of tetravalent and hexavalent neptunium can be correlated conveniently in terms of the separation factor from hexavalent uranium, i.e., the ratio of the distribution coefficient of neptunium to that of uranium.

Tetravalent neptunium. Srinivasan et al. [S18, S19] measured distribution coefficients of tetravalent and hexavalent neptunium and hexavalent uranium as functions of nitric acid and uranyl nitrate concentrations. At 45 and 60°C, the ratio of the observed [S19] separation factor for tetravalent neptunium to that of hexavalent uranium can be correlated within an average deviation of 6 percent by Eq. (10.25),

$$\frac{D_{Np(IV)}}{D_{U(VI)}} = 0.01129 \exp (0.3208 x_{NO_3^-} + 0.03636 t_{°C}) \tag{10.25}$$

At 25°C the equation is less satisfactory, with an average deviation of 18 percent from observations by Srinivasan et al. [S18].

Hexavalent neptunium. Distribution coefficients of hexavalent neptunium at 25, 45, and 60°C measured by Srinivasan et al. [S18, S19] are simply related to measured distribution coefficients for hexavalent uranium by Eq. (10.26), with an average deviation of only 5 percent,

$$\frac{D_{Np(VI)}}{D_{U(VI)}} = 0.54 \tag{10.26}$$

at all uranium concentrations and at nitric acid molarities between 1 and 4 *M*. Germain et al.'s observed [G6] Np(VI) distribution data at 22°C yield an average value of 0.47 for this ratio.

In the HA extracting and HS scrubbing sections of the Purex process, pentavalent neptunium is partially oxidized to the hexavalent state by nitrate ion,

$$2Np^VO_2^+ + NO_3^- + 3H^+ \rightarrow 2Np^{VI}O_2^{2+} + HNO_2 + H_2O$$

when nitrous acid is present to act as catalyst. For the reaction to proceed at a useful rate, the

Table 10.23 Measurements of neptunium distribution in 30 v/o TBP

Neptunium valence	Temperature, °C	M HNO_3	Uranium present	Investigators
4, 5, 6	~25	1–12	No	Flanary and Parker [F4]
6	25, 35, 50	0.6–4.5	No	Siddall and Dukes [S16]
6	25	0.2–3	Yes	Koch [K4]
4, 6	~25	1–4	Yes	Srinivasan et al. [S18]
6	45	2–4	Yes	Swanson [S25]
4, 6	45, 60	1–4	Yes	Srinivasan et al. [S19]
4, 6	22	1–3.5	Yes	Germain et al. [G6]

HNO_2 concentration of the aqueous phase must be over 0.00004 M. At equilibrium, neptunium in the aqueous phase is then divided between the hexavalent and pentavalent states. The ratio of hexavalent to pentavalent neptunium is given by Eq. (10.27), obtained from the equilibrium ratio K_{Np} defined by Eq. (10.24), and plotted in Fig. 10.30.

$$\frac{[Np(VI)]}{[Np(V)]} = K_{Np} \frac{[H^+]^{3/2}[NO_3{}^-]^{1/2}}{[HNO_2]^{1/2}} \tag{10.27}$$

The total neptunium concentration in the aqueous phase x_{Np} is

$$x_{Np} = [Np(VI)] + [Np(V)] = [Np(VI)]\left(1 + \frac{[HNO_2]}{K_{Np}[H^+]^{3/2}[NO_3{}^-]^{1/2}}\right) \tag{10.28}$$

Because pentavalent neptunium is essentially inextractable, the neptunium concentration in the organic phase y_{Np} is related to the distribution coefficient of hexavalent neptunium $D_{Np(VI)}$ by

$$y_{Np} = D_{Np(VI)}[Np(VI)] \tag{10.29}$$

The apparent equilibrium distribution coefficient of neptunium, D_{app}, defined as

$$D_{app} \equiv \frac{y_{Np}}{x_{Np}} \tag{10.30}$$

is then given by

$$D_{app} = \frac{D_{Np(VI)}}{1 + [HNO_2]^{1/2}/K_{Np}[H^+]^{3/2}[NO_3{}^-]^{1/2}} \tag{10.31}$$

Figure 10.31, calculated [G12] from Eq. (10.31), $D_{Np(VI)}$, and the observed equilibrium ratios of Fig. 10.30, shows the dependence of D_{app} on temperature and the concentrations of HNO_2 and HNO_3. Figure 10.31 is strictly valid only in the absence of nitrates other than nitric acid and traces of neptunium. When uranyl nitrate is present at appreciable molarity x_U, $D_{Np(VI)}$ is given by Eq. (10.26), and the apparent equilibrium distribution coefficient for neptunium may be estimated from

$$D_{app} = \frac{0.54\, D_U(x_H, x_U)}{1 + [HNO_2]^{1/2}/K_{Np}x_H^{3/2}(x_H + 2x_U)^{1/2}} \tag{10.32}$$

D_U is given in Figs. 10.13 and 10.15 and K_{Np} in Fig. 10.30. Complete ionization of HNO_3 and $UO_2(NO_3)_2$ is assumed.

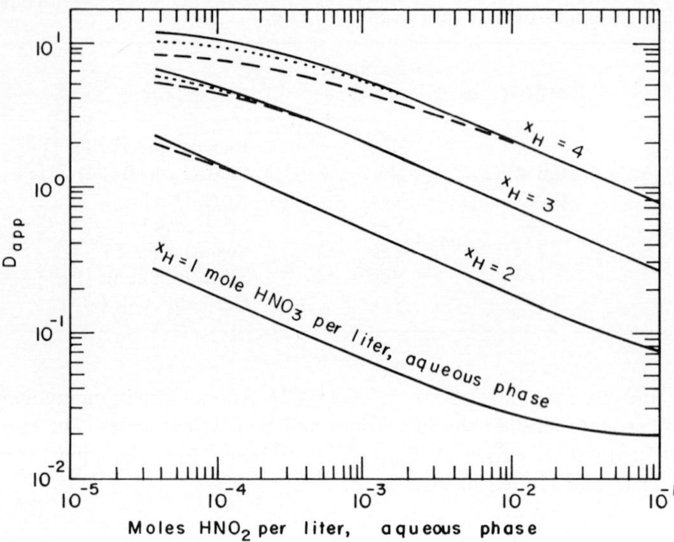

Figure 10.31 Equilibrium apparent distribution coefficient of neptunium in the system 30 percent TBP-dodecane-HNO_3-HNO_2-H_2O, from [G12]. —— 25°; ···· 35°; --- 50°C.

7.5 Kinetics of Neptunium Oxidation and Reduction

Oxidation of pentavalent neptunium by nitric acid. Oxidation of pentavalent neptunium to hexavalent by nitric acid requires catalysis by nitrous acid. The kinetics of this reaction have been studied by Siddall and Dukes [S16], Swanson [S24] and Mouline [M9]. Siddall and Dukes reported that the reaction was first order in neptunium concentration, independent of nitrous acid concentration if greater than 5×10^{-5} M, and depended on temperature T (K) and nitric acid molarity x_H as can be represented by Eq. (10.33):

$$k_N = 7.03 \times 10^{-7}\, e^{7062/T}\, x_H^{3.280} \text{ min}^{-1} \tag{10.33}$$

k_N is the specific rate constant in the equation

$$\frac{dx_5}{dt} = -k_N(x_5 - x_5^0) \tag{10.34}$$

where x_5 is the aqueous molarity of Np(V) and x_5^0 is its equilibrium molarity.

Swanson's results appear very different. He reported that the first-order reaction rate constant in Eq. (10.34) was independent of nitric acid concentration and proportional to nitrous acid concentration:

$$k_N = k_1\,[HNO_2] \tag{10.35}$$

Values of k_1 are 46 $M^{-1}\cdot\text{min}^{-1}$ at 24°C and 250 $M^{-1}\cdot\text{min}^{-1}$ at 46°C. However, at concentrations of nitric and nitrous acid used in reprocessing, values of k_N from Eqs. (10.33) and (10.35) are not far apart.

Mouline's experiments partially explain the apparent discrepancy. When the nitrous acid molarity is less than that of neptunium, the rate is proportional to nitrous acid concentration. At nitrous acid/neptunium concentration ratios above unity, the rate is independent of nitrous acid concentration. Values of the first-order rate constant observed by Mouline at 35°C for the latter condition are compared below with ones calculated from Eq. (10.33) correlating Siddall and Duke's data.

Nitric acid molarity, x_H	First-order constant k_N, min^{-1}	
	Average observed, Mouline	Eq. (10.33)
2	0.013	0.0127
3	0.059	0.048
4	0.14	0.124

These reaction rates are too low to explain the appreciable extraction of neptunium obtained in the short-residence-time HA contactors used at Hanford and elsewhere. Swanson [S24] reported that radiolysis reaction products of TBP and nitric acid present in Purex solutions increased the neptunium oxidation rate and provided a possible explanation. He found that the oxidation rate could be increased several orders of magnitude by adding a synthetic "rate-accelerating material" (RAM) produced by reacting the aciform of nitropropane, $C_2H_5(CH)(NO)(OH)$, with nitric and nitrous acids, and recommended addition of such a catalyst to Purex feed if increased neptunium extraction were desired.

Oxidation of pentavalent neptunium by pentavalent vanadium. Oxidation of pentavalent neptunium by pentavalent vanadium proceeds at a practical rate without catalyst. Dukes [D4] found that the rate of reaction could be represented by

$$\frac{dx_5}{dt} = -k_V[H^+]^2[VO_2^+]x_5 \tag{10.36}$$

with values for the specific rate constant k_V given in the second column of Table 10.24. The third column gives values for the first-order rate constant $k_V[H^+]^2[VO_2^+]$ for conditions to be recommended in the HA contactor, 2.5 M HNO_3 and 0.01 M VO_2^+. The rate is much greater than the rate with nitrous acid catalysis and is high enough for a practical process. Srinivasan et al. [S20] extracted more than 90 percent of the neptunium in laboratory mixer-settler experiments with VO_2^+ as oxidant.

Reduction of neptunium. To separate neptunium from plutonium in the Purex process, plutonium is reduced to inextractable Pu(III) while neptunium is reduced from extractable

Table 10.24 Rate of oxidation of pentavalent neptunium by pentavalent vanadium

Temperature, °C	Specific rate constant k_V, mol^{-3} min^{-1}	First-order rate constant in 2.5 M nitric acid and 0.01 M VO_2^+ min^{-1}
24	11.7	0.73
30	13.8	0.86
40	26.2	1.65
50	51.2	3.20

Source: E. K. Dukes, "Oxidation of Neptunium (V) by Vanadium (V)," Report DP-434, 1959.

Np(VI) through inextractable Np(V) to extractable Np(IV). Reduction to Np(V) is rapid, but reduction to Np(IV) is slow, probably because of need to remove oxygen from NpO_2^+. Of the three reductants considered, ferrous iron reacts most rapidly, but must be present in such great excess for complete reduction to Np(IV) that one of the stronger reductants, tetravalent uranium or hydroxylamine, is preferred.

Reduction with tetravalent uranium. Newton [N3] found the rate of reduction of hexavalent neptunium to pentavalent to be rapid and given at 25°C by

$$-\frac{d[\mathrm{Np(VI)}]}{dt_s} = 21.7[\mathrm{Np(VI)}]\,[\mathrm{U(IV)}] \tag{10.37}$$

The rate of reduction of pentavalent neptunium to tetravalent is much slower. Shastri et al. [S10] made an extensive study of the reduction of Np(V) by U(IV). In one series of experiments at 25°C, $[H^+] = 0.1$, and an ionic strength of 0.6 M, the rate of increase of Np(IV) molarity x_4 could be represented by

$$\frac{dx_4}{2\,dt_{\mathrm{min}}} = (0.039x_5 + 0.26x_4)[\mathrm{U(IV)}] \tag{10.38}$$

The rate varied inversely as $[H^+]^2$. At hydrogen ion molarity $[H^+]$ and when U(IV) is present in sufficient excess to remain effectively constant during reduction, the integrated rate equation with $x_4 = 0$ at $t = 0$ is

$$t_{\mathrm{min}} = \frac{1}{0.442[\mathrm{U(IV)}]}\frac{[H^+]^2}{(0.1)^2}\ln\left(1 + \frac{0.221x_4}{0.039x_5^0}\right) \tag{10.39}$$

In the process example to be used in Sec. 7.7, where $[H^+] = 0.06$, $[\mathrm{U(IV)}] = 0.044$ and $x_4/x_5^0 = 0.99$; $t = 35$ min.

Reduction with hydroxylamine. No comparable rate data for reduction of Np(V) by hydroxylamine are available. Barney [B2] reported that the initial rate of reduction of Pu(IV) by 0.1 M hydroxylamine nitrate (HAN) was about one-fourth the initial rate of reduction of Pu(IV) by U(IV) at 25°C. If the same ratio applies to reduction of Np(V) by HAN or U(IV), a reaction time of around (4)(35) = 140 min might be required. Use of hydroxylamine would have the advantage of not requiring reduction of uranium to U(IV) and its subsequent recycle.

7.6 Neptunium Recovery Examples

Special campaigns for recovering neptunium from Purex solutions have been run at Oak Ridge [F4], Hanford [D3], Savannah River [P7], Windscale [N1], and Marcoule [C6]. None of these sought complete recovery. A brief description will be given of the first three.

Oak Ridge [F4]. The solution obtained by dissolving irradiated, natural uranium in nitric acid was treated with 0.01 M $NaNO_2$ to convert most of the neptunium to extractable Np(VI) in 2 M HNO_3. In the first Purex cycle, 90 percent of the neptunium was extracted with the uranium and plutonium. Ferrous sulfamate used in the partitioning step reduced most of the neptunium to Np(V) and Np(IV), which followed Pu(III) into the aqueous phase, but some Np(IV) remained with the uranium. When the aqueous phase containing Pu(III) and most of the neptunium was reoxidized with HNO_3 and $NaNO_2$ and extracted with TBP in the second cycle, from one-half to two-thirds of the neptunium was recovered with the plutonium, from which the neptunium was separated by anion exchange. The process produced 99.9 percent pure neptunium, but the recovery was incomplete and very sensitive to nitrite and nitric acid concentrations.

Hanford [D3]. Nitrite concentration in feed to the HA column of a standard Purex plant was adjusted to route most of the neptunium in irradiated natural uranium into the extract from the HS scrubbing column. Sufficient ferrous sulfamate was used in the partitioning column to reduce neptunium to Np(IV), which followed uranium. This neptunium was separated from uranium by fractional extraction with TBP in the second uranium cycle. The dilute neptunium product was recycled to HA column feed, to build up its concentration. Periodically, irradiated uranium feed was replaced by unirradiated uranium, which flushed plutonium and fission products from the system. The impure neptunium remaining was concentrated and purified by solvent extraction and ion exchange.

Savannah River [P7]. At the Savannah River Purex plant, neptunium in irradiated natural uranium was recovered by the alternative method of forcing most of it into the aqueous waste stream HAW from the first extraction cycle and then recovering it from waste directly by anion exchange. Neptunium in the first extraction step was converted mostly to the inextractable pentavalent state by adding sufficient nitrite to the next-to-the-last mixer-settler stage of the HA section to make the solvent 0.007 M in HNO_2.

7.7 Neptunium Recovery Process

Process selection. The processes just described recovered neptunium only partially and in variable yield because of the difficulty in controlling the distribution of neptunium valence between 5 and 6 in the primary extraction step with nitrite-catalyzed HNO_3 and the incomplete reduction of neptunium from valence 5 to 4 in the partitioning step with ferrous ion. This section describes a modified Purex process that could be used if more complete recovery of neptunium were required. It is based on process design studies by Tajik [T1]. The principal process steps are shown in the material flow sheet Fig. 10.32. In the primary decontamination step, pentavalent vanadium oxidizes neptunium to the extractable hexavalent state. In the partitioning step, tetravalent uranium reduces plutonium to the inextractable trivalent state while converting neptunium to the still-extractable tetravalent state.

Decontamination. Prior to decontamination, nitrites in the aqueous feed must be decomposed by air sparging, to prevent them from reducing vanadium in the HA contactor. Extraction of uranium and Pu(IV) in the HA contactor and scrubbing of fission products in the HS contactor are carried out substantially as in the conventional Purex process described in Sec. 4. The neptunium oxidant, 0.054 M pentavalent vanadium in 3 M nitric acid, is fed to the extracting section two theoretical stages below the aqueous feed point. This point is selected so that most of the uranium and plutonium will have been already extracted from the aqueous stream. As the aqueous stream flows through the remaining five extracting stages, vanadium oxidizes neptunium to Np(VI) which is extracted by counterflowing solvent. Experiments by Srinivasan [S20] and Koch [K4] suggest that 87 to 97 percent of the neptunium can be recovered in this way.

Uranium, plutonium, and neptunium in the extract are returned to the aqueous phase in the HC stripping unit.

Reduction. To convert plutonium to inextractable Pu(III) and neptunium to still extractable Np(IV), 0.5 M U(IV) reductant is added to the aqueous stream from the HC unit. It is necessary to hold the reacting mixture for half an hour or more to obtain nearly complete reduction of neptunium. This is best done batchwise in a set of reactors, some of which would be reducing while others are receiving feed to be reduced. Reduced fuel is then concentrated to 1.172 M uranium in a set of batch evaporators.

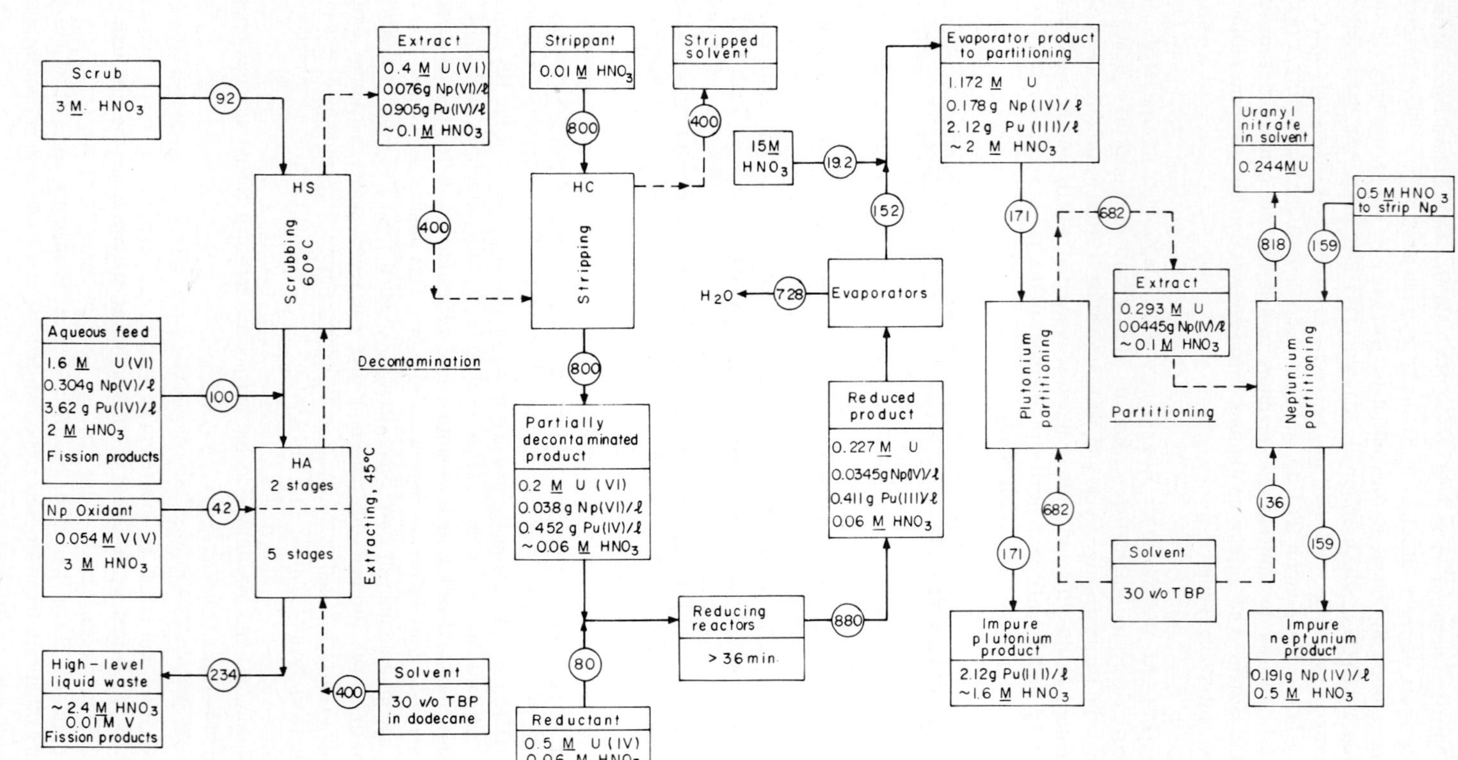

Figure 10.32 Principal steps in Purex process modified for neptunium recovery. Circles indicate relative volume flow rate; —— aqueous; ––– organic.

Partitioning of plutonium. Evaporator product is made 2 *M* in nitric acid and extracted with four volumes of 30 v/o TBP in the plutonium partitioning unit. This leaves plutonium in the aqueous raffinate and extracts the uranium and neptunium.

Partitioning of neptunium. Uranium and tetravalent neptunium in the extract are separated by fractional extraction with 0.5 *M* HNO_3. The less extractable Np(IV) is returned to the aqueous phase while uranium remains in the solvent, from which it can be stripped with 0.01 *M* HNO_3 (not shown).

The process just described has the advantage of providing nearly complete recovery of neptunium. Its principal disadvantages are addition of vanadium to first-cycle wastes and the need to recycle some uranium as U(IV).

8 PREVENTION OF CRITICALITY IN REPROCESSING PLANTS

A nuclear fission chain reaction in a reprocessing plant is an accident that must be carefully guarded against. Although such a critical reaction is not likely to generate sufficient energy to be mechanically destructive, it emits intense neutron and gamma radiation that can kill nearby plant personnel and may release radioactive fission products outside the plant.

This section outlines the methods for preventing nuclear criticality and gives some background for the conditions placed on reprocessing plant design and operation by criticality considerations. The brief discussion here and the limited examples to be cited should be used only to suggest conditions for safe design and operation. Greater detail is given in the U.S. *Nuclear Safety Guide* [T5], which is the source of the examples of this section, and in earlier reports [C3, C9, P1, P2, T4], which contain experimental data also. Nuclear criticality safety has been codified in American National Standards published by the American Nuclear Society, of which the ones most applicable to reprocessing plants are [A4, A5, and A6]. Even after using these standards, the design or operation of equipment in which fissile material is to be processed should be reviewed for criticality safety by an expert. And even after such review, some reprocessing systems may have novel features whose safety can be verified only by experiment.

8.1 Factors Affecting Criticality Safety

The principal factors that must be taken into account in assessing criticality safety are as follows:

1. Fissile nuclide (^{235}U, ^{233}U, or ^{239}Pu)
2. Proportion of fertile nuclide (^{238}U, ^{232}Th, or ^{240}Pu) diluting fissile nuclide
3. Mass of fissile nuclide
4. Geometry (shape and dimensions) of region holding fissile material
5. Volume of region holding fissile material
6. Concentration of fissile material
7. Nature and concentration of moderators
8. Nature and thickness of reflectors surrounding fissile material
9. Nature and concentration of neutron-absorbing poisons, such as nitrate ion or gadolinium nitrate
10. Homogeneity or heterogeneity of fuel-moderator-poison mixture
11. Degree of interaction between two or more regions containing fissile material

For given fuel composition (factors 1 and 2 specified), the simplest but most restrictive condition to ensure subcriticality is one of items 3, 4, 5, or 6 (limitation of mass, dimensions, volume, or concentration of fissile material). These so-called ***single-parameter limits for fissile nuclides*** are spelled out in American National Standard ANSI N16.1-1975 [A4]. They were abstracted in Table 4.11 of Chap. 4 and are amplified somewhat in Sec. 8.2, following. These single-parameter limits give the largest mass, size, volume, or concentration that will be safely subcritical no matter what other criticality-limiting conditions may be present.

Use of a single-parameter limit often leads to an inconveniently small size of batch or equipment. To permit safe operation on a larger scale, combinations of two parameters that together are safely subcritical are sometimes specified, provided that the simultaneous presence of both parameters can be assured. For example, if the maximum concentration of plutonium in aqueous solution can be limited to 20 g/liter, the maximum safe diameter of a cylinder may be increased from the single-parameter limit of 15.7 cm (Table 4.11 or 10.25) to 25 cm (Fig. 10.35).

By restricting the concentration of moderators (item 7) or the presence of reflectors (item 8), the dimensions or concentrations of safely subcritical systems may be increased further. The presence of neutron-absorbing poisons such as boron, cadmium, or gadolinium (item 9) also sometimes permits such increase. On the other hand, heterogeneity (item 10, such as lumping of fuel containing ^{238}U or interaction between two systems containing fissile material (item 11, such as adjacent pipes carrying fissile material) reduce the dimensions, mass, or concentration of safely subcritical systems.

8.2 Single-Parameter Limits for Fissile Nuclides†

Operations with fissile materials may be performed safely by complying with any one of the subcritical limits given in Sec. 8.2 provided the conditions under which it applies are maintained. A limit shall be applied only when the effects of neutron reflectors and of other nearby fissionable materials are no greater than reflection by an unlimited thickness of water.‡ The limits shall not be applied to mixtures of ^{235}U, and ^{233}U, and ^{239}Pu.

Process specifications shall incorporate margins to protect against uncertainties in process variables and against a limit being accidentally exceeded.

Uniform aqueous solutions. Any one of the limits of Table 10.25 is applicable provided a uniform aqueous solution is maintained and provided, for ^{239}Pu, at least four nitrate ions are present for each plutonium ion. The ^{239}Pu limits apply to mixtures of plutonium isotopes provided the concentration of ^{240}Pu exceeds that of ^{241}Pu and provided ^{241}Pu is considered to be ^{239}Pu in computing mass or concentration.

Uniform slurries. The limits of Table 10.25 may be used for macroscopically uniform slurries, provided:

1. There are at least four nitrate ions intimately associated with each plutonium atom, and
2. For the dimensional and volume limits, the ratio of hydrogen-to-fissionable material does not exceed that in an aqueous solution having the same concentration of fissionable material.

†Section 8.2 is taken verbatim from Sec. 5 of [A4] except for footnotes and changes in references to tables, sections, and literature citations. Extracted from American National Standard N-16.1-1975 (ANS-8.1), with permission of the publisher, the American Nuclear Society.

‡The limits do not apply to reflection by graphite, beryllium, or heavy water.

Table 10.25 Single-parameter limits for uniform aqueous solutions containing fissile nuclides

Parameter	Subcritical limit for		
	^{235}U	^{233}U	^{239}Pu provided N:Pu ⩾ 4
Mass of fissile nuclide, kg	0.76	0.55	0.51
Solution cylinder diameter, cm	13.9	11.5	15.7
Solution slab thickness, cm	4.6	3.0	5.8
Solution volume, liters	5.8	3.5	7.7
Concentration of fissile nuclide, g/liter	11.5	10.8	7.0
Areal density of fissile nuclide, g/cm^2	0.40	0.35	0.25
Uranium enrichment, wt % ^{235}U	1.00	–	–
Uranium enrichment in presence of two nitrates ions per uranium atom, wt % ^{235}U	2.07	–	–

The limit on the 1.00 wt % enrichment of uranium is valid only for the slurries in which the ratio of surface-to-volume of the particles is at least 80 cm^{-1}.

Nonuniform slurries. The limits on cylinder diameter and slab thickness in Table 10.25 may be used for nonuniform slurries provided:

1. Four nitrate ions are intimately associated with each plutonium atom,
2. The restriction on the ratio of hydrogen-to-fissionable atoms, specified in Condition 2 for uniform slurries is met everywhere throughout the system,
3. For cylinders, the concentration gradient is only along the length, and
4. For slabs, the concentration gradient is only parallel to the faces.

For ^{239}Pu in the absence of nitrate ions, but with the proviso that no localized regions of density greater than 0.25 g of $^{239}Pu/cm^3$ are permitted, limits of 15.1 and 5.4 cm on cylinder diameter and slab thickness, respectively, are applicable under Conditions 2, 3, and 4 above.

The areal densities given in Table 10.25 are valid for nonuniform slurries provided these densities are uniform.

The subcritical mass limits for ^{235}U, ^{233}U, and ^{239}Pu in nonuniform slurries are 0.70, 0.52, and 0.45 kg, respectively. Nitrate ions need not be present.

Metallic units. The enrichment limit for uranium and the mass limits given in Table 10.26 apply to a single piece having no concave surfaces. They may be extended to an assembly of smaller units provided there is no inter-unit moderation.

The ^{235}U and ^{233}U limits apply to mixutres of either isotope with ^{234}U, ^{236}U, or ^{238}U provided all isotopes except ^{238}U are considered to be ^{235}U or ^{233}U, respectively, in computing mass. The ^{239}Pu limits apply to isotopic mixtures of plutonium provided the concentration of ^{240}Pu exceeds that of ^{241}Pu, all plutonium isotopes are considered to be ^{239}Pu in computing mass, and no more than 1% ^{238}Pu is present.

Table 10.26 Single-parameter limits for metal units

Parameter	Subcritical limit for ^{235}U	^{233}U	^{239}Pu
Mass of fissile nuclide, kg	20.1	6.7	4.9
Cylinder diameter, cm	7.3	4.6	4.4
Slab thickness, cm	1.3	0.54	0.65
Uranium enrichment, wt % ^{235}U	5.0	—	—

8.3 Multiparameter, Concentration-dependent Limits for Criticality Control

This section contains examples of how limits on critical mass or dimensions can be increased from the single-parameter limits of Sec. 8.2 by limiting two or more parameters simultaneously. Many more examples than could be cited here are given in the references listed at the beginning of Sec. 8. When using multiparameter limits, the caution cited for single-parameter limits at the beginning of Sec. 8.2 is even more essential, because of the larger number of variables to be controlled.

Highly enriched uranium or plutonium. Figures 10.33, 10.34, and 10.35 show the relation between the safe diameter of a cylinder of infinite length or the safe thickness of a slab of

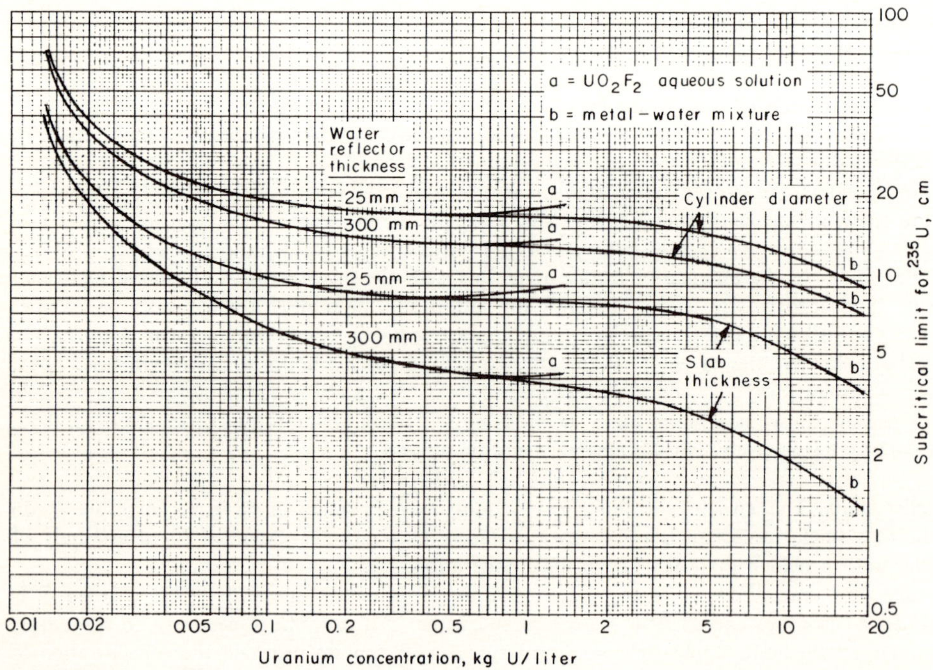

Figure 10.33 Subcritical limits for individual cylinders and slabs of homogeneous water-reflected and moderated ^{235}U.

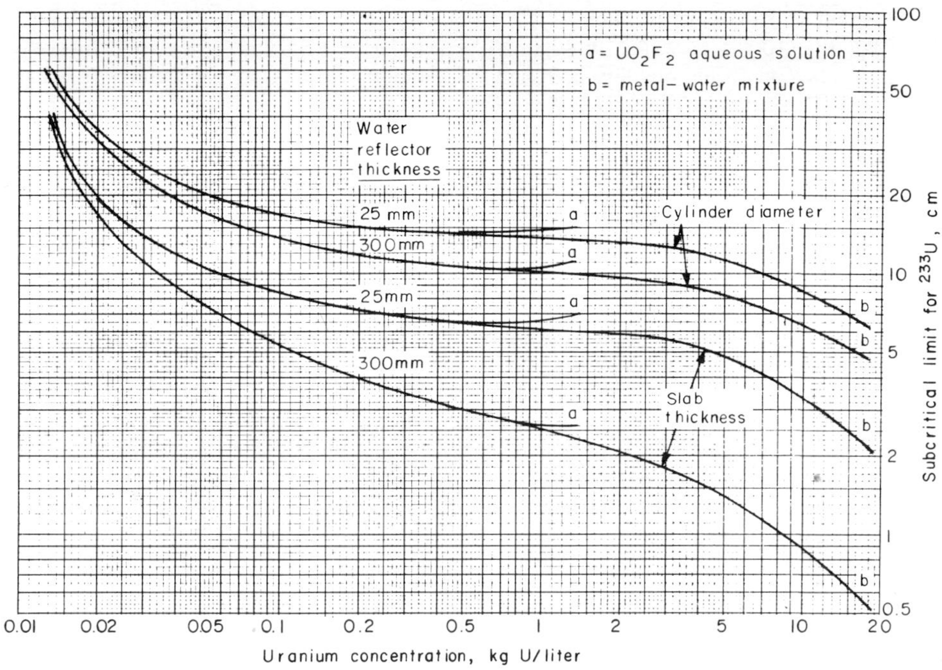

Figure 10.34 Subcritical limits for individual cylinders and slabs of homogeneous water-reflected and moderated ^{233}U.

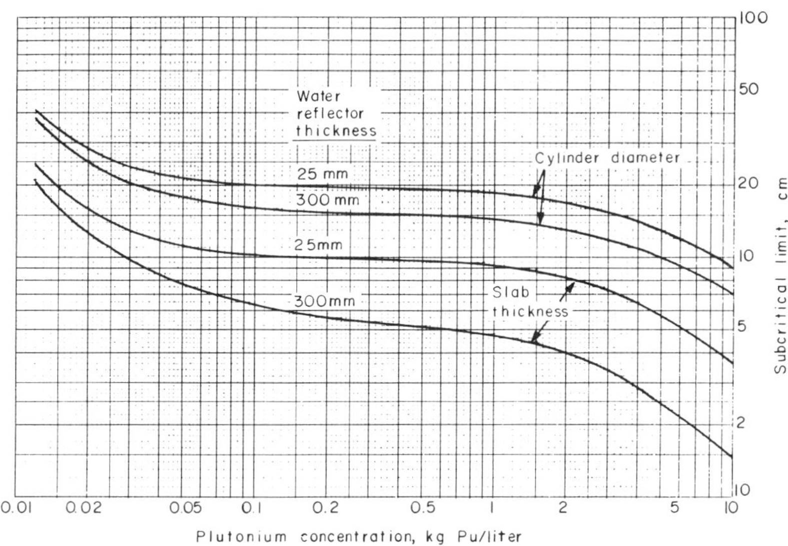

Figure 10.35 Subcritical limits for individual cylinders and slabs of homogeneous water-reflected and moderated plutonium containing at least 5 w/o ^{240}Pu and $^{240}Pu > {}^{241}Pu$.

infinite extent and the concentration of ^{235}U (Fig. 10.33), ^{233}U (Fig. 10.34), or plutonium (>5 w/o ^{240}Pu, Fig. 10.35) in water. In Figs. 10.33 and 10.34, curves are given for solutions of UO_2F_2 in water up to its solubility limit and for homogeneous mixtures of uranium metal and water up to the density of the metal. The dimensions at the minima of the UO_2F_2 curves are lower than corresponding solution values in Table 10.25 because the figures have a greater safety factor than the table.

Figure 10.35 refers to a homogeneous mixture of PuO_2 and water and is conservative for an aqueous solution of $Pu(NO_3)_4$.

Curves are given for two thicknesses of water reflector. The 25-mm curves "generally provide a sufficient margin of subcriticality to compensate for water jackets around piping and for reflection by concrete 300 mm or more distant. Limits for a 300-mm-thick water reflector are appropriate when reflector conditions cannot be rigidly controlled" [T5, p. 40]. Lower limits are required when reflection is by close-fitting concrete, uranium, tungsten, beryllium, D_2O, or plastic.

Mixtures with ^{238}U. When ^{238}U is mixed with ^{235}U, the subcritical limit for a cylinder or slab of UO_2F_2 solution given by Fig. 10.33 may be increased by the factor given in Fig. 10.36. These factors may also be applied to uniform slurries of water and UO_2 provided that the ^{235}U enrichment is greater than 6 w/o or the particle sizes are smaller than 127 μm. With ^{235}U enrichment below 6 w/o and larger particles, the factors are smaller than given in Fig. 10.36 because of reduced absorption by ^{238}U. The factors are conservative for aqueous solutions of uranyl nitrate because of neutron absorption by nitrogen.

When plutonium dioxide is mixed uniformly with uranium dioxide containing 0.71 w/o or less ^{235}U, the subcritical limits for infinite water-reflected cylinders or slabs are greater than given in Table 10.25. Table 10.27 shows the dependence of critical dimensions on w/o PuO_2 in $PuO_2 + UO_2$ and on plutonium isotopic composition.

Soluble neutron absorbers. The preceding limits on critical concentration or dimensions can be greatly relaxed when soluble neutron absorbers, such as boric acid or gadolinium nitrate, are

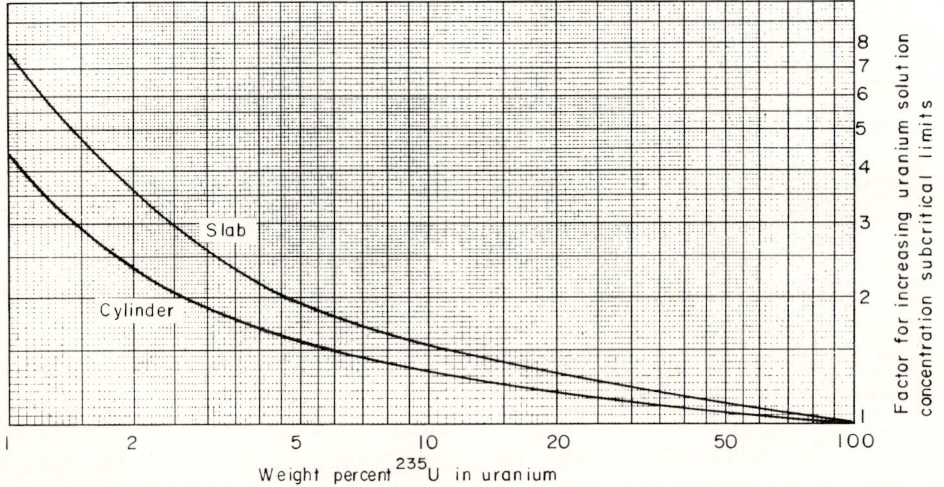

Figure 10.36 Factors by which the subcritical limits for aqueous homogenous solutions of uranium reflected by 300 mm water in Fig. 10.33 may be increased for reduced uranium enrichment.

Table 10.27 Subcritical limits for uniform aqueous mixtures of the oxides of plutonium and natural uranium†

PuO_2 in ($PuO_2 + UO_2$), w/o	3			8			15			30		
Plutonium isotopic composition‡	I	II	III	I	II	III	I	II	III	I	II	III
Mass of plutonium in oxide mixture, kg	0.73	1.35	2.00	0.61	1.06	1.53	0.54	0.94	1.28	0.50	0.87	1.16
Mass of ($PuO_2 + UO_2$), kg	27.5	51.3	75.9	8.6	15.1	21.7	4.1	7.1	9.7	1.9	3.3	4.4
Diameter of infinite cylinder, cm	24.3	30.8	34.8	19.8	24.9	27.5	17.8	22.5	24.8	16.2	21.0	23.4
Thickness of infinite slab, cm	11.0	14.9	17.4	8.2	11.2	12.9	6.9	9.6	11.0	5.9	8.7	9.9
Volume of oxide mixture, liter	23.5	44.8	63.4	14.0	25.9	34.4	11.0	20.4	26.6	8.5	16.8	21.6
Concentration of plutonium in an infinite volume, g Pu/liter	6.8§	8.1	9.3	6.9	8.2	9.4	7.0	8.2	9.4	7.0	8.1	9.3
Concentration of oxides in an infinite volume, g ($PuO_2 + UO_2$)/liter	257§	305	351	97.3	116	134	52.9	61.7	71.0	26.5	30.7	35.2
H/Pu atomic ratio	3780	3203	2780	3780	3210	2790	3780	3237	2818	3780	3253	2848
Areal density of plutonium in infinite slab, g Pu/cm^2	0.27	0.38	0.47	0.25	0.34	0.42	0.25	0.33	0.41	0.24	0.32	0.37
Areal density of oxides in infinite slab, g ($PuO_2 + UO_2$)/cm^2	10.2	14.4	17.7	3.5	4.8	5.9	1.9	2.5	3.1	0.9	1.2	1.4

† All values are upper limits except atomic ratios, which are lower limits.

‡ Plutonium isotopic composition: I—$^{240}Pu > {}^{241}Pu$. II—$^{240}Pu \geqslant 15$ w/o and $^{241}Pu \leqslant 6$ w/o. III—$^{240}Pu \geqslant 25$ w/o and $^{241}Pu \leqslant 15$ w/o. The small quantities of ^{238}Pu and ^{242}Pu expected in these isotopic mixtures are considered to have neglible effects on the limits.

§ This concentration limit is not applicable to oxide mixtures in which the $PuO_2/(PuO_2 + UO_2)$ ratio is less than 3 w/o because of the increased relative importance of ^{235}U in high-uranium-bearing materials.

Source: Extracted from American National Standard ANSI/ANS-8.12-1978, with permission of the publisher, the American Nuclear Society.

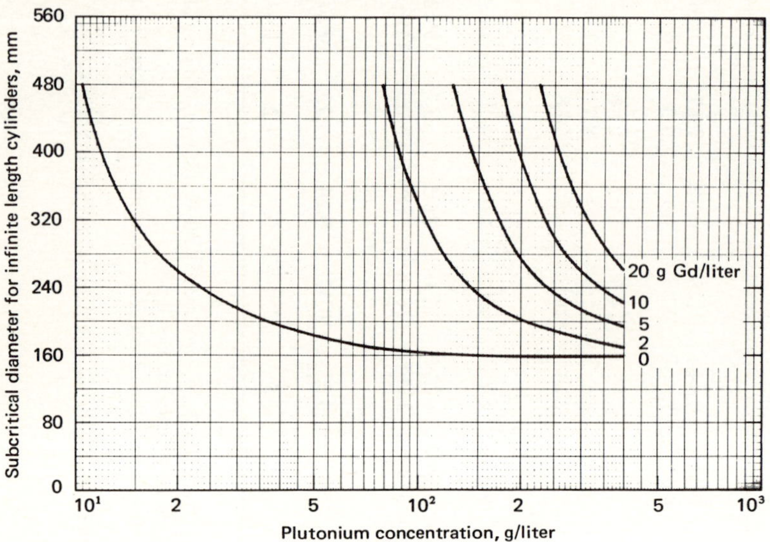

Figure 10.37 Subcritical diameter limits for thick water-reflected individual cylinders of homogeneous solutions of $Pu(NO_3)_4$ containing $Gd(NO_3)_3$.

assuredly uniformly distributed in the fissile material. As one example, one atom of natural boron per atom of ^{235}U will keep a large volume of aqueous solution subcritical for ^{235}U concentrations up to 400 g/liter. As another example, Fig. 10.37 shows how the subcritical diameter of an infinite cylinder of an aqueous solution of $Pu(NO_3)_2$ is increased by addition of $Gd(NO_3)_3$.

Solid neutron absorbers. In the disengaging sections of pulse columns and in storage vessels for solutions, it is sometimes desirable to have larger vessels than the maximums allowed in the preceding text. By packing such equipment with borosilicate glass Raschig rings, the maximum

Table 10.28 Maximum permissible concentrations of homogeneous solutions of fissile materials in vessels of unlimited size packed with borosilicate glass Raschig rings

	Maximum concentration in vessels with minimum glass content of		
Isotopic composition	24 v/o	28 v/o	32 v/o
1. 5 w/o $<$ ^{235}U $<$ 100 w/o; ^{233}U $<$ 1 w/o	270	330	400 g U/liter
2. 0.7 w/o $<$ ^{235}U $<$ 5 w/o; $^{233}U = 0$	270	330	400 g ^{235}U/liter
3. 0 $<$ ^{233}U $<$ 100 w/o	150	180	200 g U/liter
4. $^{239}Pu \geqslant 50$ w/o; $^{241}Pu \leqslant 15$ w/o; $^{240}Pu \geqslant {}^{241}Pu$			
(a) $\leqslant$5 w/o ^{240}Pu	115	140	180 g Pu/liter
(b) $>$5 w/o ^{240}Pu	140	170	200 g Pu/liter

Source: American Nuclear Society, "Proposed American National Standard, Use of Borosilicate-Glass Raschig Rings as a Neutron Absorber in Solutions of Fissile Material," Report ANS-8.5-1979, La Grange Park, Ill.

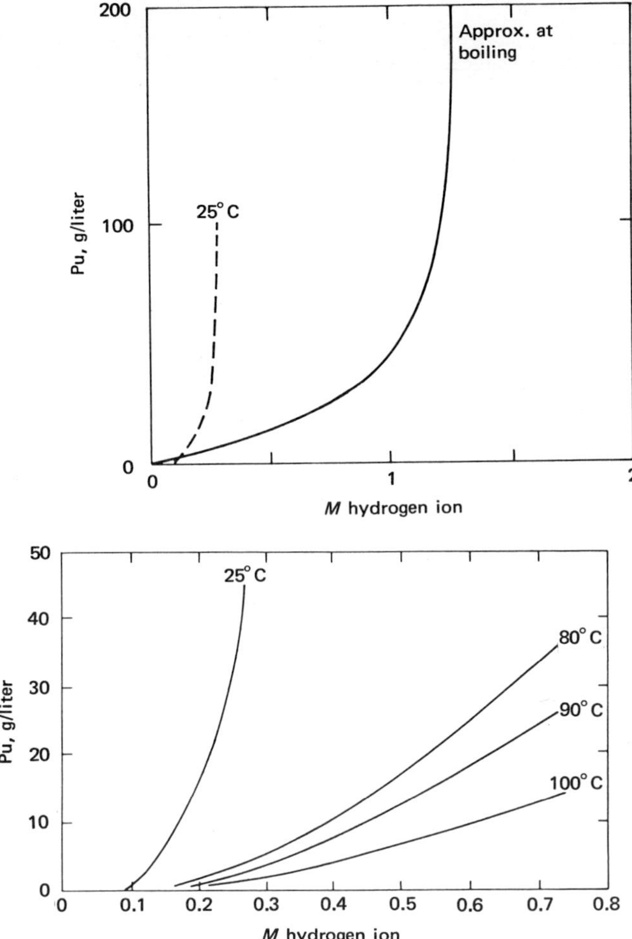

Figure 10.38 Plutonium polymer formation limits. (*From Mann and Irene [M3].*)

concentration of fissile materials that can be contained in indefinitely large vessels without becoming critical can be increased to the values given in Table 10.28. A proposed American National Standard [A7] gives specifications on the dimensions and composition of the rings.

Plutonium polymer. At low acidity and high temperature, plutonium forms a polymer that deposits as an insoluble solid film on the walls of process equipment. Polymer deposition plugs lines, fouls surfaces, and may result in unanticipated accumulation of a critical mass of plutonium. Figure 10.38 summarizes [M3] the results of investigations of the combinations of low acidity and high temperature that must be avoided if plutonium polymer formation is to be prevented.

As an additional precaution, process equipment in which plutonium polymer might form should be soaked periodically in boiling, concentrated nitric acid. If plutonium is found in solution, the presence of a polymer deposit is indicated. Complete removal may require addition of 0.01 to 0.1 M HF to the hot HNO_3.

REFERENCES

A1. Allied Chemical Company: Idaho Chemical Processing Plant, Pamphlet obtainable from Idaho Operations Office, U.S. Department of Energy, Idaho Falls, Idaho.

A2. Allied-General Nuclear Services: Barnwell Nuclear Fuel Plant Environmental Statement, Jan. 1974.

A3. Allied-General Nuclear Services: Final Safety Analysis Report, Barnwell Nuclear Fuel Plant Separations Facility, Oct. 1973.

A4. American Nuclear Society: "American National Standard, Nuclear Criticality Safety in Operations with Fissionable Materials Outside Reactors," Report ANSI N16.1-1975, La Grange Park, Ill.

A5. American Nuclear Society: "American National Standard, Nuclear Criticality Safety in the Storage of Fissile Materials," Report ANSI N16.5-1975, La Grange Park, Ill.

A6. American Nuclear Society: "American National Standard, Nuclear Criticality Control and Safety of Homogeneous Plutonium-Uranium Fuel Mixtures Outside Reactors," Report ANSI/ANS-8.12-1978, La Grange Park, Ill.

A7. American Nuclear Society: "Proposed American National Standard, Use of Borosilicate-Glass Raschig Rings as a Neutron Absorber in Solutions of Fissile Material," Report ANS-8.5-1979, La Grange Park, Ill.

A8. Anon.: Explosion in Evaporator, *Nucl. Safety* **1**(3): 78 (1960).

A9. Argonne National Laboratory: "Chemical Engineering Division Highlights, May 1963 to April 1964," Report ANL-6875, 1964.

A10. Aylward, J. R., and E. M. Whitener: "Electrolytic Dissolution of Nuclear Fuels, Part II. Nichrome in Nitrate Solutions," Report IDO-14575, Dec. 29, 1961.

A11. Aylward, J. R., and E. M. Whitener: "Electrolytic Dissolution of Nuclear Fuels, Part III. Stainless Steel (304) in Nitrate Solutions," Report IDO-14584, June 1, 1962.

B1. Barghusen, J. J., et al.: "Fluid-Bed Fluoride Volatility Processing of Spent Reactor Fuel Materials," in *Progress in Nuclear Energy,* series III, *Process Chemistry,* vol. 4, Pergamon, New York, 1970, p. 347.

B2. Barney, G. S.: "Kinetics and Mechanism of Pu(IV) Reduction by Hydroxylamine," Report ARH-SA-100, 1971.

B3. Barney, R., and F. Keneshea: "The Distribution of Pu and Fission Products between Molten Uranium and Magnesium," Report NAA-SR-1324, 1956.

B4. Baroncelli, F., and G. Grossi: "Chemical Degradation of Aromatic Diluents Exposed to Nitric Acid Attack," in *Solvent Extraction of Metals,* H. A. C. McKay (ed.), Macmillan, London, 1965.

B5. Baumgärtner, F.: "Reprocessing Problems Associated with the Increasing Burnup of Future Fuels," Report KFK-884, Dec. 1968.

B6. Baumgärtner, F., and W. Ochsenfeld: "Development and Status of LMFBR Reprocessing in the Federal Republic of Germany," Report KFK-2301, May 1976.

B7. Bebbington, W. P.: *Sci. Amer.* **235**: 30 (Dec. 1976).

B8. Becker, R., and L. Stieglitz: Report KFK-1373, 1973.

B9. Bendixsen, C. L., and F. O. German: "Operation of the ICPP Rare Gas Recovery Facility During Fiscal Year 1970," Report ICP-1001, 1971.

B10. Bennett, G. A., L. Burris, Jr., and R. C. Vogel: "Halide Slagging of Uranium-Plutonium Alloys," Report ANL-6918, 1964.

B11. Blake, C. A.: "Solvent Stability in Nuclear Fuel Processing . . . ," Report ORNL-4212, Mar. 1968.

B12. Blanco, R. E.: "Dissolution and Feed Adjustment," in *Symposium on the Reprocessing of Irradiated Fuels, Held at Brussels, Belgium, May 20–25, 1957,* USAEC Report TID-7534, book 1, pp. 22–44.

B13. Blanco, R. E., L. M. Ferris, and D. E. Ferguson: "Aqueous Processing of Thorium Fuels," Report ORNL-3219, Mar. 1962.
B14. Blanco, R. E., et al.: "Aqueous Processing of Thorium Fuels, Part II," Report ORNL-3418, June 7, 1963.
B15. Bloom, G. R.: *Proceedings of the International Conference on Sodium Technology and Large Fast Reactor Design,* Report ANL-7520, vol. 1, 1968, p. 410.
B16. Bray, L. A.: "Denitration of Purex Waste with Sugar," Report HW-76973 Rev., Apr. 1963.
B17. British Nuclear Fuels Ltd.: "Windscale Planning Application," Risley, England, May 9, 1977.
B18. Brookhaven National Laboratory: "Nuclear Engineering Department Progress Report, April 1–June 30, 1958," Report BNL-516, 1958.
B19. Brookhaven National Laboratory: "Nuclear Engineering Division Annual Report, 1966," Report BNL-50023, 1967.
B20. Brooks, L. H.: "Survey and Evaluation of Methods for Reprocessing Spent HTGR Fuel," Report GA-A12853, Dec. 1974.
B21. Buckham, J. A.: Letter to M. Benedict, Nov. 8, 1978.
B22. Bull, H., and J. E. Koonce, Jr.: "Performance of an Electrolytic Dissolver at the Savannah River Plant," Preprint No. 67D, Annual Meeting of Amer. Inst. Chem. Eng., 1970. Available from Engineering Societies Library, 345 E. 47th St., New York, N.Y. 10017.
C1. Campbell, W. M.: *Nucleonics* **14**(9): 92 (Sept. 1956).
C2. Carr, W. H.: *Chem. Eng. Progr. Symp. Series 28* **56**: 57 (1960).
C3. Carter, R. D., et al.: *Criticality Handbook,* Report ARH-600, 3 vols., June 1968, May 1969, Sept. 1971.
C4. Catlin, J. R., D. Morehouse, and R. P. P. Connop: Improvements in Apparatus for Removal of Sheaths from Nuclear Fuel Elements, British Patent 1,180,033, Feb. 1970.
C5. Chauvin, M.: "French Situation in Fuel Reprocessing and Waste Management," Statement to California Energy Resources Conservation and Development Commission, April 19, 1977.
C6. Chesné, A.: *Industries Atomiques* **7/8**: 71 (1966).
C7. Chiotti, P., and J. S. Klepfer: *Ind. Eng. Chem. Process Design and Development* **4**: 232 (1965).
C8. Chulos, L. E.: "Sodium Removal and Cleaning of Reusable Hardware," Report BNWL-637, Dec. 1967.
C9. Clark, H. K.: "Handbook of Nuclear Safety," Report DP-532, 1961.
C10. Codding, J. W., W. O. Haas, and F. K. Heumann: "Equilibrium Data for Purex Systems," Report KAPL-602, Nov. 26, 1951; *Ind. Eng. Chem.* **50**: 145 (1958).
C11. Colven, T. J., et al.: "Interim Technical Report—TNX Evaporator Incident January 12, 1953," Report DP-25, May 1953.
C12. Commissariat à l'Energie Atomique: *Irradiated Fuel Reprocessing, La Hague Center,* 1970.
C13. Commissariat à l'Energie Atomique: "Rapport Semestrial du Département de Chimie, CEN (Fontenay-aux-Roses)," Reports CEA-N-969 (1968), 1044 (1969), 1241 (1970), and 1419 (1971).
C14. Commissariat à l'Energie Atomique: *Marcoule Nuclear Industrial Center,* 1976.
C15. Couture, J.: *Chem. Eng. Progr. Symp. Ser. 94* **65**: 26 (1969).
C16. Croff, A. G.: "An Evaluation of Options Relative to the Fixation and Disposal of ^{14}C-Contaminated CO_2 as $CaCO_3$," Report ORNL/TM-5171, Apr. 1976.
C17. Cubicciotti, D.: "The Evaporation of U from Small Pieces of U Reactor Fuel," Report NAA-SR-1057, 1954.

C18. Culler, F. L., and R. E. Blanco: "Dissolution and Feed Preparation for Aqueous Radiochemical Separation Processes," *PICG(2)* **17**: 259 (1959).

D1. Detilleux, E., and S. Cao: "Recent Eurochemic Reprocessing Experiences," *Proceedings of the International Conference on Constructive Uses of Atomic Energy, Washington, D.C., 1968,* American Nuclear Society, 1969, p. 275.

D2. Duboz, M.: *Energie Nucléaire* **7**: 228 (1965).

D3. Duckworth, J. P., and L. R. Michels: *Ind. Eng. Chem. Process Design and Development* **3**: 302 (1964).

D4. Dukes, E. K.: "Oxidation of Neptunium(V) by Vanadium(V)," Report DP-434, 1959.

D5. Dwyer, O. E., et al.: "High-Temperature Processing Systems for Liquid-Metal Fuels and Breeder Blankets," *PICG(1)* **9**: 604 (1956).

E1. English, J. L.: "Thorex Pilot Plant Corrosion Studies, II," Report ORNL-2844, Jan. 1960.

F1. Ferguson, D. E., et al.: "Chemical Technology Division Annual Progress Report for Period Ending March 31, 1971," Report ORNL-4682, 1971.

F2. Fisher, F. D.: "The Sulfex Process Terminal Development Report," USAEC Report HW-66439, Aug. 22, 1960.

F3. Flagg, J. F.: "Solvent Extraction Processes Based on Hexone," in *Chemical Processing of Aqueous Fuels,* J. F. Flagg (ed.), Academic, New York, 1961, chap. 6.

F4. Flanary, J. R., and G. W. Parker: "The Development of Recovery Processes for Neptunium-237," *Progress in Nuclear Energy,* series III, vol. 2, Pergamon, London, 1958, p. 501.

G1. Galley, R.: *Trans. Inst. Chem. Eng. (London)* **36**: 401 (1958).

G2. General Electric Company: *Purex Technical Manual,* Report HW-31000, Mar. 25, 1955.

G3. General Electric Company: "Final Safety Analysis Report for Midwest Fuel Reprocessing Plant," Section 2 of Amendment 35 of Docket 50-268, Dec. 20, 1973.

G4. General Electric Company: "Design and Analysis of Midwest Fuel Recovery Plant," Report Docket 50-268, 1966.

G5. General Electric Company: "Midwest Fuel Recovery Plant Technical Report," July 5, 1974.

G6. Germain, M., D. Gourisse, and M. Sougnez: *J. Inorg. Nucl. Chem.* **32**: 245 (1970).

G7. Goldberg, S. M.: In LMFBR Fuel Cycle Studies Progress Reports. No., 23, 24, 25, 26, month, Jan. 1971, Feb. 1971, Mar. 1971, Apr. 1971, report no., and pages are, respectively, ORNL-TM-3312, 10–13; ORNL-TM-3345, 5–6; ORNL-TM-3375, 8–10; ORNL-TM-3412, 12–14.

G8. Goode, J. H. (ed.): "Volatile Fission Product Removal from LMFBR Fuels," Report ORNL-TM-3723, 1972.

G9. Goode, J. H., and S. D. Clinton: "Aqueous Processing of LMFBR Fuels—Technical Assessment and Experimental Program Definition," Report ORNL-4436, secs. 4.4 and 5.4, 1970.

G10. Goode, J. H., and J. G. Moore: "Adsorption of Protactinium on Unfired Vycor: Final Hot-Cell Experiments," Report ORNL-3950, June 1967.

G11. Gourisse, D.: "Laboratory Studies of Nitrous Acid and Neptunium Behavior in TBP Extraction Processes," *Proceedings of the International Solvent Extraction Conference,* vol. 1, 1971, p. 781.

G12. Gourisse, D.: *J. Inorg. Nucl. Chem.* **33**: 831 (1971).

G13. Gray, J. H., A. Schneider, A. F. Cermak, and A. L. Ayers: Apparatus for Electrolytic Oxidation or Reduction, Concentration and Separation of Elements in Solution, U.S. Patent 3,770,612, Nov. 6, 1973.

G14. Gresky, A. T.: "Solvent-Extraction Separation of ^{233}U and Thorium from Fission Products by Means of Tributyl Phosphate," *PICG(1)* **8**: 505 (1956).

G15. Gronier, W. S.: "Equipment for the Dissolution of Core Material from Sheared Power Reactor Fuels," Report ORNL-TM-3194, Apr. 1971.

G16. Gronier, W. S.: "Calculation of the Transient Behavior of a Dilute-Purex Solvent Extraction Process Having Application to the Reprocessing of LMFBR Fuels," Report ORNL-4746, Apr. 1972.

H1. Haas, W. O., Jr., and D. J. Smith: "Thorex Development at KAPL," Report KAPL-1306, May 1956.

H2. Hammond, V. L., and V. P. Kelly: "Low-Speed Saw Testing," Report HW-62843, June 9, 1960.

H3. Hardy, C. J., D. Scargill, and J. M. Fletcher: *J. Inorg. Nucl. Chem.* **7**: 257 (1958).

H4. Harmon, K. M., and G. Jansen, Jr.: "The Salt Cycle Process," in *Progress in Nuclear Energy,* series III, *Process Chemistry,* vol. 4, Pergamon, New York, 1970, p. 429.

H5. Healy, T. V., and B. L. Davies: "The Destruction of Nitric Acid by Formaldehyde, Parts II, III and IV," Report AERE-C-R-1739, Feb. 22, 1956.

H6. Healy, T. V., and B. L. Ford: "The Destruction of Nitric Acid by Formaldehyde, Part I," Report AERE-C-R-1339, Apr. 2, 1954.

H7. Hesson, J. C., M. J. Feldman, and L. Burris, Jr., "Description and Proposed Operation of Fuel Cycle Facility for Second Experimental Breeder Reactor," Report ANL-6605, Apr. 1963.

H8. Howells, G. R., et al.: "The Chemical Processing of Irradiated Fuels from Thermal Reactors," *PICG(2)* **17**: 3 (1958).

I1. Ishikawa, K., and S. Sato: *Chem. Eng. Progr. Symp. Ser. 94* **65**: 102 (1969).

I2. ISO-440 RD: "Process Performance of the First ^{233}U Production Campaign at the Hanford Purex Plant," Mar. 11, 1968.

J1. Jackson, R. R., and R. L. Walser: "Purex Process Operations and Performance; 1970 Thoria Campaign," Report ARH-2127, Mar. 1977.

J2. Jonke, A. A.: "Reprocessing of Nuclear Reactor Fuels by Process Based on Volatilization, Fractional Distillation and Selective Adsorption," *Atomic Energy Rev.* **3**(1) (1965).

J3. Jouannaud, C.: "Expérience de Six Années de Fonctionnement de l'Usine de Retraitement de Marcoule," *PICG(3)* **10**: 215 (1965).

J4. Joyce, A. W., L. C. Perry, and E. B. Sheldon: *Chem. Eng. Progr. Symp. Ser. 28* **56**: 21 (1960).

K1. Kaiser, G., et al.: *Kerntechnik* **20**: 550 (1978).

K2. Kanaan, Z. R., and C. R. Nash: "Removal of Sodium from Core Subassemblies with White Oil and Ultrasonics," Report APDA-142, Mar. 1961.

K3. Kelly, V. P.: "Final Report Shear Development for the Non-Production Fuels Reprocessing Program," Report HW-69667, 1961.

K4. Koch, G.: "Recovery of Actinides from Power Reactor Fuel," Report KFK-976, 1969.

K5. Koltunov, V. S., et al.: "Kinetics and Mechanisms of Some Reactions of Neptunium and Plutonium," in *4th International Transplutonium Element Symposium (Sept. 1975),* W. Muller and R. Lindner (eds.), North Holland, Amsterdam, 1976.

K6. Küchler, L., L. Schäfer, and B. Wojtech: *Kerntechnik* **12**: 327 (1970).

K7. Küchler, L., L. Schäfer, and B. Wojtech: "The Thorex Two-Stage Process for Reprocessing Thorium Reactor Fuel with High Burnup," *Kerntechnik* **13**: 319 (1971).

L1. Lawroski, S., and M. Levenson: "The Redox Process—A Solvent Extraction Processing Method for Irradiated Uranium," Report TID-7534, 1957, p. 45.

L2. Leary, J., et al.: "Pyrometallurgical Processing of Plutonium Reactor Fuels," *PICG(2)* **17**: 376 (1958).

L3. Lindauer, R. B.: "Processing of the MSRE Flush and Fuel Salts," Report ORNL-TM-2578, Aug. 1969.

M1. Malody, C. W., C. W. Pollock, and T. R. McKenzie: "Silver Reactor Reclamation," Report HW-59702, 1959.

M2. Maness, R. F., and L. L. Burger: "Laboratory Studies of Fumeless Dissolving," Report BNW/XN-136, Aug. 1973.

M3. Mann, S., and A. R. Irene: "A Study of Plutonium Polymer Formation and Precipitation as Applied to LMFBR Fuel Reprocessing," Report ORNL-TM-2806, Dec. 22, 1969.

M4. Martin, F. S., I. L. Jenkins, and N. J. Keen: "Processing of Reactor Fuels by Liquid Metals," *PICG(2)* **17**: 352 (1958).

M5. McKenzie, D. E.: *Can. J. Chem.* **34**: 515 (1956).

M6. Merriman, J. R., et al.: "Removal of Kr-85 from Reprocessing Plant Off-Gas by Selective Absorption," Report K-L-6201, 1972.

M7. Merz, E. R., G. Kaiser, and E. Zimmer: "Progress in Th-^{233}U Recycle Technology," Amer. Nuclear Soc. Topical Meeting, Gatlinburg, Tenn., May 1974.

M8. Motta, E. E.: "High Temperature Fuel Processing Methods," *PICG(1)* **9**: 596 (1956).

M9. Mouline, J. P.: "Contribution to the Study of the Oxidation Reaction of Np(V) by Nitric Acid Catalyzed by Nitrous Acid," Report CEA-R-4665, 1975.

M10. Murbach, E. W.: Personal communications to M. Benedict, Jan. 1979.

N1. Nairn, J. S., et al.: "Extraction of Actinide Elements," *PICG(2)* **17**: 216 (1958).

N2. Naylor, A.: "TBP Extraction Systems–TBP and Diluent Degradation," in Report KR-126, 1967, p. 120.

N3. Newton, J. W.: *J. Phys. Chem.* **63**: 1493 (1959).

N4. Nicholls, C. M.: *Trans. Inst. Chem. Eng. (London)* **36**: 336 (1958).

N5. Nichols, G. S.: "Decomposition of Tributyl Phosphate-Nitrate Complexes," Report DP-526, Nov. 1960.

N6. Niedrach, L. W.: "Fuel Reprocessing by Electrorefining," in *Progress in Nuclear Energy,* series III, *Process Chemistry,* vol. 2, Pergamon, New York, 1958.

N7. Notz, K. J.: "An Overview of HTGR Fuel Recycle," Report ORNL-TM-4747, Apr. 30, 1975.

N8. Nuclear Fuel Services, Inc.: Press release, Sept. 22, 1976.

O1. Oak Ridge National Laboratory: "Monthly Progress Report of Chemical Technology Division, March 1960," Report CF 60-4-36, 1960, pp. 27–28.

O2. Oak Ridge National Laboratory: "Monthly Progress Report of Chemical Technology Division, June–July 1960," Report CF-60-7-76, 1960, p. 33.

O3. Oak Ridge National Laboratory, Chemical Technology Division: "Annual Progress Report for Period Ending May 31, 1965," Report ORNL-3830, 1965, pp. 69–75.

O4. Oak Ridge National Laboratory, Chemical Technology Division: "Annual Report for Period Ending May 31, 1964," Report ORNL-3627, 1964, pp. 29–35.

O5. Oak Ridge National Laboratory, Chemical Technology Division: "Annual Progress Report for the Period Ending May 31, 1969," Report ORNL-4422, 1969.

O6. Oak Ridge National Laboratory, Chemical Technology Division: "Annual Progress Report for the Period Ending May 1970," Report ORNL-4572, 1970.

O7. Oak Ridge National Laboratory: "LMFBR Fuel Cycle Studies Progress Report for October 1970, No. 20," Report ORNL-TM-3217, Nov. 1970, pp. 15–17.

O8. Oak Ridge National Laboratory, Chemical Technology Division: "Annual Progress Report for Period Ending March 31, 1972," Report ORNL-4794, Aug. 1972.

O9. Oak Ridge National Laboratory: "LMFBR Fuel Cycle Studies Progress Report for July 1971, No. 29," Report ORNL-TM-3534, Aug. 1971.

O10. Oak Ridge National Laboratory: "LMFBR Fuel Cycle Studies Progress Report for Aug. 1971, No. 30," Report ORNL-TM-3571, Sept. 1971.

O11. Oak Ridge National Laboratory and Bechtel National Incorporated: "Hot Experimental Facility, Interim Design Report," Report ORNL/AFRP-78/6, Oct. 1978.

O12. Odom, C. H.: "Continuous or Semi-Continuous Leacher for Leaching Soluble Core Material from Sheared Spent Nuclear Fuel Tubes," *Proceedings of 20th Conference on Remote Systems Technology,* 1972.

O13. Orth, D. A., J. M. McKibben, and W. C. Scotten: "Progress in Tributyl Phosphate Technology at the Savannah River Plant," *Proceedings of the International Solvent Extraction Conference,* vol. 1, 1971, p. 514.

P1. Paxton, H. C.: "Criticality Control in Operations with Fissile Material," Report LA-3366 (Rev.), Nov. 1972.

P2. Paxton, H. C., et al.: "Critical Dimensions of Systems Containing U^{235}, Pu^{239} and U^{233}," Report TID-7028, June 1974.

P3. Pence, D. T., et al.: "Metal Zeolites–Iodine Absorption Studies," Report IN-1455, June 1971.

P4. Pence, D. T., and T. R. Thomas: "NO_x Abatement at Nuclear Processing Plants," *Second AEC Environmental Protection Conference, Albuquerque, N.M., April 16, 1974,* Report CONF-740406-18, 1974.

P5. Perry, J. H.: *Chemical Engineers Handbook,* 4th ed., McGraw-Hill, New York, 1963, pp. 3–63, 64.

P6. Pierce, R. D., and L. Burris, Jr.: "Pyroprocessing of Reactor Fuels," in *Selected Review of Reactor Technology,* Report TID-8540, 1964, chap. 8.

P7. Poe, W. L., A. W. Joyce, and R. I. Martens: *Ind. Eng. Chem. Process Design and Development* **3**: 314 (1964).

P8. Proctor, J. F.: Letter to M. Benedict, Feb. 20, 1979.

R1. Rainey, R. H., A. B. Meservey, and R. G. Mansfield: "Laboratory Development of the Thorex Process, Progress Report, Dec. 1, 1955 through Jan. 1, 1958," Report ORNL-2591, Jan. 1959.

R2. Rainey, R. H., and J. G. Moore: *Nucl. Sci. Eng.* **10**: 367 (1961).

R3. Rainey, R. H., and J. G. Moore: "Laboratory Development of the Acid Thorex Process for Recovery of Consolidated Edison Thorium Reactor Fuel," Report ORNL-3155, May 11, 1962.

R4. Rainey, R. H., and S. B. Watson: "Modification of the SEPHIS Computer Program for Calculation of the Acid Thorex Solvent Extraction System," *Amer. Nucl. Soc. Trans.* **22**: 315–317 (Nov. 1975).

R5. Rathvon, H. C., et al.: "Recovery of ^{233}U with Low ^{232}U Content," *Proceedings of the 2nd International Thorium Fuel Cycle Symposium, Gatlinburg, Tenn., May 1966,* USAEC CONF-660524, 1966, pp. 765–824.

R6. Reddick, G. W.: "Solvent Extraction in HTGR Reprocessing," Interim Development Report GA-A13835, Feb. 1976.

R7. Richardson, G. L., and J. L. Swanson: "Plutonium Partitioning in the Purex Process with Hydrazine-Stabilized Hydroxylamine Nitrate," Report HEDL-TME-75-31, June 1975.

R8. Rodger, W. A.: "Reprocessing of Spent Nuclear Fuel," Presentation to California Energy Resources Conservation and Development Commission, Mar. 7, 1977.

R9. Rosenthal, M. W., et al.: *Atomic Energy Rev.* **9**: 601 (1971).

R10. Runion, T. C., and W. H. Lewis: *Chem. Eng. Progr. Symp. Ser. 94* **65**: 53 (1969).

R11. Ryon, A. D.: "McCabe-Thiele Graphical Solution of Uranium-Thorium Partitioning from 30% TBP-Amsco Solvent," Report ORNL-3045, Jan. 1961.

S1. Salmon, L., et al.: "Tests on the CNEN Alpha Decontaminating Solvent for Final Uranium Product–Preliminary Control," Eurochemic, Mol, Belgium, IDL Report 47, Apr. 1971.

S2. Schmets, J. J.: "Reprocessing of Spent Nuclear Fuels by Fluoride Volatility Processes," *Atomic Energy Rev.* **8**(1): 3 (1970).

S3. Schuller, W., et al.: "Nuclear Reprocessing and Waste Treatment at Karlsruhe Nuclear Research Center," *Proceedings of Nuclear Power and Its Fuel Cycle,* vol. 3, International Atomic Energy Agency, Vienna, 1977, p. 579.

S4. Schultz, W. W.: "Macroreticular Anion Exchange of TBP Solvent," Report ARH-SA-129, May 15, 1972.

S5. Scotten, W. C.: "SOLVEX–A Computer Program for Simulation of Solvent Extraction Processes," Report DP-1391, Sept. 1975.

S6. Seaborg, G. T.: *Man-Made Transuranium Elements,* Prentice-Hall, Englewood Cliffs, N.J., 1963.

S7. Sethna, H. N., and N. Srinivasan: "Fuel Reprocessing Plant at Trombay," *PICG(3)* **10**: 272 (1964).

S8. Shabbir, M., and R. G. Robins: *J. Appl. Chem. (London)* **18**: 129 (1968).

S9. Shank, E. M.: "Operation of the Thorium Pilot Plant with Highly Irradiated Thorium," in *Progress in Nuclear Energy,* series III, *Process Chemistry,* vol. 2, Pergamon, New York, 1958, p. 279.

S10. Shastri, N. K., E. S. Amis, and J. O. Wear: *J. Inorg. Nucl. Chem.* **27**: 2413 (1965).

S11. Shevchenko, V. B., N. S. Povitsky, and A. S. Solovkin: "Problems in the Treatment of Irradiated Fuel Elements at the First USSR Atomic Power Station," *PICG(2)* **17**: 46 (1958).

S12. Shirin, V. M., et al.: "Use of Lead in Unloading Systems of Sodium-Cooled Facilities," in *IAEA Symposium on Progress in Sodium-Cooled Fast Reactor Engineering,* Monaco, Mar. 1970.

S13. Siddall, T. H., III: "Extraction of Thorium Nitrate from Nitric Acid by TBP-Ultrasene," Report DP-181, Oct. 1956.

S14. Siddall, T. H., III: "A Rationale for the Recovery of Irradiated Uranium and Thorium by Solvent Extraction," *PICG(2)* **17**: 339 (1958).

S15. Siddall, T. H., III: "Solvent Extraction Processes Based on TBP," in *Chemical Processing of Reactor Fuels,* J. F. Flagg (ed.), Academic, New York, 1961, chap. V.

S16. Siddall, T. H., III, and E. K. Dukes: "Kinetics of HNO_2 Catalyzed Oxidation of Np(V) by Aqueous Solutions of Nitric Acid," *J. Amer. Chem. Soc.* **81**: 790 (1959).

S17. Smith, P. W.: "The Zirflex Process Terminal Development Report," Report HW-65979, Aug. 20, 1960.

S18. Srinivasan, N., et al.: "Process Chemistry of Neptunium–Part I," Report B.A.R.C.-428, 1969.

S19. Srinivasan, N., et al.: "Process Chemistry of Neptunium–Part II," Report B.A.R.C.-736, 1974.

S20. Srinivasan, N., et al.: "Counter-Current Extraction Studies for the Recovery of Neptunium by the Purex Process, Parts I and II," Reports B.A.R.C.-734 and 735, 1974.

S21. Steinberg, M.: "The Recovery of Fission Product Xenon and Krypton by Absorption Processes," Report BNL-542, 1959.

S22. Steunenberg, R. K., R. D. Pierce, and I. Johnson: "Status of the Salt Transport Process for Fast Breeder Reactor Fuels," in *Reprocessing of Nuclear Fuels, Proceedings of the Symposium Held at Ames, Iowa, 1969,* Report CONF-690801, 1969.

S23. Stoller, S. M., and R. B. Richards (eds.): *Reactor Handbook,* vol. II, *Fuel Reprocessing,* 2d ed., Interscience, New York, 1961.

S24. Swanson, J. L.: "Oxidation of Neptunium(V) in Nitric Acid Solution–Laboratory Study of Rate Accelerating Materials (RAM)," Report BNWL-1017, Apr. 1969.

S25. Swanson, J. L.: "Neptunium and Zirconium Extraction under Purex HA Column Scrub Conditions," Report BNWL-1588, 1971.

T1. Tajik, S.: "Recovery of Neptunium in the Modified Purex Process," thesis submitted in partial fulfillment of requirements for the M.S. degree in Nuclear Engineering and Chemical Engineering, Massachusetts Institute of Technology, Cambridge, Mass., 1979.

T2. Tarnero, M., and J. Dollfus: "Le Transfer de Matière dans les Appareils d'Extraction, Centrifuges, Multistages," *Chem. Ind.-Génie Chimique* **99**(11) (June 1968).

T3. Taylor, R. F., and G. P. Wall: "Development of a Production Process for Radiokrypton Recovery by Fractional Absorption," in *Progress in Nuclear Energy,* series IV, vol. 5, Pergamon, New York, 1963, p. 307.

T4. Templin, L. J.: "Reactor Physics Constants," Report ANL-5800, 1963.

T5. Thomas, J. T. (ed.): *Nuclear Safety Guide,* TID-7016, Revision 2, Report ORNL/NUREG/CSD-6, June 1978.

U1. Uematsu, K.: Personal communication to M. Benedict, 1975.

U2. Unger, W. E., et al.: "Aqueous Fuel Reprocessing Quarterly Report, Period Ending March 31, 1973," Report ORNL-TM-4240, June 1973.

V1. Vaughen, V. C. A., Oak Ridge National Laboratory: Letters to M. Benedict, Nov. 1978–Jan. 1979.

V2. Vogel, R. C., et al.: "Chemical Engineering Division Highlights, May 1966–April 1967," Report ANL-7350, 1967.

V3. Voight, A. F., et al.: "Removal of Plutonium from Uranium by Liquid-Metal Extraction," Report IS-470, May 1962.

V4. Vondra, B. L.: "LWR Fuel Reprocessing and Recycle Program Quarterly Report for Period October 1 to December 1, 1976," Report ORNL/TM-5760, Feb. 1977.

W1. Wagner, R. M.: "Investigation of Explosive Characteristics of Purex Solvent Decomposition Products 'Red Oil,' " Report HW-27492, Mar. 1953.

W2. Warf, J. C.: *J. Amer. Chem. Soc.* **71**: 2187 (1949).

W3. Warner, B. F.: *Kerntechnik* **9**: 249 (June 1967).

W4. Watson, C. D., et al.: "Mechanical Processing of Spent Power Reactor Fuel at Oak Ridge National Laboratory," in *Proceedings of the AEC Symposium for Chemical Processing of Irradiated Fuels from Power, Test and Research Reactors,* Report TID-7583, Jan. 1960, p. 306.

W5. Watson, S. B., and R. H. Rainey: "Modifications of the SEPHIS Computer Code for Calculating the Purex Solvent Extraction System," Report ORNL-TM-5123, Dec. 1975.

W6. Weinberger, A. J., D. L. Marley, and D. A. Constanzo: "A Solvent-Extraction Study of the Thorium Nitrate, Nitric Acid, Tri-Butyl Phosphate/Dodecane System," ORNL/TM-6337, 1978.

W7. Wilhelm, J. G., et al.: "An Inorganic Absorber Material for Off-Gas Cleaning in Fuel Reprocessing Plants," *12th AEC Air Cleaning Conference, Oak Ridge, 1972,* Report CONF-720823-P2, Jan. 1973, p. 540.

W8. Wilson, E. J., and K. J. Taylor: "The Separation and Purification of Krypton-85 at the Multicurie Level," Report AERE-1/R-2673, 1958.

Y1. Young, D. T.: "Fluidized Combustion of Beds of Large, Dense Particles in Reprocessing HTGR Fuel," Report GA-A14327, Mar. 1977.

PROBLEMS

10.1 Show that the concentration proposed for the solution to be fed to the Barnwell Nuclear Fuel Plant, containing 1.21 mol of uranium, 2.9 g of plutonium and 5.6 g of gadolinium per liter, would be subcritical even with fully enriched ^{235}U. 2200 m/s cross sections are accurate enough for this check.

10.2 A Purex plant processes 1 MT of fuel from a PWR whose content of radioactivity is given in Table 8.7. Air discharged from the plant stack contains 5 percent of the tritium and 1 percent each of the krypton and iodine in the fuel. The highest radionuclide concentration to which humans are exposed is one one-thousandth the concentration in stack effluent.

Find the total volume of air in cubic meters that must be discharged through the stack with the tritium, krypton, and iodine so that human exposure does not exceed that from the radioactivity concentration limit (C) values of App. D. The total volume of air is the sum of the volumes needed to dilute each of the three radionuclides to the C values.

10.3 Table 10.7 gives the flow rates and compositions of the high-level aqueous waste HAW stream 5 leaving the extracting contactor HA of the Barnwell Nuclear Fuel Plant and the organic streams, solvent HAX (number 6), and recycle 1SP (number 60) entering this contactor. Assume that 1 SP joins the organic stream one theoretical stage upstream from the HAX feed point and that the combined organic streams have a second theoretical stage contact with the aqueous phase. Calculate the uranium, plutonium, and nitric acid molarities of the aqueous phase entering the second theoretical stage. Assume that distribution equilibrium at 40°C is obtained in each stage. Compare the calculated aqueous composition with that of the aqueous feed HAF (stream 3) entering the HA contactor.

10.4 In the IBS uranium extracting column of the Hanford acid Thorex flow sheet, Fig. 10.21, the average aqueous molarities of thorium and nitric acid are $x_{\mathrm{Th}} = 0.25$ and $x_{\mathrm{H}} = 0.33$, respectively. At the flow rates: aqueous $F = 645$ and organic $E = 220$, how many theoretical stages would be needed to reduce the aqueous uranium concentration from 0.0096 g/liter to 0.00001 g/liter? Assume that $D_{\mathrm{U}} = 14D_{\mathrm{Th}}$ and that distribution coefficients are constant throughout the column.

10.5 In the IBX thorium stripping column of the Hanford acid Thorex flow sheet, Fig. 10.21, flow quantities and concentrations adjusted to satisfy material balances are as follows:

	Stream			
	IBX	IBXF	IBXT	IBU
Name	Thorium strippant	IBX feed	Partially stripped thorium	Uranium in solvent
Phase	Aqueous	Organic	Aqueous	Organic
Flow	In	In	Out	Out
Flow rate, liters/h	638	1238	638	1238
M HNO_3	0.2	0.104	0.3514	0.026
M thorium	0	0.1392	0.270	0.00004
g uranium/liter	0	0.0549	0.0096	0.050
Mol HNO_3/h	127.6	128.8	224.2	32.2
Mol thorium/h	0	172.3	172.3	0.0
g uranium/h	0	68.0	6.1	61.9

(*a*) Starting at the organic feed end, find the number of theoretical stages needed to reduce the thorium concentration of the outflowing organic to 0.00004 M. Show that the HNO_3 molarity in the outflowing organic then is approximately 0.026.

(*b*) Show that the number of stages found in (*a*), with a uranium/thorium separation factor of 14, leads to a uranium concentration in the outflowing organic close to 0.050 g/liter.

CHAPTER
ELEVEN

RADIOACTIVE WASTE MANAGEMENT

1 INTRODUCTION

1.1 Definition

Radioactive waste is any waste material—gas, liquid, or solid—whose radioactivity exceeds certain limits. These limits have been established by governments or by local authorities, guided by the recommendations of the International Commission on Radiation Protection (ICRP). The ICRP recommendations define the *maximum permissible concentration* (MPC) for each individual radionuclide and for mixtures of radionuclides in water or air. The U.S. regulation defines such limiting concentration as the *radioactivity concentration limit* (C), which is the terminology used in this text. Values of C for selected actinides and long-lived fission products in water or air are given in App. D.

The intention of regulations limiting the release of radioactive material from nuclear installations is to keep the radioactivity concentration in ground and surface water or in air well below the levels recommended by the ICRP. The regulations may follow either one of two principles or may combine them:

Limitation of the total amount of radioactivity associated with a certain material that may be released over a given period of time

Limitation of the radioactivity concentration in the material to be released

As a consequence of these limitations, most of the radioactivity arising as waste from nuclear technology has to be isolated from the environment by some storage or final disposal technique. The first step toward this goal is usually a volume reduction, preparing the waste for interim storage as a liquid or solid. This is considered part of the waste-generating technology rather than of the waste management. Waste management is defined to include interim storage, final conditioning, and long-term storage or disposal.

1.2 Classification

The techniques of waste management depend largely on the type of waste to be dealt with. The criteria are the level of the radioactivity concentration in the waste, the nature of the radionuclides present in the waste, and the properties of the material that carries the radioactivity.

With respect to the level of radioactivity concentration, several waste classifications are in use appropriate for particular handling schemes. More basic distinctions are between waste that requires radiation shielding and that which does not and between waste that needs to be cooled and that which does not.

The radionuclides associated with the waste have to be considered in terms of the type of radiation, the half-life, and possibly the chemical nature. Long lived alpha-emitting actinides call for particular attention because of the high and long-lasting radiotoxicity typical of those alpha-emitters.

As for the material, the primary distinction is between solid, liquid, and gaseous waste. Solid waste includes any kind of contaminated or activated plant components, tools, filters, and protective clothing. Most of the liquid wastes are aqueous solutions or sludges. The term gaseous waste will be used for radioactive gases recovered from off-gas streams and contained in an appropriate form.

From these criteria a number of waste categories may be derived according to further treatment and final disposal requirements.

1. High-level waste (HLW): alpha, beta, and gamma emitters; shielding required, cooling may be required.
   (*a*) Liquid HLW concentrate
   (*b*) Solid HLW
2. Non-high-level waste (medium-level waste, MLW; low-level waste waste, LLW): shielding may be required.
   (*a*) Alpha waste (liquid and solid): alpha (beta, gamma) emitters, alpha activity dominating
   (*b*) Non-alpha waste (liquid and solid): beta, gamma emitters
3. Special radionuclide waste†
   (*a*) ^{85}Kr (gaseous beta emitter; 10-year half-life)
   (*b*) Tritiated water (weak beta emitter; 12-year half-life)
   (*c*) ^{129}I (beta, gamma emitter; 10^7-year half-life)

1.3 Sources

The main sources of radioactive waste are fuel reprocessing plants. More than 99 percent of the total radioactivity generated by nuclear technology appears eventually in wastes from reprocessing plants, most of it in HLW. In a nuclear economy doing without reprocessing, spent fuel itself would be high-level radioactive waste.

Liquid HLW is the concentrated aqueous raffinate from the liquid-liquid extraction process. It contains practically all fission products, neptunium and transplutonium elements as well as 0.5 to 1.0 percent of the uranium and plutonium fed to the extraction process. It represents a very small fraction of the total radioactive waste volume produced in nuclear installations.

Solid high-level wastes are the cladding hulls of spent fuel elements from the chop-leach

† ^{14}C being released from reprocessing plants now operating and to be released from those under construction or in the planning may become a waste in the future. Because of its long half-life, it will accumulate in the atmosphere and, in the long run, contribute significantly to the total radiation exposure from fuel-cycle operations. It may therefore become necessary to recover ^{14}C from dissolver off-gas and to treat it as a waste. There is, however, no urgent need to develop the required technology. Only if high-temperature gas-cooled reactor fuel were to be reprocessed would ^{14}C recovery be necessary.

head end of the reprocessing plant and some undissolved solids as sludges from feed clarification. Their radiation characteristics are similar to liquid HLW but on a lower concentration level.

Refabrication plants for plutonium-recycle fuel and liquid-metal fast-breeder reactor (LMFBR) fuel to be combined with reprocessing plants will be the principal sources of alpha wastes. These may be liquids (sludges) or solids, the latter being combustible or not. Combustion is an effective way to reduce the volume. The beta and gamma activity concentration of alpha waste is by orders of magnitude lower than that of corresponding HLW. Therefore only light or no shielding and no cooling are required. However, the total alpha activity is within the same order of magnitude as that of HLW, causing a long-term biological hazard potential similar to that of HLW. This is reflected in similar critieria for conditioning and disposal techniques.

Among the MLW and LLW streams from reprocessing plants are non-alpha wastes. The term non-alpha waste includes all nongaseous wastes that are not high-level wastes and whose radioactivity is due mainly to beta and gamma emitters. Usually, their alpha radiotoxicity is on the same order as that of a relatively rich uranium ore. These waste streams are generated by various operations, including decontamination of equipment. Their biological hazard potential is much smaller than that of HLW and alpha waste, and lasts for much shorter periods of time.

Tritium as well as radioactive krypton and carbon dioxide are in today's technology released to the atmosphere. Increasing fuel-cycle activities and enforced environmental protection standards, however, have resulted in the requirement for recovery and safe storage of krypton and in some countries of tritium. The design of a German 1400 MT/year reprocessing plant provides for the recovery of at least 95 percent of the krypton, which is to be recovered by cryogenic distillation and will be delivered to the waste management section pressurized in steel bottles. Iodine will be fixed on solid absorbers and tritium will be collected as tritiated water in tanks for intermediate storage. At the AGNS† plant, provision is presently made only for iodine removal (Chap. 10).

Two more LLW streams from sources other than reprocessing should be mentioned that are significant because of their large volumes. Nuclear power plants produce large volumes of non-alpha waste whose biological hazard potential is small. Techniques for treatment and disposal of this waste type will easily meet the safety requirements but will have to be optimized in terms of economics. Uranium mills generate large amounts of ore tailings with relatively high concentrations of alpha emitters, particularly radium. This is basically a naturally occurring material. However, it is moved from an underground ore deposit to the surface and therefore creates an additional health hazard. Attention given to mill tailings is surprisingly modest compared to that given waste from reprocessing.

In this chapter, the primary emphasis will be on HLW. Non-high-level alpha waste, tritium, ^{129}I, and ^{85}Kr will be treated to some extent. Volumes and radioactivity concentrations of these wastes to be expected from a 1400 MT/year reprocessing plant according to a German design are given in Table 11.1 [D2].

2 HIGH-LEVEL WASTE

Liquid HLW is the concentrate of the aqueous raffinates from the reprocessing extraction cycles. This means that up to 1 percent of the uranium and plutonium and practically all of the

†For brevity, the Barnwell Nuclear Fuel Plant is referred in this chapter as the AGNS plant, an abbreviation for the plant owner, Allied General Nuclear Services.

Table 11.1 Annual amounts of wastes ready for intermediate storage prior to final conditioning generated by a 1400 MT/year reprocessing plant†

Type of waste	Volume (m^3/yr)	Radioactivity after 1 yr collecting time, Ci/m^3	Plutonium-concentration, kg/m^3	Type of intermediate storage
High-level				
Liquid concentrate	600	$<4 \times 10^6$	<0.07	Tank with cooling
Dissolver sludge	80	$<6 \times 10^5$	<0.45	Tank
Cladding hulls	800	$<1 \times 10^4$	<0.09	Container
Medium-level				
Liquid concentrate	1500	$<2 \times 10^3$	$<10^{-3}$	Tank; 400 g/liter salt
Tritiated water	3000	<200 tritium, 0.1 others	$<10^{-5}$	Tank
Solids (noncombustible + ash)	800	<1	$<3 \times 10^{-4}$	Fixed in concrete
Krypton	2	$<8 \times 10^6$	–	Pressurized steel bottles

†Fuel elements cooled 1 year before reprocessing.

Source: Deutsche Gesellschaft für Wiederaufbereitung von Kernbrennstoffen (DKW): "Bericht über das in der Bundesrepublik Deutschland geplante Entsorgungszentrum für ausgediente Brennelemente aus Kernkraftwerken," Hannover, 1977.

fission products, neptunium, and the transplutonium elements produced in nuclear reactors end up in the HLW. It is therefore a reservoir of radioactivity that will not reach a steady state as long as nuclear power is generated, and its hazard potential will last much longer than the use of nuclear energy. Therefore, a reliable technology for long-term isolation of high-level wastes from the environment is a key to environmental protection against the consequences of nuclear power, and it is also a key to the public acceptance of nuclear power.

HLW arising in solid form, mainly cladding hulls, has activity concentrations more than two orders of magnitude lower than liquid HLW. It presents somewhat different technical problems, which will be discussed briefly in this chapter.

2.1 Characterization of Liquid HLW

The aqueous raffinate from the first extraction cycle of light-water reactor (LWR) fuel reprocessing has an original volume of up to 5 m^3/MT† of heavy metal. It is concentrated by evaporation, and the residues of the evaporated raffinates from further extraction cycles may be combined with the concentrate. The result of these operations is the HLW concentrate that will be transferred to the waste management section of the reprocessing plant.

Volume. The volume reduction factor that can be achieved depends strongly on the burnup and on the cooling time the fuel has experienced. According to the length of the cooling period, either the heat generation of the resulting HLW or its content of dissolved solids including fission products and process chemicals are the factors limiting the degree of concentration. Most of the volume reduction is achieved in the HLW evaporator. Some further

†1 MT = 1 metric ton = 1 megagram (1 Mg).

evaporation usually takes place in the storage tank, where the temperature of the solution is kept below 60°C. According to the wide range of parameters the specific volumes of the HLW concentrate may range from 0.4 to more than 1 m^3/MT of heavy metal reprocessed. A value of 0.6 m^3/MT may be used in the design of reprocessing plants to calculate tank volume requirements.

Chemical composition. In a well-designed reprocessing scheme the amount of process chemicals that would appear in wastes will be kept as small as possible. Then the bulk of dissolved solids includes fission products, uranium and plutonium losses, neptunium, and the transplutonium elements. Table 11.2 shows the amounts of fission-product elements (more than 0.1 percent contribution) present in the waste from reprocessing 1 MT of heavy metal from spent LWR fuel with 30,000 MWd/MT burnup, at the time of and 6 years after discharge from reprocessing. According to widely used design parameters, reprocessing is assumed to take place 150 days after reactor discharge. An aging time of 6 years can presently be envisaged prior to solidification of the liquid HLW. In practice, spent fuel will be aged much longer, and storage of liquid HLW may be shorter.

With a specific HLW concentrate volume of 600 liters/MT of heavy metal, the total fission product concentration will be on the order of 50 g/liter and the actinide concentration on the order of 10 g/liter.

Table 11.2 Amounts of fission-product elements and actinide elements in the waste from 1 MT LWR uranium fuel (30,000 MWd/MT burnup) at discharge from reprocessing (150 days cooled fuel elements) and 6 years after discharge (contributions of more than 0.1 percent) according to ORIGEN

	g/MT of heavy metal			g/MT of heavy metal	
Element	At discharge	After 6 years	Element	At discharge	After 6 years
Se	4.71 E + 01	4.71 E + 01	La	1.15 E + 03	1.15 E + 03
Kr†	3.36 E + 02	3.28 E + 02	Ce	2.47 E + 03	2.25 E + 03
Rb	3.00 E + 02	3.08 E + 02	Pr	1.09 E + 03	1.09 E + 03
Sr	8.04 E + 02	7.34 E + 02	Nd	3.52 E + 03	3.73 E + 03
Y	4.22 E + 02	4.19 E + 02	Pm	1.00 E + 02	2.05 E + 01
Zr	3.31 E + 03	3.37 E + 03	Sm	7.40 E + 02	8.17 E + 02
Mo	3.13 E + 03	3.15 E + 03	Eu	1.66 E + 02	1.55 E + 02
Tc	7.68 E + 02	7.68 E + 02	Gd	9.08 E + 01	1.05 E + 02
Ru	2.09 E + 03	1.97 E + 03			
Rh	3.63 E + 02	3.66 E + 02	Total F.P.‡	3.18 E + 04	3.18 E + 04
Pd	1.20 E + 03	1.31 E + 03			
Ag	5.79 E + 01	5.74 E + 01	U	4.79 E + 03	4.79 E + 03
Cd	7.72 E + 01	7.76 E + 01	Np	4.19 E + 02	4.19 E + 02
Sn	4.78 E + 01	4.74 E + 01	Pu	4.42 E + 01	5.28 E + 01
Te	5.17 E + 02	5.22 E + 02	Am	1.29 E + 02	1.30 E + 02
I†	2.48 E + 02	2.48 E + 02	Cm	3.19 E + 01	2.18 E + 01
Xe†	4.94 E + 03	4.94 E + 03			
Cs	2.50 E + 03	2.23 E + 03	Total actinides	5.42 E + 03	5.42 E + 03
Ba	1.26 E + 03	1.53 E + 03			

†Not present in the liquid waste.
‡F.P., fission products.
Source: M. J. Bell, "The ORNL Isotope Generation and Depletion Code (ORIGEN)," Report ORNL-4628, May 1973.

Corrosion products play a minor role and make up to about 1 percent of the total solid content of the HLW solution. If gadolinium is used as homogeneous poison for criticality control, it will significantly increase the total solid content of the waste.

The optimum HNO_3 concentration for tank storage is in the range of 2 to 4 M, determined by the corrosion behavior of stainless steel. The aqueous raffinate stream (HAW) leaves the extraction process with 1 to 3 M HNO_3. A good deal of the nitric acid is stripped in the HAW evaporator or destroyed by a reductant (Chap. 10), and more by subsequent water vapor distillation. The final adjustment may be performed in the storage tank, taking advantage of radiolysis, which removes HNO_3 effectively when the radiolytic products are swept by air.

Table 11.3 shows typical HAW concentrate data as they have been observed in pilot-plant operations and as they are envisaged for commercial operation.

Table 11.3 Chemical composition of HAW concentrate from reprocessing uranium discharge fuel

	Early operation	Steady-state operation
A. HAW concentrate design data for the AGNS plant		
Age, years	6.25	1.8
Heat load, W/liter	3.5	6.17
Specific volume, liters/MT heavy metal	336	1,375
Burnup, MWd/MT heavy metal	23,000	35,000
Initial enrichment, %	2.5	3.5
Fission products, g/liter	65	22.38
Uranium, 1% loss, g/liter	29.76	7.27
Plutonium, 1% loss, g/liter	0.846	0.24
Soluble poison, g/liter	66.66	24.0
Phosphate, g/liter	0.505	0.124
Free nitric acid, M	4–7	4–7
Total nitrate, 7 M HNO_3, M	10.2	8.1
Chloride, g/liter	0.06	0.015
Shear fines, g/liter	0.744	0.182
Iodine, g/liter	0.0025	0.001
Corrosion products, g/liter	0.22	0.083
Iron, g/liter	6	–
B. German HAW concentrate data		
	Analytical figures from WAK†	Design figures, 1400 MT/year plant
Age, years	5–10	6
Specific volume, liters/MT heavy metal	630	430
Burnup, MWd/MT heavy metal	30–39,000	36,000
Radioactivity concentration, Ci/liter	3×10^2	1×10^3
Uranium, g/liter	5.0‡	0.35
Plutonium, g/liter	0.15	0.07
Free nitric acid, M	4	2–5
Total salt content, g/liter		250
Density, g/liter	1.2016	

†Wiederaufarbeitungsanlage Karlsruhe (reprocessing plant Karlsruhe).

‡No effort was made to recover uranium effectively.

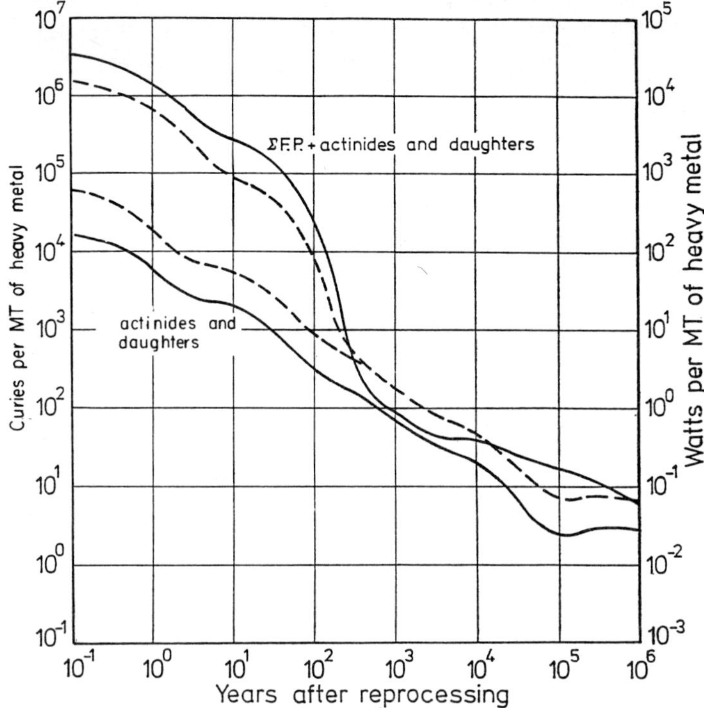

Figure 11.1 Radioactivity and thermal power of LWR uranium waste. Reprocessed, 150 days after discharge from reactor; enrichment, 3% ^{235}U; burnup, 30,000 MWd/MT heavy metal; specific power, 27.3 MW/MT heavy metal; residence time, 1100 days, uranium loss, 0.5 percent; plutonium loss, 0.5 percent; —— radioactivity; – – – thermal power.

Radioactivity and heat generation. The total radioactivity and the heat generation of the waste solution up to 100 years is essentially a function of the fission-product concentration and of their age. Then the actinides start to contribute significantly. The computer program ORIGEN [B2] permits the calculation of all relevant data such as radionuclide activities, ingestion hazards, element concentrations, heat generation, and neutron generation for fission products and actinides as a function of age. Figure 11.1 shows radioactivity concentrations and heat generation of the waste from 1 MT of heavy metal with 30,000 MWd/MT burnup up to an age of 10^6 years. Typical maximum radioactivity concentrations and specific heat generations for a freshly filled tank and reprocessing of 150-day-old fuel are of the order of 10^3 Ci and 10 W/liter, respectively. After about 500 years radioactivity and heat generation have decayed by a factor of more than 1000. The heat generation will then be insignificant.

Table 11.4 shows the contributions (more than 0.1 percent) of individual fission-product nuclides to the total fission-product activity and of individual actinide nuclides to the total actinide activity. After 6 years only nine fission products or fission-product mother/daughter pairs contribute significantly. After 100 years, $^{90}Sr/^{90}Y$ and $^{137}Cs/^{137m}Ba$ make up to 98 percent of the fission-product activity, and among the actinides ^{238}Pu, ^{241}Am, $^{243}Am/^{239}Np$, and ^{244}Cm are responsible for 90 percent of the total alpha radioactivity.

Hazard indices. The radioactivity of the waste is no direct measure of its radiotoxicity or its hazard. When we assume ingestion as the most likely path of incorporating radioactivity from

Table 11.4 Radioactivities of fission products and actinides in the waste from 1 MT LWR uranium fuel (30,000 MWd/MT burnup) at discharge from reprocessing (150 days cooled fuel elements) and 6 years after discharge (contributions of more than 0.1 percent only), according to ORIGEN

	Curies per MT of heavy metal			Curies per MT of heavy metal	
Nuclide	At discharge	After 6 years	Nuclide	At discharge	After 6 years
			^{141}Ce	5.13 E + 04	–
^{85}Kr†	9.90 E + 03	6.74 E + 03	^{144}Ce	6.98 E + 05	3.32 E + 03
^{89}Sr	8.74 E + 04	–	^{144}Pr	6.98 E + 05	3.32 E + 03
^{90}Sr	6.89 E + 04	5.94 E + 04	^{147}Pm	9.30 E + 04	1.90 E + 04
			^{151}Sm	1.17 E + 03	1.12 E + 03
^{90}Y	6.89 E + 04	5.95 E + 04	^{154}Eu	6.11 E + 03	4.71 E + 03
^{91}Y	1.45 E + 05	–	^{155}Eu	5.78 E + 03	–
^{95}Zr	2.52 E + 05	–	Total F.P.‡ activity	4.00 E + 06	3.61 E + 05
^{95}Nb	4.72 E + 05	–			
^{103}Ru	8.08 E + 04	–	^{239}Np	1.61 E + 01	1.61 E + 01
^{103m}Rh	8.08 E + 04	–	^{238}Pu	1.20 E + 01	9.13 E + 01
^{106}Ru	3.84 E + 05	6.12 E + 03	^{241}Pu	4.96 E + 02	3.73 E + 02
			^{241}Am	1.52 E + 02	1.55 E + 02
^{106}Rh	3.84 E + 05	6.12 E + 03	$^{242m/242}$Am	1.81 E + 01	1.76 E + 01
^{123}Sn	3.56 E + 03	–	^{243}Am	1.61 E + 01	1.61 E + 01
^{125}Sb	7.41 E + 03	–	^{242}Cm	1.65 E + 04	8.71 E + 00
$^{127m/127}$Te	1.13 E + 04	–	^{243}Cm	3.31 E + 00	2.91 E + 00
^{134}Cs	1.86 E + 05	2.45 E + 04	^{244}Cm	2.03 E + 03	1.61 E + 03
^{137}Cs	9.72 E + 04	8.47 E + 04	Total actinide activity	1.93 E + 04	2.30 E + 03
^{137m}Ba	9.10 E + 04	7.92 E + 04			

†Not present in the liquid waste.

‡F.P., fission products.

Source: M. J. Bell, "The ORNL Isotope Generation and Depletion Code (ORIGEN)," Report ORNL-4628, May 1973.

waste, we may characterize the hazard by the ingestion hazard index. This is defined for an individual radionuclide as radioactivity divided by the radioactivity concentration limit for drinking water (general public) and has the dimension of a volume. It may be understood as the volume of water required to dilute a given quantity of radioactive material so that drinkable water will be obtained. For a mixture of radionuclides, such as radioactive waste, the ingestion hazard index is the sum of the ingestion hazard indices of all radionuclides present. The hazard index characterizes the potential hazard rather than the actual risk associated with the waste. It does not give credit for the various barriers between waste and humans.

Figure 11.2 shows the ingestion hazard index of HLW as a function of time up to 10^6 years. After about 500 years the actinide radiotoxicity clearly dominates.

HLW from advanced fuel cycles. Advanced fuel cycles that will be considered are plutonium recycling and the LMFBR fuel cycle. There is little difference from LWR waste as far as fission products are concerned. Actinide concentrations, however, are considerably different, as shown in Chap. 8 for spent fuel.

To discuss other properties of advanced fuel-cycle waste, such as chemical composition, is of little use. As yet, the reprocessing technology for advanced fuel cycles is in a very premature state of development.

Cladding hulls. Cladding hulls as collected from the chop-leach head end are radioactive due to activation products in the zircaloy and to fission products and actinides from $(U,Pu)O_2$ adsorbed at the inside of the hulls. The principal activation products are ^{60}Co, $^{125}Sb/^{125m}Te$, and ^{63}Ni. Their total activity is on the order of 10^3 μCi/g zircaloy after 6 years and about 100 times lower after 100 years. The fission-product activity after 6 years is of the same order of magnitude and is dominated by ^{137}Cs, ^{90}Sr, and tritium. The residual $(U,Pu)O_2$ after leaching is estimated to be of the order of 0.1 percent of the charge.

A number of processes are under development for consolidation and volume reduction of the hulls. Melt densification has received much effort, resulting in volume reductions by a factor of 6. Other processes seriously considered are mechanical compaction and consolidation in concrete. In any case, final disposal of consolidated hulls will be similar to that of other HLW. Thus, hulls (and dissolver sludge) will add considerably to the volume of waste eventually to be disposed of with essentially the HLW technology.

2.2 Projections of HLW Generation

Projections of future waste generation depend strongly on projections of nuclear power generation and of reprocessing capacity. Therefore they have a high degree of uncertainty.

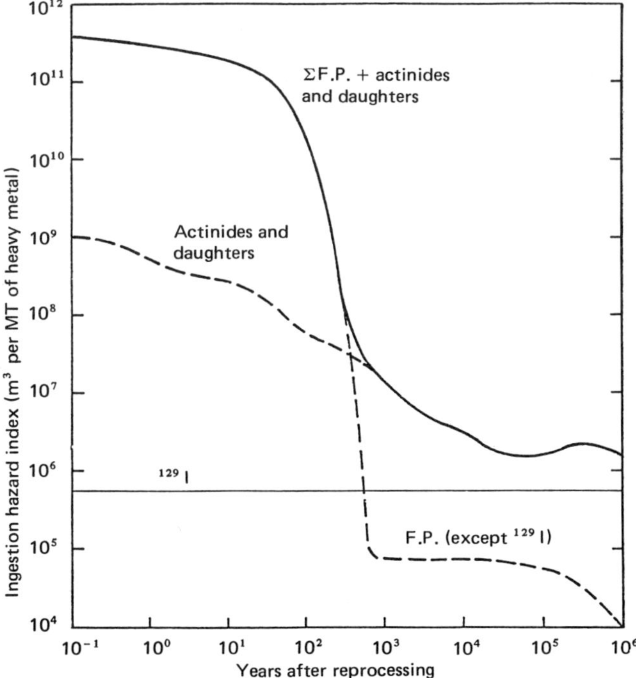

Figure 11.2 Ingestion hazard index of LWR uranium waste without ^{129}I and of ^{129}I from LWR uranium waste.

Nuclear capacity forecasts have considerably decreased over the last years, and it is presently an open question when reprocessing plants will go on stream. Table 11.5 shows the most recent estimate of the amount of solidified HLW to be accumulated at a federal repository in the United States [B4]. There may be a shift of time scale, but this will not greatly affect the general conclusion from this table: Early in the next century on the order of 10^4 m^3 of solidified waste with on the order of 10^3 MT of actinides and on the order of 10^{10} Ci of total radioactivity will be collected in the repository and probably be buried underneath the United States.

To put these numbers into perspective, they should be compared with natural radioactivity already contained in the crust of the earth. We consider a layer of soil all over the United States that is 1 m thick. Its volume is of the order of 10^{13} m^3. With an average uranium concentration of 3 ppm, its uranium content is very roughly 3×10^7 MT, corresponding to a total radioactivity of 10^{10} Ci. This corresponds to about 1500 MT of ^{234}U in the 1-m layer, a nuclide whose relative ingestion hazard resembles that of ^{239}Pu within one order of magnitude.

This simple calculation shows that a 1-m layer of the United States contains about as much long-lived radioactivity and actinides as nuclear industry will put beneath several hundred of those layers within the next 30 years. This illustration does not mean that HLW represents no serious and long-lasting hazard potential, but it emphasizes that the amounts of radioactive material dealt with in waste management are not at all alien to nature.

2.3 Alternatives for Commercial HLW Management

Any liquid HLW that has been generated either in national programs or commercially is presently contained in tanks, some of them still with artificial cooling. In the United States, there is more than 200,000 m^3 of HLW from defense programs. Most of it is not as highly radioactive as waste solutions from commercial reprocessing will be. But the volume corresponds to more than 200,000 MT of commercial fuel or about 7×10^6 MWe-years of nuclear energy, and this is about twice as much as commercial reprocessing in the United States will generate up to the year 2010.

If the storage conditions are properly chosen (which has not always been so in the past), tank storage is a perfectly safe technique. This has been proved by almost three decades of experience in the United States and Great Britain. Proper conditions mean storage of acidic waste in stainless steel tanks to minimize corrosion and to avoid formation of sludges.

However, liquid storage cannot be the ultimate solution to HLW management. A surface storage system, which is relatively vulnerable, with a large and steadily increasing inventory of

Table 11.5 Estimate of solidified HLW generated in the United States and received by a federal repository [B4]†

Year	Nuclear power capacity, GWe	Volume, $10^3 m^3$	Actinide mass, MT	Radioactivity, MCi	Thermal power, MW
1985	160	–	–	–	–
1990	285	0.28	18.2	726	2.5
1995	445	1.30	85.0	3,556	12.3
2000	625	3.21	221.0	8,550	31.9
2005		6.25	445.1	16,408	64.0
2010		10.75	772.7	27,800	108.9

†Assuming that reprocessing of spent fuel begins in 1978, capacity increases, spent fuel backlog is eliminated in 1988, shipment to federal repository 10 years after the time of waste generation.

liquid highly radioactive material, means a potential hazard that is not acceptable in the long run. Moreover, the high degree of maintenance and surveillance that is required would have to be provided for an extremely long period of time.

The general concept how to achieve permanent isolation of radioactive waste from the environment without human activity being required in the long run has been well established internationally. First, the waste solution will be immobilized by solidification after some time of intermediate storage as a liquid. The solid product should have a high degree of stability to ensure that the radionuclides contained in the waste remain immobile for a long time. With present knowledge, materials that meet this requirement are glasses and ceramics. The solidified waste contained in a steel canister will then be stored irretrievably in a stable geologic formation deep underground.

The most likely path for such radioactive material to find access to the environment is transport by groundwater. Three barriers are provided in the above concept to prevent this: (1) the inherent stability of the solidification product, particularly against attack of water, and the durability of the waste container; (2) the stability of the geologic containment, which prevents water from providing means of communication between waste and circulating groundwater; (3) the large sorption capacity provided by rock and soil on the long distance from the underground repository to the surface combined with a low groundwater velocity.

In various countries this concept may materialize in different ways. Alternative solidification processes and solidification products have been developed, and some of them are in or close to the prototype stage. As for final disposal, some countries have made a decision in favor of rock salt as the most suitable geologic formation. In West Germany all effort is focused on salt domes. In the United States much effort is devoted to bedded salt, although alternatives such as granite and shale are under consideration as well. Other countries are investigating those geologic formations to which they have easy access. Because of strong public opposition, most governments have decided not to take any foreign waste for final disposal.

Some more basic alternatives of waste disposal are studied in some countries. Seabed disposal is one that attracts some attention in the United States and in Great Britain. The idea is to drill holes into the bottom of the deep sea that will be self-sealing after they have been filled with solidified waste. The water covering the disposal site will act as an additional barrier between waste and human beings. An obvious advantage of the seabed option is the very remote location. An obvious disadvantage is the complex and not at all conventional technology required.

Another approach to HLW management takes into account that after some 400 years the radiotoxicity of the waste begins to be determined by the actinides. Consequently, if the actinides can be removed from the waste to a sufficiently high degree, the integrity of the geologic containment is required only for hundreds of years rather than for thousands. However, it will be difficult to achieve the high degree of separation required, and it will even be difficult to deal effectively with the separated actinides. The very substantial additional effort to separate actinides can be justified only if the separated actinides can be eliminated. Presently no way seems feasible other than transmutation by neutron bombardment to yield shorter-lived fission-product nuclides. As yet, the technology for both, actinide separation and actinide transmutation, is not available. Moreover, it seems doubtful whether there is a sound incentive to undertake the extra effort required.

Occasionally some more exotic alternatives to get rid of radioactive waste are mentioned, such as disposal into the earth's ice caps or into outer space. They may look promising at first sight. Studying them in greater depth, however, makes obvious that there may be insurmountable problems. Both ice-cap disposal and disposal into outer space do not seem safe as yet and have political implications that will be hard to resolve.

For the first generation of the world's commercial reprocessing plants, there is little doubt that the HLW will be solidified as a glass or as a ceramic and, after some interim storage, will be disposed of into a geologic formation deep underground.

2.4 Tank Storage

Storage of acidic liquid HLW in stainless steel tanks is the only HLW management technique presently available on a large scale. Although it is not an option for final storage, it will remain a necessary intermediate step before solidification. According to present considerations the waste should be aged for about 6 years prior to solidification. An aging period of 6 years will reduce the heat generation of waste from reprocessing 150-day-old fuel by one order of magnitude. It is obvious, however, that the desired reduction of heat generation can as well be achieved by prolonged aging of the fuel elements prior to reprocessing. Then the role of intermediate tank storage would be that of an operational buffer requiring substantially less tank capacity. As there is no strong economic incentive for early recovery of uranium and plutonium, it is merely a question to be solved by a safety analysis whether to decide in favor of early or of late reprocessing.

Because of the large backlog of spent fuel that has been accumulating for many years, the first generation of reprocessing plants will see hardly any fuel that has been aged for less than a few years. Because of lack of experience with solidification, however, a substantial buffer capacity to store liquid waste will still be desirable to ensure steady operation of the reprocessing plant.

Basic technology of tank storage. Liquid HLW enters the storage tank as a solution with up to 250 g/liter salt content and up to 7 M HNO_3. The intense radiation absorbed by the liquid causes self-heating and radiolysis of water and HNO_3. The radiolysis, however, does not lead to substantial amounts of hydrogen, as verified by Windscale and WAK experience. The liquid contains undissolved solids carrying considerable amounts of radioactivity, which may lead to hot spots at the wall if they settle.

The most important consideration in tank design is minimization of corrosion. Originally two storage philosophies were believed to be equally safe in this respect: (1) neutralized waste in mild steel tanks; and (2) acid waste in stainless steel tanks. Almost three decades of experience have proved that only the latter satisfies all safety requirements. No leakage from stainless steel tanks has become known, whereas 20 out of 183 mild steel tanks at the Hanford and Savannah River sites developed leaks [L2]. It is now generally accepted that a minimum corrosion rate can be maintained with suitable types of stainless steel and nitric acid concentrations in the range of 2 to 4 M. If the HNO_3 concentration falls below 1 M stress corrosion due to chloride ions may be promoted.

Neutralized waste may develop another problem. Sludge will be formed that will carry most of the radioactivity and will eventually settle. That happened at Hanford, Savannah River, and at the Nuclear Fuel Services plant at West Valley, New York, the first commercial reprocessing plant, which is now out of operation. Considerable problems will have to be solved there to transfer the waste entirely from the storage tanks to a final treatment facility.

Despite proper choice of tank material and chemical conditions, continuous leak monitoring is required. Two safety features have to be provided to keep leak incidents under control. The tank has to be located in a vault that is able to hold the entire tank volume, and the tank has to be connected to an empty spare tank ready to receive the content of a leaking tank.

To prevent the highly active liquid from boiling and to maintain a temperature below 60°C, a cooling system is required. As loss of coolant is a severe hazard when very active liquid is stored, the cooling system has to be sufficiently redundant. The cooling system adds considerably to the total investment costs of a storage facility. It should therefore be designed for optimum use of its capacity even when the heat rate of the waste decreases. This may be achieved by providing tanks with different cooling capacities and suitable fill/empty schedules.

Because of the great importance of the cooling systems as a safety requirement and because of its high cost, consideration should be given to storing liquid HLW at a much lower concentration level. This means storage of correspondingly greater volumes and consequently higher tank costs. On the other hand, savings will be achieved because of less cooling capacity required, and the safety philosophy with respect to overheating of the highly radioactive liquid may be greatly relaxed.

Another safety feature that is required is a pressurized air system that can be used to keep the concentration of radiolytic hydrogen below a given safety level as well as to prevent solids from settling. The air stream may also be useful to regulate in-tank evaporation and radiolytic stripping of HNO_3. The air has to be released via an elaborate off-gas system.

Design of the HAW tanks at the AGNS plant. The AGNS plant design [L2] is similar to that at Windscale [W1]. The design is based on a burnup of 27,000 MWd/MT and a heat rate of 15.5 kW/MT of heavy metal for waste from 150-day-cooled fuel and of 7.9 kW/MT for waste from 1-year-cooled fuel. The waste will arrive at the tank facility with a specific volume of 1100 liters/MT heavy metal. The volume will be reduced to 550 liters by in-tank evaporation.

Two tanks of 1100-m^3 capacity each with 10 percent freeboard are provided, one in operation and one as a spare tank. The total operational capacity of the tank is 2000 MT heavy metal. This means a waste storage capacity of not even 2 years at full-load operation. When the plant goes on stream, either a solidification plant has to be operational not much more than 1 year later or additional tanks will have to be built.

The tanks are 16.5 m in diameter and 6 m high with walls between 0.95 and 1.25 cm thick. The expected corrosion rate is smaller than 0.01 mm/year, and the total corrosion allowance is 0.3 mm. A pressure-vacuum relief system ensures that the range of operating pressures does not exceed a vacuum of 25 cm of water and a pressure of 7.5 cm of water.

The tank is contained in a stainless steel-lined concrete vault. The concrete walls are 1 m thick, the top is 1.75 m thick. The liner includes a sump at the low side of the floor which collects any liquid in the vault. An elaborate system is provided to monitor the integrity of the liner.

The cooling system is designed to remove 10^4 kW of thermal power. To keep the liquid at or below 60°C, almost 11 km of stainless steel cooling coils are connected in six valved, parallel banks. If one of them is out of service, the remaining five will maintain a full tank at or below 70°C. The primary closed cooling loop contains a heat exchanger. Three separate cooling-water supply systems are provided to ensure adequate cooling under all foreseen conditions. The first system, used during normal operation, is a closed cooling-tower loop. The second system provides well water on a once-through basis to the heat exchanger. The third system employs either of two diesel-powered pumps to circulate cooling water from a large pond directly through the tank coils. Under maximum heat rejection conditions, 14 m^3/min flow through the cooling coils with inlet and outlet temperatures of 35°C and 45°C, respectively.

Agitation is effected by two different schemes involving air-lift circulators and ballast tanks. Both are able to keep particulate matter in suspension and both are operated pneumatically. The prime difference is that the air-lift circulators intimately mix air and solution, whereas the ballast tanks always maintain an interface between air and solution inside the ballast tanks. Thus, evaporation of the waste solution as well as radiolytic stripping of HNO_3 are significantly greater when the air-lift circulator is operated. The air-lift circulators thereby provide the means for in-tank concentration and HNO_3 adjustment. Both the air-lift circulators and ballast tanks provide enough air to keep the hydrogen concentration in the tank below 3 percent if this were to be exceeded by radiolysis. The hydrogen generation rate is very conservatively based on about 0.9 standard cubic meters per hour per megawatt of decay power.

The air-lift circulator is shown in Fig. 11.3. There are 152 $\frac{1}{8}$-in-diameter holes distributed over a 6-in-diameter face plate of the diffuser cone. The vertical air supply line supports an 18-in-diameter outer pipe. Figure 11.3 demonstrates how this device works. There are 22 air-lift circulators, to provide for vertical mixing of the waste solution.

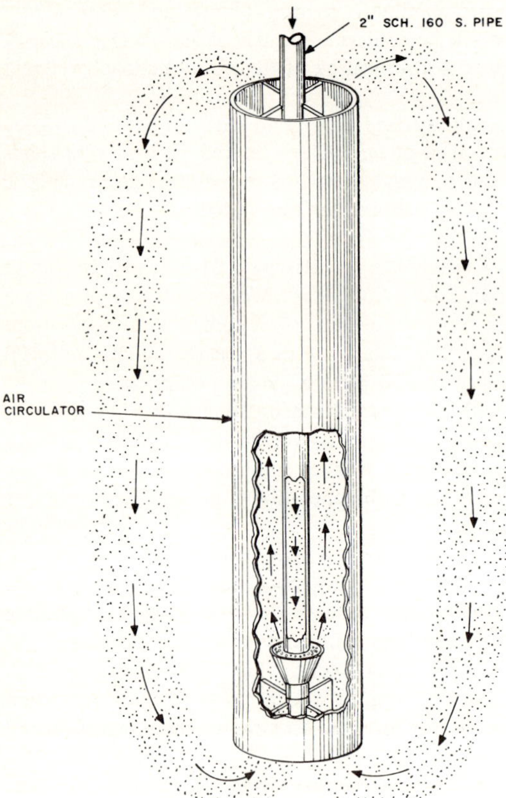

Figure 11.3 Air-lift circulator in AGNS HLW tank. (*From Legler and Bray [L2].*)

The ballast tanks as shown in Fig. 11.4 are internally mounted air tanks arranged in two groups, 9 central and 18 peripheral. Each ballast tank is 16 ft high and 3 ft in diameter. The bottom tapers to a cone and terminates in a nozzle 2 ft above the floor of the tank. Air is supplied to each tank intermittently at a controlled pressure for a specific time, resulting in a flushing action. When the air supply is off, the air in the ballast tank vents through a $\frac{3}{16}$-in hole near the top of the waste tank in the air-supply line and allows the solution to flow back into the ballast tank. For a full waste tank, air at 15 psig is supplied to a central ballast tank for about 60 s.

An off-gas system ensures that no excessive radioactivity is released together with the sparge air. Condensate from the off-gas can be transferred either back to the HLW tank or to the intermediate-level waste tank.

Figure 11.5 shows a schematic drawing of the AGNS tank.

2.5 Solidification Products

HLW solidification has to serve two purposes:

1. Immobilization of the waste
2. Long-term fixation of long-lived radionuclides

Immobilization of the waste helps to facilitate and to add safety to all handling and storage operations before the waste is in its final disposal location. Whereas liquid HLW is definitely

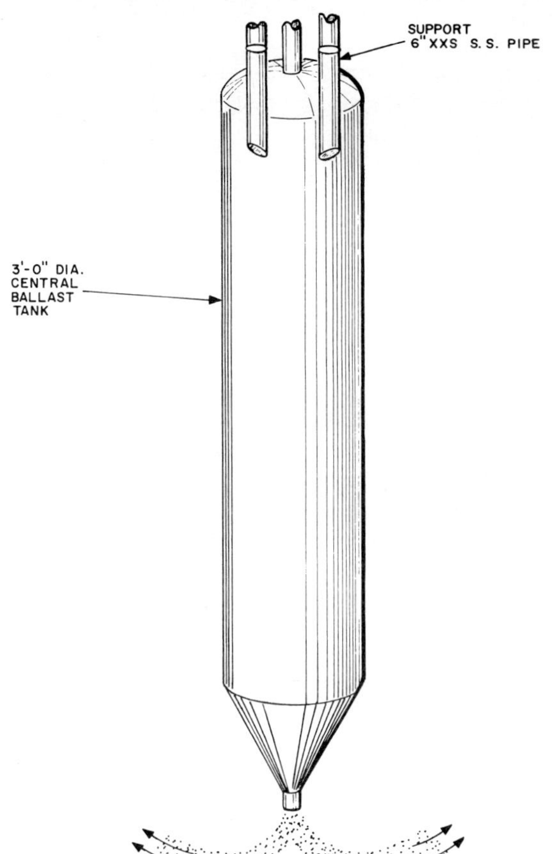

Figure 11.4 Ballast tank in AGNS HLW tank. (*From Legler and Bray [L2].*)

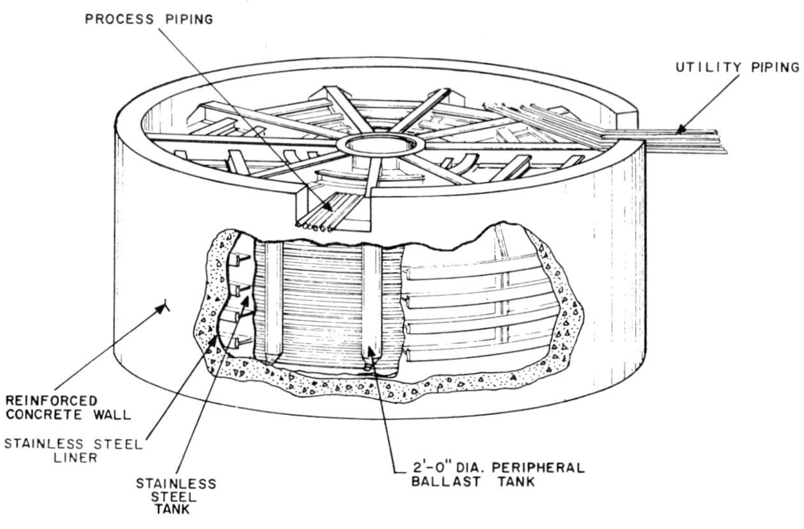

Figure 11.5 AGNS HLW tank. (*From Legler and Bray [L2].*)

not suitable for transportation on an industrial scale, solidified waste may well be when its integrity can be maintained and appropriate shielding is provided. Interim storage of solidified waste presents fewer problems because of significantly reduced container corrosion and relaxed cooling requirements. The nature of the waste solidification product is of particular importance during the period of several decades while the repository is being filled, and hence open. In the long term, the basic requirement of the geologic disposal concept is a repository located deep underground under geologic conditions that ensure permanent isolation of radioactive waste material from the environment, regardless of its particular form. The inherent stability of the solidification product is considered an additional release barrier for radioactivity within the overall safety concept, as part of a system of engineered barriers including the container and, possibly, an overpack and highly absorbing backfill materials.

Product alternatives. The general requirement of a solidification product is stability against destructive influences to which a highly radioactive solid may be exposed, i.e., irradiation stability, thermal stability, mechanical stability, and chemical stability. Although stability of the waste form will not last forever and although it will be even impossible to verify long-lasting stability of the waste form, it is highly desirable as an additional safety barrier in the geologic waste repository concept. Consequently, the solidification product will have to be made as stable as practically achievable.

The more important product alternatives, classified in terms of two principal lines—calcine and glass—are shown in Table 11.6.

Calcines are products obtained by removing the volatile components of the waste, i.e., water and nitrate, at temperatures between 400 and 900°C. The result is a mixture of oxides of fission products, actinides, and corrosion products in particulate form with a specific surface of 0.1 to 5 m^2/g. The plain calcine is not very stable chemically because of its large surface area and the chemical properties of some of the oxides, and it is highly friable. To improve the properties of calcines, advanced forms are developed. One such product is the so-called multibarrier waste form, a composite consisting of calcine particles with inert coatings, such as pyrocarbon, silicon carbide, or aluminum, embedded in a metal matrix. Another advanced calcine is the so-called supercalcine. This is essentially a ceramic obtained by adding appropriate chemicals to the HLW to form refractory compounds of fission products and actinides when fired at 1200°C. Supercalcine requires consolidation by embedding in a matrix but does not need to be coated, as the material is supposed to have inherent chemical stability.

Glasses are products obtained by melting the waste oxides together with additives such as SiO_2, B_2O_3, Al_2O_3, P_2O_5, Na_2O, and CaO. On solidification, the melt forms a glass or a near-glassy solid with good stability. Borosilicate glass is the type of solidification product most

Table 11.6 Solidification-product alternatives

	Product	
Alternative	Calcine	Glass
Basic	Fluidized-bed [L1] (particles) Pot [B3] (cake)	Borosilicate [M2, C3] (cylinder) Phosphate [H1] (cylinder)
Advanced	Supercalcine [M1] (chemical additives, high-temperature ceramic product)	Borosilicate glass ceramic [D1] (cylinder)
Composite	Multibarrier waste form [M2] (coated particles in metal matrix)	Vitromet [G1] (glass or glass ceramic particles in metal matrix)

thoroughly studied all over the world. Phosphate glass has long been abandoned in the United States, where it had been studied for the first time. It was still considered in West Germany for quite a while, but has been given up there as well. Only in the Soviet Union is work on phosphate glass kept alive.

Glasses also have certain drawbacks, such as the possibility of devitrification leading to products with less predictable properties. Advanced developments along the glass line are glass ceramics obtained by controlled crystallization of glass to avoid uncontrolled devitrification. Another advanced product is vitromet, i.e., glass or glass ceramic beads embedded in a metal matrix, with extremely high heat conductivity and mechanical strength.

It is the common principle of supercalcine and glass ceramic to have stable crystalline phases hosting fission products and actinides. Along the same line, a third group of crystalline solidification products, synrock (synthetic rock), has been developed. All crystalline waste forms contain significant quantities of glassy phases remaining from their formation at high temperatures.

Fission-product content. The solidification products may incorporate different fractions of fission-product oxides. This fraction is desired to be high on economic grounds. Costs of handling, packaging, and transportation are considerable and depend on the volume and the number of containers to be handled. On the other hand, the fission-product concentration is limited by chemical reasons or by reason of heat production. Chemical limitations are typical for glasses where either phase separations may occur or the product may not be a glass at all. For borosilicate glasses, 20 to 25 w/o (weight percent) of fission products is about an upper limit. Higher concentration may lead to the segregation of a yellow crystalline phase composed of alkaline and alkaline earth molybdates. This easily soluble phase contains long-lived fission products such as ^{90}Sr and ^{137}Cs.

No chemical limitations exist for calcines, which may consist of pure waste oxides. In this case, however, the heat production in the solid may impose a limitation depending on the burnup and the fraction of nonradioactive oxides in the waste.

Making a composite means a further reduction of the fission-product concentration in the final product. If small particles of glass or calcine are embedded in a close-packed matrix, the fission-product content is reduced by about one-third.

Recently, there has been a trend to give less priority to a high-fission-product content and to a small product volume. Economic penalties may be compensated by benefits due to more flexibility in handling, interim storage, and disposal of the waste products.

Irradiation stability. Any solidified HLW will be exposed to energetic radiation from radioactive decay of fission products and actinides. Part of the radiation energy is dissipated in elastic collisions with atoms from the solid material, thereby displacing them and causing radiation damage. This may affect macroscopic properties such as mechanical or chemical ones, and it may cause storage of energy.

The following types of radiation will occur in the waste:

Gamma radiation (average energy 2 MeV) and beta radiation (1.5 MeV) from fission-product decay

Alpha radiation (6 MeV) from actinide decay and some alpha radiation from (n,α) reactions (e.g., with boron)

Recoils from alpha decay (100 keV)

Fast neutrons from spontaneous fissions and from (α,n) reactions with light elements, and fission recoils

The energy dissipated in elastic collision is at least two orders of magnitude lower for gamma and beta radiation than for the others. The total fast-neutron and fission recoil doses

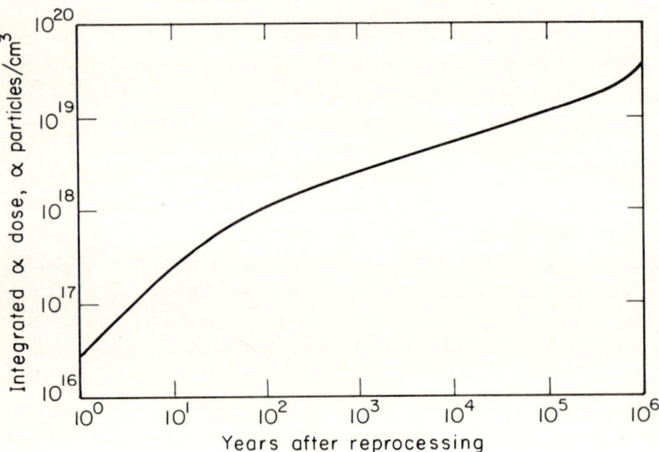

Figure 11.6 Alpha dose in LWR uranium waste from 1 MT heavy metal incorporated in 70 liters of glass.

are by several orders of magnitude lower than for alpha radiations and their recoils. Thus, mainly alpha particles and even more their recoils have to be considered. Their effect can be simulated on a reduced time scale by incorporation of an appropriate amount of Cm_2O_3 (mainly ^{244}Cm) into a synthetic waste solid. The alpha-radiation rate of Cm_2O_3 is about 1×10^{14} $min^{-1} \cdot g^{-1}$. Figure 11.6 shows the total alpha dose per metric ton of heavy metal reprocessed. The first 100 years are considered most significant for radiation damage, as the total alpha dose increases only by a factor of 10 over the next 100,000 years. The 100-year alpha and alpha-recoil dose can be simulated during 1 year with a Cm_2O_3 content of 1 w/o.

Among the possible consequences of radiation damage on solidified waste, energy storage has to be considered a potential risk. The temperature of the waste solid would suddenly rise if stored energy were released. A quite thorough experimental study by Roberts et al. [R2] on energy storage in calcines and borosilicate glass comes to the conclusion that there will be hardly more than 50 cal/g stored. With an average heat capacity of 0.2 cal/(°C·g), this corresponds to a maximum temperature rise of 250°C, which should be tolerable.

Other possible effects of radiation on the solidified waste are deterioration of mechanical properties and changes of volume due to radiation damage or as a consequence of helium formation from alpha decay.

Furthermore, one may think of radiation influencing the chemical stability of the solid. There is no experimental evidence for any of these effects [H2]. In all test procedures a long-term dose has been simulated in a short time, which will probably rather enhance the effects. The apparent radiation stability should be not too surprising, bearing in mind that an alpha dose of $10^{18}/cm^3$ over 100 years is a rather modest one compared to the fast-neutron doses to which materials in a nuclear reactor are exposed.

Thermal stability. Heat generation in the solidified waste causes it to be at an elevated temperature for more than 100 years. The specific heat generation in a solid with 20 w/o fission products is shown in Fig. 11.7. A cylindrical waste block being a homogenous heat source will have a radial temperature gradient. Given the heat generation, the temperature difference between the surface (the surface temperature is determined by the storage conditions) and the centerline is a function of the heat conductivity of the material. The maximum temperature difference in the waste cylinder is

$$\Delta T_{max} = \frac{q'(d/2)^2}{4\kappa} \tag{11.1}$$

In this equation q' is the homogenous thermal power density (W/m^3), d is the diameter of the waste cylinder (m), and κ is the thermal conductivity [W/(m·°C)].

Equation (11.1) is essentially a solution of Eq. (11.7) and is based on a few assumptions and simplifications, e.g., no axial heat conduction, constant average heat conductivity and specific heat, constant heat source, steady-state heat transfer, one-dimensional (radial) heat flux, cylindrical geometry in the waste and in the surrounding material, e.g., salt, and no heat source in the salt.

Table 11.7 shows the thermal conductivities of several solidification products. In Fig. 11.8 maximum temperature differences are plotted against the age of the waste assuming 20 w/o fission-product oxides in the solid. For the calculations typical canister diameters of about 250 and 500 mm and typical thermal conductivities of 0.25 (particulate calcine), 1.2 (glass), and 10 (vitromet) W/(m·°C) have been used [E2].

When a 6-year storage period prior to solidification is assumed, glass in canisters 25 cm in diameter and glass-metal composites in canisters 50 cm in diameter exhibit maximum temperature differences well below 100°C. However, glass in a large canister and plain calcine even in a small canister will produce centerline temperatures as much as 300°C higher than the surface temperature, thus possibly reaching 500 or 600°C. After about 200 years the temperature gradient will have essentially disappeared.

One limitation of the maximum temperature in a solidified-waste block is given by the need to maintain the immobilization of the waste. Table 11.8 shows softening and melting temperatures of various products. For high-melting materials volatilization of individual fission products below the melting temperature has to be considered.

A long-term effect promoted by high temperature is devitrification of glass, converting it

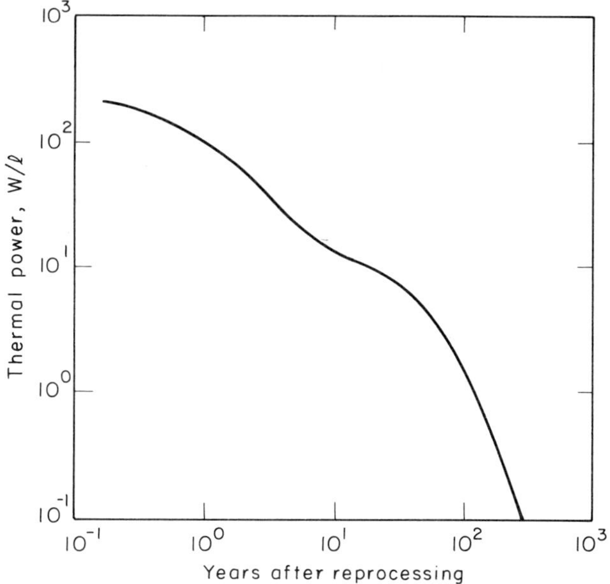

Figure 11.7 Thermal power of 1 liter of solidified LWR uranium waste (70 liters of glass per metric ton of heavy metal; 150 days aged prior to reprocessing, 30,000 MWd/MT).

Table 11.7 Thermal conductivity ranges for various classes of solidification products in the temperature range 100 to 500°C

Product	Thermal conductivity κ, W/(m·°C)
Particulate calcine	0.2–0.3
Phosphate glass	0.8–1.2
Borosilicate glass	0.9–1.3
Borosilicate glass ceramic	1.5–2.0
Particulate calcine or glass beads in metal matrix (e.g., vitromet)	~10

into a thermodynamically more stable form. This effect is supported by the presence of a great number of components and of impurities that may act as crystallization nuclei. Both conditions are present in a waste glass. The crystallization process in a multicomponent borosilicate glass with simulated fission-product oxides has been quite thoroughly investigated [H4]. It was found that partial crystallization may occur within days at temperatures above 600°C depending on the composition. Fission products are selectively enriched in certain crystalline phases. The remaining glass phase still containing fission products is enriched in boron oxide. The devitrified product may therefore be less leach-resistant than the original glass.

With full radioactivity, phosphate glasses showed strong devitrification at 500°C with deterioration of leach resistance. Under the same conditions borosilicate glasses did not devitrify within 7 months [B3]. Recent investigations show that extensive additional tests under hydrothermal conditions are required to simulate underground storage conditions.

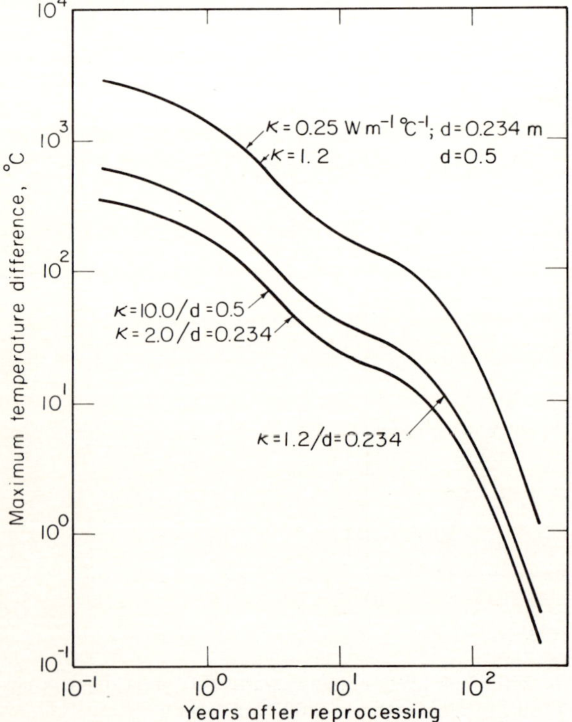

Figure 11.8 Maximum temperature difference in a cylinder of solidified waste for different diameters and thermal conductivities (70 liters/MT of heavy metals; 30,000 MWd/MT, 150 days aged prior to reprocessing).

Table 11.8 Softening and melting temperatures of HLW solidification products and associated materials

Material	Softening temperature, °C	Melting temperature, °C
Calcine†	Very high	Very high
Phosphate glass	350–450	800–1000
Borosilicate glass	500–600	1000–1200
Glass ceramic	ca. 750	ca. 1200
Lead matrix	–	327
Sodium chloride	–	801

†Usually, calcines contain significant volatile residues, limiting the storage temperature.

Ceramic-type products such as supercalcine and glass ceramics have been subject to a crystallization process and are therefore thermodynamically more stable. Consequently, as long-term structural changes are less likely with those solidification products, they may maintain their original properties at a higher temperature than glasses.

The temperature gradient will give rise to thermal stresses in any monolithic material, which in turn may cause cracking. The stress at any point of a glass sample depends on the difference in temperature between this point and the average. At the surface of a cylinder, where cracking is most likely, the longitudinal and the circumferential stresses are given by [K1]

$$\sigma = \frac{E\alpha}{1-\mu}(T_a - T_s) \tag{11.2}$$

with σ = stress (kg/cm^2), E = modulus of elasticity (kg/cm^2), α = linear expansion coefficient (1/°C), μ = Poisson's ratio, T_a = average temperature (°C), and T_s = surface temperature (°C). The maximum difference between surface and average temperature beyond which stress cracking is to be expected may be roughly estimated as a function of the expansion coefficient by inserting $E = 7 \times 10^5$ kg/cm^2, $\mu = 0.20$, and a fracture stress of 1000 kg/cm^2, which are reasonable for an average soft glass. Then this temperature difference limit is

$$\Delta T_{\text{limit}} \approx 3 \times 10^{-4}\,\frac{1}{\alpha} \tag{11.2a}$$

Table 11.9 shows expansion coefficients of certain solidification products and the corresponding temperature-difference limit. However, even if the stationary temperature differ-

Table 11.9 Expansion coefficients for various materials and temperature differences beyond which stress cracking is to be expected

Material	$\alpha \times 10^5$	ΔT_{limit}, °C
Borosilicate glass	0.75–1.80	40–15
Borosilicate glass ceramic	0.50–1.00	60–30
Phosphate glass	1.10	30
Bottle glass	0.90	35
Fused silica	0.05	600
Steel	1.00–1.40	

ence could be kept below these limits, there may be severe stresses due to fast cooling of the glass.

Thermal expansion has to be looked on under one more aspect. Different expansion coefficients of glass and canister material may cause stress in the canister wall. It has been observed that this may significantly promote corrosion of the canister.

Mechanical stability. When a block of solidified waste is crushed by mechanical impact, fragments of various size will be formed. Two consequences are to be considered: (1) The fraction of radioactivity leached in a certain period of time will be increased in proportion to the increase in surface area; (2) the formation of very small particles in the order of 100 μm and less may enable radioactive material to be spread by air. Even larger particles may be carried by water.

Nonmonolithic calcines do not have any mechanical stability, metal matrix products are the most stable ones. Glass ceramics are more stable than glasses.

Chemical stability. The only chemical attack on the solidification products deserving serious consideration is leaching by water or brine, if such exist in the repository. Chemical interaction between solid rock salt or other geologic material and any of the solidification products under consideration will not be significant unless the temperature rises above the melting point of the salt or the solidification product. Diffusion of fission products into the salt at reasonable temperatures is not a significant safety concern either.

As leaching is the most likely mechanism by which radionuclides from the waste may be remobilized, the leaching behavior is the most thoroughly studied property of solidification products. As a consequence, there is a wide spectrum of procedures and results. The samples are powders, grains, or small blocks. The leaching procedures are characterized by different temperatures, different leach liquors—e.g., pure water, seawater, saturated NaCl solutions—and different renewal schemes for the liquor. The latter is important for the result, as the average leach rate is found the smaller the longer the liquor remains in contact with the sample. With a soxhlet-type apparatus continuous renewal may be achieved. In the majority of the experiments an integral leaching rate R [g/(cm^2·day)] is determined in order to compare the leach resistance of different solidification products. In addition, a number of experiments have been designed to study the time dependence of the leaching process. They provide the information necessary to calculate the total radioactivity release upon accidental contact with water over a long period of time. To have a sound basis for extrapolation to the leach behavior of a large glass cylinder over a time period relevant for long-term disposal considerations, those experiments have to employ a suitable leach technique, e.g., leaching of fine powders. In general, the average leach rate observed decreases with increasing duration of the leach experiment. Most experimental leach rates are obtained from leach processes lasting for days or weeks.

Leach rates measured on borosilicate glasses at room temperature are in the range of 10^{-6} to 10^{-4} g/(cm^2·day), largely depending on experimental conditions. Almost identical results were obtained in some simultaneous tests with water and NaCl solution. A careful study of the spectrum of data and procedures leads to the conclusion that 10^{-5} g/(cm^2·day) is a reasonable weighted average suitable to characterize the leach resistance of borosilicate glasses in water at room temperature and atmospheric pressure [B7, T1]. Leach rates of phosphate glasses are within the range of borosilicate glasses. There is experimental evidence that devitrification increases the leaching rates, more in the case of phosphate glass than in the case of borosilicate glass.

For borosilicate glass ceramics, leaching rates of the order of 10^{-5} g/(cm^2·day) and greater have been found. Remarkably, the controlled crystallization process used in making these ceramics does not increase the leachability relative to that of the parent glass, in contast to the increased leachability occurring on spontaneous devitrification.

Calcines are well known to be readily leachable in water. Hot pressed supercalcine shows a leach resistance similar to that of the other chemically stable products.

In general, it may be concluded that a leach rate in the range of 10^{-6} g/(cm^2·day) is probably something like a lower limit unless a substantially different solidification technology is employed. Less leachable products may be high-temperature glasses or ceramics or very sophisticated composites.

It should be mentioned that leach data presently available have been obtained under standard laboratory conditions. Leaching experiments modeling conditions that may be experienced in a specific type of repository, including solutes in the leach liquor and elevated temperature and pressure, have been initiated in a number of laboratories.

Extrapolation of long-term leaching. An important task in evaluating the risk of radioactivity release by water leaching is the mathematical description of the process. All efforts to do this on the basis of a physical understanding of the process have not led very far. This leaves us with empirical approaches where much of the physics of the process is packed into coefficients and exponents obtained by fitting experimental leach curves.

It is reasonable to assume two limiting cases of leaching kinetics, dissolution of the waste form and diffusion of radioactive species from the waste form, representing upper and lower limits of the fraction leached within a given time period. The corresponding types of equation are (F = fractional release)

$$F_1 = a_1 t \tag{11.3}$$

and

$$F_2 = a_2\sqrt{t} \tag{11.4}$$

Geometric changes of the sample on leaching have not been taken into account. Equation (11.4) is the well-known type of equation approximating a solution of Fick's second law for a semiinfinite sample valid for fractional releases up to about 25 percent.

These are oversimplifications very likely indicating upper and lower limits of radioactivity release. Essentially the complex process will be composed of both dissolution and diffusion. This can be described by equations of the type

$$F_3 = a_3\sqrt{t} + a_{33} t \tag{11.5}$$

This type of equation may work well for relatively short periods. For long-term leaching, however, it assumes that corrosion becomes more and more dominating and eventually controls the process. Therefore an entirely empirical approach with the following type of equation appears to be the most appropriate:

$$F_4 = a_4 t^x \qquad (1 > x > 0.5) \tag{11.6}$$

There are examples where test runs can be best fitted with $x = \frac{2}{3}$. None of these simple equations takes into account that a piece of solidified waste will become smaller on leaching and so will the surface area. Although this simplification is on the safe side, the effect will be small for a full-size glass cylinder within the time period of interest.

Figure 11.9 shows the results of sample calculations of cumulated fractional releases from a 1.65-m-long and 0.234-m-diameter waste cylinder over a period of 10^6 years taking into account decay of radioactivity. They are obtained by fitting Eqs. (11.3), (11.4), and (11.6) to experimental data of sodium-leaching from a 144-day column leach experiment with a powdered borosilicate glass (1236 cm^2/g specific surface), recalculation to the specific surface of the glass block and coupling with the ORIGEN program for LWR uranium waste from 30,000 MWd/MT burnup fuel [E1]. The total fraction of sodium leached in the experiment was about 30 percent. The constants a_n in Eqs. (11.3), (11.4), and (11.6) include a factor a_n' from fitting experimental data and a geometric conversion factor. They are

$$a_1 = a_1' S_{\text{cyl}} = 4.45 \times 10^{-5} \text{ (yr}^{-1}\text{)}$$

$$a_2 = a_2' \frac{S_{\text{cyl}}}{S_{\text{sample}}} = 2.53 \times 10^{-5} \text{ (yr}^{-1/2}\text{)}$$

$$a_4 = a_4' \frac{S_{\text{cyl}}}{S_{\text{sample}}} = 3.24 \times 10^{-5} \text{ (yr}^{-x}\text{)}$$

with $S_{\text{cyl}} = 6.53 \times 10^{-2}$ cm^2/g (specific surface of waste cylinder)

$S_{\text{sample}} = 1.24 \times 10^3$ cm^2/g (specific surface of leach sample)

$$x = 0.666$$

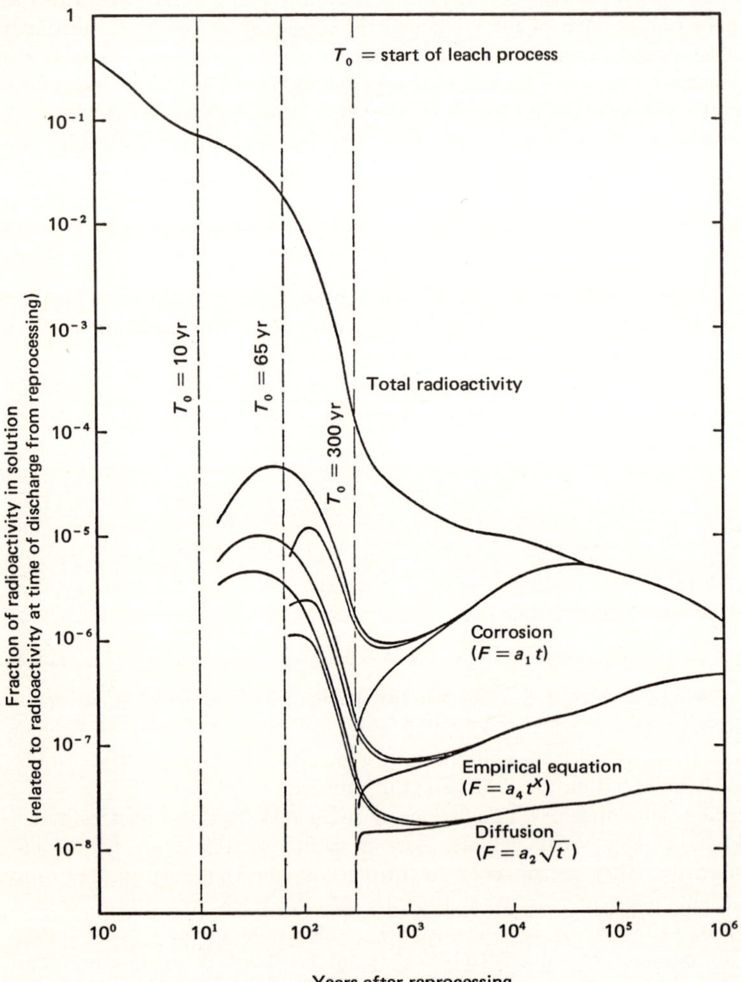

Figure 11.9 Long-term leaching curves of a vitrified waste cylinder calculated according to different kinetic models [initial leach rate 2×10^{-6} g/(cm^2·day)]. (*From Ewest [E1].*)

Table 11.10 Stability criteria of the waste form

Stability category	Individual criteria
1. Radiation	Energy storage
	Effects on other stability categories
2. Thermal	Heat conductivity (maximum temperature and thermal stresses)
	Softening or melting temperature
	Thermodynamics
3. Mechanical	Degree of fragmentation
4. Chemical	Leachability

The curves indicate the fraction of the radioactivity present in the waste at the time of reprocessing that will be in solution if the glass block has come into contact with water 10, 65, or 300 years after reprocessing. An increase with time means that the leaching process is faster than the overall decay of radioactivity and vice versa. Thus, the curves represent the fraction of the initial radioactivity available for release into the environment at any time when the geologic barrier may fail. For comparison, the top curve of Fig. 11.9 shows the total fraction of the initial radioactivity available at any time. The plot thereby demonstrates that the solidification products presently envisaged for final disposal of highly radioactive waste may in fact provide an effective release barrier.

Figure 11.9 is an example of how to extrapolate leaching data. It should be noted that this extrapolation is based on experimental data not necessarily representative of conditions expected in a geologic repository. In general such extrapolations have considerable uncertainty. They are based on the assumption that the basic properties of the glass are not significantly altered in a period of thousands of years.

Product evaluation. Table 11.10 lists the stability criteria of solidification products. The stability categories are of different relevance. Categories 3 and 4 are directly relevant for the release-barrier function of the solidification product. Radioactivity release from the final disposal site may occur by leaching or, less likely, via spreading particulate matter by air or water. Mechanical effects may also alter the leachable surface of the waste form. Categories 1 and 2 are not directly relevant for the release of radioactivity but for maintaining the original properties of the solidification product, which are to prevent radioactivity from being released.

The radiation dose to be expected is moderate and smeared out over a long period of time. Serious effects on stability are unlikely, and the experimental data presently available do not indicate such effects. Properties relating to thermal stability have been characterized in some detail. They are considered important for maintaining the product stability over intermediate periods of time.

In the final evaluation, product technology has to be considered as well. When the point is to select a product for near-future demand, technology will have a high weight. It will be better to have the waste solidified in an acceptable form even if not in the very best one. One has to keep in mind that the amount of highly radioactive waste to be solidified over the next two decades will be a small fraction of the waste that nuclear energy will produce altogether. In the long run, however, stability criteria should clearly dominate.

It should be noted that the canister may add considerably to the durability of a high-level radioactive waste package. Often it may be easier to achieve a high degree of sophistication with a nonradioactive canister material than with a highly radioactive solidification product.

Figure 11.10 is a semiquantitative approach to a rating of solidification products according to stability and technological simplicity [E2].

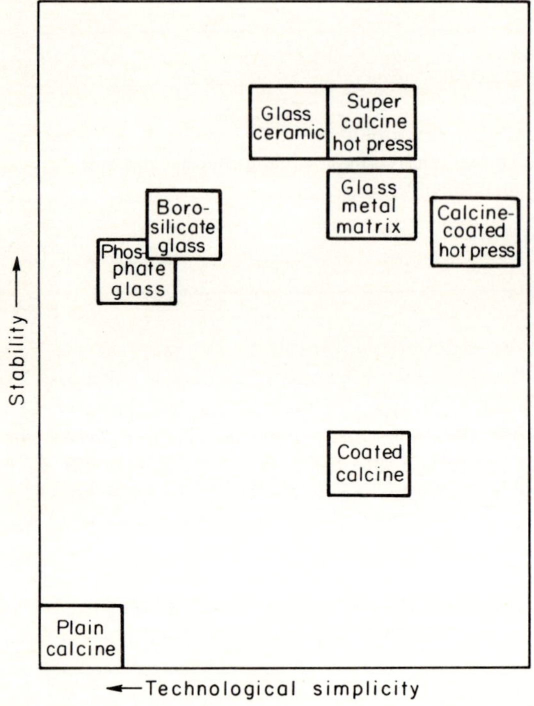

Figure 11.10 Rating of waste solidification products.

Policy considerations. Calcines as such do not seem suitable as final solidification products. They do not have any favorable properties with regard to mechanical and chemical stability, which are the categories of primary relevance for the release of radioactivity.

Coating and matrixing the calcine particles may overcome the stability drawbacks but will replace them by severe technological difficulties. Calcine may only be discussed as a nonfinal solidification product for interim storage provided that the safety concept does not require inherent stability of the product. In this case it may be an advantage that the calcine can be converted into a final product with some advanced technology possibly available at a later time.

Glass is among the solidification products with the highest mechanical, chemical, and irradiation stabilities presently known. As far as thermal stability of glass is concerned, the possibility of devitrification and the not very high softening temperature are disadvantages causing some uncertainty about its long-term performance. Glass solidification technology is in an advanced state of development and, in fact, is already available on a technical scale. Considering properties and the state of technology, glasses are presently the first choice for HLW solidification.

A ceramic product, thermodynamically more stable than a glass, will reduce the uncertainty about the state of solidified waste after some time of storage or disposal. Among alternative ceramic products, glass ceramic probably requires only slight modifications of the borosilicate glass process, whereas the supercalcine process will be quite different from ordinary calcination processes. It is therefore not unlikely that glass ceramics offer a chance to achieve with reasonable effort a ceramic product with at least the same chemical and mechanical stability as the parent glass. The technological penalty for any ceramic product with tailored crystalline host phases is a reduced flexibility toward the chemical composition of the waste.

The German/Eurochemic metal-matrix process PAMELA, originally developed for phosphate glass particles, is also suitable for borosilicate glass and for glass ceramic. It will provide a product of excellent impact stability and extremely high thermal conductivity. The latter may be useful if less aging time prior to solidification or cylinders of larger diameters are desired. Drawbacks are the lower overall melting point of the product due to the metal, uncertainties about the long-term chemical stability of the metal, and the relatively complex technology.

Compared to well-characterized solidification products carefully designed for final storage, unreprocessed spent fuel elements are a less well-defined waste form. Fuel elements are designed for operation in nuclear reactors rather than for final storage. Very little is known about their stability in final storage. However, they have at least two disadvantages: Spent fuel contains radioactive gases at rather high pressure, and it has been damaged by radiation to quite an extent. There is no doubt that spent fuel that is to be disposed of needs some processing before final storage, such as additional canning, to make it suitable for disposal.

2.6 Solidification Processes

A broad spectrum of processes to solidify HLW has been considered in various countries over the last two decades. The more urgent the need for an operational process has become, the more has this spectrum narrowed. Attention is now focused worldwide on a few types of vitrification processes with a strong preference on those for borosilicate glass, and on a fluidized-bed calcination process. The latter yields granules of calcine as the primary product. It must be consolidated for final disposal, preferably by mixing with molten glass. Other products under investigation are considered long-term developments rather than present technology.

Besides the alternatives concerning material and shape of the product discussed in the last paragraph, there are a few characteristic alternatives among the process parameters that may serve to classify solidification processes.

Glass melting. Glass melting may be performed either in a continuously fed melter with discontinuous discharge of the melt into a storage canister or directly in the storage canister (in-can melting). The continuous melter may be a joule-heated ceramic melter or a furnace-heated metallic melter.

In-can melting is the simplest choice as far as the melting device is concerned. No replacement or repair of a melter is necessary, and the potentially troublesome melt drain is avoided. On the other hand, the capacity of a canister is limited compared to a melter. Therefore parallel melting units are required with a complex technique to divert the feed from one canister to another.

Among continuous melters, the ceramic melter is favored over the metallic one, usually made of Inconel, because of its better corrosion resistance. The ceramic melter will have a longer life than the metallic one, and it may be the only practical device with sufficient corrosion resistance for processing high-temperature glasses if they are desired for the sake of improved long-term stability. On the other hand, remote replacement of a bulky ceramic melter is a more difficult task than that of a metallic crucible.

Joule heating is practically a requirement when a ceramic melter is employed. This means dissipating electrical energy in the molten glass between immersed electrodes. Joule heating has been shown to be feasible with sufficiently refractory electrode material such as molybdenum or even tin oxide. An auxiliary heating system has to be provided for initial start-up and for restarting.

A separate melter, particularly a ceramic one, is more flexible with respect to the feeding technique than a canister, mainly because of its greater surface. Furthermore, a continuous melter leaves the option either to fill a canister with the molten glass or to be coupled to a glass-shape forming device. This, for example, can produce beads to be embedded in a metal matrix.

Feed to the melter. The feed to the melter or to the canister may be liquid waste or a calcine with the glass frit either added to the waste or as a separate stream to the melt.

Liquid feeding saves the separate calcination step, which requires considerable engineering effort and which may be the source of a number of operational problems. Liquid feeding presents problems as well, such as capacity limitation due to the higher heat demand per unit weight of glass and a chance of unsteady boiling of the liquid fed onto the frozen but still very hot surface of the melt. However, liquid feeding seems feasible and, apparently, requires less sophisticated technology.

If calcination is to be performed solely to feed a melter, a fine powered calcine is desired, although not suitable as such for transport or interim storage. Granulated calcine can also be fed to a glass melter, but this will be considered a calcine consolidation treatment rather than a vitrification process.

To obtain a calcine powder, two techniques have been developed to the demonstration stage, the spray process and the rotary kiln process. Both have specific problems, such as the replacement frequency of the spray nozzle and the general reliability of a large rotating tube. Nevertheless, both have received intensive development and have proved to be feasible.

To obtain a granulated calcine a fluidized-bed process is available on a technical scale. A crucial point of this process is the treatment of dusty off-gas, which is created in large amounts by fluidizing the bed.

Denitration. Denitration may be performed thermally on calcination and/or melting. No separate denitration equipment is then required. The penalty is that the off-gas contains nitric oxides and that ruthenium volatility may be promoted by the oxidizing environment. However, there is still debate about the significance of the latter effect.

Denitration may also be achieved in solution by chemical means prior to feeding the calciner or the melter. The major part of nitric acid and nitrates will be destroyed, forming N_2/N_2O, NO, and NO_2 depending on the reducing agent and the conditions of the chemical reaction. The nitrates are converted to oxides or oxide hydrates, forming a suspension that can easily be transferred to the calciner or melter. The most common denitrating agents to be applied are formic acid, formaldehyde, and sucrose.

Off-gas purification. As a high-temperature process, any type of vitrification process will have to have a very effective off-gas cleaning system. In fact, besides the remote operation and maintenance technique, off-gas treatment will be among the most important waste-processing problems to be solved. The off-gas may contain volatile fission products, such as ruthenium and cesium, as well as aerosols and dust. Multistage systems will be required with wet and dry cleaning procedures to obtain an off-gas sufficiently clean for release to the atmosphere.

Specific vitrification processes. An installation that has to serve a 5 MT/day reprocessing plant will have to have a capacity of about 150 liters/h, corresponding to a specific HLW volume of 600 liters/t of heavy metal and 80 percent load factor. As yet, none of the vitrification processes has been operated on this scale and with full radioactivity. In fact, there is only one process that is now demonstrated on a technical scale with highly radioactive waste and about 25 percent of full capacity, the French AVM process. The others are in different stages of development and still awaiting the hot demonstration phase. Consequently, design and operation data available are preliminary, and the following discussions of individual vitrification processes will not go deeply into the details but rather emphasize the principles of the processes.

U.S. vitrification processes. In the United States [M2] development efforts are focused on two processes, a spray-calcine/in-can melting process and a ceramic melter process that may be coupled with a spray calciner, a fluidized-bed calciner, or liquid feeding (Figs. 11.11 and 11.12). No separate chemical denitration is provided in either case. Both processes are to produce borosilicate glass cylinders. Work is in progress at Battelle Pacific Northwest Laboratories in Hanford.

In the spray calciner, liquid waste is pumped to a nozzle at the top of the calciner where it is atomized by pressurized air, producing droplets with diameters less than 70 μm that are dried and calcined in-flight in the 700°C-wall-temperature spray chamber. Sintered stainless steel dust filters collect a portion of the powder with a mean diameter of 10 μm. They are periodically cleaned by a reverse pulse of air. Calcine from the spray chambers and filters drops directly into the melting canister. Frit is fed to the cone of the calciner.

Two problems typical of such a device have been largely eliminated. Spray-chamber fouling has been overcome by improved feed atomization and by use of vibrators mounted radially on

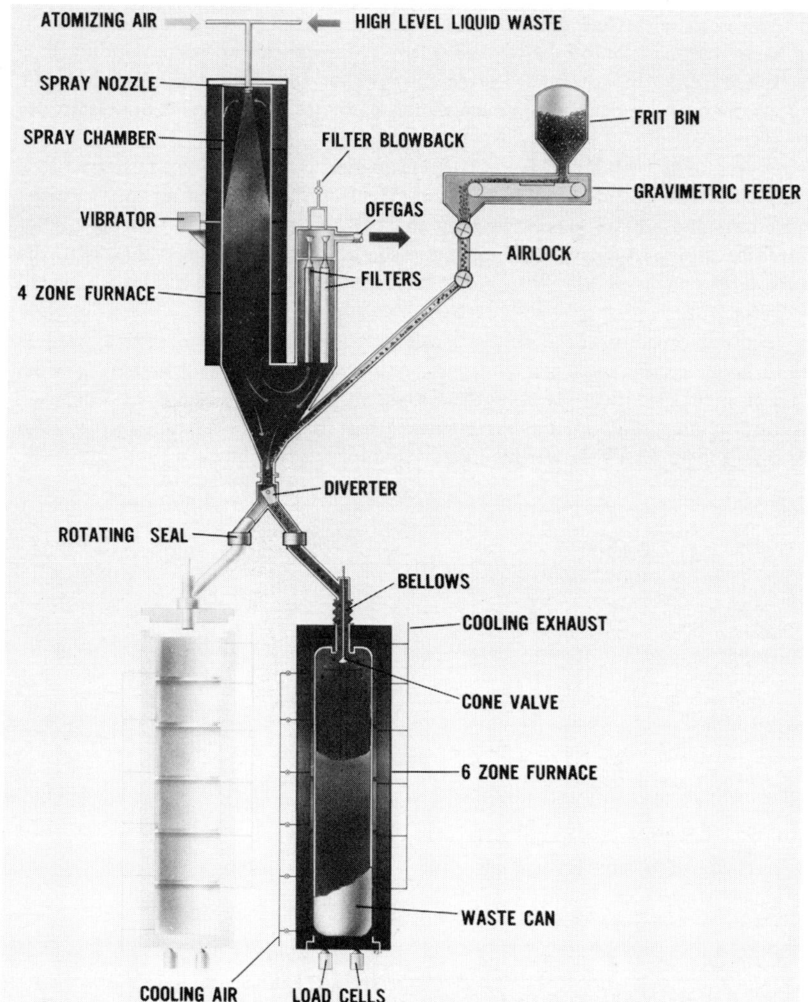

Figure 11.11 Spray calciner/in-can melter. (*Courtesy of Battelle Pacific Northwest Laboratories.*)

the spray chamber. The operating life of the nozzle has been significantly increased by using an alumina insert in the nozzle.

A spray calciner designed for a 5 MT/day reprocessing plant has an incoming waste flow rate of 118 liters/h with 75 g/liter solids to be combined with an off-gas system recycle of 8 liter/h to a total calciner feed of 126 liter/h with 70 g/liter solids. The product stream is 15.6 kg/h calcined solids. The demand on atomizing air is 0.85 m^3/min.

The canister is placed in a multizone furnace and coupled directly to the output of the calciner as shown in Fig. 11.11. The canister is heated to 1050°C and calcine and frit, which are metered continuously at a rate proportional to the calcine generation rate, are fed to the canister. Once the calcine-frit mixture in a zone has melted to a glass, heating is stopped and cooling is initiated to prevent excessive exposure of the canister to high temperatures. After the canister is full, feed is diverted to another canister in a parallel furnace.

The spray calciner is also used to feed a ceramic melter as shown in Fig. 11.12. In January 1975 an engineering-scale ceramic melter was started. The melting cavity is 0.36 m wide, 0.76 m long, and 0.30 m deep. The outside dimensions are 1.28 m wide, 1.36 m long, and 0.89 m high. This melter has a capacity of 60 kg/h glass when fed with calcine corresponding to a specific melting rate of 200 kg/m^2 surface. The ceramic melter was inspected after about 11 months of continuous operation and only minor corrosion of the refractories and the electrodes was detected.

The same melter was used with liquid feed. This means that liquid HLW is transferred to a mix tank where the frit is slurried into the waste liquid. The slurry is then fed directly into the melting cavity and covers the molten glass. Flooding the entire surface with 40 to 80 mm of the slurry is preferred because particulate entrainment in the off-gas stream is less than with the slurry falling directly on the melt surface. With liquid feed the capacity of the melter is reduced very roughly by a factor of 5.

To cover the required capacity range for full-scale operation with liquid feed as well as with calcined feed, a larger melter has been built. It has a melting cavity that is 0.86 m wide, 1.22 m long, and 0.71 m deep with a glass depth of 0.48 m. The overall size is 1.95 m wide, 2.13 m long, and 1.62 m high. This melter has a surface that is about four times larger than that of the smaller one. Figure 11.13 shows this melter equipped for liquid feed.

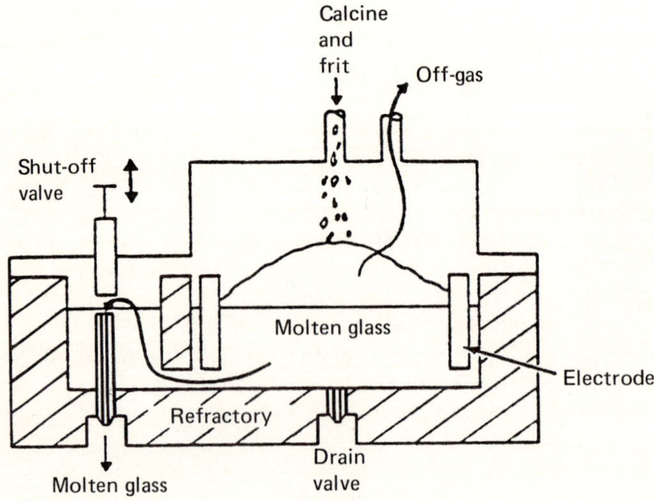

Figure 11.12 Joule-heated ceramic melter process. (*From McElroy et al. [M2].*)

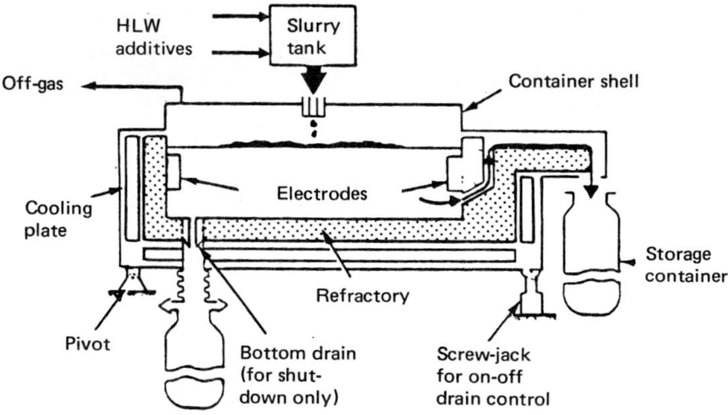

Figure 11.13 Direct liquid-fed ceramic melter. (*From McElroy et al. [M2].*)

For this new device, which has been operational since 1977, a new technique has been designed for periodic on-off drain control. The overflow is permanently open, but the melt flows only when the whole melter is tilted by a few degrees. When the canister is to be replaced, the melter is tilted back to a horizontal position, thereby interrupting the melt flow.

An alternative calcination process derived from the Idaho Waste Calcining Facility to be employed in connection with a vitrification unit has been developed to take advantage of the excellent heat transfer and solid mixing properties of fluidized beds. Silica is used as bed particles and is continuously fed into the bed at the rate needed in the final glass.

The German PAMELA process. In West Germany it has been decided to concentrate all development efforts on a modified PAMELA process, PAMELA II. The original PAMELA process, developed by Gelsenberg AG and Eurochemic [C3, G1], is a liquid-feed/ceramic melter process with chemical denitration, producing phosphate glass. From the glass, beads of about 5 mm diameter are formed and embedded in a lead matrix. The product is called vitromet. This process has been developed up to a semitechnical scale, cold as well as hot. The advantages of this process are considered to be relative simplicity of phosphate glass preparation, suitability of a particulate product for quality-control procedures, and favorable mechanical and thermal properties of vitromet as discussed before.

PAMELA II, to be built by DWK (Deutsche Gesellschaft für Wiederaufbereitung von Kernbrennstoffen) as a demonstration plant at the Eurochemic site near Mol (Belgium), differs from the original PAMELA with respect to the product. Phosphate glass will be replaced by borosilicate glass and the plant is to be operated on two product versions, glass blocks and vitromet. Construction of PAMELA II will profit a lot from two former process developments in West Germany: VERA, a spray-calcine/ceramic melter process carried out by Kernforschungszentrum Karlsruhe, and FIPS, a drum dryer/in-can melting process carried out by Kernforschungsanlage Jülich. The PAMELA II plant will vitrify the Eurochemic Purex waste (LEWC), with a specific activity of about 200 Ci/liter, and a specific heat rate of 0.7 W/liter.

According to the present design, the liquid waste is transferred from a process storage vessel by air lift or steam jet to the denitrator. The denitration is performed batchwise with one batch of 720 liter waste per day. Then formaldehyde solution (37 w/o) is metered into the waste. It will destroy the nitric acid and much of the nitrates. The effect of the denitration step has not yet proven in detail. A final decision as to whether the additional effort is justified is still pending.

The denitrated and concentrated waste is transferred to a mixing vessel where 140 g borosilicate glass frit per liter of waste is added. The slurry is fed on top of the molten glass in a ceramic melter. The ceramic melter will have a surface area of 0.8 × 0.8 m. The depth of the glass melt is 0.4 m. The continuous feed rate is 30 liters/h. There will be two alternatives for the melt drain, one to fill a storage canister and another to produce glass beads.

For the glass block production, the melt is drained periodically from the melter by means of a bottom drain. This bottom drain uses joule heating as well as medium-frequency heating. It can be frozen with air cooling. The glass is cast into storage canisters. For the production of beads continuous draining is needed. Because of the low flow rates desired (2 liters melt/h), an overflow drain will be more suitable than a bottom drain. The glass melt leaves the drain as droplets.

Beads for vitromet production are prepared by means of a slowly (0.5 to 3 r/min) rotating stainless steel disk with a diameter of 700 mm. The droplets hit the disk and solidify to beads of about 5 mm diameter and 0.08 cm^3 volume with a flat bottom. The beads are transferred to an intermediate storage vessel for product control and mass balance and then via a metering vessel to the final canister. The bead production is shown in Fig. 11.14. The canister is heated at 350 to 400°C and can be vibrated to achieve a close packing of the beads. When the canister is filled with beads, molten lead is introduced through a central pipe extending to the bottom. After some cooling the canister is sealed. Then it contains 67 v/o (volume percent) glass and 33 v/o lead.

The PAMELA II demonstration plant at Mol will have a capacity of about 30 liters HLW/h, corresponding to a scale-up factor of about 5 related to a 5 MT/day reprocessing plant. It is scheduled to be in operation in 1985.

The British HARVEST process. Another vitrification process is the HARVEST process [C3, M3], an improved version of the former FINGAL process. It is a pot process or, in the categories of this chapter, a liquid-feed/in-can melting process. A full-scale, fully radioactive plant is scheduled to be in operation at the Windscale site in 1986.

Concentrated radioactive waste solution, together with glass-forming chemicals such as silica and borax, are fed into a stainless steel vessel held at a temperature of 1050°C by a multizone, resistance-heated furnace. Evaporation, denitration, sintering, and glass formation occur steadily during the filling cycle and the feed rate is kept constant to ensure that the free liquid level rises at a rate equal to the rate at which glass is formed.

With the FINGAL process, the off-gases from the first, that is, the glass-making, vessel were passed through a second and a third vessel that contained filters to trap the particulate material and volatile ruthenium. At the end of the process cycle, when the first vessel was filled with glass, it was removed to storage and the vessel from the second position containing the primary filter was moved into the furnace and the filter was incorporated in the glass. A new vessel with a new filter was put in the middle position. The third vessel was only to provide a backup filter and did not require frequent replacement. Although this filter system was very efficient, it will not be used in the present HARVEST process. This is because of the filter size in a full-scale plant, because of problems that will arise when it fails blocking the entire off-gas system, and because of the necessity of handling additional pipe connections. The HARVEST off-gas system will rather follow the more conventional pattern of most other solidification processes.

The French AVM process. The French vitrification process at Marcoule is the first one in the world that is now effectively operating on a routine industrial basis after an exceptionally smooth period of test operation. With a team of 21 workers distributed among six shifts, AVM (Atelier de Vitrification de Marcoule) is operated continuously and produces one 150-liter glass block per day. It is used to solidify the backlog of military waste and future waste from natural uranium fuel produced at Marcoule.

AVM is a continuous rotary-kiln calciner/induction-heated melter process [B6, C3]. The

Figure 11.14 Glass-bead production device in the German PAMELA II process. (*Courtesy of DWK.*)

development of AVM was based on extensive experience with the pot process PIVER, which has been operated on a pilot-plant scale with full radioactivity for several years. Figure 11.15 shows the basic flow sheet of AVM.

The plant has two 15-m^3 tanks for receipt of the liquid to be solidified. The liquid is cooled and agitated to avoid any buildup of solid residues. It is blended with additives before being fed to the calciner, to prevent caking.

The calciner, which receives a feed of 40 liters/h, comprises a tube of wrought Uranus-65 which has been machine-finished. The ends of the tube are fitted with graphite-ring air seals. These end fittings lie on "fore-and-aft" movable trolleys. The tube lies on easily removable roller bearings, has a slight slope, rotates at 30 r/min, and is heated by four separate heaters arranged in zones.

The solution is fed through the upper end fitting and dried in the first half of the tube. The dry product, which is calcined in the second half of the tube at a temperature of 300 to 400°C, leaves by gravity through the lower end fitting and passes to the melting furnace, which is fed, through another connection, with small batches of glass frit.

The presence of a free rod inside the tube and the use of a chemical additive produces a more consistent calcined product and prevents material sticking to the inside surface of the tube. Great care was taken in the design of this component, in particular with respect to the quality of the output. It had been successfully tested in a 6000-h cold operation.

The melting furnace consists of a ceramic melting crucible heated to a temperature of around 1150°C by five induction heaters. The molten glass is allowed to build up in the furnace for a period of 8 h, and then a glass plug in the bottom of the furnace is melted through the use of two additional induction heaters and the glass is poured into the stainless steel canister. The canister is 50 cm in diameter and 1 m tall. It takes 3 h to fill with about 150 liters of glass.

In the original AVM design, an Inconel crucible was used for glass melting, which requires more frequent replacement than a ceramic one.

The off-gas system ensures that the bulk of active material escaping from the furnace is trapped in a countercurrent water-scrubbing column and recirculated directly to the calciner. Further off-gas treatment includes a condenser, an absorption column, and a washing column. The low-activity liquid from this section of the plant is recycled to the adjoining reprocessing plant.

The main cell of the vitrification plant is provided with a 2-t bridge crane, eight shielded windows, and 14 manipulator positions. Every component in the plant is designed for remote disconnection and removal to an adjoining maintenance cell.

On-site engineered storage in air-cooled underground vaults is provided for the glass canisters. The storage facility has a 10-year capacity related to the AVM output.

The AVM plant is designed to produce glass blocks with a heat rate of up to 400 W/liter. The basic design of AVM is considered appropriate for the construction of a further plant to serve the La Hague reprocessing center. The intention is to build a vitrification plant of about twice the Marcoule capacity—probably with two parallel lines—to produce glass with a heat rate of up to 100 W/liter from the waste of oxide fuel reprocessing in the present plant (UP 2). This should be available to start glassmaking by about 1983. Follow-on vitrification plants of about the same size will be built for the two new oxide-reprocessing plants planned at La Hague, UP-3A and UP-3B. As the first of them is being assigned to the reprocessing of fuel from foreign customers and with contracts that provide for the return of waste in solidified form, the availability of proven technology for vitrification has assumed special importance.

Fluidized-bed calcination. The fluidized-bed calcination process has been developed at the Idaho National Engineering Laboratory (INEL), where in 1962 the Waste Calcining Facility (WCF)

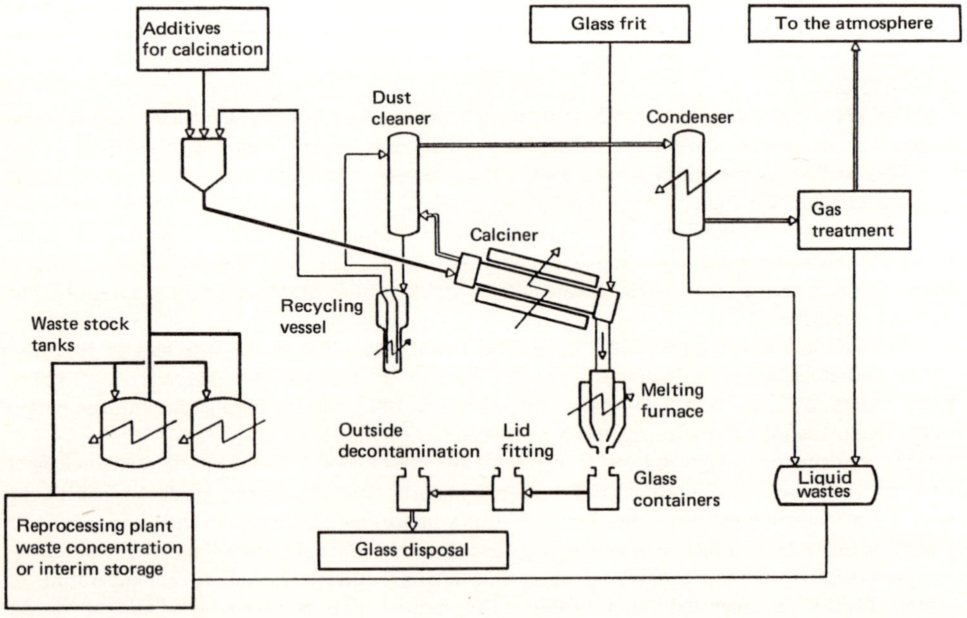

Figure 11.15 The continous process employed in the Marcoule vitrification plant (AVM). (*Courtesy of CEA.*)

started operation. Since then about 10^7 liters of liquid waste from the reprocessing of aluminum- and zirconium-alloy fuels have been calcined to produce about 1000 m^3 of granular solids.

In the fluidized-bed calcination process, exemplified by the WCF, pneumatically atomized waste solution is sprayed at a gross rate of 375 liters/h into a 1.22-m-diameter by 1.37-m-deep fluidized bed of solidified waste granules maintained at 400 to 500°C. A recycle stream of off-gas scrubbing solution representing 20 to 30 percent of the total feed rate is added to the calciner feed stream. Inlet fluidizing velocities, based on only the fluidizing air flowing through the empty cross-sectional area of the calciner vessel, of 0.18 to 0.36 m/s are generally used, and a freeboard of about 2.3 m, supplemented by a louvered baffle, is provided for deentrainment of solids from the off-gas within the calciner vessel. The bed height is maintained at a constant level above the feed-spray nozzles by adjusting the rate of withdrawal of the product. Process heat is provided by in-bed combustion of kerosene with oxygen.

During operation waste is blended with required feed additives and fed by air lift and gravity to the calciner. The feed is atomized by air through spray nozzles located on the wall of the calciner vessel. The primary solidification mechanism is the evaporation of atomized liquid droplets on the fluidized-bed particles. A portion of the atomized liquid also evaporates to a dry powder before striking the surface of a bed particle. Therefore, the calciner produces a mixture of powdery solids and granules in the size range 0.05 to 0.5 mm.

Calcination of the waste solution to granular solids is accompanied by the release of large amounts of water vapor and gaseous products. These vapors and gases, along with the air employed for fluidizing the bed, atomizing the feed, and purging connecting lines, sweep a portion of the bed material—mainly, fine particles—into the off-gas piping. The initial separation of these solids from the gas takes place in a cyclone, the collected solids being combined with the primary product from the fluidized bed and transported pneumatically to product-storage bins in an underground vault. The gas then flows to a wet-scrubbing system that includes a quench tower, a venturi scrubber, and a deentrainment cyclone. In the scrubbing system, condensing takes place, which provides a scrubbing-solution recycle flow back to the calciner feed tanks at a rate sufficient to keep the dissolved solids content of the scrubbing solution well below the saturation level. For a final treatment, the off-gas is passed through four silica-gel beds in parallel and then through three off-gas filters in parallel. The silica-gel beds were installed primarily to remove gaseous ruthenium compounds, the only fission-product compounds in the wastes, other than tritium, that volatilize at the calcination temperatures.

Four solids-storage facilities have been placed in operation. The first and second facilities have been filled, and the third is presently being filled. The bins are cooled by atmospheric air, which flows through prefilters, down an inlet duct to the bottom of the vault. Air then flows upward through the vault by natural convection and out of the vault through a 15-m cooling-air stack. A forced-air cooling system was installed in the first storage facility, but has not been needed. The cooling air can be shut off, and high-efficiency filters can be installed, should radioactivity be detected.

In 1976 pilot-scale testing with simulated commercial high- and medium-level waste feedstock composition was conducted to demonstrate the feasibility of the process for this type of waste. Currently expected commercial waste compositions do not seem to present major problems in fluidized-bed calcination [M2]. A conceptual flow sheet for fluidized-bed solidification of commercial waste is shown in Fig. 11.16.

The specific volume of calcine will be about 40 liters/MT of heavy metal for combined HLW and MLW, corresponding to that to be expected from the AGNS plant. For final disposal, the product from the fluidized-bed calcination will have to be consolidated by melting with a glass flux. If it is to be stored for extended periods directly in sealed canisters, the calcined solid will have to be stabilized (denitrated, dehydrated) at approximately 900°C.

Fluidized-bed calcination is the only solidification process where long-term operation experience is available. Thereby it is probably the most readily available solidification process.

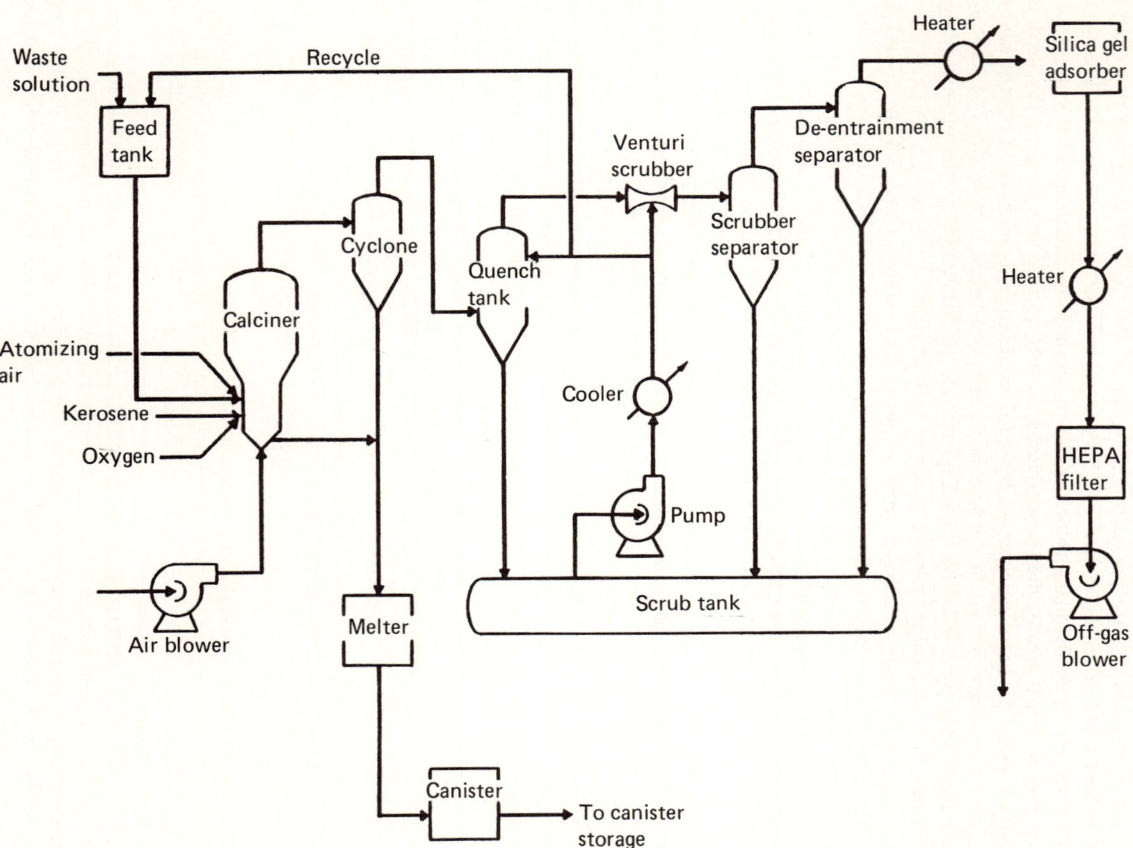

Figure 11.16 INEL fluidized-bed calciner flow sheet. (*From McElroy et al. [M2]*.)

The solidification processes–vitrification and calcination–whose principles have been described in the last two sections are summarized in Table 11.11.

2.7 Actinide Partitioning

It has been an attractive idea for some time to reduce the long-term potential hazard of the waste by chemical removal of the actinides and subsequent transmutation in a neutron flux. The overall incentive for actinide partitioning is not very great. The reduction of the ingestion hazard after recycling equilibrium has been reached will be only modest, and the technical effort will be enormous. The technology for actinide partitioning is not available as yet, and considerable development will be required to make it available. Moreover, it has to be considered that part of the actinides are transferred from the waste to the fuel cycle on recycling, where they may create an even greater hazard than in the waste.

The overall effect of actinide partitioning depends not only on the degree of chemical separation but also on the efficiency of transmutation. At present transmutation would have to be performed by recycling the separated actinides to LWRs, where it will be less effective than in a LMFBR. The reduction of the potential hazard achieved by actinide removal will decrease with repeated recycling of these actinides as a result of the buildup of the higher actinides and will eventually attain an equilibrium value. Figure 11.17 is a plot of equilibrium hazard index reduction factors in LWR uranium waste versus age of the waste for 99.5 and 99.9 percent chemical separation efficiency. Between 100 and 50,000 years, reduction factors are found of not more than 5 and 30, respectively [C2]. Any actinide separation higher than 99.9 percent makes it necessary to consider ^{99}Tc as well and seems out of reach, as presently nothing even close to 99.9 percent is technically feasible.

Figure 11.17 reflects the effect that actinide partitioning and transmutation has on the actinide hazard index of only the HLW itself. If the total quantity of actinides accumulated in the HLW and in the fuel cycle is considered, the same equilibrium reduction factor will eventually be attained provided that a constant nuclear power level is assumed, but it will take a very long time. In the fuel-cycle study performed for the American Physical Society [P2], an example with recycling the actinides to a LMFBR has been calculated that is shown in Fig. 11.18.

It should also be obvious that actinide reduction in HLW is reasonable only if an equivalent reduction of actinides in non-high-level waste, such as refabrication waste, can also be achieved. Also, ^{129}I must be considered in a long-term hazard balance.

Chemical separation. Current concepts for high-efficiency separation of actinides call for improved plutonium recovery, coextraction of uranium and neptunium with subsequent partitioning by valence control, and extraction of amercium and curium from the HAW stream. There are a number of major problems to be solved before a technically feasible process will be available.

Actinide losses to undissolved residues of fuel and to solids generated in the process have to be eliminated. To improve the recovery of plutonium, inextractable forms have to be identified and means have to be found to recover them.

For the recovery of americium and curium from the waste stream, cation-exchange and extraction processes appear most promising. The outstanding problem is a highly effective separation of actinides from lanthanides. The latter would be harmful upon transmutation in thermal reactors because of the high-neutron-capture cross sections of some of them. An actinide/lanthanide fraction would probably have to be separated first from the other fission products and waste components and then the actinides would have to be recovered with high purity. Also, by taking into account that substantial additional waste streams would have to be managed without significantly increasing the overall waste quantity, it is obvious that the recovery of americium and curium will be the most difficult task in waste partitioning [B5].

Table 11.11 Summary of principal high-level waste solidification processes being developed in the Western world

	Process characteristics				
Process	Product material	Product shape	Melter	Feed	Denitration
Spray calciner/in-can-melter (U.S.–Battelle PNL)	Borosilicate glass	Monolithic cylinder	Furnace-heated canister	Spray calcine	Thermal
Ceramic melter (U.S.–Battelle PNL)	Borosilicate glass	Monolithic cylinder	Joule-heated ceramic	Spray calcine or liquid	Thermal
Fluidized-bed calciner (U.S.–INEL, Idaho Falls)	Calcine	Granules (for disposal to be consolidated)			Thermal (incomplete)
Rotary-kiln/continuous melter (AVM) (France–Cogéma, Marcoule)	Borosilicate glass	Monolithic cylinder	Furnace-heated metallic or ceramic crucible	Rotary-kiln calcine	Thermal
Liquid-feed/ceramic melter (PAMELA II) (FRG–DWK/Eurochemic/KFK, Mol)	Borosilicate glass or glass ceramic	Monolithic cylinder	Joule-heated ceramic	Liquid	Chemical or thermal
Modification of PAMELA II (FRG–DWK/Eurochemic/KFK, Mol)	Borosilicate glass	Beads in lead matrix	Joule-heated ceramic	Liquid	Chemical or thermal
Pot vitrification (HARVEST) (UK–AERE, Harwell)	Borosilicate glass	Monolithic cylinder	Furnace-heated ceramic	Liquid	Thermal

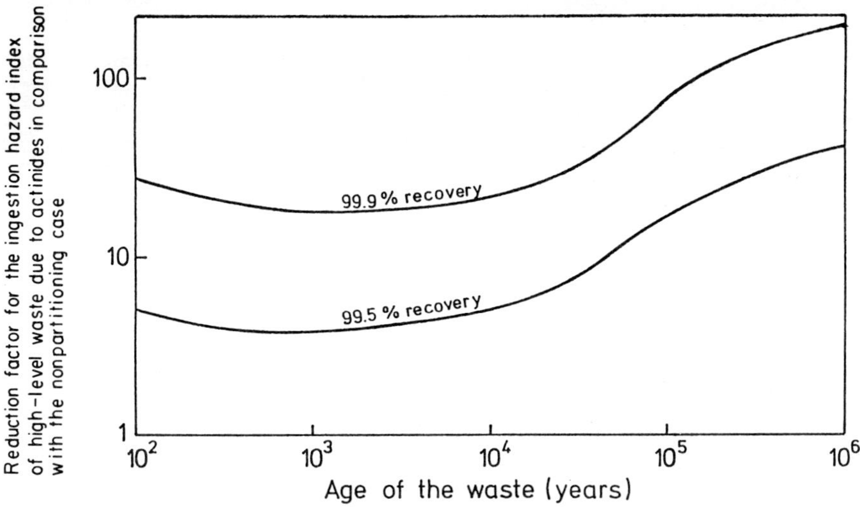

Figure 11.17 Reduction factors of the ingestion hazard index due to actinides of LWR uranium HLW by actinide partitioning and transmutation in LWRs after the twentieth cycle.

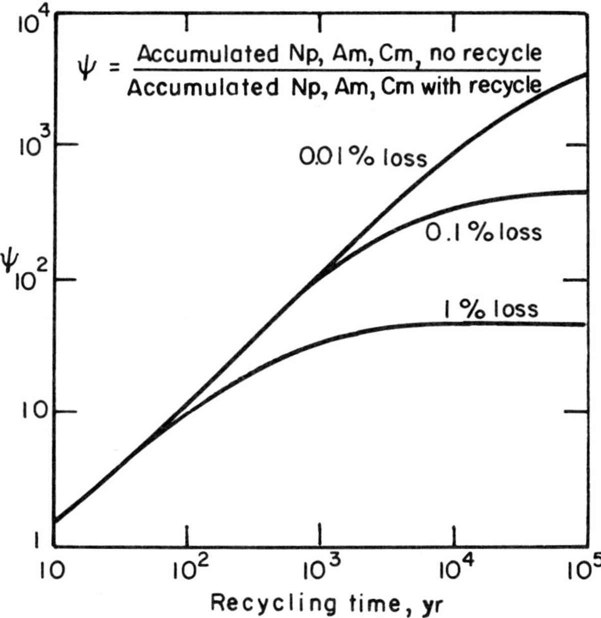

Figure 11.18 Ratio of actinide quantities accumulated in the fuel cycle (including radioactive wastes) with and without actinide partitioning and recycling as a function of time for different values of actinide loss to the waste stream. (*From Pigford and Choi [P2].*)

Transmutation. Recycling actinides to the LWRs will decrease the average material neutron multiplication factor by only 0.8 percent, provided that they are of high purity [C2]. Recycling to LMFBRs, however, will be preferred. There will be less neutron capture in impurities, such as lanthanides, and the average fission-to-capture ratio of the actinides should be higher in a fast spectrum than in a thermal one.

Recycling of actinide waste will increase radiation problems associated with processing of fuel. After a few cycles, for example, ^{252}Cf builds up to the strongest neutron source and reaches 10^{12} n/s per MT of heavy metal at 150 days after discharge.

Figure 11.19 is a schematic flow sheet for actinide recycling.

3 NON-HIGH-LEVEL WASTE

The term non-high-level waste includes low- and medium-level wastes (LLW and MLW) and covers a very large range of wastes [B1, C4]. Whereas the generation of HLW is determined by the quantity of fission products and transuranium elements inevitably generated in nuclear fission, that of non-high-level waste is rather dependent on specific process design and performance. It should be minimized in terms of volume as well as activity by appropriate process design. Recycling of waste streams in a reprocessing plant to reduce the volume of tritium waste is one example of how this can be done. A crucial point is the non-high-level alpha waste. The environmental benefit of the reprocessing fuel-cycle option depends largely on the minimization of this waste stream. Only if this minimization can be achieved will the long-term environmental impact of the fuel cycle be limited to a very small volume of solidified HLW with most of the plutonium eliminated by recycling.

The goal of non-high-level waste treatment is primarily volume reduction. This, however, does not hold for alpha-bearing waste, often called TRU waste.† Effective immobilization may

†For *transuranium.*

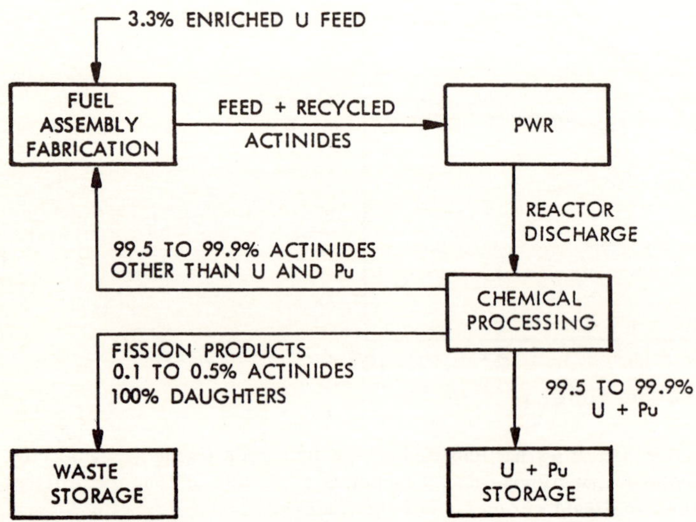

Figure 11.19 Flow sheet for actinide recycling. (*From Claiborne [C2].*)

generally be required for alpha waste. In contrast to HLW, recovery of actinides from alpha-bearing non-high-level waste may be beneficial. Because of the low fission-product concentration, actinide recovery will appreciably reduce the actual ingestion hazard of non-high-level waste. Moreover, it will be simpler from a technical point of view. As most of the actinide contamination in this waste will be plutonium, there may be even a certain economic compensation.

Non-high-level waste may be classified into three categories:

1. Process waste (aqueous solutions, slurries, ion-exchange resins, organic liquids)
2. General trash (combustible and noncombustible trash)
3. Discarded equipment (contaminated or activated items)

There are three basic steps in treating these wastes, which may be applied in sequence or individually:

1. Volume reduction
2. Actinide recovery
3. Immobilization and packaging

3.1 Volume Reduction

The methods available for volume reduction are different for liquid and solid waste. For liquid waste evaporation, ion exchange, and flocculation are used; for solid waste incineration, baling and surface decontamination are the most common processes.

Evaporation. Evaporation is a process whereby a solution or a slurry is concentrated by vaporizing the solvent, normally water. Then a residue with a high solids content, usually a sludge, will be formed that is handled as the radioactive waste concentrate.

Evaporators coupled to efficient deentrainment devices provide capability for a high degree of separation for most radioactive materials. The inherently high operating cost of evaporation limits its application to those liquids that have a high concentration of dissolved solids and require high decontamination factors.

An evaporator consists basically of a device to transfer heat to the solution and a device to separate the vapor and the liquid phases. The principal parameters involved in evaporator design are heat transfer, vapor-liquid separation, and energy utilization. Common problems in radioactive waste evaporators are foaming, severe scaling, and corrosion. To resist corrosion, evaporators are usually constructed of stainless steel and operated at as low a temperature as is practical. Scale has to be removed periodically, either mechanically or chemically. Foaming can be avoided by foam-breaking devices inside the evapoarator or by antifoam agents.

The basic types of evaporators are pot evaporators and circulation—either natural or forced—evaporators. Figure 11.20 shows a natural-circulation evaporator. To improve the economy of the process, vapor compression may be employed. Vapor-compression evaporators make the latent heat of condensation available at a higher temperature to use the energy potentials of vapors by compressing it and combining it with fresh steam input.

The wiped-film evaporator is a special type of evaporator that permits evaporation to a much higher concentration of solids than do other evaporators. Liquid is fed into a heated cylinder that contains rotating blades or wipers to reduce the liquid to a film, thereby improving the heat-transfer efficiency. Wiped-film evaporators can also be operated as dryers. Other equipment that can be used for drying and calcining non-high-level waste is the same as for HLW, e.g., spray calciners, fluidized-bed calciners, and rotary kilns.

Ion-exchange. Ion exchange is a process whereby ions from an aqueous solution are bound to a solid adsorbent. Either the ion-exchanger itself, loaded with radioactive ions, will then be handled as a waste concentrate, or it may be regenerated. In the latter case a liquid concentrate is obtained that has a volume greater than that of the ion exchanger but smaller than that of the original liquid waste. The decision as to which way will be more appropriate depends on the radioactivity concentration in the exhausted ion exchanger as well as on the price of the ion-exchanger material. In this respect, inexpensive inorganic ion-exchangers such as vermiculite are of some interest.

As the capacity of the ion-exchanger is equally exhausted by radioactive and inactive ions, this method is suitable only for waste solutions with a high radioactivity concentration relative to the total concentrations of dissolved solids.

According to the ionic nature of the radioactive contaminant, cation- or anion-exchangers have to be used, but usually the radioactive species in the waste are cations. The most efficient decontamination can be achieved by using a mixed-bed ion-exchanger as a final process stage. This is an intimate mixture of a cation-exchange resin in H^+ form and an anion-exchange resin in OH^- form in a 2:1 ratio. Its high decontamination effect is due to the favorable equilibrium of the reaction $2H^+ + OH^- \rightleftharpoons 2H_2O$. With mixed beds decontamination factors as high as 10^3 may be obtained. The product is fully demineralized water suitable as reactor coolant.

Flocculation. Flocculation is the least costly procedure to concentrate non-high-level waste. The principles are unspecific adsorption of radionuclides on a carrier, such as $Fe_2O_3(aq)$ or calcium phosphate, or cocrystallization with a suitable crystalline precipitate, such as strontium with $CaCO_3$. The sludge has to be collected by settling or filtering and is handled as the radioactive waste concentrate. This technique, because of its rather poor decontamination effect, is suitable only for LLW. Usually, the concentrate has a high water content.

Volume reduction of solid waste. Concentration of burnable solid waste can be very effectively achieved by incineration. The ashes are handled as radioactive concentrate. This is a rather costly technique because of much effort spent for off-gas filtration and safe handling of the ashes. Figure 11.21 shows an example flow sheet of an incinerator.

A much simpler though less effective technique is baling of the waste under high pressure.

If bulky equipment, which is radioactive only because of surface contamination, is to be

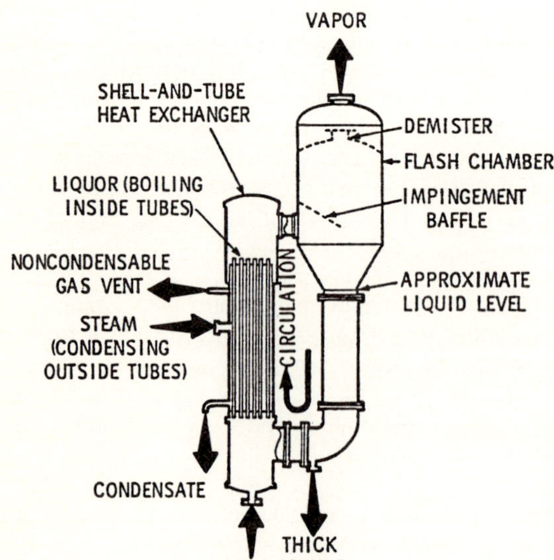

Figure 11.20 Natural circulation evaporator with external vertical-tube heat exchanger. (*From Cooley and Clark [C4].*)

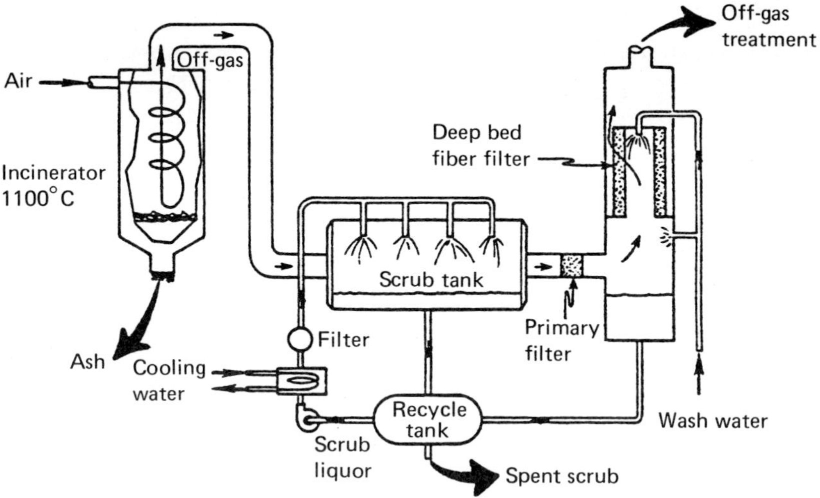

Figure 11.21 Excess-air (cyclone) incinerator (Mound Laboratory). (*From Richardson [R1].*)

discarded, the actual radioactive waste volume can be significantly reduced by complete decontamination. The techniques available include rinsing with acids or other suitable solvents, ultrasonic treatment, and sandblasting.

3.2 Recovery of Transuranium Elements

To begin with, it is necessary to measure the transuranic content of large volumes of rather heterogeneous wastes. Basically there are three ways to do so, namely, by making use of gamma- and x-ray spectra accompanying the alpha decay, of radiation produced by spontaneous fission, and of radiation produced by induced fission.

Recovery of transuranium elements is mainly of interest for solid waste from the refabrication of mixed-oxide fuel. Plutonium is the major element to be recovered, and ^{241}Am may be recovered as a by-product. Other transuranium elements are usually present in minor quantities. The treated wastes are seldom decontaminated to levels of plutonium that would permit unrestricted release.

The most rigorous recovery technique is burning of plutonium-impregnated material in a plutonium scrap-recovery incinerator followed by grinding and leaching the ash with a mixture of hot nitric and hydrofluoric acids. Undissolved plutonium in the ashes may be recovered by fusion with a suitable salt, such as a 10:1 $K_4(SO_4)_2$-NaF melt, to get a product soluble in nitric acid. Nonburnable solids can be leached directly with a HNO_3-HF mixture.

Once the plutonium is in solution, it can be recovered and purified for recycle by well-established solvent extraction and ion-exchange techniques. Aluminum nitrate is added to the feed to complex the fluoride and thus decrease its interference with plutonium recovery.

Figure 11.22 presents a scheme of typical plutonium recovery operations. The Plutonium Reclamation Facility (PRF) [R1] at Hanford incorporates many of these options. Geometrically favorable process equipment and storage tanks are used to ensure criticality safety.

The PRF also recovers ^{241}Am from the raffinate of the TBP-solvent extraction plutonium-purification process. The process employs 30 v/o dibutyl butylphosphonate in CCl_4 as the solvent to extract both americium and residual plutonium from the high-nitrate feed solution, adjusted to about pH 1 by the addition of NaOH. Americium is selectively stripped from the solvent and purified by a cation-exchange procedure.

3.3 Immobilization

Volume reduction as described above usually leads to a product that still contains considerable quantities of water or that is quite easily leached or dissolved by water. The policy as to the degree of immobilization required for final disposal varies in different countries. As yet, there is no official regulation in the United States requiring that non-high-level waste be immobilized before disposal. It is, however, practiced in many places. In West Germany, by regulation, any non-high-level waste has to be immobilized before disposal in such a way that low leachability is warranted over a sufficient period of time.

There is no doubt that immobilization at least of alpha-bearing waste must generally be required and will be in the future. It has been mentioned before that the total transuranic inventory of alpha-bearing non-high-level waste will be in the same order of magnitude as that of HLW.

The range of suitable immobilization products for non-high-level waste is broader than that for HLW because there will be no significant heat generation. It includes glasses as well as cement, bitumen, and polymers.

Hydraulic cement. Immobilization of radioactive waste by incorporation into hydraulic cements, as typified by portland cement, has been practiced for many years. The optimum proportions of cement and waste vary with the type of waste to be immobilized.

Several additives have been used to improve the setting properties, fission-product

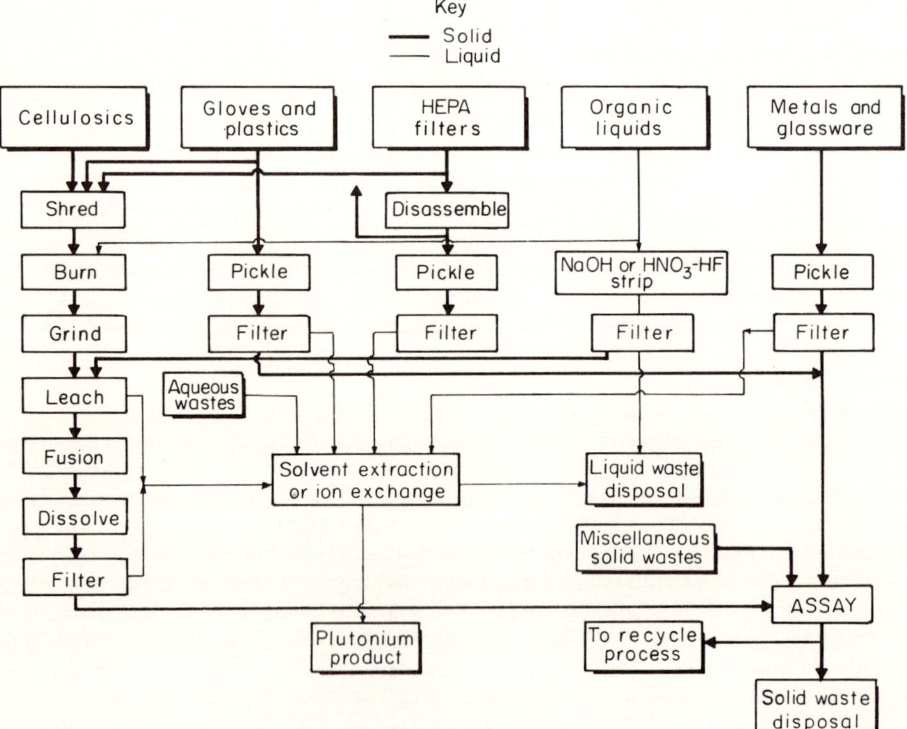

Figure 11.22 Typical plutonium-recovery operations (Hanford Engineering Development Laboratory). (*From Richardson [R1].*)

retention, or the volumetric efficiency of cement. A useful mixture is the portland cement sodium silicate system developed by United Nuclear Industries, Inc. The sodium silicate additive produces a quick set with no free water, readily solidifies pressurized-water-reactor boric acid solutions, which set poorly with cement alone, and provides a significant reduction in the solidified volume [H3].

Another way to improve cement products is polymer impregnation. The process being developed in Italy consists of preparation of the cement product, thermal dehydration of the cement (165°C), impregnation with a catalyzed organic monomer, such as styrene, and polymerization by heating at 75 to 85°C [D3].

In spite of experience, solidification with cement is still an art. Each new waste application must be considered individually because of possible interactions between cement and the waste constituents.

Bitumen. Bituminization systems for immobilizing liquid and solid wastes are used in several countries. Bitumen, or asphalt, has certain advantages for immobilizing LLW and MLW. It is highly leach-resistant, it has good coating properties, and it possesses a certain degree of plasticity. Perhaps the greatest advantage is that at the operating temperature of 150 to 250°C, 99 percent of the water evaporates, resulting in a volume reduction of up to fivefold compared with conventional cementing techniques for products made with evaporator concentrates. Typical bitumen products contain 40 to 60 w/o waste solids.

One of the major drawbacks of bitumen is its potential fire hazard, particularly if used to encapsulate oxidants such as nitrates. The combustion problem is minimized by using bitumen grades with high flash points ($\geqslant$290°C). Improved safety can also be obtained by substituting more expensive polyethylene for bitumen. Fires have occurred in bituminization facilities, but they have been readily controlled.

Another problem to be observed is the radiation resistance of bitumen. There may be some radiolysis resulting in the release of hydrogen, methane, carbon dioxide, and ethylene. In the order of 0.5 cm^3 H_2/g product has been found to be generated per 10^8 rad. This is of significance primarily for alpha-bearing products.

The bituminization process is performed basically by adding a concentrated waste slurry or even a predried waste mixture to the molten bitumen. Residual water is evaporated and the solids are evenly distributed in the bitumen. After solidification a homogeneous product is obtained. Figure 11.23 shows a flow sheet of a screw extruder plant for bituminization; Figure 11.24 is a photograph of the screw extruder evaporator.

Glass. For liquid non-high-level alpha-bearing wastes with sufficiently high activity concentration, glass may be a suitable fixation product as it is for high-level waste.

In terms of radiation stability, glass is superior to cement and particularly to bitumen. The leach rates, however, are about the same for glass and for bitumen, both being smaller than that of cement by up to three orders of magnitude depending on the type of cement. The fire hazard is a disadvantage of bitumen compared with both glass and cement. The costs of immobilization decrease in the order glass, bitumen, cement.

4 SPECIAL RADIOACTIVE WASTE

In terms of special radioactive waste three radionuclides will be discussed, which are collected separately in the reprocessing plant: tritium, ^{129}I, and ^{85}Kr. ^{14}C, as mentioned before, is presently not considered waste in the sense that attempts are made to develop techniques for recovery and final disposal.

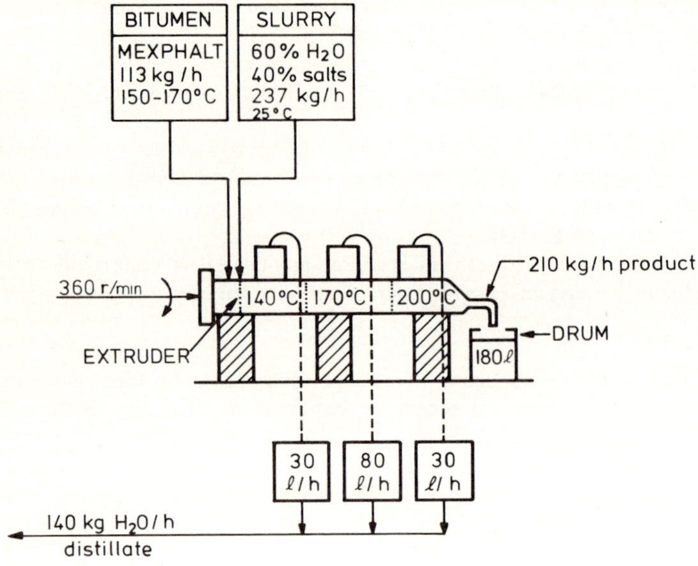

Figure 11.23 Eurochemic MLW bituminization flow sheet. (*Courtesy of Eurochemic.*)

Figure 11.24 Screw extruder evaporator used for bituminization of MLW solutions at Eurochemic in Mol. (*Courtesy of Eurochemic.*)

4.1 Tritium

Spent fuel elements contain appreciable amounts of tritium, partly produced by fission, partly by other nuclear reactions. About half of the tritium is released from the fuel upon dissolution. The rest is bound to the zircaloy of the hulls and is disposed of with them. The fraction of tritium that is released exchanges with water, forming HTO. The total annual input of tritium in a 1400 MT/year reprocessing plant is about 10^6 Ci. In West Germany a reprocessing plant of this size is supposed to retain 75 to 80 percent.

The fundamental problem of tritium waste management is that there is no simple way to reduce the volume of tritiated water. There are techniques available to minimize the volume generated in reprocessing, e.g., reuse of tritiated water to feed steam jets, and strict confinement of tritium in the first extraction cycle. These techniques, however, add complications to the process. If, therefore, an inexpensive way were available to dispose of untreated tritiated water, severe generation restrictions would not be appropriate. If, however, expensive methods were to be applied, such as solidification or even concentration by isotopic enrichment, the volume generated has to be limited as much as possible.

Another approach is a suitable head-end process in the reprocessing plant, such as voloxidation (Chap. 10, Sec. 4.3). However, such a head-end process is not yet available technology but requires several more years of development.

There are minor quantities of tritium smeared out over the whole reprocessing flow scheme that will ultimately arise as low-activity condensate with tritium concentrations of the order of 10^{-4} Ci/liter and 10^{-8} Ci/liter of other radionuclides. It is very likely that this can be released to surface waters.

Basically three options are considered to dispose of tritiated water that is stored in tanks and cannot be released.

Deep well disposal. Injection of tritium-containing liquid into isolated aquifers or depleted oil horizons is the most interesting option. This technique has been used increasingly for almost 20 years to dispose of industrial wastes. In the United States, for instance, some hundred injection wells have been drilled and are actually in operation at depths between 60 m and 3600 m. Although there are still licensing problems, this is a safe and economic way to dispose of tritiated water.

This technique will be tested for tritiated water in the neighborhood of the Karlsruhe Nuclear Research Center in West Germany. An isolated oil lens that is exhausted but located in an oil field still being exploited will be used. Thereby any migrations occurring deep underground will be detected.

Solidification. In principle, any solid that contains firmly bound water may be suitable as a solidification form for HTO. This includes drying agents, such as silica gel, molecular sieves, and calcium sulfate, as well as hydraulic cement and organic polymers. Most experience is available with cement, which has been used to solidify non-high-level waste for quite a while.

Although concrete is a monolithic solid, it is quite porous. In contact with water, about a third of the tritium will be released, mainly by isotopic exchange, in the first month. The release may be retarded by coating the cement. Because of the relatively high leachability, cemented HTO would have to be stored in gastight steel cylinders, probably in a nonaccessible geologic repository.

If it turns out that a more leach-resistant and probably more expensive solidification product has to be developed, it may well become beneficial not only to restrict the volume arising from reprocessing but also to further reduce it by isotopic enrichment prior to solidification. An enrichment process suitable for this purpose must provide a very effectively depleted waste stream.

Ocean disposal. In view of the relatively short half-life of tritium and of the enormous isotopic dilution, sea disposal is another alternative for dealing with tritium waste. Transport will be an economic drawback of this alternative, and political and administrative problems will have to be solved.

4.2 ^{129}I

All iodine isotopes except ^{129}I will have decayed prior to reprocessing as long as a large backlog of unreprocessed spent fuel exists. The ^{129}I activity per metric ton of heavy metal (30,000 MWd/MT) is only 34 mCi. However, its extremely long half-life of 17 million years makes ^{129}I a permanent contaminant if released to the atmosphere. In shorter-cooled fuel elements radioactive ^{131}I will also be present and must be recovered.

Practically all iodine from spent fuel will be released upon dissolution with the dissolver off-gas. There are several scrubbing techniques that remove iodine effectively from the off-gas but do not yield a stable product for long-term disposal.

For permanent fixation of ^{129}I adsorption on silver-loaded adsorbents, such as zeolites, silica, or alumina, will be the choice [P1, W2]. The process is simple, the bed temperature may be relatively high, the product is a dry solid, the chemisorbed iodine is highly insoluble, and the adsorbent is very efficient in removing both organic and inorganic iodine from gas streams.

The ^{129}I content of spent fuel with an average burnup of 30,000 MWd/MT heavy metal is 211 g/MT corresponding to 34 mCi. This corresponds to an annual production from a 1400 MT/year reprocessing plant of 300 kg ^{129}I. As there will be some isotopic dilution, an iodine-recovery system could conceivably be required to remove 600 kg of iodine annually. If iodine will be adsorbed on silver zeolite beds ready for final disposal, the total amount of iodine waste is then estimated to be about 5 m^3/year with a total activity of 50 Ci. The amount of silver corresponding to 600 kg iodine is about 500 kg. Even though there will be excess silver required, this does not seem an unreasonable silver consumption in view of the overall reprocessing costs. The world's silver production was almost 10^4 tons in 1976. There is, however, some research in progress on regeneration of iodine-loaded silver zeolite and reloading the iodine on a lead zeolite.

4.3 ^{85}Kr

^{85}Kr, a 10-year half-life krypton isotope, is currently released from reprocessing plants to the atmosphere. There will probably be no urgent need in terms of radiation dose to the local population to retain ^{85}Kr. However, in view of a worldwide accumulation of ^{85}Kr in the atmosphere, krypton recovery from reprocessing plants is required or will be required in the near future.

The krypton disposal problem is characterized by the fact that there is no easy way of converting it into a nongaseous form stable at ambient temperature. There are interesting experiments in progress to fix krypton in zeolites by adsorption under high pressure. In England a pilot plant for krypton implantation in metals is under construction. Nevertheless, the containment technique presently envisaged for technical use is pressurization in steel bottles.

There are a number of problems in developing efficient krypton-removal processes. The great portion of xenon present in the noble gas fraction tends to solidify at the krypton condensation temperature and to block the equipment. Small fractions of krypton adsorbed in the pretreatment steps may be lost from the main krypton streams. A mechanical problem is presented by the need to exchange steel bottles for krypton collection without significant leakage.

The annual amount of krypton from a 1400 MT/year reprocessing plant is about 500 kg with a ^{85}Kr activity of about 2×10^7 Ci. This corresponds to 50 standard bottles at 175 atm pressure with a surface dose rate of 400 rem/h. The temperature may be as high as 150°C. The krypton bottles are to be stored in an engineered facility with dry cooling.

There is some consideration of ultimately disposing of these bottles into the sea. This may be well justified because of the relatively short half-life, the low radiotoxicity, and the chemical inertness of ^{85}Kr. It may even reduce the ^{85}Kr hazard in comparison with surface storage of high-pressure bottles. At present, however, the London Convention on sea disposal of radioactive waste permits only disposal of solid waste.

5 DISPOSAL OF RADIOACTIVE WASTE

The final disposal technique depends on the type of waste. For the extremely long-lived HLW a concept has been accepted worldwide, namely, storage in a stable geologic formation deep underground, eventually nonretrievable. For non-high-level waste the storage philosophy is less uniform. Alpha waste is almost unanimously regarded a potential hazard similar to HLW and will probably be disposed of in a similar way. Other non-high-level waste will probably be handled differently. In less densely populated countries, including the United States, shallow burial is considered adequate for non-alpha, non-high-level waste. Heavily populated countries such as West Germany have rather decided to put any solidified radioactive waste eventually into a deep underground facility. However, as the design of a geologic waste repository will largely be letermined by the heat-generation rate of the waste, it will be simpler and cheaper to build a safe repository for non-high-level waste.

The United States, West Germany, and Canada have the most active programs in the field of geologic disposal. The U.S. program focuses on two pilot plants to be operational in 1986. The Canadian program has a similar time schedule. A German pilot plant in the abandoned salt mine Asse (Fig. 11.25) has been operated with LLW and MLW for about 10 years. Asse will also be used as an experimental facility for HLW. A preliminary site decision has already been made in West Germany for a full-scale repository. In other countries, e.g., the United Kingdom, France, Italy, the Netherlands, Belgium, and Spain, research programs on geologic disposal are also in progress.

As a means for intermediate storage of solidified HLW, engineered surface facilities are studied. These facilities shall be designed to contain and control liquid, solid, and gaseous waste resulting from normal and abnormal operations of the facility and from exposure to natural phenomena.

5.1 Basic Considerations on Geologic Disposal

Geologic disposal has two principal objectives:

1. Isolation of the waste from people to make incidental or intentional access to the waste highly improbable
2. Isolation of the waste from circulating groundwater, which is the most likely if not the only possible vehicle to carry radioactivity to people

This is to be achieved without maintenance or surveilance of the disposal site in the long run.

There are several types of geologic formations that in principal will be suitable to meet these objectives, e.g., argillaceous formations, crystalline rocks, and in the first place, rock salt as bedded salt or as a salt dome. To select a specific site for the repository, local conditions must be carefully evaluated.

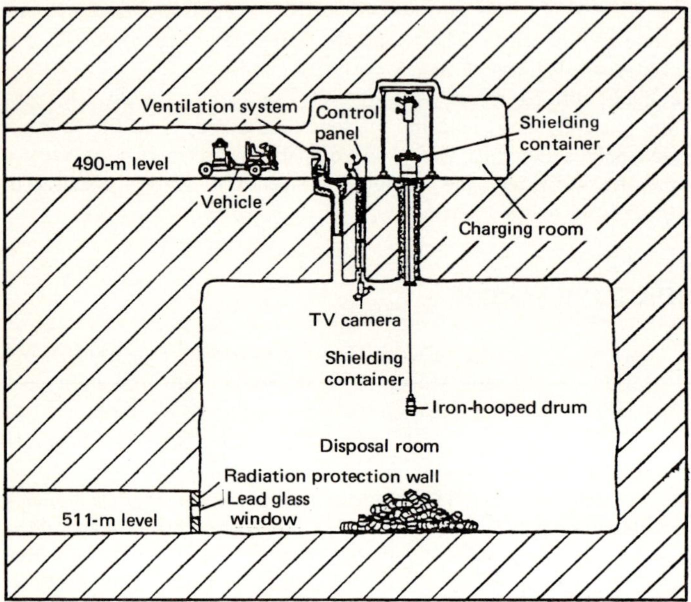

Figure 11.25 Scheme of the Asse disposal facility for intermediate-level wastes. (*From Kühn and Hamstra [K2].*)

Properties of the host rocks. The prime requirement is low porosity of the rock hosting the solidified waste. As long as the geologic formation retains its integrity, low porosity will be a reliable barrier against circulating groundwater. There is no rock with zero permeability for water, but rock salt meets this requirement to such a degree that it may be considered virtually impermeable. Salt, however, is soluble in water, so that the salt formation must be sufficiently isolated from circulating groundwater.

Plasticity of the host rock is another desirable property. It ensures that fractures that may develop and may provide access for groundwater will be self-sealing. Plasticity also permits a gradual closure of excavated openings. In rock salt, with the compaction, reconsolidation, and recrystallization of the crushed-salt backfill in the galleries and rooms, the waste will eventually be contained in a massive and solid salt formation.

Plasticity, on the other hand, complicates the design of the disposal mine. With salt, excavated openings while still in use will tend to close again by natural convergence, which is accelerated when the salt is heated by the waste. This will be especially a problem when retrievability of the waste for an extended period of time is desired.

Thermally, rapid dissipation of decay heat is most desirable to keep temperature peaks in the neighborhood of waste cylinders low and to reach as soon as possible a temperature equilibrium throughout the geologic formation. Table 11.12 summarizes heat conductivities of various rocks. Among them, rock salt is again a very favorable type of rock.

As for the physical-chemical properties, radiation stability and thermal stability are most important. Irradiation does not appear to pose great problems. Elevated temperatures may cause dehydration processes in the host rock, resulting in large quantities of free water. Rock salt formations are frequently interspersed with carnallite ($KCl \cdot MgCl_2 \cdot 6H_2O$), which will probably be dehydrated when heated above 110°C and may even develop hydrochloric acid above 165°C. In addition it is of low strength and easily soluble in water.

Table 11.12 Heat conductivities of rocks

	Heat conductivity, W/(m·°C)		
Type of rock	20°C		200°C
Argillaceous rock	2–2.7		1.3–1.6
Sandstone	2–2.7		1.3–1.5
Granite		2.85	
Rock salt	5.7		3.5
Anhydrite	9.3–11		3–3.5

Argillaceous rock may also dehydrate. The clay formation under investigation in Italy contains 20 to 25 percent water and will lose most of it at 110°C. Moreover, these formations will alter their mechanical properties by dehydration as well as by additional moisture absorption, which may even be caused by mine ventilation. An advantage of clay is its high ion-exchange capacity. This property can also be employed to add safety to the waste repository when a formation is chosen where the layers overlying the actual host rock are argillaceous.

Properties of the geologic formation. The area hosting the formation to be used for final storage of HLW must have a very low seismicity and correspondingly a high tectonic stability.

A most important criterion for the choice of a geologic repository is the specific pattern of groundwater occurrence. This includes the directions and velocities of groundwater flow, the distances between the waste-emplacement zone and water-bearing layers, and the specific conditions of these layers. Naturally, slow water flow is desired. The distance between waste zone and water zone may be smaller if the host rock has a high plasticity, but a minimum separation of a few hundred meters is required anyway. Intervening bodies of impermeable shales as well as overlying and underlying impermeable layers are good additional protection. In the United States these are frequently found in bedded salt formations. The water-bearing soil layers should have a high adsorption capability for ions, as this phenomenon will drastically

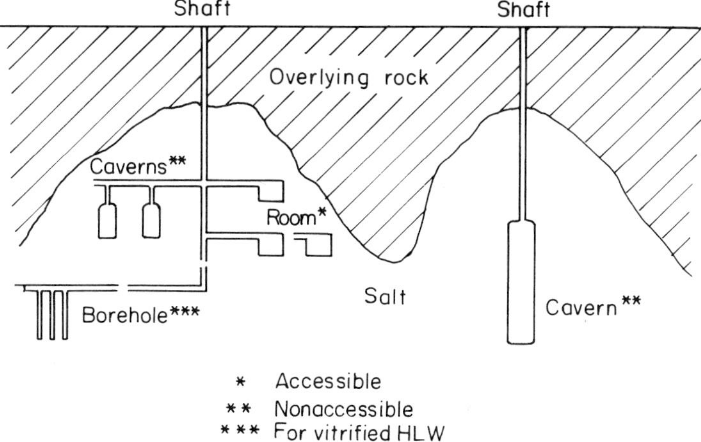

Figure 11.26 Schematic presentation of possible disposal techniques in salt formation. (*Courtesy of Nukem.*)

reduce the velocity of the transport of radioactive ions compared to that of groundwater in which they may be dissolved.

The isolation zone should be sufficiently deep underground to avoid surface phenomena such as erosion and biologic processes. A depth of about 1500 m, however, will be an upper limit for plastic rocks such as salt or shale. With respect to operating conditions the geothermal gradient will pose another limit on the depth of the waste-emplacement zone. For the German waste repository a depth between 500 and 1000 m is envisaged. Vertical and lateral dimensions of the entire host rock formation should be sufficient to allow for good heat dissipation.

5.2 Design of Repositories in Salt Formations

Figure 11.26 illustrates the three basic options for repository design:

1. Rooms, mined, accessible to store waste containers
2. Caverns, mined or leached, not accessible, to be charged through a shaft either from the surface or from a lower level
3. Galleries with storage boreholes in the floor, mined, accessible to charge the boreholes

The choice among these options depends on the type of waste and the filling technique appropriate for the type.

Accessible storage rooms are simple but useful only for waste with a low surface dose rate. Otherwise the waste would have to be stored with lost shielding, which is usually uneconomic.

Caverns may be used for wastes with higher surface dose rates because the waste container can be dropped from a shielded cask into the cavern. The heat generation, however, must be very moderate, because dropping the containers into the cavern leads to a random array not optimized in terms of heat dissipation. In West Germany an in situ solidification process for non-high-level waste is being investigated where granulated waste mixed with a binder is to be pumped into a cavern.

Single boreholes in the floors of galleries are provided to hold high-level glass cylinders. The cylinders are carried in heavily shielded casks and are then lowered into the boreholes. Single boreholes can be arranged in a way that the heat is sufficiently dissipated to maintain maximum permissible peak temperatures in the salt.

Rock mechanical stability. The main potential hazard to the integrity of an underground repository has its roots in rock mechanical failures. The stability of the repository depends on many factors, such as the volume of the rooms relative to the pillars. Convergence of rooms due to the plasticity of the salt and enhanced by the elevated temperature may cause stresses within the rock salt. It is therefore important that a repository at least for HLW should be built in a salt formation not mined before. Moreover, only the space required for a minimum number of years should be mined at the same time, and every room used up should be backfilled with crushed salt. On the other hand, convergence will help to eliminate open space in the rock salt quickly after rooms have been backfilled and will thereby be beneficial.

Another factor affecting rock mechanics is the temperature. The rock has attained a quasi-equilibrium state corresponding to the geothermal temperature gradient over millions of years. Only formations having this tectonic stability are eligible as waste repositories. Inserting high-level waste will inevitably disturb this equilibrium by raising the temperature in the salt and by creating new gradients. The natural temperature at depths of 1000 m is in the neighborhood of 40 to 45°C with gradient of a little more than 2°C per 100 m.

According to suggestions made in an Oak Ridge study [C1], the following temperature criteria are to be met:

Waste temperature should not exceed the temperature of the solidification process.
No more than 1 percent of the salt shall be at a temperature above 250°C.
No more than 25 percent of the salt shall be at a temperature above 200°C.
If carnallite interspersion is expected, the maximum temperature shall be limited to 100°C.

In West Germany 200°C is presently envisaged as an upper limit of the waste canister surface temperature.

In general, the temperature increase caused by the waste should be kept low to ensure that the quasi-equilibrium is disturbed as little as possible. It may turn out as a result of further thermomechanical analyses that it is desirable to age the solidified HLW for quite a while in engineered storage before it is put into a geologic repository.

Thermal analysis. The temperature distribution in space and time is given by the following differential equation:

$$c\rho \frac{\partial T}{\partial t} = \text{div}\,(\kappa \text{ grad } T) + q' \tag{11.7}$$

where c = specific heat
ρ = density
κ = heat conductivity
q' = heat-production rate of the source per unit volume

c, ρ, and κ are functions of space and temperature. Equation (11.7) has been solved numerically [C1].

A parametric analysis has been conducted with room width, waste package array (pitch), waste characteristics, and diameter of HLW container as parameters.

Optimization leads to a set of parameters indicated in the schematic cross sections of the repository presented in Fig. 11.27. The diameter of the glass cylinders is 6 in (15.24 cm). These parameters will permit storage of the 20-year HLW production of a 1400-MT/year reprocessing plant for 20 years in an area of about 0.5 km^2.

Figure 11.28 illustrates the temperature distribution throughout the repository. The maximum temperature rise at the hottest spot of the salt, according to this calculation, will be about 175°C and will be reached after about 50 years.

5.3 Other Disposal Techniques

Apart from the more exotic approaches to waste disposal that have been mentioned before, shallow burial and sea disposal are widely used. Disposing of liquids into isolated aquifers or exhausted oil lenses has been mentioned as a special technique for tritium waste.

Burial grounds have become quite common, mainly in those countries where nuclear activities have a long history and originated in weapons research. The safety of this disposal technique is largely dependent on the type of soil, particular groundwater occurrence, and on the type of land use. Presently, a volume of over 200,000 m^3 containing approximately 2×10^6 Ci of radioactivity including 80 kg of plutonium are disposed of in commercial burial grounds in the United States and about the same order of magnitude in burial grounds established by the former U.S. Atomic Energy Commission. Although this technique cannot be considered unsafe when properly conducted, some incidents of radionuclide migration resulting in off-site contamination have occurred. It is fair to say that shallow burial of non-high-level, non-alpha waste may be safe in remote areas, but these usually do not exist in Europe. The overall policy of establishing burial ground needs reconsideration.

The term sea disposal includes two basically different techniques, namely, disposal into coastal waters and deep-sea disposal. Deep-sea disposal may be perfectly safe if handled

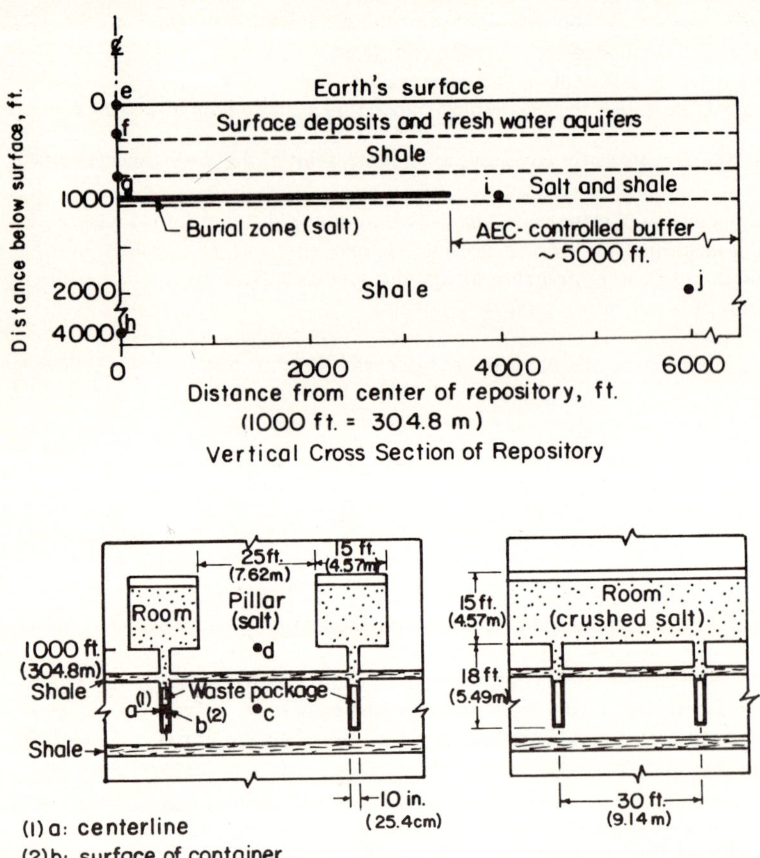

Figure 11.27 Schematic cross sections of proposed HLW repository. (*From Cheverton and Turner [C1].*)

responsibly. For certain types of waste that are difficult to deal with on land, such as bulky parts from decommissioning, it may even be the most appropriate technique. Deep-sea disposal has been practiced mainly under the supervision of international agencies. Disposal into coastal waters as practiced with non-high-level liquids from European reprocessing plants, however, is highly debated.

6 ASSESSMENT OF LONG-TERM SAFETY

The waste repository will be the final reservoir for all radioactivity generated by nuclear power. It will remain radioactive for a very long time, with some radioactivity even remaining for millions of years. As yet, complete safety analyses of waste repositories are not available. However, several attempts to approach the problem are known, and a number of systematic programs are on their way in various countries.

Proceeding from the assumption that water will be the only vehicle that possibly can carry radioactive material from the repository to people, the following processes must take place to create an actual risk:

1. The geologic containment fails and water is allowed to enter the repository and to find its way to the solidified waste, or brine present in the repository may contact the waste.
2. Radioactivity is released from the repository through contaminated water or brine entering an aquifier which is conected to circulating groundwater.

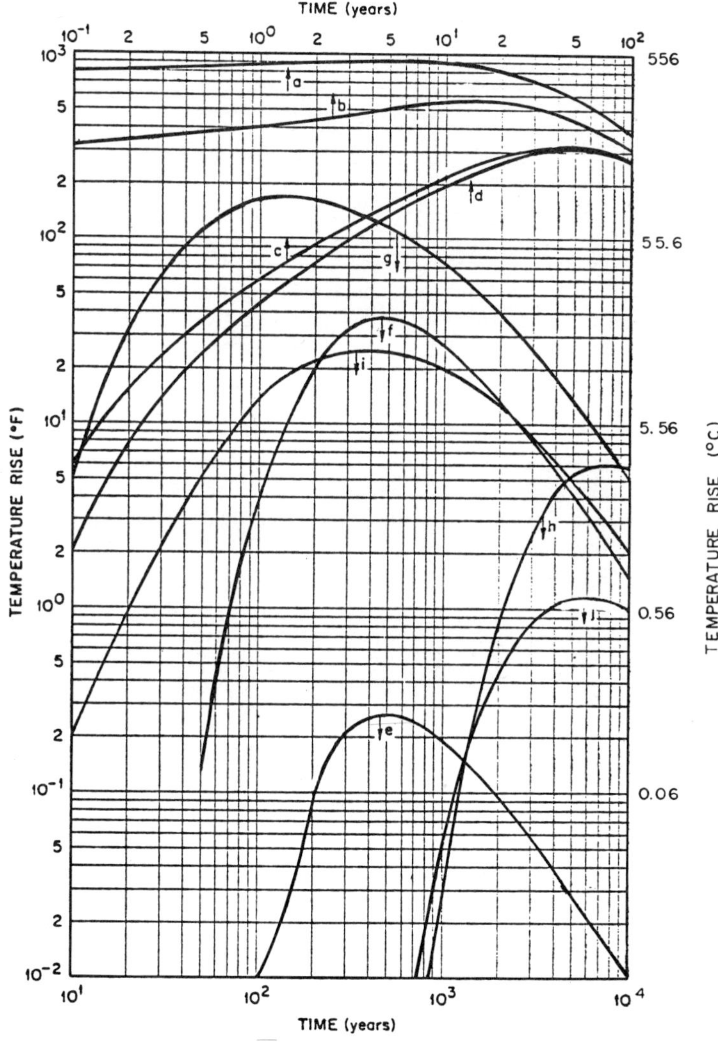

Figure 11.28 Temperature rise versus time afte burial of 10-year-old LWR calcined waste, 15-ft (4.57-m) room, 25-ft (7.62-m) pillar, single row of 6-in (15.24-cm) diameter containers on 30-ft (9.14-m) pitch, each containing 165 kg of 10-year-old waste nuclides equivalent to a power level of 4.4 kW. (See Fig. 11.27 for identification of curves). (*From Cheverton and Turner [C1].*)

3. The radioactive material released from the repository is carried to a drinking water well or to the surface by groundwater and enters the food chain, thereby causing consequences to human beings.

The magnitude of the consequences will obviously be a function of the radioactive inventory of the waste repository at the time when the sequence starts. As this inventory decreases by natural decay, the consequences will also decrease and will eventually drop below the level of significance.

6.1 Evaluation of Barriers between Waste and People

Geologic containment. The salt domes in the northern part of Germany where the tentative site for a waste repository is located are one example of a geologic containment under consideration. They are more than a hundred million years old. It was only after the formation of these salt domes that America and Europe began to separate, forming the Atlantic Ocean, and that the Alps came into being. The salt domes withstood numerous geologic catastrophes without changing their shape or location. The area was three times covered by ocean water and dried again, vulcanism developed all over Germany, and later several glaciers moved across the salt. Nothing, however, happened to the salt formations, thereby proving that they were in perfect equilibrium with the geologic environment. As long as this equilibrium is not disturbed by human activities, it is extremely unlikely that the salt domes will undergo any changes. The utilization as waste repository will influence this equilibrium mechanically by mining the salt and thermally by charging it with heat sources. Geologists and mining engineers, however, have no doubt that this can be done without serious disturbance of the equilibrium. This gives a very high degree of confidence in the long-term integrity of the salt formation.

Migration of radionuclides. Even in the event of intrusion of groundwater into the waste repository, the low solubility of the waste, the slow motion of water at depth, the sorptive capacity of the soil, and the distance from the repository to water used by people provide additional protection against contamination of the environment.

Burkholder [B8] and others [F1] have developed models for analyzing the consequences of accidental intrusion of underground water into a geologic repository for HLW. In an example calculation [B8] it is assumed that the geologic medium has sorptive properties typical of U.S. western desert subsoil, that the waste material dissolves at the slow rate of from 0.03 to 0.003 percent per year, that the solution percolates unidirectionally through 10 km of sorptive soil at rates of from 1 to 10 m/year, and that the underground water is discharged into a river used for drinking. The general result is that nuclides that are not sorbed by the soil, e.g., tritium, ^{14}C, ^{129}I, and possibly ^{99}Te, reach the river within a few thousand years. Other radionuclides that are sorbed by the soil are delayed for a much longer period, e.g., over a million years for plutonium, and are attenuated by radioactive decay and dispersion.

Procedures for evaluating the rate of migration in the event of intrusion of water into an underground repository are detailed in the foregoing references.

6.2 Significant Period of the Hazard

In analyzing the safety of a waste repository, it is crucial to know the time period under consideration. A number of geologic processes and events are relevant for the safety analysis only if the time frame exceeds a certain range. As it is obvious that the hazard of a waste repository due to the decrease of its radioactive inventory will eventually approach a level that is no longer significant, it will be feasible to estimate a time frame for the safety analysis. This time frame will be called the significant period of the waste repository hazard. The estimation of

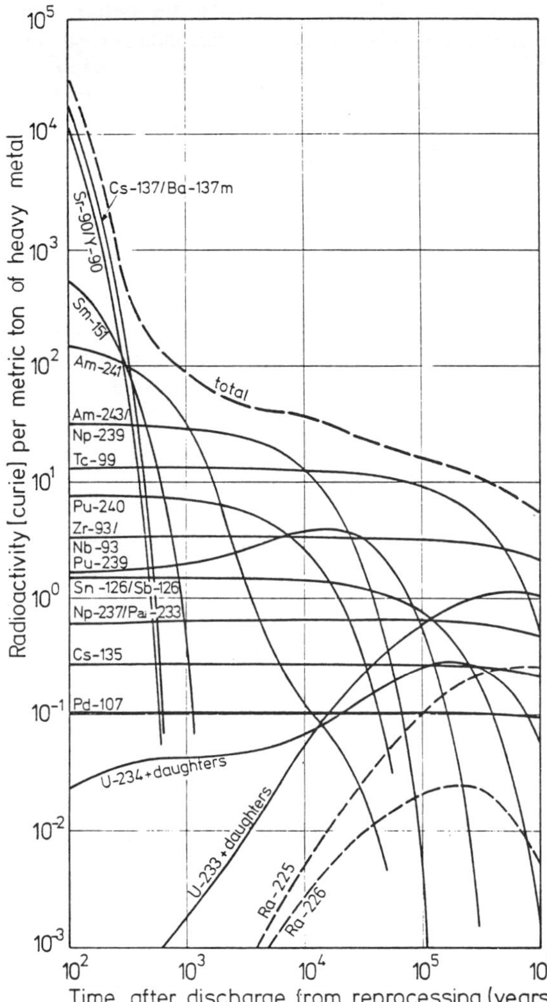

Figure 11.29 Radioactivity of individual radionuclides in HLW from the LWR uranium fuel cycle. Reprocessing, 150 days after reactor discharge; enrichment, 3% ^{235}U; burnup, 30,000 MWd/MT heavy metal; residence time, 1100 days; 0.5% uranium and 0.5% plutonium remaining in HLW.

such a significant period of hazard may be considered the first step in an iteration that may need refinement before the safety analysis is completed.

Definition of a significant level. To define a level of significance for the geologic waste repository hazard, a point of reference is required. The hazard of naturally occurring uranium in equilibrium with its daughters is frequently used as such reference. This choice implies the reasonable assumption that an artificial hazard equal to that of naturally occurring uranium is not considered significant because the natural uranium hazard is inevitable and people have been living with it all the time.

Such a comparison of hazards is meaningful only for similar chemical species and if the barriers protecting people from the hazards are at least qualitatively similar. This is true for a geologic waste repository as compared to a uranium deposit. The locations are similar, that of waste is even likely to be more favorable, and the key radionuclides involved, particularly ^{226}Ra and its parents, behave similarly.

As for the location, many uranium deposits occur considerably closer to the surface than a waste repository is supposed to be located. Therefore, radionuclides from uranium deposits may have to travel a shorter distance than those from waste repositories. Moreover, groundwater at greater depth is usually less mobile. The geologic containment of the waste repository is not taken into account as a barrier because the significant period of the hazard is supposed to be the period for which the integrity of the geologic containment is to be analyzed. Even disregarding this barrier, it is reasonable to consider the remaining barriers of a waste repository similar to those of a natural uranium deposit.

The specific radionuclides reponsible for the waste hazard are important because of their different mobilities when migrating with groundwater. Figures 11.29 and 11.30 show the long-term radioactivities and ingestion hazard indices of the most significant radionuclides in LWR uranium waste versus time. Beyond 500 years, the ingestion hazard is controlled by americium, plutonium, and eventually by radium as a uranium daughter. The neptunium itself contributes to the ingestion hazard, but less than 10 percent. The ingestion hazard of natural uranium is that of its daughter radium, and consequently over a long period of time is identical

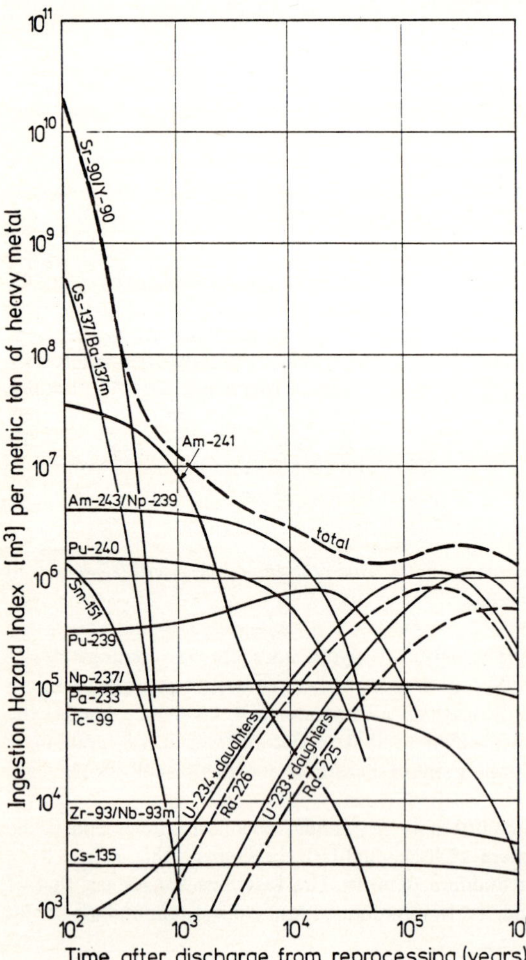

Figure 11.30 Ingestion hazard index (defined in Sec. 2.1) of individual radionuclides in HLW from the LWR uranium fuel cycle. Reprocessing, 150 days after reactor discharge; enrichment, 3% ^{235}U; burnup, 30,000 MWd/MT heavy metal; residence time, 1100 days; 0.5% uranium and 0.5% plutonium remaining in HLW.

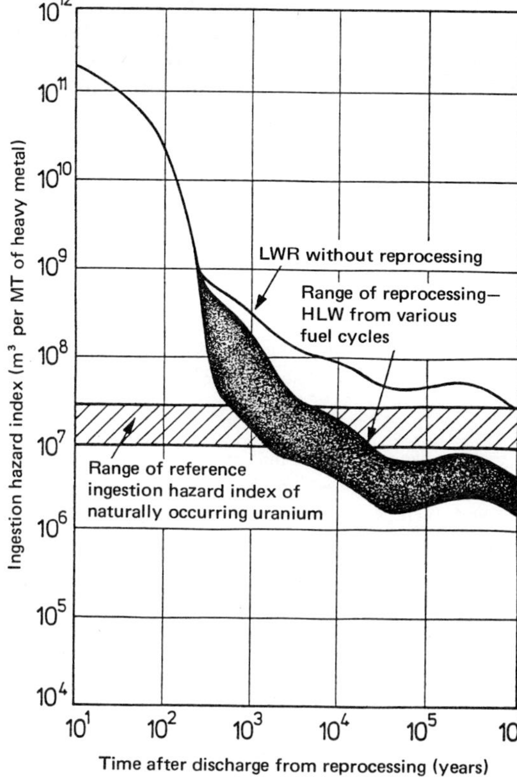

Figure 11.31 Range of ingestion hazard index of HLW and range of reference ingestion hazard index of naturally occurring uranium.

with the ingestion hazard of waste. Plutonium and americium have essentially the same mobility as uranium. The mobility of radium is correlated with that of its parent uranium. Only the not very abundant neptunium is faster by a factor of 100 [B8].

The conclusion is that a comparison of ingestion hazard indices of waste in a geologic repository and of naturally occurring uranium is a reasonable basis for the definition of a significant level of the waste hazard.

Estimation of the significant period of the waste hazard. Figure 11.31 shows a band of long-term ingestion hazard indices of HLW from various fuel cycles and a line corresponding to unreprocessed LWR fuel versus time. It shows also a horizontal band representing various reference levels [L4, L5].

Reference level means the quantity of natural uranium whose ingestion hazard index is used as a reference to which that of the waste from 1 MT of heavy metal reprocessed is compared. These quantities according to different approaches are as follows:

The quantity of natural uranium to be mined for the production of the heavy metal reprocessed. This type of reference has already been used in Chap. 8 because it is the most general one with no special assumption about the form of the natural uranium involved. Its disadvantage is the strong dependence on fuel-cycle type. With an equilibrium LMFBR fuel cycle, for instance, the quantity of uranium to be mined becomes close to zero and, consequently, the period of significance of the waste hazard becomes extremely long. To maintain its applicability, the uranium equivalent must always be calculated on the virtual basis that all power has been generated from freshly mined uranium.

The volume of natural U_3O_8 equal to the volume of solidified waste from reprocessing 1 MT of heavy metal. This volume is assumed to be 80 liters as an average. For unreprocessed fuel 120 liters have been used. U_3O_8 has been chosen as the standard uranium species because this is the radioactive concentrate in a uranium ore just as solidified waste is the radioactive concentrate in a waste repository. Moreover, it is a sufficiently generalized uranium species. This reference leads to a dependence of the significant period on the waste oxide concentration in the waste form.

The waste from 1 MT of heavy metal is assumed to be evenly distributed in that volume of rock which is required to accommodate the boreholes for the corresponding number of waste blocks, disregarding rock above and beneath the boreholes. The waste blocks are assumed to have 20 w/o waste oxides and to be arranged in a hexagonal array with 10-m distances. The ingestion hazard index of a unit volume of this homogenized disposal field is compared to the ingestion hazard index of the same volume of 0.2 percent uranium ore. This approach leads to a dependence of the significant hazard period on the density of waste in the host rock of the geologic repository.

The range of intersection between the ingestion hazard index band and the horizontal band indicates the range of significant periods of the hazard. These significant periods vary in a relatively narrow range, namely, between 500 and 10,000 years for the whole variety of waste from different fuel cycles except for unreprocessed fuel.

REFERENCES

B1. Bähr, W., et al.: "Experiences in the Treatment of Low- and Intermediate-Level Radioactive Wastes in the Nuclear Research Center, Karlsruhe," *Management of Low- and Intermediate-Level Radioactive Wastes,* International Atomic Energy Agency, Vienna, 1970.

B2. Bell, M. J.: "The ORNL Isotope Generation and Depletion Code (ORIGEN)," Report ORNL-4628, May 1973.

B3. Blasewitz, A. G., et al.: "The High Level Waste Solidification Program," *Proceedings of the Management of Radioactive Wastes from Fuel Reprocessing,* Paris, 1972, Report CONF-721107, Mar. 1973, p. 615.

B4. Blomeke, J. O., and C. W. Kee: "Projections of Waste to Be Generated," *Proceedings of the International Symposium on the Management of Wastes from the LWR Fuel Cycle,* Denver, 1976, Report CONF-76-0701, p. 96.

B5. Bond, W. D., and R. E. Leuze: "Feasibility Studies of the Partitioning of Commercial High-Level Wastes Generated in Spent Fuel Reprocessing," Report ORNL-5012, 1975.

B6. Bonniaud, R.: "Continuous Vitrification in France Taken to Industrial Plant Scale," *Nucl. Eng. Int.* **21**: 67–69 (Nov. 1976).

B7. Bradley, D. J.: "Leaching of Fully Radioactive High-Level Glass," Report PNL-2664, 1978.

B8. Burkholder, H. C., et al.: "Incentives for Partitioning High-Level Waste," *Nucl. Technol.* **31**: 202 (1976).

C1. Cheverton, R. D., and W. D. Turner: "Thermal Analysis of the National Radioactive Waste Repository," Report ORNL-4789, 1972.

C2. Claiborne, H. C.: "Neutron-Induced Transmutation of High-Level Radioactive Waste," Report ORNL-TM-3964, 1972.

C3. Clelland, D. W., et al.: "A Review of European High-Level Waste Solidification Technology," *Proceedings of the International Symposium on the Management of Wastes from the LWR Fuel Cycle,* Denver, 1976, Report CONF-76-0701, p. 137.

C4. Cooley, C. R., and D. E. Clark: "Treatment Technologies for Non-High-Level Wastes (USA)," *Proceedings of the International Symposium on the Management of Wastes from the LWR Fuel Cycle,* Denver, 1976, Report CONF-76-0701, p. 250.

D1. De, A. K., et al.: "Fixation of Fission Products in Glass Ceramics," *Proceedings of the Management of Radioactive Wastes from the Nuclear Fuel Cycle,* vol. 2, International Atomic Energy Agency, Vienna, 1976, p 63.

D2. Deutsche Gesellschaft für Wiederaufbereitung von Kernbrennstoffen (DWK): "Bericht über das in der Bundesrepublik Deutschland geplante Entsorgungszentrum für ausgediente Brennelemente aus Kernkraftwerken," Hannover, 1977.

D3. Donato, A.: "Incorporation of Radioactive Wastes in Polymer-Impregnated Cement," *Proceedings of the Management of Radioactive Wastes from the Nuclear Fuel Cycle,* vol. 2, International Atomic Energy Agency, Vienna, 1976, p. 143.

E1. Ewest, E.: "Calculation of Radioactivity Release Due to Leaching of Vitrified High-Level Waste," *Symposium on Science Underlying Radioactive Waste Management,* Boston, 1978, Plenum, New York, 1979.

E2. Ewest, E., and H. W. Levi: "Evaluation of Products for the Solidification of High-Level Radioactive Waste from Commercial Reprocessing in the Federal Republic of Germany," *Proceedings of the Management of Radioactive Wastes from the Nuclear Fuel Cycle,* vol. 2, International Atomic Energy Agency, Vienna, 1976, p. 15.

F1. Foglia, M., et al.: "The Superposition Solution of the Transport of a Radionuclide Chain Through a Sorbing Medium," Report UCB-NE-3348, Sept. 1979.

G1. Van Geel, J., et al.: "Solidification of High-Level Liquid Waste of Phosphate Glass-Metal Matrix Blocks," *Proceedings of the Management of Radioactive Wastes from the Nuclear Fuel Cycle,* vol. 1, International Atomic Energy Agency, Vienna, 1976, p. 341.

H1. Halaszowich, St., et al.: "Interim Storage and Solidification for Thorex-Type Fission-Product Solutions," *Proceedings of the Management of Radioactive Waste from Fuel Reprocessing,* Paris, 1972, Report CONF-721107, Mar. 1973, p. 705.

H2. Hall, A. R., et al.: "Development and Radiation Stability of Glasses for Highly Radioactive Wastes," *Proceedings of the Management of Radioactive Wastes from the Nuclear Fuel Cycle,* vol. 2, International Atomic Energy Agency, Vienna, 1976, p. 3.

H3. Heacock, H. W., and J. W. Riches: "Waste Solidification–Cement or Urea Formaldehyde," Paper 74-WA/NE-9, Amer. Soc. Mech. Eng., Annual Winter Meeting, New York, Nov. 12–22, 1974.

H4. Heimerl, W., et al.: "Studies on the Behaviour of Radioactive Waste Glasses," *Proceedings of the Management of Radioactive Wastes from Fuel Reprocessing,* Paris, 1972, Report CONF-721107, Mar. 1973, p. 515.

K1. Kingery, W. D.: *Introduction to Ceramics,* Wiley, New York, 1976, p. 628.

K2. Kühn, K., and J. Hamstra: "Geologic Isolation of Radioactive Wastes in the Federal Republic of Germany and the Respective Program of the Netherlands," *Proceedings of the International Symposium on the Management of Wastes from the LWR Fuel Cycle,* Denver, 1976, Report CONF-76-0701, p. 580.

L1. Lakey, L. T., and B. R. Wheeler: "Solidification of High-Level Radioactive Wastes at the Idaho Chemical Processing Plant," *Proceedings of the Management of Radioactive Wastes from Fuel Reprocessing,* Paris, 1972, Report CONF-721107, Mar. 1973, p. 731.

L2. Legler, B. M., and G. R. Bray: *Chem. Eng. Progr.* **72**: 52 (Mar. 1976).

L3. Lennemann, W. L.: "Management of Radioactive Aqueous Waste from U.S. Atomic Energy Commission's Fuel Reprocessing Operations, Experience and Planning," *Proceedings of the Management of Radioactive Wastes from Fuel Reprocessing,* Paris, 1972, Report CONF-721107, Mar. 1973, p. 357.

L4. Levi, H. W.: "Project Safety Studies Entsorgung in the Federal Republic of Germany," *Proceedings of the Underground Disposal of Radioactive Waste,* Helsinki, 1979.

L5. Levi, H. W., and E. Ewest: "Zur Frage einer zeitlichen Begrenzung der Störfallanalyse des geologischen Endlagers," Report PSE-79/1, 1979.

M1. McCarthy, G. J. L.: "Ceramics and Glass Ceramics as High-Level Waste Forms," *ERDA Workshop,* Germantown, Jan. 1977, p. 83.

M2. McElroy, J. L., et al.: "Waste Solidification Technology (USA)," *Proceedings of the International Symposium on the Management of Wastes from the LWR Fuel Cycle,* Denver, 1976, Report CONF-76-0701, p. 166.

M3. Morris, J. B., and B. E. Chidley: "Preliminary Experience with the New Harwell Inactive Vitrification Plant," *Proceedings of the Management of Radioactive Wastes from the Nuclear Fuel Cycle,* vol. 1, International Atomic Energy Agency, Vienna, 1976, p. 241.

P1. Pence, D. T., et al.: "Metal Zeolites: Iodine Absorption Studies," Project Report Jan. 1–Dec. 31, 1970, Report IN-1455, June 1971.

P2. Pigford, T. H., and J. Choi: In "Report to the APS by the Study Group on Nuclear Fuel Cycles and Waste Management," *Rev. Mod. Phys.* **50**(1), part II: S116 (Jan. 1978).

R1. Richardson, G. L.: "Technologies for the Recovery of the Transuranium Elements and Immobilization of Non-High-Level Wastes," *Proceedings of the International Symposium on the Management of Wastes from the LWR Fuel Cycle,* Denver, 1976, Report CONF-76-0701, p. 289.

R2. Roberts, F. P., et al.: "Radiation Effects in Solidified High-Level Waste, Part I, Stored Energy," Report BNWL-1944, Jan. 1976.

T1. Tymochowicz, S.: "A Collection of Results and Methods on the Leachability of Solidified High-Level Radioactive Waste Forms," Report HMI-B 241, 1977.

W1. Warner, B. F., et al.: "Operational Experience in the Evaporation and Storage of Highly-Active Fission-Product Waste at Windscale," *Proceedings of the Management of Radioactive Wastes from Fuel Reprocessing,* Paris, 1972, Report CONF-721107, Mar. 1973, p. 339.

W2. Wilhelm, J. G., et al.: "An Inorganic Absorber Material for Off-Gas Cleaning in Fuel Reprocessing Plants," *12th Air Cleaning Conference, Oak Ridge, Tenn., 1972,* Report CONF-720823-P2, Jan. 1973, p. 540.

PROBLEM

11.1 The high-level radioactive waste storage facility for a reprocessing plant is to be designed.

Plant data: Slightly enriched uranium is to be reprocessed with a burnup of 30,000 MWd/MT uranium and a specific power of 30 MW/MT uranium.

Cooling time: 150 days

Capacity: 1 MT UO_2/day

Flow rate of feed solution: 110 liters/h

Flow rate of scrub solution: 55 liters/h

The raffinate from the first extraction column contains 99.5 percent of the total fission-product activity. Most of the rest appears in the concentrate from the MLW evaporator, which is produced at a rate of 35 liters/h. Both streams are fed to the HLW evaporator, where concentration by a factor of 10 is achieved. The plant is to be equipped with a 500-m^3 HLW tank.

(*a*) What are the activity concentrations in the aqueous raffinate from the first extraction column and in the HLW concentrate to be stored?

(*b*) What is the total activity in the storage tank after it is filled?

(*c*) At what rate is heat generated in the freshly filled tank?

(*d*) What coolant flow is required (water with 35°C inlet and 45°C outlet temperature) to keep the tank content at 60°C?

CHAPTER

TWELVE

STABLE ISOTOPES: USES, SEPARATION METHODS, AND SEPARATION PRINCIPLES

Although the isotopes of an element have very similar chemical properties, they behave as completely different substances in nuclear reactions. Consequently, the separation of isotopes of certain elements, notably ^{235}U from ^{238}U and deuterium from hydrogen, is of great importance in nuclear technology. The fact that isotopes of an element have such similar gross physical and chemical properties, however, makes their separation unusually difficult and has necessitated the development of processes and concepts especially adapted to this purpose. Despite the novelty of some of these isotope separation techniques, they have features in common with distillation and other familiar separation methods, and study of isotope separation is helpful in understanding more conventional separation methods.

1 USES OF STABLE ISOTOPES

Table 12.1 lists separated isotopes that are being produced on a significant industrial scale. In addition to these, separated isotopes of practically all natural elements are being produced in research quantities by the U.S. Department of Energy (DOE) and by the atomic energy agencies of England, France, the Soviet Union, and other nations.

1.1 ^{235}U

^{235}U is the separated isotope of by far the greatest industrial importance, with the value of annual production throughout the world of the order of a billion dollars. Uranium enriched from the natural level of 0.7 percent to from around 1.5 to 4 percent is used as fuel in power reactors moderated by natural water or graphite.

^{235}U enriched to 90 percent or higher, mixed with thorium, is proposed as fuel for the high-temperature gas-cooled reactor, the light-water breeder reactor, and the thorium-fueled CANDU type of heavy-water reactor, and as an alternative fuel for light-water reactors. In these reactor systems fission of ^{235}U is supplemented by five times or more as many fissions from ^{233}U produced by neutron absorption in thorium, as outlined in Chap. 3. Highly enriched ^{235}U is used as fuel for research or testing reactors, where the highest attainable neutron flux is wanted, and in compact power reactors, where high power density is needed.

Table 12.1 Uses of separated isotopes

Isotope	Natural atom percent	Use
^{235}U	0.7205	Fuel for nuclear fission reactors
D	0.015	1. D_2O moderator for natural uranium reactors 2. Fuel for thermonuclear reactors
^{6}Li	7.56	1. Source of tritium 2. Fuel for thermonuclear reactors
^{7}Li	92.44	1. As LiOH, water conditioner for water-cooled reactors 2. As lithium metal, possible high-temperature reactor coolant
^{10}B	19.61	1. Neutron absorber in control rods and shielding 2. Neutron-capture medical therapy
^{13}C	1.107	1. Stable isotopic tracer in living systems 2. Nuclear magnetic resonance studies of molecular structure
^{15}N	0.366	
^{17}O	0.037	
^{18}O	0.204	

1.2 Deuterium

Of the three moderators that make possible a fission chain reaction in natural uranium, heavy water, graphite, or beryllium, heavy water has become the preferred material. It is used both as coolant and moderator in the CANDU line of heavy-water reactors, which are the exclusive source of nuclear power in Canada, Argentina, and Pakistan, are being used in India, and are being considered in other countries wishing to have a nuclear power system not dependent on a source of enriched uranium.

Deuterium, either mixed with tritium or in the form of ^{6}Li deuteride, ^{6}LiD, is an essential ingredient in the fuel proposed for fusion power reactors. In the magnetically confined type of fusion power system, the working substance is a plasma mixture of fully ionized deuterium and tritium. In the laser or electron beam imploded type of system, the fuel form is a small sphere containing deuterium and tritium or ^{6}LiD. Although power systems of these types have not yet been proved feasible, their successful development would create a market for deuterium and ^{6}Li as great as the current market for enriched uranium.

1.3 Lithium Isotopes

^{6}Li may be used in fusion power systems, as noted above, and is the starting material for producing tritium by neutron absorption:

$$^{6}_{3}Li + ^{1}_{0}n \rightarrow ^{3}_{1}T + ^{4}_{2}He$$

In some types of thermonuclear power systems it is desirable to use a blanket of lithium enriched in ^{6}Li to increase the volumetric rate of neutron capture to produce tritium.

^{7}Li hydroxide is now used in some water-cooled reactors to inhibit corrosion by control of hydrogen ion concentration. Because the thermal-neutron absorption cross sections of the lithium isotopes are ^{6}Li, 940 b, and ^{7}Li, 0.037 b, it is necessary to use ^{7}Li containing less than 0.01 percent ^{6}Li. ^{7}Li metal, which melts at 180°C, was proposed as coolant for an aircraft-propulsion reactor, because of its low vapor pressure at high temperature and low neutron-absorption cross section.

1.4 ^{10}B

The thermal-neutron absorption cross section of natural boron, which contains 19.61 percent ^{10}B, is 759 b, whereas that of separated ^{10}B is 3837 b. Thus, enriched ^{10}B is useful in applications where the highest volumetric rate of neutron absorption is wanted. Examples are compact shielding for thermal neutrons and control rods for fast reactors.

Neutron-capture therapy is an experimental technique for selective destruction of cancerous tissue surrounded by healthy tissue. In this technique a compound of ^{10}B that is selectively absorbed by the cancer is injected into the bloodstream, followed by irradiation of the cancerous tissue by a beam of neutrons. Energetic alpha particles, produced by the reaction

$$^{10}_{5}B + ^{1}_{0}n \rightarrow ^{4}_{2}He + ^{7}_{3}Li$$

where the neutron beam reacts with the boron compound in the cancer, destroy the cancer while leaving the neighboring healthy tissue, containing less boron, less affected.

1.5 ^{13}C

Carbon, hydrogen, oxygen, and nitrogen are the elements that occur in greatest abundance in living systems. Tracer experiments using either radioactive isotopes or separated natural isotopes are of great importance in understanding biochemical reactions. Although with carbon there is the possibility of using the short-lived radioisotope ^{11}C or the very long-lived ^{14}C, for many experiments it is preferable to avoid radioactivity and use separated stable ^{13}C. Another important use of ^{13}C is in nuclear magnetic resonance experiments on the structure of carbon compounds. By synthesizing a compound with a ^{13}C atom in a known location, it is possible to draw conclusions about the configuration of the molecule, because ^{13}C has a nuclear magnetic moment and ^{12}C has none.

1.6 ^{15}N

^{15}N can be used in very much the same way as ^{13}C, as a tracer for nitrogen compounds and in nuclear magnetic resonance experiments. The fact that the longest-lived nitrogen radioisotope, ^{13}N, has a half-life of only 10 min gives ^{15}N added significance.

An additional possible use suggested for ^{15}N is in $U^{15}N$ fuel material for a fast reactor. ^{15}N has a lower absorption and inelastic scattering cross sections for fast neutrons than the more abundant ^{14}N. Its use avoids ^{14}C production from the reaction $^{14}N + ^{1}n \rightarrow ^{14}C + ^{1}H$.

1.7 Oxygen Isotopes

Because the longest-lived oxygen radioisotope, ^{15}O, has a half-life of only 124 s, the separated isotopes ^{17}O and ^{18}O are valuable in tracer experiments. The nuclear magnetic moment of ^{17}O gives it application in determining molecular structure by nuclear magnetic resonance measurements.

2 ISOTOPE SEPARATION METHODS

2.1 ^{235}U

Table 12.2 lists methods that have been used on an industrial or large pilot-plant scale to enrich uranium in ^{235}U.

Table 12.2 Methods for enriching ^{235}U

Method	Status
Gaseous diffusion of UF_6	Three large plants operating in United States; large plants operating in the Soviet Union and China; smaller plants operating in England and France; large plant being constructed in France
Centrifugation of UF_6	Large pilot plants operating and commercial plants under construction in England and Holland; large plant to be built in United States
Thermal diffusion of UF_6	Small amount of slightly enriched UF_6 produced in United States in 1945; process abandoned
Electromagnetic separation of UCl_4	Used in United States in 1945 for first large-scale production of highly enriched ^{235}U; process abandoned in 1946
Separation nozzle process	Process demonstrated on large pilot-plant scale at Karlsruhe, Germany; semicommercial plant being built in Brazil
UCOR process	Process demonstrated in pilot plant at Valindaba, Union of South Africa; commercial plant under consideration

Gaseous diffusion process. Figure 12.1 illustrates the principle of one stage of the gaseous diffusion process. Stage feed gas, UF_6, flows past a diffusion barrier made of porous material with very fine holes, smaller than the mean free path of the UF_6 molecules. About half of the feed gas flows through the barrier to a lower-pressure region. The gas passing through the barrier is slightly richer in ^{235}U than the gas remaining on the high-pressure side, because the mean speed of $^{235}UF_6$ molecules is slightly higher than that of $^{238}UF_6$ molecules. These mean speeds are in the inverse ratio of the square roots of the molecular weights of the two molecules. Under practical operating conditions the ratio of $^{235}UF_6$ atoms to $^{238}UF_6$ atoms in the enriched UF_6 fraction passing through the barrier, $y/(1-y)$, to the corresponding ratio in the depleted UF_6 fraction remaining behind, $x/(1-x)$, is in the ratio of their mean speeds:

$$\alpha \equiv \frac{y/(1-y)}{x/(1-x)} = \sqrt{\frac{m_{^{238}UF_6}}{m_{^{235}UF_6}}} = \sqrt{\frac{352}{349}} = 1.00429 \tag{12.1}$$

The ratio $[y/(1-y)]/[x/(1-x)]$ is called the stage separation factor and is denoted by α. Analogous separation factors are used to characterize all separation processes. A value of α close to unity indicates that the separation is difficult; a value far from unity, easier. For gaseous diffusion of UF_6, α is so close to unity that the process must be repeated many times for a useful degree of separation. To do this, the low-pressure enriched UF_6 must be recompressed to the feed pressure and cooled. The depleted UF_6, which experiences some pressure loss, must also be recompressed (not shown).

Because of the small change in enrichment from a single stage, for a useful degree of enrichment, it is necessary to use many stages in series in countercurrent cascade. Figure 12.2 shows how stages are connected together in such a cascade. On each stage a motor-driven compressor takes partially depleted gas from the next higher stage and partially enriched gas from the next lower stage and recompresses them before passage through a cooler and the diffusion barrier. To separate natural uranium feed containing 0.00711 fraction ^{235}U into product containing 0.03 and tails 0.002 fraction ^{235}U requires 1272 stages. The minimum total

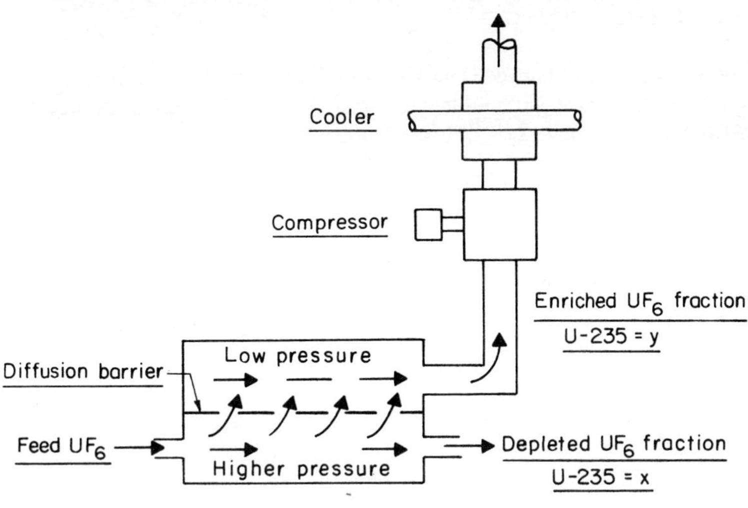

Separation factor

$$\alpha \equiv \frac{y(1-x)}{x(1-y)} = \sqrt{\frac{m_{238_{UF_6}}}{m_{235_{UF_6}}}} = \sqrt{\frac{352}{349}} = 1.00429$$

Figure 12.1 Gaseous diffusion stage.

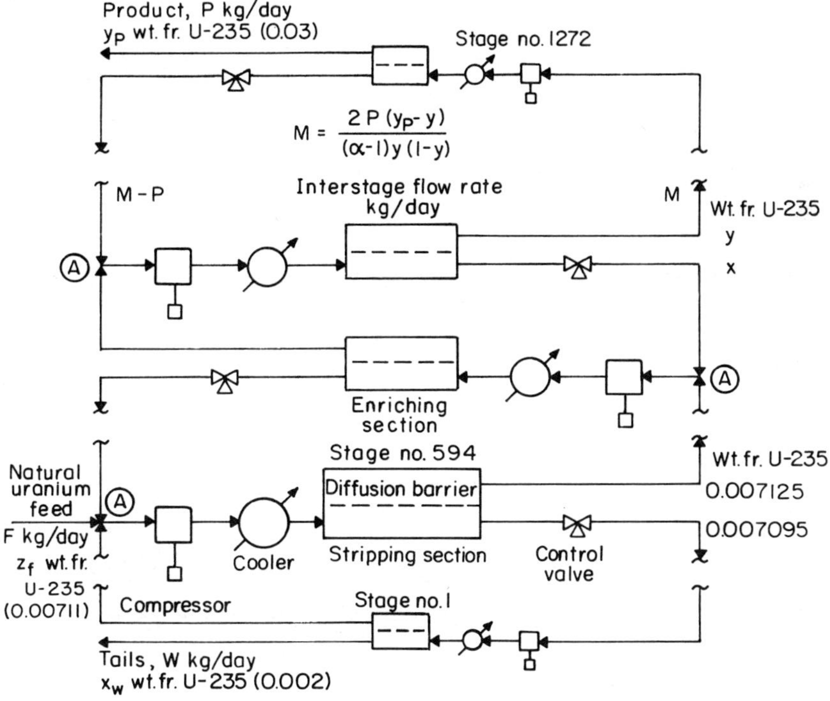

Figure 12.2 Ideal gaseous diffusion cascade.

interstage flow in such a cascade is obtained when the compositions of the streams mixed at each point A are equal. Such a cascade is called an *ideal cascade.* In such a cascade, the interstage flow rate M from a stage where the ^{235}U fraction is y is

$$M = \frac{2P(y_P - y)}{(\alpha - 1)y(1 - y)} \tag{12.2}$$

where P is the flow rate of product containing y_P fraction ^{235}U. The theory of such an ideal cascade is developed later in this chapter, and details of the gaseous diffusion process are given in Chap. 14.

Figure 12.3 is a photograph of the large gaseous diffusion plant of the U.S. Department of Energy at Portsmouth, Ohio, which use 4080 stages to enrich ^{235}U to 97 percent.

The gas centrifuge. Figure 12.4 shows the principle of the type of countercurrent gas centrifuge proposed 20 years ago by the German engineer, Gernot Zippe [Z1], and now generally adopted by groups continuing development of this promising method of isotope separation. Such a centrifuge consists of a rapidly rotating cylindrical bowl made of a material with high strength-to-density ratio. The UF_6 gas rotating inside in this bowl is subjected to centrifugal accelerations thousands of times greater than gravity. This makes the pressure at the outer radius of the bowl millions of times greater than at the axis and causes the concentration of $^{238}UF_6$ relative to $^{235}UF_6$ to be appreciably higher at the outer radius than at the axis. In a machine made of fiberglass running at the highest speed possible without mechanical failure, the ^{235}U content at the center of the bowl can be as much as 18 percent higher than at the

Figure 12.3 Gaseous diffusion plant at Portsmouth, Ohio. (*Courtesy of U.S. Department of Energy.*)

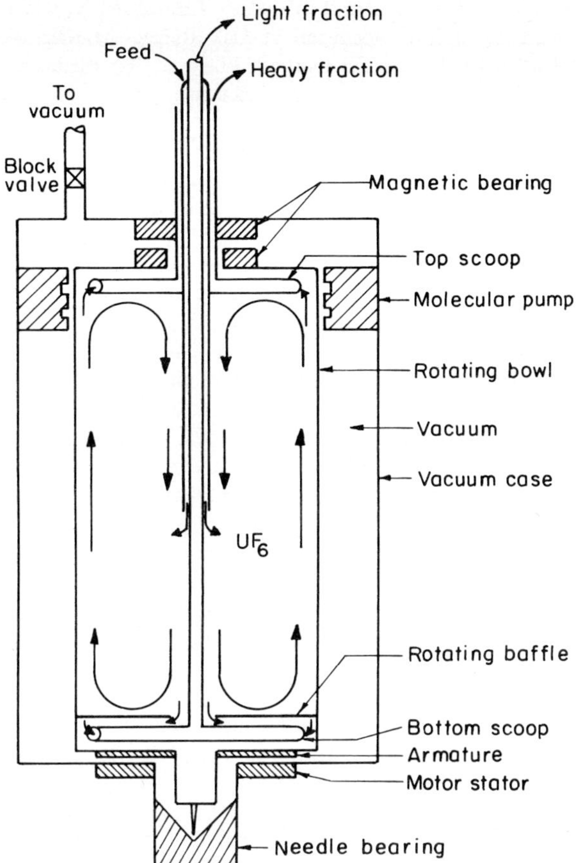

Figure 12.4 Zippe gas centrifuge schematic.

outside. In addition, longitudinal countercurrent flow of UF_6 is induced by a system of rotating baffles and stationary scoops. In Fig. 12.4, gas enriched in $^{235}UF_6$ at the center flows downward and gas enriched in $^{238}UF_6$ at the outside flows upward. Under these conditions the gas toward the bottom of the bowl becomes progressively richer in $^{235}UF_6$ and the gas at the top richer in $^{238}UF_6$. By making the bowl sufficiently long, the concentration difference between top and bottom can be made many times greater than between center and outside.

Gas centrifuges of greater capacity than described by Zippe have been developed in the United States, England, Germany, and Holland. Commercial centrifuge plants are operating in England and Holland and are planned in the United States. The power consumption of the centrifuge process is much lower than gaseous diffusion, and it is expected that separation costs will become lower. The process is described in more detail in Chap. 14.

Thermal diffusion of UF_6. The thermal diffusion process makes use of the small difference in $^{235}U/^{238}U$ ratio that is established when heat flows through a mixture of $^{235}UF_6$ and $^{238}UF_6$. The principle of the process is described in Chap. 14. The process was used [A1] in 1945 in the United States by the Manhattan Project to enrich uranium to 0.86 percent ^{235}U. This slightly enriched material was used as feed for an electromagnetic separation plant. Although the process could be put into production quickly because of the simplicity of the equipment, it was very inefficient, with very high heat consumption per unit of output. Consequently, when the more efficient gaseous diffusion plant came into operation at Oak Ridge, the thermal diffusion plant was dismantled. Thermal diffusion is a useful method, however, for separating small amounts of isotopes for research purposes. It is used, for example, at the Mound Laboratory to enrich ^{13}C from 90 to 99 percent.

Electromagnetic processes. The possibility of using electromagnetic means for separating isotopes was established by Thomson [T5] in 1911. When Thomson passed a beam of positive neon ions through electric and magnetic fields, two traces were produced on a photographic plate, one for ^{20}Ne and the other for ^{22}Ne. The modern mass spectrometer works on the same general principle. With it, the existence of naturally occurring isotopes of 61 elements has been established, and isotopic abundances and masses have been determined (App. C).

In 1940, Nier and co-workers [N2] used a mass spectrometer to separate around 0.01 μg of ^{235}U from ^{238}U, to show that ^{235}U was the fissionable isotope of uranium. Because of its demonstrated ability to separate ^{235}U, the electromagnetic method was the first one selected by the Manhattan District for large-scale production of this isotope [S5]. Under the direction of Lawrence [L1] at the University of California, mass spectrometers of greatly increased capacity were developed. The end result was the calutron† electromagnetic isotope separator used in the Y-12 plant at Oak Ridge, in which, in 1944, the first kilograms of ^{235}U were produced.

When the gaseous diffusion plant came into operation, the cost of separating ^{235}U electromagnetically was found to be higher, and in 1946, the Y-12 plant was taken off uranium-isotope separation. Some of this equipment is now being used to produce gram quantities of partially separated isotopes of most of the other polyisotopic elements, for research uses. These units have also been used to separate artificially produced isotopes, such as ^{236}U from irradiated uranium, and the various plutonium isotopes.

Large-capacity electromagnetic isotope separation equipment has also been developed in Russia [Z2], and at Harwell [S4], Amsterdam [K2], and other centers of nuclear research [K5].

Becker separation nozzle process. Recently there has been increased interest in aerodynamic processes in which partial separation of isotopes is obtained in flowing gas streams subjected to high linear or centrifugal acceleration. The aerodynamic process about which most information is available is the Becker separation nozzle process.‡ This originally employed linear acceleration of UF_6 through a divergent nozzle, but now uses a combination of linear and centrifugal acceleration through a curved slit.

Figure 12.5 is a cross section of the slit-shaped separation element used in the most fully tested form of the Becker nozzle process. Feed gas consists of a mixture of about 5 m/o (mole percent) UF_6 and 95 m/o hydrogen at a pressure of around 1 atm. This flows into a low-pressure region through a long curved slit, or "nozzle" (perpendicular to the plane of the figure), with first a convergent, then a divergent cross section. The change in cross section accelerates the gas mixture to supersonic speed, and the curved groove downstream of the slit produces a centrifugal field. This sets up a concentration gradient in the mixture, with the gas adjacent to the curved wall enriched in ^{238}U relative to ^{235}U. A knife-edge downstream from the slit divides the stream into a more-deflected light fraction and a less-deflected heavy fraction.

†From *Cal*ifornia *U*niversity Cyclo*tron*.

‡See, for example, papers presented by Dr. E. W. Becker and his associates at the International Conference on Uranium Isotope Separation of the British Nuclear Energy Society, London, March 1975.

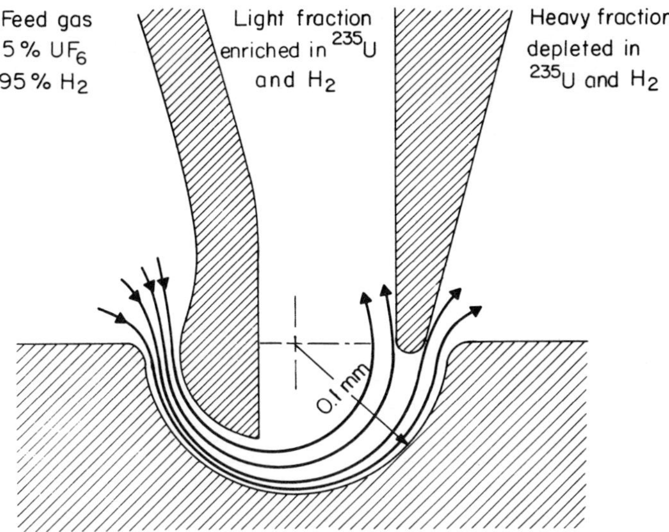

Figure 12.5 Cross section of slit used in Becker separation nozzle process.

Dilution of UF_6 with hydrogen has two beneficial effects. The mixture has a much higher sonic velocity than pure UF_6, so that much higher flow velocities are attainable, and inert gas makes the isotope separation factor greater than would be predicted for the prevailing centrifugal field. A separation factor of 1.015 can be obtained with a mixture of 5 percent UF_6-95 percent H_2 flowing through a pressure ratio of 3.5.

A more complete description of the process is given in Chap. 14. A semicommercial plant using this process is being built in Brazil.

UCOR process. The UCOR process, developed by the Uranium Enrichment Corporation of South Africa, Ltd., also makes use of high-speed flow of UF_6-hydrogen mixtures through sharply curved ducts. By using a new cascade technique, called the Helikon, in which an axial-flow compressor handles several streams simultaneously without mixing, it is expected that natural uranium can be enriched to 3 percent ^{235}U with from 90 to 115 multistage compressor modules. A partial description of a South African pilot plant using this process was given by Roux and Grant [R2].

Laser-based processes. In addition to the processes listed in Table 12.2, intensive research is being conducted on using high-intensity, tunable lasers to separate uranium isotopes by making use of the small differences in absorption spectra of ^{235}U and ^{238}U or one of their compounds. Laser-based processes have demonstrated capability for selective separation of isotopes of many elements on a small scale and are considered promising candidates for eventual large-scale economic production of enriched uranium. Letokhov and Moore [L3] provide a good review of laser isotope separation work through 1976.

2.2 Deuterium

Commercial production of deuterium has been almost universally in the form of heavy water, D_2O. Table 12.3 lists processes that have been used for production of heavy water at rates above a ton per year. These processes are divided into two classes. Parasitic processes take feed

Table 12.3 Processes for commercial production of heavy water

Method	Status
	Parasitic processes
Electrolysis of water	Used at Rjukan, Norway; Ems, Switzerland; and Nangal, India, to partially enrich deuterium for final concentration by another process
Hydrogen distillation	Tested in pilot plants in France and Germany in 1950s, used at Ems, Switzerland, in 1960s, and still used at Nangal, India, for final concentration
Water-hydrogen exchange	Used at Trail, Canada, in 1940s and still used at Rjukan, Norway, for intermediate concentration
Ammonia-hydrogen exchange	Used at Mazingarbe, France, in 1960s and planned for three plants in India to operate in late 1970s
	Self-contained processes
Water distillation	Used in three U.S. plants in 1940s for primary production; now restricted to final concentration
Hydrogen sulfide-water dual-temperature exchange	Two 500 MT/year plants built in United States in 1950s, one still operating at 69 MT/year; three plants operating in Canada with combined capacity of 1600 MT/year, more under construction

from a primary plant producing hydrogen or ammonia synthesis gas (75 percent H_2, 25 percent N_2), extract deuterium from it, and return the depleted hydrogen for commercial use, usually ammonia synthesis. Self-contained processes have heavy water as their sole product and use natural water as feed. Generally speaking, the parasitic processes produce heavy water at lower cost, but their output is limited to the deuterium contained in the feed gas, which seldom contains more than 0.013 a/o (atom percent) deuterium. Even with complete deuterium extraction, a large plant producing 1000 short tons (t) of ammonia synthesis gas per day and operating 330 days/year could yield only

$$\left(\frac{1000\ \text{t NH}_3}{\text{day}}\right)\left(\frac{330\ \text{days}}{\text{yr}}\right)\left(\frac{3\ \text{atoms H}}{\text{molecule NH}_3}\right)\left(\frac{0.00013\ \text{atoms D}}{\text{atom H}}\right)\left(\frac{20\ \text{t D}_2\text{O}}{\text{t-mol D}_2\text{O}}\right)\Big/$$

$$\left(\frac{17\ \text{t NH}_3}{\text{t-mol NH}_3}\right)\left(\frac{2\ \text{atoms D}}{\text{molecule D}_2\text{O}}\right) = 75.7\ \text{t D}_2\text{O/yr} \tag{12.3}$$

Concentration of deuterium by the electrolysis of water was proposed by Washburn and Urey [W1], used by Lewis [L4] to make the first samples of pure D_2O, and employed for the first production of heavy water on a large industrial scale by the Norsk Hydro Company, at Rjukan, Norway. The Rjukan plant makes use of cheap hydroelectric power to produce electrolytic hydrogen for ammonia synthesis and by-product heavy water.

When Germany occupied Norway in World War II, this plant was producing 1.5 MT/year of heavy water, and around 90,000 MT/year of ammonia. The water being electrolyzed contained 21 MT/year of heavy water, of which 10 could have been recovered by burning hydrogen enriched in deuterium from the higher stages of the plant and recycling the deuterium-rich water. This, however, would have reduced the ammonia output by 23,000 MT/year. The German scientists Harteck, Hoyer, and Suess [C2] conceived the ingenious idea of recovering deuterium from the hydrogen gas by absorption in water, by making use of the exchange reaction

$$HD + H_2O \rightleftharpoons HDO + H_2 \qquad K = 3.0$$

in which deuterium concentrates in the water. A nickel catalyst for carrying out this reaction in the gas phase was developed. One catalytic reactor was installed at Rjukan, and others to bring the heavy-water production up to 5 MT/year were planned, but the plant was destroyed in 1943 in a series of daring commando raids. It was rebuilt after the war and has been in operation since then.

At about the same time, a similar exchange process was developed by Urey and Taylor [M5, S2, T1], working under the Manhattan Project in the United States. The Standard Oil Development Company designed the exchange equipment [B1] and installed it in the electrolytic hydrogen plant of the Consolidated Mining and Smelting Company, at Trail, British Columbia, where it was operated until 1955. This plant produced 6 MT D_2O/year at a concentration of 2.37 w/o (weight percent) D_2O. Final concentration to 99.7 w/o D_2O was by electrolysis. The cost was \$130/kg D_2O.

A second method for the industrial production of heavy water, used by the Manhattan Project in the United States [M5], was the distillation of water. Three plants having a total capacity of 13 MT D_2O/year were built at Army Ordinance plants. Because the relative volatility for separating H_2O from HDO is only 1.03 at atmospheric pressure, the size of equipment and the heat consumption of these plants per unit of D_2O produced was very high, and the cost of heavy water was greater than in other processes. Nevertheless, the distillation of water was attractive as a wartime production method because the process needed little development work and used standard equipment. These plants were shut down after the war. More recently, distillation of water has come to be one of the most satisfactory methods for final concentration of heavy water.

Because the relative volatility for separation of deuterium by the distillation of liquid hydrogen is around 1.5 at atmospheric pressure, the size and heat consumption of a hydrogen distillation plant would be much smaller than that of a water distillation plant producing the same amount of deuterium. Plants to concentrate deuterium by the distillation of liquid hydrogen were designed by German engineers [C2] and by the Manhattan Project [M5] during World War II, and by Hydrocarbon Research, Inc. [H6], in the United States, but none of these plants was built because of uncertainty about the performance of industrial equipment operating at the very low temperatures needed to liquefy hydrogen. In 1949 a group of Soviet engineers undertook the development work necessary to ensure success of this type of plant, and in 1958 announced [M1] that a plant producing deuterium by distillation of electrolytic hydrogen had been in operation in the Soviet Union for some years. The plant consists of multiple units, each with a capacity of around 4 MT D_2O/year.

In 1958, two companies specializing in cryogenic engineering put into operation experimental plants for concentrating deuterium by distillation of ammonia synthesis gas (75 percent H_2, 25 percent N_2). Société de l'Air Liquide designed and built one at the ammonia plant of Office National Industriel de l'Azote (ONIA), at Toulouse, France, which is operated by Compagnie Française de l'Eau Lourde, jointly owned by Air Liquide and ONIA. Gesellschaft für Linde's Eismaschinen designed and built a second deuterium plant at the ammonia plant of Farbwerke Hoechst, at Hoechst, Germany. The production rates of the plants were roughly 2 and 6 MT D_2O/year, respectively. Because of the small size of these plants, the high local cost of electric power, and the less-than-natural deuterium content of the available synthesis gas, the cost of heavy water produced in these plants was high. After sufficient information had been obtained to permit design of larger plants at other locations where local conditions were more favorable, both plants were shut down in 1960.

In 1959, Sulzer Brothers designed and built a plant to distill electrolytic hydrogen enriched to six times the natural abundance of deuterium, which was available at the ammonia plant of Emswerke AG, at Ems, Switzerland [H1]. At this plant, about 2 MT/year of heavy water were produced at a cost near \$62/kg. The cost at Ems was lower than at Toulouse or Hoechst because of the higher deuterium content of feed and the low content of nitrogen and other condensable impurities in electrolytic hydrogen. This plant has been shut down because production of the electrolytic hydrogen that fed the heavy water plant has become too costly. In 1961, a 14 MT/year plant of this type was built by Linde to distill electrolytic hydrogen enriched to three times the natural abundance of deuterium, which was available at the Indian government's ammonia plant at Nangal, India.

Another process that has been used to extract deuterium from ammonia synthesis gas is the deuterium-exchange reaction between liquid ammonia and gaseous hydrogen:

$$NH_3(l) + HD(g) \rightleftharpoons NH_2D(l) + H_2(g)$$

In the presence of potassium amide, KNH_2, as catalyst dissolved in liquid ammonia, equilibrium favors concentration of deuterium in the liquid phase. A 26 MT/year plant using this process was operated at Mazingarbe, France, in the late 1960s, and three larger plants with a combined capacity over 200 MT/year are being built in India.

All of the previously mentioned plants except those employing distillation of water were parasitic to a synthetic ammonia plant. Their deuterium-production rate is limited by the amount of deuterium in ammonia synthesis gas. To produce heavy water at a sufficient rate, a growing industry of heavy-water reactors requires a deuterium-containing feed available in even greater quantity than ammonia synthesis gas. Of the possible candidates, water, natural gas, and petroleum hydrocarbons, water is the only one for which an economic process has been devised, and the dual-temperature hydrogen sulfide-water exchange process is the most economic of the processes that have been developed.

This process, invented by Spevack [S7] and developed independently by Geib [C2] in Germany, makes use of the fact that the separation factor α for exchange of deuterium between liquid water and gaseous hydrogen sulfide,

$$H_2O(l) + HDS(g) \rightleftharpoons HDO(l) + H_2S(g)$$

is

$$\alpha_c = 2.32 \text{ at } 32°\text{C} \quad \text{and} \quad \alpha_h = 1.80 \text{ at } 138°\text{C}$$

By running liquid water countercurrent to recycled gaseous hydrogen sulfide through first a cold tower and then a hot tower, as shown schematically in Fig. 12.6, water enriched in deuterium may be withdrawn from the water leaving the cold tower. The principle of the process and process flow sheets are described in detail in Chap. 13.

The first plant of this type, designed by the Girdler Corporation and operated by E. I. du Pont de Nemours and Company, built at the Wabash Ordnance Plant at Dana, Indiana, in 1952 but later shut down, gave this process the name the G-S process, for Girdler-Sulfide. Three improved units, each with a capacity of 160 MT/year, were designed, built, and operated by du Pont at Aiken, South Carolina [B2]; one is still in operation at a reduced capacity of 69 MT/year.

Figure 12.7 is a photograph of this plant. The world's principal heavy-water production capacity is found in Canada, where G-S plants with a total capacity of 4000 MT/year are in operation or under construction.

2.3 Lithium Isotopes

Many methods have been used to achieve partial separation of lithium isotopes on a small scale. Examples of processes and reported separations are listed in Table 12.4. A process somewhat similar to the last one listed in this table, involving countercurrent exchange of lithium isotopes between aqueous lithium hydroxide and lithium amalgam, is to be used in a plant being built

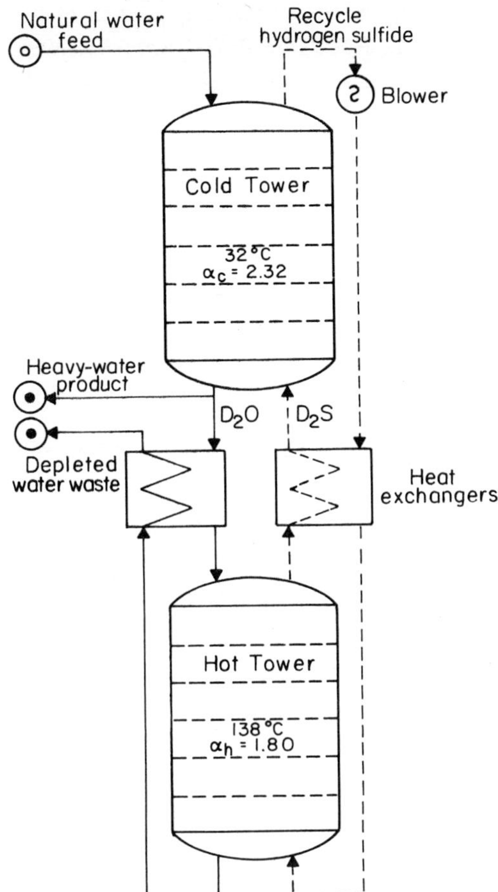

Figure 12.6 Dual-temperature water-hydrogen sulfide process.

by Eagle Picher Industries, Inc., at Quapaw, Oklahoma, to produce 1000 kg 99.99 percent ^{7}Li per year at an approximate price of \$3/g.

2.4 ^{10}B

Table 12.5 compares four processes that have been used for concentrating ^{10}B. The research that led to the first commercial production of ^{10}B was carried out by Crist and Kirshenbaum [C5] in the laboratory of H. C. Urey at Columbia University in 1943. As reported by Kilpatrick and co-workers [K1], it was concluded that the most satisfactory process consisted in the equilibrium distillation of the complex of boron trifluoride and dimethyl ether, $BF_3 \cdot (CH_3)_2O$. When this substance vaporizes, it dissociates partially according to the reaction

$$BF_3 \cdot (CH_3)_2O \rightleftharpoons BF_3 + (CH_3)_2O$$

The isotopic exchange equilibrium

$$^{10}BF_3(g) + {}^{11}BF_3 \cdot (CH_3)_2O(l) \rightleftharpoons {}^{11}BF_3(g) + {}^{10}BF_3 \cdot (CH_3)_2O(l)$$

is then established, with an equilibrium constant of 1.027 at 100°C [K2]. When the liquid is distilled at 100°C, the vapor phase is 60 percent dissociated.

Because there is no separation of boron isotopes in the equilibrium vaporization of the complex, the overall separation factor in the combined process of evaporation and dissociation is

$$(1.027)(0.6) + (1.000)(0.4) = 1.016 \tag{12.4}$$

This value has been confirmed experimentally [K1].

A semicommercial plant based on this process was built and operated for the Manhattan Project by the Standard Oil Company of Indiana [C4]. In 1953, the U.S. Atomic Energy Commission authorized construction of a larger plant at Niagara Falls, New York, with the Hooker Electrochemical Company as operating contractor [M3]. This plant produced 460 kg/year of ^{10}B at an enrichment of 92 a/o ^{10}B. The plant was shut down in January 1958. Eagle Picher Industries, Inc., has been producing ^{10}B at Quapaw, Oklahoma, by this process since 1973 and is expanding capacity to 1000 kg/year. The cost is from $5 to $15/g.

A plant producing 2 kg of ^{10}B per year by equilibrium distillation of the complex of BF_3 and diethyl ether, $BF_3 \cdot (C_2H_5)_2O$, was operated by 20th Century Electronics, Ltd., in New Addington, England [E1]. The process, developed by the U.K. Atomic Energy Authority (UKAEA), is generally similar to the U.S. process using the dimethyl ether complex. Both plants are operated at subatmospheric pressure, to minimize irreversible decomposition of the complex.

Distillation of BF_3 is another process that has been used to concentrate ^{10}B. This has the advantage over the processes using ether complexes of BF_3 that decomposition is not a problem, so that the plant can be operated at atmospheric pressure and can be scaled up without special concern about increased column pressure drop. Disadvantages of BF_3, however, are that the separation factor is only 1.0075 [N1], and the reflux condenser must be operated

Figure 12.7 Heavy-water plant at Aiken, South Carolina. (*Courtesy of U.S. Energy Research and Development Administration.*)

Table 12.4 Methods tested for separating lithium isotopes

Method	Investigated by	Reference	Separation factor	Enrichment obtained
Differential ion migration				
Fused LiCl	Klemm et al.	[K4]		^{7}Li to 97%; ^{6}Li to 16%
	Klemm	[K3]		^{7}Li to 99.974%
Fused LiBr	Lundén	[L7]		
Fused $LiNO_3$	Vallet et al.	[V1]		
Electrolysis of LiCl in H_2O	Johnston and Hutchison	[J2]	1.055	
	Perret et al.	[P1]	1.05–1.07	
Molecular distillation of Li	Trauger et al.	[T6]	1.06	^{6}Li to 9% in 8 stages
Equilibrium distillation of Li	Perret et al.	[P1]	1.03	
Chemical exchange				
Li amalgam vs. LiCl in alcohol	Lewis and MacDonald	[L5]		^{6}Li to 14%
Li amalgam vs. LiBr in DMF†	Perret et al.	[P1]	~1.05	
Ion exchange				
Aqueous LiCl vs. zeolite	Taylor and Urey	[T3]	1.022	
Aqueous LiCl vs. zeolite	Sessions et al.	[S3]	1.004–1.006	
Aqueous LiCl vs. Dowex 50 × 12	Perret et al.	[P1]	1.002	^{6}Li to 10.2%
Aqueous LiCl vs. Dowex 50	Lee and Begun	[L2]	1.0038	
Chemical exchange between lithium amalgam and aqueous solution of lithium compound	Saito and Dirian	[S1]		

†DMF, dimethyl formamide.

Table 12.5 Methods used for separating ^{10}B

Reference	Method of separation	Working substance	Operating conditions		Separation factor	^{10}B production rate, kg/year	Percent ^{10}B
			Pressure, Torr	Temperature, °C			
[M3]	Distillation + exchange	$BF_3 \cdot (CH_3)_2O$	150–275	91–104	1.016	460	92
[E1]	Distillation + exchange	$BF_3 \cdot (C_2H_5)_2O$	20–53	10–75	1.016	2	95
[N1]	Distillation	BF_3	760	−101	1.0075	26.5	95
[H2]	Distillation + exchange	$BF_3 \cdot$anisole		25	1.032	–	–

at temperatures in the inconvenient range between the melting point of BF_3 (−127°C) and its normal boiling point (−101°C). Despite these difficulties, the process was used successfully in the Soviet Union [M4] to produce 0.5 kg/year of ^{10}B enriched to 83 percent, and in England by the UKAEA [N1] to produce 26.5 kg/year enriched to 95 percent. ^{10}B concentrates in the liquid phase, as in the exchange equilibrium.

2.5 ^{13}C

Natural carbon contains 1.11 percent ^{13}C. This isotope was first produced commercially at a rate of around 1 g/day by the Eastman Kodak Company [S8], using the exchange reaction between HCN gas and NaCN solution developed in 1940 by Urey and co-workers [H5]. The separation factor is 1.013.

^{13}C has also been produced by the low-temperature distillation of carbon monoxide, in a process developed by London and co-workers [J1, L6]. A carbon monoxide distillation plant has been in operation at Harwell since 1949, producing 0.4 g/day of ^{13}C at 60 to 70 percent enrichment. Simultaneously, the plant produces 0.045 g/day of ^{18}O at 5 to 6 percent enrichment. The separation factors for these two separations are

$$^{12}C^{16}O/^{13}C^{16}O: \quad 1.011$$

$$^{12}C^{16}O/^{12}C^{18}O: \quad 1.008$$

A carbon monoxide distillation plant at Los Alamos Scientific Laboratory produces 4 kg ^{13}C/year [A2] at 90 percent enrichment.

2.6 ^{15}N

Natural nitrogen contains 0.365 percent ^{15}N. Methods that have been used for separating ^{15}N on a small scale are listed in Table 12.6.

The exchange reaction between NH_3 gas and NH_4NO_3 in aqueous solution was used by Thode and Urey in 1939 to obtain the first samples of enriched ^{15}N, and was employed by the Eastman Kodak Company to produce ^{15}N at a rate of around 1 g/day. The only production of ^{15}N in the United States at present is by distillation of NO at Los Alamos [M2].

2.7 Heavy Oxygen Isotopes

Natural oxygen contains 0.037 percent ^{17}O and 0.204 percent ^{18}O. The isotope ^{18}O was first concentrated by Huffman and Urey [H4] in 1937, by the distillation of water. Although the separation factor is very low (1.004 at 100°C), the method has been adapted to semi-commercial production by Dostrovsky [D3], who produced 11 g/day of ^{18}O at an enrichment

Table 12.6 Methods used for separating nitrogen isotopes

Process	Investigators	Reference	Separation factor	Percent ^{15}N
NH_3-NH_4^+ exchange	Thode and Urey	[T4]	1.023	72.8
N_2 thermal diffusion	Clusius	[C1]	—	99.8
NH_3-NH_4R ion exchange	Spedding et al.	[S6]	1.026	99.7
NO-HNO_3 exchange	Spindel and Taylor	[T2]	1.055	99.9
NO distillation	McInteer and Potter	[M2]	1.027	93.9

Table 12.7 Isotope separation methods

Method	Applied to
Electromagnetic	^{235}U, all others
Electrolysis	D, Li
Distillation	D, ^{10}B, ^{13}C, ^{15}N, ^{18}O
Chemical exchange	D, Li, ^{10}B, ^{13}C, ^{15}N, ^{18}O
Ion migration	Li
Diffusion methods	^{235}U
Gas centrifuge	^{235}U
Aerodynamic methods	^{235}U

of 95 percent. ^{18}O is also concentrated when water is distilled for deuterium separation, but the low separation factor for the oxygen isotopes limits the degree of enrichment.

Other methods used to concentrate ^{18}O are the distillation of CO, referred to in Sec. 2.4, the distillation of NO, and the exchange reaction between CO_2 gas and water [B4], for which the separation factor is 1.02. Boyd [B3] estimated that ^{18}O could be produced at a rate of 4 g/day at a cost of $93/g by this process.

2.8 Recapitulation of Separation Methods

The most useful methods mentioned above are recapitulated in Table 12.7.

This text is concerned primarily with methods used on a large industrial scale. Electrolysis, distillation, and chemical exchange, which are useful primarily for separating deuterium and isotopes of other light elements, will be described in Chap. 13. Diffusion methods, the gas centrifuge, and aerodynamic methods, which are used primarily for uranium but are applicable also to other heavy elements, will be described in Chap. 14.

The separation factor in all of these processes is so close to unity that production of separated isotopes requires repeated partial separations in a multistage cascade generally similar to the gaseous diffusion cascade of Fig. 12.2. The remainder of this chapter develops theoretical principles of isotope separation in such cascades.

3 TERMINOLOGY

3.1 Separating Unit, Stage, and Cascade

The smallest element of an isotope separation plant that effects some separation of the process material is called a *separating unit.* Examples of a single separating unit are one stage of a mixer-settler, one plate of a distillation column, one gas centrifuge, one calutron, or one electrolytic cell.

A group of parallel-connected separating units, all fed with material of the same composition and producing partially separated product streams of the same composition, is known as a *stage.* Often a single unit serves as a stage, like a plate of bubble-plate column. However, in some separation methods whose units have low capacity, such as an electrolytic cell or centrifuge, it is necessary to use many units in parallel.

When the degree of separation effected by a single stage is less than the degree of separation desired between product and waste, it is necessary to connect stages in series. Such a series-connected group of stages is known as a *cascade.* Examples of a cascade are a complete distillation column or a battery of solvent extraction mixer-settlers.

The relation between unit, stage, and cascade is illustrated by Fig. 12.8. Each unit of this cascade might represent, for example, an electrolytic cell. The group of parallel-connected cells, each of which separates feed of composition z_1 into a partially enriched stream of composition y_1 and a partially depleted stream of composition x_1, constitutes the first stage of this cascade. The cascade is the entire group of series- and parallel-connected cells.

A cascade that has the same number of units (i.e., the same capacity) in all stages of a group is known as a *"squared-off" cascade.* A cascade in which the number of units, or the capacity, in each stage decreases as the produce and waste ends of the cascade are approached is called a *tapered cascade.* A single multiplate distillation column is an example of a squared-off cascade; a gaseous diffusion plant for uranium separation is an example of a tapered cascade.

The engineering analysis of separation processes frequently employs the concept of an *ideal*, or *equilibrium stage.* In such a stage, the feed streams, which may be one or two in number, are acted upon to produce two product streams that are in equilibrium. The use of such a concept can be employed in the design and analysis of both stagewise and continuous contacting equipment. Determination of the number of stages in a cascade required to achieve a given separation involves the determination of the number of such ideal stages followed by application of a *stage efficiency*, which expresses the fraction of ideal transfer achieved in the actual stages employed.

3.2 Measures of Composition

In Chaps. 12, 13, and 14, dealing with isotope separation, the composition of a mixture may be expressed in terms of the *weight* (or mass) *fraction* of each component, the *mole fraction* of

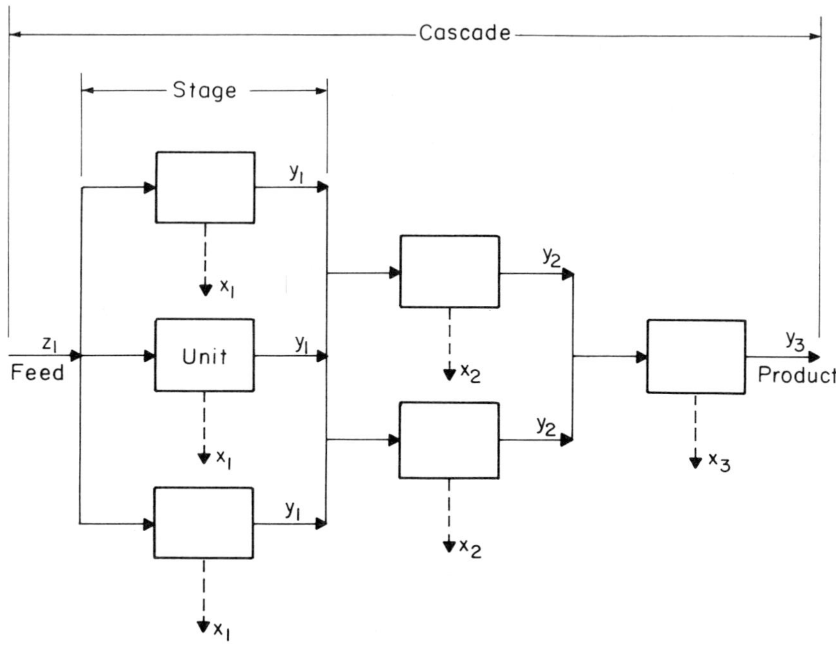

Figure 12.8 Unit, stage, and cascade.

each component, or the *atom fraction* of each isotope. The relations among these three measures of composition may be illustrated by the example of water containing 0.79 weight fraction H_2O (molecular weight 18), 0.19 weight fraction HDO (molecular weight 19), and 0.02 weight fraction D_2O (molecular weight 20). The procedure to obtain mole fractions from these weight fractions is shown below:

	Mol/g mixture	Mole fraction
H_2O	0.79/18 = 0.0439	0.0439/0.0549 = 0.800
HDO	0.19/19 = 0.010	0.010/0.0549 = 0.182
D_2O	0.02/20 = 0.001	0.001/0.0549 = 0.018
	0.0549	1.000

The atom fraction of deuterium is the ratio of the number of atoms of deuterium to the number of atoms of deuterium plus hydrogen in the mixture, or

$$\frac{(0.182 \times 1) + (0.018 \times 2)}{2} = 0.109 \tag{12.5}$$

The symbol z will be used to represent the fraction of a component in the feed stream to a unit, stage, or cascade; y the fraction in the enriched stream leaving a unit, stage, or cascade; and x the fraction in the depleted stream leaving a unit, stage, or cascade. The context will indicate whether weight, mole, or atom fractions are being dealt with. In the case of compounds containing a single atom of a polyisotopic element, such as UF_6, atom fractions and mole fractions are identical.

For mixtures of two isotopes, the symbol z, y, or x refers to the fraction of desired isotope (for example, ^{235}U in the case of uranium or D in the case of hydrogen). For mixtures of three or more isotopes, the first subscript following z, y, or x indicates the specific isotope.

The location of a stream in a unit, stage, or cascade is also designated by a subscript, standing alone for a two-component system, or standing second after a comma for a multicomponent one. For example, $z_{i,F}$ is the fraction of the ith isotope in feed.

Some relations for isotope separation plants are simpler when expressed as weight, mole, or atom ratios, defined as the ratio of the fraction of one component to the fraction of a second. These ratios are denoted by Greek letters ζ, ξ, or η for feed, depleted, or enriched stream, corresponding to z, x, or y. In a two-component mixture, these ratios are defined as the ratio of the fraction of the desired component to that of the other component. For example, in a tails stream, the weight, mole of atom ratio for a two-component mixture is

$$\xi \equiv \frac{x}{1-x} \tag{12.6}$$

For a multicomponent mixture the two components entering the ratio are designated by a double subscript, without comma, for example,

$$\xi_{58} \equiv \frac{x_5}{x_8} \tag{12.7}$$

The ratio of atom fractions is frequently termed the *abundance* ratio. For example, the abundance ratio of ^{235}U to ^{238}U in natural uranium containing 0.007205 atom fraction ^{235}U and 0.99274 atom fraction ^{238}U is 0.007205/0.99274 = 0.007258.

4 STAGE PROPERTIES

4.1 Terminology

The simplest type of separating unit or stage is one that receives one feed stream and produces one heads stream enriched in the desired component and one tails stream depleted in the desired component. Figure 12.9 shows such a stage, which is fed at rate Z with z fraction desired component and which produces a *heads* stream at rate M with y fraction desired component and a *tails* stream at rate N with x fraction desired component. Flow rate and composition should be on the same basis, e.g., weight, mole, or atom.

Because of overall material balance,

$$Z = M + N \tag{12.8}$$

Material balanced on desired component leads to

$$Zz = My + Nx \tag{12.9}$$

Hence

$$\frac{M}{Z} = \frac{z - x}{y - x} \tag{12.10}$$

and

$$\frac{N}{Z} = \frac{y - z}{y - x} \tag{12.11}$$

The ratio of heads flow rate to feed rate is known as the *cut* θ,

$$\theta \equiv \frac{M}{Z} = \frac{z - x}{y - x} \tag{12.12}$$

The fraction of a component appearing in the heads stream is known as the *recovery* r of that component. The recovery of the desired component, for example, is

$$r \equiv \frac{yM}{zZ} = \frac{y\theta}{z} \tag{12.13}$$

Alternatively,

$$r = 1 - \frac{x(1 - \theta)}{z} \tag{12.14}$$

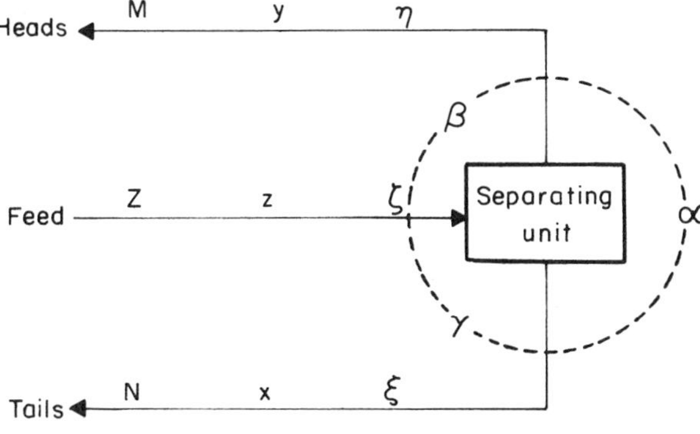

Figure 12.9 Flow rates, compositions, and separation factors.

4.2 Separation Factors

The degree of separation achieved by a single stage is known as the stage separation factor, or simply the *separation factor* α. This is defined as the weight, mole, or atom ratio in the heads stream divided by the corresponding ratio in the tails. For a two-component mixture,

$$\alpha \equiv \frac{\eta}{\xi} = \frac{y/(1-y)}{x/(1-x)} \tag{12.15}$$

The separation factor defined in this way is useful because in many isotope separation processes, it is independent of composition. The ratio y/x, on the other hand, may vary strongly with composition.

Other useful measures of the degree of separation effected by a stage are the heads *separation factor* β, defined by

$$\beta \equiv \frac{\eta}{\zeta} = \frac{y/(1-y)}{z/(1-z)} \tag{12.16}$$

and the tails separation factor γ, defined by

$$\gamma \equiv \frac{\zeta}{\xi} = \frac{z/(1-z)}{x/(1-x)} \tag{12.17}$$

The composition differences measured by α, β, and γ are indicated in Fig. 12.9 by the curved lines. From the definitions of α, β, and γ, it follows that

$$y = \frac{\beta z}{\beta z + 1 - z} = \frac{\alpha x}{\alpha x + 1 - x} \tag{12.18}$$

$$z = \frac{\alpha x}{\alpha x + \beta(1-x)} = \frac{y}{y + \beta(1-y)} \tag{12.19}$$

$$x = \frac{y}{y + \alpha(1-y)} = \frac{z}{1 + \gamma(1-z)} \tag{12.20}$$

A relation between β, α, and θ may be obtained from (12.12), (12.15), and (12.16):

$$\beta - 1 = \frac{(\alpha - 1)(1-\theta)}{1 + \theta(\alpha - 1)(1-y)} \tag{12.21}$$

4.3 Differential Stage Separation

In some stage processes, the heads and tails streams are separated in such a way that all portions of each stream have uniform composition. This occurs, for example, in a well-mixed electrolytic cell operated with steady flow of feed water and steady withdrawal of partially electrolyzed water. In other stage processes, the heads or tails stream may be withdrawn in such a way that the other stream changes progressively in composition during the separation process. This occurs, for example, when water flows through an electrolytic cell without mixing, and becomes progressively richer in deuterium, or when water is electrolyzed batchwise and becomes richer in deuterium as time goes on. These are examples of *differential stage separation*, in which successive small portions of one stream are removed from a second without mixing the second stream or giving the first stream further opportunity to exchange material with the second.

Two types of differential stage separation are illustrated in Fig. 12.10. In type A the stream being removed in small portions is *depleted* in the desired component, while the remaining stream becomes progressively enriched in this component; the concentration of

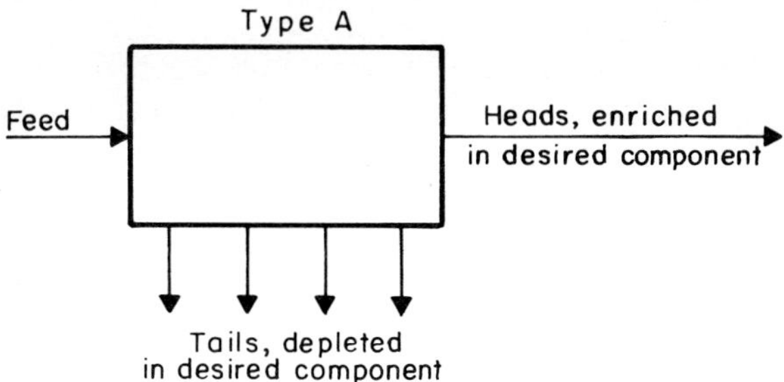

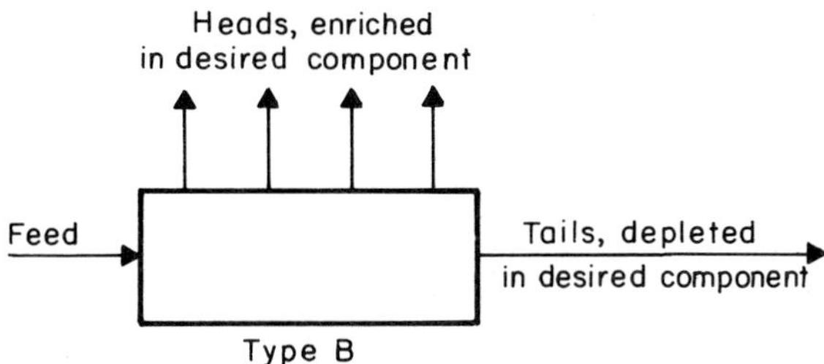

Figure 12.10 Two types of differential stage separation.

deuterium in batch electrolysis of water is an example of this type of differential stage separation.

In type B the stream being removed in small portions is enriched in the desired component, while the remaining stream becomes progressively depleted in this component. The flow of a mixture of $^{235}UF_6$ and $^{238}UF_6$ along the barrier of a gaseous diffusion stage is an example of this type of process. The small portions of gas that pass through the barrier are enriched in the desired component, $U^{235}F_6$, and the remaining gas flowing along the upstream side of the barrier becomes progressively depleted in $^{235}UF_6$.

Equations relating the flow rates and compositions of feed and product streams in differential separation processes, first derived by Lord Rayleigh [R1] for batch distillation, are often called the Rayleigh distillation equation. We shall derive some of these relationships for type B differential stage separation, using the nomenclature shown in Fig. 12.11.

At a point in the stage where a small amount of heads stream having flow rate dM' and composition y' is separated, the flow rate of the remaining depleted stream is changed by amount dN' and its composition is changed by dx'. The material balance equation on total flow is

$$dM' = -dN' \tag{12.22}$$

and the material balance equation on flow of desired component is

$$y'dM' = -d(x'N') \tag{12.23}$$

The result of elimination dM' is

$$-y'dN' = -d(x'N') \tag{12.24}$$

or

$$\frac{dN'}{N'} = \frac{dx'}{y'-x'} \tag{12.25}$$

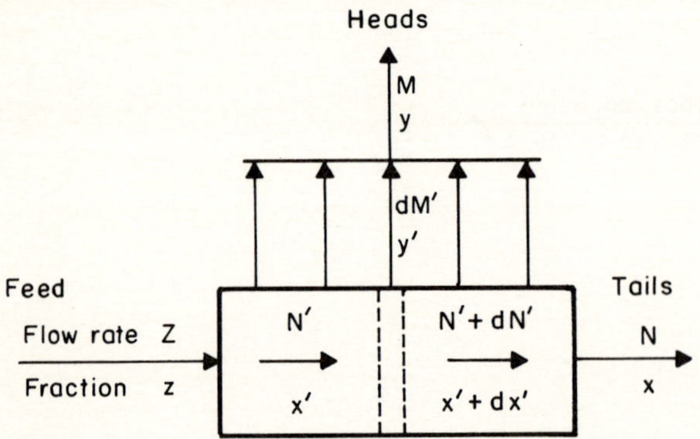

Figure 12.11 Nomenclature for type B differential stage separation.

The result of integrating this equation from the feed end of the stage at which the flow rate is Z and composition z to the tails end where the flow rate is N and the composition x is

$$\ln \frac{N}{Z} = \int_z^x \frac{dx'}{y' - x'} \tag{12.26}$$

This is the general form of the Rayleigh equation. When the relationship between y' and x' is known, the equation may be integrated graphically or numerically.

For a two-component mixture, the relationship between y' and x' may be expressed in terms of a *local separation factor* α', defined as

$$\alpha' \equiv \frac{y'/(1-y')}{x'/(1-x')} \tag{12.27}$$

in analogous fashion to the stage separation factor defined by (12.15). The result of using this equation to eliminate y' from (12.26) is

$$\ln \frac{N}{Z} = \int_z^x \left(\frac{\alpha'}{1-x'} + \frac{1}{x'} \right) \frac{dx'}{\alpha' - 1} \tag{12.28}$$

When α' is constant throughout the stage, this equation may be integrated to give

$$\ln \frac{N}{Z} = \frac{\alpha'}{\alpha' - 1} \ln \frac{1-z}{1-x} + \frac{1}{\alpha' - 1} \ln \frac{x}{z} \tag{12.29}$$

Because

$$\frac{N}{Z} = 1 - \theta \tag{12.30}$$

this may be transformed to

$$\frac{x}{z}(1-\theta) = \left[\left(\frac{1-x}{1-z} \right) (1-\theta) \right]^{\alpha'} \tag{12.31}$$

A relation between the stage separation factor α and the local separation factor α' may be obtained from Eq. (12.31) by using (12.12) to replace z by y and (12.15) to eliminate y:

$$1 + \frac{\alpha\theta}{(1-\theta)(\alpha x + 1 - x)} = \left[1 + \frac{\theta}{(1-\theta)(\alpha x + 1 - x)} \right]^{\alpha'} \tag{12.32}$$

When $\alpha' - 1 \ll 1$, as in separating uranium isotopes by gaseous diffusion, this equation reduces to

$$\alpha - 1 = -\frac{(\alpha' - 1)\ln(1-\theta)}{\theta} \approx (\alpha' - 1)\left(1 + \frac{\theta}{2} + \frac{\theta}{3} + \cdots\right) \tag{12.33}$$

In this form it can be seen that α is greater than α', and becomes much greater as θ approaches unity. Thus, differential stage separation may be used to enhance the difference in composition attainable in simple stage separation.

For type A differential stage separation, a similar derivation leads to

$$\frac{1-y}{1-z}\theta = \left(\frac{y\theta}{z}\right)^{\alpha'} \tag{12.34}$$

Because of Eq. (12.13) defining r and (12.16) defining β,

$$r = \frac{1}{\beta^{1/(\alpha'-1)}} \tag{12.35}$$

or

$$\beta = \frac{1}{r^{\alpha'-1}} \tag{12.36}$$

The equation corresponding to (12.33), applicable when $\alpha - 1 \ll 1$, is

$$\alpha - 1 = -\frac{(\alpha' - 1)\ln\theta}{1-\theta} \tag{12.37}$$

5 TYPES OF CASCADE

A cascade like Fig. 12.12, in which no attempt is made to reprocess the partially depleted tails streams leaving each stage, will be called a simple cascade. In a simple cascade the feed stream for one stage is the heads stream from the next lower stage of the cascade. This type of cascade connection is used in the lower stages of the Norsk Hydro electrolytic heavy-water plant where the tails streams have too little deuterium to warrant processing for deuterium recovery. The theory of such a cascade is developed in Sec. 6.

When partially depleted tails have sufficient value to warrant reprocessing, a countercurrent recycle cascade like Fig. 12.13 may be used. This cascade flow scheme is by far the most

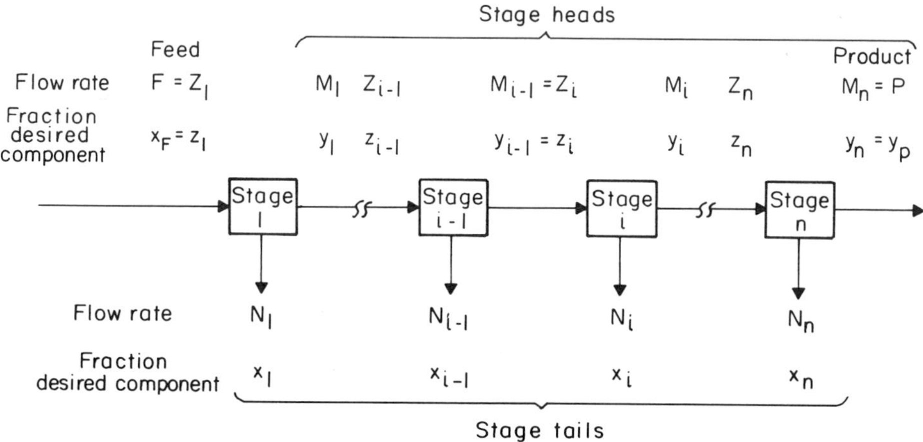

Figure 12.12 Simple cascade, no reprocessing of tails.

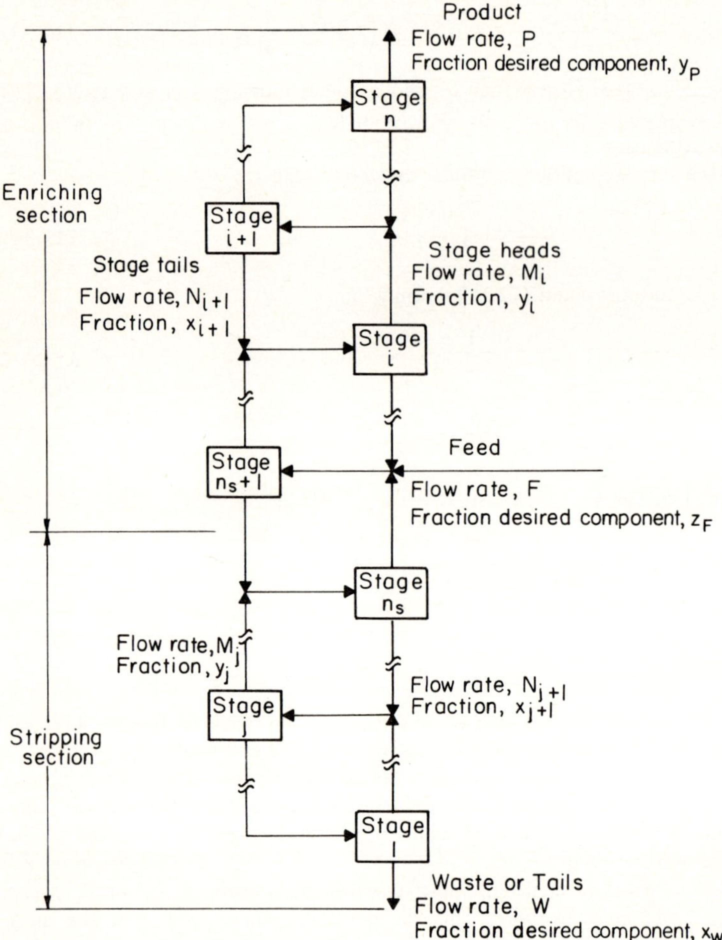

Figure 12.13 Countercurrent recycle cascade.

common. It is approached, for example, in a bubble-plate distillation column and is used in a battery of series-connected solvent extraction mixer-settlers or in the gaseous diffusion cascade of Fig. 12.2.

Such a countercurrent cascade separates *feed* containing z_F fraction of desired component flowing at rate F into *product* containing y_P fraction of desired component flowing at rate P and *waste*, or *tails*, containing x_W fraction of desired component flowing at rate W. These six compositions and flow rates are called the *external variables* of the cascade.

Feed for each stage consists of heads from the next lower stage and tails from the next higher stage. These interstage flow rates and compositions will be called the *internal variables* of the cascade.

The portion of the cascade between the feed point and product end is called the *enriching section*; the portion between the feed point and waste end is called the *stripping section.* The purpose of the enriching section is to make material of product composition; the purpose of the stripping section is to increase the recovery of desired isotope from feed. The enriching section is essential in making product of the desired grade; the stripping section is used only to reduce the amount of feed required to make a given amount of product. When feed has no value, as with water feed for a deuterium plant, the stripping section may be eliminated altogether.

Stages of the cascade are numbered consecutively from 1 at the waste end of the plant to n at the product end. The highest stage of the stripping section is numbered n_S.

The streams that move away from the ends of the cascade, that is, the tails stream in the enriching section and the heads stream in the stripping section, are known as *reflux*.

The theory of a recycle cascade is developed in Sec. 7.

6 THE SIMPLE CASCADE

Figure 12.12 illustrates flow through a simple cascade, fed at rate F with material containing z_F fraction of desired component, to produce product at rate P containing y_P fraction of desired component. Feed for one stage consists of heads from the next lower stage, so that

$$Z_i = M_{i-1} \tag{12.38}$$

and

$$z_i = y_{i-1} \tag{12.39}$$

The recovery of desired component from the ith stage, r_i, is

$$r_i = \frac{M_i y_i}{Z_i z_i} \tag{12.40}$$

but

$$\frac{M_i}{Z_i} = \frac{z_i - x_i}{y_i - x_i} \tag{12.41}$$

as in (12.10), so that

$$r_i = \frac{1-(x_i/z_i)}{1-(x_i/y_i)} \tag{12.42}$$

The result of replacing y_i and z_i by their expressions in terms of x_i, Eqs. (12.18) and (12.19), respectively, is

$$r_i = \frac{1-\{[\alpha_i x_i + \beta_i(1-x_i)]/\alpha_i\}}{1-[(\alpha_i x_i + 1 - x_i)/\alpha_i]} = \frac{\alpha_i - \beta_i}{\alpha_i - 1} \tag{12.43}$$

The recovery r from all n stages of the cascade is

$$r = r_1 \cdots r_{i-1} r_i \cdots r_n \tag{12.44}$$

The overall enrichment of a simple cascade ω may be defined as

$$\omega \equiv \frac{y_n/(1-y_n)}{z_1/(1-z_1)} \tag{12.45}$$

From the definition of heads separation factor,

$$\beta_i = \frac{y_i/(1-y_i)}{z_i/(1-z_i)} \tag{12.46}$$

and condition (12.39), it follows that

$$\omega = \beta_1 \cdots \beta_{i-1} \beta_i \cdots \beta_n \tag{12.47}$$

In a cascade in which α_i and β_i are independent of stage number, (12.43) and (12.44) reduce to

$$r = \left(\frac{\alpha - \beta}{\alpha - 1}\right)^n \tag{12.48}$$

and (12.47) to

$$\omega = \beta^n \tag{12.49}$$

The relation between recovery, overall enrichment, and number of stages then is

$$r = \left(\frac{\alpha - \omega^{1/n}}{\alpha - 1}\right)^n \tag{12.50}$$

Figure 12.14 illustrates the variation of r with ω for a simple cascade of electrolytic cells with $\alpha = 7$ and $n = 1$, 2, or 3. The recovery is greater the greater the number of stages. In the limit, as the number of stages increases indefinitely, the recovery from Eq. (12.50) approaches

$$\lim_{n \to \infty} r = \frac{1}{\omega^{1/(\alpha - 1)}} \tag{12.51}$$

The line for $n \to \infty$ is also shown in Fig. 12.14. This is the highest recovery that can be obtained in a simple cascade, with $\alpha = 7$.

Such a simple cascade, with an infinite number of stages each performing an infinitesimal amount of separation, is equivalent to type A differential stage separation. Equation (12.51) is equivalent to the form of the Rayleigh equation (12.35), when one recognizes that ω in the simple cascade is equivalent to the heads separation factor β in differential stage enrichment, and α in the simple cascade is equivalent to the local separation factor α'.

7 THE RECYCLE CASCADE

In the simple cascade of Fig. 12.12, whose performance was illustrated in Fig. 12.14, it is impossible to obtain high recovery of desired component because of losses in the tails streams leaving every stage. Desired component in these streams can be recovered by recycling these

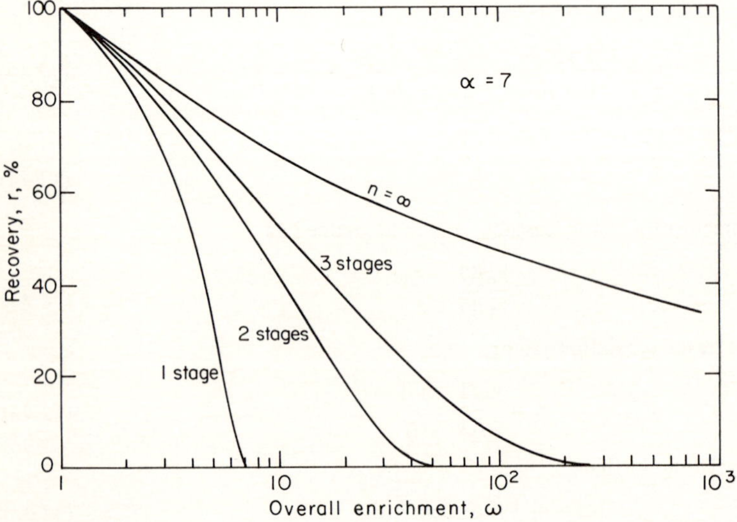

Figure 12.14 Recovery versus overall enrichment in simple cascade, $\alpha = 7$.

streams to a lower stage in the cascade. Figure 12.13 illustrates the simplest type of recycle cascade, in which the tails stream from stage $i+1$ is recycled to become part of the feed to stage i from which stage $i+1$ received part of its feed. This recycle flow scheme is by far the most common in countercurrent separation processes. It is approached, for example, in a bubble-plate distillation column and is used in a battery of series-connected solvent extraction mixer-settlers or in the gaseous diffusion cascade of Fig. 12.2. More complex recycle flow schemes will be treated in Sec. 14.

7.1 Material-Balance Relations

In a recycle cascade such as Fig. 12.13, feed, product, and tails quantities and compositions (the external variables) must satisfy the material-balance relations

$$F = P + W \tag{12.52}$$

and

$$Fz_F = Py_P + Wx_W \tag{12.53}$$

Because there are two equations and six variables, it is possible to specify four external variables independently. For example, these might be product rate and product, feed, and tails compositions. In such a case the other two variables would be given by

$$\text{Feed rate:} \quad F = \frac{P(y_P - x_W)}{z_F - x_W} \tag{12.54}$$

$$\text{Tails rate:} \quad W = \frac{P(y_P - z_F)}{z_F - x_W} \tag{12.55}$$

Two material-balance relations among internal variables may also be written for each stage. Consider the portion of the cascade from the product end down to, but not including, stage i. In this portion

$$M_i = N_{i+1} + P \tag{12.56}$$

and

$$M_i y_i = N_{i+1} x_{i+1} + Py_P \tag{12.57}$$

or

$$x_{i+1} = \left(1 + \frac{P}{N_{i+1}}\right) y_i - \frac{Py_P}{N_{i+1}} \tag{12.58}$$

In the stripping section, where the direction of net flow is reversed, stage material-balance relations are

$$M_j = N_{j+1} - W \tag{12.59}$$

and

$$M_j y_j = N_{j+1} x_{j+1} - Wx_W \tag{12.60}$$

or

$$x_{j+1} = \left(1 - \frac{W}{N_{j+1}}\right) y_j + \frac{Wx_W}{N_{j+1}} \tag{12.61}$$

A relation for the difference in composition between heads from one stage (y_i) and tails from the next higher stage (x_{i+1}) may be obtained from (12.58):

$$y_i - x_{i+1} = \frac{y_P - y_i}{N_{i+1}/P} \tag{12.62}$$

Thus, x_{i+1} is less than y_i by an amount that decreases as the reflux ratio N_{i+1}/P increases. At total reflux ($N_{i+1}/P \to \infty$), x_{i+1} and y_i are equal.

7.2 Number of Ideal Stages

If the separation factor for the system is known and the variation of the reflux ratio is specified as a function of stage number in the cascade, the number of ideal stages required to separate feed into product and tails of specified composition can be calculated. For example, starting with the known tails composition x_W, the heads composition from stage 1, y_1, is calculated using values of α and Eq. (12.18). This composition is used in Eq. (12.61) to calculate the tails composition from stage 2, x_2. Equation (12.18) is again used to calculate y_2, the heads composition from stage 2, and so on. Thus by a repetitive, stepwise, calculation involving the equilibrium expression (12.18) and the two difference equations (12.58) and (12.61), the compositions on each stage in the cascade can be calculated. Equation (12.61) is employed for compositions less than the feed composition and Eq. (12.58) for compositions greater than the feed composition. When the heads composition from a stage equals or exceeds the desired product composition, the required number of ideal stages has been calculated. This calculational method applies generally to all stage processes. Simplified or analytic methods of solution are available for special cases.

7.3 Minimum Number of Stages: Constant Separation Factor

The number of stages required to separate feed into product and tails of specified composition is a minimum at total reflux, when $N_{i+1}/P \to \infty$. Under this condition we have seen that

$$x_{i+1} = y_i \tag{12.63}$$

Abundance ratios in these two streams are also equal:

$$\xi_{i+1} = \eta_i \tag{12.64}$$

Because of the definition of separation factor (12.15), abundance ratios on adjacent stages at total reflux are related by

$$\eta_{i+1} = \alpha \eta_i \tag{12.65}$$

When applied to stage 1, this equation is

$$\eta_2 = \alpha \eta_1 \tag{12.66}$$

When applied to stage 2 it is, for constant α,

$$\eta_3 = \alpha \eta_2 = \alpha^2 \eta_1 \tag{12.67}$$

By proceeding in this way through the entire cascade, we find

$$\eta_n = \alpha^{n-1} \eta_1 \tag{12.68}$$

But

$$\eta_P = \frac{y_P}{1 - y_P} \tag{12.69}$$

and

$$\eta_1 = \alpha \xi_1 = \frac{\alpha x_W}{1 - x_W} \tag{12.70}$$

so that

$$\frac{y_P}{1 - y_P} = \frac{\alpha^n x_W}{1 - x_W} \tag{12.71}$$

or

$$n = \frac{\ln [y_P(1 - x_W)/(1 - y_P)x_W]}{\ln \alpha} \tag{12.72}$$

This is the familiar Underwood [U1]-Fenske [F1] equation for total reflux. The ratio of abundance ratios appearing in (12.72) is the *overall separation* (Ω) of the recycle cascade:

$$\Omega \equiv \frac{y_P(1 - x_W)}{(1 - y_P)x_W} \tag{12.73}$$

Equation (12.72) gives the minimum number of stages for a particular overall separation. The minimum number of stages requires that the ratio of interstage flow rate to product be infinite.

The minimum number of stages increases as the overall separation increases and as the separation factor approaches unity. Because both these conditions hold in a typical isotope separation plant, the minimum number of stages is often very large. For example, in a ^{235}U gaseous diffusion plant ($\alpha = 1.00429$) making product containing 90 percent ^{235}U and tails 0.3 percent,

$$n_{\min} = \frac{\ln [(0.90)(0.997)/(0.10)(0.003)]}{\ln 1.00429} = 1869.6 \tag{12.74}$$

7.4 Minimum Reflux Ratio

At total reflux, the difference in composition between corresponding streams on adjacent stages is a maximum. As the reflux ratio is decreased, the difference in composition decreases, and reaches zero at minimum reflux. A condition for minimum reflux thus is

$$y_{i+1} = y_i \tag{12.75}$$

From (12.20),

$$y_{i+1} - x_{i+1} = \frac{(\alpha - 1)y_{i+1}(1 - y_{i+1})}{y_{i+1} + \alpha(1 - y_{i+1})} \tag{12.76}$$

At minimum reflux this becomes

$$y_i - x_{i+1} = \frac{(\alpha - 1)y_i(1 - y_i)}{y_i + \alpha(1 - y_i)} \tag{12.77}$$

But this difference in composition is already given by the material-balance equation (12.62), so that

$$\left(\frac{N_{i+1}}{P}\right)_{\min} = \frac{(y_P - y_i)[y_i + \alpha(1 - y_i)]}{(\alpha - 1)y_i(1 - y_i)} \tag{12.78}$$

In terms of the tails composition x_{i+1}, this is

$$\left(\frac{N_{i+1}}{P}\right)_{\min} = \frac{y_P(\alpha x_{i+1} + 1 - x_{i+1}) - \alpha x_{i+1}}{(\alpha - 1)x_{i+1}(1 - x_{i+1})} \tag{12.79}$$

Several special forms of Eq. (12.78) will be useful. When $y_i \ll 1$,

$$\left(\frac{N_{i+1}}{P}\right)_{\min} \approx \frac{y_P - y_i}{y_i} \frac{\alpha}{\alpha - 1} \tag{12.80}$$

This equation is applicable to the portions of a heavy-water separation plant or ^{235}U plant near the feed point. In a *close-separation* cascade, in which $\alpha - 1 \ll 1$,

$$\left(\frac{N_{i+1}}{P}\right)_{\min} \approx \frac{y_P - y_i}{(\alpha - 1)y_i(1 - y_i)} \tag{12.81}$$

These equations all show that the minimum reflux ratio increases as the composition departs more from product or tails composition. In isotope separation cascades in which α is close to unity, the minimum reflux ratio is enormous. For example, at the feed point of a plant to produce 90 percent ^{235}U from natural uranium ($y = 0.0072$) by gaseous diffusion ($\alpha = 1.00429$), the minimum reflux ratio, from (12.81), is

$$\left(\frac{N}{P}\right)_{\min} = \frac{0.9 - 0.0072}{(0.00429)(0.0072)(0.9928)} = 29{,}114 \tag{12.82}$$

Yet, as the product end of this cascade is approached, the minimum reflux ratio approaches zero.

7.5 Practical Reflux Ratio

In any practical separation plant, the preferred reflux ratio will clearly be greater than the minimum, which would lead to an infinite number of stages, and less than the infinite reflux ratio needed for the minimum number of stages. In most nonisotopic separation plants it is customary to select a reflux ratio somewhat greater than the minimum at the feed point and to use the same value throughout the entire enriching or stripping section, even though a smaller value would suffice toward the product or waste end of the plant. In distillation this is done because the reflux ratio in an adiabatic column remains nearly constant, and it is cheaper to add or remove heat only at the ends of the column than at a number of intermediate points. In many isotope separation plants, however, so much can be saved in the way of reduced equipment size and material holdup by reducing the reflux ratio at intervals between the feed point and the product ends of a cascade that this is usually done. Investigation of the properties of such a "tapered" cascade is therefore important in isotope separation, and of interest in other separation problems because it indicates how equipment size and holdup could be reduced in cases where the increased complexity of a "tapered" plant is justified.

Properties of a cascade with constant reflux ratio over a substantial composition interval are considered in Sec. 13.

8 THE IDEAL CASCADE

One type of tapered plant that is easy to treat theoretically, which has minimum interstage flow for a specified separation, and which is approximated by all isotope separation plants designed for minimum cost, is the so-called *ideal cascade*. An ideal cascade is one in which

1. The heads separation factor β is constant.
2. The heads stream and tails stream fed to each stage have the same composition:

$$x_{i+1} = y_{i-1} = z_i \qquad (i = 2, 3, \ldots, n-1) \tag{12.83}$$

The theory of such cascades was developed by P. A. M. Dirac and R. Peierls in England and by K. Cohen and I. Kaplan in the United States and is described in *The Theory of Isotope Separation* by Cohen [C3]. The most important results are summarized in Secs. 8 through 12 of this chapter, with some changes in terminology and notation.

8.1 Heads Separation Factor

The above condition for an ideal cascade may also be expressed in terms of abundance ratios:

$$\xi_{i+1} = \eta_{i-1} = \zeta_i \tag{12.84}$$

Figure 12.15 shows three stages of an ideal cascade in which this condition holds. From the

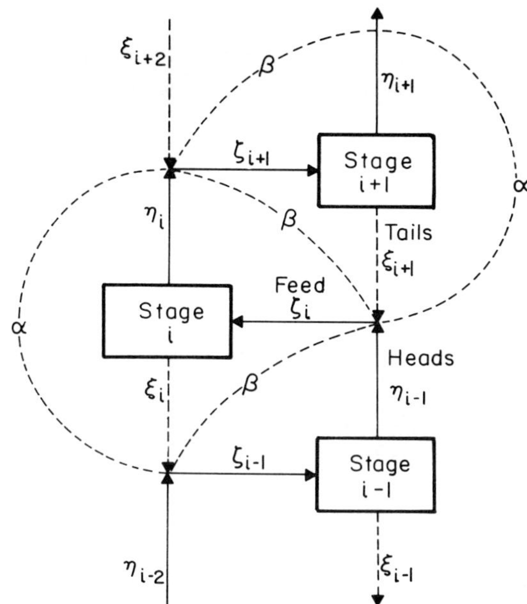

Figure 12.15 Abundance ratios in ideal cascade.

definition of the heads separation factor,

$$\eta_i = \beta\zeta_i \tag{12.85}$$

In an ideal cascade, because of (12.84),

$$\eta_i = \beta\xi_{i+1} \tag{12.86}$$

Similarly,

$$\eta_{i+1} = \beta\eta_i \tag{12.87}$$

By multiplying these two equations together,

$$\eta_{i+1} = \beta^2 \xi_{i+1} \tag{12.88}$$

But, from the definition of the separation factor,

$$\eta_{i+1} = \alpha\xi_{i+1} \tag{12.89}$$

so that

$$\beta = \sqrt{\alpha} = \gamma \tag{12.90}$$

This relation between the heads and tails separation factors and the stage separation factor is the key property of an ideal cascade.

In the close-fractionation case, in which $\beta - 1$ and $\alpha - 1$ are small compared to unity,

$$\beta - 1 \approx \frac{\alpha - 1}{2} \tag{12.91}$$

An equation for the cut θ in an ideal cascade is obtained from Eq. (12.12) by substituting for y and x their values in terms of z from Eqs. (12.18) and (12.20) and using the condition $\beta = \gamma$:

$$\theta = \frac{1 + z(\beta - 1)}{\beta + 1} \tag{12.92}$$

8.2 Number of Stages

The number of stages in an ideal cascade may be evaluated by a procedure similar to that used in deriving Eq. (12.72) for the minimum number of stages at total reflux. The result is

$$n = \frac{\ln [y_P(1-x_W)/(1-y_P)x_W]}{\ln \beta} - 1 = 2\frac{\ln [y_P(1-x_W)/(1-y_P)x_W]}{\ln \alpha} - 1 \quad (12.93)$$

Thus the number of stages required for a given separation in an ideal cascade is just twice the minimum number needed at total reflux minus 1.

By a similar procedure, the number of stages in the stripping section is found to be

$$n_S = \frac{\ln [z_F(1-x_W)/(1-z_F)x_W]}{\ln \beta} - 1 \quad (12.94)$$

and in the enriching section

$$n - n_S = \frac{\ln [y_P(1-z_F)/(1-y_P)z_F]}{\ln \beta} \quad (12.95)$$

A relation between composition and stage number may be derived by a procedure similar to that which led to (12.68):

$$\eta_n = \beta^{n-i}\eta_i \quad (12.96)$$

Because

$$\eta_n = \frac{y_P}{1-y_P} \quad (12.97)$$

and

$$\eta_i = \frac{y_i}{1-y_i} \quad (12.98)$$

(12.96) may be solved for y_i, with the result

$$y_i = z_{i+1} = x_{i+2} = \frac{\beta^i y_P}{\beta^i y_P + \beta^n(1-y_P)} \quad (12.99)$$

The corresponding equations in the stripping section are

$$x_j = z_{j-1} = y_{j-2} = \frac{\beta^{j-1} x_W}{\beta^{j-1} x_W + (1-x_W)} \quad (12.100)$$

8.3 Reflux Ratio

The reflux ratio required to bring about condition (12.83) defining an ideal cascade may be found as follows. From (12.62),

$$\frac{N_{i+1}}{P} = \frac{y_P - y_i}{y_i - x_{i+1}} \quad (12.101)$$

But $y_i = z_{i+1}$ in an ideal cascade, and z_{i+1} is given in terms of x_{i+1} by (12.19) with $\alpha = \beta^2$, so that

$$\frac{N_{i+1}}{P} = \frac{y_P(\beta x_{i+1} + 1 - x_{i+1}) - \beta x_{i+1}}{(\beta - 1)x_{i+1}(1 - x_{i+1})} = \frac{1}{\beta - 1}\left[\frac{y_P}{x_{i+1}} - \frac{\beta(1-y_P)}{1-x_{i+1}}\right] \quad (12.102)$$

This equation is the same as for minimum reflux (12.79), except that β replaces α.

Figure 12.16 is a McCabe-Thiele diagram for an ideal cascade. The *equilibrium* line, relating

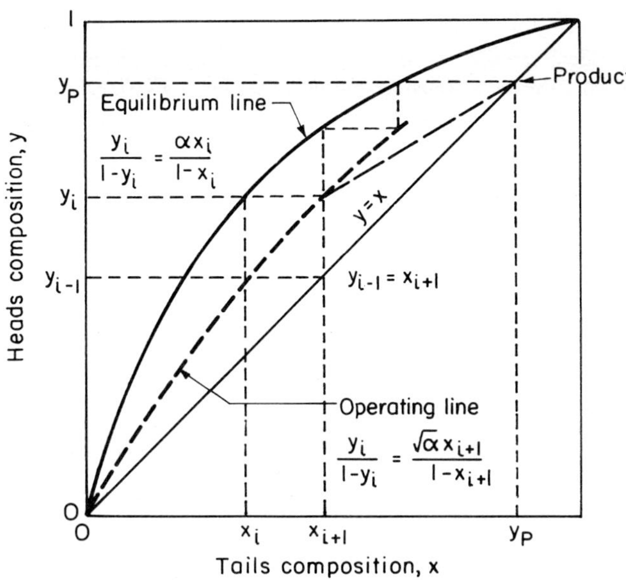

Figure 12.16 McCabe-Thiele diagram for ideal cascade.

y_i to x_i, is represented by the solid curved line, with the equation

$$\frac{y_i}{1-y_i} = \frac{\alpha x_i}{1-x_i} \tag{12.103}$$

The *operating* line, relating y_i to x_{i+1}, is represented by the dashed curved line, with the equation

$$\frac{y_i}{1-y_i} = \frac{\sqrt{\alpha} x_{i+1}}{1-x_{i+1}} \tag{12.104}$$

The graphic construction shows that with these two lines $x_{i+1} = y_{i-1}$, as required for an ideal cascade. The straight line connecting the product point (y_P, y_P) with the point on the operating line (y_i, x_{i+1}) has a slope $(y_P - y_i)/(y_P - x_{i+1})$, which equals $N_{i+1}/(N_{i+1} + P)$, the ratio of tails to heads flow at this point in the cascade. N_{i+1}/P is given by Eqs. (12.101) and (12.102).

In the stripping section, the equation corresponding to (12.102) is

$$\frac{M_j}{W} = \frac{1}{\beta - 1}\left(\frac{1-x_W}{1-y_j} - \frac{\beta x_W}{y_j}\right) \tag{12.105}$$

An equation for the reflux ratio in the enriching section as a function of stage number may be obtained by substituting x_{i+1} from (12.99) into (12.102):

$$\frac{N_{i+1}}{P} = \frac{1}{\beta - 1}[y_P(1-\beta^{i-n}) + (1-y_P)\beta(\beta^{n-i} - 1)] \tag{12.106}$$

Similarly, in the stripping section, Eqs. (12.105) and (12.100) lead to

$$\frac{M_j}{W} = \frac{1}{\beta - 1}[x_W\beta(\beta^j - 1) + (1-x_W)(1-\beta^{-j})] \tag{12.107}$$

8.4 Shape of Ideal Cascade

To illustrate the shape of a typical ideal cascade, we shall work out the variation of interstage flow with stage number for an ideal cascade to separate natural uranium ($z_F = 0.0072$) into enriched uranium with $y_P = 0.90$ and depleted uranium tails with $x_W = 0.003$ by gaseous diffusion, with $\alpha = 1.00429$.

To produce 1 mol of product, the amount of feed, from (12.54), is

$$F = \frac{0.90 - 0.003}{0.0072 - 0.003} = 213.57 \text{ mol} \tag{12.108}$$

and the amount of tails, from (12.55), is

$$W = \frac{0.90 - 0.072}{0.0072 - 0.003} = 212.57 \text{ mol} \tag{12.109}$$

The heads separation factor is given by (12.90), with

$$\beta = \sqrt{1.00429} = 1.00214 \tag{12.110}$$

The total number of stages n is twice the minimum, given by (12.74), less 1, or 3738. The number of stages in the stripping section n_S is given by (12.94), with

$$n_S = 2\,\frac{\ln\,[(0.0072)(0.997)/(0.9928)(0.003)]}{\ln 1.00429} - 1 = 410 \tag{12.111}$$

The heads flow rate in the enriching section, from (12.106), is

$$M_i = N_{i+1} + 1 = 1 + \frac{(1 - 1.00214^{i-3738})(0.90)}{0.00214} + \left(\frac{1.00214}{0.00214}\right)(1.00214^{3738-i} - 1)(0.10) \qquad (410 < i \leqslant 3738) \tag{12.112}$$

The heads flow rate in the stripping section, from (12.107), is

$$M_j = \frac{212.57}{0.00214}[(1.00214)(1.00214^j - 1)(0.003) + (1 - 1.00214^{-j})(0.997)] \qquad (0 < j \leqslant 410) \tag{12.113}$$

Figure 12.17 is a plot of these equations represented as a tapered column whose height is proportional to stage number above tails and whose width is proportional to heads flow rate.

The large interstage flow rate at the feed point ($M_{410}/P = 58{,}229$) and its rapid decrease as the product and tails ends of the plant are approached are characteristic of an ideal cascade.

8.5 Total Flow Rates

The total interstage flow rate of heads or tails is a measure of the size of the separation plant. In a distillation plant, for example, the total volume of column internals is proportional to the total interplate vapor flow rate. In a gaseous diffusion plant, the total amount of power expended in pumping gas from one stage to the next is proportional to the total heads flow rate.

An expression for the total flow rate of heads or tails in stripping or enriching section may be derived by summing the appropriate Eq. (12.106) or (12.107). For example, the total heads flow rate in the stripping section J_S is

$$J_S = \sum_{j=1}^{n_S} M_j = \frac{W}{\beta - 1}\left[x_W \beta \sum_{j=1}^{n_S} (\beta^j - 1) + (1 - x_W) \sum_{j=1}^{n_S} (1 - \beta^{-j})\right] \tag{12.114}$$

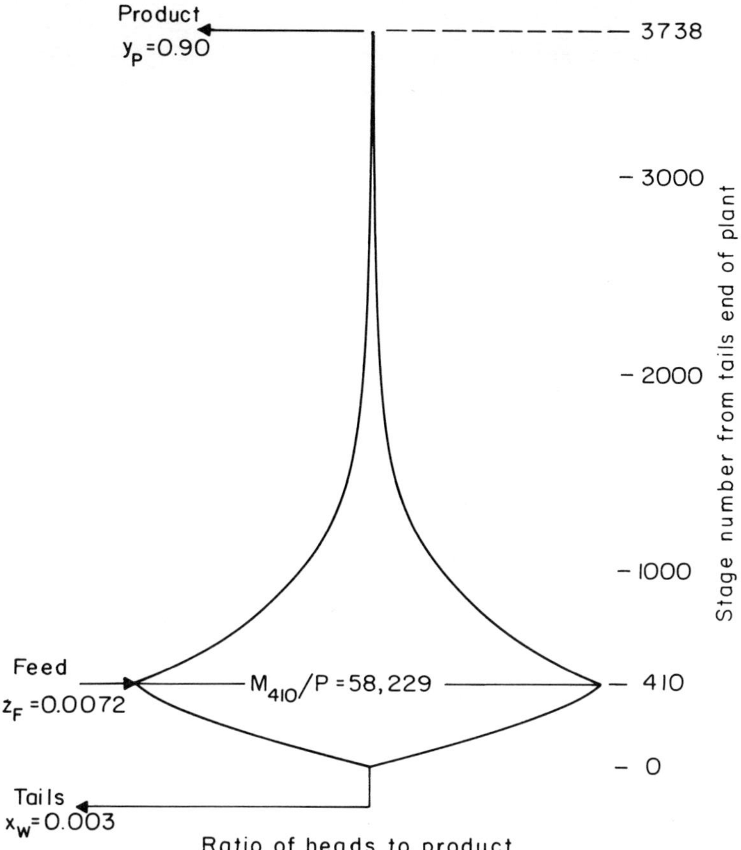

Figure 12.17 Heads flow rate versus stage number in ideal cascade. $\beta = 1.00214$.

Because
$$\sum_{j=1}^{n_S} \beta^j = \frac{\beta^{n_S} - 1}{\beta - 1}\,\beta \tag{12.115}$$

this becomes
$$J_S = \frac{W}{\beta - 1}\left[\beta x_W\left(\beta\,\frac{\beta^{n_S} - 1}{\beta - 1} - n_S\right) + (1 - x_W)\left(n_S - \frac{1 - \beta^{-n_S}}{\beta - 1}\right)\right] \tag{12.116}$$

n_S in this equation may be eliminated by (12.94), with the result

$$J_S = \frac{W}{\beta - 1}\left\{[1 - x_W(\beta + 1)]\,\frac{\ln\,[z_F(1 - x_W)/x_W(1 - z_F)]}{\ln\beta} - \frac{\beta(z_F - x_W)(1 - 2z_F)}{(\beta - 1)z_F(1 - z_F)}\right\} \tag{12.117}$$

By a similar procedure, the total tails flow rate in the stripping section is found to be

$$K_S = \sum_{j=1}^{n_S} N_j = \frac{W}{\beta - 1}\left\{[\beta - x_W(\beta + 1)]\,\frac{\ln\,[z_F(1 - x_W)/x_W(1 - z_F)]}{\ln\beta}\right. \tag{12.118}$$

(Cont. on p. 664)

$$-\frac{(z_F - x_W)[\beta^2 - (\beta^2 + 1)z_F]}{z_F(1 - z_F)(\beta - 1)}\Bigg\} \tag{12.118 (Cont.)}$$

The total tails flow rate in the enriching section is obtained from (12.106):

$$K_E = \sum_{i=n_S}^{n-1} N_{i+1} = \frac{P}{\beta - 1}\Bigg\{\frac{[y_P(\beta+1) - \beta] \ln [y_P(1 - z_F)/z_F(1 - y_P)]}{\ln \beta}$$

$$+ \frac{(y_P - z_F)}{z_F(1 - z_F)} \frac{[\beta^2 - (\beta^2 + 1)z_F]}{\beta - 1}\Bigg\} \tag{12.119}$$

and the total heads flow rate in the enriching section is

$$J_E = \sum_{i=n_S}^{n-1} M_{i+1} = \frac{P}{\beta - 1}\Bigg\{\frac{[y_P(\beta+1) - 1] \ln [y_P(1 - z_F)/z_F(1 - y_P)]}{\ln \beta}$$

$$+ \beta \frac{(y_P - z_F)(1 - 2z_F)}{z_F(1 - z_F)(\beta - 1)}\Bigg\} \tag{12.120}$$

The total flow in the entire cascade, $J + K$, is the sum of (12.117) through (12.120):

$$J + K = \frac{\beta + 1}{(\beta - 1) \ln \beta}\Bigg\{W(1 - 2x_W) \ln \left[\frac{z_F(1 - x_W)}{x_W(1 - z_F)}\right] + P(2y_P - 1) \ln \left[\frac{y_P(1 - z_F)}{z_F(1 - y_P)}\right]\Bigg\} \tag{12.121}$$

Terms in $y_P - z_F$ and $z_F - x_W$ have canceled out because of the material-balance relations (12.52) and (12.53). Also, because of these material-balance relations, (12.121) may be written

$$J + K = \frac{\beta + 1}{(\beta - 1) \ln \beta}\Bigg[\overbrace{W(2x_W - 1) \ln \left(\frac{x_W}{1 - x_W}\right) + P(2y_P - 1) \ln \left(\frac{y_P}{1 - y_P}\right)}^{\text{output}}$$

$$- \overbrace{F(2z_F - 1) \ln \left(\frac{z_F}{1 - z_F}\right)}^{\text{input}}\Bigg] \tag{12.122}$$

This result is of great importance for isotope separation plants. It states that the total flow in the plant is the product of two factors, the first a function only of the heads separation factor β, and the second a function only of the flow rates and composition of feed, product, and tails.

The first factor is a measure of the relative ease or difficulty of the separation; it is large when β is close to unity and small when β differs markedly from unity. The second factor is a measure of the magnitude of the job of separation; it is proportional to the throughput, and is large when product and tails differ substantially in composition from feed, and small when these compositions are nearly equal. The second factor has been termed the *separative capacity*, because it is a measure of the rate at which a cascade performs separation. It equals the sum of two output terms, each the product of an output flow rate and a function of the corresponding output condition, minus an input term that is the product of the feed rate and a function of the input condition. The separative capacity is discussed in more detail in Sec. 10.

9 CLOSE-SEPARATION CASCADE

A close-separation cascade is one in which $\alpha - 1 \ll 1$. In such a cascade, the condition $\beta = \sqrt{\alpha}$ for an ideal cascade, in which heads and tails fed to a stage have the same composition, may be approximated by

$$\beta - 1 = \frac{\alpha - 1}{2} \tag{12.123}$$

Equation (12.102) for the tails flow rate in an ideal cascade may be approximated by

$$N = \frac{P(y_P - x)}{(\beta - 1)x(1 - x)} \tag{12.124}$$

or

$$N = \frac{2P(y_P - x)}{(\alpha - 1)x(1 - x)} \tag{12.125}$$

because of (12.123). We shall now show that when the total tails flow rate of a close-separation cascade is a minimum, the tails flow rate at each stage is given by (12.125).

The difference in composition between stage heads and stage tails, given by (12.76), may be approximated by

$$y_{i+1} - x_{i+1} = (\alpha - 1)y_{i+1}(1 - y_{i+1}) \tag{12.126}$$

in the close-separation case. A relation for the change in heads composition between adjacent stages is obtained by combining this with the material-balance equation (12.62):

$$y_{i+1} - y_i = (\alpha - 1)y_{i+1}(1 - y_{i+1}) - \frac{P}{N_{i+1}}(y_P - y_i) \tag{12.127}$$

Because y_{i+1}, y_i, and x_i are nearly equal, this difference equation may be approximated by the differential equation

$$\frac{dx}{di} = (\alpha - 1)x(1 - x) - \frac{P}{N}(y_P - x) \tag{12.128}$$

The total tails flow rate in the enriching section is

$$K_E = \int_{n_W+1}^{n} N\,di = \int_{z_F}^{y_P} N\frac{di}{dx}\,dx \tag{12.129}$$

K_E will be minimum when the integrand

$$N\frac{di}{dx} = \frac{1}{[(\alpha - 1)x(1 - x)/N] - (P/N^2)(y_P - x)} \tag{12.130}$$

is a minimum at all x. The optimum value of N that makes this a minimum is that at which the derivative of the denominator vanishes, or at which

$$-\frac{(\alpha - 1)x(1 - x)}{N^2} + \frac{2P}{N^3}(y_P - x) = 0 \tag{12.131}$$

Thus

$$N_{\text{opt}} = \frac{2P(y_P - x)}{(\alpha - 1)x(1 - x)} \tag{12.132}$$

This is just twice the minimum tails flow rate at which $dx/di = 0$.

This is identical with (12.125). Thus it has been shown that in the close-separation case an ideal cascade may be defined in any one of the three following equivalent ways:

$$\beta - 1 = \frac{\alpha - 1}{2}$$

$$N = \frac{P(y_P - x)}{(\beta - 1)x(1 - x)}$$

N is so chosen that total interstage flow is a minimum

In such a cascade, the heads and tails fed to each stage have the same composition, and the cut θ is $\frac{1}{2}$. The last may be seen from (12.21), which becomes

$$\beta - 1 = (\alpha - 1)(1 - \theta) \qquad (\alpha - 1 \ll 1) \tag{12.133}$$

At the optimum flow rate, the change in composition per stage, from (12.128), is

$$\frac{dx}{di} = \frac{\alpha - 1}{2} x(1 - x) \tag{12.134}$$

which is just half the change at total reflux at which $P/N = 0$. The total number of stages in the enriching section is

$$n - n_S = \int_{z_F}^{y_P} \frac{di}{dx} dx = \frac{2}{\alpha - 1} \int_{z_F}^{y_P} \frac{dx}{x(1 - x)} = \frac{2}{\alpha - 1} \ln \frac{y_P(1 - z_F)}{z_F(1 - y_P)} \tag{12.135}$$

Equation (12.95) reduces to this expression, except for terms of the order of unity.

The total tails flow rate in the enriching section at the optimum flow rate is

$$K_E = \int_{z_F}^{y_P} N_{\text{opt}} \frac{di}{dx} dx \tag{12.136}$$

With N_{opt} from (12.132) and di/dx from (12.134), this becomes

$$K_E = \frac{4P}{(\alpha - 1)^2} \int_{z_F}^{y_P} \frac{y_P - x}{x^2 (1 - x)^2} dx$$

$$= \frac{4P}{(\alpha - 1)^2} \left[(2y_P - 1) \ln \frac{y_P(1 - z_F)}{z_F(1 - y_P)} + \frac{(y_P - z_F)(1 - 2z_F)}{z_F(1 - z_F)} \right] \tag{12.137}$$

Equation (12.119) gives the same result, except for terms in $1/(\alpha - 1)$, with the substitution

$$\beta = 1 + \frac{\alpha - 1}{2} \tag{12.138}$$

The total heads flow rate, from (12.120), reduces to the same expression.

The total flow rate in both stripping and enriching sections, from (12.122), becomes

$$J + K = \frac{8}{(\alpha - 1)^2} \left[W(2x_W - 1) \ln \frac{x_W}{1 - x_W} + P(2y_P - 1) \ln \frac{y_P}{1 - y_P} - F(2z_F - 1) \ln \frac{z_F}{1 - z_F} \right] \tag{12.139}$$

The total heads flow rate or total tails flow rate in both sections is one-half this value.

These formulas are extraordinarily useful in roughing out the characteristics of an isotope separation plant without the necessity of designing every one of its stages, which often number in the thousands. As an illustration, the total heads flow rate in the uranium isotope separation example considered in Fig. 12.17 is

$$J = \frac{4}{(0.00429)^2}\left\{[(2)(0.9)-1]\ln\frac{0.9}{0.1} + 212.57[(2)(0.0030)-1]\ln\frac{0.003}{0.997}\right.$$

$$\left. -213.57[(2)(0.0072)-1]\ln\frac{0.0072}{0.9928}\right\} = (217{,}343)(191.57) = 41{,}636{,}000 \tag{12.140}$$

This is the area within Fig. 12.17. Around 42 million moles of UF_6 must be pumped for every mole of 90 percent $^{235}UF_6$ separated.

10 SEPARATIVE CAPACITY, SEPARATIVE WORK, AND SEPARATION POTENTIAL

10.1 Definitions

The second factor appearing in Eqs. (12.122) and (12.139) for the total flow rate in an ideal cascade is known as the *separative capacity*, or *separative power* [C3], D. For a plant with a single tails, product, and feed stream, it is given by

$$D = W(2x_W - 1)\ln\frac{x_W}{1-x_W} + P(2y_P - 1)\ln\frac{y_P}{1-y_P} - F(2z_F - 1)\ln\frac{z_F}{1-z_F} \tag{12.141}$$

The separative capacity has the same dimensions as used for the flow rates. It is a measure of the rate at which a cascade is performing separation.

The separative capacity concept may be generalized to a plant with any number of external streams of composition x_k and molar flow rate X_k (positive when a product, negative when a feed). The total internal flow rate in such a plant, $J + K$, is†

$$J + K = \frac{8}{(\alpha - 1)^2}D = \frac{2}{(\beta - 1)^2}D \tag{12.142}$$

where the separative capacity D now is

$$D = \Sigma_k X_k \phi(x_k) \tag{12.143}$$

and

$$\phi(x_k) = (2x_k - 1)\ln\frac{x_k}{1-x_k} \tag{12.144}$$

The function ϕ defined by (12.144) is called the *separation potential*, or the *elementary value function* [C3]. It is a function only of composition and is dimensionless. It is plotted in Fig. 12.18. It is symmetrical about $x = 0.5$, at which value it vanishes. It is positive for all other x and increases without limit as x approaches zero or unity. This expresses the fact that a plant of infinite size is required to produce a pure isotope. The curve of ϕ versus x is convex downward, because

$$\frac{d^2\phi}{dx^2} = \frac{1}{x^2(1-x)^2} \tag{12.145}$$

is positive.

†This holds for a close-separation, ideal cascade. When $\beta - 1$ is not small relative to unity, the more general equation is

$$J + K = \frac{(\beta + 1)D}{(\beta - 1)\ln\beta} \tag{12.142a}$$

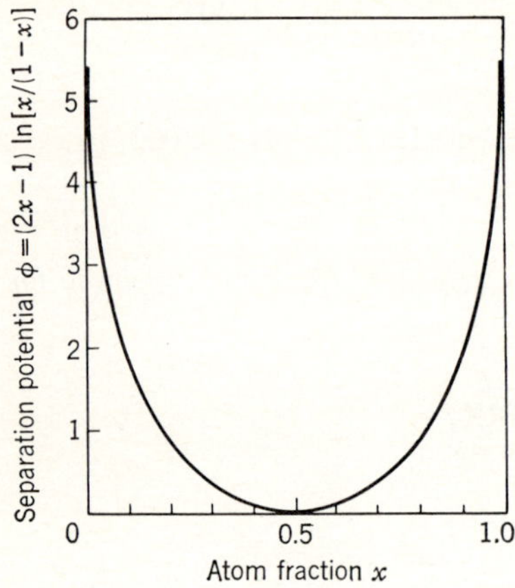

Figure 12.18 Separation potential.

Because ϕ is convex downward, D is always positive.

The importance of the separative capacity in isotope separation lies in the fact that it is a good measure of the magnitude of an isotope separation job. Many of the characteristics of the plant that make important contributions to its cost are proportional to the separative capacity. For example, in a gaseous diffusion plant built as an ideal cascade of stages operated at the same conditions, the total flow rate, the total pump capacity, the total power demand, and the total barrier area are all proportional to the separative capacity. In a distillation plant, the total column volume and total rate of loss of availability are proportional to the separative capacity.

The separative capacity is analogous to the heat duty of an evaporator or other process equipment. The separation potential is analogous to the enthalpy per mole of the streams entering or leaving an evaporator. Calculations of material balances and separative capacity in an isotope separation plant are made in similar fashion to conventional material and heat balances. A form for such calculations is illustrated in Table 12.8, which illustrates the calculation of the separative capacity of an isotope separation cascade producing 1 mol/day of ^{235}U at 0.80 mole

Table 12.8 Example of material-balance and separative-capacity calculations for ideal cascade†

				Flow rates, mol/day		
	Stream	Mole fraction x	Separation potential ϕ	Both isotopes X	Desired isotope Xx	Separative capacity $X\phi$
Out	Product P	0.800†	0.83178	1.25	1.00†	1
	Tails W	0.0036†	5.58273	275.27	0.9910	1537
	Total			276.52	1.9910	1538
In	Feed F	0.0072†	4.85551	276.52	1.9910	1343
	Total			276.52	1.9910	1343
	Net separative capacity of cascade					195

†Given: $Px_P = 1.0$ mol/day desired isotope in product; $x_P = 0.800$ mole fraction desired isotope in product; $x_F = 0.0072$ mole fraction desired isotope in feed; $x_W = 0.0036$ mole fraction desired isotope in tails. Required: net separative capacity of cascade.

fraction of ^{235}U, from normal uranium feed, with cascade waste containing 0.0036 mole fraction ^{235}U.

It is also useful to have a measure of the amount of separation performed by a cascade in making E_P moles of product and E_W moles of waste from E_F moles of feed. This measure is provided by the *separative work* S, defined in similar fashion to the separative capacity.

$$S = E_W(2x_W - 1) \ln \frac{x_W}{1 - x_W} + E_P(2y_P - 1) \ln \frac{y_P}{1 - y_P} - E_F(2z_F - 1) \ln \frac{z_F}{1 - z_F} \tag{12.146}$$

Separative work S has the same dimensions as used for the amounts of material E. Each term of the form $E\phi$ in Eq. (12.146) represents the separative work content associated with amount of material E in the corresponding stream.

Generalization to more than three streams is treated in Sec. 11.

Practical applications of Eq. (12.141) for separative capacity and Eq. (12.146) for separative work are usually expressed in terms of kilograms of uranium and weight fractions rather than moles and mole fractions. When atomic weights of the components are as close together as ^{235}U and ^{238}U, the equations on a weight basis still provide a valid measure of the magnitude of a job of separation.

10.2 Applications of Separative Capacity and Separative Work

If the rate of production of an ideal cascade at one set of feed, product, and tails compositions is known, so that its separative capacity can be evaluated, the best possible performance of the cascade for another set of compositions can be calculated by treating its separative capacity as a constant property of the cascade. This will be true if under the changed conditions the number of separating units in series and parallel are so rearranged that mixing of streams of different compositions is avoided.

The following may be cited as examples of the kinds of problems that may be solved by this means:

1. The effect of change in product rate on product purity
2. The effect of change in feed rate on product rate at constant product purity
3. The effect of providing supplementary feed of a different composition on product rate
4. The effect of withdrawing partially enriched product on product rate

Problem 12.3 illustrates how problems of this kind can be solved.

Most large isotope separation plants have so much flexibility that their separative capacity can be kept very nearly constant under moderately changed conditions.

10.3 Costs from Separative Work

In many isotope separation plants the initial cost of the plant is proportional to the separative capacity of the plant and the annual operating costs are proportional to the amount of separative work done per year. In such cases the annual charges for plant investment plus annual operating costs exclusive of feed, in dollars per year, equal Dc_S, where D is the annual separative capacity in kilograms of uranium per year and c_S is the unit cost of separative work, in dollars per kilogram of uranium of separative work units (\$/kg SWU). If F kg of feed is charged per year at a unit cost of c_F \$/kg, the total annual cost c is

$$c = Dc_S + Fc_F \tag{12.147}$$

If P kg of product is made per year, the unit cost of product, c_P, is

$$c_P = \frac{Dc_S}{P} + \frac{Fc_F}{P} \tag{12.148}$$

From (12.141),

$$\frac{D}{P} = \phi_P + \frac{W}{P}\phi_W - \frac{F}{P}\phi_F \tag{12.149}$$

where ϕ is the separation potential (12.144). By material balances,

$$\frac{W}{P} = \frac{y_P - z_F}{z_F - x_W} \tag{12.150}$$

and

$$\frac{F}{P} = \frac{y_P - x_W}{z_F - x_W} \tag{12.151}$$

Substitution of Eqs. (12.149) through (12.151) into (12.148) yields for the unit cost of product

$$c_P = \left[(\phi_P - \phi_F) - (y_P - z_F)\frac{\phi_F - \phi_W}{z_F - x_W}\right] c_S + \left(\frac{y_P - x_W}{z_F - x_W}\right) c_F \tag{12.152}$$

The first term on the right gives the separative work component of the cost of product; the second term gives the feed component.

10.4 Toll Enrichment Charges

When a power company or other customer wishes to obtain E_P kg of uranium enriched to y_P weight fraction, the usual arrangement is for the customer to purchase $[(y_P - x_W)/(z_F - x_W)]E_P$ kg of natural uranium with $z_F = 0.00711$, deliver it to a uranium enrichment plant providing toll enrichment services, and pay for an amount of separative work S calculated from

$$S = \left[(\phi_P - \phi_F) - (y_P - z_F)\frac{\phi_F - \phi_W}{z_F - x_W}\right] E_P \tag{12.153}$$

The U.S. DOE sets the tails assay x_W in transactions with its customers; in 1977 this tails assay was $x_W = 0.002$. In the future, it is likely that customers will be given some latitude in the choice of x_W so as to minimize the sum of the costs of separative work and natural uranium feed.

Substitution of these values into (12.153) and using (12.144) for ϕ yields

$$\frac{S}{E_P}\text{(ERDA, 1977)} = (2x_P - 1)\ln\frac{y_P}{1 - y_P} + 258.0964\,y_P - 6.7039 \tag{12.154}$$

This equation has been used in the U.S. "Standard Table of Enriching Services" [U2]. This text has used $x_W = 0.003$ as a more probable value of the tails assay in enrichment transactions after 1977. With $z_F = 0.00711$ and $x_W = 0.003$, Eq. (12.153) becomes

$$\frac{S}{E_P}\text{(this text)} = (2x_P - 1)\ln\frac{y_P}{1 - y_P} + 219.5666\,y_P - 6.4300 \tag{12.155}$$

The second column of Table 12.9 gives, for different values of the product weight fraction ^{235}U, y_P, the kilograms of natural uranium feed required to produce 1 kg of product, E_F/E_P,

Table 12.9 Table of uranium enriching services and unit cost of enriched uranium

	Table of enriching services Basis: Feed, $z_F = 0.00711$ Tails, $x_W = 0.003$		Unit cost of product as UF_6 Nat. UF_6, c_F = \$89.11/kg U Sep. work, c_S = \$100/SWU	
Weight percent ^{235}U, 100 y_P	kg nat. U feed/ kg U product, E_F/E_P	Sep. work units/ kg U product, S/E_P	\$/kg U	\$/g ^{235}U
0.3	0	0	–	–
0.4	0.243	−0.078	13.85	3.46
0.6	0.730	−0.064	58.65	9.78
0.711	1.000	0.000	89.11	12.53
1.0	1.703	0.269	178.65	17.87
2.0	4.136	1.697	538.26	26.91
3.0	6.569	3.425	927.86	30.93
3.2	7.056	3.787	1,007.46	31.48
5.0	11.436	7.198	1,738.86	34.78
10.0	23.601	17.284	3,831.49	38.31
20.0	47.932	38.315	8,102.72	40.51
50.0	120.925	103.353	21,110.93	42.22
80.0	193.917	170.055	34,285.44	42.86
90.0	218.248	192.938	38,741.88	43.05
94.0	227.981	202.384	40,553.79	43.14

from the material-balance relation

$$\frac{E_F}{E_P}(\text{this text}) = \frac{y_P - 0.003}{0.00711 - 0.003} \tag{12.156}$$

The third column gives the number of separative work units required to produce 1 kg of product, S/E_P, from Eq. (12.155). The units of S are kilograms of uranium, but are conventionally referred to as SWUs (for separative work units).

10.5 Cost of Enriched Uranium

The last two columns of Table 12.9 give the unit cost of product, in dollars per kilogram of uranium and dollars per gram of ^{235}U, calculated from Eq. (12.152), with $z_F = 0.00711$, the assumed transaction tails assay of $x_W = 0.003$, and the unit costs employed in Sec. 5 of Chap. 3, c_F = \$89.11/kg uranium and c_S = \$100/SWU, which lead to the equation:

$$c_P = 89.11\frac{y_P - 0.003}{0.00411} + 100\left[(2y_P - 1)\ln\frac{y_P}{1 - y_P} + 219.5666\, y_P - 6.4300\right] \tag{12.157}$$

Because costs change frequently, this equation must be considered an example rather than a permanent relation.

Figure 12.19 has been calculated from Eq. (12.157). It shows the dependence of the unit cost of enriched uranium, in dollars per gram of ^{235}U, on the weight percent ^{235}U and shows the contributions to this cost from natural uranium feed and enrichment.

10.6 Optimum Tails Composition

In the future, it is probable that the supplier of enrichment services will permit a customer to specify the assay (^{235}U content) of the tails to which feed is to be stripped so as to minimize the combined cost to the customer of natural UF_6 feed and separative work. Figure 12.20 shows qualitatively the effect of tails composition on the contributions to product cost arising from costs for feed and for separative work in stripping and enriching sections. The amount of separative work required in the enriching section is independent of tails composition. But the cost of separative work required in the stripping sections varies from zero when $x_W = z_F$ (no stripping) to infinity when $x_W = 0$. Conversely, the cost of feed varies from infinity when $x_W = z_F$ to a minimum at $x_W = 0$, as may be seen from Eq. (12.152). There is therefore an optimum tails assay x_0 between $x_W = 0$ and $x_W = z_F$, at which the sum of the cost of separative work and the cost of natural uranium feed is a minimum.

An equation for evaluating the optimum tails composition is derived by substituting explicitly into Eq. (12.152) for the unit cost of product c_P the separation potentials ϕ_F, ϕ_W, and ϕ_F expressed in terms of the corresponding weight fractions x_P, x_W, and x_F by Eq. (12.144):

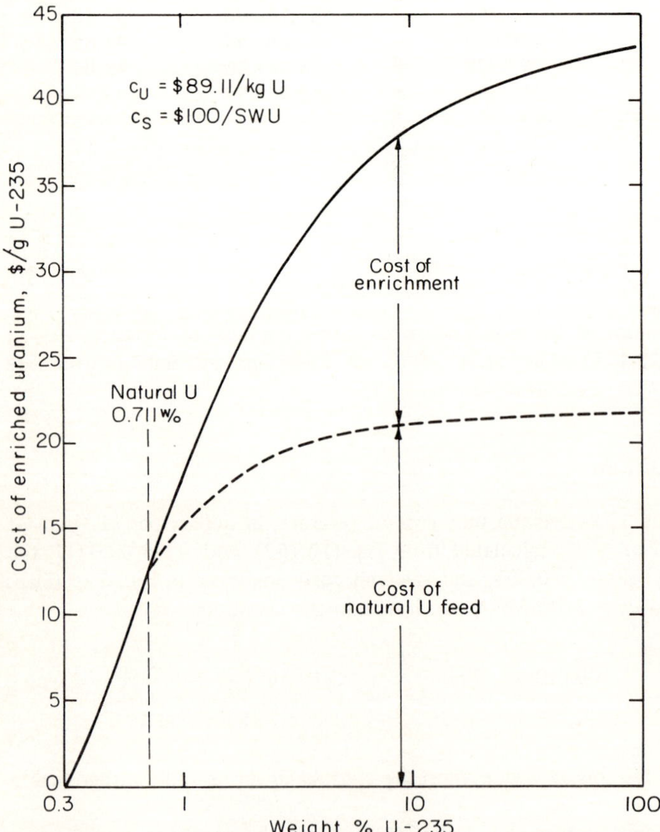

Figure 12.19 Cost of enriched uranium.

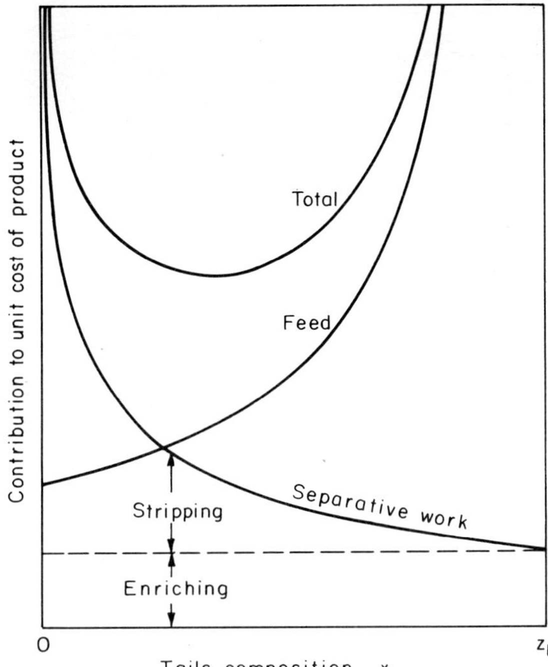

Figure 12.20 Effect of tails composition on cost of product.

$$c_P = c_F \frac{y_P - x_W}{z_F - x_W} + c_S \left[(2y_P - 1) \ln \frac{y_P}{1 - y_P} + \frac{y_P - z_F}{z_F - x_W} (2x_W - 1) \ln \frac{x_W}{1 - x_W} - \frac{y_P - x_W}{z_F - x_W} (2z_F - 1) \ln \frac{z_F}{1 - z_F} \right] \tag{12.158}$$

Optimum tails composition occurs when

$$\left(\frac{\partial c_P}{\partial x_W} \right)_{y_P, x_F} = 0 \tag{12.159}$$

When c_P from Eq. (12.158) is substituted into Eq. (12.159) and optimum tails composition x_0 is substituted for x_W, the result is

$$0 = c_F \frac{y_P - z_F}{(z_F - x_0)^2} + c_S \left\{ \frac{y_P - z_F}{(z_F - x_0)^2} (2x_0 - 1) \ln \frac{x_0}{1 - x_0} + \frac{y_P - z_F}{z_F - x_0} \left[2 \ln \frac{x_0}{1 - x_0} + \frac{2x_0 - 1}{x_0(1 - x_0)} \right] - \frac{y_P - z_F}{(z_F - x_0)^2} (2z_F - 1) \ln \frac{z_F}{1 - z_F} \right\} \tag{12.160}$$

This may be simplified to

$$\frac{c_F}{c_S} = (2z_F - 1) \ln \frac{z_F(1 - x_0)}{x_0(1 - z_F)} + \frac{(z_F - x_0)(1 - 2x_0)}{x_0(1 - x_0)} \tag{12.161}$$

Figure 12.21 shows the dependence of optimum tails composition on the feed-to-separative work cost ratio.

An interesting interpretation may be given Eq. (12.161). The optimum tails composition is

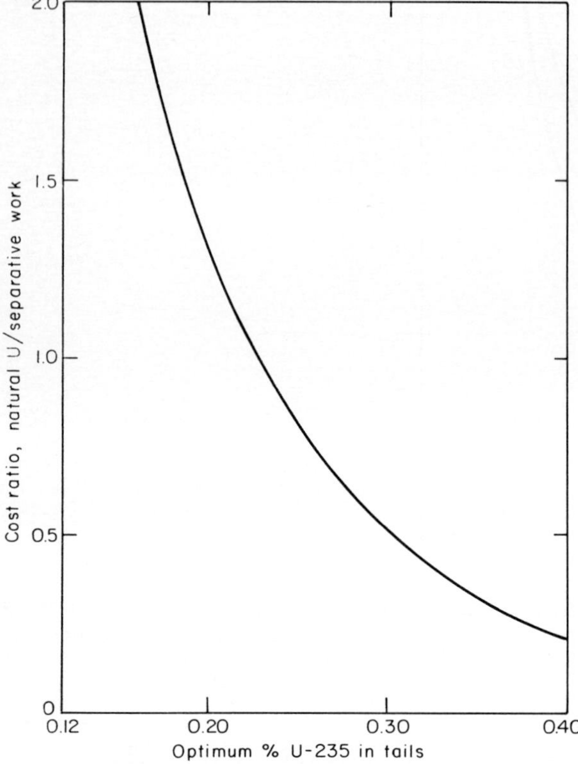

Figure 12.21 Optimum tails composition.

the composition of material from which natural uranium can be produced in an ideal cascade without stripping section for the same cost as natural uranium from an external source. This may be seen by comparing the right side of Eq. (12.161) with the term in brackets of Eq. (12.137) for the total flow rate in the enriching section of an ideal cascade. Further discussion of these equations is given by Hollister and Burrington [H3].

11 DIFFERENTIAL EQUATION FOR SEPARATION POTENTIAL

The fact that the total internal flow rate in a close-separation, ideal cascade is given by Eq. (12.142) may be derived without solving explicitly for the individual internal flow rates by the following development, due originally to P. A. M. Dirac. This procedure is valuable in showing the fundamental character of the separation potential and the separative capacity, and provides a point of departure for the treatment of multicomponent isotope separation.

We consider a close-separation, ideal cascade whose external streams have molar flow rates X_k (positive if a product, negative if a feed), and compositions x_k expressed as mole fraction. Let us look for a function of composition $\phi(x_k)$, to be called the separation potential, with the property that the sum over all external streams, to be called the separative capacity D,

$$D = \Sigma\, X_k \phi(x_k) \tag{12.162}$$

is proportional to the sum of the flow rate of all internal streams. At this point in the derivation, the nature of $\phi(x_k)$ is assumed not to be known.

Figure 12.22 represents stages $i-2$, $i-1$, i, and $i+1$ of such a cascade, with the kth product stream consisting of part of the heads stream of stage $i-1$. The total internal flow from stage i is $M_i + N_i$. The separative capacity of the ith stage, considered as an isolated plant, is

$$\Delta_i = M_i\phi(y_i) + N_i\phi(x_i) - (M_i + N_i)\phi(z_i) = (M_i + N_i)[\theta_i\phi(y_i) + (1 - \theta_i)\phi(x_i) - \phi(z_i)] \tag{12.163}$$

By expanding (y_i) and (x_i) is a Taylor's series about z_i, we obtain

$$\phi(y_i) = \phi(z_i) + (y_i - z_i)\frac{d\phi(z_i)}{dz} + \frac{(y_i - z_i)^2}{2}\frac{d^2\phi(z)}{dz^2} + \cdots \tag{12.164}$$

and

$$\phi(x_i) = \phi(z_i) + (x_i - z_i)\frac{d\phi(z_i)}{dz} + \frac{(x_i - z_i)^2}{2}\frac{d^2\phi(z_i)}{dz^2} + \cdots \tag{12.165}$$

Substitution of these expansions into (12.163) yields

$$\Delta_i = \frac{M_i + N_i}{2}[\theta_i(y_i - z_i)^2 + (1 - \theta_i)(x_i - z_i)^2]\frac{d^2\phi(z_i)}{dz^2} \tag{12.166}$$

where the term in $d\phi/dz$ has dropped out because of the material-balance relations (12.8) and (12.9).

In a close-separation cascade,

$$y_i - z_i = (1 - \theta_i)(\alpha - 1)z_i(1 - z_i) \tag{12.167}$$

as may be seen from (12.18) and (12.21), with $(\alpha - 1)$ and $(\beta - 1)$ considered small relative to unity. Similarly,

$$x_i - z_i = -\theta_i(\alpha - 1)z_i(1 - z_i) \tag{12.168}$$

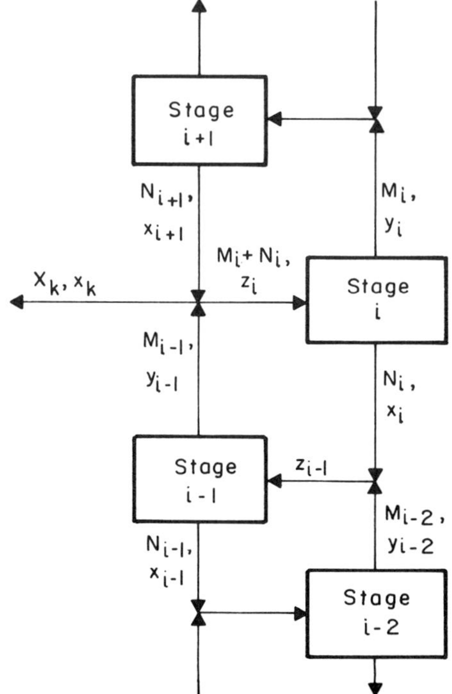

Figure 12.22 Flow in portion of ideal cascade. Molar flow rates denoted by capital letters, mole fractions by small letters.

Therefore, (12.166) becomes

$$\Delta_i = \frac{M_i + N_i}{2} (\alpha - 1)^2 \theta_i (1 - \theta_i) z_i^2 (1 - z_i)^2 \frac{d^2\phi}{dz^2} (z_i) \tag{12.169}$$

In a close-separation, ideal cascade $\theta_i = \frac{1}{2}$, so that the total flow leaving the ith stage is

$$M_i + N_i = \frac{8}{(\alpha - 1)^2} \frac{\Delta_i}{z_i^2 (1 - z_i)^2 [d^2\phi(z_i)/dz^2]} \tag{12.170}$$

where Δ_i is defined by (12.163).

The total internal flow leaving all stages is

$$J + K \equiv \sum_i^{\substack{\text{all}\\\text{stages}}} (M_i + N_i) = \frac{8}{(\alpha - 1)^2} \sum_i^{\substack{\text{all}\\\text{stages}}} \frac{\Delta_i}{z_i^2 (1 - z_i)^2 [d^2\phi(z_i)/dz^2]} \tag{12.171}$$

If and only if ϕ satisfies the differential equation

$$\frac{d^2\phi(z_i)}{dz^2} = \frac{1}{z_i^2 (1 - z_i)^2} \tag{12.172}$$

this may be reduced to

$$J + K = \frac{8}{(\alpha - 1)^2} \sum_i^{\substack{\text{all}\\\text{stages}}} \Delta_i \tag{12.173}$$

where Δ_i is defined by (12.163).

When the separation potential satisfies (12.172), the separative capacity of a single stage in a close-separation cascade operated at a cut of $\frac{1}{2}(M = N)$ from Eq. (12.170) is

$$\Delta_i = \frac{M_i(\alpha - 1)^2}{4} \tag{12.174}$$

We shall now show that the separative capacity of the entire cascade, D, is given by

$$D = \sum_i^{\substack{\text{all}\\\text{stages}}} \Delta_i = \sum_i^{\substack{\text{all}\\\text{external}\\\text{streams}}} X_k \phi(x_k) \tag{12.175}$$

Consider first the sum of the separative capacity of stages i and $i - 1$.

$$\begin{aligned}\Delta_i + \Delta_{i-1} = M_i\phi(y_i) + N_i\phi(x_i) - (M_i + N_i)\phi(z_i) + M_{i-1}\phi(y_{i-1})\\ + N_{i-1}\phi(x_{i-1}) - (M_{i-1} + N_{i-1})\phi(z_{i-1})\end{aligned} \tag{12.176}$$

The internal streams between this pair of stages, M_{i-1} and N_i, may be expressed in terms of the streams external to this pair of stages M_i, N_{i+1}, X_k, M_{i-2}, and N_{i-1} by means of the material-balance relations:

$$N_i - M_{i-1} = N_{i-1} - M_{i-2} \tag{12.177}$$

and

$$N_i - M_{i-1} = N_{i+1} - M_i - X_k \tag{12.178}$$

Because of the assumption that this is an ideal cascade,

$$x_{i+1} = z_i = y_{i-1} \;(= x_k) \tag{12.179}$$

and

$$x_i = z_{i-1} = y_{i-2} \tag{12.180}$$

By means of these four equations (12.176) may be expressed as

$$\Delta_i + \Delta_{i-1} = M_i\phi(y_i) - N_{i+1}\phi(x_{i+1}) - M_{i-2}\phi(y_{i-2}) + N_{i-1}\phi(x_{i-1}) + X_k\phi(x_k) \tag{12.181}$$

This is an example of (12.175) applied to the pair of stages i and $i-1$. If Δ_{i+1} is added to this expression, terms in M_i and N_{i+1} may be eliminated in the same way. By proceeding in this way until the separative capacity of every stage has been included in the sum, terms representing all internal streams cancel out, the only terms that remain on the right are those representing external streams, and Eq. (12.175) results.

Thus, we have shown that the separative capacity of an ideal cascade is the sum of the separative capacities of its component stages. And if the separation potential satisfies the differential equation (12.172), the total internal flow is given by

$$J + K = \frac{8}{(\alpha - 1)^2} \sum_{k}^{\substack{\text{all}\\ \text{external}\\ \text{streams}}} X_k\phi(x_k) \tag{12.182}$$

as was to have been shown.

The general solution of (12.172) is

$$\phi = (2x - 1)\ln\frac{x}{1-x} + ax + b \tag{12.183}$$

Here a and b are arbitrary constants, and the general composition variable x has been substituted for z_i. The arbitrary constants a and b do not affect the value of the right side of (12.182) because of the overall material-balance relations

$$\sum_k X_k x_k = 0 \tag{12.184}$$

and

$$\sum_k X_k = 0 \tag{12.185}$$

In Eq. (12.157) for the price of uranium, it may be noted that the term in brackets has the general form (12.183) for the separation potential, with

$$a = 219.5666 \tag{12.186}$$

and

$$b = -6.4300 \tag{12.187}$$

The separation potential may be thought of as related to the value of a mixture of isotopes, and has, in fact, been called the "value function" by Cohen [C3].

12 EQUILIBRIUM TIME FOR ISOTOPE SEPARATION PLANTS

One of the most striking aspects of plants for producing heavy water or ^{235}U is the long time they must be operated when first started before it is possible to withdraw enriched material of specified product composition from them. This is because the amount of desired isotope held up in the plant may represent many days or even months of normal production, and at start-up the plant must be run without product withdrawal for a time sufficient to produce the plant's working inventory of desired isotope. The purpose of this section is to derive approximate relations that may be used to estimate the so-called equilibrium, or start-up, time of an isotope separation plant. Exact evaluation of the equilibrium time requires numerical intergration of the partial differential equation describing the change of isotopic abundance with time and stage number, and is beyond the scope of this text. This equation has been derived by Cohen [3], p. 29.

12.1 Operating Procedure during Start-up

Figure 12.23 shows the nomenclature to be used in describing the operation of an isotope separation plant during the transient period in which it is approaching steady-state performance. Figure 12.24 represents qualitatively the way tails and product flow rates and compositions will change with time during this transient period. Compositions are represented by a scale linear in $\ln [x/(1-x)]$.

At time zero, all stages of the plant contain material of feed composition, z_F. Initially the plant is operated with no feed supply and no tails or product withdrawal. As the plant operates, the fraction of desired isotope in material at the tails end of the plant decreases and the fraction of desired isotope in material at the product end increases. At time t_1 material at the tails end of the plant reaches the desired steady-state level x_W. At this time tails withdrawal is started at such a rate $W(t)$ as to keep the composition at this point constant at x_W. Feed is supplied at a rate equal to tails withdrawal. At first, tails withdrawal is at a rate below the steady-state value W because the compositions elsewhere in the stripping section have not yet reached steady-state values. The tails rate increases and may temporarily exceed the steady-state value for a time, until product withdrawal can be started.

The fraction of desired isotope in material at the product end of the plant continues to increase, reaching the steady-state value y_P at time t_2. Product withdrawal is then started at such a rate $P(t)$ as to keep product composition constant at y_P. Feed is supplied at the rate $P(t) + W(t)$. At first, product withdrawal is at a rate below the steady-state value P because the compositions elsewhere in the enriching section have not yet reached steady-state values. As time goes on, $P(t)$ approaches P asymptotically.

The equilibrium, or start-up, time for product withdrawal t_P is defined as the number of days of equivalent production lost during the approach to steady state. In Fig. 12.24, the area of the rectangle between the vertical line at t_P and the horizontal line at unity equals the area between this horizontal line and the curve for $P(t)/P$. Mathematically,

$$t_P = \lim_{T\to\infty} \int_0^T \left[1 - \frac{P(t)}{P}\right] dt \tag{12.188}$$

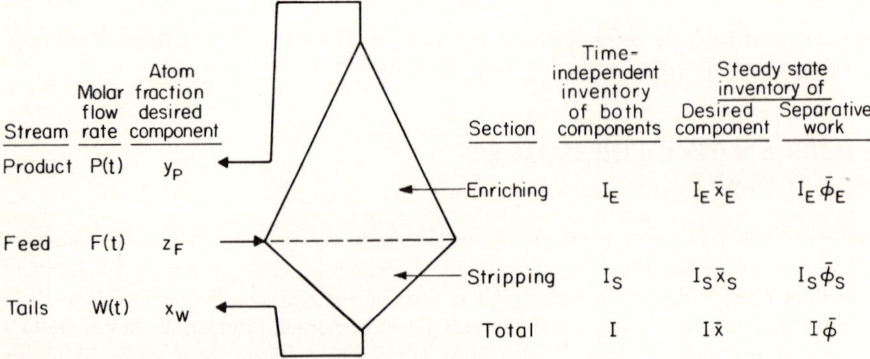

Figure 12.23 Nomenclature for start-up of cascade.

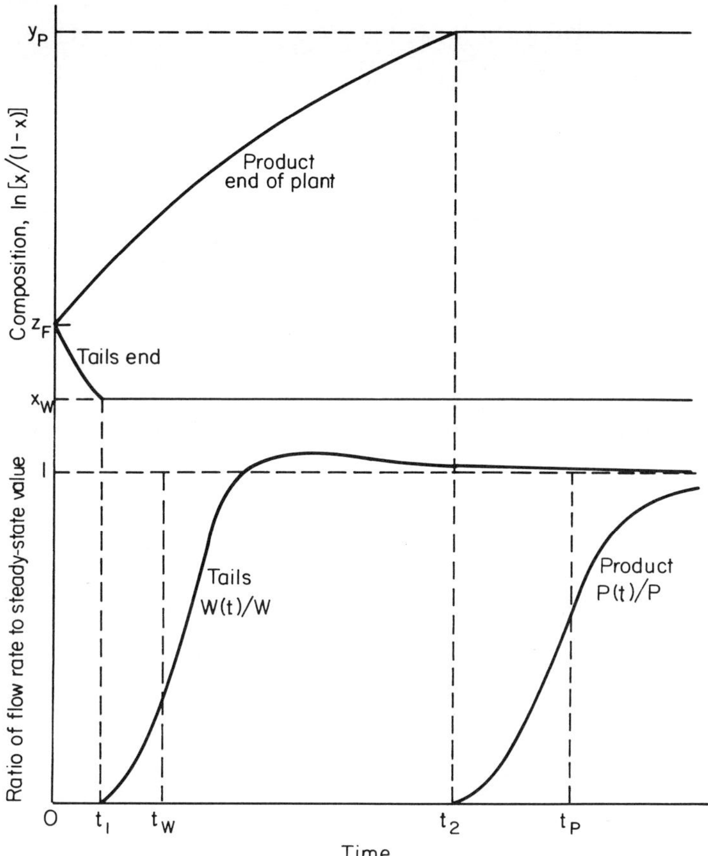

Figure 12.24 Flow rates and compositions during approach to steady state.

Similarly, the equilibrium time for tails withdrawal t_W is defined as

$$t_W = \lim_{T \to \infty} \int_0^T \left[1 - \frac{W(t)}{W} \right] dt \tag{12.189}$$

12.2 Relation between Equilibrium Time and Inventory

A simple relation between t_P, t_W, and the cascade inventory may be derived as follows. The inventory of both components in the plant is assumed to remain constant at I during the start-up period. The average fraction of desired component in the plant changes from z_F, the value throughout the plant at time zero, to $\bar{x}$ when the steady state is reached. The material-balance equation for desired component during the transient period is

$$F(t)z_F - P(t)y_P - W(t)x_W = I \frac{d\bar{x}}{dt} \tag{12.190}$$

But

$$F(t) = P(t) + W(t) \tag{12.191}$$

so that
$$(z_F - y_P)P(t) + (z_F - x_W)W(t) = I\frac{d\bar{x}}{dt} \tag{12.192}$$

At steady state
$$(z_F - y_P)P + (z_F - x_W)W = 0 \tag{12.193}$$

Subtract (12.193) from (12.192) and integrate from 0 to T:
$$(y_P - z_F)\int_0^T [P - P(t)]\,dt - (z_F - x_W)\int_0^T [W - W(t)]\,dt = I\int_0^T \frac{d\bar{x}}{dt}\,dt \tag{12.194}$$

In the limit, as $T \to \infty$,
$$(y_P - z_F)Pt_P - (z_F - x_W)Wt_W = I(\bar{x} - z_F) \tag{12.195}$$

where the limit of the integrals have been expressed in terms of the equilibrium times through (12.188) and (12.189).

Hence
$$t_P = \frac{I(\bar{x} - z_F)}{P(y_P - z_F)} + t_W \tag{12.196}$$

This equation is a consequence of material-balance relations and is exact.

Its usefulness for evaluating the equilibrium time of the enriching section t_P is diminished, however, because t_W is usually not known exactly. Nevertheless, an approximate equation for t_P can be developed by considering the result of decreasing the size of the stripping section of the plant until only the enriching section is left. The inventory of desired component at steady state then becomes $I_E\bar{x}_E$, that of the enriching section alone. The equilibrium time for waste withdrawal t_W becomes small, because tails withdrawal may be started at time zero, so that
$$t_P \approx \frac{I_E(\bar{x}_E - z_F)}{P(y_P - z_F)} \tag{12.197}$$

This is the equation usually used to estimate the start-up time of a separation cascade. In most cases, it overestimates the time somewhat, because t_W for a plant without stripping section is usually negative.

12.3 Inventory of Ideal Cascade

The total inventory I_E and the inventory of desired component $I_E\bar{x}_E$ may be evaluated if the inventory per stage is known. The stage inventory H_i may be related to the stage feed rate $M_i + N_i$ by
$$H_i = h(M_i + N_i) \tag{12.198}$$

where h is the *stage holdup time*, the time it takes material to flow through one stage. We shall assume that h is constant throughout the cascade. This will be strictly true of an ideal cascade made up of identical separating units and is often approximately true of an ideal cascade made up of stages of decreasing size.

The total inventory of the enriching section I_E then is just h times the total flow rate in the enriching section; for a close-separation, ideal cascade,
$$I_E = \frac{8Ph}{(\alpha - 1)^2}\left[(2y_P - 1)\ln\frac{y_P(1 - z_F)}{z_F(1 - y_P)} + \frac{(y_P - z_F)(1 - 2z_F)}{z_F(1 - z_F)}\right] \tag{12.199}$$

as may be seen from Eq. (12.137) and the fact that heads and tails flow rates are approximately equal in a close-separation, ideal cascade.

The inventory of desired component in the enriching section is

$$I_E \bar{x}_E = \int_{n_S+1}^{n} Hx\, di \tag{12.200}$$

Because
$$H = h(M + N) = \frac{4Ph(y_P - x)}{(\alpha - 1)x(1 - x)} \tag{12.201}$$

from (12.132) and

$$di = \frac{di}{dx}\, dx = \frac{2\, dx}{(\alpha - 1)x(1 - x)} \tag{12.202}$$

from (12.134), this inventory is given by

$$I_E \bar{x}_E = \int_{z_F}^{y_P} \frac{8Ph}{(\alpha - 1)^2} \frac{y_P - x}{x(1 - x)^2}\, dx = \frac{8Ph}{(\alpha - 1)^2}\left[y_P \ln \frac{y_P}{z_F}\frac{(1 - z_F)}{(1 - y_P)} - \frac{y_P - z_F}{1 - z_F}\right] \tag{12.203}$$

With Eqs. (12.199) and (12.203), approximate equation (12.197) for the start-up time of a close-separation, ideal cascade becomes

$$t_P = \frac{8h}{(\alpha - 1)^2}\left[\frac{(y_P - 2y_P z_F + z_F)}{y_P - z_F} \ln \frac{y_P(1 - z_F)}{z_F(1 - y_P)} - 2\right] \tag{12.204}$$

12.4 Relation between Equilibrium Time and Separative Work

A lower bound for the equilibrium time of an ideal cascade may be found by determining the length of time it would take the plant to produce its own steady-state composition gradient if at all times during the start-up period it was possible to prevent loss of separative work through mixing of streams of different composition. Conceptually, this might be done if the plant consisted of a large number of small separating units whose connection in parallel and in series could be changed continuously during the start-up period. We assume that no product is withdrawn until the steady-state composition gradient has been established. Then product withdrawal may be started at the steady-state rate.

During the start-up period prior to product withdrawal it is necessary to bring in enough feed of composition z_F and withdraw enough tails of composition x_W to provide the increase in inventory of desired component from its initial value of Iz_F to its steady-state value of $I\bar{x}$. By material balance, the required amount of feed E_F is

$$E_F = \frac{I(\bar{x} - z_F)}{z_F - x_W} \tag{12.205}$$

During the start-up period, the change in the plant's inventory of separative work is $I(\bar{\phi} - \phi_F)$, where

$$I\bar{\phi} = \sum_{i}^{\text{all stages}} I(i)\phi_i \tag{12.206}$$

If the duration of the start-up period is τ, the amount of separative work S done by a plant whose separative capacity is D is

$$S = D\tau \tag{12.207}$$

This must equal the amount of separative work done by the plant in changing $E_F + I$ mol of material of feed composition into E_F mol of tails of composition x_W and I mol of inventory having the requisite steady-state composition distribution. That is,

$$D\tau = E_F(\phi_W - \phi_F) + I(\bar{\phi} - \phi_F) \tag{12.208}$$

This may be solved for the equilibrium time τ, with E_F given by (12.205):

$$\tau = \frac{I}{D}\left[\frac{(\bar{x} - z_F)(\phi_W - \phi_F)}{z_F - x_W} + \bar{\phi} - \phi_F\right] \tag{12.209}$$

This expression gives a lower bound for the equilibrium time, which can be attained only if mixing of streams of different composition can be prevented during the entire start-up period. It provides a lower bound for the equilibrium time in somewhat the same way that consideration of a thermodynamically reversible process provides a lower bound for the amount of work needed to carry out a given change of state.

12.5 Inventory Functions

To make use of Eq. (12.209) we need expressions for the inventory of both components I, the inventory of desired component $I\bar{x}$ and the inventory of separative work $I\bar{\phi}$ in a close-separation, ideal cascade. To derive these expressions we shall assume that the stage inventory is proportional to the stage feed rate, as stated by (12.198), and that the average stage composition is that of the stage feed z_i.

Because of the first assumption, the total inventory is proportional to the total interstage flow rate, given by (12.181), so that

$$I = \frac{8h}{(\alpha - 1)^2} \sum_k X_k \phi(x_k) \tag{12.210}$$

ϕ, the separation potential, may be thought of in this connection as a function for evaluating the inventory. We have proved that ϕ satisfies differential equation (12.172) and is given by (12.144).

By a development similar to that which showed the separation potential to have these properties, it can be shown that the inventory of desired component is given by

$$I\bar{x} = \frac{8h}{(\alpha - 1)^2} \sum_k X_k \psi(x_k) \tag{12.211}$$

and the inventory of separative work

$$I\bar{\phi} = \frac{8h}{(\alpha - 1)^2} \sum_k X_j \pi(x_k) \tag{12.212}$$

The functions ϕ, ψ, and π and their second derivatives are listed in Table 12.10.

The derivation of differential equation (12.215) for the separative work inventory function π is similar to the derivation of differential equation (12.172) for the separation potential ϕ. Equation (12.170) is valid for any function of composition that can be expressed as a Taylor series. Therefore, the feed rate to stage i may be expressed in terms of π instead of ϕ as

$$M_i + N_i = \frac{8}{(\alpha - 1)^2} \frac{M_i \pi(y_i) + N_i \pi(x_i) - (M_i + N_i)\pi(z_i)}{z_i^2(1 - z_i)^2 [d^2\pi(z_i)/dz^2]} \tag{12.217}$$

Because of the assumption that the stage inventory is given by (12.198) and the assumption

Table 12.10 Inventory functions for ideal cascade

Inventory of	Second derivative	Function
Both components	$\frac{d^2\phi}{dx^2} = \frac{1}{x^2(1-x)^2}$ (12.172)	$\phi = (2x-1)\ln\frac{x}{1-x}$ (12.144)
Desired component	$\frac{d^2\psi}{dx^2} = \frac{1}{x(1-x)^2}$ (12.213)	$\psi = x\ln\frac{x}{1-x}$ (12.214)
Separative work	$\frac{d^2\pi}{dx^2} = \frac{(2x-1)\ln[x/(1-x)]}{x^2(1-x)^2}$ (12.215)	$\pi = \frac{1}{2}\ln\left(\frac{x}{1-x}\right)^2 + (1-2x)\ln\frac{x}{1-x}$ (12.216)

that the average composition† of the inventory is z_i, the inventory of separative work on the stage is

$$h(M_i + N_i)\phi(z_i) = \frac{8h}{(\alpha-1)^2}\frac{(2z_i-1)\ln[z_i/(1-z_i)]}{z_i^2(1-z_i)^2[d^2\pi(z_i)/dz^2]}[M_i\pi(y_i) + N_i\pi(x_i) - (M_i+N_i)\pi(z_i)] \tag{12.218}$$

The separative work inventory of all stages will be given by Eq. (12.212) if and only if the second factor is independent of i, that is, if

$$\frac{d^2\pi(z_i)}{dz^2} = \frac{(2z_i-1)\ln[z_i/(1-z_i)]}{z_i^2(1-z_i)^2} \tag{12.219}$$

The proof is similar to that given in Sec. 11 to establish Eq. (12.182) for the total flow rate of all stages. Differential equation (12.213) for the component inventory function ψ may be derived in similar fashion (see Prob. 12.9).

12.6 Equilibrium Time Example

To compare equilibrium times evaluated by approximate Eq. (12.204) and the lower bound Eq. (12.209), the example of an ideal cascade to perform the separation of Table 12.8 will be considered. It is assumed, in addition, that the stage holdup time h is 1 s and the stage separation factor is 1.0043, the nominal value for separating $^{235}UF_6$ from $^{238}UF_6$. For this cascade,

$$\frac{8h}{(\alpha-1)^2} = \frac{8/86{,}400}{(0.0043)^2} = 5.008 \text{ days} \tag{12.220}$$

From Eq. (12.204), the approximate equilibrium time is

$$t_E \approx 5.008\left(\frac{0.8 - 0.01152 + 0.0072}{0.8 - 0.0072}6.313 - 2\right) = 21.7 \text{ days} \tag{12.221}$$

Calculation of the inventories needed in Eq. (12.209) is shown in Table 12.11.

†This disregards the slight difference in separation potential between stage feed, heads, and tails, which does not affect the final equation.

Table 12.11 Inventories in UF_6 separation example

	Stream		
	Product	Tails	Feed
Mole fraction x	0.8000	0.0036	0.0072
Flow rate X, mol/day	1.25	275.27	−276.52
Separation potential ϕ, Eq. (12.144)	0.83178	5.58273	4.85551
Component inventory function ψ, Eq. (12.214)	1.10904	−0.020243	−0.035469
Separative work inventory function π, Eq. (12.216)	0.12917	10.2276	7.2794

$$D = \Sigma X_k \phi_k = 195 \text{ mol/day} \qquad \frac{8h}{(\alpha - 1)^2} = 5.008 \text{ days}$$

$$\Sigma X_k \psi_k = 5.54 \text{ mol/day}$$

$$\Sigma X_k \pi_k = 802 \text{ mol/day}$$

UF_6 inventory (12.210): $I = (5.008)(195) = 977$ mol

$^{235}UF_6$ inventory (12.211): $I\bar{x} = (5.008)(5.54) = 27.74$ mol

Separative work inventory (12.212): $I\bar{\phi} = (5.008)(802) = 4016$ mol

With these inventories, the lower bound for the equilibrium time may be evaluated from Eq. (12.209).

$$\tau = \frac{1}{195}\left\{\frac{[27.74 - (977)(0.0072)](5.58273 - 4.85551)}{0.0072 - 0.0036} + 4016 - (977)(4.85551)\right\}$$

$$= 17.7 \text{ days} \qquad (12.222)$$

The true value lies between 17.7 and 21.7 days.

This example shows that the equilibrium time in an ideal cascade with $\alpha - 1 \ll 1$ may be relatively long, even when the stage holdup time h is very short. In a cascade that is not tapered at the product end, the equilibrium time will be even greater, because of the increased inventory of desired component in this part of the plant. Equation (12.197) may be used to estimate the equilibrium time of such a nonideal cascade; Eq. (12.209) is restricted to ideal cascades.

13 SQUARED-OFF CASCADE

In some isotope separation plants, notably those using distillation or exchange processes, it is more economic to use a constant interstage flow rate over a considerable composition interval rather than a flow rate that decreases steadily from the feed point to the product ends, as is characteristic of an ideal cascade. Cohen [C3] has called such cascades "squared-off" cascades and has derived equations for their separation performance. This section summarizes the derivation for a close-separation, squared-off cascade.

In the enriching section of a cascade with constant tails flow rate N, the change in composition x with stage number i is given by differential equation (12.128). The number of enriching stages n_{12} needed to span the composition range between x_1 and x_2 is then obtained by integration of

$$\frac{di}{dx} = \frac{1}{(\alpha - 1)x(1 - x) - (P/N)(y_P - x)} \tag{12.223}$$

Hence
$$n_{12} = \int_{x_1}^{x_2} \frac{dx}{(\alpha - 1)x(1 - x) - (P/N)(y_P - x)} = \frac{1}{(\alpha - 1)b} \ln \frac{1 + a}{1 - a} \tag{12.224}$$

where
$$a = \frac{b(x_2 - x_1)}{(x_2 + x_1)(1 + c) - 2x_1x_2 - 2cy_P} \tag{12.225}$$

$$c = \frac{P}{N(\alpha - 1)} \tag{12.226}$$

and
$$b = [1 + 2c(1 - 2y_P) + c^2]^{1/2} \tag{12.227}$$

If a constant value of N is used for the entire enriching section spanning the composition range from z_F to y_P,

$$a = \frac{b(y_P - z_F)}{y_P + z_F - 2y_P z_F - c(y_P - z_F)} \tag{12.228}$$

In the stripping section similar equations hold, with substitution of $-W$ for P and x_W for y_P in Eqs. (12.224) through (12.227). Equation (12.228) for a square stripping section, with constant value of N on all stages, becomes

$$a = \frac{b(z_F - x_W)}{z_F + x_W - 2z_F x_W - c(x_W - z_F)} \tag{12.229}$$

14 GENERALIZED IDEAL CASCADE

14.1 Separation Factor

In the ideal cascade discussed up to this point, each stage receives as feed two streams of the same composition, a tails stream from the stage next higher in the cascade and a heads stream from the stage next lower in the cascade. In such a cascade the heads separation factor β, tails separation factor γ, and overall separation factor α are related by

$$\beta = \gamma = \alpha^{1/2} \tag{12.230}$$

The cut θ at which condition (12.230) is satisfied is given by

$$\theta = \frac{1 + (\beta - 1)z}{\beta + 1} \tag{12.231}$$

The cut thus ranges in value from $1/(\beta + 1)$ at $z = 0$ to $\beta/(\beta + 1)$ at $z = 1$. Because β for most isotope separation processes is close to unity, θ in this type of ideal cascade must be close to $\frac{1}{2}$.

In some isotope separation processes it is impractical to operate a stage at a cut of $\frac{1}{2}$ for mechanical or hydraulic reasons, and in others the separative capacity of the stage is higher at a cut substantially different from $\frac{1}{2}$. In the Becker separation nozzle process described in Chap. 14, the separative capacity of a stage producing a heads stream at a given rate is substantially higher at a cut of $\frac{1}{3}$ than at a cut of $\frac{1}{2}$.

To permit operation at a cut different from $\frac{1}{2}$ while still ensuring that the composition of heads and tails streams entering each stage be equal requires a more complex cascade

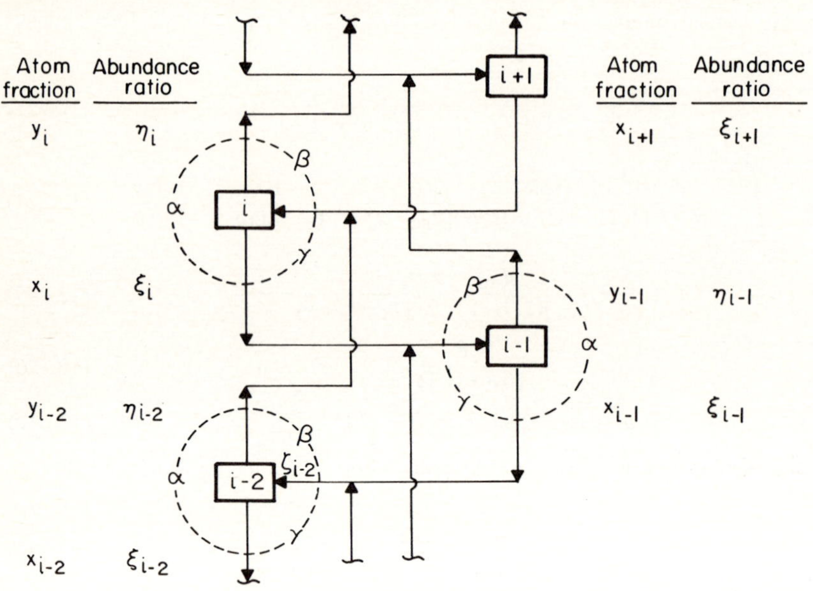

Figure 12.25 Compositions and separation factors in two-up, one-down ideal cascade.

connection scheme than the one shown in Fig. 12.13. Figure 12.25 is an example of such a more complex cascade in which the cut of each stage would be approximately $\frac{1}{3}$. In this cascade the heads stream leaving a stage is fed to the stage two stages up (at higher enrichment) in the cascade and the tails stream leaving a stage is fed to the stage one stage down in the cascade. Olander [O1] calls this a "two-up, one-down" cascade. The condition for an ideal cascade, that the streams entering a stage have the same composition, applied to this cascade, requires that

$$x_{i+1} = y_{i-2} \tag{12.232}$$

in terms of fractions, or

$$\xi_{i+1} = \eta_{i-2} \tag{12.233}$$

in terms of ratios. From the definition of separation factor α,

$$\eta_{i-2} = \alpha\xi_{i-2} \tag{12.234}$$

From the definition of tails separation factor γ,

$$\xi_{i-1} = \zeta_{i-2} = \gamma\xi_{i-2} \tag{12.235}$$

Similarly,

$$\xi_i = \gamma\xi_{i-1} = \gamma^2\xi_{1-2} \tag{12.236}$$

and

$$\xi_{i+1} = \gamma\xi_i = \gamma^3\xi_{1-2} \tag{12.237}$$

From (12.233), (12.234), and (12.237),

$$\alpha = \gamma^3 \tag{12.238}$$

or

$$\gamma = \alpha^{1/3} \tag{12.239}$$

Because

$$\alpha = \beta\gamma \tag{12.240}$$

the heads separation factor β is given by

$$\beta = \frac{\alpha}{\gamma} = \alpha^{2/3} \tag{12.241}$$

An extension of this development to the general, p-up, q-down ideal cascade shows that the heads separation factor β is

$$\beta = \alpha^{p/(p+q)} = \lambda^p \tag{12.242}$$

and the tails separation factor γ is

$$\gamma = \alpha^{q/(p+q)} = \lambda^q \tag{12.243}$$

where

$$\lambda \equiv \alpha^{1/(p+q)} \tag{12.244}$$

14.2 Cut

We have seen that the cut θ of a stage is related to the fractions in heads y, feed z, and tails x by

$$\theta = \frac{z - x}{y - x} \tag{12.245}$$

When x and y are expressed in terms of z and the heads and tails separation factors by Eqs. (12.18) and (12.20), this becomes

$$\theta = \frac{z - z/[z + \gamma(1 - z)]}{[\beta z/(\beta z + 1 - z)] - z/[z + \gamma(1 - z)]} \tag{12.246}$$

After clearing fractions this becomes

$$\theta = \frac{\gamma - 1}{\beta\gamma - 1}\,[1 + (\beta - 1)z] \tag{12.247}$$

Using (12.242) and (12.243),

$$\theta = \frac{\lambda^q - 1}{\lambda^{p+q} - 1}\,[1 + (\lambda^p - 1)z] \tag{12.248}$$

or

$$\theta = \frac{\sum_{i=0}^{q-1} \lambda^i}{\sum_{i=0}^{p+q-1} \lambda^i}\,[1 + (\lambda^p - 1)z] \tag{12.249}$$

For the stages important in deuterium and ^{235}U separation, $z \ll 1$ and Eq. (12.248) reduces to

$$\theta = \frac{\alpha^{q/(p+q)} - 1}{\alpha - 1} \tag{12.250}$$

so that

$$\frac{q}{p+q} = \frac{\ln\,[1 + \theta(\alpha - 1)]}{\ln \alpha} \tag{12.251}$$

For the close-separation case, in which $\lambda^{p+q} \approx 1$, Eq. (12.249) reduces to

$$\theta = \frac{q}{p+q} \tag{12.252}$$

and (12.243) becomes

$$\gamma - 1 = \frac{(\alpha - 1)q}{p + q} = (\alpha - 1)\theta \tag{12.252a}$$

Thus, in a process like the Becker nozzle process, in which it is desirable to design stages for a cut of $\frac{1}{3}$, the cascade might advantageously be of the two-up, one-down type shown in Fig. 12.25 with $p = 2$ and $q = 1$.

14.3 Separative Capacity

The separative capacity Δ of a stage receiving feed of atom fraction z at rate Z and producing heads of atom fraction y at rate M and tails of atom fraction x at rate N is

$$\Delta = M(2y - 1) \ln \frac{y}{1 - y} + N(2x - 1) \ln \frac{x}{1 - x} - Z(2z - 1) \ln \frac{z}{1 - z} \tag{12.253}$$

Substitution of Eq. (12.10) for M, (12.11) for N, (12.18) for y in terms of z, and (12.20) for x in terms of z into Eq. (12.253) and simplification leads to

$$\Delta = \frac{Z}{\beta\gamma - 1} \{\gamma(\beta - 1) \ln \gamma - (\gamma - 1) \ln \beta + z[(\beta + 1)(\gamma - 1) \ln \beta - (\gamma + 1)(\beta - 1) \ln \gamma]\} \tag{12.254}$$

In this general expression, the ratio of total flow to a stage to separative capacity of the stage, Z/Δ, is a function of stage feed composition z. Hence, for this general case, the total flow to all stages cannot be obtained simply as $D(Z/\Delta)$, as was done in Sec. 11. That is, the concept of separative capacity does not provide a simple, accurate way of evaluating the total flow in general for a p-up, q-down cascade. There are, however, a number of practically important special cases for which the term in braces of Eq. (12.254) is substantially or completely independent of z, in which the separative capacity may still be used.

14.4 Special Cases

Standard ideal cascade, $p = q = 1$. For the standard ideal cascade, $p = q = 1$. From Eqs. (12.242) and (12.243),

$$\beta = \gamma = \alpha^{1/2} \tag{12.255}$$

and the coefficient of z in Eq. (12.254) is identically zero. Hence

$$\frac{\Delta}{Z} = \frac{(\beta - 1)^2}{\beta^2 - 1} \ln \beta = \frac{\beta - 1}{\beta + 1} \ln \beta \tag{12.256}$$

which is equivalent to Eq. (12.142a).

Close-separation case. In many multistage isotope separation processes $\alpha - 1 \ll 1$, so that $\beta - 1 \ll 1$ and $\gamma - 1 \ll 1$. The gaseous diffusion process for separating uranium isotopes and the water distillation process for enriching deuterium are examples.

Define

$$\delta \equiv \beta - 1 \tag{12.257}$$

and

$$\epsilon \equiv \gamma - 1 \tag{12.258}$$

and expand

$$\ln \beta = \delta - \frac{\delta^2}{2} + \cdots \tag{12.259}$$

and

$$\ln \gamma = \epsilon - \frac{\epsilon^2}{2} + \cdots \tag{12.260}$$

in Eq. (12.254):

$$\Delta = \frac{Z}{\delta + \epsilon + \delta\epsilon}\left\{(1+\epsilon)\delta\left(\epsilon - \frac{\epsilon^2}{2}\right) - \epsilon\left(\delta - \frac{\delta^2}{2}\right) + z\left[(2+\delta)\epsilon\left(\delta - \frac{\delta^2}{2}\right)\right.\right.$$

$$\left.\left. - (2+\epsilon)\delta\left(\epsilon - \frac{\epsilon^2}{2}\right)\right]\right\} = \frac{Z}{\delta + \epsilon + \delta\epsilon}\left[\frac{\delta^2\epsilon + \delta\epsilon^2 - \delta\epsilon^3}{2} + \frac{z(\delta\epsilon^3 - \delta^3\epsilon)}{2}\right] \tag{12.261}$$

To the second order in δ and ϵ this reduces to

$$\Delta = \frac{Z\delta\epsilon}{2} = \frac{Z(\beta - 1)(\gamma - 1)}{2} \tag{12.262}$$

which is independent of z. Hence, the concept of separative capacity may be used to evaluate the total flow rate in a close-separation, ideal cascade for all values of p and q.

Low-enrichment case, $z \ll 1$. In the largest and most important stages of a deuterium enrichment or uranium isotope separation plant $z < 0.03$. For this low-enrichment case $z[(\beta + 1)(\gamma - 1)\ln\beta - (\gamma + 1)(\beta - 1)\ln\gamma]$ in Eq. (12.254) is small compared with $\gamma(\beta - 1)\ln\gamma - (\gamma - 1)\ln\beta$ and may be neglected for values of β and γ under 2.

14.5 Separative Capacity of Low-Enrichment, Two-Up, One-Down Ideal Cascade

This type of cascade may have practical application in a Becker nozzle plant or centrifuge plant for producing low-enriched uranium, with individual stages operated at a cut of around $\frac{1}{3}$. Figure 12.26 is a schematic diagram of stage connections showing the nomenclature to be used in solving the enrichment equations for such a cascade. Olander [O1] has solved the enrichment equations for such a cascade.

The cascade receives feed of fraction z_F at flow rate F and produces an upper product of fraction y_P at flow rate P, a lower product of fraction y_Q at flow rate Q, and tails of fraction x_W at flow rate W. For this two-up, one-down cascade, $p = 2$, $q = 1$, the heads separation factor β is

$$\beta = \alpha^{p/(p+q)} = \alpha^{2/3} \tag{12.263}$$

and the tails separation factor γ is

$$\gamma = \alpha^{q/(p+q)} = \alpha^{1/3} \tag{12.264}$$

By counting the number of stages n_S in the stripping section of Fig. 12.26, it is seen that

$$\frac{z_F}{1 - z_F} = \frac{y_{n_S - 1}}{1 - y_{n_S - 1}} = \gamma^{n_S + 1}\frac{x_W}{1 - x_W} \tag{12.265}$$

Similarly, the total number of stages n satisfies

$$\frac{y_P}{1 - y_P} = \gamma^{n+2}\frac{x_W}{1 - x_W} \tag{12.266}$$

For this low-enrichment case in which $1 - x \approx 1$,

$$n_S + 1 = \frac{\ln(z_F/x_W)}{\ln\gamma} \tag{12.267}$$

and

$$n + 2 = \frac{\ln(y_P/x_W)}{\ln\gamma} \tag{12.268}$$

Also,

$$n + 1 = \frac{\ln(y_Q/x_W)}{\ln\gamma} \tag{12.269}$$

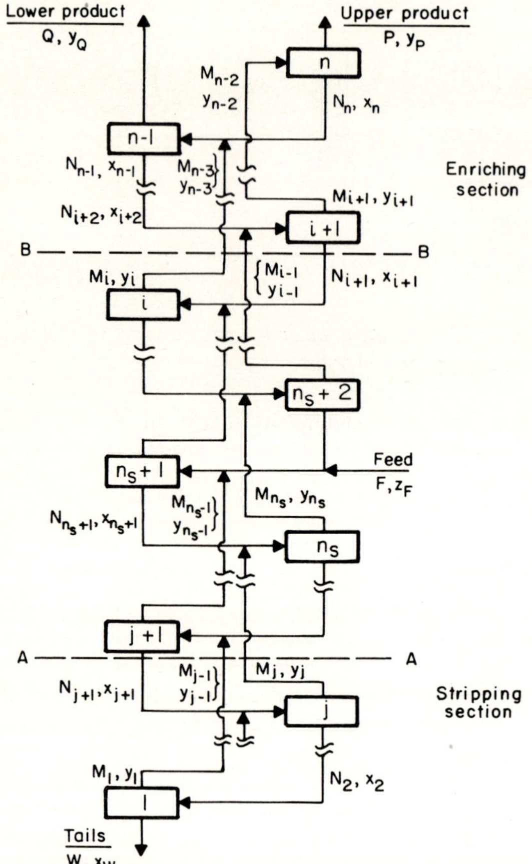

Figure 12.26 Flow rates (capital letters) and fractions (x, y, or z) in two-up, one-down cascade.

The separative capacity D of the two-up, one-down cascade is

$$D = P(2y_P - 1)\ln\frac{y_P}{1 - y_P} + Q(2y_Q - 1)\ln\frac{y_Q}{1 - y_Q}$$

$$+ W(2x_W - 1)\ln\frac{x_W}{1 - x_W} - (P + Q + W)(2z_F - 1)\ln\frac{z_F}{1 - z_F} \tag{12.270}$$

For this low-enrichment approximation,

$$D \approx -P\ln\frac{y_P}{z_F} - Q\ln\frac{y_Q}{z_F} + W\ln\frac{z_F}{x_W} \tag{12.271}$$

Substitution of (12.267), (12.268), and (12.269) into (12.271) yields

$$D = P(\ln\gamma)\left[-(n - n_S + 1) - (n - n_S)\left(\frac{Q}{P}\right) + (n_S + 1)\left(\frac{W}{P}\right)\right] \tag{12.272}$$

Although this equation is concise, it cannot be used to determine the separative capacity until the relative flow rates Q/P and W/P have been determined. This requires solution of the difference equations representing interstage flows in the stripping and enriching sections of Fig. 12.26.

Stripping section. Material balances on the section of the stripping section below line AA of Fig. 12.26 are

$$M_j + M_{j-1} + W = N_{j+1} \tag{12.273}$$

and

$$M_j y_j + M_{j-1} y_{j-1} + W x_W = N_{j+1} x_{j+1} \tag{12.274}$$

But

$$y_j = x_W \gamma^{j+2} \tag{12.275}$$

and

$$x_{j+1} = x_W \gamma^j \tag{12.276}$$

The result of eliminating N_{j+1} between (12.273) and (12.274), and expressing compositions in terms of γ by (12.275) and (12.276) is

$$M_j(\gamma^{j+2} - \gamma^j) + M_{j-1}(\gamma^{j+1} - \gamma^j) = W(\gamma^j - 1) \tag{12.277}$$

Olander [O1] has shown that the general solution of this first-order, inhomogenous difference equation is

$$\frac{M_j}{W} = k\left(-\frac{1}{\gamma+1}\right)^{j-1} - \frac{\gamma^{-j}}{(\gamma-1)(2\gamma+1)} + \frac{1}{(\gamma-1)(\gamma+2)} \tag{12.278}$$

This may be verified by direct substitution. The constant k is determined from a material balance on stage 1, which requires that

$$\frac{M_1}{W} = \frac{\gamma - 1}{\gamma^3 - \gamma} = \frac{1}{\gamma(\gamma+1)} \tag{12.279}$$

so that

$$k = \frac{1}{\gamma(\gamma+1)} + \frac{1}{\gamma(\gamma-1)(2\gamma+1)} - \frac{1}{(\gamma-1)(\gamma+2)} \tag{12.280}$$

Enriching section. Material balances above line BB in Fig. 12.26 are

$$M_i + M_{i-1} = P + Q + N_{i+1} \tag{12.281}$$

and

$$M_i y_i + M_{i-1} y_{i-1} = P y_P + Q y_Q + N_{i+1} x_{i+1} \tag{12.282}$$

but

$$y_i = \frac{y_P}{\gamma^{n-i}} \tag{12.283}$$

$$x_{i+1} = \frac{y_P}{\gamma^{n-i+2}} \tag{12.284}$$

$$y_Q = \frac{y_P}{\gamma} \tag{12.285}$$

The result of eliminating N_{i+1} between (12.281) and (12,282) and expressing compositions in terms of γ by (12.283), (12.284), and (12.285) is

$$M_i(\gamma^2 - 1) + M_{i-1}(\gamma - 1) = P(\gamma^{n-i+2} - 1) + Q(\gamma^{n-i+1} - 1) \tag{12.286}$$

Olander [O1] has shown that the solution of Eq. (12.286), with the boundary condition $M_n = P$, is

$$\frac{M_i}{P} = [-(\gamma+1)]^{n-i} - \frac{\gamma(\gamma + Q/P)}{(\gamma-1)(2\gamma+1)}\{[-(\gamma+1)]^{n-i} - \gamma^{n-i}\}$$
$$+ \frac{1 + Q/P}{(\gamma-1)(\gamma+2)}\{[-(\gamma+1)]^{n-i} - 1\} \tag{12.287}$$

This may be confirmed by substitution of (12.287) into (12.286). For the three top stages, Eq. (12.187) gives

$$\frac{M_n}{P} = 1 \tag{12.288}$$

$$\frac{M_{n-1}}{P} = \frac{Q}{P} \tag{12.289}$$

$$\frac{M_{n-2}}{P} = 1 + \gamma + \gamma^2 \tag{12.290}$$

which are also obtainable by inspection of Fig. 12.26.

External flow rates. One relation among the external flow rates W, Q, and P is obtained by equating the heads flow rate from the top stage of the stripping section M_j evaluated from Eq. (12.278) with $j = n_S$ to the heads flow rate into the bottom stage of the enriching section M_i evaluated from Eq. (12.287) with $i = n_S$. From Eqs. (12.278) and (12.280),

$$M_{j=n_S} = rW \tag{12.291}$$

where

$$r = \left[\frac{1}{\gamma(\gamma+1)} + \frac{1}{\gamma(\gamma-1)(2\gamma+1)} - \frac{1}{(\gamma-1)(\gamma+2)}\right]\left(-\frac{1}{\gamma+1}\right)^{n_S-1} - \frac{1}{(\gamma-1)(2\gamma+1)}\left(\frac{1}{\gamma}\right)^{n_S} + \frac{1}{(\gamma-1)(\gamma+2)} \tag{12.292}$$

From Eq. (12.287),

$$M_{i=n_S} = sP + tQ \tag{12.293}$$

where

$$s = \left[1 - \frac{\gamma^2}{(\gamma-1)(2\gamma+1)} + \frac{1}{(\gamma-1)(\gamma+2)}\right][-(\gamma+1)]^{n-n_S} + \frac{\gamma^{n-n_S+2}}{(\gamma-1)(2\gamma+1)} - \frac{1}{(\gamma-1)(\gamma+2)} \tag{12.294}$$

and

$$t = \left[-\frac{\gamma}{(\gamma-1)(2\gamma+1)} + \frac{1}{(\gamma-1)(\gamma+2)}\right][-(\gamma+1)]^{n-n_S} + \frac{\gamma^{n-n_S+1}}{(\gamma-1)(2\gamma+1)} - \frac{1}{(\gamma-1)(\gamma+2)} \tag{12.295}$$

The condition that $M_{j=n_S} = M_{i=n_S}$ thus is

$$rW = sP + tQ \tag{12.296}$$

where r, s, and t are functions of n, n_S, and γ. A second relation between W, P, and Q in terms of these variables may be obtained from the material-balance relations

$$F = W + P + Q \tag{12.297}$$

and

$$Fz_F = Wx_W + Py_P + Qy_Q \tag{12.298}$$

Inspection of Fig. 12.26 shows that

$$z_F = x_W \gamma^{n_S+1} \tag{12.299}$$

$$y_P = x_W \gamma^{n+2} \tag{12.300}$$

and

$$y_Q = x_W \gamma^{n+1} \tag{12.301}$$

The result of eliminating F, z_F, x_W, y_P, and y_Q from Eqs. (12.297) through (12.301) is

$$(\gamma^{n_S+1} - 1)W = (\gamma^{n+2} - \gamma^{n_S+1})P + (\gamma^{n+1} - \gamma^{n_S+1})Q \tag{12.302}$$

Equations (12.296) and (12.302) make it possible to determine the flow ratios Q/P and W/P as functions of n, n_S, and γ:

$$\frac{Q}{P} = \frac{r(\gamma^{n+2} - \gamma^{n_S+1}) - s(\gamma^{n_S+1} - 1)}{t(\gamma^{n_S+1} - 1) - r(\gamma^{n+1} - \gamma^{n_S+1})} \tag{12.303}$$

and

$$\frac{W}{P} = \frac{t(\gamma^{n+2} - \gamma^{n_S+1}) - s(\gamma^{n+1} - \gamma^{n_S+1})}{t(\gamma^{n_S+1} - 1) - r(\gamma^{n+1} - \gamma^{n_S+1})} \tag{12.304}$$

Design example. The foregoing equations will be applied to the two-up, one-down ideal cascade considered by Olander [O1] having three stripping stages ($n_S = 3$), seven total stages ($n = 7$), and a tails separation factor (γ) of 1.3027. Values of r, s, and t then are

	Equation	Value
r	(12.292)	0.592674
s	(12.294)	15.99783
t	(12.295)	−2.99973

Table 12.12 gives compositions and flow rates relative to top product calculated from the preceding equations for feed containing 0.71 percent ^{235}U.

15 THREE-COMPONENT ISOTOPE SEPARATION

Although most isotope separation problems involve only two components, it is occasionally necessary to consider the effect of one or more additional components on cascade design or performance. Examples are the effect of the 0.0058 percent ^{234}U present in natural uranium, the ^{236}U present in uranium recovered from a nuclear fuel reprocessing plant, the three isotopes found naturally in oxygen, or the five isotopes occurring in natural tungsten. de la Garza and co-workers have extended the theory of the close-separation, ideal cascade to multicomponent mixtures. In this section, their development is used to derive equations that describe the effect of small amounts of ^{236}U on the performance of a cascade designed to separate ^{235}U and ^{238}U. For extension of the theory to systems containing large amounts of a third component and to multicomponent systems, de la Garza's papers [D1, D2] and Pratt's [P2] summary of them may be used.

Table 12.12 Flow rates and compositions in example of two-up, one-down ideal cascade

		Flow rate relative to top product		
Stream	Percent ^{235}U	Symbol	Value	Equation
Tails	0.2465	W/P	12.476	(12.304)
Heads, stage 1	0.545	M_1/P	4.159	(12.278), (12.304)
Heads, stage 2	0.710	M_2/P	5.546	(12.278), (12.304)
Heads, stage 3	0.925	M_3/P	7.394	(12.287), (12.303)
Heads, stage 4	1.205	M_4/P	8.472	(12.287), (12.303)
Heads, stage 5	1.570	M_5/P	4.000	(12.287), (12.303)
Heads, stage 6	2.045	Q/P	2.868	(12.303)
Heads, stage 7	2.664	P/P	1.000	–

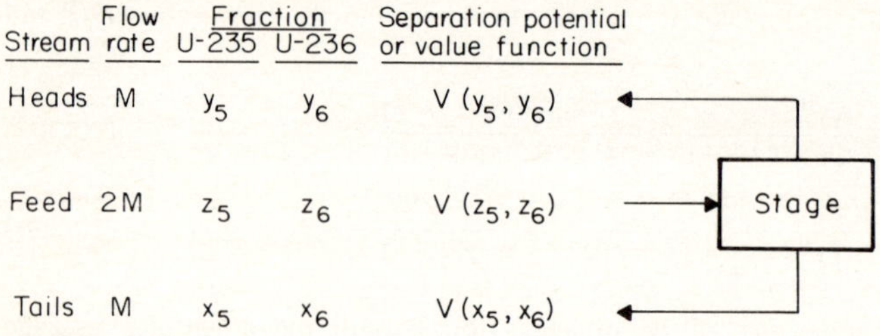

Stream	Flow rate	Fraction U-235	Fraction U-236	Separation potential or value function
Heads	M	y_5	y_6	$V(y_5, y_6)$
Feed	2M	z_5	z_6	$V(z_5, z_6)$
Tails	M	x_5	x_6	$V(x_5, x_6)$

Stage

Figure 12.27 Nomenclature for stage processing mixture of ^{235}U, ^{236}U, and ^{238}U.

15.1 Separation Factors

Figure 12.27 represents one stage of an ideal, close-separation, one-up, one-down cascade whose feed flows at rate $2M$ and contains z_5 fraction ^{235}U, z_6 fraction ^{236}U, and $z_8 = 1 - z_5 - z_6$ fraction ^{238}U. At the cut of $\frac{1}{2}$ used in such a cascade, heads flows at rate M and contains y_5 fraction ^{235}U and y_6 ^{236}U. Stage tails flows at rate M and contains x_5 fraction ^{235}U and x_6 ^{236}U. Stage separation factors are defined as

$$^{235}\text{U from }^{238}\text{U:} \qquad \alpha_{58} = \frac{y_5(1 - x_5 - x_6)}{x_5(1 - y_5 - y_6)} \tag{12.305}$$

$$^{235}\text{U from }^{236}\text{U:} \qquad \alpha_{56} = \frac{y_5 x_6}{y_6 x_5} \tag{12.306}$$

For close isotope separation processes that depend on differences in molecular weight, such as gaseous diffusion or the Becker nozzle process,

$$\frac{\alpha_{56} - 1}{\alpha_{58} - 1} = \frac{236 - 235}{238 - 235} = \frac{1}{3} \tag{12.307}$$

α_{56} and α_{58} are to be replaced by the overall enrichment factor for ^{235}U from ^{238}U, ψ, defined as

$$\psi \equiv \alpha_{58} - 1 = 3(\alpha_{56} - 1) \tag{12.308}$$

For this close-separation case, with $(y_5 - x_5)/x_5 \ll 1$ and $(y_6 - x_6)/x_6 \ll 1$, Eq. (12.305) may be approximated by

$$\psi = \frac{y_5 - x_5 - x_6 y_5 + x_5 y_6}{x_5(1 - x_5 - x_6)} \tag{12.309}$$

and Eq. (13.306) by

$$\frac{\psi}{3} = \frac{y_5 x_6 - y_6 x_5}{x_5 x_6} \tag{12.310}$$

Hence,

$$y_5 - x_5 = \psi\left[x_5(1 - x_5 - x_6) + \frac{x_5 x_6}{3}\right] \tag{12.311}$$

and

$$y_6 - x_6 = \psi\left[\frac{2x_6(1 - x_5 - x_6)}{3} - \frac{x_5 x_6}{3}\right] \tag{12.312}$$

15.2 Three-Component Value Function

We wish to find a value function V, a generalization of the separation potential ϕ for a two-component mixture, now a function of x_5 and x_6, which can be used to evaluate the separative capacity, and from it, the total flow rate. The difference equation (12.313) for V is obtained by writing a V balance for the stage, in which the difference between the separation potential carried by the stage effluents and the stage feed is equated to the separative capacity of the stage, given by Eq. (12.174) as $M\psi^2/4$:

$$MV(y_5, y_6) + MV(x_5, x_6) - 2MV(z_5, z_6) = \frac{M\psi^2}{4} \tag{12.313}$$

For a close-separation cascade with a cut of $\frac{1}{2}$,

$$z - x = \frac{y - x}{2} \tag{12.314}$$

When Eq. (12.313) is expanded in a Taylor series about x_5 and x_6, the following differential equation is obtained:

$$(y_5 - x_5)^2 \frac{\partial^2 V}{\partial x_5^2} + 2(y_5 - x_5)(y_6 - x_6) \frac{\partial^2 V}{\partial x_5 \, \partial x_6} + (y_6 - x_6)^2 \frac{\partial^2 V}{\partial x_6^2} = \psi^2 \tag{12.315}$$

Terms in V, $\partial V/\partial x_5$, and $\partial V/\partial x_6$ have dropped out because of material-balance relations. Substitution of $y_5 - x_5$ from (12.311) and $y_6 - x_6$ from (12.312) into (12.315) leads to

$$\left[x_5(1 - x_5 - x_6) + \frac{x_5 x_6}{3}\right]^2 \frac{\partial^2 V}{\partial x_5^2} + 2\left[x_5(1 - x_5 - x_6) + \frac{x_5 x_6}{3}\right]\left[\frac{2x_6(1 - x_5 - x_6)}{3} - \frac{x_5 x_6}{3}\right]$$
$$\times \frac{\partial^2 V}{\partial x_5 \, \partial x_6} + \left[\frac{2x_6(1 - x_5 - x_6)}{3} - \frac{x_5 x_6}{3}\right]^2 \frac{\partial^2 V}{\partial x_6^2} = 1 \tag{12.316}$$

We wish to find a solution of Eq. (12.316) that can be used to evaluate total flow rates, as was done for two components in Sec. 11. To do this, it is necessary to arrange that there be no loss of V when two streams are mixed. In a two-component system this was done by requiring the two streams to have the same composition. In a three-component system this is not generally possible. The mole fractions of only one component in the two streams may be made equal, or one function of the mole fractions in the two streams may be made equal. For the present derivation, we shall require that the abundance ratio R of the two principal components, ^{235}U and ^{238}U, be equal whenever two streams are mixed.

$$R \equiv \frac{x_5}{1 - x_5 - x_6} \tag{12.317}$$

de la Garza et al. [D1, D2] have shown that this leads to a cascade with nearly the minimum total internal flow as long as the fraction of other components is small, and have called such a cascade a matched R cascade. We then need to find the most general solution of Eq. (12.316) that has the property that when two streams are mixed, V is conserved.

If the streams being mixed have flow rates M' and M'' and compositions (R, x_6') and (R, x_6''), the condition that V be conserved is

$$(M' + M'')\,V(R, x_6) = M'V(R, x_6') + M''V(R, x_6'') \tag{12.318}$$

with the ^{236}U fraction in the mixed stream x_6 given by material balance

$$x_6 = \frac{M'x_6' + M''x_6''}{M' + M''} \tag{12.319}$$

To satisfy (12.318) and (12.319), $V(R, x_6)$ must be a linear function of x_6:

$$V(R, x_6) = a(R) + b(R)x_6 \tag{12.320}$$

The most general solution of (12.316) of the form (12.320) is

$$V(R, x_6) = \kappa_0 + \kappa_5 x_5 + \kappa_6 x_6 + \frac{\kappa x_6}{R^{1/3}} + (2x_5 + 4x_6 - 1) \ln R \tag{12.321}$$

κ_0, κ_5, κ_6, and κ are arbitrary constants. The fact that (12.321) satisfies the differential equation (12.316) may be verified by direct substitution.

When interstage flows are adjusted so that the abundance ratios R of ^{235}U to ^{238}U of each pair of streams being mixed are equal, the separative capacity D of an entire cascade whose feed, product, and tails are

Stream	Flow rate	$^{235}U/^{238}U$ ratio	Fraction ^{236}U
Feed	F	R_F	$z_{6,F}$
Product	P	R_P	$y_{6,P}$
Tails	W	R_W	$x_{6,W}$

is

$$D = PV(R_P, y_{6,P}) + WV(R_W, x_{6,W}) - FV(R_F, z_{6,F}) \tag{12.322}$$

This may be shown by a development similar to that of Sec. 11.

When Eqs. (12.321) for feed, product, and tails are substituted into (12.232), the coefficients of κ_0, κ_5, and κ_6 vanish because of material-balance relations. The coefficient of the remaining arbitrary constant κ in Eq. (12.321) for the separative capacity may be made to vanish by requiring that

$$\frac{Py_{6,P}}{R_P^{1/3}} + \frac{Wx_{6,W}}{R_W^{1/3}} - \frac{Fz_{6,F}}{R_F^{1/3}} = 0 \tag{12.323}$$

Equation (12.323) and the material-balance equation for ^{236}U make possible evaluation of the distribution of ^{236}U between product and tails in terms of the specified fraction of ^{236}U in feed $z_{6,F}$ and the specified abundance ratios R_P, R_W, and R_F of ^{235}U to ^{238}U in product, tails, and feed, respectively. It should be noted that it is not possible to specify in advance the distribution of the third component, ^{236}U in this case. The distribution of only two components, called *key components,* ^{235}U and ^{238}U in this case, are the only ones that can be specified in advance.

With the distribution of ^{236}U between product and tails thus determined, the separative capacity of the entire cascade, from Eqs. (12.321) and (12.322), becomes

$$D = P(2y_{5,P} + 4y_{6,P} - 1) \ln R_P + W(2x_{5,W} + 4x_{6,W} - 1) \ln R_W$$
$$- F(2z_{5,F} + 4z_{6,F} - 1) \ln R_F \tag{12.324}$$

Thus, it has been shown that the separation potential, or value function, for an ideal cascade treating a mixture of ^{235}U, ^{236}U, and ^{238}U in which the ratios of ^{235}U to ^{238}U in each pair of streams being mixed are made equal, is

$$V(x_5, x_6) = (2x_5 + 4x_6 - 1) \ln \frac{x_5}{1 - x_5 - x_6} \tag{12.325}$$

15.3 Three-Component Separation Example

As an example of the use of these equations, we shall calculate the distribution of ^{236}U and the amount of separative work expended in a matched $^{235}U/^{238}U$ cascade producing 1000 kg/day of uranium containing 3.2 w/o ^{235}U from feed containing 0.711 w/o ^{235}U and 0.4 w/o ^{236}U, while stripping tails to 0.3 w/o ^{235}U. External conditions specified for the cascade are listed in Table 12.13.

Distribution of ^{236}U is evaluated from the material-balance equation

$$(1)y_{6,P} + 6.056\, x_{6,W} - 7.056\, z_{6,F} = 0 \tag{12.326}$$

and application of Eq. (12.323):

$$\frac{(1)\, y_{6,P}}{[0.032/(0.968 - y_{6,P})]^{1/3}} + \frac{6.056\, x_{6,W}}{[0.003/(0.997 - x_{6,W})]^{1/3}} - \frac{7.056\, z_{6,F}}{[0.00711/(0.99289 - z_{6,F})]^{1/3}} = 0 \tag{12.327}$$

Values of $y_{6,P}$ and $x_{6,W}$ obtained from these equations using $z_{6,F} = 0.004$ are tabulated below, together with weight fractions of ^{235}U and ^{238}U.

	Weight fraction		
	^{235}U	^{236}U	^{238}U
Product	$y_{5,P} = 0.032$	$y_{6,P} = 0.0128575$	$y_{8,P} = 0.9551425$
Tails	$x_{5,W} = 0.003$	$x_{6,W} = 0.0025374$	$x_{8,W} = 0.9944626$
Feed	$z_{5,F} = 0.00711$	$z_{6,F} = 0.004$	$z_{8,F} = 0.98889$

The amount of separative work expended per day in making 1000 kg/day of product, from Eq. (12.324) with these values of $y_{6,P}$ and $x_{6,W}$ is 3813.4 kg SWU/day. This may be compared with the amount of separative work needed in the absence of ^{236}U with the same weight fractions of ^{235}U in product, tails, and feed and the same quantities of these streams:

$$D = 1000\, \phi(0.032) + 6056\, \phi(0.003) - 7056\, \phi(0.00711) = 3787.5 \tag{12.328}$$

using values of the separation potential ϕ from Eq. (12.144).

Thus, the presence of 0.004 weight fraction ^{236}U in the feed increases the amount of

Table 12.13 External conditions for ^{235}U, ^{236}U, ^{238}U example

	Weight fraction		Weight ratio ^{235}U:^{238}U,	Mass,
Stream	^{235}U	^{236}U	R	kg
Product	0.032	$y_{6,P}$	$\frac{0.032}{0.968 - y_{6,P}}$	$P = 1000$
Tails	0.003	$x_{6,W}$	$\frac{0.003}{0.997 - x_{6,W}}$	$F = \frac{0.032 - 0.003}{0.00711 - 0.003} P = 7056$
Feed	0.00711	$z_{6,F} = 0.004$	$\frac{0.00711}{0.99289 - z_{6,F}}$	$W = \frac{0.032 - 0.00711}{0.00711 - 0.003} P = 6056$

separative work needed for this example by $3813.4 - 3787.5 = 25.9$ kg SWU/day, or $25.9/(7056)(0.004) = 0.918$ kg SWU/kg ^{236}U in feed. This could serve as the basis for a penalty to be charged for ^{236}U present in feed to a uranium enrichment plant operating between the ^{235}U concentrations of this example. It should be noted that these results for the effect of ^{236}U on separative capacity are independent of the separation process under consideration.

15.4 Number of Stages

The development next to be given of equations for the number of stages, interstage flow rates, and fraction ^{236}U on each stage does depend on the process used. In subsequent numerical examples, the gaseous diffusion process with a $^{235}U/^{238}U$ stage separation factor of 1.00429 is assumed.

Because the $^{235}U/^{238}U$ ratio is matched, the number of stages is given by equations analogous to (12.93) and (12.94). The total number of stages is

$$n = 2\frac{\ln (y_{5,P}x_{8,W}/y_{8,P}x_{5,W})}{\ln \alpha} - 1$$

$$= 2\frac{\ln [(0.032)(0.9944626)/(0.9551425)(0.003)]}{\ln 1.00429} - 1 = 1124 \tag{12.329}$$

The number of stripping stages is

$$n_S = 2\frac{\ln (z_{5,F}x_{8,W}/z_{8,F}x_{5,W})}{\ln \alpha} - 1$$

$$= 2\frac{\ln [(0.00711)(0.9944626)/(0.98889)(0.003)]}{\ln 1.00429} - 1 = 405 \tag{12.330}$$

15.5 Interstage Flow Rates and Compositions

Equations for interstage flow rates and compositions in the enriching section are obtained by applying to the section of the cascade from the product end through stage $i + 1$ shown in Fig. 12.28 a development similar to the one used earlier for the complete cascade.

The $^{235}U/^{238}U$ ratios η_i and ξ_{i+1} for this matched-ratio cascade may be related by

$$\eta_i = \eta_P \beta^{i-n} \tag{12.331}$$

and

$$\eta_{i+1} = \eta_P \beta^{i-n-1} \tag{12.332}$$

where β is the $^{235}U/^{238}U$ heads-separation factor.

Material-balance relations are

$$^{235}\text{U:} \qquad M_i y_{5,i} - N_{i+1}x_{5,i+1} = Py_{5,P} \tag{12.333}$$

$$^{236}\text{U:} \qquad M_i y_{6,i} - N_{i+1}x_{6,i+1} = Py_{6,P} \tag{12.334}$$

$$^{238}\text{U:} \qquad \frac{M_i y_{5,i}}{\eta_i} - \frac{N_{i+1}x_{5,i+1}}{\xi_{i+1}} = \frac{Py_{5,P}}{\eta_P} \tag{12.335}$$

Interstage flow rates of ^{235}U and ^{238}U may be obtained by solving (12.333) and (12.335) for $N_{i+1}x_{5,i+1}$ and $N_{i+1}x_{5,i+1}/\xi_{i+1}$:

$$^{235}\text{U:} \qquad N_{i+1}x_{5,i+1} = Py_{5,P}\frac{(1/\eta_i) - (1/\eta_P)}{(1/\xi_{i+1}) - (1/\eta_i)} = Py_{5,P}\frac{1 - \beta^{i-n}}{\beta - 1} \tag{12.336}$$

$$^{238}\text{U:} \qquad N_{i+1}x_{8,i+1} = \frac{N_{i+1}x_{5,i+1}}{\xi_{i+1}} = \frac{Py_{5,P}}{\eta_P}\frac{\eta_P - \eta_i}{\eta_i - \xi_{i+1}} = Py_{8,P}\beta\frac{\beta^{n-i} - 1}{\beta - 1} \tag{12.337}$$

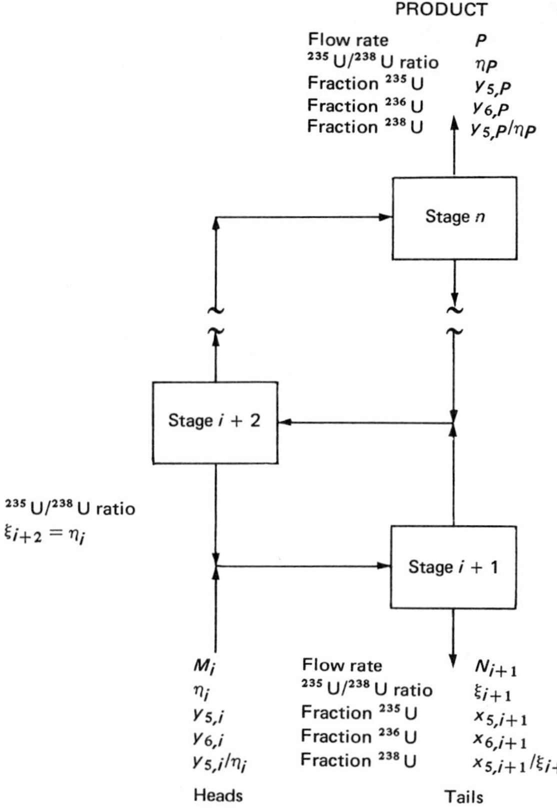

Figure 12.28 Enriching section of matched $^{235}U/^{238}U$ cascade for ^{235}U, ^{236}U, ^{238}U separation.

Interstage flow rate of ^{236}U is obtained from Eq. (12.334) and Eq. (12.338), derived in similar fashion to Eq. (13.323):

$$\frac{M_i y_{6,i}}{(\eta_i)^{1/3}} - \frac{N_{i+1} x_{6,i+1}}{(\xi_{i+1})^{1/3}} = \frac{P y_{6,P}}{(\eta_P)^{1/3}} \tag{12.338}$$

$$^{236}\text{U:} \qquad N_{i+1} x_{6,i+1} = P y_{6,P} \frac{(1/\eta_i)^{1/3} - (1/\eta_P)^{1/3}}{(1/\xi_{i+1})^{1/3} - (1/\eta_i)^{1/3}} = P y_{6,P} \frac{1 - \beta^{(i-n)/3}}{\beta^{1/3} - 1} \tag{12.339}$$

The total interstage downflow rate, N_{i+1}, is obtained as the sum of Eqs. (12.336), (12.337), and (12.339):

$$\frac{N_{i+1}}{P} = \frac{1}{\beta - 1} [y_{5,P}(1 - \beta^{i-n}) + y_{8,P}\beta(\beta^{n-i} - 1)] + \frac{y_{6,P}}{\beta^{1/3} - 1} [1 - \beta^{(i-n)/3}] \tag{12.340}$$

When $y_{6,P}$ is zero, this equation reduces to Eq. (12.106) for a two-component ideal cascade.

For the uranium isotope separation case in which

$$\beta - 1 = \frac{\psi}{2} \ll 1 \tag{12.341}$$

Eqs. (12.336), (12.337), (12.339), and (12.340) may be approximated by

$$^{235}\text{U:} \qquad \frac{N_{i+1} x_{5,i+1}}{P} = 2 y_{5,P} \frac{1 - e^{\psi(i-n)/2}}{\psi} \tag{12.342}$$

$$^{238}\text{U:} \qquad \frac{N_{i+1}x_{8,i+1}}{P} = 2y_{8,P}\,\frac{e^{\psi(n-i)/2}-1}{\psi} \tag{12.343}$$

$$^{236}\text{U:} \qquad \frac{N_{i+1}x_{6,i+1}}{P} = 6y_{6,P}\,\frac{1-e^{\psi(i-n)/6}}{\psi} \tag{12.344}$$

$$\frac{N_{i+1}}{P} = \frac{2y_{5,P}(1-e^{\psi(i-n)/2}) + 2y_{8,P}(e^{\psi(n-i)/2}-1) + 6y_{6,P}(1-e^{\psi(i-n)/6})}{\psi} \tag{12.345}$$

Hence

$$x_{5,i+1} = \frac{y_{5,P}(1-e^{\psi(i-n)/2})}{y_{5,P}(1-e^{\psi(i-n)/2}) + y_{8,P}(e^{\psi(n-i)/2}-1) + 3y_{6,P}(1-e^{\psi(i-n)/6})} \tag{12.346}$$

$$x_{6,i+1} = \frac{3y_{6,P}(1-e^{\psi(i-n)/6})}{y_{5,P}(1-e^{\psi(1-n)/2}) + y_{8,P}(e^{\psi(n-i)/2}-1) + 3y_{6,P}(1-e^{\psi(i-n)/6})} \tag{12.347}$$

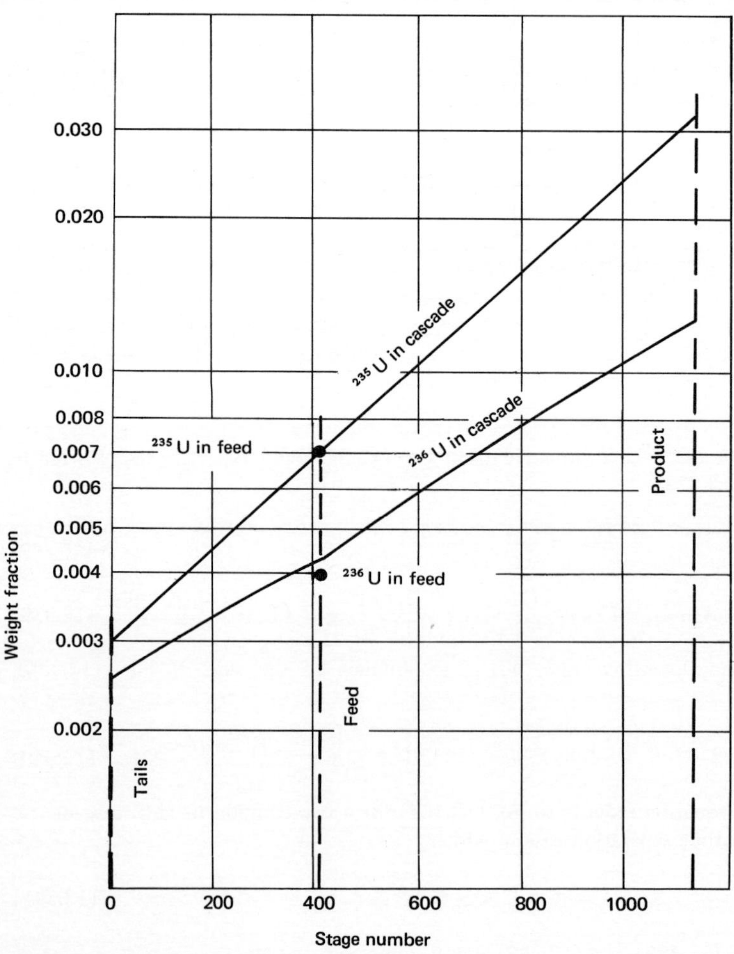

Figure 12.29 Composition versus stage number in example of matched ^{235}U/^{238}U abundance-ratio cascade for separating mixture of ^{235}U, ^{236}U, and ^{238}U.

A similar derivation leads to the following equations for the stripping section.

$$\frac{M_j}{W} = \frac{2x_{5,W}(e^{\psi j/2} - 1) + 2x_{8,W}(1 - e^{-\psi j/2}) + 6x_{6,W}(e^{\psi j/6} - 1)}{\psi} \tag{12.348}$$

$$x_{5,j} = \frac{x_{5,W}(e^{\psi j/2} - 1)}{x_{5,W}(e^{\psi j/2} - 1) + x_{8,W}(1 - e^{-\psi j/2}) + 3x_{6,W}(e^{\psi j/6} - 1)} \tag{12.349}$$

$$x_{6,j} = \frac{3x_{6,W}(e^{\psi j/6} - 1)}{x_{5,W}(e^{\psi j/2} - 1) + x_{8,W}(1 - e^{-\psi j/2}) + 3x_{6,W}(e^{\psi j/6} - 1)} \tag{12.350}$$

Figure 12.29 shows the fraction of ^{235}U and of ^{236}U as a function of stage number for the three-component separation example of Table 12.13. Characteristic features of this plot are as follows: (1) The ^{235}U composition gradient is nearly linear on a semilog scale. This feature holds only for low-^{235}U and low-^{236}U fractions. (2) The composition gradient of ^{236}U is about two-thirds that of ^{235}U on a semilog scale. (3) The ^{236}U plot has a noticeable discontinuity in slope at the feed point and is noticeably curved upward in each section. (4) The ^{236}U content of the cascade at the feed point is substantially higher than the ^{236}U content of the feed.

This buildup of ^{236}U at the feed point and the bulge in ^{236}U gradient in each section are characteristic of a third component whose molecular weight is between those of the key components. It is responsible for the increase in separative capacity caused by the presence of the third component. de la Garza [D1] gives an extreme example of buildup of concentration of a component of intermediate molecular weight.

de la Garza has shown that it would be impossible to separate ^{236}U completely from ^{235}U in the product of a cascade designed with matched ^{235}U/^{238}U ratio, no matter how many enriching stages were used. The property that determines whether a component can be completely separated from product is the arithmetic mean of the molecular weights of the key components, which de la Garza has called the *key weight*. This is 236.5 when ^{235}U and ^{238}U are key components. Only those components with molecular weight greater than the key weight can be fractionated completely out of the product. An example of a cascade that would do this for ^{236}U would be one that matched ^{235}U/^{236}U ratios.

NOMENCLATURE

a, b	constants in general Eq. (12.183) for separation potential
a, b, c	constants for enrichment equation of square cascade, defined by Eqs. (12.225) through (12.229)
c_F	unit cost of feed
c_P	unit cost of product
c_S	unit cost of separative work
D	separative capacity, Eq. (12.141) or (12.143)
E	quantity of material fed to or withdrawn from cascade
F	cascade feed flow rate
h	stage holdup time, Eq. (12.198)
H	stage inventory
i	serial number of stage in enriching section
I	cascade inventory
j	serial number of stage in stripping section
J	total heads flow rate
k	chemical equilibrium constant
K	total tails flow rate

m	molecular weight
M	stage heads flow rate
n	number of stages in cascade
n_S	number of stages in stripping section
N	stage tails flow rate
p	number of stages heads stream moves up cascade
P	cascade product flow rate
q	number of stages tails stream moves down cascade
Q	product flow rate from stage next below top of cascade
r	recovery
r	parameter for two-up, one-down cascade, Eq. (12.292)
R	gas constant
R	weight ratio, ^{235}U to ^{238}U
s	parameter for two-up, one-down cascade, Eq. (12.294)
S	separative work, Eq. (12.146) or (12.153)
t	parameter for two-up, one-down cascade, Eq. (12.295)
t	time after start-up
t_P	equilibrium time for product withdrawal, Eq. (12.188)
t_W	equilibrium time for tails withdrawal, Eq. (12.189)
V	generalized separation potential for multicomponent system
w	weight fraction ^{235}U
W	cascade tails flow rate
x	fraction desired component in tails
X	cascade external flow rate
y	fraction desired component in heads
z	fraction desired component in feed
Z	stage feed flow rate
α	stage separation factor, Eq. (12.15)
α'	local separation factor, Eq. (12.27)
β	stage heads separation factor, Eq. (12.16)
γ	stage tails separation factor, Eq. (12.17)
δ	$\beta - 1$
Δ	separative capacity of one stage
ϵ	$\gamma - 1$
ζ	weight, mole, or atom ratio in feed
η	weight, mole, or atom ratio in product
θ	cut, ratio of stage heads flow rate to feed, Eq. (12.12)
κ	constant in equation for separation potential of multicomponent mixture
λ	separation parameter defined by Eq. (12.244)
ξ	weight, mole, or atom ratio in tails
π	separative work inventory function, Eq. (12.216)
τ	lower bound for equilibrium time of ideal cascade, Eq. (12.209)
ϕ	separation potential (elementary value function), Eq. (12.144)
ψ	overall $^{235}U/^{238}U$ enrichment factor, $\alpha_{58} - 1$
ψ	component inventory function, Eq. (12.214)
ω	overall enrichment, Eq. (12.45)
Ω	overall separation, Eq. (12.73)

Superscripts

$'$	local value within a stage
$\bar{}$	average value in cascade

Subscripts

E	enriching section
F	cascade feed
i	stage number in enriching section, counting from tails end of cascade
j	stage number in stripping section, counting from tails end of cascade
k	external stream
n	highest stage number of enriching section
n_S	highest stage number of stripping section
P	cascade product
Q	heads from stage next below product stage
S	stripping section
W	cascade tails
min	minimum
opt	optimum
0	tails of optimum composition
5	^{235}U
6	^{236}U
8	^{238}U

REFERENCES

A1. Abelson, P. H., and J. I. Hoover: "Separation of Uranium Isotopes by Liquid Thermal Diffusion," *Proceedings of the International Symposium on Isotope Separation,* Interscience, New York, 1958, p. 483.

A2. Armstrong, D. E., et al.: "A Carbon-13 Production Plant Using Carbon Monoxide Distillation," Los Alamos Scientific Laboratory Report LA-4391, Apr. 10, 1970.

B1. Barr, F. T.: U.S. Patent 2,676,875, Apr. 27, 1954.

B2. Bebbington, W. P., and V. R. Thayer: *Chem. Eng. Progr.* **55**(9): 70 (Sept. 1959).

B3. Boyd, W. T.: Sc.D thesis, University of Michigan, 1951.

B4. Boyd, W. T., and R. R. White: *Ind. Eng. Chem.* **44**: 2207 (1952).

C1. Clusius, K.: *Helv. Chim. Acta* **33**: 2134 (1950).

C2. Clusius, K., et al.: "Nuclear Physics and Cosmic Rays, Part II," *FIAT Rev. Ger. Sci.,* 1948, pp. 182–188.

C3. Cohen, K.: *The Theory of Isotope Separation,* McGraw-Hill, New York, 1951.

C4. Conn, A. L., and J. E. Wolfe: "Large Scale Separation of Boron Isotopes," paper presented at 132nd Meeting of American Chemical Society, New York, Sept. 12, 1957.

C5. Crist, R. H., and I. Kirshenbaum: U.S. Patent 2,796,323, June 18, 1957.

D1. de la Garza, A.: *Chem. Eng. Sci.* **18**: 73 (1963).

D2. de la Garza, A., G. A. Garrett, and J. E. Murphy: U.S. AEC Report K-1455, July 1960; *Chem. Eng. Sci.* **15**: 188 (1961).

D3. Dostrovsky, I.: "Production and Distribution of the Heavy Isotopes of Oxygen," *PICG(2)* **4**: 605 (1958).

E1. Edmunds, A. O., and F. C. Loveless: "Production of Boron-10 and Other Stable Isotopes," *PICG(2)* **4**: 576 (1958).

F1. Fenske, M. R.: *Ind. Eng. Chem.* **24**: 482 (1932).

H1. Hänny, J.: *Schweizer Archiv. Angew. Wiss. Tech.* **26**: 115 (1960).

H2. Healy, R. M., A. A. Palko, E. F. Joseph, and J. S. Drury: "Chemical Separation of Stable Isotopes," *Proceedings of the International Symposium on Isotope Separation,* Interscience, New York, 1958, p. 199.

H3. Hollister, H., and A. J. Burrington: *Nucleonics* **16**(1): 54 (1958).
H4. Huffman, J. R., and H. C. Urey: *Ind. Eng. Chem.* **29**: 531 (1937).
H5. Hutchison, C. A., D. W. Stewart, and H. C. Urey: *J. Chem. Phys.* **8**: 532 (1940).
H6. Hydrocarbon Research, Inc.: "Low Temperature Heavy Water Plant," Report NYO-889, 1951.
J1. Johns, T. F., H. Kronberger, and H. London: *Mass Spectrometry,* Institute of Petroleum, London, 1950, pp. 141–147.
J2. Johnston, H. L., and C. A. Hutchison: *J. Chem. Phys.* **8**: 869 (1940). See also T. I. Taylor and H. C. Urey: *J. Chem. Phys.* **5**: 597 (1937); L. Holleck: *Z. Elektrochem.* **44**: 411 (1938).
K1. Kilpatrick, M., et al.: *Separation of Boron Isotopes,* National Nuclear Energy Series, vol. III-5, McGraw-Hill, New York, 1952.
K2. Kistemaker, J., C. J. Zilverschoon, and J. Schutter: *Ned. Tijdschr. Natuurk.* **20**: 5 (1954).
K3. Klemm, A.: "Ionenwanderung in Geschmolzenen Salzen," *Proceedings of the International Symposium on Isotope Separation,* Interscience, New York, 1958, p. 275.
K4. Klemm, A., M. Hintenberger, and P. Hoernes: *Z. Naturforsch.* **2a**: 245 (1947); A. Klemm: *Z. Naturforsch.* **6a**: 512 (1951).
K5. Koch, J. (ed.): *Electromagnetic Isotope Separators and Applications of Magnetically Enriched Isotopes,* Interscience, New York, 1958.
L1. Lawrence, E. O.: U.S. Patent 2,709,222, May 24, 1955; U.S. Patent 2,721,272, Oct. 18, 1955.
L2. Lee, D. A., and G. M. Begun: *J. Amer. Chem. Soc.* **81**: 2332 (1959).
L3. Letokhov, V. S., and C. J. Moore: "Laser Isotope Separation," Lawrence Berkeley Laboratory Report LBL-4904, Mar. 1976; *Sov. J. Quant. Electron.* **6**(2): 129 (Feb. 1976).
L4. Lewis, G. N.: *J. Amer. Chem. Soc.* **55**: 1297 (1933); G. N. Lewis and R. T. MacDonald: *J. Chem. Phys.* **1**: 341 (1933); *J. Amer. Chem. Soc.* **55**: 3058 (1933).
L5. Lewis, G. N., and R. T. MacDonald: *J. Amer. Chem. Soc.* **58**: 2519 (1936).
L6. London, H.: "Isotope Separation by Fractional Distillation," *Proceedings of the International Symposium on Isotope Separation,* Interscience, New York, 1958, p. 319.
L7. Lundén, A.: *Z. Naturforsch.* **11a**: 590 (1956).
M1. Malkov, M. P., A. G. Zeldovitch, A. B. Fradkov, and I. B. Danilov: "Industrial Separation of Deuterium by Low-Temperature Distillation," *PICG(2)* **4**: 491 (1958).
M2. McInteer, B. B., and R. M. Potter: "Nitric Oxide Distillation Plant for Isotope Separation," *Ind. Eng. Chem. Proc. Design and Devel.* **4**: 35 (1965).
M3. Miller, G. T., R. J. Kralik, E. A. Belmore, and J. S. Drury: "Production of Boron-10," *PICG(2)* **4**: 585 (1958).
M4. Mühlenpfordt, J., et al.: "Die Anreichung des Isotopes ^{10}B durch Fraktionierte Destillation von BF_3," *Proceedings of the International Symposium on Isotope Separation,* Interscience, New York, 1958, p. 408.
M5. Murphy, G. M. (ed.): *Production of Heavy Water,* McGraw-Hill, New York, 1955.
N1. Nettley, P. T., D. K. Cartwright, and H. Kronberger: "The Production of ^{10}Boron by Low-Temperature Distillation of Boron Trifluoride," *Proceedings of the International Symposium on Isotope Separation,* Interscience, New York, 1958, p. 385.
N2. Nier, A. O., E. T. Booth, J. R. Dunning, and A. V. Grosse: *Phys. Rev.* **57**: 546, 748 (1940).
O1. Olander, D. R.: *Nucl. Technol.* **29**: 108 (1976).
P1. Perret, L., L. Rozand, and E. Saito: "Investigation of the Separation Coefficient of Certain Processes Involving the Isotopes of Lithium," *PICG(2)* **4**: 595 (1958).
P2. Pratt, H. R. C.: *Countercurrent Separation Processes,* Elsevier, New York, 1967, chap. 10.
R1. Rayleigh, Lord: *Phil. Mag.* (6), **4**: 521 (1902).
R2. Roux, A. J. A., and W. L. Grant: "Uranium Enrichment in South Africa," European Nuclear Conference, Paris, Apr. 1975.

S1. Saito, E., and G. Dirian: Process for the Isotopic Enrichment of Lithium by Chemical Exchange, British Patent 902,755, Aug. 9, 1962.
S2. Selak, P. J., and J. Finke: *Chem. Eng. Progr.* **50**: 221 (1954).
S3. Sessions, R. I., A. H. Kibbey, J. T. Roberts, and R. E. Blanco: Oak Ridge National Laboratory Report CF-53-6-241, June 1, 1953.
S4. Smith, M. L.: "Operational Experience with Hermes, the Harwell Active Electromagnetic Separator," *Proceedings of the International Symposium on Isotope Separation,* Interscience, New York, 1958, p. 581; *Progr. in Nucl. Phys.* **6**: 162 (1957).
S5. Smyth, H. D.: *Atomic Energy for Military Purposes,* Princeton University Press, Princeton, N.J., 1945.
S6. Spedding, F. H., J. E. Powell, and H. J. Svec: *J. Amer. Chem. Soc.* **77**: 1393 (1955).
S7. Spevack, J. S.: U.S. Patent 2,787,526, Apr. 2, 1957; U.S. Patent 2,895,803, July 21, 1959.
S8. Stewart, D. W.: *Nucleonics* 2(9): 25 (1947).
T1. Taylor, H. S.: U.S. Patent 2,690,380, Sept. 28, 1954.
T2. Taylor, T. I., and W. Spindel: "Preparation of Highly Enriched Nitrogen-15 by Chemical Exchange of NO with HNO_3," *Proceedings of the International Symposium on Isotope Separation,* Interscience, New York, 1958, p. 158.
T3. Taylor, T. I., and H. C. Urey: *J. Chem. Phys.* **6**: 429 (1938).
T4. Thode, H. G., and H. C. Urey: *J. Chem. Phys.* **7**: 34 (1939).
T5. Thomson, J. J.: *Rays of Positive Electricity,* Longmans, Green, London, 1921.
T6. Trauger, D. B., et al.: "Some Experiments on the Separation of Lithium Isotopes by Molecular Distillation," *Proceedings of the International Symposium on Isotope Separation,* Interscience, New York, 1958, p. 350.
U1. Underwood, A. J. V.: *Trans. Inst. Chem. Eng.* **10**: 112 (1932).
U2. U.S. Energy Research and Development Administration: *Federal Register,* vol. 38, Aug. 9, 1973, pp. 12158–12159.
V1. Vallet et al.: "Investigation of a Cascade Element for Use in the Separation of Lithium Isotopes," *PICG(2)* **4**: 602 (1958).
W1. Washburn, E. W., and H. C. Urey: *Proc. Natl. Acad. Sci.* **18**: 493 (1932).
Z1. Zippe, G.: "The Development of Short Bowl Ultracentrifuges," University of Virginia School of Engineering and Applied Science, Report EP-4420-101-60U, July 1960.
Z2. Zolotarev, V. S., A. I. Iljin, and E. G. Komar: "Isotope Separation by Electromagnetic Separators in the Soviet Union," *PICG(2)* **4**: 471 (1958).

PROBLEMS

12.1 It is proposed that ^{10}B be concentrated by the gaseous diffusion process applied to $^{10}BF_3$ and $^{11}BF_3$. The plant is to be designed as an ideal cascade and is to separate feed containing 19 percent ^{10}B into product containing 90 percent and tails containing 8 percent. The stage-separation factor is 1.0074.

How many stages are needed in the stripping section? In the enriching section? Where in the cascade does the maximum value of the reflux ratio (tails to product) occur? What is this maximum value?

12.2 A uranium enrichment plant is to produce 200 kg of ^{235}U/day in product containing 3.2 w/o ^{235}U, while stripping tails to 0.2 w/o, from natural uranium feed containing 0.711 w/o ^{235}U.

In a gaseous diffusion plant designed as an ideal cascade with $\alpha = 1.00429$, how many stripping stages would be required? How many enriching stages?

At what rate would natural uranium be fed? What is the separative capacity of the plant in kilograms of uranium per year?

What is the total heads flow rate in the plant in kilograms of uranium per day?

12.3 A uranium isotope separation plant has been operating as an ideal cascade to produce 200 kg of ^{235}U/day in product containing 3.2 w/o ^{235}U while stripping tails to 0.2 w/o, from natural uranium feed containing 0.711 w/o ^{235}U.

Assume that the plant can be rearranged to operate as an ideal cascade at constant separative capacity when operating conditions are changed individually as described in each of the following ways:

(*a*) If tails assay were raised to 0.3 w/o, what would be the feed and product rates?

(*b*) If product assay were raised to 4.0 w/o, what would be the feed and product rates?

(*c*) If the natural uranium feed rate were changed to 45,000 kg/day, at what rate could 3.2 w/o product be produced? What would be the tails assay?

(*d*) If 3000 kg/day of uranium containing 2.0 w/o ^{235}U were also to be produced while keeping the tails assay constant at 0.2 w/o, at what rate could 3.2 w/o product be made? What would be the natural uranium feed rate?

12.4 (*a*) If natural uranium in the form of UF_6 costs \$130/kg uranium and separative work in a gas centrifuge plant costs \$100/kg SWU, what is the optimum tails assay?

(*b*) If individual centrifuges have a separation factor of 2 and a separative capacity of 50 kg SWU/year, how many centrifuges in an ideal cascade would be needed to produce 200 kg ^{235}U/day in product containing 3.1 w/o ^{235}U while stripping tails to the optimum assay? Assume 365 operating days per year.

(*c*) How many stages would be needed?

12.5 If heavy water containing 99.8 a/o D costs \$150/kg D_2O when made from natural water containing 0.015 a/o D in an ideal cascade without stripping section, what would be the cost in dollar per kilogram of contained D_2O in water containing 1 a/o D? 0.1 a/o D? Assume that natural water costs nothing and that the unit cost of separative work is constant between 0.015 and 99.8 percent D.

12.6 If natural uranium costs \$100/kg uranium and separative work \$125/kg SWU, what is the cost, in dollars per kilogram uranium product, of producing uranium enriched to 90 w/o ^{235}U from natural uranium feed while stripping tails to 0.3 w/o ^{235}U? To 0.2 w/o?

12.7 A water distillation plant to produce heavy water containing 99.8 a/o D from natural water containing 0.015 percent D is designed as an ideal cascade without stripping with a separation factor of 1.03. The plant's inventory of water is effectively all in the liquid phase. The depth of liquid on each distillation plate is 30 cm. The plate efficiency is 100 percent. The liquid downflow rate is 1 cm^3/s per cm^2 of column cross section. Using Eq. (12.204), what would be an upper bound for the equilibrium time?

12.8 The gaseous diffusion plant of Prob. 12.2 is to be designed as a square cascade with constant interstage heads flow rate on every stage. The flow ratio of heads from the feed stage, with $y = 0.00711$, to product is 1.2 times the minimum at that assay.

(*a*) What is the required heads flow rate?

(*b*) How many enriching and stripping stages are required?

(*c*) What is the total heads flow rate?

Compare with the total heads flow rate of Prob. 12.2.

12.9 Derive differential equation (12.213) for the component inventory function ψ.

12.10 In separating deuterium from hydrogen by distillation of ammonia, the separation factor is 1.043, with deuterium concentrating in the liquid. It is desired to concentrate deuterium from 0.00014 atom fraction in natural ammonia to 0.1 atom fraction while stripping to 0.00002 atom fraction.

(*a*) What is the minimum number of theoretical plates needed for this separation?

(*b*) What is the minimum ratio of reboil vapor to product enriched ammonia?

(*c*) If a reboil vapor ratio 1.2 times the minimum is used, how many theoretical stripping and enriching stages would be needed?

Assume liquid ammonia feed and constant separation factor and constant vapor flow rate throughout the entire plant.

12.11 Consider a cascade to produce 1.25 mol/day of uranium enriched to 80 m/o ^{235}U from natural uranium feed containing 0.72 m/o while stripping tails to 0.36 m/o. The separative capacity of such a cascade was evaluated in Table 12.8.

(*a*) Find the number of gaseous diffusion stages in an ideal cascade, using a $^{235}U/^{238}U$ separation factor of 1.00429.

(*b*) Assume that the feed contains 0.4 m/o ^{236}U. What would be the ^{236}U content of product?

(*c*) How many stages would be needed in a matched $^{235}U/^{238}U$ cascade operating between the above ^{235}U mole percents?

(*d*) What would be the separative capacity?

This problem illustrates the severe contamination of highly enriched ^{235}U caused by ^{236}U in feed. In practice, a cascade with fewer stages and less total interstage flow would result from using ^{235}U and ^{236}U as key components in the more highly enriched stages.

12.12 To make tritium in natural water more readily measurable for analysis, it is proposed that the water be electrolyzed batchwise until the remaining volume is one one-thousandth of its original volume. The separation factor between tritium and hydrogen is 15, with tritium concentrating in the liquid phase.

By what factor will the tritium in the remaining water be concentrated?

12.13 The stage holdup time h and separation factor α of a solvent extraction column for uranium enrichment are $h = 10$ s, $\alpha = 1.0010$. What is the minimum equilibrium time of an ideal cascade fed with natural uranium, stripping to 0.2 w/o ^{235}U and enriching to 3 w/o ^{235}U product? Repeat for 90 w/o ^{235}U product.

CHAPTER

THIRTEEN

SEPARATION OF ISOTOPES OF HYDROGEN AND OTHER LIGHT ELEMENTS

This chapter describes processes most suitable for separation of isotopes of light elements on an industrial scale. Principal emphasis is on separation of deuterium through production of heavy water, but some information on separation of isotopes of other light elements is also given. Processes to be discussed include distillation, electrolysis, and chemical exchange.

1 SOURCES OF DEUTERIUM

The most abundant source of deuterium, of course, is natural water. Other potential natural sources are natural gas and petroleum. Of these, natural water is by far the most significant. No economic method has been found for extracting deuterium from natural gas or petroleum without first converting them chemically to other materials. Industrial hydrogen and ammonia synthesis gas, produced by chemical conversion of natural gas and petroleum, are being used as sources of deuterium, but the amount of heavy water that can be produced from these industrial sources is small compared with the amount needed for heavy-water reactors. As shown in Chap. 12, a large plant producing 1000 short tons of synthetic ammonia per day could produce only around 75 short tons of heavy water per year, a small amount compared with around 500 short tons needed as the initial charge of a 600-MW heavy-water nuclear power plant.

Unlike other elements, the variability of isotopic composition of hydrogen from different sources is great enough to be a factor in the location, design, and economic performance of heavy-water plants.

The deuterium content of natural waters varies from place to place and from time to time because of isotopic fractionation which occurs when water evaporates from land or sea or is condensed from the air. The deuterium content of natural waters relative to standard water samples has been determined by a number of investigators; representative results of two workers are abstracted in Table 13.1. The percent differences from standards have been converted to atom percent deuterium by using the indicated deuterium content of the standards, which, however, are less accurately known than the differences.

Ocean water in the tropics contains around 0.0156 a/o (atom percent) deuterium. Water vapor in the air in equilibrium with the ocean has a deuterium content about 7 percent lower than seawater because H_2O has a higher vapor pressure than HDO. Consequently, water vapor over the ocean should contain about $0.0156/1.07 = 0.0146$ a/o deuterium. The first rain to fall out from this water vapor is richer in deuterium than 0.0146 percent, again because of the

Table 13.1 Deuterium content of natural waters

	Percent difference from standard	Atom ppm deuterium
A. Friedman [F2], standard contains 0.0148 a/o D		
Surface ocean waters		
Mid-Atlantic Ocean at equator	+5.41	156.0
Jacksonville, Fla.	+5.02	155.4
La Jolla, Calif.	+4.56	154.8
Bering Sea	+4.07	154.0
West coast of Greenland	+2.42	151.6
North American Rivers		
Columbia at Trail, B.C., 1943	−10.1	132.9
Missouri at Kansas City, Kan., 1948	−7.06	137.5
Colorado at Yuma, Ariz., 1948	−6.06	139.0
Connecticut, 1948	−2.15	144.8
Mississippi at Baton Rouge, La., 1948	+0.39	148.5
Red at Colbert, Okla., 1948	+3.05	152.5
Arkansas at Van Buren, Ark., 1948	+3.25	152.8
Rio Grande at Mission, Tex., 1948	+3.28	152.8
B. Craig [C13], standard (mean ocean water) contains 0.01566 a/o D according to Horibe and Kobayakawa [H6]		
1955–1956 snow, 200 mil east of Thule, Greenland	−23.62	119.5
Snow, Little America, Antarctica	−14.32	134.1
Columbia River, Hood River, Ore.	−13.64	135.2
Danube River, Regensburg, Germany	−7.76	144.4
Hudson River	−6.0	147.2
Niagara River	−5.3	148.3
Gulf of Suez, Red Sea	+1.42	158.8
White Nile, Khartoum, Sudan	+4.22	163.2
Chicago, mean precipitation	−5	149
rain, 4/10/54	+0.21	156.9
snow, 2/5/54	−16.19	131.2

lower vapor pressure of HDO. As moisture-laden air from the ocean flows away from the tropics and over the continents, it becomes steadily depleted in deuterium. Rainfall on the leeward side of mountains and snowfall in the polar regions, where most of the moisture has already been condensed from the air, will contain less than 0.0146 a/o deuterium. This is shown in Table 13.1 for the Columbia, Missouri, and Colorado rivers, and for snowfall in Greenland, Antarctica, and, in an exceptional instance, in Chicago. For the same reason, rivers whose flow is substantially reduced by evaporation during passage through arid regions will contain more than 0.0146 a/o deuterium, as is shown in Table 13.1 for the Red, Arkansas, Rio Grande, and Nile rivers. Most of the differences in deuterium content given in this table can be explained by fractionation of deuterium during evaporation and condensation of water.

The difference in deuterium content of snow and rain at Chicago is an extreme example of the change in deuterium content with changes in conditions of precipitation.

The examples of this table have been selected to illustrate the variability of the deuterium content of natural waters. Actually, over large parts of the earth where conditions of precipitation are comparatively uniform and evaporation of groundwater unimportant, the variability is much less. For example, the deuterium content of river and lake waters in the eastern United States and Canada, where most of the world's heavy water is now produced, is within 1 or 2 ppm of 148 ppm (0.0148 percent).

Because the cost of producing heavy water is roughly inversely proportional to the deuterium content of plant feed, local variations are of major economic importance. The low deuterium content of the Columbia River at Trail, British Columbia, 0.0133 percent, made the cost of producing heavy water at the U.S. Atomic Energy Commission's (AEC) plant at this location higher than if the Columbia River had been as rich in deuterium as the Niagara or the Nile, for example.

The deuterium content of natural gas and petroleum is also variable. Values as low as 0.0107 percent have been found for Texas natural gas [H11].

When natural gas or petroleum is converted to hydrogen by reforming with an excess of steam, equilibrium is established in the reactions

$$CH_4 + H_2O \rightleftharpoons CO + 3H_2$$

$$CO + H_2O \rightleftharpoons CO_2 + H_2$$

and

$$HD + H_2O \rightleftharpoons H_2 + HDO$$

The equilibrium constant for the third, deuterium exchange, reaction is around 2 at the temperature at which the second, water-gas shift, reaction is carried out. Because an excess of water is used to convert CO completely to CO_2, the deuterium content of hydrogen will be less than that of the methane and water fed, unless the excess water is fully recycled. Because water recycle is usually not practiced at ammonia synthesis plants, the deuterium content of synthesis gas at operating plants is sometimes as low as 0.009 percent [M7]. If the ammonia plant were specifically designed for deuterium recovery from its synthesis gas, the deuterium content could be increased to the average of the methane and water feeds by recycling all water and preventing losses.

2 DEUTERIUM PRODUCTION PROCESSES AND PLANTS

Table 13.2 lists all plants in the non-Communist world that have been built or are planned for production of deuterium, in the form of heavy water, at a rate of 1 t/year or more.

The following general comments may be made about these plants and processes:

1. All plants, except 18, have a different process for primary enrichment than for final concentration.
2. Those plants that for primary concentration use water distillation (WD) or the dual-temperature, water-hydrogen sulfide (GS) process are self-contained plants whose sole product is heavy water.
3. All other plants that for primary concentration use water electrolysis (WE), steam-hydrogen exchange (SH), synthesis gas distillation (SD), hydrogen distillation (HD), or ammonia-hydrogen exchange (AH) are parasitic to a synthetic ammonia plant. Heavy water is a by-product of these plants, and its production rate is limited by the amount of deuterium in the ammonia plant feed.
4. Water distillation is used for final concentration in all plants still operating, except 18.
5. The relative amount of heavy water produced by each primary concentration process up to 1975 was reported [M7] to have been

   90%, GS process
   6%, water electrolysis and steam-hydrogen exchange
   2%, hydrogen and synthesis gas distillation

Table 13.2 Deuterium production plants

Site, country	Designer, owner†	Start, shutdown	Most recent capacity, MT/yr	Concentration processes: primary, final‡
1. Rjukan & Glomfjord, Norway	Norsk Hydro, Norsk Hydro	1934, Oper.	12	WE + SH, WD
2. Morgantown, W.Va., United States	du Pont, Man. Dist.	1943, 1945	3	WD WE
3. Childersburg, Ala., United States	du Pont, Man. Dist.	1943, 1945	5	WD, WE
4. Dana, Ind., United States	du Pont, Man. Dist.	1943, 1945	8	WD, WE
5. Trail, B.C., Canada	Man. Dist., Cominco	1944, 1956	6	WE + SH, WE
6. Dana, Ind., United States	du Pont, U.S. AEC	1952, 1958	490	GS, WD, WE
7. Savannah River, S.C., United States	du Pont, U.S. DOE	1952, Oper.	Originally 480, reduced to 69	GS, WD
8. Hoechst, Germany	Linde, Farbwerke Hoechst	1958, 1960	6	SD, HD
9. Toulouse, France	Air Liquide, ONIA	1958, 1960	2	SD, HD
10. Domat Ems, Switzerland	Sulzer, Emser Werke	1960, 1967	2	WE + HD, WD
11. Nangal, India	Linde, DAE	1962, Oper.	14	WE, HD
12. Mazingarbe, France	Sulzer-Air-Liquide, SCC	1968 1972	26	AH1 AD
13. Port Hawkesbury, Canada	Lummus, AECL	1970, Oper.	400	GS, WD
14. Bruce A, Canada	Lummus, Ont. Hydro	1973, Oper.	800	GS, WD
15. Glace Bay, Canada	Canatom, AECL	1976, Oper.	400	GS, WD
16. Baroda, India	GELPRA, DAE	1979§	67	AH1, AH1
17. Kota, India	DAE, DAE	1980§	100	GS, WD
18. Tuticorin, India	GELPRA, DAE	1979§	71	AH1, AH1
19. Talcher, India	Uhde, DAE	1979§	63	AH2, WD
20. Bruce B, Canada	Lummus, Ont. Hydro	1979	800	GS, WD
21. La Prade, Canada	Canatom, AECL	Planned	800	GS, WD
22. Bruce D, Canada	Lummus, Ont. Hydro	Planned	800	GS, WD

†Organizations: AECL, Atomic Energy of Canada, Ltd.; DAE, Dept. of Atomic Energy, India; GELPRA, Groupement Eau Lourde Procédé Ammoniac; ONIA, Organisation Nationale Industrielle de l'Azote; SCC, Societe Chimique de Charbonnage; U.S. AEC, U.S. Atomic Energy Commission; U.S. DOE, U.S. Department of Energy.

‡Processes: AD, ammonia distillation; AH1, monothermal ammonia-hydrogen exchange; AH2, dual-temperature ammonia-hydrogen exchange; GS, Girdler-sulfide, dual-temperature, water-hydrogen sulfide exchange; HD, hydrogen distillation; SD, ammonia synthesis gas distillation; SH, steam-hydrogen exchange; WD, water distillation; WE, water electrolysis.

§ Scheduled start-up year.

1%, ammonia-hydrogen exchange
0.3%, water distillation

The rest of this chapter is organized according to process rather than individual plants. The simplest and most familiar process, distillation, is taken up first.

Section 3 describes the separation factors obtainable in distillation of the principal substances used in isotope separation. Section 4 describes deuterium concentration plants using distillation of hydrogen or ammonia synthesis gas. Section 5 describes use of water distillation for primary deuterium concentration, for final deuterium concentration, and for separation of oxygen isotopes.

Section 6 describes the enrichment of deuterium in electrolysis of water. Section 7 describes how steam-hydrogen exchange has been used to increase the recovery of deuterium in electrolytic hydrogen plants.

Section 8 summarizes separation factors obtainable in isotope exchange reactions and their temperature dependence. The latter is the key property in dual-temperature exchange processes. Section 9 develops equations to be used for calculating the number of theoretical stages needed in exchange separation towers.

Section 10 describes monothermal exchange processes, with principal emphasis on ammonia-hydrogen exchange.

Section 11 describes the principle of dual-temperature exchange processes with particular reference to the water-hydrogen sulfide exchange reaction and gives more detailed engineering information about plants using this, the GS process, the process of greatest commercial significance.

Dual-temperature exchange processes using ammonia and hydrogen, methylamine and hydrogen, and water and hydrogen are described in Secs. 12, 13, and 14, respectively, and are compared with the GS process in Sec. 14.

Section 15 gives a brief description of exchange processes for separating lithium isotopes, and Sec. 16 gives a limited account of exchange processes for separating isotopes of carbon, nitrogen, oxygen, and sulfur.

3 SEPARATION FACTORS IN DISTILLATION

3.1 Terminology

In analyzing processes for separating isotopes by distillation, it is desirable to select as components those species whose proportions can be varied independently. When each molecule of the mixture being processed contains only one atom of the element whose isotopes are being separated, such as $H_2{}^{16}O$ and $H_2{}^{18}O$, it is immaterial whether the components be selected as the pair ($H_2{}^{16}O$, $H_2{}^{18}O$) or (^{16}O, ^{18}O), as the mole fraction of $H_2{}^{18}O$ in ($H_2{}^{16}O$, $H_2{}^{18}O$) is identical with the atom fraction of ^{18}O in (^{16}O, ^{18}O). However, when the molecules of the mixture being processed contain two or more atoms of the element whose isotopes are being separated, such as hydrogen containing H_2, HD, and D_2, or water containing H_2O, HDO, and D_2O, it is necessary to choose as components those species whose proportions can be varied independently. In distilling a mixture of H_2, HD, and D_2, the amount of any one of the three components can be varied independently of the other two; the mixture is therefore treated as containing the three components H_2, HD, and D_2, and compositions are expressed as mole fractions of H_2, HD, and D_2. However, in distilling a mixture of H_2O, HDO, and D_2O, equilibrium is continuously maintained in the disproportionation reaction

$$2HDO \rightleftharpoons H_2O + D_2O$$

so that the amount of only two of the three components can be varied independently. In this case, separation performance equations are simplest if compositions are expressed as atom fractions of deuterium or hydrogen.

All of the processes for separating isotopes of hydrogen or other light elements dealt with in this chapter involve distribution between a liquid and a vapor phase. To remain consistent with standard chemical engineering usage, component fractions in the vapor phase are denoted by y and the liquid phase by x. For a two-component mixture, the symbol y or x will denote the fraction of desired component (e.g., atom fraction deuterium in a mixture of H_2O, HDO, and D_2O) in the vapor or liquid phase. For a mixture containing three or more components, a subscript will be used to designate the component. For example, y_{HD} denotes mole fraction HD in a vapor mixture of H_2, HD, and D_2. However, in mixtures of H_2, HD, and D_2 whose deuterium content is so low that the fraction of D_2 can be neglected, the mole fraction of HD will be denoted by y or x without subscript.

In a two-component mixture, the separation factor α is defined as the fraction of desired component in the phase in which it concentrates divided by the fraction of desired component in the other phase. Deuterium, the isotope principally discussed in this chapter, almost always concentrates in the liquid phase. For such deuterium separation processes, the deuterium separation factor α is given by

$$\alpha \equiv \frac{x/(1-x)}{y/(1-y)} \tag{13.1}$$

This is the reciprocal of the equation used to define the separation factor in Chap. 12, Eq. (12.1). This change in notation for Chap. 13 is regrettable, but is hard to avoid.

3.2 Relation of Separation Factor to Vapor Pressures

When only two isotopic compounds are present in the mixture being separated, such as a mixture of CH_4 and CH_3D or a mixture of $H_2{}^{16}O$ and $H_2{}^{18}O$, the separation factor in distillation may be estimated with sufficient accuracy for survey purposes from the ratio of the vapor pressures π of the two compounds,

$$\alpha_{AB} = \frac{\pi_A}{\pi_B} \tag{13.2}$$

where A is the compound with higher vapor pressure. Measurements of the separation factor in liquid-vapor equilibrium of many isotopic mixtures have shown that $\ln \alpha$ (measured) is within 10 percent of $\ln \alpha$ [calculated from (13.2)] except for ^{3}He-^{4}He or H_2-HD-D_2 mixtures. With the same exceptions, measured $\ln \alpha$'s vary less than 10 percent with isotopic composition at constant temperature or pressure.

For Eq. (13.2) to be strictly true, it is sufficient that the liquid and vapor phases form ideal solutions, which is usually very nearly the case for isotopic mixtures at pressures up to 1 atm.

When more than two isotopic compounds are present in the mixture being separated, such as H_2, HD, and D_2, or H_2O, HDO, and D_2, the relation between separation factor and vapor pressures becomes more involved. The situation is complicated further when the vapor pressure of a mixed isotopic compound cannot be measured, because it cannot be isolated in pure form. HDO is such a compound, because it remains in equilibrium with H_2O and D_2O:

$$2HDO \rightleftharpoons H_2O + D_2O$$

The approximate relation between separation factor for hydrogen from deuterium and the measurable vapor pressures of H_2O and D_2O is

$$\alpha^*(\mathrm{H,D}) = \sqrt{\frac{\pi_{\mathrm{H_2O}}}{\pi_{\mathrm{D_2O}}}} \tag{13.3}$$

The general rule is that in a mixture of isotopic compounds

$$XA_n, XA_{n-1}B, XA_{n-2}B_2, \ldots, XB_n$$

the separation factor for isotopes A and B may be approximated by

$$\alpha^*(\mathrm{A,B}) = \sqrt[n]{\frac{\pi_{XA_n}}{\pi_{XB_n}}} \tag{13.4}$$

The conditions required for this relation to be strictly true will be described later.

3.3 Separation Factors

Table 13.3 lists for a number of isotopic mixtures the separation factor computed from vapor pressures by this general formula. This table gives separation factors at the normal boiling point and at the triple point, the lowest temperature at which distillation is possible. As this table shows, the separation factor is greatest for compounds of elements of low atomic weight and increases as the temperature is reduced.

Table 13.3 Separation factors in distillation estimated from vapor-pressure ratios

Compounds and function of vapor pressure	Separation factor at: Triple point	Separation factor at: Normal boiling point	Triple point: Pressure, Torr	Triple point: Temperature, °C	Normal boiling point, °C	Reference
ortho-H_2/HD	3.61	1.81	54	−259.4	−252.9	[W5]
$\sqrt[3]{NH_3/ND_3}$	1.080	1.036	45.6	−77.7	−33.6	[K3]
$\sqrt{H_2O/D_2O}$	1.120	1.026	4.6	0.0	100	[K2]
$\sqrt{H_2O/T_2O}$	–	1.029	–	–	100	[P3]
CH_4/CH_3D	1.0016	0.9965	87.5	−182.5	−161.9	[A3]
$\sqrt{H_2S/D_2S}$	–	1.001	–	–	−60.7	[K4]
$^3He/^4He$	Ratio = 70.4 at 1 K, 3.08 at 3.3 K					[R1]
$^{20}Ne/^{22}Ne$	1.046	1.038	325	−248.6	−245.9	[K1]
$^{36}A/^{40}A$	1.006	–	516	−189.4	−185.7	[C4]
$^{128}Xe/^{136}Xe$	1.000	–	317	−111.8	−109.1	[C8, G5]
$^{12}CH_4/^{13}CH_4$	1.0054	–	87.5	−182.5	−161.9	[J2]
$^{12}CO/^{13}CO$	1.0113	1.0068	111.3	−205.7	−191.3	[J1, J3]
$\sqrt{^{14}N_2/^{15}N_2}$	1.006	1.004	96.4	−209.9	−195.8	[U1]
$^{14}NH_3/^{15}NH_3$	1.0055	1.0025	45.6	−77.7	−33.6	[U1]
$^{14}NO/^{15}NO$	1.033	1.027	164.4	−163.6	−151.8	[C5, C6]
$N^{16}O/N^{18}O$	1.046	1.037	164.4	−163.6	−151.8	[C5]
$C^{16}O/C^{18}O$	1.008	–	111.3	−205.7	−191.3	[J3]
$^{16}O_2/^{16}O^{18}O$	–	1.0052	–	–	−183.0	[J2]
$H_2{}^{16}O/H_2{}^{18}O$	1.010	1.0046	4.6	0.0	100	[U1]

Deuterium. The first part of Table 13.3 lists vapor-pressure ratio data for four compounds of hydrogen that are handled in large enough volumes to be possible feed materials for a plant to concentrate deuterium by distillation.

H_2 + HD is the only mixture of compounds of hydrogen that has a separation factor as favorable as in conventional industrial distillation. In this case, however, the true separation factor is less favorable than here calculated from the vapor-pressure ratio, because of nonidealities in gaseous and liquid mixtures of hydrogen and HD. Moreover, it is desirable to operate above atmospheric pressure, to preclude in-leakage of air. Under practical conditions, at 1.6 atm, the relative volatility obtainable is around 1.6 [N1]. This is the most favorable relative volatility for separation of deuterium by distillation.

Although water has a slightly less favorable relative volatility than ammonia, water makes the better working substance because it is available in unlimited quantities, whereas the amount of deuterium that could be extracted from ammonia is limited to the amount present in ammonia produced industrially.

Methane cannot be used as working substance in a distillation process because its relative volatility is so close to unity. This is regrettable in view of the large amount of natural gas that might be used as a source of deuterium.

Concentration of deuterium by distillation of hydrogen will be discussed in Sec. 14.4 and water in Sec. 15.5.

Noble gases. The second part of Table 13.3 lists vapor-pressure ratios for isotopes of the noble gases helium, neon, argon, and xenon. The vapor-pressure ratio is very high for helium, much smaller for neon, scarcely different from unity for argon, and precisely 1 for xenon. This illustrates the general rule that distillation is a possible separation method for isotopes of the lightest elements, but becomes useless at atomic weights much over 20. Distillation is the preferred method for separating helium isotopes.

Carbon, oxygen, and nitrogen. The only other compounds listed in Table 13.3 whose isotopic species have been concentrated to a significant degree by distillation are CO, NO, and H_2O (for oxygen isotope separation). Distillation becomes unattractive as a method for separating an isotope of low natural abundance when the vapor-pressure ratio is below 1.01, because the plant required for a given output becomes very large and the time required to bring the plant into steady production becomes very great. This is a consequence of the high holdup per unit separation capacity in this method in which the process fluid is liquid. Gas-phase separation processes such as gaseous diffusion are less subject to this difficulty.

Derivation of Eq. (13.3). The following derivation of Eq. (13.3) relating the deuterium separation factor in the distillation of water to the vapor pressure π of H_2O and D_2O is similar to that given by Urey [U1]. It is assumed that:

1. Liquid and vapor phases form ideal solutions.
2. The vapor pressure of HDO is the geometric mean of the vapor pressures of H_2O and D_2O.
3. Equilibrium in the reaction

$$H_2O + D_2O \rightleftharpoons 2HDO$$

   is maintained in the liquid phase.
4. The distribution of deuterium and hydrogen atoms among the three species of water is random, so that the equilibrium constant for this reaction has the value of 4.0. These assumptions are plausible, but are not subject to complete experimental confirmation because liquid HDO cannot be isolated, because it disproportionates into H_2O and D_2O. Values for the equilibrium constant calculated by statistical mechanics are around 3.8.

Water contains the three molecular species H_2O, HDO, and D_2O. In concentrating heavy water by distillation, the deuterium separation factor is defined as the ratio of the atomic ratio of deuterium to hydrogen in the liquid to the corresponding ratio in the vapor. In terms of the mole fractions of individual compounds in the liquid x and vapor y, the separation factor α^* is

$$\alpha^* = \left(\frac{x_{HDO} + 2x_{D_2O}}{2x_{H_2O} + x_{HDO}}\right)\left(\frac{2y_{H_2O} + y_{HDO}}{y_{HDO} + 2y_{D_2O}}\right) \tag{13.5}$$

Because of the ideal solution assumption 1,

$$y = \frac{\pi x}{p} \tag{13.6}$$

where p is the pressure. Because of assumption 4,

$$x_{HDO} = 2\sqrt{x_{H_2O}x_{D_2O}} \tag{13.7}$$

With these substitutions in (13.5),

$$\alpha^* = \left(\frac{2\sqrt{x_{H_2O}x_{D_2O}} + 2x_{D_2O}}{2x_{H_2O} + 2\sqrt{x_{H_2O}x_{D_2O}}}\right)\left[\frac{(2\pi_{H_2O}x_{H_2O}/p) + (2\sqrt{\pi_{H_2O}\pi_{D_2O}x_{H_2O}x_{D_2O}}/p)}{(2\sqrt{\pi_{H_2O}\pi_{D_2O}x_{H_2O}x_{D_2O}}/p) + (2\pi_{D_2O}x_{D_2O}/p)}\right]$$

$$= \frac{\sqrt{x_{D_2O}}\sqrt{\pi_{H_2O}x_{H_2O}}}{\sqrt{x_{H_2O}}\sqrt{\pi_{D_2O}x_{D_2O}}} = \sqrt{\frac{\pi_{H_2O}}{\pi_{D_2O}}} \tag{13.8}$$

All mole fractions have canceled out, and α^* is independent of composition.

The general equation (13.4) may be derived in similar fashion from analogous assumptions.

Distillation of water. Combs et al. [C11] have determined the deuterium separation factor in the distillation of water by measuring the H/D ratio in water liquid and vapor in equilibrium. The third and fourth columns of Table 13.4 compare their measured separation factors with values predicted by Eq. (13.3) from their values for the vapor pressures of pure H_2O and D_2O. The agreement in the two sets of values of ln α is within 6 percent. The agreement with Kirshenbaum's vapor-pressure ratios [K2] is somewhat poorer. Rolston et al. [R8] have proposed the equation $\ln \alpha^* = 0.0592 - 80.3/T + 25{,}490/T^2$ to correlate all data to 1976.

Distillation of ammonia. Petersen and Benedict [P2] have made similar direct measurements of the deuterium separation factor in the distillation of ammonia. Table 13.5 compares their values for ammonia containing 24 percent deuterium with those predicted by Eq. (13.9) from vapor pressures of NH_3 and ND_3 measured by Kirshenbaum and Urey [K3], Groth et al. [G4], and Taylor and Jungers [T1].

$$\alpha^* = \sqrt[3]{\frac{\pi_{NH_3}}{\pi_{ND_3}}} \tag{13.9}$$

The agreement at this deuterium content is within experimental uncertainty. However, a small but significant trend of separation factor with deuterium content was observed, as indicated in Table 13.6.

These results for water and ammonia suggest that Eq. (13.4) can be used to predict separation factors in distillation with an error in ln α^* no greater than 10 percent.

Table 13.4 Deuterium separation factors in distillation of water

		Vapor-pressure ratio $\sqrt{\pi_{H_2O}/\pi_{D_2O}}$		Separation factor	
Temperature, °C	Vapor pressure of H_2O, Torr	Kirshenbaum [K2]	Combs et al. [C11]	Measured [C11]	Correlated [R8]
0	4.58	1.12_0			1.113
10	9.21	1.08_7	1.094	1.100_3	1.098
20	17.54	1.07_4	1.082	1.087_3	1.085
30	31.8	1.06_6	1.071	1.074_8	1.074
40	55.3	1.05_9	1.063	1.062_9	1.065
50	92.5	1.052	1.055	1.051	1.056
60	149.4	1.046			1.049
70	233.7	1.040			1.043
80	355	1.035			1.037
90	526	1.030			1.032
100	760	1.026			1.027
120	1,489	1.019			1.020
140	2,711	1.013_5			1.014
160	4,636	1.009			1.010
180	7,521	1.005			1.006
200	11,661	1.002_5			1.003
220	17,400	1.000_5			
240	25,100	0.997_9			

4 DISTILLATION OF HYDROGEN

Deuterium was discovered by Urey et al. [U2] in samples of liquid hydrogen in which deuterium had been concentrated by partial evaporation. Because of the high deuterium separation factor, separation of deuterium by distillation of liquid hydrogen was studied by engineers in Germany [C3] and the United States [M8] during World War II and more recently by groups in the Soviet Union [M1], France [A1], Germany [L2], Switzerland [H3], England [D2], and the United States [B3, B4]. The main difficulties with the process have been the extremely low operating temperatures, which until recently have been without industrial precedent, and elimination of condensable impurities from the feed stream, which would foul heat exchangers and stop flow if not removed. Because these difficulties are those of low-temperature plants and are not unique to isotope separation, they will not be dealt with

Table 13.5 Deuterium separation factors in distillation of ammonia

		Separation factor	
Pressure, Torr	Temperature, °C	Measured, at 24 percent deuterium	From Eq. (13.9)
764	−32.6	1.0429 ± 0.0015	1.042
500	−40.6	1.050 ± 0.0007	1.047
250	−52.2	1.0564 ± 0.0013	1.055

Table 13.6 Effect of deuterium content on separation factor in distillation of ammonia

	Atom percent deuterium in liquid (760 Torr pressure)				
	10	24	42	58	Eq. (13.9)
Separation factor	1.0435	1.0436	1.0402	1.0383	1.042
Experimental uncertainty	0.0016	0.0010	0.0006	0.0007	

extensively here. The references cited above may be consulted for more detailed information. Plants producing deuterium by distillation of liquid hydrogen that have been built and operated are listed in Table 13.7. The process used for the primary concentration of deuterium in all of these plants is similar in principle and is illustrated in generalized fashion in Fig. 13.5. The individual plants differ in detail; some of the principal differences are noted in Table 13.7. More detail is given in the references cited in Table 13.7.

The history of these plants has been sketched in Sec. 2.2 of Chap. 12. Each plant is parasitic to an ammonia synthesis plant, taking deuterium-bearing, hydrogen-rich feed gas from the ammonia plant, and returning gas depleted in deuterium to the ammonia plant, with little loss of hydrogen (less than 5 percent). The first two plants listed in Table 13.7 used as feed ammonia synthesis gas, which contains around 75 percent H_2, 25 percent N_2, and small amounts of CH_4, A, CO_2, CO, O_2, and H_2O. The remaining plants used as feed electrolytic hydrogen, which contains as impurities only H_2O and traces of N_2 and O_2. The high content of nitrogen and the presence of other impurities in the ammonia synthesis gas used as feed in the first two plants caused their design to be more complicated, their specific energy consumption higher, and the cost of heavy water produced in them greater than in the three plants using electrolytic hydrogen and feed. In fact, the first two plants were built primarily as pilot plants rather than as economic producers of heavy water, and they have been shut down, having served their purpose.

The process used in the primary section of these plants may be understood by reference to Fig. 13.1. Where gas from the ammonia plant is available under pressure, it is fed directly to the hydrogen distillation plant; otherwise it is compressed in the feed compressor. The gas is cooled down to around −175°C by outflowing cold gas depleted in deuterium in a heat exchange system in which water is condensed and removed from the feed. Refrigeration to compensate for heat leaking into the plant may next be supplied to the feed. The gas is cooled further to about −245°C by outflowing cold gas in a second heat exchange system in which nitrogen is condensed and removed from the feed. Much nitrogen is condensed from synthesis gas; traces, from electrolytic hydrogen. Final cooling is provided by Joule-Thomson expansion through a valve, in which hydrogen is cooled to around −250°C and partially liquefied.

The hydrogen is distilled in the primary tower into a bottom product enriched in deuterium and an overhead product depleted in deuterium. Final concentration of the bottom product is effected by distillation either of liquid hydrogen or water (not shown in Fig. 13.1). The depleted hydrogen flows back through the feed exchanger system where it is warmed to room temperature. It is returned to the ammonia plant at the supply pressure, being compressed if necessary.

To provide heat to reboil the tower and to supply liquid hydrogen reflux, additional depleted hydrogen is circulated by the reflux compressor through another system of heat exchangers, to which additional refrigeration may be supplied. Cold, compressed hydrogen from this system flows through a coil at the bottom of the tower where it is condensed, supplying

Table 13.7 Hydrogen distillation heavy-water plants

Plant location	Toulouse, France	Hoechst, Germany	Soviet Union	Ems, Switzerland	Nangal, India
Designer	Compagnie Francaise	Linde	Soviet	Sulzer Bros.	Linde
Operator	de l'Eau Lourde	Farbwerke Hoechst	govt.	Emswerke AG	Indian govt.
Year production started	1958	1958	?	1960	1962
Year production ended	1960	1960	?	1967	
Feed gas (1)†					
Material	NH_3 synthesis gas	NH_3 synthesis gas	Electrolytic H_2	Electrolytic H_2	Electrolytic H_2
$nm^3\ H_2/h$	3000	6300	4000	400	5300
ppm D in H	120	105	150	970	450
D recovery, %	65	85	?	85	90
Production rate, kg D_2O/day	3.5	12	13	7	45
Energy cons., kWh/kg D_2O	17,200	8000	5000	2400	2100
Pressure, atm					
Feed (1)	230	20	2–4	3.7	5
Recycle (2)	2–16	50	6 + 70	14	40
Stripped gas (3)	2.5	1	?	1.5	1
Flow ratio, recycle/feed H_2	?	0.43	?	7.5	1.24
Method of removing H_2O	Alumina	Regenerators	Switch exchangers	Switch exchangers	Regenerators
Method of removing N_2	Adsorption	Regenerators	Adsorption	Switch exchangers	Regenerators
Refrigeration	Piston expander	Liquid N_2	Liquid NH_3 + liquid N_2	Turbine expander	Liquid NH_3 + liquid N_2
Stream applied to	Feed	Feed + recycle	Feed + recycle	Recycle	Feed + recycle
Primary tower					
Type	Double	Triple	Single	Single	Triple
Internals	l'Air Liquide	Sieve plates	Bubble caps	Kuhn, Dixon	Sieve plates
Diameter, m	?	1.2	1.05	90 tubes, 5-cm diam.	?
Packed height	85 plates	30 m	77 plates	2 m	?
% HD in bottoms (4)	2	4	7–9	60	4
Material distilled for final D conc.	Hydrogen	Hydrogen	?	Water	Hydrogen
Reference	[A1]	[L2]	[M1]	[H3]	[G1]

†Numbers are keyed to Fig. 13.1.

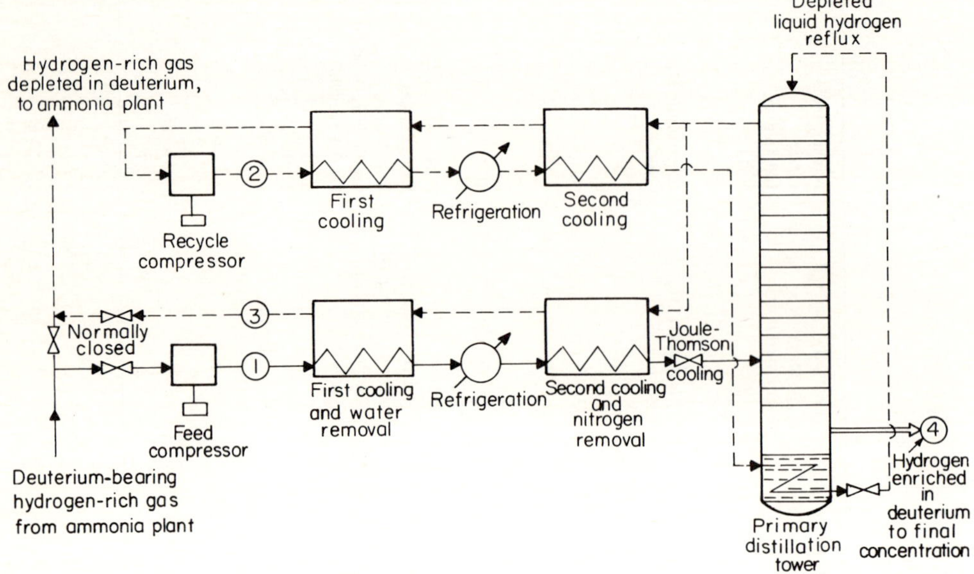

Figure 13.1 Generalized flow sheet for hydrogen distillation heavy-water plants. (For process conditions at numbered points, see Table 13.7.)

heat to reboil the tower at the same time. This liquid is then expanded to tower pressure through a valve and fed to the top of the tower for reflux.

The product of these primary plants is a stream of hydrogen containing from 2 to 60 percent HD. At Ems this hydrogen was converted to water by burning it with oxygen, and pure heavy water was produced by distilling the water. At Toulouse, Hoechst, and Nangal, the HD-rich hydrogen stream was distilled directly to produce pure deuterium, which might then be burned to make heavy water. The basic flow sheet for this final concentration of deuterium was devised by Clusius and Starke [C7], who conducted the first experimental work on the fractional distillation of liquid hydrogen and showed that a mixture of hydrogen, HD, and deuterium could be separated by fractional distillation at atmospheric pressure into relatively pure fractions of H_2, HD, and D_2 without HD undergoing disproportionation and without appreciable conversion of ortho to para modifications.

The flow sheet for final concentration of deuterium developed by Clusius and Starke, which was used in the Hoechst and Nangal plants, is shown in Fig. 13.2, together with the primary tower of Fig. 13.1.

The bottoms from the primary tower are fed into the upper half of a smaller secondary tower, where fractionation into a bottom product of nearly pure HD is completed. This HD is warmed to room temperature in a heat exchanger and passed through a catalytic exchange reactor where its disproportionation into an equilibrium mixture of H_2, HD, and D_2 is catalyzed. The product of the exchange reaction is cooled to liquid hydrogen temperatures in the heat exchanger and fed to the bottom half of the secondary tower where it is fractionated into an overhead product of $HD + H_2$ and a bottom product of pure deuterium. This is warmed to room temperature in the heat exchanger and constitutes the product of the plant. The HD and H_2 overhead from the bottom of the secondary tower is fed to the top of the secondary tower for recovery of HD.

Heat to reboil these towers is provided by a stream of compressed, HD-free hydrogen, which is condensed in reboiler coils located in the sump of these towers. The condensed HD-free hydrogen is then used as open reflux in the top of the primary tower. A Linde, double-column arrangement is used to provide reflux for the bottom of the secondary tower and reboil vapor for the top of this tower.

Of the plants listed in Table 13.7, the one at Nangal, India, may be regarded as indicating the full potentialities of this method of producing deuterium. It is a relatively large plant, producing around 14 t of heavy water per year. It uses clean electrolytic hydrogen as feed. This hydrogen has been preconcentrated by two stages of partial electrolysis of water to around three times natural abundance. Power costs at Nangal, which is the site of a large hydroelectric project, are low. These three favorable circumstances make it possible to produce heavy water at a specific energy consumption in the distillation plant of only 2.1 kWh/g D_2O. This is lower than the energy consumption at the other sites, and of course is much lower than the 468 kWh/g D_2O for electrolysis alone noted in Sec. 6. Gami et al. [G1] in 1958 predicted that heavy water would be produced at Nangal at a cost of $27.2/lb or $60/kg. Data cited by these authors in 1958 as typical of what production rate and costs might be experienced at Nangal when the plant went into operation are summarized in Table 13.8.

A special problem of hydrogen distillation plants is the need to minimize conversion of ortho to para hydrogen. At room temperature, hydrogen contains 75 percent ortho and 25 percent para hydrogen. At low, hydrogen distillation temperature, the equilibrium proportion is nearly 100 percent para hydrogen. Conversion of ortho to para hydrogen is very slow in the absence of catalysts. Conversion must be minimized in a deuterium separation plant because about 1.5 times as much heat is released in conversion of ortho hydrogen as in liquefaction; it would greatly increase power consumption if allowed to occur. Conversion is catalyzed by paramagnetic materials, such as solid oxygen, and by ferromagnetic materials, such as certain steels. These must be excluded from the plant.

Condensed oxygen is especially objectionable, both because of the heat produced in ortho-para catalysis and because of its liability to explode when in contact with cold hydrogen.

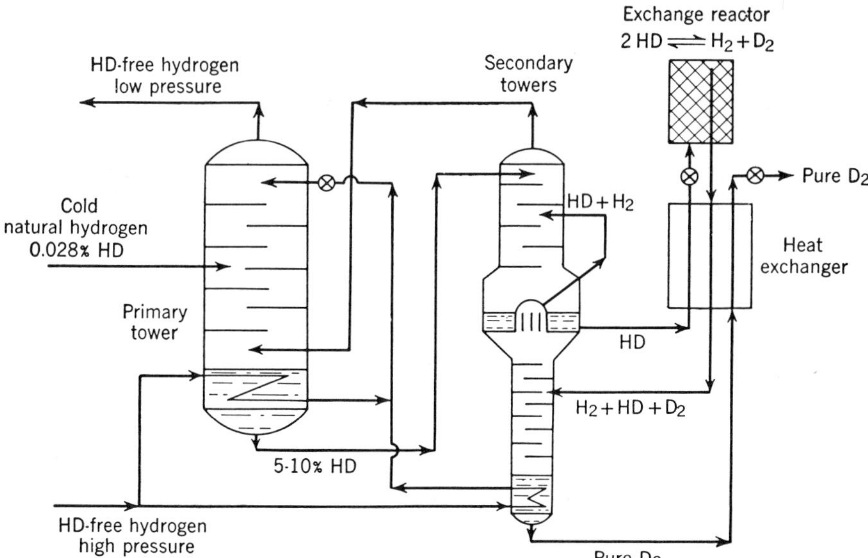

Figure 13.2 Flow sheet for final concentration of deuterium by distillation of liquid hydrogen. *[From K. Clusius and K. Starke, Z. Naturforsch. 4A:549 (1949).]*

Table 13.8 Production and cost data anticipated for Nangal heavy-water plant

Stages of electrolytic preconcentration, 2
Hydrogen production rate, 25,000 nm^3/h
Hydrogen feed rate to distillation plant, 5000 nm^3/h
Producing hours per year, 8000
D_2O production rate, 14,000 kg/yr
Erected cost of plant, $2.75 million

Production costs	$ million/yr	$/kg D_2O
Capital charges at 16.8%/yr	0.462	33.0
Power	0.130	9.3
Hydrogen loss	0.090	6.4
Labor and maintenance	0.130	9.3
Supplies	0.027	1.9
Total	0.839	59.9

5 DISTILLATION OF WATER

Distillation of water was used in the early plants of the Manhattan Project [M8] for primary concentration of deuterium. It is now the method generally used for final concentration of deuterium and for reconcentration of heavy water that has picked up light water during use. Distillation of water has been used by Dostrovsky [D4, D5] to produce ^{18}O.

5.1 Primary Concentration of Deuterium

Although water distillation is no longer used for primary concentration of deuterium because of its high energy consumption, the principal features of water distillation plants for this purpose will be described briefly because they illustrate isotope separation principles so well.

Process requirements. Distillation of water for deuterium separation differs from all other industrial distillation processes in the extremely small difference in normal boiling point between the key components, 0.7°C between H_2O and HDO. This, coupled with the very low natural abundance of deuterium, leads to an extraordinarily high reboil vapor ratio, so that the heat consumption per unit of D_2O product is enormous.

A rough idea of the requirements of the water distillation process may be derived from a representative separation factor of 1.05. The minimum number of theoretical plates (n_{min}) needed to enrich deuterium from the natural concentration of $x_F = 0.000149$ atom fraction to product concentration of $x_P = 0.998$ is

$$n_{min} = \frac{\ln [x_P(1 - x_F)/x_F(1 - x_P)]}{\ln \alpha^*} = 308 \tag{13.10}$$

The optimum number will be somewhat more than twice this, or around 700 plates.

The minimum consumption of steam per mole of heavy water produced is secured when an infinite number of plates is used, so that the outgoing steam depleted in deuterium may be in equilibrium with incoming feed.

From Eq. (12.80), the minimum molar ratio of steam flow rate G to product P is

$$\left(\frac{G}{P}\right)_{\min} = \frac{x_P - x_F}{x_F} \frac{\alpha^*}{\alpha^* - 1} = 141{,}000 \tag{13.11}$$

For a practical plant, with a finite number of stages, around 200,000 mol of steam must be provided per mole of heavy water produced. Because of the small difference in boiling point between the two products, this large amount of heat flows through a relatively small temperature difference; in fact, the principal temperature differences are due to pressure drop across the column and temperature difference across reboiler and condenser heat exchange surface, rather than differences between the boiling points of the components. Economical operation requires that the large heat demand be supplied as nearly reversibly as possible, with the minimum practicable loss in availability. Reboil heat should be supplied with good thermodynamic efficiency, and column pressure drop should be minimized.

History of process. Despite these severe requirements, the water distillation process has been of interest because of its simple, conventional equipment. For primary concentration of deuterium from natural water, it received attention in Germany, where pilot-plant work was done by I. G. Farben during World War II [C3], and in the United States [M8], where most of the heavy water used by the Manhattan District was produced in this way.

Manhattan District plants. The water distillation plants of the Manhattan District were built to provide a simple and certain way of producing heavy water, although not necessarily at minimum cost. Because speed was more important than economy, it was not possible to explore fully developments that might have permitted more economical production. These plants are described briefly in this section; more detailed information has been given by Maloney and Ray [M8] and by Selak and Finke [S3].

Plants. Three water distillation plants were designed and built for the Manhattan District by E. I. du Pont de Nemours and Company, Inc. These plants were located at Morgantown, West Virginia, Childersburg, Alabama, and Dana, Indiana. Parts of the plants were started up in June 1943, and concentrations reached steady-state values in June 1944. About 90 days were needed to reach steady state. The plants were shut down in October 1945 because of reduced demand and because of the high cost of their heavy water.

These distillation plants concentrated deuterium from 0.0143 a/o (atom percent) to 87 to 91 a/o. Further concentration to 99.8 percent was effected by electrolysis. The average recovery of D_2O from the steam fed was only 1.94 percent; 360,000 mol of steam were fed per mole of D_2O produced.

The combined capacity of the three plants was 1.15 MT/month. The total production of 99.8 percent D_2 was 20.7 MT.

The total cost of the plants was \$14.5 million. The unit investment cost was therefore

$$\frac{\$14{,}500{,}000}{(1.15)(12)(1000)} = \$1051/(\text{kg/yr})$$

The operating costs were as follows:

	Per month	Per kg D_2O
Steam	\$295,000	\$271
Other	127,000	117
Total	\$422,000	\$388

Process. A simplified flow sheet of the process used at the Morgantown plant, the smallest and most efficient of the three, is shown in Fig. 13.3. This plant produced 254 kg D_2O/month, with a deuterium recovery of 2.8 percent. The plant consists of an eight-stage cascade of distillation towers, with associated reboilers, condensers, and pumps. Summary data on the towers of each stage are given in Table 13.9.

The first stage consists of five parallel groups of two series-connected towers, of which one group, 1A and 1B, are shown in Fig. 13.3. Feed for each 1A tower consists of condensate from the reboiler of the associated 1B tower. Feed is introduced at the top of the 1A tower. Stripped vapor from the top plate is condensed in a barometric condenser, vented to a steam ejector that maintains a pressure of from 50 to 90 Torr at the top of the tower.

Slightly enriched water from the bottom of tower 1A is pumped to the top of tower 1B, and vapor from the top of 1B flows back to the bottom of 1A.

Most of the water at the bottom of 1B, now enriched to 0.117 a/o deuterium, is converted to vapor in the reboiler and returned to 1B, but around 12 percent is pumped ahead to the top of 2A. Vapor from the top of 2A is condensed in a condenser refrigerated with ammonia, to prevent loss of the now valuable water. This condenser is also vented to a steam ejector, which maintains a pressure of 130 Torr.

The second stage consists of two towers, 2A and 2B, connected in series, like each 1A and 1B pair. The third and higher stages consist of single towers, of progressively decreasing diameter. Arrangements for reboiling water at the bottom of each tower and condensing and returning vapor at the top of the next stage are the same as at the bottom of 2B and the top of 3. The progressive decrease in tower diameter from the feed point to the product end is characteristic of an isotope separation plant.

As the water flows through the stages of the plant, it is enriched progressively in deuterium, until it reaches 89 a/o in the bottom product of the eighth and last stage.

Most of the steam for the plants at Morgantown and elsewhere was generated at 165 psia† and throttled to 55 psia, the pressure at which it was used in the reboilers, even though steam at 22 psia would have sufficed to reboil the tower bottoms, where were at subatmospheric pressure. Because low-pressure, by-product steam was not available in the required amounts, it was necessary to generate steam solely for the water distillation plant. This was inefficient and added to the operating cost in these plants.

Towers. Towers of these plants over 18 in in diameter were of the plate type, with plates on 1-ft spaces. All the large towers used bubble caps, except 1A, which had tunnel caps. Towers 18 in in diameter and smaller were packed with $\frac{5}{8}$- by $\frac{5}{8}$-in ceramic rings.

Possible improvements. The designers of the Manhattan District plants recognized that two major improvements could be made in a future water distillation plant designed for economy rather than speed of construction. These were as follows:

1. More efficient utilization of heat than generating 150-psig steam solely for the distillation plant
2. The use of tower internals with greater capacity per unit volume than tunnel- or bubble-cap plates, to increase plant capacity for the same capital investment

More efficient utilization of heat. In the first-stage towers of the Manhattan District plants, where most of the heat was consumed, heat flowed from the tower bottom temperature of

†1 psia (pound force per square inch absolute) = 51.7 Torr = 0.06895 bar = 6895 Pa.

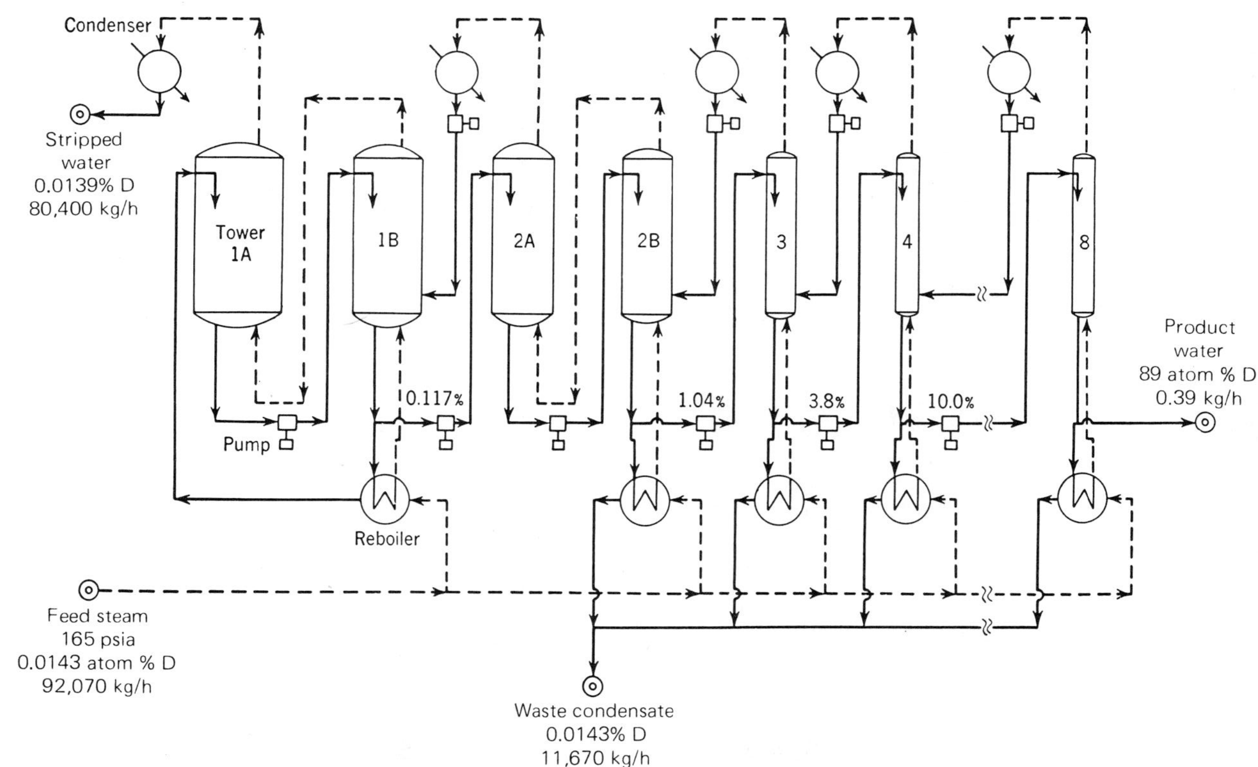

Figure 13.3 Morgantown water distillation plant.

Table 13.9 Towers of Morgantown water distillation plant

Tower	Number in parallel	Diameter†	No. of plates	kg vapor/h	Pressure, Torr Top	Pressure, Torr Bottom	a/o deuterium, bottom
1A	5	15 ft	80	(80,400)	67	238	
1B	5	12 ft	90	80,400	238	536	0.117
2A	1	10.5 ft	72	(9,620)	129	340	
2B	1	8 ft	83	9,620	340	645	1.04
3	1	3.3 ft	72	1,380	124	343	3.8
4	1	1.5 ft	72‡	330	127	440	10.0
5	1	10 in	72‡	85	127	340	11.5
6	1	10 in	72‡	85	124	328	21.2
7	1	10 in	72‡	90	124	333	56.4
8	1	10 in	72‡	90	127	308	89
Total	18		757	92,070			

†1 ft = 12 in = 30.48 cm.
‡Number of theoretical plates in packed column.

195°F (10.5 psia)† to the tower top temperature of 111°F (1.3 psia). To transfer this heat through the reboilers, steam at 233°F (22 psia) was required. Because this heat is needed only at relatively low temperatures, it is very inefficient to obtain it by burning fuel under boilers, without making use of the heat at higher temperatures first. Two possible ways of providing low-temperature heat more efficiently are these:

1. Using 22-psia exhaust steam from the turbines of a power plant.
2. Using a vapor-recompression system

Examples of these two schemes are shown in Figs. 13.4 and 13.5. The turbine-exhaust scheme of Fig. 13.4 has two advantages over the vapor-recompression scheme of Fig. 13.5.

1. The cost of the condenser and steam jet ejector is less than that of the vapor compressor and feed-water preheater.
2. The power lost in the steam turbine plant is less than the power consumed by the vapor compressor. Although the theoretical power W lost by the turbine exhausting at 22 psia instead of at 1.3 psia is exactly the same as the power consumed by a compressor taking the same amount of steam from 1.3 to 22 psia, the actual turbine efficiency of E drops the turbine power lost to WE and the compressor efficiency of E' raises the power consumed by the compressor to W/E'.

The turbine exhaust scheme has the disadvantage of making the production rate of heavy water dependent on the production rate of power from the steam turbines.

Packed towers. After these plants were built, several improved types of tower internals were developed that have higher capacity per unit volume and lower pressure drop per theoretical

†For consistency with the original references, conditions in U.S. heavy-water plants have been expressed in English units. Conversion tables to SI units are given in App. B.

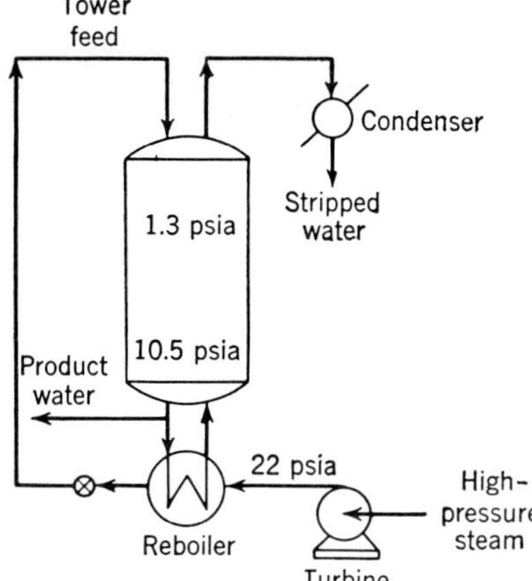

Figure 13.4 Water distillation tower reboiled by steam turbine exhaust.

plate than bubble-cap plates and are claimed to be more economical. The British Atomic Energy Research Establishment has developed a tower packing known as Spraypak [M4] for use especially in the water distillation process.

Distillation-cascade design principles. Some of the principles involved in designing an isotope separation plant for minimum cost will be illustrated by roughing out optimum conditions for a water distillation plant incorporating the two improvements noted above.

Design variables. The principal design variables in a water distillation plant are

1. The type of tower internals
2. The pressure p, Torr
3. The vapor velocity v, cm/s
4. The ratio of reboil vapor to product, G/P

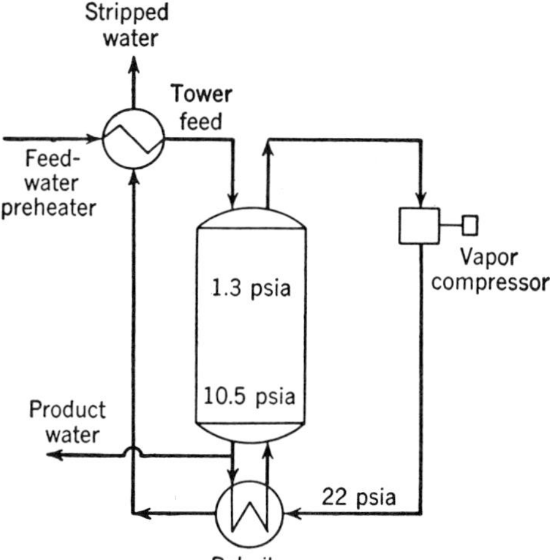

Figure 13.5 Water distillation tower reboiled by vapor recompression.

The best inernals and the optimum values of pressure, vapor velocity, and reboil vapor ratio are those that permit production of heavy water at minimum cost. The initial cost of the plant depends on a number of factors including the total number of towers, the total amount of reboiler and condenser surface, and the total volume of tower internals. The principal operating cost is for power, which is proportional to total loss in availability of steam as it flows through the towers. A complete minimum-cost analysis requires knowledge of the unit cost of all the important cost components and is beyond the scope of this book. Design for minimum volume of tower internals or minimum loss in availability due to tower pressure drop and for minimum cost of these two important contributors to total cost can be carried out without complete unit-cost data and will be discussed. Because the same choice of reboil vapor ratio minimizes the number of towers, their volume, and the loss of availability within them, this reboil vapor ratio is close to that which leads to minimum production cost. An equation for this optimum reboil vapor ratio will now be derived, and expressions will be developed for the total volume of towers and the total loss in availability in towers designed for the optimum ratio.

Enrichment equation. The differential equation for the increase in deuterium content x with distance z from the top of the tower is

$$h\frac{dx}{dz} = (\alpha^* - 1)x(1-x) - \frac{P}{G}(x_P - x) \tag{13.12}$$

This equation is derived in a manner similar to (12.128); h is the height of a transfer unit, $h\,dx/dz$ replaces dx/di, and G, the molar flow rate of steam, plays the role of the tails flow rate N.

Tower volume. At a point in the tower where the vapor velocity is v and the absolute pressure is p, the area A needed to accommodate a steam flow rate of G mol/s is

$$A = \frac{GRT}{pv} \tag{13.13}$$

where R is the gas constant and T the absolute temperature. The volume of tower dV needed to increase the deuterium content of the liquid by an amount dx is

$$\frac{dV}{dx} = \frac{A}{dx/dz} = \frac{hRT/pv}{[(\alpha^* - 1)x(1-x)/G] - (P/G^2)(x_P - x)} \tag{13.13a}$$

The steam flow rate that leads to minimum tower volume is that which makes this expression a minimum at every x, or

$$G_{\text{opt}} = \frac{2P(x_P - x)}{(\alpha^* - 1)x(1-x)} \tag{13.14}$$

This is the tails flow rate for an ideal cascade. The details are the same as in deriving (12.132). At this optimum steam rate,

$$\left(\frac{dV}{dx}\right)_{\text{min}} = \frac{4hRT}{pv(\alpha^* - 1)^2}\,\frac{P(x_P - x)}{x^2(1-x)^2} \tag{13.15}$$

In a tower in which h, T, p, v, and α^* are held constant,

$$V_{\text{min}} = \frac{4hRT}{pv(\alpha^* - 1)^2}P\int_{x_F}^{x_P} \frac{x_P - x}{x^2(1-x)^2}\,dx = \frac{4hRT}{pv(\alpha^* - 1)^2}PD_{P,F} \tag{13.16}$$

where $D_{P,F}$, the separative capacity for the enriching section of an ideal cascade per unit product rate, is given by

$$D_{P,F} = (2x_P - 1)\ln\frac{x_P(1-x_F)}{x_F(1-x_P)} + \frac{(x_P - x_F)(1-2x_F)}{x_F(1-x_F)} \tag{13.17}$$

The factor $4hRT/[pv(\alpha^* - 1)^2]$ gives the tower volume required for a plant performing 1 mol of separative work per second; it is a measure of the relative volume needed for different types of packing as a function of vapor velocity and pressure. When the design objective is to minimize tower volume, the type of packing and the velocity and pressure that minimize this factor should be selected.

Rate of loss of availability. In the scheme of Fig. 13.4 for reboiling a tower with low-pressure exhaust steam from a turbine, factors that reduce the power output of the turbine are (1) the temperature difference across the reboiler, which causes the turbine exhaust pressure to be higher than the tower bottom pressure, and (2) the steam pressure drop through the tower, which causes the tower bottom pressure to be higher than the tower top. We shall focus attention on the second of these inefficiencies and shall derive an expression for the reduction in turbine power caused by steam pressure drop through the tower.

If this were the only thermodynamic inefficiency, the loss in turbine power would equal the rate of loss of availability in the tower Q, given by

$$Q = T_0\frac{dS}{dt} \tag{13.18}$$

where dS/dt is the rate of production of entropy in the tower and T_0 is the temperature at which heat is rejected to cooling water. When liquid and vapor have the same temperature and when liquid-phase pressure changes are neglected, the rate of entropy production is simply that due to steam-pressure changes,

$$\frac{dS}{dt} = -\int_0^Z G\left(\frac{\partial s}{\partial p}\right)_T \frac{dp}{dz}\,dz \tag{13.19}$$

where Z is the height of the tower, and s is the entropy per mole of steam. If steam is treated as a perfect gas,

$$\left(\frac{\partial s}{\partial p}\right)_T = -\frac{R}{p} \tag{13.20}$$

so that

$$Q = \int_0^Z \frac{RT_0G}{p}\frac{dp}{dz}\,dz \tag{13.21}$$

The rate of availability loss per unit height is the integrand

$$\frac{dQ}{dz} = \frac{RT_0G}{p}\frac{dp}{dz} \tag{13.22}$$

and the rate of availability loss per unit increase in deuterium content is

$$\frac{dQ}{dx} = \frac{dQ/dz}{dx/dz} = \frac{(hRT_0/p)(dp/dz)}{[(\alpha^* - 1)x(1-x)/G] - (P/G^2)(x_P - x)} \tag{13.23}$$

The optimum steam rate, which makes this a minimum, is again given by (13.14), so that

$$\left(\frac{dQ}{dx}\right)_{\min} = \frac{4hRT_0}{(\alpha^* - 1)^2 p} \frac{dp}{dz} \frac{P(x_P - x)}{x^2(1-x)^2} \tag{13.24}$$

In a tower in which h, p, and α^* are held constant, the minimum rate of loss of availability is obtained in the same way as the minimum volume (13.16) and is

$$Q_{\min} = \frac{4hRT_0}{(\alpha^* - 1)^2 p} \frac{dp}{dz} P\, D_{P,F} \tag{13.25}$$

The factor

$$\frac{4hRT_0}{(\alpha^* - 1)^2 p} \frac{dp}{dz}$$

gives the loss of turbine power in a plant performing 1 mol of separative work per second; it is a measure of the relative power consumption with different types of packing as a function of vapor velocity and pressure.

Costs for tower volume and power. The contribution of tower volume and availability loss to the cost of heavy water produced by the distillation of water in an ideal cascade may be evaluated when values are assigned to

j, the fractional charge against investment per year
c_V, the unit cost of tower volume
c_Q, the unit cost of turbine work lost owing to tower pressure drop

The annual charge A for tower volume and power, then, is

$$A = \left[jc_V \frac{4hRT}{pv(\alpha^* - 1)^2} + 3.15 \times 10^7 c_Q \frac{4hRT_0}{(\alpha^* - 1)^2 p} \frac{dp}{dz} \right] P\, D_{P,F} \tag{13.26}$$

where the numerical factor is the number of seconds per year. The contribution of tower volume and power to the unit cost of heavy water, in dollars per mole, is obtained by dividing the annual cost by the number of moles of heavy water produced per year, $3.15 \times 10^7 P x_P$:

$$c_{D_2O} = \left\{ \frac{jc_V[4hRT/pv(\alpha^* - 1)^2]}{3.15 \times 10^7} + c_Q \frac{4hRT_0}{(\alpha^* - 1)^2 p} \frac{dp}{dz} \right\} \frac{D_{P,F}}{x_P} \tag{13.27}$$

$D_{P,F}$ may be obtained from (13.17). With $x_F = 0.000149$ and $x_P = 0.998$,

$$\frac{D_{P,F}}{x_P} = 6729 \tag{13.28}$$

Packing characteristics. We have shown that the optimum steam rate that leads to minimum tower volume and minimum power is that of the ideal cascade (13.14). The optimum type of packing, optimum pressure, and optimum vapor velocity is that which makes the expression in braces (13.27) a minimum. We shall not attempt to evaluate a number of types of packing, but shall use Spraypak no. 37 packing as an example of the selection of optimum vapor velocity and pressure. This is the type of packing recommended by McWilliams and co-workers [M4] for a water distillation plant.

Figure 13.6 is a plot of the height of a transfer unit in feet, h, and the pressure drop per unit transfer unit in torr, $h\, dp/dz$, versus percent of flooding velocity, obtained from the data of McWilliams and co-workers [M4]. These data are for the system H_2O-HDO at total reflux and pressures of 420, 760, or 1245 Torr. Flooding velocities v_f reported by McWilliams et al. at

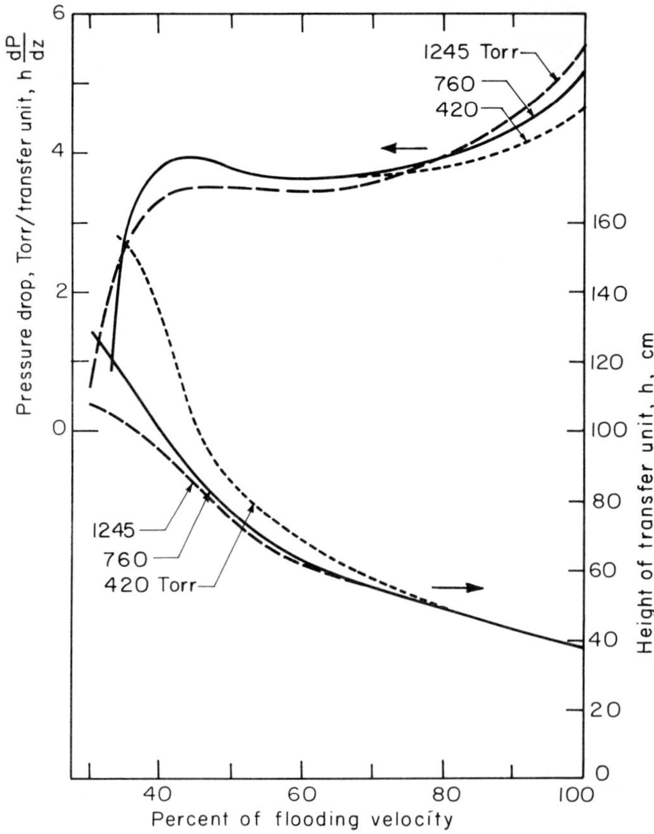

Figure 13.6 Characteristics of Spraypak no. 37 for H_2O-HDO, total reflux.

these three pressures are given in Table 13.10. For pressures below 420 Torr it will be assumed that the product $v_f^2\rho$ is constant at its value of 128.5 for 420 Torr and that h and $h\,dp/dz$ have the same values as for 420 Torr shown in Fig. 13.6.

The following values will be assigned the parameters of the cost equation (13.27):

$j = 0.20/\text{yr}$
$c_V = \$0.002/\text{cm}^3$
$c_Q = \$0.015/\text{kWh}$

Table 13.10 Flooding velocities for Spraypak no. 37, system H_2O-D_2O, total reflux

Pressure p, Torr	420	760	1245
Temperature T, K	357.4	373.2	387.6
Vapor density ρ, g/cm^3 (pM/RT)	0.000339	0.000588	0.000928
Flooding velocity, g/(cm^2·s) [M4]	0.208	0.275	0.338
v_f, cm/s (above/ρ)	614	468	364
$v_f^2\rho$, g/(cm·s^2)	127.8	128.8	123.0

$R = 62360$ (Torr·cm^3)/(g-mol·K) (first term)
$R = 0.000002310$ kWh/(g-mol·K) (second term)
T = absolute temperature, K, corresponding to p, from Table 13.4
$T_0 = 310$ K
$\alpha^* - 1$, from Table 13.4
h, cm, from Fig. 13.6
$h\,dp/dz$, Torr, from Fig. 13.6
v, vapor velocity, cm/s, and p, pressure, Torr, are independent variables

With these values, Eq. (13.27) becomes

$$c_{D_2O}(\$/\text{g-mol}) = \frac{0.0213}{(\alpha^* - 1)^2} \frac{hT}{pv} + \frac{0.289}{(\alpha^* - 1)^2} \frac{h}{p} \frac{dp}{dz} \tag{13.29}$$

The first term gives the contribution of tower volume to the cost of heavy water; the second, the contribution of power.

Figure 13.7 represents these parts of the cost of heavy water from (13.29) as a function of vapor velocity, for pressures of 200, 420, 760, and 1245 Torr. Conditions that lead to minimum cost are listed in Table 13.11.

The above loss in availability is equivalent to 8.5 kWh/g D_2O. Because this takes account only of tower pressure drop, and does not include the loss in availability associated with temperature drop across heat exchange surface in reboilers and condensers, it is apparent that power consumption in water distillation is appreciably higher than in hydrogen distillation. The cost of \$193/kg D_2O covers only the cost of tower packing and power loss associated with tower pressure drop. If account were taken of the cost of tower shells and foundations,

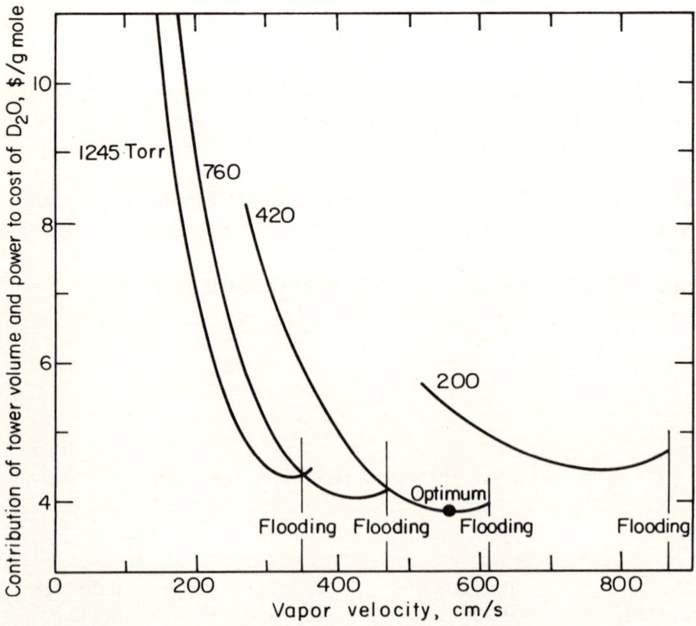

Figure 13.7 Contribution of tower volume and power to costs of heavy water made by distillation of water with Spraypak no. 37.

Table 13.11 Optimum conditions for water distillation plant using Spraypak no. 37

Pressure p	420 Torr
Vapor velocity v	552.6 cm/s
Separation factor α^*	1.033
Height transfer unit h	43 cm
Pressure drop dp/dz	0.0945 Torr/cm
Total tower volume in ideal cascade needed to produce 1 g-mol D_2O/s (1.73 Mg/day)	1.02×10^{11} cm^3
Loss in availability due to tower pressure drop	171 kWh/g-mol D_2O
Contribution to cost of D_2O	
Tower packing	\$1.30/g-mol D_2O
Power loss	2.56
Sum	\$3.86/g-mol D_2O
or	\$193/kg D_2O

reboilers, condensers, pumps, and other process equipment and other sources of power loss, the cost of producing heavy water by water distillation would be much greater than \$215/kg, U.S. Energy Research and Development Administration (ERDA)'s charge in 1977 for heavy water made by the GS process.

Holdup and start-up time. McWilliams and co-workers [M4] have found that the holdup of Spraypak no. 37 under these conditions is about 4 lb water/ft^3, or 0.064 g/cm^3. Consequently, the total water holdup of the columns of a plant producing 1 g-mol of D_2O/s would be $(0.064)(1.02 \times 10^{11})/18 = 3.63 \times 10^8$ g-mol water. From Eqs. (13.17), (12.199), and (12.203), the average deuterium content of the water inventory of this plant is

$$\bar{x}_E = \frac{\ln [x_P(1-x_F)/x_F(1-x_P)] - [(x_P - x_F)/x_P(1-x_F)]}{D_{P,F}/x_P}$$

$$= \frac{\ln [(0.998)(0.999851)/0.000149)(0.002)] - [(0.998 - 0.000149)/(0.998)(0.999851)]}{6729}$$

$$= 0.00209 \text{ atom fraction} \tag{13.30}$$

The increase in D_2O inventory during start-up of this plant would be

$$I_E(\bar{x}_E - x_F) = 3.63 \times 10^8 (0.00209 - 0.000149)$$

$$= 7.05 \times 10^5 \text{ g-mol } D_2O \tag{13.31}$$

The start-up time for this water distillation plant, evaluated from approximate Eq. (12.197), is

$$t = \frac{I_E(\bar{x}_E - x_F)}{P(x_P - x_F)} = \frac{7.05 \times 10^5 \text{ g-mol}}{1 \text{ g-mol/s}} = 7.05 \times 10^5 \text{ s} \tag{13.32}$$

or 8.17 days.

The low holdup and low start-up time is another advantage of Spraypak compared with bubble-plate columns.

Squared-off cascade. The preceding treatment of a water distillation plant as an ideal cascade operated at uniform vapor velocity has required that the steam flow rate be varied continuously as its deuterium content changes and that the number of towers in parallel, or the tower area, be changed continuously. On the other hand, a practical water distillation plant, like the Morgantown plant, will consist of a number of multiplate towers in parallel in the first group at the feed point, a smaller number in parallel in the second group at a higher deuterium content, a smaller number still or a smaller tower in the third group, and so on until at the product end a very small tower will suffice. A practical plant like this is characterized by uniform heads and tails flow over a large number of stages. Cohen [C9] has called such a plant a "squared-off" cascade and has developed general equations for it.

Figure 13.8 compares the variation of tails flow rate with stage number in a squared-off cascade with the variation in an ideal cascade performing the same job of separation in the same number of stages. Because the total flow rate in an ideal cascade is the lowest possible, the area under the stepped curve of the squared-off cascade is greater than under the smoothly tapered curve of the ideal cascade.

Consider a squared-off cascade making product containing x_P fraction deuterium at the rate P. An equation giving the number of stages n_{12} needed in a section of the plant that enriches the deuterium content of water from x_1 to x_2, with a uniform steam rate G, is obtained from Eq. (12.224):

$$n = \frac{1}{(\alpha^* - 1)b} \ln \frac{1+a}{1-a} \tag{13.33}$$

where

$$a = \frac{b(x_2 - x_1)}{(x_2 + x_1)(1+c) - 2x_1 x_2 - 2cx_P} \tag{13.34}$$

$$c = \frac{P}{G(\alpha^* - 1)} \tag{13.35}$$

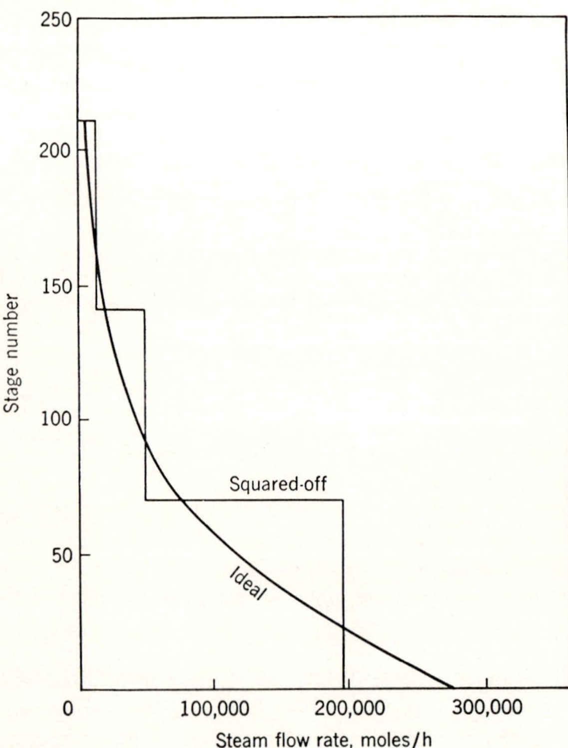

Figure 13.8 Steam flow rate versus stage number in ideal and squared-off cascades.

and $$b = [1 + 2c(1 - 2x_P) + c^2]^{1/2} \tag{13.36}$$

A much simpler result may be obtained when

$$\frac{x_1}{x_P} \ll 1 \tag{13.37}$$

and $$\frac{x_2}{x_P} \ll 1 \tag{13.38}$$

as is the case in the large columns near the feed end of a water distillation plant whose product is highly enriched. In this special but important case,

$$n = \int_{x_1}^{x_2} \frac{dx}{(\alpha^* - 1)x - (P/G)x_P} = \frac{1}{\alpha^* - 1} \ln \frac{g\omega - 1}{g - 1} \tag{13.39}$$

where $$\omega = \frac{x_2}{x_1} \tag{13.40}$$

the overall separation of the section, and

$$g = \frac{Gx_1(\alpha^* - 1)}{Px_P} \tag{13.41}$$

the ratio of the steam rate to the minimum steam rate at x_1.

Because the number of stages in the portion of an ideal cascade whose overall separation is ω is

$$n_{\text{ideal}} = \frac{2}{\alpha^* - 1} \ln \omega \tag{13.42}$$

the value of g that leads to the same number of stages in a square section as in an ideal cascade is given by

$$2 \ln \omega = \ln \frac{g\omega - 1}{g - 1} \tag{13.43}$$

or $$g = \frac{\omega + 1}{\omega} \tag{13.44}$$

This is close to the optimum value of g for a square cascade section, as will now be shown.

The volume of the section is given by

$$V = \frac{hRT}{pv} nG = \frac{hRT}{pv} \frac{G}{\alpha^* - 1} \ln \frac{g\omega - 1}{g - 1} \tag{13.45}$$

The optimum value of G for a section of a squared-off cascade is one that makes nG a minimum, with n given by (13.39).

We shall evaluate this optimum value and compare the minimum size of a section of a squared-off cascade with the portion of an ideal cascade performing the same job of separation. This will give us an idea of the penalty in increased equipment size paid by using practical towers with uniform vapor flow rates instead of the constantly changing flow rate of an ideal cascade.

The volume of the portion of an ideal cascade enriching deuterium from x_1 to x_2 is

$$V_{\text{ideal}} = \frac{hRT}{pv} \frac{4P}{(\alpha^* - 1)^2} \left[(2x_P - 1) \ln \frac{x_P(1 - x_1)}{x_1(1 - x_P)} + \frac{(x_P - x_1)(1 - 2x_1)}{x_1(1 - x_1)} \right. \tag{13.46}$$

(*Cont. on p. 736*)

$$- (2x_P - 1) \ln \frac{x_P(1-x_2)}{x_2(1-x_P)} - \frac{(x_P - x_2)(1-2x_2)}{x_2(1-x_2)} \Bigg] \qquad (13.46) \; (Cont.)$$

For the present special case, in which $x_2/x_P \ll 1$, this may be approximated by

$$V_{\text{ideal}} = \frac{hRT}{pv} \frac{4P}{(\alpha^* - 1)^2} \frac{x_P(x_2 - x_1)}{x_1 x_2} = \frac{hRT}{pv} \frac{4}{\alpha^* - 1} G_{\min} \frac{\omega - 1}{\omega} \qquad (13.47)$$

With V from (13.45), V_{ideal} from (13.47), n from (13.39), and g defined by (13.41), there results

$$\frac{V}{V_{\text{ideal}}} = \frac{\omega g}{4(\omega - 1)} \ln \frac{\omega g - 1}{g - 1} \qquad (13.48)$$

Figure 13.9 is a plot of V/V_{ideal} against g, the ratio of the steam flow rate to the minimum at x_1 for several values of the overall enrichment of the section ($\omega = x_2/x_1$). Figure 13.9 shows (1) that the optimum steam rate in a square section is less than the optimum in an ideal cascade, and (2) that the penalty in using a squared-off cascade is less than 15 percent so long as the overall enrichment of a section is under 4. At $x_2/x_1 = 4$, the optimum steam rate is 1.35 times the minimum, at point A in the figure. In practice, a somewhat lower steam rate would be used in order to reduce the size of reboilers and condensers. A point around B might be chosen, at $g = 1.25$, where the number of stages in a square section is equal to the number in the portion of an ideal cascade with the same overall enrichment [cf. Eq. (13.44)]. The tower volume and power consumption at this condition are 1.156 times those of an ideal cascade, and costs are higher by the same factor.

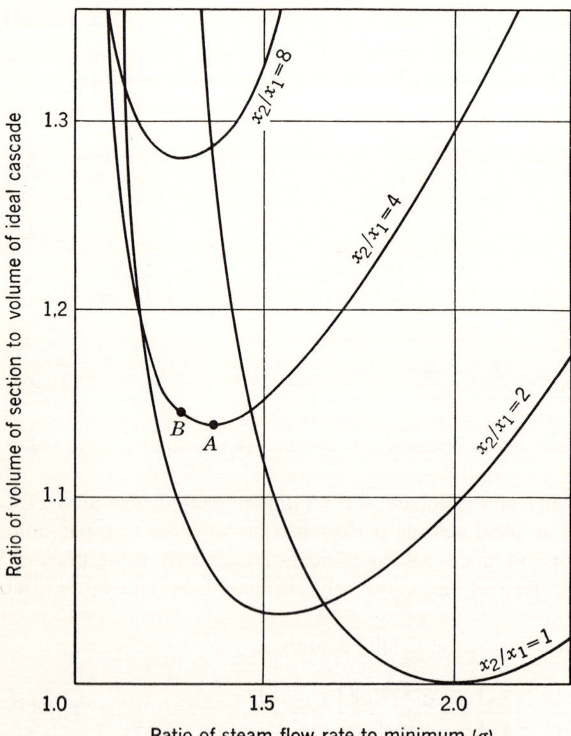

Figure 13.9 Volume of square section relative to ideal cascade.

Figure 13.10 Example of Sulzer CY packing for water distillation columns. Diameter, 250 mm.

These equations for a squared-off cascade and those previously given for an ideal cascade were used to work out the variation of steam flow rate with stage number shown in Fig. 13.8 for an overall enrichment per section $x_2/x_1 = 4$ and a steam flow rate (g) 1.25 times the minimum. The following conditions have been assigned to each cascade:

$$P = 1 \text{ mol/h}$$

$$x_P = 0.998$$

$$x_F = 0.000149$$

$$\alpha^* = 1.04$$

The portion of each cascade up to $x = 0.0093$ has been shown.

5.2 Final Concentration of Deuterium by Distillation of Water

Although water distillation is not competitive with other processes for primary concentration of deuterium from natural water, owing to the high energy consumption and the large number of towers used by this process, it is the preferred method for final concentration of deuterium from water previously enriched to several percent deuterium. In this high-enrichment range, energy consumption or equipment size is only a small fraction of that needed for primary concentration, and the reliability and simplicity of water distillation make it the process preferred for final concentration. Almost all final concentration water distillation plants installed since 1960 have been built by Sulzer Brothers, of Winterthur, Switzerland, including the final concentration sections of plants 1, 10, 12, 13, 15, 19, and 20 of Table 13.2.

Water distillation is also generally used for purifying D_2O that may have become contaminated in use by H_2O through dilution or DTO through neutron absorption.

The design of Sulzer water distillation plants for these purposes, described in references [B2], [D1], [H7], [M5], and [Z1], has evolved through several stages. The most recent plants use Sulzer CY packing made of copper, chemically treated to improve wettability. As shown in Fig. 13.10, this packing consists of parallel strips of wire mesh with oblique corrugations, arranged vertically. The slopes of adjacent strips are in opposite directions. The packing is fabricated in cylindrical cartridges about 160 mm high. Successive cartridges are turned 90° from adjacent ones. Because of the 90° displacement and the oblique corrugations, the gas stream is well mixed at each elevation, and the liquid trickles down in a zigzag motion. After flowing through 3 to 4 m of packing, the liquid is collected, mixed, and redistributed over the top of the next packed section. Because of these means to keep gas and liquid well mixed, no increase in transfer unit height has been noted in column diameters up to 2 m.

Figure 13.11 shows the principal characteristics of Sulzer CY packing for water distillation service [M6]. The optimum throughput is said to be at 75 percent of flooding, at which the F factor is 1.7. At this load, the gas-phase pressure drop is about 4 Torr/m, the liquid holdup is about 6 percent of the packed volume, and the observed height of a transfer unit (htu) has been found to be between 6.5 and 12 cm. The observed variations in htu are attributed to variations in the wetting of the packing, which is impaired by traces of oil and other hydrophobic impurities in the water.

At an F factor of 1.7, the water vapor throughput at several pressures is as follows:

Pressure, Torr	Temperature, °C	Vapor density, kg/m³	Throughput, kg/(m²·h)
60	41.5	0.0551	1437
120	55.4	0.106	1993
240	70.6	0.203	2757
360	80.3	0.296	3330
760	100.0	0.598	4733

5.3 Separation of ^{18}O by Distillation of Water

Water distillation has been used by Dostrovsky [D4] to produce 13 g/day of $D_2{}^{18}O$ containing 99.8 percent ^{18}O from natural water containing 0.204 a/o ^{18}O. Figure 13.12 shows external flows between stages in this water distillation plant; reboilers and condensers are not shown. Table 13.12 summarizes process conditions in this plant. This plant has the stepped-down tapered shape characteristic of a squared-off cascade.

The columns are packed with Dixon [D3] rings made from 100-mesh phosphor bronze wire gauze. The columns are operated at a mean pressure of approximately 0.5 atm, with a pressure drop of 130 Torr. Under these conditions, Dostrovsky and co-workers [D5] have found the ^{16}O-^{18}O separation factor to be 1.0064, and the height of a transfer unit in the larger columns to be about 2 cm.

6 ELECTROLYSIS

6.1 Electrolysis of Water

History of process. Until 1943, all the heavy water produced commercially was made by electrolysis. The largest single producer of heavy water was the Norsk Hydro Company, which operated the world's largest electrolytic hydrogen plant at Rjukan, Norway. In 1942, this plant was making about 1.5 MT of heavy water per year as a by-product of the production of 17,300 nm³ of electrolytic hydrogen per hour, used for ammonia synthesis. The average power consumption of this plant was 91,000 kW, or 5.2 kWh/nm³ of hydrogen.

The primary plant at Rjukan made water containing 15 a/o deuterium. The electrolytic cells were of the Pechkranz [M2] type, with steel cathodes and diaphragms to prevent mixing of hydrogen and oxygen. Nine stages of parallel-connected cells were used, with the number of

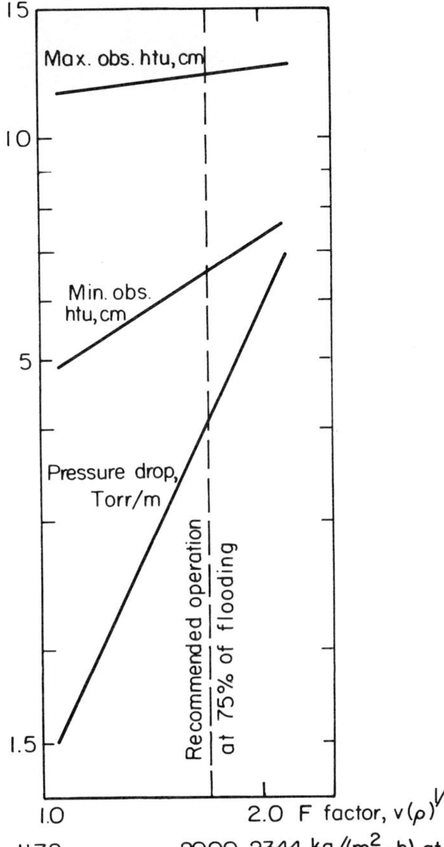

Figure 13.11 Characteristics of Sulzer CY packing for water distillation service.

cells per stage decreasing as the deuterium content increased. The stages were connected in a series cascade, without recycle of partially enriched hydrogen, and the cascade was operated in steady flow. A schematic flow sheet for this kind of plant is shown in Fig. 13.13. About 73 percent of the water fed to each stage was electrolyzed; and 27 percent was carried from the stage by the products of electrolysis as water vapor, condensed, and fed to the next higher stage of the cascade. The fraction of water fed forward was controlled by the vapor pressure of water; 27 percent forward feed requires an electrolyte temperature of 60°C.

The product of the primary plant was refined to pure D_2O in a small, nine-stage secondary plant, also operated with steady flow, but with the partially enriched hydrogen burned and recycled, as shown in Fig. 13.14. The secondary electrolytic plant has since been replaced by a water distillation plant.

During World War II heavy-water production at Rjukan was increased by addition of steam-hydrogen deuterium exchange equipment, to be described in Sec. 7. In 1943 operation was interrupted by a series of commando raids, but production was resumed after the war and was at the rate of 6.5 Mg/year in 1975 [R3]. A second electrolysis and exchange plant at Glomfjord, Norway, was then producing 5.9 Mg/year.

The steady-flow electrolytic process without recycle, shown in Fig. 13.13, was also used at the plant of Emswerke AG at Ems, Switzerland, to produce 400 nm^3/h of hydrogen enriched

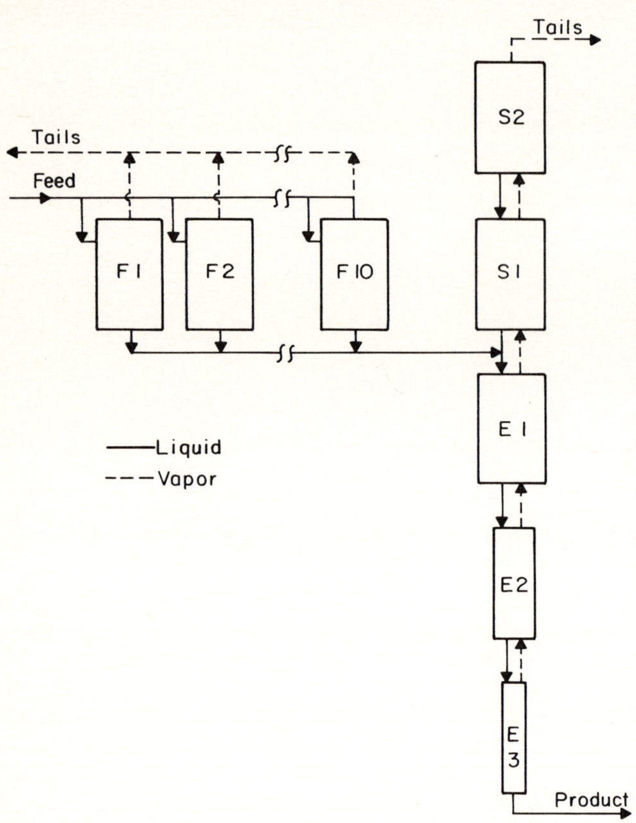

Figure 13.12 Dostrovsky's [D4] water distillation plant for concentration of ^{18}O.

sixfold in deuterium over natural abundance [H3] and is being used at the Indian government's fertilizer plant at Nangal, India, to produce 5,000 nm^3/h of hydrogen containing 3.1 times the natural abundance of deuterium [G1]. In each case the partially enriched hydrogen goes to a hydrogen distillation plant for final concentrations of deuterium, as was described in Sec. 13.4.

At the Manhattan District's heavy-water plant at Trail, British Columbia, primary concentration of deuterium was effected by the combination of electrolysis and steam-hydrogen

Table 13.12 Process conditions in ^{18}O water distillation plant

Columns	Diam., cm	Packed height, m	Packing diam., mm	Steam flow, kg/day	Percent ^{18}O at bottom
F1 to F10	10	9.5	3	175	1.6
S1 and S2	10	9.5	3	170	
E1	10	9.5	3	170	6.4
E2	6.3	9.5	4	35	60.0
E3	3.2	10	4	5	99.8

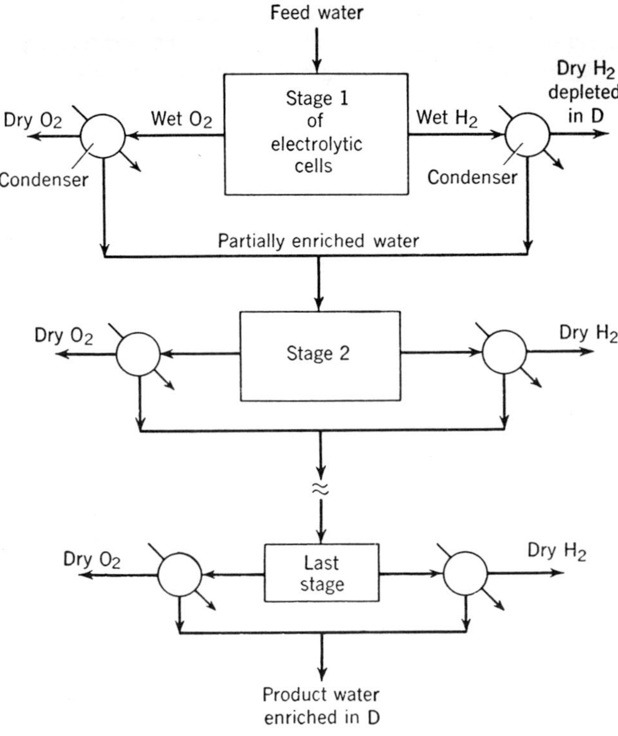

Figure 13.13 Steady-flow cascade of electrolytic cells, without recycle.

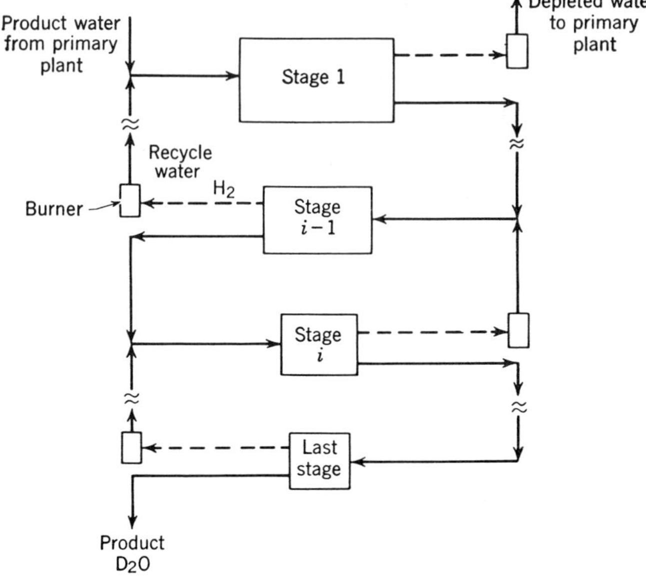

Figure 13.14 Steady-flow recycle cascade of electrolytic cells.

exchange, to be described in Sec. 7. The plant made use of hydrogen produced electrolytically by the Consolidated Mining and Smelting Company for ammonia synthesis. This was the largest electrolytic hydrogen plant in North America. In 1945 the average hydrogen production rate was 14,000 nm^3/h, almost as great as at Rjukan. The electrolytic cells used at Trail have been described by Mantell [M2]. The cells were operated with steady flow. Because the principal means for isotope separation in the primary plant at Trail was by the exchange process rather than by electrolysis, no special efforts were made to obtain a high separation factor in the primary plant.

The electrolytic process was also used by the Manhattan District, at Morgantown, West Virginia, and at Trail, British Columbia [M8], to refine crude heavy water from a primary plant where some process other than electrolysis was employed. These electrolytic plants were operated batchwise. The cells had no diaphragm, so the product was a mixture of hydrogen and oxygen. The gases were recombined in a burner, and the water was recycled to the primary plant when its deuterium content was leaner than primary-plant product or to the next batch of the electrolytic plant when its deuterium content was richer than primary-plant product.

Details of the Manhattan District's secondary electrolytic plants are given by Maloney et al. [M8].

Batch electrolysis was used to concentrate deuterium from 90 to 99.87 percent at the large Savannah River heavy-water plant of the U.S. Atomic Energy Commission, at Aiken, South Carolina [B7, B8], but this final concentration step is not needed when the plant is operated at reduced capacity.

Separation factors. Deuterium separation factors in the electrolytic plants described above, together with the types of cells used and operating conditions that may have had an effect on separation factor, are listed in Table 13.13. Separation factors of from 6 to 10 have been reported for the secondary plants, and from 3.8 to 7.0 for the primary plants. The lower values for the primary plants are attributed to their higher cell temperatures, their use of diaphragms, and the greater difficulty of keeping large equipment clean.

In a detailed laboratory investigation of the effect of cell variables on the deuterium separation factor in electrolysis of water, Brun and co-workers [B13] have found that α depends on the cathode material, electrolyte composition, and cell temperature, generally as follows. The separation factor is higher for an alkaline electrolyte than for an acid. With KOH, at 15°C, a pure iron cathode gave the highest value reported, 13.2. The separation factor for mild steel, the material used in most commercial electrolyzers, was 12.2. Values as low as 5 were reported for tin, zinc, and platinized steel. At 25°C the separation factor with a steel cathode was 10.6, and at 75°C it had dropped to 7.1.

Because the equilibrium constant for the reaction

$$H_2O(l) + HD(g) \rightleftharpoons HDO(l) + H_2(g)$$

is 3.81 at 25°C and 2.95 at 75°C, it is evident that the much higher separation factors obtained in electrolysis must be due to some mechanism other than establishment of equilibrium in this reaction at the cathode surface. One plausible explanation is that the hydrogen ion is discharged more readily at the cathode than the deuterium ion.

6.2 Analysis of Electrolysis

In this analysis of electrolysis, the somewhat optimistic assumption will be made that a separation factor of 7 can be obtained at a cell voltage of 2.1. At 95 percent current efficiency, the power consumption per gram-mole of water electrolyzed is then

$$\frac{(2)(96{,}501\text{ C/g-mol})(2.1\text{ V})}{[3{,}600{,}000\text{ (C}\cdot\text{V)/kWh}](0.95)} = 0.118\text{ kWh/g-mol} \tag{13.49}$$

Table 13.13 Separation factors in electrolytic heavy-water plants

Type of concentration	Primary (steady flow)				Secondary (batch)		
Plant	Rjukan	Ems	Nangal	Trail	Trail	Morgantown	Savannah River
Type of cell	Pechkranz	Oerlikon	De Nora	Trail	Special	Special	Special
Cell diaphragm	Yes	Yes	Yes	Yes	No	No	No
Cathode	Steel	Armco iron	Steel	Steel	Steel	Steel	Steel
Electrolyte							
Material	KOH	KOH	KOH	KOH	KOH	K_2CO_3	K_2CO_3
Initial w/o†		28	29	28	2.5	7.5	6–8
Final w/o					15	15	
Temperature, °C	60	80–82	65–75	60–70	23	40	20
Voltage	2.1	2.2	2.13	2.1	2.6	2.6–3.4	2.5–3.5
A/cm^2				0.072			0.07–0.14
Separation factor	6.6	6.5–7.0	6.36‡	3.8	6.9–9.7	6.0–8.2	8.5–10
Reference	[T6]	–	[G1]	[M2, M8]	[M8]	[M8]	[B7]

† w/o, weight percent.
‡ Predicted.

or

$$\frac{0.118 \text{ kWh/g-mol}}{0.022415 \text{ nm}^3/\text{g-mol}} = 5.3 \text{ kWh/nm}^3 \tag{13.50}$$

of hydrogen produced.

Let us first consider the production of heavy water in a simple cascade of electrolytic cells, without recycle, as in Fig. 13.13. Such a cascade, used at Ems and Nangal, preconcentrates deuterium prior to final concentration by distillation of hydrogen. With a sufficient number of stages, such a cascade could be used to produce pure heavy water in low yield from natural water.

If the heads separation factor β is constant throughout such a simple cascade, the fraction of deuterium that may be recovered depends on the number of stages n and the overall enrichment ω in accordance with

$$r_n = \left(\frac{\alpha - \omega^{1/n}}{\alpha - 1}\right)^n \tag{13.51}$$

as has been shown in Eq. (12.50).

As an example of the recovery of deuterium obtainable in a simple electrolytic cascade without recycle, production of heavy water containing 99.693 a/o deuterium from natural water containing 0.0149 a/o will be considered. The overall enrichment ω is

$$\omega = \frac{x_P(1 - x_F)}{x_F(1 - x_P)} = \frac{(0.99693)(0.99851)}{(0.000149)(0.00307)} = 2{,}176{,}168 \tag{13.52}$$

A high, but attainable, overall separation factor of $\alpha = 7$ will be used. A heads separation factor such as the one that would be used in an ideal recycle cascade of $\beta = \sqrt{\alpha} = \sqrt{7} = 2.646$ will be assumed. Then the number of stages n is given by

$$n = \frac{\ln \omega}{\ln \beta} = 2\,\frac{\ln \omega}{\ln \alpha} = 2\,\frac{\ln 2{,}176{,}168}{\ln 7} = 15.00 \tag{13.53}$$

The recovery of deuterium, from Eq. (13.51), is

$$r = \left[\frac{7 - (2{,}176{,}168)^{1/15}}{7 - 1}\right]^{15} = 0.00816 \tag{13.54}$$

The reason for this low recovery, of course, is that most of the deuterium is carried off by the hydrogen produced during electrolysis, and only a small fraction is left in the residual water.

The maximum amount of heavy water that could be produced in this way as the by-product of 10,000 g-mol of hydrogen would be

$$(10{,}000)(0.000149)(0.00816) = 0.0122 \text{ g-mol} \tag{13.55}$$

or 0.244 g. The electric power consumption is 1180 kWh, or 4836 kWh/g D_2O. Because electric power costs of the order of \$0.02/kWh, this corresponds to \$100/g D_2O. It is evident that a profitable use must be made of the hydrogen, because the value of the heavy water is only a small fraction of the cost of power.

The recovery of deuterium can be increased substantially by burning deuterium-rich hydrogen from the upper stages of the plant and recycling the water to the electrolytic cells, as in Fig. 13.14. Figure 13.15 shows a generalized flow sheet for such a plant, with hydrogen from the lower stages being used as the principal plant product and with hydrogen from the upper stages being burned and recycled to increase recovery of heavy-water by-product. The principal variables in such a flow sheet are

1. The increase in deuterium content taken per stage
2. The deuterium content y_m at and above which hydrogen is burned and recycled

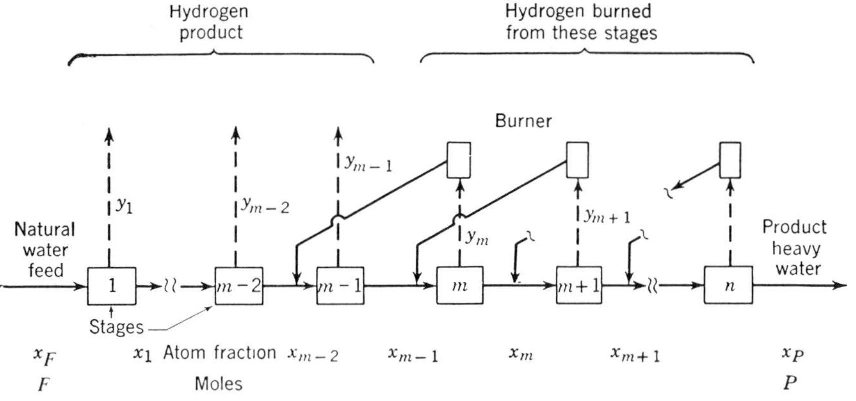

Figure 13.15 Electrolytic plant with hydrogen burned and recycled on upper stages.

The increase in deuterium content per stage is measured by the heads enrichment factor β, defined in terms of the atom fraction deuterium in the heads water leaving this stage x_m and the atom fraction deuterium in the water leaving the next lower stage x_{m-1}, as

$$\beta = \frac{x_m(1 - x_{m-1})}{x_{m-1}(1 - x_m)} \tag{13.56}$$

In stages $m - 1$ and higher, the optimum enrichment per stage is that of an ideal cascade, in which

$$x_{m-1} = y_{m+1} \tag{13.57}$$

and

$$\beta = \sqrt{\alpha} \tag{13.58}$$

We shall use condition (13.58) to set the enrichment per stage in the lower stages in which hydrogen is not recycled, also. The total number of stages n is then given by Eq. (13.53).

As a specific example, we may consider a plant designed to produce 10,000 g-mol of hydrogen per minute, while recovering as a by-product as much heavy water as can be economically justified. The atom fraction D in feed water x_0 will be taken as the natural value of 0.000149, and the atom fraction D in the heavy-water product x_P will be taken as 0.99693 (to make the number of stages exactly 15.00). The deuterium content of hydrogen and water leaving the lower stages of this cascade is given in Table 13.14.

It evidently would not pay to burn and recycle hydrogen from stages 1 and 2, because it is no richer in deuterium than feed. To determine at which stage it would pay to begin to burn and recycle hydrogen, a number of alternative flow sheets have been worked out, with the most significant results summarized in the last four columns of Table 13.14. In the first case listed, hydrogen is burned and recycled on stage 3 and all higher stages, in the second case on stage 4 and higher, etc. In each case, the unburned hydrogen production rate has been held constant at 10,000 mol. For each case there has been evaluated:

1. The deuterium recovery, from (13.51)
2. The moles of heavy water produced, P
3. The moles of hydrogen burned and recycled, H, the total tails flow of the recycle portion of the cascade, from (12.119)

Table 13.14 Electrolytic plants for production of 10,000 mol of hydrogen, with heavy water as by-product ($\alpha = 7, \beta = \sqrt{7}$, 15 stages)

Stage no. (m)	a/o deuterium in stage product: Hydrogen $100y_m$	Water $100x_m$	Deuterium recovery	Moles D_2O product, P	Moles hydrogen burned and recycled from this and higher stages, H	$\Delta H/\Delta P$	Value of hydrogen and oxygen consumed to produce incremental D_2O, \$/kg D_2O
Feed		0.0149					
1	0.00563	0.0394					
2	0.0149	0.1042					
3	0.0394	0.2752	0.439	0.655	1550		
						6539	520
4	0.1042	0.725	0.327	0.488	458		
						2469	196
5	0.2752	1.89	0.238	0.3542	127.6		
						936	74
6	0.725	4.86	0.172	0.256	35.7		
						361	28.7
7	1.89	11.88	0.124	0.1856	10.3		
						59	4.7
15	97.89	99.693	0.00816	0.0122	0.0		

To determine which of these cases is best economically, it is necessary to set a value on the hydrogen that is burned and therefore lost. A representative value for hydrogen produced commercially by reforming naphtha is \$1.50/1000 ft^3; an electrolytic plant would probably not be built unless it could sell hydrogen for this figure. Because 0.5 mol of oxygen is consumed per mole of hydrogen burned, it is necessary to set a value on this, too. A value of \$20/short ton, or \$0.80/1000 ft^3, will be used. The value of hydrogen and oxygen consumed is therefore \$1.90/1000 ft^3 or \$1.59/kg-mol of hydrogen burned.

The next-to-the-last column of Table 13.14 gives $\Delta H/\Delta P$, the ratio of the additional hydrogen that must be made to the additional heavy water produced when hydrogen from an additional stage is burned and recycled. The last column gives the value of the additional hydrogen and the associated oxygen required to produce 1 kg of additional D_2O. This is obtained as $(\$1.59/20)(\Delta H/\Delta P)$. Because of the high value of \$196/kg of incremental D_2O made by burning hydrogen from stage 4, it is evident that it would not be economical to burn hydrogen from this stage. The incremental value of \$74/kg for hydrogen from stage 5 is under the cost of heavy water in competing processes. Burning hydrogen from stage 5 to increase heavy-water production therefore might be justified.

The average cost of hydrogen and oxygen from stage 5 and higher burned to make heavy water in the most favorable case is

$$\frac{(127.6)(1.59)}{(0.3542)(20)} = \$28.6/\text{kg } D_2O \tag{13.59}$$

Although the need for efficient condensers and the complications of connecting electrolytic cells in series cascade would add something to the cost, it is evident that electrolysis provides a way of making small amounts of heavy water at a very low cost as a by-product of hydrogen and oxygen.

Figure 13.16 is a flow sheet for a plant for the case in which hydrogen from stage 5 and higher is burned and recycled. The fraction of deuterium in the feed that is recovered is only 0.238. This low recovery is characteristic of the electrolytic process when used as the sole means of concentrating deuterium. As a result, the amount of heavy water that could be produced by electrolysis alone, even at a large electrolytic plant, is small.

Although the recovery of heavy water is better than in the simple cascade without recycle,

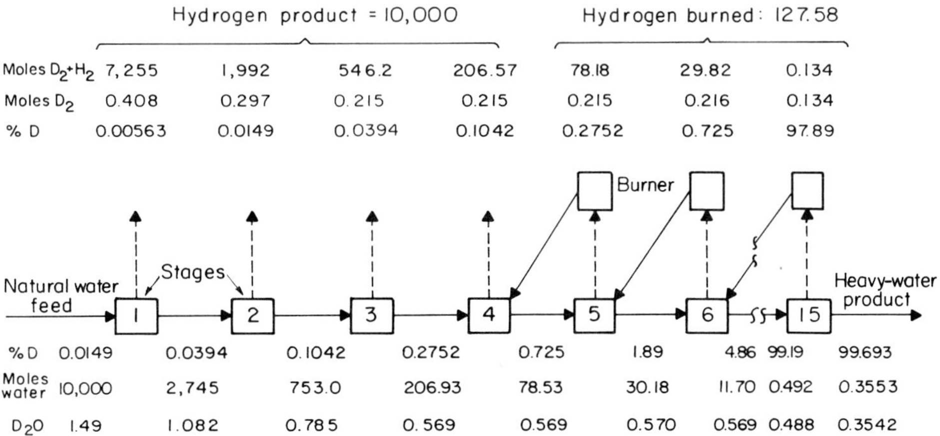

Figure 13.16 Optimum electrolytic cascade for production of 10,000 mol hydrogen and heavy-water by-product. $\alpha = 7; \beta = \sqrt{7}$.

the power consumption per unit of heavy water is still very large. At 0.118 kWh/g-mol of water electrolyzed, the plant of Fig. 13.16 would consume

$$\frac{(0.118)(10{,}127)}{(0.3542)(20)} = 169 \text{ kWh/g } D_2O \tag{13.60}$$

In Fig. 13.15, stages 1 to $m-2$ constitute a simple cascade, without recycle, and the remaining stages, from $m-1$ to n, constitute an ideal, recycle cascade. We shall show how flow quantities may be derived for this flow sheet.

The deuterium content of water heads leaving stage m is

$$x_m = \frac{\beta^m x_F}{\beta^m x_F + 1 - x_F} \tag{13.61}$$

The deuterium content of hydrogen tails from the same stage is

$$y_m = \frac{\beta^{m-2} x_F}{\beta^{m-2} x_F + 1 - x_F} \tag{13.62}$$

Compositions in Fig. 13.16 were obtained in this way.

The total amount of hydrogen formed from stages m to n is given by (12.119), for the total tails flow in the enriching section of an ideal cascade, with x_{m-1} replacing z_F. In the plant shown in Fig. 13.16, the total number of moles formed is

$$K_E = \frac{0.3553}{\sqrt{7}-1}\left\{\frac{0.997(\sqrt{7}+1)-\sqrt{7}}{\ln\sqrt{7}} \ln\left[\left(\frac{0.997}{0.00725}\right)\left(\frac{0.993}{0.00313}\right)\right] + \left[\frac{0.997-0.007}{(0.00725)(0.993)}\right]\left[\frac{7-(8)(0.00725)}{\sqrt{7}-1}\right]\right\} = 127.58 \tag{13.63}$$

Hydrogen from stage $m-1$ constitutes tails from the ideal cascade section, whose quantity relative to product is given by

$$W = \frac{P(x_P - x_{m-2})}{x_{m-2} - y_{m-1}} \tag{13.64}$$

In the plant shown in Fig. 13.16, the tails quantity is

$$W = \frac{0.3553(0.99693 - 0.002752)}{0.002752 - 0.001042} = 206.57 \tag{13.65}$$

The feed rate to stage 4 is $W + P = 206.93$.

In Fig. 13.15, the first $m-2$ stages constitute a simple cascade, operated without recycle, with constant heads separation factor β. The recovery of deuterium from a simple cascade of $m-2$ stages operated at constant β is

$$r_{m-2} = \left(\frac{\alpha-\beta}{\alpha-1}\right)^{m-2} \tag{13.66}$$

from Eq. (12.48).

In the flow sheet of Fig. 13.16, the recovery of deuterium from the three stages of the simple cascade is

$$r_3 = \left(\frac{7-\sqrt{7}}{7-1}\right)^3 = 0.38220 \tag{13.67}$$

The number of moles of natural water fed to stage 1 required to get 206.93 mol of water

Table 13.15 Separation factors in electrolysis

Isotopes	Solute in water	Electrode	Separation factor	Reference
$^{6}Li/^{7}Li$	LiCl	Hg	1.055	[J4]
$^{14}N/^{15}N$	NH_4Cl	Hg	1.008	[H10]
$^{18}O/^{16}O$	KOH	Ni	1.036	[T6]
$^{41}K/^{39}K$	KCl	Hg	1.0054	[H9]

containing 0.2752 percent deuterium from stage 3 is then

$$F = \frac{(206.93)(0.002752)}{(0.38220)(0.000149)} = 10{,}000 \tag{13.68}$$

6.3 Electrolytic Separation of Other Elements

Separation factors in electrolysis for other elements are much lower than for hydrogen. A few values that have been reported are listed in Table 13.15. These values are so low, and the cost of electric energy per unit electrolyzed is so high, that electrolysis is uneconomical for separating isotopes of any element other than hydrogen. Some concentration of ^{18}O takes place in an electrolytic deuterium plant.

7 ELECTROLYSIS AND STEAM-HYDROGEN EXCHANGE

7.1 Principle of Process

In the electrolytic cascade shown in Fig. 13.16, 76.2 percent of the deuterium in the water fed leaves with the hydrogen product at too low a concentration for economical recovery by electrolysis, even though over half of the deuterium in the hydrogen product is at or above the natural abundance. Some of the deuterium in this hydrogen may be recovered economically by the steam-hydrogen exchange process. The principle of this process may be seen through an example. Consider the effect of mixing hydrogen from stage 3 of Fig. 13.16 containing 0.0394 percent deuterium with an equal volume of steam containing the natural abundance of deuterium, 0.0149 percent, and passing the mixture over a catalyst at 80°C in which the exchange equilibrium

$$HD + H_2O \rightleftharpoons H_2 + HDO$$

is established. Because the equilibrium constant for this reaction is 2.8 at this temperature, the deuterium content of hydrogen and steam will be changed as follows:

	Percent deuterium	
	Before equilibration	After equilibration
Hydrogen	0.0394	0.0143
Steam	0.0149	0.0400

By cooling the mixture to condense the steam and separating the hydrogen and water, it is possible to transfer deuterium from hydrogen gas to water without burning the hydrogen, and thus to increase the heavy-water output from an electrolytic plant without having to sacrifice hydrogen production. This principle would be applicable not only to hydrogen slightly enriched in deuterium, as from stage 3 of Fig. 13.16, but to any hydrogen containing more than 1/2.8 of the natural abundance of deuterium, because deuterium would be transferred to natural water from such hydrogen.

The variation of this equilibrium constant with temperature is given in Table 13.16.

The exchange reaction proceeds at a negligible rate unless catalyzed, and the only catalysts available until recently lost activity in the presence of liquid water. It was therefore necessary to use a gas-phase catalytic reactor, as described in the previous example.

The recovery of deuterium from hydrogen by exchange with water could be increased over the single-stage example just cited by using a multistage countercurrent cascade. The simplest arrangement, consisting of a tower packed with catalyst through which liquid water flows down and gaseous hydrogen flows up, was not practical, because of the inactivation of catalyst mentioned above. One possible arrangement of gas-phase exchange reactors is shown in Fig. 13.17. In such a cascade each exchange reactor, with its associated evaporator, condenser, and separator, acts like a single plate of a distillation column. If the condenser condenses all the steam leaving a stage, the separation factor is simply the equilibrium constant k for the above exchange reaction, with a value around 2.8. In such a cascade, the electrolytic cell acts like a reboiler to provide hydrogen recycle for the exchange cascade. In fact, the additional enrichment of deuterium occurring in the electrolytic cell is not essential for the operation of the process, because any desired degree of enrichment could be obtained by using a sufficient number of exchange stages.

The maximum recovery of deuterium possible with such a cascade is achieved by increasing the number of stages indefinitely and reducing the reflux ratio of depleted hydrogen to product heavy water until the depleted hydrogen is in deuterium exchange equilibrium with feed. The deuterium recovery r is given by

$$r = 1 - \frac{Wy_W}{Fx_F} \tag{13.69}$$

Table 13.16 Equilibrium constant for steam-hydrogen exchange reaction $H_2O + HD \rightleftharpoons HDO + H_2$

	Equilibrium constant k	
Temperature, °C	Calculated [R8]	Observed [C1]
0	4.07	
25	3.52	
50	3.12	3.05
75	2.82	2.77
100	2.58	2.55
125	2.39	2.36
200	1.99	1.98
300	—	1.69
400	—	1.52
500	—	1.39
600	—	1.30
750	—	1.21

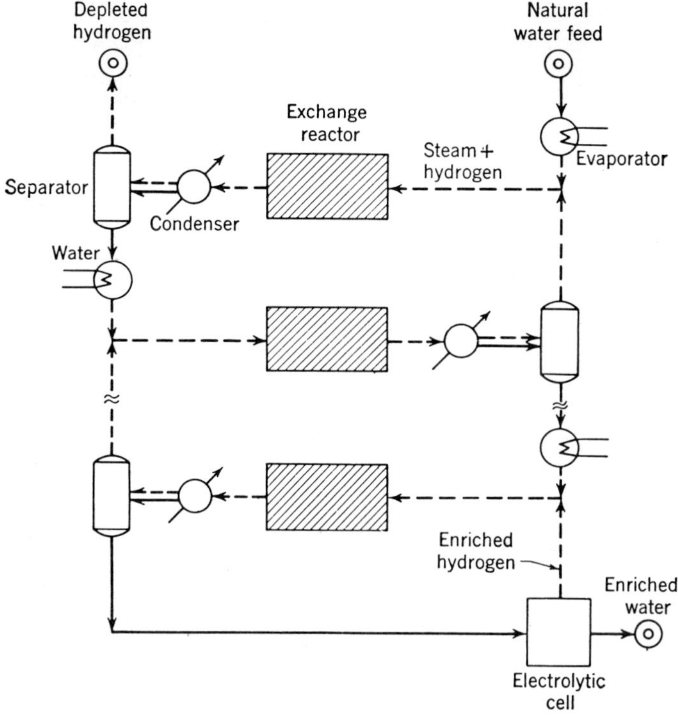

Figure 13.17 Cascade of exchange reactors.

where y_W and x_F are the atom fraction of deuterium in depleted hydrogen and feed, respectively, and W and F are the corresponding flow rates. At the low atom fractions in natural water and depleted hydrogen, W/F is very nearly unity, and y_W/x_F may be approximated by

$$\frac{y_W}{x_F} \approx \frac{1}{k} \tag{13.70}$$

Therefore, the maximum recovery of deuterium obtainable by steam-hydrogen exchange at 80°C is

$$r_{\max} \approx 1 - \frac{1}{2.8} = 0.64 \tag{13.71}$$

This is more than twice the maximum recovery attainable economically by electrolysis alone. The deuterium content of depleted hydrogen in equilibrium with natural steam is 0.0053 a/o.

7.2 History

Use of this vapor-phase deuterium exchange reaction between steam and hydrogen was proposed independently in 1941 by Harteck and co-workers in Germany and by Urey and co-workers in the United States as a means for recovering deuterium from electrolytic hydrogen. Harteck and Seuss [C3, G3] developed a supported nickel catalyst that caused the reaction to take place at an acceptable rate below 100°C, where the high value of the equilibrium constant favors high recovery. In 1942, a set of exchange reactors containing this catalyst was installed to treat hydrogen from the sixth stage of the Norsk Hydro electrolytic plant at Rjukan, Norway [S10], and additional reactors were planned for the fourth and fifth stages, but operation was interrupted by the war before this could be completed. After the war, catalytic reactors were installed at Norsk Hydro plants at Rjukan and Glomfjord, Norway, bringing their combined heavy-water production at one time to around 20 MT/year [B9].

In the United States, a similar vapor-phase, steam-hydrogen, deuterium exchange process was the first one selected for large-scale production by the Manhattan District [M8], and a heavy-water plant using this process was built at the synthetic ammonia plant of the Consolidated Mining and Smelting Company at Trail, British Columbia. Urey and co-workers at Columbia University developed a nickel-on-chromia catalyst, and Taylor and co-workers at Princeton University developed a platinum-on-charcoal catalyst, both of which were used in this plant. The exchange tower system used in the Trail plant was devised by Barr and co-workers [B5] of the Standard Oil Company of New Jersey, which was responsible for the basic design of the plant.

7.3 Trail Plant

The exchange cascade of Fig. 13.17 is impractical because a volume of steam equal to the volume of hydrogen has to be evaporated and condensed on every exchange stage. The Barr towers used in the Trail plant greatly reduced the heat load by permitting production and condensation of steam only once in an entire cascade. The principle of these towers is shown in Fig. 13.18. A gaseous mixture of steam and hydrogen flows up this column, passing alternately through a pair of bubble-plate absorption trays, a heater to vaporize entrained water, a chamber filled with catalyst, another pair of absorption trays, another catalyst chamber, and so on through the top pair of absorption trays. Water flows down the column through the top pair of absorption trays, then bypasses the catalyst chamber, and continues through the second pair of absorption trays, and so on through the bottom set of absorption trays. Each tower of the Trail plant contained 13 catalytic sections and 14 absorption sections.

In the gas flowing up through a catalyst chamber, deuterium is partially transferred from HD to HDO; as the gas next flows up through an absorption section, the HDO is partially absorbed from the gas phase by the downflowing water. The overall result is a transfer of deuterium from gas to liquid, so that as the gas flows up it is progressively depleted in HD, and as the liquid flows down it is progressively enriched in HDO.

Steam is produced and condensed only once for an entire tower, and the ratio of steam to hydrogen may be varied at will. Steam consumption is therefore only a small fraction of that of Fig. 13.17, but more catalytic stages are required for a given change in deuterium content.

The primary heavy-water plant at Trail consisted of four groups of exchange towers and electrolytic cells connected in countercurrent cascade. Figure 13.19 shows the flow through one such group. At the top of the tower, vapor is cooled in a condenser; most of the steam is condensed and combined with water from the next lower group of electrolytic cells. Hydrogen from the condenser is returned to the next lower group of towers. At the bottom of the tower, steam is generated by vaporizing part of the water in a reboiler; hydrogen is generated by dissociating part of the water in a group of electrolytic cells with diaphragms to separate hydrogen from oxygen. Upflowing vapor consists of this steam and hydrogen, together with hydrogen from the next higher tower. Water fed forward to the next higher tower is obtained as condensate from the hydrogen and oxygen gases leaving the electrolytic cells.

Towers are operated at a pressure close to atmospheric and a temperature around 70°C, at which the steam-hydrogen ratio of the vapor is optimum.

The four groups of exchange towers and electrolytic cells at Trail produced partially

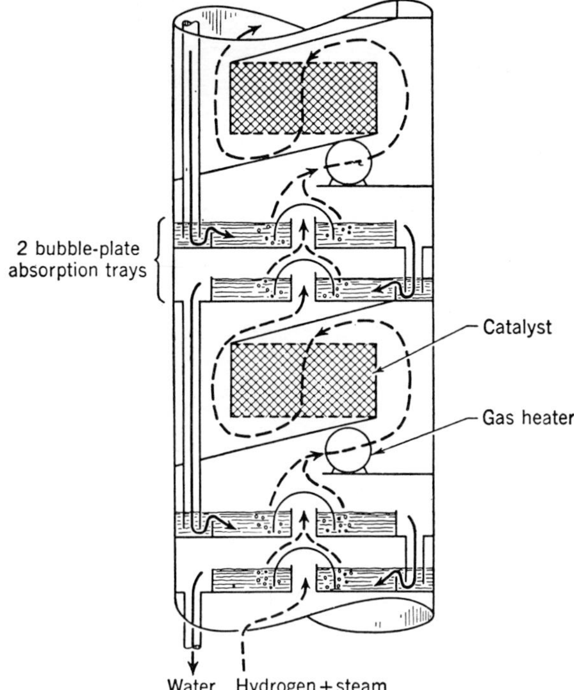

Figure 13.18 Section of exchange tower of Trail plant.

concentrated water containing 2.14 a/o deuterium. This water was concentrated further to 99.8 percent deuterium in the secondary electrolytic plant described in Table 13.13. The production rate was 6 Mg of heavy water per year.

Unit costs in 1945 were [M8]

Investment \$500/kg D_2O/yr
Operation \$60/kg D_2O

Production costs in 1954, including overhead and profit to CM & S, were \$133/kg D_2O [S3].

The Trail plant was started up in 1943 and began producing heavy water in 1944. It was shut down in 1956 because of the high cost of its heavy water compared with that produced by the GS process (Sec. 11).

Details of the Trail plant have been given by Maloney et al. in [M8].

7.4 Recovery of Deuterium from Electrolytic Hydrogen by Exchange with Liquid Water Under Pressure

The high cost of recovering heavy water from electrolytic hydrogen by exchange with steam is due largely to the cost of making and condensing steam and to the large mass of catalyst needed for this vapor-phase reaction at low pressure. These difficulties would be avoided if the reaction could be carried out at an acceptable rate in the presence of liquid water. Becker [B10] developed a colloidal platinum-on-charcoal catalyst, suspended in liquid water, which was circulated in countercurrent flow to gaseous hydrogen through a conventional sieve-plate column. This catalytic exchange system was tested on a semicommercial scale at Dortmund,

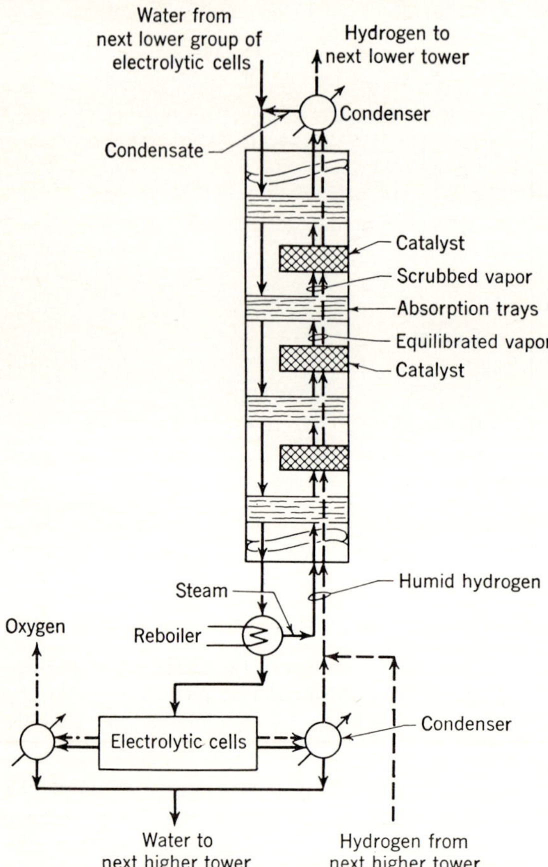

Figure 13.19 Flow sheet for one exchange tower and electrolytic cell group.

Germany [W1], in the early 1960s, using a dual-temperature flow sheet. Even with this finely divided catalyst and a pressure of 200 bar, the plate efficiency at 30°C was only around 1 percent. The resulting large column and catalyst volume made the process appear to be only marginally economical.

The low plate efficiency is due to the low solubility of gaseous hydrogen in liquid water, which results in a low mass-transfer coefficient for hydrogen to and from the catalyst surface, which is wetted by liquid water. Stevens [S8], in Canada, has recently developed a catalyst for the deuterium exchange reaction that is not wetted by liquid water and is much more active. The catalyst consists of nickel or platinum deposited on a conventional support such as silica gel and then coated with a thin layer of a water-repellent resin through which hydrogen can rapidly diffuse. Liquid water and gaseous hydrogen flow countercurrent through a column packed with particles of such a catalyst. The gas-phase deuterium exchange reaction between water and hydrogen takes place on the catalyst surface, while H_2O and HDO are transferred simultaneously between the gas and liquid phases. From experiments described in Stevens' patents, transfer-unit heights of around 1.5 m are predicted [H2] at a superficial gas velocity of 3 m/s evaluated at standard conditions of 0°C and 1 atm for actual conditions of 60°C and pressures in the range of 14 to 40 bar.

Figure 13.20 shows how exchange towers packed with such a catalyst permitting counterflow of hydrogen and water might be used to increase the recovery of deuterium from the cascade of electrolytic cells shown in Fig. 13.16. Exchange towers are used to reduce the deuterium content of hydrogen leaving electrolytic stages 2, 3, and 4 to 0.0563 a/o, the same value as in hydrogen leaving the first electrolytic stage. In this flow sheet it has been assumed that the exchange towers are so designed and operated that water and hydrogen leaving them have the same deuterium content as the corresponding streams leaving the electrolytic stages with which they are mixed. This ideal cascade condition minimizes exchange tower volume.

Comparison of Fig. 13.20 with Fig. 13.16 shows that the use of exchange towers would increase heavy-water production from 0.3553 to 0.930 mol and deuterium recovery from 23.8 to 62.2 percent.

To reduce the number of electrolytic stages to three, Hammerli and co-workers have suggested [H2] the flow sheet of Fig. 13.21. The big advantage of this flow sheet is the great simplification of interstage connections compared with Fig. 13.20. The principal disadvantage of Fig. 13.21 is its much greater catalyst volume. Hammerli [H1] estimates, however, that the cost of catalyst and catalyst towers for a flow sheet like Fig. 13.21 is only 15 percent of the cost of the electrolytic cells, so that it is cost-effective to simplify the flow sheet at the expense of increased catalyst volume. The dimensions of the catalyst towers of Fig. 13.21 for a superficial hydrogen gas velocity of 3 m/s at standard conditions are

Stage	1	2	3
Area, m^2	20.75	1.02	0.0287
Diameter, m	5.14	1.14	1.191
Packed height, m	9.7	11.1	15.8

In comparing Figs. 13.20 and 13.21, the following should be noted. Hydrogen product rates are substantially equal, 10,000 kg-mol/h. The total amounts of water electrolyzed are about the same. Use of an exchange tower on hydrogen from the first cell coupled with the closer approach to exchange equilibrium at the top of the towers of Fig. 13.21 permits reduction in deuterium content of depleted hydrogen from 0.0563 to 0.050 percent and

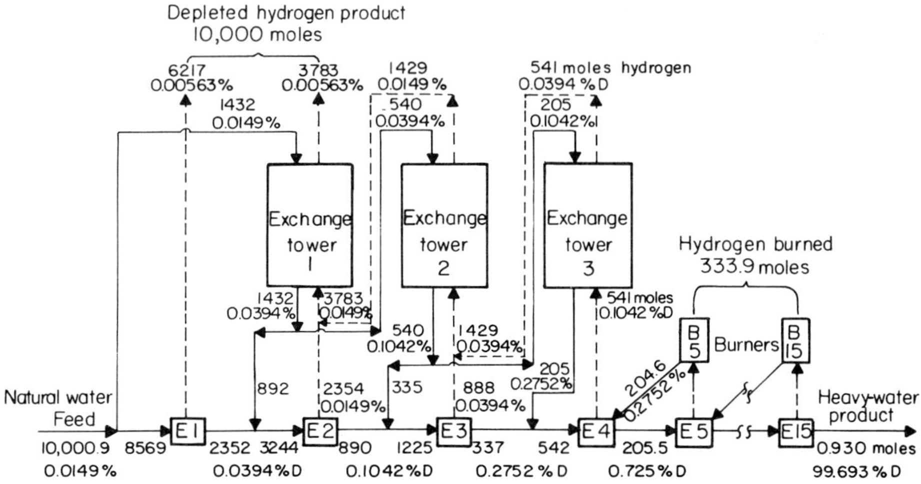

Figure 13.20 Cascade of electrolytic cells and exchange towers. E = electrolysis stage. —— water; --- hydrogen; flow units, kg-mol/h.

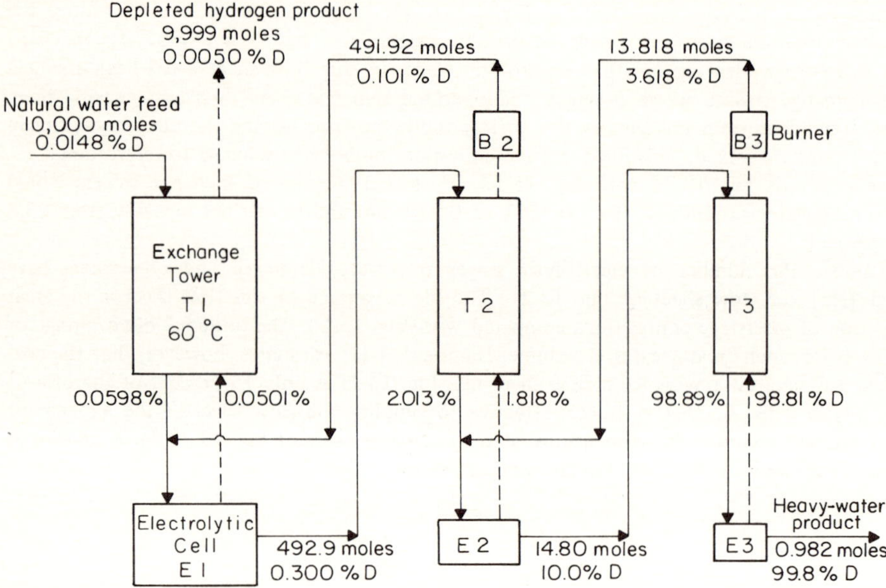

Figure 13.21 Three-stage cascade of electrolytic cells and exchange towers. —— water; --- hydrogen; flow units, kg-mol/h.

increases heavy-water production from 0.930 to 0.982 kg-mol/h. More separative work is performed in the exchange towers of Fig. 13.21 than in those of Fig. 13.20, primarily to compensate for the loss of separative work in Fig. 13.21 where the water recycled from each burner is mixed with water of quite a different composition from an exchange tower. Other factors increasing the separative work demand on the towers in Fig. 13.21 are the lower electrolytic separation factor of 6† used in that figure compared with 7 in Fig. 13.20 and the lower deuterium content of hydrogen product.

One possible difficulty with Fig. 13.21 is the much higher average deuterium content of water in the electrolytic cells compared with Fig. 13.20. This requires that cell leak rates and water holdup be kept small.

8 SEPARATION FACTORS IN DEUTERIUM EXCHANGE PROCESSES

The deuterium exchange reaction between water and hydrogen discussed in Sec. 7 is one of a group of deuterium exchange reactions that have been extensively studied and are the basis for most of the world's heavy-water production. Table 13.17 lists deuterium separation factors between liquid water and gaseous compounds of hydrogen for temperatures in the range 0 to 200°C. The ratio of the separation factor at 25°C to that at 125°C, α_{25}/α_{125}, is also given. The higher this ratio is, the greater is the fractional recovery of deuterium and the smaller is the number of stages needed in the dual-temperature exchange process to be described in Sec. 11.

†A flow sheet like Fig. 13.21 would concentrate deuterium even if electrolysis produced no separation at all.

Table 13.17 Separation factors in liquid-vapor deuterium exchange reactions involving water

Reactants			Products			Separation factor α						Ratio	
Liquid	Gas		Liquid	Gas	α/K	0°C	25°C	50°C	100°C	125°C	200°C	α_{25}/α_{125}	References
H_2O +	NH_2D	⇌	HDO +	NH_3	$\frac{3}{2}$	1.02	1.00	1.00	0.99	0.99	0.99	1.01	[K2]
H_2O +	PH_2D	⇌	HDO +	PH_3	$\frac{3}{2}$	2.71	2.44	2.27	2.04	1.96	1.78	1.24	[W4], [K2]
H_2O +	HDS	⇌	HDO +	H_2S	1	2.60	2.37	2.19	1.94	1.84	1.64	1.29	[C3]
H_2O +	DCl	⇌	HDO +	HCl	$\frac{1}{2}$	2.87	2.51	–	–	1.88	–	1.34	[K2], [U1]
H_2O +	DBr	⇌	HDO +	HBr	$\frac{1}{2}$	3.57	3.07	–	–	2.18	–	1.41	[K2], [U1]
H_2O +	DI	⇌	HDO +	HI	$\frac{1}{2}$	4.56	3.84	–	–	2.58	–	1.49	[K2], [U1]
H_2O +	HD	⇌	HDO +	H_2	1	4.53	3.81	3.30	2.65	2.43	1.99	1.57	[R8]

The reactions of Table 13.17 have been listed in order of increasing values for this ratio. Because water is one of the components of each pair in Table 13.17, processes based on these reactions could use liquid water as feed and thus would not be limited in output by limited feed availability.

Table 13.18 lists deuterium separation factors between gaseous hydrogen and liquid ammonia or methylamine, two compounds of hydrogen proposed for deuterium separation processes. The ratios of separation factors between the temperatures marked by a dagger, which have been proposed for dual-temperature processes based on these reactions, are also given. Both the separation factors and the separation factor ratios of the reactions involving hydrogen are greater than those involving water in Table 13.17. These higher values are what give the reactions of Table 13.18 their practical importance. A disadvantage of the reactions of Table 13.18 is that their deuterium production is limited to the amount present in commercially available hydrogen.

In all systems deuterium tends to concentrate in the phase normally liquid except ammonia-water at high temperature. Separation factors in chemical exchange are much higher than separation factors in distillation for the corresponding materials (cf. Table 13.3) except for ammonia-water. The high value of these separation factors and their strong dependence on temperature are what give the chemical exchange process its importance for separation of deuterium and isotopes of other light elements.

The deuterium exchange reaction between water and ammonia, water and hydrogen sulfide, or water and the hydrogen halides proceeds rapidly in the liquid phase without catalysis, because of ionic dissociation. In the case of a mixture of water and hydrogen sulfide, for example, the ionic equilibria

$$H_2O \rightleftharpoons H^+ + OH \qquad H_2S \rightleftharpoons H^+ + SH^-$$

$$HDO \rightleftharpoons H^+ + OD^- \qquad HDS \rightleftharpoons H^+ + SD^-$$

$$HDO \rightleftharpoons D^+ + OH^- \qquad HDS \rightleftharpoons D^+ + SH^-$$

permit rapid exchange of H^+ and D^+ between the two materials. Deuterium exchange between water and phosphine, water and hydrogen, ammonia and hydrogen, or methylamine and hydrogen does not proceed without catalysis. The water-phosphine reaction can be catalyzed by

Table 13.18 Separation factors in liquid-vapor deuterium exchange reactions involving hydrogen

Reactants	$NH_3 + HD$	$CH_3NH_2 + HD$
Products	$NH_2D + H_2$	$CH_3NHD + H_2$
α/K	$\frac{2}{3}$	1
Separation factor α at		
−50	6.6	7.90†
−25	5.19†	6.04
0	4.25	4.85
25	3.62	—
40	3.32	3.6†
50	3.15	—
60	2.99†	—
100	2.55	—
125	2.34	—
Ratio at †	1.74	2.19
Reference	[P1], [R4] averaged	[R7]

strong acids [W4], the water-hydrogen reaction by nickel or platinum-metal catalysts (see Sec. 7), the ammonia-hydrogen reaction by potassium amide dissolved in liquid ammonia [C2], and the methylamine-hydrogen reaction by potassium methylamide.

Solutions used in the ammonia-water, water-hydrogen, and ammonia-hydrogen processes are relatively noncorrosive and may be handled in ordinary steel equipment. Solutions used in all of the other processes are relatively corrosive, and require use of stainless steel or other expensive construction materials.

The constant-boiling mixtures formed by water and the hydrogen halides make it difficult to use these systems in a practical exchange process.

Of the reactions listed, the water-hydrogen sulfide case has the greatest practical importance because it needs no catalysis and has a fairly large change of separation factor with temperature. This case is discussed in detail in Sec. 11. The water-hydrogen reactions discussed in Sec. 7 and the ammonia-hydrogen and methylamine-hydrogen reactions, with their large separation factors and large change of separation factor with temperature, are also of practical importance.

In some cases the separation factors given in these tables have been determined experimentally from equilibrium constants K for gas-liquid reactions such as

$$H_2O(l) + HDS(g) \rightleftharpoons HDO(l) + H_2S(g)$$

In other cases, they have been derived from experimental measurements of the equilibrium constants k for gas-phase reactions such as

$$H_2O(g) + HDS(g) \rightleftharpoons HDO(g) + H_2S(g)$$

In still other cases gas-phase equilibrium constants have been computed by statistical mechanics from molecular properties. Procedures for calculating k have been described by Bigeleisen and Mayer [B12]. Varshavskii and Vaisberg [V1] have given a very extensive tabulation of values of k calculated for many deuterium exchange equilibria.

Expressions for the relation between K, k, and the chemical exchange separation factor α will now be derived. Let us consider first the exchange reaction between liquid water and gaseous hydrogen sulfide. As in distillation, the deuterium separation factor in the chemical exchange reaction is defined as the ratio of the abundance ratio of deuterium to light hydrogen in the liquid to the corresponding ratio in the vapor. In terms of the mole fractions of individual compounds in the liquid and vapor, the separation factor is†

$$\alpha = \frac{(x_{HDO} + 2x_{D_2O})/(2x_{H_2O} + x_{HDO})}{(y_{HDS} + 2y_{D_2S})/(2y_{H_2S} + y_{HDS})} \tag{13.72}$$

When the deuterium content of liquid and vapor is low, under a few percent, $x_{D_2O} \ll x_{HDO}$, $x_{HDO} \ll x_{H_2O}$, etc., so that the above equation reduces to

$$\alpha = \frac{x_{HDO}/x_{H_2O}}{y_{HDS}/y_{H_2S}} = K \tag{13.73}$$

When the deuterium content is appreciable, equilibrium constants for such reactions as

$$HDO(l) + D_2S(g) \rightleftharpoons D_2O(l) + HDS(g)$$

must be also taken into account. These do not greatly affect the value of α, however.

The equilibrium constant k for the gas-phase reaction is defined as

†In this equation the solubility of hydrogen sulfide in the liquid and the vaporization of water in the vapor have been neglected. These effects are treated in Sec. 11.

$$k = \frac{y_{HDO}/y_{H_2O}}{y_{HDS}/y_{H_2S}} \tag{13.74}$$

Because liquid-vapor exchange reaction is the resultant of vapor-phase exchange reaction and the vaporization equilibrium reaction

$$H_2O(l) + HDO(g) \rightleftharpoons HDO(l) + H_2O(g)$$

for which the equilibrium constant is the relative volatility α^*, defined by

$$\alpha^* = \frac{x_{HDO}/x_{H_2O}}{y_{HDO}/y_{H_2O}} \tag{13.75}$$

it follows that

$$K = k\alpha^* \tag{13.76}$$

so that

$$\alpha = k\alpha^* \tag{13.77}$$

In the more general exchange reaction

$$MH_m(l) + ZH_{z-1}D(g) \rightleftharpoons MH_{m-1}D(l) + ZH_z(g)$$

in which the liquid compound MH_m and the gaseous compound ZH_z contain different numbers of hydrogen atoms, the separation factor is related to the equilibrium constant by

$$\alpha = K\left(\frac{z}{m}\right) \tag{13.78}$$

Values of α/K have been listed in Tables 13.17 and 13.18.

9 NUMBER OF THEORETICAL STAGES IN EXCHANGE COLUMNS

This section derives a general equation for the dependence of stream compositions in an exchange column on stream flow rates and number of equilibrium stages. In Fig. 13.22, the vapor flow rate is V kg-mol of exchangeable element in unit time, and the liquid flow rate is L in the same units. In the present simplified derivation these flow rates are treated as constant throughout the column. Vapor compositions y and liquid compositions x are expressed as atom fraction of desired isotope of exchangeable element. To keep the derivation simple, atom fractions are to be restricted to values below 0.05, as are found in the large stages of plants to concentrate deuterium, ^{13}C, ^{15}N, or ^{18}O from the natural element.

The equilibrium relation between vapor and liquid leaving stage i is

$$x_i = \alpha y_i \tag{13.79}$$

where α is the separation factor. This convention is used to make α greater than unity for exchange separation of deuterium, the example of greatest practical importance.

The material-balance equation for the section between the top of the column and the top of stage i is

$$y_i = \frac{(x_{i-1} - x_0)L}{V} + y_1 \tag{13.80}$$

Hence

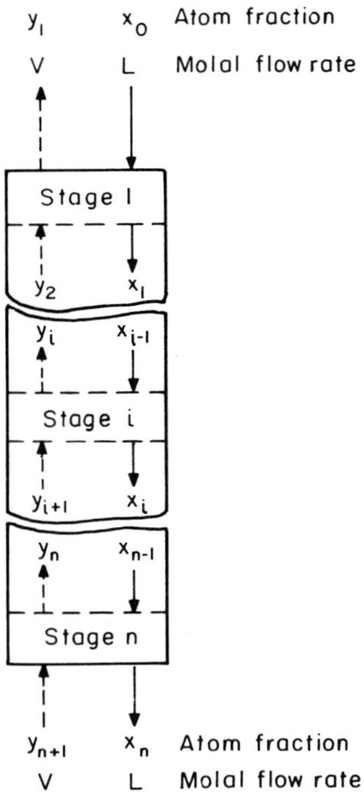

Figure 13.22 Flow rates and compositions in exchange column.

$$x_i = (x_{i-1} - x_0)\left(\frac{\alpha L}{V}\right) + \alpha y_1 \tag{13.81}$$

This is a first-order difference equation in x whose general solution is

$$x_i = AB^i + Z \tag{13.82}$$

Substitution into (13.81) gives

$$AB^i + Z = (AB^{i-1} + Z - X_0)\left(\frac{\alpha L}{V}\right) + \alpha y_1 \tag{13.83}$$

For this to hold, it is necessary that

$$B = \frac{\alpha L}{V} \tag{13.84}$$

and

$$Z = \frac{\alpha(x_0 L/V - y_1)}{\alpha L/V - 1} \tag{13.85}$$

A is obtained from (13.82) with $i = 0$,

$$x_0 = A + Z \tag{13.86}$$

so that

$$A = x_0 - Z = \frac{\alpha y_1 - x_0}{\alpha L/V - 1} \tag{13.87}$$

From (13.82),

$$x_i = \frac{(\alpha y_1 - x_0)(\alpha L/V)^i + \alpha(x_0 L/V - y_1)}{\alpha L/V - 1} \tag{13.88}$$

and

$$x_n = \frac{(\alpha y_1 - x_0)(\alpha L/V)^n + \alpha(x_0 L/V - y_1)}{\alpha L/V - 1} \tag{13.89}$$

From (13.80) and (13.89),

$$y_{n+1} = \frac{[(\alpha y_1 - x_0)(\alpha L/V)^n + x_0] L/V - y_1}{\alpha L/V - 1} \tag{13.90}$$

Hence

$$\alpha y_{n+1} - x_n = (\alpha y_1 - x_0)\left(\frac{\alpha L}{V}\right)^n \tag{13.91}$$

so that

$$n = \frac{\ln [(\alpha y_{n+1} - x_n)/(\alpha y_1 - x_0)]}{\ln (\alpha L/V)} \tag{13.92}$$

This is the general equation for the number of theoretical stages needed in exchange columns. It is a form of the Kremser [K5, S5] equation, derived originally for gas absorption.

10 MONOTHERMAL EXCHANGE PROCESSES

In addition to needing an exchange equilibrium constant different from unity, exchange processes for concentrating deuterium require a reflux-making step in which part of the deuterium in the liquid phase leaving an enriching column is transferred to the vapor phase returned to the column. This can be done either by a chemical-phase conversion operation in the monothermal exchange processes to be described in this section or by another exchange column at a higher temperature in the dual-temperature exchange processes to be described in Secs. 11 through 14.

The exchange towers of Fig. 13.21 are an example of a monothermal exchange process for concentrating deuterium, with the electrolytic cells providing reflux-making phase conversion. Because the equilibrium constant for the reaction

$$HD(g) + H_2O(l) \rightleftharpoons H_2(g) + HDO(l)$$

is 3.2 at 60°C, a flow sheet similar to this figure would permit concentration of deuterium even if no separation occurred in electrolysis.

Chemical reflux-making steps, such as the electrolysis of water in Fig. 13.21, cost more than thermal reflux-making steps such as the reboiler or condenser in distillation. Consequently, there are only a few examples of economical monothermal exchange processes for concentrating deuterium. Hydrogen-water exchange refluxed by water electrolysis as in Fig. 13.21 is one example that can be economical where electricity is cheap enough and electrolytic hydrogen valuable. The only other commercial example of production of deuterium by monothermal exchange is use of the ammonia-hydrogen exchange reaction

$$HD(g) + NH_3(l) \rightleftharpoons H_2(g) + NH_2D(l)$$

This was used in the exchange plant at Mazingarbe (item 12, Table 13.2) and is planned in plants in India (items 16 and 18). The economic attractiveness of this process comes from two factors:

1. The relative ease with which hydrogen reflux can be obtained by thermal cracking of ammonia
2. The relatively low cost of providing liquid ammonia reflux as incremental production of an existing synthetic ammonia plant

10.1 Monothermal Ammonia-Hydrogen Exchange

Figure 13.23 illustrates the principle of enriching deuterium by monothermal ammonia-hydrogen exchange in conjunction with a synthetic ammonia plant. The flow quantities have been developed from partial information reported for the Mazingarbe plant [L1, E1, N2]. Feed, at point (1), consists of ammonia synthesis gas, $3H_2/1N_2$, which has been purified to reduce its content of water, CO, and CO_2 to less than 1 ppm and compressed to the pressure of ammonia

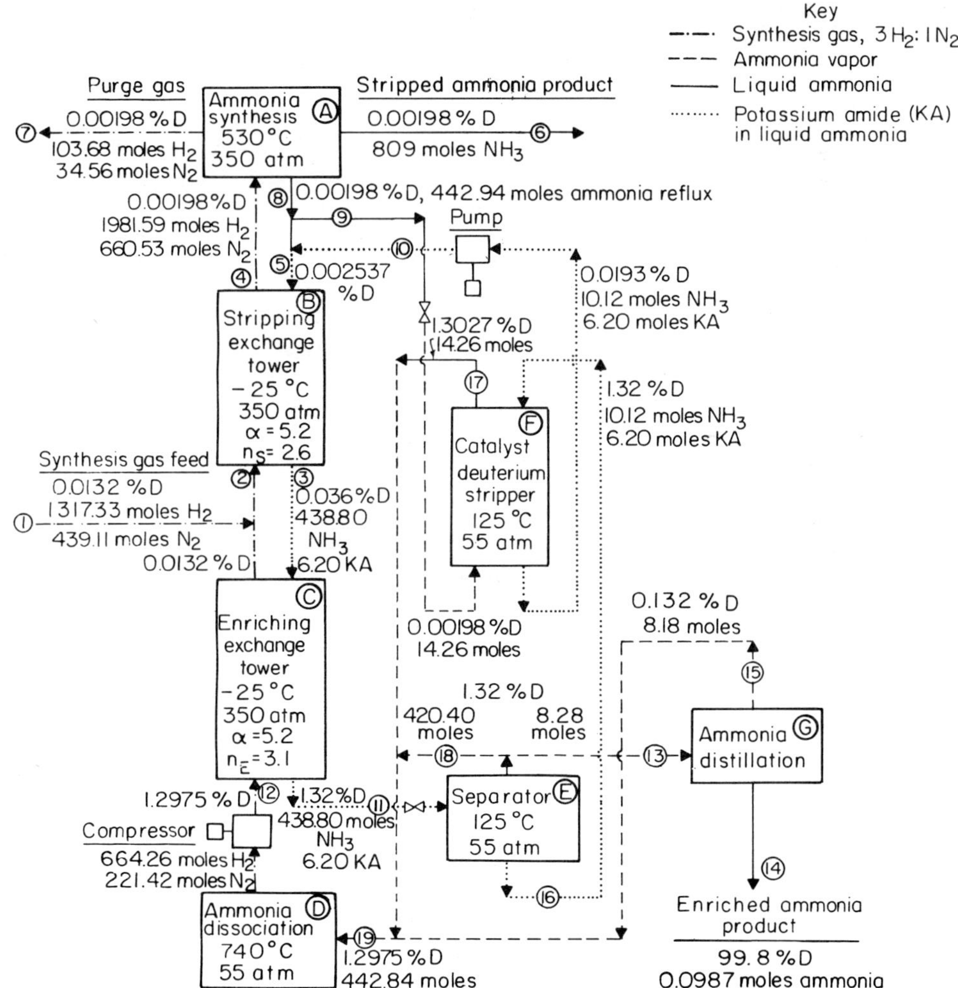

Figure 13.23 Monothermal ammonia-hydrogen exchange process. Flow quantities, kg-mol/h.

synthesis, 350 atm† in this example, as at Mazingarbe. At (2) feed joins additional synthesis gas circulating countercurrent to liquid ammonia in the stripping exchange column B and the enriching exchange column C, both operated at −25°C. At this temperature, the equilibrium constant for the foregoing deuterium exchange reaction is 5.2. At the countercurrent flow conditions of Fig. 13.23, deuterium is transferred from upflowing synthesis gas to downflowing ammonia. For the reported [E1] 85 percent recovery of deuterium, the atom percent deuterium in synthesis gas is reduced from 0.0132 percent in feed [L1] to 0.00198 percent in the gas leaving the stripping column (4). For the reported [L1] 100-fold enrichment, the ammonia leaving the enriching column (11) is enriched to 1.32 percent deuterium. Final concentration to 99.8 percent deuterium [N2] is by distillation of ammonia, G.

At Mazingarbe synthesis gas leaving the stripping column at (4) was converted to ammonia in the synthetic ammonia plant A of Houillères du Bassin du Nord et du Pas-de-Calais. In Fig. 13.23 about 5 percent of the synthesis gas is purged at (7) to remove inert impurities, mostly methane and argon, present in feed. About 65 percent of the ammonia synthesized in A is withdrawn as stripped ammonia product at (6) and the remainder is returned (8) to the exchange column as liquid reflux.

At the bottom of the enriching column C, a small portion of the partially enriched ammonia (13) is withdrawn for final concentration, and the remainder (19) is dissociated in D into partially enriched synthesis gas (12) to serve as reboil vapor for the enriching column. To obtain nearly complete dissociation of ammonia without using excessively high temperatures, the Mazingarbe plant is reported [L1] to have reduced the ammonia pressure in the cracking section to 55 atm. At this pressure and the assumed cracking temperature of 740°C, the ammonia content at equilibrium would be under 1 percent. For simplicity, the small amounts of ammonia in recycle gas (12) and hydrogen dissolved in recycle ammonia (3) have been neglected in Fig. 13.23.

To cause the deuterium exchange reaction between hydrogen and ammonia to take place at a useful rate, it is necessary to have 1 to 2 m/o (mole percent) of potassium amide, KNH_2, dissolved in the liquid phase to serve as a homogenous catalyst [C2]. Presence of potassium amide complicates the process for three reasons:

1. Potassium amide is expensive and must be recovered and recycled.
2. It remains in deuterium exchange equilibrium with ammonia:

$$KNHD + NH_3 \rightleftharpoons KNH_2 + NH_2D$$

3. It reacts, sometimes violently, with oxygen-containing impurities in synthesis gas.

To recover and recycle potassium amide, the partially enriched ammonia containing dissolved partially enriched potassium amide leaving the exchange column at (11) is depressured to 55 atm, heated to 125°C, and separated at E into ammonia vapor and a concentrated liquid solution of potassium amide (16). Because this solution contains 1.32 percent deuterium, its deuterium content must be reduced before it is recycled to the top of the stripping exchange column. In Fig. 13.23 this is done by countercurrent exchange with stripped ammonia vapor (9) in the catalyst deuterium stripper F. This is a conventional sieve-plate column in which deuterium is transferred from dissolved potassium amide to ammonia vapor in the overall reaction

$$KNHD(l) + NH_3(g) \rightleftharpoons KNH_2(l) + NH_2D(g)$$

which proceeds rapidly with an equilibrium constant near unity. The deuterium-depleted catalyst solution leaving F is repressured to 350 atm and returned (10) to the top of the

†1 atm = 1.01325 bar.

stripping exchange column B. At Mazingarbe, catalyst stripping was done with synthesis gas feed.

Potassium amide reacts with water, CO_2, CO, and oxygen, forming solid impurities that would plug the columns. With oxygen it forms potentially explosive potassium azide, KN_3. To prevent loss of exchange catalyst and formation of undesirable reaction products in the exchange system, synthesis gas feed is purified ahead of (1) by two systems not shown. It is dried by molecular sieves and then passed through a guard chamber containing sacrificial KNH_2 dissolved in liquid ammonia to remove oxygen-containing impurities [N2].

The flow quantities of synthesis gas feed (1), stripped ammonia product (6), and enriched ammonia product (14) were calculated from the net ammonia production rate of 330 MT/day [E1] reported at Mazingarbe, the reported [N2] heavy-water production rate of 26 MT of 99.8 percent D_2O per year at 85 percent recovery [E1], and the 0.0132 percent deuterium in feed [L1]. The reflux rates of ammonia (3) to column B and synthesis gas (12) to column C and the number of theoretical stages were evaluated in unpublished design studies at Massachusetts Institute of Technology which led to a requirement of $n_S = 2.6$ theoretical stages in stripping column B and $n_E = 3.1$ theoretical stages in enriching column C. These values were obtained from the Kremser equation (13.92). The number of stages in the stripping section, n_S, is

$$n_S = \frac{\ln [(\alpha y_2 - x_3)/(\alpha y_4 - x_5)]}{\ln (\alpha L_3/V_2)} \tag{13.93}$$

The number of stages in the enriching section, n_E, is

$$n_E = \frac{\ln [(\alpha y_{12} - x_{11})/(\alpha y_2 - x_3)]}{\ln (\alpha L_{11}/V_{12})} \tag{13.94}$$

In these equations y is the atom fraction deuterium in hydrogen, and x is the atom fraction deuterium in the solution of potassium amide in ammonia at the numbered points in Fig. 13.23. V is the molar flow rate of hydrogen in the vapor phase, and L is the molar flow rate of hydrogen in the liquid solution of KNH_2 in ammonia. For example, $L_3 = \frac{3}{2}(438.80) + 6.20 = 664.4$.

The small number of theoretical stages is a consequence of the high value of the separation factor, 5.2, and is the principal advantage of the monothermal ammonia-hydrogen exchange process. There are, however, a number of offsetting disadvantages. Even in the presence of KNH_2 catalyst, the rate of the exchange reaction is low, primarily because of the low solubility of hydrogen in liquid ammonia. Even at the high pressure of 350 atm used in Fig. 13.23, the hydrogen content of the liquid is only 0.85 m/o. With conventional sieve-plate or bubble-plate columns the plate efficiency would be only 1 or 2 percent, necessitating use of hundreds of plates. The Mazingarbe plant is reported [L1] to have used special ejectors for upflow of gas to entrain liquid and increase interphase transfer area. Because of the pressure drop taken by the upflowing gas, it was necessary to pump the liquid from one plate to the plate next below. Even with this enhanced contacting, it was necessary to use towers 36 m high [E1].

The Mazingarbe exchange plant produced its first heavy water in January 1968. It was taken out of service in 1972 because of an explosion in the ammonia synthesis plant, which has not been repaired. Operation of the exchange plant itself was satisfactory; the availability factor was 92 percent in 1970. Lefrancois stated [L1] that an operating temperature of $-10°C$ in the exchange towers would have been preferable to the design temperature of $-25°C$.

10.2 Monothermal Water-Hydrogen Sulfide Exchange

The deuterium exchange reaction between water and hydrogen sulfide,

$$H_2O(l) + HDS(g) \rightleftharpoons HDO(l) + H_2S(g)$$

proceeds rapidly without catalysis and has an equilibrium constant of 2.32 at 32°C. A monothermal process using this reaction, thus, could concentrate deuterium without the need for the complicated catalyst-recovery steps used in the ammonia-hydrogen exchange process, Fig. 13.23. Moreover, a water-hydrogen sulfide exchange plant can use natural water as feed and thus, unlike hydrogen-fed processes, is not limited in capacity to the amount of deuterium present in other industrial operations.

Despite these advantages of the water-hydrogen sulfide deuterium exchange reaction, it is not economical to use it in a monothermal flow sheet to produce heavy water because of the high cost of chemical reflux in this system. This may be shown by reference to Fig. 13.24.

In this monothermal, water-hydrogen sulfide flow sheet, natural water is fed at the top of a bubble-plate exchange column, and the water becomes progressively enriched in deuterium as it flows down the column in countercurrent contact with upflowing hydrogen sulfide gas. Heavy-water product is drawn off the bottom of the column and hydrogen sulfide gas depleted in deuterium is drawn off the top.

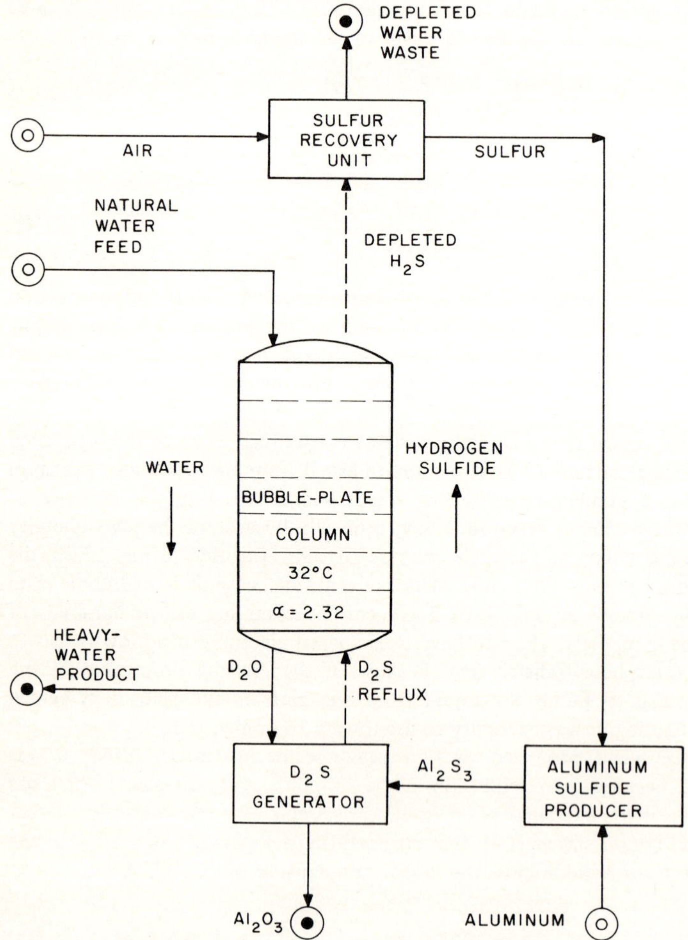

Figure 13.24 Example of reflux by chemical conversion for water-hydrogen sulfide exchange process.

The critical and essential feature of this flow sheet is the D_2S generator at the bottom of the column in which deuterium is transferred from D_2O to D_2S to provide reflux. Various means for effecting this chemical transfer can be imagined; all are costly. The means assumed here is the reaction between water and aluminum sulfide,

$$3D_2O + 2Al_2S_3 \rightleftharpoons 3D_2S + Al_2O_3$$

Aluminum sulfide may be made by reacting aluminum metal with sulfur:

$$2Al + 3S \rightarrow Al_2S_3$$

The sulfur needed for this step may be reclaimed from the depleted hydrogen sulfide leaving the top of the column by partial combustion with air:

$$H_2S + \tfrac{1}{2}O_2 \rightarrow H_2O + S$$

The overall effect is to separate natural water into D_2O and water depleted in hydrogen, with reflux provided by consumption of aluminum metal and production of aluminum oxide. Sulfur and hydrogen sulfide circulate internally and are not consumed by the process. The minimum molar ratio of D_2S reflux G to D_2O product P may be evaluated from Eq. (12.80):

$$\left(\frac{G}{P}\right)_{\min} \approx \frac{x_P - x_F}{x_F}\frac{\alpha}{\alpha - 1} = \frac{1}{0.000149}\frac{2.32}{1.32} = 11{,}800 \tag{13.95}$$

Because this reflux ratio is much lower than the reflux ratio in the distillation of water derived in Eq. (13.11), the towers of a hydrogen sulfide exchange plant could be much smaller in diameter than the towers of a water distillation plant. Because the separation factor for the exchange process (2.32) is much greater than for water distillation (~1.05), the towers could contain a much smaller number of plates.

However, the cost of providing chemical reflux is so high as to preclude the use of the flow sheet of Fig. 13.24 for heavy-water production. From the preceding chemical reactions it is seen that $\frac{2}{3}$ mol of aluminum metal is consumed for each mole of D_2S reflux. Because aluminum metal costs around \$0.50/lb, the minimum cost of aluminum (MW = 27) per pound of heavy-water product (MW = 20) is

$$\frac{(11{,}800)(\tfrac{2}{3})(27)(\$0.50)}{20} = \$5310/\text{lb } D_2O \tag{13.96}$$

Even without allowing for the additional costs of the conversion operations themselves, this is clearly prohibitive. Other possible chemical conversion schemes are similarly uneconomical.

11 DUAL-TEMPERATURE WATER–HYDROGEN SULFIDE EXCHANGE PROCESS

11.1 Principle of Process

To circumvent the high cost of chemical reflux, Geib [C3] in Germany and Spevack [S6] in the United States conceived of the dual-temperature system for providing reflux by purely physical means. The principle of the dual-temperature process using the water-hydrogen sulfide reaction is shown in Fig. 13.25. The cold tower of Fig. 13.25 performs the same function as the tower of Fig. 13.24; it operates at 32°C with a separation factor of 2.32 in this example, and it enriches deuterium from feed concentration to product concentration by exchanging deuterium from upflowing hydrogen sulfide to downflowing water.

The D_2S reflux needed for the cold tower is provided by the hot tower. This operates at a

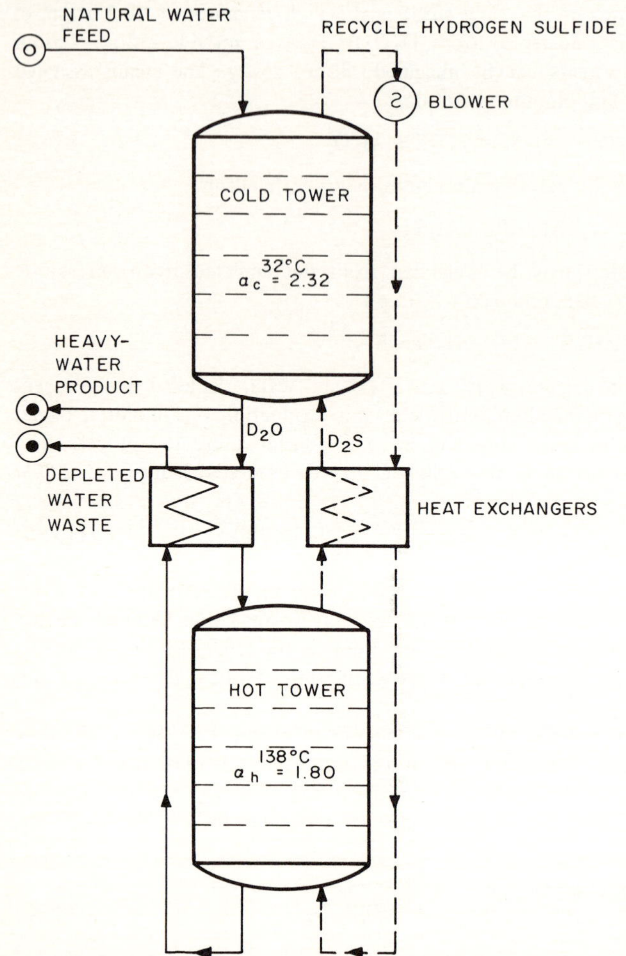

Figure 13.25 Dual-temperature reflux for water-hydrogen sulfide process.

high temperature, 138°C in this example, at which the separation factor is lower, 1.80 in this example.

With a proper flow ratio of hydrogen sulfide to water, this lower separation factor makes possible transfer of deuterium from water to hydrogen sulfide in the hot tower and thus converts the H_2S entering the hot tower into the D_2S needed for refluxing the cold tower.

Hydrogen sulfide is conserved by returning depleted H_2S from the top of the cold tower to the bottom of the hot. Heat is conserved by heat exchange between hot and cold liquid and between hot and cold vapor. In principle, no materials other than feed water are consumed in the dual-temperature system; energy consumption can be reduced by efficient heat exchange, with a lower bound set by the minimum required by thermodynamics for the separation.

The detailed manner in which the dual-temperature system effects separation will be explained in Sec. 11.3. That separation is possible can be made plausible by the simple qualitative considerations of Fig. 13.26. This represents one vessel containing cold water and a second containing hot water through which water flows in series and through which hydrogen sulfide may be recirculated.

Deuterium exchange equilibrium at the appropriate temperature is established between the hydrogen sulfide leaving each vessel and the water contained in it. The separation factor in the cold vessel α_c is greater than that in the hot vessel α_h. The deuterium-to-hydrogen abundance ratio in the water in the cold vessel is related to the abundance ratio in the gas leaving the cold vessel by

$$\xi_c = \alpha_c \eta_c \tag{13.97}$$

Similarly, the abundance ratio in the water in the hot vessel is related to the abundance ratio in the gas leaving the hot vessel by

$$\xi_h = \alpha_h \eta_h \tag{13.98}$$

Imagine that water containing the normal abundance ratio of deuterium to hydrogen ξ_F is started flowing through the system before hydrogen sulfide is charged. At this time product, waste, and feed water all have the same deuterium abundance ξ_F. Then assume that the hydrogen sulfide is charged and its circulation started. Because $\xi_c = \xi_h$ at this time, because $\alpha_c > \alpha_h$, and because of Eqs. (13.97) and (13.98), $\eta_h > \eta_c$; that is, the hydrogen sulfide leaving the hot vessel is richer in deuterium than that leaving the cold vessel. Therefore, there will be a net transport of deuterium from the hot vessel to the cold vessel; when a steady state is reached, ξ_c must be greater than ξ_F, and ξ_h must be less than ξ_F. Partial separation of the deuterium in the feed is effected. Addition of more cold and more hot contacting stages, as in Fig. 13.25, makes possible more complete separation.

11.2 History

The exchange reaction between water and hydrogen sulfide was one of a number of reactions investigated by Urey and co-workers at Columbia University from 1940 to 1943 for possible

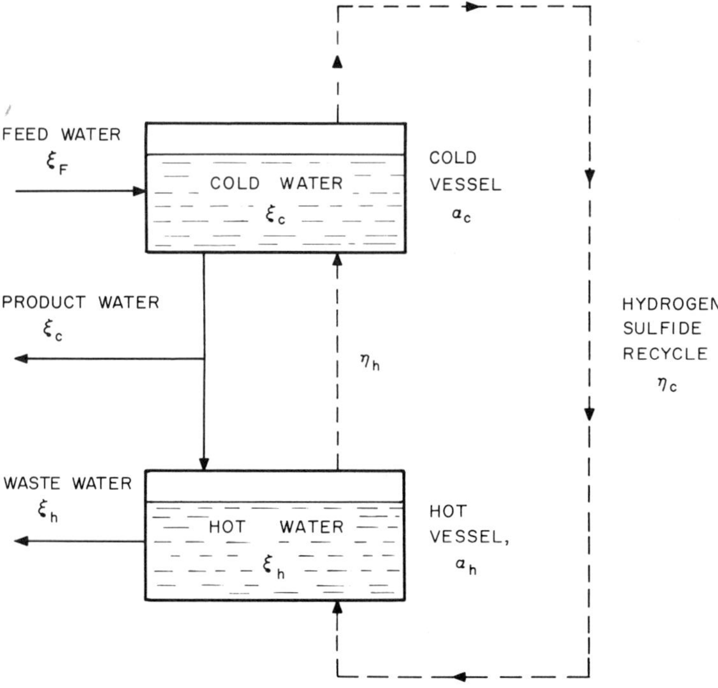

Figure 13.26 Simplified illustration of dual-temperature principle.

use by the Manhattan District for heavy-water production. During this time, Spevack [S6] conceived and patented the dual-temperature process and suggested its use with the water-hydrogen sulfide system. Because of concern about corrosion by aqueous solutions of hydrogen sulfide, the process was not used by the Manhattan District.

In 1949, when the need for large amounts of heavy water for the Savannah River reactors of the U.S. AEC was recognized, E. I. du Pont de Nemours and Company selected this process as the most economical means for producing heavy water on the large scale then required.

Spevack [S7] had developed improvements in the process that reduced its energy consumption, and corrosion research established where it was necessary to use stainless steel and where carbon steel could be used without undue corrosion by hydrogen sulfide. Under du Pont direction the Girdler Corporation designed a plant to produce heavy water at Dana, Indiana, where some of the equipment formerly used for the Manhattan District's water distillation plant was available. The process came to be known as the GS process, for Girdler-Sulfide. Lummus designed and du Pont built a second GS plant at Savannah River, of about the same capacity as the Dana plant. Both plants came into operation in 1952. By 1957, production rates were 490 MT/year at Dana and 480 MT/year at Savannah River. At this time the demand for heavy water began to decrease; the Dana plant was shut down and dismantled, and two-thirds of the GS units at Savannah River were shut down and put into standby condition. In 1977 the production rate from the operating portion of the Savannah River plant was 69 MT/year.

At both Dana and Savannah River the GS process was used for primary concentration of deuterium to 15 percent, with the remaining concentration being effected by distillation of water and electrolysis.

Pilot-plant investigations of the GS process have been carried out in France [R4] and in Sweden [E2], and a thorough analysis of the process has been published by Weiss [W3].

11.3 Simplified Analysis of Process

To show the main features of the GS process, a simplified analysis is first given, in which the complications introduced by the solubility of hydrogen sulfide in liquid water and the vaporization of water into hydrogen sulfide gas are neglected. The effects of the solubility of hydrogen sulfide and the volatility of water on the process are considered in Sec. 11.7. Figure 13.27 shows the flow of gas and liquid assumed and the nomenclature to be used. Figure 13.28 is a McCabe-Thiele diagram for the process. The analysis is formally similar to that given for solvent extraction with constant distribution coefficients in Chap. 4.

To simplify the treatment further, only low deuterium abundances are considered, so that the atom fractions of deuterium in liquid x and in vapor y in the streams leaving stage i are related by

$$y_{ci} = \frac{x_{ci}}{\alpha_c} \tag{13.99}$$

in the cold tower and

$$y_{hi} = \frac{x_{hi}}{\alpha_h} \tag{13.100}$$

in the hot tower. These are the equations for the equilibrium lines of the McCabe-Thiele diagram, which pass through the origin with slope $1/\alpha_c$ and $1/\alpha_h$.

For the cold tower, the overall deuterium material balance is

$$F(x_P - x_F) = G(y_P - y_F) \tag{13.101}$$

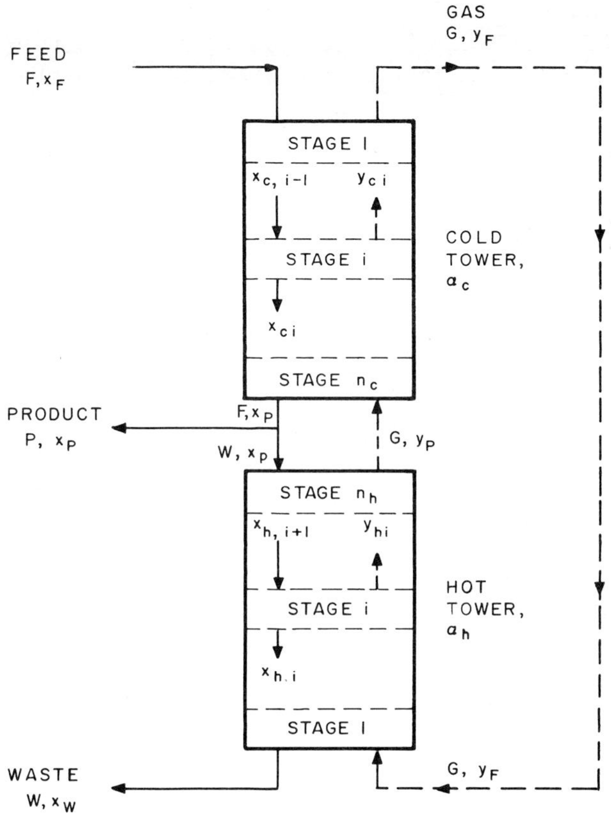

Figure 13.27 Nomenclature for simplified treatment of dual-temperature process.

The material balance above stage i is

$$F(x_{c,i-1} - x_F) = G(y_{ci} - y_F) \tag{13.102}$$

or

$$y_{ci} = y_F + \frac{F}{G}(x_{c,i-1} - x_F) \tag{13.103}$$

or because of (13.101),

$$y_{ci} = y_F + \frac{y_P - y_F}{x_P - x_F}(x_{c,i-1} - x_F) \tag{13.104}$$

This is the equation for the operating line in the cold tower, which passes through the points (y_F, x_F) and (y_P, x_P) and has the slope

$$\frac{F}{G} = \frac{y_P - y_F}{x_P - x_F} \tag{13.105}$$

Similarly, in the hot tower, the equation for the operating line is

$$y_{hi} = y_F + \frac{W}{G}(x_{h,i+1} - x_W) \tag{13.106}$$

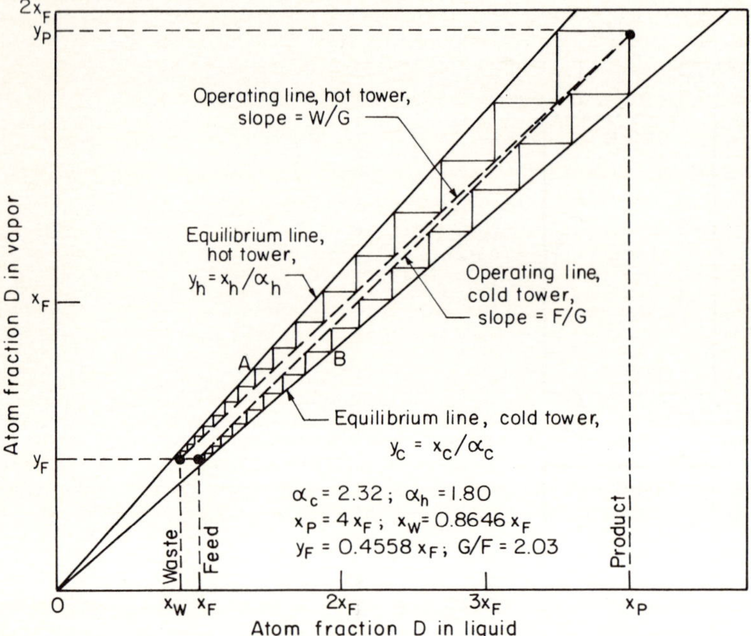

Figure 13.28 Example of McCabe-Thiele diagram for dual-temperature process.

or

$$y_{hi} = y_F + \frac{y_P - y_F}{x_P - x_W}(x_{h,i+1} - x_W) \tag{13.107}$$

because

$$\frac{W}{G} = \frac{y_P - y_F}{x_P - x_W} \tag{13.108}$$

This line passes through the points (y_F, x_W) and (y_P, x_P) and has the slope W/G given by (13.108).

Because the deuterium content of water leaving the cold tower (x_P) equals that entering the hot, and the deuterium content of hydrogen sulfide leaving the hot tower (y_P) equals that entering the cold, the two operating lines end in a common point at top right. Because the deuterium content of hydrogen sulfide leaving the cold tower (y_F) equals that entering the hot, the left end of each operating line is at the same value of y.

It is thus possible to draw the McCabe-Thiele diagram with equilibrium lines established from the separation factors α_c and α_h, and the operating lines established from specified values of feed, product, and waste compositions x_F, x_P, and x_W and assumed values of the gas-phase compositions y_F and y_P. The number of theoretical stages needed in the cold tower for a given set of conditions is then determined by the number of horizontal steps required to go from x_F to x_P; the number of stages in the hot tower, from the number of steps to go from x_W to x_P. For the separation example of Fig. 13.28, the number of stages in each tower is 16.

The McCabe-Thiele diagram can be used to demonstrate two important characteristics of a dual-temperature plant.

1. If x_F, x_W, and y_F are held constant and the number of plates in both towers is increased, the deuterium content of product x_P can be increased to any desired degree.

2. If x_P, x_F, and y_P are held constant and the number of plates is increased, y_F decreases and the lower end of each operating line approaches the corresponding equilibrium line. The maximum spread between x_W and x_F occurs with an infinite number of plates, at which

$$\left(\frac{x_W}{x_F}\right)_{\min} = \frac{\alpha_h}{\alpha_c} \tag{13.109}$$

The fractional recovery of deuterium is

$$r = \frac{Px_P}{Fx_F} = \frac{1 - x_W/x_F}{1 - x_W/x_P} \tag{13.110}$$

The maximum deuterium recovery possible is

$$r \approx 1 - \frac{\alpha_h}{\alpha_c} \tag{13.111}$$

because usually $x_W/x_P \ll 1$.

This shows the importance of using a reaction in which the separation factor in the hot tower differs substantially from that in the cold; in fact, separation is possible only because the slopes of the two equilibrium lines in Fig. 13.28 are different. For the GS process example of Fig. 13.25, the maximum recovery of deuterium possible is

$$r_{\max} = 1 - \frac{1.80}{2.32} = 0.224 \tag{13.112}$$

It is found in practice [B7] that the minimum number of stages for a given separation, or the maximum production rate for a given number of stages, is realized where two conditions are satisfied:

1. The approach to equilibrium at the top of the cold tower equals that at the bottom of the hot:

$$\frac{x_F}{\alpha_c y_F} = \frac{\alpha_h y_F}{x_W} \tag{13.113}$$

and

2. The ratio of the slope of the equilibrium line to the slope of the operating line in the hot tower equals the ratio of the slope of the operating line to the slope of the equilibrium line in the cold tower:

$$\frac{1/\alpha_h}{W/G} = \frac{F/G}{1/\alpha_c} \tag{13.114}$$

The approximate validity of these two conditions can be seen qualitatively by considering the effect on the number of stages of changing the location of the operating lines in Fig. 13.28, while keeping x_P and x_W constant.

The diameter of the towers of a GS plant, the principal heat exchanger duties, and the heat consumption are determined mainly by the ratio of gas flow rate to product rate, G/Px_P. The optimum value of G is, from (13.114),

$$G = \sqrt{FW\alpha_c\alpha_h} \tag{13.115}$$

When $x_P \gg x_F$, $F \approx W$, and

$$G \approx F\sqrt{\alpha_c\alpha_h} \tag{13.116}$$

The gas flow rate per unit product is

$$\frac{G}{Px_P} \approx \frac{F\sqrt{\alpha_c \alpha_h}}{Px_P} = \frac{\sqrt{\alpha_c \alpha_h}}{x_F r} \tag{13.117}$$

The minimum value of this ratio is obtained at maximum recovery; with r_{max} from (13.111),

$$\left(\frac{G}{Px_P}\right)_{min} \approx \frac{\sqrt{\alpha_c \alpha_h}}{x_F(1 - \alpha_h/\alpha_c)} \tag{13.118}$$

For the GS process with natural water feed,

$$\left(\frac{G}{Px_P}\right)_{min} = \frac{\sqrt{2.32 \times 1.80}}{(0.000149)(0.224)} = 61{,}100 \text{ mol gas/mol } D_2O \tag{13.119}$$

Although the minimum gas flow rate is large, it is much smaller than in the distillation of water [141,000, from Eq. (13.11)]. Moreover, the GS process can be operated at much higher pressure than water distillation, which also helps to reduce the number and diameter of towers.

Equations for the dependence of composition in the cold and hot towers on stage number are obtained by application of Eq. (13.92) to the nomenclature of Fig. 13.27. For the cold tower,

$$\alpha_c y_P - x_P = \left(\frac{\alpha_c F}{G}\right)^{n_c} (\alpha_c y_F - x_F) \tag{13.120}$$

By material balance,

$$y_P = \frac{(x_P - x_F)F}{G} + y_F \tag{13.121}$$

so that

$$x_P = \frac{(\alpha_c y_F - x_F)(\alpha_c F/G)^{n_c} + \alpha_c[(F/G)x_F - y_F]}{\alpha_c(F/G) - 1} \tag{13.122}$$

Application of Eq. (13.92) to the hot tower leads to

$$\alpha_h y_P - x_P = \left(\frac{G}{W\alpha_h}\right)^{n_h} (\alpha_h y_F - x_F) \tag{13.123}$$

By material balance,

$$y_P = \frac{(x_P - x_W)W}{G} + y_F \tag{13.124}$$

so that

$$x_P = \frac{(\alpha_h y_F - x_W)(G/W\alpha_h)^{n_h+1} + x_W - (G/W)y_F}{1 - (G/W\alpha_h)} \tag{13.125}$$

These equations may be used either to determine the number of stages required to separate feed of given composition x_F into product and waste of specified compositions x_P and x_W, at a specified flow ratio F/G, or to determine the recovery attainable from a plant of a given number of stages n_c and n_h when operated at a specified flow ratio F/G. An example of the latter application will be given.

To do this, y_F is eliminated from Eqs. (13.122) and (13.125) and the resulting equation is solved for x_W/x_F.

$$\frac{x_W}{x_F} = \frac{(G/W)[(\alpha_c F/G)^{n_c} - (\alpha_c F/G)]\,[(G/W\alpha_h)^{n_c} - 1]}{\alpha_c[(\alpha_c F/G)^{n_c} - 1]\,[(G/W\alpha_h)^{n_h+1} - 1]} \tag{13.126}$$

(*Cont. on p. 775*)

$$+\frac{(x_P/x_F)\{[\alpha_c(F/G)-1](G/W)[(G/W\alpha_h)^{n_h}-1]+\alpha_c[(\alpha_c F/G)^{n_c}-1][(G/W\alpha_h)-1]\}}{\alpha_c[(\alpha_c F/G)^{n_c}-1][(G/W\alpha_h)^{n_h+1}-1]}$$

(13.126)
(*Cont.*)

Because G/W depends on x_W/x_F through

$$\frac{G}{W}=\frac{F}{W}\frac{G}{F}=\frac{(x_P/x_F)-(x_W/x_F)}{(x_P/x_F)-1}\frac{G}{F} \quad (13.127)$$

Eq. (13.126) is implicit in x_W/x_F and must be solved by trial. Figure 13.29 shows values of x_W/x_F calculated for the specific case of $n_c = n_h = 16$; $x_P/x_F = 4$, for a range of values of G/F.

This figure brings out an important characteristic of dual-temperature exchange processes: The recovery (or production rate) of a given plant is very sensitive to the gas-to-liquid flow ratio. There is only a narrow range of flow ratios within which optimum performance is obtained. In the example of Fig. 13.29, the minimum value of x_W/x_F, 0.8563, is obtained at $G/F = 2.03$. If G/F is less than 1.85 or greater than 2.25, x_W/x_F becomes greater than 0.90, and the recovery of deuterium is decreased by 30 percent or more.

At the optimum value of $G/F = 2.03$, $G/W = 2.127$ and $y_F = 0.4558x_F$. The approach to equilibrium is

$$\text{At top of cold tower:} \quad \frac{x_F}{\alpha_c y_F}=\frac{1}{(2.32)(0.4558)}=0.946 \quad (13.128)$$

$$\text{At bottom of hot tower:} \quad \frac{\alpha_h y_F}{x_W}=\frac{(1.80)(0.4558)}{0.8563}=0.958 \quad (13.129)$$

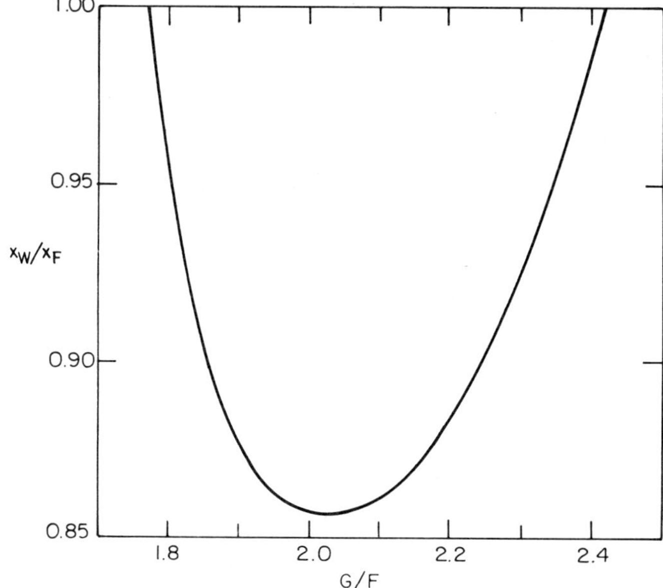

Figure 13.29 Effect of vapor-to-feed ratio on recovery in GS process example. $\alpha_h = 1.80$; $\alpha_c = 2.32$; $x_P/x_F = 4$; $n_c = n_h = 16$.

This illustrates the applicability of condition (13.113) when the plant is operated for maximum recovery.

Condition (13.114) also is approximately satisfied at the optimum flow ratio, because

$$\frac{1/\alpha_h}{W/G} = \frac{2.127}{1.80} = 1.182 \tag{13.130}$$

and

$$\frac{F/G}{1/\alpha_c} = \frac{2.32}{2.03} = 1.143 \tag{13.131}$$

The McCabe-Thiele diagram, Fig. 13.28, is drawn for this separation at the optimum flow ratio $G/F = 2.03$. At this optimum condition, the size of each step in the cold tower is approximately equal to the size of the step in the hot tower at the corresponding plate. In operating the Savannah River plant [B7], the flow ratio of gas to liquid is controlled to give optimum performance by setting it so that the deuterium content of corresponding streams at the middle of the hot and cold towers are equal. In Fig. 13.28 this is illustrated by the fact that the deuterium content of vapor flowing between the eighth and ninth plates of the hot tower (step A) is approximately equal to the deuterium content of vapor flowing between the eighth and ninth plates of the cold tower (step B). Use of this principle greatly simplified what would otherwise be a difficult problem in flow control.

11.4 Detailed Process Flow Sheet for GS Plant

The GS plant for which the most detailed information has been published is the Savannah River plant of the U.S. AEC. This section summarizes the design and operating characteristics of this plant, which has been in operation since 1955. Section 11.8 describes improvements that du Pont personnel suggested for future GS plants, some of which presumably have been adopted in the newer Canadian plants.

Figure 13.30 is a flow sheet showing the main process equipment of the Savannah River GS plant and the principal process conditions† as given by Bebbington and Thayer [B7]. The plant consists of 24 units of type shown, operated in parallel. Not shown in the figure are the feed-water deaerator, the tower to recover H_2S from purge gas, and pumps for liquid.

Natural water feed is deaerated, brought to around 32°C, and pumped to the top of the cold tower CT-1, at 292 psia. It dissolves H_2S and becomes saturated after flowing down through the first few plates, and it becomes enriched in deuterium by isotopic exchange as it flows through the entire tower. Liquid leaving the bottom of CT-1, enriched to 0.085 percent deuterium, is split into two streams. About one-fourth is pumped to the top of cold tower CT-2A; three-fourths is bypassed around CT-2A, heated to 125°C by exchange against outgoing hot waste water, and pumped to the top of hot tower HT-1. Liquid flowing down through cold towers CT-2A and 2B in series is enriched to about 15 percent deuterium by further isotopic exchange. Liquid leaving the bottom of CT-2B is heated to 120°C by exchange against outgoing hot waste water and is pumped to the top of hot tower HT-2A. In HT-2A the deuterium content of the water is reduced by exchange at the higher temperature. The deuterium content of the water is further reduced, to around 0.012 percent, by exchange in the top of hot tower HT-1; this water is drawn off above the eleventh plate above the bottom of HT-1.

Before this depleted water can be discharged from the plant, it is necessary to strip it of H_2S, down to less than 2 ppm. To do this, the water is heated to 200°C in heat exchanger

†For consistency with du Pont literature, pressures are given in pounds per square inch absolute (psia) and flow quantities in pound-moles. Conversion factors are 1 psia = 0.068046 atm = 6895 Pa; 1 lb-mol = 0.4536 kg-mol.

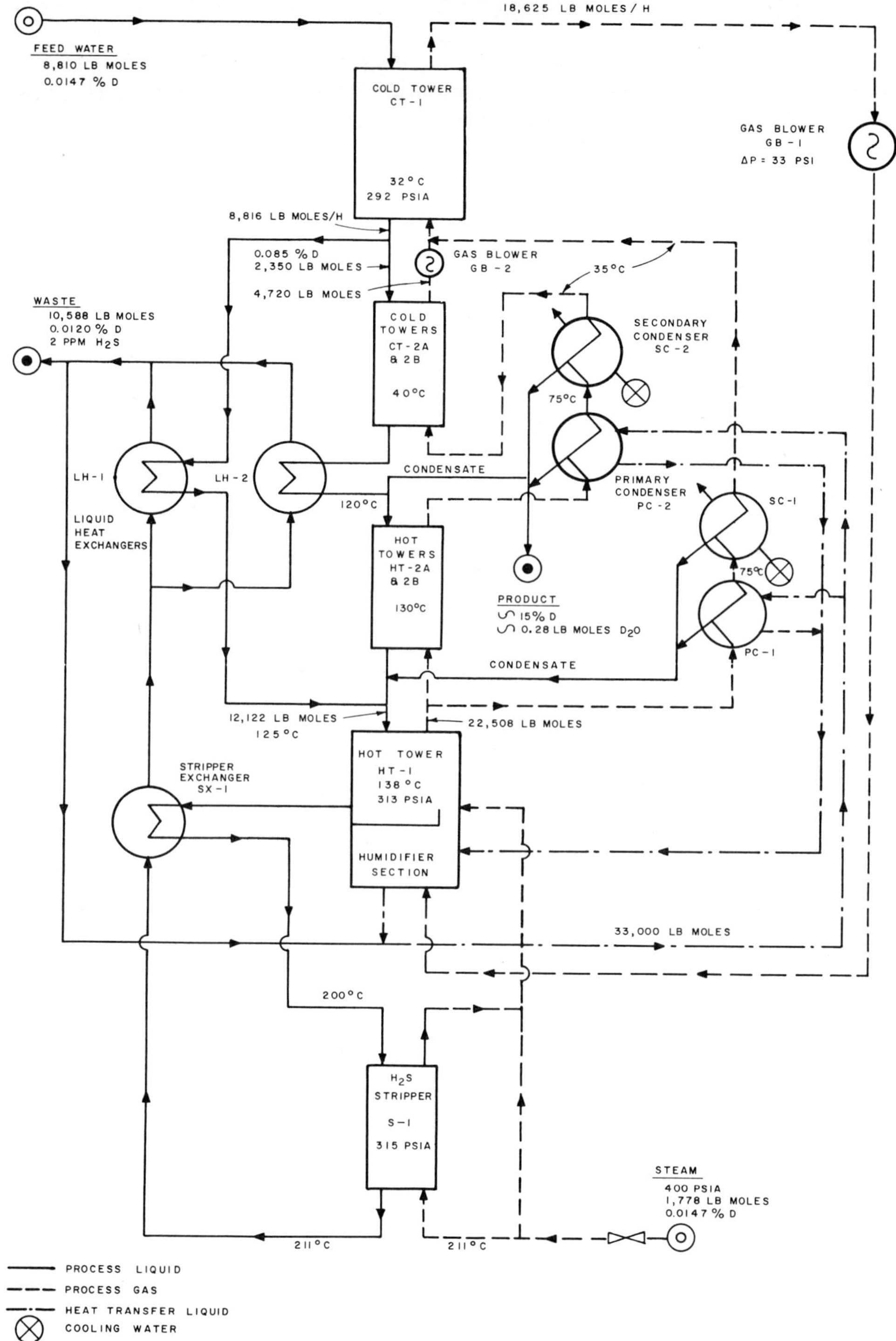

Figure 13.30 Flow diagram for unit of Savannah River GS plant. Basis, 1 h. Plant consists of 24 units.

SX-1 and fed to the top of the 12-plate H_2S stripper S-1. Heat in water leaving the bottom of this tower is recovered by heat exchange against colder water in exchangers SX-1, LH-1, and LH-2. This water, cooled, depleted in deuterium, and stripped of H_2S, leaves the plant as waste.

Depleted hydrogen sulfide at 32°C from the top of cold tower CT-1 is compressed 33 psi in gas blower GB-1 and fed to the bottom of hot tower HT-1. The bottom 11 plates of this tower are used to heat and humidify the hydrogen sulfide through direct countercurrent contacting with downflowing hot water charged to the eleventh plate. As the hot humid hydrogen sulfide from the eleventh plate of HT-1 and from the stripper S-1 flows up through the top 59 plates of HT-1, it is partially enriched by exchange of deuterium from the downflowing water. Gas leaving the top of HT-1 is split into two streams. About one-fourth goes to the second-stage hot tower HT-2B; three-fourths is bypassed around the second stage. As gas flows up through hot towers HT-2B and HT-2A in series, its enrichment in deuterium is completed. Gas leaving the top of HT-2A is dehumidified and cooled to 75°C in primary condenser PC-2 by closed heat exchange against cold water from the bottom of the humidifier section of HT-1. The gas is cooled further to 40°C in secondary condenser SC-2 by closed heat exchange against cooling water. The hot gas bypassed around the second stage is similarly dehumidified and cooled by closed heat exchange in PC-1 and SC-1. As gas flows up through cold towers HT-2B, HT-2A, and HT-1, its deuterium is transferred to cold water flowing down in these towers.

Condensate from SC-1 and PC-1 is returned to the top of hot tower HT-1, and part of the condensate from SC-2 and PC-2 is returned to the top of HT-2A. The rest of the water condensed in PC-2 and SC-2 containing around 15 percent deuterium is withdrawn as plant product. Use of this stream for product instead of water from the bottom of cold tower CT-2B, which has about the same enrichment, is preferred because the condensate is cleaner.

All heat requirements for the process are provided in the form of open steam at 400 psia. Some is used at the bottom of S-1 to strip H_2S and the rest is fed to the twelfth plate in HT-1 to control the temperature of the hot towers and to compensate for heat losses and heat exchanger inefficiencies. Steam consumption is $1778/0.28 = 6400$ mol/mol of D_2O produced. This is much less than the 200,000 mol/mol D_2O needed in water distillation. Additional energy in the amount of 680 kWh/kg D_2O is used to circulate gas and pump liquid. This, however, is much less than is used in electrolysis or hydrogen distillation (Table 13.7). The low energy consumption of the GS process is due in large measure to the efficient heat recovery obtainable in the flow sheet Fig. 13.30, which follows Spevack's patent [S7].

11.5 Materials of Construction

The principal disadvantage of the GS process is the toxic and corrosive character of aqueous solutions of hydrogen sulfide. Extensive corrosion research and experience with the Dana and Savannah River plants has shown what materials of construction can be used to withstand corrosion, without prohibitive cost. The following summary of recommendations regarding materials of construction is condensed from reference [T4].

1. Carbon steel is used for a large part of the equipment in the heavy-water plants. This includes the shells of the exchange towers, the shells of most of the heat exchangers, and practically all the process piping. These items are protected from surface corrosion by a coating of iron sulfide that forms during the first few weeks of operating, after which further corrosion of the steel is so slight as to be negligible.

2. Bubble-cap trays in the exchange towers are made of stainless steel, preferably type 304 18-8. Carbon steel is unsuitable because the impingement of spray to which the trays are subjected prevents an iron sulfide layer from forming, and under these conditions carbon steel corrodes rapidly. The same considerations apply to other parts exposed to erosion by spray

impingement or by high liquid velocity, such as heat exchanger tubes, centrifugal pumps, and throttle valves.

3. Carbon steel, "low-alloy" steel, and stainless steel must be suitably heat treated to relieve stresses that may result from fabrication procedures. If this is not done, failure may occur due to hydrogen embrittlement or to stress-corrosion cracking. Carbon steels are satisfactory if annealed according to the ASTM Boiler Code. Austenitic stainless steels that have been heavily cold worked must be quench annealed at 1800 to 2000°F.

4. Bolts (normally made of low-alloy steel) that are used at flanged joints (or for any other purpose) pose a special problem when handling H_2S-water. Even when such bolts are outside of the process equipment, they may be exposed to H_2S because of leaks. A small leak of H_2S in air will attack the surface of a bolt, causing hydrogen absorption into the metal. If such a bolt is stressed beyond a certain threshold value, dependent on its hardness, it will crack. For this reason, all bolts are heat treated, after machining, to reduce hardness below a critical maximum value and are installed to a predetermined stress level using torque wrenches. The unavoidable reduction in tensile strength resulting from the heat treatment is accepted.

5. In order to ensure that corrosion of process piping, if this should occur, would not result in a major release of H_2S, all piping above 3 in in diameter is provided with "minimum thickness holes." These holes, about $\frac{1}{8}$ in in diameter, are drilled from the outside to about half the wall thickness. Loss of metal on the inside of the pipe will result in a small but detectable leakage through the test holes, while the structural strength is still adequate to withstand the operating pressure.

11.6 Economics

The construction cost of the Savannah River heavy-water plant, built in 1951/1952, is summarized in Table 13.19 [B6]. The unit investment cost of this plant, capable of producing 454 Mg D_2O/year, then was \$163,000,000/454,000 kg/year = \$359/(kg/year).

The cost of Atomic Energy of Canada, Ltd.'s 800 Mg/year plant at La Prade, item 21, Table 13.2, was predicted [H5] to be \$300 million in 1974, for a unit investment cost of \$375/(kg/year), exclusive of escalation and interest during construction.

Heavy water from the Savannah River plant was sold by the U.S. AEC in the 1960s for a price of \$61.73/kg. Demand for heavy water subsequently decreased, and two of the three original wings of the plant were shut down. In 1976, when one wing was operating at its full

Table 13.19 Construction cost of Savannah River heavy-water plant

	\$ million
Process facilities	
H_2S exchange units	113
Water distillation plant	2.5
Electrolytic plant	1.5
Steam and electric power plant	31
Water system	8
General facilities	7
	\$163

Source: W. P. Bebbington and V. R. Thayer, *Chem. Eng. Progr.* 55(9): 70 (1959).

capacity of 177 MT/year, production costs [J5] were as summarized in Table 13.20. In 1977, when the one remaining wing was operating at reduced capacity, the price charged by U.S. ERDA for heavy water was $245/kg [F1].

Utility requirements reported for heavy-water production by the GS process are as follows:

	Heat, kWht/kg D_2O	Electricity, kWhe/kg D_2O
Savannah River [B8]	7800	680
Canada [R2]	6800	700

11.7 Detailed Analysis of Process

Separation factor. In the simplified analysis of the water-hydrogen sulfide exchange process in Sec. 11.3, the effects of the solubility of hydrogen sulfide in water and the vaporization of water into hydrogen sulfide were neglected. In the following they will be taken into account.

The deuterium separation factor α for the hydrogen sulfide exchange process is defined as

$$\alpha \equiv \frac{x(1-y)}{y(1-x)} \tag{13.132}$$

where y and x are the atom fractions of deuterium in the vapor and liquid, respectively. In terms of the molecular species H_2O, HDO, D_2O, H_2S, HDS, and D_2S that make up each phase, α is given by

Table 13.20 Heavy-water production cost at Savannah River

	Quantity per kg D_2O	Cost, \$/kg D_2O
Direct production cost		
Feed water, kg	24,000	5.07
Hydrogen sulfide	0.66	0.24
Salaries		2.89
Operating labor		4.37
Miscellaneous		1.12
		13.69
Direct maintenance cost		
Labor		4.37
Materials		7.38
		11.75
Utilities		
Electricity, kWh	604	13.76
Steam, kg (900 psig equiv.)	5,660	41.45
Cooling water, kg	125,000	1.48
Miscellaneous		0.84
		57.53
Depreciation		24.95
Administrative and general		15.01
Total cost of production		122.93

$$\alpha = \frac{(2x_{D_2O} + x_{HDO} + 2x_{D_2S} + x_{HDS})(2y_{H_2O} + y_{HDO} + 2y_{H_2S} + y_{HDS})}{(2y_{D_2O} + y_{HDO} + 2y_{D_2S} + y_{HDS})(2x_{H_2O} + x_{HDO} + 2x_{H_2S} + x_{HDS})} \tag{13.133}$$

where y refers to the mole fraction of the indicated species in the vapor and x in the liquid.

An expression will be derived for the dependence of α on the physical properties of the water-hydrogen sulfide system, temperature and pressure. The slight dependence of α on deuterium content will be neglected by considering only low deuterium abundances, at which $x_{D_2O} \ll x_{HDO}$, etc. In this limiting case, the expression for α reduces to

$$\alpha = \frac{(x_{HDO} + x_{HDS})(2y_{H_2O} + 2y_{H_2S})}{(y_{HDO} + y_{HDS})(2x_{H_2O} + 2x_{H_2S})} \tag{13.134}$$

The following properties of water, hydrogen sulfide, and their mixtures are used to evaluate α:

1. The humidity H of H_2O-H_2S vapor in equilibrium with liquid mixtures, defined as

$$H \equiv \frac{y_{H_2O}}{y_{H_2S}} \tag{13.135}$$

2. The solubility S of H_2S in liquid in equilibrium with vapor, defined as

$$S \equiv \frac{x_{H_2S}}{x_{H_2O}} \tag{13.136}$$

The dependence of H and S on temperature and pressure has been determined experimentally [S4] and is shown in Figs. 13.31 and 13.32.

3. The relative volatility α^* of H_2O to HDO, defined by Eq. (13.5).

The dependence of α^* on temperature has been given in Table 13.4; it is assumed to be unchanged by the presence of H_2S.

4. The relative volatility γ of H_2S to HDS, defined as

$$\gamma \equiv \frac{y_{H_2S} x_{HDS}}{y_{HDS} x_{H_2S}} \tag{13.137}$$

In the design of the Savannah River plant [B7] it was assumed that γ equaled α^*.

Roth et al. [R9] have determined γ for anhydrous hydrogen sulfide and have found it to be substantially equal to unity. No data are available for values of γ in aqueous solutions of hydrogen sulfide, but its value probably lies in the range 1.00 to 1.05.

5. The equilibrium constant k for the gas-phase deuterium exchange reaction,

$$H_2O(g) + HDS(g) \rightleftharpoons HDO(g) + H_2S(g)$$

defined by

$$k = \frac{y_{HDO} y_{H_2S}}{y_{H_2O} y_{HDS}} \tag{13.138}$$

The mole fractions y_{H_2O}, x_{H_2S}, x_{HDS}, y_{HDO}, and x_{HDO} occurring in Eq. (13.134) will be expressed in terms of y_{H_2S}, x_{H_2O}, and y_{HDS} by the following equations derived from those given above defining H, S, γ, k, and α^*:

$$y_{H_2O} = H y_{H_2S} \tag{13.139}$$

$$x_{H_2S} = S x_{H_2O} \tag{13.140}$$

$$x_{HDS} = \frac{\gamma S x_{H_2O} y_{HDS}}{y_{H_2S}} \tag{13.141}$$

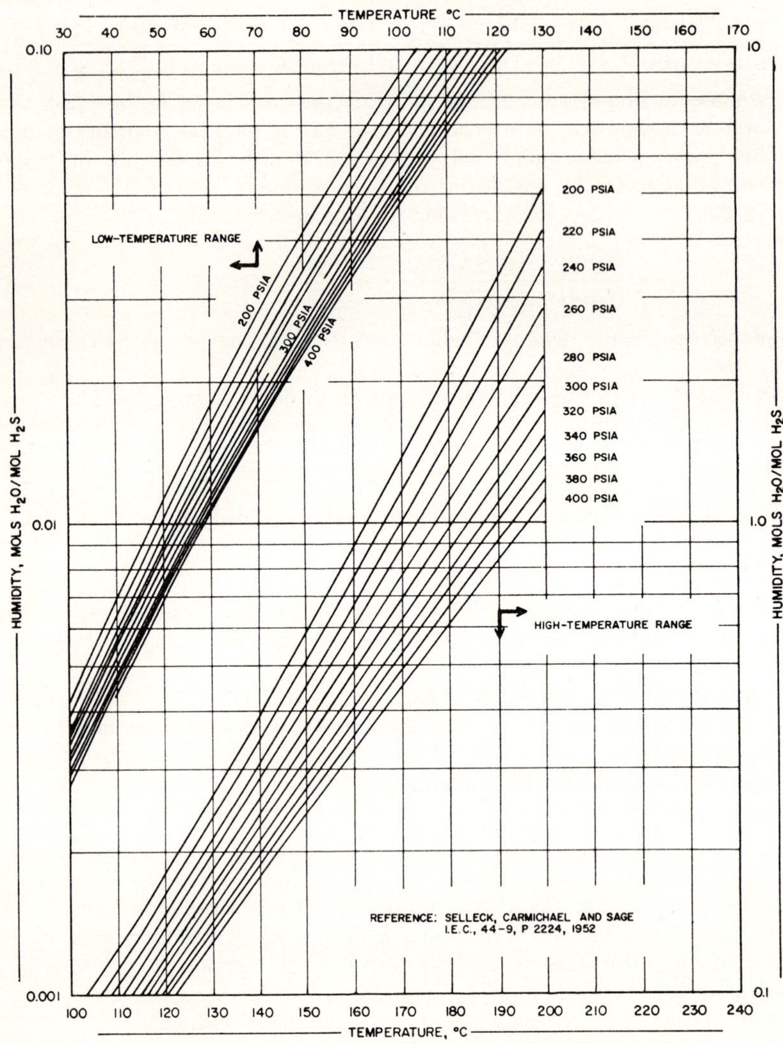

Figure 13.31 Humidity of H_2S vapor in equilibrium with liquid water.

$$y_{\text{HDO}} = kHy_{\text{HDS}} \tag{13.142}$$

$$x_{\text{HDO}} = \frac{k\alpha^* x_{\text{H}_2\text{O}} y_{\text{HDS}}}{y_{\text{H}_2\text{S}}} \tag{13.143}$$

The result of substituting Eqs. (13.139) through (13.143) into (13.134) is

$$\alpha = \frac{(k\alpha^* x_{\text{H}_2\text{O}} y_{\text{HDS}}/y_{\text{H}_2\text{S}}) + (\gamma S x_{\text{H}_2\text{O}} y_{\text{HDS}}/y_{\text{H}_2\text{S}})}{kHy_{\text{HDS}} + y_{\text{HDS}}} \frac{Hy_{\text{H}_2\text{S}} + y_{\text{H}_2\text{S}}}{x_{\text{H}_2\text{O}} + Sx_{\text{H}_2\text{O}}} = \frac{k\alpha^* + \gamma S}{kH + 1} \frac{H + 1}{1 + S} \tag{13.144}$$

The remaining mole fractions have cancelled out, and α has been expressed in terms of H, S, γ, k, and α^*.

Equation (13.144) is the exact expression for the deuterium exchange separation factor in liquid-vapor mixtures of water and hydrogen sulfide at low deuterium abundances. Values evaluated from it are customarily used without correction up to 15 percent deuterium. When the vaporization of water into H_2S is small ($H \ll 1$) and the solubility of H_2S in water is small ($S \ll 1$), Eq. (13.144) reduces to Eq. (13.77).

A number of experimental measurements and theoretical calculations have been made of the equilibrium constant k for the gas-phase reaction that have been correlated by the equation

$$k = Ae^{B/T} \tag{13.145}$$

Values of A and B given by four investigators and k at 32 and 138°C from Eq. (13.145) are listed in Table 13.21.

The equilibrium constant $k\alpha^*$ for the gas-liquid reaction has also been determined by a number of investigators. Results at several temperatures are given in Table 13.22. Data of Geib and Seuss have been computed from their equation for k given in Table 13.21 and their equation (13.146) for α^*:

$$\alpha^* = 0.8624e^{65.43/T} \tag{13.146}$$

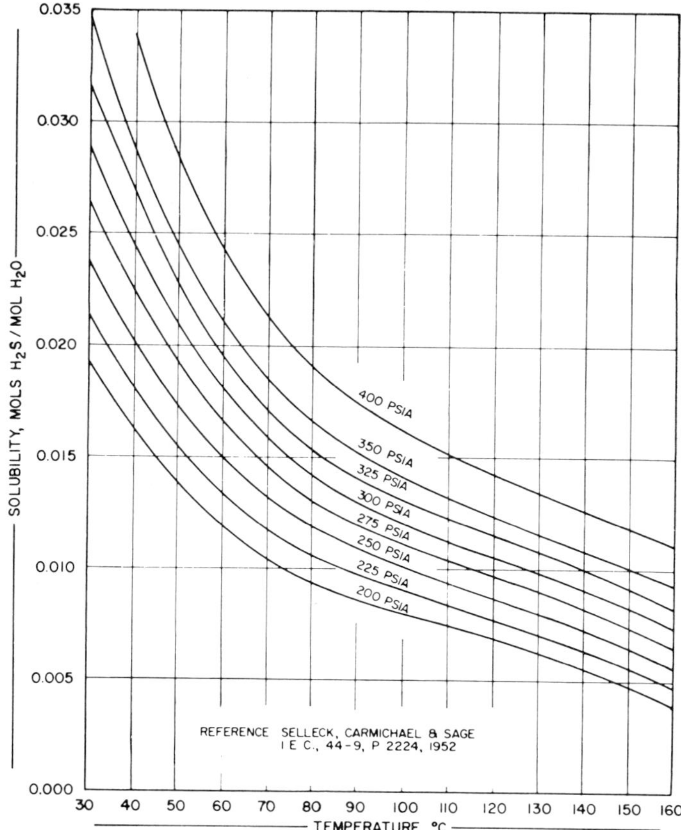

Figure 13.32 Solubility of H_2S in liquid water.

Table 13.21 Comparison of equilibrium constants for gas-phase reaction $H_2O + HDS \rightleftharpoons HDO + H_2S$†

Source	Geib and Suess	Bigeleisen	Varshavskii and Vaisberg	Roth et al.
Reference	[C3]	[B11]	[V1]	[R9]
A	1.010	1.051	1.0084	1.001
B	233	218	219.0	221.3
k at 32°C	2.167	2.147	2.067	2.067
138°C	1.780	1.786	1.718	1.715

†$k = Ae^{B/T\,(\mathrm{K})}$.

There are substantial differences among the results for k and for $k\alpha^*$ given by the various investigators. The equations of Geib and Seuss have been used by Bebbington and Thayer [B7] in the most complete published account of the Savannah River plant. Table 13.23 compares values of the separation factor α for the hot and cold towers of the Savannah River plant computed by Eq. (13.144) from the data recommended by Bebbington and Thayer with values computed from the data recommended by Roth et al. The data recommended by Bebbington and Thayer have been used in this chapter because they have been successful in interpreting the performance of the Savannah River plant.

The dependence of α on temperature and pressure, as computed from Eq. (13.144), is shown in Fig. 13.33. In the cold tower an increase in pressure decreases α because it increases the concentration of H_2S in the liquid more than it decreases the concentration of H_2O in the vapor. In the hot tower, an increase in pressure increases α because it decreases the concentration of H_2O in the vapor more than it increases the concentration of H_2S in the liquid.

Optimum operating conditions. Because the deuterium recovery increases with increasing ratio of α in the cold tower to α in the hot, it might be supposed that the optimum operating conditions would be the lowest possible cold tower temperature, the highest possible hot tower temperature, and low pressure. Other factors beside α must be considered, however.

An increase in pressure above atmospheric leads to lower costs, despite the reduced spread in α's between the hot and cold tower, because of the greater mass flow rate of gas per unit area that can be taken through the towers at higher pressure. At a pressure of 300 psig, however, there is a discontinuous increase in the cost of equipment, because of the need to

Table 13.22 Equilibrium constant $k\alpha^*$ for gas-liquid reaction $H_2O(l) + HDS(g) \rightleftharpoons HDO(l) + H_2S(g)$

	$k\alpha^*$			
Temperature, °C	Calculated from Geib and Seuss [C3]	McClure and Herrick [M3]	Haul et al. [H4]	Interpolated from Roth et al. [R9]
24	2.38	2.38		2.267
25	2.37		2.35	2.259
78	2.03	2.02		1.948
141	1.79	1.82		1.729

Table 13.23 Comparison of separation factors at conditions of Savannah River plant

	Tower			
	Cold		Hot	
Temperature, °C	32		138	
Pressure, psia	292		313	
Humidity H	0.0036		0.215	
Solubility S	0.027		0.0096	
Source of data	DP-400 [B7]	Roth [R9]	DP-400 [B7]	Roth [R9]
k	2.167	2.067	1.780	1.715
α^*	1.0686		1.0111	
$k\alpha^*$	2.316	2.218	1.800	1.737
γ	1.0686	1.000	1.0111	1.000
α	2.275	2.178	1.576	1.518

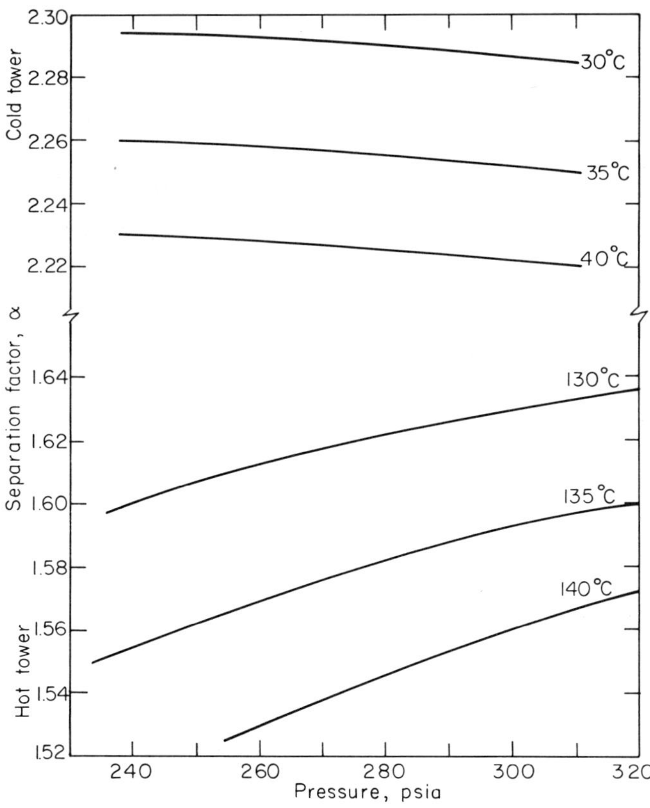

Figure 13.33 Separation factor for water-hydrogen sulfide exchange process.

change to the heavier pipe and fittings required for use in this higher pressure range. This sets the optimum pressure around 300 psi. The same pressure is used in each tower, except for pressure drop due to flow, to keep gas-recompression costs at a minimum.

The optimum temperature of the cold tower is as low as possible without risking formation of a third phase in addition to vapor and aqueous solution. Table 13.24 gives the temperatures at which solid hydrogen sulfide hydrate or liquid hydrogen sulfide form in the system H_2S-H_2O. At 300 psi, the minimum safe cold tower temperature is around 30°C. The rapid increase in condensation temperature above 300 psi is another reason for this being the optimum pressure. Before the first pilot plant for the GS process was operated, the possibility of hydrate formation was not recognized, and freeze-ups occurred until the cold tower temperature was raised above 30°C.

The optimum hot tower temperature is around 130 to 140°C and is determined by a balance between the improvement in separation at higher temperature and the increased costs for heat and for humidifying the gas entering the hot tower at higher temperature.

Effect of hydrogen sulfide solubility and water volatility on analysis of process. The solubility of hydrogen sulfide and the volatility of water introduce changes in flow rates of gas and liquid and deuterium concentrations at the top and bottom of the hot and cold towers. Figure 13.34 illustrates the flow scheme and nomenclature to be used in working out these effects.

The flow rate of liquid into the cold tower is increased from F, in feed water, to L_c leaving the top tray of the tower, owing to formation of a saturated solution of hydrogen sulfide. L_c then remains constant throughout the cold tower. Between the cold and hot tower the liquid flow rate is changed to L_h because of withdrawal of product P, addition of condensate L_a, and vaporization of some gas, G_a. L_h remains constant through the exchange section of the hot tower down to the point where liquid is drawn off to the H_2S stripper and vapor from the humidifying section is returned.

Vapor flows up through the cold tower at a constant rate G_c until in leaving the tower the rate is reduced to G_0 owing to solution of some H_2S in incoming feed water. The vapor flow rate to the hot tower is increased from G_0 to G_h by hydrogen sulfide from the stripper and by the water vapor needed to saturate the hydrogen sulfide at the temperature of the hot tower. G_h remains constant in the hot tower.

It is possible to set independently three of the nine flow rates F, L_c, P, L_h, L_a, G_c, G_0, G_h, and G_a. The other six are determined by the following material-balance equations:

Table 13.24 Equilibrium conditions for three phases in H_2O-H_2S system†

Pressure, psia	Temperature, °C	Third phase
15	1.1	Hydrate
30	7.5	Hydrate
50	12.2	Hydrate
100	18.6	Hydrate
200	25.0	Hydrate
300	28.9	Hydrate
325	29.5	Hydrate + liquid H_2S
400	38.6	Liquid H_2S
500	48.3	Liquid H_2S
600	56.1	Liquid H_2S

†Data from Bebbington and Thayer [B7].

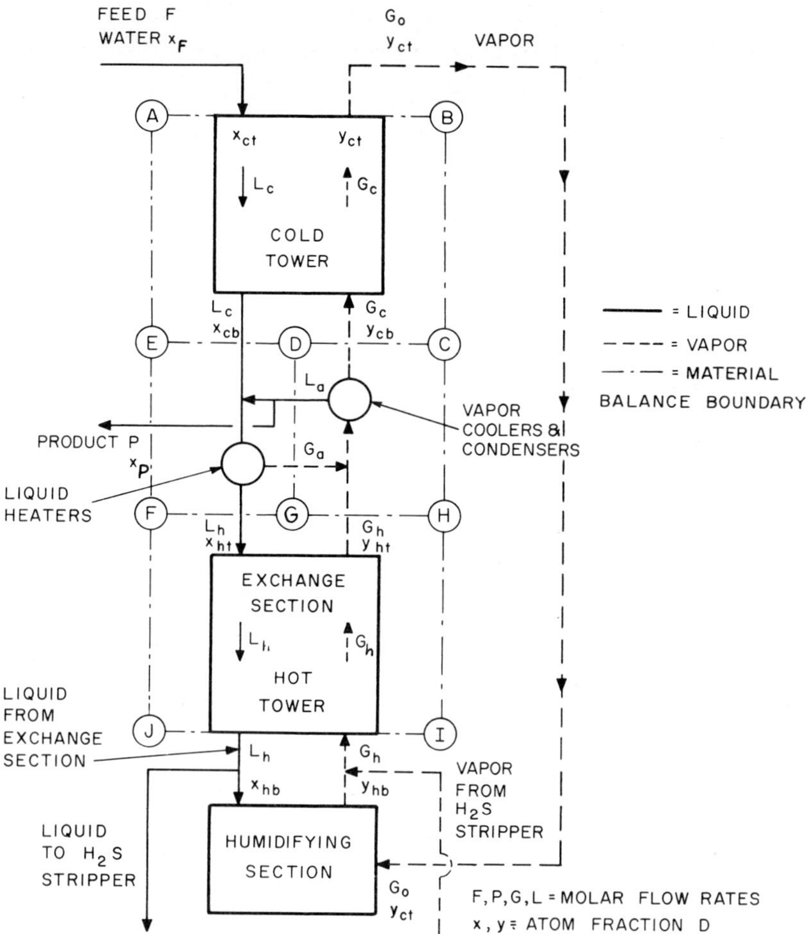

Figure 13.34 Nomenclature for detailed analysis of GS process.

At top of cold tower

$$\text{Total flow:} \quad L_c + G_0 = F + G_c \tag{13.147}$$

$$\text{Hydrogen sulfide:} \quad \frac{L_c S_c}{1+S_c} + \frac{G_0}{1+H_c} = \frac{G_c}{1+H_c} \tag{13.148}$$

Between cold and hot towers (ABCHGFE, Fig. 13.34)

$$\text{Total flow:} \quad L_c + G_h = L_h + G_c + P \tag{13.149}$$

$$\text{Hydrogen sulfide:} \quad \frac{L_c S_c}{1+S_c} + \frac{G_h}{1+H_h} = \frac{L_h S_h}{1+S_h} + \frac{G_c}{1+H_c} + \frac{P S_c}{1+S_c} \tag{13.150}$$

Around the vapor coolers and condensers (DCHG, Fig. 13.34)

$$\text{Total flow:} \quad G_h + G_a = L_a + G_c \tag{13.151}$$

$$\text{Hydrogen sulfide:} \quad \frac{G_h}{1+H_h} + \frac{G_a}{1+H_h} = \frac{L_a S_c}{1+S_c} + \frac{G_c}{1+H_c} \tag{13.152}$$

For example, if F, P, and G_0 are set, one obtains:

$$L_c = \frac{F(1+S_c)}{1-S_c H_c} \tag{13.153}$$

$$G_c = G_0 + \frac{FS_c(1+H_c)}{1-S_c H_c} \tag{13.154}$$

$$L_h = \frac{1+S_h}{1-S_h H_h}\left(F - P\frac{1-S_c H_h}{1+S_c} + G_0\frac{H_h - H_c}{1+H_c}\right) \tag{13.155}$$

$$G_h = \frac{1+H_h}{1-S_h H_h}\left(FS_h + G_0\frac{1-S_h H_c}{1+H_c} + P\frac{S_c - S_h}{1+S_c}\right) \tag{13.156}$$

$$L_a = \frac{(H_h - H_c)(1+S_c)}{1-S_c H_h}\left(\frac{G_0}{1+H_c} + \frac{FS_c}{1-S_c H_c}\right) \tag{13.157}$$

$$G_a = \frac{(S_c - S_h)(1+H_h)}{1-S_h H_h}\left[\frac{F}{1-S_c H_h} + \frac{G_0(H_h - H_c)}{(1+H_c)(1-S_c H_h)} - \frac{P}{1+S_c}\right] \tag{13.158}$$

In designing a plant, G_0 and P might first be set. At several values of F, Eqs. (13.151) through (13.154) would then be used to evaluate G_c, L_c, G_h, and L_h. The ratios $\alpha_c L_c/G_c$ and $G_h/L_h\alpha_h$ would be determined; the optimum value of F that leads to the minimum number of plates is the one at which

$$\frac{G_h}{L_h \alpha_h} = \frac{\alpha_c L_c}{G_c} \tag{13.159}$$

This is equivalent to Eq. (13.114).

With the values of the flow rates thus determined, the nine atom fractions of deuterium x_{ct}, y_{ct}, x_{cb}, y_{cb}, x_P, x_{ht}, y_{ht}, x_{hb}, and y_{hb} may be related to the composition of feed x_F and the number of plates n_c and n_h in the cold and hot towers, respectively, by the nine equations (13.161) through (13.169), derived as follows.

At the top of the cold tower, a deuterium balance on the streams above and below the point of H_2S solution gives

$$L_c x_{ct} - F x_F = \frac{(L_c - F)x_{ct}}{(k\alpha^*)_c} \tag{13.160}$$

where $x_{ct}/(k\alpha^*)_c$ is a sufficient approximation for the atom fraction of deuterium in the hydrogen sulfide transferred from gas to liquid. Because $L_c \approx F(1+S_c)$, this may be approximated by

$$x_{ct} = x_F\left[1 - S_c\left(1 - \frac{1}{k\alpha^*}\right)_c\right] \tag{13.161}$$

A deuterium balance over the cold tower gives

$$\frac{G_c}{L_c} y_{cb} - x_{cb} = \frac{G_c}{L_c} y_{ct} - x_{ct} \tag{13.162}$$

The Kremser-type equation (13.120) for the streams at the top and bottom of the cold tower, converted to the notation of Fig. 13.34, leads to

$$\alpha_c y_{cb} - x_{cb} = (\alpha_c y_{ct} - x_{ct}) \left(\frac{\alpha_c L_c}{G_c} \right)^{n_c} \tag{13.163}$$

The deuterium content of product is given by

$$x_P = \alpha_c y_{cb} \tag{13.164}$$

A deuterium balance between the hot and cold towers, on streams flowing across CDEFGH in Fig. 13.34, gives

$$L_c x_{cb} + G_h y_{ht} = P x_P + L_h x_{ht} + G_c y_{cb} \tag{13.165}$$

Similarly, a deuterium balance over the vapor coolers and condensers, on streams flowing across CDGH in Fig. 13.34, gives

$$\frac{G_a x_{ht}}{\alpha_h} + G_h y_{ht} = L_a \alpha_c y_{cb} + G_c y_{cb} \tag{13.166}$$

A deuterium balance over the hot tower, on streams flowing across FGHIJ, gives:

$$x_{ht} - \frac{G_h}{L_h} y_{ht} = x_{hb} - \frac{G_h}{L_h} y_{hb} \tag{13.167}$$

The Kremser-type equation (13.120) for the streams at the top and bottom of the hot tower, converted to the notation of Fig. 13.34, leads to

$$x_{ht} - \alpha_h y_{ht} = (x_{hb} - \alpha_h y_{hb}) \left(\frac{G_h}{\alpha_h L_h} \right)^{n_h} \tag{13.168}$$

The final equation is obtained by making a deuterium balance on the vapor stream entering the bottom of the hot tower. The hydrogen sulfide content of this stream consists of $G_0/(1 + H_c)$ mol from the top of the cold tower plus $L_h S_h/(1 + S_h)$ mol recycled by the stripper and humidifier from the liquid leaving the hot tower. The deuterium content of this latter hydrogen sulfide is approximately $x_{hb}/(k\alpha^*)_h$. The water content of this stream consists of $G_0 H_c/(1 + H_c)$ mol from the top of the cold tower plus $G_h H_h/(1 + H_h) - G_0 H_c/(1 + H_c)$ mol supplied by the humidifier and steam from the stripper. The deuterium content of this latter water vapor is approximately x_{hb}. The balance equation expressing the deuterium content of the vapor entering the exchange section of the hot tower is

$$\underbrace{G_h y_{hb}}_{\text{Vapor to exchange section}} = \underbrace{G_0 y_{ct}}_{\text{Vapor from cold tower}} + \underbrace{\frac{L_h S_h}{1 + S_h} \frac{x_{hb}}{(k\alpha^*)_h}}_{\text{Hydrogen sulfide added}} + \underbrace{\left(\frac{G_h H_h}{1 + H_h} - \frac{G_0 H_c}{1 + H_c} \right) x_{hb}}_{\text{Water vapor added}} \tag{13.169}$$

11.8 Improved GS Flow Sheets

Because of the complexity of the GS process flow sheet, there are a number of opportunities for making improvements in the process that, taken together, should increase deuterium production, reduce the number of separate pieces of equipment, improve energy utilization, and reduce costs. U.S. work on improvements in the early 1960s was described by Proctor and Thayer [P4] and has been used in the first Canadian plants. Later improvements patented by Thayer [T3] have been considered for the newer Canadian plants.

This section will describe one flow sheet improvement patented by Babcock [B1], which would increase deuterium production by providing supplementary natural water feed to the hot tower. Burgess [B14] describes computer calculations of the increased production that would be possible if additional natural water were fed to the first stage hot tower of one of Savannah River GS units. This section will derive equations for the improved deuterium production obtainable by feeding natural water to one of the stages of the hot tower of the 24-stage example used in Sec. 11.3 of this chapter in the simplified analysis of the process.

The McCabe-Thiele diagram for this process, Fig. 13.28, shows that the deuterium content of the liquid phase flowing down through the hot tower drops to feed level x_F between the third and second stages from the bottom of that tower. By feeding additional hot water at rate F' to the second or first stage, it should therefore be possible to increase deuterium production P at constant H_2S circulation rate G, although at the cost of increased tails assay x_W, reduced fractional deuterium recovery, and higher heat requirements.

Analysis of the increased deuterium production made possible through use of supplementary hot water feed will be made by reference to Fig. 13.35. Here it is assumed that the flow rate of supplementary feed F' to the top of stage number n_S of the hot tower and the product rate P are so adjusted that the deuterium content of water flowing from stage $n_S + 1$ to stage n_S equals that of natural water feed x_F, to prevent mixing loss at the supplementary

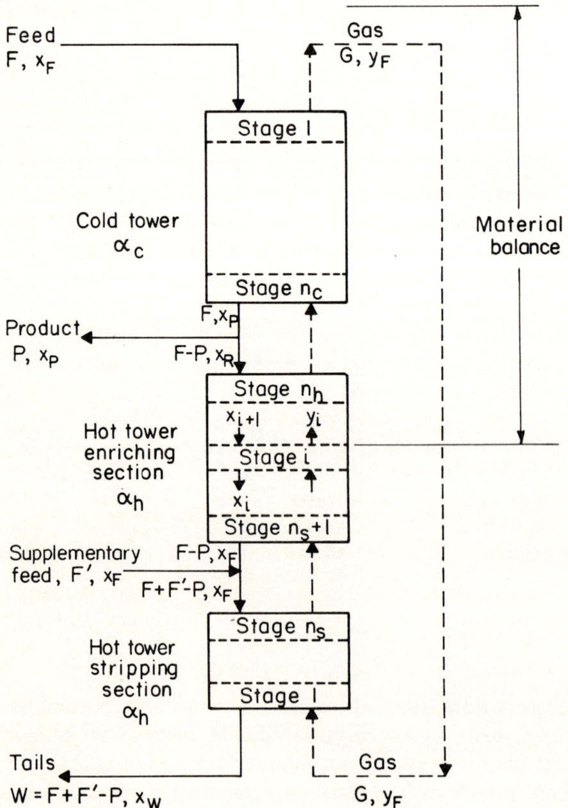

Figure 13.35 Nomenclature for dual-temperature process with supplementary feed.

feed point. Primary natural water feed rate F and gas circulation rate G are assumed to remain unchanged, with their ratio at the optimum value $G/F = 2.03$ found for the previous case with no supplementary feed. Burgess [B14] has shown that G/F should be increased slightly to obtain maximum benefit from supplementary feed, but this refinement has been neglected to simplify the subsequent analysis somewhat.

Under these assumptions, conditions in the cold tower will remain unchanged. Equation (13.122) still relates the product liquid assay x_P, natural feed liquid assay x_F, and cold-tower gas effluent assay y_F. With $n_C = 16$, $G/F = 2.03$, $x_P/x_F = 4.0$, and $\alpha_C = 2.32$, $y_F/x_F = 0.4558$.

Equation (13.170) for the separation performance of the stripping section of the hot tower is obtained by analogy with Eq. (13.125), by substituting x_F for x_P and n_S for n_h.

$$x_F = \frac{(\alpha_h y_F - x_W)(G/W\alpha_h)^{n_S+1} + x_W - (G/W)y_F}{1 - G/W\alpha_h} \tag{13.170}$$

By overall material balance,

$$W = F + F' - P \tag{13.171}$$

By deuterium material balance,

$$x_W = \frac{(F + F')x_F - Px_P}{F + F' - P} \tag{13.172}$$

An equation for the separation performance of the enriching section of the hot tower is derived by reference to Fig. 13.35. The deuterium balance for the entire plant above stage i of the hot-tower enriching section (between the dashed lines) is

$$Px_P + (F - P)x_{i+1} + Gy_F = Fx_F + Gy_i \tag{13.173}$$

At low deuterium content, the equilibrium relation for stage i is

$$y_i = \frac{x_i}{\alpha_h} \tag{13.174}$$

By eliminating y_i from (13.173) and (13.174), difference equation (13.175) for the liquid-phase deuterium atom fraction x is obtained:

$$x_{i+1} = \frac{G}{F-P}\left(\frac{x_i}{\alpha_h} - y_F\right) + \frac{Fx_F - Px_P}{F - P} \tag{13.175}$$

The solution of (13.175) for the boundary conditions $x_i = x_P$ at $i = n_h + 1$ and $x_i = x_F$ at $i = n_S + 1$ is

$$x_P = \frac{[P(x_P - x_F) + Gy_F - Gx_F/\alpha_h]\,[G/(F-P)\alpha_h]^{n_h - n_S} + Fx_F - Gy_F}{F - G/\alpha_h} \tag{13.176}$$

Equation (13.176) provides an implicit relation between the product/feed ratio P/F and the number of enriching stages $n_h - n_S$. For the present case, with $x_P/x_F = 4$, $y_F/x_F = 0.4558$, $\alpha_h = 1.80$, $G/F = 2.03$, and $n_h = 24$.

$$24 - n_S = \ln\left(\frac{0.58584}{0.20250 - 3P/F}\right) \Big/ \ln\left(\frac{1.12778}{1 - P/F}\right) \tag{13.177}$$

To complete the analysis, it is necessary to find the amount of supplementary feed for the hot tower, F', for a given number of stripping stages n_S. To do this, W and x_W are eliminated from (13.170) by means of (13.171) and (13.172), and the resulting equation is solved for $n_S + 1$:

$$n_S + 1 = \ln\left[\frac{(P/F)(x_P/x_F - 1) + (G/F)(y_F/x_F - 1/\alpha_h)}{(1 + F'/F - P/F)(\alpha_h y_F/x_F) - (1 + F'/F - Px_P/Fx_F)}\right] \Bigg/ \ln\left[\frac{G/F}{\alpha_h(1 + F'/F - P/F)}\right] \tag{13.178}$$

Substitution of the given values for $x_P/x_F = 4$, $G/F = 2.03$, $y_F/x_F = 0.4558$, and $\alpha_h = 1.80$ yields

$$n_S + 1 = \ln\left[\frac{0.20250 - 3P/F}{0.17954(1 + F'/F) - 3.17954P/F}\right] \Bigg/ \ln\left(\frac{1.1278}{1 + F'/F - P/F}\right) \tag{13.179}$$

Numerical solution of Eqs. (13.177) for P/F and (13.179) for F'/F with auxiliary feed to the top of the second stage of the hot tower ($n_S = 2$) or to the first stage ($n_S = 1$) yields the results of Table 13.25, where they are compared with the case of no supplementary feed.

Figure 13.36 compares the above results for this 24-plate case, without reoptimization of the feed rate to the cold tower, with Burgess' [B14] calculations for the Savannah River plant, in which feed to the cold tower was reoptimized for maximum production.

Determination of the economic proportion of supplementary feed to the hot tower involves balancing the advantage of increased production against the extra costs of preheating additional feed water and stripping H_2S from additional waste. In the stripping section of the hot tower, larger downcomers would be needed for the increased liquid flow, and at some value, a larger tower diameter. In a new plant designed for it, some supplementary feed to the hot tower would seem to be advantageous. It would probably be neither practical nor economical to use more than 50 percent extra feed to the hot tower.

12 DUAL-TEMPERATURE AMMONIA-HYDROGEN EXCHANGE PROCESS

The dual-temperature principle for providing reflux for the ammonia-hydrogen deuterium exchange process was proposed by the British firm Constructors John Brown [C12], has been tested in pilot-plant experiments conducted by Friedrich Uhde Gmbh at the plant of Farbwerke Hoechst in Germany [W2], and is to be used in a commercial plant at Talcher, India (item 19, Table 13.2), being constructed by Uhde.

Figure 13.37 is a material flow sheet for a dual-temperature ammonia-hydrogen exchange plant using the same amount of synthesis-gas feed and producing the same amount of enriched ammonia product as the monothermal ammonia-hydrogen exchange plant of Fig. 13.23. Comparison of these figures shows that the hot exchange column of Fig. 13.37 performs both the function of the ammonia dissociation step D of Fig. 13.23 in providing enriched

Table 13.25 Increase in heavy-water production resulting from supplementary feed to hot tower

Supplementary feed to plate number n_S =	Ratio, supplementary feed to feed, F'/F	Ratio, production/ feed, P/F	Percent production increase
None	0	0.04571	–
2	0.3137	0.04975	8.8
1	0.7685	0.05339	16.8

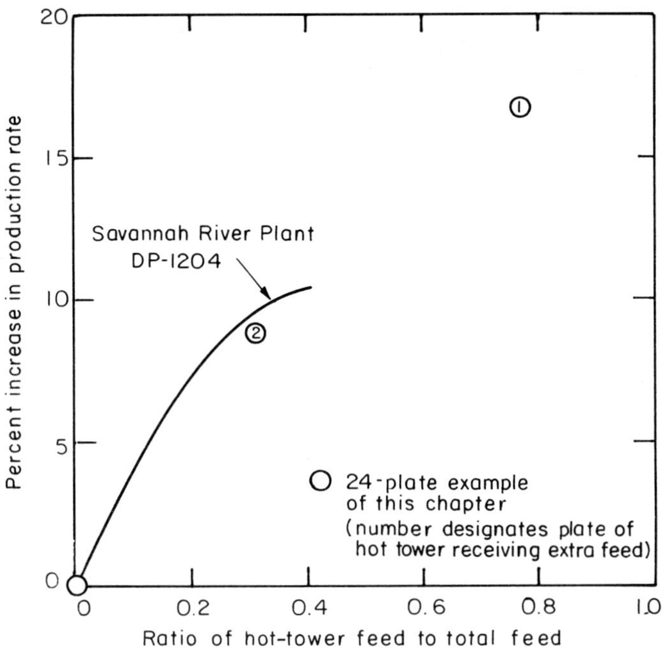

Figure 13.36 Calculated effect of extra feed to hot tower on D_2O production rate in GS plant.

synthesis-gas recycle vapor for the enriching cold column of Fig. 13.37 and the function of the ammonia synthesis step A of Fig. 13.23 in providing depleted liquid ammonia reflux for the stripping cold column of Fig. 13.37. Comparison of these figures shows the advantages of the dual-temperature system to be as follows:

1. Elimination of the ammonia dissociation step D
2. Elimination of the work of recompressing synthesis gas from 55 to 350 atm
3. Elimination of the net heat input needed to dissociate ammonia at 740°C
4. Elimination of the need to synthesize ammonia for reflux and the costs associated with this step
5. Elimination of the catalyst deuterium stripper, F, Fig. 13.23

The dual-temperature system, however, is not without its disadvantages. Because the hot exchange column of Fig. 13.37 returns synthesis gas with a much lower deuterium content than the ammonia dissociation step of Fig. 13.23 and returns liquid ammonia reflux with a much higher deuterium content than the ammonia synthesis step of Fig. 13.23, it is necessary to operate the cold columns of Fig. 13.37 with higher liquid and vapor flow rates than those of Fig. 13.23 and to run them closer to minimum reflux conditions. Consequently, a much larger number of theoretical stages is needed in the cold columns of Fig. 13.37 than in the corresponding columns of Fig. 13.23. In addition, the dual-temperature system requires a large hot exchange column. Table 13.26 compares the liquid and vapor flow rates and number of theoretical stages in the two systems.

Even though flow conditions for the dual-temperature system, Fig. 13.37, were chosen to give a minimum number of stages, the increase from 5.7 stages for the monothermal system to

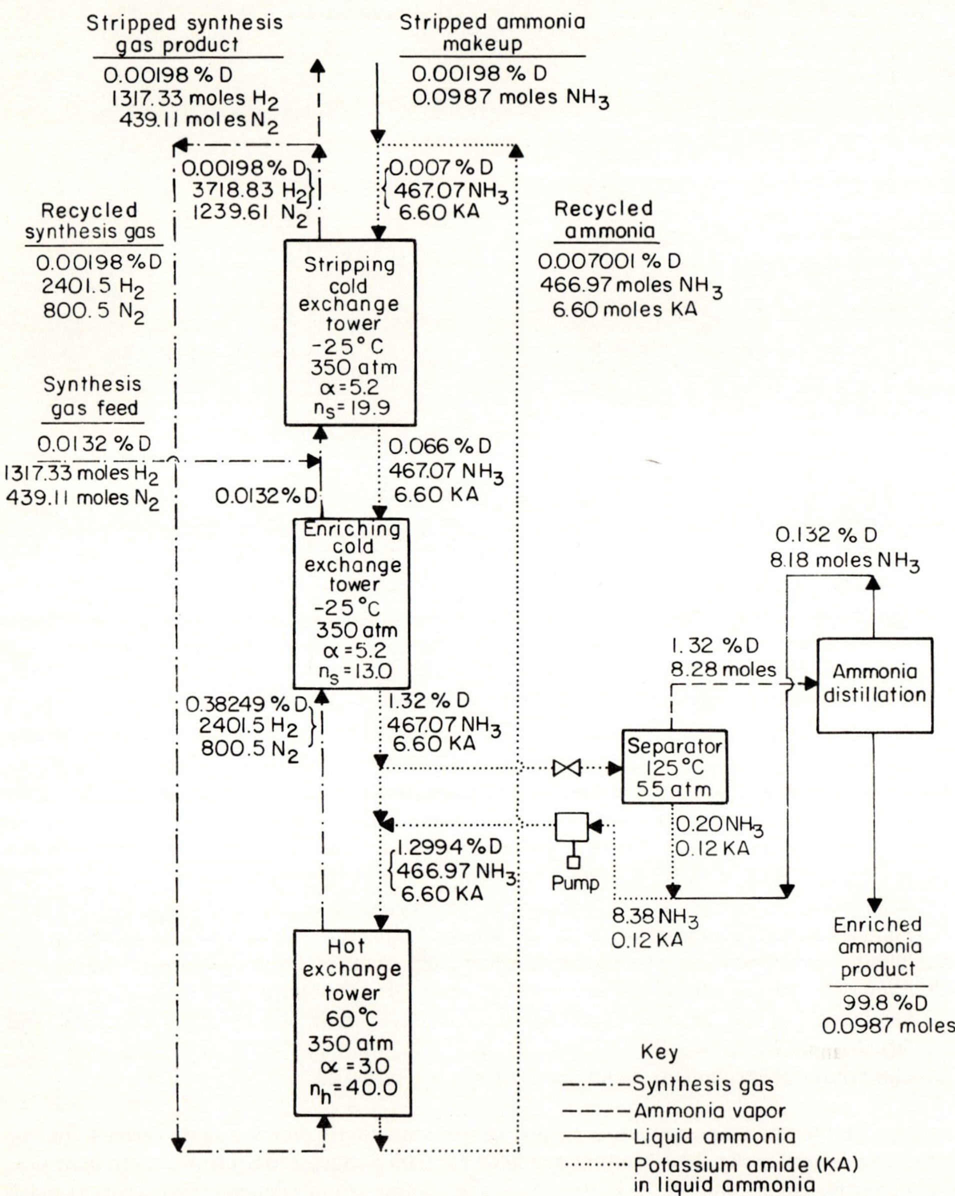

Figure 13.37 Material flow sheet for dual-temperature ammonia-hydrogen exchange process. Flow units, kg-mol/h.

Table 13.26 Comparison of monothermal and dual-temperature ammonia-hydrogen exchange processes

	System	
	Monothermal	Dual temperature
Figure	13.23	13.37
Flow rates, kg-mol/h		
Ammonia	439	467
Synthesis gas		
Stripping, cold	2642	4957
Enriching, cold	886	3202
Hot	None	3202
Number of stages in columns		
Stripping, cold	2.6	19.9
Enriching, cold	3.1	13.0
Hot	0	40.0
Total	5.7	72.9

72.9 for the dual-temperature system must be viewed as a serious disadvantage of the latter. Another disadvantage of the dual-temperature flow sheet would be the complicated heat exchange system needed for heat recovery and humidification between the hot and cold towers, which is not shown in Fig. 13.37.

Nitschke [N2] has given a partial description of the flow sheet used by Uhde for the dual-temperature ammonia-hydrogen heavy-water plant that company is building at Talcher, India (item 19, Table 13.2). Figure 13.38 is a qualitative material flow sheet for the first-stage exchange columns of that plant.

Feed for this heavy-water plant consists of synthesis gas for the ammonia plant of the Indian Department of Atomic Energy, at 190 to 200 atm. The heavy-water plant, however, operates at 300 atm. To avoid the need for compressing synthesis gas, and to isolate gas flow in the ammonia plant from gas flow in the heavy-water plant, deuterium in feed synthesis gas is transferred to a solution of potassium amide in ammonia in the transfer column A, and synthesis gas 85 percent stripped of deuterium is returned to the ammonia plant.

Ammonia for the heavy-water plant is pumped to 300 atm, cooled to −25°C, and introduced as feed between the stripping (B) and enriching (C) sections of the first-stage cold exchange tower, where it joins ammonia circulating at the rate L. In C the deuterium content of the ammonia is raised to first-stage product level x_P by exchange against synthesis gas flowing at rate $G_1 + G_2$ whose deuterium content is reduced from y_P to y_F. A portion of the ammonia is sent to the first of two higher stages for further enrichment, and an equal flow of partially depleted ammonia is returned, reducing the deuterium content of ammonia entering the hot enriching section D to x'_P. Here, because of the lower separation factor, the deuterium content of the ammonia is reduced to x'_F, somewhat below that of feed, while the deuterium content of synthesis gas is raised from y_F to y_P. The deuterium content of ammonia is further reduced to the tails level x_W in the hot stripping section E, where the gas flow rate has been reduced to G_2 because of the recycle at rate G_1 to the enriching sections C and D. The gas in E is enriched from y_W to y_F. A portion F of the tails is reenriched to feed level x_F in the transfer column A, and the remainder, L, is fed to the cold stripping column B to be reenriched to feed level while stripping synthesis gas flowing at rate G from y_F to y_W.

The function of the four exchange-column sections can be better understood by reference

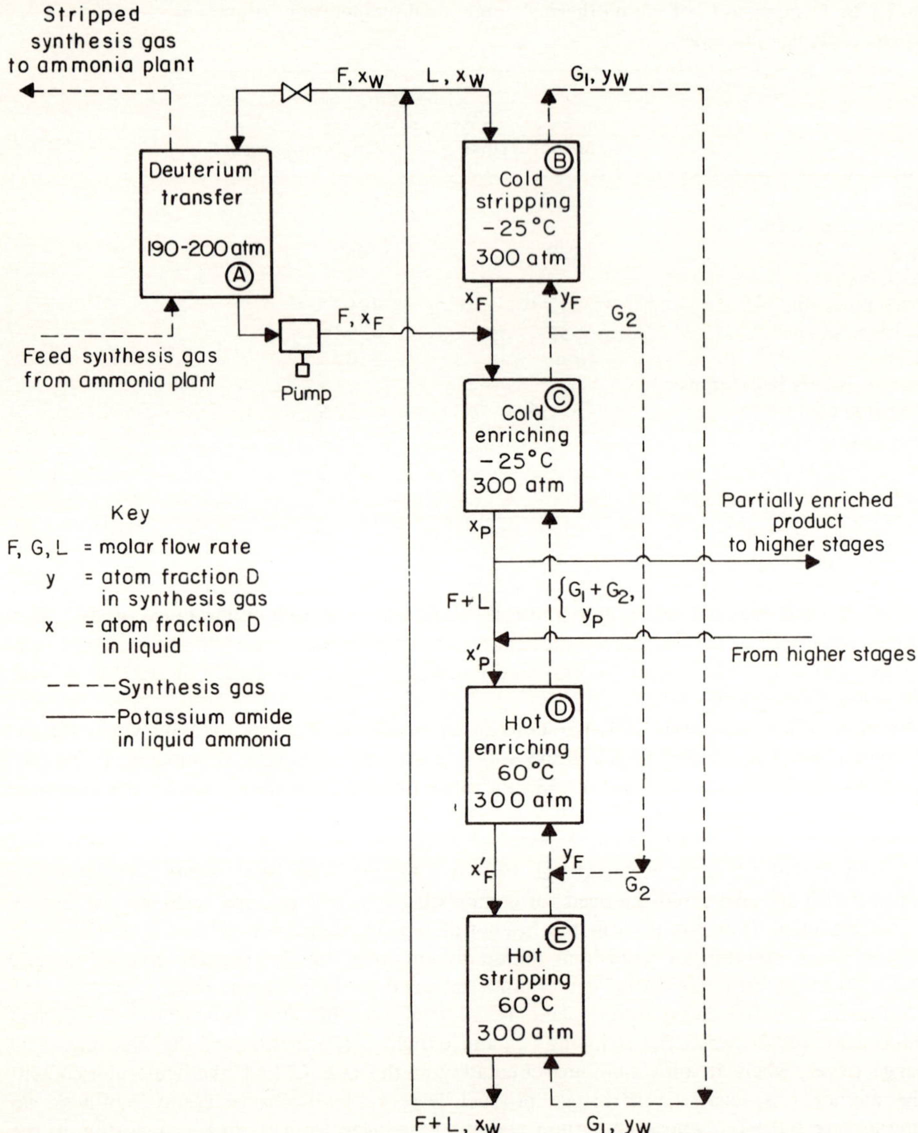

Figure 13.38 Flow scheme for first stage of Talcher dual-temperature ammonia-hydrogen exchange heavy-water plant.

to the qualitative McCabe-Thiele diagram Fig. 13.39, whose nomenclature is keyed to Fig. 13.38. The slopes of the four operating lines are

Cold stripping, L/G_1
Cold enriching, $(F+L)/(G_1+G_2)$
Hot enriching, $(F+L)/(G_1+G_2)$
Hot stripping, $(F+L)/G_1$

By providing enough stripping plates, x_W and y_W could be made as close to zero as desired. By providing enough enriching plates, x_P and y_P could be made as close to unity as desired.

13 METHYLAMINE-HYDROGEN EXCHANGE PROCESSES

The deuterium exchange reaction between liquid methylamine and gaseous hydrogen,

$$CH_3NH_2(l) + HD(g) \rightleftharpoons CH_3NHD(l) + H_2(g)$$

is catalyzed by potassium methylamide, CH_3NHK. This reaction proceeds with sufficient speed at −50°C to permit operation of a cold tower at this temperature, where the equilibrium constant, 7.9, is the highest known for any practical deuterium exchange reaction. The optimum hot-tower temperature for a dual-temperature process using this reaction is +40°C, a limit set by thermal decomposition of potassium methylamide at higher temperatures. At +40°C the deuterium exchange equilibrium constant is 3.6. The ratio of these two separation factors, 7.9/3.6 = 2.19, is also higher than the ratio for any other practical system (Tables 13.17 and 13.18). For this reason, Atomic Energy of Canada, Limited (AECL), has undertaken a development program for a dual-temperature process using methylamine and hydrogen from a synthetic ammonia plant with a flow sheet similar to Fig. 13.37.

Sulzer Brothers Canada, Ltd., working with AECL, has given a partial description [W6] of a dual-temperature flow sheet modified from Fig. 13.37 proposed for use in recovering deuterium from ammonia synthesis gas made from Alberta natural gas containing 135 ppm deuterium. Figure 13.40 is a material flow sheet for the synthesis-gas generation section and first deuterium-enrichment stage of such a heavy-water plant. Deuterium contents have been given as [xN], where x is the ratio of the deuterium content to that of Alberta water containing 135 ppm deuterium. The deuterium contents of methane, water, and hydrogen are those given by Wynn [W6]. The deuterium contents of methylamine streams have been assumed to give a plausible number of stages in the various towers. Total flow quantities

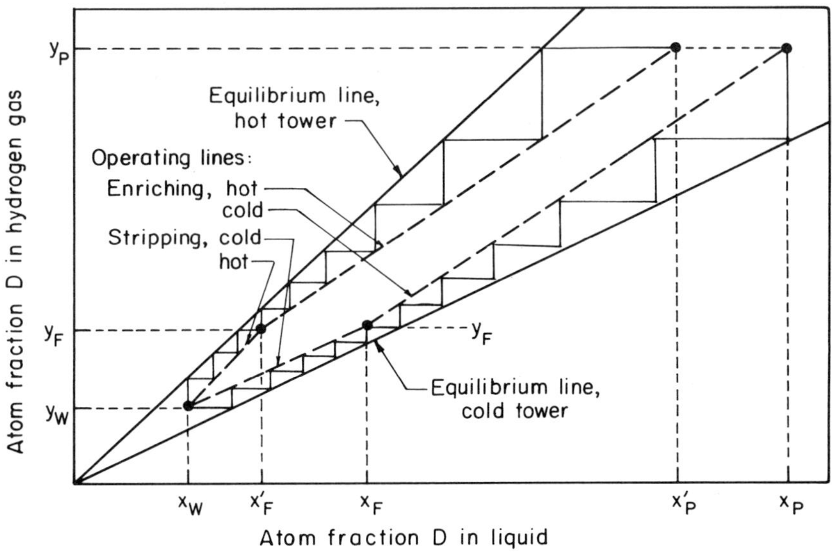

Figure 13.39 McCabe-Thiele diagram for Fig. 13.38.

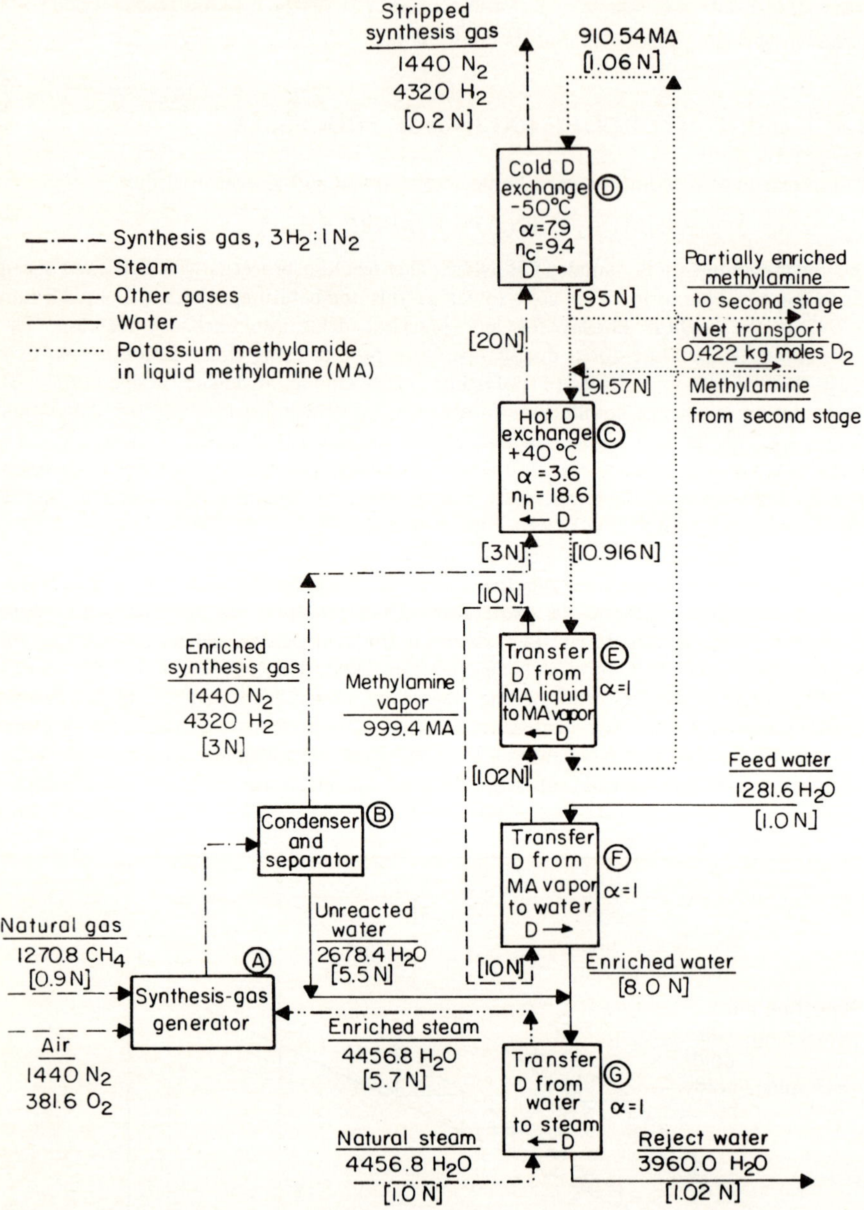

Figure 13.40 Material flow sheet for first stage of Sulzer dual-temperature methylamine-hydrogen exchange heavy-water process. $[N]$ = deuterium content of hydrogen relative to natural water containing 135 ppm. Flow quantities, kg-mol/h.

assumed for this flow sheet are those for a plant permitting production of 1150 MT ammonia/day, after allowing for losses in the ammonia plant. The net deuterium extraction of 0.422 kg-mol/h would produce 66.8 MT D_2O in 330 operating days per year.

The novel feature of this flow sheet is the production of synthesis gas enriched threefold relative to natural water to provide enriched feed for the exchange plant and thus reduce its size. In the synthesis-gas generator (A), natural gas whose deuterium content is 0.9*N* is re-formed with air and enriched steam, 5.7*N*, to produce threefold-enriched synthesis gas and unreacted enriched water, whose enriched deuterium content is recycled. Enriched synthesis gas is the vapor feed to the hot tower (C) of the first dual-temperature exchange stage. Here the deuterium content of synthesis gas is raised from 3*N* to 20*N*, while that of counterflowing methylamine and catalyst is reduced from 91.57*N* to 10.916*N*. In the cold tower (D), the deuterium content of synthesis gas is decreased from 20*N* to 0.2*N* while that of methylamine is increased from 1.06*N* to 95*N*. A portion of the 95-fold-enriched methylamine is fed to the second enriching stage, and an equal amount of partially depleted methylamine is returned; the resultant net flow of 0.422 kg-mol D/h, after further enrichment in higher stages, provides the plant's heavy-water product.

Enriched steam for the synthesis-gas generator (A) is produced in the series of sieve-plate contactors (E), (F), and (G). In (E) deuterium is transferred from methylamine liquid to methylamine vapor, reducing the deuterium content of the liquid from 10.916*N* to 1.06*N* while increasing that of the vapor from 1.02*N* to 10*N*. In (F) deuterium is transferred from methylamine vapor to water, increasing the deuterium content of the latter from 1*N* to 8*N*. This two-step transfer of deuterium from liquid methylamine leaving (C) to water leaving (F) is necessary to prevent chemical reaction between water and the catalyst dissolved in liquid methylamine.

Deuterium in enriched water (8*N*) leaving (F) and unreacted enriched water (5.5*N*) leaving (B) is transferred to steam in step G, producing enriched steam (5.7*N*) for the synthesis-gas generator from natural steam. This transfer step is used instead of simply recycling the water leaving (B) and (F) to avoid returning nonvolatile impurities to the synthesis-gas generator.

Because of the reduced rate of the deuterium exchange reaction at −50°C, the stages of the cold tower (D) are to be of the type developed by Sulzer [L1] for the ammonia-hydrogen exchange process and used in the Mazingarbe plant, Sec. 9.1. For the methylamine-hydrogen system at −50°C, a stage efficiency of 70 percent has been obtained [W6].

At the temperature of the hot tower, 40°C, potassium methylamide slowly decomposes into potassium dimethyl formamidide:

$$2CH_3NH_2 + CH_3NHK \rightarrow CH_3(NK)(CH)NCH_3 + 2H_2 + NH_3$$

This reaction is suppressed by addition of an equimolal amount of lithium methylamide, which has little catalytic activity but inhibits decomposition of the potassium compound.

The great advantage of this methylamine-hydrogen exchange process compared with the dual-temperature ammonia-hydrogen system is the much smaller number of stages needed with methylamine. Intratower flow rates relative to product D_2O with methylamine are also smaller than with ammonia. Table 13.27 compares the two processes. The lower internal flow rates with methylamine also lead to lower utility requirements. A disadvantage of this methylamine-hydrogen flow sheet is the need to operate the synthesis-gas-generating section of the ammonia plant with enriched water. This necessitates recycle and strict control of losses of unreacted deuterium-enriched steam and water.

14 DUAL-TEMPERATURE WATER-HYDROGEN EXCHANGE PROCESSES

Section 7.4 described the development in Canada [S8] of a catalyst for the deuterium exchange reaction between hydrogen and liquid water that is not inactivated when submerged in water.

Table 13.27 Comparison of dual-temperature ammonia-hydrogen and methylamine-hydrogen exchange processes

	Fig. 13.37	Fig. 13.40
	Ammonia process	Methylamine process
Deuterium content relative to feed		
First-stage product	100	95
Stripped synthesis gas	0.15	0.2
Number of stages		
Cold	32.9	9.4
Hot	40.0	18.6
Molal flow rates, relative to product D_2O		
Hydrogen feed	8,898	10,237
Hydrogen, cold tower, stripping	25,119	10,237
Liquid	3,200	2,160

Availability of this catalyst has led to interest in its possible use in dual-temperature water-hydrogen exchange processes. With liquid-water feed and recirculated hydrogen gas, this catalyst could be used in a dual-temperature process similar in principal to the GS process, with a schematic flow sheet like Fig. 13.25. With ammonia synthesis-gas feed and recirculated water, this catalyst could be used in a dual-temperature process similar to the ammonia-hydrogen process flow scheme of Fig. 13.37, provided that impurities in synthesis-gas feed that would poison the catalyst can be recovered sufficiently completely.

Miller and Rae [M7] have suggested process conditions for a dual-temperature process using this catalyst at 69 atm pressure and temperatures of 50°C for the cold tower and 170°C for the hot. These conditions have been used to estimate optimum flow rates and numbers of theoretical stages for dual-temperature water-hydrogen processes using these two flow schemes. The results are tabulated in Table 13.28 and compared with similar data for the other dual-temperature processes discussed previously.

With water feed, the water-hydrogen exchange process has the advantages of lower gas and liquid flow rates and fewer stages than the water-hydrogen sulfide process. Utility requirements would also be smaller. Disadvantages of the hydrogen process are the higher pressure and the need to use large volumes of an expensive catalyst. If the catalyst were sufficiently active and not too expensive, the hydrogen process might be economically attractive.

With synthesis-gas feed, the water-synthesis-gas exchange process appears to be at a disadvantage relative to the ammonia and methylamine exchange processes because the water process has the highest flow rates and the largest number of stages.

15 EXCHANGE PROCESSES FOR SEPARATION OF LITHIUM ISOTOPES

Saito [S1] has patented separation of lithium isotopes by countercurrent exchange between lithium amalgam and lithium chloride or bromide dissolved in dimethyl formamide or other organic solvent. Arkenbout [A2] has measured a separation factor of 1.05 for this process, with ^6Li concentrating in the amalgam phase. With countercurrent flow through a packed column, natural lithium (7.5 percent ^6Li) was separated into 5.8 percent ^6Li at the top of a 1-m column and 12 percent ^6Li at the bottom. Reflux at the bottom was obtained by making the amalgam the anode (positive electrode) of an electrolytic cell in contact with the organic solution of the lithium salt. Reflux at the top was obtained by crystallizing lithium salt from organic solvent, dissolving it in water, and electrolyzing the aqueous solution at a mercury cathode.

Saito and Dirian [S2] have patented separation of lithium isotopes by countercurrent exchange between lithium amalgam and an aqueous solution of lithium hydroxide, with ^6Li concentrating in the amalgam phase. Reflux at the bottom is obtained by making the amalgam the anode of an electrolytic cell against an aqueous solution of LiOH. Reflux at the top is obtained by the reverse reaction, which takes place spontaneously between lithium amalgam and water. The simpler cathodic process is an advantage of this system compared with the previous one using an organic solvent. A disadvantage is the spontaneous transfer of lithium from amalgam to aqueous phase by chemical reaction that takes place as amalgam flows through the column. This has to be reversed by applying a negative potential to the amalgam either continuously or at intervals. Saito and Dirian report a separation factor of 1.06 to 1.07. Collén [C10] reports 1.069 ± 0.004. A process like this was used in the Y-12 plant of the U.S. AEC.

16 EXCHANGE PROCESSES FOR OTHER ELEMENTS

16.1 Separation Factors

Table 13.29 is a partial list of separation factors for exchange of isotopes of carbon, nitrogen, oxygen, and sulfur between an aqueous solution and a gas phase containing compounds of these elements.

Table 13.28 Comparison of dual-temperature processes

	Feed				
	Water		Synthesis gas		Enriched synthesis gas
Liquid	H_2O	H_2O	H_2O	NH_3	CH_3NH_2
Gas	H_2S	H_2	Synthesis gas, $3H_2/1N_2$		
Flow scheme, Fig.	13.25	13.25	13.37	13.37	13.40
Percent deuterium recovery	18.4	30	85	85	–
Pressure, atm	20	69	69	350	60
Temperature, °C, Cold	32	50	50	−25	−50
Hot	138	170	170	60	40
Separation factor, Cold	2.32	3.3	3.3	5.2	7.9
Hot	1.80	2.1	2.1	3.0	3.6
Percent deuterium, feed	0.0149	0.0149	0.0132	0.0132	0.0405
First-stage product	1.32	1.32	1.32	1.32	1.28
First-stage molal flow rate per mole 99.8% D_2O product					
Cold gas	74,400	58,800	42,700	33,500	13,600
Liquid	36,400	22,300	9,700	3,200	2,160
Number of theoretical stages					
Cold tower	48	27	39	33	9.4
Hot tower	49	28	43	40	18.6

Table 13.29 Separation factors for isotopic exchange

Reactants	Products	Separation factor	Reference
$C^{16}O_2 + H_2{}^{18}O(l)$	$C^{16}O^{18}O + H_2{}^{16}O(l)$	1.044	[K2]
$^{13}CO_2 + H^{12}CO_3{}^-$	$^{12}CO_2 + H^{13}CO_3{}^-$	1.012	[H8, R5]
$H^{12}CN + {}^{13}CN^-$	$H^{13}CN + {}^{12}CN^-$	1.013	[H8, R6]
$^{14}NH_3 + {}^{15}NH_4{}^+$	$^{15}NH_3 + {}^{14}NH_4{}^+$	1.034	[T5]
$^{15}NO + H^{14}NO_3(aq)$	$^{14}NO + H^{15}NO_3(aq)$	1.055	[T2]
$^{34}SO_2 + H^{32}SO_3{}^-$	$^{32}SO_2 + H^{34}SO_3{}^-$	1.019	[S9]
$^{36}SO_2 + H^{32}SO_3{}^-$	$^{32}SO_2 + H^{36}SO_3{}^-$	1.040	[S9]

Although these values are close to unity, each is greater than the separation factor in distillation for compounds of the corresponding element. Processes based on these exchange reactions thus have been used for laboratory-scale separations of these isotopes described in the references of Table 13.29. In 1977, however, none was being used on a semiindustrial scale in the United States. The exchange processes for carbon, oxygen, and nitrogen isotopes have been replaced by low-temperature distillation of CO and NO, described in Chap. 12, Secs. 2.5, 2.6, and 2.7. Even though the separation factor in distillation is smaller, distillation is preferred because reflux is obtained by providing and removing heat, whereas in chemical exchange reflux must be obtained by chemical reaction. The dual-temperature principle cannot be applied to elements other than hydrogen because hot and cold separation factors are so close together that it would be impossible to control liquid-vapor flow ratio with the requisite precision.

16.2 Separation of Nitrogen Isotopes

To give an example of one of the most successful applications of chemical exchange to separation of isotopes of an element heavier than hydrogen that may have industrial application, a brief description will be given of the process and equipment used by Taylor and Spindel [T2] to product ^{15}N 99.8 percent pure. This separation depends on the exchange reaction

$$^{15}NO + H^{14}NO_3 \rightleftharpoons {}^{14}NO + H^{15}NO_3$$

which takes place in the gas phase because of the presence there of the species NO, NO_2, N_2O_3, N_2O_4, H_2O, HNO_2, and HNO_3. These interact at acceptably high rates at temperatures of 25°C or higher. The separation factor for this process, defined as the ratio of $^{15}N/^{14}N$ in the liquid phase to $^{15}N/^{14}N$ in the gas phase, was found by Taylor and Spindel to be 1.055 at 25°C in 10 M HNO_3, and to decrease with increasing acid concentration and increasing temperature. Because the value of the equilibrium constant for the foregoing reaction calculated from spectroscopic data is 1.096, it appears that isotopic exchange reactions between species other than HNO_3 and NO enter into the observed overall exchange equilibrium. This reaction, however, may be used to characterize the process.

Taylor and Spindel found that the optimum conditions for operating this process on the laboratory scale were 8 to 10 M HNO_3, 25 to 50°C, and atmospheric pressure. Although a higher temperature speeds up attainment of exchange equilibrium, α is lower, and more NO_2 is present with a lower exchange equilibrium constant.

The process used by Taylor and Spindel is illustrated in Fig. 13.41. Liquid aqueous HNO_3 flows downward through a packed column countercurrent to an upflowing gas stream consisting largely of NO with lesser amounts of other nitrogen compounds. Nitric acid containing the normal abundance of ^{15}N, 0.365 a/o, is fed at the top of the larger column, no. 1, and is enriched in ^{15}N by the foregoing exchange reaction as it flows down this column. At the foot of the column, where its ^{15}N content is around 7 percent, the reflux ratio of NO vapor to product may be substantially reduced. This is done by diverting 4 percent of the acid downflow to the smaller column, no. 2. The remaining 96 percent of the acid downflow is sent to NO reflux generator no. 1, where it is reduced to NO by reaction with SO_2:

$$H_2O + HNO_3 + \tfrac{3}{2}SO_2 \rightarrow \tfrac{3}{2}H_2SO_4 + NO$$

The NO is returned to column no. 1 as reboil vapor.

The HNO_3 flowing down through the smaller column, no. 2, countercurrent to NO is enriched further in ^{15}N to 99.8 percent at the foot of the column. At this point some of the downflowing HNO_3 is withdrawn as plant product, and the remainder of the HNO_3 is reduced to NO with SO_2 in reflux generator no. 2. This NO is used to reboil column no. 2.

NO vapor depleted in ^{15}N leaving column no. 1 at the top of the plant is converted to HNO_3 depleted in ^{15}N by mixing it with air and passing the mixture counter to downflowing water in a packed column, where the reaction

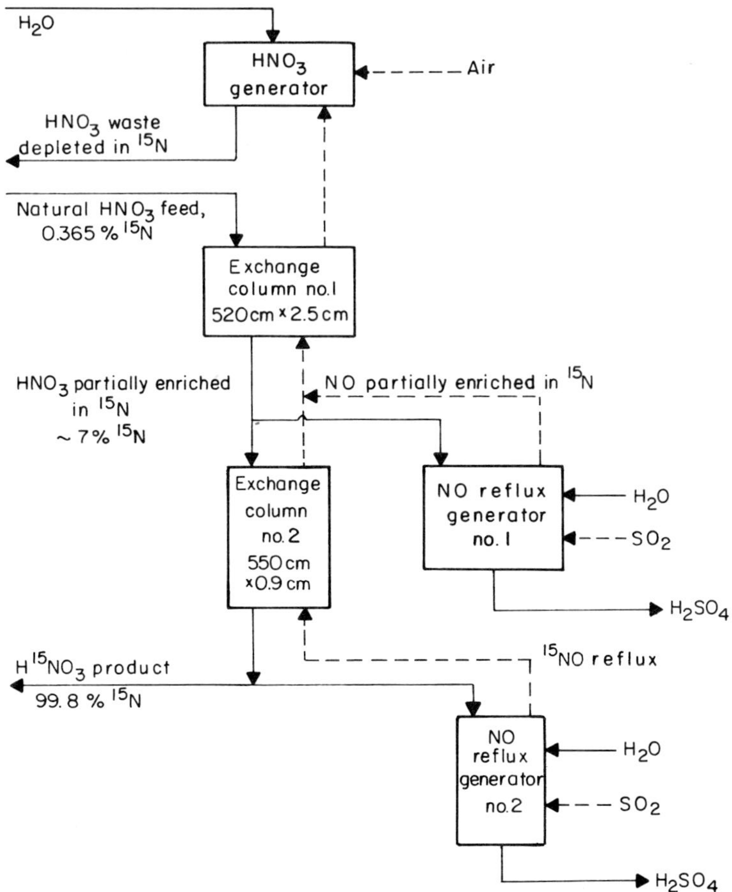

Figure 13.41 Plant used for production of ^{15}N by Taylor and Spindel.

$$NO + \tfrac{3}{4}O_2 + \tfrac{1}{2}H_2O \rightarrow HNO_3$$

takes place.

The net result of the process, then, is to separate HNO_3 containing the natural abundance of ^{15}N into product HNO_3 highly enriched in ^{15}N and waste HNO_3 slightly depleted in ^{15}N, while converting SO_2 and air to H_2SO_4. The minimum ratio of H_2SO_4 to ^{15}N is $\frac{3}{2}$ times the minimum molal reboil vapor ratio, which is given by Eq. (12.80), or

$$\frac{3}{2}\frac{x_P - x_F}{x_F}\frac{\alpha}{\alpha - 1} = \frac{3}{2}\frac{0.998 - 0.00365}{0.00365}\frac{1.055}{0.055} = 7838 \qquad (13.180)$$

This high reflux requirement is not a complete economic drain because H_2SO_4 is a more valuable material than SO_2. In this respect, Taylor and Spindel's process is in a more favorable economic position than the chemical exchange system of Fig. 13.24 to concentrate deuterium, which consumes aluminum to make less valuable Al_2O_3.

In their engineering analysis of the HNO_3-NO process, Garrett and Schacter [G2] considered a plant to produce 30.2 g-mol ^{15}N/day while simultaneously producing 239,670 g-mol H_2SO_4/day. They recommended use of substantially the same conditions employed by Taylor and Spindel and estimated that ^{15}N could be produced at a cost of \$4/g. This relatively low cost is due to the credit for converting SO_2 to H_2SO_4.

It is important to note that the use of a cascade of columns of decreasing size, such as in Fig. 13.41, does not affect the consumption of chemicals for reflux, because this depends on the interstage flow required at the feed point. The cascade of columns of decreasing size does, however, reduce the total volume and the holdup of desired isotope. If the cascade of columns were not used for the ^{15}N separation example, with its low feed concentration and separation factor close to unity, the holdup would be so great that product concentration would not reach 99.8 percent in any practical time.

NOMENCLATURE

a defined by Eq. (13.34)
A annual cost, \$/year
A tower cross-sectional area
b defined by Eq. (13.36)
c defined by Eq. (13.35)
c unit cost
D separative capacity
E efficiency
F molar feed rate
F' molar flow rate of supplementary feed to hot tower
g ratio of steam rate to minimum rate
G vapor molar flow rate
h height of transfer unit
H moles of hydrogen
H humidity, mol water/mol noncondensible gas
I inventory, mol
j annual charge against investment
k equilibrium constant for gas-phase exchange reaction
K total tails flow rate
K equilibrium constant for gas-liquid exchange reaction

L liquid molar flow rate
M molecular weight
n number of stages
p pressure
P mol product (or molar product flow rate)
P kg D_2O/year
Q rate of loss of availability
r fractional recovery
R gas constant
s entropy per mole
S entropy
S solubility, mol dissolved gas/mol water
t time
T absolute temperature
T_0 absolute temperature at which heat is rejected
v vapor velocity, cm/s
V tower volume
V vapor molar flow rate
W moles of tails (or molar tails flow rate)
W power
x atom fraction or mole fraction in liquid
y atom fraction or mole fraction in vapor
z distance from top of tower
Z height of tower
α stage separation factor
α^* relative volatility, separation factor in distillation
β heads separation factor
γ relative volatility of H_2S to HDS
η relative abundance in vapor
ξ relative abundance in liquid
π vapor pressure
ρ density
ω overall separation factor

Subscripts

a stream produced in heating liquid or cooling gas
b bottom of tower
c cold tower
F feed stream
h hot tower
i stage number
m stage number
P product stream
Q turbine work
S supplementary feed point
t top of tower
V tower volume
W tails stream
0 gas stream from cold tower to hot tower, Fig. 13.34

REFERENCES

A1. Akar, P., and G. Simonet: *PICG(2)* **4**: 522 (1958).
A2. Arkenbout, G. J.: *PICG(3)* **12**: 478 (1964).
A3. Armstrong, G. T.: "A Compilation of Vapor Pressure Data of Deuterium Compounds," Natl. Bur. Std. (U.S.) Report 2306, 1953.
B1. Babcock, D. F.: U.S. Patent 3,549,323, Dec. 22, 1970.
B2. Baertschi, P., and W. Kuhn: *PICG(1)* **8**: 411 (1956).
B3. Bailey, B. M.: *PICG(2)* **4**: 556 (1958).
B4. Banikiotes, G. C., E. Cimler, and M. C. Sze: *Chem. Eng. Progr. Symp. Ser. 39* **58**: 17 (1962).
B5. Barr, F. T.: U.S. Patent 2,676,785, Apr. 27, 1954.
B6. Bebbington, W. P., and V. R. Thayer: *PICG(2)* **4**: 527 (1958).
B7. Bebbington, W. P., and V. R. Thayer: Report DP-400, July 1959.
B8. Bebbington, W. P., and V. R. Thayer: *Chem. Eng. Progr.* **55**(9): 70 (1959).
B9. Becker, E. W.: "Production of Heavy Water," International Atomic Energy Agency, Vienna, 1961.
B10. Becker, E. W., R. P. Hübener, and R. W. Kessler: *Chem.-Ing.-Tech.* **30**: 288 (1958).
B11. Bigeleisen, J.: *Proceedings of the International Symposium on Isotope Separation*, Interscience, New York, 1958, p. 130.
B12. Bigeleisen, J., and M. G. Mayer: *J. Chem. Phys.* **15**: 261 (1947).
B13. Brun, J., and T. Varberg: *Kgl. Norske Videnskab. Selskabs. Forh.* **26**(6): 19 (1953); J. Brun, T. Varberg, W. Gundersen, and R. Solli: ibid. **29**(2): 5 (1956); J. Brun, W. Gundersen, and T. Varberg: ibid. **30**(5): 30 (1957).
B14. Burgess, M. P.: Report DP-1204, Aug. 1969.
C1. Cerrai, E., et al.: *Chem. Eng. Progr. Symp. Ser. 11* **50**: 271 (1954).
C2. Clayes, Y., J. Dayton, and W. K. Wilmarth: *J. Chem. Phys.* **18**: 759 (1950).
C3. Clusius, K., et al.: *FIAT Rev. Ger. Sci.*, 1939–1946, Physical Chemistry.
C4. Clusius, K., and H. Meyer: *Helv. Chim. Acta* **36**: 2045 (1953).
C5. Clusius, K., and K. Schleich: *Helv. Chim. Acta* **41**: 1342 (1958).
C6. Clusius, K., K. Schleich, and M. Vecchi: *Helv. Chim. Acta* **42**: 2654 (1959).
C7. Clusius, K., and K. Starke: *Z. Naturforsch.* **4A**: 549 (1949).
C8. Clusius, K., L. Stavely, and G. Dickel: *Z. Phys. Chem. B* **50**: 403 (1941).
C9. Cohen, K.: *The Theory of Isotope Separation*, McGraw-Hill, New York, 1951.
C10. Collén, B.: *Acta Chem. Scand.* **18**: 805 (1964).
C11. Combs, R. L., J. M. Googin, and H. A. Smith: *J. Chem. Phys.* **58**: 1000 (1954).
C12. Constructors John Brown: Circular distributed at Industrial Exposition, Geneva, Sept. 1958.
C13. Craig, H.: Personal communication to M. Benedict, Aug. 30, 1960.
D1. Damiani, M., R. Winkler, and M. Huber: *Sulzer Tech. Rev.*, Nuclex 75 Issue, 92 (1975).
D2. Denton, W. H., B. Shaw, and D. E. Ward: *Trans. Inst. Chem. Eng.* **36**: 179 (1958).
D3. Dixon, O. G.: *J. Soc. Chem. Ind.* **68**: 88 (1949).
D4. Dostrovsky, I.: *PICG(2)* **4**: 605 (1958).
D5. Dostrovsky, I., J. Gillis, D. R. Llewellyn, and B. H. Vromen: *J. Chem. Soc.* 3517 (1952).
E1. Elwood, P.: *Chem. Eng.*, July 1, 1968, pp. 56–58.
E2. Erikson, B. J.: *Nuclear Eng.* **9**: 409 (1960); *Chem. Process Eng.* **2**: 53 (1960).
F1. *Federal Register*, **42**: 14768 (Mar. 16, 1977).
F2. Friedman, I.: *Geochim. Cosmochim. Acta* **4**: 89 (1953).
G1. Gami, D. C., D. Gupta, N. B. Prasad, and K. C. Sharma: *PICG(2)* **4**: 534 (1958).
G2. Garrett, G. A., and J. Schacter: *Proceedings of the International Symposium on Isotope Separation*, Interscience, New York, 1958, p. 17.

G3. Groth, W.: *Z. Elektrochem.* **54**: 5 (1950).
G4. Groth, W., H. Ihle, and A. Murrenhoff: *Angew. Chem.* **68**: 605 (1956).
G5. Groth, W., and P. Harteck: *Z. Elektrochem.* **47**: 167 (1940).
H1. Hammerli, M., Letter to M. Benedict, Mar. 16, 1977.
H2. Hammerli, M., W. H. Stevens, W. J. Bradley, and J. P. Butler: Report AECL-5512, Apr. 1976.
H3. Hänny, J.: *Kältetechnik* **12**(6): 158 (1960).
H4. Haul, R., H. Behnke, and H. Dietrich: *Angew. Chem.* **71**: 64 (1959).
H5. Haywood, L. R., and P. B. Lumb: *Chem. Can.* **27**: 19 (Mar. 1975).
H6. Horiba, Y., and M. Kobayakawa: *Bull. Chem. Soc. Japan* **33**: 116 (1960).
H7. Huber, M., and A. Sperandio: *Sulzer Tech. Rev.* **46**(4): 177 (1970).
H8. Hutchison, C. A., D. W. Stewart, and H. C. Urey: *J. Chem. Phys.* **8**: 532 (1940).
H9. Hutchison, D. A.: *J. Chem. Phys.* **14**: 401 (1946).
H10. Hutchison, D. A.: *Phys. Rev.* **75**: 1303 (1949).
H11. Hydrocarbon Research, Inc.: "Low-Temperature Heavy Water Plant," Report NYO-889 to USAEC, Mar. 15, 1951.
J1. Johns, T. F.: *Proc. Phys. Soc.* **B66**: 808 (1953).
J2. Johns, T. F.: *Progr. Nucl. Phys.* **6**: 1 (1957).
J3. Johns, T. F., H. Kronenberger, and H. London: *Mass Spectrometry*, Institute of Petroleum, London, 1950, pp. 141–147.
J4. Johnston, H. L., and C. A. Hutchison: *J. Chem. Phys.* **8**: 869 (1940).
J5. Jones, D. W.: "Deuterium," in *Encyclopedia of Chemical Processing and Design*, vol. 1, Dekker, New York, 1976.
K1. Keesom, W. H., and J. Haantjes: *Physica* **2**: 986 (1935).
K2. Kirschenbaum, I.: *Physical Properties of Heavy Water*, McGraw-Hill, New York, 1951.
K3. Kirschenbaum, I., and H. C. Urey: *J. Chem. Phys.* **10**: 712 (1942).
K4. Kiss, I., and L. Matus: *Magy. Tud. Akad. Kosp. Fiz. Kut. Int. Koslemen.* **10**: 61 (1962).
K5. Kremser, A.: *Natl. Petroleum News* **22**(21): 42 (May 21, 1930).
L1. Lefrancois, B.: *Proceedings of Conference on Techniques and Economy of Production of Heavy Water, Turin, Italy, 1970*, Comitato Nazionale Energia Nucleare, Rome, 1971, pp. 197–208.
L2. Lehmer, W., A. Sellmaier, and W. Baldus: *Linde Ber. Tech. Wissensch.* **5**: 3 (1959).
M1. Malkov, M. P., A. G. Zeldovitch, A. B. Fradkov, and I. B. Danilov: *PICG(2)* **4**: 491 (1958).
M2. Mantel, C. L.: *Industrial Electrochemistry*, 3d ed., McGraw-Hill, New York, 1950, pp. 462–466.
M3. McClure, D. S., and C. E. Herrick, Jr.: Report A-582, Apr. 8, 1943.
M4. McWilliams, J. A., H. R. C. Pratt, F. R. Dell, and D. A. Jones: *Trans. Inst. Chem. Eng.* **34**: 17 (1956).
M5. Meier, W.: *Sulzer Tech. Rev.* **52**(3): 147 (1970).
M6. Meier, W., et al.: "Sulzer Experience with DW Systems," Paper presented at AECL Symposium on Heavy Water Distillation, Apr. 1976.
M7. Miller, A. I., and H. K. Rae: *Chem. Can.* **27**: 25 (Mar. 1975).
M8. Murphy, G. M. (ed.): *Production of Heavy Water*, McGraw-Hill, New York, 1955.
N1. Newman, R. B., thesis, Bristol University, Oct. 1954, quoted by H. London, *Separation of Isotopes*, Newnes, London, 1961, p. 85.
N2. Nitschke, E.: *Atomwirt.*, June 1973, pp. 274–280.
P1. Perlman, M. L., J. Bigeleisen, and N. Elliott: *J. Chem. Phys.* **21**: 70 (1953).
P2. Petersen, G. T., and M. Benedict: *Nucl. Sci. Eng.* **15**: 90 (1963).
P3. Popov, M. M., and F. I. Tazetdinov: *Atom. Energ.* **8**: 420 (1960).
P4. Proctor, J. F., and V. R. Thayer: *Chem. Eng. Progr.* **58**(4): 53 (1962).

R1. Rabinovich, I. B.: *Influence of Isotopy on the Physicochemical Properties of Liquids*, translated by Consultants Bureau, New York, 1970.
R2. Rae, H. K.: "Selecting Heavy Water Processes," paper presented at Joint Canadian Institute of Chemistry and American Chemical Society Meeting, Montreal, May 31, 1977.
R3. Rafn, I., Norsk Hydro Co.: Personal communication to M. Benedict, Dec. 1976.
R4. Ravoire, J., P. Grandcollot, and G. Dirian: *J. Chem. Phys.* **60**: 130 (1963).
R5. Reid, A. F., and H. C. Urey: *J. Chem. Phys.* **11**: 403 (1943).
R6. Roberts, I., H. G. Thode, and H. C. Urey: *J. Chem. Phys.* **7**: 137 (1939).
R7. Rolston, J. H., J. P. Butler, and J. den Hartog: *J. Phys. Chem.*, to be published.
R8. Rolston, J. H., J. den Hartog, and J. P. Butler: *J. Phys. Chem.* **80**: 1064 (1976).
R9. Roth, E., et al.: *PICG(2)* **4**: 499 (1958).
S1. Saito, E.: U.S. Patent 3,105,737, Oct. 1, 1963.
S2. Saito, E., and G. Dirian: British Patent 902,755, Jan. 19, 1960.
S3. Selak, P. J., and J. Finke: *Chem. Eng. Progr.* **50**: 221 (1954).
S4. Selleck, F. T., L. T. Carmichael, and B. H. Sage: *Ind. Eng. Chem.* **44**: 2219 (1952).
S5. Sherwood, T. K., and R. L. Pigford: *Absorption and Extraction*, 2d ed., McGraw-Hill, New York, 1952, pp. 146, 406.
S6. Spevack, J. S.: U.S. Patent 2,787,526, Apr. 2, 1957.
S7. Spevack, J. S.: U.S. Patent 2,895,803, July 21, 1959.
S8. Stevens, W. H.: U.S. Patents 3,888,974, June 10, 1975; 3,981,976, Sept. 21, 1976.
S9. Stewart, D. W., and K. Cohen: *J. Chem. Phys.* **8**: 904 (1940); **11**: 403 (1943).
S10. Suess, H.: Personal communication to M. Benedict, Oct. 1953.
T1. Taylor, H. S., and J. C. Jungers: *J. Chem. Phys.* **2**: 373 (1934).
T2. Taylor, T. I., and W. Spindel: *Proceedings of the International Symposium on Isotope Separation*, Interscience, New York, 1958, p. 158. See also Spindel and Taylor: *J. Chem. Phys.* **23**: 981 (1955); **24**: 626 (1956).
T3. Thayer, V. R.: U.S. Patents 3,685,966 and 3,685,967, Aug. 22, 1972; and 3,692,477, Sept. 19, 1972.
T4. Thayer, V. R., and W. B. DeLong: *Chem. Eng. Progr. Symp. Ser. 39* **58**: 86 (1962).
T5. Thode, H. G., and H. C. Urey: *J. Chem. Phys.* **8**: 904 (1940).
T6. Tronstad, L., and J. Brun: *Trans. Faraday Soc.* **34**: 766 (1938).
U1. Urey, H. C.: *J. Chem. Soc.* 562 (1947).
U2. Urey, H. C., F. G. Brickwedde, and G. M. Murphy: *Phys. Rev.* **40**: 1 (1932).
V1. Varshavskii, J. M., and F. E. Vaisberg: *J. Phys. Chem. (USSR)* **29**: 523 (1955).
W1. Walter, S., et al.: *Chem.-Ing.-Tech.* **34**: 7 (1962).
W2. Walter, S., and V. Schindewolf: *Chem.-Ing.-Tech.* **37**: 1185 (1965).
W3. Weiss, G.: *Chem.-Ing.-Tech.* **30**: 433 (1958).
W4. Weston, R. E., and J. Bigeleisen: *J. Chem. Phys.* **20**: 1400 (1952).
W5. Wooley, H., R. B. Scott, and F. G. Brickwedde: *J. Res. Natl. Bur. Std.* **41**: 379 (1948).
W6. Wynn, N. P.: "The AECL-Sulzer Amine Process for Heavy Water," paper presented at Joint Canadian Institute of Chemistry and American Chemical Society meeting, Montreal, May 31, 1977.
Z1. Zmasek, R.: *Sulzer Tech. Rev.* **54**(3): 199 (1972).

PROBLEMS

13.1 How many theoretical plates are required to produce heavy water containing 99 percent deuterium from natural water in a no-mixing distillation cascade? Assume that distillation is carried out at a temperature of 50°C. What is the minimum reboil rate for a plant producing

1000 kg D_2O/day? How many columns 3 m in diameter must be used in parallel on the feed stage if the maximum vapor velocity based on the empty tower is 1 m/s?

13.2 The bubble plates of the towers of the Morgantown water distillation plant were set on 0.3-m spaces. At a pressure of 234 Torr, the maximum operable steam velocity in these towers was 2 m/s. At this condition, the enrichment per plate was 75 percent of that attainable in one equilibrium contact, and the pressure drop per foot was 3.5 Torr.

Consider an ideal cascade of towers of this type producing 99.8 percent D_2O from natural water containing 0.0149 percent deuterium. What is the total volume of towers required to produce 100 t D_2O/year? What is the rate of loss of availability due to pressure drop of steam in such a plant, in kilowatts? Compare both results with a distillation plant using Spraypak no. 37, at a pressure of 420 Torr and a velocity of 560 cm/s.

13.3 Deuterium is to be produced at a rate of 30 kg/day by the distillation of ammonia at −52.2°C and a pressure of 250 Torr. Atom fractions deuterium are

Feed: 0.00014
Product: 0.10
Tails: 0.00007

A cascade of distillation columns arranged like an ideal cascade is used. Pressure drop is 0.05 percent per theoretical plate.

(*a*) How many theoretical plates are required?

(*b*) What is the reboil vapor ratio at the feed plate?

(*c*) What is the total rate of loss of availability in the cascade, in kilowatts?

(*d*) If the allowable vapor velocity is 0.6 m/s, how many 2-m-diameter columns must be used in parallel at the feed point?

13.4 A plant is to be built to produce 1 MT/day of D_2O containing 99.8 percent deuterium by the distillation of natural water containing 0.0149 percent deuterium. No stripping section is required. The plant will use Spraypak no. 37 tower packing, and the towers will operate at an average pressure of 1 atm and a vapor velocity of 5 m/s. The columns of the plant fed with natural water are 4 m in diameter and operate at a reboil vapor 1.5 times the minimum. The latent heat of evaporation of water at 1 atm is 539 kcal/kg.

(*a*) How many towers in parallel are needed in the largest section of the plant?

(*b*) How much heat, in kilocalories per hour, must be supplied to the towers at the feed point?

(*c*) What fraction of the deuterium in the feed is recovered?

(*d*) The towers of the largest section are 50 m high. What is the percent deuterium in the water at the bottom of these towers?

13.5 One hundred tons of 99.8 percent D_2O per year is to be produced by distillation of natural water containing 0.0149 percent deuterium. The towers are to be packed with Spraypak no. 37, operating at a pressure of 420 Torr, at the velocity of 560 cm/s. Each tower is to be designed to provide an overall enrichment of 4 and to use a reboil vapor ratio 1.25 times the minimum at the top.

(*a*) What is the packed height required in each tower?

(*b*) The low-concentration portion of the plant is to be made up of towers 4 m in internal diameter. Work out a schedule of the number of towers in parallel needed in each stage to enrich deuterium up to the point at which one column is sufficient.

13.6 Partially enriched heavy water containing 15 a/o deuterium is to be concentrated to 99.8 a/o deuterium, while stripping to 1 a/o deuterium in a water distillation plant using Sulzer CY packing operating at an effective pressure of 120 Torr and at a throughput of 2000 kg/(m^2·h). The product rate is 400 MT/year. If designed as an ideal cascade, (*a*) How many towers 2 m in internal diameter would be needed to handle the vapor load at the feed point?

(*b*) How many transfer units would be needed to span the composition range between 99.8 a/o and 1 a/o deuterium?

(*c*) What total packing volume would be needed?

(*d*) What would be the rate of loss of availability of the vapor due to pressure drop, in kilowatts?

13.7 An electrolytic hydrogen plant fed with natural water containing 0.0149 percent deuterium is operated as a simple cascade, without recycle, to produce water containing 0.142 percent deuterium. The stage separation factor α has the value 7.0, and the heads separation factor β is independent of stage number.

(*a*) How many moles of water must be electrolyzed per mole of D_2O product when the cascade has two stages? Four stages? An infinite number of stages?

(*b*) How many moles of steam are generated (to serve as stage feed) per mole of D_2O product when the cascade has two stages? Four stages? An infinite number of stages?

In Fig. 13.13, the first two stages illustrate the type of cascade to which this problem refers.

13.8 ^{7}Li is to be separated from ^{6}Li in a no-mixing recycle cascade of electrolytic cells with LiOH electrolyte and a mercury cathode. Seven kilograms per day of ^{7}Li at 99.99 a/o ^{7}Li is to be produced from natural lithium containing 92.48 a/o. No stripping section is used. How many kilograms of feed is consumed per day? How many stages are required? What is the total electric power consumption of the cascade? Assume that $\alpha = 1.055$, that ^{6}Li concentrates in the amalgam, that the cell drop is 5 V, and that the current efficiency is 75 percent.

13.9 In the reaction $H_2O + DCl \rightleftharpoons HDO + HCl$, the equilibrium constant is 5 and the separation factor for enrichment of deuterium by chemical exchange is 2.5. Explain qualitatively why these are not equal.

13.10 A hydrogen sulfide-water dual-temperature exchange plant is required to produce 1000 kg-mol/day of water containing 1.2 a/o deuterium from natural water containing 0.0144 a/o deuterium with waste containing 0.012 a/o deuterium. A single cold tower at 32°C and a single hot tower at 138°C are to be used.

For the approximate optimum values of hydrogen sulfide flow rate G given by Eq. (13.115) and intertower atom fraction deuterium in hydrogen sulfide y_F given by Eq. (13.113), find the number of plates in the hot and cold towers that will perform this separation.

Neglect the solubility of H_2S in water and the vaporization of water into gaseous H_2S. Use equations valid for low D/H ratios.

13.11 Verify that the number of theoretical stages for the stripping and enriching sections of Fig. 13.23 are 2.6 and 3.1, respectively.

13.12 The dual-temperature, methylamine-hydrogen exchange process described in Sec. 13 could also be used to concentrate deuterium from ammonia synthesis gas produced from natural gas and steam containing the normal abundance of deuterium instead of the enriched steam used in the Sulzer flow sheet, Fig. 13.40. Figure 13.42 is a flow sheet for such a process giving the deuterium content of each stream in the first stage of the plant.

(*a*) For an assumed hydrogen feed rate of 4320 kg-mol/h (the same as in Fig. 13.40), by deuterium material balance find:

(1) Heavy-water production rate P
(2) Methylamine circulation rate L
(3) Hydrogen recycle rate G

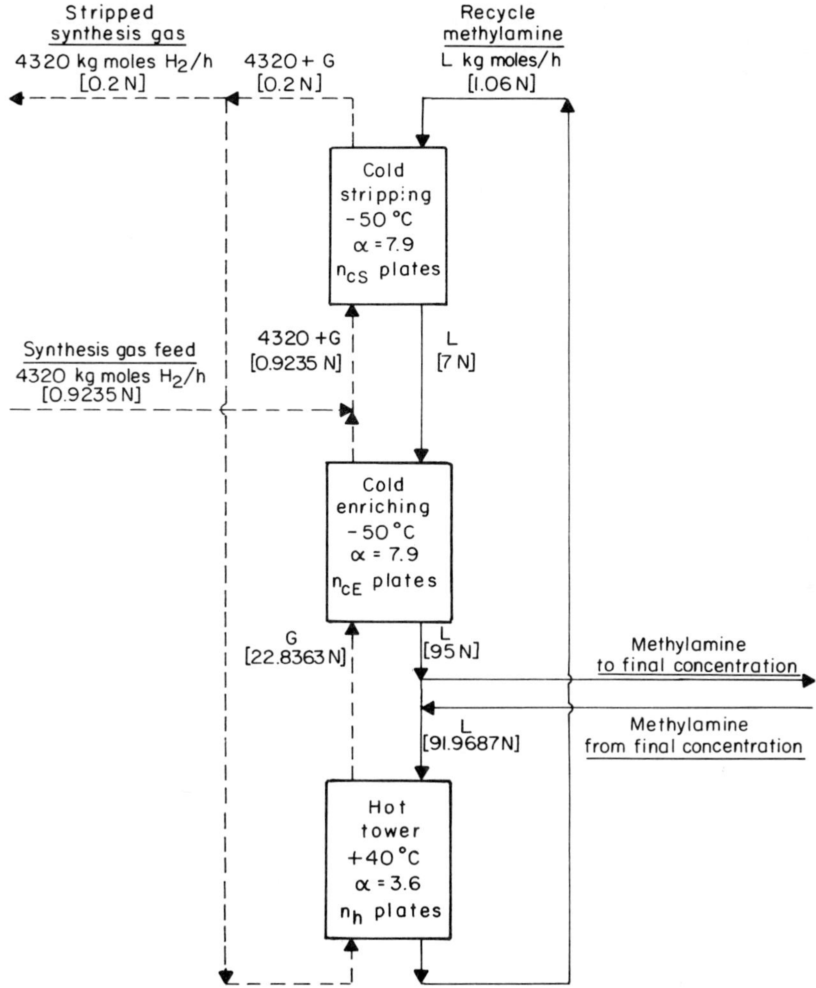

Figure 13.42 Primary concentration step in dual-temperature methylamine-hydrogen exchange process fed with synthesis gas made from normal water. Flow rates G and L in kg-mol/h. $[1N]$ = 135 parts deuterium per million parts deuterium + hydrogen.

(*b*) Find the number of theoretical plates in the cold stripping section n_{cS}, the cold enriching section n_{cE}, and the hot tower n_h.

(*c*) Compare this process with the methylamine process of Fig. 13.40 and the ammonia process of Fig. 13.37 with respect to:

(1) Number of cold-tower plates
(2) Number of hot-tower plates
(3) Flow ratio, hydrogen to D_2O
(4) Flow ratio, liquid to D_2O

CHAPTER
FOURTEEN

URANIUM ISOTOPE SEPARATION

1 INTRODUCTION

Because of the importance of ^{235}U at compositions above natural abundance, originally for military purposes and more recently for nuclear electricity generation, great effort has gone into investigation and development of many processes for enriching ^{235}U. This chapter deals only with those processes that have been used on an industrial scale, those that seem likely to become of future industrial importance, and those that illustrate the shortcomings of the processes used industrially for separating the isotopes of light elements when applied to heavy elements such as uranium.

Discussion of processes for industrial separation of uranium isotopes cannot be as complete as the discussion of deuterium separation in Chap. 13. The detailed technology of the most economical and most promising processes is subject to security classification and to proprietary restrictions. Nevertheless, processes for enriching uranium can be described in sufficient detail to make their principles clear and to illustrate the similarities and differences between them and processes for separating isotopes of light elements.

For a more detailed discussion of uranium isotope separation than is possible in this chapter, reference may be made to papers on this subject presented at the four International Conferences on the Peaceful Uses of Atomic Energy sponsored by the United Nations at Geneva, to the *Proceedings of the International Conference on Uranium Isotope Separation* sponsored by the British Nuclear Energy Society in London in April 1975 [B20], to the *Chemical Engineering Progress Symposium Series* volume on uranium enrichment [B14], the articles on diffusion separation methods [H11, S3] in the *Encyclopedia of Chemical Technology*, informative reports ORO-684, 685, 690, and 694 on uranium enrichment activities of the U.S. Atomic Energy Commission (AEC), and the authoritative monograph on uranium enrichment edited by Villani [V1a].

The processes used most extensively for separating isotopes of light elements, distillation and chemical exchange, become progressively less selective as the atomic weight increases and are ineffective for uranium.

The processes used most extensively for separating uranium isotopes, gaseous diffusion and the gas centrifuge, are much less efficient than distillation for light elements, but are impaired less by an increase in molecular weight, so that they are the preferred methods for uranium. Table 14.1 compares the separation factors for these four processes when applied to mixtures

Table 14.1 Representative separation factors for isotope separation processes

Process	Property†	Separation factor for isotopic mixture:		
		H_2-HD	^{14}NO-^{15}NO	$^{235}UF_6$-$^{238}UF_6$
Distillation	α^*	1.5	1.033	1.00002 [O2]
Monothermal chemical exchange	K	3.6‡	1.055§	1.0016¶ [S1]
Gaseous diffusion	$\sqrt{m_2/m_1}$	1.225	1.017	1.00429
Gas centrifuge	$\exp \dfrac{(m_2 - m_1)v_a^2}{2RT}$	1.056	1.056	1.162

†α^*, relative volatility. K, exchange equilibrium constant: ‡for HD-H_2O exchange; §for ^{14}NO-$H^{15}NO_3$ exchange; ¶for $^{235}UF_6$-$^{238}UF_5$ NOF exchange. m_2, m_1, molecular weight of heavy, light component. v_a, peripheral speed, 500 m/s. R, 8314 J/(kg-mol·K). T, 300 K.

of H_2 and HD, ^{14}NO and ^{15}NO, and $^{235}UF_6$ and $^{238}UF_6$. Although many features besides separation factor enter into choice of the preferred process, it is clear that the higher values for gaseous diffusion and the gas centrifuge give these processes a substantial advantage over distillation and chemical exchange for uranium isotope separation.

Section 2 of this chapter deals with the isotopic content of uranium. Section 3 lists the principal processes for separating uranium isotopes on an industrial scale and describes briefly projects using these processes. Section 4 gives a detailed description of the gaseous diffusion process, which until now has produced almost all of the world's enriched uranium. Section 5 is a parallel treatment of the gas centrifuge process, which is emerging as an effective competitor of gaseous diffusion. Section 6 describes aerodynamic processes that separate uranium isotopes through composition differences developed when mixtures of $^{235}UF_6$ and $^{238}UF_6$ are subjected to high linear or centrifugal accelerations in flowing gas streams. Such processes are in an advanced stage of development and are to be used industrially in Brazil and possibly South Africa. The remainder of this chapter discusses in less detail other processes not yet ready for industrial use. Mass diffusion (Sec. 7) and thermal diffusion (Sec. 8) are clearly not economical for uranium isotope separation but are described briefly because they illustrate isotope separation principles in the comparison with gaseous diffusion, and have been used to advantage for other elements. Laser-based processes (Sec. 9) appear very promising and may, with sufficient development, become the most economical means of separating uranium isotopes.

2 ISOTOPIC CONTENT OF URANIUM

Uranium isotope separation plants may be fed either with natural uranium, which contains only the isotopes ^{234}U, ^{235}U, and ^{238}U in nearly fixed proportions; or uranium discharged from a nuclear reactor, which contains the above three isotopes in many possible proportions, together with ^{236}U from neutron capture in ^{235}U, some ^{233}U from neutron capture in thorium present in the irradiated uranium, and traces of ^{232}U from fast-neutron irradiation of thorium or decay of ^{236}Pu.

Until recently it has been assumed that natural uranium from all sources had exactly the same content of ^{234}U, ^{235}U, and ^{238}U. As lately as 1977, U.S. Energy Research and Development Administration (ERDA) used 0.711 w/o (weight percent) as the ^{235}U content of all natural uranium feed supplied to U.S. ERDA plants for enrichment. However, accurate measurement of the $^{235}U/^{238}U$ ratios of uranium from various minerals and various locations has shown significant variations. Cowan and Adler [C9] have summarized measurements of the weight percent of ^{235}U in 90 samples of natural uranium stated by the measuring laboratories to have a relative error of 0.0003 or less at the 95 percent confidence level. Average values of the weight percent ^{235}U in different classes of samples and the standard deviation as reported by Cowan and Adler are listed in Table 14.2.

The difference between sandstone-type minerals and high-temperature minerals is considered to be significant. It is attributed probably to isotopic fractionation that occurred when uranium initially deposited at high temperatures from magmas was dissolved by water at lower temperature and reprecipitated in sandstones. The difference between non-U.S. and U.S. samples is explained in the same way, as most non-U.S. samples were of magmatic origin and most U.S. samples were of the sandstone type.

Another possible cause of lower ^{235}U content more dramatic than isotopic fractionation during mineral deposition is possible occurrence of a critical fission chain reaction in a uranium deposit subsequent to primary mineralization, which would deplete ^{235}U relative to ^{238}U. One such deposit has been found and extensively studied at Oklo in the Republic of Gabon, West Africa. One uranium sample from this mine contained only 0.3 percent ^{235}U [N1], and much of the ore contains substantially less than 0.711 w/o ^{235}U. Extensive nuclear chemical research reported in the proceedings containing [N1] has found higher than normal concentrations of fission-product nuclides such as ^{143}Nd and ^{145}Nd in regions where the ^{235}U content of uranium is lower than normal. The evidence is conclusive that a fission chain reaction operated for many years in this deposit about 2 billion years ago. At that time the ^{235}U content of natural uranium would have been around 3 percent, compared with today's 0.711 percent, because of the shorter half-life of ^{235}U compared with ^{238}U. In portions of the Oklo deposit where the uranium content was high and neutron absorbers were scarce, water made its way into the ore in sufficient concentration to establish a low-power fission chain reaction that persisted for thousands of years and used up a substantial fraction of the ^{235}U present at the start of the reaction. Cowan [C8] has summarized salient findings about this dramatic natural event and has given reasons for anticipating future discoveries of other one-time natural uranium reactors where the present $^{235}U/^{238}U$ ratio would also be less than normal.

Because of the possibility of natural depletion of ^{235}U and because of the availability of tails from isotope separation plants that might become mixed with natural uranium, it is important that natural uranium feed for an isotope separation plant be analyzed for its ^{235}U content.

The $^{235}U/^{238}U$ ratio of natural uranium is generally assumed to be the same as the ratio of the half-lives of these elements, 2.47×10^5 years/4.51×10^9 years = 0.000055.

Table 14.2. Average w/o ^{235}U in natural uranium from different sources

Source	Number of samples	Weight percent ^{235}U	Standard deviation
All independent samples	88	0.7107	0.0002_2
High-temperature minerals	33	0.7108	0.0001_5
Non-U.S. samples	26	0.7108	0.0001_0
U.S. samples	62	0.7106	0.00025
Sandstone-type minerals	54	0.7106	0.0002_3

Source: G. A. Cowan and H. H. Adler, *Geochim. et Cosmochim. Acta* **40**:1487 (1976).

3 URANIUM ENRICHMENT PROJECTS

3.1 Processes Developed by Manhattan Project

During the period from 1943 to 1947 in the United States, the Manhattan Project carried four uranium enrichment processes through the large pilot stage and into production to the extent noted below.

The electromagnetic process, using the Calutron isotope separator in the Y-12 plant at Oak Ridge, Tennessee, produced the first kilogram quantities of highly enriched uranium in 1944. Because costs proved to be higher than in the gaseous diffusion process, separation of uranium isotopes by this method was terminated in 1946, with some of the equipment being converted to separating isotopes of other elements.

The thermal diffusion process, in the Oak Ridge S-50 plant, enriched natural uranium to 0.86 percent ^{235}U, which was fed to the Y-12 plant to increase slightly the ^{235}U production rate of the latter. Its heat source was steam from the steam-electric power plant built to provide electricity for the K-25 gaseous diffusion plant. As the thermal diffusion process makes much less efficient use of energy for uranium enrichment than gaseous diffusion, the S-50 plant was shut down in 1945 when enough of the K-25 gaseous diffusion plant was operating to use productively the full electric output of the power plant. This process will be described briefly in Sec. 8.

The Oak Ridge K-25 gaseous diffusion plant was completed in sections in 1945 and 1946. When partially completed, its partially enriched ^{235}U product was fed to the Y-12 plant to increase the output of fully enriched uranium from the latter. After all sections of the K-25 plant were in operation, the Y-12 plant was shut down in 1946 because of the lower cost and more efficient energy use of the gaseous diffusion process. Later, Section K-27, containing larger gaseous diffusion stages than K-25, was brought into operation at Oak Ridge. By 1977 all of these Manhattan Project stages at Oak Ridge had been retired from operation because of the later construction of the more efficient gaseous diffusion stages of the K-29, K-31, and K-33 Sections at Oak Ridge and the Paducah and Portsmouth gaseous diffusion plants.

The gas centrifuge process was developed by the Manhattan Project through the construction and operation in 1944 at the Bayway, New Jersey, refinery of Standard Oil Company (N.J.) of a pilot plant of centrifuges 4 m long. After the gaseous diffusion process proved to be reliable, work on the gas centrifuge was suspended because of the low separative capacity of the individual centrifuges and the mechanical complexity of the machines then under development. With the advent of the simpler Zippe [Z2] centrifuge to be described in Sec. 5, development of the gas centrifuge for uranium enrichment was resumed in the 1960s, leading to its current industrial use.

3.2 Current Industrial Uranium Enrichment Projects

Gaseous diffusion. Table 14.3 lists gaseous diffusion plants in operation in 1977 and those then under construction, planned, or under consideration. Part 1 of Table 14.3 lists plants in operation at that time. The three large plants of the U.S. Department of Energy (DOE) had a capacity of over 17 million kg separative work units (SWU) per year when supplied with the maximum amount of electric power, 6100 MW, they could then utilize.

The U.S.S.R. plant is rumored to have an annual capacity of from 7 to 10 million units, of which 3 million are thought available for export. The existing plants of the French Commissariat à l'Energie Atomique (CEA) and British Nuclear Fuels, Ltd. (BNFL) are too small to be a major source. Little is known about the Chinese plant.

Table 14.3 Gaseous diffusion projects

Owner	Location	Capacity, million kg separative work units per year	
1. Now operating			
U.S. DOE	Oak Ridge, Tenn.	4.73	
	Paducah, Ky.	7.31	
	Portsmouth, Ohio	5.19	
Total, U.S.		17.23	
Soviet Union	Siberia	7–10	
CEA	Pierrelatte, France	0.4–0.6	
BNFL	Capenhurst, England	0.4–0.6	
Peoples' Republic of China	Lanchow, China	?	
2. Under construction			Scheduled operation
Improvement and uprating of U.S. DOE Plants–Adds		10.5	1975–1985
Eurodif (CEA, Iran, Belgium, Italy, Spain)	Tricastin, France	10.8	1978–1981
3. To be built			
Coredif (Eurodif, CEA, Iran)	France, Belgium, or Italy	5.4	Late 1980s
4. Under consideration			
Coredif expansion	France, Belgium, or Italy	5.4	?

Part 2 of Table 14.3 lists additional separative capacity by gaseous diffusion under construction. U.S. DOE is improving the barrier in its three existing plants and increasing the power input to the stage compressors to increase capacity by 10.5 million units per year. The Eurodif combination of French, Belgian, Italian, Spanish, and Iranian interests is building a 10.8 million unit per year plant in France, using French-developed technology, to start operation in 1978.

Parts 3 and 4 list additional gaseous diffusion enrichment projects likely to be built. The Coredif project uses French diffusion technology, and appears to be committed to construction of 5.4 million units of additional diffusion capacity at a European site still to be selected. Possible expansion of capacity of this plant by another 5.4 million units per year is under consideration.

Gas centrifuge projects. Table 14.4 lists gas centrifuge projects. The Urenco-Centec Organization, a combination of British, Dutch, and German interests, has been operating three pilot units at Capenhurst, England, and Almelo, Holland, since 1972. By 1982 these plants will have been expanded to an annual capacity of 2 million units. This group is seeking additional orders with intention of increasing capacity to 10 million units by 1985 if orders materialize.

President Carter announced on April 20, 1977, that the United States would expand its uranium enrichment facilities and would shortly reopen its order book for sale of additional units of separative work. After the cascade uprating and cascade improvement programs have been completed, all new separative capacity would be provided by the gas centrifuge, whose much lower energy demand and greater flexibility were perceived as decisive advantages. U.S. DOE is building a centrifuge enrichment plant with capacity of 2.2 million kg SWU/year at Portsmouth, Ohio, for operation in the late 1980s. Expansion of 8.8 million kg SWU/year is possible.

Japan is building a 7000-machine centrifuge pilot plant to operate in 1979 and is considering a 6 million SWU/year production plant to start operation in 1985.

Aerodynamic processes. Two projects have developed to industrial-scale processes for separating uranium isotopes by causing a mixture of UF_6 and hydrogen to flow at high speed in a sharply curved path and thus experience centrifugal acceleration large enough to effect partial separation of $^{235}UF_6$ and $^{238}UF_6$. The separation nozzle process developed by Becker and his associates at the Karlsruhe Nuclear Research Center in Germany and adapted for industrial use by Steag, A.G., and Gesellschaft für Kernforschung is being used in a plant with a capacity of 180,000 SWU/year being built in Brazil for operation in 1982. The UCOR process, developed by Roux, Grant, and their associates of the Uranium Enrichment Corporation of South Africa, has been demonstrated in a 6000 SWU/year pilot plant at Valindaba, South Africa; in 1978 a decision was to be made whether to build a commercial plant based on this process. These processes will be described in Sec. 6.

3.3 Processes Under Development

Laser-based processes. Laser-based processes, which use intense, narrow-frequency radiation to cause atoms or molecules containing ^{235}U to undergo selectively a different physical or chemical process than those containing ^{238}U, are under intensive development in many countries, but have not yet advanced to industrial use.

The principal U.S. projects of this type are research at U.S. DOE's Los Alamos Laboratory, which uses UF_6 vapor, and work by U.S. DOE's Livermore Laboratory and a joint venture of Avco Everett Research Laboratory, Inc., and Exxon Nuclear Company, which use uranium metal vapor. The two groups [J2, T3] using uranium metal vapor reported production of milligram quantities of partially enriched uranium in 1975. Avco and Exxon applied for a license to build a pilot plant to demonstrate their process in the mid-1980s.

Improved electromagnetic processes. Developments in plasma physics and magnet design in the 30 years since the Y-12 plant was taken off uranium isotope separation have caused many groups to reexamine electromagnetic processes for separating uranium isotopes, some of which reported at the London Conference on Uranium Isotope Separation [B20]. In the United States

Table 14.4 Gas centrifuge projects

Owner	Location	Capacity, million separative work units per year	Scheduled operation
1. Now operating			
Urenco-Centec (United Kingdom, Holland, Germany)	Capenhurst, England; Almelo, Holland	0.120	Since 1975
2. Under construction			
Urenco-Centec	Capenhurst, Almelo	0.4–2.0	1977–1985
3. To be built			
U.S. DOE	Portsmouth, Ohio	2.2–8.8	1986–1988
4. Under consideration			
Urenco-Centec	Capenhurst; W. Germany	Add 8	Late 1980s
Japan		6	1985

a company, Phrasor, Inc., has been formed to continue development of an improved process of this general type. Dawson and associates [D3] have given a partial description of a process using ion-cyclotron resonance to ionize selectively and separate K-40. This process is being investigated for ^{235}U with funding by U.S. DOE and TRW Defense and Space Systems.

Solvent extraction. At the 1977 International Atomic Energy Agency (IAEA) Conference on Atomic Energy at Salzburg, Austria, Commissioner Giraud of the French CEA announced development of a new process for producing uranium enriched sufficiently for reactor fuel, but impractical for producing more highly enriched weapons-grade material because it has too high a specific inventory. At the same conference, Dr. Frejacques and colleagues of the CEA [F4] said that "a new process using crown compounds of uranium is currently under study." Such a process could involve complexing and fractional solvent extraction of ^{235}U from an aqueous solution with a crown ether dissolved in an immiscible organic solvent.

4 GASEOUS DIFFUSION

4.1 Principle

The gaseous diffusion process makes use of the phenomenon of molecular effusion to effect separation. In a vessel containing a mixture of two gases, molecules of the gas of lower molecular weight have higher speeds and strike the walls of the vessel more frequently, relative to their concentration, than do the molecules of the gas with higher molecular weight. If the walls of the vessel have holes just large enough to allow passage of molecules one by one without permitting flow of the gas as a continuous fluid, more of the lighter molecules flow through the wall, relative to their concentration, than the heavier molecules. The flow of individual molecules through minute holes is known as *molecular effusion*. The possibility of separating gases by effusion through porous media was discovered experimentally by Graham over a hundred years ago. Maxwell showed that this separation was due to the fact that the relative frequency with which molecules of different species enter a small hole is inversely proportional to the square root of their molecular weights. For a mixture of $^{235}UF_6$ and $^{238}UF_6$ this ratio, the ideal separation factor for gaseous diffusion α_0, is

$$\alpha_0 = \sqrt{\frac{m_{^{238}UF_6}}{m_{^{235}UF_6}}} = \sqrt{\frac{352}{349}} = 1.00429 \tag{14.1}$$

Because this value is so close to unity, to obtain a useful degree of separation the process must be repeated many times in a countercurrent cascade of gaseous diffusion stages, such as was shown in Fig. 12.2.

4.2 History

The first use of gaseous diffusion for isotope separation was by Aston [A4], who in 1920 effected a slight separation of the isotopes of neon in a single stage of gaseous diffusion through a porous clay tube. Hertz [H2, H5, H6] greatly increased the separation obtainable by this method by using a countercurrent recycle cascade of from 24 to 50 stages of the type shown in Fig. 12.2. This apparatus effected practically complete separation of the neon isotopes of mass 20 and 22 and completely separated hydrogen and deuterium. With a 34-stage cascade, Wooldridge, Jenkins, and Smythe [W3, W4] enriched $^{13}CH_4$ from 1 to 16 percent.

When World War II created a demand for ^{235}U, the proved ability of gaseous diffusion to effect isotope separation and the existence of a stable, volatile compound of uranium, UF_6, led to intensive development of this process in England and the United States. Because of greater security against attack and more abundant energy supplies, the two governments decided that the first gaseous diffusion uranium enrichment plant would be built in the United States. The Manhattan Project, under the leadership of General Leslie R. Groves, built the first gaseous diffusion plant, the K-25 plant, at Oak Ridge, Tennessee, which began operation in 1945. Partial descriptions of this plant and the demanding development effort that led to its successful operation have been given by Smyth [S6], Keith [K1], Hogerton [H10], Groves [G5], and Groueff [G4], and the official U.S. history by Hewlett and Anderson [H7]. The development effort in England and the construction of the British gaseous diffusion plant at Capenhurst in the 1950s has been described by Jay [J3]. The independent development of the gaseous diffusion process in France in the 1950s and the construction of the first French plant at Pierrelatte in 1964-1967 has been described by CEA [C7].

4.3 U.S. Process Equipment

Partial descriptions of the type of equipment used in the gaseous diffusion plants of the U.S. DOE are given in references [U1] and [U2]. Figure 14.1 is a schematic plan view of three gaseous diffusion stages. The separating unit on each stage, called a converter, contains

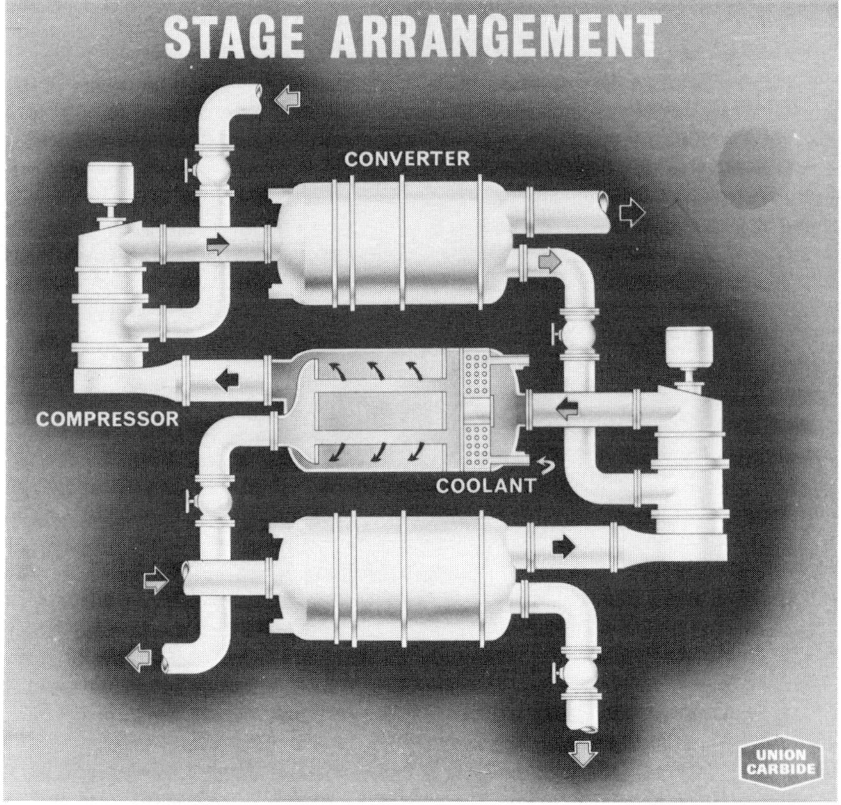

Figure 14.1 Arrangement of gaseous diffusion stages. (*Courtesy of U.S. Energy Research and Development Administration.*)

thousands of tubes of diffusion barrier supported by tube sheets at each end. As UF_6 gas at the highest process pressure flows along the inside of these tubes, about one-half of it effuses through the tubes into the region at the lowest process pressure outside of the tubes and is thereby slightly enriched in $^{235}UF_6$. This low-pressure, slightly enriched gas, the stage heads stream, is compressed to an intermediate pressure in the first stage of a horizontally mounted, two-stage, axial-flow compressor of the next higher stage of the cascade. Here it is joined by an equal amount of UF_6 at the same pressure and ^{235}U content representing the tails stream from the second higher stage of the cascade. The combined streams are compressed by the second stage of the compressor to the highest process pressure. The compressed gas flows through a cooler, where the heat of compression is removed by heat exchange against coolant $C_2F_4Cl_2$, chosen because it will not react with UF_6 should a leak occur. The compressed and cooled gas then flows through the tubes of the converter on the next higher stage of the cascade.

The tails stream from each converter, the gas that has not effused through the holes in the barrier tubes, flows through a control valve and into the intermediate pressure inlet of the compressor on the next lower stage of the cascade. The valve position is adjusted so as to control the pressure level of the converter upstream at the desired level.

In some stages of the U.S. plants the flow sequence is modified with the stage cooler inserted between the converter outlet for the heads stream and the compressor inlet. This permits the converter to operate at the compressor outlet temperature rather than the lower inlet temperature, and improves somewhat the separation performance of the barrier.

Figure 14.2 is a photograph of the process equipment used in the largest stages of the U.S.

Figure 14.2 View of converters and compressor. (*Courtesy of U.S. Energy Research and Development Administration.*)

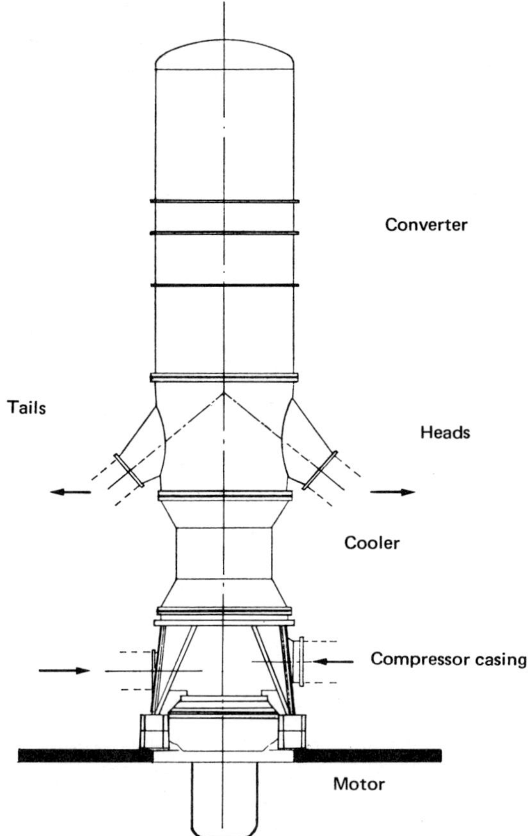

Figure 14.3 Eurodif gaseous diffusion stage.

diffusion plants. The large drums in the foreground are the converters, each of which contains a cooler and thousands of barrier tubes. The two-stage axial-flow compressor that recompresses the UF_6 that has passed through the barrier and circulates the undiffused gas is at the back of the figure. From 8 to 16 stages such as these are grouped into cells, housed in steel enclosures heated to around 60°C to prevent condensation of UF_6. Each cell is the smallest independently operable unit, and is equipped with block and bypass valves to permit shutdown for maintenance.

As Fig. 12.2 showed, about 1270 stages are needed to separate natural uranium into product containing 3 w/o ^{235}U and tails containing 0.2 w/o. The Portsmouth plant of U.S. DOE, which produces uranium enriched to 97 percent ^{235}U, contains 4080 stages.

The large plants of U.S. DOE have operated for 20 years at a capacity factor over 99 percent and attest to the reliability of the gaseous diffusion process.

4.4 French Process Equipment

Stages in the Eurodif gaseous diffusion plant contain the same components in the same process sequence as Fig. 14.1, but they are arranged more compactly, as shown in Fig. 14.3, with converter, cooler, compressor, and motor mounted vertically on the same axis. This arrangement greatly reduces the length of interconnecting piping and the required floor area and building space.

Charpin et al. [C3] and Massignon [M5] have described several types of diffusion barrier developed in France and given examples of their characteristics. Materials from which these barriers were made include sintered alumina, oxidized aluminum, Teflon, and nickel. Pore radii were in the range of 0.01 to 0.05 μm. Barriers developed in Sweden have been described by Mártensson et al. [M3].

4.5 Flow of Gases through Diffusion Barrier

Pure gases. A typical diffusion barrier consists of a thin sheet of material perforated by a very large number of small holes of nearly uniform diameter. If the diameter of the holes and the thickness of the sheet are smaller than the mean free path of UF_6 at the pressure upstream of the barrier, individual molecules of UF_6 will flow through the holes without colliding with other molecules in what is known as *molecular flow*. The rate of molecular flow through a circular capillary is given by Knudsen's law [K3]:

$$G_{\text{mol}} = \frac{8r(p'' - p')}{3l\sqrt{2\pi mRT}} \tag{14.2}$$

where G = molar velocity, kg-mol/(m$^2\cdot$s)
r = capillary radius, m
l = capillary length, m
m = molecular weight, 349 for $^{235}UF_6$ and 352 for $^{238}UF_6$
R = gas constant, 8314 (Pa$\cdot$m^3)/(kg-mol$\cdot$K)†
T = absolute temperature, K
p'' = upstream pressure, Pa
p' = downstream pressure, Pa

The fact that G is different for $^{235}UF_6$ than for $^{238}UF_6$ is what makes separation by gaseous diffusion possible.

If the pressure is sufficiently high or the holes sufficiently large to cause the gas molecules to collide with each other a number of times during flow through the barrier, laminar or viscous flow obtains. The rate of viscous flow through a circular capillary is given by Poiseuille's law:

$$G_{\text{vis}} = \frac{r^2(p''^2 - p'^2)}{16l\mu RT} \tag{14.3}$$

where μ is the viscosity. For UF_6 [D5],

$$\mu = 1.67(1 + 0.0026t) \times 10^{-5} \text{ kg/(m}\cdot\text{s)} \qquad t = \text{temperature, °C} \tag{14.4}$$

The principal differences from molecular flow are as follows:

1. The flow law is the same for $^{235}UF_6$ as for $^{238}UF_6$, so no separation takes place during viscous flow.
2. The flow rate is inversely proportional to the viscosity instead of to the square root of the molecular weight.
3. The flow rate is proportional to the difference in the *square* of the pressures instead of the first power.

†1 Pa = 0.007500 Torr = 0.000750 cmHg = 9.87 × 10^{-6} atm.

The openings in a diffusion barrier are neither circular, straight, nor of uniform diameter, but its flow characteristics approach molecular flow at low pressures, in the form

$$G_{\mathrm{mol}} \propto \frac{p'' - p'}{\sqrt{m}} \tag{14.5}$$

and approaches viscous flow at high pressure, in the form

$$G_{\mathrm{vis}} \propto \frac{p''^2 - p'^2}{\mu} \tag{14.6}$$

In the intermediate-pressure region, in which flow has some features of both molecular and viscous flow, experiments reported by Present and de Bethune [P3] have shown that the flow may be expressed as a linear combination of Eqs. (14.5) and (14.6):

$$G = \frac{a(p'' - p')}{\sqrt{m}} + \frac{b(p''^2 - p'^2)}{\mu} \tag{14.7}$$

where a and b are properties of the barrier.

For several different models of barrier structure, the constants a and b in Eq. (14.7) can be related to dimensions of holes in the barrier. For straight circular holes of uniform radius r occupying ϵ fraction of a barrier of uniform thickness l, Present and de Bethune assign to the constant a the value it would have for molecular flow and to b the value it would have for viscous flow, so that for this "mixed flow" model,

$$G(\text{mixed flow}) = \left[\frac{8r(p'' - p')}{3\sqrt{2\pi mRT}} + \frac{r^2(p''^2 - p'^2)}{16\mu RT}\right]\frac{\epsilon}{l} \tag{14.8}$$

For a barrier consisting of two sizes of straight circular holes, with ϵ_{mol} fraction occupied by small holes of radius r_{mol} through which pure molecular flow takes place and ϵ_{vis} fraction occupied by larger holes of radius r_{vis} through which viscous flow takes place, the molar velocity for this "viscous leak" model would be

$$G(\text{viscous leak}) = \frac{8r_{\mathrm{mol}}(p'' - p')\epsilon_{\mathrm{mol}}}{3l\sqrt{2\pi mRT}} + \frac{r_{\mathrm{vis}}^2(p''^2 - p'^2)\epsilon_{\mathrm{vis}}}{16l\mu RT} \tag{14.9}$$

Real barriers contain crooked, noncircular holes distributed in size about a mean radius in the range of 0.005 to 0.03 μm. Molar velocity through most barrier materials is found experimentally to depend on pressures as in Eq. (14.10):

$$\Gamma \equiv \frac{G}{p'' - p'} = \Gamma_0\left[1 + S\left(\frac{p'' + p'}{2}\right)\right] \tag{14.10}$$

Here Γ is known as the permeability, Γ_0 is interpreted as the permeability for molecular flow, and S is sometimes called the "slope factor." Comparison of Eq. (14.10) with (14.8) and (14.9) shows that a physical interpretation can be given to the parameters Γ_0 and S in terms of pore radius and void fraction for the mixed flow and viscous leak models:

$$\Gamma_0 = \underset{\text{(Mixed flow)}}{\frac{8\epsilon r}{3l\sqrt{2\pi mRT}}} = \underset{\text{(Viscous leak)}}{\frac{8\epsilon_{\mathrm{mol}} r_{\mathrm{mol}}}{3l\sqrt{2\pi mRT}}} \tag{14.11}$$

$$S = \underset{\text{(Mixed flow)}}{\frac{3r}{64\mu}\sqrt{\frac{2\pi m}{RT}}} = \underset{\text{(Viscous leak)}}{\frac{3r_{\mathrm{vis}}^2\epsilon_{\mathrm{vis}}}{64\mu r_{\mathrm{mol}}\epsilon_{\mathrm{mol}}}\sqrt{\frac{2\pi m}{RT}}} \tag{14.12}$$

It is desirable to have a high value of Γ_0, to reduce the barrier area needed for a given gas flow, and a low value of S, to reduce the fraction of flow that is nonseparating.

Another parameter used to characterize flow through a barrier is the specific permeability γ, defined as the ratio of the actual flow through unit barrier area to the flow by molecular effusion alone through a hole of unit area. Because the latter is

$$\frac{\bar{v}(p''-p')}{4RT}$$

where $\bar{v}$ is the mean molecular speed,

$$\bar{v} = \sqrt{\frac{8RT}{\pi m}} \tag{14.13}$$

thus

$$\gamma \equiv \frac{G\sqrt{2\pi mRT}}{p''-p'} \tag{14.14}$$

The limiting value of γ as the pressures p'' and p' approach zero has simple physical significance. In the mixed flow model,

$$\gamma_0 \equiv \lim_{p'',p' \to 0} \gamma = \frac{8r\epsilon}{3l} \tag{14.15}$$

and in the viscous leak model,

$$\gamma_0 \equiv \lim_{p'',p' \to 0} \gamma = \frac{8r_{\text{mol}}\epsilon_{\text{mol}}}{3l} \tag{14.16}$$

Gas mixtures. Nomenclature to be used in describing the flow of a binary gas mixture through a diffusion barrier is shown in Fig. 14.4. The problem is to determine how the molar velocities of light and heavy components, G_1 and G_2, respectively, depend on upstream and downstream

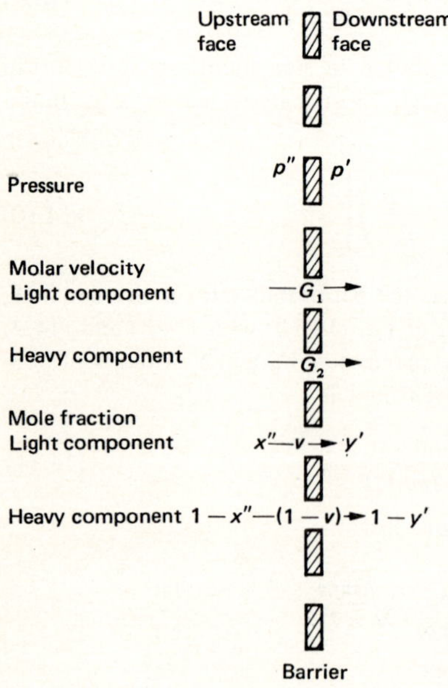

Figure 14.4 Flow of binary mixture through diffusion barrier.

pressures and compositions. We shall also be interested in the composition of the net flow through the barrier, expressed as mole fraction of light component υ, defined by

$$\upsilon = \frac{G_1}{G_1 + G_2} \tag{14.17}$$

Ideal separation. When the upstream pressure is so low that only molecular flow takes place and when the downstream pressure is negligible ($p'/p'' \to 0$), Eq. (14.7) shows that the molar velocity of each component is proportional to its partial pressure on the upstream faces and inversely proportional to $\sqrt{m}$:

$$G_1 = \frac{ap''x''}{\sqrt{m_1}} \tag{14.18}$$

$$G_2 = \frac{ap''(1-x'')}{\sqrt{m_2}} \tag{14.19}$$

The composition of the net flow through the barrier in this case is

$$\upsilon = \frac{G_1}{G_1 + G_2} = \frac{x''\sqrt{m_2/m_1}}{x''\sqrt{m_2/m_1} + (1-x'')} \tag{14.20}$$

$$\frac{\upsilon(1-x'')}{x''(1-\upsilon)} = \sqrt{\frac{m_2}{m_1}} = \alpha_0 \tag{14.21}$$

$\sqrt{m_2/m_1}$ is known as the ideal barrier separation factor α_0. For $^{235}UF_6$-$^{238}UF_6$ mixtures,

$$\alpha_0 = \sqrt{\frac{352}{349}} = 1.00429 \tag{14.22}$$

The composition of the upstream gas x_0 that would give a net flow of composition υ under these ideal conditions is

$$x_0 = \frac{\upsilon}{\upsilon + \alpha_0(1-\upsilon)} \tag{14.23}$$

When α_0 is as close to unity, as it is for UF_6, many equations are simpler when expressed in terms of

$$\delta \equiv \alpha_0 - 1 \approx \frac{m_2 - m_1}{m_2 + m_1} \tag{14.24}$$

When $\delta \ll 1$, Eq. (14.23) may be approximated by

$$\upsilon - x_0 = \delta\upsilon(1-\upsilon) \tag{14.25}$$

This approximation will be used in all subsequent derivations.

Barrier separation efficiency. In practice, the difference in composition between gas on the upstream face of the barrier x'' and gas flowing through the barrier is less than under ideal conditions for the following reasons, among others:

1. Downstream pressure p' is not negligible, and some molecular flow takes place from downstream to upstream faces, partially offsetting separation achieved by flow in the forward direction.
2. Some of the flow through the barrier is of a nonseparating type, such as viscous flow of the gas acting as a continuous fluid.

It is convenient to express the reduction in separation from these two causes in terms of a barrier separation efficiency E_B, defined as

$$E_B \equiv \frac{\upsilon - x''}{\upsilon - x_0} \tag{14.26}$$

Because of (14.25) this is

$$E_B = \frac{\upsilon - x''}{\delta\upsilon(1-\upsilon)} \tag{14.27}$$

First, we consider the effect of a finite back pressure p', but with both p'' and p' so low that only molecular flow takes place. In this case, the net flow through the barrier equals the difference between the molecular flow from the upstream face and the molecular flow from the downstream face:

$$G_1 = \frac{a}{\sqrt{m_1}}(p''x'' - p'y') \tag{14.28}$$

$$G_2 = \frac{a}{\sqrt{m_2}}[p''(1-x'') - p'(1-y')] \tag{14.29}$$

The composition of the net flow is

$$\upsilon = \frac{G_1}{G_1 + G_2} = \frac{\alpha_0(x'' - qy')}{\alpha_0 x'' + 1 - x'' - q(\alpha_0 y' + 1 - y')} \tag{14.30}$$

where q is the ratio of downstream to upstream pressures,

$$q \equiv \frac{p'}{p''} \tag{14.31}$$

and α_0 is the ideal separation factor given by (14.21). This may be solved for x'':

$$x'' = \frac{\upsilon + q(\alpha_0 y' - \alpha_0 \upsilon y' - \upsilon + \upsilon y')}{\upsilon + \alpha_0(1-\upsilon)} \tag{14.32}$$

The barrier separation efficiency, from (14.27), is

$$E_B = 1 - q\frac{\alpha_0 y'(1-\upsilon) - \upsilon(1-y')}{(\alpha_0 - 1)\upsilon(1-\upsilon)} \tag{14.33}$$

For the special but practically important case in which the composition of the downstream gas equals that of the net flow of gas through the barrier ($y' = \upsilon$),

$$E_B = 1 - q \tag{14.34}$$

When $y' \neq \upsilon$, to the first order in $\delta \equiv \alpha_0 - 1$ and $y' - \upsilon$,

$$E_B = 1 - q - \frac{q(y' - \upsilon)}{\delta\upsilon(1-\upsilon)} \tag{14.35}$$

Effect of nonseparating flow. The effect of nonseparating viscous flow on the barrier separation efficiency depends on the detailed structure of the barrier. Because the theoretical derivation of this effect for the viscous leak model can be worked out simply and completely, this model will be dealt with first. Then results for the mixed flow model derived by Present and de Bethune [P3] will be summarized and both models will be compared with empirical correlations of separation performance suggested by experimental investigators of barrier performance.

Barrier separation efficiency, viscous leak model. In the viscous leak model, flow through the small holes of radius r_{mol} is of the separating, molecular type dealt with in deriving Eqs.

(14.33) through (14.35), and flow through the large holes of radius r_{vis} is of the nonseparating, viscous type at the rate given by Poiseuille's law (14.3).

In this model, net flows for each component, from (14.7), are

$$G_1 = \frac{a}{\sqrt{m_1}}(p''x'' - p'y') + \frac{b}{\mu}x''(p''^2 - p'^2) \tag{14.36}$$

$$G_2 = \underbrace{\frac{a}{\sqrt{m_2}}[p''(1-x'') - p'(1-y')]}_{\text{Molecular}} + \underbrace{\frac{b}{\mu}(1-x'')(p''^2 - p'^2)}_{\text{Viscous}} \tag{14.37}$$

where

$$a = \frac{8r_{\mathrm{mol}}\epsilon_{\mathrm{mol}}}{3l\sqrt{2\pi RT}} \tag{14.38}$$

and

$$b = \frac{r_{\mathrm{vis}}^2\epsilon_{\mathrm{vis}}}{16lRT} \tag{14.39}$$

The composition of the net flow is

$$v = \frac{\alpha_0(x'' - qy') + x''(p'' + p')(1-q)/p_c}{\alpha_0 x'' + 1 - x'' - q(\alpha_0 y' + 1 - y') + (p'' + p')(1-q)/p_c} \tag{14.40}$$

with $q = p'/p''$, α_0 from (14.21), and the characteristic pressure p_c given by

$$p_c = \frac{\mu a}{b\sqrt{m_2}} = \frac{128\mu r_{\mathrm{mol}}\epsilon_{\mathrm{mol}}}{3r_{\mathrm{vis}}^2\epsilon_{\mathrm{vis}}}\sqrt{\frac{RT}{2\pi m}} \tag{14.41}$$

For this viscous leak model, from Eq. (14.12),

$$p_c = \frac{2}{S} \tag{14.42}$$

Equation (14.42) suggests that the separation parameter p_c for a mixture could be evaluated from measurement of the slope factor S obtained from the pressure dependence of the permeability for a pure gas, Eq. (14.10). For real barrier materials it is found that the separation parameter p_c is appreciably smaller than would be predicted from the slope factor in Eq. (14.10).

Equation (14.40) may be solved for x'':

$$x'' = \frac{v + v(p'' + p')(1-q)/p_c + q(\alpha_0 y' - \alpha_0 v - v + vy')}{v + \alpha_0(1-v) + (p'' + p')(1-q)/p_c} \tag{14.42a}$$

The barrier separation efficiency, from (14.26), is

$$E_B = \left[1 - q\frac{\alpha_0 y'(1-v) - v(1-y')}{(\alpha_0 - 1)v(1-v)}\right]\left[\frac{v + \alpha_0(1-v)}{v + \alpha_0(1-v) + (p'' + p')(1-q)/p_c}\right] \tag{14.43}$$

To the first order in $\alpha_0 - 1 \equiv \delta$ and $y' - v$, Eq. (14.43) reduces to

$$E_B = \frac{1 - q - q(y' - v)/\delta v(1-v)}{1 + (\pi'' + \pi')(1-q)} \tag{14.44}$$

where

$$\pi \equiv \frac{p}{p_c} \tag{14.45}$$

is a dimensionless pressure and

$$q = \frac{\pi'}{\pi''} \tag{14.46}$$

When the composition of the downstream gas y equals that of the net flow v,

$$E_B = \frac{1-q}{1+(\pi''+\pi')(1-q)} \tag{14.47}$$

Note that $E_B = 0.500$ when $p' = 0$ and $p'' = p_c$.

Barrier separation efficiency, mixed flow model. Present, Pollard, and de Bethune [P3, P4] have worked out the transport equations for each component of a two-component mixture flowing through a circular capillary of radius r and length l under conditions in which both molecular and viscous flows are taking place in the same capillary. They find that separation is impaired over what would be predicted from the slope factor by the viscous leak model because of an effect important at pressures below the pure viscous flow regime, in which occasional collisions between faster-moving lighter molecules and slower-moving heavier molecules slow down the former and speed up the latter and thus reduce separation. Their derivation is limited to the practically important case in which the composition of the gas downstream of the barrier equals that of the net flow through the barrier ($y' = v$). They give a rather complex set of equations for the case in which α_0 differs appreciably from unity, which reduce for the close-separation case of interest in uranium isotope separation to Eq. (14.48) for the barrier separation efficiency E_B.

$$E_B = \frac{\int_{\phi'}^{\phi''} \exp\,[(1+X)\phi + (X/2)\phi^2]\,d\phi}{\phi''\exp\,[(1+X)\phi'' + (X/2)\phi''^2]} \tag{14.48}$$

Here

$$\phi \equiv \frac{3rp}{16\mu}\sqrt{\frac{\pi m}{8RT}} \tag{14.49}$$

and

$$X \equiv \frac{256}{9\pi}\,\frac{\mu}{\rho D} \tag{14.50}$$

where ρ is the density and D is the diffusion coefficient. Ney and Armistead [N2] have found that $\rho D/\mu$ for mixtures of $^{235}UF_6$ and $^{238}UF_6$ is close to $\frac{4}{3}$. With this value,

$$X(UF_6) = \frac{64}{3\pi} \tag{14.51}$$

Numerical inversion of Eq. (14.48) shows that $E_B = 0.500$ when $\phi' = 0.00$ and $\phi'' = 0.1834$. To provide an equation that may be compared with (14.47), the characteristic pressure p_c is defined by

$$p_c = \frac{(0.1834)(16\mu)}{3r}\sqrt{\frac{8RT}{\pi m}} \tag{14.52}$$

and the dimensionless pressure π is

$$\pi = \frac{p}{p_c} \tag{14.53}$$

Hence

$$\pi = \frac{\phi}{0.1834} \tag{14.54}$$

In Eq. (14.48), substitution of (14.51) for X and change of variable from ϕ to π through (14.54) results in

$$E_B(\pi'', \pi') = \frac{\int_{\pi'}^{\pi''} \exp\,(1.430\pi + 0.1142\pi^2)\,d\pi}{\pi''\exp\,(1.430\pi'' + 0.1142\pi''^2)} \tag{14.55}$$

Comparison of Eqs. (14.52) and (14.12) shows that for this mixed flow model,

$$p_c = \frac{0.1834}{S} \tag{14.56}$$

Comparison with Eq. (14.42) shows that p_c evaluated from the slope factor with the mixed flow model is only 9.17 percent the value of p_c evaluated from the slope factor with the viscous leak model. Equation (14.56) comes closer to representing the characteristics of actual barriers.

Empirical equations for barrier separation efficiency. Even Present and de Bethune's development does not represent accurately conditions in an actual diffusion barrier because gas flow paths are neither straight, circular, nor of uniform cross section. Consequently, a number of empirical equations have been suggested to characterize the separation performance of barriers. Bilous and Counas of the French CEA [B17] have proposed the empirical equation

$$E_B = (1-q)\left(1 - \frac{\pi''}{2}\right) \tag{14.57}$$

valid for a limited range of values of π''. C. H. Bosanquet [K5], of the British gaseous diffusion project, proposed Eq. (14.58), a modification of the viscous leak formula (14.47):

$$E_B = \frac{1-q}{1+\pi''(1-q)} = \frac{1-p'/p''}{1+(p''-p')/p_c} \tag{14.58}$$

which brings its results closer to the Present and de Bethune formula (14.55). Table 14.5 compares the barrier separation efficiencies predicted by the Bilous and Counas Eq. (14.57), the viscous leak Eq. (14.47), Bosanquet's Eq. (14.58), and Present and de Bethune's Eq. (14.55). For gaseous diffusion process analysis, this text will use Bosanquet's Eq. (14.58) because of its comparatively simple form and its fairly close correspondence with the theoretically based Eq. (14.55) of Present and de Bethune. As Table 14.5 shows, all four equations give a barrier efficiency of 0.500 at upstream condition $\pi'' = 1.00$ and downstream condition $\pi' = 0.00$.

Diffusion barrier characteristics. Because of security classification, quantitative information on barrier characteristics is scarce. The most comprehensive report in the open literature was made

Table 14.5 Comparison of equations for barrier separation efficiency

			Barrier efficiency given by			
$\pi'' = p''/p_c$	$\pi' = p'/p_c$	$q = p'/p''$	Bilous & Counas Eq. (14.57)	Viscous leak Eq. (14.47)	Bosanquet Eq. (14.58)	Present & de Bethune Eq. (14.55)
1.00	0.00	0.00	0.500	0.500	0.500	0.500
0.72	0.144	0.20	0.512	0.473	0.508	0.528
0.72	0.18	0.25	0.480	0.448	0.487	0.507
0.72	0.24	0.333	0.427	0.406	0.450	0.469
0.4	0.10	0.25	0.600	0.545	0.577	0.604
0.5	0.125	0.25	0.563	0.511	0.545	0.572
0.72	0.18	0.25	0.480	0.448	0.487	0.507
1.0	0.25	0.25	0.375	0.387	0.429	0.436
1.2	0.30	0.25	0.300	0.353	0.395	0.391

Table 14.6 Characteristics of French diffusion barriers

Method of preparation	Mean pore radius $\bar{r}$, μm	Permeability $\Gamma \times 10^5$, g-mol air/ $(\text{cm}^2 \cdot \text{cmHg} \cdot \text{min})$	Pressure for 50% separation efficiency p_c, Torr	Permeability $\gamma \times 10^5$
Dissolving silver from gold-silver alloy	0.03	8	500	20.8
Anodic oxidation of aluminum	0.01	6	1500	15.6
Sintering alumina	0.025	4	600	10.4
Sintering nickel	0.020	2.5	750	6.5
Rolling Teflon powder into nickel gauze	0.015	2.5	1000	6.5

by Frejacques et al. [F3] in 1958. The first two columns of Table 14.6 give properties reported by these workers for five different barrier types developed by the French CEA. These reported properties have been converted to the units given in the last two columns as follows. Frejacques et al. state that the barrier separation efficiency depends on upstream pressure p'', downstream pressure p', and mean pore radius $\bar{r}$ as

$$E_B = \left(1 - \frac{p'}{p''}\right)\left(1 - \frac{\bar{r}p''}{A}\right) \tag{14.59}$$

Bilous and Counas [B17] recommend for the parameter A a value of 3 (μm$\cdot$cmHg) as providing an adequate correlation between their pore size measurements and barrier separation performance on UF_6 at temperatures between 35 and 85°C. Hence the upstream pressure $p'' = p_c$ in torr at which the barrier would have an efficiency of $E_B = 0.500$ at a downstream pressure $p' = 0.00$ is

$$p_c(\text{Torr}) = \frac{(1 - 0.5)3(\mu\text{m}\cdot\text{cmHg})10(\text{Torr/cmHg})}{\bar{r}(\mu\text{m})} = \frac{15}{\bar{r}(\mu\text{m})} \tag{14.60}$$

The relation between the observed permeability Γ reported in units of gram-moles air per square centimeter per cmHg pressure difference per minute and the dimensionless permeability γ defined earlier is

$$\Gamma[\text{g-mol air}/(\text{cm}^2\cdot\text{min}\cdot\text{cmHg})]$$

$$= \frac{\gamma\bar{v}_{\text{air}}(\text{cm/s})60(\text{s/min})}{(4)82.06[(\text{cm}^3\cdot\text{atm})/(\text{g-mol}\cdot\text{K})]\,293(\text{K})76(\text{cmHg/atm})} = 8.2 \times 10^{-6}\gamma\bar{v} \tag{14.61}$$

Here 82.06 $(\text{cm}^3\cdot\text{atm})/(\text{g-mol}\cdot\text{K})$ is the gas constant and 293 K is the test temperature. Because the mean speed of air molecules at 293 K is

$$\bar{v} = \sqrt{\frac{8RT}{\pi m}} = \sqrt{\frac{(8)(8.31434 \times 10^7)(293)}{\pi(28.8)}} = 46{,}411 \text{ cm/s} \tag{14.62}$$

$$\gamma = \frac{\Gamma}{(8.2 \times 10^{-6})(46{,}411)} = 2.63\Gamma \tag{14.63}$$

In subsequent analysis of the gaseous diffusion process, the diffusion barrier will be assumed to have the properties of the French barrier listed second in Table 14.6, made by anodic oxidation of aluminum, with $p_c = 1500$ Torr and $\gamma = 15.6 \times 10^{-5}$. Over the range of operating conditions of economic interest, the specific permeability γ will be treated as independent of pressure and temperature.

4.6 Mixing Efficiency

Because the atom fraction of light component in the net transport of gas through the barrier, v, is greater than the atom fraction of light component in the gas at the high-pressure face of the barrier, x'', there must be a difference between the average composition of the gas flowing past the high-pressure side of the barrier, x_i, and the gas at the barrier face, x'', to maintain the required transport of light component to the barrier surface. The local barrier mixing efficiency is defined as

$$E_M = \frac{v - x_i}{v - x''} \tag{14.64}$$

The purpose of this section is to show how this mixing efficiency depends on conditions on the high-pressure side of the barrier.

Figure 14.5 is a transverse section of a circular barrier tube of diameter d with high-pressure flow along the inside of the tube. With turbulent flow of gas inside the tube, molar velocity is practically uniform at a value slightly above the average, H, over most of the tube diameter, but drops to zero at the tube wall. Atom fraction light component is practically constant at a value slightly above the average, x_i, over most of the tube diameter owing to turbulent mixing where the velocity is uniform, but drops to a lower value of x'' adjacent to the tube wall to provide the required transport through the poorly mixed gas adjacent to the

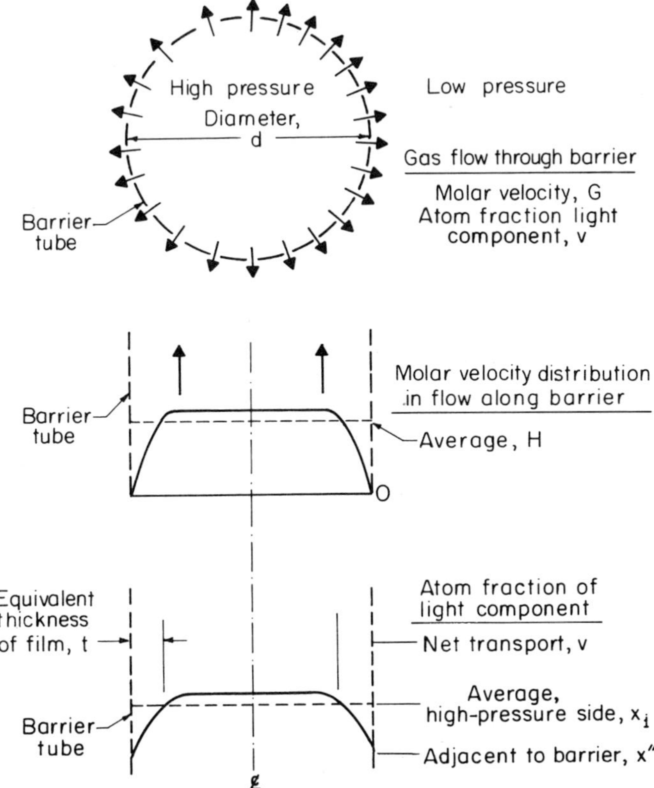

Figure 14.5 Velocity and composition distribution in flow through barrier.

tube wall. The molar velocity of gas flow through the barrier is G, with v atom fraction light component.

Bilous and Counas [B17] have used an equation derived originally for wetted-wall gas-absorption towers to evaluate E_M as a function of the molar velocities G and H. The basic assumption is that the actual gas flow pattern behaves as if there were a stagnant film of thickness t adjacent to the tube wall, through which light component is transported by molecular diffusion, with diffusion coefficient D. As will be shown later, E_M in this model is given by

$$E_M \equiv \frac{v - x_i}{v - x''} = \exp\left(\frac{-352Gt}{D\rho}\right) \tag{14.65}$$

Here 352 is the molecular weight of UF_6 and ρ is the mass density. The empirical correlation for the thickness of the stagnant film t, obtainable from standard chemical engineering texts such as [S4], is

$$t = 43d(\text{Re})^{-0.83}\left(\frac{\rho D}{\mu}\right)^{0.44} \tag{14.66}$$

Here $\rho D/\mu$ has the value $\frac{4}{3}$ for UF_6, and Re is the Reynolds number on the high-pressure side of the barrier:

$$\text{Re} \equiv \frac{352Hd}{\mu} \tag{14.67}$$

μ is the viscosity of UF_6, Eq. (14.4).

Equation (14.65) for the composition gradient in mass transfer through a stagnant film of thickness t may be derived with the aid of Fig. 14.6. At a distance t_f into the film, where the atom fraction of light component is x_f, the required net transport of Gv mol of light component per unit area per second is the resultant of that due to flow, Gx_f, and that due to diffusion:

$$Gv = Gx_f - \frac{D\rho}{352}\frac{dx_f}{dt_f} \tag{14.68}$$

The solution of this equation, with boundary condition $x_f = x_i$ at $t_f = 0$, is

$$\ln \frac{v - x_i}{v - x_f} = -352Gt_f D\rho \tag{14.69}$$

At the barrier surface, where $t_f = t$ and $x_f = x''$,

$$\frac{v - x_i}{v - x''} = \exp\left(\frac{-352Gt}{D\rho}\right) \tag{14.70}$$

which is (14.65).

The molar velocity H of the gas along the high-pressure side of the barrier may be obtained from the dimensions of the barrier as follows: The total number of moles of gas flowing through a barrier tube d in diameter and L long is πdGL. In a well-designed diffusion stage, one-half of the gas entering the stage is diffused. The molar velocity of gas at the inlet end of each tube of the stage then is

$$H_{\text{in}} = \frac{2\pi dGL}{\pi d^2/4} = \frac{8GL}{d} \tag{14.71}$$

and the molar velocity at the outlet end is one-half of this value, or

$$H_{\text{out}} = \frac{4GL}{d} \tag{14.72}$$

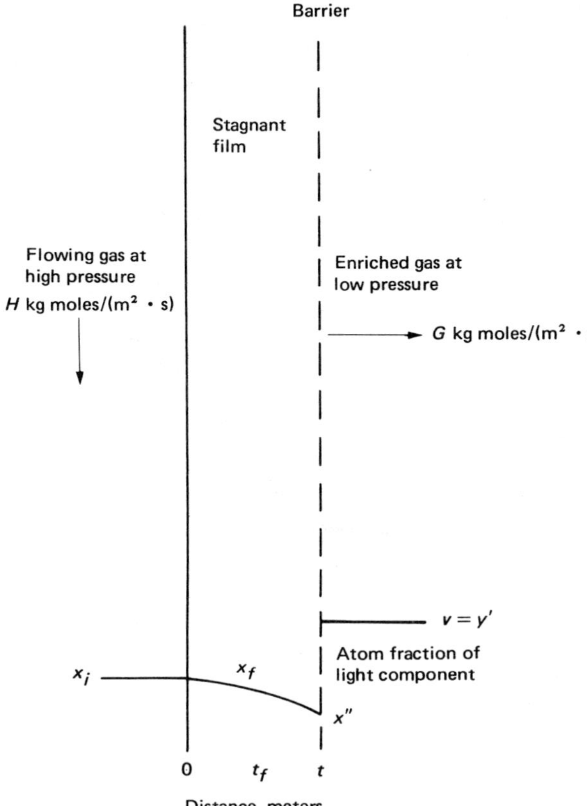

Figure 14.6 Nomenclature for deriving Eq. (14.65) for mixing efficiency.

For a rough estimate of average mixing efficiency, the value of H at midlength of the tube may be used, at which

$$H_{\text{ave}} = \frac{6GL}{d} \tag{14.73}$$

For a barrier tube of given diameter and permeability, the mixing efficiency is higher the longer the barrier tube, because the molar velocity along the tube is proportional to the length. However, the pressure drop experienced by the gas flowing along the tube is greater the longer the tube, both because of the increased flow path and the increased molar velocity. This pressure drop is detrimental for three reasons: The barrier separation efficiency is decreased, more barrier area is needed, and more energy must be expended to restore the pressure drop. Determination of optimum tube length requires an economic balance among the gain in mixing efficiency, the loss of barrier efficiency, and the cost of increased energy input. Such detailed balance is beyond the scope of this text. Instead, calculations will be given for the mixing efficiency and pressure drop for several tube lengths, and an arbitrary choice of length will be made for subsequent design examples.

The pressure gradient in a circular tube of diameter d through which turbulent flow at molar velocity H is taking place is

$$\frac{dp}{dz} = \frac{0.046m^2H^2}{(\text{Re})^{0.2}\, d\rho} \tag{14.74}$$

Here ρ is the gas density:

$$\rho = \frac{pm}{RT} \tag{14.75}$$

Tube lengths of 2, 4, and 6 m will be considered. A specific high-side inlet pressure of 1 atm (101,325 Pa) and low-side pressure of 0.25 atm (25,331 Pa) will be used.

For the example of the French aluminum barrier with $\gamma = 15.6 \times 10^{-5}$, the molar velocity G of UF_6 through the barrier at 358 K, an upstream pressure of $p'' = 1.0$ atm (101,325 Pa), and a downstream pressure of $p' = 0.25$ atm (25,331 Pa), from (14.14), is

$$G = \frac{\gamma(p'' - p')}{\sqrt{2\pi mRT}} = \frac{15.6 \times 10^{-5}(101{,}325 - 25{,}331)}{\sqrt{(2\pi)(352)(8314)(358)}} = 14.61 \times 10^{-5} \text{ kg-mol/(m}^2\cdot\text{s)} \tag{14.76}$$

For this barrier at 358 K, in general, when pressures p'' and p' are expressed in atmospheres,

$$G = 14.61 \times 10^{-5} \frac{p'' - p'}{1 - 0.25} = 19.48 \times 10^{-5} (p'' - p') \text{ kg-mol/(m}^2\cdot\text{s)} \tag{14.77}$$

Table 14.7 gives the mixing efficiency and pressure gradient at the inlet, midlength, and outlet of barrier tubes of these three lengths, and the overall pressure drop in the direction of flow down the tube. The overall pressure drop is approximated by multiplying the average of the pressure gradient at the three calculated points by the length of the tube.

Table 14.7 shows that increasing the tube length from 2 to 4 m increases the mixing efficiency at midlength by 9 percent with an increase in pressure drop under 1 percent. Further increase in tube length to 6 m increases mixing efficiency by less than 4 percent, with an increase in pressure drop of 2 percent. Determination of the optimum tube length would require an economic balance that is beyond the scope of this text. A length of 4 m will be used

Table 14.7 Variation of local mixing efficiency and pressure drop with length of barrier tube†

Tube length L, m	2	4	6
Molar velocity along tube, kg-mol/($m^2\cdot$s)			
Inlet	0.1669	0.3339	0.5008
Midlength	0.1252	0.2504	0.3756
Outlet	0.0834	0.1669	0.2504
Mixing efficiency			
Inlet	0.823	0.896	0.925
Midlength	0.781	0.870	0.906
Outlet	0.708	0.823	0.870
Pressure gradient, Pa/m			
Inlet	114	396	821
Midlength	68	236	489
Outlet	33	114	236
Overall pressure drop			
Pa	143	995	3092
Fraction of inlet	0.0014	0.0098	0.0305

†Tube diameter d, 0.014 m; barrier specific permeability γ, 15.6×10^{-5}; temperature T, 358 K; high-side pressure p'', 101,325 Pa; low-side pressure p', 25,331 Pa.

in Sec. 4.7 in examining the effect of various combinations of high-side and low-side pressures on stage design and plant requirements.

Table 14.7 also shows that the mixing efficiency at midlength is close to the average value over the tube length. To simplify the calculations to be made in Sec. 4.7, the mixing efficiency will be treated as if constant at its value at midlength, and the pressure drop along the tube will be neglected. In accurate design calculations, point-by-point calculations along the barrier tube should be made of pressure drop, flow through the barrier, flow along the tube, and mixing efficiency, refinements that are neglected in the remaining treatment of gaseous diffusion.

4.7 Stage Characteristics

Stage separation efficiency. Figure 14.7 illustrates the nomenclature to be used in describing flow rates, compositions, and degree of separation in a cross-flow gaseous diffusion stage, with $v = y'$. The stage separates feed containing x_F mole fraction light component into a light fraction containing y mole fraction and a heavy fraction containing x mole fraction.

$$x < x_F < y$$

The separation factor of the stage α is defined as

$$\alpha \equiv \frac{y(1-x)}{x(1-y)} \tag{14.78}$$

A stage separation efficiency E, analogous to the overall Murphree plate efficiency in distillation, may be defined as

$$E \equiv \frac{y-x}{y-x_0} \tag{14.79}$$

where x_0 is the composition of gas that, on the high-pressure side of an ideal barrier, would give low-pressure gas of composition y. From (14.23) it follows that

$$y - x_0 = \frac{(\alpha_0 - 1)y(1-y)}{y + \alpha_0(1-y)} \tag{14.80}$$

In the close separation case,

$$\alpha - 1 = \frac{y-x}{x(1-x)} \tag{14.81}$$

$$E = \frac{y-x}{(\alpha_0 - 1)x(1-x)} \tag{14.82}$$

and

$$\alpha - 1 = E(\alpha_0 - 1) \tag{14.83}$$

Our problem is to determine the relationship between the stage separation efficiency, given by (14.82), the barrier separation efficiency E_B, defined by (14.26), and the local mixing efficiency E_M, defined by (14.64). The stage separation efficiency depends on the relative direction of flow of the high-pressure and low-pressure streams and the degree of mixing of these streams.

No mixing, cross flow. In a common type of gaseous diffusion stage, high-pressure gas flows along the inside of a number of barrier tubes in parallel without significant mixing in the direction of flow, and the low-pressure gas that has passed through the barrier is removed in cross-flow paths approximately perpendicular to the barrier tubes. With cross flow on the low-pressure side, the composition of the gas at each point on the low-pressure side of the barrier, y', equals the composition of the net flow through the barrier at the point, v. This

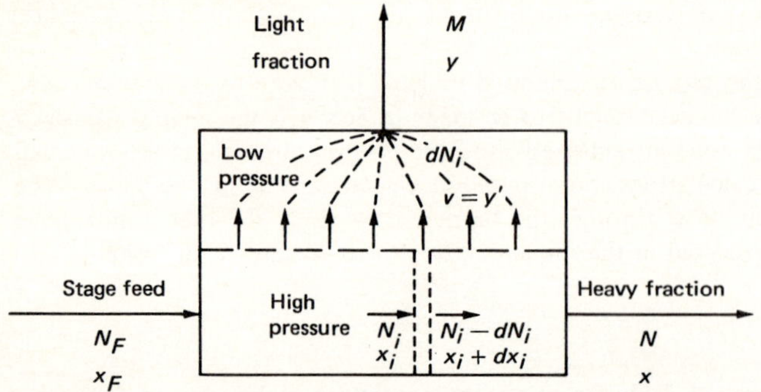

Figure 14.7 Cross-flow gaseous diffusion stage. Stage separation factor $\alpha \equiv y(1-x)/x(1-y)$; stage efficiency $E \equiv (y-x)/(y-x_0)$; N_F, N_i, N, M = molar flow rates; x_F, x_i, x, y, v = mole fraction light component.

practically important condition obtains in most barrier testing experiments and is a condition for Eqs. (14.47) and (14.55) for barrier separation efficiency. When $y' = v$, there is no mixing efficiency correction on the downstream side of the barrier.

Figure 14.7 shows the nomenclature to be used in deriving an equation relating the stage separation efficiency to the barrier efficiency E_B and the local mixing efficiency E_M for the above kind of cross-flow diffusion stage. A material balance on light component over the portion of the high-pressure side of the stage in which the flow rate decreases by dN_i may be expressed as

$$v\,dN_i = N_i x_i - (N_i - dN_i)\,(x_i + dx_i) \tag{14.84}$$

This leads to the differential equation

$$N_i \frac{dx_i}{dN_i} = v - x_i \tag{14.85}$$

In the close-separation case, from (14.65),

$$v - x_i = E_M(v - x'') \tag{14.86}$$

and

$$v - x_i = E_M E_B (v - x_0) = E_M E_B \delta v(1 - v) \tag{14.87}$$

from (14.27). In the close-separation case v changes so little from point to point in a diffusion stage that v in (14.87) may be replaced by x, the outlet heavy fraction composition. With this substitution, Eqs. (14.87) and (14.85) become

$$dx_i = (\alpha_0 - 1)x(1 - x)E_M E_B \frac{dN_i}{N_i} \tag{14.88}$$

The integral of this equation between $x_i = x_F$ at $N_i = N_F$ and $x_i = x$ at $N_i = N$ is

$$x_F - x = (\alpha_0 - 1)x(1 - x)E_M E_B \ln \frac{N_F}{N} \tag{14.89}$$

The fraction diffused is the stage cut θ:

$$\theta \equiv \frac{M}{N_F} = 1 - \frac{N}{N_F} \tag{14.90}$$

By material balance,

$$y - x = \frac{x_F - x}{\theta} \tag{14.91}$$

Hence
$$y - x = -(\alpha_0 - 1)x(1 - x)E_M E_B \frac{\ln(1-\theta)}{\theta} \tag{14.92}$$

From (14.82), the stage separation efficiency is

$$E = -E_M E_B \frac{\ln(1-\theta)}{\theta} \tag{14.93}$$

Because $-[\ln(1-\theta)]/\theta$ is greater than unity, the stage efficiency in cross flow exceeds the local efficiency $E_M E_B$. A similar result is found in distillation, where the plate efficiency is greater than the point efficiency when there is cross flow of liquid without mixing across the plate.

In an ideal cascade in which $\theta = \frac{1}{2}$,

$$E = 1.386 E_M E_B \tag{14.94}$$

Stage performance equations for a mixture in which α_0 differs substantially from unity have been derived by Weller and Steiner [W1].

Stage design variables. The principal independent variables involved in designing a gaseous diffusion stage to serve in an ideal cascade are as follows:

1. The product rate P and product composition y_P of the cascade of which the stage is a member
2. The fraction of light component y in the stage of interest
3. The quality of the barrier selected, as measured by its characteristic pressure p_c and its permeability γ
4. The diameter d and length L of barrier tubes
5. The high-side pressure p'' and low-side pressure p'
6. The barrier absolute temperature T

The principal stage characteristics that depend on the choice of the above independent variables are as follows:

1. The stage separation factor α
2. The heads flow rate M
3. The stage separative capacity Δ
4. The compressor volumetric capacity V
5. The stage barrier area A
6. The stage power requirement Q
7. The initial cost of the stage C_0
8. The annual cost charged to the stage C
9. The unit cost of separative work $c_S \equiv C/\Delta$

The objectives of this section are as follows:

1. To develop equations for the dependence of these stage characteristics on the independent variables
2. To show how the stage design may be optimized for various criteria
3. To work out the stage design conditions that lead to minimum unit cost of separative work for a specific barrier type

Stage separation factor. For a cross-flow gaseous diffusion stage with a cut θ of $\frac{1}{2}$, the stage separation factor α is given by

$$\alpha - 1 = 1.386(\alpha_0 - 1)E_M E_B \tag{14.95}$$

from (14.83) and (14.94).

For a barrier whose separation performance on UF_6 is given by the Bosanquet equation (14.58),

$$\alpha - 1 = 1.386(\alpha_0 - 1)E_M \frac{1 - p'/p''}{1 + (p'' - p')/p_c} \tag{14.96}$$

where

$$\alpha_0 - 1 = 0.00429 \tag{14.97}$$

and E_M is given by (14.65) and (14.66).

Heads flow rate. The flow rate M of stage heads of composition y in the enriching section of a close-separation ideal cascade producing product at rate P and composition y_P is

$$M = \frac{2P(y_P - y)}{(\alpha - 1)y(1 - y)} \tag{14.98}$$

as can be seen by a development similar to Eq. (12.125).

Stage separative capacity. The separative capacity Δ of a stage in a close-separation cascade with a cut of $\frac{1}{2}$ is

$$\Delta = \frac{M(\alpha - 1)^2}{4} \tag{14.99}$$

from Eq. (12.174).

Compressor volumetric capacity. The volumetric capacity V of the compressor for the heads stream at pressure p' and absolute temperature T flowing at molar rate M is

$$V = \frac{MRT}{p'} \tag{14.100}$$

The ratio of compressor capacity to separative capacity is

$$\frac{V}{\Delta} = \frac{4RT}{(\alpha - 1)^2 p'} = \frac{4RT}{(1.386)^2(\alpha_0 - 1)^2 E_M^2} \left\{ \frac{[1 + (p'' - p')/p_c]^2}{p'(1 - p'/p'')^2} \right\} \tag{14.101}$$

This ratio is independent of isotopic composition. Because E_M is only slightly dependent on p' and p'', the pressures p' and p'' that would result in minimum compressor capacity for a given separative capacity are close to those that minimize the term in braces in (14.101). These are found to be

$$p'' = 2p_c \quad \text{and} \quad p' = p_c \tag{14.102}$$

at which the term in braces in (14.101) has the value $16/p_c$, so that the minimum compressor capacity is

$$\frac{V_{\min}}{\Delta} = \frac{33.3RT}{(\alpha_0 - 1)^2 E_M^2 p_c} \tag{14.103}$$

Barrier area. The barrier area A required for a heads stream flowing at molar rate M between pressures of p'' and p' is

$$A = \frac{M}{G} = \frac{\sqrt{2\pi mRT}\,M}{\gamma(p'' - p')} \tag{14.104}$$

from (14.14). Hence the ratio of barrier area to separative capacity is

$$\frac{A}{\Delta} = \frac{M}{G\Delta} = \frac{4\sqrt{2\pi mRT}}{(\alpha - 1)^2\gamma(p'' - p')} = \frac{4\sqrt{2\pi mRT}}{(1.386)^2(\alpha_0 - 1)^2 E_M^2 \gamma}\left\{\frac{[1 + (p'' - p')/p_c]^2}{(p'' - p')(1 - p'/p'')^2}\right\} \tag{14.105}$$

The pressures p' and p'' that would result in minimum barrier area for a given separative capacity are close to those that minimize the term in braces in (14.105). These are

$$p'' = p_c \quad \text{and} \quad p' = 0 \tag{14.106}$$

at which the term in braces in (14.105) has the value $4/p_c$, so that the minimum barrier area is

$$\frac{A_{\min}}{\Delta} = \frac{8.33\sqrt{2\pi mRT}}{(\alpha_0 - 1)^2 E_M^2 p_c \gamma} \tag{14.107}$$

This indicates the desirability of having a high value of the characteristic pressure p_c (hence small pores) and a high value of γ (hence many pores per unit area).

Power. Flow of gas through the barrier at rate M is accompanied by loss of availability at rate

$$Q = MRT_0 \ln\left(\frac{p''}{p'}\right) \tag{14.108}$$

where T_0 is the temperature of heat rejection to the environment. This represents the minimum net power needed to recompress the gas from p' to p'' when the heat of compression at temperatures above T_0 is converted to work in a reversible heat engine. Hence the ratio of this net power to separative capacity is

$$\frac{Q}{\Delta} = \frac{4RT_0 \ln(p''/p')}{(\alpha - 1)^2} = \frac{4RT_0}{(1.386)^2(\alpha_0 - 1)^2 E_M^2}\left\{\frac{[1 + (p'' - p')/p_c]^2}{(1 - p'/p'')^2}\ln\left(\frac{p''}{p'}\right)\right\} \tag{14.109}$$

Minimum power results when both p' and p'' approach zero, with their ratio $q = p'/p''$ selected to make $[\ln(1/q)]/(1 - q)^2$ a minimum. This occurs at $q = 0.285$, at which the term in braces in (14.109) has the value 2.455. E_M at zero pressure equals 1.0. Hence the minimum power per unit separative capacity is

$$\left(\frac{Q}{\Delta}\right)_{\min} = \frac{5.11RT_0}{(\alpha_0 - 1)^2} \tag{14.110}$$

This important result is independent of the type of barrier and isotopes being separated. For $^{235}UF_6$ and $^{238}UF_6$. with $T_0 = 300$ K,

$$\left(\frac{Q}{\Delta}\right)_{\min} = \frac{(5.11)[8314\ \text{J/(kg-mol}\cdot\text{K)}](300\ \text{K})}{(0.00429)^2(3.6 \times 10^6\ \text{J/kWh})(8760\ \text{h/yr})(238\ \text{kg U/kg-mol})}$$
$$= 0.0923\ \text{kW/(kg SWU/yr)} \tag{14.111}$$

The pressure conditions that minimize compressor capacity, barrier area, and power consumption are listed in part 1 of Table 14.8. Part 2 of Table 14.8 gives for a diffusion plant with a separative capacity of 1 kg uranium/year, using anodized aluminum barrier tubes 0.014 m in diameter and 4 m long, the compressor capacity, barrier area, and power for the conditions that minimize these three plant requirements. Since these conditions are different, no one design can minimize simultaneously compressor capacity, barrier area and power.

The appropriate criterion to optimize the design of a gaseous diffusion stage is that the

Table 14.8 Optimum pressure conditions in gaseous diffusion and corresponding plant requirements

	Minimum compressor capacity	Minimum barrier area	Minimum power
1. Any barrier			
Optimum pressure conditions			
Low-side pressure p'/p_c	1.0	0.0	0.0
High-side pressure p''/p_c	2.0	1.0	0.0
Pressure ratio p'/p''	0.5	0.0	0.285
Barrier efficiency E_B	0.25	0.5	0.715
Factor in braces in equation for			
Compressor capacity (14.101)	$16/p_c$	∞	∞
Barrier area (14.105)	$16/p_c$	$4/p_c$	∞
Power (14.109)	11.1	∞	2.46
2. Anodized aluminum barrier, UF_6†			
Optimum low-side pressure p', atm	1.974	0.0	0.0
Optimum high-side pressure p'', atm	3.948	1.974	0.0
Mixing efficiency E_M	0.849	0.849	1.000
Stage separation factor $\alpha - 1$	0.000991	0.00297	0.00425
Plant requirements (capacity basis 1 kg U SWU/yr)			
Compressor capacity, m^3/s	0.00498	∞	∞
Barrier area, m^2	0.870	0.217	∞
Power, kW	0.578	∞	0.0923

†Diameter, 0.014 m; length, 4 m; permeability γ, 15.6×10^{-5}; p_c, 1.974 atm; T, 358 K.

unit cost of separative work produced by the stage be a minimum. To illustrate how Eqs. (14.101), (14.105), and (14.109) may be used to select optimum values of the stage pressures p' and p'' that minimize the unit cost of separative work, specific assumptions will be made about the unit cost of the principal stage characteristics on which the cost of separative work depends. The unit costs assumed for this purpose are listed below:

Direct capital costs	
Compressors and piping	\$10,000/(m^3/s)
Converters and barrier	\$50/m^2 barrier area
Electrical equipment and cooling system	\$100/kW
Indirect capital costs	50 percent of direct capital costs
Capital charge rate	20 percent/year
Electric power	\$0.02/kWh
Ratio of actual power to power for isothermal compression at T_0	2

Other costs that make small additional contributions to the cost of separative work, which are to be disregarded in this example, include costs of operation, maintenance, and supervision; fixed stage costs for such components as instruments; and the cost of enriched UF_6 inventory.

With the above assumptions, the unit cost of separative work c_S is

$$c_S(\$/\text{kg SWU}) = (0.2)(1.5)\left[10{,}000\left(\frac{V}{\Delta}\right) + 50\left(\frac{A}{\Delta}\right) + (2)(100)\left(\frac{Q}{\Delta}\right)\right] + (2)(0.02)(8760\ \text{h/yr})\left(\frac{Q}{\Delta}\right) \tag{14.112}$$

where the separative capacity basis for V/Δ, A/Δ, and Q/Δ is 1 kg SWU/year.

On this separative capacity basis, from (14.101),

$$\frac{V}{\Delta}\ \frac{\text{m}^3/\text{s}}{\text{kg SWU/yr}} = \frac{(4)[0.08206\,(\text{m}^3\cdot\text{atm})/(\text{kg-mol}\cdot\text{K})](358\ \text{K})}{(3.154\times 10^7\ \text{s/yr})(238\ \text{kg U/kg-mol})(1.386)^2(0.00429)^2 p' E_M^2 E_B^2} = \frac{0.0004428}{p' E_M^2 E_B^2} \tag{14.113}$$

where

$$E_B = \frac{1 - p'/p''}{1 + (p'' - p')/p_c} \tag{14.114}$$

and E_M is given by (14.65) and (14.66).

Because the required barrier area $A = M/G$ and $\Delta = M(\alpha - 1)^2/4$,

$$\frac{A}{\Delta}\ \frac{\text{m}^2}{\text{kg SWU/yr}} = \frac{4}{(3.154\times 10^7\ \text{s/yr})(238\ \text{kg U/kg-mol})[G\ \text{kg-mol}/(\text{m}^2\cdot\text{s})](\alpha - 1)^2} \tag{14.115}$$

With G for this barrier from Eq. (14.77) and $(\alpha - 1)$ from (14.95),

$$\frac{A}{\Delta}\ \frac{\text{m}^2}{\text{kg SWU/yr}} = \frac{4}{(3.154\times 10^7)(238)(19.48\times 10^{-5})(p'' - p')(1.386)^2(0.00429)^2 E_M^2 E_B^2} = \frac{0.07737}{(p'' - p')E_M^2 E_B^2} \tag{14.116}$$

On this separative capacity basis of 1 kg separative work unit per year, from (14.109),

$$\frac{Q}{\Delta}\ \frac{\text{kW}}{\text{kg SWU/yr}} = \frac{(4)[8.314\,(\text{kW}\cdot\text{s})/(\text{kg-mol}\cdot\text{K})]\,[300\ \text{K}\ \ln(p''/p')]}{(3.154\times 10^7\ \text{s/yr})(238\ \text{kg U/kg-mol})(1.386)^2(0.00429)^2 E_M^2 E_B^2} = \frac{0.03759\ \ln(p''/p')}{E_M^2 E_B^2} \tag{14.117}$$

Figure 14.8 shows the dependence of the unit cost of separative work, evaluated from Eq. (14.112), with V/Δ from (14.113), A/Δ from (14.115), and Q/Δ from (14.117), on the high-side pressure p'' and the pressure ratio p'/p''. Minimum unit cost is in the neighborhood of an upstream pressure of 0.55 atm and a pressure ratio of 0.32. There is a considerable range of pressures around this optimum in which the unit cost of separative work changes but little.

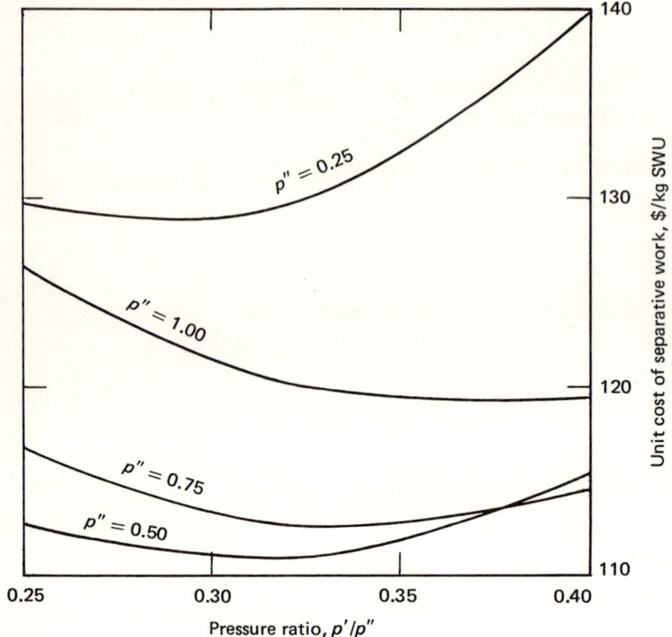

Figure 14.8 Effect of high-side pressure p'' and low-side pressure p' on unit cost of separative work from gaseous diffusion.

Table 14.9 gives the characteristics of a diffusion stage using these optimum conditions of $p'' = 0.55$ atm and $p'/p'' = 0.32$ for anodized aluminum barrier tubes 0.014 m in diameter and 4 m in length, with $p_c = 1.974$ atm and $\gamma = 15.6 \times 10^{-5}$.

The unit cost of \$110/kg SWU is not far from the value of \$100 anticipated for 1980 delivery. The power consumption of (2)(0.16776) = 0.336 kW/(kg SWU/year) may be compared with the power of 6,060,000 kWe consumed by U.S. ERDA's diffusion plants when operating at their full capacity of 17,230,000 kg SWU/year [U1]: 6,060,000/17,230,000 = 0.352 kW/(kg SWU/year).

After the cascade improvement and cascade operating programs planned by U.S. DOE have been completed, their power consumption will be increased to 7,380,000 kW and their separative capacity to 27,700,000 kg SWU/year, equivalent to a specific power consumption of 0.266 kW/(kg SWU/year).

4.8 Minimum Power Requirement of Gaseous Diffusion Process

Equation (14.111) for the minimum power of 0.0923 kW to produce 1 kg of separative work per year in uranium isotope separation was derived for cross flow on the low-pressure side of the barrier, with the composition of gas on that side y' equal to the composition of the net flow v. The purpose of this section is to show that the minimum power requirement could be reduced further by having v greater than y' by an appropriate amount and to derive an expression for the optimum difference between v and y' and the corresponding power consumption per unit separative capacity. For this minimum-power case, pressures on the high-pressure and low-pressure sides of the barrier must be so low the only flow through the barrier is of the separating, molecular type, and the mixing efficiency on each side of the barrier is unity.

A stream containing x mole fraction light component flowing at molar rate N carries separative work at the rate

$$N(2x-1)\ln\left(\frac{x}{1-x}\right)=(N_1-N_2)\ln\left(\frac{N_1}{N_2}\right) \tag{14.118}$$

where N_1 and N_2 are the molar flow rates of light and heavy components, respectively.

Consider the small element of barrier area dA shown in Fig. 14.9, at which the flow rates of light and heavy components are as follows:

		Molar flow rates			
	Molar velocity through barrier	High-pressure side		Low-pressure side	
Component		To dA	From dA	To dA	From dA
Light	G_1	N_1	N_1-dN_1	M_1	M_1+dM_1
Heavy	G_2	N_2	N_2-dN_2	M_2	M_2+dM_2

Table 14.9 Design conditions and characteristics of gaseous diffusion stage designed for minimum unit cost of separative work

Barrier type	Anodized aluminum
Permeability γ	15.6×10^{-5}
Characteristic pressure p_c	1.974 atm
Tube diameter d	0.014 m
Tube length L	4 m
Operating conditions	
Barrier temperature T	358 K
High-side pressure p''	0.55 atm
Pressure ratio $q=p'/p''$	0.32
Low-side pressure p'	0.176 atm
Stage properties	
Barrier efficiency E_B, Eq. (14.58)	0.57168
Mixing efficiency E_M, Eq. (14.65)	0.88388
Separation factor $\alpha-1$, Eq. (14.95)	0.003004
For a separative capacity of 1 kg SWU/yr	
Compressor capacity V/Δ, Eq. (14.113)	0.0098539 m^3/s
Barrier area A/Δ, Eq. (14.115)	0.81024 m^2
Loss of availability Q/Δ, Eq. (14.117)	0.16776 kW
Contributions to capital cost, \$/(kg SWU/yr)	
Compressors and piping, 10,000 V/Δ	\$ 98.54
Converters and barrier, 50 A/Δ	40.51
Electrical equipment and cooling system, (2)(100) Q/Δ	33.55
Direct capital costs	\$172.60
Indirect capital costs @ 50%	86.30
Total capital costs	\$258.90
Unit costs, \$/kg SWU	
Capital charges, (0.2)(258.90)	\$ 51.78
Power 2[Q/Δ) kW/(kg SWU/yr)] (8760 h/yr) (0.02 \$/kWh)	58.78
Total	\$110.56

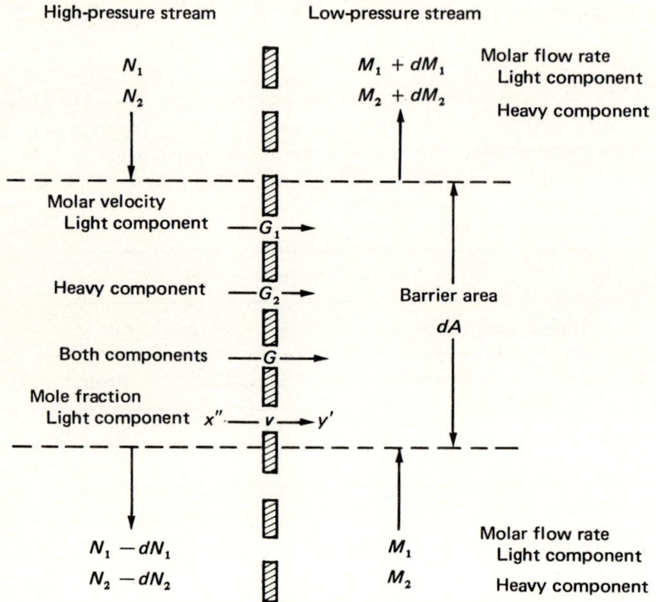

Figure 14.9 Nomenclature for deriving Eq. (14.144) for minimum power consumption in gaseous diffusion.

Material-balance equations are

$$G_1 dA = dN_1 = dM_1 \tag{14.119}$$

$$G_2 dA = dN_2 = dM_2 \tag{14.120}$$

The rate of production of separative work $d\Delta$ by the element dA is the difference between the rate of output separative work carried by the streams leaving dA and the rate of input carried by the streams entering dA.

$$\begin{aligned} d\Delta = (M_1 + dM_1 - M_2 - dM_2) \ln\left(\frac{M_1 + dM_1}{M_2 + dM_2}\right) \\ + (N_1 - dN_1 - N_2 + dN_2) \ln\left(\frac{N_1 - dN_1}{N_2 - dN_2}\right) \\ - (M_1 - M_2) \ln\left(\frac{M_1}{M_2}\right) - (N_1 - N_2) \ln\left(\frac{N_1}{N_2}\right) \end{aligned} \tag{14.121}$$

To the first order in dN_1 and dN_2, expansion of (14.121) yields

$$\begin{aligned} (dM_1 - dM_2) \ln\left(\frac{M_1}{M_2}\right) + (M_1 - M_2)\left(\frac{dM_1}{M_1} - \frac{dM_2}{M_2}\right) \\ - (dN_1 - dN_2) \ln\left(\frac{N_1}{N_2}\right) - (N_1 - N_2)\left(\frac{dN_1}{N_1} - \frac{dN_2}{N_2}\right) \end{aligned} \tag{14.122}$$

Introduction of the mole fractions

$$y' = \frac{M_1}{M_1 + M_2} \tag{14.123}$$

and

$$x'' = \frac{N_1}{N_1 + N_2} \tag{14.124}$$

and use of G_1 and G_2 from (14.119) and (14.120) in (14.122) yields

$$\frac{d\Delta}{dA} = (G_1 - G_2) \ln\left(\frac{y'}{1 - y'}\right) + (2y' - 1)\left(\frac{G_1}{y'} - \frac{G_2}{1 - y'}\right)$$

$$- (G_1 - G_2) \ln\left(\frac{x''}{1 - x''}\right) - (2x'' - 1)\left(\frac{G_1}{x''} - \frac{G_2}{1 - x''}\right) \tag{14.125}$$

In terms of the molar velocity G of both components through the barrier and the mole fraction υ of light component in the net flow through the barrier,

$$G_1 = \upsilon G \tag{14.126}$$

$$G_2 = (1 - \upsilon)G \tag{14.127}$$

Substitution of G and υ for G_1 and G_2 in Eq. (14.125) results in

$$\frac{d\Delta}{G\,dA} = (2\upsilon - 1) \ln\left[\frac{y'(1 - x'')}{x''(1 - y')}\right] + (y' - x'')\left[\frac{\upsilon}{x''y'} - \frac{1 - \upsilon}{(1 - x'')(1 - y')}\right] \tag{14.128}$$

With

$$\frac{y'(1 - x'')}{x''(1 - y')} = 1 + \epsilon \tag{14.129}$$

Eq. (14.128) becomes

$$\frac{d\Delta}{G\,dA} = (2\upsilon - 1) \ln(1 + \epsilon) + \left(\frac{\epsilon}{1 + \epsilon}\right)\left[1 - 2\upsilon + \frac{\upsilon - x''}{x''(1 - x'')} - \frac{\epsilon x''(1 - \upsilon)}{1 - x''}\right] \tag{14.130}$$

For uranium isotope separation, $\epsilon \ll 1$ and $\upsilon - x'' \ll 1$. Hence, to the second order in ϵ and $\upsilon - x''$, Eq. (14.130) reduces to

$$\frac{d\Delta}{G\,dA} = \frac{\epsilon(\upsilon - x'')}{x''(1 - x'')} - \frac{\epsilon^2}{2} \tag{14.131}$$

For the present assumption of pure molecular flow, $\upsilon - x''$ from Eq. (14.30) is

$$\upsilon - x'' = \frac{\delta x''(1 - x'') + q(x'' - y') - \delta q y'(1 - x'')}{1 + \delta x'' - q(1 + \delta y')} \tag{14.132}$$

where

$$\delta \equiv \alpha_0 - 1 \tag{14.133}$$

To the first order in δ and $y' - x''$, Eq. (14.132) reduces to

$$\upsilon - x'' = \delta x''(1 - x'') - \frac{q}{1 - q}(y' - x'') \tag{14.134}$$

To the first order in ϵ, Eq. (14.129) reduces to

$$y' - x'' = \epsilon x''(1 - x'') \tag{14.135}$$

With Eqs. (14.134) and (14.135), Eq. (14.131) becomes

$$\frac{d\Delta}{GdA} = \delta\epsilon - \frac{\epsilon^2(1+q)}{2(1-q)} \tag{14.136}$$

The optimum value of ϵ is the one that makes (14.136) a maximum, at which

$$\frac{d}{d\epsilon}\left(\frac{d\Delta}{GdA}\right) = \delta - \frac{\epsilon(1+q)}{1-q} = 0 \tag{14.137}$$

Hence

$$\epsilon_{\text{opt}} = \frac{\delta(1-q)}{1+q} \tag{14.138}$$

and

$$\left(\frac{d\Delta}{GdA}\right)_{\max} = \frac{\delta^2(1-q)}{2(1+q)} \tag{14.139}$$

From (14.129), (14.134), and (14.138), it is found that

$$(v-y')_{\text{opt}} = \frac{\delta x''(1-x'')q}{1+q} \tag{14.140}$$

The minimum power required to recompress GdA moles flowing through pressure ratio q is

$$dQ_{\min} = (GdA)RT_0 \ln\left(\frac{1}{q}\right) \tag{14.141}$$

From (14.139) and (14.141), the ratio of minimum power to maximum separative capacity is

$$\left(\frac{dQ}{d\Delta}\right)_{\min} = \frac{2RT_0}{(\alpha_0-1)^2} \frac{(1+q)\ln(1/q)}{1-q} \tag{14.142}$$

where $(\alpha_0 - 1)$ has been substituted for δ. Values of $2[(1+q)\ln(1/q)]/(1-q)$ are tabulated below.

Pressure ratio q	0.2	0.3	0.4	0.6	0.8	1.0
$2[(1+q)\ln(1/q)]/(1-q)]$	4.83	4.47	4.28	4.09	4.02	4.00

Hence the minimum value occurs at a ratio of 1.0, as the low-side pressure becomes equal to the high-side pressure. At this limiting condition, the minimum ratio of power to separative capacity is

$$\lim_{q\to 0}\left(\frac{dQ}{d\Delta}\right)_{\min} = \frac{4RT_0}{(\alpha_0-1)^2} \tag{14.143}$$

The coefficient 4 in Eq. (14.143) for optimum counterflow is to be compared with 5.11 in Eq. (14.110) for cross flow. The minimum possible power input to produce 1 kg of separative work per year is

$$\frac{4RT_0}{m(\alpha_0-1)^2} = \frac{(4)[0.002310 \text{ kWh/(kg-mol}\cdot\text{K)} (300 \text{ K})}{[238 \text{ kg U/(kg-mol)}]\ (0.00429)^2(8760 \text{ h/yr})}$$

$$= 0.0722 \text{ kW/(kg SWU/yr)} \tag{14.144}$$

This result is to be compared with 0.0923 kW for the minimum with cross flow.

Because this minimum value with counterflow is obtained in the limit of zero pressure difference across the barrier, it would require use of an infinite amount of barrier surface. This condition is analogous to the familiar thermodynamic condition that the loss in availability in a heat exchanger is a minimum when an infinite amount of surface is used.

The foregoing tabulation shows, however, that even at a practical pressure ratio of 0.3, the coefficient of $RT_0/(\alpha_0 - 1)^2$ would be 4.47, substantially less than 5.11 with cross flow. However, this result would be somewhat offset by mixing inefficiency on the low-pressure side when v differs from y', and by the need to use a counterflow, p-up, one-down cascade to obtain the optimum difference at as many points as possible in the cascade.

5 THE GAS CENTRIFUGE

5.1 Principle

The principle of the countercurrent gas centrifuge is shown in Fig. 14.10. The device consists of a long, thin, vertical cylinder made of material with high strength-to-density ratio, rotating in an evacuated casing about its axis with high peripheral speed. The gas rotating inside the cylinder is subject to centrifugal acceleration thousands of times greater than gravity. This makes the pressure at the outer radius of the cylinder millions of times greater than at the center and causes the relative abundance of the heavier isotope to be appreciably greater at the outer radius than at the center. For UF_6 at 300 K, for example, in a centrifuge rotating at a peripheral speed of 500 m/s, the abundance ratio of ^{238}U to ^{235}U at the outer radius is greater

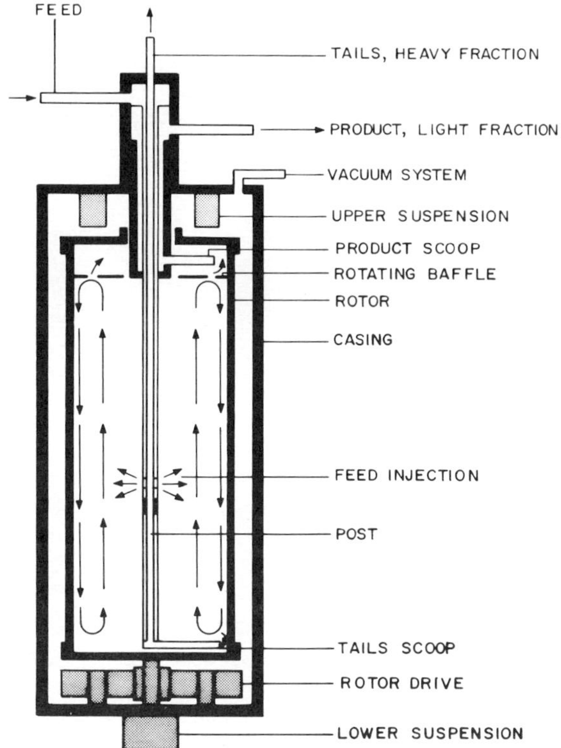

Figure 14.10 Countercurrent gas centrifuge with internal circulation.

than at the center by a factor of 1.162 and the pressure at the outside is greater than at the center by a factor of 46 million. By inducing countercurrent flow between the ^{235}U-depleted stream near the outer radius and the ^{235}U-enriched stream near the axis, the difference in composition between the top and bottom can be made much greater than between the two streams at one elevation. Three general methods have been used for inducing countercurrent flow: (1) by the system of internal scoops and baffles shown in Fig. 14.10, (2) by convection currents set up by heating one end and cooling the other or establishing a temperature gradient along the wall, or (3) by flow induced by pumps external to the machine, as shown in Fig. 14.11. The last gives greater operating flexibility, but is much more complex mechanically.

5.2 History

The concept of separating isotopes in a centrifugal field was first suggested by Lindemann and Aston [L3] in 1919. The first successful use of this method was by Beams and co-workers, who developed vacuum ultracentrifuges with the high peripheral speed needed for measurable isotope separation [B1] and applied it in 1938 to partial separation of the isotopes of chlorine in CCl_4 [B4] and other elements. In 1939, Urey [U4] suggested use of countercurrent flow to multiply single-stage enrichment by heating the bottom of the rotor and cooling the top. In

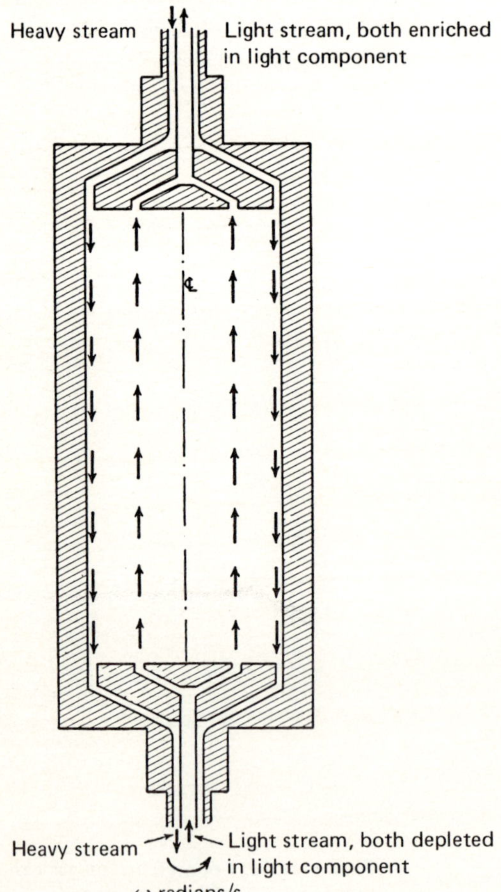

Figure 14.11 Countercurrent gas centrifuge, with externally pumped streams.

Table 14.10 Gas centrifuges built by Groth et al. [G3]

Machine	Length L, cm	Radius a, cm	L/a	Peripheral speed v_a, m/s	Separative capacity kg SWU/yr
UZI	40.0	6.0	6.7	302	0.582
UZIIIB	63.5	6.7	9.5	302	0.935
ZG3	66.5	9.25	7.2	302	0.97
ZG5	113	9.25	12.2	302	1.64
ZG6	240	20	12.0	340	5.32
ZG7	316	22.5	14.0	340	7.25

1940, Bramley and Brewer [B19] and Martin and Kuhn [M4] proposed alternative thermal convective means of internal circulation. The system of internal scoops and baffles shown in Fig. 14.10 was an important simplification introduced by Zippe [Z2].

The first reported enrichment of ^{235}U by the gas centrifuge was obtained by Beams [B2] and co-workers at the University of Virginia in 1941, when 1.2 g of uranium enriched in ^{235}U by 4 percent were produced. Development of larger machines was undertaken by Westinghouse Electric Company with separation performance measurements made by the Standard Oil Development Company under direction of E. V. Murphree. The largest centrifuge tested on UF_6 consisted of a duraluminum tube 18.29 cm in internal diameter, 1.27 cm thick, and 335.3 cm long. Externally driven counterflow was used, as in Fig. 14.11. When operated at a peripheral speed of 206 m/s, a separative capacity around 1.0 kg SWU/year was obtained [B3]. When the success of the gaseous diffusion process was demonstrated, this centrifuge work was terminated.

The German engineer, G. Zippe, devised the simple method of inducing counterflow by internal scoops and baffles shown in Fig. 14.10 during World War II. After the war Zippe continued his work, first in the Soviet Union and later with Beams at the University of Virginia [Z2] from 1958 to 1960. The highest separative capacity there reported was 0.3 kg SWU/year for a rotor 7.41 cm in diameter and 30.2 cm long run at a peripheral speed of 350 m/s.

Groth [G3] and co-workers, in Germany, in the 1950s developed and built a series of progressively larger gas centrifuges whose principal features are summarized in Table 14.10. Tests were reported for centrifuge ZG3 on UF_6 and ZG5 on argon. Separative capacities of the larger machines were predicted on the assumption that they would have the same separation efficiency, 75 percent, as ZG3. These machines used thermal convection to provide controlled internal counterflow, and hence had great operating flexibility. However, their mechanical construction was much more complex than the Zippe-type machine.

In the 1960s, when the gas centrifuge began to appear competitive for large-scale uranium enrichment, the nations then mainly responsible for its development, the United States, Great Britain, the Netherlands, and West Germany, agreed to place security restrictions on description of the technology. Consequently, no details are available of the characteristics of the machines being used either by the Urenco-Centec tripartite organization of English, Dutch, and German interests in their plants at Capenhurst, England, and Almelo, Holland, or by the U.S. DOE for its proposed plant at Portsmouth, Ohio. Trade gossip [N5] is to the effect that the capacity of the Urenco-Centec machines is around 5 kg SWU/year and that their peripheral speed is around 400 m/s.

U.S. machines are presumed to be larger. Figure 14.12 is a photograph of centrifuges of German manufacture in the Urenco/Centec pilot plant at Almelo, Holland; Fig. 14.13 is a photograph of machines under development in the United States.

Figure 14.12 Urenco/Centec pilot plant of German centrifuge machines at Almelo, Netherlands. (*Courtesy of Urenco, Ltd.*)

5.3 Description of Centrifuges

The two types of centrifuge whose features have been described most completely are the Groth and the Zippe machines. Figure 14.14 is a schematic drawing of Groth's ZG5 machine [G3].

The aluminum alloy rotor R is suspended and driven from the top by an electric motor, not shown, within the vacuum case C. UF_6 gas V is fed into the center of the rotor through the stationary tube R_1. Heavy fraction is removed at the top through scoop S_1, a stationary tube concentric with the feed tube, and outlet Z_1. Light fraction is removed in similar fashion through scoop S_2 and outlet Z_2 at the bottom. Circulation of gas within the rotor is shielded from interference from the scoops by baffles B_1 and B_2. Controlled countercurrent circulation of gas is effected by heating the top end cap by induction from the electromagnet E and cooling the bottom end cap by radiation to cooling coil K. Temperatures are measured by thermocouples Th_1 and Th_2. Pressure at the axis is measured through connection M. The rotor is connected to hollow shafts at top and bottom, which rotate within oil-lubricated bearings, not shown. To keep oil and UF_6 from mixing, labyrinth seals D_1, D_2, D_3, and D_4 are used on the top and bottom shafts. These are fed with hydrogen and discharge a mixture of hydrogen and UF_6 to cold traps through P_1 and P_2, and a mixture of hydrogen and oil to other outlets, not shown.

A significant advantage of Groth's machine is its control of internal circulation by convective heating and cooling; this permitted attainment of 75 percent of the maximum theoretical separative capacity, at least in the smaller machines. A serious disadvantage is the very complex construction associated with the oil-lubricated bearings and hydrogen-fed seals at top and bottom, which makes the machine expensive and complicates operation. The Zippe-type machine, free of these complications but with less flexibility in controlling internal circulation, is less costly and easier to operate.

Figure 14.15 is a cross section of one of the centrifuges tested by Zippe at the University of Virginia [Z2]. The rotor is a duraluminum cylinder 7.62 cm in diameter and 38 cm long. It rotates inside a vacuum casing and is closed at the bottom by an end cap, which rests on a flexible steel needle. The needle spins in a bottom bearing supported by springs and oil-filled vibration dampers. The top of the rotor is covered by an end cap fitting with small clearance around a center post that carries three concentric tubes for withdrawing light fraction from the bottom of the rotor, admitting feed to the center, and withdrawing heavy fraction from the top. Leakage of UF_6 between the top cap of the rotor and the center post is small because of the low pressure maintained at the axis by the centrifugal field. Any UF_6 that leaks is kept out of the vacuum casing by the spiral grooves of a molecular pump that surrounds the top of the rotor. The rotor is positioned at the top by a magnet rotating on the top cap below a stationary magnet supported by flexible plastic strips and steel wires to provide positioning and damping.

Countercurrent circulation of UF_6 is provided by the top scoop, which also serves to

Figure 14.13 U.S. gas centrifuge pilot plant. (*Courtesy of U.S. Energy Research and Development Administration.*)

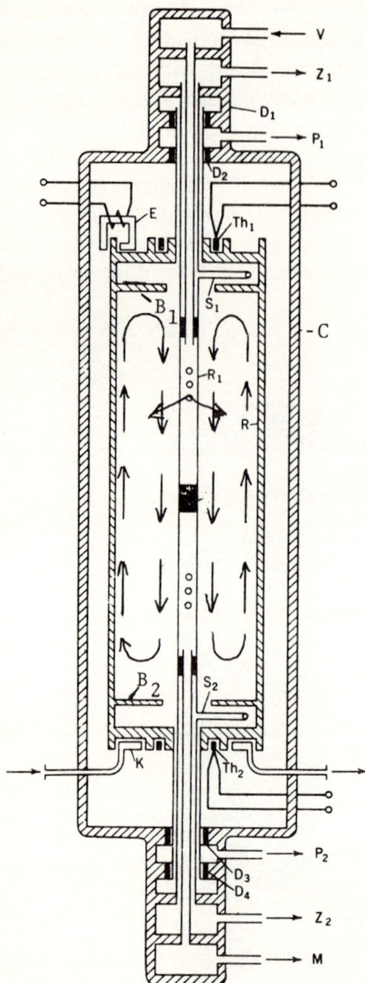

Figure 14.14 Schematic of Groth ZG5 centrifuge. *(Adapted from Shacter et al. [S3].)*

remove heavy fraction. Light fraction is removed by the bottom scoop, which is prevented from disturbing circulation within the rotor by the bottom baffle.

The rotor is driven by a planar induction motor whose armature plate is attached to the bottom end cap and whose stator is a flat winding with pole pieces outside the vacuum case. The motor is provided with cooling coils and a speed pickup.

5.4 Mechanical Performance of Centrifuges

As will be shown in Sec. 5.5, the separative capacity of a countercurrent gas centrifuge is proportional to its length L and increases rapidly as the peripheral speed v_a increases. Hence it is advantageous to run at the highest practical speed and to use centrifuges of the greatest practical length. An absolute limit to the speed is reached when tangential stresses caused by centrifugal forces equal the tensile strength of the rotor material. Limitations on practical values of the length are set by the need to avoid combinations of length, radius, and speed at which the rotor experiences resonant vibrations. These two factors limiting centrifuge mechanical performance, which have the greatest effect on separation performance, will be discussed in this section. Many other relevant mechanical topics, such as design of bearings, motor drives, and damping mechanisms, are beyond the scope of this text.

Maximum peripheral speed. Consider a cylindrical shell of radius r and thickness dr, made of material of density ρ and rotating at angular velocity ω rad/s. Figure 14.16 represents a volume element of the shell of height dz subtending an angle $d\theta$. The mass of the element

$$dm = \rho r dr dz d\theta \tag{14.145}$$

experiences a centrifugal force

$$\omega^2 r dm = \rho \omega^2 r^2 dr dz d\theta \tag{14.146}$$

in the outward r direction. This must be balanced by the components in the opposite direction, $\sigma_\theta \sin(d\theta/2)$, of the tangential stresses σ_θ acting on the two surfaces $drdz$ offset by angle $d\theta$.

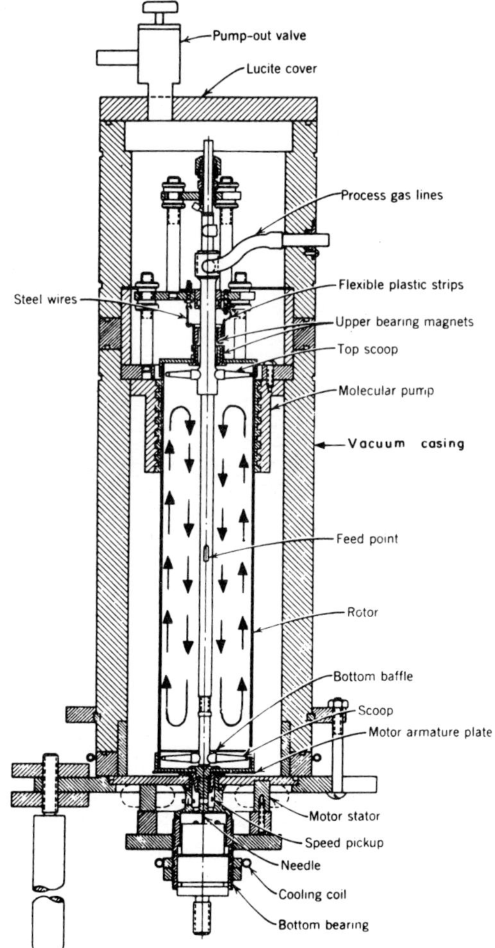

Figure 14.15 The Zippe centrifuge. *(Adapted from Shacter et al. [S3.])*

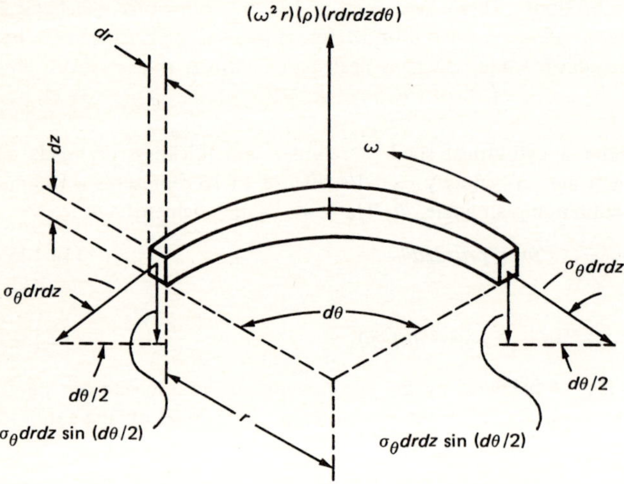

Figure 14.16 Forces on volume element of rotating cylindrical shell.

$$2\sigma_\theta \sin\left(\frac{d\theta}{2}\right) drdz = \rho\omega^2 r^2 drdzd\theta \tag{14.147}$$

To the first order in $d\theta$,

$$\sigma_\theta = \rho\omega^2 r^2 \tag{14.148}$$

Because ωr is the tangential speed v,

$$v_{\text{max}} = \sqrt{\frac{\sigma}{\rho}} \tag{14.149}$$

where v_{max} is the maximum tangential speed, at which the tangential stress reaches the tensile strength σ of the material.

Table 14.11 gives the density, tensile strength, and modulus of elasticity E of six possible high-tensile materials for centrifuge rotors. These properties are given in metric units and SI units.

The maximum tangential speed ranges from 400 m/s for aluminum alloy to 720 m/s for a carbon fiber-resin composite.

Conditions for resonant vibrations. Certain angular velocities ω_i cause a thin, hollow cylinder to go into resonant longitudinal vibrations. If a centrifuge rotor is driven for any length of time at or near one of these angular velocities, rotational energy is used to increase the amplitude of longitudinal vibrations until the rotor or its bearings may be wrecked. Consequently, it is important to avoid tangential speeds at which a rotor of given length and radius will be in resonance. Texts on mechanical vibrations such as [D4] show that the longitudinal vibration frequency ω_i of a thin hollow cylinder of radius a, modulus of elasticity E, and length L, unrestrained at the ends, in the ith mode is

$$\omega_i = \lambda_i \sqrt{\frac{E}{2\rho}}\, \frac{a}{L^2} \tag{14.150}$$

where the eigenvalues λ_i are

i	1	2	3	4	5
λ_i	22.0	61.7	121.0	200.0	298.2

i is the number of loops into which the profile of the cylinder is displaced. Because

$$v_i = \omega_i r \tag{14.151}$$

the length-to-radius ratio at which rotors of each of the materials run at maximum tangential speed v_{max} would be in resonant vibration is

$$\left(\frac{L}{a}\right)_i = \sqrt{\frac{\lambda_i}{v_{max}}}\sqrt[4]{\frac{E}{2\rho}} \tag{14.152}$$

With v_{max} from (14.149),

$$\left(\frac{L}{a}\right)_i = \sqrt{\lambda_i}\sqrt[4]{\frac{E}{2\sigma}} \tag{14.153}$$

The last part of Table 14.11 gives values of L/a for the first five resonances in cylinders of the five materials operated at the maximum speed, at which tangential stress equals the tensile strength of the material. At lower speeds v, the critical L/a ratio is obtained by multiplying the values of Table 14.11 by $\sqrt{v_{max}/v}$.

Rotors that are shorter than the first critical length are said to be *subcritical.* Such rotors do not need special means to avoid resonant speeds. Rotors that are longer than the first critical length are called *supercritical.* They must be operated at speeds away from resonance

Table 14.11 Physical properties and operating limits of possible centrifuge materials

Material	Aluminum alloys	High-tensile steel	Titanium	Maraging steel	Glass fiber	Carbon fiber/resin
Density						
g/cm^3†	2.8	7.8	4.6	7.8	1.8	1.6
kg/m^3 (ρ)	2,800	7,800	4,600	7,800	1,800	1,600
Tensile strength						
kg/cm^2†	4,570	14,080	9,150	19,700	5,000	8,450
MPa ($10^{-6}\ \sigma$)	448	1,381	897	1,932	490	829
Modulus of elasticity						
Mg/cm^2	724	2,110	1,160	2,110	738	
MPa ($10^{-6}\ E$)	71,000	207,000	114,000	207,000	72,400	
Max. tangential speed, $v_{max} = \sqrt{\sigma/\rho}$, m/s	400	421	442	498	522	720
Length-to-radius ratio at v_{max}, Eq. (14.153)						
First resonance	14.0	13.8	13.2	13.8	13.8	
Second resonance	23.4	23.1	22.2	23.1	23.0	
Third resonance	32.8	32.4	31.1	32.4	32.2	
Fourth resonance	42.2	41.6	39.9	41.6	41.4	
Fifth resonance	51.5	50.8	48.8	50.8	50.6	

†From Avery and Davis [A5], p. 44.

and must be provided with drives of sufficient power to accelerate them quickly through resonant speeds and brakes of sufficient power dissipation to decelerate them quickly. All of Groth's rotors listed in Table 14.10 have length-to-diameter ratios below the first critical at the listed peripheral speeds. However, if ZG7 had been made of titanium and operated at its maximum peripheral speed of 442 m/s, it would have run between the first and second resonance.

Power consumption. Because the separation performed by the gas centrifuge is a thermodynamically reversible process, the minimum energy input necessary to separate an isotopic mixture is merely the small difference in free energy between the separation products and the feed. The actual energy input is thousands of times greater because it is dominated by the work necessary to overcome mechanical friction in bearings, aerodynamic drag, and pressure drops in gas circulation. These energy inputs are specific to details of centrifuge and plant design and cannot be estimated from principles of the separation process, as was possible for gaseous diffusion. The U.S. DOE stated [U3] that the power consumption of a centrifuge plant per unit separative capacity would be around 4 percent of the power consumption of a gaseous diffusion plant.

The comparatively low power consumption of a gas centrifuge plant is its greatest advantage over competing processes. The relatively low separative capacity of a single centrifuge is its greatest disadvantage.

Means for estimating the separative capacity of a centrifuge will be developed in Sec. 5.5.

5.5 Separation Performance of Gas Centrifuge

Notation. As used for isotope separation, the gas centrifuge is a cylinder of radius a and length L, rotating about a vertical axis with angular velocity ω rad/s. Cylindrical polar coordinates are used, with the following notation for position and velocity components:

Direction	Position	Velocity, relative to solid cylinder rotating about axis with angular velocity ω
Radial, out from axis	r	u
Tangential	θ (angle)	v
Axial, up from midplane	z	w

Properties of the light component of a binary mixture are denoted by subscript 1; heavy component by subscript 2.

Equilibrium separation. When a gas mixture in a centrifuge rotates as a solid body without motion relative to the wall of the cylinder, its pressure and composition are independent of θ and z and vary with r according to the equations for equilibrium in a centrifugal field.

In a centrifuge rotating at ω rad/s, gas of density ρ at radius r is subjected to centrifugal force of $\omega^2 r\rho$ per unit volume, which equals the pressure gradient at that point.

$$\frac{dp}{dr} = \omega^2 r\rho \tag{14.154}$$

Because

$$\rho = \frac{pm}{RT} \tag{14.155}$$

$$\frac{1}{p}\frac{dp}{dr} = \frac{m\omega^2 r}{RT} \tag{14.156}$$

This equation is analogous to the equation for the change in barometric pressure B with altitude h under gravitational acceleration g:

$$\frac{1}{B}\frac{dB}{dh} = -\frac{mg}{RT} \tag{14.157}$$

By integration, the pressure ratio or density ratio between an interior radius r and the outer wall of the centrifuge at radius a is

$$\frac{p(r)}{p_a} = \frac{\rho(r)}{\rho_a} = \exp\left[-\frac{mv_a^2}{2RT}\left(1 - \frac{r^2}{a^2}\right)\right] \tag{14.158}$$

where v_a is the speed of rotation ωa at the outer wall, termed the peripheral speed.

Table 14.12 illustrates pressure ratios for UF_6 gas ($m = 352$) at several values of r/a for peripheral speeds of 400, 500, and 700 m/s at 300 K. Most of the gas is in a thin shell near the wall.

In a binary mixture of gases of molecular weights m_1 and m_2, an equation like (14.157) describes the partial pressure ratio of each component,

$$\frac{px}{p_a x_a} = \exp\left[-\frac{m_1 v_a^2}{2RT}\left(1 - \frac{r^2}{a^2}\right)\right] \tag{14.159}$$

$$\frac{p(1-x)}{p_a(1-x_a)} = \exp\left[-\frac{m_2 v_a^2}{2RT}\left(1 - \frac{r^2}{a^2}\right)\right] \tag{14.160}$$

where x is the mole fraction of light component. The local separation factor $\alpha(a,r)$ between radii r and a, obtained by dividing (14.159) by (14.160), is

$$\alpha(a,r) \equiv \frac{x(1-x_a)}{(1-x)x_a} = \exp\left[\frac{\Delta m v_a^2}{2RT}\left(1 - \frac{r^2}{a^2}\right)\right] \tag{14.161}$$

where

$$\Delta m \equiv m_2 - m_1 \tag{14.162}$$

The separation factor in the gas centrifuge thus depends on the difference between molecular weights, whereas in gaseous diffusion it depends on their ratio. Table 14.13 gives local separation factors for mixtures of $^{235}UF_6$ and $^{238}UF_6$ ($\Delta m = 3$) for the same speeds and radial locations as Table 14.12. Because most of the gas is in a thin shell adjacent to the wall, the more significant values are those for r/a near unity. Even with this restriction, the separation factor for the centrifuge is much more favorable than $\alpha_0 = 1.00429$ for gaseous diffusion.

Table 14.12 Pressure ratios for UF_6 in a centrifugal field

	Pressure ratio $p(r)/p_a$ for speed v_a of		
r/a	400 m/s	500 m/s	700 m/s
0	1.25E-5	2.2E-8	1E-15
0.5	2.1E-4	1.8E-6	5E-12
0.8	1.7E-2	1.7E-3	4E-6
0.9	0.12	3.5E-2	1.4E-3
1.0	1	1	1

Table 14.13 Local separation factors for $^{235}UF_6$-$^{238}UF_6$ in a centrifugal field

	Local separation factors for speed v_a of		
r/a	400 m/s	500 m/s	700 m/s
0	1.101	1.162	1.343
0.5	1.075	1.119	1.247
0.8	1.035	1.056	1.112
0.9	1.018	1.029	1.058
1.0	1.0	1.0	1.0

When concentration equilibrium is established in a centrifugal field, the composition gradient is given by the derivative of Eq. (14.161):

$$\left\{\frac{d \ln [x/(1-x)]}{dr}\right\}_{\text{equil}} = \frac{-\Delta m v_a^2 r}{RTa^2} \tag{14.163}$$

Hence

$$-\left(\frac{dx}{dr}\right)_{\text{equil}} = \frac{\Delta m \omega^2 rx(1-x)}{RT} \tag{14.164}$$

Transport equations. When centrifugal equilibrium is disturbed, as by establishment of counterflow or injection of feed and removal of effluents, flow of the gas mixture and of its individual components takes place relative to the rotating centrifuge. The analysis to be given has the following restrictions.

1. All gas is rotating at angular velocity ω so that there is no angular motion relative to the rotating centrifuge. In a coordinate system rotating with angular velocity ω, $v = 0$.
2. Analysis is to be limited to the case of no radial motion of the gas as a whole, $u = 0$. This condition cannot hold at the top and bottom of the centrifuge, but may be nearly correct away from the ends, in the so-called long bowl development.
3. The change of ρD with temperature and pressure, and thermal diffusion effects, are neglected.

Transport of light component is to be described in terms of its mass velocity, the vector **J**, with component J_r in the radial direction and J_z in the axial. In the coordinate system rotating at angular velocity ω, the angular component J_θ is zero.

When the radial composition gradient $\partial x/\partial r$ differs from the gradient at equilibrium $(\partial x/\partial r)_{\text{equil}}$, transport of light component against the composition gradient takes place with radial mass velocity

$$J_r = -D\rho\left[\frac{\partial x}{\partial r} - \left(\frac{\partial x}{\partial r}\right)_{\text{equil}}\right] = -D\rho\left[\frac{\partial x}{\partial r} + \frac{\Delta m \omega^2 rx(1-x)}{RT}\right] \tag{14.165}$$

The axial mass velocity J_z is the sum of a convective term ρwx and a diffusive term $-D\rho(\partial x/\partial z)$:

$$J_z = \rho wx - D\rho\frac{\partial x}{\partial z} \tag{14.166}$$

Differential enrichment equation. Under steady-state conditions, the differential equation for conservation of light component, in cylindrical polar coordinates, is

$$\frac{1}{r}\frac{\partial (rJ_r)}{\partial r} + \frac{\partial J_z}{\partial z} + \frac{1}{r}\frac{\partial^2 J_\theta}{\partial \theta^2} = 0 \tag{14.167}$$

With J_r from (14.165), J_z from (14.166), and $J_\theta = 0$, Eq. (14.167) becomes

$$-\frac{1}{r}\frac{\partial}{\partial r} rD\rho\left[\frac{\partial x}{\partial r} + \frac{\Delta m \omega^2 rx(1-x)}{RT}\right] - \frac{\partial}{\partial z} D\rho\frac{\partial x}{\partial z} + \frac{\partial}{\partial z}\rho wx = 0 \tag{14.168}$$

Cohen [C6] made the following assumptions to simplify solution:

1. $x(1-x)$ is treated as a constant.
2. $\partial^2 x/\partial z^2$ is neglected.

3. $\partial x/\partial z$ is independent of r.
4. ρw is independent of z.

Then, multiplying (14.168) by r and integrating with respect to r' from $r' = 0$ to $r' = r$ yield

$$r\frac{\partial x}{\partial r} = -\frac{\Delta m\omega^2 r^2 x(1-x)}{RT} + \frac{1}{D\rho}\frac{\partial x}{\partial z}\int_0^r \rho w r' dr' \tag{14.169}$$

because $r(\partial x/\partial r) = 0$ at $r' = 0$.†

Integration of (14.169) requires use of boundary conditions for the net flow. In the enriching section, the net flow P is

$$P = 2\pi \int_0^a \rho w r dr \tag{14.170}$$

The net flow of light component is

$$Px_P = 2\pi \int_0^a J_z r\, dr = 2\pi \int_0^a x\rho w r\, dr - 2\pi D\rho \int_0^a \frac{\partial x}{\partial z} r\, dr \tag{14.171}$$

Integrate the first term of the right side by parts:

$$Px_P = 2\pi x(a) \int_0^a \rho w r\, dr - 2\pi \int_0^a \frac{\partial x}{\partial r} dr \int_0^r \rho w r'\, dr' - 2\pi D\rho \int_0^a \frac{\partial x}{\partial z} r\, dr \tag{14.172}$$

Replace the first integral by Eq. (14.170) and substitute for $\partial x/\partial r$ from Eq. (14.169):

$$\frac{P[x_P - x(a)]}{2\pi} = \int_0^a \frac{\Delta m\omega^2 r x(1-x)}{RT} dr \int_0^r \rho w r'\, dr' - \int_0^a \frac{dr}{rD\rho}\left(\int_0^r \rho w r'\, dr'\right)^2 \frac{\partial x}{\partial z} - D\rho \int_0^a \frac{\partial x}{\partial z} r\, dr \tag{14.173}$$

Because of assumption (3), this may be solved for $\partial x/\partial z$:

$$\frac{\partial x}{\partial z} = \left\{\int_0^a \frac{\Delta m\omega^2 r x(1-x)}{RT} dr \int_0^r \rho w r'\, dr' - \frac{P[x_P - x(a)]}{2\pi}\right\} \Big/ \left[\int_0^a \frac{dr}{rD\rho}\left(\int_0^r \rho w r'\, dr'\right)^2 + \frac{D\rho a^2}{2}\right] \tag{14.174}$$

†This condition and the lower integration limit of 0 are strictly correct only when a tube at the axis of the centrifuge is not present. In most centrifuges, with such a tube, the lower limit of integration should be the outer radius of the tube. However, at peripheral speeds of 400 m/s or higher, the density of gas at the central tube is so low that use of 0 for the lower limit of integration introduces no significant error, and $r(dx/dr)$ at the lower limit is much smaller than at the upper limit.

This solution was first given by Cohen [C6]. It has become conventional to define a *flow function* $F(r)$† by

$$F(r) = 2\pi \int_0^r \rho w r' \, dr' \tag{14.175}$$

Cohen used the notation

$$C_1 = \frac{\Delta m \omega^2}{RT} \int_0^a F(r) r \, dr \tag{14.176}$$

$$C_2 = \pi D \rho a^2 \tag{14.177}$$

$$C_3 = \frac{1}{2\pi D\rho} \int_0^a [F(r)]^2 \frac{dr}{r} \tag{14.178}$$

$$C_5 = C_2 + C_3 \tag{14.179}$$

In terms of these functions, the differential enrichment equation (14.174) becomes

$$\frac{dx}{dz} = \frac{C_1}{C_5} x(1 - x) - \frac{P(x_P - x)}{C_5} \tag{14.180}$$

Here the variation of x with r has been neglected, as it is small compared to its change with z in a long centrifuge.

For the same reason, it is permissible to write a similar equation for the composition y of the enriched stream:

$$\frac{dy}{dz} = \frac{C_1}{C_5} y(1 - y) - \frac{P(y_P - y)}{C_5} \tag{14.181}$$

Because the coefficients C_1 and C_5 are to be evaluated for the velocity distribution with zero net flow, (14.181) is as valid an approximation as (14.180). When feed is added to the enriched stream, as in a centrifuge with feed introduced by a tube at the axis, Eq. (14.181) is easier to use than (14.180).

The equation corresponding to (14.181) for the stripping section is

$$\frac{dy}{dz} = \frac{C_1}{C_5} y(1 - y) - \frac{W(y - y_W)}{C_5} \tag{14.182}$$

In an exact treatment, values of C_1 and C_5 in the stripping section would differ slightly from the enriching section because of the slightly different flow profile. In the present approximate treatment, the constants are to be evaluated for the total reflux case in which the flow patterns in both sections are the same. If the net flow rate is a small fraction of the circulation rate, studies by Parker [P1] and others have shown that the effect on C_1 and C_5 of the changed flow pattern with net flow is small.

Equation (14.181) may be compared with the corresponding differential equation for the enriching section of a two-stream, close-separation, countercurrent column like a distillation column:

$$h\frac{dy}{dz} = (\alpha - 1)\, y\, (1 - y) - \frac{P(y_P - y)}{N} \tag{14.183}$$

†Physically, $F(r)$ is the total mass upflow rate between the center and radius r.

where h is the height of a transfer unit, α is the local separation factor, and N is the flow rate of the stream moving from the product end of the column. Comparison of (14.181) and (14.183) shows that C_5 may be interpreted as

$$C_5 = hN \tag{14.184}$$

and C_1/C_5 as

$$\frac{C_1}{C_5} = \frac{\alpha - 1}{h} \tag{14.185}$$

Thus

$$h = \frac{C_5}{N} \tag{14.186}$$

and

$$\alpha - 1 = \frac{C_1}{N} \tag{14.187}$$

In the countercurrent centrifuge N is the depleted stream flow rate

$$N = 2\pi \int_{r_1}^{a} \rho w r \, dr \tag{14.188}$$

where r_1 is the internal radius at which the axial velocity w changes sign. In the present approximation, in which the centrifuge parameters are evaluated for the velocity profile at total reflux, the flow rates of enriched stream and depleted streams are equal and an equivalent equation is

$$N = 2\pi \int_{0}^{r_1} \rho w r \, dr = F(r_1) \tag{14.189}$$

Local separative capacity. The separative capacity of a gas centrifuge per unit length, $d\Delta/dz$, may be derived from Eq. (14.181) for the composition gradient, dy/dz. In the enriching section of a gas centrifuge the net flow rate of light component toward the product end is Py_P and of heavy component is $P(1 - y_P)$. As these flows make their way through gas of composition y against a composition gradient dy/dz, the rate of production of separative work per unit height $d\Delta/dz$ is

$$\frac{d\Delta}{dz} = Py_P \frac{d}{dz}\left(\frac{\partial S}{\partial n_1}\right)_{n_2} + P(1 - y_P)\frac{d}{dz}\left(\frac{\partial S}{\partial n_2}\right)_{n_1} \tag{14.190}$$

where S is the separative work associated with n_1 mass of component 1 and n_2 mass of component 2.

From Eq. (14.118):

$$S \equiv (n_1 - n_2)\ln\left(\frac{n_1}{n_2}\right) \tag{14.191}$$

Because

$$\left(\frac{\partial S}{\partial n_1}\right)_{n_2} = \ln\left(\frac{n_1}{n_2}\right) + \frac{n_1 - n_2}{n_1} = \ln\left(\frac{y}{1-y}\right) + \frac{2y-1}{y} \tag{14.192}$$

and

$$\left(\frac{\partial S}{\partial n_2}\right)_{n_1} = -\ln\left(\frac{n_1}{n_2}\right) - \frac{n_1 - n_2}{n_2} = -\ln\left(\frac{y}{1-y}\right) + \frac{1-2y}{1-y} \tag{14.193}$$

$$\frac{d\Delta}{dz} = Py_P\left[\frac{1}{y(1-y)} + \frac{1}{y^2}\right]\frac{dy}{dz} + P(1-y_P)\left[-\frac{1}{y(1-y)} - \frac{1}{(1-y)^2}\right]\frac{dy}{dz}$$

$$= \frac{P(y_P - y)}{y^2(1-y)^2}\frac{dy}{dz} \tag{14.194}$$

Using (14.181) for dy/dz,

$$\frac{d\Delta}{dz} = \frac{C_1}{C_5}\frac{P(y_P - y)}{y(1-y)} - \frac{1}{C_5}\left[\frac{P(y_P - y)}{y(1-y)}\right]^2 \tag{14.195}$$

The optimum value of the group of variables

$$\phi \equiv \frac{P(y_P - y)}{y(1-y)} \tag{14.196}$$

is the value that maximizes $d\Delta/dz$, at which

$$\frac{d}{d\phi}\left(\frac{d\Delta}{dz}\right) = 0 \tag{14.197}$$

Because

$$\frac{d}{d\phi}\left(\frac{d\Delta}{dz}\right) = \frac{C_1}{C_5} - \frac{2\phi}{C_5} \tag{14.198}$$

$$\phi_{\text{opt}} = \frac{C_1}{2} \tag{14.199}$$

and

$$\left(\frac{d\Delta}{dz}\right)_{\max} = \frac{C_1^2}{2C_5} - \frac{C_1^2}{4C_5} = \frac{C_1^2}{4C_5} \tag{14.200}$$

In a centrifuge with axial flow independent of height, C_1 and C_5 are constant and condition (14.199) can be satisfied at only one height.

In terms of the parameters $\alpha - 1$, h, and N,

$$\left[\frac{P(y_P - y)}{y(1-y)}\right]_{\text{opt}} = \frac{N(\alpha - 1)}{2} \tag{14.201}$$

and

$$\left(\frac{d\Delta}{dz}\right)_{\max} = \frac{N(\alpha - 1)^2}{4h} \tag{14.202}$$

Equation (14.201) is analogous to condition (12.125) for an ideal cascade, and (14.202) is the separative capacity of a stage of an ideal cascade divided by h.

The parameters C_1, C_5, and N, and from them h, $\alpha - 1$, and the separative capacity, depend on the radial distribution of mass velocity $\rho w(r)$. These parameters will be evaluated for two velocity distributions:

1. Mass velocity independent of radius
2. Berman-Olander distribution

Mass velocity independent of radius. The optimum radial distribution of longitudinal mass velocity $w\rho(r)$ that leads to the highest possible separative capacity per unit length when condition (14.199) is satisfied is a distribution in which the mass velocity in one direction is independent of radius r up to the outer radius a, with all countercurrent flow in the opposite direction occurring in a cylindrical shell of infinitesimal thickness at the outer radius a. Under the total reflux conditions to be used in evaluating C_1 and C_5,

$$\rho w = \frac{N}{\pi a^2} - N\delta(a) \tag{14.203}$$

where $\delta(a)$ is the delta function in cylindrical coordinates defined by

$$\delta(r) = 0 \quad (r \neq a) \tag{14.204}$$

and

$$2\pi \int_0^a r\delta(r)\, dr = 1 \tag{14.205}$$

From (14.175), the flow function $F(r)$ for this case is

$$F(r) = \frac{Nr^2}{a^2} \quad (r \neq a) \tag{14.206}$$

$$F(a) = 0 \tag{14.207}$$

From (14.176),

$$C_1 = \frac{\Delta m N \omega^2 a^2}{4RT} = \frac{\Delta m N v_a^2}{4RT} \tag{14.208}$$

From (14.178),

$$C_3 = \frac{N^2}{8\pi D\rho} \tag{14.209}$$

Hence

$$\alpha - 1 = \frac{\Delta m v_a^2}{4RT} \tag{14.210}$$

from (14.187), and

$$h = \frac{N}{8\pi D\rho} + \frac{\pi D\rho a^2}{N} \tag{14.211}$$

from (14.186).

The maximum separative capacity per unit length, from (14.200), (14.208), (14.209), (14.177), and (14.179), is

$$\left(\frac{d\Delta}{dz}\right)_{\max} = \frac{\pi D\rho(\Delta m)^2 v_a^4}{8(RT)^2} \quad \frac{1}{1 + 8(\pi D\rho a)^2/N^2} \tag{14.212}$$

Cohen [C6] has shown that the first factor represents the maximum possible separative capacity per unit length for a centrifuge operating at peripheral speed v_a, in the absence of axial back diffusion. The second factor, termed the circulation efficiency E_c, takes into account reduction in separative capacity caused by axial back diffusion. It approaches unity as the circulation rate N increases or the radius a decreases. Values of the first factor for separating ^{235}U from ^{238}U ($\Delta m = 3$) at 300 K, using the value of $D\rho = 2.161 \times 10^{-4}$ g UF_6/(cm·s) recommended by May [M6],† are

v_a, m/s	400	500	700
First factor, kg UF_6 SWU/(yr · m)	9.912	24.198	92.959
kg U SWU/(yr · m)	6.702	16.361	62.853

†This value, 7 percent lower than 2.32×10^{-4} g/(cm·s) given by $D\rho = 4\mu/3$, with μ from Eq. (14.4), is used so that results will be consistent with May's.

These values are much higher than can be obtained in an actual centrifuge because most of the gas flow, up or down, actually occurs in a thin shell near the outer wall, the only region in which the gas density is appreciable.

Berman-Olander velocity distribution. Even if the optimum radial distribution of mass velocity (14.203) could be established at one elevation in a countercurrent centrifuge, it could not persist over any distance because of the great shear force at the velocity discontinuity between the counterflowing streams. Determination of a more stable countercurrent radial velocity distribution, which would persist over a substantial length of centrifuge, requires solution of the hydrodynamic equations for motion of a compressible fluid in a centrifugal field. Even for the simplified "long-bowl" case considered here, in which the radial velocity u is zero and the longitudinal velocity w is a function only of r, the solution procedure is difficult and the equations complex. During the past 20 years, long-bowl solutions of progressively increasing rigor have been given by Parker and Mayo [P1], Soubbaramayer [S7], Berman [B16], and others.† Olander [O1] showed that Berman's solution could be approximated for the large values of

$$A^2 \equiv \frac{mv_a^2}{2RT} \tag{14.213}$$

met in centrifuges of practical importance by

$$\frac{w(r)}{w_0} = 1 - \frac{a^2}{r^2}\left[2A^2\left(1 - \frac{r^2}{a^2}\right) + 1\right] \exp\left[-A^2\left(1 - \frac{r^2}{a^2}\right)\right] \tag{14.214}$$

w_0 is an adjustable parameter proportional to the circulation rate N. The mass velocity at radius r is the product of Eqs. (14.214) and (14.158):

$$\frac{w\rho(r)}{w_0\rho(a)} = \left\{1 - \frac{a^2}{r^2}\left[2A^2\left(1 - \frac{r^2}{a^2}\right) + 1\right] \exp\left[-A^2\left(1 - \frac{r^2}{a^2}\right)\right]\right\} \exp\left[-A^2\left(1 - \frac{r^2}{a^2}\right)\right] \tag{14.215}$$

Equations (14.214) and (14.215) approach zero as $r \to a$ and thus properly represent the condition of no slip at the outer wall. They fail to represent exactly conditions as $r \to 0$ because $w(r)$ should be finite for a centrifuge without a central tube or should be zero as $r \to r_0$ when the centrifuge has a central tube of radius r_0. However, for a practical centrifuge with peripheral speed over 400 m/s, for which A^2 for $UF_6 > 11.3$, $w\rho(r)/w_0\rho(a)$ from (14.215) at $r/a = 0.1$ is less than 1.2×10^{-5} times its maximum value, so that little error is made in replacing $w\rho$ from (14.215) by zero at $r/a < 0.1$.

In Fig. 14.17 the curve marked "mass velocity" is a plot of Eq. (14.215) for $A = 11.3$, corresponding to peripheral speed $v_a = 400$ m/s. Most of the flow occurs in the outer half of the cross-sectional area, with flow reversal taking place at $r^2/a^2 = 0.88$, so that all heavy-fraction flow occurs in the outer 12 percent of the area. At 700 m/s this area shrinks to only 3.6 percent of the total cross section.

The curve marked "flow function," obtained from

$$f\left(\frac{r}{a}\right) \equiv 2\pi \int_0^{r/a} \frac{w\rho(r')}{w_0\rho(a)} \frac{r'}{a} \frac{dr'}{a} \tag{14.216}$$

†Theoretical analyses of flow and separation in a gas centrifuge published too late for inclusion in this text may be found in [S2a] and [V1a].

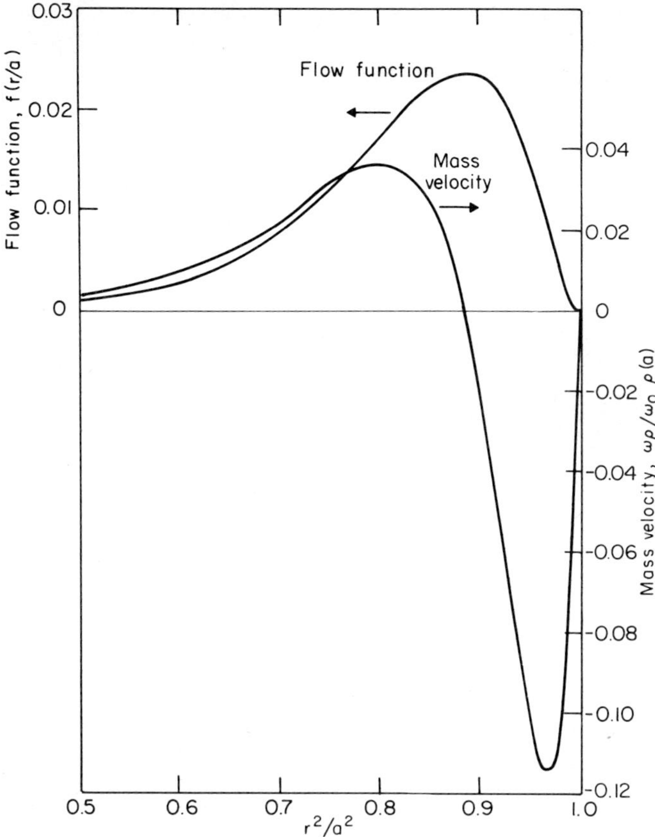

Figure 14.17 Dependence of mass velocity and flow function on radius for UF_6 in centrifuge with peripheral speed of 400 m/s, at 300 K.

is proportional to the total mass flow in the direction of light fraction through the area between the center of the centrifuge and radius r.

For physical interpretation, it is instructive to cast these equations into a form that contains the internal circulation rate N. N is given by

$$N = 2\pi \int_0^{r_1} w\rho r \, dr$$

$$= a^2 f\left(\frac{r_1}{a}\right) w_0 \rho(a) \tag{14.217}$$

where r_1 is the radius at which w changes sign and

$$f\left(\frac{r_1}{a}\right) = 2\pi \int_0^{r_1/a} \frac{w\rho}{w_0 \rho(a)} \frac{r}{a} \frac{dr}{a} \tag{14.218}$$

Hence

$$w_0 \rho(a) = \frac{N}{a^2 f(r_1/a)} \tag{14.219}$$

From (14.175), (14.176), (14.216), and (14.219),

$$C_1 = \frac{\Delta m \omega^2 a^2 N \int_0^1 f(r/a)\,(r/a)\,d(r/a)}{RTf(r_1/a)} \tag{14.220}$$

From (14.175), (14.178), (14.216), and (14.219),

$$C_3 = \frac{N^2 \int_0^1 [f(r/a)]^2\,d(r/a)/(r/a)}{2\pi D\rho\,[f(r_1/a)]^2} \tag{14.221}$$

In terms of the dimensionless integrals,

$$I_1 \equiv \int_0^1 f\left(\frac{r}{a}\right)\left(\frac{r}{a}\right)d\left(\frac{r}{a}\right) \tag{14.222}$$

and

$$I_3 \equiv \int_0^1 \left[f\left(\frac{r}{a}\right)\right]^2 \frac{d(r/a)}{r/a} \tag{14.223}$$

$$C_1 = \frac{\Delta m v_a^2 N I_1}{RTf(r_1/a)} \tag{14.224}$$

and

$$C_3 = \frac{N^2}{2\pi D\rho}\,\frac{I_3}{[f(r_1/a)]^2} \tag{14.225}$$

From Eq. (14.200) the maximum value of the separative capacity per unit length, for a given value of N and velocity profile, is

$$\left(\frac{d\Delta}{dz}\right)_{\max} = \frac{\pi D\rho(\Delta m)^2 v_a^4}{8(RT)^2}\,\frac{4I_1^2}{I_3}\,\frac{1}{1 + [8(\pi D\rho a)^2/N^2]\,[f(r_1/a)]^2/4I_3} \tag{14.226}$$

Equation (14.226) has been written in this form to facilitate comparison with Eq. (14.212) for the maximum separative capacity obtainable from the optimum, constant mass velocity profile. The first factor in these two equations is the same and is the maximum separative capacity per unit length obtainable for a given v_a. The second factor,

$$E_F \equiv \frac{4I_1^2}{I_3} \tag{14.227}$$

is called the *flow pattern efficiency* and represents the reduction in separative capacity caused by departure of the mass velocity profile in an actual centrifuge from the optimum, constant mass velocity profile.

The third factor,

$$E_C \equiv \frac{1}{1 + [8(\pi D\rho a)^2/N^2]\,[f(r_1/a)]^2/4I_3} \tag{14.228}$$

is the circulation efficiency, which takes longitudinal back diffusion into account. It approaches unity as the radius a decreases or the circulation rate N increases. It is written in this form to facilitate comparison with Eq. (14.212) for the optimum, constant mass velocity profile. The two expressions differ by the factor $[f(r_1/a)]^2/4I_3$ in the denominator of (14.228).

Table 14.14 gives the results of calculation of these centrifuge parameters for the Berman-Olander velocity profile (14.214) for UF_6 at 300 K and peripheral speeds of 400, 500,

Table 14.14. Functions of Berman-Olander velocity distribution (14.214) for UF_6 at 300 K

Peripheral speed $v_a = \omega a$, m/s	400	500	700
$A^2 = \exp\left(\frac{352\, v_a^2}{RT}\right)$	11.3	17.6	34.6
$\frac{r_1}{a}$, at which $w = 0$	0.938	0.963	0.982
$f\left(\frac{r_1}{a}\right)$, Eqs. (14.215) and (14.218)	0.02253	0.01638	0.00861
$I_1 = \int_0^1 \frac{r}{a} f\left(\frac{r}{a}\right) d\left(\frac{r}{a}\right)$	2.4945E-3	1.1841E-3	0.2904E-3
$I_3 = \int_0^1 \left[f\left(\frac{r}{a}\right)\right]^2 \frac{d(r/a)}{r/a}$	44.24E-6	14.60E-6	1.736E-6
Flow pattern efficiency, $E_F = \frac{4I_1^2}{I_3}$	0.5626	0.3841	0.1943
$\frac{[f(r_1/a)]^2}{4I_3}$	2.868	4.594	10.68
Separation factor, $\alpha - 1 = \frac{\Delta m v_a^2}{RT} \frac{I_1}{f(r_1/a)}$	0.02131	0.02174	0.01988

and 700 m/s. These parameters have been cast in the present dimensionless form from numerical integrations carried out by May [M6].

As this table shows, concentration of counterflow near the outer wall of the centrifuge in the Berman-Olander profile has these principal effects compared with the optimum uniform mass velocity distribution:

1. The flow pattern efficiency decreases from 0.56 at $v_a = 400$ m/s to 0.19 at 700 m/s. When combined with the v_a^4 in the first factor of Eq. (14.226), the overall effect is to cause the separative capacity per unit length to vary as $v_a^{2.02}$ over this range of v_a, instead of as v_a^4.

2. In the circulation efficiency, Eq. (14.228), the factor $[f(r_1/a)]^2/4I_3$, which has the value unity for the uniform mass velocity profile, increases rapidly with v_a, thus reducing the circulation efficiency.

3. The radial separation factor, expressed as $\alpha - 1$, is nearly independent of v_a over this range, instead of varying as v_a^2 as would be the case for a uniform mass velocity profile, and is much smaller than the local separation factor between the center and outer radius of the centrifuge, given in Table 14.13.

Overall separation performance. To evaluate the overall separative performance of a gas centrifuge from the preceding results for the local separative performance at particular height z and composition y, it is necessary to integrate the differential enrichment equations (14.181) for the enriching section and (14.182) for the stripping section. Because the parameters C_1 and C_5 are functions of the circulation rate N for a given axial velocity profile, it is necessary to know how N varies with z before these integrations can be carried out. A qualitative description can be given of the dependence of N on z for the principal means used to drive the circulation.

Scoop and baffle. With an unbaffled scoop at one end and a rotating baffle at the other, such as shown in Figs. 14.10 and 14.15, the circulation rate will decrease exponentially from the driving end of the centrifuge to the end with the rotating baffle. When the unbaffled scoop is at the bottom, as in Fig. 14.10, and vertical distance z is measured up from the midplane,

$$N_s = N_s^0 \exp\left[-\lambda_s\left(z + \frac{L}{2}\right)\right] \tag{14.229}$$

where L = length of centrifuge
N_s^0 = circulation rate at bottom

λ_s is the decay constant for the circulation rate, whose dependence on the peripheral speed v_a and centrifuge radius a can be estimated from the aerodynamics of the centrifuge. Qualitatively, λ_s is higher the larger v_a and the smaller a.

End cap thermal drive. When flow is induced by heating the top and cooling the bottom of the centrifuge, as in Groth's machine (Fig. 14.14), and the lateral wall is isothermal, the circulation rate decays exponentially from both ends and can be represented qualitatively by

$$N_e = N_e^0 \cosh \lambda_e z \tag{14.230}$$

Here N_e^0 is the circulation rate at the midplane ($z = 0$) and λ_e is the decay constant, which is higher the larger v_a and the smaller a.

Wall thermal drive. When a linear temperature gradient is imposed on the lateral wall of the centrifuge, Durivault and Louvet [D7] have shown that the circulation direction at the wall is in the direction of increasing wall temperature. The rate is highest at the midplane and decreases to zero at the top and bottom. Hence the circulation rate for this type of drive can be modeled approximately by

$$N_w = N_w^0 \frac{\cosh(\lambda_w L/2) - \cosh \lambda_w z}{\cosh(\lambda_w L/2) - 1} \tag{14.231}$$

Here N_w^0 is the circulation rate at the midplane and λ_w is the decay constant.

Combination of drives. In an actual centrifuge driven by a motor at the bottom, motor inefficiency introduces heat at the bottom end cap. The longitudinal variation of circulation rate then depends on where this heat is removed and whether other heat sources are present. Examples of centrifuge separation performance will be given for two cases:

1. Constant circulation rate, independent of z
2. Optimized circulation rate, varied for maximum separative capacity at every elevation

Centrifuge considered. The centrifuge example whose separation performance is to be evaluated has the dimensions of the centrifuge tested by the Standard Oil Development Company in 1944 and described by Beams et al. [B3]:

Length, L = 335 cm
Radius, a = 9.15 cm

which was tested at a peripheral speed of 206 m/s. May [M6] has evaluated separation parameters for a centrifuge of the foregoing dimensions for peripheral speeds of 400, 500, and 700 m/s possibly obtainable with more modern materials. Dimensionless integrals used in separation performance equations have been given in Table 14.14 for these speeds.

Circulation rate independent of height. The case of circulation rate independent of height is analogous to distillation at constant reflux ratio. For this case, explicit equations can be given for overall separation performance of the centrifuge. Conceptually, a constant circulation rate might be realized by a proper combination of end cap thermal drive, Eq. (14.230), and wall thermal drive, Eq. (14.231). Specifically, if

$$\lambda_e = \lambda_w = \lambda \tag{14.232}$$

and

$$N_w^0 = N_e^0 \left(\frac{\cosh \lambda L}{2} - 1 \right) \tag{14.233}$$

the circulation rate has the constant value

$$N = N_e + N_w = N_e^0 \frac{\cosh \lambda L}{2} \tag{14.234}$$

When N and the radial velocity profile are constant, the separation parameters C_1 and C_5 in the differential equations (14.181) for the enriching section and (14.182) for the stripping section are independent of position z. For the low-enrichment case ($y \ll 1$) to be treated here, the equations may be linearized to

$$\text{Enriching:} \qquad C_5 \frac{dy}{dz} = (C_1 + P)y - Py_P \tag{14.235}$$

$$\text{Stripping:} \qquad C_5 \frac{dy}{dz} = (C_1 - W)y + Wy_W \tag{14.236}$$

The integral of (14.235) between $y = y_F^E$ at the feed point, $z = 0$, and $y = y_P$ at the top, $z = L_E$, is

$$\frac{y_P}{y_F^E} = \frac{P + C_1}{P + C_1 \exp\left[-(P + C_1)L_E/C_5\right]} \tag{14.237}$$

The integral of (14.236) between $y = y_W$ at the bottom, $z = -L_S$, and $y = y_F^S$ at $z = 0$ is

$$\frac{y_W}{y_F^S} = \frac{W - C_1}{W - C_1 \exp\left[-(W - C_1)L_S/C_5\right]} \tag{14.238}$$

The material-balance equation on light component at the feed point where feed rate $P + W$ joins enriched stream at rate $N - W$ is

$$(N - W)y_F^S + (P + W)y_F = (N + P)y_F^E \tag{14.239}$$

Overall material balance on light component is

$$Wy_W + Py_P = (W + P)y_F \tag{14.240}$$

For given values of y_F, L_E, P, W, and N, these equations are sufficient to determine the product composition y_P, tails composition y_W, and the heads compositions y_F^S leaving the stripping section and y_F^E entering the enriching section.

To avoid mixing losses at the feed point, the solution for which

$$y_F^S = y_F \tag{14.241}$$

is desired. When this is true, y_F^E also equals y_F, from (14.239). When L_E, L_S, y_F, N, and the feed rate $F \equiv P + W$ are given, it is necessary to find by trial the value of P (or W) at which the preceding five equations are satisfied. The condition for this is obtained by substitution into the overall material-balance equation (14.240) for y_P/y_F^E from (14.237) and y_F^S/y_W from

(14.238):

$$\frac{W}{y_F^S/y_W} + P\left(\frac{y_P}{y_F^E}\right) = (W + P) \tag{14.242}$$

The separative capacity Δ for this low-enrichment case, from Eq. (12.141), is

$$\Delta = -P \ln y_P - W \ln y_W + F \ln y_F$$

$$= -P \ln\left(\frac{y_P}{y_F}\right) + W \ln\left(\frac{y_F}{y_W}\right) \tag{14.243}$$

When the no-mixing-loss condition (14.241) is satisfied, the separative capacity, from (14.243), (14.237), and (14.238), is

$$\Delta = -P \ln \left\{\frac{P + C_1}{P + C_1 \exp\left[-(P + C_1)L_E/C_5\right]}\right\} + W \ln \left\{\frac{W - C_1}{W - C_1 \exp\left[-(W - C_1)L_S/C_5\right]}\right\} \tag{14.244}$$

This separative capacity is lower than the product of the maximum separative capacity per unit length $(d\Delta/dz)_{\max}$, given by Eq. (14.200), and the length $L = L_E + L_S$. The ratio

$$E_I \equiv \frac{\Delta}{(C_1^2/4C_5)L} \tag{14.245}$$

is termed the *ideality efficiency* E_I. In terms of the three efficiencies: ideality efficiency E_I, Eq. (14.245), circulation efficiency E_C, Eq. (14.228), and flow pattern efficiency E_F, Eq. (14.227), the overall separative capacity is

$$\Delta = E_I E_C E_F \frac{\pi D \rho (\Delta m)^2 v_a^4}{8(RT)^2} L \tag{14.246}$$

Centrifuge example. Tables 14.15 and 14.16 summarize calculations of the separative capacity of a centrifuge 335.3 cm long, 18.29 cm in diameter, run at a peripheral speed of 400 m/s at 300 K, with circulation rate independent of height. Some of these were given by May [M6]. In Table 14.15, the circulation rate for all cases is 0.1884 g UF_6/s, and the feed rate is varied. The separative capacity has a maximum of 10.05 kg uranium SWU/year at a feed rate of 0.038052 g UF_6/s (1200 kg UF_6/year. Δ remains close to 10 with a variation in feed rate of ± 20 percent, but decreases considerably at feed rates outside of this range. The axial separation factor is 1.67 at the lowest feed rate, decreases steadily with increasing feed rate, and equals 1.37 at optimum. The height of a transfer unit is 12.39 cm. The maximum ideality efficiency, at the optimum feed rate, from (14.245), is 0.8147. The circulation efficiency, from (14.228), is $E_C = 0.9757$. The overall efficiency $E \equiv E_F E_C E_I$ has a maximum value of 0.4472.

In Table 14.16, the feed rate is held constant at 0.03171 g UF_6/s (1000 kg UF_6/year) and the circulation rate is varied. The separative capacity has a maximum of 10.03 kg uranium SWU/year at an optimum circulation rate N = 0.1884 g UF_6/s and decreases rather rapidly with changes from this rate. The axial separation factor has a maximum of 1.41 at the optimum circulation rate. The height of a transfer unit increases almost proportionally with circulation rate. The circulation efficiency increases from 0.9095 at the lowest circulation rate of 0.0942 g/s to practically unity at the highest, showing the decreasing influence of axial back diffusion as circulation rate increases.

In Tables 14.15 and 14.16, the cut (ratio of product flow rate to feed flow rate) at conditions that lead to maximum separative capacity is 0.45. Because the centrifuge is

Table 14.15 Effect of feed rate on separation performance of gas centrifuge

Length: stripping, $L_S = 167.65$ cm; enriching, $L_E = 167.65$ cm
Radius: $a = 9.145$ cm
Temperature: 300 K
Peripheral speed: $v_a = 40{,}000$ cm/s; $A^2 = 11.3$
Circulation rate: $N = 0.1884$ g UF_6/s
Centrifuge parameters: $C_1 = 0.00402$ g UF_6/s
$C_5 = 2.3347$ (g $UF_6 \cdot$cm)/s
Radial enrichment factor: $\alpha - 1 = 0.02131$
Height transfer unit: $h = 12.39$ cm
Efficiencies: flow pattern, $E_F = 0.5626$; circulation, $E_C = 0.9757$

UF_6 flow rate, g/s						
Feed	0.006342	0.019026	0.031710	0.038052†	0.044394	0.06342
Product	0.002749	0.008409	0.014239	0.017200†	0.020200	0.02928
Tails	0.003593	0.010617	0.017471	0.020852†	0.024194	0.03414
Axial separation factors						
Heads, β	1.29638	1.23601	1.19165	1.17394	1.15851	1.12318
Tails, γ	1.29314	1.22993	1.18510	1.16752	1.15246	1.11812
Overall, $\beta\gamma$	1.67640	1.52020	1.41222	1.37060	1.33514	1.25585
Cut, θ	0.433	0.442	0.449	0.452	0.455	0.462
Separative capacity, kg U SWU/yr	4.484	8.859	10.03	10.05	9.830	8.749
Efficiencies						
Ideality, E_I	0.3635	0.7182	0.8132	0.8147	0.7970	0.7093
Overall, $E = E_F E_C E_I$	0.1996	0.3943	0.4464	0.4472	0.4375	0.3894

†Optimum.

connected in a cascade, cascade conditions may require the centrifuge to operate at a somewhat different cut. This would violate the no-mixing-loss condition at the feed point (14.241) (unless the feed location can be changed) and reduce the separative capacity.

May [M6] has calculated the effect of varying the circulation rate N on the separative capacity of this centrifuge example operated at peripheral speeds of 400, 500, and 700 m/s, at a feed rate of 0.03171 g UF_6/s, using the parameters of Table 14.14, with results shown in Fig. 14.18. The optimum heavy-stream flow rate N increases with increasing speed. The separative capacity at optimum N increases as $v_a^{2.02}$.

Optimum distribution of circulation rate. When the circulation rate N is independent of height, the parameters C_1 and C_5 are constant. It is thus possible to satisfy condition (14.199) for the composition variable and (14.200) for the maximum separative capacity gradient at only one elevation and one value of y in each of the enriching and stripping sections. This is what causes the ideality efficiency for the constant N condition to be less than unity. We shall now give an example of a centrifuge in which the heavy-fraction flow rate N is varied so as to have its optimum value at every height in the centrifuge, and will find by how much the overall height of a centrifuge of a given capacity could be reduced compared with one with uniform N.

The composition gradient dy/dz in the enriching section may be expressed as a function of the composition y and the circulation rate N by Eqs. (14.181) and (14.179):

$$\frac{dy}{dz} = \frac{N(C_1/N)y(1-y) - P(y_P - y)}{N^2(C_3/N^2) + C_2} \tag{14.247}$$

Table 14.16 Effect of circulation rate on separation performance of gas centrifuge

Length: stripping, $L_S = 167.65$ cm; enriching, $L_E = 167.65$ cm
Radius: $a = 9.145$ cm
Temperature: 300 K
Peripheral speed: $v_a = 40{,}000$ cm/s; $A^2 = 11.3$
Radial enrichment factor: $\alpha - 1 = 0.02131$
Feed rate: $F = 0.03171$ g UF_6/s (1000 kg UF_6/yr)

UF_6 flow rate, g/s					
Circulation rate, N	0.0942	0.1884†	0.2200	0.3768	0.5652
Product	0.01431	0.014239†	0.014345	0.014796	0.01511
Tails	0.01740	0.017471†	0.017365	0.016914	0.01660
Centrifuge parameters					
C_1, g/s	0.00201	0.00402	0.00469	0.00804	0.01205
C_5, (g · cm)/s	0.6262	2.3347	3.1630	9.169	20.56
Height transfer unit h, cm	6.65	12.39	14.38	24.3	36.4
Axial separation factor					
Enriching, β	1.1384	1.19165†	1.1856	1.1366	1.0966
Stripping, γ	1.1285	1.18510†	1.1810	1.1357	1.0964
Overall, $\beta\gamma$	1.2847	1.41222†	1.4002	1.2908	1.2023
Cut, θ	0.451	0.449†	0.452	0.467	0.477
Separative capacity, kg U SWU/yr	5.29	10.03†	9.53	5.50	2.87
Efficiencies					
Flow pattern, E_F	0.5626	0.5626	0.5626	0.5626	0.5626
Circulation, E_C	0.9095	0.9757	0.9821	0.9938	0.9972
Ideality, E_I	0.4601	0.8132	0.7676	0.4378	0.2277
Overall, $E = E_F E_C E_I$	0.2354	0.4464	0.4241	0.2448	0.1277

†Optimum.

For a given speed v_a, C_1/N is a constant,

$$B_1 \equiv \frac{C_1}{N} \tag{14.248}$$

independent of N and y, as shown by Eq. (14.224). Similarly, C_3/N^2 is a constant,

$$B_3 \equiv \frac{C_3}{N^2} \tag{14.249}$$

independent of N and y, as shown by Eq. (14.225). In these terms,

$$\frac{dy}{dz} = \frac{NB_1 y(1-y) - P(y_P - y)}{N^2 B_3 + C_2} \tag{14.250}$$

The optimum value of N for a given y is the value at which

$$\frac{\partial \ln(dy/dz)}{\partial N} = \frac{B_1 y(1-y)}{NB_1 y(1-y) - P(y_P - y)} - \frac{2NB_3}{N^2 B_3 + C_2} = 0 \tag{14.251}$$

or

$$-N^2 B_1 B_3 y(1-y) + 2NPB_3(y_P - y) + B_1 C_2 y(1-y) = 0 \tag{14.252}$$

Hence,
$$N_E^{\text{opt}} = \frac{PY_E}{B_1} + \sqrt{\left(\frac{PY_E}{B_1}\right)^2 + \frac{C_2}{B_3}} \tag{14.253}$$

where
$$Y_E \equiv \frac{y_P - y}{y(1-y)} \tag{14.254}$$

In the low-enrichment case ($y \ll 1$),

$$Y_E = \frac{y_P}{y} - 1 \tag{14.255}$$

In a centrifuge with the optimum value of N at every y, the length dz needed for enrichment dy from (14.250) is

$$y(1-y)\left(\frac{dz}{dy}\right)_{\min} = \frac{(N_E^{\text{opt}})^2 B_3 + C_2}{(N_E^{\text{opt}}) B_1 - PY_E} = \frac{2B_3}{B_1} N_E^{\text{opt}} \tag{14.256}$$

The last expression results from using (14.253) to eliminate Y_E. In the low-enrichment case,

$$\left(\frac{dz}{d \ln y}\right)_{\min} = \frac{2B_3}{B_1} N_E^{\text{opt}} \tag{14.257}$$

The minimum length of enriching section z_E necessary to increase y from the feed value y_F to a higher value y_E is

$$z_E(y_E) = \int_0^{z_E} dz_{\min} = \frac{2B_3}{B_1} \int_{y_F}^{y_E} N_E^{\text{opt}} \, d \ln y \tag{14.258}$$

for the low-enrichment case.

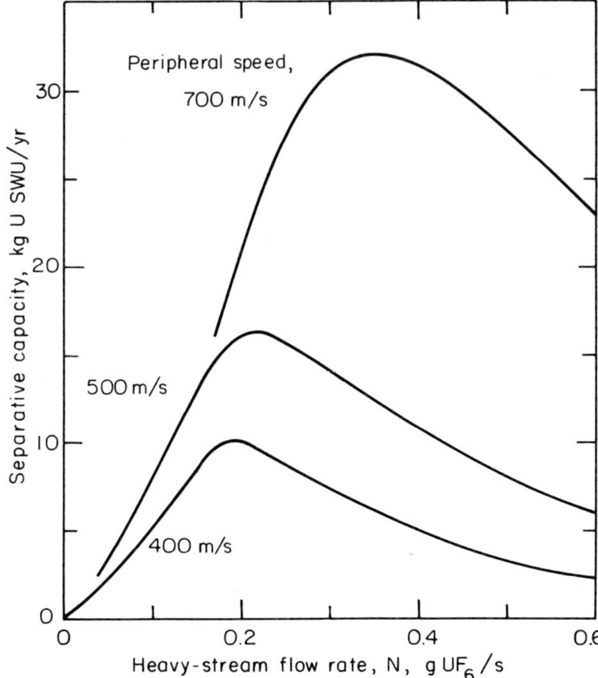

Figure 14.18 Effect of heavy-stream flow rate on separative capacity of centrifuge at peripheral speeds of 400, 500, and 700 m/s. Length 335.3 cm; diameter 18.29 cm; feed rate 1000 kg UF_6/year. *(From May [M6].)*

A similar development for the stripping section leads to

$$N_S^{\text{opt}} = \frac{PY_S}{B_1} + \sqrt{\left(\frac{PY_S}{B_1}\right)^2 + \frac{C_2}{B_3}} \tag{14.259}$$

where

$$Y_S \equiv \frac{y - y_W}{y(1-y)} \tag{14.260}$$

or

$$Y_S = 1 - \frac{y_W}{y} \tag{14.261}$$

for the low-enrichment case. The minimum length of stripping section z_S necessary to decrease y from y_F to a lower value y_S is

$$z_S(y_S) = \int_{-z_S}^{0} dz_{\min} = \frac{2B_3}{B_1} \int_{y_S}^{y_F} N_S^{\text{opt}}\, d \ln y \tag{14.262}$$

To give an example of the reduction in centrifuge height for a given separative capacity that could be obtained if it were possible to use an optimized, variable heavy-stream flow rate instead of the uniform flow rate employed in Table 14.15, a centrifuge with optimized, variable heavy-stream flow rate was designed for the conditions of Table 14.15 marked with a dagger. These led to maximum separative capacity in a centrifuge operated with uniform heavy-stream flow rate. Centrifuge characteristics for the optimum flow-rate distribution are shown in Fig. 14.19 and compared with the uniform-flow-rate case in Table 14.17.

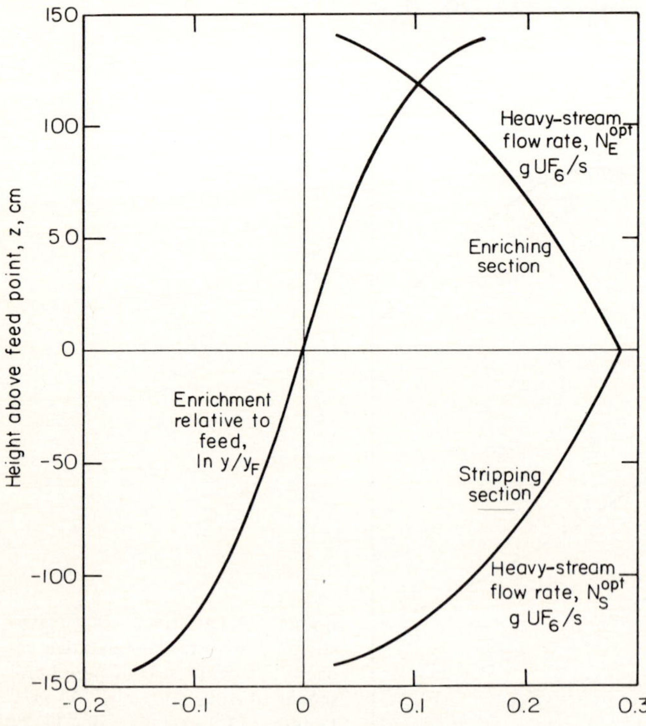

Figure 14.19 Variation of optimum heavy-stream flow rate and enrichment with height in centrifuge at 300 K, 400 m/s, and 1000 kg UF_6/year feed rate.

Table 14.17 Comparison of centrifuges with uniform and optimized variable heavy-stream flow rates

	Conditions common to both cases	
UF_6 flow rate, g/s		
Feed	0.03171	
Product	0.014239	
Tails	0.017471	
Separation factor		
Enriching	1.19165	
Stripping	1.18510	
Overall	1.41222	
Separative capacity, kg SWU/yr	10.03	
	Conditions differing in optimized case	
Heavy-stream flow rate	Uniform	Optimized
g UF_6/s at feed point	0.1884	0.2837
at top	0.1884	0.0297
at bottom	0.1884	0.0297
Height, cm, enriching	167.65	140.41
stripping	167.65	142.02
total	335.3	284.43
Circulation efficiency		
Feed point	0.9757	0.9891
Top or bottom	0.9757	0.5000

In Fig. 14.19, height above feed point is plotted vertically to correspond with orientation of an operating centrifuge. Optimum heavy-stream flow rate has a maximum of 0.2837 g UF_6/s at the feed location ($z = 0$) and decreases to 0.0297 at the top and bottom. These are to be compared with 0.1884 g/s in the uniform-flow-rate case. This decrease in flow rate from feed location to withdrawal ends of the centrifuge is qualitatively similar to that of the ideal cascade discussed in Chap. 12. However, the tails flow rate at the top, product end of the centrifuge cannot drop to zero, as it would in an ideal cascade, because dy/dz would become zero at $N = 0$, as can be seen from Eq. (14.250).

In Fig. 14.19 composition is plotted horizontally as $\ln y/y_F$, to bring out another difference from an ideal cascade. A plot of distance versus $\ln y$ in an ideal cascade with constant height of a transfer unit (htu) would be a straight line. In this centrifuge with variable circulation rate, the htu from Eq. (14.186) varies from 18.4 cm at the feed elevation to 3.8 cm at the top and bottom. This causes $\ln y$ to change more rapidly with z at the top and bottom than at the feed elevation.

As Table 14.17 shows, optimization of flow distribution permits reduction in centrifuge length for the stated separation performance from 335.3 to 284.43 cm. The ratio of these lengths, 0.8483, is somewhat greater than the ideality efficiency of the uniform-flow case, 0.8132 from Table 14.16. The reason for this may be seen by comparing the circulation efficiencies for the two cases. With variable flow rate, the circulation efficiency ranges from 0.9891 at the feed location to 0.5000 at top and bottom, compared with a constant efficiency of 0.9757 for the uniform-flow-rate case. Thus, the average circulation efficiency with variable flow rate is lower than with constant, a disadvantage that partially cancels the use of optimum flow rate at every height.

6 AERODYNAMIC PROCESSES

6.1 Introduction

Processes in which isotopic composition changes are produced when a flowing gas mixture experiences large linear or centrifugal acceleration are termed aerodynamic processes. Of the many aerodynamic processes that have been proposed or investigated experimentally, only two have been carried through large-scale pilot-plant experiments to intended commercial deployment. These are the separation nozzle process, developed by Becker and his associates of the Nuclear Research Center at Karlsruhe, West Germany, and the UCOR process, developed by the Uranium Enrichment Corporation of South Africa. The separation nozzle process has passed through a number of development stages, which have been described in detail by Becker and his associates [B5–B12, G1]. These will be summarized in Sec. 6.2. The South African process is subject to considerable industrial secrecy; a brief summary of three published articles on this process [G2, H1, R3] will be given in Sec. 6.3.

Numerous other schemes for separating isotopes in flowing gas streams have been conceived and subjected to small-scale test, but none has appeared sufficiently promising to enlist the major development support given the nozzle and UCOR processes. Summary descriptions of other aerodynamic processes are in references [T2] and [M2].

6.2 The Separation Nozzle Process

Evolution of process. The separation nozzle process has evolved through a number of forms. The first process tested experimentally by Becker [B6] is illustrated schematically in Fig. 14.20, with dimensions for one of the devices tested on UF_6. UF_6 feed at a pressure p of around 20 Torr flows through a slit-shaped nozzle 0.045 mm wide into a region at much lower pressure p', where a fraction θ, about 0.2, of the feed diverges from the feed jet and is somewhat enriched in the light isotope. The remaining fraction of the feed jet, $1 - \theta$, somewhat enriched in the heavy isotope, passes through a wider separator slit, where its pressure p'' is somewhat higher than p' because of deceleration. London [L4] gives examples of the separation factor, cut, and UF_6 feed rate observed by Becker [B6]. Optimum pressure conditions at which power consumption, compressor capacity, and nozzle length per unit separative capacity were smallest are listed in the first column of Table 14.18, together with the minimum values of these performance indices.

Comparison with corresponding performance indices for gaseous diffusion, taken from

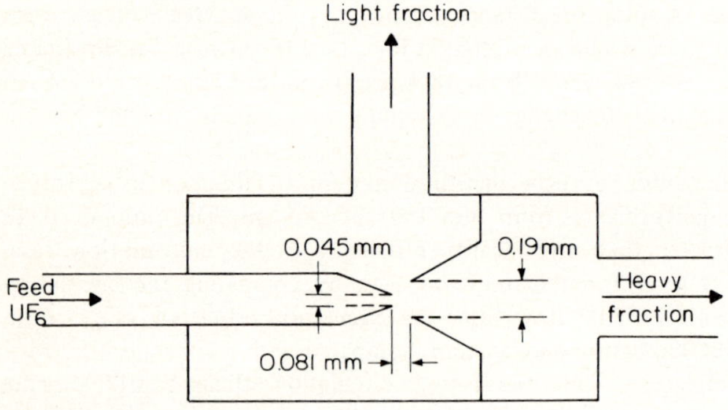

Figure 14.20 First form of separation nozzle process.

Table 14.18 Comparison of operating conditions and performance indices of two forms of nozzle process and gaseous diffusion process

	Nozzle process		Gaseous diffusion process
	Early	Improved	
Reference	[B6]	[G1]	Table 14.9
Operating pressures, Torr			
Feed p	20	290	418
Light fraction p'	0.5	138	138
Heavy fraction p''	2.8	138	~ 200
Mole fraction UF_6 in feed	1.0	0.042	1.0
Feed rate, kg UF_6/h·m)	3.2	3.96	–
Separation factor $\alpha - 1$	0.0037	0.0148	0.0030
Cut θ	0.2	0.25	0.5
Per meter slit length			
Separative capacity Δ, kg SWU/(yr·m)	0.0208	0.48	–
Power (rate of loss of availability) Q, kW/m	0.0146	0.138	–
Compressor volumetric capacity V, $m^3/(s \cdot m)$	0.0324	0.0108	–
Performance indices per unit separative capacity			
Slit length, m/(kg SWU/yr)	48	2.08	–
Power Q/Δ, kW/(kg SWU/yr)	0.70	0.287	0.168
Compressor capacity V/Δ, (m^3/s)/(kg SWU/yr)	1.5	0.0225	0.00985
Relative number of stages	2.0	0.4	1.0

Table 14.9, shows that in this early version of the separation nozzle process, the separation factor was slightly better than for gaseous diffusion, but the power consumption, Q/Δ, the rate of loss of availability, was four times as great as in gaseous diffusion, and the compressor capacity, V/Δ, was 150 times as great. The high power consumption was a consequence of the high pressure ratio through which both light and heavy fractions were expanded in this early version of the nozzle process, and the very high compressor capacity was caused both by the high pressure ratio and the low operating pressure level.

Two modifications of the process developed by Becker and his associates have greatly improved these process characteristics. (1) Dilution of UF_6 feed with a gas of lower molecular weight, helium in early developments [B7] and hydrogen in later developments [G1], has had two beneficial effects. Sonic velocity in the nozzle is increased, with accompanying increase in separation factor, and diffusion rates are increased, permitting operation at higher pressure and higher uranium throughout without impairment of separation. (2) The radical change in nozzle geometry illustrated in Fig. 14.21 adds the relatively large separation caused by centrifugal acceleration to the smaller separation accompanying expansion through the slit.

Improved nozzle process. In Fig. 14.21, a dilute mixture of f mole fraction UF_6 in hydrogen at upstream pressure p is expanded through a convergent-divergent slit with a throat spacing s into a curved groove of radius a. After being deflected through 180° by the wall of the curved groove, the gas stream at lower pressure p' traveling at high speed is separated by a flow divider set at radius c into an outer heavy fraction depleted in $^{235}UF_6$ and hydrogen and an inner light fraction enriched in these components. The cut θ is determined by the position of the flow divider. The separation factor α (1) is higher the higher the speed attained by the gas, which is higher the higher the pressure ratio p/p' and the lower the UF_6 content of the feed gas; (2) has a maximum value at an optimum pressure level, which is inversely proportional to the dimensions s and a; and (3) is higher the lower the cut θ.

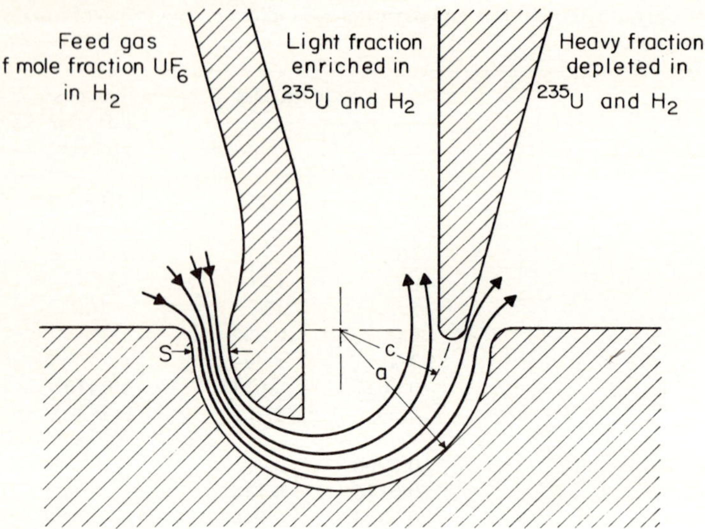

Figure 14.21 Cross section of slit used in separation nozzle process.

Figure 14.22 shows the dependence of separation factor on cut. The lower lines show the separation factor calculated by assuming that the $^{235}UF_6$ and $^{238}UF_6$ density distribution in the curved groove reaches centrifugal equilibrium at the indicated peripheral speed v, using the theory to be derived later in this section. The top line shows the highest values of the separation factor reported in Becker's papers [B10], at a pressure ratio of 8 and a low UF_6 content, 1.6 m/o (mole percent) in hydrogen, at which the calculated outlet gas velocity for reversible expansion is 1042 m/s. Because these extreme conditions result in gas-compression energy consumption per unit of separative work produced much greater than optimum, they are not recommended for a commercial plant. They do indicate, however, that values of $\alpha - 1$ in the current version of the nozzle process can be 10 times as high as in the early, linear nozzle process of Fig. 14.20 or in the gaseous diffusion process of Table 14.9.

Design studies for a commercial plant by Geppert and associates [G1] indicate that optimum conditions are feed composition $f = 0.042$ mole fraction UF_6 in hydrogen, pressure ratio $p/p' = 2.1$, and a cut $\theta = \frac{1}{4}$, at which $\alpha - 1 = 0.0148$, still four times that in gaseous diffusion, and somewhat higher than what would be predicted for centrifugal equilibrium at the speed attainable from expansion through this pressure ratio. The cut of $\frac{1}{4}$ necessitates use of a three-up, one-down cascade, as shown in Sec. 14.2 of Chap. 12.

Attainment of separation factors higher than predicted for equilibrium in a centrifugal field have been explained by Becker and associates [B10] as follows. Before the mixture of hydrogen, $^{235}UF_6$, and $^{238}UF_6$ enters the curved groove, the concentration of each is spatially uniform. While undergoing linear and centrifugal acceleration, the heaviest component, $^{238}UF_6$, experiences the highest forces and migrates more rapidly toward the outer wall than the lighter component, $^{235}UF_6$. Thus, there is a transient time during flow along the curved wall when the $^{238}UF_6/^{235}UF_6$ concentration ratio is a maximum, after which the ratio decreases toward the limiting, equilibrium value. This transient phenomenon is enhanced by high dilution by hydrogen, which reduces the frequency of collisions between $^{235}UF_6$ and $^{238}UF_6$ molecules, which otherwise would speed attainment of centrifugal equilibrium between these species. Maximum benefit from this transient phenomenon for a given pressure ratio is obtained at an optimum pressure level for a given set of nozzle dimensions. At a pressure level lower than

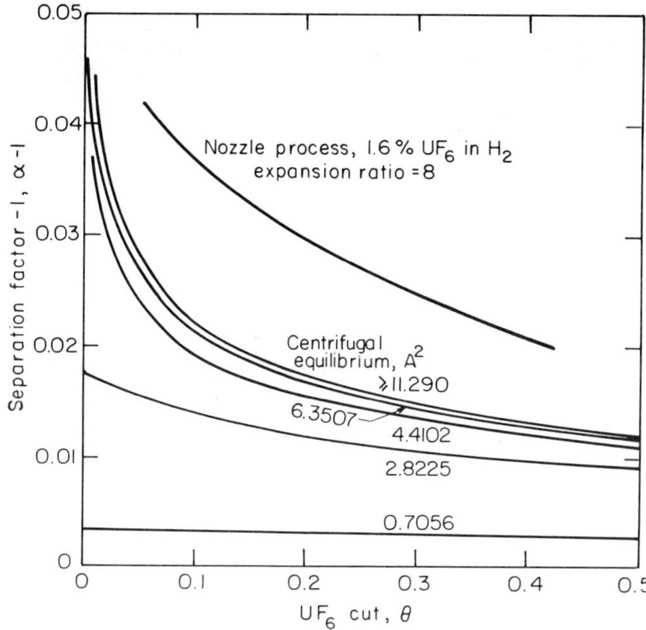

Figure 14.22 Comparison of highest reported separation factors in nozzle process with calculated values for equilibrium in centrifugal field.

optimum, diffusion rates, which are inversely proportional to pressure, cause attainment of centrifugal equilibrium before the gas mixture reaches the flow divider. At a pressure level higher than optimum, diffusion rates are too slow to permit the initially spatially uniform $^{238}UF_6/^{235}UF_6$ ratio to reach its maximum transient value.

The left half of Fig. 14.23 shows the dependence of separation factor, expressed as $\alpha - 1$, on pressure ratio p/p' and upstream pressure p, for a cut $\theta = \frac{1}{4}$ and $f = 0.04$ mole fraction UF_6 in hydrogen, as reported by Becker et al. [B10]. At each pressure ratio p/p' there is an

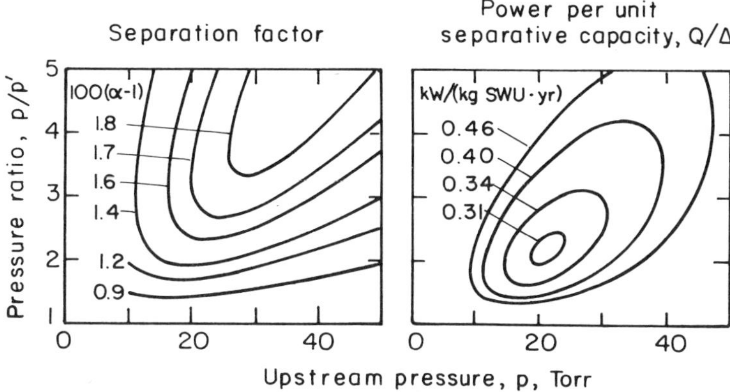

Figure 14.23 Separation factor and power consumption per unit separative capacity in nozzle process. 4 m/o UF_6 in hydrogen, cut $= \frac{1}{4}$.

optimum inlet pressure p at which the separation factor is a maximum. At high pressure ratios, the separation factor is higher than predicted by Fig. 14.22 for centrifugal equilibrium at a cut of $\frac{1}{4}$, for any speed. At each pressure ratio, i.e., at each speed, there is an inlet pressure at which the separation factor is a maximum; this inlet pressure is higher the higher the pressure ratio and the higher the speed.

The right half of Fig. 14.23 shows the dependence of power consumption per unit separative capacity Q/Δ on the same pressure variables. The power consumption has been calculated as the rate of loss of availability, so that Q/Δ is given by

$$\frac{Q}{\Delta} = \frac{2RT_0 \ln p/p'}{(\alpha - 1)^2 \theta(1 - \theta)} \tag{14.263}$$

Optimum pressure conditions for minimum Q/Δ are inlet pressure $p = 22$ Torr, and pressure ratio $p/p' = 2.1$, at which $\alpha - 1 = 0.0148$ and the power consumption is 0.31 kW/(kg SWU/year).

The nozzle dimensions a and s with which the pressure level of Fig. 14.23 was associated were not stated in reference [B10]. Dimensions and related operating pressures reported [V1] as optimum for UF_6-helium mixtures are

Throat spacing s, mm	0.4	0.2	0.03
Groove radius a, mm			0.1
Downstream pressure p', Torr	12	20	150
Upstream pressure p, Torr	48	80	600

Because the diffusion coefficient of UF_6 into hydrogen is about 20 percent higher than into helium, optimum pressures for UF_6-hydrogen mixtures would be about 20 percent higher than the foregoing values. The inference then is that the data of Fig. 14.23 were obtained with a nozzle with a throat spacing around 0.4 mm.

Operation at the highest feasible pressure is economically desirable because the volumetric flow rate is lower and compressors and piping are smaller. Later design studies for a commercial plant by Geppert et al. [G1] selected optimum outlet and inlet pressures of 138 and 290 Torr, respectively. These are for 4.2 m/o UF_6 in hydrogen feed, presumably with nozzle dimensions of

Throat spacing $s = 0.03$ mm
Groove radius $a = 0.1$ mm

the smallest dimensions reported [V1]. This lower pressure ratio of 2.1 was chosen to reduce the specific power consumption and to permit operation with a single stage of compression without intercooling.

The second column of Table 14.18 summarizes characteristics of the improved nozzle plant whose design was described by Geppert et al. [G1]. The slit length, power, and compressor capacity per unit separative capacity are greatly improved over the early process because of the much higher separation factor and operating pressures. However, the last two are still not as small as those for the gaseous diffusion process, restated from Table 14.9 in the third column. The higher compressor capacity and power consumption of the nozzle process compared with gaseous diffusion results from the 24-fold dilution of UF_6 with hydrogen and the need to recompress both light and heavy fractions through the full pressure ratio in the nozzle process. However, the much higher separation factor of the nozzle process causes the number of stages it requires to be only 40 percent of those needed by gaseous diffusion for the same separation, despite the smaller cut used in the nozzle process. When all sources of process inefficiency, such as pressure drops and compressor inefficiency, are taken into account, Geppert [G1] has estimated that the actual power consumption of a complete nozzle plant with capacity of 5,045,000 kg SWU/year would be 2520 MW, for a specific power consumption of 0.50 kW/(kg SWU/year). This may be compared with the capacity of the gaseous diffusion plants of the U.S. DOE, 17,230,000 kg SWU/year and their power consumption of 6,060 MW, for a specific power consumption of 0.352 kW/(kg SWU/year). These actual power consumptions are in approximately the same ratio as the values of Q/Δ in Table 14.18.

Equipment of nozzle plants. Becker [B11] has described two types of separating elements with the cross-section contour shown in Fig. 14.21. The more fully developed type, produced by mechanical means by Messerschmidt-Bölkow-Blohm Gmbh, Munich, is illustrated in Fig. 14.24. This consists of a cylindrical aluminum tube 2 m long, whose outer surface carries 10 semicircular longitudinal grooves, through each of which a portion of the feed gas flows circumferentially. The convergent-divergent nozzle contour and flow divider are provided by properly shaped strips fitted into 10 dovetail-shaped notches cut into the aluminium tube. The aluminum tube is divided into 10 radial sectors which carry, alternately, inflowing feed gas and outflowing heavy fraction. The light fraction flows into the space outside the tube through a slot between the dovetail strips, which are held in position by small spherical spacers at regular intervals. A complete separation stage contains 80 or more of these separating tubes mounted vertically, with appropriate headers for admitting feed and withdrawing light and heavy fractions. Their predicted separating capacity when operated on 4.2 percent UF_6 in hydrogen and pressures of 290 and 138 Torr is 0.48 kg SWU per year per meter slit length [G1]. A separating element of this type has been run on UF_6 for over 30,000 h without change in measured separation factor [B11]. The cost of mass-produced separating tubes of this type predicted in 1971 [B9] was less than $16/(kg SWU/year).

A second type of separating element, developed by Siemens AG, is fabricated by photoetching of metal foils by techniques used in miniaturizing electronic circuits. The left side of Fig. 14.25 is an enlarged contact print of such an etched foil. The middle of Fig. 14.25 shows how these foils are stacked into chips held by cover plates pierced with holes in register with the feed and heavy fraction passages. The right side of Fig. 14.25 shows assembly of chips into a tube.

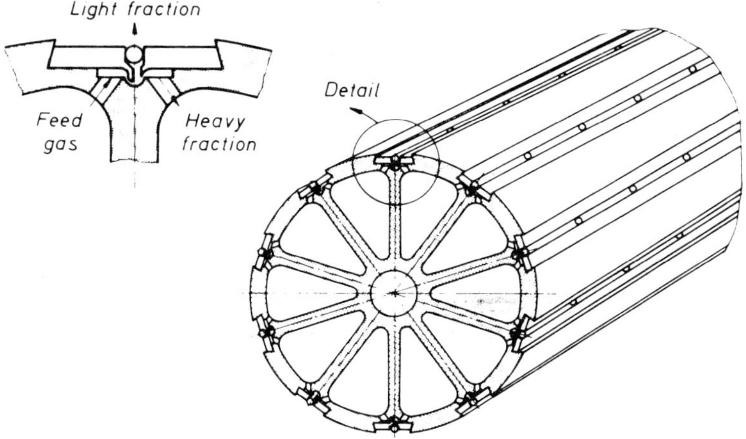

Figure 14.24 Tubular separation element for nozzle process. *(Courtesy of Dr. E. W. Becker.)*

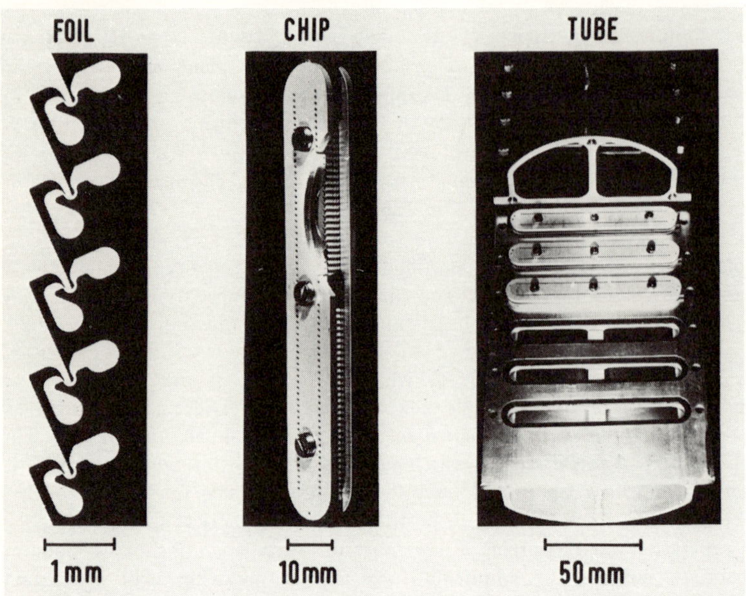

Figure 14.25 Separation nozzle element made by stacking photoetched metal foils. (*Courtesy of Dr. E. W. Becker. Reproduced with permission of the copyright holder, American Institute of Chemical Engineers.*)

Figure 14.26 is a partially cutaway side view of a small prototype separation nozzle stage that has been run [G1] on total recycle with UF_6 and hydrogen. The stage contains 54 of the 10-sector elements 1 m long. The separating elements are mounted vertically inside a metal tank from which is suspended a two-stage gas cooler and a two-stage radial centrifugal compressor. The two-stage arrangement was necessitated by design for a compression ratio of 4. Stages for a larger production plant, based on later designs, will use a compression ratio of 2.1, and a single-stage cooler and axial-flow compressor.

Theory of separation. Theoretical analysis of the current form of the separation nozzle process is very difficult because of the presence of three components of widely different molecular weight, the complex flow geometry, and the importance of transient diffusion effects during the brief exposure of the mixture to centrifugal acceleration. A simplified, approximate analysis of the effect of cut and gas velocity on separation factor, separative capacity, and power consumption will be given by assuming (1) that $^{235}UF_6$ and $^{238}UF_6$ attain their equilibrium concentration distribution at the end of the 180° rotation the expanded gas undergoes, and (2) that gas motion is in "wheel flow" at uniform angular velocity ω. Finally, the effect of factors neglected in this simplified treatment will be discussed qualitatively. Malling and Von Halle [M2] made similar assumptions in their simplified analysis of the nozzle process.

The flow geometry assumed is illustrated in Fig. 14.21. The gas mixture is assumed to be rotating at uniform angular velocity ω in a semicircular groove of radius a. Centrifugal equilibrium is established where the mixture is separated by the flow divider at radius c into an inner, light fraction enriched in hydrogen and $^{235}UF_6$ and an outer, heavy fraction depleted in these components relative to $^{238}UF_6$.

From the treatment of the gas centrifuge in Sec. 5.5, the dependence of concentration of light isotope (e.g., $^{235}UF_6$) on radius r at centrifugal equilibrium is

$$\rho_1(r) = \rho_1(0) \exp\left(\frac{m_1\omega^2 r^2}{2RT'}\right) \tag{14.264}$$

where $\rho_1(0)$ is the density of component (1) of molecular weight m_1 at the center of rotation ($r = 0$). A similar equation for the density of component 2 (e.g., $^{238}UF_6$) is

$$\rho_2(r) = \rho_2(0) \exp\left(\frac{m_2\omega^2 r^2}{2RT'}\right) \tag{14.265}$$

T' is the absolute temperature of the mixture after acceleration to angular velocity ω. If the flow divider is set at $r = c$, the mass flow rate of component 1 in the light fraction per unit length is

$$\mathfrak{M}_1 = \int_0^c \omega r \rho_1(r)\,\mathrm{d}r = \int_0^c \omega r \rho_1(0) \exp\left(\frac{m_1\omega^2 r^2}{2RT'}\right) dr = \frac{RT'\rho_1(0)}{m_1\omega}\left[\exp\left(\frac{m_1\omega^2 c^2}{2RT'}\right) - 1\right] \tag{14.266}$$

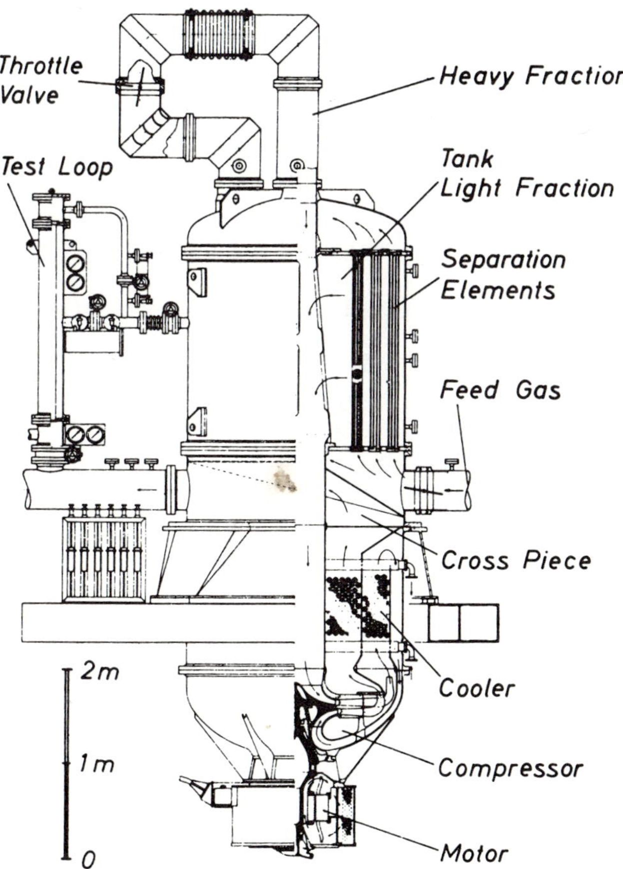

Figure 14.26 Cross section of separation nozzle stage with planned capacity of 2000 kg SWU/year. *(Courtesy of Dr. E. W. Becker.)*

Similarly, the mass flow rate per unit length of component 2 in the light fraction is

$$\mathfrak{M}_2 = \int_0^c \omega r \rho_2(r)\, dr = \frac{RT'\rho_2(0)}{m_2\omega}\left[\exp\left(\frac{m_2\omega^2 c^2}{2RT'}\right) - 1\right] \tag{14.267}$$

The mass flow rate of component 1 in the heavy fraction flowing between radius c and the outer wall at radius a is

$$\mathfrak{N}_1 = \int_c^a \omega r \rho_1(r)\, dr = \frac{RT'\rho_1(0)}{m_1\omega}\left[\exp\left(\frac{m_1\omega^2 a^2}{2RT'}\right) - \exp\left(\frac{m_1\omega^2 c^2}{2RT'}\right)\right] \tag{14.268}$$

and that of component 2 in the heavy fraction is

$$\mathfrak{N}_2 = \int_c^a \omega r \rho_2(r)\, dr = \frac{RT'\rho_2(0)}{m_2\omega}\left[\exp\left(\frac{m_2\omega^2 a^2}{2RT'}\right) - \exp\left(\frac{m_2\omega^2 c^2}{2RT'}\right)\right] \tag{14.269}$$

Let

$$\frac{m_2\omega^2 a^2}{2RT'} \equiv A^2 \tag{14.270}$$

as in Sec. 5.5. In the low-enrichment case, when $\mathfrak{M}_1 \ll \mathfrak{M}_2$ and $\mathfrak{N}_1 \ll \mathfrak{N}_2$, the cut θ is

$$\theta \equiv \frac{\mathfrak{M}_2}{\mathfrak{M}_2 + \mathfrak{N}_2} = \frac{\exp(A^2 c^2/a^2) - 1}{\exp A^2 - 1} \tag{14.271}$$

Hence, the fraction of the total flow area used for the light fraction to provide a cut of θ is

$$\frac{c^2}{a^2} = 1 + \frac{1}{A^2}\ln\left[\theta + (1-\theta)\exp(-A^2)\right] \tag{14.272}$$

The fraction of the flow area used for the light fraction has a lower limit of θ when the speed is low ($A \to 0$) and approaches unity as the speed increases ($A \to \infty$), as in the countercurrent centrifuge, because all flow is compressed against the outer wall.

The separation factor α is

$$\alpha \equiv \frac{\mathfrak{N}_2\mathfrak{M}_1}{\mathfrak{M}_2\mathfrak{N}_1} = \frac{1-\theta}{\theta}\frac{\mathfrak{M}_1}{\mathfrak{N}_1} = \frac{1-\theta}{\theta}\,\frac{\exp(A^2c^2/\alpha_0^2 a^2) - 1}{\exp(A^2/\alpha_0^2) - \exp(A^2c^2/\alpha_0^2 a^2)} \tag{14.273}$$

where

$$\alpha_0^2 \equiv \frac{m_2}{m_1} \tag{14.274}$$

This notation is used to facilitate comparison with gaseous diffusion, for which the ideal separation factor is

$$\alpha_0 = \sqrt{\frac{m_2}{m_1}} \tag{14.275}$$

With c^2/a^2 from (14.272),

$$\alpha = \frac{1-\theta}{\theta}\left\{\frac{[\theta + (1-\theta)\exp(-A^2)]^{1/\alpha_0} - \exp(-A^2/\alpha_0^2)}{1 - [\theta + (1-\theta)\exp(-A^2)]^{1/\alpha_0^2}}\right\} \tag{14.276}$$

At low speed ($A \to 0$), α approaches unity. At high speed,

$$\lim_{A \to \infty} \alpha = \frac{1-\theta}{\theta} \frac{\theta^{1/\alpha_0^2}}{1-\theta^{1/\alpha_0^2}} \tag{14.277}$$

When $\alpha_0 - 1 \ll 1$, as in uranium isotope separation,

$$\lim_{A \to \infty} (\alpha - 1) = -\frac{2(\alpha_0 - 1) \ln \theta}{1-\theta} \tag{14.278}$$

The corresponding expression for a cross-flow gaseous diffusion stage, from Eqs. (14.92) and (14.93), is

$$(\alpha - 1)_{\text{diff}} = -\frac{(\alpha_0 - 1)E_M E_B \ln(1-\theta)}{\theta} \tag{14.279}$$

Hence, in the nozzle process at high speed, the separation factor at cut θ is $2/E_M E_B$ times as great as in gaseous diffusion at cut $1 - \theta$.

In Fig. 14.22 the curves of separation factor versus cut for centrifugal equilibrium were calculated from Eq. (14.276) for $^{235}UF_6$ ($m_1 = 349$) and $^{238}UF_6$ ($m_2 = 352$). $\alpha_0^2 = 1.008596$.

The temperature T' and peripheral speed $\omega a \equiv v$ occurring in the definition of A^2, Eq. (14.270), are for the mixture of UF_6 and hydrogen *after* expansion to speed v. The nozzle process ordinarily is operated at a known constant temperature T *before* expansion. T, T', and v are related by the enthalpy balance

$$C_p(T - T') = \frac{mv^2}{2} \tag{14.280}$$

where C_p is the molar specific heat at constant pressure and m is the molecular weight. At $T =$ 313 K, assumed [G1] as the temperature at which the mixture of UF_6 and H_2 enters the nozzle separator,

$$C_p(H_2) = 6.874 \text{ cal/(g-mol·K) [P2]}$$

$$C_p(UF_6) = 31.3 \text{ cal/(g-mol·K) [D6]}$$

and

$$C_p(\text{mixture}) = 6.874(1-f) + 31.3f \text{ cal/(g-mol·K)} \tag{14.281}$$

where f is the mole fraction of UF_6. In dealing with gas expansion processes, it is conventional to use the heat capacity ratio

$$\gamma \equiv \frac{C_p}{C_v} = \frac{1}{1 - R/C_p} \tag{14.282}$$

The molecular weight m of a mixture of UF_6 and hydrogen is

$$m = 2.016(1-f) + 352.02f \tag{14.283}$$

On the assumption that $v = \omega a$, from Eqs. (14.270), (14.280), and (14.282),

$$v^2 = \frac{2RT\,A^2}{352 + mA^2(\gamma - 1)/\gamma} \tag{14.284}$$

Values of v calculated from Eq. (14.284) for the values of A^2 shown in Fig. 14.22 are tabulated at the bottom of Fig. 14.27 for several mole fractions of UF_6, f. The equilibrium separation factor increases rapidly between 100 and 250 m/s and approaches a limiting value above 300 m/s. At the higher speeds, the separation factor increases substantially as the cut is reduced.

The separative capacity in kg SWU/kg uranium fed to a nozzle stage is

$$\frac{\Delta}{Z} = \frac{\theta(1-\theta)(\alpha-1)^2}{2} \tag{14.285}$$

from Eqs. (12.169) and (12.172). Figure 14.27 shows the dependence of this separative capacity on cut for the peripheral speeds used in Fig. 14.22. The important point to note is that the cut at which separative capacity is highest for a given speed shifts from $\theta = \frac{1}{2}$ at low speed to $\theta = \frac{1}{5}$ at the highest speeds. Because both the light and heavy fractions have to be recompressed in this version of the nozzle process, the cut at which the separative capacity is highest is the cut at which power consumption is lowest for a given speed.

Power requirement. In the separation nozzle elements shown in Figs. 14.24 and 14.25, the kinetic energy of the expanded gas is dissipated after separation. Then the minimum net power to recompress the gases leaving the separator at pressure p' to the feed pressure p is

$$Q = \frac{ZRT_0 \ln (p/p')}{238\, f} \tag{14.286}$$

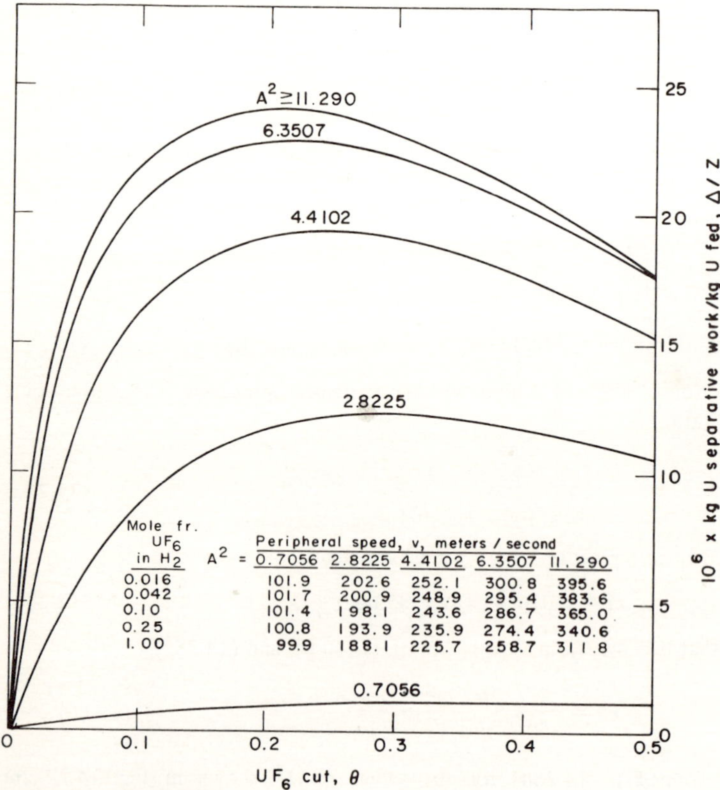

Figure 14.27 Variation of separative capacity with peripheral speed and cut for centrifugal equilibrium at constant angular velocity.

where Z = feed rate in kg uranium/yr
$R = 2.6365 \times 10^{-7}$ (kW·yr)/(kg-mol·K)
$T_0 = 300$ K

$$Q(\text{kW}) = \frac{3.3233 \times 10^{-7} Z \ln (p/p')}{f} \tag{14.287}$$

Under the most favorable possible conditions of reversible, adiabatic expansion through the nozzle,

$$\frac{p}{p'} = \left(\frac{T}{T'}\right)^{\gamma/(\gamma - 1)} \tag{14.288}$$

From (14.280), (14.282), and (14.284),

$$\frac{T}{T'} = 1 + \frac{mA^2(\gamma - 1)}{352\gamma} \tag{14.289}$$

Hence
$$Q(\text{kW}) = \frac{3.3233 \times 10^{-7} Z\gamma \ln [1 + mA^2(\gamma - 1)/352\gamma]}{f(\gamma - 1)} \tag{14.290}$$

The minimum power consumption per unit separative capacity is obtained from (14.290) and (14.285):

$$\frac{Q}{\Delta} \text{ kW/(kg SWU/yr)} = \frac{6.6466 \times 10^{-7} \gamma \ln [1 + mA^2 (\gamma - 1)/352\gamma]}{f(\gamma - 1)\, \theta(1 - \theta)(\alpha - 1)^2} \tag{14.291}$$

The dependence of α on A^2 and θ is given by (14.276).

For every feed composition f and cut θ, there will be an optimum value of A^2, because the numerator of (14.291) increases continuously with A^2, whereas the denominator approaches a limit. As a practical matter, values of A^2 are limited to those corresponding to the speed of sound because expansion through the curved nozzle becomes very irreversible at higher speeds. Because the sonic speed is

$$v_s^2 = \frac{RT_1\gamma}{m} = \frac{2RT\gamma}{m(1 + \gamma)} \tag{14.292}$$

$$A_s^2 = \frac{352 v_s^2}{2RT_1} = \frac{176\gamma}{m} \tag{14.293}$$

and
$$\left(\frac{Q}{\Delta}\right)_s = \frac{6.6466 \times 10^{-7} \gamma \ln [(1 + \gamma)/2]}{f(\gamma - 1)\, \theta(1 - \theta)(\alpha_s - 1)^2} \tag{14.294}$$

The lower curve of Fig. 14.28 is a plot of $(Q/\Delta)_s$ versus mole fraction UF_6 in feed, f, for $\theta = \frac{1}{4}$. The minimum value of $(Q/\Delta)_s$ is 0.072 kW/(kg SWU/year) at a feed composition of 0.18 mole fraction UF_6. This is to be contrasted with the optimum value of 0.31 kW/(kg SWU/year) reported by Geppert et al. [G1] for experiments with a feed composition of 0.04 mole fraction UF_6, and a design value of 0.287 kW/(kg SWU/year) for a commercial plant with a feed composition of 0.042 mole fraction (Table 14.18).

Part of the lack of agreement can be explained by the fact that flow in the curved groove in which separation takes place is quite different from the wheel flow assumed in the foregoing derivation. Instead of v at the wall ($r = a$) being a maximum as assumed, v actually drops to zero there because of wall friction. Also, flow through the curved nozzle cannot be perfectly reversible, so that the speed of the mixture after expansion will be lower than calculated for reversible expansion through a given pressure ratio. Justification for the choice of a feed

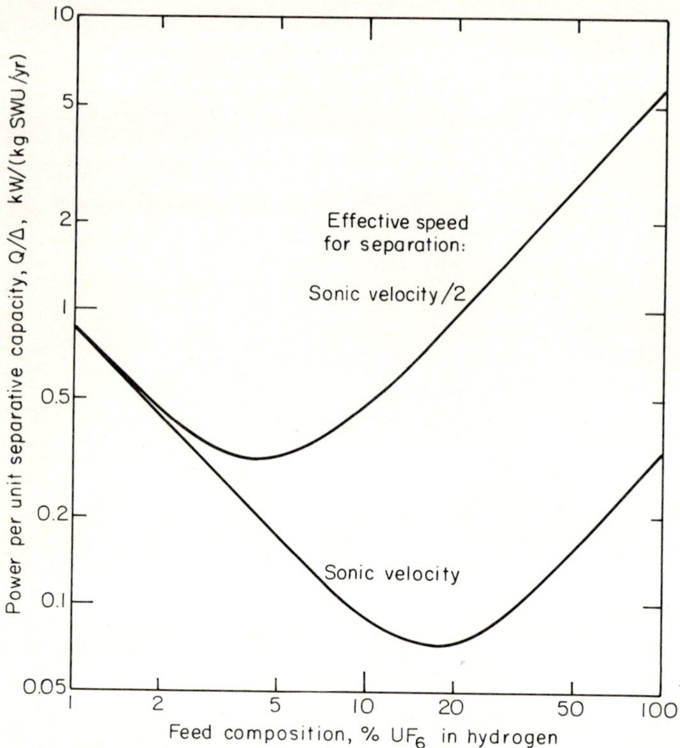

Figure 14.28 Power per unit separative capacity for nozzle process with UF_6-hydrogen mixtures expanded through critical pressure ratio. Cut $= \frac{1}{4}$.

composition of 0.042 fraction UF_6 and agreement with Geppert's reported Q/Δ of 0.31 kW/(kg SWU/year) can be obtained by assuming that the effective peripheral speed v of the gas after expansion through the pressure ratio corresponding to sonic speed is one-half the sonic speed. The upper curve of Fig. 14.28 was calculated for this condition. The minimum value of 0.308 at a feed composition of 0.042 mole fraction UF_6 in hydrogen is close to the values cited by Geppert [G1].

6.3 The South African UCOR Process

History. The UCOR process, developed by the Uranium Enrichment Corporation of South Africa, has been operated on a large pilot-plant scale at Valindaba, Union of South Africa. Partial information on the process, its separation factor and specific power demand, and its projected economics was given by Roux and Grant [R3]. The ingenious Helikon cascade technique developed for this process, in which a single axial-flow compressor handles several process streams simultaneously, was described by Grant et al. [G2] and analyzed theoretically by Haarhoff [H1]. Cost estimates, prepared in 1974 and converted to dollars with the purchasing power of that year, predicted that the capital cost of the 5000 MT/year plant would be $1,350 million, and that the cost of separative work from it, using electricity priced at 6 mills/kWh, would be $74/kg SWU. This cost was close to the price then charged by the U.S. Atomic Energy Commission.

The UCOR project is a major effort. In 1975, some 1200 persons were employed, and $150 million had already been spent on development. Extensive experiments had confirmed the separation performance and power consumption of individual stages. A "prototype module" with design separative capacity of 6000 kg SWU/year had been built and tested. The design of a full-scale prototype, expected to have a capacity of 50,000 kg SWU/year, was well advanced. On February 14, 1978, S. P. Botha, South African Minister for Mines and Industry, announced [B18] that South Africa would expand the pilot enrichment plant to meet domestic needs, but had abandoned plans to build a full-scale plant.

Description of process. Because many features of the process, including details of the separating element, have not been disclosed, this description is necessarily incomplete. The following partial description has been given by Roux and Grant [R3]:

> The South African–or UCOR–process is of an aerodynamic type. It has been possible to develop a separating element which in effect is a high performance stationary-walled centrifuge using UF_6 in hydrogen as process fluid. All process pressures throughout the system will be comfortably above atmospheric and depending on the type of "centrifuge" used, the maximum process pressure will be in a range of up to 600 kPa (6 bar). The UF_6 partial pressure will however be sufficiently low to eliminate the need for process heating during plant operation, and the maximum temperature at the compressor delivery will not exceed 75°C.
>
> The process is characterised by a high separation factor over the element, namely from 1.025 to 1.030 depending on economic considerations. Furthermore it has a high degree of asymmetry with respect to the UF_6 flow in the enriched and depleted streams, which emerge at different pressures. The feed to enriched stream pressure ratio is typically 1.5 whereas the feed to depleted stream pressure ratio is typically only 1.12.
>
> To deal with the small UF_6 cut, a new cascade technique was developed, the so-called "helikon" technique, based on the principle that an axial flow compressor can simultaneously transmit several streams of different isotopic composition without there being significant mixing between them. The UCOR process must therefore be regarded as a combination of the separation element and this technique, which makes it possible to achieve the desired enrichment with a relatively small number of large separation units by fully utilising the high separation factor available. . . .
>
> The theoretical lower limit to the specific energy consumption of the separation element can be shown to be about 0.30 MWh/kg USW. The minimum figure we have been able to obtain with laboratory separating elements is about 1.80 MWh/kg USW, based on adiabatic compression and ignoring all system inefficiencies. Although we do not believe that the present energy consumption can, in the short term be drastically reduced, the discrepancy between the above figures illustrates that the UCOR process still has a large development potential.

In discussion following presentation of the above information, the actual power consumption of a complete UCOR plant, allowing for pressure drops, and other process inefficiencies, was given as 3.5 MWh/kg SWU, or 0.40 kW/(kg SWU/year). This is to be compared with 0.50 estimated by Geppert [G1] for a complete nozzle plant and 0.266 for the improved U.S. gaseous diffusion plants.

An additional important bit of process information, from Grant et al. [G2], is: "For the UCOR process, the cut is typically 0.045 to 0.055." Figure 14.29 is a flow sheet for one stage of the UCOR process on which the preceding information has been represented, with a particular cut of θ = 0.050. This cut requires use of a 19-up, 1-down cascade. The only important process variable not stated in published information is the UF_6 content of the mixture with hydrogen fed to the stage. As will be shown in the next section, a UF_6 feed composition of 0.032 mole fraction is consistent with the reported process information.

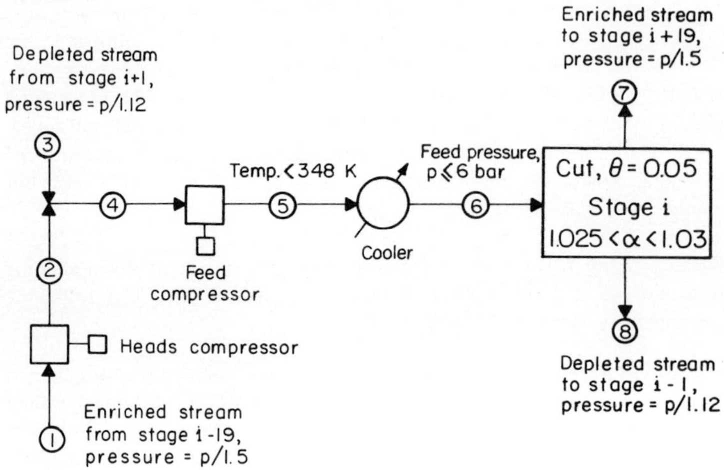

Figure 14.29 Stage conditions in UCOR process.

Cut: $0.045 < \theta < 0.055$
Separation factor: $1.025 < \alpha < 1.030$
Specific power: $Q/\Delta = 1.80$ MWh/kg SWU
Temperatures $< 75°$C

Theoretical analysis of UCOR process. Because the UCOR process has been characterized [R2] as a "stationary-wall centrifuge," its performance for $^{235}UF_6/^{238}UF_6$ separation can be represented by Eq. (14.276). The speed parameter A^2 is related to the given pressure ratio $p/p' = 1.5$ by

$$A^2 = \frac{352\,\gamma}{m(\gamma - 1)}\left[\left(\frac{p}{p'}\right)^{(\gamma - 1)/\gamma} - 1\right] \tag{14.295}$$

as may be seen from Eqs. (14.288) and (14.289).

In the UCOR process, unlike the separation nozzle process, the depleted stream is recompressed through a smaller pressure ratio (1.12) than the enriched stream (1.5). Hence, to evaluate the energy used in compression it is necessary to know the hydrogen cut θ_H, the fraction of hydrogen fed that leaves in the enriched stream, and the composition of the enriched stream represented by the mole fraction f_1 of UF_6 in it. A development analogous to the one that led to Eq. (14.271) for the UF_6 cut results in Eq. (14.296) for the hydrogen cut:

$$\theta_H = \frac{\exp\,(A_H^2 c^2/a^2) - 1}{\exp A_H^2 - 1} \tag{14.296}$$

where

$$A_H^2 = \frac{m_H v^2}{2RT'} = \frac{2}{352}A^2 = \frac{A^2}{176} \tag{14.297}$$

because the molecular weight of hydrogen is $m_H = 2$. Because the fraction of flow area used by the enriched stream, c^2/a^2, is given by (14.272), the hydrogen cut θ_H is related to the UF_6 cut θ by

$$\theta_H = \frac{[\theta + (1-\theta)\exp\,(-A^2)]^{1/176} - \exp\,(-A^2/176)}{1 - \exp\,(-A^2/176)} \tag{14.298}$$

The mole fraction UF_6 in the enriched stream 1, Fig. 14.29, is

$$f_1 = \frac{\theta f}{\theta f + \theta_H(1 - f)} \tag{14.299}$$

and the moles of UF_6 (M) plus hydrogen (M_H) in the enriched stream per mole of UF_6 fed ($M + N$) is

$$\frac{M + M_H}{M + N} = \theta + \theta_H \frac{1 - f}{f} \tag{14.300}$$

Similarly, the mole fraction UF_6 in the depleted stream, f_3, is

$$f_3 = \frac{(1 - \theta)f}{(1 - \theta)f + (1 - \theta_H)(1 - f)} \tag{14.301}$$

and the moles of UF_6 (N) plus hydrogen (N_H) in the heavy fraction per mole UF_6 fed is

$$\frac{N + N_H}{M + N} = 1 - \theta + \frac{(1 - \theta_H)(1 - f)}{f} \tag{14.302}$$

Equations for gas temperatures and compression energy are obtained by reference to Fig. 14.29. The temperature T_2 of the light fraction after adiabatic, reversible compression from point 1 to point 2 is

$$T_2 = T_1 \left(\frac{p_2}{p_1}\right)^{(\gamma_1 - 1)/\gamma_1} \tag{14.303}$$

The temperature of the mixed gases from points 2 and 3 entering feed compressor at point 4 is

$$T_4 = \frac{T_2(M + M_H)\ [\gamma_1/(\gamma_1 - 1)] + T_3(N + N_H)\ [\gamma_3/(\gamma_3 - 1)]}{(M + M_H + N + N_H)\ [\gamma_4/(\gamma_4 - 1)]} \tag{14.304}$$

where $\gamma/(\gamma - 1)$ is C_p/R for the stream designated by the subscript. The temperature of feed after compression is

$$T_5 = T_4 \left(\frac{p_5}{p_4}\right)^{(\gamma_4 - 1)/\gamma_4} \tag{14.305}$$

Temperatures at points 1, 3, 6, 7, and 8 are assumed to equal T, the temperature of feed to the separating element. Then, the power input from compression is

$$K = (M + M_H + N + N_H) \frac{\gamma R}{\gamma - 1} (T_5 - T) \tag{14.306}$$

because the heat capacity $\gamma R/(\gamma - 1) = C_p$ is a linear function of mole fraction, Eq. (14.281). The energy input in joules per kilogram of UF_6 fed, K/Z, is

$$\frac{K}{Z} = \frac{K}{238(M + N)} = \frac{\gamma R}{238(\gamma - 1)f} (T_5 - T) \tag{14.307}$$

with $R = 8314$ J/(kg-mol·K). Because the kilogram separative capacity of the stage is

$$\Delta = \frac{Z\theta(1 - \theta)(\alpha - 1)^2}{2} \tag{14.308}$$

the adiabatic, reversible energy input in joules per kilogram separative work is

$$\frac{K}{\Delta} = \frac{2\gamma R(T_5 - T)}{238(\gamma - 1)f\ \theta(1 - \theta)(\alpha - 1)^2} \tag{14.309}$$

From an assumed feed temperature $T = 313$ K, a UF_6 cut $\theta = 0.05$, and the stated expansion pressure ratios of 1.12 and 1.5 for the heavy and light fractions, a value for the mole fraction of UF_6 in feed of $f = 0.03225$ was found by trial to lead to the value of 1.80 MWh/kg SWU given by Roux and Grant [R2] for the energy per kilogram uranium separative work.

Table 14.19 summarizes the steps in the calculation of compositions, properties, and flow rates of the numbered streams in Fig. 14.29, and from them, the energy per kilogram of uranium fed, the separation factor, and the separative work.

The following should be noted:

1. The high hydrogen cut, 0.73, coupled with the low UF_6 cut, 0.05, causes the mole fraction UF_6 in the enriched stream, 0.0029, to be much lower than in the feed, 0.032, and the mole fraction UF_6 in the depleted stream, 0.105, to be much higher.
2. For every mole of UF_6 fed, 21.9 mol of enriched stream and 9.1 mol of depleted stream are processed.
3. The maximum calculated temperature, 340.35 K, provides margin below the 75°C (348 K) maximum temperature cited by Roux and Grant, to allow for process inefficiencies.
4. The heavy fraction containing 0.105 mole fraction UF_6 would start to condense at a pressure of 3.8 bar at 313 K. Hence the pressure of the heavy stream must be below this value and the feed pressure, p, must be below (1.12)(3.8) = 4.3 bar. This pressure is much higher than the subatmospheric pressures reported for the nozzle process and would result in much lower volumetric flow rates in a UCOR plant than in a nozzle plant of the same separative capacity.

Table 14.19 Steps in calculating separation performance of UCOR process

Variable	Symbol	Equation	Value
Mole fraction UF_6 in feed	f	Assumed	0.03225
Temperatures to stage	T_1 and T_3	Assumed	313 K
Molecular weight feed	m	(14.283)	13.3036
R/C_p of feed	$(\gamma - 1)/\gamma$	(14.281 & 2)	0.259363
Speed parameter	A^2	(14.295)	11.31261
UF_6 cut	θ	Given	0.05
Hydrogen cut	θ_H	(14.298)	0.728919
Mole fraction UF_6 in enriched stream	f_1	(14.299)	0.0022807
R/C_p of enriched stream	$(\gamma_1 - 1)/\gamma_1$	(14.281 & 2)	0.286761
Mole fraction UF_6 in depleted stream	f_3	(14.301)	0.104573
R/C_p of depleted stream	$(\gamma_3 - 1)/\gamma_3$	(14.281 & 2)	0.210766
Moles enriched stream†	$(M + M_H)/(M + N)$	(14.300)	21.9232
Moles depleted stream†	$(N + N_H)/(M + N)$	(14.302)	9.0845
Compression ratio, heads compressor	p_1/p_1	Given	1.5/1.12
Temperature from heads compressor	T_2	(14.303)	340.3507 K
Temperature to feed compressor	T_4	(14.304)	330.4904 K
Compression ratio, feed compressor	p_5/p_4	Given	1.12
Temperature from feed compressor	T_5	(14.305)	340.3488 K
Energy, MWh/kg U fed	$K/3.6 \times 10^9 Z$	(14.307)	3.1728E-5
Separation factor	α	(14.276)	1.027239
kg U separative work/kg U fed	Δ/Z	(14.308)	1.7622E-5
MWh/kg SWU	$K/3.6 \times 10^9 \Delta$	(14.309)	1.8005

†Per mole UF_6 fed.

5. The calculated separation factor of 1.0272 is in the range 1.025 to 1.03 cited by Roux and Grant and is higher than optimum in the nozzle process.
6. The value of $A^2 = 11.31$ calculated for wheel flow is sufficiently high that even if the effective gas speed were well below that corresponding to the stated expansion ratio of 1.5, the separation factor would not be much below the calculated 1.027 value.
7. The specific power of 1.80 MWh/kg SWU, with no allowance for process inefficiencies, is equivalent to 0.205 kW/(kg SWU/year). This may be compared with 0.168 for gaseous diffusion (Table 14.9), and 0.31 for the nozzle process (Fig. 14.23). The higher value for the nozzle process may be due to its expanding the heavy stream through the full pressure ratio.

UCOR process equipment. The low cut, $\theta = 0.045$ to 0.055, selected for the UCOR process requires use of more stages than the gaseous diffusion or nozzle process, despite the higher UCOR separation factor. To reduce the number of independent items of process equipment, the UCOR process uses the ingenious Hilikon technique to consolidate as many as 20 stages in a single independently operable unit. Figures 14.30, 14.31, and 14.32, adapted from UCOR publications [G2, H1], provide a partial description of the Helikon principle and the process equipment used in it.

Each Helikon module uses two axial-flow compressors, one for the enriched streams (point 1, Fig. 14.29) and a second for the feed streams (point 4). The nature of flow through this type of compressor is such that there is rather little mixing of material fed into the barrel at one angular position with material of another composition fed at another angular position. Such streams of different composition flow through the compressor in helical paths and leave the compressor still relatively unmixed.

Figure 14.30 shows how the inlet end of the compressor would be divided into sectors to handle the streams fed to three stages with ^{235}U fractions increasing in the order $z_1 < z_2 < z_3$. Each feed stream is divided into two halves which are introduced symmetrically about plane AA through the axis into sectors formed by radial partitions. In this way, composition differences between adjacent streams are minimized. The partitions stop at the inlet rotor blades and begin again after the outlet blades. To deal with possible helical displacement during compression, the

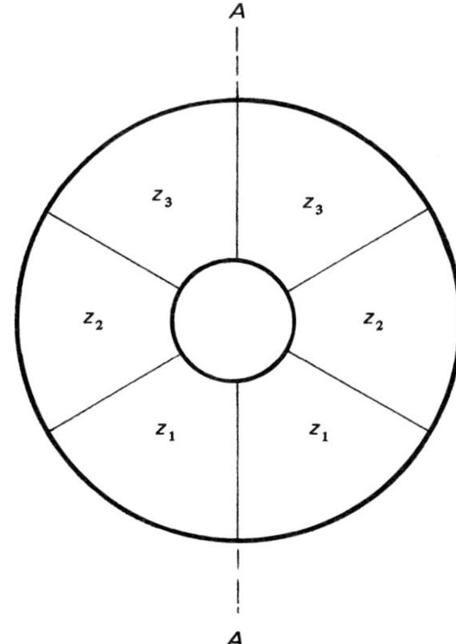

Figure 14.30 Introduction of three streams of different ^{235}U content $z_1 < z_2 < z_3$ into axial-flow compressor.

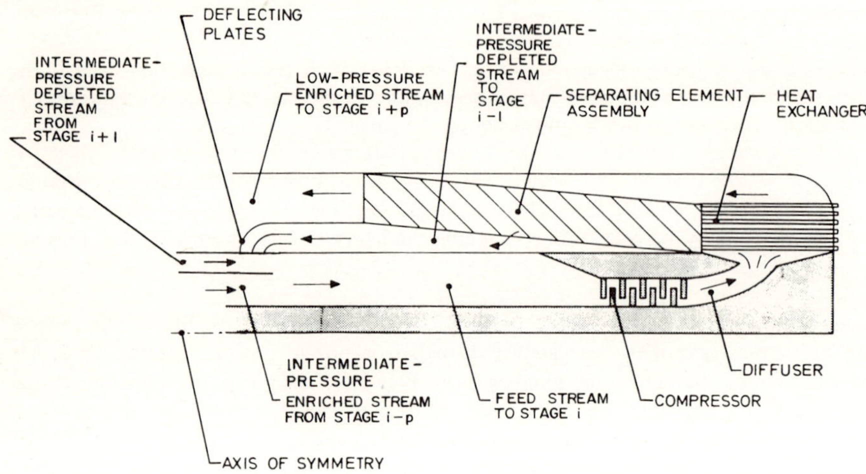

Figure 14.31 Schematic representation of flow through stage i of p-up, one-down Helikon module. *(Reproduced with permission of the copyright holder, American Institute of Chemical Engineers, and Dr. W. L. Grant.)*

outlet partitions may be displaced through an appropriate angle. In the UCOR plant with a cut of $\frac{1}{20}$, 38 (2 × 19) sectors would be used.

The flow path through one sector of a Helikon module, containing all equipment of stage i except the light-stream compressor, is shown in Fig. 14.31. Depleted stream from stage $i + 1$ and enriched stream from stage $i - p$, both at intermediate pressure, are mixed and fed into one sector at the compressor inlet. At the compressor discharge the compressed feed is

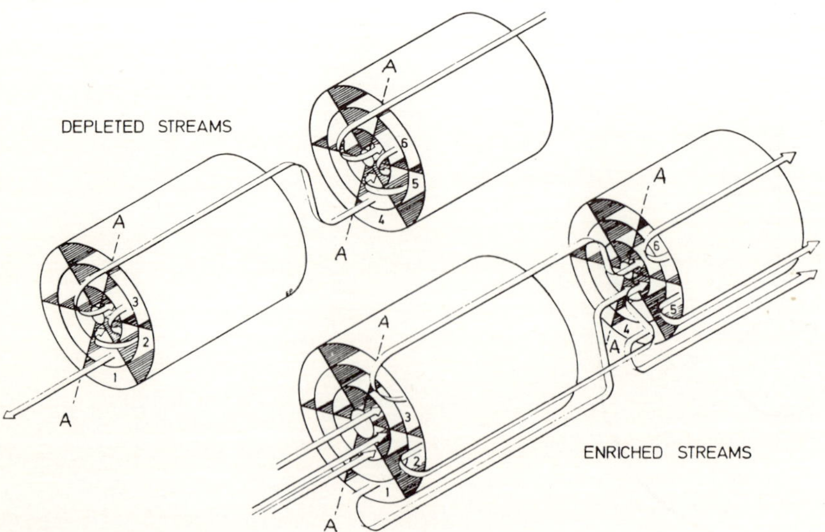

Figure 14.32 Flow between modules of three-up, one-down Helikon cascade. *(Reproduced with permission of the copyright holder, American Institute of Chemical Engineers, and Dr. W. L. Grant.)*

collected in the appropriate sector, passed first through a stage cooler, and then through the separating element where it is divided into the low-pressure enriched stream and the intermediate-pressure depleted stream.

The enriched stream from each sector is transported to the enriched stream compressor for stage $i + p$ in the module handling the next higher enrichment. The depleted stream is rotated by deflecting plates into the feed stream of stage $i - 1$ of the same module, or if from the least enriched stage, is sent to the highest stage of the module handling the next lower enrichment.

To illustrate the Helikon principle, flow between two adjacent modules of a three-up, one-down Helikon cascade is shown schematically in Fig. 14.32. The upper half shows the flow of depleted streams from one stage to the next lower stage; the lower half shows the flow of enriched streams from a sector of one module to the corresponding sector of the module of next higher enrichment. Because the figures are symmetric about the plane AA, the other half of the flow paths are not shown.

To permit construction of a complete plant with one size, or at most a few sizes, of compressor, while providing the variation in stage throughput desirable in an ideal cascade, it is proposed that the number of sectors in a module be varied to provide a smaller number of large sectors near the feed point and a larger number of small sectors toward the product and tails ends of the cascade.

Experiments reported by Grant et al. [G2] have shown that mixing of streams of different composition in an axial flow compressor can be kept acceptably low.

The number of stages needed for a given overall enrichment is inversely proportional to $\theta(\alpha - 1)$. Because of its low cut the UCOR process needs more stages than the separation nozzle or gaseous diffusion process, despite its higher separation factor. This potential disadvantage is dealt with by the Helikon technique, which combines a number of stages into a single module. Table 14.20 compares the gaseous diffusion process design of Table 14.9, the improved separation nozzle process of Table 14.18, and the UCOR process of Table 14.19 with respect to cut, separation factor, number of stages in an ideal cascade producing product containing 3 percent ^{235}U and tails containing 0.25 percent ^{235}U, and the number of modules for such a UCOR plant cited by Grant et al. [G2].

7 MASS DIFFUSION

In mass diffusion, separation of isotopes occurs through diffusion of the light isotope of a gas mixture into a condensible vapor at higher rate than diffusion of the heavy isotope. Mass diffusion separation has been carried out in a cascade of individual mass diffusion stages and in a mass diffusion column.

Table 14.20 Comparison of gaseous diffusion, nozzle, and UCOR processes

	Process		
	Gaseous diffusion	Separation nozzle	UCOR
Cut θ	$\frac{1}{2}$	$\frac{1}{4}$	$\frac{1}{20}$
Separation factor α	1.0030	1.0148	1.0272
Number of stages	1675	679	1848
Number of modules			~ 100

7.1 Mass Diffusion Stage

The stage type of mass diffusion was patented by Hertz [H4], who used this method to separate the isotopes of neon [H5, H6]. The means by which separation is effected in a mass diffusion stage are shown in Fig. 14.33, which illustrates the type of equipment used by Maier [M1] to separate hydrogen from other gases.

The heart of this apparatus is the mass diffusion stage, in this case of cylindrical cross section, which is divided into two annular chambers by the mass diffusion screen. Feed gas is brought to the top of the inner compartment by a riser. As this stream flows down through the inner chamber, it gives up a portion of the feed, which diffuses through the screen into the outer chamber against the inward-diffusing separating agent. Because the light component of the feed diffuses at a higher speed than the heavy, the stream in the inner chamber is progressively depleted in the light component relative to the heavy.

Steam or other separating-agent vapor is admitted at the bottom of the outer chamber. As this stream flows upward it gives up separating agent to the heavy stream and picks up from it a portion of the feed, enriched in the light component.

After the light and heavy streams leave the diffusion stage, they are cooled to condense the separating agent. After separation of the condensate, they leave the apparatus as the light and heavy fractions.

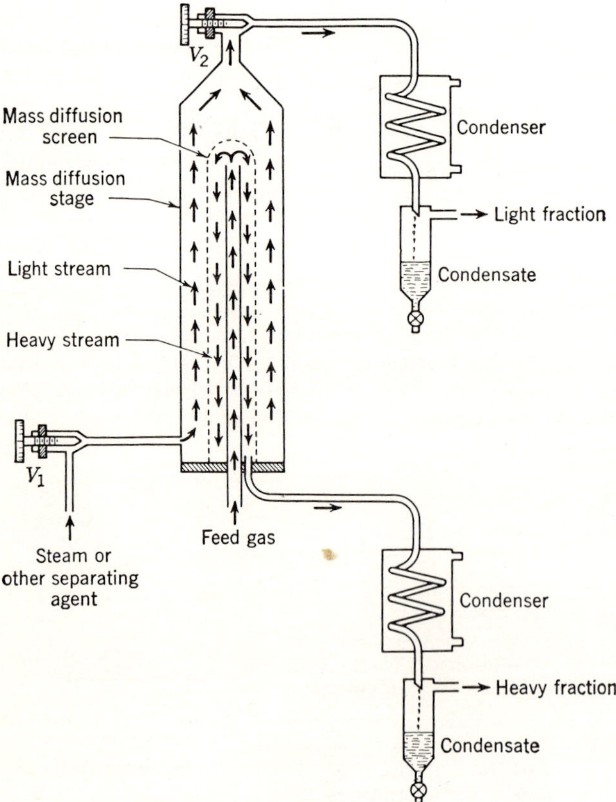

Figure 14.33 Flow in mass diffusion stage.

Table 14.21. Isotopes separated by cascade of mass diffusion stages

Working substance	Rare isotope concentrated	m/o Feed	m/o Product	Investigator	Year
Neon	^{22}Ne	9.7	50	Hertz	1934
Argon	^{36}A	0.23	50	Kopferman and Kruger	1937
Nitrogen	^{15}N	1.9	20	Kruger	1938
Methane	^{13}C	1.06	50	Capron and Hemptinne	1939

For the mass diffusion screen, Maier used a variety of materials, such as plates perforated with 0.4-mm holes, fine-mesh wire screen, or alundum filter plates. Very fine holes, such as is needed in gaseous diffusion, are not required, although holes with diameter under 10 μm are preferred because control of mass flow through the screen is easier. In the uranium isotope separation design example to be given in Sec. 7.4, electroformed nickel screen with holes 6.76 μm in diameter and 30 percent free area was specified.

The main requirements of the separating agent are that it be selective, be readily separable from the components of the mixture to be separated, and be chemically inert to it. For isotopic mixtures in the form of a permanent gas, such as neon or methane, a readily condensible vapor such as steam or mercury has been used. For UF_6 feed neither of these can be used because of chemical reactivity, and fluorocarbon vapor is specified. Selectivity is enhanced by using a separating agent of appreciably higher molecular weight than the components to be separated.

Table 14.21 gives examples of isotope separation reported for this type of apparatus.

7.2 Mass Diffusion Column

Because the separation obtainable in a mass diffusion stage is even smaller than in a gaseous diffusion stage, a practical degree of separation requires either a multistage cascade, such as the 48-stage cascade used by Hertz [H3] to separate neon isotopes, or a mass diffusion column.

The mass diffusion column, devised by Benedict [B13, B15] to provide a greater degree of separation than attainable from a single mass diffusion stage, is shown in Fig. 14.34. The main differences from a mass diffusion stage are as follows: (1) Separating agent is charged to the light stream at a uniform rate over the entire length of the column, instead of at one end of a stage. (2) Separating agent is condensed at a uniform rate over the entire length of the column instead of just from the streams leaving the column. To permit uniform charging and condensation of separating agent, this mass diffusion column contains four compartments instead of the two used in the stage type. These may take the form of cylindrical shells (Fig. 14.34) or parallel ducts.

In Fig. 14.34 the innermost chamber carries separating agent and distributes it at uniform rate over the length of the column; the second chamber carries the light stream; the third chamber, the heavy stream; and the fourth chamber, cooling water. Separating agent vapor flows radially through holes in the central tube and diffuses through the chambers carrying light and heavy streams to the cooling surface, where it is condensed.

The mass diffusion screen divides the second chamber, carrying the light stream, from the third chamber, carrying the heavy. As the light stream flows up, it is progressively enriched in the light isotope, which diffuses through the screen against the separating agent. As the heavy stream flows down, it is progressively enriched in the heavy isotope, which is carried through the screen with the separating agent.

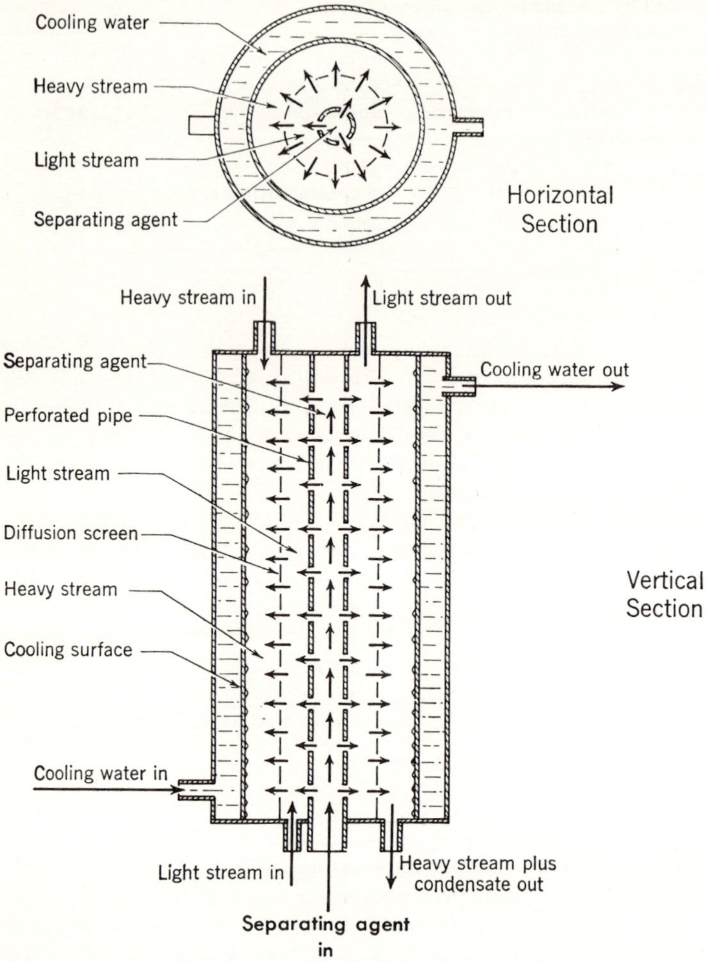

Figure 14.34 Flow in mass diffusion column.

By proper control of flow rates of separating agent and light- and heavy-stream feed rates, it is possible to make the molar velocity of light component inward just equal to the molar velocity of heavy component outward, a preferred condition for isotope separation in this equipment.

7.3 Sweep Diffusion

Sweep diffusion is a form of mass diffusion column in which the screen separating the counterflowing light and heavy streams is not present. Cichelli et al. [C4] developed the theory of such a column and used it to separate hydrogen from natural gas and to enrich air. Table 14.22 gives examples of partial separation of isotopes by this process.

Because it has no screen, the sweep diffusion column is simpler to construct and has a lower transfer-unit height than the mass diffusion column. A disadvantage is the greater difficulty of maintaining undisturbed counterflow over a long column.

Both the sweep and mass diffusion columns have two advantages over the stage process. The columns are more efficient because it is possible to maintain optimum conditions over the entire screen area. In the stage, this is not possible because separating agent concentrations change from point to point. The other main advantage of the column is its ability to achieve the equivalent of many stages of separation in a single piece of equipment.

One disadvantage of the column type is its more complex construction compared with the stage, which makes scale-up more difficult. The column type has high mixing losses when the gas mixture to be separated is appreciably soluble in condensed separating agent. This is not a problem when separating isotopes of permanent gases such as nitrogen or argon, but it practically precludes use of the column type for UF_6, because UF_6 is readily soluble in all known separating agents that do not react with it chemically.

7.4 Separation of Uranium Isotopes by Cascade of Mass Diffusion Stages

Process description. Mass diffusion has one potential advantage over gaseous diffusion for separation of uranium isotopes, in that a special diffusion membrane with ultrafine holes is not required. Also, the energy input needed to effect separation can be provided by conventional refrigerant compressors rather than special compressors for UF_6. To evaluate mass diffusion for uranium isotope separation, Forsberg [F2] carried out a detailed design and cost estimate of separation of uranium isotopes in a cascade of mass diffusion stages. Perfluorotributylamine, $N(C_4F_9)_3$, trade name N_{43}, was specified as separating agent. This compound was chosen as having the highest molecular weight of any commercially available substance that was thermally stable and not reactive with UF_6 in the vapor phase at a pressure of 1200 Torr. This operating pressure was chosen as being sufficiently above the triple-point pressure of UF_6, 1142 Torr, to prevent freezing during separation of N_{43} from UF_6 by partial condensation at this pressure. The stage temperature was 194°C, slightly above the condensation temperature of any process mixture of N_{43} and UF_6 at this pressure.

The mass diffusion screen used in the study was the electroformed nickel sheet referred to earlier, 3 μm thick, with holes 6.76 μm in diameter through 30 percent of its surface.

Because this method of separating uranium isotopes was found not to be economically competitive with gaseous diffusion, only a summary of Forsberg's principal results will be given here.

Figure 14.35 shows the flow through a mass diffusion stage and the nomenclature to be used in characterizing its performance. Figure 14.36 shows how three stages are connected into a simple, ideal cascade and the location of equipment used to separate the mixture of UF_6 and N_{43} leaving each stage into a UF_6-rich stream and an N_{43}-rich stream.

Physical separation. Figure 14.37 is a schematic flow sheet for the system used for the physical separation stage that prepares feed for each mass diffusion stage. The mixture of UF_6 and N_{43}

Table 14.22. Isotopes separated by sweep diffusion

Working substance	Rare isotope concentrated	Column length, cm	Sweep vapor	Overall separation	Reference
Hydrogen	HD	90	Methanol	1.39	[H8]
Nitrogen	^{14}N, ^{15}N	90	Methanol	1.04	[H8]
Neon	^{22}Ne	100	Xylene	6.50	[G6]
Neon	^{22}Ne	Three in series	Mercury	8.8 to 97%	[N4]
Argon	^{36}Ar	90	Methanol	1.17	[H8]

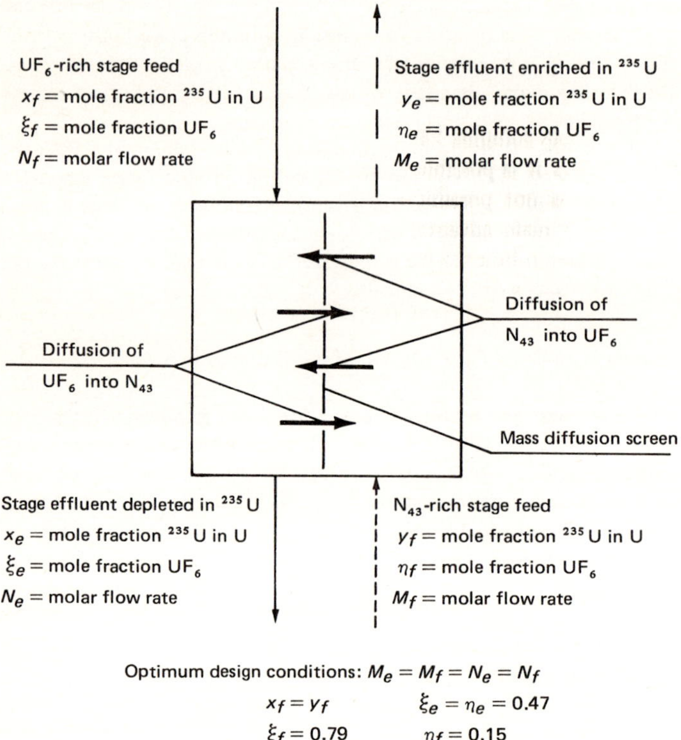

Figure 14.35 Nomenclature for mass diffusion stage.

to be fed to a stage, at the top center of Fig. 14.37, is cooled and partially condensed in the principal heat exchanger by separated UF_6-rich vapor and N_{43}-rich liquid. Final cooling to condense the required fraction of feed is by heat exchange against boiling liquid refrigerant-114 (R-114, $C_2Cl_2F_4$). The UF_6-rich vapor and N_{43}-rich liquid are reheated part way to process temperature in the principal heat exchanger. Heating of UF_6 and evaporation of N_{43} is completed by heat exchange against condensing R-114 vapor in the final heaters.

Energy to drive the process is provided by the R-114 vapor compressor, which pumps heat from the lower temperature of the final cooler to the higher temperature of the final heaters. Most of this energy is removed in the R-114 condenser as heat to cooling water. Because the heat-pump circuit operates between temperatures of 100 to 200°C, some energy is recovered in the power-recovery turbine.

Equations for separation performance of mass diffusion stage. At the bottom of Fig. 14.35, the stage conditions found to lead to minimum unit cost of separative work are tabulated. The condition that the UF_6 content of the two effluent streams be equal,

$$\xi_e = \eta_e \tag{14.310}$$

is set so that the enriched stream leaving stage $1 + i$ can be mixed with the depleted stream from stage $i - 1$ without loss of thermodynamic efficiency before these are separated into the UF_6-rich stream and the N_{43}-rich stream to be fed to stage i.

The four molar flow rates M_f, M_e, N_e, and N_f are equal for the following reasons. They are necessarily related by the overall material balance

$$M_f + N_f = M_e + N_e \tag{14.311}$$

The ideal cascade condition of a cut of $\frac{1}{2}$ requires that

$$N_e \xi_e = M_e \eta_e \tag{14.312}$$

Hence N_e and M_e are equal. Forsberg found that a condition of zero net flow through the screen led to minimum unit cost of separative work, so that

$$M_f = M_e \tag{14.313}$$

Hence N_f and N_e are equal.

The condition that the UF_6 mole fraction of the effluent streams be

$$\xi_e = 0.47 \tag{14.314}$$

is the result of a detailed optimization study [F2]. With a lower value (less UF_6), more N_{43} would have to be circulated, heated, and cooled, with higher processing costs. With a higher

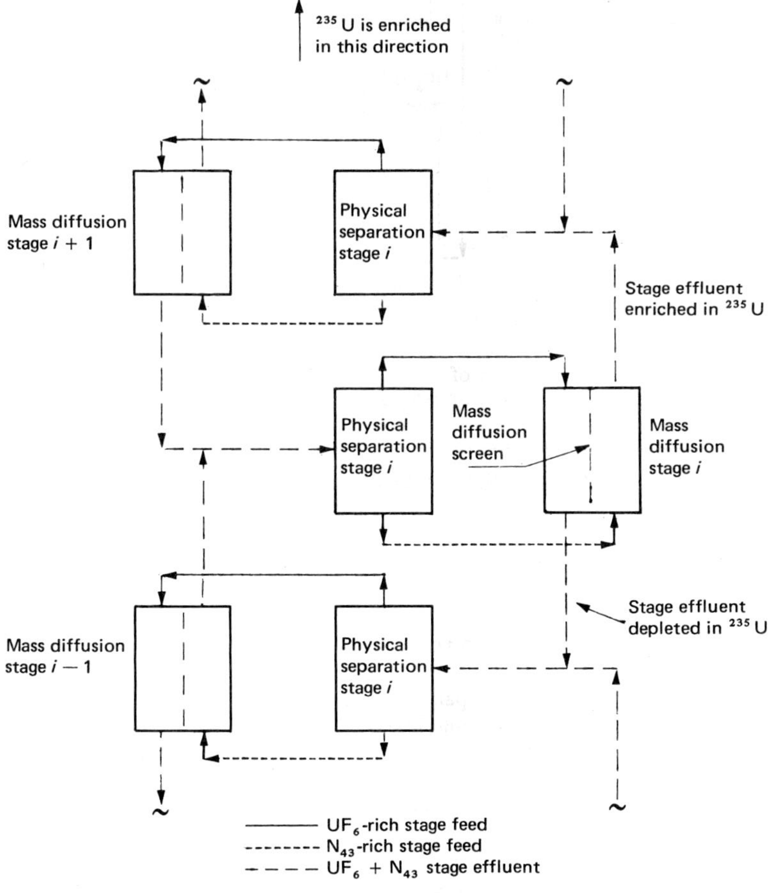

Figure 14.36 Cascade of mass diffusion stages.

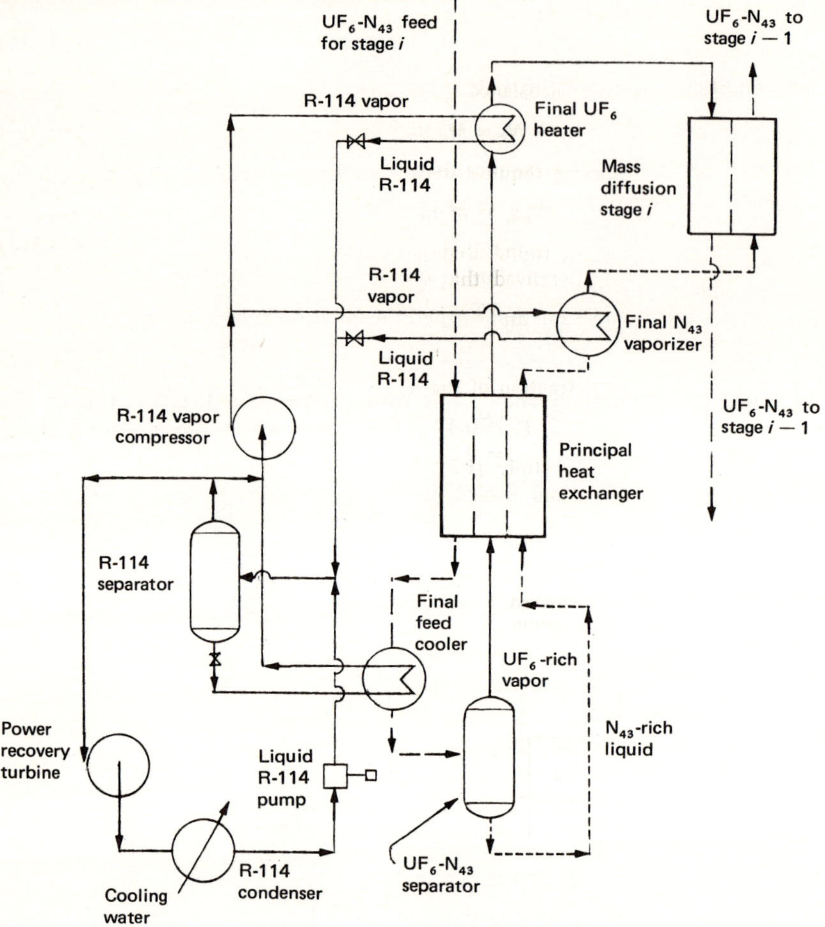

Figure 14.37 Flow sheet for physical separation of UF_6 and N_{43}.

value (less separating agent), the stage separation factor drops off rapidly, so that more stages and more UF_6 reflux would be required.

After the above conditions are set, the UF_6 content of the UF_6-rich stream

$$\xi_f = 0.79 \tag{14.315}$$

and of the N_{43}-rich stream

$$\eta_f = 0.15 \tag{14.316}$$

are equilibrium compositions that result from separation of the effluent stream containing 0.47 mole fraction UF_6 into equimolal amounts of liquid and vapor by partial condensation at 1200 Torr [F2].

The mole fractions of ^{235}U in UF_6 in the two stage feeds are equal,

$$x_f = y_f \tag{14.317}$$

because these two streams came from the same physical separation stage, in which no isotope separation occurs. Because of the ideal cascade requirement that the cut be $\frac{1}{2}$, the mole fractions of ^{235}U in the feed and effluent uranium are related by

$$x_f - x_e = y_e - y_f \tag{14.318}$$

Equations (14.310) through (14.318) provide nine independent relations among the 12 variables M_e, M_f, N_e, N_f, ξ_e, ξ_f, η_e, η_f, x_e, x_f, y_e, and y_f. For a given stage ^{235}U content (e.g., x_e), the two remaining relations needed to specify completely all 12 variables are the interstage UF_6 flow rate (e.g., $N_e\xi_e$) and the stage separation factor

$$\alpha \equiv \frac{y_e(1 - x_e)}{x_e(1 - y_e)} \tag{14.319}$$

The optimum interstage UF_6 flow rate is that of an ideal cascade. For example, in the enriching section of a close-separation ideal cascade making product of composition y_P at rate P, the UF_6 flow rate in stage tails is given by Eq. (12.132). In the notation of this section,

$$N_e\xi_e = \frac{2P(y_P - x_e)}{(\alpha - 1)x_e\,(1 - x_e)} \tag{14.320}$$

Forsberg [F2] has solved the differential equations for diffusion of $^{235}UF_6$, $^{238}UF_6$, and N_{43} through the holes of a mass diffusion screen and for material balances in countercurrent flow of the two streams in the mass diffusion stage of Fig. 14.35 to obtain Eq. (14.321) for the stage separation factor α. When there is no net flow through the screen,

$$\alpha - 1 = \frac{2(\xi_f - \xi_e)\,\{\exp\,[(\xi_e - \eta_f)(1 - D_0/D_{12})] - 1\}\;\gamma}{\xi_e(1 - D_0/D_{12})\,\{\xi_f - \eta_f \exp\,[(\xi_e - \eta_f)(1 - D_0/D_{12})]\}} \tag{14.321}$$

where γ, the *separability*, is

$$\gamma \equiv \frac{D_{01} - D_{02}}{D_{01} + D_{02}} \tag{14.322}$$

and

$$D_0 \equiv \frac{D_{01} + D_{02}}{2} \tag{14.323}$$

The three diffusion coefficients are

D_{01}, light component into separating agent
D_{02}, heavy component into separating agent
D_{12}, light component into heavy component

For an isotopic mixture with nearly equal molecular weights m_1 and m_2,

$$\gamma = \frac{m_1 - m_2}{m_1 + m_2}\,\frac{1}{1 + (m_1 + m_2)/2m_0} \tag{14.324}$$

Because

$$\frac{m_1 - m_2}{m_1 + m_2} = \alpha_0 - 1 \tag{14.325}$$

where α_0 is the ideal separation factor in gaseous diffusion, the separability in mass diffusion is necessarily smaller than $\alpha_0 - 1$ in gaseous diffusion by the factor $1/[1 + (m_1 + m_2)/2m_0]$. This factor is closer to unity, the larger the molecular weight of the separating agent m_0 is relative to the molecular weights of the isotopic compounds m_1 and m_2. This shows the advantage of using mercury or N_{43} as separating agent. For separating $^{235}UF_6$ from $^{238}UF_6$ with N_{43},

$$\gamma = \frac{352 - 349}{352 + 349} \frac{1}{1 + (352 + 349)/(2)(671)} = (0.00428)(0.657) = 0.00281 \quad (14.326)$$

Thus, in separating uranium isotopes by mass diffusion with N_{43}, the separability is less than two-thirds of $\alpha_0 - 1$ in gaseous diffusion.

For the ratio of diffusion coefficients in (14.321), for UF_6-N_{43} mixtures, Forsberg recommended

$$\frac{D_0}{D_{12}} = 0.5 \quad (14.327)$$

With the optimum values of 0.15, 0.47, and 0.79 given previously for η_f, ξ_e, and ξ_f,

$$\alpha - 1 = 0.770\gamma = 0.00216 \quad (14.328)$$

The coefficient of γ in Eqs. (14.321) and (14.328) is analogous to the stage efficiency E in gaseous diffusion, Eq. (14.94).

The rate of production of separative work in this mass diffusion stage, with a cut of $\frac{1}{2}$, is

$$\Delta = \frac{(M_e\eta_e + N_e\xi_e)(\alpha - 1)^2}{8} \quad (14.329)$$

For the optimum conditions of Fig. 14.35,

$$\Delta = \frac{0.47M_e(\alpha - 1)^2}{4} \quad (14.330)$$

The mixing of a UF_6-rich stream with a separating agent-rich stream as occurs in a mass diffusion stage results in an irreversible loss of availability analogous to the irreversible pressure drop in a gaseous diffusion stage. The rate of loss of availability in the adiabatic mass diffusion stage is

$$Q = T_0(S_{out} - S_{in}) \quad (14.331)$$

where S is the rate at which entropy is carried by the indicated streams. Treating UF_6-N_{43} mixtures as ideal solutions,

$$Q = RT_0 \{N_f[\xi_f \ln \xi_f + (1 - \xi_f) \ln (1 - \xi_f)] + M_f[\eta_f \ln \eta_f + (1 - \eta_f) \ln (1 - \eta_f)]$$
$$- N_e[\xi_e \ln \xi_e + (1 - \xi_e) \ln (1 - \xi_e)] - M_e[\eta_e \ln \eta_e + (1 - \eta_e) \ln (1 - \eta_e)]\} \quad (14.332)$$

For the optimum conditions of Fig. 14.35,

$$Q = RT_0M_e[0.79 \ln 0.79 + 0.21 \ln 0.21 + 0.15 \ln 0.15 + 0.85 \ln 0.85$$
$$- 2(0.47 \ln 0.47 + 0.53 \ln 0.53)] = 0.446\, RT_0M_e \quad (14.333)$$

The rate of loss of availability per unit separative capacity is the ratio of (14.333) to (14.330),

$$\frac{Q}{\Delta} = \frac{3.80RT_0}{(\alpha - 1)^2} \quad (14.334)$$

With $\alpha - 1$ from (14.328),

$$\frac{Q}{\Delta} = \frac{(3.80)[8.314\ (\text{kW}\cdot\text{s})/(\text{kg-mol}\cdot\text{K})(300\ \text{K})}{(3.154 \times 10^7\ \text{s/yr})(238\ \text{kg U/kg-mol})(0.00216)^2} = 0.271\ \text{kW/(kg SWU/yr)} \quad (14.335)$$

This is to be compared with 0.168 kW/(kg SWU/year) for the rate of loss of availability associated with pressure drop through the diffusion barrier of the optimized gaseous diffusion stage of Table 14.9.

In addition to this loss of availability resulting from mixing UF_6 and N_{43} in each mass diffusion stage of Fig. 14.36, flow of heat through temperature differences of the heat exchanger, cooler, vaporizer, and heater of each physical separation stage of Fig. 14.37 results in even greater availability losses. Forsberg's [F2] heat balances for an optimized plant predict an additional availability loss of 0.49 kW/(kg SWU/year) for a physical separation system consisting of two stages of partial condensation and evaporation, each with a minimum temperature difference of 5.5°C. With an allowance of 0.12 kW/(kg SWU/year) for other thermodynamic inefficiencies such as pressure drops, the total available energy consumption of an economically optimized mass diffusion plant was estimated to be 0.88 kW/(kg SWU/year). This is to be compared with 0.266 kW/(kg SWU/year) for the U.S. gaseous diffusion plants and 0.366 for the gaseous diffusion design of Table 14.9.

This poor energy utilization compared with gaseous diffusion is inherent in the mass diffusion process. Using a thermodynamic argument similar to Sec. 4.8 for gaseous diffusion, Forsberg showed that the minimum ratio of availability loss rate to rate of production of separative work at any point in a mass diffusion screen is

$$\left(\frac{Q}{\Delta}\right)_{\min} = \frac{4RT_0}{\gamma^2} \tag{14.336}$$

For the UF_6-N_{43} system, with $\gamma = 0.00281$,

$$\left(\frac{Q}{\Delta}\right)_{\min} = \frac{0.168 \text{ kW}}{(\text{kg SWU/yr})} \tag{14.337}$$

This is to be compared with 0.0722 for gaseous diffusion. The minimum value of 0.168 is a theoretical lower limit, attained only by letting the UF_6 content of the mixture with N_{43} approach zero, a condition that would require that the flow rate of N_{43}, the amount of mass diffusion screen, and the size of physical separation equipment increase without limit.

Table 14.23 summarizes the foregoing comparison of gaseous diffusion and mass diffusion for uranium isotope separation. A mass diffusion plant would need about 1.4 times as many stages as a gaseous diffusion plant performing the same separation, and would consume about three times as much available energy. For this reason, and because the mass diffusion plant needs the complex physical separation system besides its diffusion separation stages, Forsberg [F2] has estimated that the capital cost of such a mass diffusion plant would be almost five times that of a gaseous diffusion plant of the same capacity.

Table 14.23 Comparison of mass diffusion with gaseous diffusion for uranium isotope separation

	Process	
	Mass diffusion	Gaseous diffusion
Ideal separation factor	$\gamma = 0.00281$	$\alpha - 1 = 0.00429$
Separation factor in optimized plant	0.00216	0.00300
Power per unit separative capacity, kW/(kg SWU/yr)		
Theoretical minimum	0.168	0.0722
Optimized plant		
For separating elements	0.271	0.168
Complete plant	0.88	0.266

Despite this unfavorable conclusion for uranium isotopes, mass diffusion does appear to have favorable features for small-scale separation of isotopes of heavy elements such as argon, for which the column type of separator can be used.

8 THERMAL DIFFUSION

8.1 General Description

When heat flows through a mixture initially of uniform composition, small diffusion currents are set up, with one component transported in the direction of heat flow, and the other in the opposite direction. This is known as the thermal diffusion effect. The existence of thermal diffusion was predicted theoretically in 1911 by Enskog [E1, E2] from the kinetic theory of gases and confirmed experimentally by Chapman [C1, C2] in 1916. It is not surprising that the effect was not discovered sooner, because it is very small. For example, when a mixture of 50 percent hydrogen and 50 percent nitrogen is held in a temperature gradient between 260 and 10°C, the difference in composition at steady state is only 5 percent. In isotopic mixtures the effect is even smaller.

Thermal diffusion column. Thermal diffusion remained a scientific curiosity until 1938, when Clusius and Dickel [C5] developed their thermal diffusion column, which made possible useful separations in simple equipment. In the Clusius-Dickel column the mixture to be separated is confined in a long, vertical tube, cooled externally and heated internally by a hot wire at the axis of the tube. Other workers have used the annular type of equipment shown in Fig. 14.38. In both types, the mixture to be separated is confined in a narrow space between an inner heated and an outer cooled surface. The outward flow of heat sets up a small difference in isotopic composition through the thermal diffusion effect, with the light isotope usually concentrating in the inner zone at the higher temperature. At the same time, convection currents are set up, as indicated by the arrows, with the lighter heated fluid adjacent to the inner wall moving upward and the heavier cooled fluid adjacent to the outer wall moving downward. This counterflow multiplies the small composition difference obtained from the thermal diffusion effect and makes possible substantial degrees of separation in a practical length of column. For example, in a column 36 m long, Clusius and Dickel were able to separate the isotopes of chlorine, producing HCl containing 99.6 percent ^{35}Cl at one end of the column and HCl containing 99.4 percent ^{37}Cl at the other.

For most isotopes it is preferable to work with gases rather than liquids, because the higher diffusion coefficients result in higher separative capacity. The optimum pressure is usually near atmospheric. However, when ^{235}U was first found to be fissionable, Nier [N3] attempted to separate it by thermal diffusion of UF_6 vapor at low pressure without success, so that it was necessary to work with the liquid at high pressures [A1] to obtain useful separation. The optimum spacing between hot and cold surfaces is a few millimeters for gases and fraction of a millimeter for liquids.

The degree of separation obtainable in thermal diffusion (the difference in composition between hot and cold walls) is much less than in other diffusion processes, so that use of a column to multiply the composition difference is practically essential. The stage type of thermal diffusion has been used only to measure the thermal diffusion coefficient and is never used for practical separations. In some thermal diffusion columns, htu's are as low as 1.5 cm, and as many as 800 stages of separation have been obtained from a single column. Even with such a great increase in separation, it is often necessary to use a tapered cascade of thermal diffusion columns for isotopic mixtures, to minimize hold-up of partially enriched isotopes and to reduce equilibrium time.

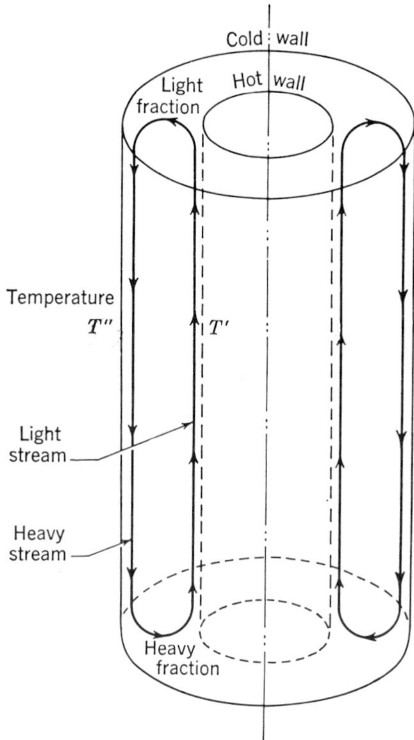

Figure 14.38 Thermal diffusion column.

Isotopes separated. Table 14.24 gives examples of some of the highest reported concentrations of separated isotopes that have been obtained by thermal diffusion. Most of these separations were on a small laboratory scale. The high purity to which scarce isotopes such as ^{13}C, ^{15}N, and ^{18}O have been concentrated is a notable feature of these examples of thermal diffusion. The feasibility of concentrating rare isotopes of intermediate mass, such as ^{21}Ne and ^{38}A, by thermal diffusion is also noteworthy. These separations are facilitated by the large number of stages obtainable from a single thermal diffusion column.

Thermal diffusion is a convenient way of separating isotopes on a small scale. It is a very inefficient process for large-scale use because of its high energy consumption.

8.2 Partial Separation of Uranium Isotopes

Abelson and Hoover [A1], working in the U.S. Naval Research Laboratory, found that thermal diffusion in UF_6 at pressures above the critical (4.6 MPa) resulted in small but measurable enrichment of ^{235}U at the hot wall. Because of the simplicity of thermal diffusion equipment compared with the advanced technology needed for the gaseous diffusion process, the Manhattan District in the United States in 1944–1945 used thermal diffusion of UF_6 to raise its ^{235}U content to 0.86 percent, to serve as partially enriched feed for the Y-12 electromagnetic plant. Energy for the S-50 thermal diffusion plant was obtained from steam which later powered the 150-MW electric generating station which drove the compressors of the K-25 gaseous diffusion plant.

The thermal diffusion plant [A1] contained 2100 columns, each with an effective height of 14.6 m. Each column consisted of three concentric tubes, the innermost being made of nickel,

Table 14.24 Examples of isotope separation by thermal diffusion

Working fluid	Isotope separated	m/o in product	Phase	Single column (S) or cascade (C)	Investigator	Year
HCl	^{35}Cl	99.6	Gas	S	Clusius and Dickel	1939
	^{37}Cl	99.4				
Kr	^{84}Kr	98.2	Gas	S	Clusius and Dickel	1941
	^{86}Kr	99.5				
O_2	^{17}O	0.5	Gas	C	Clusius and Dickel	1944
	^{18}O	99.5				
UF_6	^{235}U	0.86	Liquid	C	Manhattan Dist.	1945
N_2	^{15}N	99.8	Gas	S	Clusius & Dickel	1950
Xe	^{134}Xe	1	Gas	C	Clusius et al.	1956
	^{136}Xe	99				
He	^{3}He	10	Gas	C	Bowring and Davies	1958
A	^{36}A	99.8	Gas	C	ORNL†	1961
	^{38}A	23.2				
Ne	^{20}Ne	99.99	Gas	C	ORNL	1961
	^{22}Ne	99.99				
Kr	^{78}Kr	10	Gas	C	ORNL	1961
	^{86}Kr	96.1				
He§	^{3}He	99	Gas	C	Mound Lab.‡	1962
Ne	^{21}Ne	33.9	Gas	C	ORNL	1963
CH_4	^{13}C	90	Gas	C	Mound Lab.	1963
Xe	^{124}Xe	4.4	Gas	C	ORNL	1964

†Oak Ridge National Laboratory, U.S. AEC, Oak Ridge, Tennessee.
‡Mound Laboratory, U.S. AEC, Miamisburg, Ohio.
§Feed not of normal abundance, contained 1 percent ^3He from nuclear reaction.

the middle of copper, and the outer of iron. The inner tube, about 5 cm in diameter, carried condensing steam at temperatures that could be varied from 188 to 286°C. The annular space (about 0.025-cm gap) between nickel and copper was filled with UF_6 at a pressure of 6.7 MPa, well above the critical. The outer annular space between copper and iron carried cooling water at 63°C, slightly above the freezing point of UF_6. The columns were operated batchwise, with periodic removal of slightly enriched UF_6 from a header connected to the top of a group of columns and slightly depleted UF_6 from a larger reservoir connected to the bottom. Operation of the complete plant of 2100 columns was affected by frequent leaks and freezeups, so that its performance is less representative than that of tests made in individual columns, which are summarized in Table 14.25.

Their separation performance was characterized by two parameters. Y is ln y_P/y_F, the overall separation between top and bottom when equilibrium is attained at total reflux. ϕ is a parameter that was inferred from the rate at which product composition at total reflux approached equilibrium. The theory of the time-dependent separation performance of a thermal diffusion column developed by Cohen [C6] and others shows that ϕ is given by

$$\phi = \frac{C_1^2 L}{C_5} \tag{14.338}$$

where C_1 and C_5 are the parameters in the differential equation for the steady-state separation performance of a countercurrent column:

Table 14.25 Separative capacity and power consumption of UF_6 thermal diffusion column†

Operating conditions												
Pressure, MPa		Temperature, K					kW					
			UF_6									
UF_6	Steam	Steam, T	Hot, T'	Cold, T''	Annular spacing Δr, cm	UF_6 inventory, g	Heat, H	Availability, Q	$Y = C_1 L/C_5$	$\phi = C_1^2 L/C_5$, g UF_6 SWU/day	Max. sep. capacity, Δ_{max}, kg U SWU/yr	Power/sep. cap., Q/Δ_{max} kW/(kg U SWU/yr)
6.7	1.1	461	438	340	0.0273	2040	109	38	0.50	13.6	0.67	57
6.7	4.0	527	497e	341e	0.0256	1720	172	74	0.53	27.3	1.34	55
6.7	6.7	559	517	342	0.0248	1600	201	93	0.6	50	2.46	38
6.7	1.1	461	438	340	0.0253	1860	117	41	0.6	13.3	0.66	62
6.7	5.3	544	504e	341e	0.0230	1500	214	96	0.65	44.1	2.18	44
6.7	1.1	461	438	340	0.0225	1600	131	46	>0.7	6.2	0.31	148
6.7	5.0	540	500e	341e	0.0200	1320	216	96	0.8	26.4	1.30	74
6.7	6.7	559	517	342	0.025	1600	203	94	0.6	44.4	2.19	43
10	6.7	559	517	342	0.025	1700	198	92	0.77	31.5	1.55	59
20	6.7	559	517	342	0.025	1800	188	87	–	31.2	1.54	56

†e, estimated, $Q = H(1 - 300/T)$. $L = 1460$ cm. $Y = \ln (y_P/y_F)$ at steady state at total reflux.
Δ kg U SWU/yr = (0.238 kg U/352 g UF_6) (365 day/yr) (0.80 $C_1^2 L/4C_5$) (g UF_6 SWU/day) = 0.0494 ϕ.

$$\frac{dy}{dz} = \frac{C_1 y(1-y)}{C_5} - \frac{P(y_P - y)}{C_5} \tag{14.339}$$

An equation of the same form (14.181) was derived for the gas centrifuge treated as a countercurrent column. L is the active length of the column, 1460 cm.

The maximum separative capacity, Δ_{max}, and the power consumed per unit separative capacity, Q/Δ_{max}, given in the last two columns of Table 14.25 have been calculated from Abelson's parameters Y and ϕ to permit comparison with the other processes for enriching uranium treated in this chapter. Because the thermal diffusion column operates with constant reflux ratio, its steady-state separation performance as an enricher is given by Eq. (14.237), expressed here in the form

$$\frac{y_P}{y_F} = \frac{(P/C_1) + 1}{(P/C_1) + \exp\{-[(P/C_1) + 1](C_1 L/C_5)\}} \tag{14.340}$$

Its separative capacity Δ, for $y \ll 1$, is

$$\Delta = \lim_{y_W \to y_F} \left(-P \ln \frac{y_P}{y_F} - W \ln \frac{y_W}{y_F}\right) \tag{14.341}$$

Because

$$W = \frac{P(y_P - y_F)}{y_F - y_W} \tag{14.342}$$

$$\Delta = -P\left(\ln \frac{y_P}{y_F} - \frac{y_P}{y_F} + 1\right) \tag{14.343}$$

With y_P/y_F from (14.340) and the separative capacity Δ from (14.343), the maximum value of Δ at $C_1 L/C_5$ around 0.6 is found to be

$$\Delta_{max} = 0.80 \frac{C_1^2 L}{4C_5} \tag{14.344}$$

At P/C_1 around 1.8, 0.80 is the maximum value of the ideality efficiency E_I for this thermal diffusion column considered as a square enriching cascade.

The next-to-the-last column of Table 14.25 gives maximum values of the separative capacity of this thermal diffusion column if operated at the optimum product rate for each set of the operating conditions given in the first six columns. The last column gives the ratio of the power loss from heat input to separative capacity. The optimum set of operating conditions are those in the third row of Table 14.25, with a UF_6 pressure of 6.7 MPa, a steam temperature of 559 K, and an annular spacing of 0.0248 cm. At these conditions this column would have a separative capacity of 2.46 kg uranium SWU/year and would consume heat equivalent to a power loss of 38 kW/(kg uranium SWU/year). The separative capacity of 2.46 kg uranium SWU/year of this thermal diffusion column 1460 cm high may be compared with the centrifuge of Tables 14.15 and 14.16, which had a higher separative capacity of 10 kg uranium SWU/year in a lower height of 335.3 cm. The specific power consumption of 38 kW/(kg uranium SWU/year) may be compared with 0.266 kW/(kg uranium SWU/year) for the U.S. gaseous diffusion plants. The much greater specific power of thermal diffusion was the principal reason that the Manhattan District's thermal diffusion plant was shut down as soon as the K-25 gaseous diffusion plant began operation.

Although its very poor power utilization compared with gaseous diffusion and the gas centrifuge precludes use of thermal diffusion for large-scale uranium isotope separation, the simplicity of the equipment, the absence of moving parts, and the large separation attainable in a convenient height have led to its use for small-scale separation of many isotopes, as suggested by Table 14.24.

8.3 Theory of Thermal Diffusion Separation

Theoretical prediction of the constants C_1 and C_5 of the UF_6 thermal diffusion column would be very difficult because of the great difference in properties of UF_6 between the liquid at the cold wall and the dense gas at the hot wall. For other gases at pressures around atmospheric, at temperature differences between hot and cold walls small enough so that separation performance can be characterized by gas properties at a mean temperature, closed expressions can be given for the separation parameters C_1 and C_5. Quantities involved are

Hot wall temperature T'
Cold wall temperature T''
Density at mean temperature ρ
Viscosity μ
Thermal diffusion constant γ
Diffusion coefficient D

Thermal diffusion effect. When a composition gradient $\partial y/\partial r$ and a temperature gradient $\partial T/\partial r$ occur together in a stationary gas mixture, the usual diffusion mass velocity $-D\rho\ \partial y/\partial r$ is modified by the thermal diffusion effect so that the mass velocity J_r in the r direction becomes

$$J_r = -D\rho\left[\frac{\partial y}{\partial r} - \gamma y(1-y)\frac{\partial \ln T}{\partial r}\right] \tag{14.345}$$

γ is known as the *thermal diffusion constant*. When γ and $\partial T/\partial r$ have the same sign and are large enough, this thermal diffusion effect can cause transport of an isotope *against* a composition gradient and thus produce separative work. If a steady state has been established with zero transport, the mole fraction gradient is related to the temperature gradient by

$$\left(\frac{\partial y}{\partial r}\right)_{\text{zero transport}} = \gamma y(1-y)\frac{\partial \ln T}{\partial r} \tag{14.346}$$

Integration of Eq. (14.346) between (y', T') and (y'', T'') leads to Eq. (14.347) for the separation factor α:

$$\ln \alpha \equiv \ln \frac{y'(1-y'')}{y''(1-y')} = \gamma \ln \frac{T'}{T''} \tag{14.347}$$

Measurement of steady-state compositions y' and y'' after thermal diffusion equilibrium has been established between temperatures T' and T'' is the most accurate way of determining the thermal diffusion constant. The next-to-the-last column of Table 14.26 gives values of the measured thermal diffusion constant for several binary isotopic mixtures. In all these cases, γ is positive, which means that the light isotope concentrates at the higher temperature under the experimental conditions listed.

Values of the thermal diffusion constant can be calculated by the kinetic theory of gases if the intermolecular potential energy is known. Because the calculation is quite sensitive to the detailed intermolecular interaction, calculated values of the thermal diffusion constant are in less satisfactory agreement with experiment than other transport properties.

Hirschfelder et al. [H9] gives a generalized relation for the variation of the thermal diffusion constant with temperature for gases whose molecules interact with the so-called Lennard-Jones potential function, the difference between a repulsion energy inversely

Table 14.26. Thermal diffusion constants for isotopic mixtures

Isotopes	Log mean reciprocal temperature, K	Thermal diffusion constant, γ		Reference
		Theory	Measured	
^{3}He-^{4}He	398	0.079	0.059	[M7]
^{20}Ne-^{22}Ne	408	0.026	0.023	[M9]
^{36}A-^{40}A	450	0.023	0.020	[S2]
^{83}Kr-^{84}Kr	449	0.0019	0.0013	[M9]
^{131}Xe-^{132}Xe	448	0.0011	0.00085	[M9]
$^{16}O_2$-$^{16}O^{18}O$	443	0.015	0.013	[W2]
$^{12}C^{16}O$-$^{13}C^{16}O$	431	0.0085	0.0070	[M9]
$^{12}CH_4$-$^{13}CH_4$	407	0.011	0.0073	[D1]

proportional to the twelfth power of the intermolecular distance and an attraction energy inversely proportional to the sixth power. Shacter et al. [S3] have used this theory to predict the thermal diffusion constants in the third column of Table 14.26. The agreement with experimental values is only semiquantitative.

Figure 14.39, based on this theory, may be used to predict the magnitude and temperature dependence of the thermal diffusion constant for 15 isotopic mixtures. The quantity plotted, k_T^*, is the ratio of the calculated thermal diffusion constant to the thermal diffusion constant

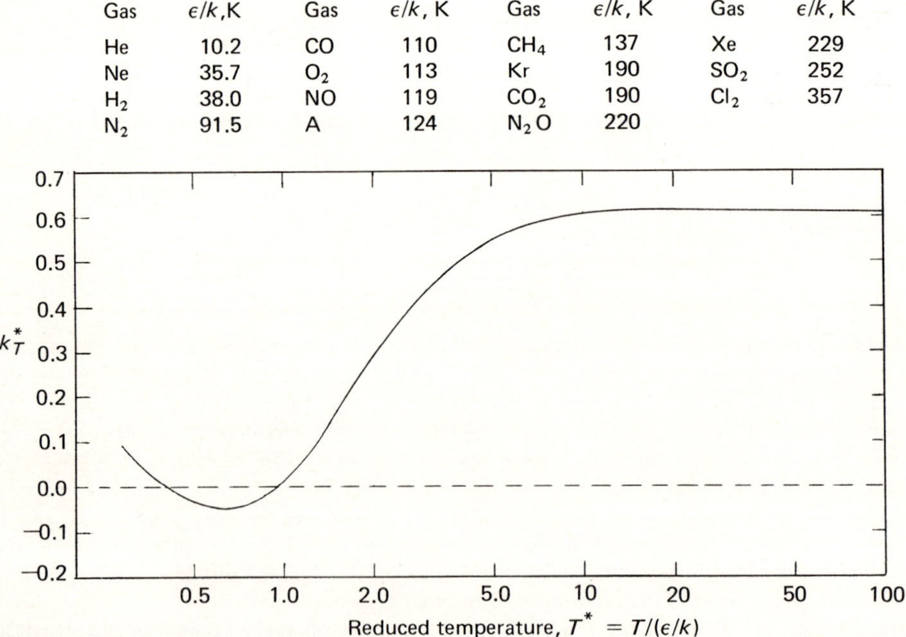

Figure 14.39 The function k_T^* for calculating thermal diffusion constants of isotopic mixtures from Lennard-Jones 6-12 potential function.

γ^* for rigid, spherical, nonattracting isotopic molecules, for which

$$\gamma^* = \frac{105}{108} \frac{m_2 - m_1}{m_2 + m_1} \tag{14.348}$$

Because the maximum theoretical value of k_T^* for this intermolecular potential is 0.627, γ in thermal diffusion is smaller than $\alpha_0 - 1$ in gaseous diffusion, Eq. (14.24). k_T^* becomes negative at temperatures near the normal boiling point and changes back to positive at still lower temperatures.

Figure 14.40 shows the most accurate measurements of the thermal diffusion effect in UF_6 vapor at low pressure, by Kirch and Schütte [K2]. Results are plotted both as k_T^*, for comparison with other gases in Fig. 14.39, and as the thermal diffusion constant γ. The very low values, under 0.00005, explain Nier's [N3] inability to detect a thermal diffusion effect in UF_6 vapor. The thermal diffusion coefficient γ is so much smaller than the analogous parameter in gaseous diffusion, $\alpha_0 - 1 = 0.0043$, that vapor-phase thermal diffusion cannot compete economically with gaseous diffusion for uranium enrichment.

Equations for thermal diffusion column. Equations for the separation performance of a thermal diffusion column can be derived in somewhat similar fashion to the countercurrent gas centrifuge of Sec. 5.5. The results will be summarized for the simplest case to treat theoretically, that of an annular column in which the spacing d between the heated and cooled

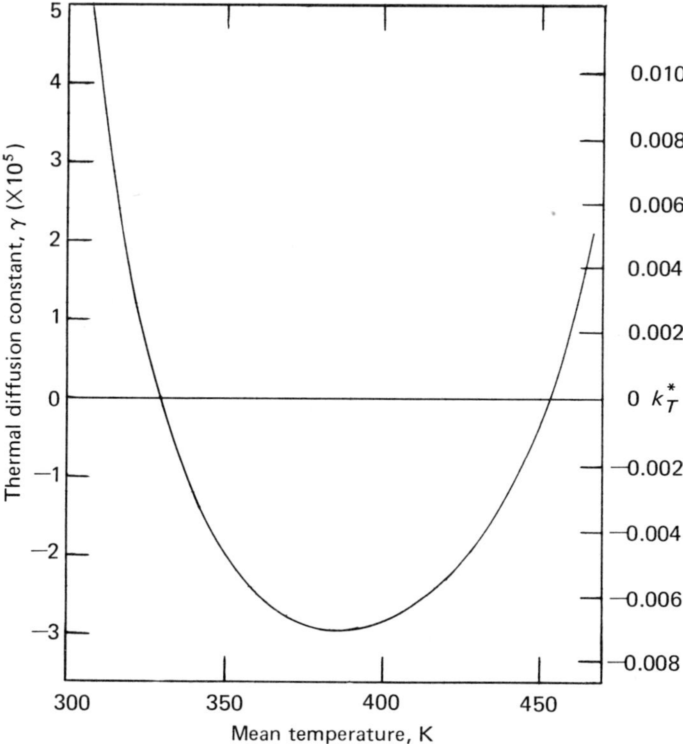

Figure 14.40 Thermal diffusion constant of UF_6 vapor at subatmospheric pressure. *(From Kirch and Schütte [K2].)*

tubes is much smaller than the log mean radius $\bar{r}$, and in which the temperature difference between heated and cooled walls, $\Delta T = T' - T''$, is small enough so that the gas properties can be evaluated at the log mean temperature $\bar{T}$. This theory was developed first by Jones and Furry [J5], but using different notation.

Thermal convection between the hot and cold walls under gravitational acceleration g induces longitudinal countercurrent mass flow at the rate

$$N = \frac{2\pi\bar{r}g\rho^2\, d^3}{384\, \mu} \frac{\Delta T}{\bar{T}} \tag{14.349}$$

Because longitudinal velocity is zero at the heated and cooled walls, the logarithm of the effective separation factor is found to be $\frac{8}{15}$ that of Eq. (14.347). For $\gamma \ll 1$ and $\Delta T/\bar{T} \ll 1$, this becomes

$$\alpha - 1 = \frac{8}{15}\gamma \frac{\Delta T}{\bar{T}} \tag{14.350}$$

The differential enrichment equation for the countercurrent thermal diffusion column is of the same form as for the countercurrent gas centrifuge, Eq. (14.181), here written as

$$\frac{dy}{dz} = \frac{C_1 y(1-y) - P(y_P - y)}{C_2 + C_3} \tag{14.351}$$

Form (14.187),

$$C_1 = N(\alpha - 1) = \frac{2\pi\bar{r}g\rho^2 d^3 \gamma}{720\, \mu}\left(\frac{\Delta T}{\bar{T}}\right)^2 \tag{14.352}$$

For the annular geometry presently considered, the parameter C_2, representing longitudinal back diffusion, is

$$C_2 = 2\pi\bar{r}D\rho d \tag{14.353}$$

Jones and Furry's development leads to Eq. (14.354) for C_3:

$$C_3 = \frac{128\, dN^2}{315\, \rho D(2\pi\bar{r})} = \frac{2\pi\bar{r}g^2\rho^4 d^7}{9!\, \mu^2 \rho D}\left(\frac{\Delta T}{\bar{T}}\right)^2 \tag{14.354}$$

Generalizations of these equations for larger $\Delta T/\bar{T}$ and for wider annuli ($d \approx \bar{r}$) have been given by McInteer and Reisfeld [M8].

9 LASER ISOTOPE SEPARATION

9.1 Introduction

The possibility of using the slight differences that exist in the absorption spectra of isotopes of an element for isotope separation has been recognized ever since isotopes were discovered. The first reported successful photochemical separation of isotopes was that of Kuhn and Martin [K4], who dissociated $CO^{35}Cl_2$ molecules in natural phosgene by light of 281.618-nm wavelength from an aluminum spark, which happened to be the correct wavelength. The first photochemical separation of isotopes on a practical scale was that of mercury isotopes. In one example [Z1], light from a mercury arc containing a preseparated mercury isotope was used to excite the same isotope in natural mercury vapor and cause it to form HgO with water vapor also present. This method is not generally applicable to other elements because it makes use of the especially simple character of the mercury spectrum, with few, widely spaced lines.

Invention of the laser provided the intense, monochromatic, tunable light source needed to make photochemical isotope separation applicable to all elements, at least on a laboratory scale. The promise of this method was recognized as early as 1965 by Robieux and Auclair [R1], who were issued the first patent on it. Since the pioneering experiments of Tiffany et al. [T1] on bromine isotopes in 1966, an enormous amount of work has been done with lasers, with small-scale separation reported for most elements.

This text can describe only briefly the incomplete information publicly available on laser separation of uranium isotopes. For a more detailed discussion of the history and principles of laser isotope separation, reference may be made to the review articles of Letokhov and Moore [L1] and Aldridge et al. [A2], and to Farrar and Smith's report on uranium [F1].

Two general methods have been proposed for separating uranium isotopes. In the photoionization method to be discussed in Sec. 9.2, ^{235}U in uranium metal vapor is ionized selectively and then separated from unionized ^{238}U by deflection in electric or magnetic fields. In the photochemical method, to be described in Sec. 9.3, $^{235}UF_6$ in UF_6 vapor is excited selectively and caused to react chemically to produce a solid lower fluoride, which is then separated from unreacted $^{238}UF_6$ vapor.

9.2 Laser Isotope Separation of Uranium Metal Vapor

Absorption spectrum of uranium metal vapor. The absorption spectrum of uranium metal vapor is very complex, with over 300,000 lines at visible wavelengths. However, many of these absorption lines are very sharp, with sufficient displacement between a ^{238}U absorption line and the ^{235}U absorption line for the corresponding transition, and without overlap of the ^{235}U line with the ^{238}U line for a different transition, to permit selective excitation of the ^{235}U atoms. However, choice of the wavelength most suitable for a practical process is made difficult by the large number of possibilities. Janes et al. [J2] discuss some of the alternatives.

History. In the United States, laser isotope separation (LIS) with uranium metal vapor has been investigated experimentally by the Lawrence Livermore Laboratory (LLL) of the U.S. DOE at Livermore, California, and by Jersey-Nuclear-Avco Isotopes, Inc. (JNAI), a joint venture of Exxon Nuclear Company of Bellevue, Washington, and Avco-Everett (Massachusetts) Research Laboratory, which holds a number of patents on this method, of which the most significant are those of Levy and Janes [J1, L2].

Workers at LLL [T3, D2] have reported production of milligram quantities of uranium enriched to 2.5 percent ^{235}U by this method. In the LLL work, the source of uranium metal vapor was a uranium-rhenium alloy, chosen to reduce attack by the hot metal on the containing crucible. Deflection of ^{235}U ions was by an electric field.

In the JNAI work, solid uranium metal is vaporized by an electron beam impinging on its surface, and deflection of ^{235}U ions is by either space charge expansion, a magnetic field, an electric field, or a combination. The JNAI process, as described in patents [J1, L2] and a 1977 article [J2], has evolved through several stages. The next section describing the uranium metal LIS process follows the 1977 article.

Process description. Figure 14.41 is a schematic assembly drawing of one module of the JNAI uranium metal laser isotope separator. Figure 14.42 is a transverse section of this module. In Fig. 14.41, separation takes place inside a vacuum chamber about 1 m long. The uranium vapor source at the bottom consists of a charge of uranium metal, held in a water-cooled crucible, whose top surface is heated to 3000 K by a sheet of high-energy electrons curved and focused in a line on the uranium by a magnetic field of 100 to 200 gauss. Uranium vapor atoms diverge radially upward from the heated line source and travel in straight lines because of the absence of collisions in the high vacuum. These atoms flow upward between longitudinal, cooled,

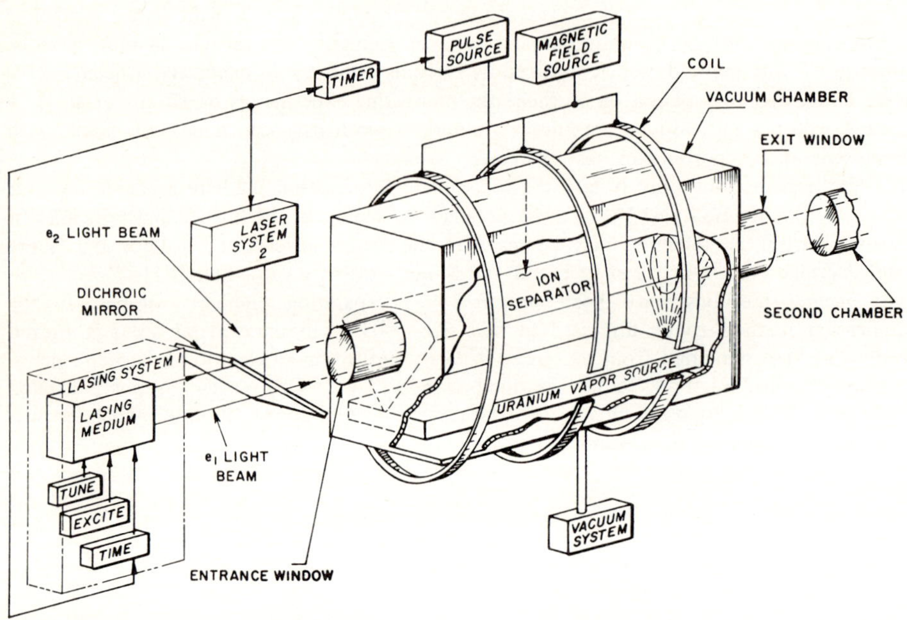

Figure 14.41 Schematic diagram illustrating basic elements of the JNAI atomic LIS process. *(Reproduced with permission of the copyright holder, American Institute of Chemical Engineers.)*

product collector plates, oriented so that the atoms move parallel to them and do not impinge. The space between the plates is illuminated by light from a system of lasers, to be described later, which ionize most of the ^{235}U atoms selectively, while leaving most of the ^{238}U atoms un-ionized. The ^{235}U ions, being electrically charged, can be deflected from the outward flowing uranium vapor and caused to impinge on and adhere to the product collector plates. Three possible methods for deflecting the ^{235}U ions are (1) expansion with energetic electrons released when the uranium is ionized, (2) motion in circular orbits around longitudinal magnetic field lines, or (3) deflecting by electric fields produced by giving adjacent collector plates alternative positive and negative charges. Un-ionized ^{238}U atoms move outward beyond the product collector plates and condense on the upper cooled tails collection surface.

For maximum capacity, the uranium vapor density should be as high as possible. An upper limit is around 10^{13} atoms/cm^3, because at higher density collisions between ^{235}U ions and ^{238}U atoms, or charge exchange between them, would occur too frequently, resulting in too high deflection of ^{238}U. Assuming a plate height of around 4 cm, a flow area 4 cm wide by 100 cm long, and a uranium vapor thermal velocity of 40,000 cm/s, the uranium feed rate per module would be

$$\frac{(100)(4)(40{,}000)(10^{13})(238)}{6.025 \times 10^{23}} = 0.063 \text{ g uranium/s} \qquad (14.355)$$

which represents a maximum daily ^{235}U production rate of

$$(0.063 \text{ g uranium/s})(0.00711 \text{ g } ^{235}\text{U/g uranium})(86{,}400 \text{ s/day}) = 39 \text{ g } ^{235}\text{U/day} \qquad (14.356)$$

per module 1 m long.

The lasers used to ionize the ^{235}U should be pulsed sufficiently often to irradiate all ^{235}U atoms passing between the plates. With a plate height of 4 cm and a vapor velocity of 40,000 cm/s, this requires a pulse repetition rate of 10,000 Hz. This, and other requirements to be described below, require development of lasers more advanced than any now available.

The light path through the module is limited to around 1 m to prevent the uranium metal vapor from itself becoming a laser, with consequent loss of selectivity. This length limitation prevents full utilization of laser photons in a single module and makes desirable connecting several physically separate modules optically in series as suggested by the second chamber shown in Fig. 14.41.

Laser requirements. In order to utilize photons efficiently, absorption by ^{235}U should be selective. A ^{235}U absorption line should be found that (1) occurs at a frequency at which ^{238}U does not absorb, and (2) has a high absorption cross section, to reduce the light path needed for efficient use of photons. Because the isotope shift between ^{235}U and ^{238}U absorption frequencies is of the order of 1 in 50,000, the first requirement calls for use of a very narrow ^{235}U absorption line. Because the absorption lines for transitions in which uranium is ionized are very broad, it is necessary to ionize the ^{235}U atoms in two or more steps, in which the first step is selective excitation of ^{235}U to an energy level below the uranium ionization potential of 6.18 eV. This would be followed by less selective absorption of one or more additional photons of sufficient energy to ionize the excited ^{235}U atoms but of too little energy to ionize the unexcited ^{238}U atoms. One of the JNAI patents [L2] suggests use of a narrow-frequency laser supplying visible light at 502.74 nm to excite ^{235}U, followed by ultraviolet light at 262.5 nm to carry the excited atoms over the 6.18 eV ionization level. At 502.74 nm, the ^{235}U absorption line, of half-width around 0.001 nm, is displaced 0.01 nm from the ^{238}U absorption line, so that the required selectivity is obtained.

The energy E imparted to the ^{235}U atom by absorption of a photon of wavelength λ is evaluated from Planck's law,

$$E(\mathrm{J}) = h\nu = \frac{hc}{\lambda} \tag{14.357}$$

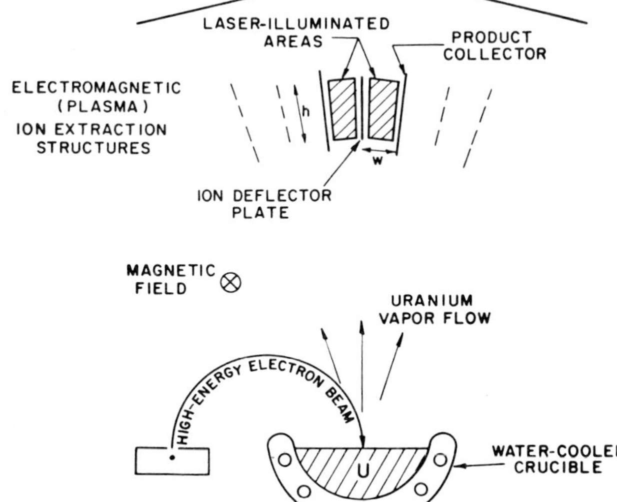

Figure 14.42 Cross-sectional view of the interior of the module shown in Fig. 14.41. *(Reproduced with permission of the copyright holder, American Institute of Chemical Engineers.)*

h is Planck's constant, 6.62559×10^{-34} (J · s). c is the velocity of light, 2.997925×10^8 m/s. The energy in electron volts V is

$$V = \frac{E(\mathrm{J})}{e} \tag{14.358}$$

where e is the electron charge, 1.60210×10^{-19} C, so that

$$V = \frac{hc}{e\lambda} = \frac{(6.62559 \times 10^{-34})(2.997925 \times 10^{8})}{1.60210 \times 10^{-19}\ \lambda\ (\mathrm{m})} = \frac{1.23981 \times 10^{-6}}{\lambda\ (\mathrm{m})} \tag{14.359}$$

Hence the energy given the ^{235}U atom by successive absorption of photons of wavelength 502.74 and 262.5 nm is

λ (nm)	E (eV)
502.74	2.466
262.5	4.723
	7.189

Since the total 7.189 eV absorbed by ^{235}U exceeds its ionization potential of 6.18 eV, this two-step photon absorption process imparts sufficient energy to ionize ^{235}U. But since ^{238}U absorbs only the 262.5-nm photon, ^{238}U receives only 4.723 eV and is not ionized. Other possible combinations of two or more photons are described by Janes et al. [J2].

Even though the foregoing photon absorption process selectively ionizes ^{235}U, charge exchange between ^{235}U ions and neutral ^{238}U atoms and atomic collisions deflect enough ^{238}U atoms to the collector plates to limit the heads enrichment factor to around 10. For example, a product content of 6 percent ^{235}U is the highest value that has been obtained from natural uranium. At the same time, however, very complete stripping of ^{235}U from tails is claimed. Three advantages cited for this kind of separation performance are as follows:

1. A single stage of separation suffices to produce uranium of high enough enrichment for light-water reactors.
2. More complete stripping is achieved than is economical in gaseous diffusion or the gas centrifuge.
3. This LIS process can be used to produce uranium containing 2 to 3 percent ^{235}U from tails from these other processes.

The lasers for the process just described have three requirements more exacting than any yet developed:

1. They must deliver pulses with a frequency of 10,000 Hz.
2. To be economical they must last for a year or more to deliver over 3×10^{11} pulses before replacement.
3. They must deliver far more energy per pulse than any high-frequency lasers now available.

Development problems. Despite the promise apparent in this laser enrichment process, it has a number of development problems. As just stated, lasers with higher repetition rate, longer life, and higher energy must be developed. Optical windows that do not lose transparency or mechanical integrity from deposition of uranium or intense illumination must be developed. Materials problems associated with handling corrosive uranium metal at high temperatures must be solved. Perhaps most important of all, convenient means must be developed for charging uranium to the high-vacuum, high-temperature system and for collecting and removing the separated uranium product and tails fractions. This LIS process appears to have one of the disadvantages of the Y-12 electromagnetic process, of having feed material deposit all over the vacuum chamber, necessitating troublesome interruptions for disassembly and clean-out.

Economic estimates. Despite these problems, JNAI was sufficiently optimistic about the ultimate economics to go ahead with pilot-plant construction. At this stage of development, however, economics are very uncertain. This is illustrated by Table 14.27, which compares estimates of process characteristics and costs made by JNAI and a Japanese group. The specific power estimate of Janes et al. is around that predicted for the gas centrifuge. The estimate of Ozaki et al., although 10 times higher, is lower than that of gaseous diffusion (Sec. 4.7). The unit investment costs predicted by both groups, although very different, are much lower than for gaseous diffusion or the gas centrifuge and are the principal reason for the interest being shown in this process.

Two features that make separative work cost estimates very uncertain are uncertainty about laser energy efficiencies and ignorance of operating and maintenance costs, which can be obtained only by completing the development and making life tests on plant equipment.

9.3 Laser Isotope Separation of UF_6

The absorption spectrum of UF_6 is far more complex even than that of uranium metal, because the spectrum of the UF_6 molecule involves transitions between many vibrational and rotational states that are absent in the uranium atom. Absorption bands of the $^{235}UF_6$ molecule overlap those of $^{238}UF_6$, so that highly selective absorption by one isotope is seldom found. This is illustrated by Fig. 14.43, which shows the absorption by $^{235}UF_6$ and $^{238}UF_6$ at four different pressures at room temperature at an infrared wavelength around 16 μm at which the difference between the spectra of the two compounds is greatest. The peak in the $^{235}UF_6$ absorption band is displaced 0.55 cm^{-1} from the peak in the $^{238}UF_6$ band at a wave number (reciprocal wavelength) of 625 cm^{-1}, about 1 part in 1000. However, the absorption by $^{238}UF_6$ at the peak absorption by $^{235}UF_6$ is so great as to preclude selective absorption under these conditions.

It has been predicted theoretically by Sinha et al. [S5] and observed experimentally by Jensen and Robinson [J4, R2] that if UF_6 is cooled to 55 K and its absorption spectra measured with high resolution, wavelengths can be found at which selective absorption by $^{235}UF_6$ takes place with relatively little absorption by $^{238}UF_6$. The reason for this is as follows.

Uranium in the UF_6 molecule is at the center of an octahedron, with the six fluorine atoms equally spaced at the corners. Such a molecule can vibrate in six different modes, of which the uranium atom moves in only two, the only ones with an isotopic shift. In the ν_3 mode to which Fig. 14.43 is attributed, the uranium and two opposite fluorine atoms move up and down together out of the plane of the other four fluorine atoms. The absorptions of Fig. 14.43 are caused by transitions in which the vibrational quantum number increases by unity, while the rotational quantum numbers change by plus or minus unity. If all transitions were

Table 14.27. Estimated characteristics of uranium metal laser isotope separation plants

	Source of estimate	
	Osaki et al. [O3]	Janes et al. [J2]
Capacity, million kg SWU/yr	8.75	3
Specific electric power, kW/(kg SWU/yr)	0.20	0.02
Unit investment cost, $/(kg SWU/yr)	36	195

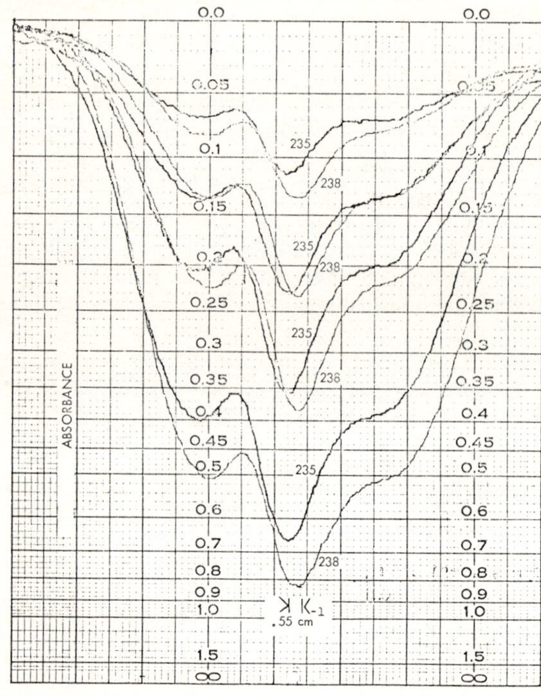

Wavelength, 15.8 μm 16.0 μm 16.2 μm

Wave number, 632.9 cm^{-1} 625 cm^{-1} 617.3 cm^{-1}

Figure 14.43 Absorption spectra of $^{235}UF_6$ and $^{238}UF_6$ around 16 μm (625 cm^{-1}).

from the lowest vibrational level, the fine structure of the absorption bands would be as shown qualitatively in Fig. 14.44, where the individual peaks are due to transitions from different rotational levels. Under such conditions, a $^{235}UF_6$ absorption maximum might be found that occurred at a $^{238}UF_6$ absorption minimum, as shown in the figure. Then a tuned laser beam with a frequency spread narrower than the line spacing of 0.1 cm^{-1} might be able to excite $^{235}UF_6$ to the first vibrational level without exciting $^{238}UF_6$.

Such selective absorption is not possible at room temperature. There, only about 1 percent of the UF_6 molecules are in their lowest vibrational state, so that the observed absorption spectrum is a composite of vibrational transitions from the ground state and many excited states, in each case to the next higher vibrational quantum number. These excited-state absorption frequencies are displaced somewhat from the ground state, so that $^{238}UF_6$ lines from an excited state overlap $^{235}UF_6$ lines from the ground state, thus destroying selectivity.

There is another difficulty with working at room temperature. In the photochemical method, a second light beam would be used to dissociate vibrationally excited $^{235}UF_6$ molecules into a physically separable, nonvolatile lower fluoride and fluorine, while leaving unexcited $^{238}UF_6$ molecules with too little energy to be dissociated. However, because most of the $^{238}UF_6$ molecules at room temperature are already in excited states, many of these would necessarily also be dissociated.

For these two reasons, two-step photochemical dissociation of UF_6 at room temperature would yield only very partial separation and would make very inefficient use of laser energy. At very low temperatures, the fraction of UF_6 in the lowest vibrational state increases, reaching 69 percent at 77 K and 85 percent at 55 K. However, the vapor pressure of UF_6 at 77 K, extrapolated from measurements at higher temperature, is only 5×10^{-25} Torr.

Jensen and Robinson [J4] have described an experiment at Los Alamos in which a dilute mixture of UF_6 in hydrogen was cooled to 30 K by expansion to high speed through a hypersonic nozzle. In this experiment, subcooled UF_6 molecules remained uncondensed long enough to assume the low-temperature energy distribution and display an absorption spectrum in which $^{235}UF_6$ lines and $^{238}UF_6$ lines were separate and did not overlap.

In the proposed separation process, this high-speed, subcooled gas mixture would be irradiated first by 16-μm light of a frequency absorbed by $^{235}UF_6$ and not by $^{238}UF_6$ and then by additional light of sufficient energy to dissociate the excited $^{235}UF_6$, but insufficient to dissociate the unexcited $^{238}UF_6$. The dissociated lower fluoride of ^{235}U and undissociated $^{238}UF_6$ would then be separated in one of several possible ways. If condensation of subcooled UF_6 could be delayed long enough, the solid lower fluoride of ^{235}U might be separated mechanically from the still gaseous $^{238}UF_6$. Or both might be condensed and the $^{238}UF_6$ leached with water from the insoluble lower fluoride of ^{235}U. In either method, transfer of a fluorine atom from undissociated $^{238}UF_6$ to the possibly unstable lower fluoride of ^{235}U would impair selectivity. Because of classification, information is not available on how successful this postirradiation separation step has been.

A possible simplification of the photochemistry of this process is afforded by the discoveries of multistep photon absorption by Lyman et al. [L5] and by Ambartzumian et al. [A3]. An intense laser beam of the frequency absorbed by $^{235}UF_6$ will deliver a sufficient number of photons successively to a $^{235}UF_6$ molecule to dissociate it, while hopefully leaving $^{238}UF_6$ unaffected. This would permit a single laser to do the job.

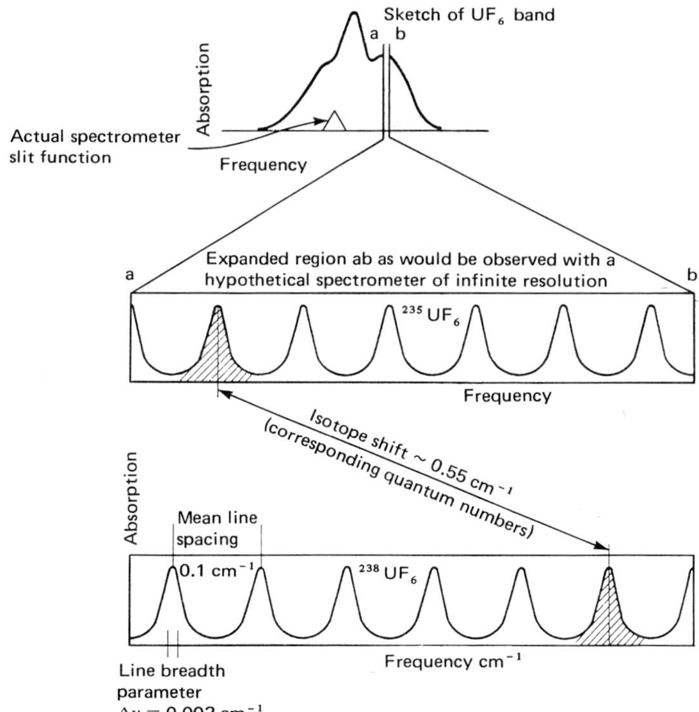

Figure 14.44 Schematic representation of unresolved structure of absorption spectra of $^{235}UF_6$ and $^{238}UF_6$.

NOMENCLATURE

a	molecular flow parameter, Eq. (14.38)
a	outer radius
A	barrier area
A	barrier parameter in Eq. (14.59)
A^2	peripheral speed parameter in gas centrifuge, Eq. (14.213)
A^2	peripheral speed parameter for UF_6 in nozzle, Eq. (14.270)
A_H^2	peripheral speed parameter for hydrogen, $A^2/176$
b	viscous flow parameter, Eq. (14.39)
B_1	countercurrent column parameter, C_1/N
B_3	countercurrent column parameter, C_3/N^2
c	radial position of flow divider in nozzle process
c	velocity of light, Sec. 9
c_S	unit cost of separative work
C	annual cost, \$/year
C_0	initial cost, \$
C_1	countercurrent column parameter, Eq. (14.176) or (14.352)
C_2	countercurrent column parameter, Eq. (14.177) or (14.353)
C_3	countercurrent column parameter, Eq. (14.178) or (14.354)
C_5	$C_2 + C_3$
C_p	constant pressure heat capacity, per mole
d	diameter
D	self-diffusion coefficient
D_0	mean diffusion coefficient, Eq. (14.323)
D_{01}	diffusion coefficient of light component in separating agent
D_{02}	diffusion coefficient of heavy component in separating agent
D_{12}	diffusion coefficient of light component in heavy component
e	electron charge, 1.60210×10^{-19} C
E	ion energy, J, Sec. 9
E	modulus of elasticity, Sec. 5
E	overall efficiency of diffusion stage, Eq. (14.93)
E	overall efficiency of centrifuge, $E_C E_F E_I$
E_B	diffusion barrier separation efficiency, Sec. 4
E_C	centrifuge circulation efficiency, Eq. (14.228)
E_F	centrifuge flow pattern efficiency, Eq. (14.227)
E_I	centrifuge ideality efficiency, Eq. (14.245)
E_M	local mixing efficiency
f	dimensionless flow function of centrifuge, Eq. (14.216)
f	mole fraction UF_6 in mixture with hydrogen, Sec. 6
F	flow function of centrifuge, Eq. (14.175)
g	acceleration of gravity, 9.80665 m/s^2
G	molar velocity
h	Planck's constant, Sec. 9
h	height of transfer unit
H	mass velocity of flow along diffusion barrier, Sec. 4
H	heat flow rate, Sec. 8
I_1	dimensionless separation parameter of centrifuge, Eq. (14.222)
I_3	dimensionless separation parameter of centrifuge, Eq. (14.223)
J	mass velocity
k_T^*	γ/γ^*, thermal diffusion ratio

K	power in isentropic compression
l	length of barrier pore
L	overall length
m	mass, Sec. 5.4
m	molecular weight
M	in gaseous diffusion, flow rate of light stream
M	in centrifuge, mass flow rate of light stream
M	in UCOR process, molar flow rate of UF_6 in light stream
M	in mass diffusion, molar flow rate of N_{43}-rich stream
$\mathfrak{M}$	mass flow rate of UF_6 in light stream per unit slit length in nozzle process
M_H	in UCOR process, molar flow rate of H_2 in light stream
N	in gaseous diffusion, flow rate of heavy stream
N	in centrifuge and thermal diffusion, mass flow rate of heavy stream
N	in UCOR process, molar flow rate of UF_6 in heavy stream
N	in mass diffusion, molar flow rate of UF_6-rich stream
$\mathfrak{N}$	mass flow rate of UF_6 in heavy stream per unit slit length in nozzle process
N_H	in UCOR process, molar flow rate of H_2 in heavy stream
p	absolute pressure
p_c	characteristic barrier pressure
P	product flow rate
q	pressure ratio across diffusion barrier, p'/p''
Q	rate of loss of availability
r	barrier pore radius, Sec. 4
r	radial position
r_1	radial position in centrifuge at which longitudinal velocity changes sign
$\bar{r}$	log mean radius in annular column
R	gas constant
Re	Reynolds number
s	throat spacing
S	barrier slope factor, Eq. (14.10)
S	entropy flow rate, in Eq. (14.331)
S	separative work
t	equivalent film thickness, Eq. (14.66)
t	temperature, °C, in Eq. (14.4)
T	absolute temperature, K
T_0	absolute temperature at which heat is rejected to environment
T'	in aerodynamic processes, absolute temperature after isentropic expansion
T'	in thermal diffusion, absolute temperature of fluid at hot wall
T''	in thermal diffusion, absolute temperature of fluid at cold wall
$\bar{T}$	log mean absolute temperature
ΔT	$T' - T''$
u	radial velocity
v	in gaseous diffusion, mole fraction light component in net flow through barrier
v	in centrifuge, tangential velocity relative to coordinates rotating at angular velocity ω
v	in aerodynamic processes, peripheral speed
$\bar{v}$	mean molecular speed
v_a	in centrifuge, peripheral speed
V	compressor volumetric capacity
V	ion energy, eV, Sec. 9
w	longitudinal velocity

w_0	scale factor in Eq. (14.214) for velocity profile
W	tails flow rate
x	fraction of light component in stream in which heavy component concentrates
x_i	mole fraction of light component in high-pressure stream within diffusion stage
x_0	mole fraction of light component defined by Eq. (14.23)
x''	mole fraction of light component at high-pressure surface of barrier
X	barrier parameter, Eq. (14.50)
y	fraction light component in stream in which light component concentrates
y'	mole fraction of light component at low-pressure surface of barrier
Y	in thermal diffusion, C_1L/C_5
Y_E	defined by Eq. (14.254) or (14.255)
Y_S	defined by Eq. (14.260) or (14.261)
z	in gaseous diffusion, distance
z	longitudinal distance
Z	stage UF_6 feed rate
α	separation factor
α_0	ideal separation factor in gaseous diffusion, $\sqrt{m_2/m_1}$
β	heads separation factor
γ	tails separation factor
γ	specific permeability, Eq. (14.14)
γ	in Sec. 6, ratio of heat capacities
γ	in mass diffusion, separability, Eq. (14.322) or (14.324)
γ	thermal diffusion constant, Sec. 8
γ^*	thermal diffusion constant for rigid spheres, Eq. (14.348)
Γ	barrier permeability, Eq. (14.10)
Γ_0	permeability for molecular flow
δ	$\alpha_0 - 1$
Δ	separative capacity
ϵ	fraction of barrier area open for flow
ϵ	measure of composition difference across diffusion barrier, Eq. (14.129) and (14.135)
ϵ/k	characteristic temperature for estimating thermal diffusion constant from Fig. 14.39
η	mole fraction UF_6 in N_{43}-rich stream
θ	in Sec. 5, angular position
θ	UF_6 cut
θ_H	hydrogen cut in UCOR process
κ	thermal conductivity
λ_i	in Sec. 5.4, eigenvalue of rotating cylindrical shell, Eq. (14.150)
λ	in Sec. 5.5, decay constant for velocity profile
λ	in Sec. 9, wavelength
μ	viscosity
ν	frequency
ξ	mole fraction UF_6 in UF_6-rich stream
π	pressure ratio, p/p_c
ρ	mass density
σ	stress
ϕ	barrier parameter, Eq. (14.49)
ϕ	C_1^2L/C_5, in thermal diffusion
ω	angular velocity
ω_i	ith resonant angular velocity

Subscripts

a	at outer wall
e	end cap thermal drive in gas centrifuge
e	stage effluent in mass diffusion
equil	at centrifugal equilibrium
E	enriching section
f	in film near diffusion barrier
f	mass diffusion stage feed
F	in feed
F	at centrifuge feed point
in	at inlet
max	maximum
min	minimum
opt	optimum
out	at outlet
P	product
r	radial
s	in Sec. 5, scoop-and-baffle drive
s	in Sec. 6, sonic
S	stripping section
vis	viscous flow
w	wall thermal drive
W	tails
z	longitudinal component
θ	tangential component
0	separating agent
1	light component
2	heavy component
1, 2, . . . , 8	numbered points in Fig. 14.29

Superscripts

E	enriching section
opt	optimum
S	stripping section
$'$	stream in which light isotope concentrates
$''$	stream in which heavy isotope concentrates
$\bar{}$	average

REFERENCES

A1. Abelson, P. H., and J. I. Hoover: "Separation of Uranium Isotopes by Liquid Thermal Diffusion," *Proceedings of the International Symposium on Isotope Separation,* Interscience, New York, 1958, p. 483.

A2. Aldridge, J. P., et al. "Experimental and Theoretical Studies of Laser Isotope Separation," in *Physics of Quantum Electronics,* vol. 4, S. F. Jacobs et al. (eds.), Addison-Wesley, Reading, Mass., 1976.

A3. Ambartzumian, R. V., et al.: *Chem. Phys. Lett.* **27**: 87 (1975).

A4. Aston, J. G.: *Phil. Mag.* **39**: 449 (1920).

A5. Avery, D. G., and E. Davies: *Uranium Enrichment by Gas Centrifuges,* Mills and Boon, London, 1973.

B1. Beams, J. W.: *Rev. Mod. Phys.* **10**: 245 (1938).
B2. Beams, J. W.: *Early History of the Gas Centrifuge Work in the United States,* University of Virginia, Charlottesville, Va., May 1975.
B3. Beams, J. W., A. C. Hagg, and E. V. Murphree: "Developments in Centrifuge Separation," Report TID-5230, U.S. AEC, 1951.
B4. Beams, J. W., and C. W. Skarstrom: *Phys. Rev.* **56**: 266 (1939).
B5. Becker, E. W., et al.: "Physics and Technology of Separation Nozzle Process," in *Nuclear Energy Maturity, Proceedings of the European Nuclear Conference, Paris, 21-25 April, 1975, Invited Sessions,* Pergamon, New York, 1975, p. 172.
B6. Becker, E. W., and R. Schütte: *Z. Naturforsch.* **15A**: 336 (1960).
B7. Becker, E. W., et al.: *Angew. Chem., Int. Ed.* **6**: 507 (1967).
B8. Becker, E. W., et al.: *Z. Naturforsch.* **26a**: 1377 (1971).
B9. Becker, E. W., et al.: "The Separation Nozzle Process for Enrichment of Uranium 235," *PICG (4)* **9**: 3 (1972).
B10. Becker, E. W., et al.: Paper no. 1, *Proceedings of the International Conference on Uranium Isotope Separation,* British Nuclear Energy Society, London, Mar. 5-7, 1975.
B11. Becker, E. W., et al.: Paper no. 2, *Proceedings of the International Conference on Uranium Isotope Separation,* British Nuclear Energy Society, London, Mar. 5-7, 1975.
B12. Becker, E. W., et al.: *AIChE Symp. Ser. 169* **73**: 25 (1977).
B13. Benedict, M.: U.S. Patent 2,609,059, Sept. 2, 1952.
B14. Benedict, M. (ed.): "Developments in Uranium Enrichment," *AIChE Symp. Ser. 169* **73**: (1977).
B15. Benedict, M., and A. Boas: *Chem. Eng. Progr.* **41**: 51, 111 (1951).
B16. Berman, A.: U.S. AEC Report K-1535, 1963.
B17. Bilous, O., and G. Counas: "Determination of the Separation Factor of the Uranium Isotopes Produced by Gaseous Diffusion," *PICG(2)* **4**: 405 (1958).
B18. Botha, S. P.: Reported by *Wall Street J.*, Feb. 14, 1978.
B19. Bramley, A., and K. Brewer: *Science* **92**: 427 (1940).
B20. British Nuclear Energy Society: *Proceedings of the International Conference on Uranium Isotope Separation,* British Nuclear Energy Society, London, Mar. 5-7, 1975.
C1. Chapman, S.: *Phil. Trans. Roy. Soc. London, Ser. A* **216**: 279 (1916); **217**: 115 (1917).
C2. Chapman, S., and F. W. Dootson: *Phil. Mag.* **33**: 248 (1917).
C3. Charpin, J., P. Plurien, and S. Mommejac: "Application of General Methods of Study of Porous Bodies to the Determination of Characteristics of Barriers," *PICG (2)* **4**: 380 (1958).
C4. Cichelli, M. T., W. D. Weatherford, and J. R. Bowman: *Chem. Eng. Progr.* **47**: 63, 123 (1951).
C5. Clusius, K., and G. Dickel: *Z. Phys. Chem.* **B44**: 397, 451 (1939).
C6. Cohen, K.: *The Theory of Isotope Separation,* McGraw-Hill, New York, 1951.
C7. Commissariat à l'Energie Atomique: *Pierrelatte, Usine de Separation des Isotopes de l'Uranium,* Brun, Paris, 1973.
C8. Cowan, G. A.: *Sci. Amer.* **235**(1): 36 (July 1976).
C9. Cowan, G. A., and H. H. Adler: *Geochim. et Cosmochim. Acta* **40**: 1487 (1976).
D1. Davenport, A. N., and E. R. S. Winter: *Trans. Faraday Soc.* **47**: 1160 (1951).
D2. Davis, J. I., and R. W. Davis: *AIChE Symp. Ser. 169* **73**: 69 (1977).
D3. Dawson, J. M., et al.: *Phys. Rev. Lett.* **37**: 1547 (Dec. 6, 1976).
D4. Den Hartog, J. P.: *Mechanical Vibrations,* 4th ed., McGraw-Hill, New York, 1956.
D5. DeWitt, R.: "Uranium Hexafluoride: A Survey of the Physico-Chemical Properties," Report GAT-280, Aug. 12, 1960, p. 39.
D6. DeWitt, R.: "Uranium Hexafluoride: A Survey of the Physico-Chemical Properties," Report GAT-280, Aug. 12, 1960, p. 63.

D7. Durivault, J., and P. Louvet: "Theoretical Study of Flow in Thermal Countercurrent Centrifuges," Report CEA-R-4714, 1976.

E1. Enskog, D.: *Phys. Z.* **12**: 56, 533 (1911).

E2. Enskog, D.: *Ann. Phys.* **38**: 731 (1912).

F1. Farrar, R. L., Jr., and D. F. Smith: "Photochemical Isotope Separation as Applied to Uranium," Report K-L-3054, rev. 1, U.S. AEC, 1972.

F2. Forsberg, C. W.: "A Technical and Economic Study of Uranium Enrichment by Mass Diffusion," D. Sc. thesis submitted to Massachusetts Institute of Technology, Cambridge, Mass., Nov. 1973.

F3. Frejacques, C., et al.: *PICG(2)* **4**: 418 (1958).

F4. Frejacques, C., M. Gelee, D. Massignon, and P. Plurien: Paper 36/FR/257, *Proceedings of the International Conference on Nuclear Energy and Its Fuel Cycle,* International Atomic Energy Agency, Salzburg, 1977.

G1. Geppert, H., et al.: "The Industrial Implementation of the Separation Nozzle Process," *Proceedings of the International Conference on Uranium Isotope Separation,* British Nuclear Energy Society, London, Mar. 5–7, 1975.

G2. Grant, W. L., J. J. Wannenburg, and P. C. Haarhoff: *AIChE Symp. Ser. 169* **73**: 20 (1977).

G3. Groth, W., et al.: *PICG (2)* **4**: 439 (1958).

G4. Groueff, S.: *Manhattan Project,* Little, Brown, Boston, 1967.

G5. Groves, L. R.: *Now It Can Be Told,* Harper & Row, New York, 1962.

G6. Gverdtsiteli, L. G., R. Y. Kucherov, and V. K. Tskhakaya: "Isotope Separation by Diffusion in a Current of Steam," *PICG (2)* **4**: 608 (1958).

H1. Haarhoff, P. C.: "The Helikon Technique for Isotope Enrichment," Report VAL 1, Uranium Enrichment Corp. of South Africa, Pretoria, Nov. 1976.

H2. Harmsen, H., G. L. Hertz, and W. Schutze: *Phys. Z.* **90**: 703 (1934).

H3. Hertz, G. L.: *Phys. Z.* **91**: 810 (1934).

H4. Hertz, G. L.: U.S. Patents 1,486,521, Mar. 11, 1924; 1,498,097, June 17, 1924.

H5. Hertz, G. L.: *Phys. Z.* **79**: 108, 700 (1932).

H6. Hertz, G. L.: *Naturwiss.* **21**: 884 (1933).

H7. Hewlett, R. G., and O. E. Anderson, Jr.: *The New World, 1939–1946,* vol. I, *A History of the United States Atomic Energy Commission,* Pennsylvania State University Press, University Park, Pa., 1962.

H8. Heymann, D., and J. Kistemaker: *J. Chem. Phys.* **24**: 165 (1956).

H9. Hirschfelder, J. O., C. F. Curtis, and R. B. Bird: *Molecular Theory of Gases and Liquids,* Wiley, New York, 1954, chap. 8.

H10. Hogerton, J. F.: *Chem. Eng.* **52**(12): 98 (1945).

H11. Hoglund, J. S., J. Shacter, and E. Von Halle: "Diffusion Separation Methods," in *Encyclopedia of Chemical Technology,* 3d ed., vol. 7, Wiley, New York, 1979, pp. 639–723.

J1. Janes, G. S.: U.S. Patents 3,935,451, Jan. 27, 1976; 3,939,354, Feb. 17, 1976.

J2. Janes, G. S., H. K. Forsan, and R. H. Levy: *AIChE Symp. Ser. 169* **73**: 62 (1977).

J3. Jay, K. E. B.: *Britain's Atomic Factories,* Her Majesty's Stationery Office, London, 1954.

J4. Jensen, R. J., and C. P. Robinson: *AIChE Symp. Ser. 169* **73**: 76 (1977).

J5. Jones, R.C., and W. H. Furry: *Rev. Mod. Phys.* **18**: 151 (1946).

K1. Keith, P. C.: *Chem. Eng.* **53**(2): 112 (1946).

K2. Kirch, P., and R. Schütte: *Z. Naturforsch.* **22A**: 1532 (1967).

K3. Knudsen, M.: *Ann. Phys.* **28**: 75 (1909).

K4. Kuhn, W., and H. Martin: *Naturwiss.* **20**: 772 (1932); *Z. Phys. Chem.* **21B**: 93 (1933).

K5. Kynch, G. J.: "Differential Diffusion Through a Capillary," UKAEA Reports MS 118, MS 119.

L1. Letokhov, V. S., and C. B. Moore: "Laser Isotope Separation," Report LBL-4904, Mar. 1976; *Sov. J. Quant. Electron.* **6**(2): 129 (Feb. 1976).

L2. Levy, R. S., and G. S. Janes: U.S. Patents 3,772,519, Nov. 13, 1973; 3,944,825, Mar. 16, 1976.

L3. Lindemann, F. A., and F. W. Aston: *Phil. Mag.* **37**: 350 (1919).

L4. London, H.: *Separation of Isotopes,* Newnes, London, 1963, p. 366.

L5. Lyman, J., et al.: *Appl. Phys. Lett.* **27**: 87 (1975).

M1. Maier, C. G.: *J. Chem. Phys.* **7**: 854 (1939).

M2. Malling, G. F., and E. Von Halle: "Aerodynamic Isotope Separation Processes for Uranium Enrichment," U.S. ERDA Report K/OA-2872, Oct. 7, 1976.

M3. Mårtensson, M., et al.: "Some Types of Membranes for Isotope Separation by Gaseous Diffusion," *PICG(2)* **4**: 395 (1958).

M4. Martin, H., and W. Kuhn: *Z. Phys. Chem.* **A189**: 219 (1941).

M5. Massignon, D.: "Characteristics of Barriers That Can Be Used for Isotope Separation by Gaseous Diffusion," *PICG (2)* **4**: 388 (1964).

M6. May, W. G.: *AIChE Symp. Ser. 169* **73**: 30 (1977).

M7. McInteer, B. B., L. T. Aldrich, and A. O. Nier: *Phys. Rev.* **72**: 510 (1947).

M8. McInteer, B. B., and M. J. Reisfeld: *J. Chem. Phys.* **33**: 570 (1960).

M9. Moran, T. I., and W. W. Watson: *Phys. Rev.* **109**: 1184 (1958).

N1. Naudet, R., and C. Renson: *Proceedings of an International Symposium on the Oklo Phenomenon,* Libreville, Gabon, June 23–27, 1975, IAEA-SM-240/23, pp. 265–291.

N2. Ney, E. P., and F. C. Armistead: *Phys. Rev.* **71**: 14 (1947).

N3. Nier, A. O. C.: "The Determination of the Coefficient of Thermal Diffusion of Uranium Hexafluoride," U.S. Report A-41, May 28, 1941.

N4. Nikolaev, B. I., et al.: *Isotopenpraxis* **6**: 417 (1970).

N5. *Nuclear Fuel,* Oct. 31, 1977, p. 15.

O1. Olander, D. R.: "Technical Basis of the Gas Centrifuge," *Adv. Nuclear Sci. Tech.* **6**: 105 (1972).

O2. Oliver, G. D., and J. W. Grisard: "Separation of Uranium Isotopes by Distillation," Report K-829, Oct. 29, 1951.

O3. Ozaki, N., et al.: *Appl. Energy* **2**: 279 (1976).

P1. Parker, H. M., and T. T. Mayo, IV: "Countercurrent Flow in a Semi-Infinite Gas Centrifuge," Report UVA-279-63U, University of Virginia, Richmond, Va., 1963.

P2. Perry, R. H., and C. H. Chilton (eds.): *Chemical Engineers' Handbook,* 5th ed., McGraw-Hill, New York, 1973, pp. 3–120.

P3. Present, R. D., and A. J. de Bethune: *Phys. Rev.* **75**: 1050 (1949).

P4. Present, R. D., and W. G. Pollard: *Phys. Rev.* **73**: 762 (1948).

R1. Robieux, J., and J. M. Auclair: French Patent 1,391,738, Mar. 12, 1965.

R2. Robinson, C. P., and R. J. Jensen: "Some Developments in Laser Enrichment at Los Alamos," Report LA-UR-76-91, Feb. 1975.

R3. Roux, A. J. A. and W. L. Grant: "Uranium Enrichment in South Africa," in *Nuclear Energy Maturity, Proceedings of the European Nuclear Conference, Paris, 21–25 April, 1975, Invited Sessions,* Pergamon, New York, 1975, p. 167.

S1. Saraceno, A. J., and C. F. Trivisonno (eds.): "Uranium Isotope Separation by Chemical Exchange Reactions Between UF_6 and UF_6-Nitrogen Oxide," Report GAT-674, Feb. 4, 1972.

S2. Saxena, S. C., J. G. Kelley, and W. W. Watson: *Phys. Fluids* **4**: 1216 (1961).

S2a. Scuricini, G. B. (ed.): *Proceedings of the Third Workshop on Gases in Strong Rotation*, Comitato Nazionale Energia Nucleare, Rome, 1979.

S3. Shacter, J., E. Von Halle, and J. S. Hoglund: *Encyclopedia of Chemical Technology,* 2d ed., vol. 7, Wiley, New York, 1965, p. 91.

S4. Sherwood, T. K., and R. L. Pigford: *Absorption and Extraction,* 2d ed., McGraw-Hill, New York, 1952, p. 79.

S5. Sinha, M. P., A. Schultz, and R. N. Zare: *J. Chem. Phys.* **58**: 549 (1973).

S6. Smyth, H. D.: *Atomic Energy for Military Purposes,* Princeton University Press, Princeton, N.J., 1945.

S7. Soubbaramayer: "Approximate Solutions to Thermal Countercurrent Centrifuge Problems," Report CEA-R-4186, July 1971.

T1. Tiffany, W. B., N. W. Moos, and A. L. Schawlow: *Science* **157**: 40 (1967).

T2. Touryan, K. J., E. P. Muntz, L. Talbot, and E. Von Halle: "Gas Dynamic Problems in Isotope Separation," Sandia Laboratories' Report, SAND 75-0121, Mar. 1975.

T3. Tuccio, S. A., R. J. Foley, J. W. Dubrin, and O. Krikorian: *IEEE J. Quant. Elect.* **QE-11**: 101D (1975).

U1. U.S. Atomic Energy Commission: "AEC Gaseous Diffusion Plant Operations," Report ORO-684, Jan. 1972.

U2. U.S. Atomic Energy Commission: "Data on New Gaseous Diffusion Plants," Report ORO-685, Apr. 1972.

U3. U.S. Department of Energy: *United States Gas Centrifuge Program for Uranium Enrichment,* 1978.

U4. Urey, H. C.: *Rep. Prog. Phys.* **6**: 48 (1939).

V1. Villani, S.: *Isotope Separation,* American Nuclear Society, La Grange Park, Ill., 1976, p. 266.

V1a. Villani, S. (ed.): *Uranium Enrichment,* Topics in Applied Physics, vol. 35, Springer-Verlag, New York, 1979.

W1. Weller, S., and W. A. Steiner: *J. Appl. Phys.* **21**: 279 (1950).

W2. Whalley, E., and E. R. S. Winter: *Trans. Faraday Soc.* **45**: 1091 (1949).

W3. Wooldridge, D. E., and F. A. Jenkins: *Phys. Rev.* **49**: 404, 704 (1936).

W4. Wooldridge, D. E., and W. R. Smythe: *Phys. Rev.* **50**: 233 (1936).

Z1. Zelikoff, M., L. M. Aschenbrand, and P. H. Wykoff: *J. Chem. Phys.* **21**: 376 (1953).

Z2. Zippe, G.: "The Development of Short Bowl Ultracentrifuges," Report EP-4420-101-6OU, University of Virginia, Charlottesville, Va., July 1960.

PROBLEMS

14.1 A gaseous diffusion plant is to be designed for a capacity of 10.8 million kg uranium separative work units per year with the following feed, product, and tails compositions (w/o = weight percent).

Feed: 0.711 w/o ^{235}U
Product: 3.0 w/o ^{235}U
Tails: 0.2 w/o ^{235}U

The plant is to use the barrier prepared by anodic oxidation of aluminum whose properties are given in Table 14.6, in tubes 0.014 m in diameter and 4 m long. The plant is to be built as an ideal cascade of stages operating at the optimized conditions of Table 14.9.

(*a*) What are the annual feed, product, and tails flow rates, in kilograms of uranium?
(*b*) How many stripping and enriching stages are required?
(*c*) How much total barrier area is required?
(*d*) How many electric kilowatts are required?
(*e*) For the largest stage, what is
  (1) The compressor capacity?
  (2) The barrier area?

(3) The electric power demand?

(4) The heads flow rate, in kilograms of uranium per second?

14.2 For the optimized gaseous diffusion stage of Table 14.9, what is the ratio of UF_6 inventory inside the barrier tubes to UF_6 stage feed rate?

Assume that the stage holdup time h, Eq. (12.198), is three times the above ratio. What is the minimum equilibrium time τ, Eq. (12.209), of the ideal cascade of Prob. 14.1?

14.3 In qualitative terms, how would the optimum design of the gaseous diffusion stage of Sec. 4.7 be changed if better barrier, with higher p_c, were specified?

14.4 How does the separative capacity of a cross-flow gaseous diffusion stage vary with the cut? At what cut is the separative capacity highest? What is the ratio of separative capacity at a cut of $\frac{1}{2}$ to the maximum separative capacity?

14.5 A gaseous diffusion plant is to be designed to separate 10,000 kg UF_6/year whose isotopic content is 40 w/o ^{235}U and 60 w/o ^{236}U into product containing 90 w/o ^{235}U and tails containing 10 w/o ^{235}U. The plant is to use the barrier prepared by anodic oxidation of aluminum whose properties are given in Table 14.6, in tubes 0.014 m in diameter and 4 m long. The plant is to be built as an ideal cascade.

(*a*) What are annual feed, product, and tails rates, in kilograms of uranium?

(*b*) Justify the selection of the same optimized conditions as used in Table 14.9.

(*c*) With these conditions, how many stripping and enriching stages are needed?

(*d*) How much total barrier area is needed?

(*e*) How many electric kilowatts are needed?

14.6 Assume that the centrifuge whose dimensions are given in Table 14.16 is made of glass fiber composite, whose properties are given in Table 14.11, and that the centrifuge is run at a peripheral speed v_a of 500 m/s.

(*a*) What is the tangential stress in kilograms per square centimeter?

(*b*) Through how many longitudinal vibration resonances will the rotor pass while being brought to operating speed?

14.7 A gas centrifuge with the dimensions of Table 14.16, running at a peripheral speed of 400 m/s, is to be operated at total reflux, with circulation rate N the same at all values of the length, z. For the Berman-Olander velocity distribution whose separation parameters f, I_1, and I_3 have the values given in Table 14.14, find the circulation rate N at which the overall separation y_P/y_W is a maximum and the value of this maximum. Compare with the values at maximum separative capacity. $D\rho = 2.161 \times 10^{-4}$ g/(cm·s).

14.8 A gas centrifuge 300 cm long and 40 cm in diameter is to be run at 300 K and 500 m/s peripheral speed. It is fed at the midplane with UF_6 at a rate of 0.03 g UF_6/s. The longitudinally uniform heavy-stream flow rate is 0.20 g UF_6/s. Heads and tails flow rates are set so that there is no mixing loss at the point of feed injection. Find the heads separation factor, tails separation factor, and separative capacity. Note: I_1, I_3 and $f(r_1/a)$ are given in Table 14.14.

14.9 Repeat Prob. 14.8, for a cut of $\frac{1}{2}$.

14.10 This problem shows how much poorer the separation factor in the nozzle process would be if pure UF_6 were used instead of a dilute mixture with hydrogen.

Pure UF_6 is expanded isentropically from 313 K through a Becker nozzle to sonic speed. Assuming centrifugal equilibrium in wheel flow and a cut of $\frac{1}{4}$, find the separation factor. C_p for UF_6 = 31.3 cal/(K·g-mol).

14.11 Natural methane containing 1.1 percent $^{13}CH_4$ is to be enriched to 10 percent $^{13}CH_4$ in an ideal cascade of mass diffusion stages, without stripping, using steam as separating agent. Each stage is operated at 100°C and 1.0 atm, under balanced pressure conditions such that there is no net flow through the diffusion screen. Steam fed to the stage contains no methane;

methane feed contains 10 m/o (mole percent) H_2O. Flow ratio is steam/(CH_4 + H_2O) = 1.9. The methane/steam ratio of all stage effluent streams are equal. Diffusion coefficients at operating conditions are

CH_4—H_2O	0.39 cm^2/s
$^{12}CH_4$—$^{13}CH_4$	0.33 cm^2/s

(*a*) How many stages are required?

(*b*) How many moles of water must be vaporized per mole of 10 percent $^{13}CH_4$ produced?

14.12 Suppose that all 2100 columns of the S-50 thermal diffusion plant, with individual column characteristics as given in row 3 of Table 14.25, were operated in parallel as an enriching section, at very high natural UF_6 feed rate, without stripping section. At the product rate at which separative capacity is a maximum, what would be the ^{235}U content of product? The product flow rate in kilograms of uranium per year?

14.13 An annular thermal diffusion column with a large reservoir at the top is to be used to enrich ^{13}C by thermal diffusion of natural methane containing 1.1 percent $^{14}CH_4$. The column is operated at total reflux until a steady state is reached. Dimensions, operating conditions, and properties of methane are as follows:

Radius of inner heated tube, 0.559 cm
Radius of outer cooled tube, 0.864 cm
Spacing d, 0.305 cm
Log mean radius $\bar{r}$, 0.700 cm
Length L, 213 cm
Pressure p, 1 atm
Temperatures, K
  Inner wall, T', 572 K
  Outer wall, T'', 300 K
Properties of methane at $\bar{T}$ = 421.5 K
  Density ρ, 4.63×10^{-4} g/cm^3
  Viscosity μ, 1.5×10^{-4} g/(cm·s)
  $\rho D/\mu$, 1.33
  Thermal diffusion constant γ, 0.0074

What is the percent $^{13}CH_4$ at the bottom of the column at steady state?

14.14 Estimate the energy in kilowatt-hours consumed in the uranium metal LIS process to produce 1 kg uranium enriched to 6 percent ^{235}U product and 0.1 percent ^{235}U tails from natural uranium feed under the following assumptions:

Energy imparted to product atoms, 7 eV
Conversion efficiency, photon energy absorbed/input energy to laser system, 1/1000
Heat of vaporization of uranium, 550,000 J/g-atom
Conversion efficiency, vaporization energy/electric energy input, $\frac{1}{4}$.

APPENDIX

A

FUNDAMENTAL PHYSICAL CONSTANTS

Symbol	Quantity	Value[†] This text[‡]	More recent value[§]	Units
c	Speed of light in vacuum	$2.997925(1) \times 10^{8}$	$2.99792458(1.2) \times 10^{8}$	m/s
e	Electron charge	$1.602101(23) \times 10^{-19}$	$1.6021892(46) \times 10^{-19}$	C
$\mathfrak{F}$	Faraday's constant (Ne)	9.64868×10^{4}	$9.648456(27) \times 10^{4}$	C/g-mol
h	Planck's constant	$6.62559(16) \times 10^{-34}$	$6.626176(36) \times 10^{-34}$	J·s
M_e	Electron rest mass	$5.48597(3) \times 10^{-4}$	$5.4858026(21) \times 10^{-4}$	amu
M_{H}	Hydrogen atom rest mass	1.00782519(8)	$1.007825050(11) \times 10^{-4}$	amu
M_n	Neutron rest mass	1.00866520(10)	1.008665012(37)	amu
N	Avogadro's constant	$6.02252(9) \times 10^{23}$	$6.022045(31) \times 10^{23}$	$(\text{g-mol})^{-1}$
R	Molar gas constant	8.31434(35)	8.31441(26)	J/(g-mol·K)

[†] Numbers in parentheses are the standard deviation uncertainty in the last digits.

[‡] Principal source, E. R. Cohen and J. W. M. Dumond, *Rev. Mod. Phys.* **37**: 537 (1965).

[§] Reprinted with permission from *Handbook of Chemistry and Physics*, 58th ed. 1977–1978, copyright the Chemical Rubber Co., CRC Press, Inc., Boca Raton, Fla.

APPENDIX

B

CONVERSION FACTORS

B.1 BASIC EQUIVALENTS

Length: One foot (ft) = 12 inches (in) = 0.3048 meter (m)
Mass: One pound (lb) = 453.59237 grams (g)
One short ton (ST) = 0.91718474 megagrams (Mg)
One metric ton (MT) = one megagram (Mg)
Temperature: T, Kelvin (K) = t, Celsius (°C) + 273.15
T, Rankine (°R) = 1.8T, Kelvin (K)
t, Fahrenheit (°F) = 1.8t, Celsius (°C) + 32

B.2 MASS AND ENERGY EQUIVALENTS†

				Calorie	
Per	MeV	amu	Joule, J	(USNBS)‡	(IST)§
MeV	1.00000	1.07356E-3	1.60210E-13	3.82911E-14	3.82655E-14
amu	9.31480E2	1.00000	1.49232E-10	3.56674E-11	3.56435E-11
J	6.24180E12	6.70096E9	1.00000	2.39006E-1	2.38846E-1
cal (USNBS)	2.61157E13	2.80368E10	4.18400	1.00000	9.99331E-1
cal (IST)	2.61332E13	2.80556E10	4.18680	1.00067	1.00000
Btu (IST)	6.58545E15	7.06988E12	1.05506E4	2.52164E2	2.51996E2
hp•h	1.67562E19	1.79889E16	2.68452E6	6.41616E5	6.41187E5
kWh	2.24705E19	2.41234E16	3.60000E6	8.60421E5	8.59845E5
MWd	5.39292E23	5.78963E20	8.64000E10	2.06501E10	2.06363E10
g	5.60985E26	6.02252E23	8.98755E13	2.14808E10	2.14664E13
lb	2.54459E29	2.73177E26	4.07669E16	9.74352E15	9.73700E15

(See footnotes on page 936.)

Per	British thermal unit, Btu (IST)§	Horsepower-hour, hp•h	Kilowatt-hour, kWh	Megawatt-day, MWd	Gram, g	Pound, lb
MeV	1.51850E-16	5.96792E-20	4.45028E-20	1.85428E-24	1.78258E-27	3.92991E-30
amu	1.41445E-13	5.55900E-17	4.14535E-17	1.72723E-21	1.66043E-24	3.66063E-27
J	9.47817E-4	3.72506E-7	2.77778E-7	1.15741E-11	1.11265E-14	2.45297E-17
cal (USNBS)	3.96567E-3	1.55856E-6	1.16222E-6	4.84259E-11	4.65533E-14	1.02632E-16
cal (IST)	3.96832E-3	1.55961E-6	1.16300E-6	4.84583E-11	4.65844E-14	1.02701E-16
Btu (IST)	1.00000	3.93015E-4	2.93071E-4	1.22113E-8	1.17391E-11	2.58802E-14
hp•h	2.54443E3	1.00000	7.45700E-1	3.10708E-5	2.98693E-8	6.58506E-11
kWh	3.41214E3	1.34102	1.00000	4.16667E-5	4.00554E-8	8.83070E-11
MWd	8.18914E7	3.21845E4	2.40000E4	1.00000	9.61330E-4	2.11937E-6
g	8.51856E10	3.31792E7	2.49654E7	1.04023E3	1.00000	2.20462E-3
lb	3.86395E13	1.51859E10	1.13241E10	4.71839E5	4.53592E2	1.00000

†Conversion factors from joules into MeV, amu, and grams are based on the 1965 fundamental physical constants of App. A.

‡The calorie defined by the U.S. National Bureau of Standards as 4.18400 J and used in thermochemical data tables.

§The calorie and Btu used in International Steam Tables, from 4.18680 J/cal (IST).

B.3 PRESSURE EQUIVALENTS

Per	Pascal, Pa	kg_f/m^2	Torr	Inches, H_2O, in H_2O
Pa	1.00000	1.01972E-1	7.50062E-3	4.01464E-3
kg_f/m^2	9.80665	1.00000	7.35559E-2	3.93701E-2
Torr	1.33322E2	1.35951E1	1.00000	5.35240E-1
in H_2O	2.49089E2	2.54000E1	1.86827	1.00000
psia	6.89476E3	7.03070E2	5.17149E1	2.76799E1
bar	1.00000E5	1.01972E4	7.50062E2	4.01464E2
atm	1.01325E5	1.03323E4	7.60000E2	4.06782E2
Mpa	1.00000E6	1.01972E5	7.50062E3	4.01464E3

Per	Pounds force per in^2, psia	Bar	Atmosphere, atm	Megapascal, MPa
Pa	1.45038E-4	1.00000E-5	9.86923E-6	1.00000E-6
kg_f/m^2	1.42233E-3	9.80665E-5	9.67841E-5	9.80665E-6
Torr	1.93368E-2	1.33322E-3	1.31579E-3	1.33322E-4
in H_2O	3.61273E-2	2.49089E-3	2.45832E-3	2.49089E-4
psia	1.00000	6.89476E-2	6.80460E-2	6.89476E-3
bar	1.45038E1	1.00000	9.86923E-1	1.00000E-1
atm	1.46959E1	1.01325	1.00000	1.01325E-1
MPa	1.45038E2	1.00000E1	9.86923	1.00000

APPENDIX

C

PROPERTIES OF THE NUCLIDES

This appendix lists some of the properties of nuclides and natural elements that are useful in nuclear engineering. Sources of data and explanations of notation are given below.

C.1 NUCLEAR SPECIES

The nuclear species (or nuclides) listed in this appendix are limited to naturally occurring nuclides and the following classes of synthetic radionuclides: products of neutron reaction with naturally occurring nuclides, fission products, and a few positron-emitting radionuclides important in radioisotope applications, such as ^{11}C, ^{13}N, ^{15}O, and ^{18}F. Very short-lived products of successive reaction of two or more neutrons with naturally occurring nuclides have not been listed. The nuclides listed thus include almost all those important in nuclear engineering applications, but many radionuclides that decay by positron emission or electron capture and some very short-lived negative beta emitters are not included, as these have no present nuclear engineering application.

Nuclear species are characterized in the first two columns of the table by the name of the element and the mass number of the nuclide. The complete symbol notation for a nuclide can be obtained from these two columns. For example, the symbol notation for the boron isotope of atomic number 5 and mass number 10 is $^{10}_{5}B$.

Metastable nuclides, of higher energy than the ground state, are characterized by the letter m following the mass number. For the few nuclides with two listed metastable states, the one with higher energy is designated m2 and the lower m1 (for example, $^{192m2}Ir$ and $^{192m1}Ir$).

For elements with two or more naturally occurring isotopes, the first-row entry refers to the mixture of isotopes with the atomic percentages given in the fourth column headed "Abundance (a/o)."

Several of the naturally occurring radioisotopes of elements 81 through 92 have synonymous names different from the element name. These synonymous names, or the chemical symbols for them, are listed in column 1, enclosed in square brackets. For example, ^{210}Pb is also known as radium-D [RaD].

C.2 MASS

The third column gives the mass of each nuclide, or naturally occurring mixed element, in atomic mass units (amu), relative to ^{12}C, whose mass is defined as 12.0000000. Atomic masses are those listed in the *Handbook of Physics* [C1].

C.3 ABUNDANCE OR HALF-LIFE

To conserve space, the fourth column gives either the relative abundance of a nuclide in the naturally occurring element, expressed in atomic percent and designated "a/o," or the half-life of a radionuclide designated by one of the following abbreviations for a unit of time; y, year; d, day; m, minute; or s, second. For a naturally occurring radionuclide, such as $^{40}_{19}K$, two rows are required, with the first giving the percent abundance and the second the half-life. Relative abundances are those listed in the *Handbook of Physics* [C1]. Half-lives are those given in table II of *Table of Isotopes* [L1].

To conserve space, numbers greater than 99,999 are written in the exponential notation used in computer printouts. For example, the half-life of ^{40}K, 1.26×10^9 y, is written 1.26E9y.

C.4 DECAY MODE

The fifth column gives the type of radioactive decay experienced by a radionuclide, and the sixth column gives the percent of decays of the type listed. These data are from reference [L1]. Types of decay are designated as follows: —, negative beta emission; α, alpha-particle emission, +, positron emission; EC, orbital electron capture; IT isomeric transition, accompanied by gamma-ray emission; n, neutron emission; SF, spontaneous fission. Addition of the symbol m means that the reaction product is in its metastable state. For example, when $^{99}_{42}Mo$ decays by negative beta emission, 86.7 percent of the reaction product is metastable $^{99m}_{43}Tc$ and 13.3 percent is the ground-state $^{99}_{43}Tc$. Addition of the symbol g means that the reaction product is in the ground state. For example, when $^{192m2}Ir$ decays by internal transition, 100 percent of the product is the ground-state ^{192}Ir.

C.5 DECAY ENERGY

The seventh column, headed "Total," gives the effective energy in million electron volts emitted per decay. This comprises all forms of radiation that eventually degrade into heat, which include alpha particles, negative and positive electrons, gamma photons, x-rays from orbital electron capture and bremsstrahlung, neutrons, recoil nuclei, and fission fragments. Only neutrinos are excluded. The eighth column, headed "γ," gives the energy per decay that is in the form of gamma photons whose individual energy is 0.2 MeV or greater. A "0*" in the eighth column means that the energy of such gamma photons is less than 0.001 MeV per decay, but that the total energy of weaker gamma photons is greater than 0.001 MeV/decay. A "0" in the eighth column means that the total energy of all gamma photons, excluding bremsstrahlung, is less than 0.001 MeV.

The energies per decay in columns 7 and 8 have been taken from Report ORNL-4628 [B1] for all nuclides for which these energies are given and are not clearly inconsistent with reference [L1]. When inconsistencies were noted, values were computed from reference [L1]. It was then assumed that the ratio of the average beta-particle energy (used in this appendix) to the maximum beta-particle energy (the only property given by reference [L1] for most beta emitters) was 0.400. Values of the total effective energy per decay computed from this assumption, or energy values judged to be uncertain for other reasons, are enclosed in parentheses, such as (1.40) for the total decay energy for ^{11}C.

For nuclides decaying by electron capture (EC), the effective energy per decay (other than gamma rays from isomeric transitions of the daughter) was taken as the energy of the K_α x-ray of the daughter. This differs from Report ORNL-4628, which apparently includes neutrino energy in the effective decay energy of nuclides decaying by electron capture. For example, the effective decay energy of ^{37}Ar, which decays 100 percent by electron capture, is listed in this appendix as 0.003 MeV, the K_α energy of chlorine, whereas Report ORNL-4628 gives 0.814 MeV. Of this 0.814 MeV, 0.811 MeV is emitted in the form of neutrinos and is not considered as effective decay energy.

For more detailed information about decay schemes or product energies, reference [L1] should be used.

C.6 NEUTRON REACTIONS

All cross-section data given in the last three columns are from vol. 1 of the 3d edition of BNL-325 [M1].

The column headed "Prod." identifies the reaction in which the neutron takes part. Reactions in which the heavier product is in its ground state are designated by the nature of the lighter product, viz:

Designation	Lighter product
α	^4_2He
γ	Gamma ray
p	^1_1H

In these cases, the heavier product nuclide can be inferred from the conservation of mass numbers and atomic numbers. For example, when $^{10}_5B$ reacts with a neutron and emits an alpha particle, the heavier reaction product must be ^7_3Li:

$$^{10}_5B + ^1_0n \rightarrow ^4_2He + ^7_3Li$$

In an (n, γ) reaction in which the product nuclide is in an isomeric state, the column "Prod." lists the designation of the isomer. For example, the reaction

$$^{115}_{49}In + ^1_0n \rightarrow ^{116m2}_{49}In + \gamma$$

is designated "116m2."

When absorption of a neutron may result in one of two or more reactions, the sum of the cross sections for all reactions, the absorption cross section, is denoted by "a." For example, when ^{115}In absorbs a neutron, the absorption cross section is 202 b. The reaction cross section to form $^{116m2}In$ is 92 b, and the cross section to form $^{116m1}In$ is 65 b. By inference, the cross section to form ^{116}In is $202 - 92 - 65 = 45$ b.

Fission cross sections are denoted by "f." For fissionable isotopes of thorium and elements of higher atomic number, the average number of neutrons produced per fission is listed in the same row as the fission cross section, in the same column as the mass, to conserve space in the table. The average number of prompt and delayed neutrons produced by fission with a thermal neutron is denoted by "ν." The average number of prompt neutrons produced by fission with a thermal neutron is denoted by "ν_p." The average number of neutrons emitted per spontaneous fission is denoted by "ν_{sp}."

The last column gives the resonance integral I, defined in terms of the reaction cross section $\sigma(E)$ at energy E(eV) by

$$I = \int_{0.5}^{\infty} \frac{\sigma(E)\, dE}{E}$$

This is the reaction cross section for a dilute absorber in 1/E neutron spectrum down to the cadmium cutoff energy of 0.5 eV.

C.7 CREDITS

Atomic masses are extracted from *Handbook of Physics*, 2d ed., edited by E. U. Condon and H. Oldishaw, copyright 1967 by McGraw-Hill Book Company, and are used with permission of the publisher.

Half-lives are extracted by permission of John Wiley & Sons, Inc., from *Table of Isotopes*, 6th ed. by C. M. Lederer, J. M. Hollander, and I. Perlman, copyright 1967 by General Manager of U.S. Atomic Energy Commission.

REFERENCES

B1. Bell, M. J.: "ORIGEN–The ORNL Isotope Generation and Depletion Code," Report ORNL-4628, May 1973.

C1. Condon, E. U., and Oldishaw, H. (eds.): *Handbook of Physics*, 2d ed., McGraw-Hill, New York, 1967.

L1. Lederer, C. M., Hollander, J. M., and Perlman, I.: *Table of Isotopes*, 6th ed., Wiley, New York, 1967.

M1. Mughabghab, S. F., and Garber, D. I.: *Neutron Cross Sections*, vol. 1, *Resonance Parameters*, BNL-325, 3d ed., vol. 1, June 1973.

Element (Symbol) At. no., Z	Mass no., A	Mass, amu	Abundance (a/o) or half-life	Decay		Effective energy per decay, MeV		Neutron reactions		
									Barns	
				Type	Percent	Total	γ	Prod.	2200 m/s cross section	Resonance integral
Neutron	1	1.00866520	12 m	–	100	(0.31)	0			
Hydrogen		1.00797								
(H)	1	1.00782519	99.985 a/o					γ	0.332	
	2	2.01410222	0.015 a/o					γ	0.00053	
	3	3.01604971	12.3 y	–	100	0.0057	0			
Helium		4.0026								
(He)	3	3.01602973	0.000137 a/o					p	5327	
2								γ		2390
	4	4.00260312	99.999863 a/o						0	
	6	6.0188927	0.8 s	–	100	1.58	0			
Lithium		6.939						a	70.7	
(Li)	6	6.0151247	7.5632 a/o					α	940	
3	7	7.0160039	92.4368 a/o					γ	0.037	
	8	8.0224871	0.84 s	–, 2α	100	8.49	0			
Beryllium	9	9.0121855	100 a/o					γ	0.0092	0.004
(Be)	10	10.0135344	2.5E6y	–	100	(0.22)	0	γ	< 0.001	
4										
Boron		10.811						a	759	341
(B)	10	10.0129388	19.61 a/o					α	3837	1722
5	11	11.0093053	80.39 a/o					γ	0.0055	
	12	12.0143537	0.0203 s	–	100	6.381	0.058			
Carbon		12.01115						γ	0.0034	0.0015
(C)	11	11.0114317	20.3 m	+	100	(1.40)	1.022			
6	12	12.0000000	98.893 a/o					γ	0.0034	0.0015
	13	13.003354	1.107 a/o					γ	0.0009	0.0013
	14	14.00324197	5730 y	–	100	0.045	0		0	

Element (Symbol) At. no., Z	Mass no., A	Mass, amu	Abundance (a/o) or half-life	Decay		Effective energy per decay, MeV		Neutron reactions		
									Barns	
				Type	Percent	Total	γ	Prod.	2200 m/s cross section	Resonance integral
Nitrogen		14.0067						*a*	1.85	0.90
(N)	13	13.0057384	10.0 m	+	100	1.499	1.022			
7	14	14.00307439	99.6337 a/o					γ	0.075	0.90
								p	1.81	
	15	15.0001077	0.3663 a/o					γ	0.000024	
	16	16.0061033	7.2 s	−	100	7.168	4.615			
Oxygen		15.9994						γ	0.00027	0.00031
(O)	15	15.0030703	124 s	+	100	(1.72)	1.022			
8	16	15.9949150	99.759 a/o					γ	0.000178	0.00027
	17	16.9991329	0.0374 a/o					α	0.235	0.105
	18	17.9991600	0.2039 a/o					γ	0.00016	0.00081
	19	19.0035779	29 s	−	100	2.622	1.040			
Fluorine	18	18.0009366	109.7 m	+	97	(1.24)	0.991			
(F)	19	18.9984046	100 a/o					γ	0.0095	0.0176
9	20	19.999987	11.4 s	−	100	4.146	1.63			
Neon		20.183						γ	0.038	
(Ne)	19	19.0018809	17.5 s	+	100	(1.91)	1.022			
10	20	19.9924405	90.92 a/o					γ	0.037	
	21	20.9938486	0.257 a/o					γ	0.692	
	22	21.9913847	8.82 a/o					γ	0.048	
	23	22.9944729	37.6 s	−	100	2.058	0.159			
Sodium	22	21.9944366	2.60 y	+	90	2.448	2.194	γ	29,000	
(Na)	23	22.9897707						γ	0.530	0.311
11	24	23.9909623	15.0 h	−	100	4.727	4.122			
	25	24.989955	60 s	−	100	1.909	0.381			

Magnesium		24.312						γ	0.063	0.038
(Mg)	23	22.994125	12.1 s	+	100	(2.26)	1.062			
12	24	23.9850417	78.7 a/o					γ	0.052	0.030
	25	24.985839	10.13 a/o					γ	0.180	0.111
	26	25.982593	11.17 a/o					γ	0.0382	0.025
	27	26.9843447	9.5 m	−	100	2.068	0.895	γ	0.04	
	28	27.983875	21 h	−	100	(1.56)	1.35			
Aluminum	26	25.9868909	7.4E5y	+	85	(3.12)	2.724			
(Al)	27	26.9815389	100 a/o					γ	0.230	0.17
13	28	27.9819047	2.31 m	−	100	3.022	1.780			
	29	28.980442	6.6 m	−	100	2.339	1.349			
Silicon		28.086						γ	0.16	
(Si)	28	27.9769292	92.21 a/o					γ	0.17	0.078
14	29	28.9764958	4.70 a/o					γ	0.28	
	30	29.9737628	3.09 a/o					γ	0.107	0.106
	31	30.975349	2.62 h	−	100	0.594	0.001	γ	0.48	
	32	31.974020	650 y	−	100	(0.08)	0			
Phosphorus	31	30.9737647	100 a/o					γ	0.180	0.08
(P)	32	31.9739095	14.3 d	−	100	0.695	0			
15	33	32.9717282	25 d	−	100	0.076	0			
Sulfur		32.064						γ	0.520	
(S)	32	31.9720737	95.0 a/o					γ	0.53	
16	33	32.9714619	0.76 a/o					α	0.14	
	34	33.9678646	4.22 a/o					γ	0.24	
	35	34.9690308	88 d	−	100	0.048	0			
	36	35.96709	0.0136 a/o					γ	0.15	
	37	36.97101	5.06 m	−	100	3.564	2.781			
Chlorine		35.452						γ	33.2	12
(Cl)	35	34.9688511	75.770 a/o					γ	43	17
17	36	35.9683089	3.1E5y	−	100	0.314	0	γ	<10	
	37	36.9658985	24.229 a/o					γ	0.428	0.310
	38	37.968005	37.3 m	−	100	3.101	1.628			

Element (Symbol) At. no., Z	Mass no., A	Mass, amu	Abundance (a/o) or half-life	Decay		Effective energy per decay, MeV		Neutron reactions		
									Barns	
				Type	Percent	Total	γ	Prod.	2200 m/s cross section	Resonance integral
Argon		39.948						γ	0.678	0.42
(Ar)	36	35.9675445	0.337 a/o					γ	5	2.5
18	37	36.9667722	35 d	EC	100	0.003	0*			
	38	37.9627278	0.063 a/o					γ	0.8	0.4
	39	38.964317	269 y	—	100	0.234	0	γ	600	
	40	39.9623842	99.6 a/o					γ	0.66	0.41
	41	40.9645	1.83 h	—	100	3.355	1.283	γ	0.5	
Potassium		39.102						γ	2.10	1.0
(K)	39	38.9637101	93.1 a/o					γ	1.96	0.9
19	40	39.9639998	0.01181 a/o	EC	11	(0.63)	0.161	γ	30	
			1.26E9y	—	89			p	4.4	
	41	40.9618323	6.88 a/o					γ	1.46	1.42
	42	41.962406	12.4 h	—	100	1.677	0.281			
	43	42.960730	22.4 h	—	100	1.296	0.993			
Calcium		40.08						γ	0.43	0.20
(Ca)	40	39.9625889	96.97 a/o					γ	0.40	0.18
20	41			EC	100	0.004	0*			
	42	41.9586252	0.64 a/o					γ	0.65	0.29
	43	42.9587796	0.145 a/o					γ	6.2	5.5
	44	43.9554905	2.06 a/o					γ	1.0	0.56
	45	44.9561895	165 d	—	100	0.103	0			
	46	45.953689	0.0033 a/o					γ	0.7	0.32
	47	46.954538	4.53 d	—	100	(1.45)	1.091			
	48	47.952531	0.185 a/o					γ	1.1	
	49	48.955675	8.8 m	—	100	4.057	3.17			

Scandium	45	44.955918	100 a/o					γ	26.5	11.3
(Sc)	46	45.9551726	83.9 d	–	100	2.368	2.01	γ	8.0	
21	47	46.9524129	3.43 d	–	100	0.491	0.117			
	48	47.952221	1.83 d	–	100	3.578	3.36			
Titanium		47.90						γ	6.1	
(Ti)	45	44.958129	3.09 h	+	84	(1.21)	0.86			
22	46	45.952631	7.93 a/o					γ	0.6	0.4
	47	46.951768	7.28 a/o					γ	1.7	1.8
	48	47.947950	73.94 a/o					γ	7.8	3.7
	49	48.947870	5.51 a/o					γ	2.2	1.5
	50	49.944785	5.34 a/o					γ	0.179	0.118
	51	50.946603	5.8 m	–	100	1.227	0.356			
Vanadium		50.942						γ	5.04	2.7
(V)	49	48.9485225	330 d	EC	100	0.005	0*			
23	50	49.947163	0.24 a/o	EC	70			γ	70	67
			6E15y	–	30	(0.27)	0.235			
	51	50.943961	99.76 a/o					γ	4.88	2.7
	52	51.944780	3.76 m	–	100	2.624	1.447			
	53	52.943980	2.0 m	–	100	1.892	1.00			
Chromium		51.996						γ	3.1	1.7
(Cr)	49	48.951271	41.9 m	+	94	(1.56)	1.00			
24	50	49.946054	4.35 a/o					γ	15.9	7.6
	51	50.9447682	27.8 d	EC	100	0.035	0.029			
	52	51.940513	83.76 a/o					γ	0.76	0.60
	53	52.940652	9.51 a/o					γ	18.2	8.85
	54	53.938881	2.38 a/o					γ	0.36	0.18
	55	54.940833	3.5 m	–	100	1.230	0			
Manganese	54	53.940362	303 d	EC	100	0.842	0.835	γ	<10	
(Mn)	55	54.9380503	100 a/o					γ	13.3	14.0
25	56	55.9389102	2.576 h	–	100	2.636	1.763			
Iron		55.847						γ	2.55	1.4
(Fe)	53	52.945572	8.5 m	+	98	(2.29)	1.166			
26	54	53.939617	5.82 a/o					γ	2.25	1.2
	55	54.9382986	2.6 y	EC	100	0.007	0*			

Element (Symbol) At. no., Z	Mass no., A	Mass, amu	Abundance (a/o) or half-life	Decay: Type	Decay: Percent	Effective energy per decay, MeV: Total	Effective energy per decay, MeV: γ	Neutron reactions: Prod.	Neutron reactions, Barns: 2200 m/s cross section	Neutron reactions, Barns: Resonance integral
Iron (*Cont.*)	56	55.934936	91.66 a/o					γ	2.63	1.4
	57	56.935397	2.19 a/o					γ	2.48	1.3
	58	57.933282	0.33 a/o					γ	1.15	1.19
	59	58.9348778	45 d	−						
	60	59.933964	3E5y	−	100	(0.06)	0*			
Cobalt	58	57.935761	71.3 d	+	15	(1.01)	0.981	γ	1880	6890
(Co)	59	58.933189	100 a/o					a	37.2	75.5
27								60m	2.0	
	60m		10.5 m	IT	0.997	0.064	0.005			
	60	59.9338134	5.26 y	−	100	2.637	2.505	γ	2.0	4.3
	61	60.932440	99 m	−	100	0.556	0*			
Nickel		58.71						γ	4.43	2.2
(Ni)	58	57.935342	67.88 a/o					γ	4.6	2.2
28	59	58.9343423	8×10^4 y	EC	100	0.008	0*	γ	92	138
								α	12	
	60	59.930787	26.23 a/o					γ	2.8	1.5
	61	60.931056	1.19 a/o					γ	2.5	1.6
								α	0.047	
	62	61.928342	3.66 a/o					γ	14.2	6.8
	63	62.929664	92 y	−	100	0.027	0	γ	23	
	64	63.927958	1.08 a/o					γ	1.49	1.1
	65	64.930072	2.56 h	−	100	1.207	0.584	γ	24	11
Copper		63.54						γ	3.79	3.2
(Cu)	63	62.929592	69.1739 a/o					γ	4.5	4.9
29	64	63.929759	12.8 h	+	19					
				EC	43	(0.34)	0.201			
				−	38					

	65	64.927786	30.8261 a/o					γ	2.17	2.4
	66	65.928871	5.1 m	−	100	1.163	0.098	γ	135	
Zinc		65.37						γ	1.10	2.3
(Zn)	64	63.929145	48.89 a/o					γ	0.78	1.4
30	65			+	1.7	0.563	0.564			
	66	65.926052	27.81					γ	0.85	0.8
	67	66.927145	4.11 a/o					γ	6.9	20
	68	67.924857	18.57 a/o					*a*	1.07	3.3
								69m	0.072	
	69m		13.8 h	IT	100	0.439	0.439			
	69	68.926541	57 m	−	100	0.350	0			
	70	69.925334	0.62 a/o					*a*	0.092	
								71m	0.0087	
	71m		4.0 h	−	100	1.900	1.292			
	71	70.92751	2.4 m	−	100	1.110	0.111			
Gallium		69.72						γ	2.9	18.7
(Ga)	69	68.925574	60.4 a/o					γ	1.68	15.6
31	70	69.926035	21.1 m	−	100	0.690	0.055			
	71	70.924706	39.6 a/o					γ	4.86	31.2
	72	71.926372	14.10 h	−	100	3.641	3.16			
	73	72.925126	4.9 h	−	100	1.006	0.333			
	74	73.927190	7.9 m	−	100	4.010	2.88			
	75		2 m	−	100	1.472	0.021			
Germanium		72.59						γ	2.3	6.1
(Ge)	70	69.924251	20.52 a/o					γ	3.43	2.4
32	71	70.924956	11.4 d	EC	100	0.011	0*			
	72	71.922081	27.43 a/o					γ	0.98	0.88
	73	72.923462	7.76 a/o					γ	15	65
	74	73.921180	36.54 a/o					*a*	0.34	0.43
								75m	0.143	
	75m		48 s	IT	100	0.139	0*			
	75	74.922883	82 m	−	100	0.468	0.037			
	76	75.921405	7.76 a/o					*a*	0.142	2.0
								77m	0.092	

Element (Symbol) At. no., Z	Mass no., A	Mass, amu	Abundance (a/o) or half-life	Decay		Effective energy per decay, MeV		Neutron reactions		
									Barns	
				Type	Percent	Total	γ	Prod.	2200 m/s cross section	Resonance integral
Germanium (*Cont.*)	77m		54 s	IT	24	(0.095)	0.045			
				—	76					
	77	76.9236	1.3 h	—		1.724	1.032			
	78		1.47 h	—		0.500	0.278			
Arsenic	75	74.9215964	100 a/o					γ	4.3	60
(As)	76	75.922397	26.5 h	—	100	2.327	0.384			
33	77	76.920646	38.7 h	—	100	0.240	0.100			
	78	77.92190	91 m	—	100	2.343	0.902			
	79	78.92089	9.0 m	—	100	(0.99)	0.031			
	80	79.92297	15.3 s	—	100	2.932	0.490			
	81		33 s	—	100	1.670	0			
Selenium		78.96						γ	11.7	13
(Se)	74	73.922476	0.87 a/o					γ	51	565
34	75	74.9225249	120.4 d	EC	100	0.404	0.1			
	76	75.919207	9.02 a/o					γ	85	44
	77	76.919911	7.58 a/o					γ	42	34
	78	77.917313	23.52 a/o					γ	0.4	4.7
	79m		3.9 m	IT	100	0.096	0*			
	79	78.9184943	6.5E4y	—	100	0.064	0			
	80	79.916527	49.82 a/o					*a*	0.610	1.7
								81m	0.080	
	81m		57 m	IT	100	0.103	0			
	81	80.917984	18.6 m	—	100	0.552	0.006			
	82	81.916707	9.19 a/o					*a*	0.045	
								83m	0.0058	
	83m		70 s	—	100	2.712	1.03			

	83		25 m	—	100	2.842	2.17			
	84		3.3 m	—	100	1.280	(0)			
Bromine (Br) 35		79.909						γ	6.8	90
	79	78.918329	50.6864 a/o					a	11.1	132.5
								80m	2.6	34.5
	80m		4.38 h	IT	100	0.086	0*			
	80	79.9185357	17.6 m	EC	6					
				+	2.6	(0.76)	0.076			
				—	91.4					
	81	80.916292	49.3136 a/o					a	2.69	51
								82m	2.43	
	82m		6.1 m	IT	99	0.047	0.001			
				—	0.17					
	82	81.916802	35.34 h	—	100	2.777	2.648			
	83	82.915168	2.41 h	—	100	0.394	0.007			
	84	83.91655	31.8 m	—	100	2.862	1.428			
	85	84.91553	3.0 m	—	100	1.050	0			
	86	85.9182	54 s	—	100	3.230	0*			
	87		55 s	—	98	5.676	3.81			
				—, *n*	~2					
Krypton (Kr) 36		83.8						γ	25.0	53
	78	77.920403	0.354 a/o					a	4.71	5.3
								79m	0.21	
	79m		55 s	IT	100	0.127	0*			
	79	78.920068	34.9 h	+	8	(0.24)	0.217			
	80	79.916380	2.27 a/o					a	14.0	56.1
								81m	4.55	
	81m		13 s	IT	100	0.190	0*			
	81	80.916610	2.1E5y	EC	100	0.014	0*			
	82	81.913482	11.56 a/o					a	45	200
								83m	20	
	83	82.914131	11.55 a/o					γ	200	230
	84	83.911503	56.90 a/o					a	0.130	2.7
								85m	0.090	

Element (Symbol) At. no., Z	Mass no., A	Mass, amu	Abundance (a/o) or half-life	Decay		Effective energy per decay, MeV		Neutron reactions		
									Barns	
				Type	Percent	Total	γ	Prod.	2200 m/s cross section	Resonance integral
Krypton (*Cont.*)	85m		4.4 h	IT	23	(0.44)	0.070			
				—	77					
	85	84.912523	10.76 y	—	100	0.274	0.005	γ	1.66	1.8
	86	85.910615	17.37 a/o					γ	0.060	0.03
	87	86.913365	76 m	—	100	2.738	1.374			
	88	87.91427	2.80 h	—	100	2.097	1.715			
	89	88.9166	3.2 m	—	100	3.213	1.774			
	90	89.91972	33 s	—	100	3.091	2.065			
	91		10 s	—	100	1.570	(0)			
	92		3.0 s	—	100	2.250	(0)			
	93		2.0 s	—	100	3.680	(0)			
	94		1.4 s	—	100	2.910	(0)			
Rubidium		85.47						γ	0.37	6.0
(Rb)	85	84.911800	72.15 a/o					*a*	0.46	7.5
37								86m	0.050	
	86m		1.04 m	IT	100	0.56	0.56			
	86	85.911193	18.66 d	—	100	0.797	0.095			
	87	86.909186	27.85 a/o					γ	0.12	2.0
			4.7E10y	—	100	0.110	0			
	88	87.911270	17.8 m	—	100	2.694	0.571	γ	1.0	
	89	88.91165	15.4 m	—	100	3.059	2.459			
	90	89.91482	2.9 m	—	100	4.635	3.180			
	91	90.91607	1.2 m	—	100	2.451	0.400			
	92		5.3 s	—	100	4.558	1.071			
	93		5.6 s	—	100	2.750	(0)			
	94		2.9 s	—	100	4.170	(0)			

Strontium (Sr) 38		87.62						γ	1.21	11
	84	83.913430	0.56 a/o					*a*	0.81	10.6
								85m	0.55	
	85m		70 m	EC IT	14 86	0.206	0.204			
	85	84.912989	64 d	EC	100	0.530	0.514			
	86	85.909285	9.86 a/o					87m	0.84	4
	87m		2.83 h	IT	99	0.38	0.38			
	87	86.908892	7.02 a/o					γ	16	120
	88	87.905641	82.56 a/o					γ	5.8	0.05
	89	88.907442	52 d	−	100	0.607	0	γ	0.42	
	90	89.907747	28.1 y	−	100	0.221	0	γ	0.9	
	91	90.910161	9.67 h	− *m* −	59 41	(1.38)	0.743			
	92	91.910980	2.71 h	−	100	(1.50)	1.233			
	93	92.91471	8 m	−	100	2.587	1.363			
	94	93.91538	1.3 m	−	100	2.253	1.42			
Yttrium (Y) 39	89	88.905871	100 a/o					γ	1.28	1.0
	90	89.907163	64 h	−	100	0.993	0.00035	γ	< 6.5	
	91m		50 m	IT	100	0.551	0.551			
	91	90.907295	58.8 d	−	100	0.642	0.0036	γ	1.4	
	92	91.908926	3.53 h	−	100	1.635	0.244			
	93	92.909552	10.2 h	−	100	1.249	0.095			
	94	93.91168	20.3 m	−	100	2.434	0.638			
	95	94.91254	10.9 m	−	100	2.782	1.185			
	96	95.91569	2.3 m	−	100	3.207	1.70			
Zirconium (Zr) 40		91.22						γ	0.185	1.10
	90	89.904699	51.46 a/o					γ	0.10	0.20
	91	90.905642	11.23 a/o					γ	1.03	6.5
	92	91.905030	17.11 a/o					γ	0.26	0.54
	93	92.90645	1.5E6y	−	100	(0.04)	0			
	94	93.906313	17.40 a/o					γ	0.056	0.30
	95	94.908035	65 d	− *m* −	2 98	0.883	0.725			

Element (Symbol) At. no., Z	Mass no., A	Mass, amu	Abundance (a/o) or half-life	Decay		Effective energy per decay, MeV		Neutron reactions		
									Barns	
				Type	Percent	Total	γ	Prod.	2200 m/s cross section	Resonance integral
Zirconium	96	95.908286	2.80 a/o					γ	0.017	5.0
(*Cont.*)	97	96.910966	17.0 h	$-m$	96	1.575	0.789			
				$-$	4					
Niobium	93	92.906382	100 a/o					γ	1.15	8.5
(Nb)	94	93.907303	2E4y	$-$	100	(1.77)	1.573	γ	13.6	125
41	95m		90 h	IT	100	0.235	0.235			
	95	94.9068318	35 d	$-$	100	0.812	0.765	γ	<7	
	96	95.908056	23.4 h	$-$	100	(2.73)	2.450			
	97m		1.0 m	IT	100	0.747	0.747			
	97	96.908096	72 m	$-$	100	1.159	0.672			
	98	97.91035	51 m	$-$	100	3.634	1.50			
	99	98.91105	2.4 m	$-$	100	1.411	0.260			
	100		11 m	$-$	100	3.645	1.196			
Molybdenum		95.94						γ	2.65	22
(Mo)	92	91.906810	15.84 a/o					γ	0.0045	0.52
42	93	92.90683	>100 y	EC	100	0.019	0*			
	94	93.905090	9.04 a/o					γ	0.016	0.57
	95	94.905839	15.72 a/o					γ	14.5	105
	96	95.904673	16.53 a/o					γ	1.0	20
	97	96.906021	9.46 a/o					γ	2.2	13
	98	97.905408	23.78 a/o					γ	0.13	6.2
	99	98.90772	67 h	$-m$	86.7	0.687	0.149			
				$-$	13.3					
	100	99.907474	9.63 a/o					γ	0.199	3.75
	101	100.910353	14.6 m	$-$	100	2.080	1.666			
	102	101.91025	11 m	$-$						

Technetium	96	95.90783	4.3 d	EC	100	2.52	2.50			
(Tc)	97	96.90634	2.6E6y	EC	100	0.020	0*			
43	98	97.90711	1.5E6y	—	100	(1.53)	1.42	γ	2.6	
	99m		6.0 h	IT	100	0.1427	0*			
	99		2.12E5y	—	100	0.114	0	γ	19	340
	100	99.90784	17 s	—	100	1.460	(0)			
	101	100.907326	14.0 m	—	100	0.824	0.350			
	102m		4.5 m	—		1.305	0.470			
	102	101.90918	5 s	—	100	1.920	(0)			
	103	102.90883	50 s	—	100	1.128	0.069			
	104	103.91171	18 m	—	100	4.139	3.122			
Ruthenium		101.07						γ	2.56	42
(Ru)	96	95.907598	5.51 a/o					γ	0.25	6.6
44	97	96.90763	2.9 d	EC	100	0.25	0.23			
	98	97.905288	1.87 a/o					γ	8	
	99	98.905935	12.72 a/o					γ	5	195
	100	99.904218	12.62 a/o					γ	5.8	11.2
	101	100.905576	17.07 a/o					γ	3.1	85
	102	101.904347	31.61 a/o					γ	1.3	4.1
	103	102.906306	39.6 d	— *m*	100	0.556	0.490			
	104	103.905430	18.58 a/o					γ	0.47	4.61
	105	104.907679	4.44 h	— *m*	89 }	1.074	0.677	γ	0.20	
				—	11 }					
	106	105.907322	367 d	—	100	0.010	0	γ	0.146	2.6
	107	106.91013	4.2 m	—	100	1.508	0.217			
	108	107.9101	4.5 m	—	100	0.547	0.046			
Rhodium	103m		57 m	IT	100	0.040	0*			
(Rh)	103	102.905511	100 a/o					*a*	150	1100
45								104m	11	
	104m		4.41 m	IT	100	0.129	0*	γ	800	
	104	103.906659	43 s	EC	0.5 }	1.010	0.013	γ	40	
				—	99.5 }					
	105m		45 s	IT	100	0.129	0*			
	105	104.905671	35.9 h	—	100	0.223	0.076	*a*	16,000	15,800
								106m	4,700	

Element (Symbol) At. no., Z	Mass no., A	Mass, amu	Abundance (a/o) or half-life	Decay Type	Decay Percent	Effective energy per decay, MeV Total	Effective energy per decay, MeV γ	Neutron reactions Prod.	Barns 2200 m/s cross section	Barns Resonance integral
Rhodium	106m		130 m	—	100	2.661	2.374			
(*Cont.*)	106	105.907279	30 s	—	100	1.773	0.340			
	107	106.906753	22 m	—	100	(0.79)	0.307			
	108	107.9087	17 s	—	100	2.505	0.699			
	109		< 1h	—	100	0.728	(0)			
Palladium		106.4						γ	6.9	90
(Pd)	102	101.905609	0.96 a/o					γ	4.8	
46	103	102.906107	17 d	EC *m*	100	0.07	0*			
	104	103.904011	10.97 a/o							
	105	104.905064	22.23 a/o					γ	14	90
	106	105.903479	27.33 a/o					*a*	0.305	5.73
								107m	0.013	
	107m		22 s	IT	100	0.21	0.21			
	107	106.9051316	7E6y	—	100	0.014	0			
	108	107.903891	26.71 a/o					*a*	12	250
								109m	0.2	
	109m		4.7 m	IT	100	0.188	0*			
	109	108.905954	13.47 h	— *m*	100	0.413	0			
	110	109.905164	11.81 a/o					*a*	0.22	
								111m	0.020	
	111m		5.5 h	IT	75 }	0.373	0*			
				—	25 }					
	111	110.90767	22 m	— *m*	100	0.848	0			
	112	111.907386	21 h	—	100	(0.11)	0			
	113		1.4 m	— *m*	10 }	(1.250)	0			
				—	90 }					
	114		2.4 m	—	100	(1.250)	0			

	115		45 s	— *m*	28	(2.00)	0			
				—	72					
Silver		107.87						γ	63.6	747
(Ag)	107	106.905094	51.829 a/o					*a*	37.2	94
47								108m	3.0	
	108m		127 y	EC	90	1.626	1.593			
				IT	10					
	108	107.905949	2.42 m	EC	1.7	0.738	0.018			
				—	98					
	109	108.904756	48.170 a/o					*a*	91	1450
								110m	4.5	
	110m		253 d	IT	1.3	2.850	2.761	γ	82	
				—	98.7					
	111m		74 s	IT	100	0.07	0*			
	111	110.905316	7.5 d	—	100	0.428	0.023	γ	3	105
	112	111.907064	3.2 h	—	100	1.777	0.437			
	113m		1.2 m	—	100	1.421	0.591			
	113	112.906556	5.3 h	—	100	0.942	0.592			
	114	113.908300	4.5 s	—	100	2.268	0.558			
	115m		20 s	—	100	(1.96)	(0.60)			
	115	114.90893	20 m	— *m*	9	2.065	1.210			
				—	91					
	116	115.91131	2.5 m	—	100	(3.361)	(0.66)			
	117		1.1 m	—	100	(1.25)	(0)			
Cadmium		112.40						γ	2450	
(Cd)	106	105.906462	1.215 a/o							
48	108	107.904186	0.875 a/o					γ	1.1	
	109		453 d	EC	100	0.114	0	γ	650	
	110	109.903011	12.39 a/o					*a*	11	40
								111m	0.1	
	111m		48.6 m	IT	100	0.396	0.247			
	111	110.904188	12.75 a/o					γ	24	51
	112	111.902762	24.07 a/o					γ	2.2	15
	113m		14 y	—	99+	0.223	0.0002			
	113	112.904408	12.26 a/o					γ	19,910	

Element (Symbol) At. no., Z	Mass no., A	Mass, amu	Abundance (a/o) or half-life	Decay: Type	Decay: Percent	Effective energy per decay, MeV: Total	Effective energy per decay, MeV: γ	Neutron reactions: Prod.	Neutron reactions, Barns: 2200 m/s cross section	Neutron reactions, Barns: Resonance integral
Cadmium	114	113.903360	28.86 a/o					a	0.336	20
(*Cont.*)								115m	0.036	
	115m		43 d	—	100	0.617	0.033			
	115	114.905431	53.5 h	— *m*	100	0.564	0.196			
	116	115.904761	7.58 a/o					a	0.077	
								117m	0.027	
	117m		3.4 h	— *m*	44 }	2.065	1.342			
				—	56 }					
	117	116.907239	2.4 h	— *m*	100	1.367	0.868			
	118	117.906970	49 m	—	100	0.266	0			
	119m		2.7 m	— *m*	100	1.500	0			
	119	118.909740	10 m	— *m*	100	1.601	0.10			
	120		(1 m)	— *m*	(50) }	(1.250)	(0)			
				—	(50) }					
	121		13 s	— *m*	100					
Indium		114.82						γ	193.5	3200
(In)	113	112.904089	4.28 a/o					a	11.4	282
49								114m	4.4	
	114m		50.0 d	EC	3.5 }	0.231	0.045			
				IT	96.5 }					
	114	113.904905	72 s	EC	2 }	0.78	0.003			
				—	98 }					
	115m		4.5 h	IT	95 }	0.333	0.318			
				—	5 }					
	115	114.903871	95.72 a/o					a	202	3300
			6E14y	—	100	0.198	0	116m2	92	
								116m1	65	

	116m2		2.16 s	IT*m*1	100	0.16	0*			
	116m1		54 m	—	100	2.785	2.49			
	116	115.905317	14 s	—	100	1.400	0.017			
	117m		1.93 h	IT	47	0.590	0.148			
				—	53					
	117	116.904534	44 m	—	100	1.140	0.57			
	118m		4.4 m	—	100	3.241	2.62			
	118	117.90641	5 s	—	100	1.920	0.184			
	119m		18 m	IT	5	1.174	0.061			
				—	95					
	119	118.90599	2.1 m	— *m*	5	1.487	0.775			
				—	95					
	120m		3.2 s	—	100	2.550	0.175			
	120	119.90800	46 s	—	100	4.854	3.66			
	121m		3.1 m	—	100	1.530	0			
	121		30 s	—	100	2.074	0.94			
	122		8 s	—	100	3.662	1.684			
	123m		36 s	—	100	2.029	1.10			
	123		10 s	— *m*	100	(2.029)	(1.10)			
	124		4 s	—	100	4.449	2.13			
Tin		118.69						γ	0.63	8.5
(Sn)	112	111.904835	0.96 a/o					*a*	1.15	27
50								113m	0.35	
	113m		20 m	IT	91	0.074	0*			
				EC	9					
	113	112.905187	115 d	EC	100	0.426	0.398			
	114	113.902773	0.66 a/o					γ	0.0632	1.2
	115	114.903346	0.35 a/o					γ	50	23
	116	115.901744	14.30 a/o					117m	0.006	11
	117m		14.0 d	IT	100	0.317	0*			
	117	116.902598	7.61 a/o					γ	2.6	16
	118	117.901605	24.03 a/o					119m	0.016	7
	119m		250 d	IT	100	0.089	0*			
	119	118.903313	8.58 a/o					γ	2.3	3.5
	120	119.902198	32.85 a/o					*a*	0.141	1.5
								121m	0.001	

Element (Symbol) At. no., Z	Mass no., A	Mass, amu	Abundance (a/o) or half-life	Decay		Effective energy per decay, MeV		Neutron reactions		
									Barns	
				Type	Percent	Total	γ	Prod.	2200 m/s cross section	Resonance integral
Tin (*Cont.*)	121m		76 y	−	100	0.214	0*			
	121	120.904227	27 h	−	100	0.112	0			
	122	121.903441	4.72 a/o					*a*	0.181	0.60
								123m	0.0010	
	123m		40 m	−	100	0.842	0.160			
	123	122.905738	125 d	−	100	0.597	(0.021)			
	124	123.905272	5.94 a/o					*a*	0.134	6.9
								125m	0.13	
	125m		9.7 m	−	100	1.182	0.350			
	125	124.907746	9.4 d	−	100	1.072	0.144			
	126	125.907640	1E5y	− *m*	100	(0.182)	0			
	127m		4 m	−	100	1.616	0.490			
	127	126.910260	2.1 h			(1.745)	(0.893)			
	128	127.91047	59 m	− *m*	97	(0.744)	(0.448)			
				−	3					
	129m		1.0 h	−	100	(1.980)	(1.15)			
	129		9 m	−	100	(2.570)	(1.74)			
	130		2.6 m	− *m*	90	(3.50)	(2.34)			
				−	10					
	131		3.4 m	−	100	(2)	(0)			
	132		2.2 m	−	100	(2)	(0)			
Antimony		121.75						γ	5.4	175
(Sb)	121	120.903816	57.25 a/o					*a*	6.25	200
51								122m	0.055	
	122m		4.2 m	IT	100	0.162	0*			
	122	121.905183	2.80 d	EC	3	1.02	0.413			
				−	97					

	123	122.904212	42.75 a/o					*a*	4.33	140
								124m	0.035	
	124m		93 s	IT	80	(0.45)	0.350			
				—	20					
	124	123.905973	60 d	—	100	(2.29)	1.864	γ	6.5	
	125	124.905232	2.7 y	— *m*	7.2	(0.561)	0.411			
				—	92.8					
	126m		19 m	IT	99	(1.14)	(1.08)			
				—	1					
	127	126.906927	93 h	— *m*	22	(1.064)	(0.61)			
				—	78					
	128m		11 m	—	100	(2.764)	(1.65)			
	128	127.90907	9 h	—	100	(2.313)	(1.81)			
	129	128.90926	4.3 h	— *m*	16	(1.782)	(1.02)			
				—	84					
	130m		7 m	—	100	(3.347)	(2.12)			
	130	129.91204	33 m	—	100	(3.347)	(2.12)			
	131		25 m	— *m*	15	(1.599)	(0.69)			
				—	85					
	132		2.1 m	—	100	(3.36)	(0)			
	133		4.2 m	— *m*	72	(2.76)	(0)			
				—	28					
Tellurium		127.60						γ	4.7	54
(Te)	120	119.904023	0.089 a/o					*a*	2.34	
52	121	120.905199	17 d	EC	100	0.591	0.560			
	122	121.903066	2.46 a/o					*a*	2.8	80
								123m	1.1	
	123m		117 d	IT	100	0.248	0*			
	123	122.904277	1.2E13y	EC	100	0.032	0	γ	406	5630
			0.87 a/o							
	124	123.902842	4.61 a/o					*a*	6.8	7
								125m	0.04	
	125m		58 d	IT	100	0.145	0*			
	125	124.904418	6.99 a/o					γ	1.55	20
	126	125.903322	18.71 a/o					*a*	1.04	10
								127m	0.135	

Element (Symbol) At. no., Z	Mass no., A	Mass, amu	Abundance (a/o) or half-life	Decay		Effective energy per decay, MeV		Neutron reactions		
									Barns	
				Type	Percent	Total	γ	Prod.	2200 m/s cross section	Resonance integral
Tellurium (*Cont.*)	127m		109 d	IT	99.2	0.091	0*			
	127	126.905209	9.4 h	—	100	0.272	0.001			
	128	127.904476	31.79 a/o					*a*	0.215	1.5
								129m	0.015	
	129m		34 d	IT	64	0.335	0.041			
				—	36					
	129	128.906575	69 m			0.613	0.093			
	130	129.906238	34.48 a/o					*a*	0.29	
								131m	0.02	
	131m		30 h	IT	18	(1.88)	1.465			
				—	82					
	131		25 m	—		(1.11)	0.238			
	132	131.908523	78 h	—	100	0.383	0.230			
	133m		50 m	IT	13	(3.038)	(2.12)			
				—	87					
	133		12.5 m	—	100	(2.736)	(2.12)			
	134		42 m	—	100	(2.8)	(0)			
	135		29 s	—	100	(2.63)	(0)			
Iodine (I) 53	127	126.904469	100 a/o					γ	6.2	147
	128	127.905838	25.0 m	EC	6.3	0.867	0.076			
				—	93.7					
	129	128.904987	1.7E7y	—	100	0.111	0*	*a*	27	36
								130m	18	
	130m		550 s	IT	100	(0.1)	0*			
	130	129.906676	12.4 h	—	100	2.410	2.156	γ	18	
	131	130.9061271	8.05 d	— *m*	0.6	0.589	0.383	γ	0.7	
				—	99.4					

	132	131.907981	2.3 h	—	100	2.907	2.375			
	133	132.907750	21 h	—	100	(1.177)	(0.68)			
	134	133.90985	52 m	—	100	3.141	2.534			
	135	134.91002	6.7 h	— *m*	30	2.112	1.747			
				—	70					
	136	135.91474	83 s	—	100	(4.527)	(2.485)			
	137		23 s	— *n*	6	(0.176)	(0)			
				—	94					
	138		5.9 s	—	100	(4.919)	(1.5)			
	139		2 s	—	100	(2)	(0)			
Xenon		131.30						γ	24.5	
(Xe)	124	123.906120	0.096 a/o					*a*	128	3600
54	125	124.90662	17 h	EC	100					
	126	125.904288	0.090 a/o					*a*	4.0	38
	127	126.90522	36.4 d	EC	100	0.317	0.28			
	128	127.903540	1.919 a/o					γ	5	12
	129	128.904784	26.44 a/o					γ	18	250
	130	129.903509	4.08 a/o					*a*	6.4	<14
								131m	0.42	
	131m		11.8 d	IT	100	0.164	0*			
	131	130.905085	21.18 a/o					γ	90	870
	132	131.904161	26.89 a/o					*a*	0.385	0.8
								133m	0.025	
	133m		2.26 d	IT	100	0.233	0.233			
	133	132.905815	5.27 d	—	100	0.264	0*	γ	190	
	134	133.905397	10.44 a/o					*a*	0.253	0.32
								135m	0.003	
	135m		15.6 m	IT	100	0.527	0.527			
	135	134.90702	9.2 h	—	100	0.567	0.261	γ	2.65E6	7634
	136	135.907221	8.87 a/o					γ	0.16	
	137	136.9111	3.9 m	—	100	2.057	0.305			
	138	137.91381	17 m	—	100	(3.227)	(2.104)			
	139	138.91784	43 s	—	100	(2.730)	(0.73)			
	140		16 s	—	100	(1.69)	(0)			
	141		2 s	—	100	(2.67)	(0)			

Element (Symbol) At. no., Z	Mass no., A	Mass, amu	Abundance (a/o) or half-life	Decay		Effective energy per decay, MeV		Neutron reactions		
									Barns	
				Type	Percent	Total	γ	Prod.	2200 m/s cross section	Resonance integral
Xenon (*Cont.*)	142		1.5 s	—	100	(3)	(0)			
	143		1.0 s	—	100	(3.09)	(0)			
Cesium	133	132.905355	100 a/o					*a*	29.0	415
(Cs)								134m	2.5	
55	134m		2.90 h	IT	100	0.138	0*			
	134	133.906823	2.05 y	—	100	1.787	1.560	γ	140	
	135m		53 m	IT	100	1.621	1.621			
	135	134.90577	3E6y	—	100	0.082	0	γ	8.7	
	136	135.90734	13 d	—	100	2.317	2.159	γ	1.3	76
	137	136.90677	30.0 y	— *m*	93.5	0.276	0	γ	0.11	
				—	6.5					
	138	137.9108	32.2 m	—	100	3.227	2.141			
	139	138.9129	9.5 m	—	100	2.077	0.955			
	140	139.91711	66 s	—	100	3.415	1.53			
	141		24 s	—	100	(2.04)	(0)			
	142		2.3 s	—	100	(3)	(0)			
	143		2.0 s	—	100	(2.895)	(0.43)			
Barium		137.34						γ	1.2	7.5
(Ba)	130	129.906245	0.101 a/o					*a*	13.5	
56								131m	2.5	
	131m		15 m	IT	100	0.18	0*			
	131	130.906716	12 d	EC	100	0.536	0.38			
	132	131.905120	0.097 a/o					*a*	8.5	
								133m	0.68	
	133m		38.9 h	IT	100	0.288	0.276			
	133	132.905879	7.2 y	EC	100	0.462	0.382			

	134	133.904612	2.42 a/o					*a*	2.0	23
								135m	0.158	
	135m		28.7 h	IT	100	0.268	0.268			
	135	134.905550	6.59 a/o					γ	5.8	100
	136	135.904300	7.81 a/o					*a*	0.4	1.3
								137m	0.010	
	137m		2.55 m	IT	100	0.662	0.662			
	137	136.905500	11.32 a/o					γ	5.1	4
	138	137.905000	71.66 a/o					γ	0.35	0.2
	139	138.9086	82.9 m	—	100	0.935	0.007	γ	6	
	140	139.910565	12.8 d	—	100	0.563	0.223	γ	1.6	13.6
	141	140.91405	18 m	—	100	2.327	0.959			
	142	141.91635	11 m	—	100	1.479	0.552			
	143		12 s	—	100	2.722	0.852			
Lanthanum		138.91								
(La)	138	137.906910	0.089 a/o	EC	70 }	1.288	1.235	γ	172	413
57			1.12E11y	—	30 }					
	139	138.906140	99.911 a/o					γ	9.0	12.2
	140	139.909438	40.22 h	—	100	2.976	2.125	γ	2.7	
	141	140.910828	3.9 h	—	100	1.010	0.027			
	142	141.91398	92 m	—	100	3.279	2.345			
	143	142.91587	14.0 m	—	100	2.115	0.852			
	144		4 s	—	100	(2.61)	(0)			
Cerium		140.12						γ	0.63	3.0
(Ce)	136	135.907100	0.193 a/o					*a*	7.2	70
58	137	136.90733	9.0 h	EC	100	0.062	0.012			
	138	137.905830	0.25 a/o					*a*	1.1	
	139	138.90643	140 d	EC	100	0.206	0*			
	140	139.905392	88.48 a/o					γ	0.57	0.47
	141	140.908219	33 d	—	100	0.332	0*	γ	29	0.48
	142	141.909140	11.07 a/o					γ	0.95	0.73
	143	142.912327	33 h	—	100	0.840	0.377	γ	6.0	
	144	143.913591	284 d	—	100	0.138	0*	γ	1.0	2.6
	145	144.91727	3.0 m	—	100	(1.67)	(0.6)			
	146	145.91867	14 m	—	100	(0.58)	(0.32)			

Element (Symbol) At. no., Z	Mass no., A	Mass, amu	Abundance (a/o) or half-life	Decay		Effective energy per decay, MeV		Neutron reactions		
									Barns	
				Type	Percent	Total	γ	Prod.	2200 m/s cross section	Resonance integral
Cerium (*Cont.*)	147		65 s	—	100	(2)	(0)			
	148		43 s	—	100	(2)	(0)			
Praseodymium	141	140.907596	100 a/o					*a*	11.5	14.1
(Pr)								142m	3.9	
59	142m		14.6 m							
	142	141.909978	19.2 h	—	100	0.915	0.058	γ	20	
	143	142.91099	13.6 d	—	100	0.366	0	γ	89	190
	144	143.913248	17.3 m	—	100	1.307	0.031			
	145	144.914476	5.98 h	—	100	0.749	0.056			
	146	145.91759	24 m	—	100	2.734	1.703			
	147	146.9188	12.0 m	—	100	1.585	0.75			
	148	147.92129	2.0 m	—	100	(2.088)	(0.3)			
	149		2.3 m	—	100	(2)	(0)			
Neodymium		144.24						γ	50.5	45
(Nd)	142	141.907663	27.11 a/o					γ	18.7	9
60	143	142.909779	12.17 a/o					γ	325	140
	144	143.910039	23.85 a/o					γ	3.6	5
			2.4E15y	α	100					
	145	144.912538	8.30 a/o					γ	42	240
	146	145.913086	17.22 a/o					γ	1.4	3.2
	147	146.916074	11.1 d	—	100	0.548	0.096			
	148	147.916869	5.73 a/o					γ	2.5	19
	149	148.920122	1.8 h	—	100	0.966	0.438			
	150	149.920915	5.62 a/o					γ	1.2	14
	151	150.92377	12 m	—	100	1.459	0.24			

Promethium	147	146.915108	2.62 y	—	100	0.087	0	a	181	2300
(Pm)								148m	85	1026
61	148m		42 d	IT	7	2.091	1.900	γ	22,000	3600
				—	93					
	148	147.917421	5.4 d	—	100	1.368	0.622	γ	2000	
	149	148.91833	53.1 h	—	100	0.424	0.011	γ	1400	
	150	149.92096	2.7 h	—	100	2.182	1.410			
	151	150.921198	28 h	—	100	0.536	0.185	γ	<700	
	152	151.92351	6 m	—	100	(2.317)	(1.3)			
	153	152.92403	5.5 m	—	100	(0.825)	(0)			
Samarium		150.35						γ	5800	1400
(Sm)	144	143.911989	3.09 a/o					γ	0.7	
62	145	144.913394	340 d	EC	100	0.100	0*	γ	110	
	146	145.912992	7E7y	α	100	2.54				
	147	146.914867	14.97 a/o					γ	64	714
			1.05E11y	α	100	2.314		α	0.00075	
	148	147.914791	11.24 a/o					γ	2.7	27
	149	148.917180	13.83 a/o					γ	41,000	3183
	150	149.917276	7.44 a/o					γ	102	310
	151	150.919919	87 y	—	100	(0.03)	0*	γ	15,000	3300
	152	151.919756	26.72 a/o					γ	206	3000
	153	152.922102	47 h	—	100	0.375	0.004			
	154	153.922282	22.71 a/o					γ	5.5	30
	155	154.924701	23 m	—	100	0.668	0.03			
	156	155.925569	9.4 h	—	100	0.327	0.144			
	157		0.5 m	—	100	(0.92)	(0.57)			
Europium		151.96						γ	4600	2430
(Eu)	151	150.919838	47.82 a/o					a	9200	3300
63								152m2	4	
								152m1	3300	
	152m2		96 m	ITg	100	0.10	0*			
	152m1		9.3 h	EC	23	(0.84)	0.264			
				—	77					
	152	151.921749	12 y	EC	72	(1.27)	0.096	γ	2300	
				—	28					

Element (Symbol) At. no., Z	Mass no., A	Mass, amu	Abundance (a/o) or half-life	Decay		Effective energy per decay, MeV		Neutron reactions		
									Barns	
				Type	Percent	Total	γ	Prod.	2200 m/s cross section	Resonance integral
Europium	153	152.921242	52.18 a/o					γ	390	1635
(*Cont.*)	154	153.923053	16 y	—	100	1.385	1.132	γ	1500	
	155	154.922930	1.81 y	—	100	0.142	0*	γ	4040	
	156	155.924802	15 d	—	100	1.678	1.250			
	157	156.92539	15.2 h	—	100	0.648	0.240			
	158	157.92794	46 m	—	100	(2.678)	(1.75)			
	159	158.92884	18 m	—	100	(1.395)	(0.57)			
	160	159.931	2.5 m	—	100	(1.49)	(0)			
Gadolinium		157.25						γ	49,000	390
(Gd)	152	151.919794	0.2 a/o					γ	1100	3000
64			1.1E14y	α	100	2.1				
	153	152.921503	242 d	EC	100	0.150	0*			
	154	153.920929	2.15 a/o					γ	85	215
	155	154.922664	14.73 a/o					γ	61,000	1550
	156	155.922175	20.47 a/o					γ	1.5	95
	157	156.924025	15.68 a/o					γ	254,000	730
	158	157.924178	24.87 a/o					γ	2.5	61
	159	158.926368	18.0 h	—	100	0.430	0.069			
	160	159.927115	21.90 a/o					γ	0.77	7.0
	161	160.92972	3.7 m	—	100	1.030	0.402	γ	31,000	
	162		(10.4 m)	—	100	(0.574)	(0)			
Terbium	159	158.925351	100 a/o					γ	25.5	430
(Tb)	160	159.927146	72.1 d	—	100	1.433	1.178	γ	525	
65	161	160.927572	6.9 d	—	100	0.274	0.001			
	162m		7.5 m	—	100	(2.276)	(1.144)			
	162		2.2 h	—	100	(2.222)	(1.093)			
	163m		7 m	—	100	(0.749)	(0.18)			

	163	162.930560	6.5 h	—	100	(0.809)	(0.24)			
	164		23 h	—	100	(1.58)	(0)			
Dysprosium		162.50						γ	930	1600
(Dy)	156	155.92393	0.0524 a/o					γ	33	960
66	157	156.92527	8.1 h	EC	100	0.358	0.306			
	158	157.924449	0.0902 a/o					γ	43	120
	159	158.925759	144 d	EC	100	0.067	0*			
	160	159.925202	2.294 a/o					γ	61	1160
	161	160.926945	18.88 a/o					γ	585	1060
	162	161.926803	25.53 a/o					γ	180	2730
	163	162.928755	24.97 a/o					γ	130	1680
	164	163.929200	28.18 a/o					*a*	2700	377
								165m	1700	
	165m		1.26 m	IT	97.5	0.126	0.012			
				—	2.5					
	165	164.931816	139.2 m	—	100	0.532	0*	γ	3900	22,000
	166	165.932807	81.5 h	—	100	0.147	0.004			
Holmium	165	164.930421	100 a/o					*a*	66.5	700
(Ho)								166m	3.5	
67	166m		1,200 y	—	100	1.814	1.795			
	166	165.932289	26.9 h	—	100	0.756	0.017			
Erbium		167.26						γ	162	740
(Er)	162	161.92874	0.136 a/o					γ	19	480
68	163	162.930065	75 m	EC	100	0.056	0*			
	164	163.929287	1.56 a/o					γ	13	105
	165	164.930819	10.3 h	EC	100	0.056	0*			
	166	165.930307	33.41 a/o					*a*	20	100
								167m	15	
	167m		2.3 s	IT	100	0.208	0.208			
	167	166.932060	22.94 a/o					γ	670	2970
	168	167.932383	27.07 a/o					γ	1.95	36
	169	168.93461	9.4 d	—	100	(0.139)	0*			
	170	169.935560	14.88 a/o					γ	5.7	20
	171	170.93813	7.52 h	—	100	(0.87)	0.31	γ	280	

Element (Symbol) At. no., Z	Mass no., A	Mass, amu	Abundance (a/o) or half-life	Decay Type	Decay Percent	Effective energy per decay, MeV Total	Effective energy per decay, MeV γ	Neutron reactions Prod.	Barns 2200 m/s cross section	Barns Resonance integral
Thulium	169	168.934245	100 a/o					a	103	1720
(Tm)	170	169.93606	130 d	—	100	(0.40)	0*	γ	92	460
69	171	170.93653	1.92 y	—	100	(0.040)	0*	γ	4.5	118
Ytterbium		173.04						γ	36.6	182
(Yb)	168	167.934160	0.135 a/o					γ	3470	31,000
70	169	168.93553	32 d	EC	100	0.447	0.154			
	170	169.935020	3.03 a/o					γ	10	300
	171	170.936430	14.31 a/o					γ	50	332
	172	171.936360	21.82 a/o					γ	1.3	25
	173	172.938060	16.13 a/o					γ	19	390
	174	173.938740	31.84 a/o					γ	65	33
	175	174.94114	101 h	—	100	(0.214)	0.038			
	176	175.942680	12.73 a/o					γ	2.4	6
	177	176.94541	1.9 h	—	100	(0.659)	0.112			
Lutecium		174.97						γ	77	900
(Lu)	175	174.940640	97.41 a/o					a	23.4	890
71								176m	16.4	
	176m		3.7 h	—	100	(0.556)	0*			
	176	175.942660	2.59 a/o					a	2100	1160
			3E10y	—	100	(0.766)	0.508	177m	7	
	177m		155 d	IT	22 }	0.274	0.141			
				— m	78 }					
	177	176.94343	6.7 d	—	100	(0.216)	0.014			
Hafnium		178.49						γ	102	2000
(Hf)	174	173.940360	0.18 a/o					γ	390	465
72			2E15y	α	100	2.55				

	175	174.94161	70 d	EC	100	0.419	0.347			
	176	175.941570	5.20 a/o					γ	38	700
	177	176.943400	18.50 a/o					*a*	365	7260
								178m	1.1	
	178m		4.3 s	IT	100	1.148	0.966			
	178	177.943880	27.14 a/o					*a*	86	1950
								179m	53	
	179m		18.6 s	IT	100	0.378	0.218			
	179	178.946030	13.75 a/o					*a*	45	600
								180m	0.34	
	180m		5.5 h	IT	100	1.142	1.006			
	180	179.946820	35.24 a/o					γ	12.6	43
	181	180.949105	42.5 d	—	100	(0.778)	(0.45)	γ	40	
	182	181.9507	9E6y	—	100	(0.34)	(0.23)			
Tantalum		180.948								
(Ta)	180	179.947544	0.0123 a/o					γ	700	600
73	181	180.948007	99.9877 a/o					*a*	21	710
								182m	0.00103	
	182m		16.5 m	IT	100	0.503	0.035			
	182	181.950167	115 d	—	100	1.205	1.078	γ	8200	1000
Tungsten		183.85						γ	18.5	352
(W)	180	179.947000	0.135 a/o					γ	3.5	200
74	181	180.948211	140 d	EC	100	0.070	0*			
	182	181.948301	26.41 a/o					γ	20.7	590
	183	182.950324	14.40 a/o					γ	10.2	345
	184	183.951025	30.64 a/o					*a*	1.8	14
								185m	0.002	
	185m		1.6 m	IT	100	0.368	0*			
	185	184.953519	75 d	—	100	0.160	0			
	186	185.954440	28.41 a/o					γ	37.8	500
	187	186.957244	23.9 h	—	100	0.677	0.391	γ	64	2760
	188	187.958816	69 d	—	100	(0.14)	0.002			
Rhenium		186.2						γ	88	830
(Re)	185	184.953059	37.07 a/o					γ	112	1730
75	186	185.95502	90 h	EC	5 }	0.451	0.001			
				—	95 }					

Element (Symbol) At. no., Z	Mass no., A	Mass, amu	Abundance (a/o) or half-life	Decay		Effective energy per decay, MeV		Neutron reactions		
									Barns	
				Type	Percent	Total	γ	Prod.	2200 m/s cross section	Resonance integral
Rhenium	187	186.955833	62.93 a/o					a	74.6	300
(*Cont.*)								188m	73	
	188m		18.7 m	IT	100	0.172	0*			
	188	187.958353	16.7 h	—	100	(0.87)	0.012			
Osmium		190.2								
(Os)	184	183.952750	0.018 a/o					γ	3000	
76	185	184.954113	94 d	EC	100	0.760	0.669			
	186	185.953870	1.59 a/o							
	187	186.955832	1.64 a/o					γ	336	890
	188	187.956081	13.3 a/o					γ	4.3	135
	189	188.958300	16.1 a/o					a	23	750
								190m	0.00026	
	190m		10 m	IT	100	1.706	1.706			
	190	189.958630	26.4 a/o					a	13.0	29
								191m	9.1	
	191m		13 h	IT	100	0.074	0*			
	191	190.96097	15 d	— *m*	99+	(0.056)	0*			
	192	191.961450	41.0 a/o					γ	2.0	5.4
	193	192.964227	31 h	—	100	(0.50)	0.076	γ	1540	
Iridium		192.2						γ	426	2250
(Ir)	191m		4.9 s	IT	100	0.171	0*			
77	191	190.960640	37.3 a/o					a	924	3750
								192m2	0.38	
								192m1	300	
	192m2		75 y	IT*g*	100	0.161	0*			
	192m1		1.4 m	IT	100	0.058	0*			

	192	191.9627	74.2 d	EC	4.4	(1.08)	0.857	γ	1100	
				—	95.6					
	193	192.963012	62.7 a/o					γ	112.5	1350
	194	193.965125	17.4 h	—	100	(0.91)	(0.03)			
Platinum		195.09						γ	10	140
(Pt)	190	189.959950	0.0127 a/o					γ	150	
78			6E11y	α	100	3.25				
	191	190.96145	3.0 d	EC						
	192	191.961150	0.78 a/o					a	< 14	83
			1E15y	α	100	2.3		193m	2.2	
	193m		4.3 d	IT	100	0.148	0*			
	193	192.96306	< 500 y	EC	100	< 0.05	0*			
	194	193.962725	32.9 a/o					a	1.2	4
								195m	0.090	
	195m		4.1 d	IT	100	0.259	0*			
	195	194.964813	33.8 a/o					γ	27	355
	196	195.964967	25.3 a/o					a	7.4	8
								197m	0.050	
	197m		80 m	IT	97	0.396	0.336			
				— *m*	3					
	197	196.967347	18 h	—	100	(0.36)	0.020			
	198	197.967895	7.21 a/o					a	3.7	56
								199m	0.027	
	199m		14.1 s	IT	100	0.425	0.393			
	199	198.97058	30 m	—	100	(0.81)	0.235	γ	15	
Gold	197m		7.2 s	IT	100	0.410	0.279			
(Au)	197	196.966541	100 a/o					γ	98.8	1560
79	198	197.968231	2.698 d	—	100	0.74	0.419	γ	25,800	
	199	198.968773	3.15 d	—	100	(0.28)	0.042	γ	30	
	200	199.97070	48.4 m	—	100	(1.08)	0.341			
Mercury		200.59						γ	375	73
(Hg)	196	195.965820	0.146 a/o					a	3200	472
80								197m	120	58.9
	197m		24 h	IT	94	0.305	0*			
				EC*m*	6					

Element (Symbol) At. no., Z	Mass no., A	Mass, amu	Abundance (a/o) or half-life	Decay: Type	Decay: Percent	Effective energy per decay, MeV: Total	Effective energy per decay, MeV: γ	Neutron reactions: Prod.	Neutron reactions, Barns: 2200 m/s cross section	Neutron reactions, Barns: Resonance integral
Mercury	197	196.96736	65 h	EC	100	0.181	0.004			
(*Cont.*)	198	197.966756	10.02 a/o					*a*	1.9	70
								199m	0.018	
	199m		43 m	IT	100	0.533	0.375			
	199	198.968279	16.84 a/o					γ	2000	153
	200	199.968327	23.13 a/o					γ	<60	
	201	200.970308	13.22 a/o					γ	<60	
	202	201.970642	29.80 a/o					γ	4.9	4.9
	203	202.97288	46.9 d	–	100	(0.364)	0.279			
	204	203.973495	6.85 a/o					γ	0.43	
	205	204.97621	5.5 m	–	100		>0			
Thallium		204.37						γ	3.4	12
(Tl)	203	202.972353	29.50 a/o					γ	11	39.5
81	204	203.973865	3.8 y	EC –	2.1 97.9	(0.30)	0*	γ	21.6	86
	205	204.974442	70.50 a/o					γ	0.1	0.7
	206	205.976104	4.19 m	–	100	(0.61)	0			
[AcC″]	207	206.97745	4.79 m	–	100	0.510	0.001			
[ThC″]	208	207.982013	3.10 m	–	100	3.929	3.414			
	209	208.985296	2.2 m	–	100	(2.87)	2.01			
[RaC″]	210	209.990054	1.3 m	–	100	(3.62)	2.370			
Lead		207.19						γ	0.17	0.16
(Pb)	204	203.973044	1.48 a/o					γ	0.661	1.7
82	205	204.97448	3E7y	EC	100	0.035	0*			
	206	205.974468	23.6 a/o					γ	0.0305	0.2
	207	206.975903	22.6 a/o					γ	0.709	0.4
	208	207.976650	52.3 a/o					γ	0.000487	

	209	208.981082	3.30 h	—	100	0.194	0			
[RaD]	210	209.984187	21 y	—	100	0.047	0*	γ	0.5	
[AcB]	211	210.988742	36.1 m	—	100	0.564	0.066			
[ThB]	212	211.991905	10.64 h	—	100	(0.44)	0.21			
	213	212.99629	10 m	—	100					
[RaB]	214	213.999766	26.8 m	—	100	(0.60)	0.296			
Bismuth	209	208.980394	100 a/o					*a*	0.00033	0.19
(Bi)			2E18y	α	100	3.12		210m	0.00019	
83	210m		3E6y	α	99.6 }	5.0	0.303	γ	0.054	
				—	0.4 }					
[RaE]	210	209.984121	5.01 d	—	100	0.444	0			
[AcC]	211	210.98730	2.15 m	α	99.7 }	6.73	0.056			
				—	0.28 }					
[ThC]	212	211.991279	60.6 m	α	36.0 }	2.929	0.29			
				—	64.0 }					
	213	212.994317	47 m	α	2.2 }	1.037	0.43			
				—	97.8 }					
[RaC]	214	213.998686	19.7 m	α	0.021 }	2.349	1.57			
				—	99.979 }					
Polonium										
(Po)	208	207.981243	2.93 y	α	100	5.21	0			
84	209	207.982426	103 y	α	99.5	4.96	0.006			
[RaF]	210	209.982876	138.4 d	α	100	5.408	0	γ	< 0.03	
	211	210.986657	0.52 s	α	100	7.592	0.008			
[ThC′]	212	211.988866	304 ns	α	100	8.954	0			
	213	212.992825	4.2 μs	α	100	8.54	0			
[RaC′]	214	213.995201	164 μs	α	100	7.835	0			
[AcA]	215	214.999423	1.78 ms	α	100	7.524	0			
[ThA]	216	216.001922	0.15 s	α	100	6.906	0			
	217	217.00606	< 10 s	α	100	6.67	0			
[RaA]	218	218.00893	3.05 m	α	99.981 }	6.11	0			
				—	0.019 }					
Astatine	209		5.5 h	α	5 }	1.85	1.47			
(At)				EC	95 }					
85	210		8.3 h	α	0.17 }	2.96	2.87			
				EC	99.83 }					

Element (Symbol) At. no., Z	Mass no., A	Mass, amu	Abundance (a/o) or half-life	Decay: Type	Decay: Percent	Effective energy per decay, MeV: Total	Effective energy per decay, MeV: γ	Neutron reactions: Prod.	Neutron reactions, Barns: 2200 m/s cross section	Neutron reactions, Barns: Resonance integral
Astatine	211		7.21 h	α	40.9	2.5	0*			
(*Cont.*)				EC	59.1					
	217	217.004648	0.032 s	α	100	7.199	0			
Radon										
(Rn)										
86										
[Actinon]	219	219.009481	4.0 s	α	100	6.944	0.033			
[Thoron]	220	220.011401	55 s	α	100	6.405	0	γ	< 0.2	
	221	221.01523	25 m	α	20					
				—	80					
	222	222.017531	3.821 d	α	100	5.587	0.0004	γ	0.72	
Francium										
(Fr)	221	221.014183	4.8 m	α	100	6.457	0.036			
87	222	222.01763	14.8 m	—	100					
[AcK]	223	223.019736	22 m	—	100	0.395	0.004			
Radium										
(Ra)										
88	222		38 s	α	100	6.68	0.013			
[AcX]	223	223.018501	11.43 d	α	100	5.977	0.011	γ	130	
[ThX]	224	224.020218	3.64 d	α	100	5.787	0.014	γ	12	
	225	225.023528	14.8 d	—	100	0.111	0*			
	226	226.02536	1,602 y	α	100	4.869	0*	γ	11	222
	227	227.027753	41.2 m	—	100					
[MsTh1]	228	228.031139	6.7 y	—	100	0.013	0	γ	36	
Actinium	225	225.023153	10.0 d	α	100	5.930	0.001			
(Ac)	226	226.02616	29 h	EC	20	(0.72)	0.134			
89				—	80					

	227	227.027753	21.6 y	α	1.4 }	0.085	0*	γ	515	
				—	98.6 }					
[MsTh2]	228	228.03108	6.13 h	—	100	(1.48)	(1.02)			
Thorium										
(Th)										
90										
[RdAc]	227	227.027706	18.2 d	α	100	6.145	0.13	*f*	200	
[RdTh]	228	228.02875	1.910 y	α	100	5.521	0.001	γ	123	1013
	229	229.031652	7,340 y	α	100	5.167	>0	γ	54	1000
		$\nu = 2.14$						*f*	30.5	464
[Ionium]	230	230.033087	8E4y	α	100	4.767	>0	γ	23.2	1010
[UY]	231	231.036291	25.5 h	—	100	(0.21)	0*			
	232	232.038124	100 a/o					γ	7.40	85
			1.41E10y	α	100	4.08	0*	*f*	0.000039	
	233	233.041469	22.2 m	—	100	0.427	0.009	γ	1500	400
								f	15	
[UX1]	234	234.043583	24.1 d	— *m*	100	0.060	0*	γ	1.8	
Protactinium	231	231.035877	3.25E4y	α	100	5.148	0.037	γ	210	1500
(Pa)	232	232.038612	1.31 d	—	100	1.289	0.95	γ	760	
91								*f*	700	
	233	233.040132	27.0 d	—	100	0.228	0.15	*a*	41	895
								234m	21	
[UX2]	234m		1.17 m	IT	0.13 }	0.868	0.009	*f*	< 500	
				—	99.87 }					
[UZ]	234	234.043298	6.75 h	—	100	1.533	1.21	*f*	< 5000	
Uranium		238.03								
(U)	231	231.03627	4.3 d	EC	100			*f*	400	
92	232	232.037168	72 y	α	100	5.414	0*	γ	73.1	280
		$\nu_p = 3.13$						*f*	75.2	320
	233	233.039522	1.62E5y	α	100	4.909	0*	γ	47.7	140
		$\nu = 2.492$						*f*	531.1	764
[UII]	234	234.040904	0.0056 a/o					γ	100.2	630
			2.47E5y	α	100	4.856	0*			
[AcU]	235	235.043915	0.7205 a/o					γ	98.6	144
		$\nu = 2.418$	7.1E8y	α	100	4.681	0.067	*f*	582.2	275

Element (Symbol) At. no., Z	Mass no., A	Mass, amu	Abundance (a/o) or half-life	Decay: Type	Decay: Percent	Effective energy per decay, MeV: Total	Effective energy per decay, MeV: γ	Neutron reactions: Prod.	Neutron reactions, Barns: 2200 m/s cross section	Neutron reactions, Barns: Resonance integral
Uranium	236	236.045637	2.39E7y	α	100	4.573	0*	γ	5.2	365
(*Cont.*)	237	237.048608	6.75 d	—	100	0.112	0.008	γ	411	290
[UI]	238	238.05077	99.274 a/o					γ	2.70	275
			4.51E9y	α	100	4.268	0*			
	239	239.05430	23.5 m	—	100	0.400	0*	γ	22	
								f	14	
	240	240.056594	14.1 h	— m	100	0.211	0*			
Neptunium	236m		1.29E8y					f	2500	
(Np)	236	236.046624	22 h	EC	50 }	(0.17)	0*			
93				—	50 }					
	237	237.048056	2.14E6y	α	100	4.956	0.003	γ	169	660
								f	0.019	
	238	238.050896	2.1 d	—	100	0.839	0.57	f	2070	880
	239	239.052924	2.35 d	—	100	(0.69)	0.23	a	50	
								240m	31	
	240m		7.3 m	—	100	1.065	0.34			
	240	240.05608	63 m	—	100					
Plutonium	236	236.04607	2.85 y	α	100	5.868	0*	f	165	
(Pu)	237	237.048298	45.6 d	EC	100	>0.12	0*	f	2400	
94	238	238.049511	86 y	α	100	5.592	0*	γ	547	141
		$\nu_p = 2.90$			100			f	16.5	24
	239	239.052146	24,400 y	α	100	5.243	0*	γ	268.8	200
		$\nu = 2.871$						f	742.5	301
	240	240.053882	6,580 y	α	100	5.255	0*	γ	289.5	8013
	241	241.056737	13.2 y	α	0.0023 }	0.007	0	{ γ	368	162
		$\nu_p = 2.927$		—	99.9977 }			{ f	1009	570

	242	242.058725	3.79E5y	α	100	4.98	0*	γ	18.5	1130
								f	<0.2	5
	243	243.061972	4.98 h	—	100	0.239	0*	γ	60	
								f	196	
	244	244.0641	8E7y	α	100	4.66		γ	1.7	43
	245	245.06783	10 h	—	100	0.4	0	γ	150	220
Americium	241	241.056714	458 y	α	100	5.640	0.002	a	832	1477
(Am)								242m	83.8	202
95		$\nu_p = 3.219$						f	3.15	
	242m		152 y	α	0.48 }	0.075	0*	{ γ	1400	7000
		$\nu_p = 3.264$		IT	99.52 }			{ f	6600	1570
	242	242.059502	16.0 h	EC	16 }	0.225	0*	f	2900	
				—	84 }					
	243	243.061367	7,950 y	α	100	5.439	0*	a	79.3	1820
								244m	75.2	1709
	244m		26 m	—	100	(0.6)	0*	f	1600	
	244	244.064355	10.1 h	—	100	1.256	0.788	f	2300	
	245	245.066340	2.1 h	—	100	0.313	0			
Curium	242	242.058788	163 d	α	100	6.217	0*	γ	16	150
(Cm)	243	243.06137	32 y	EC	0.3 }	6.15	0.22	{ γ	225	2345
96		$\nu_p = 3.430$		α	99.7 }			{ f	600	1860
	244	244.062821	17.6 y	α	100	5.902	0*	γ	13.9	650
								f	1.2	12.5
	245	245.065371	9,300 y	α	100	5.624	0.02	γ	345	101
		$\nu_p = 3.832$						f	2020	750
	246	246.067202	5,500 y	α	100	5.476	0*	γ	1.3	121
								f	0.17	10
	247	247.07028	1.6E7y	α	100	5.3	0	γ	60	800
								f	90	880
	248	248.0722	4.7E5y	α	89 }	21.41		{ γ	4	275
		$\nu_{sp} = 3.157$		SF	11 }			{ f	0.34	13.2
	249	249.07581	64 m	—	100	0.3		γ	1.6	
	250		1.7E4y	SF						
Berkelium	247	247.07026	1,400 y	α	100	5.86	0.17			
(Bk)	248	248.07296	16 h	EC	30 }	(0.22)	0*			
97				—	70 }					

Element (Symbol) At. no., Z	Mass no., A	Mass, amu	Abundance (a/o) or half-life	Decay Type	Decay Percent	Effective energy per decay, MeV Total	Effective energy per decay, MeV γ	Neutron reactions Prod.	Barns 2200 m/s cross section	Barns Resonance integral
Berkelium (*Cont.*)	249	249.074883	314 d	α	0.0022	0.042	0	a	1300	1240
				—	99.9978					
	250	250.07827	3.22 h	—	100	1.507	0.898	f	960	
Californium	248	248.072262	350 d	α	100	6.37	0*			
(Cf)	249	249.074749	360 y	α	100	6.295	0.363	γ	465	765
98								f	1660	2114
	250	250.076384	13 y	α	100	6.128	0*	γ	2030	11,600
								f	< 350	
	251	251.079260	800 y	α	100	5.94	0*	γ	2850	1600
								f	4300	5900
	252	252.0815	2.65 y	α	96.9	12.260	0*	γ	20.4	43.5
		$\nu = 3.74$		SF	3.1			f	32	110
	253	253.08502	17.6 d	α	0.31	(0.3)	0*	γ	17.6	
				—	99.69			f	1300	
	254		60.5 d	α	0.2	194.4		a	90	
		$\nu_{sp} = 3.93$		SF	99.8					
Einsteinium	253	253.08473	20.47 d	α	100	6.747	0*	a	< 158	7300
(Es)								254m	155	3000
99	254m			EC	0.08	(0.79)	0.53	γ	1.3	
				—	99.92			f	1840	
	254	254.08759	276 d	α	100	6.623	0*	γ	< 40	
								f	2900	2190
Fermium	254	254.086839	3.24 h	α	99.945	7.42	0*	γ	76	
(Fm)		$\nu_{sp} = 3.96$		SF	0.055					
100	255	255.08964	20.1 h	α	100	7.244	0*	γ	36	
								f	3400	
	256		2.7 h	α	3	(202)		γ	45	
		$\nu_{sp} = 3.73$		SF	97					
	257		80 d	α	100			a	6100	
								f	2950	

APPENDIX

D

RADIOACTIVITY CONCENTRATION LIMITS FOR SELECTED RADIONUCLIDES

(For Radionuclides in Effluents in Unrestricted Areas)

This table is taken from U.S. Nuclear Regulatory Commission, Rules and Regulations 10CFR20, Appendix B, Table II, *Federal Register*, Dec. 1, 1978, except for ^{225}Ra. "Sub" refers to external radiation resulting from submersion in an infinite half-space of air containing the listed gaseous radionuclide. Other quantities listed are for the radionuclide in soluble form, unless denoted as insoluble by (I). The concentration limit for the insoluble form is listed only if its value in air or water is less than that for the soluble form. An individual who continuously breathes air containing a radionuclide at the listed concentration, or whose entire potable water intake contains a radionuclide at the listed concentration, or who is continuously surrounded by an extended volume of air containing a radionuclide at the listed submersion concentration, will receive a yearly radiation dose of 500 mrem.

	Air μCi/ml	Water μCi/ml
^{241}Am, ^{242m}Am, ^{243}Am	2E-13	4E-6
^{242}Am	1E-9	1E-4
^{14}C	1E-7	8E-4
$^{14}CO_2$	1E-6 (sub)	–
^{134}Cs	1E-9	9E-6
	4E-10 (I)	4E-5 (I)
^{137}Cs	2E-9	2E-5
	5E-10 (I)	4E-5 (I)
^{242}Cm	4E-12	2E-5
^{243}Cm	2E-13	5E-6
^{244}Cm	3E-13	7E-6

	Air μCi/ml	Water μCi/ml
^{3}H (tritium)	2E-7	3E-3
	4E-5 (sub)	–
^{129}I	2E-11	6E-8
^{131}I	1E-10	3E-7
^{133}I	4E-10	1E-6
^{85m}Kr	1E-7 (sub)	–
^{85}Kr	3E-7 (sub)	–
^{140}La	5E-9	2E-5
^{237}Np	1E-13	3E-6
^{93m}Nb	4E-9	4E-4
^{95}Nb	2E-8	1E-4
	3E-9 (I)	1E-4 (I)
^{238}Pu	7E-14	5E-6
^{239}Pu, ^{240}Pu	6E-14	5E-6
^{241}Pu	3E-12	2E-4
^{242}Pu	6E-14	5E-6
^{231}Pa	4E-14	9E-7
^{233}Pa	2E-8	1E-4
	6E-9 (I)	1E-4 (I)
^{224}Ra	2E-10	2E-6
	2E-11 (I)	5E-6 (I)
$^{225}Ra^{\dagger}$	–	5E-7
^{226}Ra	3E-12	3E-8
	2E-12 (I)	3E-5 (I)
^{228}Ra	2E-12	3E-8
	1E-12 (I)	3E-5 (I)
^{220}Rn	1E-8	–
^{222}Rn	3E-9	–
^{103m}Rh	3E-6	1E-2
	2E-6 (I)	1E-2 (I)
^{103}Ru	2E-8	8E-5
	3E-9 (I)	8E-5 (I)
^{106}Ru	3E-9	1E-5
	2E-10 (I)	1E-5 (I)
^{89}Sr	3E-10	3E-6
^{90}Sr	3E-11	3E-7
^{99}Tc	7E-8	3E-4
	2E-9 (I)	2E-4 (I)
^{129}Te	2E-7	8E-4
	1E-7 (I)	8E-4 (I)
^{131m}Te	1E-8	6E-5
	6E-9 (I)	4E-5 (I)
^{228}Th	3E-13	7E-6
	2E-13 (I)	1E-5 (I)

(*See footnote on page 981.*)

	Air μCi/ml	Water μCi/ml
^{230}Th	8E-14	2E-6
^{231}Th	5E-8	2E-4
	4E-8 (I)	2E-4 (I)
^{232}Th	1E-12	2E-6
^{234}Th	2E-9	2E-5
	1E-9 (I)	2E-5 (I)
Natural thorium	2E-12	2E-6
^{232}U	3E-12	3E-5
^{233}U	2E-11	3E-5
^{234}U	2E-11	3E-5
	4E-12 (I)	3E-5 (I)
^{235}U	2E-11	3E-5
	4E-12 (I)	3E-5 (I)
^{236}U	2E-11	3E-5
	4E-12 (I)	3E-5 (I)
^{238}U	3E-12	4E-5
Natural uranium	5E-12	3E-5
^{133}Xe	3E-7 (sub)	–
^{90}Y	4E-9	2E-5
	3E-9 (I)	2E-5 (I)
^{93}Zr	4E-9	8E-4
^{95}Zr	4E-9	6E-5
	1E-9 (I)	6E-5 (I)
Mixture, if it is known that ^{129}I, ^{226}Ra, and ^{228}Ra are not present	–	1E-7

†Estimated by M. E. Laverne. Unpublished results quoted by H. C. Claiborne, "Neutron-induced Transmutation of High-Level Radioactive Waste," Report ORNL-TM/3964, Dec. 1972.

INDEX